Springer Tracts in Civil Engineering

Springer Tracts in Civil Engineering (STCE) publishes the latest developments in Civil Engineering - quickly, informally and in top quality. The series scope includes monographs, professional books, graduate textbooks and edited volumes, as well as outstanding PhD theses. Its goal is to cover all the main branches of civil engineering, both theoretical and applied, including:

- Construction and Structural Mechanics
- Building Materials
- Concrete, Steel and Timber Structures
- Geotechnical Engineering
- Earthquake Engineering
- Coastal Engineering; Ocean and Offshore Engineering
- Hydraulics, Hydrology and Water Resources Engineering
- Environmental Engineering and Sustainability
- Structural Health and Monitoring
- Surveying and Geographical Information Systems
- Heating, Ventilation and Air Conditioning (HVAC)
- Transportation and Traffic
- Risk Analysis
- Safety and Security

Indexed by Scopus

To submit a proposal or request further information, please contact:
Pierpaolo Riva at Pierpaolo.Riva@springer.com (Europe and Americas) Wayne Hu at wayne.hu@springer.com (China)

Shenghong Chen

Advanced Computational Methods and Geomechanics

Springer

Shenghong Chen
School of Water Resources
and Hydropower Engineering
Wuhan University
Wuhan, China

ISSN 2366-259X ISSN 2366-2603 (electronic)
Springer Tracts in Civil Engineering
ISBN 978-981-19-7426-7 ISBN 978-981-19-7427-4 (eBook)
https://doi.org/10.1007/978-981-19-7427-4

This Springer imprint is published by the registered company Springer Nature Singapore Pte Ltd.
The registered company address is: 152 Beach Road, #21-01/04 Gateway East, Singapore 189721, Singapore

Shenghong Chen
School of Water Resources
and Hydropower Engineering
Wuhan University
Wuhan, China

ISSN 2366-259X ISSN 2366-2603 (electronic)
Springer Tracts in Civil Engineering
ISBN 978-981-19-7426-7 ISBN 978-981-19-7427-4 (eBook)
https://doi.org/10.1007/978-981-19-7427-4

This Springer imprint is published by the registered company Springer Nature Singapore Pte Ltd.
The registered company address is: 152 Beach Road, #21-01/04 Gateway East, Singapore 189721, Singapore

Shenghong Chen

Advanced Computational Methods and Geomechanics

Springer

To my intimate friends and my family for the time, support and love they gave to me.

Chen Shenghong

Preface

Since the 1980s, the author of this book has been continuously engaged in the education, research and consultant of "hydraulic engineering" that is partially overlapped with "geotechnical engineering". Most of my works are tightly related to the "computational methods (CMs)", "geomechanics" and "hydraulics" in three application levels: the routine design and construction guided by the local failure control and global stability countermeasures within the stipulated safety margin; the deeper study on particular difficulties arise from specific engineering settings (e.g. engineering geology, material availability, etc.); and the hazardous estimation when a hydraulic structure exceeding serviceability and collapse limit states. In different levels, the key issues concerned are rather different, too.

- Routine design and construction. The knowledge and roadmap for hydraulic structures (inclusive dams, spillways, tunnels, cut slopes, sluices, head and water measuring works, ship locks and lifts, fish ways, etc.) are demanded. The works are accomplished by the type selection and configuration setup; the local failure and overall stability calibrations and engineering countermeasures; the flood releasing arrangements and scouring protections; as well as the operation and maintenance schemes. We are faced with the challenges of learning and extending the knowledge related to hydraulic structures from traditions to new frontiers, for the purposes to design, construct and manage them more safely, durably, economically and environment friendly within serviceability and collapse limit states. Among a large variety of analysis tool kits, dominant ones such as the "limit equilibrium method", "gravity method", "trial load method", "hydraulics method", etc., are primarily stipulated in the design codes and handbooks since the 1950s. These contents have been published in the opening book entitled "Hydraulic Structures" in 2015.
- Deeper study towards particular difficulties. Attributable to the impetus from the engineering demands and the progress in computer sciences, computational methods or computational mechanics (CMs) are continuously exerting profound influences on the design and construction of giant hydraulic structures, particularly when we are encountered with the strength/stability difficulties arise from

complex engineering geological setting and material availability. Among a large variety of fundamental CMs, the "finite element method (FEM)", "finite difference method (FDM)", "boundary element method (BEM)", "block element analysis (BEA)", "composite element method (CEM)", etc., are prevalently exercised in hydraulic structures and, step-by-step, recommended in the design codes and handbooks since the 1990s. They are characterized as systematic and sophisticated toolkits that comprise widely accepted material models, routinely evaluated input parameters, well-performed and standardized computation codes with competent pre-process module, certain ability for coupling physical/mechanical phenomena, vivid result interpretation techniques, rigorous safety calibration criteria and, more important, sufficient accumulation of engineering experiences. These contents with respect to some typical fundamental CMs have been published in the follow-up book entitled *Computational Geomechanics and Hydraulic Structures* in 2018.

– Hazardous estimation after the structure exceeding serviceability and collapse limit states. In the construction and service periods, blamed for the incomplete understanding of engineering geology setting and sometimes the mistake of construction quality control, incidents/accidents are inevitable. For example, concrete cracking due to the inappropriate material design and temperature control scheme often emerges as common dam incidents, the reservoir bank instability due to the defects in geological investigation and parameter evaluation frequently brings about serious landslide accidents. In the former, the most concerns are the evaluation of the rest safety margin and further crack growth risk, whereas in the latter, the critical questions are the potential damage of landslide deposits and surge waves. These issues are basically beyond the reach of fundamental CMs; hence, the advanced CMs become the call of a new era because many of them are good at handling the post-failure phenomena after the loss of structure integrity, either the local cracking or overall instability accompanied with the phenomena of discontinuous/finite deformation, and the violent fluid–structure interaction accompanied with the strong distortion/breakage of water surfaces. This is the author's initial intention of dipping into the advanced CMs since the 2000s; now, the study experiences and gains are summarized in this sequel book.

This book gives the general overview with respect to the state of the arts of advanced CMs (Chap. 1) firstly, then the preparative knowledge of material properties, mathematics and mechanics are collectively briefed (Chaps. 2–4). In the main body of the book, the basic formulations and algorithms of two typical mesh-based "general finite element methods" (XFEM and NMM), two typical "block-based/particle-type methods" (DEM and DDA) and two typical "mesh-free methods" (EFGM and SPH) are elucidated (Chaps. 5–10). Finally, use is made of hybrid DEM-SPH where special attention is focused on the landslide and generated surge waves, the author shows how two advanced CMs may be practically coupled (Chap. 11). By this unique layout of the book, the author hopes our readers will be able to clarify the strong points and weak points of different formalisms, discretization schemes and basic algorithms. After being aware of that the differences in these advanced

CMs are not so dread as at the first glance, readers may possess certain degree of freedom in the selection of appropriate methods, in the reliability/applicability evaluation of computation outcomes, even in the method improvements, for their specific purposes.

As the author has been emphasized in the precursor book entitled *Computational Geomechanics and Hydraulic Structures*, a successful analysis for engineering structure should be accomplished by the team work of experts who are experienced in engineering practices and computation techniques and supported by the comprehensive knowledge in the fields of engineering geology, material science, construction or/and operation management and even society or human sciences. Only in this manner, analysis results may be confidently presented to support the decision makers in charge of engineering construction and operation. In addition, a successful analysis for engineering structure also strongly relies on the availability of systematic and sophisticated computation toolkits that are equipped with reliable constitutive models and input parameters, standardized codes, well-accepted safety and risk criteria, etc. From this point of view, however, the advanced CMs exemplified in this book are rather far from as competent as their fundamental counterparts. Anyway, they provide promising frameworks to handle the strong discontinuous/finite deformation in the dislocation motion of solid and the violent free surface break/distortion in the high-speed flow of fluid. Without the help of these advanced CMs, the basic features of post-failure (e.g. the dam breakage, landslide) phenomena will be mostly lost. Hence, another intention of this book is to encourage our readers, as successors in hydraulic and geotechnical engineering, to pay higher attention to enhance the practice level of these charming methods.

The works presented in this book are accomplished with the collaboration of my students. Among many, they are Dr. Wu Aiqing, Dr. Peng Hua, Dr. Li Wodong, Dr. Xu Qing, Dr. Xu Ying, Dr. Tan Hai, Dr. Wang Weimin, Dr. Fu Shaojun, Dr. Qiang Sheng, Dr. Li Yongming, Dr. He Ji, Dr. Li Xinxin, Dr. Dong Zhihong. They have helped either in equation deriving, algorithm programming or validation exemplifying.

Special commendation is due Dr. Xu Qing for drawing and improving many figures of this thick volume and providing very useful editorial comments. The author is grateful to Wuhan University for the peaceful time during my 35 service years. The author is also indebted to three successive deans of the School of Water Resources and Hydropower Engineering, Mrs. Cheng Lianzhen, Mr. Huang Jiesheng and Mr. Zhou Wei, for leniently alleviating my appurtenant workload during recent ten years.

Wuhan, Hubei, China
August 2022

Chen Shenghong

Contents

Chapter 1
Introduction

Abstract This chapter gives an overview of advanced "computational methods (CMs)" that is a branch of modern mathematical modeling. The history and state-of-the-arts of the advanced CMs are examined with special focus on those that are most promising in geotechnical and hydraulic engineering. This introductory chapter is concluded with comments and suggestions on the healthy development and successful application of advanced CMs.

1.1 General

Originated from ancient Greece, "mechanics" is a branch of science concerned with the behaviors (e.g. displacement, time, velocity, acceleration, etc.) of physical bodies subjected to actions (e.g. force, temperature, etc.). During the early modern period after the Renaissance, great scientists (e.g. Galileo, Newton, et al.) laid the laws for what is now known as "classical mechanics" that describes the motion of macroscopic objects with velocities significantly smaller than the speed of light. By the principles of classical mechanics, if the present state of an object is detected, we are able to predict how it has moved in the past (reversibility) and how it will move in the future (determinism).

Modern science and engineering activities need rational, reliable and efficient calculation, which give rise to "numerical analysis", a branch of mathematics concerning the study on algorithms that use numerical approximations (as opposed to symbolic manipulations) for mathematical problems. Thanks to the sustained efforts of outstanding pioneers in more than half a century, and thanks to the ever more powerful computers today, most of the classical mathematical problems arise from science and engineering can be numerically solved to high efficiency and accuracy. Today, the application of a large stock of computers and software packages has become a routine practice of scientists and engineers to provide supportive solutions within an acceptable time. Very often, it is hard to rigorously distinguish among the terms "computational methods", "numerical analysis" and "numerical methods".

Concerned with the use of "computational methods (CMs)" to study the phenomena governed by the principles of mechanics, the "computational mechanics"

S. Chen, *Advanced Computational Methods and Geomechanics*, Springer Tracts in Civil Engineering, https://doi.org/10.1007/978-981-19-7427-4_1

is a branch of the "computation science". In this book, the author uses the abbreviation "CMs" in parallel for both the terms "computational methods" and "computational mechanics", where there is no risk of misleading.

Important specializations in which CMs are widely exercised are the "computational fluid dynamics (CFD)" and "computational solid mechanics (CSM)". Typical examples in geotechnical and hydraulic engineering may be given by the concrete dam subject to high speed or/and high pressure water flow, and the landslide in the process of large/discontinuous deformation and slip motion. Scientists and engineers working in the field of CMs for the mechanical process of geotechnical and hydraulic structures are customarily guided by the following task list (Chen 2018):

- Pre-process for a real world entity representing the structure domain is carried out. This is normally accomplished by modeling a water body or/and solid entity to build correspondent discrete system that is composed of grids/meshes/cells or nodes/particles.
- Mathematical model simulating the mechanical phenomena is constructed. This usually involves the expression of the model in terms of "ordinary differential equations (ODEs)" and "partially differential equations (PDEs)" (strong form) or "integration equations (IEs)" (weak form). The constitutive equations (relations) for the matters (e.g. water, soil, rock, concrete) and inherent parameters are important factors dominating the success of computation.
- Mathematical model is converted to a form suitable for digital computation. Typically, this operation transfers PDEs or IEs into an algebraic equation set, and is called as "discretization" or "approximation" because it produces an approximated discrete model from the original PDEs model, although the latter is not exactly equal to the prototype physical model.
- Computer program is compiled (or revised) to solve the approximately discretized model using either direct methods (single step schemes to obtain the solution) or iterative methods (starting with a trial solution and arriving at the convergent solution by successively repeated refinement). The most widely exercised programming languages in the science and engineering communities are the "Fortran" and "C++". Although they all possess well known popularity, yet the latter is becoming more and more prevalent because of its built-in visualization functions. The proprietary language/environment "MATLAB" is also widely employed, especially for the purpose of rapid application and model verification.
- Depending on the nature of the problem, eligible machine ranging from PC to supercomputers or parallel computers, may be demanded to run the program for the solution within acceptable time.
- Post-process and visualization. In order to help practitioners to understand, debug, and analyze computation outcomes conveniently, as well as to add some fun to their works, it is extremely welcome to develop/use well performed post-process software for vividly presenting the spatial-time distribution of physical and mechanical fields.
- Safety calibration. For a geotechnical or hydraulic structure, the run of computation program provides physical and mechanical fields such as the temperature,

hydraulic potential and flow rate in the water body, as well as the displacement, deformation/stress and damage/plastic zone in the structure body concerned. These outcomes should be carefully and comprehensively assessed to assure the safety and economy of the engineering structure (Chen 2015, 2018).

1.2 History of the Computation

1.2.1 From Ancient Calculation to Modern Computation

The history of calculation may date back to the dawn of human civilization (Klein 1972) when there were only basic tools such as tablet or abacus. If we look at these ancient calculation rules as the pioneering works of modern CMs supported by computers in general meaning, the author of this book divided the history of CMs into 6 major periods according to the availability of mathematics, mechanics, and computation tools (Chen 2018), which is briefly summarized in Table 1.1.

1.2.2 Classification of Modern CMs

In the precedent books, the author has respectively presented the classical analysis methods for solid deformation/stress and fluid flow that for more than eight decades have been the principal design toolkits (Chen 2015), and the fundamental CMs supported by modern computers that for more than half a century have found their wide applications in the detailed study of hydraulic structures under the normal/extreme work situations featured by small/continuous deformation (Chen 2018).

Since around the 1990s, there has been growing interest in the new families of advanced CMs towards the simulation of complex problems in science and engineering. Attributable to the accumulated works over the last decades, today we have a wide spectrum of advanced CMs for geotechnical and hydraulic structures related to high speed fluid flow and discontinuous geomaterials beyond serviceability state or even collapse limit state (Belytschko et al. 1996; Jing 2003; Oñate et al. 2004; MacLaughlin and Doolin 2006; Bobet et al. 2009; Yazid et al. 2009; Liu and Liu 2010; Gary et al. 2014; Coetzee 2017; Pan et al. 2018). They have been categorized under diverse headings such as the "general finite element method (GFEM)", the "finite volume methods (FVM)", the "extended finite element method (XFEM)", the "numerical manifold method (NMM)", the "element free method (EFM)" or "meshfree method (MFM)" or "meshless method (MLM)", the "finite particle method or finite point method (FPM)", the "discrete element method or distinct element method (DEM)", the "discontinuous deformation analysis (DDA)", the "smoothed particle hydraulics (SPH)", etc.

Table 1.1 Major period of CMs

Period	Important events
3000 BC–300 BC	• A dawn period featured by arithmetic or elementary algebraic operations consisting of the study on numbers and basic operations between them • Complex calculations needed the assistance of tools such as the earliest mathematical writings on a Babylonian tablet "YBC 7289"(1830–1531 BC) and the first appearance of the Sumerian abacus
300 BC–1600AD	• A period initiated from the Hellenistic civilization of ancient Greece whose studies on mathematics overlapped with their philosophical and mystical beliefs • The milestone work was founded by the ancient Hindu-Arabic people who incorporated a calculation system with the use of zero (0) as a separate number
1600s–1800s	• Various higher performance calculating tools were created, including the slide rule invented around 1620–1630AD, the Schickard's calculator invented in 1623AD and the Pascal's calculator (Pascaline) invented in 1642AD • The solid mechanics was initiated by the Hooke's law in 1676 • The rapid progress in fluid mechanics began with da Vinci, Newton and Pascal, and was continued until Bernoulli who established mathematical fluid dynamics • In 1715, Bernoulli systematized the virtual work principle and made explicit the infinitesimal displacement to solve problems for both rigid bodies as well as fluids
1800s–1940s	• Fast development in classical mechanics and mathematics under the impetus of the fast development in machinery manufacture and civil engineering • Leading scientists had been involved in the studies and applications of the continuum mechanics and related theory of partial differential equations • Great mathematicians had been preoccupied by the issues of numerical analysis, and invented important algorithms with their big names like the Newton's method, Lagrange interpolation, Fourier series and transform, Lagrange multiplier method, numerical quadratures of Newton-Cote/Simpson's/Gaussian rules, Euler's method, iterative methods of Jacobi/Gauss–Seidel and Gaussian elimination method for the solution of linear algebraic equation sets • The technology of mechanical analog computing reached its summit by Hazen and Bush at MIT in 1927
1940s–1970s	• The appearance of digital electronic computer • The mathematical basis of approximation theory had been built encompassing classical questions of interpolation, approximation, series expansion, harmonic analysis, semi classical questions of polynomial and rational minimax approximation, as well as much newer topics including splines, radial basis functions, and wavelets • The principle of modern computer was proposed by Turing (1937) concerning a simple device called as "universal computing machine"—now known as "universal Turing machine" • Most of algorithms that make CMs powerful were invented since 1950, of which the fundamental versions of the finite difference method (FDM), the finite element method (FEM), and the boundary element method (BEM), are without doubt the zenith

(continued)

Table 1.1 (continued)

Period	Important events
1970s–Today	• The appearance of high performance computers and expansions of fundamental FDM, FEM and BEM for various field problems, of which FEM has held dominant position in the design of super engineering structures (e.g. high dams) • Various new and advanced CMs including "general finite element methods (GFEMs)", "mesh-free methods (MFMs)", "particle-based methods (PBM)" for the issues of high-speed fluid flow with strong free surface distortion, large/discontinuous solid deformation with strong change of structure configuration, have been booming propelled by engineering and other industry demands • Computer industry has been developing at a speed normally predicted by the "Moore's law" that states a doubling law every year in the number of components per integrated circuit. Computers have to be designed in a multi-core manner to keep up with the Moore's law. There is the most remarkable hotspot to make computers out of many promising new types of technology, such as optical computers, DNA computers, neural computers, and quantum computers

Most of advanced CMs possesses one common feature: they do not rely on traditional mesh (grid) concepts, which enable them to handle the issues of fluid flow with high speed and violent free surface distortion/breakage, as well as of solid motion/deformation with strong discontinuous deformation and dislocation. Insofar, there are a number of classification methods for these existing advanced CMs, for example:

- Level one. From the viewpoint of spatial descriptions, they may be classified as Eulerian and Lagrangian methods (formalisms, descriptions).
- Level two. From the viewpoint of their spatial discretization schemes, they may be classified as mesh-based, mesh-free (meshless, element free), and particle-based methods.
- Level three. From the viewpoint of material assumptions in macroscale, they may be further illustrated as intrinsically continuous and intrinsically discontinuous methods.

It is necessary to emphasize that a method flagged by "intrinsically continuous" means in the formulation of its governing equations, the matter is postulated as a continuum. In many cases, however, these governing equations may be extended to handle discontinuum via either equivalent constitutive relation containing fractures, or special elements, or local enrichments. But such extended operations may lead to complexity/difficulty in computation algorithm and programming. On the contrary, a method flagged by "intrinsically discontinuous" means in the formulation of its governing equations, the matter is directly postulated as a discontinuum. These governing equations also may be extended to handle continuum via the input of artificial parameters of virtual structural planes/cracks/boundaries to reduce their influences as minor as possible. But such operations may lead to ill-conditioning of the equation set correspondent to the discretized system, or may give rise to perceivable errors due to improper artificial parameters.

In Fig. 1.1, a simplified classification-tree for the representative advanced CMs is constructed. It is notable that although each member of CMs may be put simultaneously in several (not always) classes mentioned above, yet it normally belongs to the most suitable one only. Where a method is applicable to both the solid and fluid, its position in the classification-tree is assigned by the solid problem only.

In Table 1.2, the representative CMs are collected in the sequence of history. We dub a method with prefix "general" to highlight that it remarkably extends the "original" or "fundamental" counterpart whereas the basic framework and features are kept. It is notable that in Table 1.2, a method is referred to the literature that is one of the earliest publications in a form of paper or Ph.D. thesis. It is also emphasized that most methods were developed by a number of scientists in a certain period, some

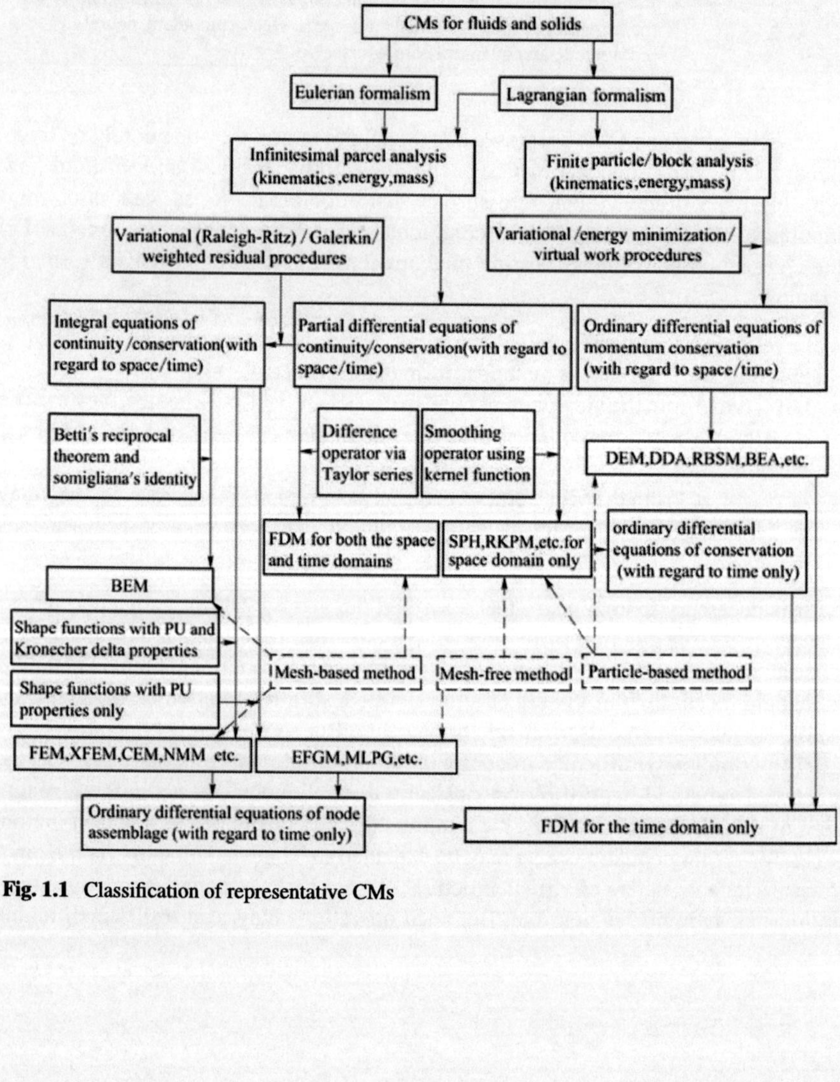

Fig. 1.1 Classification of representative CMs

even earlier than the representative author/creator, but the official creator usually summarized the previous works and gave the formal nomination of the method.

1.3 State-of-the-Arts of Fundamental and Advanced CMs

For a geotechnical and hydraulic structure (e.g. dam, landslide, tunnel, etc.), the simulation of discontinuities (e.g. structural planes and concrete cracks) in geomaterials is particularly laborious. To achieve a stipulated safety degree, it is primarily demanded that the effect of discontinuities should be defined in a quantitative manner (Chen 2018).

Modelling of geomaterials can be categorized as equivalent continuum approaches and discontinuum approaches according to their treatment w.r.t. displacement compatibility. The preference for an equivalent continuum or discontinuum approach depends on the scale of the problem and the complexity of the discontinuity system. The former demonstrates remarkable advantage of its higher efficiency in handle large-scale problems accompanied with complex fracture networks, whereas the latter is able to simulate fractures and fragmentation processes more straightforwardly and in more detail.

To provide a sufficient background for our readers, the commonly exercised fundamental and advanced CMs promising in geomechanics and hydraulics will be briefly reviewed hereinafter. In doing so, a three-class system comprising mesh-based, mesh-free and particle-based will be helpful. However, it is worth to mention that such a classification catalog is not intended to be absolute, since the boundary of many advanced CMs is rather vague.

1.3.1 Mesh-Based Methods

"Mesh-based methods (MBMs)" are very suitable for an object (solid body or water body) that its properties can be looked at as smoothly varying fields in a continuum. The basic strategy is to discretize the object into smaller sub-domains that share nodes and edges/surfaces pre-defined within the model. These nodes, edges and surfaces construct a mesh (grid) to provide the framework within which the computation for field variables and their derivatives are undertaken. The computational procedure normally runs in a series of time-marching steps. At each step, the mesh is deflected by pre-defined loads and/or displacements at the model boundaries, this change will propagate via adjacent nodes and elements until throughout the whole mesh governed by a system of mathematical equations to maintain the equilibrium condition or conservation law.

Table 1.2 Typical modern CMs

Method	Classification	Features
Fundamental finite difference method (FDM) (Thom 1933)	Eulerian formalism/mesh-based/intrinsically continuous	• It is the oldest member of CMs to solve PDEs using the approximation of nodal derivatives • It directly approximates the PDEs (strong form) of the problem by transforming them into a set of algebraic equations with unknown variables at grid points
Fundamental finite element method (FEM) (Clough 1960)	Lagrangian formalism/mesh-based/intrinsically continuous	• It solves integral/variational equations using the approximation of nodal variables by the shape functions with the properties of PU and Kronecker Delta • It is generally better suited for non-linear problems with irregular geometries and complex boundary conditions
Fundamental boundary element method (BEM) (Jaswon and Ponter 1963)	Lagrangian formalism/mesh-based/intrinsically continuous	• It solves boundary integral equations applicable to the problems for which the Green's functions can be established • It introduces isoparametric elements on boundaries using shape functions in the same fashion as that in FEM
Finite volume method (FVM) (Varga 1962)	Eulerian formalism/mesh-based/intrinsically continuous	• It is formulated with basic variables at the centers of cells (elements) or at the nodes (grid points) of unstructured grid
Generalized boundary element method (GBEM) (Rizzo and Shippy 1968)	Lagrangian formalism/mesh-based/intrinsically continuous	• It handles layered media in which the material property is constant for each material zone • Elasto-plastic material may be taken into account
Distinct element method (DEM) (Cundall 1971)	Lagrangian formalism/block-based/intrinsically discontinuous	• It uses an explicit integration scheme in the time-domain to solve the equations of motion for rigid discrete bodies • It simulates the contact deformation on discontinuities

(continued)

Table 1.2 (continued)

Method	Classification	Features
Generalized finite difference method (GFDM) (Jensen 1972)	Lagrangian formalism/mesh-based and mesh-free/intrinsically continuous	• It combines the Taylor series expansion and the moving-least squares (MLS) approximation • It uses explicit formulae for the partial derivatives of unknown functions • It may be implemented as a mesh-based method with irregular grid to form six-or eight-node stars, or as a mesh-free method
Smoothed particle hydrodynamics (SPH) (Gingold and Monaghan 1977; Lucy 1977)	Lagrangian formalism/particle-based/intrinsically continuous	• It divides the fluid into a set of discrete particles, each of them possesses a spatial length known as the "smoothing length" over which particle properties are "smoothed" by the kernel function • It is able to be extended to solid mechanics • It is the first mesh-free particle method for continuum
Rigid body-spring element method (RBSM) (Kawai 1978)	Lagrangian formalism/block-based/intrinsically discontinuous	• The rock block is looked as a rigid body • It put some springs between blocks to describe the shear and tensile behaviors of structural planes
General finite difference method/meshless finite difference method (GFDM/MFDM) (Liszka and Orkisz 1980)	Eulerian formalism/mesh-free/intrinsically continuous	• It uses general or irregular clouds of points • It uses the moving least squares (MLS) approximation to obtain the explicit difference formulae in PDEs
Discontinuous deformation analysis (DDA) (Shi and Goodman 1985)	Lagrangian formalism/block-based/intrinsically discontinuous	• It postulates the point-to-point or point-to-edge contact between blocks • It adopts linear displacement interpolation within a block • It uses the principle of minimum potential energy to get a set of equations to solve the deformation/motion process of rock block system

(continued)

Table 1.2 (continued)

Method	Classification	Features
Fast Lagrangian analysis of continua (FLAC) (Marti and Cundall 1982)	Lagrangian formalism/mesh-based/intrinsically continuous	• It is a kind of GFDM • It uses an explicit integration scheme in the time-domain • It considers large strains (geometric nonlinearities)
Block element analysis (BEA) (Chen 1987)	Lagrangian formalism/block-based/intrinsically discontinuous	• It postulates face-to-face contact between blocks only • It uses the principle of virtual work to get a set of equations to solve the deformation/motion process of rock block system
Generalized FEM (GFEM) (Wan 1990)	Lagrangian formalism/mesh-based/intrinsically continuous	• It globally enriches the approximation space in which shape functions are determined by the PU functions • It locally enriches the local function spaces for singularities • The treatment of discontinuity is at element level • It inspires the research of variant methods (e.g. XFEM, PUM, hp-Cloud Method, CEM, etc.) as its family members
Numerical manifold method (NMM) (Shi 1991)	Lagrangian formalism/mesh-based/intrinsically discontinuous	• It uses a node-based covering star for the trial functions • A covering star can be a set of finite elements or can be generally shaped • It is similar to the GFEM except for the treatment of discontinuities and discrete blocks by the truncated discontinuous shape functions in a unified form

(continued)

Table 1.2 (continued)

Method	Classification	Features
Diffusive element method (DEM) (Nayroles et al. 1992)	Lagrangian formalism/mesh-free/intrinsically continuous	• It is the first mesh-free method based on the Galerkin formulation (weak form) • It uses the weight technique among a set of scattered points without rigorous topological connection • It uses the MLS technique to generate global smooth approximation but does not pass the patch test and fails on consistency requirements • The derivatives of unknown function are evaluated only approximately • The Dirichlet boundary conditions are not accurately enforced • A background mesh is required for numerical quadrature
Element-free Galerkin method (EFGM) (Belytschko et al. 1994)	Lagrangian formalism/mesh-free/intrinsically continuous	• It is an extension of diffusive element method using the technique of MLS by a data fitting algorithm to construct trial functions • It utilizes Lagrange multipliers to enforce the Dirichlet boundary conditions
Reproducing kernel particle method (RKPM) (Liu et al. 1995)	Lagrangian formalism/mesh-free/intrinsically continuous	• It uses reproducing kernel particle interpolant as an advanced version of the MLS interpolant • It is an enhanced SPH
H-p clouds (Duarte and Oden 1995)	Lagrangian formalism/mesh-free/intrinsically continuous	• It multiplies a PU by polynomials or the other class of functions to produce h-p clouds that retain good properties of MLS function • It implements h-p adaptivity with the same remarkable features of h-p FEM but without the burden of mesh

(continued)

Table 1.2 (continued)

Method	Classification	Features
Partition of unity method (PUM) (Melenk and Babuška 1996)	Lagrangian formalism/mesh-free/intrinsically continuous	• It can be regarded as a generalization of the fundamental h-p FEM if the local approximation spaces are polynomial spaces • It provides a general approximation technique in some mesh-free methods: the approximation properties of the trial space are essentially given by the local approximation properties of polynomials
Finite point method (FPM) (Oñate et al. 1996)	Lagrangian formalism/mesh-free/intrinsically continuous	It is a point collocation method
Galerkin boundary element method (GBEM) (Nicolazzi et al. 1996)	Lagrangian formalism/mesh-free/intrinsically continuous	• A scattered set of points is used to approximate the solution • The MLS technique is employed to build a PU on the boundary that is further used to construct the trial functions for Galerkin approximations • These Galerkin approximations are then applied to the boundary element formulations
Boundary node method (BNM) (Mukherjee and Mukherjee 1997)	Lagrangian formalism/mesh-free/intrinsically continuous	• It is a hybrid between the Boundary Integral Equation and generalized MLS for interpolation functions • It retains the mesh-free attribute of the EFGM and the dimensionality advantage of the BEM
The meshless local Petrov–Galerkin method (MLPG) (Atluri and Zhu 1998)	Lagrangian formalism/mesh-free/intrinsically continuous	• It adopts test functions (Heaviside, Dirac delta) from different approximation spaces to offer high flexibility to deal with different boundary value problems • The approximations are based on the radial basis functions or MLS • It works with a local weak form instead of a global weak form • Numerical quadrature is performed over local subdomains

(continued)

Table 1.2 (continued)

Method	Classification	Features
Local boundary integral equation (LBIE) (Zhu et al. 1998)	Lagrangian formalism/mesh-free/intrinsically continuous	• It is a mesh-free method based on the local boundary integral equations and the MLS approximation • The companion solution is employed to simplify the formulation and to reduce the computation effort
Natural element method (NEM) (Sukumar et al. 1998)	Lagrangian formalism/mesh-free/intrinsically continuous	• The main difference between NEM and the MFEM is that in the former the non-Sibsonean shape functions are applied to the whole domain, whereas in the latter they are applied to each polyhedral element only
Extended finite element method (XFEM) (Belytschko and Black 1999)	Lagrangian formalism/mesh-based/intrinsically continuous	• It is a numerical technique based on the GFEM and the PUM • It enriches the approximation space to naturally reproduce weak/strong discontinuities
Finite sphere method (FSM) (De and Bathe 2001)	Lagrangian formalism/mesh-free/intrinsically continuous	• It is a truly mesh-free method • Three-dimensional domain is completely covered by a set of spheres whose centers must be within the domain and no sphere can be completely included in any other sphere • Discretization depends only on the position vectors and radii of spheres • The interpolation scheme is based on the PU

(continued)

Table 1.2 (continued)

Method	Classification	Features
General boundary element method (GBEM) (Chen et al. 2001)	Lagrangian formalism/mesh-based/intrinsically continuous	• Boundary integral equations are derived in which the traction kernel includes the full nonhomogeneous elasticity tensor • Domain integral involves the first order derivatives of the displacement kernel • It uses radial basis functions to approximate the domain integrand
Boundary particle method (BPM) (Chen 2002)	Lagrangian formalism/mesh-free/intrinsically continuous	• It applies either high-order nonsingular general solutions or singular fundamental solutions as the radial basis function • It does not require any inner nodes for inhomogeneous problems • It is a mesh-free, integration-free, and symmetric method
Radial point interpolation meshless method (Radial PIM) (Wang and Liu 2002)	Lagrangian formalism/mesh-free/intrinsically continuous	• It is a mesh-free BEM that combines the fundamental solutions and the dual reciprocity method • Using multi-quadric or Gaussian as radial basis functions, it overcomes the possible singularity associated with polynomial basis
Meshless finite element method (MFEM) (Idelsohn et al. 2003)	Lagrangian formalism/mesh-free/intrinsically continuous	• The mesh-free ideas are generalized to take into account the finite element approximations • It uses the extended Delaunay tessellation to build a mesh with the elements of different polygons
Boundary cloud method (BCM) (Li and Aluru 2003)	Lagrangian formalism/mesh-free/intrinsically continuous	• It uses a varying polynomial basis to eliminate the singularity encountered in the BNM with the MLS technique when all the points lie along a straight line • When a weighted function is centered at a point, a cloud for that point is defined

(continued)

Table 1.2 (continued)

Method	Classification	Features
Composite element method (CEM) (Chen et al. 2004)	Lagrangian formalism/mesh-based/intrinsically discontinuous	• It employs finite elements to cover the segments of discontinuities, bolts, draining holes and cooling pipes embedded in the structure domain • It is able to conveniently and explicitly describe a large amount of above meso-scale components
Gradient smoothing method (GSM) (Liu et al. 2008)	Eulerian and Lagrangian formalism/mesh-based/intrinsically continuous	• It uses gradient smoothing operation exclusively in nested fashion to develop the first-and second-order derivative approximations for a node of interest • Using the approximated derivatives, a collocation procedure is applied to governing PDEs at each node • It can be easily applied to irregular meshes
Smoothed point interpolation method (S-PIM) (Liu 2010)	Lagrangian formalism/mesh-free/intrinsically continuous	• Shape functions are constructed with the point interpolation method (PIM) to guarantee the Kronecker delta property • They belong to the G-space with the first derivative of square integrable rather than to the standard Hilbert space
Smoothed finite element method (S-FEM) (Liu et al. 2010)	Lagrangian formalism/mesh-based/intrinsically continuous	• It is a particular type of GFEM through the carefully designed combination of fundamental FEM and some techniques from mesh-free methods • It is softer than the fundamental FEM counterparts with identical mesh

1. **FDMs**

(1) **Fundamental FDM**

The fundamental "finite difference method (FDM)" is the oldest member in the family of mesh-based methods. It transforms the original "partial differential equations (PDEs)" into a set of algebraic equations w.r.t. unknown variables at grid points. It was firstly proposed by Thom in the 1920s under the title of "the method of square" to solve nonlinear hydrodynamic equations. As with the "finite element method (FEM)", the solution of the algebraic equation set is obtained after imposing initial and boundary conditions (Forsythe and Wasow 1960). FDM is excellent for static problems of heat flow and temperature modeling, and have some strong adherents for lithospheric-scale viscous models (Gerya 2010). It is prevalent in CFD, too.

(2) **FVM**

The fundamental FDM with regular grid suffers from the restraint, most of all, due to its inflexibility in dealing with fractures, complex boundaries and material heterogeneity. These make the fundamental FDM unfavorable for the solution of structural problems related to geomaterials. Today however, significant progress has been made in the FDM with irregular meshes (e.g. triangular or Voronoi), which leads to the so-called "finite volume method (FVM)".

As early as in 1953, MacNeal exploited the idea to formulate a FDM algorithm on a distorted grid (MacNeal 1953). The first apperance of FVM is in the famous book by Varga (1962). However, The term FVM was coined only in the 1970s in the lieratures of, for example, McDonald (1971), Rizzi and Inouye (1973).

In FVM, Voronoi polygons grow from points to fill the space, as opposed to tessellations where the polygons are formed by lines cutting the plane or by building up a mosaic from pre-existing polygons (Le Veque 2002). FVM can be formulated with basic field variables (e.g. displacement, flow velocity) at the cell centers (element centers) or at the nodes (grid points) of an unstructured grid. One of the most prevalent FVM codes for the "computational fluid dynamics (CFD)" is FLUENT (ANSYS Inc 2011).

FVM algorithm for the stress analysis can be traced back to the early work of Wilkins (1963) who used a vertex-centered scheme with a quadrilateral grid (Patankar 1980). Being possible to consider different material properties in different cells, FVM is therefore as flexible as FEM in handling material heterogeneity, grid generation, and treatment of boundary conditions. Today, FVM has been employed to study the mechanism of fracturing processes, such as the shear-band formation/evolution in rock and the fault system formation/propagation in the Earth's crust as a result of tectonic movement (Fang 2001). However, explicit representation of discontinuities is still not so easy in FVM because it demands the function continuity between the neighborhood grid points. In addition, it is not able to accommodate special "joint elements" as with FEM.

FVM possesses many similarities with FEM and is hence regarded as a bridge between FDM and FEM. Some authors liked to view FVM as being part of the Petrov–Galerkin family of FEM, whereas Heinrich (1987) regarded FVM as a generalization of FDMs with unstructured grids.

(3) FLAC

A commonly used FDM code for geomaterials is the "Fast Lagrangian Analysis of Continua (FLAC)" firstly released by ITASCA (1986). It utilizes an explicit integration scheme of finite volume to capture the nonlinear behaviors such as the yield/failure over large areas, or even total collapse (Marti and Cundall 1982).

FLAC employs dynamic relaxation method in the solution for time domain. At a certain step, the unbalance force at node is used to solve the motion equation for nodal acceleration according to the Newton's second law. This acceleration is further integrated to get the velocity and displacement. Afterwards, the strain and stress increments are calculated, and the correspondent unbalance force for the next iterative time-marching step is updated. This iterative recursion continues until the static equilibrium is reached or the structural collapse occurs.

An important advantage of explicit integration scheme is that it provides a very simple solution algorithm. The major drawback is the restriction of small time-marching step increment from the numerical stability, which results in the large amount of calculation time-marching steps. FLAC overcomes this drawback to some extent by utilizing automatic inertia scaling and automatic damping. Examples of its successful applications in solving structural problems can be found in many literatures (Strayer and Hudleston 1997; Detournay and Hart 1999; Strayer et al. 2004; Sheldon et al. 2006; Gerya 2010).

It is evident that the creation of FLAC was benefitted from the ideas of FVM and FEM.

(4) GSM

The "gradient smoothing method (GSM)" was initially developed for the field of CFD (Liu et al. 2008). It is similar to FVM, but it employs gradient smoothing operator exclusively in nested fashions to get the gradient approximation of field variables as well as their first-and-second order derivatives. GSM works well with unstructured grids and is very stable. It is equivalent to the central difference scheme in FDM when nodes are uniformly distributed (Mao and Liu 2018; Mao et al. 2019; Mao et al. 2019).

In addition to CDF problems, GSM has been extended to handle solid mechanics problems, too (Liu et al. 2008; Zhang et al. 2008), and fluid–structure interaction problems (Wang et al. 2013) as well. Recently, inspired by the good performance of GSM in Eulerian framework, the GSM gradient operator has been extended in Lagrangian framework to produce L-GSM. As a Lagrangian particle method, it may be used to solve solid (Mao and Liu 2018) and fluid (Mao et al. 2019) flows governed by Navier–Stokes equations. The paramount advantage of L-GSM is its excellent adaptability in simulating extremely large deformation problems. In addition, L-GSM is free from the trouble arise from tensile instability.

(5) Generalized FDM

The idea of the "generalized finite difference method (GFDM)" was stemmed from the use of irregular grids with six-point stars in the fundamental FDM (Jensen 1972). But in the early stage it was trapped by the frequent singularity or ill-conditioning of the star. Later on in 1980s, Liszka and Orkisz (Liszka 1984; Liszka and Orkisz 1980) put forward the usage of eight-node stars and weight functions to obtain finite difference formulae with irregular grids using "moving least squares (MLS)" interpolation. Benito and his co-workers (Benito et al. 2001, 2003; Ureña et al. 2005) made great contribution to the GFDM for second-order PDEs, including the study on the key parameters involving the node number of star, the weighting function, the selection criterion of nodes, and the stability condition. An h-adaptive algorithm was proposed by these pioneers, too. After years of efforts, the numerical stability of the advanced version has been greatly improved, which enables GFDM to be successfully applied to solve various types of PDEs arise from mechanical problems, such as (incomplete list):

- 2-D viscous incompressible Navier–Stokes equations around moving solid bodies (Chew et al. 2006);
- Advection–diffusion equations for fluid dynamics (Prieto et al. 2011);
- Third-and fourth-order PDEs for thin and thick elastic plates (Ureña et al. 2012);
- Vibration PDEs for the dynamic analysis of beams and plates (Gavete et al. 2013);
- Seismic wave propagation equations (Salete et al. 2017);
- Non-linear elliptic PDEs related with heat transfer, acoustics or diffusion (Gavete et al. 2017);
- Problems with cracks in general anisotropic materials (Lei et al. 2019).

GFDM also can be put in the other category of mesh-free methods.

2. **FEMs**

(1) **Fundamental FEM**

The fundamental "finite element method (FEM)", in its practical application often known as the "finite element analysis (FEA)", is a numerical method for finding the approximate solutions of "integral equations (IEs)" or PDEs (less often) by the piecewise quadrature technique (Zienkiewicz and Cheung 1967; Desai and Abel 1972; Gallagher 1975; Roy et al. 1976). It uses a similar meshing strategy of fundamental FDM for the entire domain, although the mesh in the former is not so rigorously regular. The primary distinction between these two methods is that FEM normally solves equivalent weighted IEs, or the weak form of problem using piecewise quadrature techniques, whereas FDM directly approximates PDEs, or the strong form of problem using finite difference techniques.

In structural mechanics, the fundamental FEM is customarily based on a variational principle. For the discretization, the continuous unknown variables are approximated by "trial functions" that are the linear combination of "shape functions" identical to the "test functions". The resulting equation is then integrated element by element over the whole domain concerned.

The distinctive history of FEM is its powerful impetus from structural engineers who endeavored to the practical solution techniques, in addition to the theoretical foundations laid by the scientists in mathematics and mechanics w.r.t. the existence, stability, and uniqueness of the solution and error analysis. In FEM, it is often to separate the time derivative that will be treated in a finite difference manner, and to discretize only the space derivatives with finite elements.

Today, there are many well-developed and verified academic and commercial finite element codes with high performances in terms of computing power and material complexities.

(2) **SEM**

The "spectral element method (SEM)" was introduced by Patera (1984). By expanding the solution in trigonometric series, a notable advantage of the SEM is its high order. This approach relies on the fact that the trigonometric polynomial set, usually composed of orthogonal "Chebyshev polynomials" or high order "Legendre polynomials" over non-uniformly spaced nodes, is an orthonormal basis. The computation error is reduced exponentially as the upgrade of polynomial order, hence a fast convergence to the exact solution may be achieved with fewer "degree of freedoms (DOFs)" in comparison with the fundamental FEM.

SEM uses a tensor-product space spanned by nodal basis functions that are associated with "Gauss–Lobatto points". In contrast, the p-refinement FEM spans a space of high order polynomials by nodeless basis functions (Chen 2018).

(3) **Generalized FEM**

In an attempt to develop general algorithms to overcome the deficits of traditional joint elements, the "generalized FEM (GFEM)" with discontinuous shape functions for the crack onset and growth through bifurcation theory, was proposed by Texas School (Wan 1990; Melenk 1995; Duarte and Oden 1996; Melenk and Babuška 1996). The first work in the GFEM involved the global enrichments of approximation space. The generalized finite element shape functions are determined by the enrichment of "partition of unity (PU)" functions.

The local enrichment for the singularities at sharp corners was also established by Duarte, Babuška and Oden (2000). It uses local function spaces to account for the available information on the unknown solution, in this way to guarantee a good local approximation. A PU technique is employed to "bond" these spaces together and to form approximating sub-spaces. Such a strategy creates the conforming approximations which are improved by a nodal enrichment scheme. The treatment of crack is at the element level, namely, it is defined by a distance function so that its representation demands nodal function values solely by additional DOFs in the trial functions—a kind of jump functions along the crack-tip. Crack growth is simulated using the level set technique hence no pre-defined joint elements are needed. It has been validated in the problem with domains entailed by complicated boundaries (Belytschko and Black 1999; Belytschko et al. 2001; Stolarska et al. 2001; Strouboulis et al. 2000; Strouboulis et al. 2000, 2001; Duarte et al. 2007).

GFEM performs well with both structured and unstructured finite element meshes. Structured meshes are appealing for many studies in the material science at test sample level, where the interest is directed to determine the material properties. Unstructured meshes, on the other hand, tend to be widely exercised for the analysis of engineering structures at the prototype level since it is often desirable to conform the meshes to the external boundaries, although some methods under the development are able to tackle complicated geometries with structured meshes (Belytschko et al. 2009).

GFEM inspired a generation of researchers to develop variant methods as its family members, such as the "extended finite element method (XFEM)" (Belytschko and Black 1999) and later on as a hybrid approach with the mesh-free method (Amiri et al. 2014), the "partition of unity method (PUM)" employing a set of PU functions to guarantee inter element continuity (Melenk and Babuška 1996), the mesh-free formulation in the "hp-cloud method" (Liszka et al. 1996; Ma et al. 2015) and later on as a hybrid approach with the FEM, the "numerical manifold method (NMM)" (Shi 1991), the "composite element method (CEM)" (Chen 2018), etc.

(4) **XFEM**

The "extended finite element method (XFEM)" is a numerical technique based on GFEM and PUM (Belytschko and Black 1999; Moës et al. 1999). It also may be looked at as a variant of FEM combined with some mesh-free aspects. It is normally recognized that the development of XFEM is an outgrowth of the extensive research in mesh-free methods.

The key idea of XFEM is to locally enrich the standard FE shape functions with "Heaviside functions" in the cracked element allowing for discontinuous displacements, and with "asymptotic functions" at the tip element incorporating stress singularity. The computation mesh can be completely independent of the morphology of the domain cut by cracks without the need for special elements (Yazid et al. 2009). Hence it naturally overcomes the difficulty associated with the problems of, for example, the growth of crack, the evolution of dislocation, and the evolution of phase boundary (Sukumar et al. 2000).

(5) **S-FEM**

The "smoothed finite element method (S-FEM)" is a particular member of GFEM through a carefully designed combination of fundamental FEM and some techniques from the mesh-free methods (Liu and Nguyen-Thoi 2010). The essential purpose in S-FEM is to use a finite element mesh (in particular, triangular or tetrahedral) to construct a well performed algorithm, which is achieved by modifying the compatible strain field or by constructing a strain field with displacements only. Such a modification/construction can be undertaken within, but more often, beyond element (i.e. mesh-free concepts) to bring the information from adjacent elements. Naturally, the modified/constructed strain field have to satisfy certain conditions, and the standard Galerkin weak form needs to be modified accordingly to ensure the stability and convergence of solutions.

S-FEM has been proven to be softer than the fundamental FEM counterparts with identical mesh structure. It often produces more accurate solutions with higher convergence rate and is much less sensitive to mesh distortion (Zeng and Liu 2018).

(6) NMM

The "numerical manifold method (NMM)" (Shi 1991, 1992a, b; Chen et al. 1998) also may be looked at as the combination of FEM and DDA to naturally bridge over the continuum and discontinuum representations of rock-like materials: in handling discontinuities, the NMM takes over the advantages of DDA; whereas for the strain–stress analysis in continuous rock blocks, it is as powerful as FEM.

NMM is formulated using a cover system of node-based stars for constructing trial functions. A covering star can be a set of standard finite elements associated with the node or be generated using a least-square kernel technique in the general shaped domain. The quadrature is performed analytically using simplex integration technique.

In the application of NMM, meshes also can be independent of the morphology of the domain concerned. Therefore, mesh generation is greatly simplified and re-meshing is not demanded for the simulation of crack propagation (Salami and Banks 1996; Wang and Ge 1997; Wang et al. 1997; Amadei 1999; Ohnishi and Chen 1999).

It should be noted that although NMM is able to tackle cracks/structural planes without enriched functions in most cases, yet the concept of enriched functions in XFEM may be integrated into NMM to simulate complex crack growth problems (Ma et al. 2009), such an extended NMM furnished with enriched functions for crack-tip is more powerful. Since NMM is developed on the basis of DDA, it preserves all the characteristics of "discrete element methods (DEMs)" such as the kinematics constraints, contact detection, etc. (Ma et al. 2010). However, the DOFs in NMM is usually much larger than DDA because there is more than one manifold element for each rock block in the most cases. The benefit from this cost is that NMM provides more accurate displacement and stress fields in rock blocks. If every rock block is a manifold element with linear displacement field, then NMM will be degenerated exactly into DDA.

(7) CEM

From the viewpoint of practitioners in geotechnical and hydraulic engineering, main obstacle in the explicit simulation of a large amount of discontinuities (i.e. geology structural planes, concrete cracks, etc.) and stabilization countermeasures (e.g. bolts, drainage holes, etc.) lies in the pre-process to discretize the calculation domain. The "composite element method (CEM)" employs standard finite elements to cover the segments of discontinuities, bolts, cooling pipes and draining holes embedded in the engineering structure concerned. The shape functions may be hierarchical for the p-refinement to get desired computation accuracy. The sub-elements representing rock blocks as well as the segments of bolt/cooling pipe/draining hole are assigned with independent nodal variables. Across the interfaces (structural plane patches) of these sub-elements, a jump in the gradients of essential variables (derivatives) is emerged naturally without the help of Heaviside functions. These nodal variables may be solved from the governing equations similar to that of fundamental FEM observing virtual work or variational principle.

CEM may be regarded as a special member of GFEM in its initial work to explicitly simulate passive, fully-grouted rock bolts using simple meshes. The most remarkable feature of CEM is to explicitly locate the structural planes, bolts, cooling pipes and drainage holes within finite elements. In this manner much less restraint is imposed on the mesh generation for complicated engineering structures (Chen and Qiang 2004; Chen et al. 2004a, b, 2008a, b, 2010, 2011, 2012, 2015; Chen and Feng 2006; Chen and Shahrour 2008).

CEM may be incorporated in FEM software easily with p-version enrichment for shape functions. The extension of CEM to the problems with the stress singularity at crack-tip can be implemented with the help of enriched asymptotic functions at element level in a similar manner of XFEM.

3. **BEMs**

(1) **Fundamental BEM**

"Boundary integral equations (BIEs)" are classical tools for the solution of boundary value problems based on PDEs. The "boundary element method (BEM)" is a numerical method of solving BIEs applicable to problems for which the Green's functions can be established (Jaswon and Ponter 1963; Rizzo 1967; Cruse and Rizzo 1968; Brebbia et al. 1984; Thomas 1993). It differs from other continuum methods in that only certain "boundaries" surrounding or/and within the model are discretized.

BEM initially sought a weak solution at the global level through the numerical solution of an integral equation derived by Betti's reciprocal theorem and Somigliana's identity. The introduction of isoparametric elements on boundary surfaces in the same fashion as that in FEM (Lachat and Watson 1976; Watson 1979), greatly enhanced the BEM's applicability for structural problems. The fundamental BEM is more suitable for solving the cracking problem in homogeneous and linearly elastic bodies because it possesses remarkable advantages over the fundamental FEM or FDM that:

- Only the boundary of domain needs to be discretized. This reduces the model dimension by one and consequently allows for much simpler mesh generation and data input as well as smaller data storage.
- Exterior problems with unbounded domains are handled as easily as interior problems.
- In some cases, we are not focused on the interior solution but rather on the boundary values or its derivatives of the solution. These data can be obtained directly from the solution of BEM with high accuracy.
- Solutions inside the domain are continuously approximated with a rather high convergence rate and moreover, the same rate of convergence holds for all the solution derivatives of any order in the domain.

Main difficulties with the BEM are that:

- Boundary integral equations require the explicit knowledge of fundamental solution or Green's functions which are often problematic to obtain, as they are based on the solution of system equations subject to a singularity action.
- For a given boundary value problem there exist different BIEs, and to each of them there are several numerical approximations. Thus every BEM application requires to made choice.

- The fundamental theory of integral equations and their numerical solution concentrates on ordinary Fredholm integral equations of the second kind with regular kernel. However, the BIEs frequently encountered may be of the first kind whose kernels are singular in general.
- It is not as efficient as FEM in dealing with material heterogeneity because it cannot handle as many sub-domains (elements).
- It is not as efficient as FEM in simulating non-linear behaviors (e.g. plasticity, damage) because domain integrals often emerge in these problems.

The use of BEM in geomaterials has been popularized largely through the outstanding works since the 1990s (Thomas 1993; Savage and Cooke 2004; Maerten et al. 2006). Their works took the geological faults as key boundaries within a model in addition to the domain boundaries. This type of solution is well suited to modeling the elastic interactions between faults or other structural planes within the Earth's crust, as long as strains remain relatively small.

(2) **General BEM**

The advanced version of BEM for non-linear and nonhomogeneous materials is tagged by the prefix "generalized" in this book.

An early elasto-plastic BEM was proposed by Swedlow and Cruse (1971) but various difficulties such as the strongly singular domain integrals and the stability of system equations hindered its development. Approaches to get rid of (or regularize) domain integrals were proposed afterwards (Banerjee and Henry 1989), among many others, there are two particularly efficient ones to isolate these singularities (Dallner and Kuhn 1993) and to transform them into boundary integrals (Gao and Davies 2000). The latter possesses some significant advantages in terms of generality and simplicity. Near-singular integrals also require a careful treatment in order to entail computation efforts.

Equally crucial is the solution strategy for the system equation set. For example, direct solution algorithms (Telles and Brebbia 1979) operate on boundary variables only. The system equation set is thus of moderate size but solution convergence (by iteration) is poor and even can be failed. Implicit solution algorithms (Telles and Carrer 1991; Bonnet and Mukherjee 1996; Gao 1999) that are based on the consistent tangent operator, converge rapidly but the system equation set is formulated in terms of the domain strain (or stress) tensors which means that they are very large. Anyway, this problem can be overcome by the use of variable stiffness approach (Banerjee et al. 1989; Chopra and Dargush 1994) where the domain variables are replaced by scalar variables (plastic multipliers). Davies and Gao (2006) extended this idea so that the system equation set is cast in terms of plastic multipliers to yield a remarkably compact equation set, which is then solved by a robust and rapidly convergent Newton–Raphson algorithm.

Towards the issue of nonhomogeneous materials, Rizzo and Shippy (1968) published the first work on layered media in which the material property was modeled as constant in each zone, and the interface continuity was enforced through interior boundary constraints. For more general nonhomogeneous problems, Ghosh and

Mukherjee (1984) handled the uncoupled thermos-elasticity problem in which the temperature gradient term was also taken into account as a body force. A BEM formulation free of both singular domain integral and nearly-singular boundary integral was presented by Sladek et al. (1993), in which a domain discretization into triangular or quadrilateral elements is necessary to perform the domain quadrature.

The "dual reciprocity boundary element method (DRBEM)" was proposed to solve material nonhomogeneous problems (Chen et al. 2001), too, in which a "radial basis function (RBF)" is employed to approximate the domain integral and the particular solution satisfying Navier's equation is derived in a general form. This method converts the domain integral into a boundary integral, thus avoiding the need for a domain mesh.

For heat transfer problems, Kassab and Divo (1996) successfully constructed a fundamental solution and associated singular non-symmetrical force function. They derived a generalized BIE, in which the major part of the domain contribution from the material non-homogeneity is absorbed into boundary integral, and the domain integral appearing in the BIE is often negligible. They formulated an anti-divergence scheme to deal with this domain integral.

In solid mechanics, it is possible to develop a similar approach to deal with the domain integral arise from material non-homogeneities. For example, a formulation of GBEM for nonhomogeneous, isotropic and elastic media that simplifies the BIE formulation and the approximation of domain integral, is remarkable.

1.3.2 Mesh-Free Methods

Fundamental CMs such as FDM, BEM and FEM are defined on the mesh (grid) of data points (nodes) with certain connection requirements. In such a mesh, each node has a fixed number of pre-defined neighborhood nodes, and their connectivity with each other can be employed to define mathematical operators (e.g. derivatives). These operators are, in turn, used to construct the discretized equation set, normally in the form of algebraic (linear or non-linear), to simulate the physical or mechanical fields.

Where the matter being simulated moves around or changes its domain morphology such as in the issues of free surface water flow and solid crack/fragmentation, the connectivity of mesh nodes could be difficult to maintain. If the mesh becomes strongly tangled/distorted or even degenerated during the simulation process, the operators defined upon are no longer able to provide correct solutions. Although the domain may be re-meshed following the computation, yet this introduces additional error, because all the existing data points must be mapped onto a new set of data points (Hinton and Campbell 1974; Loubignac et al. 1977; Perić et al. 1996). This barrier hampers the desired solution accuracy significantly.

Researchers came to realize that the so-called "mesh-free methods (MFMs)" are needed to remove the difficulties arise from the "mesh-based methods (MBMs)". Although these MFMs still require a set of discrete nodes to be "sprinkled/scattered" through the computation domain, yet they do not require either pre-specified connectivity of the nodes nor locally regular topological structure, namely, these nodes can be deployed automatically using a variety of schemes. Hence MFMs possess higher potentiality leading to truly automatic analysis.

The first attempt to formulate the approximation for the solution of PDEs on "randomly" scattered nodal points appeared in the early 1970s (Jensen 1972; Boresi and Lynn 1974; Perrone and Kao 1975). The basic idea was latterly interpreted as a mesh-free approximation (Lancaster and Salkauskas 1981; Liszka 1984). Although a number of applications had been presented at that time (Kaczkowski and Tribillo 1975; Tribillo 1976; Liszka and Orkisz 1977; Szmelter and Kurowski 1977; Pavlin and Perrone 1979; Liszka and Orkisz 1980), the method did not gain much attention mainly because of the overwhelming success in FEM. In the 1990s, the mesh-free technology was rediscovered and a number of MFMs towards the numerical solution of PDEs were booming (Belytschko et al. 1994; Babuška and Melenk 1995, 1997) under the pressure of slow progress in automatic mesh generation. Most of these methods are based on the "moving least-square (MLS)" technique (Lancaster and Salkauskas 1981, 1986) and the "partition of unity (PU)" technique, in which the "element-free Galerkin method (EFGM)" (Belytschko et al. 1994) is a typical example.

MFMs are more and more appealing for the researchers in the BEM community, too. Among many others, it may be incompletely cited as the "boundary node method" (Mukherjee and Mukherjee 1997), "local boundary integral equation" (Zhu et al. 1998), "boundary particle method" (Chen 2002), "radial point interpolation meshless method" (Wang and Liu 2002; Li et al. 2002; Mai-Duy and Tran-Cong 2002), and "boundary cloud method" (Li and Aluru 2003).

It is notable that in fact, MFMs are difficult to precisely define because the condition of "non-existence of mesh" is rather diffused. Generally speaking, a mesh-free (element free, meshless) method does not require any prescribed connections between the nodes in a domain because the definition of shape function depends only on the nodal positions, hence it should be an algorithm based on the interaction of each node with all its neighborhood nodes (Idelsohn et al. 2003). As a consequence, properties such as kinetic energy are no more assigned to the elements but rather to the nodes. A significant advantage of MFMs is that it greatly simplifies the preprocess works. Nevertheless, this definition is not enough to justify the use of MFMs because they are actually useless without a fast evaluation of nodal connectivity although rigorous connectivity is no longer demanded.

1. **SPH**

As one of the earliest MFMs for the solution of Navier–Stokes equations (strong form), the "smoothed particle hydrodynamics (SPH)" was proposed in 1977 (Gingold and Monaghan 1977; Lucy 1977) initially for astrophysical problems. It divides the fluid into a set of discrete elements (cells) referred to as "particles". A spatial distance is defined as the "smoothing length" over which the properties of particles are smoothed by a kernel function (Liu and Liu 2003). This means that the physical properties of any particle can be get by weighted averaging the relevant properties of all the particles that lie within the smoothing length according to their distance from the particle of interest and their density as well. SPH is able to simulate hydrodynamic flows efficiently, and latterly it has been further extended to solid mechanics (Libersky and Petschek 1990; Libersky et al. 1993; Libersky et al. 1997).

The main advantage of SPH is its flexibility with regard to local distortion because the mesh dependence is naturally avoided. The main disadvantages of the original SPH version are inaccurate results near boundaries and tension instability (Swegle et al. 1995). Over the past decades, different corrections have been put forward (Belytschko et al. 1996; Johnson and Beissel 1996; Bonet and Lok 1999; Vila 1999; Belytschko et al. 2000; Bonet and Kulasegaram 2000; Dilts 1999, 2000; Johnson et al. 2000; Rabczuk et al. 2004; Xiao and Belytschko 2005).

Nowadays, SPH is prevalent for the mechanics of continuum media (Hoover 2006; Liu and Liu 2010; Violeau 2012) undergoing violent deformation accompanied with torsional and disrupted free surface, in addition to the much wider applications to astrophysics, ballistics, volcanology, and oceanography (Monaghan 2012).

2. **EFGM**

Unlike the SPH that is based on the strong form, other MFMs developed in the 1990s are mostly based on the weak form.

The pioneer work of MFMs based on the weak form was given by Nayroles et al. (1992). They introduced a promising method called as the "diffuse element method (DEM)". In their method, only a mesh of nodes and a boundary description is needed. Polynomials are employed to fit the nodal values by a least-squares approximation. The finite element mesh is totally unnecessary.

The "element-free Galerkin method (EFGM)" (Belytschko et al. 1994) made a further advancement by the refinement and modification of DEM. It is one of the first MFMs based on a global weak form. Being consistent and quite stable, although substantially more expensive, it is a significant achievement after the invention of SPH.

The remarkable features of EFGM are:

- "Moving least square (MLS)" approximation is employed for the construction of nodal shape functions.
- Galerkin weak form is employed to develop the discretized system equation.
- Background mesh of cells is required to carry out the numerical quadrature for the integration of relevant matrices (vectors).

EFGM intrinsically differs from FEM in that the spatial discretization is undertaken by scattering an array of nodes in the computation domain and on its boundaries. The interaction of nodes can be changed by the evolution of domain configuration (e.g. crack growth), which enables the method to dynamically treat crack growth in arbitrary directions. Even more promising is its potential in adaptive refinement without the help of finite element mesh, because in an interactive mode, a large number of nodes may be simply dropped into the specific portion of domain where it looks like higher accuracy.

A number of similar methods boomed in the following years. For example, the "reproducing kernel particle method (RKPM)" (Liu et al. 1995; Liu et al. 1995) was developed one year later. In contrast to RKPM and EFGM that use a so-called intrinsic basis, other methods made use of an extrinsic basis and the PU concept. The extrinsic basis was initially used to increase the approximation order similar to the p-refinement in hp-cloud method (Duflot 2006). Melenk and Babuška (1996) found the similarities between MFMs and FEM and developed the so-called "partition of unity finite element method (PUFEM)". It is very similar to hp-cloud method, the only difference lies in the shape functions that are based on Lagrange polynomials in the PUFEM, whereas the MLS approximation is employed in the general form of hp-cloud method.

The challenges with EFGM are:

- Removal of background cells for integration.
- Formulation of shape functions that possess Kronecker delta property.

EFGM still requires a background mesh of regular cells for the integration of matrices/vectors in discretized nodal system, hence it is not truly mesh-free. This is blamed for the use of weak form. Pursuit of truly MFMs therefore continues to be challenged by the question that: is it possible not to use weak form? The answer is "yes", a good example is provided by SPH.

Another major challenge to MFMs is the non-interpolatory characteristics of approximation, that is, for example, the MLS interpolants do not pass through the nodal data because the interpolation functions are not equal to the unity at nodes unless the weighting functions are singular (Dirac-delta). This complicates the imposition of essential boundary conditions (i.e. Dirichlet boundary) and the application of point loads. To confront the awkwardness, several approaches (Belytschko et al. 1994; Chu and Moran 1995; Krongauz and Belytschko 1995) prevalently employed are:

- Lagrange multipliers;
- Penalty factors;
- Perturbed Lagrangian;
- Hybrid with other mesh-based methods (e.g. FEM).

The disadvantage of the Lagrange multiplier method is that the discrete equation set for a linear self-adjoint PDE are not positive definite nor banded. However, since it is the most precise approach for imposing Dirichlet boundary conditions, it is quite useful for smaller problems where the cost for the solution of equations is unimportant.

To avoid the difficulties inherent in MFMs, a number of MFMs have been coupled successfully to FEM (Belytschko et al. 1995; Huerta and Fernández-Méndez 2000; Fernández-Méndez and Huerta 2002; Fernández-Méndez et al. 2003; Huerta et al. 2004; Fernández-Méndez and Huerta 2004), in which finite elements are placed around the boundary of domain and essential boundary conditions are applied to finite element nodes. For the practical purposes in geotechnical and hydraulic structures, such a hybrid EFGM-FEM appears to be the most promising. This is particularly useful when finite elements are used as a background mesh for the quadrature because a FE mesh is already available.

3. **PUM**

In mathematics, a "partition of unity (PU)" in topological space R^3 is a set of continuous functions ϕ from R^3 to the unit interval [0,1] such that for every point $\mathbf{x} \in R^3$, there is a neighbor of $\mathbf{x}$ where all, but a finite number, of the functions are $\varphi(\mathbf{x}) = 0$, and the sum of all the function values at $\mathbf{x}$ is a unity, i.e., $\sum_{\phi \in R} \phi(\mathbf{x}) = 1$. A PU is useful because it allows us to extend local constructions to the whole space. A PU can be employed to the integral operation (with respect to a volume) of the function defined over a manifold: first, the integral of the function whose support is entailed in a single coordinate patch of the manifold is defined; then the PU is employed to define the integral of an arbitrary function; finally the independence of the selected PU is validated. The PU is also important in the interpolation of data, the process of signal, and the study of spline functions.

The "partition of unity method (PUM)" (Melenk and Babuška 1996) allows for the construction of conforming trial spaces with the local properties determined by the user. The method was motivated by the increasing need for new techniques to solve the problem where the fundamental FEM fails or is prohibitively expensive. It starts from a variational formulation and followed by the design of trial spaces in view of the problem under consideration. The main characteristics of PUM are:

- It permits us to include a priori knowledge of PDEs in trial spaces. This is attributable to the local approximation properties of the space constructed by PUM.
- It allows us to easily construct trial spaces of desired regularity, in this manner the trial spaces used in the variational formulations of higher-order differential equations are available. This is attributable to the conformity of these spaces.

The relationship of PUM and the other CMs, particularly various MFMs, are summarized as follows:

- PUM can be looked at as a generalization of the fundamental h or/and p versions of FEM if the local approximation spaces are chosen to be polynomial spaces. It also provides a general approximation in some MFMs, for example, h-p clouds method can be viewed as a mixture of EFGM and PUM if the local approximations are constructed by polynomials.
- Shape functions of all MFMs are derived from some form of local polynomial fitting. Therefore, all these methods are very similar to PUM if the local approximation spaces in PUM are constructed by polynomials. In fact, the analysis of Oden and Duarte indicated that under such circumstances, PUM is more efficient than EFGM.
- Since the performance of MFMs rests on the local approximation of polynomials, it may be suspected that these methods are poorly performed whenever polynomials poorly approximate the solution, as is the case of fundamental FEM meeting the problems with highly oscillatory solutions. In contrast, PUM possesses much

higher flexibility in the choice of local approximation and allows us to locally approximate functions in a custom-tailored fashion to the problem concerned. Therefore, PUM is able to deal with many difficult problems that probably cannot be coped with well by the fundamental FEM or other advanced MFMs.

4. **h-p clouds method**

Duarte and Oden (1995, 1996) shown that MLS may constitute PU functions and developed a new member of MFMs called h-p clouds. The basic idea of the method is to multiply a PU by polynomials or other appropriate class of functions. The resulting functions, called h-p clouds, retain good properties of MLS such as the high regularity and compactness. The linear combinations of these h-p clouds can replicate polynomials of any degree. These properties allow for the implementation of an adaptive h-p clouds method with the same remarkable features of the h-p FEM but without the burden of meshing.

5. **MLPGM**

Governing equations in weighted residual form—a kind of weak forms, need to be integrated over the entire problem domain as one seamless and monolithic piece. However, if we turn to satisfy the governing equation point-by-point using the information of local sub-domains related to points, the integral can be implemented locally. As a member of MFMs that is based on the local weak form of the Petrov–Galerkin residual formulation, the "meshless local Petrov–Galerkin method (MLPGM)" (Atluri and Zhu 1998, 2000; Atluri and Shen 2002) is very typical. Another well-known member mainly applied in fluid mechanics is the "moving point method (MPM)" (Oñate et al. 1996; Oñate and Idelsohn 1998; Loehner et al. 2002).

The main difference in MLPGM and EFGM is that for the former the local weak form is generated on overlapping sub-domains rather than using the global weak form. It has been fine-tuned, improved, and extended over the years by Ouatouati and Johnson (1999), Gu and Liu (2001), and many others.

It is noted that the point-by-point procedure in MLPGM is also very similar to that of the "collocation method". But the former is more stable due to the combination of MLS shape functions, which is reproductive with the local Petrov–Galerkin approach.

In MLPGM, the problem domain is discretized by a set of arbitrarily distributed nodes, but a background grid is still required for numerical quadrature. The weighted residual method in the integral form is employed to create the discrete equation set. The implementation of quadrature is confined to very small local sub-domain in a local integral sense. This is made possible by the usage of the Petrov–Galerkin formulation, in which one has the freedom to select weight and trial functions independently.

The quadrature domain can be theoretically arbitrary, of course very simple regularly shaped sub-domains, such as circles and rectangles for 2-D problems, bricks and spheres for 3-D problems, are wise choices. MLPGM can be referred to as one of truly MFMs, or at least close to the ideal of MFMs.

It is emphasized that the shapes and dimensions of quadrature sub-domains do not have to be the same for all nodes. This is important when the local quadrature domain encounters the global boundary of the problem. In addition, as long as the support domain is compact, MLPGM will produce sparse matrices for the discrete system.

The major drawback of MLPGM is the asymmetry of the system matrices arise from Petrov–Galerkin formulation. Another drawback is that the local quadrature can be very tricky due to the complexity of the integrand produced by Petrov–Galerkin approach, especially for the sub-domains that intersect with the boundary of the problem domain.

6. **MFDM**

In addition to the SPH, there are a number of MFMs that operate on the strong forms with the help of finite differential representation (Taylor series) for the function.

The "meshless finite difference method (MFDM)" is evolved from the fundamental FDM. It can be applied over general or irregular clouds surrounding scattered points (Liszka and Orkisz 1980). It was initially nominated as "general finite difference method (GFDM)", the author of this book gives this method a new abbreviated term "MFDM" to prevent it from the confusion with the other GFDMs featured by the irregular grids with stars.

The basic idea of MFDM is to obtain explicit difference formulae by the MLS approximation (Benito et al. 2001, 2003, 2007, 2008, 2009; Gavete et al. 2003; Ureña et al. 2011).

However, this kind of methods is generally not very stable, especially with an arbitrarily distributed node system. In addition, the computation results are less accurate, especially when the derivatives of field variables (strain, hydraulic potential gradient, etc.) are of interest. Efforts are still ongoing to improve the numerical stability, of which a promising direction is the usage of radial functions.

7. **MFEM**

Hybrid/coupled methods are particularly attractive where they exploit the advantages of both the mesh-free methods and mesh-based methods (Idelsohn et al. 2003, 2004; Huerta et al. 2004; Hao and Liu 2006; Rabczuk et al. 2006): mesh-based shape functions fulfill the Kronecker delta property while simultaneously exploit the smoothness and higher-order continuity of mesh-free shape functions. It needs to be explained that in the community of CMs, the distinction of terms "hybrid" and "couple" is rather vague, in this book we tend to use the former for the combination of different methods whereas the latter for the interaction between different phases or types of substances/materials as far as possible, although sometimes these terms have to be alternatively dubbed.

The "meshless finite element method (MFEM)" (Edelsbrunner and Mucke 1994) is a kind of FEMs with special shape functions. The domain is divided into polyhedral elements so the mesh may be built quickly using the extended Delaunay tessellation independent of node distribution. Non-Sibsonean shape functions are defined within each polyhedron. The main difference between MFEM and "natural element method

(NEM)" (Sukumar et al. 1998) is that in the latter the non-Sibsonean shape functions are applied to the whole domain, while in the former they are applied only to each polyhedral element.

The name of MFEM comes from its shape functions because they possess both the main characteristics of FEM and MFMs (Idelsohn et al. 2003):

- The space is divided into elements with the continuity of shape functions but with the discontinuity of derivatives, and
- The shape functions towards nodal interpolation depend only on node positions, the evaluation of nodal connectivity is bonded and linear with the total number of nodes in the domain.

A full description of MFEM may be found in the literatures (Idelsohn et al. 2003).

8. **PFEM**

The best known particle-based method is of course the SPH in which the shape functions are generated using a kernel approximation. The "particle finite element method (PFEM)" (Idelsohn and Oñate 2006; Idelsohn et al. 2008) is a hybrid method comprising remarkable features such as:

- The information is particle-based, i.e., all the geometrical and mechanical information is attached to particles;
- MFEM is used to compute the forces exerting on each individual particle; and
- Boundary contours of domain are defined by the alpha-shape method.

PFEM combines the ingredients of particle-based methods with a background mesh on which special finite element shape functions are used to evaluate the interacting forces between particles. The combination of particle-based method with MFEM and alpha-shape method, enables to avoid the difficulties in mesh-based methods and mesh-free methods.

PFEM is therefore one of the few advanced CMs that can avoid all the essential difficulties in constructing shape functions.

9. **S-PIMs**

One of the main drawbacks in most standard MFMs is the lack of Kronecker delta property, which requires a special treatment (e.g. Lagrange multipliers, penalty factors) for the imposition of essential boundary conditions (Wu and Plesha 2002; Fernández-Méndez and Huerta 2004). Liu proposed a family of MFMs called the "smoothed point interpolation methods (S-PIMs)" (Liu and Gu 2001; Liu 2003, 2010; Liu and Zhang 2013), in which the shape functions are constructed by the "point interpolation method (PIM)" to guarantee the Kronecker delta property. The price for a simpler imposition of essential boundary conditions is the presence of incompatible shape functions, which may present discontinuities in the problem domain. Such functions belong to the so-called "G-space" rather than to the standard "Hilbert space" with square integrable functions and corresponding first derivatives in FEM. Due to this incompatibility, the S-PIMs require a weakened-weak form using a strain smoothing technique for the continuum problem (Liu 2008). The smoothing operation is a sort of generalization of stabilized conforming nodal integration technique that was originally proposed by Chen and his co-workers (Chen et al. 2001).

S-PIMs have been applied to the elasto-plastic problem and the scalar damage problem (Gori et al. 2019a, b), in the latter case the method exhibits effectiveness for the localization phenomenon of continuum with a moderate soften property.

1.3.3 Particle-Based (Block-Type) Methods

Another important class in CMs is the so called "particle-based methods (PBMs)" or "block-type methods (BTMs)", in which particle dynamics is adopted directly (Walton 1984; Allen and Tildesley 1987; Place and Mora 1999; Coetzee 2017). In geotechnical and hydraulic engineering, these terms are narrowly referred to the methods that represent a physical problem of geological body with a collection of rock particles or blocks cut by structural planes (discontinuities), although many MFMs for continuum may be categorized as the sub-class of PBMs.

The narrowly defined PBMs (or BTMs) in this book comprise the best known family of "discrete element methods (DEMs)" (Zhao et al. 2012), in which a particle (or block) may be the physical part of domain (soil particles, rock blocks, etc.) or the specific part of continuous domain previously defined. Each particle moves accordingly with its own mass and the internal or/and external forces exerted on it. External forces on a particle are evaluated by its interaction with neighbor particles.

The broad family of DEMs includes the "distinct element method (DEM)" (Cundall 1971, 1976), the lattice-solid method (Mora and Place 1993), etc., of which the first is most well-known, and many researchers tend to use the abbreviation "DEM" particularly for the distinct element method.

In DEMs, particles are distinct from one another, and the continuity between particles/blocks is not required. Particles can be circular or polygonal (in 2-D), spherical or polyhedral (in 3-D), or other shapes (elliptical, angular, super-quadric) (Allen and Tildesley 1987; Hart et al. 1988; Williams and Pentland 1991; Rothenburg and Bathurst 1992; Ting et al. 1993; Lin and Ng 1997), but circular discs and spherical balls allow for the simplest contact detection. Each particle/block is assigned with certain material properties (e.g. size, density, Young's modulus, etc.) and inter-particle properties (e.g. stiffness coefficients, shear and tensile strengths, etc.). The overall behavior of particle system is determined by the inter-particle action in the form of force–displacement law. Particle interaction and resultant force are calculated in a pairwise fashion and updated throughout the whole simulation procedure. Particle acceleration, velocity, and displacement are calculated using the Newton's equation of motion.

Being advantageous to cope with discrete problems (e.g. granular matter flows) as well as continuous problems (e.g. possibilities of internal separation), DEMs are very prevalent in the areas related to geological structures (Saltzer and Pollard 1992; Strayer and Suppe 2002; Finch et al. 2004a, b) where structural planes are able to evolve naturally during a model run (collapse behavior). In addition, particle/block displacements and associated strains are unlimited, because the mesh distortion is no more an issue. Although particle properties and interactions are distinct, yet DEMs are able to calculate a composite continuum of any volume for the comparison with natural materials (Cundall and Strack 1983; Morgan and McGovern 2005a, b).

1. **DLMs**

The so-called "discrete lattice models (DLMs)" can be traced back to the 1940s (Pan et al. 2018). To solve the elasticity problem, Hrennikoff (1941) proposed a framework methodology that may be considered the prototype of lattice model. In the 1970s and 1980s, the lattice model was applied to rock mechanics (Kawai 1978) and theoretical physics (Ziman 1979; Roux and Guyon 1985; Sahimi and Goddard 1986; Louis and Guinea 1987; Herrmann et al. 1989; Zhao and Zhao 2012) for the study of fracture process in disordered media.

In the fundamental central-force lattice model (Ashurst and Hoover 1976), an isotropic and elastic medium with Poisson's ratio equal to 0.25 was represented by means of spring elements capable of transferring normal forces solely. Later on, shear spring elements were also set by which different values of Poisson's ratio can be replicated. In the beam problem (Schlangen and Garboczi 1997), nodal masses were connected either by Bernoulli or Timoshenko beams, allowing for the transfer of moments between nodal masses.

In the "rigid body spring model (RBSM)" (Kawai 1978; Hamajima et al. 1985; Kikuchi et al. 1992), a domain was discretized by polyhedral elements representing rock blocks. They are rigid bodies interconnected through their contact faces by put some springs in between to describe the shear or tension of joints.

Another different member of DLMs known as the "lattice spring models (LSM)" has been applied to the analysis of intact rock and rock mass, too. The main assumption of LSM is that a continuum medium can be represented by a collection of nodal masses interconnected by spring elements. Although simple, LSM has been proved itself sound and efficient to, for example, the analysis of hydro-fracturing and blasting problems (Sellers et al. 2012; Damjanac et al. 2016).

Among the family of DLMs, the "rigid body spring network (RBSN)" (Bolander and Saito 1998) adopts the Voronoi tessellation for both the mesh design and the scaling of spring stiffness. In its initial version, elastic, homogeneous and isotropic materials under uniform straining conditions with zero Poisson's ratio may be simulated rather well. Heretofore, it has been further developed to allow for any value of Poisson's ratio (Bolander et al. 1999; Asahina et al. 2015) and transverse isotropy (Kim et al. 2017; Rasmussen et al. 2018). Recently, the RBSN simulation for the laboratory test of sedimentary rock has succeeded (Asahina et al. 2017), in which the rock was postulated as a homogeneous, isotropic, and elastic-brittle material, and the Mohr–Coulomb strength criterion with a tension cut-off was employed.

2. **DEM**

Since the "distinct element method (DEM)" (Cundall 1971) is the most famous family member of DEMs (Burman 1971; Chappel 1972; Byrne 1974), it has been widely accepted as an effective computation tool to address engineering problems related to granular and discontinuous materials such as granular flows, powder mechanics, and rock mechanics (Sharma et al. 2018).

The key idea of DEM is that the domain of interest is treated as an assemblage of rigid or deformable particles/blocks, the contacts among them need to be identified and continuously updated during the whole deformation/motion process (Cundall 1971; Cundall and Hart 1985; Cundall 1988; Hart et al. 1988; Cundall and Hart 1992). Intrinsically, DEM is a force method that employs an explicit time-marching step scheme to directly solve the motion equations of Lagrangian formalism. Unbalanced forces drive the solution process, and an artificial damping should be introduced to quickly dissipate energy for static state. To avert the distortion of real vibration process in dynamic problems, the amount and type of damping should be very carefully specified with the help of experiments. Nevertheless, if only a quasi-static solution is desired where the intermediate results are not of interest, the amount of damping and the type of relaxation scheme can be deliberately selected to obtain the highest solution efficiency.

DEM is able to capture the stress–strain characteristics of intact rocks, the opening/shearing of pre-existing joints, and the interactions among rock blocks via joints. Combined with DFNs, it has been widely applied to the study of rock mechanical properties (Zhang and Sanderson 1998; Min and Jing 2003, 2004; Havaej et al. 2016), and the study of stress effects on interfacial fluid flow (Min et al. 2004).

The fundamental DEM formulation is not able to simulate the fracture growth in rocks driven by stress concentrations, although the plastic yielding zone can capture some aspects of rock failure process. Such a weakness has been addressed by introducing a Voronoi or Trigon discretization within rock blocks, which allows for fracturing along the internal grain boundaries governed by tensile and shear failure criteria (Kazerani and Zhao 2010; Kazerani et al. 2012; Gao and Stead 2014; Gao and Kang 2016). The DFN representation also can be integrated into such a DEM model using a dual-scale tessellation, in which the primary grid represents natural fractures whereas the secondary grid simulates microstructures (Ghazvinian et al. 2014). However, it is notable that the uncertainty of Voronoi/Trigon-DEM model arise from mesh dependency indicates a high necessity to represent natural grain shapes and textures (Mayer and Stead 2017).

Another development towards the simulation of fracture growth in rock makes use of finite element discretization in rock block matrix, which is termed as "finite-discrete element method (FDEM)" (Munjiza et al. 1995; Jing 1998; Munjiza et al. 1999; Munjiza and Andrews 2000; Munjiza 2004). It is rather important in the application because it is able to consider the fracturing and fragmentizing process of rocks (Ghaboussi 1988; Barbosa and Ghaboussi 1990, 1992; Shyu 1993; Chang 1994). To handle the non-linearity within and between blocks, a material non-linearity model using strain hardening/softening laws has been implemented (Ma 1999). Coupling of fluid flow across rock joints has been taken into account (Kim et al. 1999; Jing et al. 2001), too.

The representative DEM computer codes for simulating jointed rocks are UDEC and 3DEC (ITASCA 2013a, b) for 2-D and 3-D problems, respectively.

3. **PFC**

To simulate the progressive failure mechanism of granular materials such as soils/sands, Cundall and Strack (1979) established a "particle flow code (PFC)" for the motion and interaction of circular-or spherical-particles using the principle of DEM. A similar algorithm was also developed by Walton et al. (Walton 1982; Heuze et al. 1990).

In PFC, discrete rigid particles bonded together to form an assemblage are capable of collapse by the progressive rupture of contact bonds due to the shear and tension between these particles. The failure behavior of a jointed rock can therefore be captured either through the rupture of intact material or along joints, or through their combination. PFC has been widely exercised in diverse fields such as soil/rock mechanics, non-metal material science, processing industry, and defense industry.

Well known PFC computer codes are the dynamic materials corporation (DMC) (Taylor and Preece 1989, 1990), PFC2D and PFC3D (ITASCA 1995, 2014). An open source and particle-based modelling platform named as YADE (Kozicki and Donzé 2008, 2009) is available, too, which is able to analyze the stability of fractured rock slope controlled by the strength of fractures and intact rocks (Scholtès and Donzé 2012; Bonilla-Sierra et al. 2015).

As a special type of DEM, PFC calculates the inertial forces, velocities and displacements of interacting particles by solving the Newton's second law through an explicit time-marching scheme. The discrete particles are postulated to be rigid discs (in 2-D) or balls (in 3-D), and can have variable sizes but are usually much larger than the natural physical grain blamed for the computer capacity. The interaction between particles can be characterized by contact bond model and parallel bond model. The former serves as a linear elastic spring to transmit forces via the contact point between two particles, whereas the latter possesses certain normal stiffness and shear stiffness between two particles to resist the separation under tension, shear and rotation, and is more suitable for simulating rock-like materials.

To overcome the deficiency of the original PFC in reproducing realistic rock deformability and strength, a hierarchical bonding structure termed as "bonded particle model (BPM)" may be employed using a cluster logic (Potyondy and Cundall 2004) or clump logic (Cho et al. 2007). BPM simulates the interlocking effect of irregular grains by defining a higher value of intra-cluster bonding strength between particles than that between physical cluster boundaries. With the help of BPM, intact rocks are represented as the assemblages of cemented rigid particles, in which the macroscopic fracture is simulated as a result of the breakage and coalescence of inter-particle bonds. A fundamental challenge for BPM is its need to independently constrain the bulk constitutive properties of particle assemblage, because these are not defined a priori as for continuum models, namely, the overall behavior of particle assemblage should be determined by the aggregate effects of all particle interactions on bulk material properties, including mechanical strength, elastic modulus, contractive/dilative behavior, etc., under a certain range of stress paths and strain histories. These bulk properties should be related and tuned to prototype natural materials.

4. **DDA**

The "discontinuous deformation analysis (DDA)" was proposed by Shi and Goodman (1985, 1989) for the deformation and motion of multi-block system in which the rock block is postulated as "simply deformable", namely, first-order polynomials are employed for the displacement functions. As a result, the stress and strain fields within each rock block are constant. The stress-displacement problems are solved by taking into account of the interaction of blocks through structural planes (discontinuities).

The discretization of a DDA system is quite similar to DEM, namely, the medium is discretized into blocks by intersecting discontinuities. The fundamental difference between DDA and DEM lies in their computational frameworks: the latter treats the kinematics of each block separately based on an explicit time-marching scheme, whereas the former is basically an implicit method using displacements as essential variables solved by a set of linearized equations that should be updated following each time-marching step.

DDA can be derived by either the "principle of minimum potential energy" or "Hamilton's principle" (MacLaughlin and Doolin 2006). The kinematic constraints of no tension and no penetration between blocks are imposed by either the penalty factor method, the Lagrange multiplier method, etc. The governing equation set so derived guarantees that equilibrium condition is held at all times. Once the equation set is discretized, a step-wise time-marching scheme in the family of, Newmark (1959) for example, may be employed for the solution. DDA does not require any artificial damping term to dissipate energy for quasi-static problems, and is able to easily incorporate the mechanism of energy loss in a form of structure or/and material damping towards dynamic response problems. DDA possesses an important advantage of fast convergence with unconditional numerical stability compared to the fundamental DEM, because the latter requires a time-marching step length smaller than a critical threshold (Jing 2003).

Important extensions of the fundamental DDA include the more comprehensive representation of discontinuities, the sub-block technique (similar to Voronoi DEM) for fracturing processes, the development of 3-D models (Lin et al. 1996; Shi 2001; Jiang and Yeung 2004), etc.

After years of development, DDA has found it important role in the simulation of various topics related to rocks, including the fracturing process, wave propagation, bulk material property. The mechanical properties of jointed rock mass also may be investigated by hybrid DDA-DFN (Hatzor et al. 2004; Wu et al. 2004; Lin et al. 1996).

The applications of DDA to geotechnical and hydraulic structures such as dams, tunnels, caverns, landslides, masonry structures, etc., have been continuously exercised since the 1990s (Lin et al. 1996; Te-Chin 1997; Hatzor and Benary 1998; Ohnishi and Chen 1999; Pearce et al. 2000; Hsiung and Shi 2001; Hatzor et al. 2004; Jiang and Yeung 2004; Moosavi and Grayeli 2006).

5. **BEA**

Attributable to its extreme simplicity and rich experiences accumulated from engineering applications, the "limit equilibrium method (LEM)" is very useful in the stability analysis for rock foundations, dam abutments and cut slopes dominated by structural planes (discontinuities). It has, however, certain limitations: the deformation of rocks cannot be considered; the "factor of safety (FOS)" will be overestimated when the slip occurs on multi-discontinuities simultaneously, and additional postulations concerning the stress states on the slip surface have to be employed to render the problem statically determinate (Londe 1965).

In the early 1980s, to improve LEM for the stability analysis of rock wedges in cut slopes, the deformation characteristics of discontinuities were introduced (Chen 1984). Later on, the idea was generalized to the multi-blocky system giving rise to the "block element method or block element analysis (BEA)" (Chen 1987). With the assumption of face-to-face contact between blocks, the governing equation set of BEA is formulated by the force and moment equilibrium condition, the deformation compatibility condition, and the elasto-viscoplastic constitutive relation on discontinuities. In the time domain, the governing equation set is implicitly and stepwisely solved to present the quasi-static/dynamic and non-linear response of the whole block system. BEA had been expanded further into the areas of reinforcement and stochastic analysis (Chen 1993a, b, 1994). After the displacements in each block being interpolated by a set of polynomials of any order, and later by the overlap technique using fundamental finite elements with p-refinement, BEA is able to tackle the complicated deformation pattern in rock blocks as well as on discontinuities (Chen et al. 2004). The progress also had been achieved towards more systematical and practical algorithms, such as the unconfined seepage in the DFN embedded with drainage and grouting curtain systems, the automatic identification of multi-block system for complicated domain with irregular ground surface, the engineering applications under complicated geological setting, the dam/foundation interaction and seismic response, etc. (Chen et al. 2003, 2010).

BEA may be looked at as one family member of DEMs. Attributable to the assumption of infinitesimal deformation and related face-to-face contact of blocks through discontinuities, the contact updating during the block deflection/rotation process may be neglected. This enables us to handle various practical issues easily, such as the coupling of fluid flow across rock joints, the simulation of reinforcement components, the structural dynamic response under the action of seismic shakes, etc. It is also beneficial from its clear and accessible mechanical parameters which permit routine experimental evaluation for a specific engineering case.

It should be reminded that, however, the present form of BEA postulating infinitesimal deformation entails its applicability only towards the deformation, seepage, dynamic response, and safety margin within serviceability and collapse limit states. Although the extension of BEA to the problem of block detachment process can be implemented without essential difficulties by, for example, the contact-updating technique, yet it is wise to let the post-failure phenomenon be handled by well performed counterparts such as DEM and DDA.

1.4 Further Reflections on Advanced CMs

1.4.1 Formalisms in Continuum Mechanics

Basically, there are two formalisms in the classical theory for continuum mechanics (Spencer 1980; Hutter and Jöhnk 2004), namely, the Eulerian formalism and the Lagrangian formalism.

Eulerian formalism focuses on the current matter configuration, and what is occurring at any fixed points in the space along with the time evolution, in lieu of giving attention to individual particles w.r.t. how they move through the space and time. This is conveniently applied in the study of heat/fluid flow or solid deformation where the main interest is directed to the magnitude and rate at which the changes manifest, or to the field function gradient rather than the shape of the body at a reference time. For not very strong "unsmoothed" field problems related to thermal and fluid flow, Eulerian formalism might be more convenient and easier to incorporate the constitutive laws (relations) (e.g. Fourier's law, Darcy's law, etc.) and physical/mechanical properties observed and summarized by our predecessors.

Lagrangian formalism seems a natural development from Lagrange mechanics established in 1788, namely, a reformulation of the classical mechanics for discrete particle system with a finite "degrees of freedom (DOF)". To expand the Lagrangian formalism into the classical field theory of continuum which has an infinite number of DOF, the function of generalized coordinates—"Lagrangian", is replaced by the "Lagrangian density" that is a function of the field variable and its derivatives, and possibly of the space and time coordinates themselves. The independent variables are replaced by a spatial-time event, or still more generally, by a point on a "manifold". Often, the "Lagrangian density" is simply referred to as the "Lagrangian". In Lagrangian formalism, an observer standing in the reference frame records the changes in the position and physical properties as the continuous body moves through the space with the ongoing time. When CMs are expanded into the mechanics with large deformation accompanied with crack/separation scenarios, Lagrangian formalism starts to exhibit higher ability, although other intrinsic difficulties (e.g. counting, geometric shaping and interacting of particles) will manifest.

In this book, only the representative advanced CMs using Lagrangian formalism are elaborated in detail, this is because of their higher ability to handle the issues related to large (finite) and discontinuous deformation (flow) with strong distortion of configuration (surface) and detachment of sub-domain.

1.4.2 Spatial Discretization Schemes

In engineering practices, a computation mission is commonly accomplished by three major steps inclusive pre-processing, structural response analysis, and post-processing. In this respect, the pre-processing towards the geometrical description and domain discretization is one of the prerequisites for the analysis of complex geotechnical and hydraulic structures. If a computation program has not been equipped with well performed pre-processing software, it will be difficult to be widely accepted by practitioners.

The domain discretization scheme on the one hand, is tightly linked to the formalisms of continuum mechanics, and on the other hand, establishes the typical algorithm features in advanced CMs.

1. **Meshing**

Although the mesh (grid) generation towards irregular or complicated material domain is not easy, the fundamental CMs are normally mesh-based methods (MBMs). A number of advanced CMs, such as XFEM, CEM, and NMM, are mesh-based, too. All the meshes (if needed) in advanced CMs may be distinguished as Eulerian mesh or Lagrangian mesh, corresponding to the selected formalism illustrated above.

(1) **Eulerian mesh**

It is fixed on the space, in which the simulated object matter moves across mesh (grid) cells (elements). Therefore, all the nodes and cells remain spatially unchanged in the entire process of computation while the matter flows across the mesh. The flux of mass, momentum and energy across mesh cell boundaries are used to get the distribution of mass, velocity, energy, etc., in the domain.

Since large deformation in the object matter do not cause deformation in the mesh itself, Eulerian mesh is prevalent in the area of CFD, where the flow of fluid dominates. In principle, all CFD problems can be numerically solved in Eulerian mesh. Early simulation works w.r.t. large deformation issues such as explosion and high velocity impact were customarily performed using some kind of Eulerian mesh. However, the disadvantages associated with Eulerian mesh are nontrivial:

- It is very difficult to get the history of field variables at fixed points on the moving body, because the motion of body cannot be tracked using a fixed mesh. One can record the history of field variables at fixed points only.
- The position of free surfaces, deformable boundaries and moving material interfaces are difficult to be pinpointed accurately.
- Eulerian mesh should be sufficiently large to cover the entire area to which the object body can possibly move in and out. Sometimes, we have to use a much coarse mesh reluctantly for the computation efficiency at the expenses of low resolution and low accuracy in solution.

(2) **Lagrangian mesh**

Being fixed to or attached on the moving body in the entire computation process, a Lagrangian mesh possesses remarkable advantages such as:

- Since no convective term of flow exists in the related PDEs, the algorithm and program is simpler and faster.

- The entire history of all the field variables at a matter point can be easily tracked.
- The nodes can be placed along boundaries and material interfaces. As a result, the boundary conditions at free surfaces, moving boundaries and material interfaces may be easily imposed, tracked and determined.
- Since the mesh allows for being created only within the problem domain, namely, no additional nodes beyond the problem domain is required, as a result, it gives rise to higher computation efficient.

The major difficulty with the Lagrangian mesh is encountered in the event of extremely large deformation where the mesh is heavily distorted. In addition, the small timc-marching step length controlled by the smallest element size can lead to the high effort and even the breakdown of computation. A possible option is to re-zone or re-mesh the problem domain involving the overlay of a new and undistorted mesh on the old and distorted one, so that the following-up computation can be performed on a mesh of good quality. Adaptive re-meshing techniques are quite prevalent for the problems of impact, penetration, explosion, fragmentation, turbulence flow, and violate fluid–structure interaction. However, re-meshing procedure can be tedious and very time-consuming. Moreover, since the physical properties in the new mesh are approximated from old one through the interpolation of mass, momentum and energy transport. Hence, with each re-meshing operation, diffusion (pollution) occurs and material histories may be distorted or lost.

2. **Mesh-free point clusters**

Suppose there is a point clusters scattered in the space, and with these points we want to build a conforming mesh that must be in a good shape and the boundary contours must be respected. In order to meet these requirements, the mesh generator need large computing time and, in many cases, much human supervision is demanded.

Topologically, any complex domain can be discretized automatically with tetrahedral elements, but these elements only provide relative low accuracy. In addition, for many engineering problems the geometry/topology of domain changes with time. For instance, in fluid dynamics problem, the free surface can be broken into many free surfaces, or a number of free-surfaces can be joined together. This kind of problem needs a continuous update of node connection because two points, which are close to each other in a time-marching step, may be very far from each other in the next step. The continuous and quick regeneration of mesh is normally demanded. This explains why the MFMs that use point clusters directly are looked for.

The most important events since the 1980s are the invention of advanced CMs that comprise two larger families of mesh-free and particle-based (block-type). Most of them are based on the Lagrangian formalism and therefore possess high potentialities towards the solutions of crack propagation/dislocation and particle detachment/flow for geotechnical and hydraulic structures.

MFMs do not need a conforming mesh, but just the connection between neighbor nodes in order to build approximation functions. However, it must be noted that finding the node connection for mesh-free solutions could be as difficult as mesh generating for mesh-based solutions, in some cases, the time for creating the nodal connection is of the same order for meshing. From this point of view, MFMs are not so useful and efficient in the most cases where the same problem can be solved with MBMs, except in specific dynamic process exhibiting separation/crack/crash phenomena.

The discrete nature of particle system (e.g. in SPH) also poses unique challenges. One difficulty is the tremendous computation efforts to track all particles and their multiple interactions, which typically limits the particle size, consequently, the resolution of simulation, although this issue has been steadily overcome (Plimpton 1995; Plimpton and Hendrickson 1996; Munjiza 2004). Another difficulty is the small time-marching step length (interval) to maintain system stability. Since the critical length of time-marching step for particle translation is proportional to the square root of the ratio of particle mass to particle stiffness (Cundall and Strack 1979; Walton 1984), the dynamics of the smallest particle in the system greatly influences the run-time. The length of time-marching step also depends on expected particle velocities, because their displacements are integrated linearly. Time-marching step length must be so chosen to ensure that particle displacement increments are small enough, otherwise instability can occur. This imposes a crucial limit on the time-marching step length that can only be overcome via the advance in computer speed.

It is worthwhile to reminded that the main difficulties in building the nodal connectivity for MFMs are precisely the advantages of MBMs.

3. **Block matrix/particle clusters**

(1) **Block matrix**

In geotechnical and hydraulic engineering, the dominant factor influencing the deformation and stability of rock foundations, dam abutments, underground caverns and cut slopes is the structural plane (discontinuity) system that traverse rock masses into blocks of various sizes, shapes and positions. Hence there have being strong demand for the so-called "particle-based (block-type) methods (PBMs)" since the 1950s. Nowadays, a large family of CMs for simulating the motion of rock particles (blocks) is available that may be dubbed as "discrete element methods (DEMs)", in which the single term DEM is specially meant for the "distinct element method" proposed by Cundall (1971, 1976).

As the most important portion of pre-processor, the work towards identifying sub-domains (rock particles/blocks) in the rock mass for DEMs began from the mid1980s. Geometrical identification algorithms for the rock block system using topological techniques were established by Lin et al. (1987), Cundall (1988), Jing and Stephansson (1994). Their methods are rather sophisticated, but unlikely to identify a system including concave blocks at that time. Heliot (1988) proposed a new approach to produce more realistic rock block system, in which a rock block is always decomposed into two convex blocks. Ikegawa and Hudson (1992) made a breakthrough by the "direct body" concept to cope with both convex and concave blocks in a unified identification procedure. After a check on the various existing methods, the author of this book realized that the direct body concept could be a solid base to establish a robust identification and pre-processing algorithm for the analysis of seepage, deformation and stability of complicated rock block system. During the implementation of the method, further improvements taking into account of the existence of irregular ground surfaces (curved faults and dam surfaces as well), the existence of the grouting and drainage curtains, etc., were made (Wang and Chen 1998). This procedure, on the one hand, may be directly employed for the discrete approaches by PBMs such as the DDA, DEM and BEA elaborated in this book, on the other hand, may offer a well-organized database for the further discretization of standard mesh or mesh-free point cluster. In the latter case, in addition to natural discontinuities and interfaces, artificial structural planes also need to be additionally introduced to facilitate the discretization by transforming a complex domain into a set of simple blocks (Chen 2018).

(2) **Particle clusters**

Landslides (slopes) with soil-rock mixture body may be looked at as the particle assemblage system and simulated by PFC in which all the granular materials are packed within a number of unit periodic cells, each cell can be regarded as one fraction of the real slope. The particle with its centroid moving out of one periodic cell is mapped into another cell. The particle with only partially lying outside the cell can interact with the particle surrounding its originator cell.

There are a number of algorithms for particle cluster generation. For example, Zhao et al. (2016) proposed a "hopper discharge" technique to generate a dense slope mass, by which solid particles are used to fill the space bounded by the upper slope surface, the lower failure surface and the periodic boundaries. The "random aggregate structure (RAS)" technique also may be employed to generation the PFC model but the voids in the model should be looked at as the porosity without mortar matrix. The particles may possess the shapes of spherical, elliptical and polyhedral (Li et al. 2016; Xu and Chen 2016).

1.4.3 Material Assumptions

Substances such as solids, liquids and gases, are composed of molecules separated by voids at micro-scale. Rock-like materials (inclusive rocks and concretes) are embedded with structural planes (faults, joints, etc.) and cracks. Hence all the materials are actually discontinuous (Zhao et al. 2012).

However, often, although not always, from the standpoint view of human, certain physical phenomena of materials can be conceptually modeled by continuum, namely, the matter in a body is continuously distributed and fills the entire region of space it occupies. Consequently, a continuum is a material body that can be continually sub-divided into infinitesimal elements with properties being identical to that of the bulk material.

Continuum mechanics deals with the properties of solid and fluid that are independent of any particular coordinate system in which they are observed. These properties are customarily represented by the mathematical objects—"tensors".

The validity of continuum assumption may be verified by the theoretical analysis, in which either some clear periodicity is identified or the statistical homogeneity exists. More specifically, the continuum hypothesis hinges on the concept of a "representative elementary volume (REV)" and the separation of scales to provide a link between the viewpoints of experimentalist and theoretician on constitutive equations, as well as a way towards spatial and statistical average over different (micro-, meso-, and macro-) scale levels. These scale levels greatly dominate the selection of appropriate methods dubbed by different formalisms and spatial discretization schemes.

1. **Fluids**

For fluids, the "Knudsen number (K_n)" named after Danish physicist Martin Knudsen, is widely employed to assess to what extent the approximation of continuum works well. It is a dimensionless number defined as the ratio of the mean free path length of molecules to a representative physical length scale, i.e.

$$K_n = \lambda / L, \tag{1.1}$$

where: λ—mean free path; L—physical length scale (e.g., the radius of a body in fluid).

Knudsen number helps determine whether the statistical or continuous postulation of fluid should be used for the problem encountered. If the Knudsen number is near or greater than one, the continuum assumption is no longer a good one and statistical methods have to be used. Laurendeau (2005) suggested that $K_n \geq 10$ would be a suitable criterion for distinguishing molecular flow from continuum flow.

Airflow around an aircraft has a low Knudsen number, making it firmly in the realm of continuum mechanics. The flow of water through a geotechnical or hydraulic structure has a low Knudsen number, too. Hence we take it for granted that the hydraulic issues encountered in geotechnical and hydraulic structures may be well governed by the Navier–Stokes equations that may be further numerically solved by any advanced CMs of intrinsically continuous. For example, if the free surface will undergo severe change, such as in the case of landslide induced surge waves (tsunamis), the SPH featured by Lagrangian formalism, mesh-free, and intrinsically continuous, is one of the first considerations.

2. **Solids**

The properties of solids, particularly the geomaterials (rock-like materials and soils) concerned in this book, are strongly dependent on their micro-or/and meso-structures, but they are usually described as phenomenological (conceptual) continuum in macro-scale level referring to the large size of geotechnical and hydraulic structures. It means a mathematical expression that relates several observed phenomena by "naked-eye" or/and instruments to each other in a way consistent with fundamental physical theory only, namely, without the necessity to be directly derived from microscopic or mesoscopic mechanism. In the thermodynamics of continuum, a geomaterial body may be thought to be composed of infinitesimal "representative elementary volume (REV)" or "representative volume element (RVE)". The magnitude of REV is normally defined as the minimum volume beyond which any matter element behaves essentially homogeneous like the bulk material. Suppose REV is cubic, it side length l_{REV} may be employed to define the "representative length", then we can re-define a Knudsen number as

$$K_n = l_{\text{REV}}/L_{\text{RSV}}, \tag{1.2}$$

In which L_{RSV} is the representative length of "representative structure volume (RSV)", for example, the dam base width and span of tunnel. The author of this book suggest that:

- $K_n \leq 0.1$. It is suitable for continuous (equivalent) material model.
- $K_n \geq 1.0$. It is suitable for discontinuous material model.
- $1.0 \geq K_n \geq 0.1$. It is in between of discontinuous and continuous material models.

In the sense of phenomenological model, the properties of an elementary system are assigned to the bulk elements. The phenomena appearing at micro-or meso-level in material are represented in terms of state variables and conjugate thermodynamic forces, both attached to the elementary system. Cross-effects between phenomena on the micro-or meso-level and their macroscopic consequences may be determined from standard laboratory tests at bulk material samples. The phenomenological models most familiar to geotechnical and hydraulic engineers relate, for example, strain versus stress, temperature gradient versus heat flow rate, hydraulic gradient versus seepage flow rate.

The concept of REV has drawn high attention since the early work by Hill (1963). As is common in continuum mechanics, several other definitions of REV have been proposed by scientists for different purposes (Hashin 1983; Evesque 2000). The existence and size of REV have been studied by many scholars (Kanit et al. 2003; Stroeven et al. 2004; Gitman et al. 2007; Al-Raoush and Papadopoulos 2010; Skarzynski and Tejchman 2012), too.

(1) **Concrete**

Various studies have been carried out for a better understanding of concrete REV, based on these studies it may be rationally stipulated that the representative length of the REV of concrete is 3–5 times the maximum aggregate size (Robert 1998; Van Mier and Van Vliet 2003; Sebsadji and Chouicha 2012), namely,

$$l_{\text{REV}} = 3d_{\max} \sim 5d_{\max}, \tag{1.3}$$

in which $d_{\max}$ is the maximum size of concrete aggregate. This relation is valid not only for the deformation, but also the thermal and permeability characteristics of concrete (Keskin et al. 2011; Zhou et al. 2013; Li et al. 2016, 2017; Xu et al. 2017; Wang et al. 2019, 2020).

It is necessary noted that other scholars proposed different but similar representative lengths l_{REV}. Huet (1999) suggested that the concrete REV should be not only dependent on the maximum aggregate size, but also on the other factors such as sample shape and aggregate content. Van Vliet and Van Mier (2000) conducted a series of uniaxial tension experiments to study the size effect on the strength and fracture energy of concrete, they showed that the REV should be as large as 6–7 times the maximum aggregate size. A study using quantitative image analysis and computer simulation by Stroeven and Stroeven (2001) indicated that the concrete REV should exceed 4–5 times the maximum aggregate size for structure insensitive properties (such as stiffness) while it must be even larger for structure sensitive properties (such as fracture). Kim, Lutif and Allen (2009) indicated that for typical dense-graded asphalt concrete mixtures with a nominal maximum aggregate size of 19 mm, their effective non-damaged properties could be characterized by a REV size of 50 mm.

Where the geotechnical and hydraulic structures within serviceability and collapse limit states are the major concern, fundamental CMs of Lagrangian formalism and mesh-based such as FEM and CEM are advisable towards the spatial-time distribution of deformation/stress, hydraulic potential, and temperature under various environmental actions following the dynamic construction and operation agenda/schedule. This is mainly attributable to the Knudsen number much smaller than 0.1. Take a concrete gravity dam of 45 m high and 30 m base wide (i.e. $L_{RSV} = 30{,}000$ mm) for example, since the maximum aggregate size is 150 mm in fully-graded concrete (i.e. $l_{REV} = 5 \times 150$ mm), hence we get $K_n = 5 \times 150/30{,}000 = 0.025$ according to Eq. (1.2). Consequently, the detailed micro-and meso-structures of concrete material may be neglected for the overall response of concrete dam, and the consistent mechanics basis in the conventional computation tool kits for routine design is guaranteed.

Advanced CMs, such as the family of MBMs (XFEM, NMM, etc.) or MFMs (EFGM, etc.) featured by the finite and dislocation deformation, have to be attempted when we are encountered with macro-scale failure processes such as the localized crack propagation in concrete. However, it is emphasized that under such circumstances, the constitutive relations, physical and mechanical parameters are certainly not identical to the situations when the integrity of engineering structure is maintained within the serviceability and collapse limit states. This issue is however, not well cared and addressed insofar within the knowledge scope of the author.

(2) **Rock**

Rock masses are divided into four major types, namely blocky, layered (laminate), fractured and granular. One of the REV definitions relates rock masses to the sample size dependent on their stochastic behaviors. On the basis of crack tensor concept, Oda (1988) suggested that rock REV must be at least three times the typical length of fracture (joint) traces. This is normally applicable to the jointed rock where joint trace is smaller than 5 m. According to Müller (1974), the minimum size of rock REV should be at least ten times the joint space that is the distance between joints. Since the latter has been validated w.r.t. the deformation and permeability characteristics by the numerical material tests using hybrid DFN-CEM for jointed rocks (Chen et al. 2008, 2012), therefore it may be postulated that

$$l_{\mathrm{REV}} = 10/\lambda, \tag{1.4}$$

in which λ is the joint (fracture) frequency that is reciprocally related to joint space (Hudson and Priest 1983).

Another issue is the permissible representative length l_{REV} w.r.t. the representative length of engineering structure L_{RSV}, to justify whether the equivalent continuum assumption is valid for the macroscopic phenomenological model of jointed rock. This is more difficult and less addressed until now. Wilson and Witherspoon (1970) suggested a permissible ratio 1/50 of the maximum joint space to the minimum structure boundary, to guarantee the validity of equivalent continuum assumption. Louis (1974) believed that where there are 1000-plus joints in the domain concerned, REV does exist and the equivalent continuum assumption to support a phenomenological model is valid for seepage analysis. From these limited studies, we may postulate that the permissible representative length l_{REV} based on Knudsen number in Eq. (1.2) is applicable.

Generally, if REV exists (i.e. $K_n \leq 0.1$) for a component or sub-domain of engineering structure and its deformation is within the limit of collapse, fundamental CMs of Lagrangian formalism, mesh-based and intrinsically continuous, such as the most prevalent FEM where the joint system is simulated implicitly by equivalent constitutive relations, may be employed successfully. In doing so however, the large scale structural planes should be explicitly simulated by special finite elements (Goodman et al. 1968; Zienkiewicz et al. 1970; Desai et al. 1984; Buczkowski and Kleiber 1997).

Unfortunately, the jointed rock masses encountered in geotechnical and hydraulic structures do not often, although not always, meet the permissible REV requirement. Take the Xiaowan Arch Dam (H = 294.5 m, China) for example (Chen 2018): its dam base width varies from the maximum $L_{\mathrm{RSV}} = 72.91$ m (crown cantilever base) to the minimum $L_{\mathrm{RSV}} = 12$ m (crest arch abutments); the average space of three dominant joint sets is 0.23–0.43 m hence $l_{\mathrm{REV}} = 4.3$ m according to Müller. This means that the physical and mechanical parameters should be evaluated by test samples larger than 4.3 m for rocks. We now easily realize that it is nearly impossible to meet. Even if we success in doing so, the correspondent Knudsen number is $K_n = 4.3/72.91 = 0.059$ for the crown cantilever base and $K_n = 4.3/12 = 0.36$ for the crest arch abutments. Namely, the dam foundation rock should be looked at as a medium between discontinuous and continuous. This situation vividly shows the real difficulty we are encountered in rocks, and explains why we should make compromise among the diverse applications of computation algorithms, material models, parametric evaluations, and, safety margins. Actually, the design of dam foundation and abutment of this giant arch dam was successfully accomplished based on the comprehensive study using FEM for equivalent continuous approach and BEA for intrinsic discontinuous approach, in additional to other conventional tools such as LEM (Chen 2018). Hence, rocks need frequently be handled as both intrinsically continuous and discontinuous geomaterials, this fact explains the phenomenon that a large variety of advanced CMs illustrated in this book has been constructed in recent decades in an attempt to simulate rocks with a better balance among computation effort paid out and solution resolution paid off.

It is also worthwhile to indicate that since special finite element models are inherently formulated on continuum assumptions, therefore large-scale opening and sliding and complete detaching of elements, are not permitted even the existence of REV guarantees the fundamental CMs of Lagrangian formalism, mesh-based and intrinsically continuous. Under such circumstances, advanced CMs of Lagrangian formalism, either mesh-based (e.g. NMM) or mesh-free (e.g. EFGM) or particle-based (e.g. DEM, DDA), are demanded.

In the study of dam break, landslide, and their generated flood/surge waves, the advanced CMs of Lagrangian formalism, mesh-free (e.g. SPH) or particle-based (DEM, DDA) comes up as the first choice.

1.5 Concluding Remarks

1.5.1 Motivations of This Book

In fluid dynamics with high velocity and severe surface distortion, and solid dynamics with finite deformation (even long runout motion) and separation/crack/crash, the existing fundamental CMs find arduousness and sometimes, even helplessness. Typical examples are the dam break process and consequent flood (Ritter 1892; Alcrudo and Mulet 2007); the rock blasting process in open air/under water with crack growth, large deformation, fragmentation, or even solid/fluid flow (Fakhimi and Lanari 2014; Li and Rong 2011; Hu et al. 2015; Bohloli and Hovén 2007); the landslide process with induced surge waves in water body (Panizzo et al. 2005; Crosta et al. 2016).

The invention of advanced CMs is one of the most significant events in science and engineering. In recent decades they start to exhibit higher power although other intrinsic difficulties manifest. Most of them are based on Lagrangian formalism and therefore possess a high potentiality towards the simulation of crack growth, sub-domain dislocation and particle detachment for geotechnical and hydraulic structures.

In the prequel book (Chen 2018), a number of typical fundamental CMs such as the FEM, BEA, CEM have been elaborated. They are accepted as the important participators in the analysis of engineering structures mainly due to: the evaluation of material models and corresponding parameters have been well defined and may be routinely carried out; the permissible "factors of safety (FOS)" have been well stipulated in the design specifications to help engineers make convincible judgment concerning the safety of engineering structures within serviceability and collapse limit states.

In this follow-up book, the interest will be directed to advanced CMs. Although several members of them start to find their positions in the safety calibration and hazard estimation of geotechnical and hydraulic structures, yet insofar they mainly play an auxiliary role-set in the qualitative study where the problems are beyond the scope of fundamental CMs (e.g. FEM, BEM, FDM). This is mainly blamed for the fact that most advanced CMs are in short of systematic toolkits comprising sophisticated pre-process subroutines, standard computer software, reliable constitutive relations and parameters via acceptable test procedure, affordable computer resources for the solution within acceptable time, as well as the safety indexes for decision makers.

However, it is more and more clear that there is a great expectation for these advanced CMs attributable to their amazing performance in handling the failure-process and post-failure scenarios of engineering structure beyond collapse limit states, such as the dam break and released floods in excess of the maximum probable flood (MPF), the landslide and reservoir surge waves (tsunamis) over the check flood level (CFL), etc. In view of the demand from the hazard estimation for the disaster reduction/prevention, this book is focused on the philosophies, principles, and algorithms of a number of typical advanced CMs that possess high potential or already partially find their role in the practices of geotechnical and hydraulic engineering.

1.5.2 Layout of This Book

This book is layout as follows: This chapter reviews the state-of-the-arts of advanced CMs, Chapter 2 gives the basic material properties and Chap. 3 summarizes the classical mechanics. After the elucidation of prevalent numerical methods in Chap. 4, two mesh-based methods (XFEM, NMM), two block-based/particle-type methods (DEM, DDA), and two mesh-free methods (EFGM, SPH) are exemplified in Chaps. 5–10. Chapter 11 presents a roadmap concerning how two advanced CMs may be coupled by the example of hybrid DEM-SPH, where special attention is focused on the landslide and its generated surge waves.

Such a layout is intended, through parallel expounding, to let our readers possess the comprehensive knowledge and grasp the intrinsic features of typical advanced CMs. Therefore, most of discussions, although not always, are concentrated at 2-D formulation to facilitate the understanding. Hopefully, this will help to clarify the crucial issues such as the formalism of governing PDEs, the discretization scheme, the interpolation or approximation of field variables, the nodal connectivity and derivative calculation of field variables, of course the basic algorithms. In addition to exactly catch the strong points and weak points, readers also will be aware of that the differences in these methods are not so large as at the first glance, and be able to select an appropriate method, to improve the method for their specific purpose, and to evaluate the applicability of outcomes.

In addition to the background knowledge w.r.t. fundamental CMs, readers are presumed to have a postgraduate-level on mathematics and mechanics. Certainly, experience in engineering geology and rock mechanics, engineering hydraulics, engineering structures, etc., would be a plus.

1.5.3 Reminders and Suggestions

Fully automated analysis with adequate resolution allows for the analyst/engineer to focus on engineering issues. However, the questions such as "finer computations always lead to better understanding of the structure performances?" "is there universal method which would crown all the existing methods?" "to mesh or not to mesh the structure?" etc., are not correctly raised or cannot be appropriately answered before the fully understanding of the problem concerned.

As the author has been emphasized previously (Chen 2018), a successful and helpful computation actually relies on the team work by those who are the experts of both engineering structure and modern computation, in addition to comprehensive knowledge and practical training in the fields of engineering geology, material science, construction or/and operation management. Only in this manner, we may feel sufficient confidence to present our computation results to decision makers.

Advanced CMs featured by discontinuous and finite deformation, have to be attempted when we are encountered with post-failure processes such as dam break, landslide and generated surge waves. Unfortunately, insofar many of them are able to limitedly provide quantitative solutions only, because:

- Sometimes, the mesh generation is not an important issue for the solution, namely, they are in an unfair position to compete with fundamental CMs.
- They must use algorithms to generate and update the connectivity between scattered nodes, which has no clear and linear relation between the amount of operations and the amount of nodes involved. Namely, standardization of computation algorithm and software to overcome the bewildering that significant difference may manifest, is far from engineering expectation.
- The physical parameters and constitutive relations are certainly not identical to the situation when the integrity of geotechnical and hydraulic structures is maintained within the serviceability and collapse limit states. Namely, the matchup between the computation method and input parameters has not been well addressed.
- The safety margin, reliability or risk are not well defined and stipulated. Insofar there are no stipulated indexes in the risk-based design and hazard alarm for geotechnical and hydraulic structures concerning post-failure consequences.

For the advanced CMs elaborated in this book, there is a long way to be covered from laboratory manifestations to engineering challenges. To close this chapter, the author feels like to give three basic suggestions w.r.t. the method selection toward practical problems:

- In situations where the same problem (e.g. fracture analysis w.r.t. limited number of cracks/discontinuities) can be solved with a mesh-based method and mesh-free methods, the former is more efficient and well performed, some of them even may cope with complicated cracking issue of concrete dam convincingly (viz. Chapter 5);
- When the problem demands a step by step updating w.r.t. node/particle connectivity, the most important consideration in the selection of a mesh-free or particle-based method is to obtain a linear relation between the number of nodes/particles and the amount of operations to generate their connectivity;
- Where computer and human resources are permitted, a comprehensive study using at least two typical CMs is encouraged. This is because each method possesses unique advantages in modeling some but not all important aspects of the specific engineering problem.

References

Alcrudo F, Mulet J. Description of the Tous Dam break case study (Spain). J Hydraulic Res. 2007;45(SI):45–57.

Allen MP, Tildesley DJ. Computer simulation of liquids. Oxford: Oxford University Press; 1987.

Al-Raoush R, Papadopoulos A. Representative elementary volume analysis of porous media using X-ray computed tomography. Powder Technol. 2010;200(1–2):69–77.

Amadei B (ed.). Proceedings of the third International Conferences on Analysis of Discontinuous Deformation—From Theory to Practice (ICADD-III). Vail, Colorado: American Rock Mechanics Association; 1999.

Amiri F, Anitescu C, Arroyo M, Bordas SPA, Rabczuk T. XLME interpolants, a seamless bridge between XFEM and enriched meshless methods. Comput Mech. 2014;53(1):45–57.

ANSYS Inc. ANSYS fluent theory guide. https://kargosha.com/file/attach/20175/2812.pdf, 2011.

Asahina D, Aoyagi K, Kim K, Birkholzer JT, Bolander JE. Elastically-homogeneous lattice models of damage in geomaterials. Comput Geotech. 2017;81:195–206.

Asahina D, Ito K, Houseworth JE, Birkholzer JT, Bolander JE. Simulating the Poisson effect in lattice models of elastic continua. Comput Geotech. 2015;70:60–7.

Ashurst WT, Hoover WG. Microscopic fracture studies in 2-dimensional triangular lattice. Phys Rev B. 1976;14(4):1465–73.

Atluri SN, Shen S. The meshless local Petrov-Galerkin (MLPG) method: a simple and less-costly alternative to the finite element and boundary element methods. Comput Model Eng Sci. 2002;3:11–51.

Atluri SN, Zhu T. A new meshless local Petrov-Galerkin (MLPG) approach in computational mechanics. Comput Mech. 1998;22(2):117–27.

Atluri SN, Zhu T. The meshless local Petrov-Galerkin (MLPG) approach for solving problems in elasto-statics. Comput Mech. 2000;25(2):169–79.

Babuška I, Melenk JM. The partition of unity finite element method. Technical Report BN-1185. Maryland: Inst for Phys Sci and Tech University of Maryland; 1995.

Babuška I, Melenk JM. The partition of unity method. Int J Numer Meth Eng. 1997;40(4):727–58.

Banerjee PK, Henry DP. Advanced applications of BEM inelastic analysis of solids. In: Banerjee PK, Wilson RB, editors. Developments in boundary element methods (vol. 5). London: Elsevier Applied Science Publishers; 1989, Chapter 2.

Banerjee PK, Henry DP, Raveendra ST. Advanced inelastic analysis of solids by the boundary element method. Int J Mech Sci. 1989;31(4):309–22.
Barbosa R, Ghaboussi J. Discrete finite element method for multiple deformable bodies. Finite Elem Anal Des. 1990;7(2):145–58.
Barbosa R, Ghaboussi J. Discrete finite element method. Eng Comput. 1992;9(2):253–66.
Belytschko T, Black T. Elastic crack growth in finite elements with minimal re-meshing. Int J Numer Meth Eng. 1999;45(5):601–20.
Belytschko T, Gracie R, Ventura G. A review of extended/generalized finite element methods for material modeling. Model Simul Mater Sci Eng. 2009;17(4). featured article.
Belytschko T, Guo Y, Liu WK, Xiao SP. A unified stability analysis of meshfree particle methods. Int J Numer Meth Eng. 2000;48(9):1359–400.
Belytschko T, Krongauz Y, Organ D, Fleming M, Krysl P. Meshless methods: an overview and recent developments. Comput Meth Appl Mech Eng. 1996;139(1–4):3–47.
Belytschko Y, Lu YY, Gu L. Element-free Galerkin methods. Int J Numer Meth Eng. 1994;37(2):229–56.
Belytschko T, Moës N, Usui S, Parimi C. Arbitrary discontinuities in finite elements. Int J Numer Meth Eng. 2001;50(4):993–1013.
Belytschko T, Organ D, Krongauz Y. A coupled finite element—element free Galerkin method. Comput Mech. 1995;17(3):186–95.
Benito JJ, Ureña F, Gavete L. Influence of several factors in the generalized finite difference method. Appl Math Model. 2001;25(12):1039–53.
Benito JJ, Ureña F, Gavete L. Leading-edge applied mathematical modelling research. New York: Nova Science Publishers; 2008. Chapter 7.
Benito JJ, Ureña F, Gavete L, Alvarez R. An h-adaptive method in the generalized finite difference. Comput Methods Appl Mech Eng. 2003;192(5–6):735–59.
Benito JJ, Ureña F, Gavete L, Alonso B. Solving parabolic and hyperbolic equations by generalized finite difference method. J Comput Appl Math. 2007;209(2):208–33.
Benito JJ, Ureña F, Gavete L, Alonso B. Application of the generalized finite difference method to improve the approximated solution of PDEs. Comput Model Eng Sci. 2009;38(1):39–58.
Bobet A, Fakhimi A, Johnson S, Morris J, Tonon F, Ronald YM. Numerical models in discontinuous media: review of advances for rock mechanics applications. J Geotech Geoenviron Eng ASCE. 2009;135(11):1547–61.
Bohloli B, Hovén E. A laboratory and full-scale study on the fragmentation behavior of rocks. Eng Geol. 2007;89(1):1–8.
Bolander J, Saito S. Fracture analyses using spring networks with random geometry. Eng Fract Mech. 1998;61(5):569–91.
Bolander J, Yoshitake K, Thomure J. Stress analysis using elastically homogeneous rigid-body-spring networks. J Struct Mech Earthqu Eng. 1999;16(2):125–32.
Bonilla-Sierra V, Scholtès L, Donzé FV, Elmouttie MK. Rock slope stability analysis using photogrammetric data and DFN-DEM modelling. Acta Geotech. 2015;10(4):497–511.
Bonet J, Kulasegaram S. Correction and stabilization of smooth particle hydrodynamics methods with applications in metal forming simulations. Int J Numer Meth Eng. 2000;47(6):1189–214.
Bonet J, Lok T. Variational and momentum preservation aspects of smooth particle hydrodynamic formulations. Comput Meth Appl Mech Eng. 1999;180(1–2):97–115.
Bonnet M, Mukherjee S. Implicit BEM formulations for usual and sensitivity problems in elasto-plasticity using the consistent tangent operator concept. Int J Solids Struct. 1996;33(30):4461–80.
Boresi AP, Lynn PP. Elasticity in engineering mechanics. Englewood Cliffs, NJ: Prentice-Hall; 1974.
Brebbia CA, Telles JCF, Wrobel LC. Boundary element techniques: theory & applications in engineering. Berlin: Springer-Verlag; 1984.
Buczkowski R, Kleiber M. Elasto-plastic interface model for 3D-frictional orthotropic contact problems. Int J Numer Meth Eng. 1997;40(4):599–619.

Burman BC. A numerical approach to the mechanics of discontinua. Ph.D. Thesis. James Cook University of North Queensland (Australia); 1971.
Byrne RJ. Physical and numerical model in rock and soil-slope stability. Ph.D. Thesis. James Cook University of North Queensland (Australia); 1974.
Chang QT. Nonlinear dynamic discontinuous deformation analysis with finite element meshed block systems. Ph.D. Thesis. University of California, Berkeley (USA); 1994.
Chappel BA. The mechanics of blocky material. Ph.D. Thesis. Australia National University, Canberra (Australia); 1972.
Chen G, Ohnishi Y, Ito T. Development of high-order manifold method. Int J Numer Meth Eng. 1998;43(4):685–712.
Chen JS, Wu CT, Yoon S, You Y. A stabilized conforming nodal integration for Galerkin mesh-free methods. Int J Numer Meth Eng. 2001a;50(2):435–66.
Chen L, Kassab AJ, Nicholson DW, Chopra MB. Generalized boundary element method for solids exhibiting nonhomogeneities. Eng Anal Boundary Elem. 2001b;25(6):407–22.
Chen SH. The stability of a rock wedge in the slope. M.Sc. Thesis. WUHEE, Wuhan (China); 1984 (in Chinese with English summary).
Chen SH. The elasto-plastic and elasto-viscoplastic analysis of the discontinuous rock masses. Ph.D. Thesis. WUHEE, Wuhan (China); 1987 (in Chinese with English summary).
Chen SH, et al. Numerical analysis and model test of rock wedge in slope. In: Pasamehmetoglu AG, et al., editors. Proceedings of International Symposium on Assessment and Prevention of Failure Phenomena in Rock Engineering. Rotterdam: AA Balkema; 1993. p. 425–9.
Chen SH. Analysis of reinforced rock foundation using elastic-viscoplastic block theory. In: Ribeiro e Sousa L, Grossmann NF, editors. Proceedings of 1993 ISRM International Symposium-EUROCK 93. Rotterdam: AA Balkema; 1993. p. 45–51.
Chen SH. Hydraulic structures. Berlin: Springer; 2015.
Chen SH. Computational geomechanics and hydraulic structures. Berlin: Springer; 2018.
Chen SH, Feng XM. Composite element model for rock mass seepage flow. J Hydrodynamics (Ser B). 2006;18(2):219–24.
Chen SH, Feng XM, Shahrour I. Numerical estimation of REV and permeability tensor for fractured rock masses by composite element method. Int J Numer Anal Meth Geomech. 2008a;32(12):1459–77.
Chen SH, He J, Shahrour I. Estimation of elastic compliance matrix for fractured rock masses by composite element method. Int J Rock Mech Min Sci. 2012;49(1):156–64.
Chen SH, Li YM, Wang WM, Shahrour I. Analysis of gravity dam on a complicated rock foundation using an adaptive block element method. J Geotech Geoenviron Eng ASCE. 2004a;130(7):759–63.
Chen SH, Qiang S. Composite element model for discontinuous rock masses. Int J Rock Mech Min Sci Geomech Abstr. 2004;41(7):865–70.
Chen SH, Qiang S, Chen SF, Egger P. Composite element model of the fully grouted rock bolt. Rock Mech Rock Eng. 2004b;37(3):193–212.
Chen SH, Qiang S, Shahrour I, Egger P. Composite element analysis of gravity dam on a complicated rock foundation. Int J Geomech ASCE. 2008b;8(5):275–84.
Chen SH, Shahrour I. Composite element method for the bolted discontinuous rock masses and its application. Int J Rock Mech Min Sci. 2008;45(3):384–96.
Chen SH, Shen BK, Huang MH. Stochastic elastic-viscoplastic analysis for discontinuous rock masses. Int J Numer Meth Eng. 1994;37(14):2429–44.
Chen SH, Su PF, Shahrour I. Composite element algorithm for the thermal analysis of mass concrete: simulation of lift joint. Finite Elem Anal Des. 2011;47(5):536–42.
Chen SH, Wang WM, Zheng HF, Shahrour I. Block element method for the seismic stability of rock slopes. Int J Geotech Geoenviron Eng ASCE. 2010a;136(12):1610–7.
Chen SH, Xu MY, Shahrour I, Egger P. Analysis of arch dams using coupled trial load and block element methods. J Geotech Geoenviron Eng ASCE. 2003;129(11):977–86.

Chen SH, Xu Q, Hu J. Composite element method for seepage analysis of geo-technical structures with drainage hole array. J Hydrodynamics (Ser B). 2004c;16(3):260–6.

Chen SH, Xue LL, Xu GS, Shahrour I. Composite element method for the seepage analysis of rock masses containing fractures and drainage holes. Int J Rock Mech Min Sci. 2010b;47(5):762–70.

Chen SH, Zhang X, Shahrour I. Composite element model for the bonded anchorage head of stranded wire cable in tension. Int J Numer Anal Meth Geomech. 2015;39(12):1352–68.

Chen W. Meshfree boundary particle method applied to Helmholtz problems. Eng Anal Boundary Elem. 2002;26(7):577–81.

Chew CS, Yeo KS, Shu C. A generalized finite-difference (GFD) ALE scheme for in compressible flows around moving solid bodies on hybrid meshfree-Cartesian grids. J Comput Phys. 2006;218(2):510–48.

Cho N, Martin CD, Sego DC. A clumped particle model for rock. Int J Rock Mech Min Sci. 2007;44(7):997–1010.

Chopra MB, Dargush GF. Development of BEM for thermoplasticity. Int J Solids Struct. 1994;31(12–13):1635–56.

Chu YA, Moran B. A computational model for nucleation of solid-solid phase transformations. Model Simul Mater Sci Eng. 1995;3:455–71.

Clough RW. The finite element in plane stress analysis. J Struc Div ASCE (Proc 2nd ASCE Conf on Electronic Computation). 1960;23:345–78.

Coetzee CJ. Review: calibration of the discrete element method. Powder Technol. 2017;310:104–42.

Crosta GB, Imposimato S, Roddeman D. Landslide spreading, impulse water waves and modelling of the Vajont rockslide. Rock Mech Rock Eng. 2016;49(6):2413–36.

Cruse TA, Rizzo FJ. A direct formulation and numerical solution of the general transient elastodynamic problem. Int J Math Anal Appl. 1968;22(1):244–59.

Cundall PA. A computer model for simulating progressive large scale movements in blocky rock systems. In: Proceedings of the Symposium of International Society of Rock Mechanics (vol. 1). Nancy (France): ISRM; 1971. P. 129–136 (Paper II-8).

Cundall PA. Explicit finite difference methods in geomechanics. In: Proceedings of the EF Conference on Numerical Methods in Geomechanics, ASCE; 1976. p. 132–50.

Cundall PA. Formulation of a three-dimensional distinct element model–Part I. A scheme to detect and represent contacts in a system composed of many polyhedral blocks. Int J Rock Mech Min Sci. 1988;25(3):107–16.

Cundall PA, Hart RD. Development of generalized 2-D and 3-D distinct element programs for modelling jointed rock. Minneapolis: US Army Corps of Engineers; 1985, paper SL-85-1.

Cundall PA, Hart RD. Numerical modelling of discontinua. Eng Comput. 1992;9(2):101–13.

Cundall PA, Strack ODL. A discrete numerical model for granular assemblies. Geotechnique. 1979;29(1):47–65.

Cundall PA, Strack ODL. Modeling of microscopic mechanisms in granular materials. In: Jenkins JT, Satake M, editors. Mechanics of granular materials: new models and constitutive relations. New York: Elsevier Science; 1983. p. 137–49.

Dallner R, Kuhn G. Efficient evaluation of volume integrals in boundary element method. Comput Meth Appl Mech Eng. 1993;109(1–2):95–109.

Damjanac B, Detournay C, Cundall PA. Application of particle and lattice codes to simulation of hydraulic fracturing. Comput Part Mech. 2016;3(2):249–61.

Davies TG, Gao XW. Three-dimensional elasto-plastic analysis via the boundary element method. Comput Geotech. 2006;33(3):145–54.

De S, Bathe KJ. Towards an efficient meshless computational technique: the method of finite spheres. Eng Comput. 2001;18(1–2):170–92.

Desai CS, Abel JF. An introduction to the finite element method. New York: Van Nostrand Reinhold; 1972.

Desai CS, Zamman MM, Lightner JG, Siriwardane HJ. Thin layer element for interfaces and joints. Int J Numer Anal Meth Geomech. 1984;8(1):19–43.

Detournay C, Hart R. FLAC and numerical modelling in geomechanics. In: Proceedings of the International FLAC Symposium on Numerical Modelling in Geomechanics. Rotterdam: AA Balkema; 1999.

Dilts GA. Moving least square particle hydrodynamics. I. Consistency and stability. Int J Numer Meth Eng. 1999;44(8):1115–55.

Dilts GA. Moving least square particle hydrodynamics. II. Conservation and boundaries. Int J Numer Meth Eng. 2000;48(10):1503–24.

Duarte CA, Babuška I, Oden JT. Generalized finite element methods for three-dimensional structural mechanics problems. Comput Struct. 2000;77(2):215–32.

Duarte CA, Oden JT. Hp clouds—a meshless method to solve boundary-value problems. Technical Report 95-05. Texas: Texas Institute for Computational and Applied Mechanics (TICAM), the University of Texas at Austin; 1995.

Duarte CA, Oden JT. An hp adaptive method using clouds. Comput Meth Appl Mech Eng. 1996;139(1–4):237–62.

Duarte CA, Reno LG, Simone A. A high-order generalized FEM for through-the-thickness branched cracks. Int J Numer Meth Eng. 2007;72(3):325–51.

Duflot M. A meshless method with enriched weight functions for three-dimensional crack propagation. Int J Numer Methods Eng. 2006;65(12):1970–2006.

Edelsbrunner H, Mucke EP. Three-dimensional alpha-shape. ACM Trans Graph. 1994;13(1):43–72.

Evesque P. Fluctuations, correlations and representative elementary volume (REV) in granular materials. Poudres Grains. 2000;11:6–17.

Fakhimi A, Lanari M. DEM-SPH simulation of rock blasting. Comput Geotech. 2014;55(2):158–64.

Fang Z. A local degradation approach to the numerical analysis of brittle fracture in heterogeneous rocks. Ph.D. Thesis. Imperial College of Science, Technology and Medicine, University of London (UK); 2001.

Fernández-Méndez S, Díez P, Huerta A. Convergence of finite elements enriched with meshless methods. Numer Math. 2003;96(1):43–59.

Fernández-Méndez S, Huerta A. Coupling finite elements and particles for adaptivity: an application to consistently stabilized convection-diffusion. In: Griebel M, Schweitzer MA, editors. Meshfree methods for partial differential equations, Lecture Notes in Computational Science and Engineering (vol. 26). Berlin: Springer; 2002. p. 117–29.

Fernández-Méndez S, Huerta A. Imposing essential boundary conditions in mesh-free methods. Comput Meth Appl Mech Eng. 2004;193(12–14):1257–75.

Finch E, Hardy S, Gawthorpe R. Discrete-element modelling of extensional fault-propagation folding above rigid basement fault blocks. Basin Res. 2004a;16(4):467–88.

Finch E, Hardy S, Gawthorpe R. Discrete-element modelling of contractional fault-propagation folding above rigid basement fault blocks. J Structural Geol. 2004b;25(4):515–28.

Forsythe GE, Wasow WR. Finite-difference methods for partial differential equations. New York: Wiley; 1960.

Gallagher RH. Finite element analysis fundamentals. NJ: Prentice Hall, Inc; 1975.

Gao FQ, Kang HP. Effects of pre-existing discontinuities on the residual strength of rock mass—insight from a discrete element method simulation. J Struct Geol. 2016;85:40–50.

Gao FQ, Stead D. The application of a modified Voronoi logic to brittle fracture modelling at the laboratory and field scale. Int J Rock Mech Min Sci. 2014;68:1–14.

Gao XW. 3D non-linear and multi-region boundary element stress analysis. Ph.D. Thesis. University of Glasgow (UK); 1999.

Gao XW, Davies TG. An effective boundary element algorithm for 2D and 3D elastoplastic problems. Int J Solids Struct. 2000;37(36):4987–5008.

Gary GG, Morgan JK, Sanz PF. Overview of continuum and particle dynamics methods for mechanical modeling of contractional geologic structures. J Struct Geol. 2014;59:19–36.

Gavete L, Gavete ML, Benito JJ. Improvements of generalized finite difference method and comparison other meshless method. Appl Math Model. 2003;27(10):831–47.

Gavete L, Ureña F, Benito JJ, García A, Ureña M, Salete E. Solving second order non-linear elliptic partial differential equations using generalized finite difference method. J Comput Appl Math. 2017;318:378–87.
Gavete L, Ureña F, Benito JJ, Salete E. A note on the dynamic analysis using the generalized finite difference method. J Comput Appl Math. 2013;252:132–47.
Gerya T. Introduction to numerical geodynamic modelling. Cambridge: Cambridge University Press; 2010.
Ghaboussi J. Fully deformable discrete element analysis using a finite element approach. Int J Comput Geotech. 1988;5(3):175–95.
Ghazvinian E, Diederichs MS, Quey R. 3D random Voronoi grain-based models for simulation of brittle rock damage and fabric-guided micro-fracturing. J Rock Mech Geotech Eng. 2014;6(6):506–21.
Ghosh S, Mukherjee S. Boundary element method analysis of thermoelastic deformation in nonhomogeneous media. Int J Solids Struct. 1984;20(9–10):829–43.
Gingold RA, Monaghan JJ. Smoothed particle hydrodynamics—theory and application to non-spherical stars. Mon Not R Astron Soc. 1977;181(3):375–89.
Gitman IM, Askes H, Sluys LJ. Representative volume: existence and size determination. Eng Fract Mech. 2007;74(16):2518–34.
Goodman RE, Taylor R, Brekke TL. A model for the mechanics of jointed rock. J Soil Mech Found Div ASCE. 1968;94(SM3):637–60.
Gori L, Penna SS, Pitangueira RLD. Smoothed point interpolation methods for the regularization of material instabilities in scalar damage models. Int J Numer Meth Eng. 2019a;117(7):729–55.
Gori L, Penna SS, Pitangueira RLD. G-space theory and weakened-weak form for micropolar media: application to smoothed point interpolation methods. Eng Anal Boundary Elem. 2019b;101:318–29.
Gu YT, Liu GR. A meshless local Petrov-Galerkin (MLPG) formulation for static and free vibration analyses of thin plates. Comput Model Eng Sci. 2001;2(4):463–76.
Hamajima R, Kawai T, Yamashita K, Kusabuka M. Numerical analysis of cracked and jointed rock mass. In: Kawamoto T, Ichikawa Y, editors. Proceedings of the 5th International Conference on Numerical Meth Geomech. Rotterdam: AA Balkema; 1985. p. 207–14.
Hao S, Liu WK. Moving particle finite element method with superconvergence: nodal integration formulation and applications. Comput Meth Appl Mech Eng. 2006;195(44–47):6059–72.
Hart RD, Cundall PA, Lemos J. Formulation of three-dimensional distinct element model. Part 2. Mechanical calculations for motion and interaction of a system composed of many polyhedral blocks. Int J Rock Mech Min Sci Geomech Abstr. 1988;25(3):117–25.
Hashin Z. Analysis of composite materials—a survey. J Appl Mech. 1983;50(3):481–505.
Hatzor YH, Arzi AA, Zaslavsky Y, Shapira A. Dynamic stability analysis of jointed rock slopes using the DDA method: King Herod's Palace, Masada, Israel. Int J Rock Mech Min Sci. 2004;41(5):813–32.
Hatzor YH, Benary R. The stability of a laminated Voussoir beam: back analysis of a historic roof collapse using DDA. Int J Rock Mech Min Sci. 1998;35(2):165–81.
Havaej M, Coggan J, Stead D, Elmo D. A combined remote sensing-numerical modelling approach to the stability analysis of Delabole Slate Quarry, Cornwall, UK. Rock Mech Rock Eng. 2016;49(4):1227–45.
Heinrich B. Finite difference methods on irregular networks-a generalized approacg to second order elliptic problem. Basel: Birkhäuser; 1987.
Heliot D. Generating a blocky rock mass. Int J Rock Mech Min Sci Geomech Abstr. 1988;25(3):127–39.
Herrmann HJ, Hansen A, Roux S. Fracture of disordered, elastic lattices in two dimensions. Phys Rev b: Condens Matter. 1989;39(1):637–48.
Heuze FE, Walton OR, Maddix DM, Shaffer RJ, Butkovich TR. Analysis of explosions in hard rocks: the power of discrete element modeling. In: Rossmanith HP, editor. Mechanics of jointed and faulted rocks. Vienna: AA Balkema; 1990. p. 21–8.

Hill R. Elastic properties of reinforced solids: some theoretical principles. J Mech Phys Solids. 1963;11(5):357–72.

Hinton E, Campbell JS. Local and global smoothing of discontinuous finite element functions using a least squares method. Int J Numer Meth Eng. 1974;8(3):461–80.

Hoover WG. Smooth particle applied mechanics: the state of the art. Singapore: World Scientific Publishing; 2006.

Hrennikoff A. Solution of problems of elasticity by the frame-work method. ASME J Appl Mech. 1941;8(4):A619–715.

Hsiung SM, Shi G. Simulation of earthquake effects on underground excavations using discontinuous deformation analysis (DDA). In: Elworth D, Tinucci JP, Heasley KA, editors. Rock Mechanics in the National Interest, Proceedings of 38th US Rock Mech Symp. Washington DC: Swets & Zeitlinger Lisse; 2001. p. 1413–20.

Hu YG, Lu WB, Chen M, Yan P, Zhang YZ. Numerical simulation of the complete rock blasting response by SPH-DAM-FEM approach. Simul Model Pract Theor. 2015;56:55–68.

Hudson JA, Priest SD. Discontinuity frequency in rock masses. Int J Rock Mech Min Sci Geomech Abstr. 1983;20(2):73–89.

Huerta A, Belytschko T, Fernández-Méndez S, Rabczuk T. Meshfree Methods. In: Stein E, De Borst R, Hughes TJR, editors. Encyclopedia of computational mechanics (vol. 1, Chap. 10). New York: Wiley; 2004a. p. 279–309.

Huerta A, Fernández-Méndez S. Enrichment and coupling of the finite element and meshless methods. Int J Numer Meth Eng. 2000;48(11):1615–36.

Huerta A, Fernández-Méndez S, Liu WK. A comparison of two formulations to blend finite elements and mesh-free methods. Comput Meth Appl Mech Eng. 2004b;193(12–14):1105–17.

Huet C. Coupled size and boundary-condition effects in viscoelastic heterogeneous and composite bodies. Mech Mater. 1999;31(12):787–829.

Hutter K, Jöhnk K. Continuum methods of physical modeling: continuum mechanics, dimensional analysis, turbulence. Berlin: Springer; 2004.

Idelsohn SR, Marti J, Limache A, Oñate E. Unified Lagrangian formulation for elastic solids and incompressible fluids: application to fluid-structure interaction problems via the PFEM. Comput Meth Appl Mech Eng. 2008;197(19–20):1762–76.

Idelsohn SR, Oñate E. To mesh or not to mesh. That is the question.... Comput Meth Appl Mech Eng. 2006;195(37–40):4681–96.

Idelsohn SR, Oñate E, Calvo N, Del Pin F. The meshless finite element method. Int J Numer Meth Eng. 2003a;58(6):893–912.

Idelsohn SR, Oñate E, Del Pin F. A Lagrangian meshless finite element method applied to fluid–structure interaction problems. Comput Struct. 2003b;81(8–11):655–71.

Idelsohn SR, Oñate E, Del Pin F. The particle finite element method: a powerful tool to solve incompressible flows with free-surfaces and breaking waves. Int J Numer Meth Eng. 2004;61(7):964–84.

Ikegawa Y, Hudson JA. A novel automatic identification system for three-dimensional multi-block systems. Eng Comput. 1992;9(2):169–79.

ITASCA. FLAC—Fast Lagrangian analysis of Continua, User's guide. Minneapolis: Itasca Consulting Group Inc.; 1986.

ITASCA. PFC3D—particle flow code in 3-dimensions (version 1.1), User's manual. Minneapolis: Itasca Consulting Group Inc.; 1995.

ITASCA. UDEC (version 5.0) manual. Minneapolis: Itasca Consulting Group Inc.; 2013a.

ITASCA. 3DEC (version 5.0) manual. Minneapolis: Itasca Consulting Group Inc.; 2013b.

ITASCA. PFC (version 5.0) manual. Minneapolis: Itasca Consulting Group Inc.; 2014.

Jaswon MA, Ponter AR. An integral equation solution of the torsion problem. Proc Roy Soc Londn (Ser A). 1963;273 (1):237–46.

Jensen PS. Finite difference technique for variable grids. Comput Struct. 1972;2(1–2):17–29.

Jiang QH, Yeung MR. A model of point-to-face contact for three-dimensional discontinuous deformation analysis. Rock Mech Rock Eng. 2004;37(2):95–116.

Jing L. Formulation of discontinuous deformation analysis (DDA)—an implicit discrete element model for block systems. Eng Geol. 1998;49(3–4):371–81.

Jing L. A review of techniques, advances and outstanding issues in numerical modelling for rock mechanics and rock engineering. Int J Rock Mech Min Sci. 2003;40(3):283–353.

Jing L, Ma Y, Fang Z. Modelling of fluid flow and solid deformation for fractured rocks with discontinuous deformation analysis (DDA) method. Int J Rock Mech Min Sci. 2001;38(3):343–55.

Jing L, Stephansson O. Topological identification of block assemblages for jointed rock masses. Int J Rock Mech Min Sci Geomech Abstr. 1994;31(2):163–72.

Johnson GR, Beissel SR. Normalized smoothing functions for sph impact computations. Int J Numer Meth Eng. 1996;39(16):2725–41.

Johnson GR, Beissel SR, Stryk RA. A generalized particle algorithm for high velocity impact computations. Comput Mech. 2000;25(2–3):245–56.

Kassab AJ, Divo E. A generalized boundary integral equation for isotropic heat conduction with spatially varying thermal conductivity. Eng Anal Bound Elem. 1996;18(4):273–86.

Kaczkowski Z, Tribillo R. A generalization of the finite difference formulas. Int J Numer Meth Eng. 1975;3(9):581–99.

Kanit T, Forest S, Galliet I, Mounoury V, Jeulin D. Determination of the size of the representative volume element for random composites: statistical and numerical approach. Int J Solids Struct. 2003;40(13–14):3647–79.

Kawai T. New discrete models and their application to seismic response analysis of structures. Nucl Eng Des. 1978;48(1):207–29.

Kazerani T, Yang ZY, Zhao J. A discrete element model for predicting shear strength and degradation of rock joint by using compressive and tensile test data. Rock Mech Rock Eng. 2012;45(5):695–709.

Kazerani T, Zhao J. Micromechanical parameters in bonded particle method for modelling of brittle material failure. Int J Numer Anal Meth Geomech. 2010;34(18):1877–95.

Keskin RSO, Hover KC, Grigoriu M. Size effects in modeling diffusivity of hardened mortar. Comput Struct. 2011;89(9):713–23.

Kikuchi A, Kawai T, Suzuki N. The rigid bodies-spring models and their applications to three-dimensional crack problems. Comput Struct. 1992;44(1–2):469–80.

Kim Y, Amadei B, Pan E. Modelling the effect of water, excavation sequence and rock reinforcement with discontinuous deformation analysis. Int J Rock Mech Min Sci Geomech Abstr. 1999;36(7):949–70.

Kim YR, Lutif J, Allen D. Determining representative volume elements of asphalt concrete mixtures without damage. Transp Res Rec: J Transp Res Board. 2009;2127(2):52–9.

Kim K, Rutqvist J, Nakagawa S, Birkholzer J. ARMA-2017-0382 (Ch. Discrete modeling of fluid-driven fracture processes in anisotropic rock formations). San Francisco, California: American Rock Mechanics Association; 2017.

Klein M. Mathematical thought from ancient to modern time. New York: Oxford University Press; 1972.

Kozicki J, Donzé FV. A new open-source software developed for numerical simulations using discrete modeling methods. Comput Meth Appl Mech Eng. 2008;197(49–50):4429–43.

Kozicki J, Donzé FV. YADE-OPEN DEM: an open-source software using a discrete element method to simulate granular material. Eng Comput. 2009;26(7–8):786–805.

Krongauz Y, Belytschko T. Enforcement of essential boundary conditions in meshless approximations using finite elements. Comput Meth Appl Mech Eng. 1995;131(1–2):133–45.

Lachat JC, Watson JO. Effective numerical treatment of boundary integral equations: a formulation for three-dimensional elastostatics. Int J Numer Meth Eng. 1976;10(5):991–1005.

Lancaster P, Salkauskas K. Surfaces generated by moving least-squares methods. Math Comput. 1981;37(155):141–58.

Lancaster P, Salkauskas K. Curve and surface fitting, an introduction. San Diego: Academic Press; 1986.

Laurendeau NM. Statistical thermodynamics: fundamentals and applications. New York: Cambridge University Press; 2005.

Le Veque JR. Finite volume methods for hyperbolic problems. Cambridge: Cambridge University Press; 2002.

Lei J, Xu YJ, Gu Y, Fan CM. The generalized finite difference method for in-plane crack problems. Eng Anal Boundary Elem. 2019;98:147–56.

Li G, Aluru NR. A boundary cloud method with a cloud-by cloud polynomial basis. Eng Anal Boundary Elem. 2003;27(1):57–71.

Li J, Rong JL. Bubble and free surface dynamics in shallow underwater explosion. Ocean Eng. 2011;38(17–18):1861–8.

Li JC, Hon YC, Chen CS. Numerical comparisons of two meshless methods using radial basis functions. Eng Anal Boundary Elem. 2002;26(3):205–25.

Li XX, Chen SH, Xu Q, Xu Y. Modeling the three-dimensional unsaturated water transport in concrete at the mesoscale. Comput Struct. 2017;190:61–74.

Li XX, Xu Y, Chen SH. Computational homogenization of effective permeability in three-phase mesoscale concrete. Constr Build Mater. 2016;121:100–11.

Libersky LD, Petschek AG. Smooth particle hydrodynamics with strength of materials. In: Trease HE, Crowley WP, editors. Advances in the free-lagrange method (Lecture Notes in Physics, vol.395). Berlin: Springer; 1990. p. 248–57.

Libersky LD, Petschek AG, Carney TC, Hipp JR, Allahdadi FA. High strain Lagrangian hydrodynamics: a three-dimensional SPH code for dynamic material response. J Comput Phys. 1993;109(1):67–75.

Libersky LD, Randles PW, Carney TC, Dickinson DL. Recent improvements in SPH modeling of hypervelocity impact. Int J Impact Eng. 1997;20(6–10):525–32.

Lin CT, Amadei B, Jung J, Dwyer J. Extensions of discontinuous deformation analysis for jointed rock masses. Int J Rock Mech Min Sci Geomech Abstr. 1996;33(7):671–94.

Lin D, Fairhurst C, Starfield AM. Geometrical identification of three dimensional rock block systems using topological techniques. Int J Rock Mech Min Sci Geomech Abstr. 1987;24(6):331–8.

Lin XS, Ng TT. A three-dimensional discrete element model using arrays of ellipsoids. Geotechnique. 1997;47(2):319–29.

Liszka T. An interpolation method for an irregular net of nodes. Int J Numer Meth Eng. 1984;20(9):1599–612.

Liszka T, Duarte CAM, Tworzydlo WW. hp-Meshless cloud method. Comput Meth Appl Mech Eng. 1996;139(1–4):263–88.

Liszka T, Orkisz J. Finite difference method for arbitrary irregular meshes in nonlinear problems of applied mechanics. In: IV SMiRt. San Francisco (USA); 1977.

Liszka T, Orkisz J. The finite difference method at arbitrary irregular grids and its application in applied mechanics. Comput Struct. 1980;11(1–2):83–95.

Liu GR. A generalized gradient smoothing technique and the smoothed bilinear form for Galerkin formulation of a wide class of computational methods. Int J Comput Meth. 2008;5(2):199–236.

Liu GR. Meshfree methods: moving beyond the finite element method. New York: CRC Press-Taylor & Francis Group; 2003.

Liu GR. A G space theory and a weakened weak (W2) form for a unified formulation of compatible and incompatible methods: Part I, theory. Int J Numer Meth Eng. 2010;81(9):1093–126.

Liu GR, Gu YT. A point interpolation method for two-dimensional solids. Int J Numer Methods Eng. 2001;50:937–95.

Liu GR, Liu MB. Smoothed particle hydrodynamics: a meshfree particle method. New Jersey: World Scientific Publishing Co, Pte Ltd; 2003.

Liu GR, Nguyen-Thoi T. Smoothed finite element methods. Boca Raton: CRC Press; 2010.

Liu GR, Zhang GY. Smoothed point interpolation methods—G space theory and weakened weak forms. 1st ed. New Jersey: World Scientific; 2013.

Liu GR, Zhang J, Lam KY, Li H, Xu G, Zhong ZH, Li GY, Han X. A gradient smoothing method (GSM) with directional correction for solid mechanics problems. Comput Mech. 2008;41(3):457–72.

Liu MB, Liu GR. Smoothed particle hydrodynamics (SPH): an overview and recent developments. Arch Comput Methods Eng. 2010;17(1):25–76.

Liu WK, Jun S, Li SF, Adee J, Belytschko T. Reproducing kernel particle methods for structural dynamics. Int J Numer Meth Eng. 1995a;38(10):1655–79.

Liu WK, Jun S, Zhang YF. Reproducing kernel particle methods. Int J Numer Meth Fluids. 1995b;20(8–9):1081–106.

Loehner R, Sacco C, Oñate E, Idelsohn S. A finite point method for compressible flow. Int J Numer Meth Eng. 2002;53(8):1765–79.

Londe P. Une methode d'analyse a trois dimensions de la stabilite d'une rive rocheuse. Ann Ponts Chaussees. 1965;135(1):37–60.

Loubignac G, Cantin G, Touzot G. Continuous stress fields in finite element analysis. AIAA J. 1977;15(11):1645–7.

Louis C. Rock hydraulics. In: Müller L, editor. Rock mechanics. Wien: Springer; 1974. p. 299–387.

Louis E, Guinea F. The fractal nature of fracture. EPL (eur Lett). 1987;3(8):871–7.

Lucy LB. A numerical approach to the testing of the fission hypothesis. Astron J. 1977;82(12):1013–24.

Ma GW, An XM, He L. The numerical manifold method: a review. Int J Comput Meth. 2010;7(1):1–32.

Ma GW, An XM, Zhang HH, Li LX. Modeling complex crack problems with numerical manifold method. Int J Fract. 2009;156(1):21–35.

Ma MY. Development of discontinuous deformation analysis, the first ten years: 1986~1996. In: Amadei B, editors. ICADD-3: third international conference on analysis of discontinuous deformation, Vail, Colorado (USA). Rotterdam: American Rock Mechanics Association, AA Balkema; 1999. p. 17–32.

Ma ZH, Wang H, Pu SH. A parallel meshless dynamic cloud method on graphic processing units for unsteady compressible flows past moving boundaries. Comput Methods Appl Mech Eng. 2015;285:146–65.

MacLaughlin MM, Doolin DM. Review of validation of the discontinuous deformation analysis (DDA) method. Int J Numer Anal Meth in Geomech. 2006;30(4):271–305.

MacNeal RH. An asymmetrical finite difference network. Q Appl Math. 1953;11(3):295–310.

Maerten L, Gillespie P, Daniel J. Three-dimensional geomechanical modeling for constraint of subseismic fault simulation. Am Assoc Pet Geol Bull. 2006;90(9):1337–58.

Mai-Duy N, Tran-Cong T. Mesh-free radial basis function network methods with domain decomposition for approximation of functions and numerical solution of Poisson's equations. Eng Anal Boundary Elem. 2002;26(6):133–56.

Mao ZR, Liu GR. A Lagrangian gradient smoothing method for solid-flow problems using simplicial mesh. Int J Numer Meth Eng. 2018;113(5):858–90.

Mao ZR, Liu GR, Dong XW, Lin T. A conservative and consistent Lagrangian gradient smoothing method for simulating free surface flows in hydrodynamics. Comput Part Mech. 2019a;6(4):781–801.

Mao ZR, Liu GR, Huang Y. A local Lagrangian gradient smoothing method for fluids and fluid-like solids: a novel particle-like method. Eng Anal Boundary Elem. 2019b;107:96–114.

Marti J, Cundall P. Mixed discretization procedure for accurate solution of plasticity problems. Int J Numer Anal Meth Geomech. 1982;6(1):129–39.

Mayer JM, Stead D. Exploration into the causes of uncertainty in UDEC grain boundary models. Comput Geotech. 2017;82:110–23.

McDonald PW. The computation of transonic flow through two-dimensional gas turbine cascades. In: ASME Proceedings of the New York (USA): ASME; 1971. Paper 71-GT-89.

Melenk JM. On generalized finite element methods. Ph.D. Thesis. University of Maryland (USA); 1995.

Melenk JM, Babuška I. The partition of unity finite element method: basic theory and applications. Comput Meth Appl Mech Eng. 1996;139(1–4):289–314.

Min KB, Jing L. Numerical determination of the equivalent elastic compliance tensor for fractured rock masses using the distinct element method. Int J Rock Mech Min Sci. 2003;40(6):795–816.

Min KB, Jing L. Stress dependent mechanical properties and bounds of Poisson's ratio for fractured rock masses investigated by a DFN-DEM technique. Int J Rock Mech Min Sci. 2004;41(Supplement 1):390–5.

Min KB, Rutqvist J, Tsang CF, Jing L. Stress-dependent permeability of fractured rock masses: a numerical study. Int J Rock Mech Min Sci. 2004;41(7):1191–210.

Moës N, Dolbow J, Belytschko T. A finite element method for crack growth without remeshing. Int J Numer Meth Eng. 1999;46(1):131–50.

Monaghan JJ. Smoothed particle hydrodynamics and its diverse applications. Annu Rev Fluid Mech. 2012;44:323–46.

Moosavi M, Grayeli R. A model for cable bolt-rock mass interaction: integration with discontinuous deformation analysis (DDA) algorithm. Int J Rock Mech Min Sci Geomech Abstr. 2006;43(4):661–70.

Mora P, Place D. A lattice solid model for the non-linear dynamics of earthquakes. Int J Modern Phys C. 1993;4(6):1059–74.

Morgan JK, McGovern PJ. Discrete element simulations of gravitational volcanic deformation: 1. Deformation structures and geometries. J Geophys Res Solid Earth. 2005a;110(B5):B05402.

Morgan JK, McGovern PJ. Discrete element simulations of gravitational volcanic deformation: 2. Mechanical analysis. J Geophys Res Solid Earth. 2005b;110(B5):B05403.

Mukherjee YX, Mukherjee S. The boundary node method for potential problems. Int J Num Meth Eng. 1997;40(5):797–815.

Müller L, editor. Rock mechanics. Wien: Springer; 1974.

Munjiza AA. The combined finite-discrete element method. London: Wiley; 2004.

Munjiza A, Andrews KRF. Penalty function method for combined finite-discrete element systems comprising large number of separate bodies. Int J Numer Meth Eng. 2000;49(11):1377–96.

Munjiza A, Andrews KRF, White JK. Combined single and smeared crack model in combined finite-discrete element analysis. Int J Numer Meth Eng. 1999;44(1):41–57.

Munjiza A, Owen DRJ, Bićanić N. A combined finite-discrete element method in transient dynamics of fracturing solid. Int J Eng Comput. 1995;12(2):145–74.

Nayroles B, Touzot G, Villon P. Generalizing the FEM: Diffuse approximation and diffuse elements. Comput Mech. 1992;10(12):307–18.

Newmark NM. A method of computation for structural dynamics. J Eng Mech Div ASCE. 1959;85(EM3):67–94.

Nicolazzi LC, Duarte CA, Fancello EA, de Barcellos CS. hp clouds–a meshless method in boundary elements. Part II: implementation. In: First Brazilian seminar on the boundary element method in engineering, Rio de Janeiro (Brazil, 1996). Int J Boundary Elem Methods Commun. 1997;8:83–5.

Oda M. A method for evaluating the representative elementary volume based on joint survey of rock masses. Can Geotech J. 1988;25(3):440–7.

Ohnishi Y, Chen G. Simulation of rock mass failure with discontinuous deformation analysis. J Soc Mater Sci Jpn. 1999;48(4):329–33.

Oñate E, Idelsohn S. A mesh-free finite point method for advective-diffusive transport and fluid flow problems. Comput Mech. 1998;21(4/5):283–92.

Oñate E, Idelsohn SR, Del Pin F, Aubry R. The particle finite element method: an overview. Int J Comput Meth. 2004;1(2):267–307.

Oñate E, Idelsohn SR, Zienkiewicz OC, Taylor RL. A finite point method in computational mechanics. Applications to convective transport and fluid flow. Int J Numer Meth Eng. 1996a;39(22):3839–86.

Oñate E, Idelsohn SR, Zienkiewicz OC, Taylor RL, Sacco C. A stabilized finite point method for analysis of fluid mechanics problems. Comput Meth Appl Mech Eng. 1996b;139(1–4):315–46.

Ouatouati AE, Johnson D. A new approach for numerical modal analysis using the element-flee method. Int J Numer Meth Eng. 1999;46(1):1–27.
Pan ZC, Ma RJ, Wang DL, Chen AR. A review of lattice type model in fracture mechanics: theory, applications, and perspectives. Eng Fract Mech. 2018;190:382–409.
Panizzo A, De Girolamo P, Di Risio M, Maistri A, Petaccia A. Great landslide events in Italian artificial reservoirs. Nat Hazards Earth Syst Sci. 2005;5(5):733–40.
Patankar SV. Numerical heat transfer and fluid flow. Washington DC: Hemisphere; 1980.
Patera AT. A spectral element method for fluid dynamics—laminar flow in a channel expansion. J Comput Phys. 1984;54(3):468–88.
Pavlin V, Perrone N. Finite-difference energy techniques for arbitrary meshes applied to linear plate problems. Int J Numer Meth Eng. 1979;14(5):647–64.
Pearce CJ, Thavalingam A, Liao Z, Bićanić N. Computational aspects of the discontinuous deformation analysis framework for modelling concrete fracture. Eng Fract Mech. 2000;65(2–3):283–98.
Perić D, Hochard CH, Dutko MD, Owen DRJ. Transfer operators for evolving meshes in small strain elasto-plasticity. Comp Meth Appl Mech Eng. 1996;137(3–4):331–44.
Perrone N, Kao R. A general finite difference method for arbitrary meshes. Comput Struct. 1975;5(23):45–58.
Place D, Mora P. The lattice solid model to simulate the physics of rocks and earthquakes: incorporation of friction. J Comp Physiol. 1999;150(2):332–72.
Plimpton S. Fast parallel algorithms for short-range molecular dynamics. J Comp Physiol. 1995;117(1):1–19.
Plimpton S, Hendrickson BA. A new parallel method for molecular dynamics simulation of macromolecular systems. J Comput Chem. 1996;17(3):326–37.
Potyondy DO, Cundall PA. A bonded-particle model for rock. Int J Rock Mech Min Sci. 2004;41(8):1329–64.
Prieto FU, Benito Muñoz JJ, Corvinos LG. Application of the generalized finite difference method to solve the advection-diffusion equation. J Comput Appl Math. 2011;235(7):1849–55.
Rabczuk T, Belytschko T, Xiao SP. Stable particle methods based on Lagrangian kernels. Comput Metd Appl Mech Eng. 2004;193(12–14):1035–63.
Rabczuk T, Xiao SP, Sauer M. Coupling of meshfree methods with finite elements: basic concepts and test results. Commun Numer Meth Engnf. 2006;22(10):1031–65.
Rasmussen LL, de Farias MM, de Assis AP. Extended rigid body spring network method for the simulation of brittle rocks. Comput Geotech. 2018;99:31–41.
Ritter A. Die fortpflanzung der wasserwellen. Z Ver Dtsch Ing. 1892;36(24):947–54 (in German).
Rizzi AW, Inouye M. Time split finite volume method for three dimensional blunt-body flows. AIAA J. 1973;11(11):1478–85.
Rizzo FJ. An integral equation approach to boundary value problems of classical elastostatics. Quart Appl Math. 1967;25:83–95.
Rizzo FJ, Shippy DJ. A formulation and solution procedure for the general nonhomogeneous elastic inclusion problems. Int J Solids Struct. 1968;4(12):1161–79.
Robert A. Dielectric permittivity of concrete between 50 Mhz and 1 Ghz and GPR measurements for building materials evaluation. J Appl Geophys. 1998;40(1–3):89–94.
Rothenburg L, Bathurst RJ. Micromechanical features of granular assemblies with planar elliptical particles. Geotechnique. 1992;42(1):79–95.
Roux S, Guyon E. Mechanical percolation: a small beam lattice study. J De Physique Lett. 1985;46(21):999–1004.
Roy R, Bathe KJ, Wilson EL. Numerical methods in finite analysis. New Jersey: Prentice-Hall Inc; 1976.
Sahimi M, Goddard JD. Elastic percolation models for cohesive mechanical failure in heterogeneous systems. Phys Rev B. 1986;33(11):7848–51.
Salami MR, Banks D, editors. Discontinuous Deformation Analysis (DDA) and simulations of discontinuous media. Albuquerque, NM: TSI Press; 1996.

Salete E, Benito JJ, Ureña F, Gavete L, Ureña M, García A. Stability of perfectly matched layer regions in generalized finite difference method for wave problems. J Comput Appl Math. 2017;312:231–9.

Saltzer SD, Pollard DD. Distinct element modeling of structures formed in sedimentary overburden by extensional reactivation of basement normal faults. Tectonics. 1992;11(1):165–74.

Savage HM, Cooke ML. The effect of non-parallel thrust fault interaction on fold patterns. J Struct Geol. 2004;26(5):909–17.

Schlangen E, Garboczi EJE. Fracture simulations of concrete using lattice models: computational aspects. Eng Fract Mech. 1997;57(2):319–32.

Scholtès L, Donzé FV. Modelling progressive failure in fractured rock masses using a 3D discrete element method. Int J Rock Mech Min Sci. 2012;52:18–30.

Sebsadji SK, Chouicha K. Determining periodic representative volumes of concrete mixtures based on the fractal analysis. Int J Solids Struct. 2012;49(21):2941–50.

Sellers E, Furtney J, Onederra I, Chitombo G. Improved understanding of explosive—rock interactions using the hybrid stress blasting model. J South Afr Inst Min Metall. 2012;112(8):721–8.

Sharma VM, Saxena KR, Woods RD, editors. Distinct element modelling in geomechanics. Rotterdam: Routledge; 2018.

Sheldon HA, Barnicoat AC, Ord A. Numerical modelling of faulting and fluid flow in porous rocks: an approach on critical state soil mechanics. J Struct Geol. 2006;28(8):1468–82.

Shi GH. Manifold method of material analysis. In: Transactions of the 9th Army conference on applied mathematics and computing. Report No. 92-1. Minneapolis: US Army Research Office; 1991. p. 57–262.

Shi GH. Modeling rock joints and blocks by manifold method. In: Tillerson JR, Wawersik WR, editors. Proceedings of 33rd US Symposium on Rock mechanics, New Mexico (USA). Rotterdam: AA Balkema; 1992a. p. 639–48.

Shi GH. Discontinuous deformation analysis: a new numerical model for the statics and dynamics of deformable block structures. Eng Comput. 1992b;9(2):157–68.

Shi GH. Three dimensional discontinuous deformation analysis. In: Elworth D, Tinucci JP, Heasley KA, editors. Rock Mechanics in the National Interest, Proceedings of the 38th US Rock Mechanics Symposium. Washington DC: Swets & Zeitlinger Lisse; 2001. p. 1421–8.

Shi GH, Goodman RE. Two dimensional discontinuous deformation analysis. Int J Numer Anal Meth Geomech. 1985;9(6):541–56.

Shi GH, Goodman RE. Generalization of two-dimensional discontinuous deformation analysis for forward modelling. Int J Numer Anal Meth Geomech. 1989;13(4):359–80.

Shyu K. Nodal-based discontinuous deformation analysis. Ph.D. Thesis. University of California, Berkeley (USA); 1993.

Skarzynski L, Tejchman J. Determination of representative volume element in concrete under tensile deformation. Comput Concrete. 2012;9(1):35–50.

Sladek V, Sladek J, Markechova I. An advanced boundary element method for elasticity problems in nonhomogeneous media. Acta Mech. 1993;97(1–2):71–90.

Spencer AJM. Continuum mechanics. London: Longman Group Limited; 1980.

Stolarska M, Chopp DL, Moës N, Belytschko T. Modeling crack growth by level sets in the extended finite element method. Int J Numer Meth Eng. 2001;51(8):943–60.

Stroeven M, Askes H, Sluys LJ. Numerical determination of representative volumes for granular materials. Comput Meth Appl Mech Eng. 2004;193(30–32):3221–38.

Stroeven P, Stroeven M. Size of representative volume element of concrete assessed by quantitative image analysis and computer simulation. Image Anal Stereol. 2001;20(Suppl 1):216–20.

Strayer LM, Erickson SG, Suppe J. Influence of growth strata on the evolution of fault-related folds—distinct-element models. In: McClay KR, editors. Thrust tectonics and hydrocarbon systems. American Association of Petroleum Geologists Memoir. 2004;82:413–37.

Strayer LM, Hudleston PJ. Numerical modeling of fold initiation at thrust ramps. J Struct Geol. 1997;19(3–4):551–66.

Strayer LM, Suppe J. Out of plane motion of a thrust sheet during along strike propagation of a thrust ramp: a distinct element approach. J Struct Geol. 2002;24(4):637–50.
Strouboulis T, Babuška I, Copps K. The design and analysis of the generalized finite element method. Comput Meth Appl Mech Eng. 2000a;181(1–3):43–69.
Strouboulis T, Copps K, Babuška I. The generalized finite element method: an example of its implementation and illustration of its performance. Int J Numer Meth Eng. 2000b;47(8):1401–17.
Strouboulis T, Copps K, Babuška I. The generalized finite element method. Comput Meth Appl Mech Eng. 2001;190(32–33):4081–193.
Sukumar N, Moës N, Moran B, Belytschko T. Extended finite element method for three-dimensional crack modelling. Int J Numer Meth Eng. 2000;48(11):1549–70.
Sukumar N, Moran B, Belytschko T. The natural element method in solid mechanics. Int J Numer Meth Eng. 1998;43(5):839–87.
Swedlow JL, Cruse TA. Formulation of boundary integral equations for three-dimensional elasto-plastic flow. Int J Solids Struct. 1971;7(12):1673–83.
Swegle JW, Hicks DL, Attaway SW. Smoothed particle hydrodynamics stability analysis. J Comput Phys. 1995;116(1):123–34.
Szmelter J, Kurowski Z. A complete program for solving systems of linear partial differential equations in plain domains. In: Proceedings of the 3rd conference on computational methods in structural mechanics. Opole (Poland); 1977. p. 237–47 (in Polish).
Taylor LM, Preece SD. DMC—a rigid body motion code for determining the interaction of multiple spherical particles. Research Report SAND-88-3482. Albuquerque, New Mexico: Sandia National Laboratories; 1989.
Taylor LM, Preece SD. Simulation of blasting induced rock motion using spherical element models. Eng Comput. 1990;9(2):243–52.
Te-Chin K. Improved modeling of rock bolting in DDA. In: Yuan JX, editors. Comp Meth Adv Geomech. Rotterdam: AA Balkema; 1997.
Telles JCF, Brebbia CA. On the application of the boundary element method to plasticity. Appl Math Model. 1979;3:466–70.
Telles JCF, Carrer JAM. Implicit procedures for the solution of elastoplastic problems by the boundary element method. Math Comput Model. 1991;15(3–5):303–11.
Thom A. The flow past circular cylinder at low speeds. Proc R Soc Londn Ser A. 1933;141:651–66.
Thomas AL. Poly3D: a Three-dimensional, polygonal element, displacement discontinuity boundary element computer program with applications to fractures, faults, and cavities in the Earth's crust. M.Sc. Thesis. Stanford University, California (USA); 1993.
Ting JM, Khwaja M, Meachum L, Rowell J. An ellipse-based discrete element model for granular materials. Int J Numer Anal Meth Geomech. 1993;17(9):603–23.
Tribillo R. Application of algebraic structures to a generalized finite difference method. Poland: Politechnika Bialostocka; 1976. (in Polish).
Turing AM. On computable numbers, with an application to the entscheidungs problem. In: Proceedings of the London Mathematical Society (Ser. 2). 1937;42:230–65.
Ureña F, Benito JJ, Alvarez R, Gavete L. Computational error approximation and H-adaptive algorithm for the 3-D generalized finite difference method. Int J Comput Methods Eng Sci Mech. 2005;6(1):31–9.
Ureña F, Benito JJ, Gavete L. Application of the generalized finite difference method to solve the advection-diffusion equation. J Comput Appl Math. 2011;235(7):1849–55.
Ureña F, Salete E, Benito JJ, Gavete L. Solving third-and fourth-order partial differential equations using GFDM: application to solve problems of plates. Int J Comput Math. 2012;89(3):366–76.
Van Mier JGM, Van Vliet MRA. Influence of microstructure of concrete on size/scale effects in tensile fracture. Eng Fract Mech. 2003;70(16):2281–306.
Van Vliet MRA, Van Mier JGM. Experimental investigation of size effect in concrete and sandstone under uniaxial tension. Eng Fract Mech. 2000;65(2–3):165–88.
Varga RS. Matrix iterative analysis. London: Prentice-Hall Int; 1962.

Vila LP. On particle weighted methods and smooth particle hydrodynamics. Math Models Meth Appl Sci. 1999;9(2):161–209.

Violeau D. Fluid mechanics and the SPH method. Oxford: Oxford University Press; 2012.

Walton OR. Explicit particle dynamics model for granular materials. Proc 4th Int Conf Numer Meth Geomech. Edmonton (Canada). 1982;1261–8.

Walton OR. Numerical simulation of inelastic, frictional particle-particle interactions. In: Roco MC, editor. Particulate two-phase flow. Boston: Butterworth-Heinemann; 1984. p. 884–911.

Wan RC. The numerical modeling of shear bands in geological materials. Ph.D. Thesis. University of Alberta, Edmonton (Canada); 1990.

Wang JG, Liu GR. On the optimal shape parameters of radial basis functions used for 2-D meshless methods. Comput Meth Appl Mech Eng. 2002;191(23–24):2611–30.

Wang S, Khoo BC, Liu GR, Xu GX. An arbitrary Lagrangian-Eulerian gradient smoothing method (GSM/ALE) for interaction of fluid and a moving rigid body. Comput Fluids. 2013;71:327–47.

Wang SL, Ge XL. Application of manifold method in simulating crack propagation. Chinese J Rock Mech Eng. 1997;16(5):405–10 (in Chinese with English abstract).

Wang WM, Chen SH. Automatic identification method for three-dimensional rock block systems. J Wuhan Univ Hydr Elec Eng (WUHEE). 1998;31(5):51–5 (in Chinese with English abstract).

Wang YF, Xu Q, Chen SH. Approaches of concrete creep using mesomechanics: numerical simulation and predictive model. Modell Simul Mater Sci Eng. 2019;27(5):55012.

Wang YF, Xu Q, Chen SH, Zhou XS, Liu HY. The influence of persistent curing-loading temperature on the basic creep of early-age concrete: experimental investigation. Constr Build Mater. 2020;261:119950.

Wang Z, Wang S, Yang Z. Manifold method in analysis of large deformation for rock. Chin J Rock Mech Eng. 1997;16(5):399–404 (in Chinese with English abstract).

Watson JO. Advanced implementation of the boundary element method for two-and three-dimensional elastostatics. In: Banerjee PK, Butterfield R, editors. Developments in boundary element methods (vol. 1). London: Applied Science Publishers; 1979. p. 31–63.

Wilkins ML. Calculation of elasto-plastic flow. Research report UCRL-7322. University of California (USA): Lawrence Radiation Laboratory; 1963.

Wilson CR, Witherspoon PA. An investigation of laminar flow in fractured porous rocks. Beikalay: University of California; 1970.

Williams JR, Pentland AP. Superquadrics and model dynamics for discrete elements in concurrent design. Technical Report. Massachusetts: Intelligent Engineering Systems Laboratory, Massachusetts Institute of Technology; 1991.

Wu CKC, Plesha M. Essential boundary condition enforcement in meshless methods: boundary flux collocation method. Int J Numer Meth Eng. 2002;53(3):499–514.

Wu JH, Ohnishi Y, Nishiyama S. Simulation of the mechanical behavior of inclined jointed rock masses during tunnel construction using Discontinuous Deformation Analysis (DDA). Int J Rock Mech Min Sci. 2004;41(5):731–43.

Xiao SP, Belytschko T. Material stability analysis of particle methods. Adv Comput Math. 2005;23(1–2):171–90.

Xu Y, Chen SH. A method for modeling the damage behavior of concrete with a three-phase mesostructure. Constr Build Mater. 2016;102:26–38.

Xu Y, Xu Q, Chen SH, Li XX. Self-restraint thermal stress in early-age concrete samples and its evaluation. Constr Build Mater. 2017;134:104–15.

Yazid A, Abdelkader N, Abdelmadjid H. State-of-the-art review of the X-FEM for computational fracture mechanics. Appl Math Model. 2009;33(12):4269–82.

Zeng W, Liu GR. Smoothed finite element methods (S-FEM): an overview and recent developments. Arch Comput Methods Eng. 2018;25(2):397–425.

Zhang J, Liu GR, Lam KY, Li H, Xu G. A gradient smoothing method (GSM) based on strong form governing equation for adaptive analysis of solid mechanics problems. Finite Elem Anal Des. 2008;44(15):889–909.

Zhang X, Sanderson DJ. Numerical study of critical behaviour of deformation and permeability of fractured rock masses. Mar Pet Geol. 1998;15(6):535–48.

Zhao J, Ohnishi Y, Zhao GF, Sasaki T, editors. Advances in discontinuous numerical methods and applications in geomechanics and geoengineering. Boca Raton: CRC Press; 2012.

Zhao SF, Zhao GF. Implementation of a high order lattice spring model for elasticity. Int J Solids Struct. 2012;49(18):2568–81.

Zhao T, Utili S, Crosta GB. Rockslide and impulse wave modelling in the Vajont reservoir by DEM-CFD analyses. Rock Mech Rock Eng. 2016;49(6):2437–56.

Zhou C, Huang B, Shu X. Micromechanical model for predicting coefficient of thermal expansion of concrete. J Mater Civ Eng. 2013;25(9):1171–80.

Zhu T, Zhang JD, Atluri SN. A local boundary integral equation (LBIE) method in computational mechanics, and a meshless discretization approach. Comput Mech. 1998;21:223–35.

Zienkiewicz OC, Best B, Dullage C, Stagg K. Analysis of nonlinear problems in rock mechanics with particular reference to jointed rock systems. In: Proceedings of 2nd ISRM Congress (vol. 3). Belgrade: ISRM; 1970. p. 501–9.

Zienkiewicz OC, Cheung YK. The finite element method in structural and continuum mechanics. London: McGraw-Hill; 1967.

Ziman JM. Models of disorder: the theoretical physics of homogeneously disordered systems. Cambridge: Cambridge University Press; 1979.

Chapter 2
Preparative Knowledge of Material Properties

Abstract Basic properties of water and solid applicable to advanced CMs will be summarizes in this chapter. Termed as "rock-like materials", rocks and concrete are mostly consumed in geotechnical and hydraulic structures, and their physical and mechanical properties are highly dependent on the embedded micro-or meso-components. Instead of being prevalently described by the phenomenological (conceptual) models on macro-scale level in fundamental CMs such as FEM and BEM, the advanced CMs that will be elaborated in this book endeavor to explicitly simulate these micro-or/and meso-components, hence the basic properties (attributes, deformability and failure resistance) related to the geology structural planes and concrete crack-tips are particularly focused on. The basic properties of water are briefly collected, too. It is notable that the basic properties summarized in this chapter on one hand, provides preparative knowledge of materials and on the other hand, highlights the difficulties in the practical application of CMs intended to get a quantitative "true solution" w.r.t. the real-world process of engineering structures.

2.1 General

Classical fluid mechanics and solid mechanics are concerned with macroscopic (bulk) phenomena rather than microscopic (molecular-scale) ones. The molecular makeup of a fluid or solid is studiously ignored, namely, the matter is looked at as a smoothly varying continuum, rather than a swarm of very fine, discrete molecules. As a result, it may be assumed that the basic physical and mechanical properties such as the mass density, viscosity, deformation modulus, strength, etc., are in principle definable at every point in our matter.

The space occupied by a body will be called as the "domain". Solids are materials that emerge more or less intrinsic configurations (shapes), whereas fluids in liquid state do not have any intrinsic shapes and will totally or partially fill the container to form a free surface in the presence of gravity.

S. Chen, *Advanced Computational Methods and Geomechanics*, Springer Tracts in Civil Engineering, https://doi.org/10.1007/978-981-19-7427-4_2

Hydraulics is a branch of fluid mechanics concerning the motion of water, whereas geomechanics is a branch of solid mechanics relevant to the deformation and failure of soil, rock and concrete. According to the similarity in physical and mechanical properties, concrete and rock are customarily dubbed as "rock-like" materials.

1. **Water**

Covering 71% of the Earth's surface, water is the main constituent of streams, lakes, and oceans. It is transparent and nearly colorless, and is vital for all known forms of life. Water moves continually through the water cycle of evaporation, transpiration, condensation, precipitation, and runoff reaching oceans or lakes.

The diameter of water molecule is approximately $d = 4 \times 10^{-10}$ m. At normal indoor temperature, the space between water molecules is $s = 3 \times 10^{-10}$ m. Under a standard pressure of 1 atm (101.325 kPa) and temperature of $T = 3.98$ °C, the bulk density of water is $\rho_w = 1\ \text{g/cm}^3$.

2. **Concrete**

Concrete is an artificial composite material that essentially consists of binding media within which are embedded aggregates. In geotechnical and hydraulic engineering, binder is normally formed from a mixture of hydraulic cement, flying ash, water, and several kinds of admixtures (Mehta and Monteiro 2006).

Most natural mineral aggregates, such as sands and gravels, produce the normal-weight concrete with an approximate density of $\rho_c = 2400\ \text{kg/m}^3$, which is subjected to a slight variation within the range of 1.5% following the concrete grade—it is generally higher for higher grade concrete. This variation should be carefully studied, particularly for gravity dams, since the dam size and corresponding expenditures will be significantly influenced by the concrete density.

Aggregates are granular materials, such as the sand, gravel, or construction/demolition waste that are mixed with a cementing medium to produce either the concrete or mortar. According to the aggregate size, the concrete may be dubbed by "one-graded" with aggregate size of 5–20 mm, "two-graded" by the additional aggregate size of 20–40 mm, "three-graded" by the additional aggregate size of 40–80 mm, and "four-graded (fully aggregated)" by the additional aggregate size of 80,120 mm (or 150 mm in dam works). In slim concrete structures such as the tunnel lining and sluice and spillway pavement, the one-or two-graded concrete is dominantly employed, this is mainly required by the construction technology, whereas in massive concrete structures such as the dams, the fully-graded concrete is desirable for the reason of economy and temperature control.

Cement is a finely pulverized artificial material which may develop the binding effect as a result of hydration. A cement is tagged by "hydraulic" when its hydration products are stable in the aqueous environment. The most commonly consumed hydraulic cement is the Portland cement that consists of reactive calcium silicates forming the "calcium silicate hydrates (C–S–H)" primarily responsible for its adhesive characteristics.

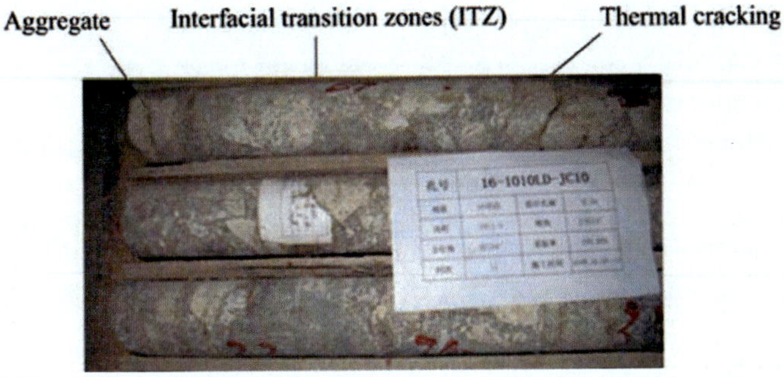

Fig. 2.1 Fully-graded concrete core extracted from the arch dam (Xiaowan Project, China)

Apart from aggregates, cement, and water, admixtures are commonly added to the concrete batch immediately before or during mixing. The use of admixtures in concrete may offer a variety of benefits such as to modify the setting and hardening characteristics of cement paste by chemically influencing the rate of cement hydration, to plasticize fresh concrete mixtures through cutting the surface tension of water by water-reducing admixtures, to improve the durability of concrete exposed to cold weather by air-entraining admixtures, and to reduce the thermal cracking in mass concrete by mineral admixtures such as pozzolan.

As an intrinsically heterogeneous and multiphase material, there are various factors such as the volume fraction and characteristics of principal constituents, the characteristics of "interfacial transition zone" (ITZ) (viz. Fig. 2.1), etc., affecting concrete properties significantly. In general, capillary voids and micro-cracks and oriented calcium hydroxide crystals are relatively more common in the ITZ than in the bulk cement paste matrix, therefore the ITZ often, although not always, provides an important clue in the exploration of concrete properties.

Although in a real-world geotechnical or hydraulic structure, most concrete is simultaneously subjected to a combination of compressive, shearing and tensile actions in two-or three-directions, yet the uniaxial compressive test is prevalently stipulated in design specifications. In China, Europe and many other countries, the compressive strength of cement and concrete determined by the standard uniaxial test using the cubes ($150 \times 150 \times 150$ mm^3) cured for 28 days is accepted as a primary index for grading. In the architecture industry of China, the strength grading is normally termed in a series of C15, C20, C25, C30, C35, C40, C45, C50, C55 and C60 where the subsequence digit denotes the uniaxial compressive strength (in MPa) of standard cubes at the age of 28 days.

Disregarding the effects such as the crack formation (matter of early age) and the chemical change (due to "alkali-aggregate reaction (AAR)" or aggressive water), growth of concrete age is in general beneficial to overall structural performance. An increase in the time-dependent elastic modulus leads to the reduction in elastic deflection, and an increase in the time-dependent strength leads to the improvement in structural safety. Attributable to the longer construction phase of hydraulic structures, the standards of China (GB50010-2010; SL191-2008) stipulate that in the design of concrete dams, the concrete grade may be specially the uniaxial compressive strength of 85% guarantee rate that is tested by the standard cube at the age of 90 days. Where the construction period is much longer, the design age may be equal to or even longer than 180 days subject to comprehensive justification. Under such circumstances, concrete strength grade is denoted as $C_{age(day)}$ Strength(MPa). For example, grade $C_{180}40$ means that standard concrete cubes (150 mm × 150 mm × 150 mm) possess uniaxial compressive strength 40 MPa tested at the age of 180 days.

3. **Rocks**

At a micro-or/and meso-scale level, rocks are composed of mineral grains held together by chemical bonds. Many rocks contain silica (SiO_2) forming crystals with other compounds, and the silica proportion in rock minerals is a major factor in determining their name and properties. Most brittle rocks comprise aggregates of crystal and amorphous particle bonded by varying amount of cementitious materials (Goodman 1989; Hudson 1997; Jaeger et al. 2007).

Over the geologic history, rocks can be transformed from one type to another, which is called the "rock cycle". These events produce three general classes of rocks, namely, "igneous rocks", "sedimentary rocks", and "metamorphic rocks":

- Igneous rocks are formed through the cooling and solidification of magma or lava;
- Sedimentary rocks are formed at the Earth's surface by the accumulation and cementation of fragments of earlier rocks, minerals, and organisms in water (sedimentation);
- Metamorphic rocks are formed by any rock types (i.e. sedimentary rocks, igneous rocks or another older metamorphic rocks) subjecting to the conditions of higher temperature or/and higher pressure than those in which the original rock was formed.

The rock density varies from the lowest 2.10 g/cm^3 (sandstone and limestone) to the highest 3.1 g/cm^3 (granite and basalt) depending on its type and weathering degree.

A widely recognized basic peculiarity of rocks is that they are fractured in the form of, for instance, faults and joints and fissures, when sufficiently strong actions are applied to them through natural or artificial processes. Being the weak links in the engineering structure, these fractures come under the term of "geological discontinuities". Significant geological discontinuities plus bedding planes and weak interlayers are commonly termed as "structural planes" that have certain shapes and sizes as well as orientations (attitudes). The overall geometrical characteristic of the major structural planes in a rock is termed as the "rock structure". A sufficiently large portion of rock entity with certain rock structure is termed as the "rock mass". Typical rock structures encountered are: blocky structure (Fig. 2.2a), layered (laminate) structure (Fig. 2.2b), fractured (fragmented) structure (Fig. 2.2c), and granular structure. The last one is actually looked at as a kind of soil or rock-fill in geotechnical and hydraulic engineering.

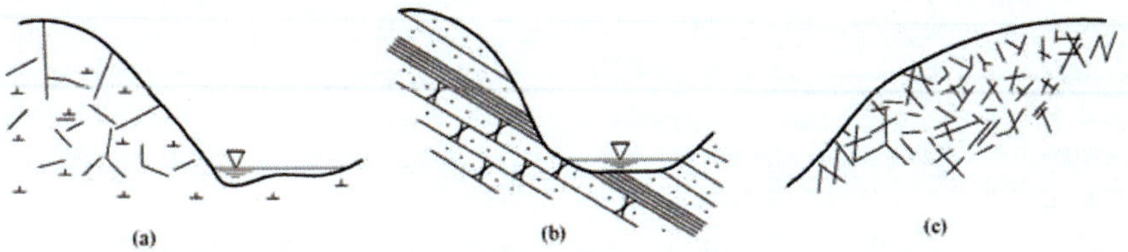

Fig. 2.2 Typical rock structures encountered in geotechnical and hydraulic engineering. **a** Blocky structure; **b** layered (laminate) structure; **c** fractured (fragmented) structure

Another important basic peculiarity of rocks is the initial in situ geo-stresses abbreviated as the "initial stresses", "in situ stresses" or "geo-stresses" that significantly affect rock characteristics and engineering performances. As an important geological setting of cut slopes, tunnels, and dam foundations, in situ stresses are the virgin natural stresses in the ground prior to any engineering activities (excavation and filling-up). They determine the initial conditions for the deformation/stress analysis and give rise to the stress adjustments accompanied with deformation when an opening/exposure surface is created (Goodman 1989; Jaeger et al. 2007).

The stability of geotechnical and hydraulic structures (e.g. dams) near or on ground surface are mainly governed by the structural planes in rock but the geo-stresses play another important role in the stability of geotechnical and hydraulic structures (e.g. tunnels) deeply embedded.

In the modern geotechnical and hydraulic engineering, rocks are generally described in the bore log by the following sequence of terms: drilling information, rock type, weathering, color, structure, rock quality designation (RQD), strength, and defects (Look 2007; Ulusay and Hudson 2007; Ulusay 2015). Since the first appearance of rational classification method (Terzaghi 1946; Lauffer 1958) for rock tunneling support, engineers had developed various classification schemes for rock masses such as, among many others, the "rock quality designation (RQD)" index (Deere et al. 1967) and the "rock structure rating (RSR)" index (Wickham et al. 1972).

All the existing rock mass classification schemes consider a few of key features (quantitative and qualitative), and assign digital values to the classes (grades). Hence they are basically the compromises between the use of a theory completely and the overlook of rock properties entirely. These schemes may provide a short-cut to the rock mass properties that are more difficult to assess (e.g. deformability), and provide direct guidance for the preliminary engineering design (e.g. type and amount of supporting for a tunnel).

Insofar, there are two prevalent classification schemes in the geotechnical engineering community essentially intended to estimate the supporting necessity of tunnels: one is the Q method by Barton and his co-workers (Barton et al. 1974; Barton 2002), another is the RMR system by Bieniawski (1973, 1978, 1989). Attempts also have been made to extend these classification systems, such as to the cut slopes (Romana 1985). In order to estimate the mechanical parameters according to rock properties and rock mass structures, Hoek and Brown proposed their famous strength criterion and the quantitative index GSI (Hoek and Brown 1980; Hoek 1994; Hoek and Brown 1997). As a quantitative description method of rock mass structural characteristics, GSI method has been continuously revised and improved (Sonmez and Ulusay 1999; Cai et al. 2004; Hoek et al. 2005).

In China, "Standards for Engineering Classification of Rock Mass (GB/T50218-2014)" stipulated in geotechnical and hydraulic engineering is supported by the test data of more than 100 engineering samples. In this standard, correlation analysis, cluster analysis and regression analysis are employed to find out that: for the basic quality classification of rock mass, the most appropriate quantitative factors are the uniaxial saturated compressive strength index and the rock mass integrity coefficient. Based on these factors, the formula relating them to the basic quality (BQ) is given in five levels (I–V grade) (Wu and Wang 2014). In addition, according to the 192 shear strength tests, 897 field pressure plate tests in 65 projects, and 350 shear strength tests for different types of structural planes in 40 projects, the range of mechanical parameters of rock mass corresponding to BQ level are recommended (Zhang et al. 2011).

2.2 Structural Planes

One of the principal factors that differentiates rocks from other materials is the presence of significant structural planes. These structural planes possess many geometrical and mechanical properties governing the overall performances of rock, particularly the anisotropy and heterogeneity in deformation, strength and permeability.

There are three fundamental types of structural plane according to their originations, namely, those that have been simply opened and come under the name of "joints", those that have been certain lateral movement and are termed as "shear zones" or "faults", those that emerge in sedimentary and layered volcanic rocks tightly linked with lithological stratigraphy and are dubbed as "bedding planes", "weak interlayers" or "intercalations".

In geotechnical literatures, the term "fracture" is employed to denote the first and second types of structural plane that are actually the separation in a geologic formation dividing the rock mass into several pieces. Fractures cause the rock to lose cohesion along its weakest plane and provide permeability networks for fluid movement. From the point view of fracture mechanics, a rock fracture can be created through three mechanisms: by pulling, by shearing, and by tearing. Accordingly, the terms "mode I, mode II and mode III" are respectively used to distinguish fractures. The mode of fracture is able to give a rough idea w.r.t. its likely mechanical properties.

However, it should be reminded that in many literatures of CMs, the term "fracture" is frequently, although not always, used for the geological discontinuity in rock and the crack in concrete, where no misunderstanding will be led to. This book will follow suit.

2.2.1 Particularizations of Structural Planes

Detailed features and origins of structural planes may be easily found in many textbooks on engineering geology, hence only the general description w.r.t. their major properties will be summarized below to provide the necessary background knowledge for our readers.

1. **Bedding planes**

Being highly persistent, bedding planes represent the interruptions in the course of rock deposition (Figs. 2.3 and 2.4). They may contain materials of different grain size, and may have been partially healed by low-order metamorphism. In either of these two cases, there would be some cohesion between bedding planes. Due to the depositional process, there may be a preferred orientation of particles in rock, giving rise to the weakness in parallel to the bedding plane.

Intercalations, also come under the name of "weak or soft interlayers", are large scale structural planes primarily emerging in sedimentary and layered volcanic rocks (Fig. 2.5). They are tightly linked with the lithological stratigraphy of rock due to the deposition of weak materials into soft and sensitive stratum layers. They are further disturbed in a manner of interlayer shearing deformation due to tectonic folding movements and followed by groundwater actions.

2. **Faults**

A fault is a large scale geological fracture in rock, across which there had been significant displacement as a result of rock movement (Fig. 2.6). It may be recognized

Fig. 2.3 Bedding in the limestone at the right abutment of RCC gravity dam (Guangzhao Project, China)

Fig. 2.4 Bedding in the metamorphic gneiss at the right abutment of arch dam (Xiaowan Project, China)

Fig. 2.5 Interlayer in the Basalt at the left abutment of arch dam (Xiluodu Project, China)

by the relative displacement of the rocks on the opposite sides of fault plane. Large faults within the Earth's crust result from the action of plate tectonic forces, and the energy release associated with the rapid rock movement is the cause of most earthquakes.

A fault plane is normally represented by its mean fracture surface, and the fault trace or fault line is the intersection of fault plane with ground/exposure surface that is commonly plotted on geologic maps. Since the geology fault usually do not consist of a single and clean fracture, geologists tend to use the term "fault zone" when referring to the zone of complex deformation associated with the fault plane.

Fig. 2.6 Fault in the metamorphic gneiss at the right abutment of arch dam (Xiaowan Project, China)

Fault thickness may vary from meters in regional geology structures to millimeters in local geology structures. Within the fault thickness, weak materials such as the fault gouge (clay), fault breccia (re-cemented), rock flour or angular fragment, may be embedded. Frequently, the wall of fault is slickenside and coated by minerals such as the graphite and chlorite with low frictional coefficients. The rock adjacent to a fault may be disturbed and weakened by associated local geological structures such as drag folding or secondary faulting.

Coming under the term of "shear zone", it is another type of large geology fractures up to several meters thick containing weak materials, in which the local shear failure of rock had previously taken place (Figs. 2.7, 2.8 and 2.9). Produced by the stress relief process or weathering, the fractured surfaces in the shear zone may be slickenside or coated by low-friction materials. Like faults, shear zones possess low shear strength but they may be much more difficult (not always) to be identified visually.

Faults may be pervasive large structural planes traversing the whole project area or may be limited within a portion of the project. In geotechnical and hydraulic engineering, a large fault may give rise to significant consequences on the adverse deformation and instability of, for example, the tunnel, dam foundation, or cut-slope.

3. **Joints**

As the most common and significant structural planes in rocks, joints are fractures of geological origin (Figs. 2.10 and 2.11) that occur with some degree of "clustering" around preferred orientations associated with their formation mechanisms. Hence, it is sometimes convenient to consider the concept of "joint set" that consists of parallel or nearly parallel joints.

Joints may be open, filled or healed. Sedimentary rocks often contain two orthogonal joint sets that are approximately perpendicular to bedding planes. These joints sometimes end at bedding planes, but others, called master joints, may cross several bedding planes.

Fig. 2.7 Shear zone in the metamorphic gneiss at the right abutment of arch dam (Xiaowan Project, China)

Fig. 2.8 Shear zone in the metamorphic gneiss of the headrace tunnel (Xiaowan Project, China)

Fig. 2.9 Shear zone in the Basalt at the left abutment of arch dam (Xiluodu Project, China)

Fig. 2.10 Orthogonal joint system in the metamorphic gneiss at the right abutment of arch dam (Xiaowan Project, China)

Fig. 2.11 Pillar joint system in the Basalt at the right abutment of arch dam (Xiluodu Project, China)

Joints have a significant influence on the mechanical and hydraulic properties of rock. Insofar, it is not completely successful to quantitatively answer the question as how the roughness, mechanical properties and boundary conditions of joints influence their frictional behavior, hence there exists numerous empirical and semi-theoretical models in an attempt to make the prediction (Bandis 1980; Bandis et al. 1981; Swam and Zongqi 1985; Leong and Randolph 1992; Haberfield 1995; Maksimovic 1996).

It is a very rare case where the spacing of joint planes is appreciably greater than the dimensions of a geotechnical and hydraulic structure. Very often, large-scale structural planes (e.g. faults) delineate rock blocks from a rock mass, and, within these blocks, lead to a further suite of small-scale structural planes (e.g. joints) that only partially delineate the rock mass. Such a hierarchical system of structural planes may significantly complicate the application of CMs. The fundamental FEM with implicit (equivalent) approach simply takes into account of the joint influences on the rock yield/failure criteria and compliance/permeability tensors meanwhile neglects their exact positions (Snow 1969; Pande and Gerrard 1983; Oda et al. 1984; Chen and Egger 1999), but it might be probably suffered from the insufficient joint amount for the existence condition of REV. On the contrary, the explicit (distinct) approach uses special elements (Mahtab and Goodman 1970; Desai et al. 1984; Chen and Egger 1997) to exactly simulate the geological and mechanical properties of joints, but it could meet difficulties in the pre-process to discrete the calculation domain into standardized mesh (grid). Actually, this dilemma is one of the strongest impetus in the development of advanced CMs since the 1980s.

To characterize jointed rocks using outcrop analogues, the key attributes of joints such as the center position, dip or strike direction, dip angle, density, trace length, and aperture, need to be individually and collectively specified (ISRM 1988). Although it is well-known that these joint attributes remarkably affect the mechanical properties of jointed rock, yet they are unfortunately difficult to precisely determine in engineering practice, which leads to the so called "uncertainties" that in turn, remind our engineering practitioners the risk of unrealistic results arise from a too delicate model with advances CMs without the competent input w.r.t. geological and material properties.

Although the uncertainty in rock mass properties is a challenge not having been well solved insofar, yet it is encouraging that the stochastic characteristics of joint attributes have been preliminarily studied by many scholars (Deere 1964; Deere and Miller, 1966; Baecher et al. 1977; Cruden 1977; Baecher 1983; Hudson and Priest 1983; Dershowitz 1984; Kulatilake and Wu 1984a, b; Chilès 1988; Hakami and Larsson 1990; Kulatilake et al. 1993a, b; Kulatilake et al. 1996). They assumed some a-priori probabilistic models of joint data that allow to define confidence intervals and to test statistical hypothesis. Their results provided selective probability distribution types and density functions for the joint attributes.

However, it is notable that aforementioned stochastic characteristics may be more confidently exercised in the fundamental CMs that in the advanced CMs, because the former is more stable w.r.t. the disturbance due to the uncertainty of joint attributes.

(1) Spacing and frequency

Spacing is the distance between adjacent joint planes and reciprocally related to joint frequency λ (Hudson and Priest 1983) defined as the number per unit distance. The definition of joint frequency can be further generalized in terms of joint density or intensity, the former represents the number of joints per unit volume, area or length, while the latter represents the total joint persistence per unit volume, area or length.

When a sufficiently large sample of these individual spacing data is plotted in histogram, a negative exponential distribution often emerges, indicating that the spacing distribution is most likely associated with the Poisson process as the result of a suite of superimposed random geological events (Snow 1970; Baecher et al. 1977), each of random geological event produces fracturing of a given distribution. In addition to the negative exponential distribution, lognormal or normal distribution depending on the degree of joint saturation in the network also may be applicable.

It is notable that the joint frequency is expected to vary with the direction of sampling line relative to the orientation of joints (Buyer and Schubert 2017).

(2) **Orientation**

Assuming a joint is planar, its strike/dip direction and dip angle uniquely define its orientation.

The orientation data can be processed using a rosette or stereogram so that joints can be grouped into different sets according to their orientations characterized by an appropriate distribution of uniform, normal or Fisher (Einstein and Baecher 1983).

(3) **Trace length**

The distribution of trace length will significantly depend on the orientation of the exposure surface, the associated orientation of the scanline, and the measurement treatments that are either truncated (values below a certain length are omitted) or censored (large values are either unobtainable because of limited rock exposure or because of equipment limitations). In engineering practice, truncation and censorship always occur that contribute to measurement bias. There are a considerable number of literatures (Baecher et al. 1977; Priest and Hudson 1981) concerning whether the distribution of trace length has a negative exponential distribution or log-normal distribution. It is rather likely that some of the differences are resulted from trace length sampling bias. At the moment, there is no clear and coherent measuring method for trace length, despite its crucial importance.

(4) **Persistence and shape**

Incorporating factors such as the shape and associated characteristic joint dimension (size) of joint plane, persistence is defined as the extent of a joint in its own plane (Shang et al. 2018). In practice, persistence is almost always measured by the one-dimensional extent of the trace lengths on the exposure surface.

Most prevalent postulations w.r.t. joint shape are circle and ellipse. According to a number of literatures (Cruden 1977; Dershowitz 1984; Davy 1993; Bonnet et al. 2001), joint size observes the distribution laws of negative exponential, lognormal, gamma or power.

(5) **Block size**

In terms of the excavation and support for tunnel and cut-slope works, it is helpful to have an estimation of average block size and block size distribution, which may be (in situ) analogous to the particle size distribution used in, for example, DEM. The block size is directly related to the factors such as the spacing distribution of joints, the persistence and intersection between joint sets, etc.

(6) **Roughness**

Although joints are postulated to be planar, the joint surface is actually not smooth. The term "roughness" is defined as the deviation of a joint surface from perfect plane, which may be indexed by either geometrical amplitude of asperities from joint profile length reference to standard charts, or mathematical procedure utilizing surface characterization techniques inclusive geo-statistics, Fourier series, noise wave forms, and fractals, etc. The roughness of rock joint is intimately related to its various mechanical and hydraulic properties.

(7) **Aperture**

Being both mechanical and hydraulic important, joint aperture is defined as the perpendicular distance between joint wall surfaces. It will be constant for parallel and planar joints and completely variable for rough joints. Joint apertures may be determined from a detailed field mapping (Jourde et al. 2002) and may be further calibrated by water flow simulation w.r.t. in-situ measurements (Taylor et al. 1999) to get "hydraulic aperture". Current research outcomes in rock hydraulics indicate that a joint cannot be exactly approximated by two parallel plates because of the phenomenon of channel flow, namely, the water mainly flows through certain channels on the joint surface created by the tracks of larger local apertures.

A distribution of joint apertures is expected, more commonly, to follow an a-priori statistical distribution (sometimes correlated with trace lengths). Joint apertures usually observe the lognormal distribution or power law distribution (Snow 1970; Barton and Zoback 1992; Hooker et al. 2009), and may be related to joint size by a power law with a linear or sublinear scaling relationship (Pollard and Segall 1987; Bonnet et al. 2001; Olson 2003).

2.2.2 *Measuring and Mapping of Rock Joint Attributes*

Natural joints are formed under self-organized dynamics mechanism, where breakage and fragmentation can occur at all scales. They are subject to in-situ stress actions at certain depth and can exhibit intricate topologies, such as cross-cutting, abutting, branching, terminating, bending, and clustering. Field data are usually collected from lower dimensional and limited exposures, such as one-dimensional (1-D) borehole logging and two-dimensional (2-D) outcrop mapping (Einstein Baecher 1983). The direct observation of detailed 3-D joint networks deep in the Earth's crust is impossible because the current technology can hardly detect widely-spreading medium and joints with sufficient resolution, although seismological surveys are able to locate large-scale structural planes (e.g. faults).

Traditionally, joint patterns can be mapped from the exposure surface of rock outcrops or man-made excavations (e.g. borehole, quarry, tunnel and road-cut) by the combination of visual and manual technologies. For example, the most common tool to directly measure the dip angle and dip direction of joint is the geological clino-compass. The geotechnical data acquisition is nonetheless limited in accessibility, geological or geotechnical knowledge, time and scale. The results are often subjective rather than objective and therefore not exactly reproducible.

To reduce the inherent bias of a single measurement, four main sampling strategies for collecting joint data are widely exercised (Watkins et al. 2015): the linear scanline sampling (Priest and Hudson 1981), the rectangular window sampling (Priest 1993), the area sampling (Wu and Pollard 1995) and the circular scanline sampling (Mauldon 1998; Mauldon et al. 2001; Rohrbaugh et al. 2002). It is notable that terrestrial remote sensing techniques have been quickly developed and are able to better capture the large-scale trace features of structural plane (Sturzenegger and Stead 2009a, b; Sturzenegger et al. 2011; Jones et al. 2016).

The measured data may be biased under truncation and censoring effects, hence they need to be amended for a better determination of underlying statistical distributions (Laslett 1982; Pickering et al. 1995).

A general description of measuring techniques for joint attributes will be given below. The details w.r.t. the measuring operations, the topological parameter calculations, as well as the use of joint rosette or stereographic projections, are not addressed here since they may be easily found in relevant textbooks, if necessary.

1. **Linear scanline sampling**

Figure 2.12 shows a linear scanline L for fast recording a wide range of joint attributes. This involves laying a tape on an outcrop and measuring the attributes of each joint that intersects the tape (Cox and Lewis 1966). Normally, there are a number of such scanlines of different orientations so as to best represent joint sets. Each scanline is preferably set up perpendicular to the strike of each joint set on the outcrop, and joint attributes are measured for this joint set only (Odling et al. 1999).

Although the linear scanline method allows a lot of joint attribute data to be collected quickly, yet it gives rise to orientation and length bias, and is sensitive

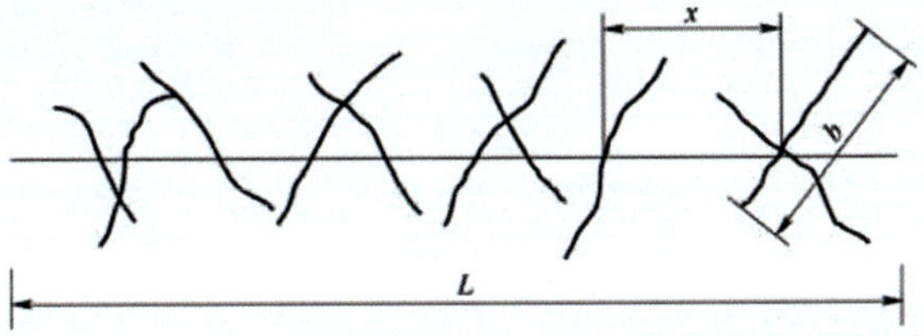

Fig. 2.12 Quantification of joint occurrence along a sample line

to censoring operation. Joints striking at a low angle to the scanline will be under-represented, giving remarkable underestimation for joint intensity and overestimation for joint spacing, unless these biases are corrected.

2. **Rectangular window sampling**

Rectangular window sampling utilizes a rectangle placed on an outcrop, and selected joint attributes are measured within the rectangle. Compared with the linear scanline sampling, this method may reduce orientation bias because all the joints within the rectangle area are measured. It allows for a simple estimation of mean trace length, too. The method can be very time consuming where many attributes are to be measured for each joint within the window area. Similar to the scanline sampling method, the measured data from window sampling are also affected by censoring due to the limitation of outcrop size and exposure quality. For example, mean trace length estimation involves analyzing joint end points, but if the outcrop has heavy vegetation cover, the estimated data may be unreliable.

For the tail water surge chamber in Guandi Project (Fig. 2.13), the geological sketch by rectangular window sampling and the correspondent block combinations at crest and upstream wall are presented in Fig. 2.14.

3. **Area sampling**

Instead of the perpendicular distance between two neighboring joints, the total length of all joints is measured on the area of a counting square. The measured joint spacing has a single value for a given outcrop area and is not affected by joint distribution within that area. This method may be useful with borehole data from a subsurface layer.

4. **Circular scanline sampling**

Fig. 2.13 Tail water surge chamber under the excavation operation (Guandi Project, China)

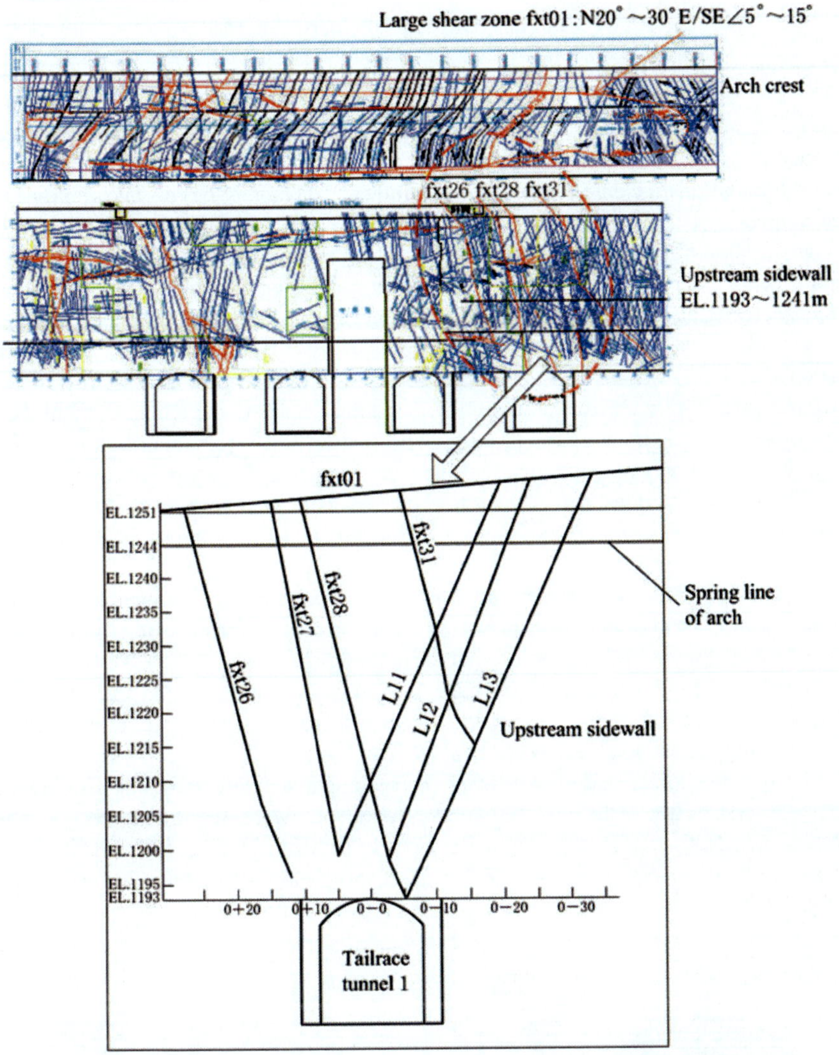

Fig. 2.14 Geological sketch and block combination at the crest and upstream wall of the tail water surge chamber (Guandi Project, China)

Instead of directly measuring joint attributes, the circular scanline method involves counting the number of joint intersections with the edge of a circular line placed on an outcrop area, together with the number of joint terminations within the circle. These values are input into a series of equations, from which the joint density, intensity and mean trace length within the circular area can be calculated. The remarkable merit of this method is insusceptible to length censoring.

2.2.3 Geometrical Representation of Joint System

There hardly exists a rock mass without any joint sets. Hence, it is of crucial importance in geotechnical and hydraulic engineering to have a profound knowledge concerning the 3-D joint system in rocks.

Although seismic data can be used to build the 3-D map of large-scale geological structures, yet the low resolution often obscures their detailed features (e.g. the segmentation of fault) and impedes the detection of small geological fractures (e.g. joint sets). Therefore, the major difficulty in modelling the jointed rock mass is the geometrical representation of 3-D joint system based on relatively reliable 1-D or/and 2-D sampling data by stereological analysis (Warburton 1980; Berkowitz and Adler 1998). As a result, the description of natural joint geometry characteristic have to largely rely on extrapolations from small samples to the whole domain. For example, borehole imaging can provide an estimation of 3-D joint distribution, but its confidence can only be guaranteed for the area close to the borehole (Wu and Pollard 2002). Hence, the question of "how to create realistic joint networks?" remains an unresolved issue, although continuous advancement is in progress (Wilson et al. 2011).

For the moment, the representation of orientation data and the identification of joint sets are commonly performed using the hemispherical projection of joint poles (i.e. unit vectors with direction normal to joint planes) (Priest, 1985). The grouping of joints into joint sets and the characterization of their orientations are important aspects of rock mass characterization for engineering applications (Priest, 1993; Starzec and Andersson 2002; Park et al. 2005).

From 1849 joints undertaken from the rectangle window sampling in Guandi Project (viz. Figs. 2.13 and 2.14), six joint sets were identified whose attitudes are: N30°–50°W/NE∠75°–85°; N5°–15°E/SE∠10°–20°; EW/ S(N)∠60°–75°; SN/W∠55°–60°; N30°–55°W/SW∠35°–50°; N60°–70°E/SE∠70°–80° (viz. Fig. 2.15).

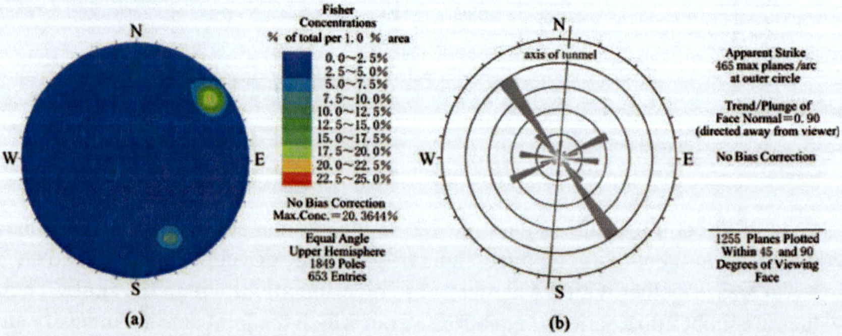

Fig. 2.15 Joint polar projection and strike direction of the tail water surge chamber (Guandi Project, China). **a** Polar isodensity diagram; **b** strike direction

Table 2.1 Dominant joint sets in tail water surge chamber (Guandi Project, China)

Joint set sequence	Description	Attitudes
1	Dislocation zone fxt01	N20°–30° E/SE∠5°–15°
2	Dislocation zone fxt26, fxt27,fxt28, fxt31	N35° W/NE∠80°–85°
3	Large fracture L11–L13	N65°–70° E/SE∠75°–80°

For the classification and grouping of joint sets, both the method assuming an initial probabilistic model of joint orientation data (Shanley and Mahtab 1976; Dershowitz et al. 1996; Marcotte and Henry 2002) and the method without a-priori probabilistic model (Hammah and Curran 1999; Zhou and Maerz 2002) are widely exercised. In addition, the method based on "artificial neural networks (ANN)" (Sirat and Talbot 2001) may be employed, too.

These geologically-mapped joint networks are widely used to understand the process of joint formation (Pollard and Segall 1987), to interpret the history of tectonic actions (Engelder and Geiser 1980; Olson and Pollard 1989), and to derive the statistics and scaling of joint populations (La Pointe 1988; Bour et al. 2002). It is more important for engineering practitioners that they, being well trained and experienced, are already able to qualitatively prognose the local failure modes and scales around an excavation face, and even further to quantitatively predict their safety margin by the principle of limit equilibrium to help the reinforcement design or evacuation early-warning.

The dominant joint sets in Guandi Project (viz. Fig. 2.15) are listed in Table 2.1.

Figure 2.16 shows the joint system on the upstream wall of the underground power house (Guandi Project, China). Figure 2.17 shows the typical failure modes emerged during the excavation, these failure modes are closely related to the joint network attributes mapped previously.

2.3 Discrete Fracture Networks (DFNs)

Rock joints often construct complex networks and dominate the mechanical behaviors of rock mass (e.g. heterogeneous stress fields and channelized fluid flow pathways) in geological formations. For the quantitative solution of deformation/seepage/stability issues using fundamental and advanced CMs, a comprehensive understanding of joint system is indispensable. Although the situation is still far from satisfaction, yet several significant advancements have been achieved, particularly on the aspects of mathematical tools such as the stochastic analysis based "discrete fracture network (DFN)". In geotechnical literature, a "fracture" is any separation in a geologic formation, such as the joint or the fault that divides the rock into pieces. A

Fig. 2.16 Joint system on the upstream wall of the underground power house (Guandi Project, China)

Fig. 2.17 Typical failure modes occurred in the cavern group of the underground power house (Guandi Project, China). **a** Block failure; **b** collapse at the crest due to slightly dipped angle fault; **c** failure along the highly dipped joints at side wall

DFN refers to a geological model that explicitly represents the attributes (e.g. orientation, size, position, shape) of each individual joint, and the topological relationships between joints as well. It can be generated from geological mapping, geomechanical simulation, digitized outcrop analogue, and stochastic realization, the last one will be elucidated below.

2.3.1 Concepts and State-of-the-Arts

Due to the difficulty in performing a complete 3-D measurement of natural joint systems, the stochastic DFN method using statistics for limited sampling data was proposed in the 1980s for the percolation of finite-sized joint populations (Balberg and Binenbaum 1983; Robinson 1983, 1984) and the fluid flow in complex joint networks (Long et al. 1982; Long et al. 1985; Andersson and Dverstorp 1987; Lin et al. 1987; Long and Billaux 1987; Dershowitz and Einstein 1988; Heliot 1988). It postulates that rock joints are straight lines (in 2-D) or planar discs/polygons (in 3-D) (Baecher 1983), and their geometrical properties (e.g. position, frequency, size, orientation, aperture) are random variables observing certain probability distributions derived from the field measurement w.r.t. outcrop traces (Zhang and Einstein 2000).

The stochastic DFN model, in essence, treats problems in a probabilistic framework and regards the real physical system as one possibility among simulated realizations that share identical statistics. Hence, a sufficient number of realizations based on a Monte-Carlo process are demanded to predict the bounded range. In practice, a balance exists between the benefits of collecting detailed information to create more representative DFNs and the increased efforts of field measurements. The uncertainty can be reduced by constraining the random process with some deterministic data, for example, to enforce the 3-D DFN generator reproducing a 2-D trace pattern exposed on tunnel walls (Dverstorp and Andersson 1989). Outcrop-based DFNs have also been constructed by many other researchers (Zhang and Sanderson 1995; Griffith et al. 2009). Advantages of such a DFN model include the preservation of natural features of joints (e.g. curvature and segmentation) and the unbiased characterization of complex topologies (e.g. intersection, truncation, arrest, spacing, clustering and hierarchy).

In the creation of stochastic DFN, joints can be randomly located and their geometrical attributes are sampled from the corresponding probability density functions (Dershowitz and Einstein 1988). Such a random fracture network approach is normally termed as the conventional "Poisson DFN model" or "Baecher model" and prevalently employed to study the connectivity, fragmentation, deformability, permeability and transport properties of jointed rocks (Zimmerman and Bodvarsson 1996a, b; Bour and Davy 1997; Bour and Davy 1998; de Dreuzy et al. 2001a, b; de Dreuzy et al. 2004; Baghbanan and Jing 2007, 2008; Zhang and Lei 2013, 2014) (only a few among many others).

However, the conventional Poisson DFN model tends to result in large uncertainties due to its assumptions in homogeneous spatial distribution, simplifications in joint shape, negligence of the correlations between different geometrical properties, and disregard of diverse topological relations (Odling and Webman 1991; Odling 1992, 1994; Berkowitz and Hadad 1997; Belayneh et al. 2009; Lei et al. 2014). It is particularly emphasized that an alternative to the Poisson DFN model may make use of the spacing distribution to locate joints in the stochastic generator (Lu and Latham 1999), which will be more suitable for highly persistent joint systems. Other improvements on the Poisson DFN model having been made include the correlation between joint attributes, unbroken areas inside individual joint planes, characterization of joint intersection types, mechanical interaction between neighboring joints, etc. (Allègre et al. 1982; Hentschel and Procaccia 1983; Long and Billaux 1987; Chilès 1988; Pollard and Aydin 1988; Billaux et al. 1989; Bour and Davy 1999; Ehlen 2000; Horgan and Young 2000; Bonnet et al. 2001; Josnin et al. 2002; Darcel et al. 2003a, b; Jourde et al. 2007; Masihi and King 2008; Neuman 2008; Xu and Dowd 2010; Shekhar and Gibson 2011; Davy et al. 2013; Lei et al. 2014; Lei et al. 2015; Sanderson and Nixon 2015).

It is emphasized that the calibration and validation of stochastic DFNs are highly necessary, which should be conducted using the in-situ data from field mapping, mechanical and hydraulic tests, etc. (Dverstorp and Andersson 1989; Cacas et al. 1990a, b; Sarda et al. 2002; Casciano et al. 2004; Lang et al. 2014).

For more detailed messages w.r.t. DFN models, readers are referred to the relevant reviews and textbooks (Dershowitz and Einstein 1988; Adler and Thovert 1999; Adler et al. 2012; Fadakar-Alghalandis 2014).

2.3.2 Generation of Stochastic DFNs by Monte-Carlo Method

"Monte-Carlo method (MCM)" is widely employed to generate DFNs in the region termed as "numerical sampling window". Normally, it is a square in 2-D cases and cube in 3-D cases.

The created DFNs can serve as important inputs for modelling the behaviors of jointed rocks. To reduce the numerical difficulties whilst to capture the key characteristics, simplifications need to be undertaken. In conventional studies, inactive joints (i.e. they are dead-end, isolated or very small) are often removed from original DFNs because of their negligible contribution to the overall permeability/deformability. Such a treatment may be applicable for low permeability/deformability rocks with well-connected joint sets. However, when all the joints are poorly connected, the interaction effects of fracture-matrix system could be significant. Under such circumstances, attentions should be drawn in setting a truncation threshold for the deletion of small-sized joints, because several studies (Berkowitz et al. 2000) have suggested that the connectivity of joint networks is ruled by joints much smaller than the system size and thus the removal of very small joints may lead to important biases. For example, from the dead-end tips of pre-existing joints under stress concentrations, new fractures may grow and link with each other to form critical pathways for fluid flow and solid failure.

1. **Pseudo-random number generator**

In a pseudo-random number generator, the random number for any distribution is transformed from the uniform random number in the interval [0, 1] according to Table 2.2. The associated algorithms fall into two categories:

Table 2.2 Random number generations for different probability distributions

Probability distribution	Density function	Mean	Standard deviation	Random number	Applicability
Uniform	$f(x) = \begin{cases} \frac{1}{b-a}, a \le x \le b, \\ 0, \ b < x < a. \end{cases}$	$\frac{1}{2}(a+b)$	$\frac{1}{2\sqrt{3}}(b-a)$	$x = a + (b-a)R$	Center position, Dip direction
Negative exponential	$f(x) = \begin{cases} \lambda \exp(z), x > 0, \\ 0, \ x \le 0. \end{cases}$ $z = -\lambda x, \lambda > 0.$	$\frac{1}{\lambda}$	$\frac{1}{\lambda}$	$x = -\frac{\ln(1-R)}{\lambda}$	Trace length
Normal	$f(x) = \frac{1}{\sqrt{2\pi}\sigma} \exp(z),$ $z = -\frac{1}{2}(\frac{x-\mu}{\sigma})^2.$	μ	σ	$u = \begin{cases} \cos(2\pi R_1)\sqrt{-2\ln(R_2)} \\ R_3 \ge 0.5 \\ \sin(2\pi R_1)\sqrt{-2\ln(R_2)} \\ R_3 < 0.5 \end{cases}$ $x = \mu + \sigma \cdot u$	Aperture

(continued)

Table 2.2 (continued)

Probability distribution	Density function	Mean	Standard deviation	Random number	Applicability
Log-normal	$f(x) = \frac{1}{\sqrt{2\pi}\sigma x}\exp(z)$, $z = -\frac{1}{2}(\frac{\ln x - \mu}{\sigma})^2$.	$\exp(\mu + \frac{\sigma^2}{2})$	$\sqrt{\exp(\gamma)(\exp(\sigma^2) - 1)}$, $\gamma = 2\mu + \sigma^2$.	$u = \begin{cases} \cos(2\pi R_1)\sqrt{-2\ln(R_2)} & R_3 \geq 0.5 \\ \sin(2\pi R_1)\sqrt{-2\ln(R_2)} & R_3 < 0.5 \end{cases}$ $x = \exp(\mu + \sigma \cdot u)$	Trace length, Aperture
Fisher	$f(x) = \frac{\eta \sin\theta \cdot \exp(\eta\cos\theta)}{\exp(\eta) - \exp(-\eta)}$.			$x = \arccos\left(\frac{\ln(1-R)}{\eta} + 1\right)$	Dip angle
Poisson	$P(x = k) = \frac{\lambda^k e^{-\lambda}}{k!}$, $k = 0, 1, \ldots$, $\lambda > 0$.	λ	λ	$F(n) = \sum_{k=0}^{n} \frac{\lambda^k e^{-\lambda}}{k}$, if $F(n) \leq R < F(n+1)$, $x = n$.	Density

- Continuous probability distribution. The random number is obtained by the inverse of its cumulative distribution function (2.1)

$$x = F^{-1}(R) \tag{2.1}$$

in which x is the random number for continuous probability distribution, F is the cumulative distribution function and F^{-1} denotes its inverse, R is the uniform random number in the interval [0, 1].

- Discrete probability distribution. The random number is obtained by trial-and-error methods on the cumulative distribution function (2.2)

$$F(n) \le R < F(n+1) \tag{2.2}$$

in which n is the random number (positive integer) for discrete probability distribution.

2. **Mainstream procedure**

The Monte-Carlo method for the generation of DFN is implemented by following mainstream steps:

① Investigation and record of the joint system for a sampling window in the geology unit concerned;
② Statistics analysis of the measured data;
③ Monte-Carlo simulation and production of random joints;
④ Construction of the stochastic NFN in the sampling window.

In general, the attributes of rock joints are correlated (e.g. length versus aperture). These correlations request extra field measuring data that are not easy to be accessed by most engineering practices. Without loss of generality, the independent assumption of these attributes is adopted in this book to simplify the generation of stochastic parent DFN.

In the following discussion, N_{sjoint} denotes the amount of joint sets, N_{joint} denotes the amount of joints in joint set I_{sjoint}, N denotes the amount of stochastic parent DFNs generated in sampling window. By the assumption that the geometrical attributes (mid position, dip direction, dip angle, trace length, aperture) of each joint are independent, the DFN generating algorithm is flow charted in Fig. 2.18.

Towards the REV study and elastic compliance tensor identification of the arch dam abutment rock (Xiaowan Project, China), one of 2-D stochastic DFNs extracted by the rock sample (10 m × 10 m) is given in Fig. 2.19. Figure 2.20 present another 3-D stochastic DFN extracted from the right bank slope of intake tunnel in the same project.

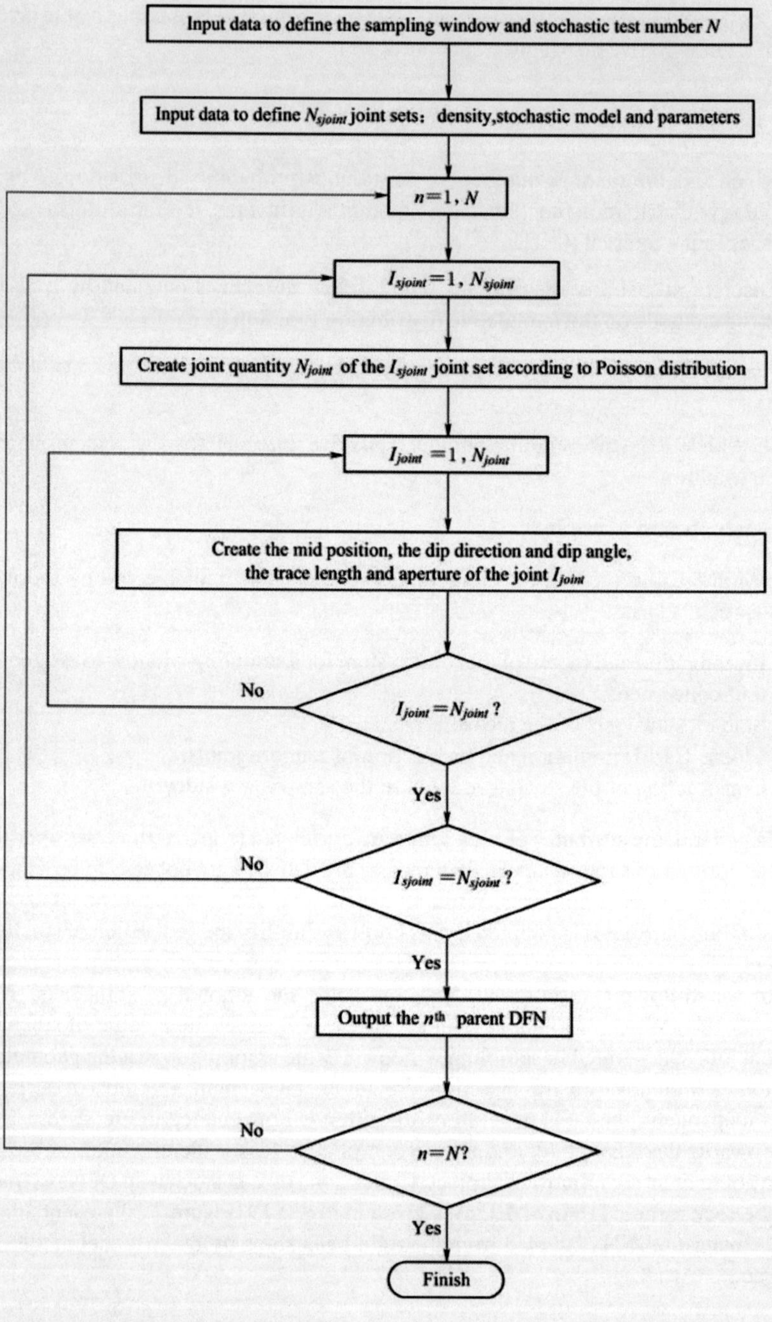

Fig. 2.18 Flowchart for the DFN generation

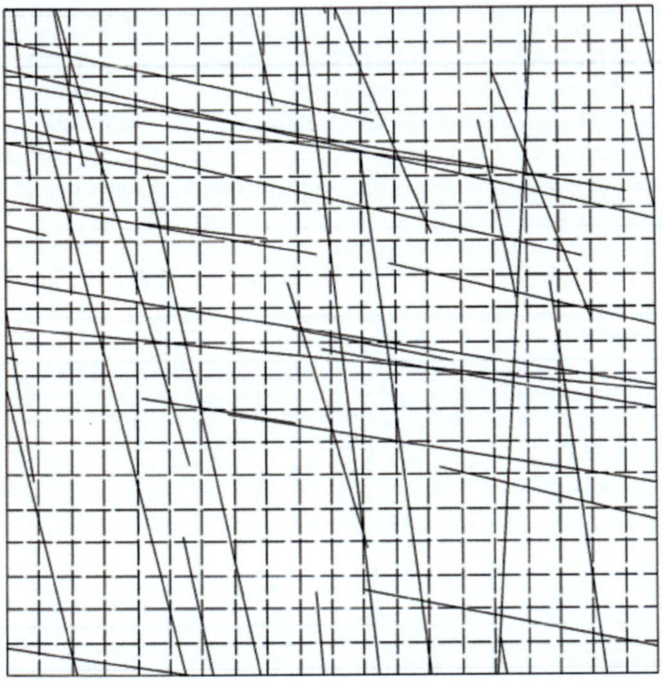

Fig. 2.19 2-D stochastic DFN at the $x - z$ plane (Xiaowan Project, China)

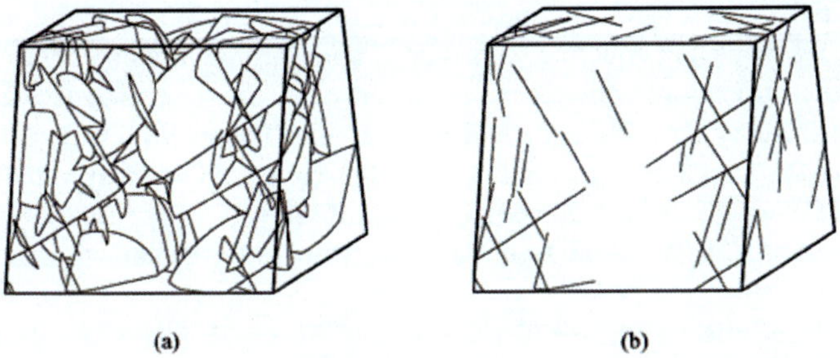

Fig. 2.20 3-D stochastic DFN (Xiaowan Project, China). **a** Perspective view; **b** surface trace view

2.3.3 Application of DFNs

The fundamental issue in modelling jointed rock mass is the evaluation of its characteristics of seepage (Louis 1969; Snow 1969; Long 1996; Hall and Hoff 2012) and deformation/failure (Zienkiewicz and Pande 1976; Pande and Gerrard 1983; Sitharam et al. 2001) under complex boundary conditions.

Extensive studies have been conducted to interpret the geological history and the formation mechanism behind natural DFNs (Engelder and Geiser 1980; Pollard and Aydin 1988; Olson and Pollard 1989). In addition to analytical solutions (Renshaw and Pollard 1994) or fundamental CM solutions (Olson 1993; Tang et al. 2006; Paluszny and Zimmerman 2011), advanced CM solutions using, for example, lattice-based rupture model (Cowie et al. 1993, 1995) and DEM (Spence and Finch 2014), have been widely exercised.

Very often, different scales (e.g. mesoscale, macroscale) w.r.t. the size of a geotechnical or hydraulic structure should be elected carefully, because they dominantly implicate whether an implicit (equivalent continuum) approach or explicit (discrete) approach for rocks could be preferentially used with the help of DFN. The former takes the influences of joints into account by means of the permeability and deformability/strength but neglects their exact positions; whereas the latter considers the joints deterministically. The DFN may be very useful both at mesoscale level to get equivalent parameters and corresponding REV towards the implicit approach (Chen et al. 2008; Chen et al. 2012), and at macroscale level to simulate dominated joints directly towards the explicit approach of prototype engineering structure (Chen et al. 2003).

1. **Mesoscale computation**

The stochastic DFN models, which may capture some mechanical characteristics of natural joint system, are sensitive to the assumed/measured joint properties and attributes. With the advance in computer technology, multi-scale or multi-level computation based on CMs termed as the "Numerical testing (NT)" or "Numerical material (NM)", has become acceptable towards the study on the mechanical behaviors of rocks and concrete. Take the solution of flow field for individual joint in the DFN for example, the closed-form formulas, the pipe models (Cacas et al. 1990a, b) and the channel lattice models (Tsang and Tsang 1987) were developed in the 1980s. They are actually a kind of special discrete methods that initially considered fluid flow and transport processes in rocks through joint system (Schwartz et al. 1983; Long et al. 1985; Andersson and Dverstop 1987; Dershowitz and Einstein 1987).

Since the 1990s, more skillful CMs (e.g. FEM, DEM, CEM) are exercised for stochastic rock samples containing complicated joint system to get the permeability tensor (Wei et al. 1995; Zhang et al. 1996; Kulatilake and Panda 2000; Min et al. 2004) and the elastic compliance matrix (Wei and Hudson 1986; Kulatilake et al. 1993a, b; Min and Jing 2003) as well as the corresponding REV size. Taking the elastic compliance matrix for example, the algorithm using "composite element method (CEM)" (Chen et al. 2012) is implemented by the following mainstream steps (viz. Fig. 2.21).

① A DFN in a sampling window is generated which will be used as the parent stochastic DFN;

② A series of rock blocks with different sizes and orientations are defined as test samples for each stochastic DFN;

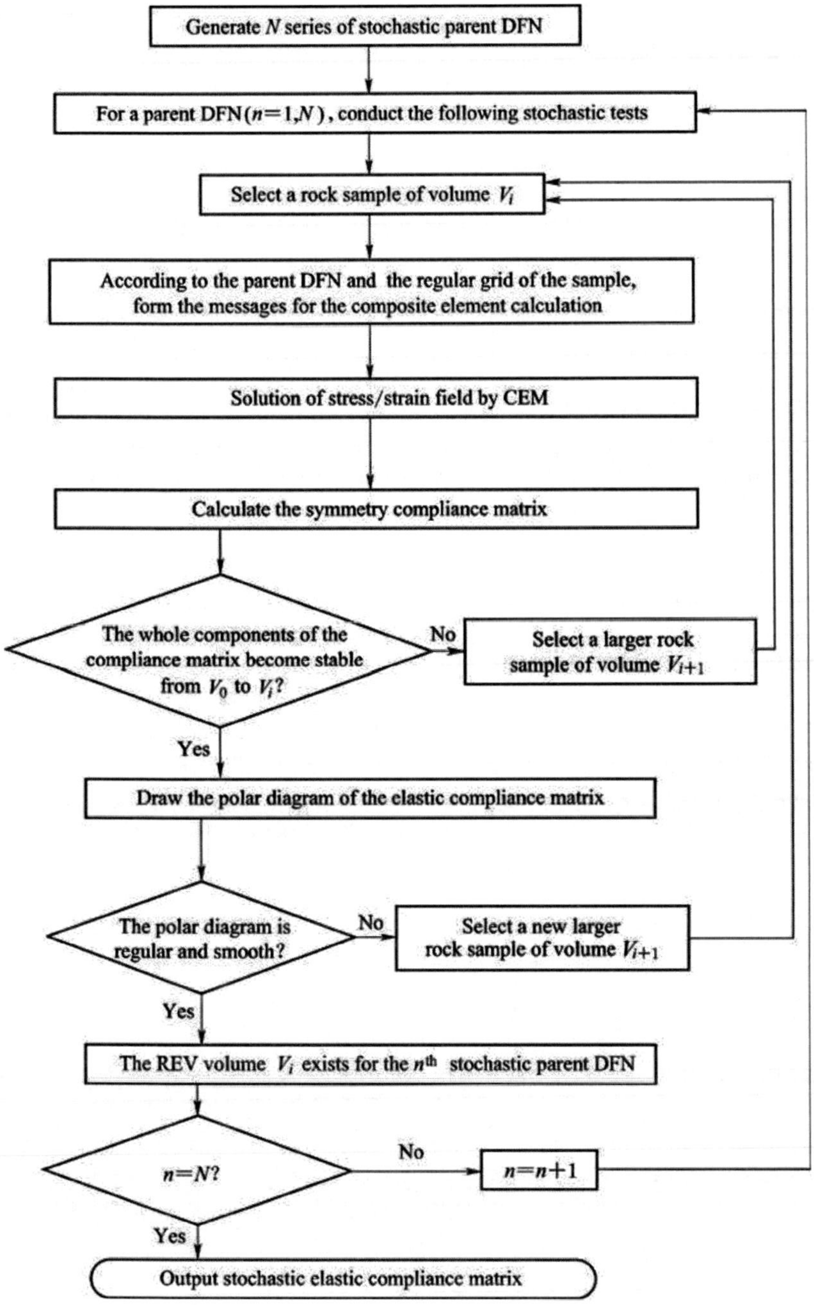

Fig. 2.21 Flowchart for the identification of elastic compliance tensor and REV

③ The numerical methods are applied to the jointed rock samples to evaluate their deformation fields;
④ The elastic compliance matrices of the samples are obtained and the existence of REV is verified.

Towards studying on the elastic compliance matrix of the rock at arch dam abutment (Xiaowan Project, China), altogether $N = 10$ stochastic 2-D parent DFNs are generated in a square region (60 m × 60 m). For each stochastic DFN, a series of rock samples are extracted whose sizes are variable ranging from 2 m × 2 m to 20 m × 20 m. The sample at the $x - z$ plane extracted from the first stochastic parent DFN ($n = 1$) is given in Fig. 2.19, in which the regular composite element mesh is also plotted (in broken lines).

The strain–stress responses are get by the hybrid DFN-CEM (Chen et al. 2012). Fluctuation tests show that when the sample size exceeds 10 m, the maximum fluctuation amplitude of the elastic compliance matrix is lower than 10%. From the practice standpoint, it may be approximately estimated that REV ≈ 10 m × 10 m.

Figure 2.22 displays the polar diagram of the elastic compliance C_α when the sample size = 10 m × 10 m, which is averaged from altogether 10 stochastic tests.

Where the performance of rock after plastic yield, particularly the ultimate strength and residual strength, are concerned, CMs need be selected appropriately in which joint pattern and evaluation can be simulated by an iterative fashion (Renshaw and Pollard 1994; Paluszny and Matthäi 2009). To be specific, the growth of a joint is influenced by its interaction with others, and can conversely generate stress perturbations in rock (Olson 2004, 2007; Renshaw 1996). The simulation of joint growth can be generally implemented by the following mainstream steps:

① Calculation of the perturbed stress field in rock caused by the presence and evolution of joints under imposed boundary conditions;

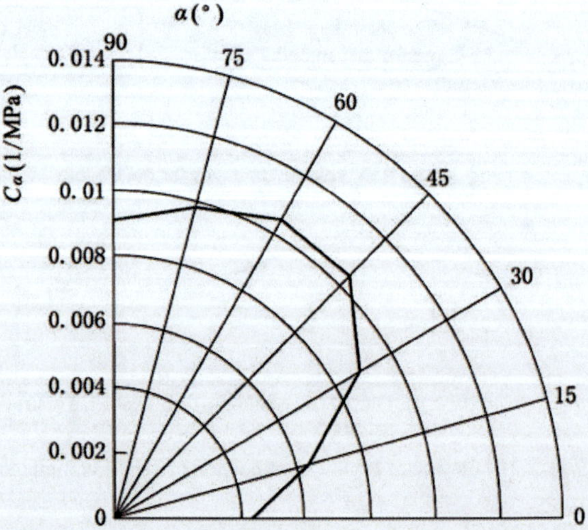

Fig. 2.22 Average polar diagram of elastic compliance matrix (sample size 10 m × 10 m)

② Derivation of "stress intensity factors (SIFs)" at the tip of each joint; and
③ Extension of joints according to the criteria w.r.t. onset, growth direction and length (Atkinson, 1984);
④ Recursion starting from step ① until the stability state of DFN or/and the collapse of numerical sample.

2. **Macroscale computation**

The procedure is nearly identical to that of mesoscale computation, but the simulation of DFNs should be more skillful, due to the complexity of prototype engineering structure. In addition, the growth dynamics of structural planes (joints and cracks) is directly related to the complexity of the engineering structures concerned.

The mesh-based methods, mesh-free methods, or particle-based methods may be selectively employed. Normally, use of hybrid methods or multi-scale methods in which the random structural planes (e.g. small joints) and located structural planes (e.g. large faults) are tackled by different methods is a wise strategy. This is because the former enables to simulate the equivalent macro-scale properties of REV containing random joints, whereas the latter may handle large faults explicitly. Figure 2.23 is the flowchart towards the analysis for the rock mass related to the prototype structures in geotechnical and hydraulic engineering.

A recent review of DFNs in rock mass modeling is given by (Lei et al. 2017). More details of advanced CMs coupled with DFNs can be found in Bobet et al. (2009), Lisjak and Grasselli (2014). In-depth discussions about stress effects on fluid flow in jointed rocks can be found in Zhang and Sanderson (2002).

It is emphasized that the identification of rock blocks as sub-domains is paramount in the construction of block assemblage for the computation using, for example, the DEM, DDA and DEA (Chaps. 7 and 8). The study to develop a sophisticated description of rock blocks began from the mid of 1980s. Lin et al. (1987), Cundall (1988), Jing and Stephansson (1994) introduced geometrical identification algorithms for rock block system using topological techniques. Their methods are rather skillful, but unlikely to identify a system including concave blocks at that time. Heliot (1988) proposed a new approach to produce more realistic rock block system, in which a rock block is always decomposed into two convex blocks. In 1992, Ikegawa and Hudson made a breakthrough by the "direct body" concept to cope with both convex and concave blocks in the same identification procedure. In 2012, Zhang and his co-workers (Zhang et al. 2012) proposed an identification method using mesh gridding technique.

After a check on the various existing methods, the author of this book realized that the direct body concept could be a solid base to establish a robust identification and pre-processing algorithm for the analysis of seepage, deformation and stability of complicated block system. During the implementation of the method, further improvements taking into account of the existence of irregular ground surfaces (curved faults and dam surfaces as well), the existence of the grouting and drainage curtains, etc., was made (Wang and Chen1998; Chen 2015). These enable us to carry out the pre-processing work more easily and more reliably. On one hand, it may be

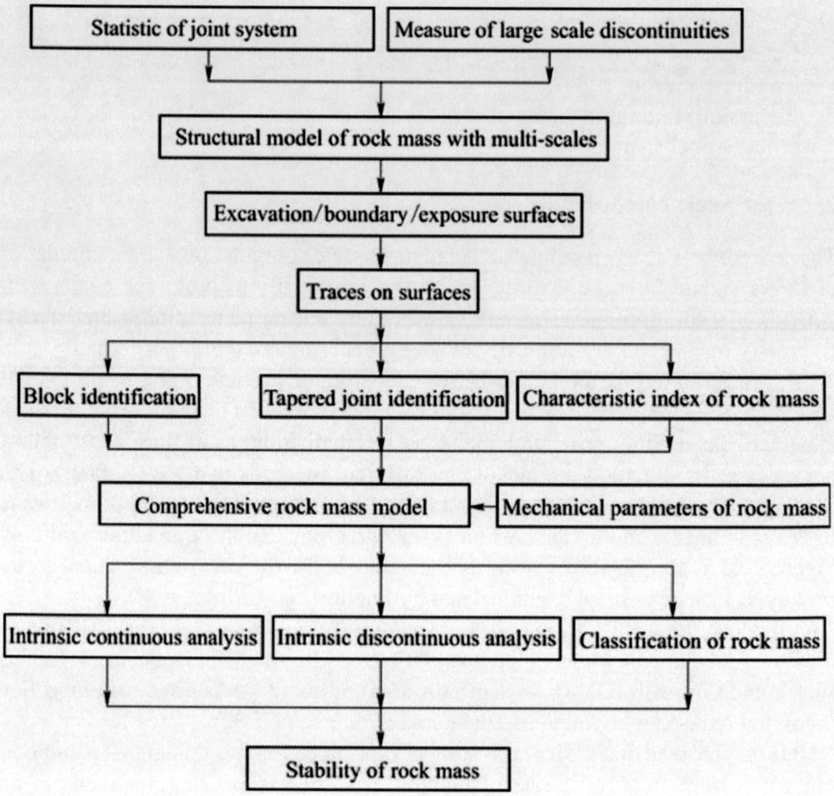

Fig. 2.23 Flowchart for the rock mass analysis in prototype engineering structure

directly employed for the block-based methods elaborated later in this book, on the other hand it may offer a well-organized database for the further FE mesh generation, by discretizing each block into a group of standard elements correctly connected.

A directed body is defined as the block enclosed by a number of directed faces (viz. Fig. 2.24b). A directed face has an external vector $\mathbf{F}^{ext}$ and an internal vector $\mathbf{F}^{\text{int}}$ perpendicular to the face and starting from the centroid of the face. The magnitude of the face vector is identical to the face area (Ikegawa and Hudson 1992).

A directed face is formed by a set of vertices arranged to make a right-handed system, with circular permutation, in terms of the external face vector (viz. Fig. 2.24a).

An edge has an external vector $\mathbf{E}^{ext}$ and an internal edge vector $\mathbf{E}^{\text{int}}$ perpendicular to both the external face vector and the edge, and starting from the middle of the edge. The magnitude of edge vector is identical to the edge length.

The directed body has a quite important geometrical invariant property that the resultant of all its external or internal face vectors is a null vector, i.e.

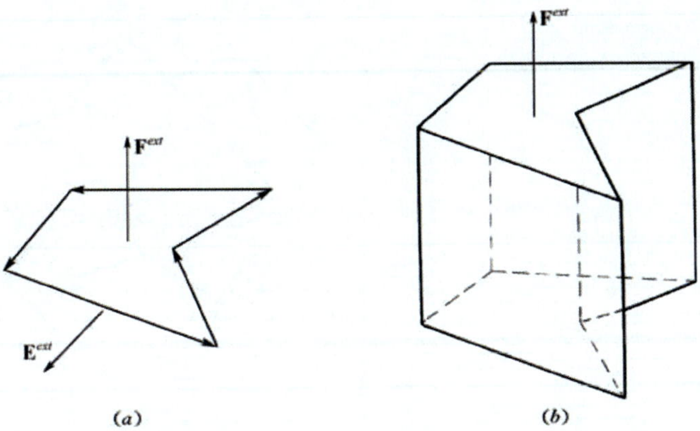

Fig. 2.24 Diagram showing directed body and directed face. **a** Directed face; **b** directed body

$$\sum_{i=1}^{n_f} \mathbf{F}_i^{ext} = 0 \tag{2.3}$$

in which i is the face sequence and n_f is thc amount of all faces bounding the body. This is called the "body completion theorem" (Ikegawa and Hudson 1992).

The directed face also possesses a similar property that the resultant of all its external or internal edge vectors is a null vector, i.e.

$$\sum_{i=1}^{n_e} \mathbf{E}_i^{ext} = 0 \tag{2.4}$$

in which i is the edge sequence and n_e is the amount of all edges bounding the face. This is called the "face completion theorem" (Ikegawa and Hudson 1992).

Based on the above concepts, the automatic identification algorithm can be formulated for the blocky system containing both the convex and concave blocks. To give a general view, Fig. 2.25 shows the axonometric drawing of a simply integrated block system, and Fig. 2.26 shows the axonometric drawing of corresponding decomposed blocks. Due to complicatedly distributed discontinuities, there are a variety of convex and concave blocks in the block system.

For more details, readers are referred to the relevant literature (Chen 2018).

2.4 Particle Assemblages

Generation of particle assemblage is crucial to a successful pre-process for particle-based methods towards either the meso-scale computation for the properties of

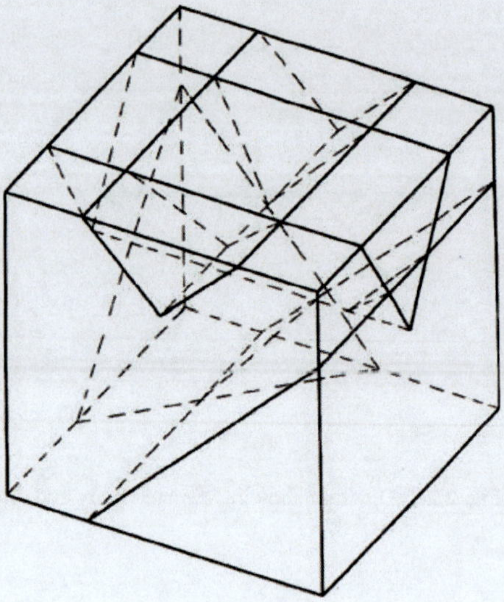

Fig. 2.25 Axonometric drawing of integrated block system

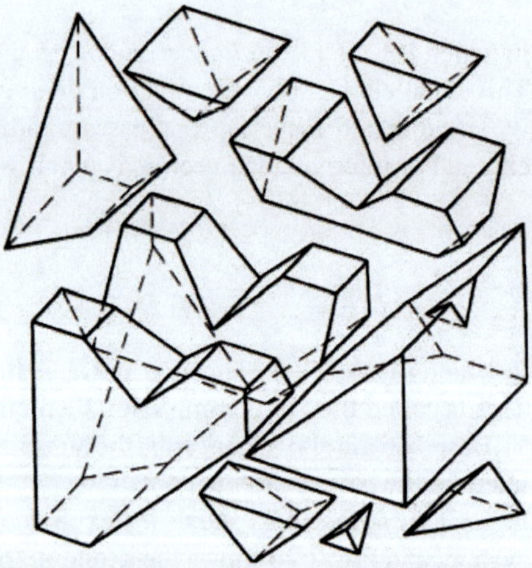

Fig. 2.26 Axonometric drawing of decomposed blocks

concrete and granular rock mass at sample level, or the macro-scale analysis for the performances of engineering structure (e.g. landslide) at prototype level.

Take the "numerical concrete (NC)" for example (Roelfstra et al. 1985), it numerically builds meso-scale samples of concrete as a three-phase composite material consisting of coarse aggregates (larger than 4.75 mm in size), the mortar matrix, and between them the "interfacial transition zone (ITZ)". Then the computations emulating the physically tested samples w.r.t. the performance of stress/deformation, heat conductivity, and water seepage/absorption, are carried out systematically to explore their response mechanisms, equivalent parameters, and of course, the correspondent REV size (Zhou et al. 2013; Abyaneh et al. 2014; Li et al. 2016; Xu and Chen 2016; Li et al. 2017; Xu et al. 2017; Wang et al. 2019; Wang et al. 2020).

The major difference in the particle assemblages for the granular rock and concrete are that in the former, the pores among grains are either empty or filled by soil/water, whereas in the latter, these pores are filled by mortar matrix.

2.4.1 Random Aggregate Techniques

In the mesoscopic study of concrete, the 3D "random aggregate structure (RAS)" (Roelfstra et al. 1985) should be established firstly for the successive modeling. This technique also may be employed to generation the PFC model but the void in the model is normally defined as the porosity without mortar matrix.

RAS is a commonly exercised mesoscale model for the numerical simulation to investigate the composite mechanical behavior of concrete, to which many significant progresses have been achieved over the last decades. In addition to numerous 2-D studies (Wang et al. 2015a, b; Wang et al. 2015), more and more researches are directed to 3-D issues. For example, Wriggers and Moftah (2006) generated a 3-D RAS model with the aggregates of spherical shape, and provided a good prediction concerning the compressive damage behavior of concrete with the help of scalar damage model (Mazars 1986), Qu and Chen (2008) developed a 3-D RAS model with spherical and polyhedral aggregates, and studied the crack growth of concrete with the maximum tensile strain fracture criterion.

It is notable that many of above 3-D models assume two-phase composite without the existence of ITZ. Hence, for better numerical test results, an exquisite mesoscale model of concrete must be generated with three-phase composite material consisting of the randomly distributed coarse aggregates, homogeneous mortar matrix and ITZ with certain thickness (viz. Fig. 2.27). The key factors for such a RAS-based concrete model are the configuration parameters of coarse aggregates inclusive their size and shape, volume fraction and spatial distributions (Abyaneh et al. 2014; Li et al. 2016; Xu and Chen 2016).

1. **Generation of aggregate particles**

(1) **Mathematical description of particle shape**

Coarse aggregates generally occupy 40–50% volume of concrete and play an important role in concrete properties. The shape of coarse aggregates depends on their types, i.e., gravels have a rounded shape while crushed stones have an angular or irregular shape. As the most prevailing exercise, spherical and ellipsoidal particles for gravel aggregates and convex polyhedral particles for crushed stones are postulated (viz. Fig. 2.28).

Only four parameters are demanded to describe a spherical body in 3-D space, namely the center (x_c, y_c, z_c) and the radius r (viz. Fig. 2.28a). While an ellipsoidal

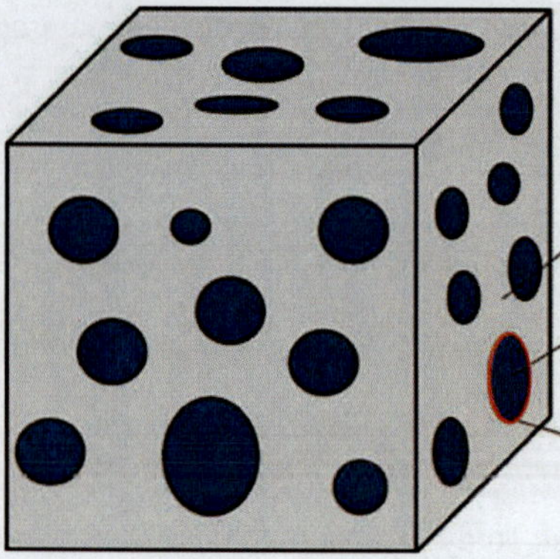

Fig. 2.27 Three-phase meso-scale model for concrete

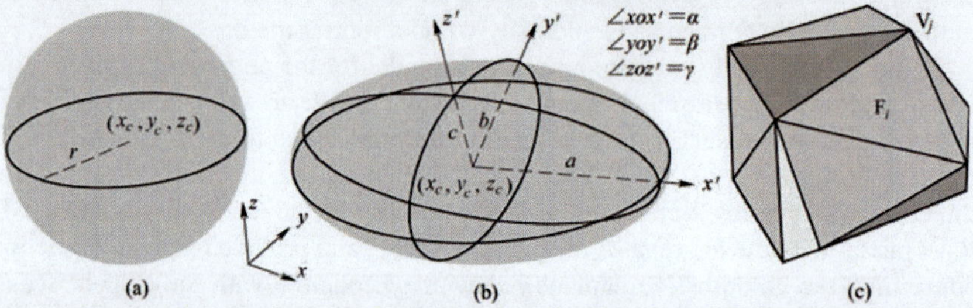

Fig. 2.28 Typical particle shapes of coarse aggregates. **a** Sphere; **b** ellipsoid; **c** polyhedron

particle is more complicated: to describe its position and orientation in 3-D space (viz. Fig. 2.28b), it requires nine parameters inclusive the ellipsoid center (x_c, y_c, z_c), the semi-principal axes (a, b, c) and the three Euler angles (α, β, γ). The geometrical model of an ellipsoid is expressed as

$$\mathbf{X}^{\mathrm{T}}\mathbf{Q}\mathbf{X} = 0 \tag{2.5}$$

in which $\mathbf{X}^{\mathrm{T}} = [x, y, z, 1]$, $\mathbf{Q}$ is the 4×4 coefficient matrix of the ellipsoid including its axes, position and orientation, which may be derived from nine parameters.

All the faces of a convex polyhedron are assumed to be triangular (viz. Fig. 2.28c), so that it can be described as a set of triangles (F_i) and a set of vertices (V_j). A convex polyhedron is customarily postulated.

(2) **Particle size distribution**

Aggregate size distribution is an important factor in the concrete mix design and optimization. As one of the well-known size distribution curves, "Fuller distribution" postulates that

$$P(d) = 100\sqrt{d/d_{\max}} \tag{2.6}$$

where: $P(d)$—cumulative percentage of aggregate particles with the size smaller than d; $d_{\max}$—maximum particle size of the aggregate.

Once the aggregate size distribution is determined by Eq. (2.6), the volume content of aggregates in different size ranges and aggregate amount may be calculated for the generation of particle assemblage (Wriggers and Moftah 2006).

(3) **Particle generation**

In the generation of particle assemblage, poorly shaped aggregate particles (e.g. needle or plate shaped) should be avoided. For this purpose, the shape characteristics of aggregate particles are specified. A widely used parameter "sphericity" (Quiroga 2003) may be adopted to characterize the shape of an aggregate particle, which is defined as

$$s = \sqrt[3]{V/V_s} \tag{2.7}$$

where: V—volume of the particle; V_s—volume of the minimum circumscribed sphere.

Figure 2.29 shows a distribution graph of sphericity, in which P_i means the distribution proportion of particles in the interval (s_i, s_{i+1}).

The composite shape index I of all aggregate particles is further expressed as

$$I = \sum \frac{1}{2}(s_i, s_{i+1})P_i \tag{2.8}$$

The index I is normally used to characterize the overall shape of aggregates. Reasonable values for the sphericity and shape index should be controlled in the generation of aggregate particles.

2. **Arrangement of particles**

The generated aggregate particles are arranged and placed in the NC sample occupying a cubic or cylindrical shaped region, which depends on the type of concrete specimen to be tested. The basic requirements for the arrangement of particles are (Wriggers and Moftah 2006):

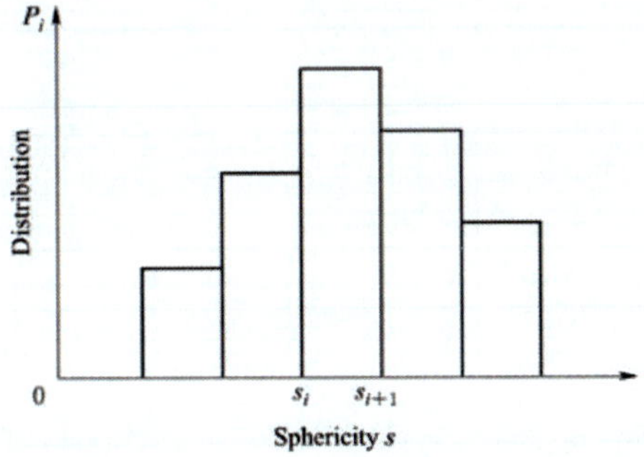

Fig. 2.29 Distribution of sphericity s

- Aggregates must be wholly located within the boundaries of the specimen;
- Any overlapping or intersection between two aggregates must be avoided;
- A minimum distance b between any two aggregates or between aggregates and boundaries must be kept, which equally means that there must be a bonding layer (i.e. ITZ) around the particle.

Several useful methods are available for the arrangement of aggregate particles. For example, the take-and-place method for low volume content aggregates (Roelfstra et al. 1985; Kwan et al.1999; Wang et al. 1999), the divide-and-fill method for random aggregates (De Schutter and Taerwe 1993), the placement and re-generation method to achieve very high volume content with spherical aggregates (Wriggers and Moftah 2006). However insofar, there are still a number of difficulties in the arrangement of ellipsoidal and irregular aggregates.

To improve the situation, a grid-based technique termed as "occupation and removal method" (Ma et al. 2005) was proposed for the aggregate arrangement of variable shapes with high volume content, in which some algebraic algorithms from analytic geometry are incorporated, so that separation checks between aggregates or between aggregates and boundaries may be accomplished rigorously.

(1) **Grid-based placement**

In the grid-based technique, aggregate particles are placed in the following procedure:

① Discretize the specimen by cubic elements smaller than the minimum particle size.
② Select any point according to the uniform probability distribution within the specimen.
③ For the particle to be placed, put its center at the selected point.
④ Separation checks are conducted to detect whether there is any overlapping with previously placed particles or boundaries. If there is no overlapping, go to step ⑤; otherwise, repeat step ② and step ③ until there is no overlapping.
⑤ Remove the elements occupied by the particle (at least one node of the element is inside the particle). When placing the rest particles, all of the removed elements are not selected again.
⑥ Store the information of the present particle, then return to step ⑤ for the rest particle until all particles have been placed.

(2) **Separation check**

Separation checks for two spheres i and j can be simply evaluated by comparing among their radii (r_i, r_j), the ITZ thickness (b) and the distance (D) between their centers, namely

$$\text{If } D = \sqrt{(x_{ci} - x_{cj})^2 + (y_{ci} - y_{cj})^2 + (z_{ci} - z_{cj})^2} \geq r_i + r_j + b \quad \text{separated} \tag{2.9}$$

$$\text{If} \quad D = \sqrt{(x_{ci} - x_{cj})^2 + (y_{ci} - y_{cj})^2 + (z_{ci} - z_{cj})^2} < r_i + r_j + b \quad \text{overlapped} \tag{2.10}$$

The algebraic approach (Wang et al. 2001; Häfner et al. 2006) is able to provide a solution for the separation check of two ellipsoids expressed as $\mathbf{X}^{\mathrm{T}}\mathbf{Q}_i\mathbf{X} = 0$ and $\mathbf{X}^{\mathrm{T}}\mathbf{Q}_j\mathbf{X} = 0$, in which $\mathbf{Q}_i$ and $\mathbf{Q}_j$ are the coefficient matrices of these two ellipsoids (viz. Eq. (2.5)), respectively. The characteristic equation can be defined as

$$f(\lambda) = \det(\lambda\mathbf{Q}_i + \mathbf{Q}_j) = 0 \tag{2.11}$$

These two ellipsoids are separated if and only if the characteristic equation has two distinct positive roots. So the problem turns out to be a question that how to know the number of positive distinct roots of the quartic equation.

To improve the computational efficiency, the algorithm for separation checks between two ellipsoids is designed as follows:

① If the central distance D between these two ellipsoids satisfies

$$D = \sqrt{(x_{ci} - x_{cj})^2 + (y_{ci} - y_{cj})^2 + (z_{ci} - z_{cj})^2} \geq a_i + a_j + b \tag{2.12}$$

no overlapping occurs.

② If the central distance D between two ellipsoids satisfies

$$D = \sqrt{(x_{ci} - x_{cj})^2 + (y_{ci} - y_{cj})^2 + (z_{ci} - z_{cj})^2} < c_i + c_j + b \tag{2.13}$$

overlapping definitely occurs.

③ If the central distance D between two ellipsoids satisfies

$$c_i + c_j + b \leq D < a_i + a_j + b \tag{2.14}$$

further effort is needed by the positive distinct roots derived from Eq. (2.11).

Separation checks between polyhedral particles begin with face-to-face intersection tests. Two polyhedrons are looked at as separated if any face of one polyhedron doesn't intersect with the faces of the other. A fast and robust 3-D algorithm for triangle-to-triangle overlapping test proposed by Guigue and Devillers (2003) may be employed to accomplish the separation checks. To accelerate the process, separation checks between the circumscribed spheres of two polyhedrons are performed firstly, namely, two polyhedrons are definitely separated if their circumscribed spheres are separated.

With the algorithms described above, the distribution of various shaped particles in different concrete specimens is presented in Figs. 2.30 and 2.31 (Li et al. 2016).

It is reminded that these NC sampling techniques are also applicable to the rock-fill (granular particle assemblage) of landslide body, after the adjustments w.r.t. filling

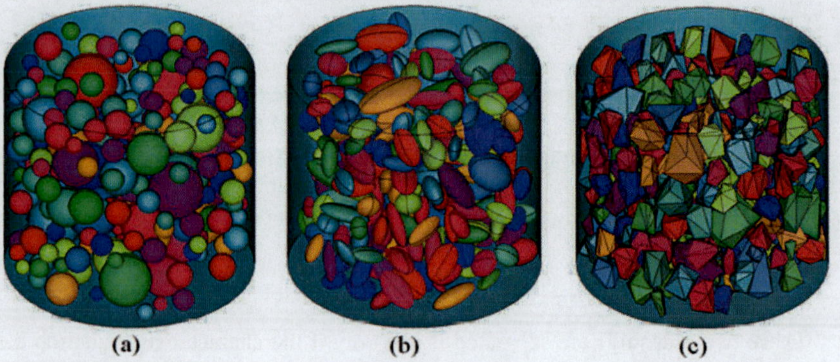

Fig. 2.30 Cylinder specimens with different aggregate shapes. **a** Spherical aggregates; **b** ellipsoidal aggregates; **c** polyhedral aggregates

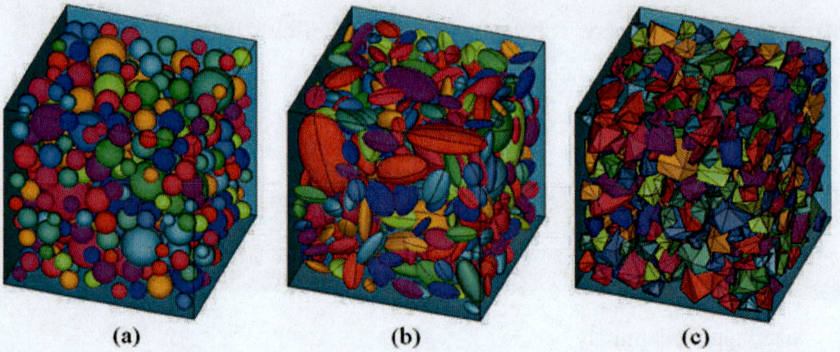

Fig. 2.31 Cubic specimens with different aggregate shapes. **a** Spherical aggregate; **b** ellipsoidal aggregates; **c** polyhedral aggregates

matter in aggregate void, eliminating the thickness or change the properties of ITZ, etc.

2.4.2 *Hopper Discharge Techniques*

In simulating the granular assemblage of landslide using particle-based methods, it is customarily to pack all the granular materials within a series of periodic cells which can be regarded as the fractions of the whole landslide body. Any particle with its centroid moving out of the periodic cell through one specific face is mapped into the cell on the opposite side of the face. Particles with only one part of the volume lying outside of the present cell can interact with particles near the cell face.

The "hopper discharge technique" (Zhao et al. 2016) is applicable to generate a dense granular rock assemblage in, for example, natural or cut slope, by which the rock particles are used to fill the space enclosed by the land surface, the lower slip (failure) surface and the periodic boundaries. Blamed for the tremendous effort, 2-D cases using discs are prevalently, although reluctantly, exercised for the moment. The procedure is implemented in five successive steps (viz. Fig. 2.32):

① Circumscribe rectangle of one periodic polygon is defined to distribute random discs one by one;
② If the center of a disc is not in the polygon, this disc is discarded;
③ If the disc intersects a boundary face (judged by the radium of disc), this disc is discarded;

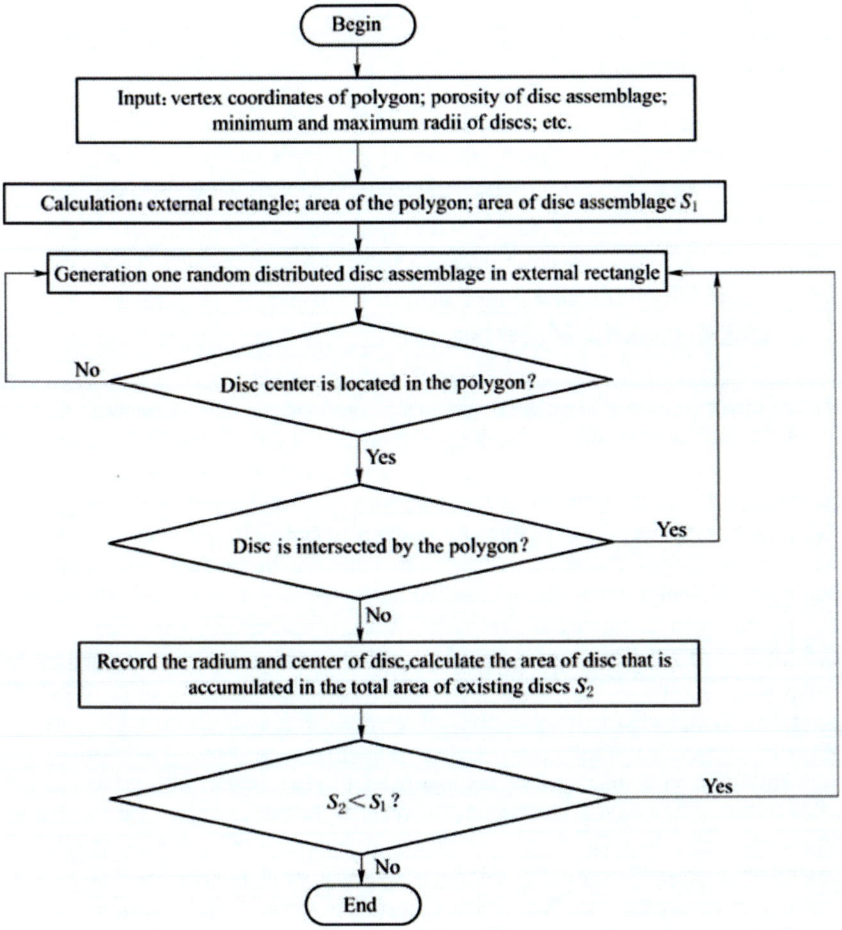

Fig. 2.32 Flowchart for filling arbitrary convex polygon with disc particles

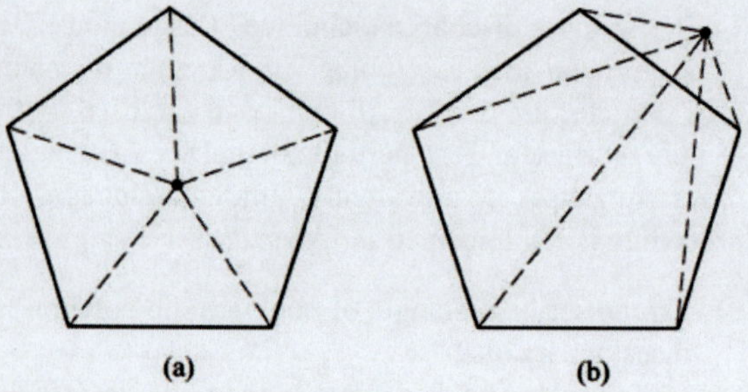

Fig. 2.33 Relationship of disc and polygon. **a** Disc is located within polygon; **b** disc is located outside of polygon

④ If the disc overlaps other discs within polygon, it is eliminated by a technique of "particle repulse".
⑤ The qualified disc is added to the tree-link list;
⑥ If the porosity of rock-fill is not met, this procedure is repeated from step ① until the porosity is met.

The judgement w.r.t. the relation of disc and polygon is depicted by Fig. 2.33: If the area summation of all triangles formed by the center of disc and the vertices of polygon is equal to the area of polygon, the disc is located within the polygon.

2.5 Basic Physical and Mechanical Properties of Water and Rock-Like Materials

Before the detailed works on the properties of fluid and solid, it is essential that the terms "stress" and "strain" are defined without ambiguity. Typically, two types of definitions are applicable. The first type deals with matters that only undergo small (infinitesimal) strains, whereas the second type deals with matters that undergo large (finite) strains. In the routine design of geotechnical and hydraulic structures, the first type is normally postulated, particularly when they are within the serviceability and collapse limit states. Under such circumstances, the use of "Cauchy stress tensor" corresponding to "infinitesimal strain tensor" is appropriate and convenient (Landau and Lipshitz 1970). This is also useful when finite strain issues are handled by the "upgraded Lagrange (UL)" scheme through successively revision of point coordinates (using displacement increments) and correspondent strain/stress transformation w.r.t. the distortion and rigid motion of the domain occupied by the substance.

If the force on a cube surface is proportional to the area of the surface, as will often be the case, then it is appropriate to consider the force per unit area, called the stress (viz. Fig. 2.34). The SI units of stress are Newton per square meter, which is commonly represented by a derived unit, the Pascal, or abbreviated as "Pa".

The component of stress tensor that is normal to the upper surface of the cube in Fig. 2.34 is denoted as σ_{zz} or simply σ_z in engineering. A normal stress can be either

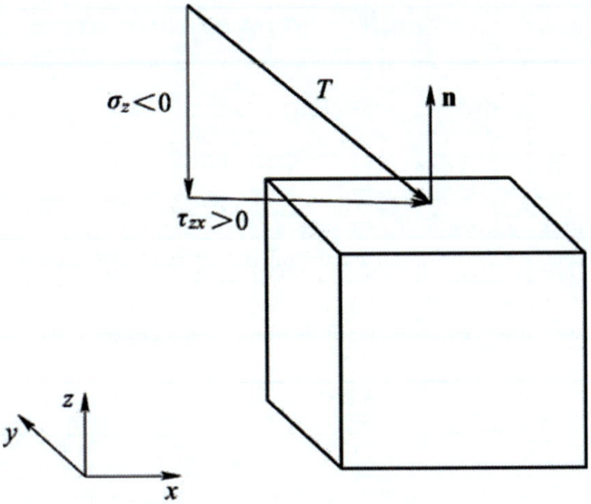

Fig. 2.34 Orthogonal triad of unit force vector at infinitesimal substance cube

a compression, namely $\sigma_{zz} \leq 0$, or a tension if $\sigma_{zz} \geq 0$. The systematic description of strain and stress tensors will be given in Chap. 3.

How the stress is related to the physical and mechanical properties and the motion of a substance body are the questions of first importance. To start, Table 2.3 gives several basic physical and mechanical properties of fluids and solids, from this table a large difference in the properties of these substances may be found.

One of the important properties of any substance is its response to an applied force. Fluids are basically different from solids in their response to tensile normal stress resisted by many solids, especially metals, that possess almost the same tensile and compressive resistance. In contrast, gases do not resist tensile stress at all, while liquids do so only very weakly in comparison to compressive stress. As a result, if a fluid volume is compressed along one dimension but is free to expand in orthogonal directions, its volume remains nearly constant. Hence in fluid, the most important compressive normal stress is almost always due to the pressure rather than viscous effects, in other words, when the discussion is limited to the compressive normal stress only we will identify σ_{zz} with the pressure p.

Table 2.3 Physical properties of air, sea water and rock (granite)

Material	Density ρ (kg m^{-3})	Heat capacity C_p (J kg^{-1} C^{-1})	Bulk modulus B (Pa)	Sound speed c (m s^{-1})	Shear modulus K(Pa)	Coefficient of dynamic viscosity μ (Pa s)
Air	1.2	1000	1.3×10^5	330	N/A	18×10^{-6}
Sea water	1025	4000	2.2×10^9	1500	N/A	1×10^{-3}
Granite	2800	2800	4×10^{10}	60,000	2×10^{10}	$\geq 10^{22}$

2.5.1 Mechanical Properties of Water

1. **Response to pressure**

Every substance will undergo certain volume change as the ambient pressure is changed, but the amount varies quite widely from gases to liquids and solids. To make a quantitative measure of volume change, an important class of phenomenon may be described by a simplified equation of state that neglects the temperature

$$\rho = \rho(p) \tag{2.15}$$

in which p is the pressure and ρ is the density.

A fluid such as water described by Eq. (2.15) is said to be "barotropic" in that the gradient of density will be everywhere parallel to the gradient of pressure, and hence the contours of constant density will be parallel to the contours of constant pressure.

A convenient measure of the stiffness or inversely the compressibility in one dimension is (viz. Fig. 2.35)

$$B = K = \frac{\sigma_{zz}}{\varepsilon_{zz}} = -V_0\frac{\Delta p}{\Delta V} = \rho_0\frac{\Delta p}{\Delta \rho} \tag{2.16}$$

in which B or K is the bulk modulus that is similar to a normalized spring constant.

With large B in Table 2.3, water can be looked at as an incompressible fluid for the most purposes in geotechnical and hydraulic structures under ordinary conditions (Fine and Millero 1973). Even in oceans at 4 km depth, where the pressure is 400 atm, water suffers only a slightly decrease (1.8%) in volume.

2. **Response to shear stress**

Fluids are remarkably different from solids in their response to the shear stress τ_{zx} (viz. Fig. 2.36). Ordinary fluids such as air and water have no intrinsic configuration, hence they are not able to maintain static balance by restoring force against shear stress. In addition, since there is no volume change associated with a pure shear deformation γ_{zx}, hence there is no coupling to the bulk modulus. As a result, there is no meaningful shear modulus for a fluid. Rather, a fluid will move or flow in response to shear stress, and the fluid flow will continue so long as the shear stress is present.

Fig. 2.35 One-dimensional case of fluid in container

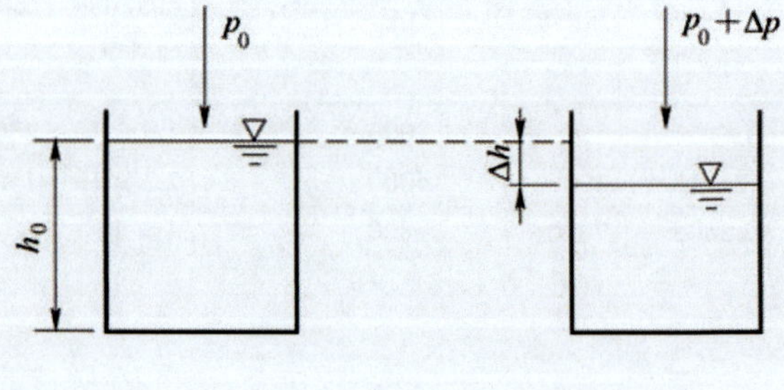

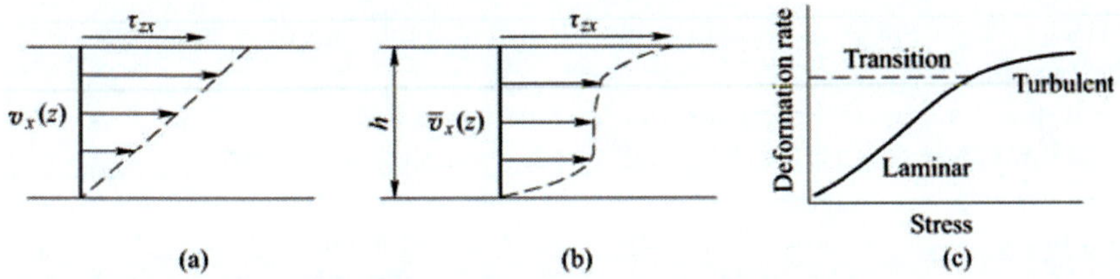

Fig. 2.36 Shear stress and corresponding shear strain rate—liquid. **a** Laminar flow; **b** turbulent flow; **c** shear stress versus shear deformation

When the shear stress is held steady, and if the geometry does not interfere, the shear deformation rate may also be steady or have a meaningful time-average. In analogy with the shear modulus, a dynamic viscosity μ as the ratio of shear stress to shear deformation rate may be defined as

$$\mu = \frac{\tau_{zx}}{h^{-1}\mathrm{d}u_x/\mathrm{d}t} \tag{2.17}$$

in which u_x is the displacement of column surface (viz. Fig. 2.36). This ratio will depend upon the kind of fluid and also upon the flow itself such as the speed. In fluid dynamics, the transformation from the coefficient of dynamic viscosity to the coefficient of kinematic viscosity is defined by $\nu = \mu/\rho$, where ρ is the fluid density.

If the flow depicted in Fig. 2.36a is set up carefully and at small Reynolds number, it may happen that the fluid velocity v_x will be steady, with velocity vectors lying smoothly, one on the top of another, in layered or laminar flow. The ratio

$$\mu = \frac{\tau_{xz}}{\partial\gamma_{xz}/\partial t} = \frac{\tau_{xz}}{\partial v_x/\partial z} \tag{2.18}$$

is then a property of the fluid alone dubbed as the "coefficient of dynamic viscosity". For turbulent flow at large Reynolds number depicted in Fig. 2.36b, Eq. (2.18) should be replaced by some non-linear ones (Davis 1952; Morris and Wiggert 1972; Featherstone and Nalluri 1995).

"Newtonian" fluids are those that their viscosity in laminar flow is a thermodynamic property of the fluid alone and not dependent upon the shear stress magnitude. To be an excellent approximation, air and water are Newtonian fluids. If a fluid is Newtonian, it is empirically found that the condition for laminar flow must satisfy the criterion

$$\mathrm{Re} = \frac{\rho v_x h}{\mu} \le \mathrm{Re}_c \tag{2.19}$$

in which Re is the non-dimensional parameter called the "Reynolds number", v_x is the speed of the upper (moving) surface relative to the lower (fixed and no-slip) surface, Re_c is the lower limit of critical Reynolds number. For open channel flow, $\text{Re}_c = 500$.

In practice, Eq. (2.19) means that the flow speed must be very low or the fluid column thickness must be very small to maintain laminar flow state. The laminar flow velocity of a Newtonian fluid will vary linearly with z and the velocity shear at each point will be equal to the overall shear deformation rate.

Assuming that we know the fluid viscosity and its dependence upon temperature, density, etc., then Eq. (2.18) relating shear stress and shear strain rate (velocity of shear deformation) may be used to estimate the viscous shear stress. This is the way that the viscous shear stress will be incorporated into the momentum balance equation of a fluid parcel (Chap. 3). In many water flows however, the equivalent Reynolds number is much larger than the upper limit for laminar flow given in Eq. (2.19), consequently, they are much more likely to be turbulent and unsteady. Hence it frequently happens that flow states, rather than physical properties of the fluid alone, determine the viscous shear stress.

2.5.2 Mechanical Properties of Concretes

For the deformation and failure analysis of structural components, scientists mainly rely on the mean value and standard deviation of relevant parameters derived from axial loaded laboratory experiments, even though in most cases the operational loads are three-dimensional. Although only the mean value is communicated to the engineers for routine design, yet it is notable that the standard deviation is also a very important parameter since it helps to design structural components appropriately with a low failure probability.

The concrete and rock in many aspects perform similarly, hence they are frequently fallen under the term of "rock-like materials".

1. **Response to pressure**

The conventional compressive test with or without confining pressure, in which a short, right cylinder is loaded axially, is one of the most widespread laboratory testing activities (viz. Fig. 2.37) towards the studies on the elastic behavior and strength, as well as on the long term creep of rock-like materials.

The most useful single description for the mechanical properties of rock-like material is the complete stress–strain curve plotted in Fig. 2.37, using a right cylinder of material sample being confined by lateral stress $\sigma_3(=\sigma_2=\sigma_{xx}=\sigma_{yy})$ and compressed in axial direction by $(\sigma_1=\sigma_a=\sigma_{zz})$, where the abscissa axis exhibits axial strain $\varepsilon_a(=\varepsilon_1=\varepsilon_{zz})$ and the ordinate axis denotes deviate stress $(\sigma_1-\sigma_3)$. If $\sigma_3 = 0$, the test is uniaxial. This type of resultant curve, known as the "strain-controlled complete stress–strain curve", was firstly obtained in 1966 for the purpose to illustrate the very significant effect of the microstructure and the history on the mechanical behaviors of rock (Jaeger et al. 2007).

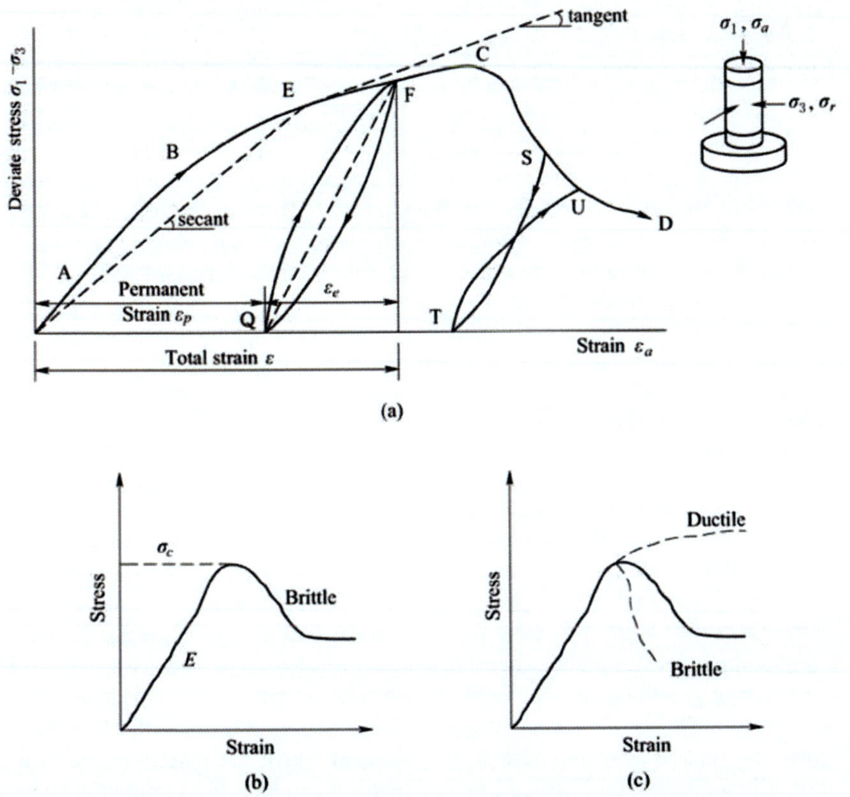

Fig. 2.37 Complete stress–strain curve of rock-like materials (concrete and rock) under constant confining pressure and quasi-static axial loading. **a** Complete stress–strain curve; **b** brittleness; **c** transition from brittle to ductile

Although there exist various factors related to loading conditions and loading rates which influence the shape of the complete stress–strain curve, yet there are always several features of importance to be found from Fig. 2.37. The first one is the Young's modulus E of the material, either tangent or secant. The next one is the characteristic strength σ_s. The third one is the steepness of the descending portion (post-failure) which is an index of material brittleness. If the post-failure strain steadily progresses at the same stress level, the material is ductile, and a residual stress may be identified; if a sharp drop in the stress level down to zero occurs at a nearly same strain value, the material is brittle and there is no residual stress. In fact, the real situation is more complicated than these two extraordinary cases because a rock-like material usually exhibits obvious hardening phase under high confining pressure while softening phase may be completed abruptly under low confining pressure.

(1) **Elastic modulus**

The stress–strain relation for linear elasticity is commonly known as the "Hooke's law". In its simplest form, this law defines the spring constant (or elasticity constant) in a scalar equation, stating the tensile/compressive force is proportional to the extended/contracted displacement and the material recovers its initial shape after unloading. At the very beginning of loading in Fig. 2.37, the curve exhibits an initial portion (O → A) which could be concave upwards for rock, but also could be convex for concrete. This is mainly due to the deficit in the rock specimen preparation—the ends of the cylinder being non-parallel, and/or the closing of micro-cracks within the intact rock. After this initial portion, there is a portion of essentially linear behavior (A → B). This portion may be used to define a Young's modulus as

$$E = \frac{\sigma_{zz}}{\varepsilon_{zz}} \tag{2.20}$$

There are various definitions of elastic moduli in addition to Young's modulus, such as the shear modulus and the bulk modulus. They are all the indexes of the inherent elastic properties as the resistance to the deformation under an applied load. These moduli are corresponding to different kinds of deformation. For instance, the Young's modulus applies to extension/compression of a solid, whereas the shear modulus applies to its torsion. There are several prevalent expressions for the definition of Young's modulus E which significantly influences the rigidity of geotechnical and hydraulic works: the "tangent modulus" is given by the slope of a line drawn tangent to the stress–strain curve at any point, of which the initial tangent is the slope at initial stress; the "elastic tangent modulus" which is conventionally named as "elastic modulus", is the slope at any specified linear point (or near linear) on the stress–strain curve, but usually at a specified stress level (such as 50% for rock) of the maximum or peak stress; the "chord modulus" is given by the slope of a line drawn between two points on the stress–strain curve; the "deformation modulus" is the slope of such a line between zero and a specified stress level with respect to the maximum or peak stress (e.g. 40% for concrete, 50% for rock, one-third for soil), which belongs to "secant modulus"; the "recovery modulus" is the slope of unloading route.

Since a correlation between the elastic modulus and strength of concrete does exist because they are both significantly affected by the porosity of the constituent phases (although not to the same degree) (Shen et al. 2016; Silva et al. 2016), therefore the Young's modulus used in preliminary phases of design may be estimated from empirical formulas that assume a direct dependence of modulus on the strength and density of concrete, or simply on concrete grade. Most design specifications provide equations predicting the Young's modulus of concrete in terms of compressive strength. For example, according to Chinese design specifications "GB50010-2010" and "SL191-2008", the elasticity modulus of concrete can be estimated by

$$E \frac{10^5}{2.2 + 34.7/\sigma_c} \tag{2.21}$$

where: E—Young's modulus, MPa; σ_c—uniaxial compression strength (28d) of standard cubic sample, MPa.

The elastic modulus E of concrete under compression varies from 14 to 40 GPa. In the initial phases of design, specifications "GB50010-2010" and "SL191-2008" also suggest its reference values in relation to the concrete grade. In the later design phases for an important engineering project, the elastic modulus of concrete should be evaluated by laboratory experiments.

It is unique that the primary factors affecting thermal stress for young concrete, apart from the structural restraint and temperature change, is the growth of Young's modulus $E(t)$ with the ongoing of age in a basic form of exponent formula

$$E(t) = E_o(1 - e^{-at}) \tag{2.22}$$

where: t—concrete age, day; E_0—finial modulus when $t \to \infty$, MPa; a—constant.

For a sample subjected to simple axial load, i.e., the confining pressure $\sigma_3 = 0$ in Fig. 2.37, the ratio ν of the lateral strain ε_r $(= \varepsilon_2 = \varepsilon_3 = \varepsilon_{xx} = \varepsilon_{yy})$ to axial strain ε_a $(= \varepsilon_1 = \varepsilon_{zz})$ within the elastic range is defined as the "Poisson's ratio". This parameter is generally not needed for the conventional design of concrete gravity dams. However, it is demanded for the analysis of tunnels, arch dams, and other statically indeterminate structures. Although there appears no consistent relationship between the Poisson's ratio and concrete characteristics (e.g. W/C, curing age, aggregate gradation, etc.), yet it is generally known to be lower for high strength concrete, higher for saturated concrete, and higher for dynamically loaded concrete.

The values of Poisson's ratio for dam concrete generally vary within 0.15–0.20, for ordinary concrete $\nu = 0.167$ may adopted in the initial phases of design.

(2) **Strength**

In Fig. 2.37, point B may be defined as the "yield point" pinpointed by the "yield strength" or "elastic limit" σ_y, after this point plastic strain ε^p will manifest along with the gentler mounting of the curve, which is termed as the "hardening phase". In the state of plasticity, the applied force will induce non-recoverable (irreversible) deformations in the solid. The significance of elastic limit in structural design lies in the fact that it represents the maximum allowable stress before the material undergoes permanent deformation. The plastic strain may be detected by an unloading–reloading circle, F → Q → F for example, where the total strain is divided into the components of elastic and plastic, i.e. $\varepsilon = \varepsilon^e + \varepsilon^p$. After the peak point C tagged by the "peak strength" σ_p or the "ultimate strength" σ_u, the complete stress–strain curve enters a descending post-peak region (C → D), which is named as the "softening phase". In this phase much more plastic strain may be found through unloading–reloading circle, S → T → U for example. For ductile materials or brittle materials under sufficient confine pressure, point D may be sustained at a stable value following the continuous increase of strain, and finally the "residual strength" σ_r is declared. All the above defined characteristic points may be uniformly dubbed as the "strength" σ_s where there is no risk of misleading.

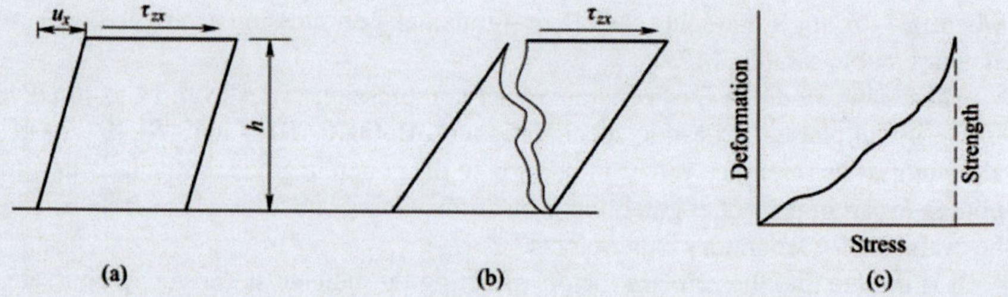

Fig. 2.38 Shear stress and corresponding shear strain—solid. **a** Elastic deformation; **b** non-elastic deformation; **c** shear stress related to shear deformation

Loading conditions include stress level and duration, application rate of load, the shape and size of test specimens, the test methods (direct tension/compression, splitting tension, bending), etc. They significantly affect the advent and degree of nonlinearity in the complete stress–strain curve as well as characteristic strengths.

Due to the most concern over the serviceability limit state and the difficulties in obtaining hardening/softening parameters, yield strength is dominantly used for the design of dam structures, and perfect plasticity behavior is postulated. Peak strength is sometimes employed where there exist difficulties to pinpoint yield strength, under such circumstances the allowable factor of safety (FOS) should be somewhat higher than that uses yield strength (Chen, 2018). Residual strength is only useful in the stability analysis for the geotechnical or hydraulic structure where its deformation is of minor concern (e.g. a landslide far from the vicinity of the water retaining dam).

2. **Response to shear stress**

For the column surface in Fig. 2.38, shear stress τ_{zx} is a single component of the stress tensor. For a small value of τ_{zx}, the shear deformation (also called the shear strain) may be measured as $\gamma_{zx} = u_x/h$.

It is fairly common that homogeneous solid materials exhibit a roughly linear stress–strain relationship for small deformations (Fig. 2.38a, c). But if the stress exceeds the shear strength, the material may break resulting in an irreversible transition (Fig. 2.38b, c).

(1) **Elastic modulus**

"Hooke's law" w.r.t. shear action leads to the definition of "shear modulus" as follows

$$G = \frac{\tau_{zx}}{\gamma_{zx}} \tag{2.23}$$

which has the unit of pressure. The magnitude of G indicates the shear stress required to achieve a unity shear deformation. For isotropic and linear elasticity, shear modulus may be related to Young's modulus and Poisson's ratio by

$$G = \frac{E}{2(1+\nu)} \tag{2.24}$$

(2) **Strength**

For concrete structures, ultimate compressive and tensile stresses are customarily used as basic strength parameters. This is due to the fact that the damage/failure of most engineering concrete is in the form of cracking and crashing. The shear strength, if demanded in engineering design, may be tested and evaluated by the similar manner of rock.

Although the shape of complete stress–strain curve, the elastic modulus and Poisson's ratio of concrete under uniaxial tension are similar to those under uniaxial compression, yet there is significant difference in its yield (and failure as well) mechanism: as the uniaxial tension state tends to arrest cracks much less frequently than the compressive state, the interval of stable crack growth is anticipatorily much shorter.

There is evidence that the compressive and tensile strength parameters are closely related but there is no fixed proportion: as the compressive strength of concrete increases, tensile strength also increases but at a lower rate. It appears that the ratio of tensile strength to compressive strength is approximately 10–11% for low-strength concrete, 8–9% for moderate-strength concrete, and 7% for high-strength concrete.

Most slim structural components are designed under the assumption that concrete would resist the compression only, and use is made of steel bars to pick up tensile loads. With massive concrete structures (e.g. dams), however, tensile stresses cannot be ignored because concrete cracking is frequently the consequence of local tensile failure due to restrained shrinkages resulted either from lowering of concrete temperature or/and from drying of concrete moist. And it is impractical to pick up all of tensile loads by the reinforcement with steel bars. Therefore, an estimation of tensile strength (or ultimate tensile strain) is demanded, especially for the concrete dam under seismic actions. In addition, a combination of tensile, compressive, and shearing stresses jointly determines the strength when concrete is in 3-D loading state, such as in the arch dam.

There exists a fundamental inverse relationship between concrete porosity and its strength. Since natural aggregates are generally dense and strong, therefore the porosity of cement paste matrix as well as ITZ will overwhelmingly determine the strength of normal-weight concrete.

3. **Response to cracking**

In modern material science, fracture mechanics is an important branch concerning the study on the crack growth in brittle materials (e.g. concrete and rock).

Fracture mechanics was developed during World War I by Griffith in 1920 to explain the failure of brittle materials (Hahn et al. 2010). His work was motivated by two contradictory facts: the stress needed to break bulk glass is around 100 MPa but the theoretical stress needed to break the atomic bond of glass is approximately 10,000 MPa. Also, experiments on glass fibers conducted by Griffith himself suggested that the strength increases as the fiber diameter decreases. Hence the uniaxial tensile strength, which had been used extensively to predict material failure, could not be a specimen-independent material property. Accordingly, Griffith came to mind that the presence of microscopic flaws in bulk material is blamed of these phenomena. By introducing artificial surface crack in his experimented glass specimens, Griffith showed what is termed today as "Griffith's criterion" in the form of

$$\sigma_f \sqrt{a} \approx C \tag{2.25}$$

where: a—flaw length; σ_f—stress at fracture; C—constant.

An explanation of this relation in terms of linear elasticity theory is problematic because it predicts that the stress (and hence the strain) at the tip of a sharp flaw is infinite. Griffith solved the elasticity problem of a finite crack in an elastic plate and he found an expression for the constant C in terms of surface energy

$$C = \sqrt{\frac{2E\gamma}{\pi}} \tag{2.26}$$

where: E—Young's modulus of the material; γ—surface energy density in the material.

Assuming $E = 62$ GPa and $\gamma = 1$ J/m^2 for glass, excellent agreement between the predicted failure stress with the experimental one was obtained.

(1) **Fracture toughness**

Griffith's work was largely ignored by engineering community until the early 1950s, when a group led by Irwin at "US Naval Research Laboratory (NRL)" found a method of calculating the amount of energy in terms of the asymptotic stress and displacement fields around a crack front (Irwin 1957).

There are three ways to apply a driving force to activate the crack (viz. Fig. 2.39): opening mode I where a tensile stress perpendicularly exerts on the crack plane; sliding mode II where a shear stress acts in parallel to the crack plane but perpendicularly to the crack front; and tearing mode III where a shear stress acts both in parallel to the crack plane and to the crack front.

According to the fracture mechanics, "fracture toughness" is one of the most important properties for many structural/mechanical design applications, because it quantitatively describes the ability of material to resist brittle fracture when a

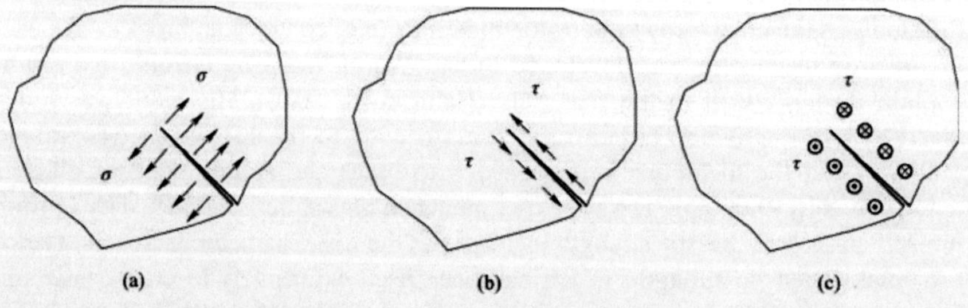

Fig. 2.39 Edge crack under loading. **a** Mode I; **b** mode II; **c** mode III

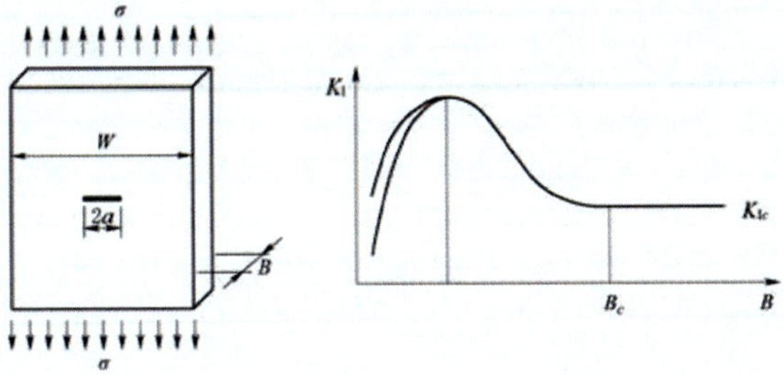

Fig. 2.40 Plate sample containing a central crack loaded with mode I stress

crack is present (Jelitto and Schneider 2019). The linear-elastic fracture toughness coefficients of a material are derived from the "stress intensity factors (SIFs)" $K_{\rm I}$, $K_{\rm II}$ and $K_{\rm III}$ correspondent to the modes at which a crack in the material begins to grow. They are denoted by $K_{\rm Ic}$, $K_{\rm IIc}$ and $K_{\rm IIIc}$ (Pa$\sqrt{\rm m}$) and are considered to be material constants w.r.t. cracking failure.

When concrete works within the limit of linear elasticity prior to its failure or the plastic zone is much smaller compared to the specimen dimension, since the most cracking failures in concrete structures are in the form of mode I, the evaluation of fracture toughness $K_{\rm Ic}$ is paramount. The actual value of SIF $K_{\rm I}$ by the closed-form (Chap. 3) or computation solution (Chap. 5) is compared to the fracture toughness $K_{\rm Ic}$, then the condition $K_{\rm I} < K_{\rm Ic}$ guarantees the stability of crack.

If the concrete is completely linear elastic, the value of $K_{\rm Ic}$ may be theoretically related to the uniaxial tension strength σ_t. Take the plate sample with a central crack of length $2a$ under tensile stress σ for example (viz. Fig. 2.40), we have

$$K_{\rm Ic} = \sigma_t\sqrt{\pi a} \tag{2.27}$$

It is normally postulated that small specimens or thin sections fail under plane stress conditions whereas plane strain fracture occurs in thick sections. Since the 1960s, size dependency on test sample has been extensively explored, and a number of size effect mechanism of rock-like materials have been revealed (Bažant and Prat 1988; Bažant and Kazemi 1990, 1991; RILEM 1990; Hu and Wittmann 1992; Bažant 2000, 2004; Chen and Liu 2004; Bažant and Yavari 2005; Ayatollahi, 2013; Sim et al. 2013). Over the years, it has been taken as an indisputable fact that fracture toughness decreases with the increase of plate thickness until a plateau is reached (viz. Fig. 2.40), this is due to the plastic zone at crack tip. The "ASTM E 399" test method reflects this viewpoint, and the critical thickness B_c for stable $K_{\rm Ic}$ is presumably calculated by

$$B_c = 2.5\left(\frac{K_{\rm Ic}}{\sigma_t}\right)^2 \tag{2.28}$$

This specimen size requirement is intended to ensure that the measured $K_{\rm Ic}$ is corresponding to the plane strain plateau.

However, there is another key issue in the test for K_{Ic}: How to ensure the specimen fracturing under nominally linear elastic conditions, or in other words, is the plastic zone sufficiently smaller than the specimen cross section? Consequently, the crack length a and the ligament width W are important specimen dimensions in addition to the specimen thickness B (viz. Fig. 2.40). Specimens for the test of K_{Ic} are usually fabricated with the specimen width W nearly equal to twice the specimen thickness B, namely, the crack length/width ratio (a/W) lies between 0.45 and 0.55.

In case the critical thickness B_c is very large, such an experimental setup would be unfeasible and therefore other experiment procedures are designed and normalized. These procedures must be validated to ensure the results are meaningful: the specimen size should be fixed, and it must be large enough to ensure the plane strain condition at crack-tip. Typical tests for fracture toughness are collected below (Gdoutos, 1993):

- Single edge crack round bar bending (SECRBB);
- Chevron bend (CB) that is one of the ISRM suggested methods;
- Semi-circular bend (SCB) that was designated in 2014 as an ISRM method for determining mode I fracture toughness;
- Straight notch disk bend (SNDB);
- Brazilian disc test (BDT);
- Flattened Brazilian disc (FBD).

It is notable that the tested fracture toughness shows significant test-method dependency, and there is a maximum difference of 42% between various test procedures. The highest and lowest values are observed in CB and SCB tests, respectively.

Brittle failure is very characteristic for materials with low fracture toughness. On the contrary, if a material possesses high fracture toughness, it will probably undergo ductile failure. Concrete and rock composites are, however, elastic–plastic materials with quasi-plastic deformations in the region around crack front. The formation of some small plastic strain before crack propagation is allowed for when applying the elasto-plastic theory. One of the basic parameters in fracture mechanics that takes into account of elasto-plastic deformation is the "crack tip opening displacement (CTOD)", and its corresponding resistance index is $CTOD_c$. This coefficient, known as one of the most important parameters in fracture mechanics w.r.t. crack driving force, is defined as the "opening at the apex of the crack". Historically, it is the first parameter for the determination of fracture toughness in the elasto-plastic region. $CTOD_c$ is considered to be the major fracture characteristics of concrete composites (Aliha et al. 2017).

Most laboratory tests of $CTOD_c$ have been made on the edge-cracked specimens loaded in three-point bending. Towards the measurement of $CTOD_c$, early experiments employed a flat paddle-shaped gage that was inserted into the crack, as the crack opened, the paddle gage rotated, and an electronic signal was sent to an $x-y$ plotter. This method is not accurate enough because it is difficult to reach the crack-tip with the paddle gage. Today, the displacement at the crack mouth is alternatively measured, and $CTOD_c$ is inferred by assuming the specimen halves are rigid and rotate about the hinge point at crack-tip. For more details of elastic–plastic fracture

Table 2.4 Fracture toughness for concrete containing FA

Addition of FA (%)	0	25	45	55
K_{Ic} (MPa $m^{1/2}$)	1.19	1.19	1.18	1.19
$CTOD_c$ (mm $\times 10^{-2}$)	1.39	1.11	1.25	1.84

mechanics, readers are referred to relevant literatures (Lemaitre and Chaboche 1990; Whittaker et al. 1992; Gdoutos 1993; Anderson 1995; Erdogan 2000; Wei 2010).

In Table 2.4, the fracture toughness for concrete containing FA (mostly employed in dam engineering) are listed according to the works of Lam et al. (1998).

The fracture toughness of mode II crack can be expressed in the fracture toughness of mode I crack by

$$K_{\prod c} = \sqrt{\frac{3}{4}} K_{Ic} \tag{2.29}$$

The tear test (e.g. Kahn tear test) may provide a semi-quantitative measurement of fracture toughness in terms of tear resistance. This type of test requires only a smaller specimen, and can therefore be used in a wider range of product forms with ductile aluminum alloys, but seldom exercised in geotechnical and hydraulic engineering.

(2) **Energy release rate**

Griffith's theory provides excellent agreement with experimental data for brittle materials such as glass. For ductile materials such as steel, although Eq. (2.25) still holds, but the surface energy (γ) predicted by Griffith's theory is usually unrealistically high. Irwin was the first to realize that the plastic zone develops at the crack-tip must play a significant role in the failure of ductile materials and even in the materials that appear to be brittle. He defined the total energy G by

$$G = 2\gamma + G_p \tag{2.30}$$

in which γ is surface energy and G_p is the plastic dissipation (and from other sources) per unit area of crack growth. The modified version of Griffith's energy criterion (2.25) can then be written as

$$\sigma_f \sqrt{a} = \sqrt{\frac{EG}{\pi}} \tag{2.31}$$

For brittle materials such as the glass, the surface energy term dominates and $G \approx 2\gamma$, hence Irwin pointed out that (Erdogan 2000): if the size of the plastic zone around crack-tip is small compared to the size of crack, the energy required to grow the crack will not be remarkably dependent on the plastic zone at crack-tip. In other words, a purely elastic solution may be used to calculate the amount of energy available for fracture. Further, he postulated that the size and shape of energy dissipation zone remains approximately constant during brittle fracture, which is equally to suggest that the energy needed to create a unit fracture surface is a constant depending only on material itself. With above assumptions, Irwin stated that the strain energy release rate for crack growth, as the change in the elastic strain energy per unit area of crack growth, also may be employed as crack growth criterion, namely, if

$$G > G_c \tag{2.32}$$

the crack will start to growth in an unstable manner.

In Eq. (2.32) the new material property G_c is designated by the name "fracture toughness", too, although different units are used by G_{Ic} and K_{Ic}. Irwin showed that for a mode I crack, the strain energy release rate and the stress intensity factor are related by (Broek 1983; Santos and Rodrigues 2003)

$$G_{\mathrm{I}} = \begin{cases} \dfrac{K_{\mathrm{I}}^2}{E} & \text{plane stress} \\ \dfrac{(1-\nu^2)K_{\mathrm{I}}^2}{E} & \text{plane strain} \end{cases} \tag{2.33}$$

where: E—Young's modulus; ν—Poisson's ratio; K_{I}—SIF of mode I.

Irwin also showed that for a planar crack in a linear elastic body, the strain energy release rate can be expressed in terms of mode I, mode II (sliding mode), and mode III (tearing mode) SIFs for the most general loading conditions.

For ductile materials (most often, metals), energy associated with plastic deformation have to be taken into account. When there is plastic deformation at crack-tip, the energy demand to activate crack may be boosted by several orders of magnitude, as the work related to plastic deformation may be much larger than the surface energy. This mechanism means a higher value for the critical value G_c. To be specific, for ductile metals G_{Ic} is around 50–200 J/m^2, for brittle metals it is usually 1–5 J/m^2, for glasses and brittle polymers it is almost always smaller than 0.5 J/m^2.

(3) J-integral

Conservation laws in elasticity theory have received considerable attention attributable to their wide applications to engineering problems. A general mathematical form of the conservation laws states that: in an elastically deformed body, the integral of a function of certain field variables over the surface enclosing a sub-region, vanishes.

Application of conservation laws to fracture mechanics leads to the establishment of so-called "path-independent integrals". The well-known Eshelby-Rice J-integral (Rice 1968) is one of these kinds and has been widely used in the engineering problem related to fracture mechanics (Chen and Shield 1977),

$$\begin{cases} J = \int\limits_{\Gamma} \left(w.n_x - t_\alpha \dfrac{\partial u_\alpha}{\partial x} \right) d\Gamma \quad (\alpha = 1, 2, 3) \\ w = \sigma_{\alpha\beta}\varepsilon_{\alpha\beta} \quad (\alpha, \beta = q, 2, 3) \end{cases} \tag{2.34}$$

where: Γ—arbitrary path around crack-tip; w—density of strain energy; t_α—components of traction vectors; u_α—components of displacement vectors; n_x—first component of the unit normal vector at Γ.

Normally, unloaded crack face is postulated, namely $t_\alpha = 0$ in Eq. (2.34), and J-integral represents the change rate of net potential energy with respect to crack growth (per unit thickness of crack front). J-integral also can be thought of as the energy flow into crack-tip, thus it is a measure of the singularity strength at crack-tip in elastic–plastic material response. For a mode I crack, it follows immediately that J-integral is equivalent to the energy release rate G_{I}. For the co-linear extension of a crack in elastic body, the integral value from Eq. (2.34) is equivalent to Irwin's energy release rate,

$$J = G_{\mathrm{I}} + G_{\mathrm{II}} \tag{2.35}$$

in which G_{I} and G_{II} are the energy release rates associated with mode-I and mode-II fractures.

The J-integral value also may be related to SIFs in both linear and nonlinear elastic solids with the help of Eq. (2.33).

2.5.3 Mechanical Properties of Intact Rocks

The mechanical performance of intact rock is, in many aspect, similar to that of concrete.

The Young's modulus of rock differs considerably for different rock types and at various stages of deformation. In compact igneous rocks, it is nearly constant until their failure. The elastic modulus of fresh intact rock with low porosity such as granite and basalt is in the range of 70–140 GPa, whereas that of sandstone, limestone, and gravel varies from 21 to 49 GPa. The secant Young's modulus E_{50} defined at 50% of the peak stress is normally employed as the "deformation modulus" and is ranged from 3.0GPa (shale, siltstone, conglomerate) to 80GPa (granite porphyry, diorite, basalt porphyrite). Correspondingly, the Poisson's ratio of rock is ranged from 0.15 (basalt porphyrite) to 0.33 (shale, siltstone, conglomerate). Generally speaking, the higher of the Young's modulus, the lower would be the Poisson's ratio (Hoek and Bray 1981).

The uniaxial compressive strength of an intact rock is related to its type ranging from the lowest 40 MPa (phyllites) to the highest 170 MPa (rhyolites).

Mostly (>70%) but not always, the tensile strength tested by splitting $\sigma_{t,s}$ is larger than that by direct axial tension $\sigma_{t,d}$, the ratio $\sigma_{t,d}/\sigma_{t,s}$ is ranged within 0.60–1.43.

Instead of compressive and tensile strengths, the most important rock resistance index for geotechnical and hydraulic structures is shear strength, particularly in the design of the abutment slope and foundation of high concrete dam. Shear strength is a result of friction and interlocking of particles, in addition to possibly cementation or bonding at particle contacts. It should be comprehensively evaluated using experiments, back analyses, and engineering analogues (empirical data). Laboratory tests employ shear box (direct shear test), tri-axial and unconfined (uniaxial) compressive apparatus, whereas in situ tests fall into vane shear and direct shear (Commission on Standardization of Laboratory and Field Tests (ISRM) 1974).

In a definite stress range, prior to the postulation of the principle of effective stress, the shear strength of rock-like material on a particular plane may be expressed as a linear function of normal stress. Therefore, in the stability analysis, it is generally assumed that the shear failure of rock-like material may be calibrated using the Mohr–Coulomb criterion, in which the shear strength is expressed in terms of parameters c (cohesion) and φ (angle of shearing friction) by the formula

$$\tau_f = c + \sigma \tan \varphi \tag{2.36}$$

in which c and φ represent the intercept at the ordinate axis and the slope of Mohr–Coulomb shear envelope corresponding to the shear failure at yield, peak or residual, which is selectively employed in the stability analysis taking into account of the type of geotechnical and hydraulic structures under the stipulated work situations.

Although Mohr–Coulomb criterion is not exactly true to rock-like materials, yet it offers a simple way to represent their failure under the combined stress state of compressive-shear from which an estimation of shear strength can be easily obtained. In the preliminary phases of structural design, shear strength parameters—the friction coefficient $\mu = \tan \varphi$ and cohesion c, may be roughly estimated by the compressive strength and tensile strength using Mohr rupture diagram.

It should be reminded that laboratory tests are always not able to simulate exact natural settings of rock-like material, particularly the rock mass with structural planes. Therefore, the testing data are normally not applied directly. They are subjected to revision taking into account of experimental conditions, sample representatives, computing methods, and natural conditions.

Assuming fresh to slightly weathered rock, peak shear strength parameters range from the lowest $c_p = 1\text{–}20$ MPa and $\varphi_p = 25°\text{–}35°$ with sedimentary (triassic, coal, chalk) to the highest $c_p = 30 - 50$ MPa and $\varphi_p = 45° - 55°$ with Igneous (granite). Strongly weathered rock will lead to significantly reduction in shear strength (Look 2007).

With regard to the fracture toughness, in addition to the test methods and test sample sizes that have been previously discussed for concrete (Hoek and Martin 2014), confining stress exhibits remarkable influence (Kataoka et al. 2017), and it is clear that loading rate dependency need to be further explored (Atkinson 1980; Nara and Kaneko 2005; Oha et al. 2019).

To give a general concept, the fracture toughness of mode I crack obtained from different test methods for Gabbro rock are listed in Table 2.5.

Table 2.5 Values of mode I fracture toughness obtained from Gabbro rock

Test method	SECRBB	BDT	CB	FBD	SCB	SNDB
Fracture toughness K_{1c}(MPa $\sqrt{\text{m}}$)	2.29	2.11	2.72	2.71	1.58–2.08	1.97

2.5.4 Mechanical Properties of Rock Joints

In most rock stability problems, conditions for slip on major pervasive features are governed by the shear resistance of rock joint. In addition, the shear stiffness and normal stiffness of joint can manifest a significant influence on the distribution of stress and displacement within the jointed rock mass.

The actual mechanism of joint deformation is complex. Various minerals and gouges may fill the joint affecting both its slip and compaction, or dilation. The presence of pore water alters the joint deformation process, too.

The most useful description for the mechanical behavior of joint is the complete strain–displacement curve of direct shear under certain pressure (normal stress), as shown in Fig. 2.41.

1. **Response to pressure**

The deformation test of a rock joint under normal loading is also called as "joint closure test". At laboratory scale, samples of cylindrical under uniaxial compression are customarily used (viz. Fig. 2.41a). Displacement sensors are positioned around the sample in order to measure the closure of joint when a normal load σ_n is applied. In order to correctly estimate the joint closure, it is necessary to deduct the intact rock deformation (Gentier 1986).

In compression, the tested joint is gradually pushed close with an apparent limit. The joint closure exhibits strong nonlinearity tightly linked to the applied normal stress and associated with the strength of intact rock/concrete, as depicted in Fig. 2.41c in which the normal stiffness is defined as $k_n = \frac{\Delta\sigma_n}{\Delta u_n}$. It is important to distinguish two types of joint closure tests: when the upper and lower surfaces match and when they do not match, because an unmatched joint can deform more than a matched one.

Shehata (1971) proposed the following equation to estimate the joint closure Δu_n,

$$\Delta u_n = \frac{\log \sigma_1 - \log \sigma_2}{C} \tag{2.37}$$

in which σ_1 and σ_2 are the reference normal stress and actual normal stress, respectively, C is a constant that depends on the elastic property of joint.

Goodman (1974) found that joint closure tends to approach a maximum value at high normal stresses, and proposed the following equation to estimate the closure of joint

$$\sigma_n = \left(\frac{\Delta u_n}{\Delta u_{n,\max} - \Delta u_n}\right)\sigma_0 + \sigma_0 \tag{2.38}$$

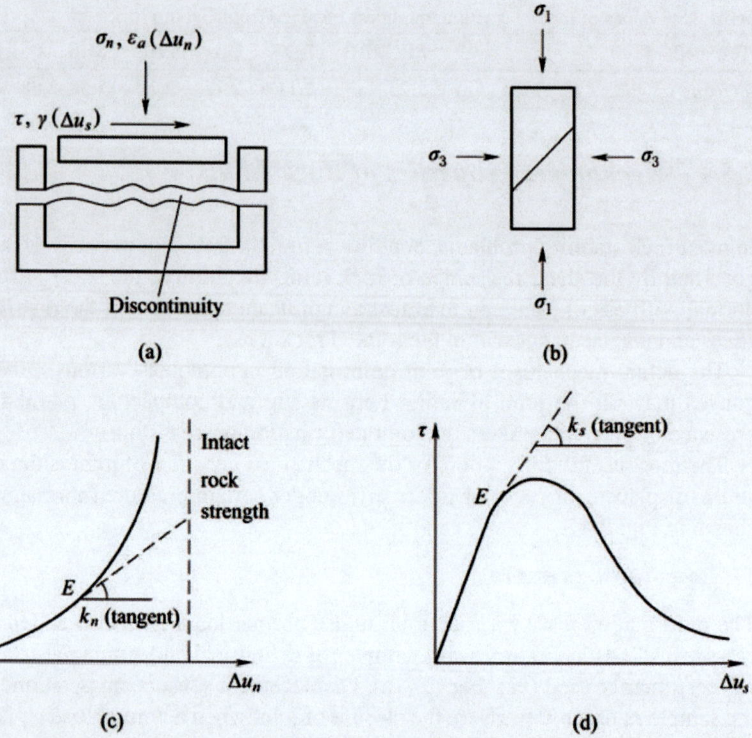

Fig. 2.41 Complete stress–strain curve of joint under constant normal pressure and quasi-static direct shear loading. **a** Schematic diagram to direct shear test; **b** schematic diagram to tri-axial test; **c** compressive behavior; **d** shear behavior

in which σ_0 and σ_n are the reference normal stress and actual normal stress, respectively, Δu_n and $\Delta u_{n,\max}$ are the actual joint closure and maximum joint closure, respectively.

In an effort to consider the influence of joint roughness on joint closure behavior, Bandis, Lumsden and Barton (1983) proposed following formula

$$\Delta u_{n,\max} = A + B(JRC) + C(JCS/a_j)^D \tag{2.39}$$

where: A, B, C and D—constants; a_j—maximum opening (mechanical aperture) of joint; JRC—joint roughness coefficient; JCS—joint compressive strength.

The corresponding normal stiffness of joint is estimated by

$$k_n = \frac{\Delta\sigma_n}{\Delta u_n} \approx 0.02(JCS/a_j) + 2JRC - 10 \tag{2.40}$$

It is notable that there are a number of other important joint closure models (Brown and Scholz 1986; Hopkins, 1990, 2000; Marache, 2002).

2. **Response to shear stress**

The shear strength and shear stiffness of rock joint can be measured by one single shear test. There exist various types of shear test, of which the most commonly exercised one is the direct shear test as shown in Fig. 2.41a, where the mean plane of joint surface is aligned parallel to the direction of applied shear force. Two walls of the specimen are fixed inside the shear box using a suitable encapsulating material (e.g. epoxy resin or plaster). Direct shear tests in the configuration of Fig. 2.41a are usually carried out at constant compressive force (normal stress). During the test, the tangential force applied is measured by a load-cell, and the vertical displacement of the upper portion of sample, also called as "dilation", is registered by the displacement sensor. During the test, the upper and lower portions of sample do not rotate.

Tri-axial cell is sometimes needed, too, which is well suited in the presence of water. Specimens are prepared from cores containing joints inclined at 25°–40° to the specimen axis. The specimen is set up in the tri-axial cell as shown in Fig. 2.41b, then cell pressure and axial load are applied successively. Tests may be either drained or undrained, preferably with a known level of water pressure being imposed and maintained throughout the test.

Figures 2.42 and 2.43 present the test results for a joint under low confining pressure in both the relations of τ versus Δu_s through which the shear stiffness may be defined as $k_s = \frac{\Delta\tau}{\Delta u_s}$, and Δu_n versus Δu_s (Archambault et al. 1997; Flamand 2000). Figure 2.44 depicts two examples of dilation curves (Marache 2002).

Archambault et al. (1997) divided the curves τ versus Δu_s and Δu_n versus Δu_s in five phases. However, from the practice point of view, a simple division in three phases is already rather appropriate to get a convincible shear model of joint:

Fig. 2.42 Relation of τ versus Δu_s of rock joint

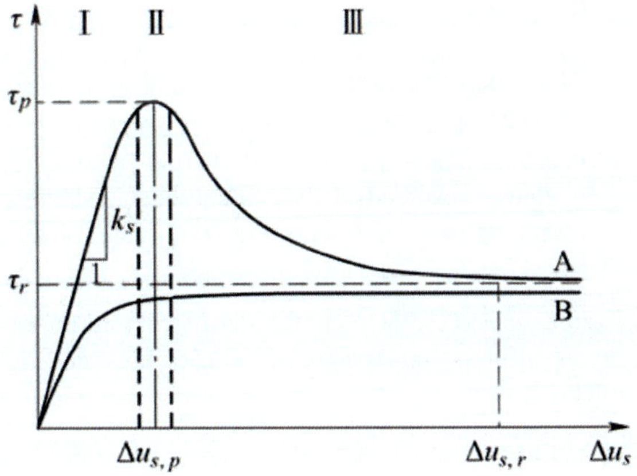

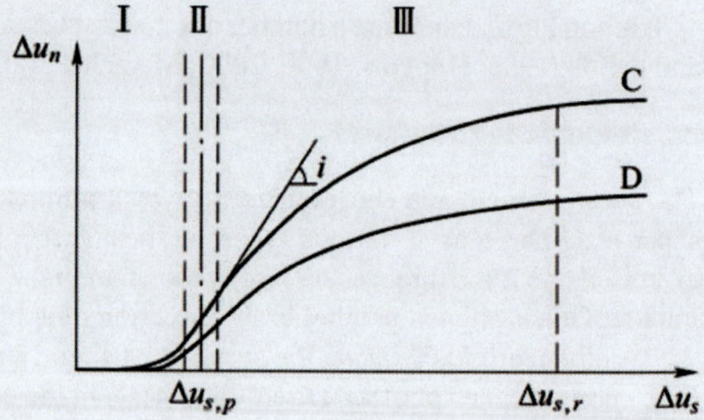

Fig. 2.43 Relation of Δu_n versus Δu_s of rock joint

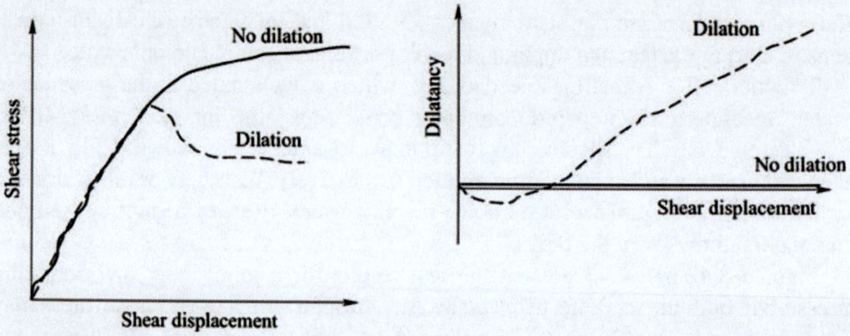

Fig. 2.44 Dilation effect

- Phase I. Joint deformation is elastic and pseudo-linear. The tangential (shear) stiffness k_s of the joint is estimated during this phase. The deformation increases progressively until the joint slip starts, leading to a dilation behavior.
- Phase II. A non-linear relation between τ and Δu_s is remarkably observed, corresponding to the behavior in the vicinity of peak shear load that coincides with the higher angle of dilation (curve C in Fig. 2.43). For a planar joint (curve D in Fig. 2.43) there is no such peak of shear load.
- Phase III. The shear load and dilation angle lower down, tending to be constant (residual shear load). This residual shear load/stress is reached much more quickly for planar joint (curve B in Fig. 2.42) than for rough joint (curve A in Fig. 2.42).

Factors such as the degree of matching between joint walls, the type of infilling, the roughness of joint walls, the size of test sample, etc., bring about nontrivial effects on the stiffness and strength of joint (Ladanyi and Archambault 1970, 1977; Bandis et al. 1981).

(1) **Stiffness**

When a joint is subjected to shear stress, the shear load vs. shear displacement curve (viz. Figs. 2.41d and 2.42) is rather like the complete stress–strain curve of intact rock under uniaxial compression (viz. Fig. 2.37), except that the slip failure is localized along the joint. In addition to the shear stiffness k_s, there are a peak shear strength and a post-peak failure region with residual strength (curve A in Fig. 2.42), but normally the yield strength is obscure.

Rock joints may be filled or unfilled. An unfilled joint can be represented by two parallel plates in contact through asperities. These asperities may be looked at as a thin layer of equivalent granular material of high porosity clipped between the plates. Accordingly, Chen et al. (1989) proposed a "filled model", in which the asperities are replaced by an evenly "filled" virtual medium with certain deformation and permeability characteristics. In this manner, they derived the non-linear and normal stress-dependent expressions of joint stiffness

$$\begin{cases} k_s = k_{s0} \exp(-\xi \sigma_{nj}) \\ k_n = k_{n0} \exp(-\xi \sigma_{nj}) \end{cases} \tag{2.41}$$

where: ξ—coupling coefficient; k_{n0} and k_{s0}—initial stiffness coefficients, MPa/m; σ_{nj}—normal stress on joint (negative for compression), MPa.

The parameters in Eq. (2.41) under the normal pressure <10 MPa may be estimated by the routine coupling tests for rock joints. Iwai (1976), Raven and Gale (1985) performed laboratory tests on the mechanical and hydraulic properties of joints in basalt, granite, and marble. Cylindrical samples of intact rocks, 0.15 m in diameter, were diamond-cored from rock blocks, and a horizontal tension fracture was created in each sample using a modified Brazilian splitting method. A central hole, 0.022 m in diameter, provided access for outward radial flow of water. Use was made of their test data, Chen et al. (1989) evaluated the parameters for their "filled model" in Table 2.6.

(2) **Strength**

Friction is a very complex phenomenon controlled by the parameters such as roughness, wear, scale effect, gouge material, boundary condition (dilation allowed for or not), speed of sliding, degree of matching, etc.

When studying the stability of rock mass, it is of capital importance to know the friction behavior of rock joints. It is customarily assumed that the strength of joint is represented by the Mohr–Coulomb criterion in Eq. (2.36). To give a general concept, unfilled rock joints possess friction angles ranging from $\varphi = 20° - 27°$ (schists, shale) to $\varphi = 34° - 40°$ (basalt, granite, limestone, conglomerate).

Table 2.6 Coupling parameters of joints (laboratory)

Parameter	Granite (Raven and Gale 1985)	Granite (Iwai 1976)	Basalt (Iwai 1976)	Marble (Iwai 1976)
k_{n0}(MPa/m)	1.977×10^4	1.449×10^4	5.495×10^4	3.571×10^4
ξ	0.22	0.3	0.35	0.8

Fig. 2.45 Evolution of shear peak stress and residual shear strength as function of the normal stress

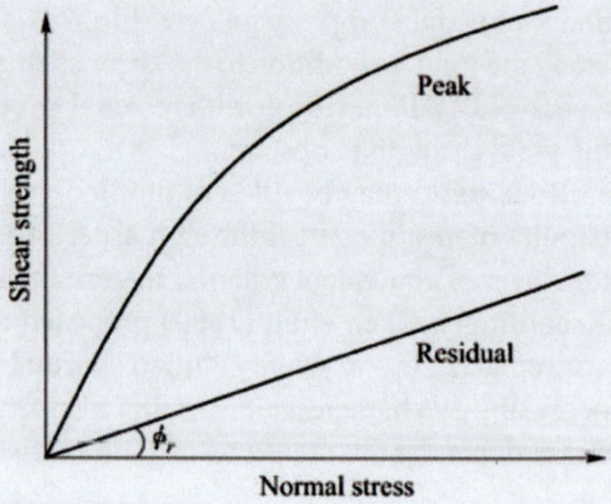

The curve A of shear load versus shear displacement in Fig. 2.42 is typical for clean and rough joints. Under constant normal stress, the peak of shear stress is reached after a small shear displacement, and the post-peak regime exhibits a residual strength after sufficiently large shear displacement. Moreover, if the shear tests are performed under various normal stresses, the peak and residual shear strengths usually exhibit their relations to normal stresses in Fig. 2.45.

To the author's knowledge, being not able to include all the variables involved in friction, there is currently no complete model able to precisely predict the shear behavior of rock joint. However, there exist some approximate models practically applicable to the behavior of joint friction in terms of, for example, surface roughness.

A simple model shown in Fig. 2.46 was introduced by Patton (1966). He proposed a bilinear model for rough surfaces that accounts for two phenomena observed in experiment: overriding of asperities at low normal stress levels, and shearing through asperities at higher normal stress levels.

$$\begin{cases} \tau_p = \sigma_n \tan(\varphi + i) & \text{when } \sigma_n < \sigma_t \\ \tau_p = c + \sigma_n \tan \varphi_r & \text{when } \sigma_n \geq \sigma_t \end{cases} \tag{2.42}$$

For most practical purposes, $\varphi_r = \varphi$.

The effects of surface roughness on joint deformation and strength were further explored by Barton and his co-workers (Barton 1973, 1976; Barton and Choubey

Fig. 2.46 Joint model by Patton (1966). **a** Rough joint inclined at an angle i; **b** rough joint with sinusoidal asperities

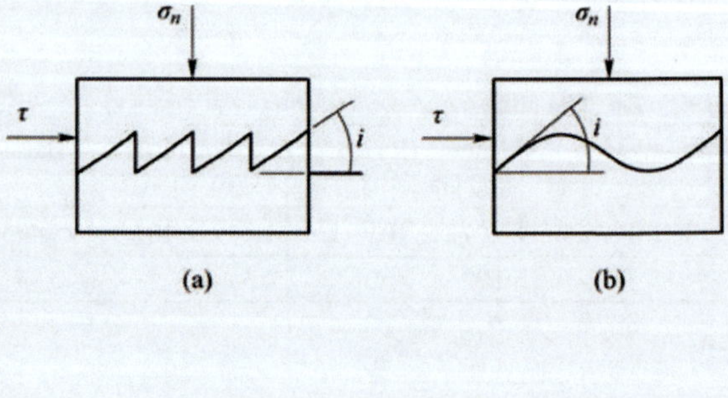

1977; Barton et al. 1985; Barton and Bandis 1990). They proposed a nonlinear model for the shear strength of joint

$$\tau = \sigma_n \tan(JRC \log_{10}(JCS/\sigma_n) + \varphi_r) \tag{2.43}$$

where: σ_n—effective normal stress; JRC—joint roughness coefficient on a scale of 1 for the smoothest to 20 for the roughest surfaces; JCS—joint compressive strength; φ_r—residual friction angle.

Equation (2.43) is termed as the "Barton-Bandis model" that comprises three components of shear strength: a basic frictional component given by φ_r, a geometrical component controlled by surface roughness JRC and an asperity failure component controlled by the ratio (JCS/σ_n).This model explicitly takes account of more characteristics for joint strength and deformation than both the elementary Mohr–Coulomb model and Patton model. However, its practical application may present some difficulties. In particular, the derivation of constitutive relations w.r.t. the mobilization and degradation of surface roughness may lead to irregularities in the numerical simulation.

Other advanced models since the 1990s such as that by Saeb and Amadei (1990), Leong and Randolph (1992), Maksimovic (1996), etc., are seldom to find their applications in geotechnical and hydraulic structures yet.

2.5.5 Mechanical Properties of Jointed Rock Masses

It has been widely recognized that joints are significantly influential to the gross response of the jointed rock mass encountered in geotechnical and hydraulic structures (e.g. dams, tunnels, cut-slopes, etc.). While the intact rock may remain effectively elastic and undergo only small strain, irreversible slip at joints changes the overall stress distribution (Müller 1974; Brown 2011).

There was a period when scholars endeavored to conduct laboratory tests for the deformation and strength properties of jointed rock mass, through which the models of gross response taking into account of the effects of joint interaction and block interlock were correlated with experimental data. These experimental works were largely carried out with specimens of plaster materials containing multiple intersecting joint sets in order to simulate rock fracture and joint slip at low stress levels (Einstein et al. 1969). However, it is not clear that even the qualitative response of jointed rock mass has been reproduced by such type of tests (Barton 1972).

1. **Elastic modulus**

"Modulus of deformation" or "deformation modulus" (E) that is a secant Young's modulus comprising both elastic and permanent deformation, signifies that its value is calculated from the data of the loading portion of stress–strain curve. Nowadays, as far as the deformation modulus (and the strength as well) of the jointed rock mass is concerned, three approaches are exercised, namely, analyses using the properties of intact rock and joints respectively, direct in-situ tests, and empirical formulae.

For the highly jointed rock masses in which the joint spacing is much smaller than the characteristic dimension of the engineering structure concerned, it is very convenient to regard them as "equivalent" continua. Along this line, many works can be found from, for example, Salamon (1968), Morland (1974), Zienkiewicz and Pande (1976), Amadei and Goodman (1981), Gerrard (1982a, b), Amadei (1983), Pande and Gerrard (1983), Yoshinaka and Yamabe (1986), Chen and Pande (1994). Although various theoretical models are available for deformation modulus (and strength as well), yet they are all subjected to the verification and adjustment using the data from in-situ tests, engineering analogues, and back analyses (if possible).

To avoid the effects of sampling disturbance, it is widely demanded to determine the deformation modulus E and Poisson's ratio ν (or shear modulus G) by the data from in-situ (field) testing. Although field testing is most straight forward to obtain the equivalent elastic deformation parameters, yet it is costly and often involves uncertain errors related to the effects of hidden factors, the control of boundary conditions, the equipment installation, the test site preparation, the blasting damage, and the interpretation of results. These errors lead to greatly inaccurate and scattered test data of rock mass (Bieniawski 1978; Palmstrom Singh 2001; Hoek and Diederichs 2006).

Empirical formulae establish certain relationships on the basis of rock mass classification (Hoek and Brown 1980; Hoek 1994; Hoek and Brown 1997; Barton 2002; Cai et al. 2004; Sonmez et al. 2004; Hoek and Diederichs 2006; The National Standards Compilation Group of People's Republic of China 2014). Although they have gained extensive popularity for the practical applications, especially in tunneling engineering, yet they often elicit too conservative estimations. This is largely because they make use of categorized parameters from case histories. Another drawback of empirical formulae is that they lack a proper mathematical platform to establish constitutive relations. Furthermore, the anisotropic deformation cannot be properly represented in the tensor form that is a base of continuum mechanics.

Through the statistical analysis conducted for 54 hydraulic projects in China, the BQ indexes and the design values of deformation modulus are correlated in Fig. 2.47 and fitted in Eq. (2.44).

Fig. 2.47 Correlation of deformation modulus E versus rock mass BQ

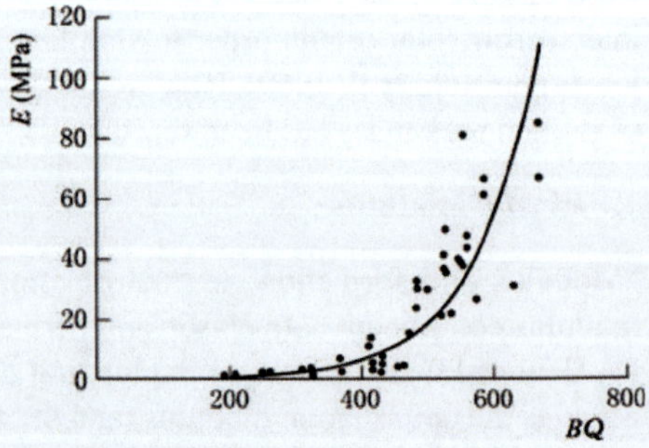

$$E = 0.0986 \times \exp(0.0105 \times BQ) \text{ with } R^2 = 0.8335 \tag{2.44}$$

When there are no sufficient test data, the recommended deformation modulus of rock mass based on the rock classification standard (GB/T50218-2014) also may be selectively referred.

It should be emphasized again that at present engineering practices, all the Young's moduli estimated by in-situ tests or/and engineering analogues can only define isotropic elastic matrices. Further endeavors are desperately desirable to access a reliable anisotropic elastic matrix with both respect to the component values and principal directions. A tight collaboration with well-experienced experts in geological exploration, theoretical/numerical analysis, and laboratory/field test, is highly encouraged.

2. **Strength**

The determination of rock mass strength can be approached either via the theoretical or/and computational analyses using the strengths of intact rocks and joints, or via in-situ tests directly. For a weathered rock, there is approximately ten-fold reduction in the deformation modulus and uniaxial strength cross-referenced to fresh one. The geologic age may considerably affect the deformation and strength parameters for sedimentary rocks, too.

In the first approach, researchers initially devoted themselves to the problems linked to stratified rock w.r.t. elastic response and failure. The seminal work of Salamon (1968) on stratified materials consisting of isotropic layers was preceded by the works of Wardle and Gerrard (1972) who examined some restrictions to the values of equivalent elastic constants. In a work by Gerrard (1982a, b), orthorhombic constituents were considered. Other works tackling the same problem with numerical approaches can be found in, for example, Adhikary and Dyskin (1998), Guz and Soutis (2001). In 2003, Bourne presented an interesting discussion on global versus local stress states. Triantafyllidis and Gerolymatou (2014) formulated a solution covering both elasticity and failure in three dimensions.

Due to the fact that real-world rock masses are discontinuous in most cases, tests conducted on small specimens in laboratory generally do not yield appropriate strength and deformation parameters (Bieniawski, 1968a, b): the smaller of the specimen, the fewer joints are presented and hence the stronger will be the specimen. A large specimen may provide a better estimation of these characteristics. Therefore, in-situ tests are highly desirable for the strength evaluation of rock mass although difficulties arise from the extreme demand of testing load capacity and resources (human, money, time).

There are many types of large scale in-situ test under compression and shear. They are exercised throughout the world for various purposes and in various applications. For detailed messages our readers are referred to the book by, for example, Dunnicliff (1988). Insofar, difficulties still exist in the evaluation of elastic matrix components and strength parameters for rock masses mainly due to:

- Stochastic characteristics. Joints are dominant factors influencing rock mass parameters, and their spatial distribution pattern is actually random. In addition, rock blocks are composed of crystals and grains in a matrix that includes cracks and fissures. The in-situ test may only comprise a portion of them and lack of full representation. The limited number of test samples and strongly fluctuated data obstruct us to conduct meaningful statistical analysis. How to explore the limited testing data from points to faces and further to a geological body is an unsolved issue, despite carefully layout testing locations.
- Scale effects. Smaller sample yields unnaturally higher deformation modulus and strength. Take the granite in the cut slope (170 m high) of the ship lock in Three-gorges Project (China) for example, the laboratory tests provided a deformation modulus of 69.6 GPa, the in-situ tests lowered it down to 48.90 GPa, and the late phase back analyses gave a much lower deformation modulus of 32.10 GPa. Coal, altered granite, shale, etc., with networks of fractures exhibit greater size effects—the ratio of in-situ tested strength to laboratory tested strength sometimes attains as high as of tens. Rather large sample size is hence demanded to get statistically complete collection of all the components influential to the deformation and strength. Ideally, a sample with at least 10 joints intersecting any of its edge might guarantee the REV with stable components of elastic matrix, but it would be very expensive, and under many circumstances, even not realistic to accomplish such in-situ testing due to the lack of technique capability.
- Dynamic evolution due to relaxation (EDZ). Again, take the granite in Three-gorges Project (China) for example, the deformation modulus is deteriorated by 60% in the strongly relaxed zone within the depth of 5 m. Under the base of Goupitan arch dam (China), the deformation modulus is deteriorated by 40% in the strongly relaxed zone within the depth of 2 m.
- Environmental effects. Conventional test codes do not take into account of factors such as high in-situ stress, high hydraulic pressure, and high geo-thermal for deeply embedded rock masses.

It is also suggestible that laboratory or in-situ testing amount is minimized until the decision on the site of geotechnical and hydraulic structure has been made. In the early phases of design, the deformation modulus, uniaxial compression strength σ_c, uniaxial tensile strength σ_t, and shear strength parameters c and $\tan\phi$ of the rock mass (weathered or intact) may be estimated using the rock classification indexes of RMR, Q, GSI, and BQ.

2.6 Concluding Remarks

The basic properties of water and rock-like materials are briefly summarized in this chapter for the elaboration of advanced CMs.

In the construction of geotechnical and hydraulic structures, concrete and rocks are largely consumed. These materials are rather difficult to be fully learned and simulated by CMs for the purpose of structural design, construction, and management. Take the jointed rock masses for example, because they are made of nature and beneath the Earth's surface, the structural plane system is difficult to be exactly established mainly blamed for the limitation from investigation and test capability. In addition, their mechanical properties are also difficult to be fully understood and interpreted.

Bearing above background knowledge in mind, it will help us to understand the motivation, state-of-the-arts, and applicability of the advanced CMs presented in this book. For our readers from, or potentially will be engaged in the research and application of advanced CMs for engineering structures, the following endeavors are all valuable:

- Enable the CMs to possess a strong ability in the simulation of meso-and macro-scale features (e.g. aggregate and ITZ in concrete, joint and fault in rock) and large deformation accompanied with distortion, crack and detachment;
- Develop the tools for more powerful geology investigation, and the facilities for more detailed experiment w.r.t. the meso-and macro-scale features;
- Select the appropriate methods for specific engineering problem, and try the best to keep the competency of input data w.r.t. the method. This is important to get meaningful outcomes, particularly the safety margin or/and accident hazard, for the engineering design and management.

We have to admit that aforementioned works are far from satisfactorily completed. Normally, many advanced CMs might be validated at the level of laboratory test or latter-witness rather well, but not be very convincible at the level of prediction when practitioners make a crucial decision under the pressure from field situations. This also may be clarified by the validation examples and engineering applications enumerated in the Chaps. 5, 6, 7, 8, 9, 10 and 11 of this book.

References

Abyaneh SD, Wong HS, Buenfeld NR. Computational investigation of capillary absorption in concrete using a three-dimensional mesoscale approach. Comput Mater Sci. 2014;87:54–64.

Adhikary D, Dyskin A. A continuum model of layered rock masses with non-associative joint plasticity. Int J Numer Anal Meth Geomech. 1998;22(4):245–61.

Adler PM, Thovert JF. Fractures and fracture networks. Dordrecht (Netherlands): Kluwer Academic Publishers; 1999.

Adler PM, Thovert JF, Mourzenko VV. Fractured porous media. Oxford (UK): Oxford University Press; 2012.

Aliha MRM, Sarbijan MJ, Bahmani A. Fracture toughness determination of modified HMA mixtures with two novel disc shape configurations. Constr Build Mater. 2017;155:789–99.

Allègre CJ, Le Mouel JL, Provost A. Scaling rules in rock fracture and possible implications for earthquake prediction. Nature. 1982;297(5861):47–9.

Amadei B. Rock anisotropy and the theory of stress measurements. Berlin (Germany): Springer-Verlag; 1983.

Amadei B, Goodman RE. A 3-D Constitutive relation for fractured rock masses. In: Selvadurai APS, editor. Proceedings of International symposium mechanical behaviour structured media, Ottawa (Canada). Amsterdam (Netherlands): Elsevier; 1981. p. 249–68.

Anderson TL. Fracture mechanics: fundamentals and applications. 3rd ed. Boca Raton, Florida (USA): CRC Press; 1995.

Andersson J, Dverstorp B. Conditional simulation of fluid flow in three-dimensional networks of discrete fractures. Water Resour Res. 1987;23(10):1876–86.

Archambault G, Gentier S, Riss J, Flamand R. The evolution of void spaces (permeability) in relation with rock joint shear behavior. Int J Rock Mech Min Sci. 1997;34(3–4). Paper No.014.

Atkinson BK. Stress corrosion and the rate-dependent tensile failure of a fine-grained quartz rock. Technopysics. 1980;65(3–4):281–90.

Atkinson BK. Subcritical crack growth in geological materials. J Geophys Res Solid Earth. 1984;89(B6):4077–114.

Ayatollahi MRAJ. Size effects in mode II brittle fracture of rocks. Eng Fract Mech. 2013;112–113(11):165–80.

Baecher GB. Statistical analysis of rock mass fracture. J Int Associ Math Geol. 1983;15(2):329–48.

Baecher GB, Lanney NA, Einstein HH. Statistical description of rock properties and sampling. In: Wang FD, Clark GB, editors. Proceeding of 18th US symposium rock mechanics. Colorado (USA): Colorado School of Mines Press; 1977. p. 5c1–8.

Baghbanan A, Jing L. Hydraulic properties of fractured rock masses with correlated fracture length and aperture. Int J Rock Mech Min Sci. 2007;44(5):704–19.

Baghbanan A, Jing L. Stress effects on permeability in a fractured rock mass with correlated fracture length and aperture. Int J Rock Mech Min Sci. 2008;45(8):1320–34.

Balberg I, Binenbaum N. Computer study of the percolation threshold in a two-dimensional anisotropic system of conducting sticks. Phys Rev B. 1983;28(7):3799–812.

Bandis S. Experimental studies of scale effects on shear strength and deformation of rock joints. PhD Thesis. University of Leeds (UK); 1980.

Bandis S, Lumsden AC, Barton BR. Experimental studies of scale effects on the shear behavior of rock joints. Int J Rock Mech Min Sci. 1981;18(1):1–21.

Bandis S, Lumsden AC, Barton NR. Fundamentals of rock joint deformation. Int J Rock Mech Min Sci & Geomech Abstr. 1983;20(2):249–68.

Barton CA, Zoback MD. Self-similar distribution and properties of macroscopic fractures at depth in crystalline rock in the Cajon Pass Scientific Drill Hole. J Geophys Res. 1992;97(B4):5181–200.

Barton NR. A model study of rock-joint deformation. Int J Rock Mech Min Sci. 1972;9(5):579–82.

Barton NR. Review of a new shear-strength criterion for rock joints. Eng Geol. 1973;7(4):287–332.

Barton NR. Rock mechanics review, the shear strength of rock and rock joints. Int J Rock Mech Min Sci. 1976;13(9):255–79.

Barton NR. Some new Q-value correlations to assist in site characterisation and tunnel design. Int J Rock Mech Min Sci. 2002;39(2):185–216.

Barton NR, Choubey V. The shear strength of rock joints in theory and practice. Rock Mech. 1977;10(1–2):1–54.

Barton NR, Bandis S. Review of predictive capabilities of JRC-JCS model in engineering practice. International Symposium on Rock Joints. Loen (Norway); 1990. p. 603–10.

Barton NR, Bandis S, Bakhtar K. Strength, deformation and conductivity coupling of rock joints. Int J Rock Mech Min Sci & Geomech Abstr. 1985;22(3):121–40.

Barton NR, Lien R, Lunde J. Engineering classification of rock masses for the design of tunnel support. Rock Mech. 1974;6(4):189–236.

Bažant ZP. Size effect. Int J of Solids Struct. 2000;37(1–2):69–80.

Bažant ZP. Probability distribution of energetic-statistical size effect in quasibrittle fracture. Probabilistic Eng Mech. 2004;19(4):307–17.

Bažant ZP, Kazemi MT. Determination of fracture energy, process zone length and brittleness number from size effect, with application to rock and concrete. Int J Fract. 1990;44(2):111–31.

Bažant ZP, Kazemi MT. Size dependence of concrete fracture energy determined by RILEM work-of-fracture method. Int J Fract. 1991;51(1):121–38.
Bažant ZP, Prat PC. Measurement of mode III fracture energy of concrete. Nuclear Engineering and Design.1988;106(1):1–8.
Bažant ZP, Yavari A. Is the cause of size effect on structural strength fractal or energetic-statistical? Eng Fract Mech. 2005;72(1):1–31.
Belayneh MW, Matthäi SK, Blunt MJ, Rogers SF. Comparison of deterministic with stochastic fracture models in water-flooding numerical simulations. Am Assoc Pet Geol Bull. 2009;93(11):1633–48.
Berkowitz B, Adler PM. Stereological analysis of fracture network structure in geological formations. J Geophys Res Solid Earth. 1998;103(B7):15339–60.
Berkowitz B, Bour O, Davy P, Odling N. Scaling of fracture connectivity in geological formations. Geophys Res Lett. 2000;27(14):2061–4.
Berkowitz B, Hadad A. Fractal and multifractal measures of natural and synthetic fracture networks. J Geophys Res Solid Earth. 1997;102:12205–18.
Bieniawski ZT. Fracture dynamics of rock. Int J Fract Mech. 1968a;4(4):415–30.
Bieniawski ZT. The effect of specimen size on compressive strength of coal. Int J Rock Mech Min Sci. 1968b;5(4):325–35.
Bieniawski ZT. Engineering classification of jointed rock mass. Trans S Afr Inst Civ Eng. 1973;15(12):335–44.
Bieniawski ZT. Determining rock mass deformability: experience from case histories. Int J Rock Mech Min Sci & Geomech Abstr. 1978;15(5):237–47.
Bieniawski ZT. Engineering rock mass classification: a complete manual for engineers and geologists in mining, civil, and petroleum engineering. New York (USA): Wiley-Interscience Publication; 1989.
Billaux D, Chilès JP, Hestir K, Long J. Three-dimensional statistical modeling of a fractured rock mass–an example from the Fanay-Augères Mine. Int J Rock Mech Min Sci & Geomech Abstr. 1989;26(3–4):281–99.
Bobet A, Fakhimi A, Johnson S, Morris J, Tonon F, Ronald YM. Numerical models in discontinuous media: review of advances for rock mechanics applications. J Geotech Geoenviron Eng. 2009;135(11):1547–61.
Bonnet E, Bour O, Odling NE, Davy P, Main I, Cowie P, Berkowitz B. Scaling of fracture systems in geological media. Rev Geophys. 2001;39(3):347–83.
Bour O, Davy P. Connectivity of random fault networks following a power law fault length distribution. Water Resour Res. 1997;33(7):1567–83.
Bour O, Davy P. On the connectivity of three-dimensional fault networks. Water Resour Res. 1998;34(10):2611–22.
Bour O, Davy P. Clustering and size distributions of fault patterns: theory and measurements. Geophys Res Lett. 1999;26(13):2001–4.
Bour O, Davy P, Darcel C, Odling N. A statistical scaling model for fracture network geometry, with validation on a multiscale mapping of a joint network (Hornelen Basin, Norway). J Geophys Res Solid Earth. 2002;107(B6):2113.
Bourne SJ. Contrast of elastic properties between rock layers as a mechanism for the initiation and orientation of tensile failure under uniform remote compression. J Geophys Res. 2003;108(B8):2395.
Broek D. Elementary engineering fracture mechanics. Boston (USA): Martinus N Publishers; 1983.
Brown ET. Fifty years of the ISRM and associated progress in rock mechanics. In: Qian Q, Zhou XY, editors. Proc 12th ISRM Congress. Beijing (China): ISRM; 2011. p. 29–45.
Brown SR, Scholz CH. Closure of rock joints. J Geophys Res. 1986;91(B5):4939–48.
Buyer A, Schubert W. Calculation the spacing of discontinuities from 3D point clouds. In: Konicek P, Soucek K, Konecny P, editors. Procedia engineering, vol. 191 (EUROCK 2017). Ostrava (Czech Republic): ISRM; 2017. p. 270–78.

Cacas MC, Ledoux E, de Marsily G, Barbreau A, Calmels P, Gaillard B, Margritta R. Modeling fracture flow with a stochastic discrete fracture network: calibration and validation: 2. The Transport Model Water Resour Res. 1990a;26(3):491–500.

Cacas MC, Ledoux E, de Marsity G, Tillie B, Barbreau A, Durand E, Feuga B, Peaudecerf P. Modeling fracture flow with a stochastic discrete fracture network: calibration and validation, 1. The Flow Model Water Resour Res. 1990b;26(3):479–89.

Cai M, Kaiser PK, Uno H, Tasaka Y, Minami M. Estimation of rock mass deformation modulus and strength of jointed hard rock masses using the GSI system. Int J Rock Mech Min Sci. 2004;41(1):3–19.

Casciano C, Ruvo L, Volpi B, Masserano F. Well test simulation through discrete fracture network modelling in a fractured carbonate reservoir. Pet Geosci. 2004;10(4):331–42.

Chen B, Liu JY. Effect of aggregate on the fracture behavior of high strength concrete. Construct Build Mater. 2004;18(8):585–90.

Chen FHK, Shield RT. Conservation laws in elasticity of the J-integral type. J Appl Math Phys (ZAMP). 1977;28:1–22.

Chen SH. Hydraulic structures. Berlin (Germany): Springer-Verlag; 2015.

Chen SH. Computational geomechanics and hydraulic structures. Berlin (Germany): Springer; 2018.

Chen SH, Egger P. Elasto-viscoplastic distinct modelling of bolt in jointed rock masses. In: Yuan JX editor. Proceedings of computer method and advances in geomechanics. Wuhan(China). Rotterdam (Netherlands): AA Balkema; 1997. p. 1985–90.

Chen SH, Egger P. Three dimensional elasto-viscoplastic finite element analysis of reinforced rock masses and its application. Int J Numer Anal Meth Geomech. 1999;23(1):61–78.

Chen SH, Feng XM, Shahrour I. Numerical estimation of REV and permeability tensor for fractured rock masses by composite element method. Int J Numer Anal Meth Geomech. 2008;32(12):1459–77.

Chen SH, He J, Shahrour I. Estimation of elastic compliance matrix for fractured rock masses by composite element method. Int J Rock Mech Min Sci. 2012;49(1):156–64.

Chen SH, Pande GN. Rheological model and finite element analysis of jointed rock masses reinforced by passive, fully-grouted bolts. Int J Rock Mech Min Sci Geomech Abstr. 1994;31(3):273–7.

Chen SH, Wang HR, Xiong WL. Study of the seepage characteristics of joint. J Wuhan University of Hydraulic and Electric Engineering (WUHEE).1989;22(1):51–60. (in Chinese with English abstract).

Chen SH, Xu MY, Shahrour I, Egger P. Analysis of arch dams using coupled trial load and block element methods. J Geotech Geoenviron Eng, ASCE. 2003;129(11):977–86.

Chilès JP. Fractal and geostatistical methods for modelling a fracture network. Math Geol. 1988;20(6):631–54.

Commission on Standardization of Laboratory and Field Tests. Suggested Methods for Determining Shear Strength (Document No. 1). Lisbon (Portugal): ISRM, 1974.

Cowie PA, Sornette D, Vanneste C. Multifractal scaling properties of a growing fault population. Geophys J Int. 1995;122(2):457–69.

Cowie PA, Vanneste C, Sornette D. Statistical physics model for the spatiotemporal evolution of faults. J Geophys Res. 1993;98(B12):21809–21.

Cox DR, Lewis PA. The statistical analysis of series of events. London (UK): Springer; 1966.

Cruden DE. Describing the size of discontinuities. Int J Rock Mech Min Sci Geomech Abstr. 1977;14(3):133–7.

Cundall PA. Formulation of three-dimensional distinct element model. Part 1. A scheme to detect and represent contacts in system composed of many polyhedral blocks. Int J Rock Mech Min Sci Geomech Abstr. 1988;25(3):107–16.

Darcel C, Bour O, Davy P. Stereological analysis of fractal fracture networks. J Geophys Res Solid Earth. 2003a;108(B9):2451.

Darcel C, Bour O, Davy P. Cross-correlation between length and position in real fracture networks. Geophys Res Lett. 2003b;30(12):1650.

Davis CD. Handbook of applied hydraulics. New York (USA): McGraw-Hill; 1952.
Davy P. On the frequency-length distribution of the San Andreas fault system. J Geophys Res Solid Earth. 1993;98(B7):12141–51.
Davy P, Le Goc R, Darcel C. A model of fracture nucleation, growth and arrest, and consequences for fracture density and scaling. J Geophys Res Solid Earth. 2013;118(4):1393–407.
de Dreuzy JR, Darcel C, Davy P, Bour O. Influence of spatial correlation of fracture centers on the permeability of two-dimensional fracture networks following a power law length distribution. Water Resour Res. 2004;40(1):W01502.
de Dreuzy JR, Davy P, Bour O. Hydraulic properties of two-dimensional random fracture networks following a power law length distribution: 1. Effective Connectivity Water Resour Res. 2001a;37(8):2065–78.
de Dreuzy JR, Davy P, Bour O. Hydraulic properties of two-dimensional random fracture networks following a power law length distribution: 2. Permeability of networks based on lognormal distribution of apertures. Water Resour Res. 2001b;37(8):2079–95.
Deere UD. Technical description of rock cores for engineering purposes. Rock Mech Eng Geol. 1964;1(1):17–22.
Deere DU, Hendron AJ, Patton FD, Cording EJ. Design of surface and near surface construction in rock. In: Proceedings of 8th US symposium rock mechanics. New York (USA): AIME; 1967. p. 237–302.
Deere D, Miller RD. Engineering classification and index properties for intact rock. Tech Rept No. AFWL-TR-65–116, University of Illinois (USA), 1966.
Dershowitz WS. Rock joint systems. PhD Thesis. Massachusetts Institute of Technology, Boston (USA); 1984.
Dershowitz W, Busse R, Geier J, Uchida M. A stochastic approach for fracture set definition. In: Aubertin M, Hassani F, Mitri S editors. Rock mechanics tools and techniques, Proceedings second North Amer rock mechanics symposium (NARMS'96, a regional conference of ISRM), vol. 2. Rotterdam (Netherlands): AA Balkema; 1996. p. 1809–13.
Dershowitz WS, Einstein HH. Three dimensional flow modeling in jointed rock masses. In: Herget G, Vongpaisal S editors. Proceedings of 6th international ISRM congress, vol. 1. Rotterdam (Netherlands): AA Balkema; 1987. p. 87–92.
Dershowitz WS, Einstein HH. Characterizing rock joint geometry with joint system models. Rock Mech Rock Eng. 1988;2(1):21–51.
Desai CS, Zaman MM, Lightner JG, Siriwardane HJ. Thin-layer element for interfaces and joints. Int J Numer Anal Meth Geomech. 1984;8(1):19–43.
De Schutter G, Taerwe L. Random particle model for concrete based on Delaunay triangulation. Mater Struct. 1993;26(2):67–73.
Dunnicliff J. Geotechnicul instrumentation for monitoring field performance. Chichester (UK): Wiley; 1988.
Dverstorp B, Andersson J. Application of the discrete fracture network concept with field data: possibilities of model calibration and validation. Water Resour Res. 1989;25(3):540–50.
Ehlen J. Fractal analysis of joint patterns in granite. Int J Rock Mech Min Sci. 2000;37(6):909–22.
Einstein HH, Baecher GB. Probabilistic and statistical methods in engineering geology. Rock Mech Rock Eng. 1983;16(1):39–72.
Einstein HH, Nelson RA, Bruhn RW, Hirschfield RC. Model studies of jointed rock behaviour. In: Proceeding of 11th US symposium rock mechanics, Chapter 6. Berkeley, California (USA): Port City Press; 1969.
Engelder T, Geiser P. On the use of regional joint sets as trajectories of paleostress fields during the development of the Appalachian Plateau New York. J Geophys Res. 1980;85(NB11):6319–41.
Erdogan E. Fracture mechanics. Int J Solids Struct. 2000;37(1–2):171–83.
Fadakar-Alghalandis Y. Stochastic modeling of fractures in rock masses. Adelaide (Australia): University of Adelaide; 2014.
Featherstone RE, Nalluri C. Civil engineering hydraulics—Essential theory with worked examples, 3rd edn. Oxford (UK): Blackwell Science Ltd.; 1995.

Fine RA, Millero FJ. Compressibility of water as a function of temperature and pressure. J Chem Phys. 1973;59(10):5529–36.

Flamand R. Validation d'un modèle de comportement mécanique pour les fractures rocheuses en cisaillement. PhD Thesis. Université du Québec à Chicoutimi (Canada); 2000 (in French).

Gdoutos EE. Fracture mechanics—an introduction. Dordrecht (Netherlands): Kluwer Academic Publishers; 1993.

Gentier S. Morphologie et comportement hydromécanique d'une fracture naturelle dans le granite sous contrainte normale—etude expérimentale et théorique. PhD Thesis. Université d'Orléans(France); 1986. (in French).

Gerrard CM. Equivalent elastic moduli of a rock mass consisting of orthorhombic layers. Int J Rock Mech Min Sci & Geomech Abstr. 1982a;19(1):9–14.

Gerrard CM. Elastic models of rock masses having one, two and three sets of joints. Int J Rock Mech Min Sci & Geomech Abstr. 1982b;19(1):15–23.

Goodman RE. Les propriétés mécaniques des joints. In: 3 ème Congrès de la Société Internationale de Mécanique des Roches. Denver (USA): ASCE, vol. 1, Tome A; 1974.

Goodman RE. Introduction to rock mechanics. 2nd ed. New York (USA): Wiley; 1989.

Griffith AA. The phenomena of rupture and flow in solids. Philos Trans R Soc Lond Ser A. 1920;221:163–98.

Griffith WA, Sanz PF, Pollard DD. Influence of outcrop scale fractures on the effective stiffness of fault damage zone rocks. Pure Appl Geophys. 2009;166(10–11):1595–627.

Guigue P, Devillers O. Fast and robust triangle-triangle overlap test using orientation predicates. Journal of Graphics Tools. 2003;8(1):25–32.

Guz IA, Soutis C. A 3-D stability theory applied to layered rocks undergoing finite deformations in biaxial compression. Eur J Mech—A/Solids. 2001;20(1):139–53.

Haberfield JP. The application of energy principles to the determination of the sliding resistance of rock joints. Rock Mech Rock Eng. 1995;28(4):211–26.

Häfner S, Eckardt S, Luther T, Könke C. Mesoscale modeling of concrete: geometry and numerics. Comput Struct. 2006;84(7):450–61.

Hahn M, Wallmersperger T, Kroplin BH. Discrete element representation of continua: proof of concept and determination of the material parameters. Comput Mater Sci. 2010;50(2):391–402.

Hakami E, Larsson E. Aperture measurement and flow experiments on a single natural fracture. Int J Rock Mech Min Sci & Geomech Abstr. 1990;33(5):395–404.

Hall C, Hoff WD. Water transport in brick, stone and concrete. London (UK): Spon Press; 2012.

Hammah RE, Curran JH. On distance measures for the fuzzy K-means algorithm for joint data. Rock Mech Rock Eng. 1999;32(1):1–27.

Heliot D. Generating a block rock mass. Int J Rock Mech Min Sci Geomech Abstr. 1988;25(3):127–38.

Hentschel HGE, Procaccia I. The infinite number of generalized dimensions of fractals and strange attractors. Phys D. 1983;8(3):435–44.

Hoek E. Strength of rock and rock masses. ISRM News J. 1994;2(2):4–16.

Hoek E, Bray JW. Rock slope engineering (3rd edn.). London (UK): Institution of Mining and Metallurgy; 1981.

Hoek E, Brown ET. Underground excavations in rock. London (UK): Inst Min Metall Stephen Austin and Sons; 1980.

Hoek E, Brown ET. Practical estimates of rock mass strength. Int J Rock Mech Min Sci. 1997;34(8):1165–86.

Hoek E, Diederichs MS. Empirical estimation of rock mass modulus. Int J Rock Mech Min Sci. 2006;43(2):203–15.

Hoek E, Marinos PG, Marinos VP. Characterisation and engineering properties of tectonically undisturbed but lithologically varied sedimentary rock masses. Int J Rock Mech Min Sci. 2005;42(2):277–85.

Hoek E, Martin CD. Fracture initiation and propagation in intact rock—A review. J Rock Mech Geotech Eng. 2014;6(4):287–300.

Hooker JN, Gale JFW, Gomez LA, Laubach SE, Marrett R, Reed RM. Aperture-size scaling variations in a low-strain opening-mode fracture set, Cozzette Sandstone. Colorado J Struct Geol. 2009;31(7):707–18.

Hopkins DL. The effect of surface roughness on joint stiffness, aperture, and acoustic wave propagation. PhD Thesis. University of California, Berkeley (USA); 1990.

Hopkins DL. The implications of joint deformation in analysing the properties and behavior of fractured rock masses, underground excavations, and faults. Int J Rock Mech Min Sci. 2000;37(1–2):175–202.

Horgan GW, Young IM. An empirical stochastic model for the geometry of two-dimensional crack growth in soil (with Discussion). Geoderma. 2000;96(4):263–76.

Hu XZ, Wittmann FH. Fracture energy and fracture process zone. Material Structure. 1992;25(150):319–26.

Hudson JA, Harrison JP. Engineering rock mechanics: an introduction to the principles. 4th ed. Oxford (UK): Pergamon; 1997.

Hudson JA, Priest SD. Discontinuity frequency in rock masses. Int J Rock Mech Min Sci & Geomech Abstr. 1983;20(2):73–89.

Ikegawa Y, Hudson JA. A novel automatic identification system for three-dimensional multi-block systems. Eng Comput. 1992;9(2):169–79.

Irwin G. Analysis of stresses and strains near the end of a crack traversing a plate. J Appl Mech. 1957;24:361–4.

ISRM. Suggested methods for determining the fracture toughness of rock. Int J Rock Mech Min Sci & Geomech Abstr. 1988;25(2):71–96.

Iwai K. Fundamental Studies of the Fluid Flow through a Single Fracture. PhD Thesis. University of California, Berkeley (USA); 1976.

Jaeger JC, Cook NGW, Zimmerman R. Fundamentals of Rock Mechanics (4th edn.). MA (USA): Wiley-Blackwell; 2007.

Jelitto H, Schneider GA. Fracture toughness of porous materials—experimental methods and data. Data Brief. 2019;23: 103709.

Jing L, Stephansson O. Topological identification of block assemblages for jointed rock masses. Int J Rock Mech Min Sci Geomech Abstr. 1994;31(2):163–72.

Jones RR, Pearce MA, Jacquemyn C, Watson FE. Robust best-fit planes from geospatial data. Geosphere. 2016;12(1):196–202.

Josnin J, Jourde H, Fénart P, Bidaux P. A three-dimensional model to simulate joint networks in layered rocks. Can J Earth Sci. 2002;39(10):1443–55.

Jourde H, Fénart P, Vinches M, Pistre S, Vayssade B. Relationship between the geometrical and structural properties of layered fractured rocks and their effective permeability tensor. A Simulation Study J Hydrol. 2007;337(1–2):117–32.

Jourde H, Flodin EA, Aydin A, Durlofsky LJ, Wen XH. Computing permeability of fault zones in eolian sandstone from outcrop measurements. Am Assoc Pet Geol Bull. 2002;86(7):1187–200.

Kataoka M, Mahdavi E, Funatsu T, Takehara T, Obara Y, Fukui K, Hashiba K. Estimation of mode I fracture toughness of rock by semi-circular bend test under confining pressure condition. Proced Eng (EUROCK 2017). 2017;191:886–93.

Kulatilake PHSW, Chen J, Teng J, Shufang X, Pan G. Discontinuity geometry characterization in a tunnel close to the Three Gorges Dam Site in China. Int J Rock Mech Min Sci & Geomech Abstr. 1996;33(3):255–77.

Kulatilake PHSW, Panda BB. Effect of block size and joint geometry on jointed rock hydraulics and REV. J Eng Mech, ASCE. 2000;126(8):850–8.

Kulatilake PHSW, Wang S, Stephansson O. Effects of finite-size joints on the deformability of jointed rock in three dimensions. Int J Rock Mech Min Sci. 1993a;30(5):479–501.

Kulatilake PHSW, Wathugala DN, Stephansson O. Joint network modeling, including a validation to an area in Stripa Mine, Sweden. Int J Rock Mech Min Sci Geomech Abstr. 1993b;30(5):503–26.

Kulatilake PHSW, Wu TH. The density of discontinuity traces in sampling windows. Int J Rock Mech Min Sci Geomech Abstr. 1984a;21(6):354–347.

Kulatilake PHSW, Wu TH. Estimation of mean trace length of discontinuities. Rock Mech Rock Eng. 1984b;17(4):215–213.

Kwan AKH, Wang ZM, Chan HC. Mesoscopic study of concrete II: nonlinear finite element analysis. Comput Struct. 1999;70(5):545–56.

Ladanyi B, Archambault G. Simulation of shear behavior of a jointed rock mass. In Proceeding 11th US Symposium Rock Mechanics Berkeley, CA (USA): AIME; 1970:105–25.

Ladanyi B, Archambault G. Shear strength and deformability of filled indented joints. In Proceeding 1st international symposium on geotechnics of structurally complex formations. Capri (Italy): Italian Geotech Assoc; 1977. p. 317–26.

Lam L, Wong YL, Poon CS. Effect of fly ash and silica fume on compressive and fracture behaviors of concrete. Cem Concr Res. 1998;28(2):271–83.

Landau LD, Lipshitz EM. Theory of elasticity. 3rd ed. Oxford (UK): Pergamon Press; 1970.

Lang PS, Paluszny A, Zimmerman RW. Permeability tensor of three dimensional fractured porous rock and a comparison to trace map predictions. J Geophys Res Solid Earth. 2014;119(8):6288–307.

La Pointe PR. A method to characterize fracture density and connectivity through fractal geometry. Int J Rock Mech Min Sci Geomech Abstr. 1988;25(6):421–9.

Laslett GM. Censoring and edge effects in areal and line transect sampling of rock joint traces. J Int Assoc Math Geol. 1982;14(2):125–40.

Lauffer H. Geobirgsklassifizierung für den stollenbau. Geol Bauwesen. 1958;24(1):46–51 (in German).

Lei QH, Latham JP, Tsang CF. The use of discrete fracture networks for modelling coupled geomechanical and hydrological behaviour of fractured rocks. Comput Geotech. 2017;85:151–76.

Lei QH, Latham JP, Tsang CF, Xiang J, Lang P. A new approach to upscaling fracture network models while preserving geostatistical and geomechanical characteristics. J Geophys Res Solid Earth. 2015;120(7):4784–807.

Lei QH, Latham JP, Xiang J, Tsang CF, Lang P, Guo L. Effects of geomechanical changes on the validity of a discrete fracture network representation of a realistic two-dimensional fractured rock. Int J Rock Mech Min Sci. 2014;70:507–23.

Lemaitre J, Chaboche JL. Mechanics of solid materials. Cambridge (UK): Cambridge University Press; 1990.

Leong EC, Randolph MF. A model for rock interfacial behavior. Rock Mech Rock Eng. 1992;25(3):187–206.

Li XX, Chen SH, Xu Q, Xu Y. Modeling the three-dimensional unsaturated water transport in concrete at the mesoscale. Comput Struct. 2017;190:61–74.

Li XX, Xu Y, Chen SH. Computational homogenization of effective permeability in three-phase mesoscale concrete. Constr Build Mater. 2016;121:100–11.

Lin D, Fairhurst C, Starfield AM. Geometrical identification of three-dimensional rock block systems using topological techniques. Int J Rock Mech Min Sci Geomech Abstr. 1987;24(6):331–8.

Lisjak A, Grasselli G. A review of discrete modeling techniques for fracturing processes in discontinuous rock masses. J Rock Mech Geotech Eng. 2014;6(4):301–14.

Long JCS. Rock fractures and fluid flow. Washington DC (USA): National Academy Press; 1996.

Long JCS, Billaux DM. From field data to fracture network modeling: an example incorporating spatial structure. Water Resour Res. 1987;23(7):1201–16.

Long JCS, Gilmour P, Witherspoon PA. A method for steady fluid flow in random three-dimensional networks of disc-shaped fractures. Water Resour Res. 1985;21(8):35–40.

Long JCS, Remer JS, Wilson CR, Witherspoon PA. Porous media equivalents for networks of discontinuous fractures. Water Resour Res. 1982;18(3):645–58.

Look BG. Handbook of geotechnical investigation and design tables. London (UK): Taylor & Francis Group; 2007.

Louis CA. Study of groundwater flow in jointed rock and its influence on the stability of rock masses (Rock Mechanics Research Report 10). London (UK): Imperial College; 1969.

Lu P, Latham JP. Developments in the assessment of in-situ block size distributions of rock masses. Rock Mech Rock Eng. 1999;32(1):29–49.
Ma HF, Chen HQ, Li BK. Influence of meso-structure heterogeneity on bending strength of concrete. J Hydraul Eng. 2005;36:846–52 (in Chinese with English abstract).
Mahtab M, Goodman RE. Three dimensional analysis of joint rock slope. In Proceedings of 2nd ISRM congress, vol. 3. Beograd (Yugoslavia): Privredni Pregled; 1970, 353–60.
Maksimovic M. The shear strength components of a rough rock joint. Int J Rock Mech Min Sci. 1996;33(8):769–83.
Marache A. Comportement mécanique d'une fracture rocheuse sous contraintes normale et tangentielle. PhD Thesis. École Centrales des Arts et Manufactures, École Centrale Paris (France); 2002. (in French).
Marcotte D, Henry E. Automatic joint set clustering using a mixture of bivariate normal distributions. Int J Rock Mech Min Sci. 2002;39(3):323–34.
Masihi M, King PR. Connectivity prediction in fractured reservoirs with variable fracture size: analysis and validation. SPE J. 2008;13(1):88–98.
Mauldon M. Estimating mean fracture trace length and density from observations in convex windows. Rock Mech Rock Eng. 1998;31(4):201–16.
Mauldon M, Dunne WM, Rohrbaugh MB Jr. Circular scanlines and circular windows: new tools for characterizing the geometry of fracture traces. J Struct Geol. 2001;23(2–3):247–58.
Mazars J. A description of microscale and macroscale damage of concrete structures. Eng Frac Mech. 1986;25(5–6):729–37.
Mehta PK, Monteiro PJM. Concrete: microstructure, properties, and materials. 3rd ed. New York (USA): McGraw-Hill; 2006.
Min KB, Jing LR. Numerical determination of the equivalent elastic compliance tensor for fractured rock masses using the distinct element method. Int J Rock Mech Min Sci. 2003;40(6):795–816.
Min KB, Rutqvist J, Tsang CF, Jing LR. Stress-dependent permeability of fractured rock masses: a numerical study. Int J Rock Mech Min Sci. 2004;41(7):1191–210.
Ministry of Housing and Urban-Rural Development of the People's Republic of China. (GB50010–2010) Code for Design of Concrete Structures. Beijing (China): China Construction Industry Press; 2010. (in Chinese).
Ministry of Water Resources of the People's Republic of China. (SL191–2008) Code for Design of Hydraulic Concrete Structures. Beijing (China): China Water Power Press; 2008. (in Chinese).
Morland LW. Continuum model of regularly jointed mediums. J Geophysi Res. 1974;79(2):357–62.
Morris HM, Wiggert JM. Applied hydraulics in engineering. 2nd ed. New York (USA): Ronald Press; 1972.
Müller L, editor. Rock mechanics. Wien (Austria): Springer; 1974.
Nara Y, Kaneko K. Study of subcritical crack growth in andesite using the double torsion test. Int J Rock Mech Min Sci. 2005;42(4):521–30.
National standards compilation group of People's Republic of China. (GB/T50218-2014) standard for engineering classification of rock mass. Beijing China): China Planning Press, 2014 (in Chinese).
Neuman SP. Multiscale relationships between fracture length, aperture, density and permeability. Geophys Res Lett. 2008;35(22):L22402.
Oda M, Suzuki K, Maeshibu T. Elastic compliance for rock-like materials with random cracks. Soils Found. 1984;24(3):27–40.
Odling NE. Network properties of a two-dimensional natural fracture pattern. Pure Appl Geophys. 1992;138(1):95–114.
Odling NE. Natural fracture profiles, fractal dimension and joint roughness coefficients. Rock Mech Rock Eng. 1994;27(3):135–53.
Odling NE, Gillespie P, Bourgine B, Castaing C, Chilès JP, Christensen NP, Fillion E, Genter A, Olsen C, Thrane L, Trice R, Aarseth E, Walsh JJ, Watterson J. Variations in fracture system geometry and their implications for fluid flow in fractured hydrocarbon reservoirs. Petrol Geosci. 1999;5(4):373–84.

Odling NE, Webman I. A "conductance" mesh approach to the permeability of natural and simulated fracture patterns. Water Resour Res. 1991;27(10):2633–43.
Oha SW, Mina GJ, Parka SW, Kima MS, Obarab Y, Choa SH. Anisotropic influence of fracture toughness on loading rate dependency for granitic rocks. Eng Fract Mech. 2019;221: 106677.
Olson JE. Joint pattern development: effects of subcritical crack growth and mechanical crack interaction. J Geophys Res. 1993;98(B7):12251–65.
Olson JE. Sublinear scaling of fracture aperture versus length: an exception or the rule? J Geophys Res. 2003;108(B9):2413.
Olson JE. Predicting fracture swarms—the influence of subcritical crack growth and the crack-tip process zone on joint spacing in rock. In: Cosgrove JW, Engelder T (eds.), Initiation, propagation, and arrest of joints and other fractures, vol. 231, Special Publications. London (UK): Geological Society, London. 2004, 73–88.
Olson JE. Fracture aperture, length and pattern geometry development under biaxial loading: a numerical study with applications to natural, cross-jointed systems. In: Lewis H, Couples GD editors. Relationship Between Damage Localization, vol. 289, Special Publications. London (UK): Geological Society, London; 2007:123–42.
Olson JE, Pollard D. Inferring paloestresses from natural fracture patterns: a new method. Geology. 1989;17:345–8.
Palmstrom A, Singh R. The deformation modulus of rock masses—comparisons between in situ tests and indirect estimates. Tunn Undergr Space Technol. 2001;16(2):115–31.
Paluszny A, Matthäi SK. Numerical modeling of discrete multi-crack growth applied to pattern formation in geological brittle media. Int J Solids Struct. 2009;46(18–19):3383–97.
Paluszny A, Zimmerman RW. Numerical simulation of multiple 3D fracture propagation using arbitrary meshes. Comput Meth Appl Mech Eng. 2011;200(9–12):953–66.
Pande GN, Gerrard CM. The behaviour of reinforced jointed rock masses under various simple loading states. In: Proceeding 5th ISRM congress. Melbourne (Australia): Brown Prior Anderson Pty Ltd; 1983, F217–23.
Park HJ, West TR, Woo I. Probabilistic analysis of rock slope stability and random properties of discontinuity parameters, Interstate Highway 40, Western North Carolina, USA. Eng Geol. 2005;79(3–4):230–50.
Patton FD. Multiple modes of shear failure in rock. In 1er Congrès de la Société Internationale de Mécanique des Roches. Lisbonne (Portugal): ISRM 1966, 509–13.
Pickering G, Bull JM, Sanderson DJ. Sampling power-law distributions. Tectonophysics. 1995;248(1–2):1–20.
Pollard DD, Aydin A. Progress in understanding jointing over the past century. Geol Soc Am Bull. 1988;100(8):1181–204.
Pollard DD, Segall P. Theoretical displacements and stresses near fractures in rock: with applications to faults, joints, veins, dikes, and solution surfaces. In: Atkinson BK, editor. fracture mechanism of rock. San Diego (USA): Academic Press; 1987. p. 277–349.
Priest SD. Hemispherical projection methods in rock mechanics. London (UK): Allen & Unwin (1985).
Priest SD. Discontinuity analysis for rock engineering. 1st ed. London (UK): Chapman & Hall; 1993.
Priest SD, Hudson JA. Estimation of discontinuity spacing and trace length using scanline surveys. Int J Rock Mech Min Sci and Geomech Abstr. 1981;18(3):183–97.
Qu HC, Chen GR. Numerical simulation of different-shaped random aggregates' influence on concrete's compression strength. Adv Mater Res. 2008;33:623–30.
Quiroga ON. The effect of the aggregates characteristics on the performance of portland cement concrete. PhD Thesis. University of Texas, Austin (USA), 2003.
Raven KG, Gale JE. Water flow in a natural rock fracture as a function of stress and sample size. Int J Rock Mech Min Sci Geomech Abstr. 1985;22(4):251–61.
Renshaw CE. Influence of subcritical fracture growth on the connectivity of fracture networks. Water Resour Res. 1996;32(6):1519–30.

Renshaw CE, Pollard DD. Numerical simulation of fracture set formation: a fracture mechanics model consistent with experimental observations. J Geophys Res Solid Erath. 1994;99(B5):9359–72.

RILEM Recommendation. Size-effect method for determining fracture energy and process zone size of concrete. Materials and Structures; 1990. 23(6):461–5.

Rice JR. A path-independent integral and approximate analysis of strain concentrations by notches and cracks. J Appl Mech. 1968;35:379–86.

Robinson PC. Connectivity of fracture systems—a percolation theory approach. J Phys A Math Gen. 1983;16(3):605–14.

Robinson PC. Numerical calculations of critical densities for lines and planes. J Phys A Math Gen. 1984;17(14):2823–30.

Roelfstra PE, Sadouki H, Wittmann FH. Le béton numérique. Mater Struct. 1985;18(5):327–35.

Rohrbaugh MB, Dunne WM, Mauldon M. Estimating fracture trace intensity, density, and mean length using circular scan lines and windows. AAPG Bull. 2002;86(12):2089–104.

Romana MR. New adjustment ratings for application of Bieniawski classification to slopes. In Proceeding international symposium on the role of rock mechanics Zacatecas (Mexico): 1985, p. 49–53.

Saeb S, Amadei B. Modelling joint response under constant or variable normal stiffness boundary conditions. Int J Rock Mech Min Sci Geomech Abstr. 1990;27(3):213–7.

Salamon MDG. Elastic moduli of a stratified rock mass. Int J Rock Mech Min Sci. 1968;5(6):519–27.

Sanderson DJ, Nixon CW. The use of topology in fracture network characterization. J Struct Geol. 2015;72:55–66.

Santos SF, Rodrigues JA. Correlation between fracture toughness, work of fracture and fractal dimensions of Alumina-mullite-zirconia composites. Mater Res. 2003;6(2):219–26.

Sarda S, Jeannin L, Basquet R, Bourbiaux B. Hydraulic characterization of fractured reservoirs: simulation on discrete fracture models. SPE Reserv Eval Eng. 2002;5(2):154–62.

Schwartz FW, Smith WL, Crowe AS. A stochastic analysis of microscopic dispersion in fractured media. Water Resour Res. 1983;19(5):1253–65.

Shang J, West LJ, Hencher SR. Zhao Z. Geological discontinuity persistence: implications and quantification. Eng Geol. 2018;241:41–54.

Shanley RJ, Mahtab MA. Delineation and analysis of clusters in orientation data. J Math Geol. 1976;8(1):9–23.

Shehata WM. Geohydrology of mount Vernon canyon area. PhD Thesis. Colorado School of Mines (USA); 1971.

Shekhar R, Gibson RL. Generation of spatially correlated fracture models for seismic simulations. Geophys J Int. 2011;185(1):341–51.

Shen DJ, Shi X, Zhu SS, Duan XF, Zhang JY. Relationship between tensile Young's modulus and strength of fly ash high strength concrete at early age. Constr Build Mater. 2016;123:317–26.

Silva RV, de Brito J, Dhir RK. Establishing a relationship between modulus of elasticity and compressive strength of recycled aggregate concrete. J Cleaner Prod. 2016;112(Part 4):2171–86.

Sim J, Yang KH, Jeon JK. Influence of aggregate size on the compressive size effect according to different concrete type. Constr Build Mater. 2013;44:716–25.

Sirat M, Talbot C. Application of artificial neural networks to fracture analysis at the Äspö HRL, Sweden: fracture sets classification. Int J Rock Mech Min Sci. 2001;38(5):621–39.

Sitharam TG, Sridevi J, Shimizu N. Practical equivalent continuum characterization of jointed rock masses. Int J Rock Mech Min Sci. 2001;38(3):437–48.

Snow DT. Anisotropic permeability of fractured media. Water Resour Res. 1969;5(6):1273–89.

Snow DT. The frequency and apertures of fractures in rock. Int J Rock Mech Min Sci Geomech Abstr. 1970;7(1):23–40.

Sonmez H, Gokceoglu C, Ulusay R. Indirect determination of the modulus of deformation of rock masses based on the GSI system. Int J Rock Mech Min Sci. 2004;41(5):849–57.

Sonmez H, Ulusay R. Modifications to the geological strength index (GSI) and their application to stability of slopes. Int J Rock Mech Min Sci. 1999;36(6):743–60.

Spence GH, Finch E. Influences of nodular chert rhythmites on natural fracture networks in carbonates: an outcrop and two-dimensional discrete element modelling study. In Spence GH, Redfern J, Aguilera R, Bevan TG, Cosgrove JW, Couples GD, Daniel JM (eds) Advanced study in fractured reservoirs, vol. 374, Special Publications. London (UK): Geological Society, London; 2014. p. 211–49.

Starzec P, Andersson J. Probabilistic predictions regarding key blocks using stochastic discrete fracture networks—example from a rock cavern in south-east Sweden. Bull Eng Geol Environ. 2002;61(4):363–78.

Sturzenegger M, Stead D. Quantifying discontinuity orientation and persistence on high mountain rock slopes and large landslides using terrestrial remote sensing techniques. Nat Hazards Earth Syst Sci. 2009a;9(2):267–87.

Sturzenegger M, Stead D. Close-range terrestrial digital photogrammetry and terrestrial laser scanning for discontinuity characterization on rock cuts. Eng Geol. 2009b;106(3–4):163–82.

Sturzenegger M, Stead D, Elmo D. Terrestrial remote sensing-based estimation of mean trace length, trace intensity and block size/shape. Eng Geol. 2011;119(3–4):96–111.

Swam G, Zongqi S. Prediction of shear behavior of joints using profiles. Rock Mech Rock Eng. 1985;18(3):183–212.

Tang CA, Zhang YB, Liang ZZ, Xu T, Tham LG, Kou SQ, Liu HY. Fracture spacing in layered materials and pattern transition from parallel to polygonal fractures. Phys Rev E. 2006;73(5): 056120.

Taylor WL, Pollard DD, Aydin A. Fluid flow in discrete joint sets: field observations and numerical simulations. J Geophys Res. 1999;104(B12):28983–9006.

Terzaghi K. Rock defects and loads on tunnel support. In: Proctor RV, White TL editors. Rock tunnelling with steel supports (1). Youngstown, OH (USA): Commercial Shearing and Stamping Co. 1946; p. 15–99.

Triantafyllidis T, Gerolymatou E. Estimation of the strength of stratified rock mass. Rock Mech Rock Eng. 2014;47(2):535–47.

Tsang YW, Tsang CF. Channel model of flow through fractured media. Water Resour Res. 1987;23(3):467–79.

Ulusay R, editor. The ISRM suggested methods for rock characterization, testing and monitoring: 2007–2014. Cham (Switzerland): Springer; 2015.

Ulusay, R., Hudson, J.A. (eds.). The complete ISRM suggested methods for rock characterization, testing and monitoring: 1974–2006. Ankara (Turkey): ISRM Turkish National Group. 2007.

Wang W, Wang J, Kim MS. An algebraic condition for the separation of two ellipsoids. Comput Aided Geometric Des. 2001;18(6):531–9.

Wang WM, Chen SH. Automatic identification method for three-dimensional rock block systems. J Wuhan Univ Hydr Elec Eng (WUHEE). 1998;31(5):51–5 (in Chinese with English abstract).

Wang XF, Yang ZJ, Jivkov AP. Monte Carlo simulations of mesoscale fracture of concrete with random aggregates and pores: a size effect study. Constr Build Mater. 2015a;80:262–72.

Wang XF, Yang ZJ, Yates JR, Jivkov AP, Zhang C. Monte Carlo simulations of mesoscale fracture modelling of concrete with random aggregates and pores. Constr Build Mater. 2015b;75:35–45.

Wang YF, Xu Q, Chen SH. Approaches of concrete creep using mesomechanics: numerical simulation and predictive model. Modell Simul Mater Sci Eng. 2019;27(5):055012.

Wang YF, Xu Q, Chen SH, Zhou XS, Liu HY. The influence of persistent curing-loading temperature on the basic creep of early-age concrete: experimental investigation. Const Build Mater. 2020;61:119950.

Wang ZM, Kwan AKH, Chan HC. Mesoscopic study of concrete I: generation of random aggregate structure and finite element mesh. Comput Struct. 1999;70(5):533–44.

Warburton PM. Stereological interpretation of joint trace data: influence of joint shape and implications for geological surveys. Int J Rock Mech Min Sci Geomech Abstr. 1980;17(6):305–16.

Wardle LJ, Gerrard CM. The "'equivalent'" anisotropic properties of layered rock and soil masses. Rock Mech Rock Eng. 1972;4(3):155–75.

Watkins H, Bond CE, Healy D, Butler RWH. Appraisal of fracture sampling methods and a new workflow to characterise heterogeneous fracture networks at outcrop. J Struct Geol. 2015;72:67–82.

Wei RP. Fracture mechanics: integration of mechanics, materials science and chemistry. Cambridge (UK): Cambridge University Press; 2010.

Wei ZQ, Egger P, Descoeuders F. Permeability prediction for jointed rock masses. Int J Rock Mech Min Sci Geomech Abstr. 1995;32(3):251–61.

Wei ZQ, Hudson JA. The influence of joints on rock modulus. In: Tan ZY, editor. Proceedings of international symposium engineering in complex formations. Beijing (China): Science Press; 1986. p. 54–62.

Whittaker BN, Singh RN, Sun G. Rock fracture mechanics: principles, design, and applications. In Developments in geotechnical engineering, vol. 71. Amsterdam (Netherlands): Elsevier; 1992.

Wickham GE, Tiedemann HR, Skinner EH. Support determination based on geologic predictions. In Proceedings rapid excavation and tunneling conference. New York (USA): AIME 1972; 43–64.

Wilson CE, Aydin A, Karimi-Fard M, Durlofsky LJ, Sagy A, Brodsky EE, Kreylos O, Kellogg LH. From outcrop to flow simulation: constructing discrete fracture models from a LIDAR survey. Am Assoc Pet Geol Bull. 2011;95(11):1883–906.

Wriggers P, Moftah SO. Mesoscale models for concrete: homogenisation and damage behavior. Finite Elem Anal Des. 2006;42(7):623–36.

Wu AQ, Wang B. Engineering rock mass classification method based on rock mass quality index BQ for rock slope. Chin J Rock Mechan Eng. 2014;33(4):699–706 (in Chinese with English abstract).

Wu H, Pollard DD. An experimental study of the relationship between joint spacing and layer thickness. J Struct Geol. 1995;17(6):887–905.

Wu H, Pollard DD. Imaging 3-D fracture networks around boreholes. Am Assoc Pet Geol Bull. 2002;86(4):593–604.

Xu C, Dowd P. A new computer code for discrete fracture network modelling. Comput Geosci. 2010;36(3):292–301.

Xu Y, Chen SH. A method for modeling the damage behavior of concrete with a three-phase mesostructure. Constr Build Mater. 2016;102:26–38.

Xu Y, Xu Q, Chen SH, Li XX. Self-restraint thermal stress in early-age concrete samples and its evaluation. Constr Build Mater. 2017;134:104–11.

Yoshinaka R, Yamabe T. Jointed stiffness and the deformation behavior of discontinuous rock. Int J Rock Mech Min Sci Geomech Abstr. 1986;23(1):295–303.

Zhang LY, Einstein HH. Estimating the intensity of rock discontinuities. Int J Rock Mech Min Sci. 2000;37(5):819–37.

Zhang QH, Xiao M, Ding XL, Wu AQ. Improvement of methodology for block identification using mesh gridding technique. Tunn Undergr Space Technol. 2012;30:217–29.

Zhang X, Sanderson DJ. Anisotropic features of geometry and permeability in fractured rock masses. Eng Geol. 1995;40(1–2):65–75.

Zhang X, Sanderson DJ. Numerical modelling and analysis of fluid flow and deformation of fractured rock masses. Oxford (UK): Pergamon; 2002.

Zhang X, Sanderson DJ, Harkness RM, Last NC. Evaluation of the 2D permeability tensor for fractured rock masses. Int J Rock Mech Min Sci Geomech Abstr. 1996;33(1):17–37.

Zhang YH, Zhou HM, Wu AQ, Zhong ZW, Qin L. Statistical analysis of deformation moduli of rock masses based on quality classification. Chin J Rock Mechan Eng. 2011;30(3):486–92 (in Chinese with English abstract).

Zhang ZX, Lei QH. Object-oriented modeling for three-dimensional multi-block systems. Comput Geotech. 2013;48:208–27.

Zhang ZX, Lei QH. A morphological visualization method for removability analysis of blocks in discontinuous rock masses. Rock Mech Rock Eng. 2014;47(4):1237–54.

Zhao T, Utili S, Crosta GB. Rockslide and impulse wave modelling in the Vajont reservoir by DEM-CFD analyses. Rock Mech Rock Eng. 2016;49(6):2437–56.

Zhou C, Huang B, Shu X. Micromechanical model for predicting coefficient of thermal expansion of concrete. J Mater Civil Eng. 2013;25(9):1171–80.
Zhou W, Maerz NH. Implementation of multivariate clustering methods for characterizing discontinuities data from scanlines and oriented boreholes. Comput Geosci. 2002;28(7):827–39.
Zienkiewicz OC, Pande GN. Time dependent multi-laminated model of rocks – a numerical study of deformation and failure of jointed rock masses. Int J Numer Anal Meth Geomech. 1976;1(3):219–47.
Zimmerman RW, Bodvarsson GS. Effective transmissivity of two-dimensional fracture networks. Int J Rock Mech Min Sci Geomech Abstr. 1996a;33(4):433–8.
Zimmerman RW, Bodvarsson GS. Hydraulic conductivity of rock fractures. Transp Porous Media. 1996b;23(1):1–30.

Chapter 3
Preparative Knowledge of Classical Mechanics

Abstract The principles and governing equations of classical mechanics are summarized in this chapter. Starting from the classical dynamics of particles, the conservation laws in continuum mechanics and the constitutive equations for water and rock-like materials are elucidated. These contents are indispensable for the further study on advanced CMs.

3.1 General

The classical mechanics is often referred to as "Newtonian mechanics". It consists of the physical concepts found and the mathematical methods invented by Newton, Leibniz and others in the seventeenth century to describe the motion of substance body under the action of force (Abraham and Marsden 1978). It provides nearly accurate results when handling objects that are not extremely massive or smaller, and move much slower than the speed of light. In case that objects become extremely massive, the theory of "general relativity" takes the place. When the objects being examined possess about the size of an atom, it becomes necessary to introduce the "quantum mechanics". To describe the object with a velocity approaching the speed of light, the theory of "special relativity" is needed.

The continuum mechanics is a branch of classical mechanics that deals with the substance as a continuum rather than as discrete particles. The French mathematician Cauchy was the first to formulate continuum mechanics in the nineteenth century. Modeling an object as a continuum basically assumes that the substance completely fills an object space, namely, the fact that any substance is made of atoms and actually not continuous, may be ignored. On the length scale much greater than that of inter-atomic distance, such a continuum model is sufficiently accurate.

Towards the description of the behavior of any continuum body, fundamental physical laws should be established to derive the governing "partial differential equations (PDEs)". In principle, the solution is straight forward: first we erect a coordinate system that is suitable for describing motion, then we derive the mathematical form for physical laws. Normally, four fundamental physical laws are demanded in the governing equations for continuum mechanics:

S. Chen, *Advanced Computational Methods and Geomechanics*, Springer Tracts in Civil Engineering, https://doi.org/10.1007/978-981-19-7427-4_3

- Conservation of mass;
- Conservation of momentum;
- Conservation of energy;
- Constitutive or/and state equation.

Constitutive or/and state equation is additionally demanded as a bridge to link system variables, which reflects the intrinsic dynamics or physical properties of the substance (e.g. viscosity, deformability, yield and failure, etc.).

The determination of stress pattern, one of the general problems to predict the field state inside single fluid/solid body, emerged as a challenge in the early time. Euler introduced the stress tensor in fluid in 1757. Later on, Navier (1824) and Cauchy (1828, 1829) described the stress tensor in solid. In 1864, Airy gave a definition of "Airy function" for the 2-D stress field in shell structures. Although these beginnings with simple examples were very fruitful, yet it became more and more obvious, that the behavior of real engineering structure or its component was not easy to predict. For the statically indeterminate structure, Weighardt (1906) put forward the idea that it should be possible to transform the stress/strain conditions from well-known continuum examples to equivalent framework structures with identical shape and boundary conditions. In 1941, Hrennikoff employed a method to explain the stress/strain distribution of a continuum by the calculation in equally supported framework system. The methods of both Weighardt and Hrennikoff had little success blamed for the restrained computer capacity. Finally, in 1956, the "finite element method (FEM)" was created (Turner et al. 1956) which soon became a great success (Zienkiewicz 1977; Bathe 2002; Knothe and Wessels 2008).

Due to the great success of fundamental CMs represented by FEM, engineers and scientists paid less attention to the other computational methods before the 1980s. Nevertheless, over the past three decades, the advanced CMs have gained more and more impetus, due to their better performance in handling finite/discontinuous deformation accompanied with dislocation/distortion motion and high speed flow accompanied with free surface distortion/break.

This chapter starts from the classical dynamics of particles, afterwards the physical laws for water and rock-like materials are elucidated for the further study on the advanced CMs in this book.

3.2 Classical Mechanics of Particles

For the detailed works on the classical mechanics of particles, readers are referred to the books of Arnold (1978) and Goldstein (1980).

1. **Virtual work principle**

The standard form of dynamic equation governing the motion of a particle system is basically expressed by the Newton's second law

$$\mathbf{F}_i + \mathbf{R}_i = m_i \ddot{\mathbf{x}}_i \quad (i = 1, 2, 3, \ldots, n) \tag{3.1}$$

where: $\mathbf{F}_i$—active force at particle i; $\mathbf{R}_i$—restraint (constraint) force at particle i; $\mathbf{x}_i$, $\ddot{\mathbf{x}}_i$—position and acceleration of particle i, respectively; m_i—mass of particle i; n—amount of particles.

In Eq. (3.1), the top-mark $(\cdot)$ is used to denote the time derivative. This notation will be employed throughout the whole book unless there is particular consideration.

D'Alembert regarded $(-m_i \ddot{\mathbf{x}}_i)$ as an equivalent inertia force, in this manner the dynamic issue was replaced by an equivalent static issue with governing equation

$$\mathbf{F}_i + \mathbf{R}_i + (-m_i \ddot{\mathbf{x}}_i) = 0 \quad (i = 1, 2, 3, \ldots, n) \tag{3.2}$$

Later on, more abstract methods were developed, leading to the reformulations of classical mechanics known as "Lagrangian mechanics" and "Hamiltonian mechanics". These advances substantially extend beyond Newton's work, particularly through their use of "analytical mechanics". With some modification, these extensions are also employed in many areas of modern physics including continuum mechanics.

Lagrange combined the virtual work principle and D'Alembert principle to establish the D'Alembert-Lagrange equation that states: under the ideal restraint, the virtual work by the active force and inertia force is null, i.e.

$$\sum_{i=1}^{n} (\mathbf{F}_i - m_i \ddot{\mathbf{x}}_i) \cdot \delta \mathbf{x}_i = 0 \tag{3.3}$$

where: $\delta \mathbf{x}_i$—virtual displacement of particle i.

2. **Variational principle**

(1) **Euler-Lagrange equations**

Suppose n mass particles are subject to k holonomic constraints, hence the freedom of the system is $s = 3n - k$ in 3-D space, and we have $\mathbf{x}_i = \mathbf{x}_i(q_1, q_2, \ldots, q_s, t)$ in which q_α is the "generalized coordinate". Lagrange push further to get the famous equation bearing his name: in a complete system, the "general external force" is equal to the difference of "general momentum rate" and "Lagrange force", i.e.

$$\frac{\mathrm{d}}{\mathrm{d}t} \frac{\partial T}{\partial \dot{q}_\alpha} - \frac{\partial T}{\partial q_\alpha} = Q_\alpha = \sum_{i=1}^{n} \mathbf{F}_i . \frac{\partial \mathbf{x}_i}{\partial q_\alpha} \quad (\alpha = 1, 2, \ldots, s) \tag{3.4}$$

where: Q_α—general external force; q_α and $\dot{q}_\alpha$—Lagrange variables (generalized coordinate and velocity); $T(q_\alpha, \dot{q}_\alpha, t) = \frac{1}{2} \sum_{i=1}^{n} m_i \dot{\mathbf{x}}_i^2$—kinetic energy of the particle system.

General external force Q_α may be distinguished as the potential force and common force. If only the potential force is exerted, the "Lagrange function" $L(\mathbf{q}, \dot{\mathbf{q}}, t)$ may be defined as the difference between kinetic energy T and potential energy V,

$$\begin{cases} L(\mathbf{q}, \dot{\mathbf{q}}, t) = T - V \\ \mathbf{q} = \left[q_1\ q_2\ \ldots\ q_s\right]^{\mathrm{T}} \\ \dot{\mathbf{q}} = \left[\dot{q}_1\ \dot{q}_2\ \ldots\ \dot{q}_s\right]^{\mathrm{T}} \end{cases} \tag{3.5}$$

in which $V = V(\mathbf{x}_1, \mathbf{x}_2, \ldots, \mathbf{x}_n) = V(q_1, q_2, \ldots, q_s)$ is the potential energy. Since

$$\begin{cases} Q_\alpha = -\frac{\partial V}{\partial q_\alpha} \\ \frac{\partial V}{\partial \dot{q}_\alpha} = 0 \end{cases} \tag{3.6}$$

the dynamics of such a particle system can be described by an action function that is the integration of L along the curve $\mathbf{q}(t)$. The variational operation undertaken for this action function leads to the "Euler-Lagrange equations" as follows (Marsden and West 2001)

$$\frac{\mathrm{d}}{\mathrm{d}t}\frac{\partial L}{\partial \dot{q}_\alpha} - \frac{\partial L}{\partial q_\alpha} = 0 \quad (\alpha = 1, 2, \ldots, s) \tag{3.7}$$

(2) **Hamilton equations**

In 1833, Hamilton kept the equalized general coordinate $q_\alpha \in R^s$ but used conjugate generalized momentum $p_\alpha = \frac{\partial L}{\partial \dot{q}_\alpha} \in R^s$ as another basic variable to replace the general velocity $\dot{q}_\alpha$ in the description of mechanics state. Then he defined the "Hamilton function (Hamiltonian)" $H(\mathbf{p}, \mathbf{q}, t)$ to replace the Lagrange function $L(\mathbf{p}, \dot{\mathbf{q}}, t)$ such that

$$H(\mathbf{p}, \mathbf{q}, t) = \mathbf{p}.\dot{\mathbf{q}} - L(\mathbf{p}, \dot{\mathbf{q}}, t) \tag{3.8}$$

Note that $H = T + V$ is the Hamiltonian representing the total energy.

Suppose a function defined as

$$J[y(x)] = \int_{x_1}^{x_2} f(y, y', x)\mathrm{d}x \tag{3.9}$$

will reach extremum when stationary condition $\delta J = 0$ is met where δ represents the variational operator. Since mathematically operators δ and d are interchangeable, hence the stationary condition leads to the famous "Euler equation"

$$\frac{\mathrm{d}}{\mathrm{d}x}\frac{\partial f}{\partial y'} - \frac{\partial f}{\partial y} = 0 \tag{3.10}$$

If the independent variable x is replaced by the time t, y is replaced by the general coordinate $\mathbf{q}$, and $f(y, y', x)$ is replaced by the Lagrange function $L(\mathbf{p}, \dot{\mathbf{q}}, t)$,

Eq. (3.10) becomes well known Euler-Lagrange equation (3.7). It states that within t_1 and t_2, if $\mathbf{q}(t_1)$ and $\mathbf{q}(t_2)$ is identical, in all the permissible restrained movements $\mathbf{q}(t)$, the real-world one is given by the minimization of $S = \int_{t_1}^{t_2} L(\mathbf{p}, \dot{\mathbf{q}}, t)\mathrm{d}t$, i.e.

$$\delta S = \delta \int_{t_1}^{t_2} L(\mathbf{p}, \dot{\mathbf{q}}, t)\mathrm{d}t = 0 \tag{3.11}$$

in which S is the "Hamilton actuating variable", and Eq. (3.11) is the "Hamilton theorem".

For non-conservative system, we have the revised Hamilton theorem

$$\delta S = \int_{t_1}^{t_2} [\delta L(\mathbf{p}, \dot{\mathbf{q}}, t) - \mathbf{F}.\delta\mathbf{q}]\mathrm{d}t = 0 \tag{3.12}$$

Use is made of Eq. (3.8), Eq. (3.11) becomes

$$\delta \int_{t_1}^{t_2} (\mathbf{p}.\dot{\mathbf{q}} - H)\mathrm{d}t = 0 \tag{3.13}$$

Recalling that $H = H(\mathbf{q}, \mathbf{p}, t)$ and $\delta t = 0$, Eq. (3.13) is identical to

$$\int_{t_1}^{t_2} (\mathbf{p}.\delta\dot{\mathbf{q}} + \dot{\mathbf{q}}.\delta\mathbf{p} - \frac{\partial H}{\partial \mathbf{q}}.\delta\mathbf{q} - \frac{\partial H}{\partial \mathbf{p}}.\delta\mathbf{p})\mathrm{d}t = 0 \tag{3.14}$$

Since

$$\mathbf{p}.\delta\dot{\mathbf{q}} = \frac{\mathrm{d}}{\mathrm{d}t}(\mathbf{p}.\delta\mathbf{q}) - \dot{\mathbf{p}}.\delta\mathbf{q} \tag{3.15}$$

Equation (3.14) becomes

$$\mathbf{p}.\delta\mathbf{q}|_{t_1}^{t_2} + \int_{t_1}^{t_2} [(\dot{\mathbf{q}} - \frac{\partial H}{\partial \mathbf{p}}).\delta\mathbf{p} - (\dot{\mathbf{p}} + \frac{\partial H}{\partial \mathbf{q}}).\delta\mathbf{q}]\mathrm{d}t = 0 \tag{3.16}$$

Recalling the conditions $\delta\mathbf{q}|_{t_1} = 0$ and $\delta\mathbf{q}|_{t_2} = 0$, and the independence of variations $\delta\mathbf{q}$ and $\delta\mathbf{p}$ as well as the arbitrariness of $\mathrm{d}t$, the well-known "Hamilton equation" naturally outcomes from Eq. (3.16) as follows

$$\begin{cases} \dot{\mathbf{q}} = \frac{\partial H}{\partial \mathbf{p}} \\ \dot{\mathbf{p}} = -\frac{\partial H}{\partial \mathbf{q}} \end{cases} \tag{3.17}$$

For the dynamic system with non-conservative force **F**, instead of Eq. (3.17) we have

$$\begin{cases} \dot{\mathbf{q}} = \frac{\partial H}{\partial \mathbf{p}} \\ \dot{\mathbf{p}} = -\frac{\partial H}{\partial \mathbf{q}} + \mathbf{F} \end{cases} \tag{3.18}$$

Being widely employed in the analytical mechanics, Hamilton equation is skew-symmetric with respect to variables **q** and **p**, hence it comes under the name of "canonical movement equation" meanwhile **q** and **p** are termed as the "canonical variables". Although the amount of equations is doubled compared to the original Euler-Lagrange Eq. (3.7), yet the differential order is reduced from 2 to 1 in the "ordinary differential equations (ODEs)", as a result, the calculi is much more simplified. In addition, Hamilton equation enables to carry out a more geometrical analysis. For example, "phase space" is the original object in the study of "symplectic geometry".

3. **Hamiltonian system and symplectic geometry**

More generally, the Hamilton Eq. (3.17) will be defined in terms of "phase space" $\boldsymbol{\xi} = (\mathbf{p}, \mathbf{q})^{\mathrm{T}} \in R^{2s}$. A dynamic system is called as canonically Hamiltonian if

$$\dot{\boldsymbol{\xi}} = \mathbf{J}\nabla_{\boldsymbol{\xi}} H \tag{3.19}$$

in which $\nabla_{\boldsymbol{\xi}}$ is the general "Hamilton (gradient) operator" and $\mathbf{J} \in R^{2s}$ is the skew-symmetric matrix

$$\nabla_{\boldsymbol{\xi}} = \left[\frac{\partial}{\partial p_1}, \frac{\partial}{\partial p_2}, \ldots, \frac{\partial}{\partial p_s}, \frac{\partial}{\partial q_1}, \frac{\partial}{\partial q_2}, \ldots, \frac{\partial}{\partial q_s}\right]^{\mathrm{T}} \tag{3.20}$$

$$\mathbf{J} = \begin{bmatrix} 0 & \mathbf{I}_s \\ -\mathbf{I}_s & 0 \end{bmatrix} \tag{3.21}$$

in which $\mathbf{I}_s$ is the unit matrix in s-dimensional space.

In the Hamilton canonical equation (3.19), the matrix **J** possesses "symplectic characteristics" of

$$\mathbf{J}^{\mathrm{T}} = \mathbf{J}^{-1} = -\mathbf{J} \tag{3.22}$$

$$\det(\mathbf{J}) = 1 \tag{3.23}$$

Symplectic characteristics are tightly linked to Hamilton system. For any vector $\boldsymbol{\xi}$ in the phase space, we always have $\boldsymbol{\xi}^{\mathrm{T}}\mathbf{J}\boldsymbol{\xi} = 0$ due to the skew-symmetry of **J**. If we express the Hamilton canonical equation in the form of

$$\dot{\mathbf{v}} = \mathbf{J}\frac{\partial H}{\partial \mathbf{v}} \tag{3.24}$$

in which $\mathbf{v}$ is another $2s$-dimensional vector in the phase space, then following conventional canonical transformation exists

$$\boldsymbol{\xi} = \boldsymbol{\xi}(\mathbf{v}) \tag{3.25}$$

and the inverse transformation may be expressed by

$$\mathbf{v} = \mathbf{v}(\boldsymbol{\xi}) \tag{3.26}$$

Equation (3.24) may be transformed as follows

$$\left(\frac{\partial \mathbf{v}}{\partial \boldsymbol{\xi}}\right)^{\mathrm{T}} \dot{\boldsymbol{\xi}} = \mathbf{J}\frac{\partial \boldsymbol{\xi}}{\partial \mathbf{v}}\frac{\partial H}{\partial \boldsymbol{\xi}} \tag{3.27}$$

Left-time Eq. (3.27) using $(\frac{\partial \boldsymbol{\xi}}{\partial \mathbf{v}})^{\mathrm{T}}$ leads to

$$\left(\frac{\partial \boldsymbol{\xi}}{\partial \mathbf{v}}\right)^{\mathrm{T}} \left(\frac{\partial \mathbf{v}}{\partial \boldsymbol{\xi}}\right)^{\mathrm{T}} \dot{\xi} = \left(\frac{\partial \boldsymbol{\xi}}{\partial \mathbf{v}}\right)^{\mathrm{T}} \mathbf{J}\frac{\partial \boldsymbol{\xi}}{\partial \mathbf{v}}\frac{\partial H}{\partial \boldsymbol{\xi}} \tag{3.28}$$

Since

$$\frac{\partial \mathbf{v}}{\partial \boldsymbol{\xi}}\frac{\partial \boldsymbol{\xi}}{\partial \mathbf{v}} = \mathbf{I}_{2s} \tag{3.29}$$

in which $\mathbf{I}_{2s}$ is the unit matrix in $2s$-dimensional space, it follows that

$$\dot{\boldsymbol{\xi}} = (\frac{\partial \boldsymbol{\xi}}{\partial \mathbf{v}})^{\mathrm{T}}\mathbf{J}(\frac{\partial \boldsymbol{\xi}}{\partial \mathbf{v}})\frac{\partial H}{\partial \boldsymbol{\xi}} = \mathbf{S}^{\mathrm{T}}\mathbf{J}\mathbf{S}\frac{\partial H}{\partial \boldsymbol{\xi}} \tag{3.30}$$

in which $\mathbf{S}$ is termed as "symplectic matrix"

$$\mathbf{S} = \frac{\partial \boldsymbol{\xi}}{\partial \mathbf{v}} \tag{3.31}$$

The condition that the canonical transformation should keep Eq. (3.19) unchanged is equally to demand

$$\mathbf{S}^{\mathrm{T}}\mathbf{J}\mathbf{S} = \mathbf{J} \tag{3.32}$$

Condition (3.32) is very useful in the selection of available algorithms or the construction of new algorithms for the solution of ODEs using "finite difference method (FDM)". In 1983, Ruth published the first paper concerning the differential scheme for Hamilton equation using symplectic algorithm (geometric integration). By taking into account of the qualitative and global characteristics of the problem, the symplectic algorithm possesses high stability and long term trace ability. This issue will be briefly discussed in Chap. 4. For more details, our readers are referred to the relevant literatures (Budd and Iserles 1999; Marsden and Ratiu 1999; Hairer et al. 2006; Gagarina et al. 2016).

3.3 Kinematics of Continuum

In the most fundamental respect, a fluid parcel is not different from a solid particle, hence a term “matter parcel” or “substance parcel” or “parcel” will be frequently employed in this chapter, if no misunderstanding would be led to. Suppose now that a matter parcel is able to move in response to an applied force, acceleration emerges exactly as expected from the second Newton’s law.

In the study of continuum mechanics, the mathematical tool of “tensor” is dominant in the society of theoretical research. In the society of engineering community however, another tool of Gibson vector {.} or matrix [.] is more welcome. In this book the latter is preferentially used with boldface notations whereas the Gibson notations are sometimes employed for the vector and matrix with detailed components, where there is no risk of misleading. To be specific, tensor index contraction rule and comma notation within the indices for partial differential operator, are normally invalid in our notation system expect for particularly stated. The cost is that the summation notation $\sum$ is very frequently appeared in our formulas.

3.3.1 Two Ways to Observe the Motion of Continuum

1. **Transformation of configurations**

Every parcel participating in substance motion is pushed or/and pulled by all of the surrounding parcels via shear stress and normal stress. Hence, we cannot predict the motion of a given parcel in isolation from its surrounding substance parcels (Landau and Lifshitz 1959; Munson et al. 1998; Kundu and Cohen 2001). The practical consequence of complex kinematics is that an appropriate description must be able to define the motion and acceleration on arbitrarily small spatial scales, i.e., the coordinates of a material theory or model must vary continuously (Hirt et al. 1974; Donéa et al. 1976, 1977).

Suppose a continuum object moving within a 3-D domain R^3, there are two quite different ways to observe its motion: either by tracking specific and identifiable parcels that are carried about with the motion (flow)—the Lagrangian formalism (material description); or by recording the motion velocity at specific locations that are fixed in space—the Eulerian formalism (spatial description). Although both of them may be successfully used in the continuum mechanics (Benson 1992; Yeung 2002; Bennett 2006), yet the material description is more natural for many observational techniques and for the statement of fundamental conservation laws in solid mechanics. On the other hand, almost all of the theories in fluid mechanics have been developed in the spatial description. This is because previously, the PDEs of fluid with Lagrangian formalism are quite difficult to solve using mesh-based method when pressure gradients are included. Nowadays however, Lagrangian formalism is more and more prevalent in the CFD attributable to the fact that the corresponding PDEs may be solved by mesh-free or/and particle-based methods without intrinsic difficulties.

In Fig. 3.1 a Lagrangian cell (parcel) dV is carried along with the motion. Then the followed question is how to identify this specific parcel. One possible scheme that can serve to label and identify the parcel throughout a domain at arbitrarily fine spatial resolution makes use of the position of parcel at some specified time, normally the initial time $t_0 = 0$, that is denote as the initial position $\mathbf{X}_0$ with Cartesian components (X_0, Y_0, Z_0). For example, if we know the density as the function of position, we can just as well write this as $\rho = \rho(X_0, Y_0, Z_0)$.

In Fig. 3.2, initial and current configurations are illustrated by Lagrangian formalism. Initial configuration is defined at $t_0 = 0$ with configuration V_0 and fixed coordinate system $\mathbf{X}$. A point P represents the Lagrangian cell whose initial position is denoted as $\mathbf{X}_0$. $\mathbf{X}_0$ plus t fixed to the initial configuration are independent variables (Lagrangian variables) for the motion and deformation of matter in Lagrangian formalism. Hereinafter the zero subscript of $\mathbf{X}_0$ will be neglected where there is no risk of misleading.

At time t, the matter domain moves to a new position with current configuration V termed as the "deformation state" and described by a new coordinate system $\mathbf{x}$. The point P moves a distance $\mathbf{u}$ to point p with the coordinate $\mathbf{x} = (x, y, z)$ termed as the "spatial coordinate", $\mathbf{x}$ plus t are independent variables (Eulerian variables) for the motion and deformation of a matter in Eulerian formalism. Hence, the spatial coordinate system is related to the current configuration.

Figure 3.3 shows the motion and deformation of a material volume in Lagrangian formalism (solid boundary) and a control volume in Eulerian formalism (broken boundary) within a velocity field. The material volume is made up of specific parcel

Fig. 3.1 Infinitesimal cell in Lagrange description

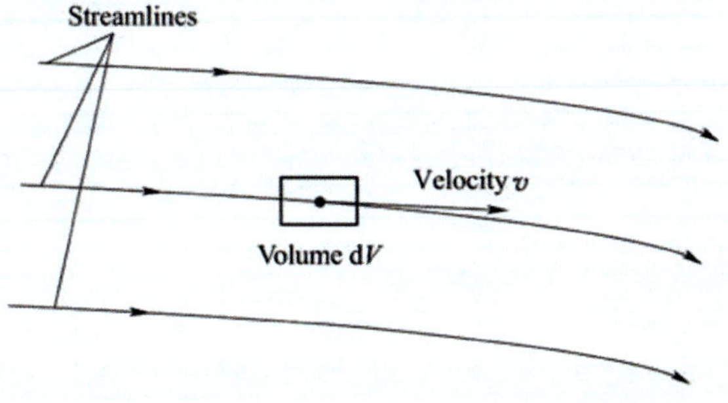

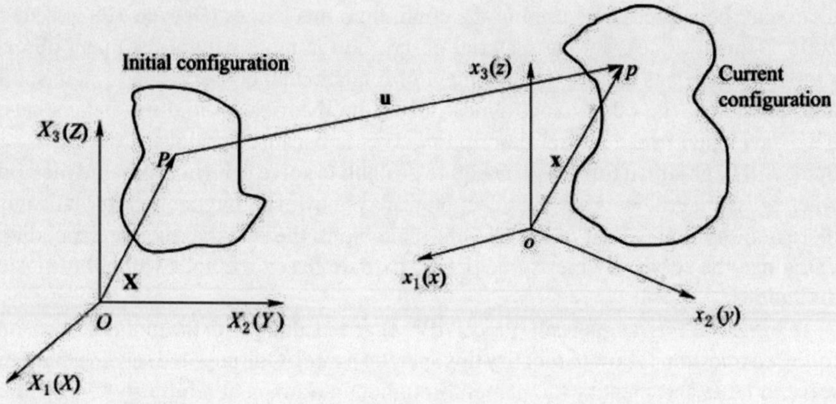

Fig. 3.2 Configurations and coordinate systems in Lagrangian formalism

that is carried along with the flow, whereas the control volume with imaginary and invisible sides is fixed in space so far as the flow is concerned. The essence of Lagrangian formalism is that we observe and seek to describe the position, pressure, velocity and other properties of moving material volumes; whereas the essence of Eulerian formalism is that we observe and seek to describe the pressure, velocity and other properties inside fixed control volumes. The continuum mechanics assumes that either material volumes or control volumes may be made as small as necessary to resolve the performances of substance, this equally requires that the substance must be a smooth continuum down to arbitrarily small spatial scales.

In solid dynamics, the focus is the specific material points, so the material description is normally adopted. In fluid dynamics, although material description possesses

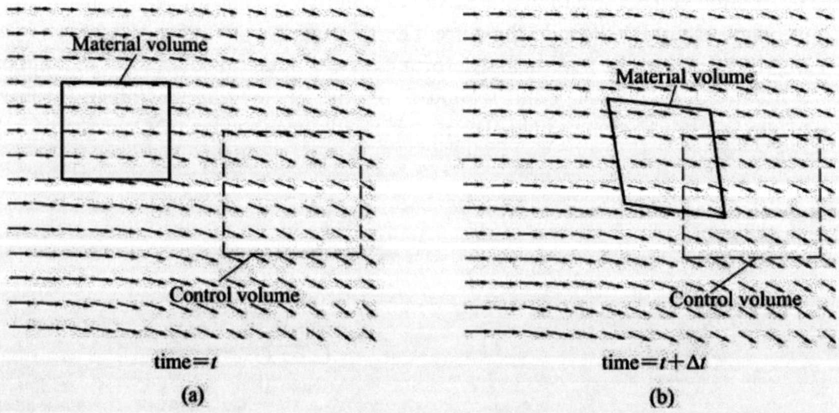

Fig. 3.3 Motion and deformation of a matter volume (in Lagrange formalism) and a control volume (in Euler formalism) within velocity field. **a** Material and control volumes at t; **b** material and control volumes at $t + \Delta t$

remarkable advantage, yet it has an inherent problem: the spatial sampling of parcels is more or less uncontrolled, since the parcels will go wherever the flow takes them, and that may not be the position our interest lies. Another problem of material description in fluid dynamics is the inability to follow the large distortion of computational domain without recourse to frequent re-meshing or re-connecting operations. As a result, in fluid dynamics the spatial description is prevalently adopted and the focus is normally the properties inside the control volume at specific spatial position. For example, most of the customarily applications of engineering fluid dynamics require the fluid properties at fixed sites such as the pressure along the surface of a spillway.

The major disadvantage of spatial description is that as the fluid moves, material interfaces lose their sharp definitions and local regions with fine resolution are difficult to achieve. This is why in some advanced CMs (e.g. SPH), Lagrangian formalism is implemented by particle-type methods to handle engineering problems such as the landslide-induced tsunami.

A hybrid approach combining the advantages of both the Lagrangian and Eulerian formalisms may be achieved by the technique of "arbitrary Lagrangian-Eulerian (ALE)" (Hirt et al. 1974; Hughes et al. 1981; Donéa et al. 1982; Fernandez and Moubachir 2002; Duarte et al. 2004).

In Fig. 3.4, there are three configurations and coordinate systems, i.e. "reference configuration", "initial configuration" and "current configuration". Denoting the motion velocity of the domain by **w**, in the Eulerian formalism **w** is zero, whereas in the Lagrangian formalism **w** is equal to the velocity of parcel. In the ALE formalism, **w** is neither equal to zero nor to the velocity of parcels, but varies smoothly and arbitrarily between both of them. This arbitrary velocity keeps the motion of computation mesh under control by specific mechanical problem.

The ALE approach uses two superimposed meshes. The Eulerian mesh remains fixed and orthogonal, whereas the Lagrangian mesh is actually a moving imaginary mesh which follows the fluid/solid domain corresponding to the reference configuration and allows to undergo quite extreme changes in shape. In 1985, SHALE—the first 2-D finite difference code for simulating stress wave propagation and fracture growth in brittle solids based on the ALE approach, was released (Demuth et al. 1985). It was originally written to investigate the chemical-explosive loading of oil shale in mine blasting process.

However, the hybrid description gives rise to additional complexity in algorithm.

2. Transformation between coordinate systems

Now that we have learned how to observe the motion of parcel, we start to survey the entire domain in order to construct a representation of motion. One helpful way to think of motion is that it maps a parcel from one position to the next, e.g., from a starting position **X** into the positions **x** at some later time. Namely, the starting position is carried along with the parcel, and thus it may serve to identify the parcel.

(1) Material coordinate system

Suppose the coordinate system for **X** and the coordinate system for **x** is coincident, and the mapping from **X** to **x** is continuous and unique, namely, adjacent parcels will

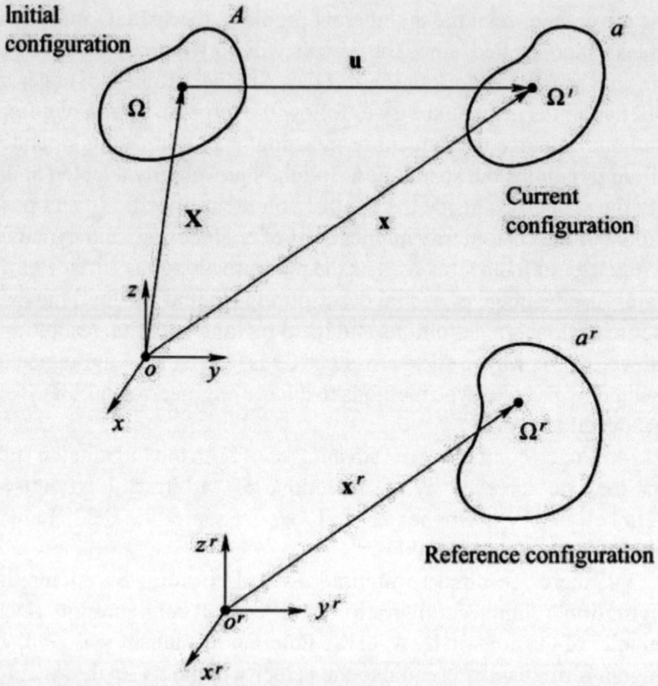

Fig. 3.4 Kinematic model in arbitrary Eulerian-Lagrangian formalism

never be split apart, and neither will one parcel be forced to occupy the same position of another. It also may be somewhat blithely assumed that the position of parcels at all later times t can be determined to form the "parcel trajectory", also called the "pathline", namely

$$x_\alpha = x_\alpha(X_\beta, t) \quad (\alpha, \beta = 1, 2, 3) \tag{3.33}$$

or

$$\mathbf{x} = \mathbf{x}(\mathbf{X}, t) \tag{3.34}$$

It is noted that the loop subscripts $\alpha, \beta = 1, 2, 3$ are customarily used in this book to indicate the coordinate components of x, y, z for the simplification of notations.

Each trajectory $\mathbf{x}$ that we observe or construct must be labelled with a unique $\mathbf{X}$, thus for a given trajectory, $\mathbf{X}$ is a constant. Since our coordinate system must be able to describe a continuum defined over the domain, hence $\mathbf{X}$ must vary continuously over the entire domain. $\mathbf{X}$ is thus independent variable in Lagrangian coordinate system. This kind of coordinate system in which parcel position $\mathbf{x}$ is the fundamental dependent variable is sometimes appropriately called as "material coordinate system". According to Eq. (3.34) it is obviously that

$$\mathbf{X} = \mathbf{x}(\mathbf{X}, 0) \tag{3.35}$$

(2) **Spatial coordinate system**

If the motion is described by **x** as independent variable at arbitrary position in spatial coordinate system, then the motion equation is

$$\mathbf{X} = \mathbf{X}(\mathbf{x}, t) \tag{3.36}$$

(3) **Transformation**

With above conventional assumptions in place, the mapping of parcels from initial to subsequent positions by Eq. (3.34) can be inverted to Eq. (3.36), namely, a Lagrangian formalism can be inverted to yield an Eulerian formalism:

$$\underset{\text{Lagrangian formalism}}{\mathbf{x} = \mathbf{x}(\mathbf{X}, t)} \quad \Leftrightarrow \quad \underset{\text{Eulerian formalism}}{\mathbf{X} = \mathbf{X}(\mathbf{x}, t)} \tag{3.37}$$

In the Lagrangian formalism we are presumably to know the starting position **X**, the independent variable, and treat the subsequent position **x** as the dependent variable. On the contrary, in the Eulerian formalism we take the fixed position **x** as the independent variable (spatial coordinate) and ask where was the initial position of the parcel now appears at this position, namely, **X** is looked at as the dependent variable. As we have been elucidated previously that in the study of solid mechanics, it does make sense to identify the initial position **X** of a parcel, but it seldom (but not always) makes sense in the study of fluid dynamics.

3.3.2 Observation of Motion

1. **Displacement vector and its gradient tensor**

According to Fig. 3.4, the displacement of a parcel is defined by

$$x_\alpha = X_\alpha + u_\alpha(X_1, X_2, X_3, t) \quad (\alpha = 1, 2, 3) \tag{3.38}$$

or

$$\mathbf{x} = \mathbf{X} + \mathbf{u} \tag{3.39}$$

Equation (3.38) may be arranged in two manners

$$u_\alpha = x_\alpha(X_1, X_2, X_3, t) - X_\alpha \quad \text{material description} \tag{3.40}$$

or

$$u_\alpha = x_\alpha - X_\alpha(x_1, x_2, x_3, t) \quad \text{spatial description} \tag{3.41}$$

For finite deformation issues such as the landslide and its induced tsunami waves, the definition and solution of corresponding PDEs may employ the "update Lagrangian (UL)" scheme (e.g. FLAC, DEM, DDA, SPH) using incremental displacement $\Delta\mathbf{u}$ obtained by time-marching step algorithms, which will be specified in the Chaps. 5–11 of this book.

Derivative to the both sides of Eq. (3.40), we get the displacement gradient tensor under material description

$$\frac{\partial u_\alpha}{\partial X_\beta} = \frac{\partial x_\alpha}{\partial X_\beta} - \delta_{\alpha\beta} \quad (\alpha, \beta = 1, 2, 3) \tag{3.42}$$

in which $\delta_{\alpha\beta}$ is the Kronecker delta symbol.

Derivative to the both sides of Eq. (3.41), we get the displacement gradient tensor under spatial description

$$\frac{\partial u_\alpha}{\partial x_\beta} = \delta_{\alpha\beta} - \frac{\partial X_\alpha}{\partial x_\beta} \quad (\alpha = 1, 2, 3) \tag{3.43}$$

2. Observation of velocity

(1) Material description

Suppose a finite collection of fluid or solid particles with its identity being represented by a subscript appended to **x**, then the derivative w.r.t. time will be an ordinary one because there is no independent variable **X**, i.e.

$$\frac{\mathrm{d}}{\mathrm{d}t} = \frac{\partial}{\partial t}|_{\mathbf{X}=\text{constant}} \tag{3.44}$$

When this derivative is applied to the Lagrangian variable **x** that depends upon **X** and *t*, the velocity $\mathbf{v}_L$ of a parcel, often termed as the "Lagrangian velocity", is just the time rate of the parcel position holding fixed **X**

$$\mathbf{v}_L(\mathbf{X}, t) = \frac{\mathrm{d}\mathbf{x}(\mathbf{X}, t)}{\mathrm{d}t} = \frac{\partial \mathbf{x}(\mathbf{X}, t)}{\partial t} \tag{3.45}$$

The Lagrangian velocity of a fluid/solid parcel in continuum is exactly identical to the velocity in the classical dynamics of particle.

(2) Spatial description

Where tracking parcels is impractical, then we may choose to observe the velocity at fixed positions **x**. Regardless of the mechanical details, the velocity $\mathbf{v}_E$ observed in this way is termed as the "Eulerian velocity" that is the velocity of the parcel within the fixed control volume (viz. Fig. 3.3).

Eulerian velocity $\mathbf{v}_E$ is normally employed as the independent variable in the spatial description of fluid, in which the partial derivative with respect to time represents the rate observed at a fixed point in space. This is not equal to the Lagrangian time rate in which $\mathbf{X}$ is held constant (i.e. parcel identity is held constant) except in the special motion/flow case where there is no spatial variation.

(3) **Transformation of velocity**

To learn how to transform time derivatives we will write the time rate following a parcel—the Lagrangian time rate, in terms of Eulerian (or field) variables. Given that parcel trajectories can be inverted to yield $\mathbf{x}(\mathbf{X}, t)$, the Eulerian velocity is defined by

$$\mathbf{v}_E(\mathbf{x}, t) = \mathbf{v}_L(\mathbf{x}(\mathbf{X}, t), t) \tag{3.46}$$

in which $\mathbf{x}$ is fixed and $\mathbf{X}$ is the initial position of fluid parcel. In other words, the Eulerian velocity at a fixed position $\mathbf{x}$ on the left side, is the Lagrangian velocity of the parcel on the right side that happens to be at that spatial position and time instant.

One way to appreciate the difference between the $\mathbf{v}_L$ and $\mathbf{v}_E$ is to understand that $\mathbf{x}$ in Lagrangian velocity (3.45) is the position of a moving parcel, whereas $\mathbf{x}$ in Eulerian velocity (3.46) is the arbitrary and fixed observation position of current configuration. Aside from the initial position, parcel position is a result of substance motion rather than our choice. As time runs, the position of any specific parcel will change, in contrast, the velocity observed at a fixed position is the velocity of sequential parcels (each having different $\mathbf{X}$).

3. **Observation of acceleration**

(1) **Material description**

According to Eq. (3.45), the acceleration is directly derived as

$$\mathbf{a}_L(\mathbf{X}, t) = \frac{d\mathbf{x}^2}{dt^2} = \frac{\partial \mathbf{v}_L(\mathbf{X}, t)}{\partial t} = \frac{\partial^2 \mathbf{x}(\mathbf{X}, t)}{\partial t^2} \tag{3.47}$$

(2) **Spatial description**

According to Eq. (3.46), the correspondent acceleration is derived as

$$\begin{aligned}\mathbf{a}_E(\mathbf{x}, t) &= \frac{d\mathbf{v}_E(\mathbf{x}, t)}{dt} = \frac{\partial \mathbf{v}_L(\mathbf{x}(\mathbf{X}, t), t)}{\partial t} + \frac{\partial \mathbf{v}_L(\mathbf{x}(\mathbf{X}, t), t)}{\partial \mathbf{x}}\frac{\partial \mathbf{x}}{\partial t} \\ &= \frac{\partial \mathbf{v}_E(\mathbf{x}, t)}{\partial t} + \frac{\partial \mathbf{v}_E(\mathbf{x}, t)}{\partial \mathbf{x}}\mathbf{v}_E(\mathbf{x}, t)\end{aligned} \tag{3.48}$$

In which the operator $\frac{d}{dt}$ is often said to be the time derivative "following the flow" in addition to a profusion of different names such as "convective derivative", "substantive (or substantial) derivative", "Stokes derivative" and "material derivative". In 3-D domain with Cartesian coordinates (x, y, z) and velocity (v_x, v_y, v_z), material derivative is specified as

$$\frac{\mathrm{d}}{\mathrm{d}t} = \frac{\partial}{\partial t} + v_x \frac{\partial}{\partial x} + v_y \frac{\partial}{\partial y} + v_z \frac{\partial}{\partial z} \tag{3.49}$$

or more compacted vector notation

$$\frac{\mathrm{d}}{\mathrm{d}t} = \frac{\partial}{\partial t} + \mathbf{v}.\nabla \tag{3.50}$$

It is noted that the subscript E for the velocity vector in Eq. (3.50) has been dropped. The notation (.) between two vectors means "dot product", and the Hamilton "∇" (or gradient "**grad**") operator expanded in Cartesian coordinates is a gradient vector

$$\nabla = \mathbf{grad} = \begin{Bmatrix} \frac{\partial}{\partial x} \\ \frac{\partial}{\partial y} \\ \frac{\partial}{\partial z} \end{Bmatrix} \tag{3.51}$$

The advection term $\mathbf{v}.\nabla$ in Eq. (3.50) is, in effect, a scalar that is multiplied onto the variable being differentiated. Advection represents the transport of fluid properties that may cause the deformation and rotation of fluid parcel.

Equation (3.50) may be generalized into the material derivate w.r.t. any function $f(\mathbf{x}, t)$

$$\frac{\mathrm{d}f}{\mathrm{d}t} = \frac{\partial f}{\partial t} + \mathbf{v}.\mathbf{grad}(f) \tag{3.52}$$

3.3.3 Observation of Deformation

1. **Deformation and strain tensors**

From Eq. (3.34), in the current configuration a linear element $\mathbf{dX}$ after deformation is

$$\mathbf{dx} = \sum_{\beta=1}^{3} \frac{\partial \mathbf{x}}{\partial X_\beta} \mathrm{d}X_\beta \tag{3.53}$$

or

$$dx_\alpha = \sum_{\beta=1}^{3} \frac{\partial x_\alpha}{\partial X_\beta} dX_\beta \quad (\alpha = 1, 2, 3) \tag{3.54}$$

in which the Greek superscripts α is employed to indicate the coordinate direction for the simplification of notations $x = x_1$, $y = x_2$ and $z = x_3$. Use is made of Eq. (3.38), the deformation gradient tensor $\mathbf{F}$ is defined as

$$\mathbf{F} = \frac{\partial x_\alpha}{\partial X_\beta} = \mathbf{I} + (\nabla \mathbf{u}^{\mathrm{T}})^{\mathrm{T}} = \begin{bmatrix} 1\,0\,0 \\ 0\,1\,0 \\ 0\,0\,1 \end{bmatrix} + \begin{bmatrix} \partial u_x/\partial X & \partial u_x/\partial Y & \partial u_x/\partial Z \\ \partial u_y/\partial X & \partial u_y/\partial Y & \partial u_y/\partial Z \\ \partial u_z/\partial X & \partial u_z/\partial Y & \partial u_z/\partial Z \end{bmatrix} \tag{3.55}$$

in which $\mathbf{I}$ is the unit tensor and $(\nabla \mathbf{u}^{\mathrm{T}})^{\mathrm{T}}$ is the mechanical strain tensor

$$(\nabla \mathbf{u}^{\mathrm{T}})^{\mathrm{T}} = \begin{bmatrix} \partial u_x/\partial X & \partial u_x/\partial Y & \partial u_x/\partial Z \\ \partial u_y/\partial X & \partial u_y/\partial Y & \partial u_y/\partial Z \\ \partial u_z/\partial X & \partial u_z/\partial Y & \partial u_z/\partial Z \end{bmatrix} \tag{3.56}$$

The deformation gradient tensor $\mathbf{F}$ transforms the linear element $\mathbf{dX}$ in the initial configuration into $\mathbf{dx}$ in the current configuration, hence it may be employed to describe the deformation of substance.

Similarly, the linear element $\mathbf{dX}$ the in initial configuration may be expressed by the deformed $\mathbf{dx}$ in the current configuration as

$$\mathbf{dX} = \sum_{\beta=1}^{3} \frac{\partial \mathbf{X}}{\partial x_\beta} dx_\beta \tag{3.57}$$

or

$$dX_\alpha = \sum_{\beta=1}^{3} \frac{\partial X_\alpha}{\partial x_\beta} dx_\beta \quad (\alpha = 1, 2, 3) \tag{3.58}$$

in which $\frac{\partial X_\alpha}{\partial x_\beta}$ is the element of the inverse tensor $\mathbf{F}^{-1}$.

Equations (3.53) and (3.57) may be re-written in the forms

$$\mathbf{dx} = \mathbf{F}\mathbf{dX} \tag{3.59}$$

$$\mathbf{dX} = \mathbf{F}^{-1}\mathbf{dx} \tag{3.60}$$

In fluid dynamics, the mechanical strain tensor (3.56) may be directly employed; whereas in solid dynamics, deformation tensors need to be further defined.

- Green deformation tensor or right Cauchy-Green tensor.

$$\mathbf{C} = \mathbf{F}^{\mathrm{T}}\mathbf{F} \tag{3.61}$$

- Cauchy deformation tensor.

$$\mathbf{c} = \mathbf{F}^{-\mathrm{T}}\mathbf{F}^{-1} \tag{3.62}$$

- Left Cauchy–Green tensor.

$$\mathbf{B} = \mathbf{F}\mathbf{F}^{\mathrm{T}} \tag{3.63}$$

Then, three types of strain tensor may be defined as follows.

- Green strain tensor.

$$\mathbf{E} = \frac{1}{2}(\mathbf{C} - \mathbf{I}) = \frac{1}{2}(\mathbf{F}^{\mathrm{T}}\mathbf{F} - \mathbf{I}) \tag{3.64}$$

Its component form is given using index contraction rule (seldom used in this book)

$$E_{\alpha\beta} = \frac{1}{2}\left(\frac{\partial x_\kappa}{\partial X_\alpha}\frac{\partial x_\kappa}{\partial X_\beta} - \delta_{\alpha\beta}\right) = \frac{1}{2}\left(\frac{\partial u_\beta}{\partial X_\alpha} + \frac{\partial u_\alpha}{\partial X_\beta} + \frac{\partial u_\kappa}{\partial X_\alpha}\frac{\partial u_\kappa}{\partial X_\beta}\right) \quad (\alpha, \beta, \kappa = 1, 2, 3) \tag{3.65}$$

- Almansi strain tensor.

$$\mathbf{A} = \frac{1}{2}(\mathbf{I} - \mathbf{c}) = \frac{1}{2}(\mathbf{I} - \mathbf{F}^{-\mathrm{T}}\mathbf{F}^{-1}) \tag{3.66}$$

Its component form is given using index contraction rule

$$A_{\alpha\beta} = \frac{1}{2}\left(\delta_{\alpha\beta} - \frac{\partial X_\kappa}{\partial x_\alpha}\frac{\partial X_\kappa}{\partial x_\beta}\right) = \frac{1}{2}\left(\frac{\partial u_\beta}{\partial x_\alpha} + \frac{\partial u_\alpha}{\partial x_\beta} - \frac{\partial u_\kappa}{\partial x_\alpha}\frac{\partial u_\kappa}{\partial x_\beta}\right) \quad (\alpha, \beta, \kappa = 1, 2, 3) \tag{3.67}$$

- Henchy strain tensor.

$$\begin{cases} \mathbf{H} = \frac{1}{2}\ln\mathbf{C} \\ \mathbf{h} = \frac{1}{2}\ln\mathbf{B} \end{cases} \tag{3.68}$$

Suppose the deformation gradient is infinitesimal, namely

$$\left|\frac{\partial u_\alpha}{\partial x_\beta}\right| \ll 1, \left|\frac{\partial u_\alpha}{\partial X_\beta}\right| \ll 1 \quad (\alpha, \beta = 1, 2, 3) \tag{3.69}$$

from Eqs. (3.53) and (3.55) we have

$$\mathbf{dx} = \mathbf{F}.\mathbf{dX} \approx \mathbf{IdX} = \mathbf{dX} \tag{3.70}$$

It means that if the deformation gradient is small, the configurations before and after deformation are nearly identical. Hence, Green tensor (3.65) and Almansi tensor (3.67) are degenerated into the "Cauchy strain tensor" ("engineering strain tensor", or "nominal strain tensor") in solid mechanics

$$\boldsymbol{\varepsilon} = \varepsilon_{\alpha\beta} = E_{\alpha\beta} = A_{\alpha\beta} = \frac{1}{2}\left(\frac{\partial u_\beta}{\partial X_\alpha} + \frac{\partial u_\alpha}{\partial X_\beta}\right) = \frac{1}{2}\left(\frac{\partial u_\beta}{\partial x_\alpha} + \frac{\partial u_\alpha}{\partial x_\beta}\right) \quad (\alpha, \beta = 1, 2, 3) \tag{3.71a}$$

or

$$\boldsymbol{\varepsilon} = \frac{1}{2}\left[(\nabla\mathbf{u}^{\mathrm{T}})^{\mathrm{T}} + \nabla\mathbf{u}^{\mathrm{T}}\right] \tag{3.71b}$$

The same is applied to the Henchy strain tensor.

Equation (3.69) also may be reasonably postulated in the finite deformation issues using "update Lagrangian (UL)" scheme w.r.t. each time-marching step, hence the Cauchy strain tensor (3.71) will be used for solid throughout this book.

2. **Stress tensor**

Stress tensor represents inner force and is the counterpart tensor w.r.t. the strain tensor defined in the same configuration. Basically, there are two definitions of stress tensor—"Green stress tensor" or "Cauchy stress tensor" $\boldsymbol{\sigma}$, and "Lagrange stress tensor" or the "first Piola-Kirchhoff stress tensor". In current configuration, Cauchy stress tensor that is the counterpart of Cauchy strain tensor (Crisfield 2000) defined in Eq. (3.71), will be used throughout this book.

At the infinitesimal element in Fig. 3.5, there are, in general, nine components of stress tensor. On each surface, there are a normal stress component plus two shear stress components. The sign convention for the subscript is that the first letter represents the surface on which the stress is acting, and the second letter represents the direction of the stress. These stresses form a Cauchy stress tensor such that

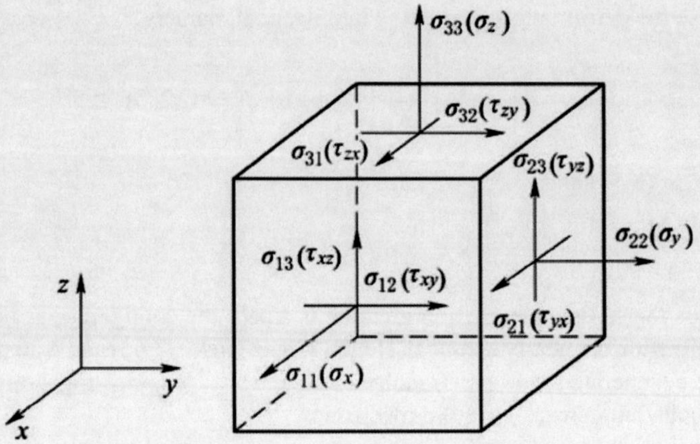

Fig. 3.5 Infinitesimal cubic element and Cauchy stress tensor

$$\boldsymbol{\sigma} = [\sigma] = \sigma_{\alpha\beta} = \begin{bmatrix} \sigma_{xx} & \tau_{xy} & \tau_{xz} \\ \tau_{yx} & \sigma_{yy} & \tau_{yz} \\ \tau_{zx} & \tau_{zy} & \sigma_{zz} \end{bmatrix} \quad (3.72)$$

In solid dynamics, the equilibrium equations w.r.t. the moments $\sum M_x = 0$, $\sum M_y = 0$ and $\sum M_z = 0$ give rise to the symmetry property of Cauchy stress tensor—the reciprocal theorem of shear stress stating $\tau_{zx} = \tau_{xz}$, $\tau_{yz} = \tau_{zy}$ and $\tau_{xy} = \tau_{yx}$. As a result, we have $\boldsymbol{\sigma}^{\mathrm{T}} = \boldsymbol{\sigma}$.

3. **Velocity gradient and stress of fluids**

(1) **Velocity gradient**

In fluid dynamics, the essential variables are assembled in a velocity vector

$$\mathbf{v} = \{v\} = \{\dot{u}\} = \begin{bmatrix} \dot{u}_x & \dot{u}_y & \dot{u}_z \end{bmatrix}^{\mathrm{T}} \quad (3.73)$$

and the velocity gradient may be directly employed in the study on constitutive (or state) equation

$$(\nabla \mathbf{v}^{\mathrm{T}})^{\mathrm{T}} = \begin{bmatrix} \partial v_x/\partial x & \partial v_x/\partial y & \partial v_x/\partial z \\ \partial v_y/\partial x & \partial v_y/\partial y & \partial v_y/\partial z \\ \partial v_z/\partial x & \partial v_z/\partial y & \partial v_z/\partial z \end{bmatrix}^{\mathrm{T}} \quad (3.74)$$

in which ∇ is the Hamilton operator.

(2) **Stress**

A fluid that sustains only pressures p is often called the "ideal fluid" or "Euler fluid". For such a fluid we have $\sigma_{xx} = \sigma_{yy} = \sigma_{zz} = -p$ and the stress tensor in Eq. (3.72) is simplified as an isotropic tensor **p** representing net pressure force

$$\mathbf{p} = -\frac{\sigma_{xx} + \sigma_{yy} + \sigma_{zz}}{3}\mathbf{I} = -p\mathbf{I} = \begin{bmatrix} -p & 0 & 0 \\ 0 & -p & 0 \\ 0 & 0 & -p \end{bmatrix} \tag{3.75}$$

For general fluid such as the water, the total stress tensor (3.72) is normally subtracted by Eq. (3.75) to get the stress tensor $\boldsymbol{\tau}$ attributable to fluid deformation only

$$\begin{aligned} \boldsymbol{\sigma} = \sigma_{\alpha\beta} &= -p\delta_{\alpha\beta} + \tau_{\alpha\beta} = [p] + [\tau] \\ &= -p\mathbf{I} + \boldsymbol{\tau} = \begin{bmatrix} -p & 0 & 0 \\ 0 & -p & 0 \\ 0 & 0 & -p \end{bmatrix} + \begin{bmatrix} \tau_{xx} & \tau_{xy} & \tau_{xz} \\ \tau_{yx} & \tau_{yy} & \tau_{yz} \\ \tau_{zx} & \tau_{zy} & \tau_{zz} \end{bmatrix} \end{aligned} \tag{3.76}$$

where: $\delta_{\alpha\beta}$—Kronecker delta symbol; $\tau_{\alpha\beta}$—shear stress in the β direction exerted on a plane perpendicular to the α axis.

4. **Cauchy strain and stress tensors of solids**

(1) **Strain**

In solid mechanics the essential variables are assembled in a displacement vector **u** or its increment $\Delta\mathbf{u}$

$$\begin{cases} \mathbf{u} = \{u\} = \begin{bmatrix} u_x & u_y & u_z \end{bmatrix}^{\mathrm{T}} \\ \Delta\mathbf{u} = \{\Delta u\} = \begin{bmatrix} \Delta u_x & \Delta u_y & \Delta u_z \end{bmatrix}^{\mathrm{T}} \end{cases} \tag{3.77}$$

Suppose the deformation gradient is infinitesimal, or the deformation is small, the configuration before and after deformation is nearly identical. Under such circumstances, the mechanical tensor defined in Eq. (3.56) may be resolute into a symmetrical portion dubbed as "strain (stretch) tensor" and a skew symmetrical portion termed as "rotation (curl) tensor"

$$(\nabla\mathbf{u}^{\mathrm{T}})^{\mathrm{T}} = \boldsymbol{\varepsilon} + \mathbf{w} = \begin{bmatrix} \varepsilon_{xx} & \frac{1}{2}\gamma_{xy} & \frac{1}{2}\gamma_{xz} \\ \frac{1}{2}\gamma_{yx} & \gamma_{yy} & \frac{1}{2}\gamma_{yz} \\ \frac{1}{2}\gamma_{zx} & \frac{1}{2}\gamma_{zy} & \varepsilon_{zz} \end{bmatrix} + \begin{bmatrix} 0 & -\frac{1}{2}\omega_z & \frac{1}{2}\omega_y \\ \frac{1}{2}\omega_z & 0 & -\frac{1}{2}\omega_x \\ -\frac{1}{2}\omega_y & \frac{1}{2}\omega_x & 0 \end{bmatrix} \tag{3.78}$$

or

$$\begin{cases} \boldsymbol{\varepsilon} = \frac{1}{2}\left[(\nabla\mathbf{u}^{\mathrm{T}})^{\mathrm{T}} + \nabla\mathbf{u}^{\mathrm{T}}\right] \\ \mathbf{w} = \frac{1}{2}\left[(\nabla\mathbf{u}^{\mathrm{T}})^{\mathrm{T}} - \nabla\mathbf{u}^{\mathrm{T}}\right] \end{cases} \tag{3.79}$$

or

$$\begin{cases} \varepsilon_{\alpha\beta} = \frac{1}{2}\left(\frac{\partial u_\alpha}{\partial x_\beta} + \frac{\partial u_\beta}{\partial x_\alpha}\right) \\ w_{\alpha\beta} = \frac{1}{2}\left(\frac{\partial u_\alpha}{\partial x_\beta} - \frac{\partial u_\beta}{\partial x_\alpha}\right) \end{cases} \quad (\alpha, \beta = 1, 2, 3) \tag{3.80}$$

If the rotation tensor is null, namely $\omega_x = \omega_y = \omega_z = 0$, the existing condition of "total differentiation" is met and the deformation comes under the term of "pure deformation", namely, there is a potential function $\Phi(x, y, z)$ that

$$\Phi(x, y, z) = u_x dx + u_y dy + u_z dz \tag{3.81}$$

or

$$\begin{Bmatrix} u_x \\ u_y \\ u_z \end{Bmatrix} = \begin{Bmatrix} \frac{\partial \Phi}{\partial x} \\ \frac{\partial \Phi}{\partial y} \\ \frac{\partial \Phi}{\partial z} \end{Bmatrix} \tag{3.82}$$

hence

$$\begin{cases} \frac{\partial u_z}{\partial y} = \frac{\partial u_y}{\partial z} \\ \frac{\partial u_x}{\partial z} = \frac{\partial u_z}{\partial x} \\ \frac{\partial u_y}{\partial x} = \frac{\partial u_x}{\partial y} \end{cases} \tag{3.83}$$

Equation (3.83) gives rise to the symmetrical property of Cauchy strain tensor (the reciprocal theorem of shear strain) that $\gamma_{zx} = \gamma_{xz}, \gamma_{yz} = \gamma_{zy}, \gamma_{xy} = \gamma_{yx}$. In the community of geotechnical and hydraulic engineering it is customarily to express six independent strain components in a vector form

$$\begin{aligned} \boldsymbol{\varepsilon} = \varepsilon_{\alpha\beta} = \{\varepsilon\} &= \left[\varepsilon_{xx}\ \varepsilon_{yy}\ \varepsilon_{zz}\ \gamma_{yz}\ \gamma_{zx}\ \gamma_{xy} \right]^{\mathrm{T}} \\ &= \left[\varepsilon_x\ \varepsilon_y\ \varepsilon_z\ \gamma_{yz}\ \gamma_{zx}\ \gamma_{xy} \right]^{\mathrm{T}} \end{aligned} \tag{3.84}$$

The strain-displacement relation under the condition of small deformation without rotation can be written in the following matrix form

$$\boldsymbol{\varepsilon} = \mathbf{L}\mathbf{u} \tag{3.85}$$

where $\mathbf{L}$ is a differential operator matrix given by

$$\mathbf{L} = \begin{bmatrix} \frac{\partial}{\partial x} & 0 & 0 \\ 0 & \frac{\partial}{\partial y} & 0 \\ 0 & 0 & \frac{\partial}{\partial z} \\ 0 & \frac{\partial}{\partial z} & \frac{\partial}{\partial y} \\ \frac{\partial}{\partial z} & 0 & \frac{\partial}{\partial x} \\ \frac{\partial}{\partial y} & \frac{\partial}{\partial x} & 0 \end{bmatrix} \tag{3.86}$$

It is reminded that for the strain tensor (3.84), there exist three "principle strains" $\varepsilon_i (i = 1, 2, 3)$ representing three uniaxial extensions or contractions along the mutually perpendicular axes corresponding to the "principal directions" in strain space.

(2) Stress

In engineering computation, it is prevalent to write six independent stress components in a vector form

$$\begin{aligned}\boldsymbol{\sigma} = \{\sigma\} &= \left[\sigma_{xx}\ \sigma_{yy}\ \sigma_{zz}\ \tau_{yz}\ \tau_{zx}\ \tau_{xy}\right]^{\mathrm{T}} \\ &= \left[\sigma_x\ \sigma_y\ \sigma_z\ \tau_{yz}\ \tau_{zx}\ \tau_{xy}\right]^{\mathrm{T}}\end{aligned} \tag{3.87}$$

As in the case of strains, it is possible to find three directions such that the shear stresses perpendicular to these directions are null, hence for the stress tensor (3.87), there exist three "principle stresses" $\sigma_i (i = 1, 2, 3)$ representing three uniaxial extensions or contractions along the mutually perpendicular axes corresponding to the "principal directions" in the stress space. It is reminded that for isotropic materials within the limit of linear elasticity, the "principal directions" of strain tensor and stress tensor are coincident.

(3) Invariants of strain and stress tensors and tensor resolutions

Towards the expression of a yield or/and failure criterion for rock-like materials, the following invariants with regard to strain and stress tensors are customarily employed (Lubliner 1990).

The invariants of strain tensor are defined as

$$\begin{cases} I_1' = \varepsilon_x + \varepsilon_y + \varepsilon_z = \varepsilon_1 + \varepsilon_2 + \varepsilon_3 \\ I_2' = -[\varepsilon_x\varepsilon_y + \varepsilon_y\varepsilon_z + \varepsilon_z\varepsilon_x + \frac{1}{4}(r_{xy}^2 + r_{yz}^2 + r_{zx}^2)] = -(\varepsilon_1\varepsilon_2 + \varepsilon_2\varepsilon_3 + \varepsilon_3\varepsilon_1) \\ I_3' = \varepsilon_x\varepsilon_y\varepsilon_z + \frac{1}{4}r_{xy}r_{yz}r_{zx} - \frac{1}{4}(\varepsilon_x r_{yz}^2 + \varepsilon_y r_{zx}^2 + \varepsilon_z r_{zy}^2) = \varepsilon_1\varepsilon_2\varepsilon_3 \end{cases} \tag{3.88}$$

in which $(\varepsilon_1, \varepsilon_2, \varepsilon_3)$ are the principal strains.

Further, the average strain ε_m (or volumetric strain $\varepsilon_v = 3\varepsilon_m$) and general shear strain $\overline{\gamma}$ defined by

$$\begin{cases} \varepsilon_m = I_1'/3 \\ \overline{\gamma} = \frac{2}{\sqrt{3}}\sqrt{J_2'} \end{cases} \tag{3.89}$$

are customarily used in the study of yield/failure criterion and corresponding constitutive relation in classical plasticity theory, particular for soils.

Which the help of strain invariants, a strain tensor may be resolved into spherical strain tensor $\boldsymbol{\varepsilon}_m$ and deviatoric strain tensor $e_{\alpha\beta}$ in a manner of

$$
\boldsymbol{\varepsilon} = \boldsymbol{\varepsilon}_m + \mathbf{e} = \begin{bmatrix} e_m & 0 & 0 \\ 0 & e_m & 0 \\ 0 & 0 & e_m \end{bmatrix} + \begin{bmatrix} \varepsilon_x - e_m & \frac{1}{2}\gamma_{xy} & \frac{1}{2}\gamma_{xz} \\ \frac{1}{2}\gamma_{yx} & \varepsilon_y - e_m & \frac{1}{2}\gamma_{yz} \\ \frac{1}{2}\gamma_{zx} & \frac{1}{2}\gamma_{zy} & \varepsilon_z - e_m \end{bmatrix}
$$

$$
= \begin{bmatrix} e_m & 0 & 0 \\ 0 & e_m & 0 \\ 0 & 0 & e_m \end{bmatrix} + \begin{bmatrix} e_x & e_{xy} & e_{xz} \\ e_{yx} & e_y & e_{yz} \\ e_{zx} & e_{zy} & e_z \end{bmatrix} \tag{3.90}
$$

The invariants of stress vector correspondent to that of strain are

$$
\begin{cases} I_1 = \sigma_x + \sigma_y + \sigma_z = \sigma_1 + \sigma_2 + \sigma_3 \\ I_2 = -\sigma_x\sigma_y - \sigma_y\sigma_z - \sigma_z\sigma_x + \tau_{xy}^2 + \tau_{yz}^2 + \tau_{zx}^2 = -\sigma_1\sigma_2 - \sigma_2\sigma_3 - \sigma_3\sigma_1 \\ I_3 = \sigma_x\sigma_y\sigma_z + 2\tau_{xy}\tau_{yz}\tau_{zx} - \sigma_x\tau_{yz}^2 - \sigma_y\tau_{zx}^2 - \sigma_z\tau_{xy}^2 = \sigma_1\sigma_2\sigma_3 \end{cases} \tag{3.91}
$$

in which $(\sigma_1, \sigma_2, \sigma_3)$ are the principal stresses.

Defining a hydrostatic stress $\sigma_m = p = \frac{1}{3}(\sigma_x + \sigma_y + \sigma_z) = \frac{1}{3}I_1$, the stress tensor resolution may be get from Eq. (3.90) as

$$
\boldsymbol{\sigma} = \boldsymbol{\sigma}_m + \mathbf{s} = \begin{bmatrix} \sigma_m & 0 & 0 \\ 0 & \sigma_m & 0 \\ 0 & 0 & \sigma_m \end{bmatrix} + \begin{bmatrix} \sigma_x - \sigma_m & \tau_{xy} & \tau_{xz} \\ \tau_{yx} & \sigma_y - \sigma_m & \tau_{yz} \\ \tau_{zx} & \tau_{zy} & \sigma_z - \sigma_m \end{bmatrix}
$$

$$
= \begin{bmatrix} \sigma_m & 0 & 0 \\ 0 & \sigma_m & 0 \\ 0 & 0 & \sigma_m \end{bmatrix} + \begin{bmatrix} s_{xx} & s_{xy} & s_{xz} \\ s_{yx} & s_{yy} & s_{yz} \\ s_{zx} & s_{zy} & s_{zz} \end{bmatrix} \tag{3.92}
$$

in which σ_m is the spherical stress and $s_{\alpha\beta}$ is the deviatoric stress tensor. It follows that there are three invariants w.r.t. the deviatoric stress vector

$$
\begin{cases} J_1 = 0 \\ J_2 = \frac{1}{6}[(\sigma_x - \sigma_y)^2 + (\sigma_y - \sigma_z)^2 + (\sigma_z - \sigma_x)^2 + 6(\tau_{xy}^2 + \tau_{yz}^2 + \tau_{zx}^2)] \\ \quad = \frac{1}{3}(I_1^2 + 3I_2) \\ J_3 = \frac{1}{27}(2I_1^3 + 9I_1I_2 + 27I_3) \end{cases} \tag{3.93}
$$

The hydrostatic stress p, the generalized shear stress q, and the Lode angle θ_σ are further defined as

$$
\begin{cases} \sigma_m = p = \frac{1}{3}(\sigma_x + \sigma_y + \sigma_z) = \frac{1}{3}I_1 \\ q = \sqrt{3J_2} \\ \theta_\sigma = \frac{1}{3}\arccos\left(\frac{3\sqrt{3}}{2}\frac{J_3}{J_2^{3/2}}\right) \end{cases} \tag{3.94}
$$

in which p and q are the counterparts of ε_m (or ε_v) and $\overline{\gamma}$ that are customarily employed in the study of yield/failure criterion and corresponding constitutive relation in classical plasticity theory, particular for soils. θ_σ is useful towards the study of yield surface on its π plane in the principal stress space.

3.4 Constitutive Relations

An important character of substance is its response to an applied action in the form of displacement or velocity, strain or strain rate, stress or stress rate, as well as their gradients. In Chap. 2, the basic physical and mechanical properties of fluid and solid have been summarized. However, they are mostly get from 1-D performances under the actions of uniaxial compression/tension/shear. Towards 2- or 3-D performances of continuum, multi-dimensional "yield/failure criteria" and "constitutive relations" are indispensable.

In continuum mechanics, "constitutive relation (equation, law)" or "equation of state" relates physical variables for a specific matter. They have everything to do with the response to a given force, and so to appreciate how or why one matter is different from another.

Although constitutive relations may be derived from the first principle in physics, yet most of them used in the CMs for geotechnical and hydraulic structures are simply phenomenological. They are combined with other governing equations (e.g. equilibrium, conservation, continuity) to solve mechanical/physical field problems. Taking the water percolation problem for instance, constitutive equation relates the seepage flow rate of water to the water head (hydraulic potential) gradient (e.g. Darcy's law). In the thermal conduction problem, it links the rate of heat flow with the gradient of temperature (e.g. Fourier's law). Towards fluids and solids, the connection between applied stresses/forces and resulted strains/deformations (e.g. Hooke's law) often comes under the term of "stress-strain relation".

3.4.1 Water

The constitutive equation of water is established with regard to pressure and shear respectively, similar to that of soils.

Since it is symmetric as usual in linear elasticity (Batchelor 1967), for compressible fluids the stress tensor can be expressed in terms of two scalar Lamé parameters—bulk viscosity λ and dynamic viscosity μ, hence it follows that

$$\begin{cases} \boldsymbol{\tau} = \lambda \mathrm{div}(\mathbf{v})\mathbf{I} + \mu\dot{\boldsymbol{\varepsilon}} \\ \dot{\boldsymbol{\varepsilon}} = \left[(\nabla\mathbf{v}^{\mathrm{T}})^{\mathrm{T}} + \nabla\mathbf{v}^{\mathrm{T}}\right] \\ \mathrm{div}(\mathbf{v}) = \nabla.\mathbf{v} = \frac{\partial v_x}{\partial x} + \frac{\partial v_y}{\partial y} + \frac{\partial v_z}{\partial z} \end{cases} \tag{3.95}$$

where: **I**—identity tensor.

Introducing the second viscosity related to the effect of volume viscosity (Landau and Lifshitz 1959)

$$\mu' = \lambda + \frac{2}{3}\mu \tag{3.96}$$

the constitutive equation becomes

$$\boldsymbol{\tau} = \mu' \text{div}(\mathbf{v})\mathbf{I} + \mu\dot{\boldsymbol{\varepsilon}} \tag{3.97}$$

$$\dot{\boldsymbol{\varepsilon}} = \left((\nabla \mathbf{v}^{\mathrm{T}})^{\mathrm{T}} + \nabla \mathbf{v}^{\mathrm{T}} - \frac{2}{3}\text{div}(\mathbf{v})\mathbf{I} \right) \tag{3.98}$$

The second viscosity μ' infers that the mechanical pressure is not equivalent to the thermodynamic pressure. But normally, this difference is neglected by assuming $\mu' = \mu$, sometimes even directly assuming $\mu' = 0$. With this simplification, the general constitutive Eq. (3.97) becomes

$$\boldsymbol{\tau} = \mu\dot{\boldsymbol{\varepsilon}} \tag{3.99}$$

It is notable that however, such a simplification could be influential in the issues of sound absorption/attenuation and shock wave.

The dynamic viscosity μ need not to be constant—in general, they depend on the density, pressure and temperature. Any equation that makes explicit one of these transport coefficients in the conservation variables is called as the "equation of state".

For incompressible and isotropic fluids, because $\text{div}(\mathbf{v}) = 0$, the constitutive Eq. (3.99) is budded as the "Stokes's stress constitutive equation" in term of deviatoric stress tensor

$$\begin{cases} \boldsymbol{\tau} = \mu\dot{\boldsymbol{\varepsilon}} \\ \dot{\boldsymbol{\varepsilon}} = \left[(\nabla \mathbf{v}^{\mathrm{T}})^{\mathrm{T}} + \nabla \mathbf{v}^{\mathrm{T}} \right] \end{cases} \tag{3.100}$$

Incompressibility rules out waves of density and pressure (e.g. sound and shock wave), so the incompressibility flow assumption is not applicable if these phenomena are of interest, although this assumption typically holds well with all fluids at low Mach numbers (say, up to about Mach 0.3).

For incompressible flows, the actual governing equation will result in prohibitive time-marching step lengths that are extremely small. This is true for all the CMs such as FDM, FEM and SPH. The fact that a theoretically incompressible flow is practically compressible leads to a concept of "artificial compressibility", which enables us to employ a quasi-incompressible equation of state for the purpose of producing the time derivative of pressure. For example, to model free surface water flows, Monaghan (1994) applied a state equation

$$p = B\left(\left(\frac{\rho}{\rho_0}\right)^{\gamma} - 1\right) \tag{3.101}$$

where: γ—constant, $\gamma = 7$ is exercised under most circumstances; ρ_0—reference density; B—problem dependent parameter setting a limit for the maximum change of density, under most circumstances B can be taken as the initial pressure, i.e. $B = p_0$ (Morris et al. 1997; Schlatter 1999).

From Eq. (3.101) it can be found that a small oscillation in density may give rise to a large variation in pressure. Another possible choice of the artificial equation of state is

$$p = c^2\rho \tag{3.102}$$

where: c—sound speed.

3.4.2 Rock-Like Materials

A comprehensive discussion on the constitutive relations for rock-like materials may be found in our precedent book (Chen 2018). For more detailed fundamental knowledge of constitutive theory, our readers are also referred to classical literatures (Oden 1972; Owen and Hinton 1980; Miyoshi 1985; Simo and Hughes 1988; Doltsinis 1989; Inoue et al. 1990; Lemaitre and Chaboche 1990; Lubliner 1990; Griffiths and Gioda 2000).

1. **Elasticity**

It is reminded that perfect elasticity is only an approximation of real-world solid, and few rock-like materials stay purely elastic even after very small deformations.

(1) **General form**

To be consistent with the theory of plasticity and viscoplasticity, elastic constitutive relations will be expressed in the form of increment. In addition, according to the prevalent convention in engineers, the matrix notation for the fourth-order tensor of elasticity $\mathbf{D}^e = [D^e]$, the vector notations for the second-order tensors of stress and strain increments $\Delta\boldsymbol{\sigma} = \{\Delta\sigma\}$ and $\Delta\boldsymbol{\varepsilon} = \{\Delta\varepsilon\}$, are employed.

The 6×6 elastic matrix $\mathbf{D}^e$ linearly links incremental stress and strain vectors by

$$\Delta\boldsymbol{\sigma} = \mathbf{D}^e\Delta\boldsymbol{\varepsilon}^e \tag{3.103}$$

Elastic matrix in Eq. (3.103) is normally considered as orthotropic symmetrical, namely, there exist three mutual perpendicular directions of symmetry, referred to as principle symmetry directions (e.g. rock with three perpendicularly joint sets). Under such circumstances the elastic matrix is specifically expressed by the inverse of compliance matrix $\mathbf{C}^e$, namely

$$\mathbf{D}^e = (\mathbf{C}^e)^{-1} \tag{3.104}$$

$$\mathbf{C}^e = \begin{bmatrix} \frac{1}{E_x} & -\frac{\nu_{xy}}{E_y} & -\frac{\nu_{xz}}{E_z} & 0 & 0 & 0 \\ -\frac{\nu_{yx}}{E_y} & \frac{1}{E_y} & -\frac{\nu_{yz}}{E_z} & 0 & 0 & 0 \\ -\frac{\nu_{zx}}{E_z} & -\frac{\nu_{zy}}{E_z} & \frac{1}{E_z} & 0 & 0 & 0 \\ 0 & 0 & 0 & \frac{1}{G_{yz}} & 0 & 0 \\ 0 & 0 & 0 & 0 & \frac{1}{G_{zx}} & 0 \\ 0 & 0 & 0 & 0 & 0 & \frac{1}{G_{xy}} \end{bmatrix} \tag{3.105}$$

$$\nu_{\alpha\beta} = \nu_{\beta\alpha}. \tag{3.106}$$

Nine independent constants in Eq. (3.105) are further reduced to five if the material is isotropic within a plane (e.g. transversely isotropic rock with laminated structure). This is normally the case where two types of rocks are regularly interlayered. It also comes about when flat minerals (e.g. mica, chlorite, graphite, serpentine, etc.) are arrayed in parallel orientation, or when long minerals (e.g. amphiboles) are oriented with their long axes randomly pointed within parallel planes.

The number of elastic constants is eventually reduced to two for isotropic material:

$$\mathbf{C}^e = \begin{bmatrix} \frac{1}{E} & -\frac{\nu}{E} & -\frac{\nu}{E} & 0 & 0 & 0 \\ -\frac{\nu}{E} & \frac{1}{E} & -\frac{\nu}{E} & 0 & 0 & 0 \\ -\frac{\nu}{E} & -\frac{\nu}{E} & \frac{1}{E} & 0 & 0 & 0 \\ 0 & 0 & 0 & \frac{1}{G} & 0 & 0 \\ 0 & 0 & 0 & 0 & \frac{1}{G} & 0 \\ 0 & 0 & 0 & 0 & 0 & \frac{1}{G} \end{bmatrix} \tag{3.107}$$

or

$$\mathbf{D}^e = \begin{bmatrix} \lambda + 2G & \lambda & \lambda & 0 & 0 & 0 \\ \lambda & \lambda + 2G & \lambda & 0 & 0 & 0 \\ \lambda & \lambda & \lambda + 2G & 0 & 0 & 0 \\ 0 & 0 & 0 & G & 0 & 0 \\ 0 & 0 & 0 & 0 & G & 0 \\ 0 & 0 & 0 & 0 & 0 & G \end{bmatrix} \tag{3.108}$$

in which shear modulus and Lamé coefficient are expressed by

$$\begin{cases} G = \frac{E}{2(1+\nu)} \\ \lambda = \frac{E\nu}{(1+\nu)(1-2\nu)} \end{cases} \tag{3.109}$$

In this book, only the two-dimensional formulation of representative CMs will be detailed, hence it is necessary to distinguish the plane stress formulation and plane strain formulation.

Plane stress state is defined by the condition that the stress vector is zero across a particular plane perpendicular to, for example, z-axis. This is often the case for thin plates, and as a result, stress analysis is considerably simplified because the stress tensor can be represented by a 2×2 matrix or a 3×1 vector. For isotropic material, it follows that

$$\mathbf{D}^e = \frac{E}{1-\nu^2}\begin{bmatrix} 1 & \nu & 0 \\ \nu & 1 & 0 \\ 0 & 0 & \frac{1-\nu}{2} \end{bmatrix} \tag{3.110}$$

$$\begin{cases} \Delta\boldsymbol{\sigma} = \begin{bmatrix} \Delta\sigma_x & \Delta\sigma_y & \Delta\tau_{xy} \end{bmatrix}^{\mathrm{T}} \\ \Delta\boldsymbol{\varepsilon} = \begin{bmatrix} \Delta\varepsilon_x & \Delta\varepsilon_y & \Delta\gamma_{xy} \end{bmatrix}^{\mathrm{T}} \end{cases} \tag{3.111}$$

It is reminded that the normal strain along z-axis is not zero and it may be calculated independently by

$$\Delta\varepsilon_z = -\frac{\nu}{E}(\Delta\sigma_x + \Delta\sigma_y) \tag{3.112}$$

If one dimension, for example along z-axis, is very large compared to the others, the principal strain in this direction is constrained in zero. This yields a plane strain condition, and the principal stress in the direction of the longest dimension can be disregarded for calculations allowing a two dimensional analysis of stress state (e.g. gravity dam monolith loaded by reservoir). The elastic matrix (3.110) may be used subject to the replacement operation of

$$\begin{cases} E \rightarrow \frac{E}{1-\nu^2} \\ \nu \rightarrow \frac{\nu}{1-\nu} \end{cases} \tag{3.113}$$

In addition, where the failure or yield criterion concerns the effect of intermediate principal stress σ_z, it may be calculated by

$$\Delta\sigma_z = \nu(\Delta\sigma_x + \Delta\sigma_y) \tag{3.114}$$

(2) Special cases

For a structural plane *sp* with thickness a_{sp} (e.g., fault, interlayer) (viz. Fig. 3.6a), its strain incremental vector will be contracted with only three components in the local Cartesian coordinate system (viz. Fig. 3.6b) such that

$$\Delta\boldsymbol{\varepsilon}_{sp} = [\,\Delta\gamma_{zx,sp}\ \Delta\gamma_{zy,sp}\ \Delta\varepsilon_{z,sp}\,]^{\mathrm{T}} = \frac{1}{a_{sp}}[\,\Delta u_{x,sp}\ \Delta u_{y,sp}\ \Delta u_{z,sp}\,]^{\mathrm{T}} \tag{3.115}$$

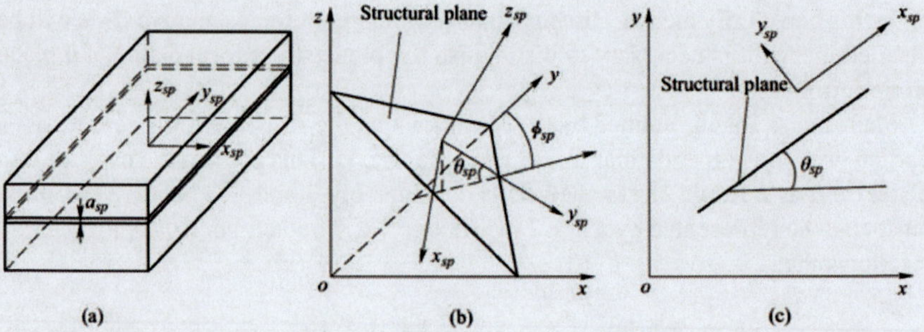

Fig. 3.6 Global and local Cartesian coordinate systems for structural plane. **a** Joint in rock and local Cartesian coordinate systems; **b** three-dimensional coordinate transformation; **c** two-dimensional coordinate transformation

$$\Delta\boldsymbol{\sigma}_{sp} = [\,\Delta\tau_{zx,sp}\ \Delta\tau_{zy,sp}\ \Delta\sigma_{z,sp}\,]^{\mathrm{T}} \tag{3.116}$$

in which a_{sp} is the aperture (thickness) of discontinuity *sp*.

For two-dimensional case (viz. Fig. 3.6c), Eqs. (3.115) and (3.116) become

$$\Delta\boldsymbol{\varepsilon}_{sp} = [\,\Delta\gamma_{yx,sp}\ \Delta\varepsilon_{y,sp}\,]^{\mathrm{T}} = \frac{1}{a_{sp}}[\,\Delta u_{x,sp}\ \Delta u_{y,sp}\,]^{\mathrm{T}} \tag{3.117}$$

$$\Delta\boldsymbol{\sigma}_{sp} = [\,\Delta\tau_{yx,sp}\ \Delta\sigma_{y,sp}\,]^{\mathrm{T}} \tag{3.118}$$

The elastic matrix $\mathbf{D}^e_{sp} = [D^e]_{sp}$ linearly linking above incremental stress and strain vectors is defined by

$$\Delta\boldsymbol{\sigma}_{sp} = \mathbf{D}^e_{sp}\Delta\boldsymbol{\varepsilon}^e_{sp} \tag{3.119}$$

$$\begin{cases} \mathbf{D}^e_{sp} = \begin{bmatrix} G_{sp} & 0 & 0 \\ 0 & G_{sp} & 0 \\ 0 & 0 & \lambda_{sp} + 2G_{sp} \end{bmatrix} & \text{for 3-D problem} \\ \mathbf{D}^e_{sp} = \begin{bmatrix} G_{sp} & 0 \\ 0 & \lambda_{sp} + 2G_{sp} \end{bmatrix} & \text{for 2-D problem} \end{cases} \tag{3.120}$$

where: G_{sp} and λ_{sp}—shear modulus and Lamé coefficients of the intercalated matter within the interlayer, MPa.

For a structural plane *sp* without thickness (e.g. joint), the normal and shear forces applied across it can be scaled by the nominal area of the joint to respectively define normal and shear stresses, which give rise to normal and shear displacements. These displacements are related to the corresponding stresses by various models and parameters extensively studied (Barton et al. 1985), of which the model initially proposed by Goodman and his co-workers (Goodman et al. 1968; Mahtab and Goodman 1970) is the most prevalent in the routine design of geotechnical and hydraulic structures. In the formulation of constitutive relation, the strain increment will be replaced by the displacement difference of two joint walls in the local Cartesian coordinate system (viz. Fig. 3.6) as follows.

$$\Delta \mathbf{u}_{sp} = [\,\Delta u_{x,sp}\ \Delta u_{y,sp}\ \Delta u_{z,sp}\,]^{\mathrm{T}} \tag{3.121}$$

The so called "Goodman joint element" model simply links the stress increments in conjugate with the relative displacement increments by the elastic matrix $\mathbf{D}^e_{sp}$

$$\Delta \boldsymbol{\sigma}_{sp} = \mathbf{D}^e_{sp} \Delta \mathbf{u}^e_{sp} \tag{3.122}$$

$$\begin{cases} \mathbf{D}^e_{sp} = \begin{bmatrix} k_{s,sp} & 0 & 0 \\ 0 & k_{s,sp} & 0 \\ 0 & 0 & k_{n,sp} \end{bmatrix} & \text{for 3-D problem} \\ \mathbf{D}^e_{sp} = \begin{bmatrix} k_{s,sp} & 0 \\ 0 & k_{n,sp} \end{bmatrix} & \text{for 2-D problem} \end{cases} \tag{3.123}$$

where: $k_{n,sp}$ and $k_{s,sp}$—coefficients of normal stiffness and tangential (shear) stiffness of structural plane *sp*, MPa/m.

There exist several nonlinear stiffness models relating joint compression and shear stresses with the closure and shear deformations of joint walls. Among various approximations, Chen et al. (1989) had established a unified stiffness and permeability model which looks at the contacted asperities on joint walls as a thin layer of evenly "filled" virtual medium (viz. Chap. 2).

In CMs, the transformation w.r.t. the displacement and velocity between global and local coordinate systems (Cartesian) may be demanded, which is defined as follows

$$\begin{cases} \mathbf{x}_{sp} = \mathbf{T}_{sp}\mathbf{X} \\ \mathbf{X} = \mathbf{T}^{\mathrm{T}}_{sp}\mathbf{x}_{sp} \end{cases} \tag{3.124}$$

where

$$\begin{cases} \mathbf{T}_{sp} = \begin{bmatrix} \cos\phi_{sp} & -\sin\phi_{sp} & 0 \\ \sin\phi_{sp}\cos\theta_{sp} & \cos\phi_{sp}\cos\theta_{sp} & -\sin\theta_{sp} \\ \sin\phi_{sp}\sin\theta_{sp} & \cos\phi_{sp}\sin\theta_{sp} & \cos\theta_{sp} \end{bmatrix} & \text{for 3 - D problem} \\ \mathbf{T}_{sp} = \begin{bmatrix} \cos\theta_{sp} & \sin\theta_{sp} \\ -\sin\theta_{sp} & \cos\theta_{sp} \end{bmatrix} & \text{for 2 - D problem} \end{cases} \tag{3.125}$$

It is noted that Eqs. (3.121)–(3.125) are valid for concrete crack, too.

2. **Plasticity**

The most advanced constitutive laws should retain the strain history of materials in all parts of the model. Sedimentary rocks, in particular, can undergo spectacular increases in strength as they age, largely due to a decreasing volume during consolidation (tracked by porosity), but also due to physical and chemical changes. In addition, concrete also exhibits age-dependent properties, particularly at its early age due to hydration process.

Materials for which a permanent deformation remains after loading and unloading cycle are called plastic materials (Hill 1998). Many materials exhibit an elastic behavior up to their yield stress, then undergo plastic deformation when loaded beyond the yield stress, hence they are dubbed as "elasto-plastic materials". As will be described in more detail below, geo-materials are often modeled by the laws with pressure sensitive elasto-plasticity.

Mathematically, a traditional plastic model needs three major ingredients:

- A yield function or criterion for the onset of plastic deformation;
- A flow rule and plastic potential function to determine plastic strain rate; and
- A hardening/softening law that governs the evolution of yield function.

Perzyna (1963, 1966) introduced a viscosity coefficient that is temperature and time dependent. His phenomenological formulation was supported by the thermodynamics of irreversible processes. The ideas presented in his works have been the basis for the most subsequent researches in rate-dependent viscoplasticity. By Perzyna formulation, if an elasto-viscoplastic computation reaches a stationary condition w.r.t. displacements, the solution to an equivalent elasto-plasticity problem is obtained. The elasto-viscoplastic approach thus provides an alternative technique to solve elasto-plastic problems, which is found to possess considerable merits over other iterative processes (Zienkiewicz and Cormeau 1972, 1974). In particular, non-associated flow and strain softening rules can be handled with rather well without requiring specific numerical artifices. Hence, Perzyna formulation will be employed throughout this book, in which the total strain increment appearing in a loaded material is divided into elastic and viscoplastic ports such that

$$\Delta\boldsymbol{\varepsilon} = \Delta\boldsymbol{\varepsilon}^{e} + \Delta\boldsymbol{\varepsilon}^{vp} \tag{3.126}$$

in which the viscoplastic strain increment is elaborated with the help of yield and load functions, flow rule, and hardening/softening law.

(1) **Yield criteria**

Strength theory (inclusive yield criterion and failure criterion), as one of the most important components in constitutive relations, has been widely exercised in various CMs.

From the uniaxial and direct shear tests illustrated in Chap. 2, it is justified that a yield or ultimate stress for 1-D problem can readily be determined. What if, however, there are bi-axial or tri-axial stresses exerting on a sample? Towards this question it is necessary to extend the definition from the uniaxial concept of yield (failure) stress to a general 3-D state of yield (or failure) criterion (Michino and Findley 1976; Brebbia 1985; Yu and Li 2012).

"Yield criterion" is a function of strain/stress state and material parameters defined either in the strain or stress space, the latter is more prevalent, for deciding what combination of multi-axial stresses will bring about yield. It is expressed mathematically by a "yield function" F being equal to zero, i.e. $F = 0$. The material will perform elasticity where the yield function $F < 0$. The suitability of any proposed yield criteria must be validated by experiments and preferably, be supported by clear physical mechanism.

Yield criteria are usually in the function forms of principal stresses or stress invariants

$$\begin{cases} F(\sigma_1, \sigma_2, \sigma_3) = 0 \\ F(I_1, I_2, I_3) = 0 \\ F(I_1, J_2, J_3) = 0 \\ F(\sigma_m, J_2, \theta_\sigma) = 0 \\ F(p, q, \theta_\sigma) = 0 \end{cases} \tag{3.127}$$

A visual representation termed as the "yield surface" may be constructed in the principal stress space $(\sigma_1, \sigma_2, \sigma_3)$ using yield criterion. Inside the yield surface the material is elastic whereas reaching the surface means the initiation of plastic deformation. It is physically impossible for a material to stay, but may temporarily go, beyond its yield surface.

For an isotropic material, yield function F should be invariant in the interchange operator w.r.t. principal coordinate axes, this means that the yield function is a symmetric function of principal stresses $(\sigma_1, \sigma_2, \sigma_3)$, namely, the yield function is tri-fold symmetric.

Nowadays, there have been numerous isotropic yield (failure) criteria available for rock and concrete normally regarded as cohesive-frictional materials. Most geologists are familiar with the Mohr-Coulomb failure criterion. This model, first proposed to predict soil strength by Coulomb in 1773, defines the shear strength of material as the linear function of the effective normal stress acting on the failure surface. Today, it is widely used for describing the yield and failure of rock-like materials.

In the principal stress space, the "Mohr-Coulomb (M-C)" criterion is expressed as

$$F = \frac{1}{3} I_1 \sin\varphi + \left(\cos\theta_\sigma - \frac{1}{\sqrt{3}} \sin\theta_\sigma \sin\varphi\right)\sqrt{J_2} - c\cos\varphi = 0 \tag{3.128}$$

in which φ and c are the internal friction angle and cohesion; the Lode angle θ_σ varies within $-\frac{\pi}{6} \le \theta_\sigma \le \frac{\pi}{6}$.

Equation (3.128) may be expressed in an alternative form

$$F = p\sin\varphi + \frac{1}{\sqrt{3}} q\left(\cos\theta_\sigma - \frac{1}{\sqrt{3}} \sin\theta_\sigma \sin\varphi\right) - c\cos\varphi = 0 \tag{3.129}$$

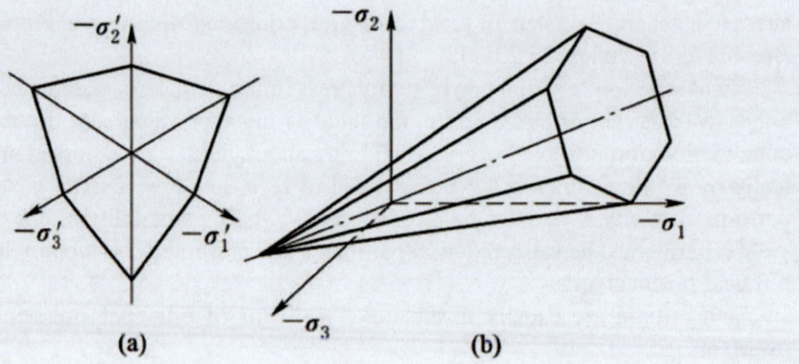

Fig. 3.7 Mohr-Coulomb yield surface. **a** On π-plane; **b** in principal stress space

in which q and p are defined in Eq. (3.94).

The M-C yield surface in the principal stress space circumscribes an irregular hexagonal prism around the hydrostatic axis (viz. Fig. 3.7).

This criterion has been equipped with different flow rules and hardening/softening laws. The main weakness of M-C model, compared to more sophisticated constitutive laws, is its inability to capture compact plastic deformation. The corners within the yield surface in three dimensions, however, make it less straightforward to implement. Moreover, it does not account for the influence of intermediate principal stress (Jaeger et al. 2007; Yu and Li 2012).

The M-C criterion with tension cut-off also may be employed to describe the yield of structural plane sp using the joint element model without thickness (viz. Fig. 3.6)

$$\begin{cases} F_{sp} = (\tau_{zx,sp}^2 + \tau_{zy,sp}^2)^{1/2} + \sigma_{z,sp} tg\varphi_{sp} - c_{sp} = 0 \text{ for } \sigma_{z,sp} - \sigma_{T,sp} < 0 \\ F_{sp} = \sigma_{z,sp} - \sigma_{T,sp} = 0 \qquad\qquad\qquad\qquad \text{for } \sigma_{z,sp} - \sigma_{T,sp} \geq 0 \end{cases} \tag{3.130}$$

in which c_{cp}, φ_{cp} and $\sigma_{T,sp}$ are the cohesion, friction angle, and tension strength, respectively.

Drucker and Prager (1952) proposed a generalized von Mises criterion abbreviated as "D-P criterion" in which the material friction is taken into account. It is a two invariant failure criterion considering deviatoric and mean effective stresses in a manner of

$$F = aI_1 + \sqrt{J_2} - k = 0 \tag{3.131}$$

$$\begin{cases} a = \sin\varphi/[3(3 + \sin^2\varphi)]^{1/2} \\ k = \sqrt{3}c\cos\varphi/\sqrt{3 + \sin^2\varphi} \end{cases} \tag{3.132}$$

in which φ and c are the internal friction angle and cohesion.

This two-invariant yield surface is a smooth circular cone centered on the isotropic line in the principal stress space (viz. Fig. 3.8), as opposed to the angular yield surface

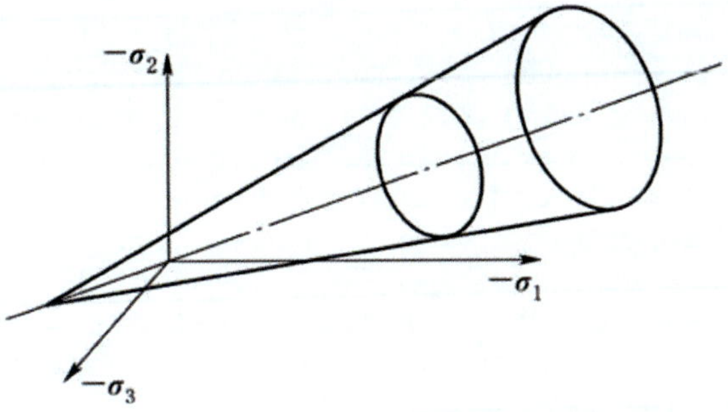

Fig. 3.8 Drucker-Prager yield surface

for M-C in Fig. 3.7. This single smooth function is differentiable everywhere due to the absence of corners.

D-P criterion is widely implemented in many commercial FEM codes. Although substantial departures from the prediction that would give a rather poor approximation to the real failure load, had been observed (Humpheson and Naylor 1975; Zienkiewicz and Pande 1977; Chen and Baladi 1985), yet this criterion are prevalent in the computation for geotechnical and hydraulic structures in China due to its simplicity, well defined and easily accessible parameters, and engineering experiences.

(2) **Loading functions**

One consequence of yield is that as plastic deformation proceeds, an increase in stress is needed to push additional plastic strain. This is known as "strain/work hardening". Plasticity materials with hardening necessitate increasingly higher stresses to get further plastic deformation. On the contrary, "perfect plasticity" is a property of material to undergo irreversible deformation without any increase in stress or load.

Except perfect plasticity materials, the yield surface, as the plastic work increases, expands or shrinks until the current stress point is reached. Under such circumstances, the "loading function" ψ is defined as the subsequent yield function criterion

$$\psi(\sigma_{\alpha\beta}, H\alpha) = 0 \tag{3.133}$$

in which H_α is "internal variable", and plastic work $w^p = \int \boldsymbol{\sigma}^{\mathrm{T}} \mathrm{d}\boldsymbol{\varepsilon}^p$ is more commonly used to construct H_α.

If

$$H_\alpha \equiv 0 \tag{3.134}$$

the material is "ideal plasticity" or "perfect plasticity".

(3) **Potential functions and flow rules**

Flow plasticity is a theory characterized by the assumption that a flow rule exists to determine the amount of plastic deformation. For the hardening material, the work done in a loading-unloading cycle is positive or zero, which is called the "Drucker stability postulation". Although it eliminates the possibility of strain softening behavior that actually exists, yet if applicable, the "normality condition"—associated flow rule that the plastic strain increment is normal to the yield surface, may be inferred. The use of associated flow rule is a cornerstone in the so-called mathematical theory of plasticity which was formulated around 1950 by, among others, Hill, Drucker and Prager. This mathematical theory of plasticity contains some very attractive results such as the necessity of convex yield surface and the existence of limit theorem which have been exercised extensively in engineering computations (Lin and Bažant 1986).

Experimental data, however, indicate that the associated flow rule often fails to describe the plastic response of rock-like materials that exhibit shear dilation characterized by nonlinear volume change and associated with shear distortion, namely, they contract at low loading levels and dilate at higher loading levels. Under such circumstances, some non-associated flow rules in which the loading (yield) and plastic potential functions are not identical, would be employed. Nowadays, associated and non-associated plastic strain flow rules are well-established concepts (Pivonka et al. 2003). The assumption of generalized associativity (or generalized normality) further relates the flow directions of all internal variables to a given loading surface (Rockafellar 1969; Jirásek and Bažant 2001; Frémond 2002).

Towards the plastic strain increment, we can, without loss of generality, assume that it can be derived from a plastic potential Q as

$$\Delta \boldsymbol{\varepsilon}^p = \lambda \nabla_\sigma Q = \lambda \frac{\partial Q}{\partial \boldsymbol{\sigma}} \tag{3.135}$$

$$\lambda = \frac{1}{A}(\nabla_\sigma \psi)\Delta \boldsymbol{\sigma} \tag{3.136}$$

in which λ is a positive scalar of proportionality and $Q = \psi$ giving rise to the so called associated flow rule.

(4) **Constitutive equations**

In CMs, we usually apply a load increment to produce a displacement increment, and thus a total strain increment. Afterwards, the stress increment can be determined by the constitutive relation as follows

$$\begin{cases} \Delta \boldsymbol{\sigma} = (\mathbf{D}^e - \mathbf{D}^p)\Delta \boldsymbol{\varepsilon} = \mathbf{D}^{ep}\Delta \boldsymbol{\varepsilon} \\ \mathbf{D}^p = \frac{\mathbf{D}^e(\nabla_\sigma Q)(\nabla_\sigma \psi)^{\mathrm{T}}\mathbf{D}^e}{A+(\nabla_\sigma \psi)^{\mathrm{T}}\mathbf{D}^e(\nabla_\sigma Q)} \end{cases} \tag{3.137}$$

in which $\mathbf{D}^p$ is "plastic matrix" and $\mathbf{D}^{ep}$ is "elasto-plastic matrix".

The elasto-plastic constitutive Eq. (3.137) was firstly derived for FEM by Zienkiewicz et al. (1969) that defines the stress increment uniquely once the total strain increment and the current state of stress are known, on the contrary, the strain increment cannot be determined uniquely on the basis of stress increment, i.e. $\mathbf{D}^{ep}$ is singular.

It also should be emphasized that the positivity of $\mathbf{D}^{ep}$ is only guaranteed for hardening materials. Following a material approaches to softening, its elasto-plastic constitutive law presents a descending branch which implies a local instability of the material. This local instability can progressively lead to irreversible strain localization, and eventually to global instability. When this behavior is introduced in a standard time-independent (e.g. elasto-plastic) constitutive relation, some physically meaningless consequences arise for both initial and boundary value problems (Bažant et al. 1984), which in turn, causes the ill-posed algorithm. When solved by means of CMs (e.g. FEM), such an ill-posed algorithm generally shows numerical instability and mesh dependence of the solution. This behavior may also emerge in perfectly plastic materials in the presence of non-associated flow rule (Rudnicki and Rice 1975). Associated to the formation of concentrated strain zones of limited thickness within the structures, the softening materials should be carefully handled with by taking into account of the conditions w.r.t. the onset, the process of spread, and the coalescence of localization zones (Ortiz et al. 1987; Leroy and Ortiz 1989; Zienkiewicz et al. 1995; Desai and Toth 1996; Chen et al. 2007).

The classical elasto-plastic theory treats the solid as "black box", which leads to the difficulties in incorporating the internal structure of material into the constitutive equation. Towards the improvement, generalized plasticity theories (Eisenberg and Phillips 1971; Zienkiewicz and Mróz 1984; Lubliner et al. 1993), endochronic models (Valanis 1971; Bažant and Krizek 1976; Watanabe and Atluri 1986; Auricchio and Taylor 1995; Erlicher and Point 2006), continuum damage mechanics (CDM) (Rice 1971; Dragon and Mróz 1979; Kachanov 1982a, b, c; Krajcinovic 1983; Oda et al. 1984; Pijaudier-Cabot and Bažant 1987; Kawamoto et al. 1988; Prat and Bažant 1997; Lemaitre 1990), etc., offer the appealing frameworks for rock-like materials because they possess indicators of internal structures.

3. **Viscoplasticity**

According to the principle of Perzyna formulation, at time t, the implicit constitutive equation will take the form of (Owen and Hinton 1980)

$$\begin{aligned}\Delta\boldsymbol{\sigma}(t) &= \overline{\mathbf{D}}(t)(\Delta\boldsymbol{\varepsilon}(t) - \dot{\boldsymbol{\varepsilon}}^{vp}(t)\Delta t) = \overline{\mathbf{D}}(t)(\Delta\boldsymbol{\varepsilon}(t) - \Delta\boldsymbol{\varepsilon}^{vp}(t)) \\ &= \overline{\mathbf{D}}(t)\Delta\boldsymbol{\varepsilon}(t) - \Delta\boldsymbol{\sigma}^{vp}(t)\end{aligned} \tag{3.138}$$

or

$$\Delta\boldsymbol{\varepsilon}(t) = \left(\overline{\mathbf{D}}(t)\right)^{-1}\Delta\boldsymbol{\sigma}(t) + \dot{\boldsymbol{\varepsilon}}^{vp}(t)\Delta t = \left(\overline{\mathbf{D}}(t)\right)^{-1}\Delta\boldsymbol{\sigma}(t) + \Delta\boldsymbol{\varepsilon}^{vp}(t) \tag{3.139}$$

in which the implicit elasticity matrix $\overline{\mathbf{D}}(t)$ and elasto-viscoplastic strain rate $\dot{\boldsymbol{\varepsilon}}^{vp}(t)$ are expressed by

$$\begin{cases} \overline{\mathbf{D}}(t) = ((\mathbf{D}^{\mathbf{e}})^{-1} + \mathbf{C}(t))^{-1} \\ \mathbf{C}(t) = \Theta \Delta t \mathbf{H}(t) \\ \mathbf{H}(t) = \frac{\partial \dot{\boldsymbol{\varepsilon}}^{vp}(t)}{\partial \boldsymbol{\sigma}(t)} = \nabla_{\boldsymbol{\sigma}} \dot{\boldsymbol{\varepsilon}}^{vp}(t) \\ \dot{\boldsymbol{\varepsilon}}^{vp}(t) = \gamma . H_1(\psi) . \psi . (\nabla_{\boldsymbol{\sigma}} Q) \\ \Delta \boldsymbol{\varepsilon}^{vp}(t) = \dot{\boldsymbol{\varepsilon}}^{vp}(t) \Delta t \end{cases} \tag{3.140}$$

where: Δt—time-marching step length, s; $\mathbf{H}(t)$—implicit matrix; Θ—implicit parameter; γ—fluidity parameter, 1/(MPa s).

$H_1(\psi)$ is a kind of "Heaviside jump function" defined as

$$H_1(\psi) = \begin{cases} 1 \text{ for } \psi > 0 \\ 0 \text{ for } \psi \leq 0 \end{cases} \tag{3.141}$$

in which ψ is the loading function. For perfect plasticity materials, it is identical to the yield function F, i.e. $\psi = F$.

It is particularly noted that for the structural plane without thickness, the strain increment in Eq. (3.138) should be replaced by the relative displacement increment $\Delta \mathbf{u}$, namely

$$\Delta \boldsymbol{\sigma}(t) = \overline{\mathbf{D}}(t)(\Delta \mathbf{u}(t) - \dot{\mathbf{u}}^{vp}(t)\Delta t) = \overline{\mathbf{D}}(t)(\Delta \mathbf{u}(t) - \Delta \mathbf{u}^{vp}(t)) = \overline{\mathbf{D}}(t)\Delta \mathbf{u}(t) - \Delta \boldsymbol{\sigma}^{vp}(t) \tag{3.142}$$

or

$$\Delta \mathbf{u}(t) = \left(\overline{\mathbf{D}}(t)\right)^{-1} \Delta \boldsymbol{\sigma}(t) + \dot{\mathbf{u}}^{vp}(t)\Delta t = \left(\overline{\mathbf{D}}(t)\right)^{-1} \Delta \boldsymbol{\sigma}(t) + \Delta \mathbf{u}^{vp}(t) \tag{3.143}$$

in which the implicit elasticity matrix $\overline{\mathbf{D}}$ and elasto-viscoplastic strain rate $\dot{\mathbf{u}}^{vp}(t)$ are expressed by

$$\begin{cases} \overline{\mathbf{D}}(t) = ((\mathbf{D}^{e})^{-1} + \mathbf{C}(t))^{-1} \\ \mathbf{C}(t) = \Theta \Delta t \mathbf{H}(t) \\ \mathbf{H}(t) = \nabla_{\boldsymbol{\sigma}} \dot{\mathbf{u}}^{vp}(t) \\ \dot{\mathbf{u}}^{vp}(t) = \gamma . H_1(\psi) . \psi . (\nabla_{\boldsymbol{\sigma}} Q) \\ \Delta \mathbf{u}^{vp}(t) = \dot{\mathbf{u}}^{vp}(t) \Delta t \end{cases} \tag{3.144}$$

The implicit matrix in Eqs. (3.140) and (3.144) may be uniformly written as

$$\begin{cases} \mathbf{H}(t) = \gamma [\psi (\nabla_{\sigma} \mathbf{q}(t))^{\mathrm{T}} + \mathbf{a}(t) \mathbf{q}^{\mathrm{T}}(t)] \\ \mathbf{a}^{\mathrm{T}}(t) = \nabla_{\sigma} \psi \\ \mathbf{q}^{\mathrm{T}}(t) = \nabla_{\sigma} Q \end{cases} \tag{3.145}$$

If $\Theta = 1$ in Eqs. (3.140) and (3.144), we obtain a "fully implicit" or "backward" time-marching step scheme with the strain increment being determined from the strain rate corresponding at t, i.e. the end of time-marching step. $\Theta = 0.5$ yields "implicit trapezoidal" scheme, and $\Theta \geq 0.5$ guarantees a time-marching step scheme of unconditionally stable. It is emphasized that the numerically stability of time-marching step scheme does not guarantee the accuracy of solution. To keep the balance between the time-marching accuracy and the computation efforts, the adaptive time-marching technique based on the error estimation may be helpful (Zhang and Chen 1996, 1997).

On the other hand, $\Theta = 0$ gives "fully explicit" or "forward stepping" scheme, where

$$\overline{\mathbf{D}} = \mathbf{D}^e \tag{3.146}$$

Attributable to the fact that the strain increment is completely determined from the strain rate corresponding to the start of time-marching step, the explicit scheme is the simplest. However, it is conditionally stable and the numerical time integration can only be proceeded for a time interval Δt smaller than some critical values (Owen and Hinton 1980).

If the fluidity parameter γ may be evaluated by experiments or in-situ back analyses, then both the history and steady-state concerning the deformation and stress of engineering structure may be obtained. However, in many cases it is not easy to get the appropriate fluidity parameter or it is thought that only the elasto-plastic solution is of interest. Under such circumstances, we can simply assume that $\gamma = 1$, in this manner the calculated history are not applicable, but the steady-state w.r.t. deformation and failure are identical to the corresponding elasto-plastic solution (Zienkiewicz and Cormeau 1974), subject to appropriate time-marching step scheme (implicit or explicit) and step length to guarantee the computation accuracy and to avert numerical escalation.

It is also noted that hereinafter the time notation t will be ignored in most cases if there is no risk of misleading.

3.5 Dynamics of Continuum: Strong Forms

3.5.1 Concepts

Depending on the specific circumstances, different forms of governing equations can be employed in the dynamics of continuum. As having been discussed previously, there are basically two formalisms—Eulerian formalism and Lagrangian formalism, for describing the physical laws. Hereinafter in this book (Chaps. 5–11), only the Lagrangian formalism will be employed in the elaboration of representative advanced CMs.

The PDEs accompanied with boundary and initial equations are normally looked at as the strong form of governing equations for continuum. Obtaining the exact (analytical, closed-form) solution for a governing equation of strong form is ideal but, unfortunately, it is usually very difficult for practical engineering problems. The "finite difference method (FDM)", which uses the finite differential representation (Taylor series) of the field function and its derivatives in a local domain, can be used to solve strong form PDE to obtain an approximated solution. However, the classical FDM requires a regular grid, hence it usually works only for problems with simple geometry and boundary conditions. One improvement for approximately solving system equations of strong form is to use arbitrary grids based on Taylor series expansions and least square (or moving least square) techniques. Such an improved formulation is very simple but less stable, and the accuracy of solution often depends on the selection of nodes for constructing differential approximation.

3.5.2 *Finite Volume and Infinitesimal Cell*

In vector calculus, the "divergence theorem", also known as the "Gauss theorem" or "Ostrogradsky's theorem" (Katz 1979), relates the flow (flux) of a vector field through a enclosed surface to the behavior of this vector field inside the surface. It is widely employed to transform the volume integral into a surface integral. More precisely, the divergence theorem states that: the outward flux of a vector field through a closed surface is equal to the volume integral of its divergence over the region inside the surface. Intuitively, it also means that the sum of all sources gives the net flux out of (or into) a region. The theorem was first discovered by Lagrange in 1762 whose article is reprinted by Serret (1867), and independently rediscovered by Gauss in 1813. In 1861 Ostrogradsky gave the first proof.

Suppose the domain Ω with volume V is a compact subset of R^3 enclosed by the piecewise smooth boundary Γ with area S (also indicated as $\partial\Omega = \Gamma$). If $\mathbf{F}$ is a continuously differentiable vector field defined on V, then the Gauss theorem is expressed as (Spiegel et al. 2009)

$$\oiiint_{\Omega} \nabla.\mathbf{F}\mathrm{d}V = \oiint_{\Gamma} \mathbf{F}.\mathbf{n}\mathrm{d}\Gamma \tag{3.147}$$

in which the left hand side is an integral over the volume V, the right hand side is the integral over the boundary area S, ∇ is Hamilton operator and the notation "." between two vectors means "dot product", $\mathbf{n}$ is the outward unit normal vector of $\mathrm{d}\Gamma$.

It should be noted that, although in two-dimension space there is an similar counterpart called "Green's theorem" or "Green formula", yet Eq. (3.147) may be used in its dimension-reduction form, too

$$\oiint_{\Omega} \nabla.\mathbf{F}\mathrm{d}\Omega = \oint_{\Gamma} \mathbf{F}.\mathbf{n}\mathrm{d}\Gamma \tag{3.148}$$

in which the left hand side is an integral over the domain Ω of unit thickness, the right hand side is the integral over the boundary line Γ enclosing Ω.

In continuum mechanics, there are a set of equations that describe the conservation of mass, energy, momentum, or other quantities. The divergence theorem states that any of them can be written either in the integral form (in term of flux) or in the differential form (in term of divergence).

In Lagrangian formalism, the matter volume moves such that the same matter is always staying inside volume V (viz. Fig. 3.3). Therefore, although the matter may be expanded, compressed and sheared, yet the mass contained in the matter volume remains unchanged. This volume may be reasonably large with finite dimensions in the continuum and the conservation laws can be directly applied to the matter inside V. Applying the conservation laws to such a finite matter volume can result in the governing equations of integral form.

Another approach in Lagrangian formalism to obtain governing equations makes use of infinitesimal matter cell (parcel). The infinitesimal cell can be regarded as a very small clump of matter associated with a very small volume ΔV and a very small surface ΔS surrounding ΔV. This infinitesimal cell, on one hand, is large enough so that the assumptions of continuum are valid, on the other hand, is small enough so that the field property inside can be regarded as homogeneous throughout the entire cell. The infinitesimal cell can move along a streamline with a velocity vector $\mathbf{v} = [\, v_x \; v_y \; v_z \,]^{\mathrm{T}}$. Applying the conservation laws to this cell, the governing equations in the form of PDEs can be established.

3.5.3 D'Alembert Principle for Continuum

Specification of the force in D'Alembert principle (3.1) for continuum is proceeded in two steps: the first step is to manipulate the stress tensor acting on any face of cell (parcel); the second step is to specify the stress components.

Equation (3.33) shows that any parcel in a velocity field with initial time and position will be found at some later time and updated position. If our goal is to measure the force applied to the matter, then by tracking these parcels in time, it may be straightforwardly to get the acceleration $\frac{\partial^2 \mathbf{x}}{\partial t^2}$ according to the classical dynamics of particle, i.e.

$$\mathbf{F}_\sigma + \mathbf{F}_{damping} + \mathbf{F} = \rho \frac{\partial^2 \mathbf{x}}{\partial t^2} \tag{3.149}$$

where: $\mathbf{F}$—vector of external body (active) forces; $\mathbf{F}_\sigma$—inner (passive) force due to stress gradient; $\mathbf{F}_{damping}$—inner (passive) force due to damping; ρ—density of substance, in most geotechnical and hydraulic engineering issues, it remains nearly constant, i.e. $\rho = \rho_0$.

The external surface (active) forces, if any, are handled by the definition of boundary conditions. Equation (3.149) are important and common in most cases of solid and fluid dynamics. For infinitesimal deformation issues, or the UL is employed for the solution of finite deformation issues, Eq. (3.149) may be rewritten in which displacements **u** (or further their increments $\Delta\mathbf{u}$) are independent variables

$$\mathbf{F}_\sigma + \mathbf{F}_{damping} + \mathbf{F}_{body} = \rho\frac{\mathrm{d}^2\mathbf{u}}{\mathrm{d}t^2} \tag{3.150}$$

in which Lagrangian acceleration $\frac{\mathrm{d}^2\mathbf{u}}{\mathrm{d}t^2}$ may be denoted as $\ddot{\mathbf{u}}$ or $\dot{\mathbf{v}}$.

The D'Alembert principle (3.150) for continuum is equivalent to the momentum conservation law.

1. **Inner forces due to stress gradients**

For the infinitesimal cubic element (cell) representing a matter parcel in continuum that is under the action of active forces and restraint forces (viz. Fig. 3.9), its equilibrium conditions of forces

$$\sum F_x = 0, \tag{3.151}$$

$$\sum F_y = 0 \tag{3.152}$$

$$\sum F_z = 0 \tag{3.153}$$

lead to "Navier equations"

$$\mathbf{F}_\sigma = \nabla\boldsymbol{\sigma} \tag{3.154}$$

where ∇ is the Hamilton (gradient) operator. Equation (3.154) may be written in the component form along the coordinate axes of x, y and z

$$\begin{cases} F_{\sigma_x} = \frac{\partial\sigma_{xx}}{\partial x} + \frac{\partial\tau_{yx}}{\partial y} + \frac{\partial\tau_{zx}}{\partial z} \\ F_{\sigma_y} = \frac{\partial\tau_{xy}}{\partial x} + \frac{\partial\sigma_{yy}}{\partial y} + \frac{\partial\tau_{zy}}{\partial z} \\ F_{\sigma_z} = \frac{\partial\tau_{xz}}{\partial x} + \frac{\partial\tau_{yz}}{\partial y} + \frac{\partial\sigma_{zz}}{\partial z} \end{cases} \tag{3.155}$$

The stress tensor $\boldsymbol{\sigma}$ is related to the strain tensor $\boldsymbol{\varepsilon}$ (in solid dynamics) or its rate $\dot{\boldsymbol{\varepsilon}}$ (in fluid dynamics) via the constitutive equation or state equation. $\boldsymbol{\varepsilon}$ or $\dot{\boldsymbol{\varepsilon}}$ is, in turn, related to the gradient of displacement vector **u** or velocity vector **v**.

(1) **Fluids**

$\mathbf{F}_\sigma$ in Eq. (3.154) is normally decomposed into two portions corresponding to Eqs. (3.75) and (3.76).

$$\mathbf{F}_\sigma = \mathbf{F}_{pressure} + \mathbf{F}_{vis\cos ity} = \nabla\mathbf{p} + \nabla\boldsymbol{\tau} \tag{3.156}$$

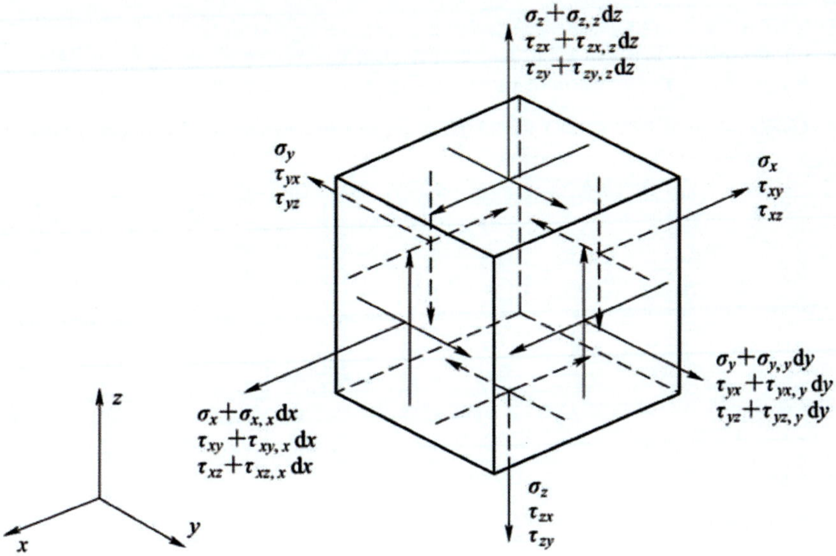

Fig. 3.9 Equilibrium of a cubic element

in which

$$\mathbf{F}_{pressure} = -\nabla p = -\begin{bmatrix} \frac{\partial p}{\partial x} & \frac{\partial p}{\partial y} & \frac{\partial p}{\partial z} \end{bmatrix}^{\mathrm{T}} \tag{3.157}$$

$$\mathbf{F}_{viscous} = \nabla \boldsymbol{\tau} = \mu \nabla \left(\left[(\nabla \mathbf{v}^{\mathrm{T}})^{\mathrm{T}} + \nabla \mathbf{v}^{\mathrm{T}} \right] - \frac{2}{3} \mathrm{div}(\mathbf{v}) \mathbf{I} \right) \quad \text{for compressible fluids} \tag{3.158}$$

$$\mathbf{F}_{viscous} = \nabla \boldsymbol{\tau} = \mu \nabla \left[(\nabla \mathbf{v}^{\mathrm{T}})^{\mathrm{T}} + \nabla \mathbf{v}^{\mathrm{T}} \right] \quad \text{for imcompressible fluids} \tag{3.159}$$

$$\mathbf{F}_{viscous} = \nabla \boldsymbol{\tau} = 0 \quad \text{for Euler fluids} \tag{3.160}$$

(2) **Solids**

There are two solution strategies in solid dynamics:

- Stress function method. The stress gradient $\mathbf{F}_\sigma$ in Eq. (3.154) is employed directly, and Eq. (3.150) is transformed to a differential equation in the "Airy stress function" Ψ. The stress components are related to the derivatives of Ψ, such that they satisfy the equilibrium equations.
- Displacement function method. A transformation of Eq. (3.150) to the differential operator form w.r.t. displacement components is undertaken in two steps. First, the substitution of constitutive equation, such as Eq. (3.138), into Eq. (3.150) leads to

three differential equations containing strain components. Next, the substitution of strain-displacement relations, namely Eq. (3.71), results in three differential equations for the solution of three displacement components.

Use is made of Eqs. (3.85) and (3.103), $\mathbf{F}_\sigma$ in Eq. (3.154) for elastic body may be expressed by

$$\mathbf{F}_\sigma = \mathbf{D}^e \nabla \mathbf{L} \mathbf{u} \tag{3.161}$$

After a series deduction, for the isotropic and homogeneous elasticity issue we get

$$\mathbf{F}_\sigma = \begin{Bmatrix} (\lambda + G)\frac{\partial \theta}{\partial x} + G\nabla^2 u_x \\ (\lambda + G)\frac{\partial \theta}{\partial y} + G\nabla^2 u_y \\ (\lambda + G)\frac{\partial \theta}{\partial z} + G\nabla^2 u_z \end{Bmatrix} \tag{3.162}$$

In which $\theta = \varepsilon_v = 3\varepsilon_m = \frac{\partial u_x}{\partial x} + \frac{\partial u_y}{\partial y} + \frac{\partial u_z}{\partial z}$ is the volumetric strain.

2. **Damping**

(1) **Fluids**

For fluid dynamics, damping is normally neglected, or in other term, has been represented by the viscous force, hence

$$\mathbf{F}_{damping} = 0 \tag{3.163}$$

(2) **Solids**

For solid dynamics related to the structural vibration problems under periodical actions, it is normally postulated as

$$\mathbf{F}_{damping} = -C\mathbf{v} = -C\dot{\mathbf{u}} \tag{3.164}$$

in which C is the damping factor (N s/m).

Normally in the structural problems such as the seismic response analysis of dams during earthquake, damping is nontrivial hence it should be carefully handled (Chen 2018). Nevertheless, in some formulations of advanced CMs intended to the phenomenon of large deformation with detachment of subdomain, damping is customarily neglected, namely

$$\mathbf{F}_{damping} = 0 \tag{3.165}$$

This because on one hand, damping factor is difficult to define and evaluate, on the other hand, damping is not so important (not always) as friction in the structure performance with large and dislocated deformation. It is notable that for some advanced

CMs (e.g. DEM), artificial damping needs to be introduced for the purpose to stabilize time recursion procedure and to get quasi-static solution with less computation efforts.

3. **Body forces**

(1) **Fluids**

Volumetric force $\mathbf{F}_{body}$ is normally the unit weight of fluid

$$\mathbf{F}_{body} = \mathbf{b} = \rho\mathbf{g} \tag{3.166}$$

In which $\mathbf{g}$ is the gravity acceleration.

(2) **Solids**

It is identically defined in Eq. (3.166). The surface/edge/point forces exerted on solid domain are handled by the definition of boundary conditions.

4. **Detailed forms of momentum equations**

(1) **Fluids**

Introducing Eqs. (3.156)–(3.158) into Eq. (3.150), the "Navier-Stokes (N-S)" equations where the second viscosity is neglected are

$$\rho\dot{\mathbf{v}} = -\nabla p + \mu\nabla\left(\left[(\nabla\mathbf{v}^{\mathrm{T}})^{\mathrm{T}} + \nabla\mathbf{v}^{\mathrm{T}}\right] - \frac{2}{3}\mathrm{div}(\mathbf{v})\mathbf{I}\right) + \rho\mathbf{g} \tag{3.167}$$

in which $\mathbf{g} = \begin{bmatrix} g_x & g_y & g_z \end{bmatrix}^{\mathrm{T}}$ is the vector of gravity acceleration.

For incompressible fluids, Eq. (3.167) becomes

$$\rho\dot{\mathbf{v}} = -\nabla p + \mu\left[(\nabla\mathbf{v}^{\mathrm{T}})^{\mathrm{T}} + \nabla\mathbf{v}^{\mathrm{T}}\right] + \rho\mathbf{g} \tag{3.168}$$

This incompressible N-S equations with conservative external field is the fundamental equation in hydraulics.

Where the viscosity is totally neglected, we can get "Euler equations" such that

$$\rho\dot{\mathbf{v}} = -\nabla p + \rho\mathbf{g} \tag{3.169}$$

Momentum Eqs. (3.167)–(3.169) contain pressure gradient terms that characterize fluid flows. Since pressure is a scalar and its gradient appears in each of the component equations, a localized pressure perturbation is likely to induce motion in all directions. The pressure gradient thus acts to couple different component equations, and is generally the physical process that allows for (or enforces) mass conservation. As fluid converges into a given volume, the pressure will rise to produce a compensating pressure-driven divergence such as in the cases of acoustic and gravity waves.

(2) **Solids**

- Stress function solution. Substitution of Eqs. (3.155) and (3.166) into Eq. (3.150) leads to

$$\begin{cases} \frac{\partial\sigma_{xx}}{\partial x} + \frac{\partial\tau_{yx}}{\partial y} + \frac{\partial\tau_{zx}}{\partial z} + b_x = \rho\ddot{u}_x \\ \frac{\partial\tau_{xy}}{\partial x} + \frac{\partial\sigma_{yy}}{\partial y} + \frac{\partial\tau_{zy}}{\partial z} + b_y = \rho\ddot{u}_y \\ \frac{\partial\tau_{xz}}{\partial x} + \frac{\partial\tau_{yz}}{\partial y} + \frac{\partial\sigma_{zz}}{\partial z} + b_z = \rho\ddot{u}_z \end{cases} \tag{3.170}$$

For plane stress problems, the dimension-reduction of above equations leads to

$$\begin{cases} \frac{\partial\sigma_{xx}}{\partial x} + \frac{\partial\tau_{yx}}{\partial y} + b_x = \rho\ddot{u}_x \\ \frac{\partial\tau_{yx}}{\partial x} + \frac{\partial\sigma_{yy}}{\partial y} + b_y = \rho\ddot{u}_y \end{cases} \tag{3.171}$$

- Displacement function solution. Substitution of elastic constitutive Eqs. (3.103) and (3.110) into Eq. (3.170) leads to well-known "Lamé equations"

$$\begin{cases} (\lambda + G)\frac{\partial\theta}{\partial x} + G\nabla^2 u_x + b_x = \rho\ddot{u}_x \\ (\lambda + G)\frac{\partial\theta}{\partial y} + G\nabla^2 u_y + b_y = \rho\ddot{u}_y \\ (\lambda + G)\frac{\partial\theta}{\partial z} + G\nabla^2 u_z + b_z = \rho\ddot{u}_z \end{cases} \tag{3.172}$$

For plane strain problems, the dimension-reduction of above equations leads to

$$\begin{cases} (\lambda + G)\frac{\partial\theta}{\partial x} + G\nabla^2 u_x + b_x = \rho\ddot{u}_x \\ (\lambda + G)\frac{\partial\theta}{\partial y} + G\nabla^2 u_y + b_y = \rho\ddot{u}_y \end{cases} \tag{3.173}$$

3.5.4 *Well-Posed Problems for Fluids*

In fluid dynamics, a set of quasilinear hyperbolic PDEs governing adiabatic and inviscid flow, the "Euler equations" named after Leonhard Euler, was formulated in 1757. Further mathematical justification was provided by Navier and Stokes. N-S equations may be established using either spatial description or material description. These equations are also of great interest in a purely mathematical sense, because it has not yet been proven that their 3-D solutions always exist, or that if exist, the solutions are smooth without any mathematical singularities. These problems are dubbed as the "Navier-Stokes existence and smoothness".

In Eq. (3.167), the shear stress tensor is linked to the velocity gradient derived from three essential variables of velocity vector **v**. The fourth essential variable—pressure p, is related to the fifth essential variable—density ρ, through the state Eq. (3.101).

ρ is in turn, related to the velocity vector $\mathbf{v}$ by the continuity equation that is equal to the conservation law of mass.

For a Lagrangian infinitesimal fluid cell (parcel) δV, the mass contained in the material volume is

$$\delta m = \rho \delta V \tag{3.174}$$

In which m and ρ are mass and density, respectively. Since the mass is conserved in the parcel, the rate of mass change is zero. Therefore, we have

$$\frac{\mathrm{d}(\delta m)}{\mathrm{d}t} = \frac{\mathrm{d}(\rho \delta V)}{\mathrm{d}t} = \frac{\mathrm{d}\rho}{\mathrm{d}t}\delta V + \rho \frac{\mathrm{d}(\delta V)}{\mathrm{d}t} = \dot{\rho}\delta V + \rho \delta \dot{V} = 0 \tag{3.175}$$

The movement of the fluid inside δV leads to the change of surface δS. The change of the surface, in turn, changes the volume $\mathrm{d}(\delta V) = \mathrm{d}t.\mathbf{v}.\mathbf{n}.\mathrm{d}(\delta S)$. The total volume change is therefore the integral over the control surface, namely, $\mathrm{d}(\delta V) = \oint_{\delta S} \mathrm{d}t.\mathbf{v}.\mathbf{n}.\mathrm{d}(\delta S)$ or $\delta \dot{V} = \frac{\mathrm{d}(\delta V)}{\mathrm{d}t} = \oint_{\delta S} \mathbf{v}.\mathbf{n}.\mathrm{d}(\delta S)$. According to Gauss theorem (3.147), we get

$$\delta \dot{V} = \oiiint_{\delta V} \nabla.\mathbf{v}.\mathrm{d}(\delta V) = \nabla.\mathbf{v} \oiiint_{\delta V} \mathrm{d}(\delta V) = \nabla.\mathbf{v}.(\delta V)$$

or

$$\nabla.\mathbf{v} = \frac{1}{\delta V}\frac{\mathrm{d}(\delta V)}{\mathrm{d}t} = \frac{\delta \dot{V}}{\delta V} \tag{3.176}$$

in which ∇ is the Hamilton (gradient) operator.

Equation (3.176) states that the velocity divergence can be physically interpreted as the time rate of volume change per unit volume. From Eqs. (3.175) and (3.176), the conservation law of mass can be mathematically expressed by continuity equation

$$\frac{\mathrm{d}\rho}{\mathrm{d}t} = \dot{\rho} = -\rho \nabla.\mathbf{v} \tag{3.177}$$

If the fluid is incompressible, it follows that

$$\nabla.\mathbf{v} = 0 \tag{3.178}$$

The conservation of energy, in the continuum mechanics, is a representation of the first law of thermodynamics. It states that the rate of energy change inside an infinitesimal fluid parcel should be equal to the net heat flux into that parcel plus the rate of work done by the body and surface forces exerting on the same parcel. If neglecting the heat flux and the body force, the rate of the internal energy change of this parcel consists of two parts: the work done by the isotropic pressure multiplying the volumetric strain + the energy dissipation due to the viscous shear forces.

For 2-dimensional problems, N-S equations of flux form consist of following equation set:

- Continuity Eq. (3.177), one PDE;
- Momentum Eq. (3.167), two PDEs;
- State Eq. (3.101), one constitutive equation.

Note that we have four independent variables ρ, $\mathbf{v}$, and p to be solved, the equation system is closed: it contains four equations for these independent variables. However, Eqs. (3.177), (3.167) and (3.101) have an infinity of solutions, hence to restrict the number of solutions, initial conditions (ICs) and boundary conditions (BCs) have to be appropriately specified.

- Initial conditions.

$$\rho|_{t=0} = \rho_0(\mathbf{x})\,,\ \mathbf{v}|_{t=0} = \mathbf{v}_0(\mathbf{x})\,,\ p|_{t=0} = p_0(\mathbf{x}) \text{ in } \Omega \tag{3.179}$$

- Boundary conditions. Let $\Gamma = \Gamma_{in} \cup \Gamma_w \cup \Gamma_{out} \cup \Gamma_{free}$ (viz. Fig. 3.10).

On inlet boundary $\Gamma_{in} = \{\mathbf{x} \in \Gamma : \mathbf{v}.\mathbf{n} < 0\}$, three conditions, namely, mass inlet, velocity inlet and pressure inlet may be selectively used:

$$\begin{cases} \rho = \rho_{in} & \text{(prescribed mass)} \\ \mathbf{v} = \mathbf{v}_{in}(\mathbf{x}) & \text{(prescribed velocity)} \\ p = p_{in}(\mathbf{x}) & \text{(prescribed pressure)} \end{cases} \tag{3.180}$$

On solid wall $\Gamma_w = \{\mathbf{x} \in \Gamma : \mathbf{v}.\mathbf{n} = 0\}$, non-slip condition is customarily used:

$$\mathbf{v}\big|_{\Gamma_w} = 0 \text{ (non-slip)} \tag{3.181}$$

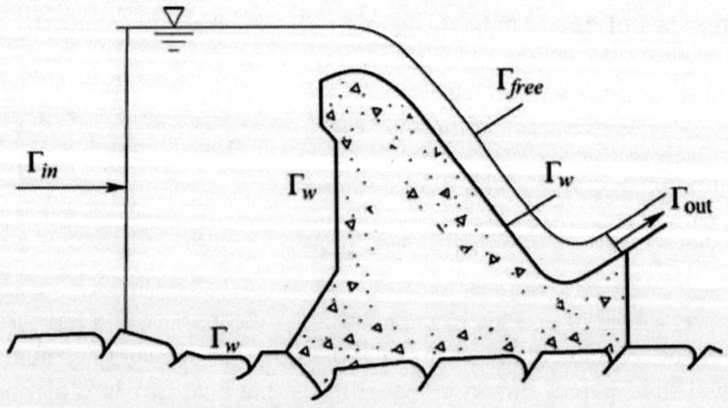

Fig. 3.10 Boundary conditions for spillway dam section (fluid domain)

On outlet $\Gamma_{out} = \{\mathbf{x} \in \Gamma : \mathbf{v.n} > 0\}$, vanish pressure condition or prescribed velocity condition may be selectively used

$$\begin{cases} \mathbf{v.n} = v_n \text{ and } \mathbf{v.s} = v_s & \text{(prescribed velocity)} \\ -p - \mathbf{n.\tau.n} = 0 \text{ and } \mathbf{s.\tau.n} = 0 & \text{(prescribed pressure or vanish pressure)} \end{cases} \tag{3.182}$$

or

$$-p - \mathbf{n.\tau.n} = 0 \text{ and } \mathbf{s.\tau.n} = 0 \quad \text{(prescribed pressure or vanish pressure)} \tag{3.183}$$

It is emphasized that the prescribed velocity inlet cannot to be used simultaneously with the prescribed pressure outlet and vice-versa. In the mesh-based methods using Eulerian formalism, the dynamic free surface $\Gamma_{free} = \{\mathbf{x} \in \Gamma : \mathbf{v.n} = 0; -p - \mathbf{n.\tau.n} = 0\}$ is normally handled as the interface of liquid domain and gas domain that may be traced by a number of algorithms such as the interface tracking methods (Harlow and Welch 1965; Donéa et al. 1977; Hughes et al. 1981) and interface capturing methods (Hirt and Nichols 1981; Osher and Sethian 1988).

More complicated boundary conditions may be found in a number of literatures (Glowinski 1984; Bristeau et al. 1985, 1987; Pironneau 1989). For example, in solving "Sturm-Liouville problems" which appear in many science and engineering fields, "Robin boundary condition" that is a weighted combination of "Dirichlet boundary conditions" and "Neumann boundary conditions", is commonly exercised (Gustafson and Abe 1998a, b; Gustafson 1998; Akin 2005). It is also called "impedance boundary conditions" from their application in electromagnetic problems, or "convective boundary conditions" from their application in heat transfer problems (Hahn and Özisik 2012). Other equivalent names in use are "Fourier-type conditions" and "radiation conditions" (Logan 2001).

The problem is well-posed if the existing solution is unique and depends continuously on ICs and BCs.

3.5.5 Well-Posed Problems for Solids

Similar to the fluid dynamics, a well-posed problem demands to appropriately define initial conditions and boundary conditions.

- Initial conditions.

$$\mathbf{u}|_{t=0} = \mathbf{u}_0(\mathbf{x}), \boldsymbol{\sigma}|_{t=0} = \boldsymbol{\sigma}_0(\mathbf{x}) \text{ in } \Omega \tag{3.184}$$

- Boundary conditions. Let $\Gamma = \Gamma_t \cup \Gamma_u$ (viz. Fig. 3.11).

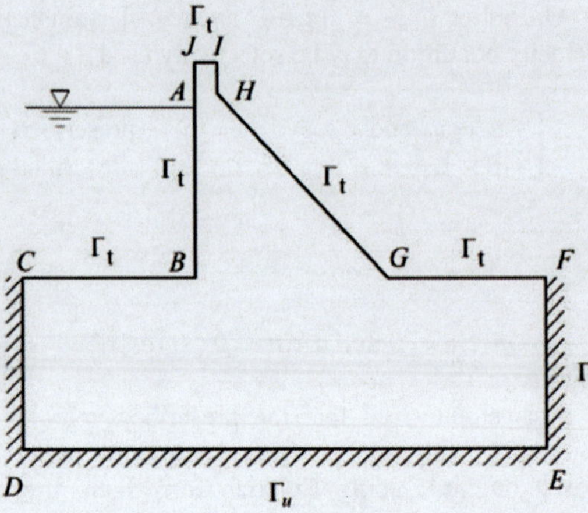

Fig. 3.11 Boundary conditions for non-overflow dam section (solid domain)

On traction (stress) boundary Γ_t such as the dam surface A-B-C, prescribed stresses are exerted (i.e. water pressure in the case of dam), whereas on displacement boundary Γ_u, prescribed displacements are imposed. It is emphasized that the dam surface A-J-I-H-G-F is a traction (stress) boundary, too, on which the prescribed stress vanishes.

Equation (3.170) supplemented by initial and boundary conditions for solid is well-posed because the solution does exist and is unique

$$\begin{cases} \nabla\boldsymbol{\sigma} + \mathbf{b} = \rho\ddot{\mathbf{u}}(\dot{\mathbf{v}}) \text{ in } \Omega \\ \boldsymbol{\sigma}.\mathbf{n} = \bar{\mathbf{t}} \quad \text{on } \Gamma_t \\ \mathbf{u} = \bar{\mathbf{u}} \quad \text{on } \Gamma_u \\ \boldsymbol{\sigma}(0) = \boldsymbol{\sigma}_0 \quad \text{in } \Omega \\ \mathbf{u}(0) = \mathbf{u}_0 \quad \text{in } \Omega \end{cases} \tag{3.185}$$

In which $\mathbf{n}$ is the outward unit normal vector of surface boundary.

Equation (3.185) is the general dynamic equation for solids used in, for example, the FLAC code with FDM. If the loads applied are static, and the concern is only the static status of solid, then the static equilibrium equation can be obtained simply by eliminating the dynamic terms from Eq. (3.185).

If the acceleration, damping and body force can be neglected, we get the governing equations for linear elastic continuum as

$$\begin{cases} \nabla\boldsymbol{\sigma} = 0 \text{ in } \Omega \\ \boldsymbol{\sigma}.\mathbf{n} = \bar{\mathbf{t}} \text{ on } \Gamma_t \\ \mathbf{u} = \bar{\mathbf{u}} \quad \text{on } \Gamma_u \end{cases} \tag{3.186}$$

Equation (3.186) may be solved by stress method. Particularly, for linear elasticity problem, the strain components are expressed in the derivatives of stress function Ψ

first, then these strain components are substituted in the compatibility equations. For 2-D problems, these operations result in a fourth-order differential equation termed as the "bi-harmonic equation". It can be written in short form using the Laplace operator ∇^2 as follows

$$\nabla^2(\nabla^2\Psi) = 0 \tag{3.187}$$

When the stress function is known, the stresses can be calculated rather straightforwardly

$$\sigma_{\alpha\beta} = -\frac{\partial^2\Psi}{\partial x_\alpha \partial x_\beta} + \delta_{ij}\nabla^2\Psi \quad (\alpha, \beta = 1, 2) \tag{3.188}$$

Afterwards, the strains and displacements may be calculated in sequence without difficulties. One of the famous achievements by the application of Airy equations is the deduction of "stress intensity factors (SIFs)" for crack-tip in elastic body.

More general solution should employ displacement method which makes use complex constitutive relations and Cauchy strain to transform Eq. (3.185) by displacements (or its increments) as essential variables. Particularly, for the elasticity problem using linear elastic (Hooke's) law, following "Lamé equations" plus boundary and initial conditions are well-posed

$$\begin{cases} (\lambda + G)\nabla\theta + G\nabla^2\mathbf{u} + \mathbf{b} = \rho\ddot{\mathbf{u}} & \text{in } \Omega \\ \boldsymbol{\sigma}.\mathbf{n} = \bar{\mathbf{t}} & \text{on } \Gamma_t \\ \mathbf{u} = \bar{\mathbf{u}} & \text{on } \Gamma_u \\ \mathbf{u}(0) = \mathbf{u}_0 & \text{in } \Omega \end{cases} \tag{3.189}$$

After the solution of Eq. (3.189), strain and stress tensors may be calculated by Eqs. (3.85) and (3.103), respectively.

It is noted that with displacement method, the weak form is much more prevalent in the study of advanced CMs for solid dynamics, rather than using strong form directly.

3.5.6 *Application of Strong Form in Elastic Fracture Mechanics*

Although reluctantly, yet we have to admit that the closed-form solution of Eqs. (3.185) and (3.189) can only be get for a few of very simple cases. Such a complete solution that is mostly of static linear elasticity, if founded, may be very valuable in the progress of continuum mechanics. A good example may be given by the fracture mechanics.

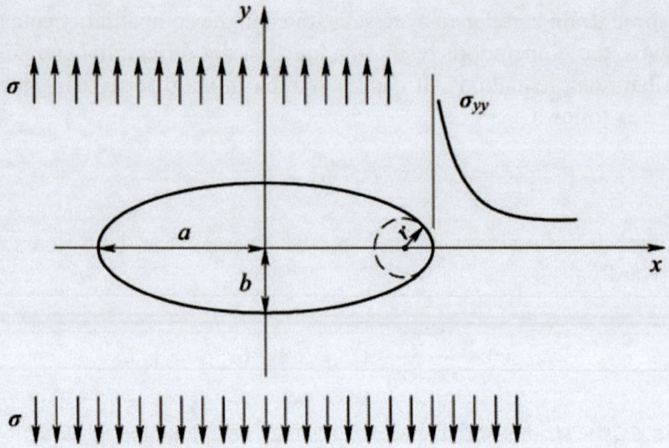

Fig. 3.12 Elliptical hole loaded with a uniform stress σ

Fracture mechanics started from the description of stress distribution in the neighborhood of a flat elliptical crack in the thin plate exerted with a uniform far-field tension σ (viz. Fig. 3.12) (Kirsch 1898; Inglis 1913; Westergaard 1939):

$$\sigma_y = (x = a, y = 0) = \sigma(1 + 2\frac{a}{b}) = \sigma(1 + 2\sqrt{a/r}) \approx \sigma\sqrt{a/r} \tag{3.190}$$

The solution reveals that at the location with the smallest radius r, the stress concentration becomes infinite when this radius approaches zero. Hence, the crack-tip is a singular point where the stress state cannot be directly evaluated and it is clear that indirect indexes are looked for towards the crack onset calibration.

A further significant achievement of Irwin (1957) and his colleagues was to find a method to calculate the amount of energy available for fracturing a linear elastic material in terms of the asymptotic stress and the displacement field around the crack front. This field was obtained by Williams (1957) using the technique of complex Airy function and variable separation technique established by Muskhelishvili in 1953 (viz. Fig. 2.39).

For 2-D problems, only mode I and mode II fractures are tackled. When a crack is loaded, stress components at the crack-tip can be calculated with the Airy stress function method. To handle singular point, the application of complex variables and functions may successfully lead to the asymptotic expression for the 2-D stress field around a crack-tip (viz. Fig. 3.13) as follows

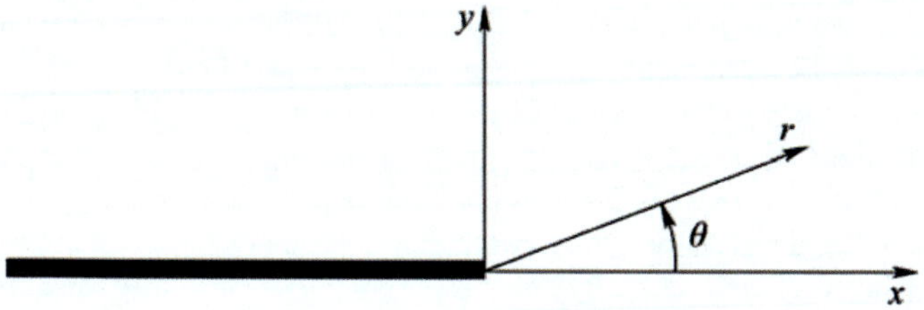

Fig. 3.13 Crack with local coordinate system origination in the crack-tip

$$\begin{cases} \sigma_x = \frac{K_I}{\sqrt{2\pi r}} \cos\frac{\theta}{2}\left(1 - \sin\frac{\theta}{2}\sin\frac{3\theta}{2}\right) + \frac{-K_{II}}{\sqrt{2\pi r}} \sin\frac{\theta}{2}\left(2 + \cos\frac{\theta}{2}\cos\frac{3\theta}{2}\right) \\ \sigma_y = \frac{K_I}{\sqrt{2\pi r}} \cos\frac{\theta}{2}\left(1 + \sin\frac{\theta}{2}\sin\frac{3\theta}{2}\right) + \frac{K_{II}}{\sqrt{2\pi r}} \cos\frac{\theta}{2}\sin\frac{\theta}{2}\cos\frac{3\theta}{2} \\ \tau_{xy} = \frac{K_I}{\sqrt{2\pi r}} \cos\frac{\theta}{2}\sin\frac{\theta}{2}\cos\frac{3\theta}{2} + \frac{K_{II}}{\sqrt{2\pi r}} \cos\frac{\theta}{2}\left(1 - \sin\frac{\theta}{2}\sin\frac{3\theta}{2}\right) \\ u_x = \frac{K_I}{8G}\sqrt{\frac{2r}{\pi}}\left[(2K-1)\cos\frac{\theta}{2} - \cos\frac{3\theta}{2}\right] + \frac{K_{II}}{8G}\sqrt{\frac{2r}{\pi}}\left[(2K+3)\sin\frac{\theta}{2} + \sin\frac{3\theta}{2}\right] \\ u_y = \frac{K_I}{8G}\sqrt{\frac{2r}{\pi}}\left[(2K+1)\sin\frac{\theta}{2} - \sin\frac{3\theta}{2}\right] + \frac{K_{II}}{8G}\sqrt{\frac{2r}{\pi}}\left[-(2K-3)\cos\frac{\theta}{2} - \cos\frac{3\theta}{2}\right] \end{cases} \tag{3.191}$$

where: G—Shear modulus; $K = (3-\nu)/(1+\nu)$ (plane stress state) or $K = (3-4\nu)$ (plane strain state); ν—Poisson's ratio; K_I and K_{II}—crack-tip stress intensity factors (SIFs) of mode I and II.

For a single fracture mode (e.g. mode I), the crack growth criterion may be expressed by

$$K_I \geq K_{Ic} \tag{3.192}$$

in which K_{Ic} is the "toughness of fracture" of mode I that is one of extremely important strength parameters determining the cracking resistance properties of a given material.

Typical values for K_{Ic} are 150 MN/m$^{3/2}$ for ductile (very tough) metals, 25 MN/m$^{3/2}$ for brittle ones and 1–10 MN/m$^{3/2}$ for glasses and brittle polymers.

For a comprehensive discussion w.r.t. history and milestone works, readers are referred to the paper by Rossmanith (1997).

3.6 Dynamics of Continuum: Weak Forms

3.6.1 Concept

In many areas of geotechnical and hydraulic structure related to the issues of solid deformation, thermal conduction and water permeability, the integral equations (via variational or weight residual technique)—the weak form of system equations for continuum, may be constructed to replace the counterpart PDEs.

In contrast to strong forms, weak forms require weaker consistency on the approximation functions of essential variables (e.g. displacements). For a $2k$th-order PDE w.r.t. field variables, the weak form formulation requires a consistency of kth-order only. As a result, the formulation of continuum mechanics based on the weak form can produce a more stable set of algebraic equations, hence the weak form is normally more preferred in advanced CMs.

There are a number of operators for establishing system equations in weak forms, of which the prevalent and often-used ones are the "variational method" and "weighted residual method".

There are different formulations of variational methods, of which the "Galerkin method" may be the most prevalent one. The "Hamilton's principle" is often employed for dynamic problems. The "minimum principle" of total potential energy is a convenient tool to derive the discrete system equations for FEM and also for many other CMs.

Weighted residual method is a more general and powerful mathematical tool that can be used for creating discretized system equations for many types of science and engineering problems, even without the existence of variational function.

In the elaboration of weak form governing equations for continuum hereinafter, the matrix $\mathbf{D} = [D]$ will be uniformly used, which could be the elastic matrix $\mathbf{D}^e$ or elasto-plastic matrix $\mathbf{D}^{ep}$, dependent on the postulated constitutive relations.

3.6.2 *Variational Principle*

Traditional discretization procedure for a differential equation is that a continuous field function is approximated by its discrete values at grid points. Although amenable to numerical analysis, this approach rests on the obvious infelicity of comparing unlike with unlike. The alternative example is to approximate the function space of infinite-dimension by a sub-space of finite-dimension. As a result, both the exact solution and its approximation inhabit the same space and the discretization error can be measured in the natural topology of ambient function space. It is without exaggeration to say that the functional analysis provides the universal language and rigor for discretizing differential and integral operators (Dahmen 1997).

The "Ritz method" replaces a differential problem by a variational problem, which is a pedigree stretching to Newton, Euler and Lagrange in the opposite direction. In a variational setting, the original PDEs are reduced to the global optimization in the underlying (infinite-dimension) Hilbert space H, which is further approximated by a finite-dimension sub-space of H. Much mathematical work has been done in this area by Vainberg (1964), Tonti (1969), Oden (1969a, b) and many others. Theorems such as the "minimization of total potential energy" in mechanical system, the "least energy dissipation principles" in viscous flow, etc., are considered as the cornerstone for the formulation of many modern CMs. Variational principles of this kind are "natural and beautiful", but unfortunately, they do not always exist for all continuum problems of which PDEs have been well-defined.

There is an extended category of variational principles dubbed as "contrived" ones. They can always be constructed for any differentially specified problem, either by extending the number of unknown functions with additional variables known as "Lagrange multipliers", or by imposing a higher degree of continuity requirements such as in the method of "penalty factors".

1. **Definitions of energy and work for continuum**

For elastic materials, the strain energy Π_s that is identical to the potential energy V of the system can be expressed as

$$\Pi_s = V = \frac{1}{2}\iiint\limits_{\Omega} \boldsymbol{\varepsilon}^{\mathrm{T}}\boldsymbol{\sigma}\mathrm{d}\Omega \tag{3.193}$$

In which Ω stands for the whole domain.

The kinetic energy is defined by

$$T = \frac{1}{2}\iiint\limits_{\Omega} \rho\left(\frac{\partial \mathbf{u}}{\partial t}\right)^{\mathrm{T}}\frac{\partial \mathbf{u}}{\partial t}\mathrm{d}\Omega = \frac{1}{2}\iiint\limits_{\Omega} \rho(\dot{\mathbf{u}})^{\mathrm{T}}\dot{\mathbf{u}}\mathrm{d}\Omega \tag{3.194}$$

The work done by the external forces can be calculated as

$$\begin{aligned} U = W_f &= \frac{1}{2}\iiint\limits_{\Omega} (\mathbf{b})^{\mathrm{T}}\mathbf{u}\mathrm{d}\Omega + \frac{1}{2}\iint\limits_{\Gamma_t} (\bar{\mathbf{t}})^{\mathrm{T}}\mathbf{u}\mathrm{d}\Gamma + \frac{1}{2}\sum_{\zeta=1}^{n_\zeta}(\mathbf{T}_\zeta)^{\mathrm{T}}\mathbf{u}_\zeta \\ &= \frac{1}{2}\iiint\limits_{\Omega} (b_x u_x + b_y u_y + b_z u_z)\mathrm{d}\Omega + \frac{1}{2}\iint\limits_{\Gamma_t} (\bar{t}_x u_x + \bar{t}_y u_y + \bar{t}_z u_z)\mathrm{d}\Gamma \\ &\quad + \frac{1}{2}\sum_{\zeta=1}^{n_\zeta}(T_{x,\zeta}u_x + T_{y,\zeta}u_y + T_{z,\zeta}u_z) \end{aligned} \tag{3.195}$$

in which Γ_t stands for the boundary surfaces of the solid domain on which traction forces are prescribed.

The work done by the external forces and internal non-adiabatic actions is

$$W = W_f + Q \tag{3.196}$$

if the deformation process is adiabatic, we have

$$Q = 0 \tag{3.197}$$

2. **Virtual work principle**

In computational mechanics, particularly in FEM, the weak form is often referred to as the "principle of virtual work". It states that for a solid body being in equilibrium state, the total virtual work arises from the stresses in the body and the external forces due to a consistent virtual displacement vector should be identical.

Suppose a virtual displacement $(\delta u_x, \delta u_y, \delta u_z)$ emerges in an equilibrium solid body, the virtual work (displacement) principle may be mathematically expressed as

$$\delta V = \delta W \tag{3.198a}$$

$$\begin{cases} \delta V = \iiint\limits_{\Omega} \delta \boldsymbol{\varepsilon}^{\mathrm{T}} \boldsymbol{\sigma} \mathrm{d}\Omega \\ \quad = \iiint\limits_{\Omega} (\sigma_{xx}\delta\varepsilon_{xx} + \sigma_{yy}\delta\varepsilon_{yy} + \sigma_{zz}\delta\varepsilon_{zz} + \tau_{xy}\delta\gamma_{xy} + \tau_{yz}\delta\gamma_{yz} + \tau_{zx}\delta\gamma_{zx})\mathrm{d}\Omega \\ \delta W = \iiint\limits_{\Omega} (b_x\delta u_x + b_y\delta u_y + b_z\delta u_z)\mathrm{d}\Omega \\ \quad + \iint\limits_{\Gamma_t} (\bar{t}_x\delta u_x + \bar{t}_y\delta u_y + \bar{t}_z\delta u_z)\mathrm{d}\Gamma + \sum\limits_{\zeta=1}^{n_\zeta} \delta \mathbf{u}^{\mathrm{T}}(\mathbf{x}_\zeta)\mathbf{T}_\zeta(\mathbf{x}_\zeta) \end{cases} \tag{3.198b}$$

where: $\mathbf{T}_\zeta(\mathbf{x}_\zeta)$—point (concentrated) force; n_ζ—amount of point force; $\mathbf{x}_\zeta$—location on which the point force exerted.

For dynamic problem, Eq. (3.198) may be further extended as follows

$$\iiint\limits_{\Omega} \delta\mathbf{u}^{\mathrm{T}}\rho\ddot{\mathbf{u}}\mathrm{d}\Omega + \iiint\limits_{\Omega} \delta\boldsymbol{\varepsilon}^{\mathrm{T}}\boldsymbol{\sigma}\mathrm{d}\Omega = \iiint\limits_{\Omega} \delta\mathbf{u}^{\mathrm{T}}\mathbf{b}\mathrm{d}\Omega + \iint\limits_{\Gamma_t} \delta\mathbf{u}^{\mathrm{T}}\bar{\mathbf{t}}\mathrm{d}\Gamma + \sum_{\zeta=1}^{n_\zeta} \delta\mathbf{u}^{\mathrm{T}}(\mathbf{x}_\zeta)\mathbf{T}_\zeta(\mathbf{x}_\zeta) \tag{3.199}$$

The virtual work principle can be regarded as an alternative statement of equilibrium condition. The first left hand item in Eq. (3.199) is the virtual work done by the inertial forces in problem domain Ω. We will see later that the virtual work statement is precisely the Galerkin weak (integral) form of equilibrium equations and is valid for non-linear constitutive relations.

3.6.3 Hamilton's Principle

1. **Unconstrained Hamilton's principle**

Hamilton's principle is one of the variational principles based on the energy conservation law. It states that, of all possible consistent displacement histories meeting three admissible conditions:

- Compatibility conditions;
- Essential (displacement) boundary conditions;
- Conditions at initial (t_1) and final time (t_2).

The unique history corresponding to the actual solution makes the Lagrangian function a minimum. This principle is mathematically formulated as follows

$$\delta \int_{t_1}^{t_2} L\mathrm{d}t = 0 \tag{3.200}$$

in which L is Lagrangian function. For a solid body, it can be defined as

$$L = T - V + W_f \tag{3.201}$$

where: T—kinetic energy; V—strain energy; W_f—work done by the external forces.

Equation (3.201) may be further specified such that

$$L = \frac{1}{2}\iiint_{\Omega} \rho(\dot{\mathbf{u}})^{\mathrm{T}}\dot{\mathbf{u}}\mathrm{d}\Omega - \frac{1}{2}\iiint_{\Omega} \boldsymbol{\varepsilon}^{\mathrm{T}}\boldsymbol{\sigma}\mathrm{d}\Omega + \iiint_{\Omega} \mathbf{u}^{\mathrm{T}}\mathbf{b}\mathrm{d}\Omega + \iint_{\Gamma_t} \mathbf{u}^{\mathrm{T}}\,\bar{\mathbf{t}}\mathrm{d}\Gamma$$
$$+ \sum_{\zeta=1}^{n_\zeta} \mathbf{u}^{\mathrm{T}}(\mathbf{x}_\zeta)\mathbf{T}_\zeta(\mathbf{x}_\zeta) \tag{3.202}$$

Use is made of the integral by parts, a set of PDEs identical to the equilibrium Eq. (3.185) will be led to.

The advantage of using Hamilton's principle is that we need not to know the PDEs in strong form for the mechanical and physical system. But in order to use Hamilton's principle, a functional has to be obtained, which may not be always possible.

2. **Constrained Hamilton's principle**

There are cases when the approximated field function does not satisfy essential boundary condition on Γ_u [viz. the third formula in Eq. (3.185)], for example, the problem domain including parts of boundaries (inner boundaries). Under such circumstances, Hamilton's principle has to be modified or constrained.

Consider a set of m_c conditions that the approximated field function should satisfy

$$\mathbf{C}(\mathbf{u}) = \left[C_1(\mathbf{u})\ C_2(\mathbf{u}) \ldots C_{m_c}(\mathbf{u}) \right]^{\mathrm{T}} = \{0\} \tag{3.203}$$

in which $\mathbf{C}$ is a given coefficient matrix.

Typical example of such constraint is the prescribed boundary displacement (essential boundary) condition,

$$\mathbf{u} - \bar{\mathbf{u}} = 0 \tag{3.204}$$

Our purpose now is to seek the stationary point of the Lagrangian functional subjected to the constraint (3.203). There are basically two methods often employed for the modification of Lagrangian function to accommodate these constraints. They are the method of Lagrange multipliers and the method of penalty factors.

(1) Method of Lagrange multipliers

The modified Lagrangian function is written as

$$\widetilde{L} = L + \iiint_{\Omega} \boldsymbol{\lambda}^{\mathrm{T}} \mathbf{C}(\mathbf{u}) \mathrm{d}\Omega \tag{3.205}$$

in which $\boldsymbol{\lambda}$ is a vector of Lagrange multipliers

$$\boldsymbol{\lambda} = \left[\lambda_1 \; \lambda_2 \ldots \lambda_{m_c} \right]^{\mathrm{T}} \tag{3.206}$$

These Lagrange multipliers are unknown functions of independent coordinates in the domain Ω. The modified Hamilton's principle seeks the stationary point of

$$\delta \int_{t_1}^{t_2} \widetilde{L} \mathrm{d}t = 0 \tag{3.207}$$

Note that since the Lagrange multipliers are unknown, in the process of formulating discretized system equation, they must be approximated in a manner similar to the field variables. Therefore, the amount of nodal unknowns in the discretized system equation set will be expanded. This method is, however, able to rigorously enforce the constraints.

(2) Method of penalty factors

The modified Lagrangian function is written as

$$\widetilde{L} = L + \frac{1}{2} \iiint_{\Omega} \mathbf{C}(\mathbf{u})^{\mathrm{T}} (\boldsymbol{\alpha}) \mathbf{C}(\mathbf{u}) \mathrm{d}\Omega \tag{3.208}$$

in which $\boldsymbol{\alpha}$ is a diagonal matrix given by

$$\boldsymbol{\alpha} = \begin{bmatrix} \alpha_1 & \ldots & 0 & \ldots & 0 \\ \ldots & \ldots & \ldots & \ldots & \ldots \\ 0 & \ldots & \alpha_\zeta & \ldots & 0 \\ \ldots & \ldots & \ldots & \ldots & \ldots \\ 0 & \ldots & 0 & \ldots & \alpha_{m_c} \end{bmatrix} \tag{3.209}$$

The penalty factor $\alpha_\zeta(\zeta = 1, 2, \ldots, m_c)$ can be given as the function of coordinates, too, but usually it is simply assigned with a positive and sufficiently large constant α.

The important difference between the penalty factor $\boldsymbol{\alpha}$ and the Lagrange multiplier $\boldsymbol{\lambda}$ is that the former is a given constant (no variation is allowed), whereas the latter is a variable. Because $\boldsymbol{\alpha}$ is a preset constant, there will be no expansion of unknowns in the discrete equation set. However, a new question raised up is how to appropriately choose the penalty factor. Theoretically, it must be infinite, which is not permitted in practical computation. Therefore, in the penalty factor method the constraints in Eq. (3.203) cannot be exactly satisfied, but only approximated. If the penalty factor is too small, the constraints will not be properly enforced, whereas if it is too large, ill-conditioned equation will be encountered. The choice of penalty factor is actually the compromise art between physical accuracy and numerical stability, and trial-and-error is suggestable for the proper penalty factor. For FEM analysis, it is normally estimated by

$$\alpha = \text{constant} \times (1/h)^n \tag{3.210}$$

in which h is the characteristic length that can be the ratio of element size to the dimension of problem domain, and n is the shape function order of finite element.

It is also widely accepted in practical computations that

$$\begin{cases} \alpha = 1.0 \times 10^{5\text{--}8} \times \text{Young's modulus, or} \\ \alpha = 1.0 \times 10^{4\text{--}13} \times \max(\text{diagonal elements in the stiffness matrix}) \end{cases} \tag{3.211}$$

3.6.4 Galerkin Weak Form

Galerkin method seeks a finite-dimensional approximation to the PDEs in the weak sense. Suppose H is a Hilbert space of suitable functions obeying the imposed boundary conditions, whereas H_0 is the same Hilbert space, except that boundary conditions are set to zero. Instead of searching for a function $\mathbf{u}$ that exactly satisfies

$$\mathbf{Au} = 0 \tag{3.212}$$

in which $\mathbf{A}$ is differential operator imposed by suitable boundary conditions, we search a trial function $\mathbf{u} \in H$ and a test function $\mathbf{v} \in H_0$, such that the inner product meets

$$(\mathbf{v}, \mathbf{Au}) = 0 \tag{3.213}$$

It is actually to approximate the solution from a finite-dimensional sub-space of H.

In the fundamental FEM, test function $\mathbf{v}$ is identical to the shape function that is simultaneously employed to construct the approximate interpolation of $\mathbf{u}$ (trial function), this is well understood and described in many monographs (Strang and Fix 1973; Ciarlet 1978).

1. **Unconstrained formulation**

In solid dynamics, Galerkin weak form can be directly derived from Hamilton's principle. Introducing the Lagrange function (3.202) into Eq. (3.200) yields

$$\delta\int_{t_2}^{t_1}\left[-\frac{1}{2}\iiint_{\Omega}\boldsymbol{\varepsilon}^{\mathrm{T}}\boldsymbol{\sigma}\mathrm{d}\Omega+\iiint_{\Omega}\mathbf{u}^{\mathrm{T}}\mathbf{b}\mathrm{d}\Omega+\iint_{\Gamma_t}\mathbf{u}^{\mathrm{T}}\bar{\mathbf{t}}\mathrm{d}\Gamma+\sum_{\zeta=1}^{n_\zeta}\mathbf{u}^{\mathrm{T}}(\mathbf{x}_\zeta)\mathbf{T}_\zeta(\mathbf{x}_\zeta)\right.$$
$$\left.+\frac{1}{2}\iiint_{\Omega}\rho(\dot{\mathbf{u}})^{\mathrm{T}}(\dot{\mathbf{u}})d\Omega\right]\mathrm{d}t=0 \tag{3.214}$$

Since changing the order of variational operation and integral operation does not affect the results because all these operations are linear, therefore it does not matter which one operates first. Moving the variational operator into integral operators, it follows that

$$-\frac{1}{2}\iiint_{\Omega}\left[\delta\int_{t_2}^{t_1}\boldsymbol{\varepsilon}^{\mathrm{T}}\boldsymbol{\sigma}\mathrm{d}t\right]\mathrm{d}\Omega+\iiint_{\Omega}\left[\delta\int_{t_2}^{t_1}\mathbf{u}^{\mathrm{T}}\mathbf{b}\mathrm{d}t\right]\mathrm{d}\Omega+\iint_{\Gamma_t}\left[\delta\int_{t_2}^{t_1}\mathbf{u}^{\mathrm{T}}\bar{\mathbf{t}}\mathrm{d}t\right]\mathrm{d}\Gamma$$
$$+\delta\sum_{\zeta=1}^{n_\zeta}\mathbf{u}^{\mathrm{T}}(\mathbf{x}_\zeta)\mathbf{T}_\zeta(\mathbf{x}_\zeta)+\frac{1}{2}\iiint_{\Omega}\left[\delta\int_{t_2}^{t_1}\rho(\dot{\mathbf{u}})^{\mathrm{T}}\dot{\mathbf{u}}\mathrm{d}t\right]\mathrm{d}\Omega=0 \tag{3.215}$$

Use is made of the chain rule of variation, the integrand in the first integral term becomes

$$\delta(\boldsymbol{\varepsilon}^{\mathrm{T}}\boldsymbol{\sigma})=\delta\boldsymbol{\varepsilon}^{\mathrm{T}}\boldsymbol{\sigma}+\boldsymbol{\varepsilon}^{\mathrm{T}}\delta\boldsymbol{\sigma} \tag{3.216}$$

Because two right hand side terms in Eq. (3.216) are all scalars, so their transposes are still themselves. For the last right hand side term in Eq. (3.216), it follows that

$$\boldsymbol{\varepsilon}^{\mathrm{T}}\delta\boldsymbol{\sigma}=\left(\boldsymbol{\varepsilon}^{\mathrm{T}}\delta\boldsymbol{\sigma}\right)^{\mathrm{T}}=\delta\boldsymbol{\sigma}^{\mathrm{T}}\boldsymbol{\varepsilon} \tag{3.217}$$

Due to the symmetrical property of elastic matrix $\mathbf{D}$ in the constitutive Eq. (3.103), Eq. (3.217) becomes

$$\boldsymbol{\varepsilon}^{\mathrm{T}}\delta\boldsymbol{\sigma}=\delta\boldsymbol{\sigma}^{\mathrm{T}}\boldsymbol{\varepsilon}=\delta(\mathbf{D}\varepsilon)^{\mathrm{T}}\boldsymbol{\varepsilon}=\delta\boldsymbol{\varepsilon}^{\mathrm{T}}\mathbf{D}^{\mathrm{T}}\boldsymbol{\varepsilon}=\delta\boldsymbol{\varepsilon}^{\mathrm{T}}\mathbf{D}\varepsilon=\delta\boldsymbol{\varepsilon}^{\mathrm{T}}\boldsymbol{\sigma} \tag{3.218}$$

Therefore, Eq. (3.216) is equal to

$$\delta(\boldsymbol{\varepsilon}^{\mathrm{T}}\boldsymbol{\sigma}) = 2\delta\boldsymbol{\varepsilon}^{\mathrm{T}}\boldsymbol{\sigma} \tag{3.219}$$

Let's now look at the last integral term in Eq. (3.215). We first move the time integration into the spatial integration to get

$$\frac{1}{2}\delta\int_{t_2}^{t_1}\left[\iiint_{\Omega}\rho(\dot{\mathbf{u}})^{\mathrm{T}}\dot{\mathbf{u}}\mathrm{d}\Omega\right]\mathrm{d}t = \frac{1}{2}\iiint_{\Omega}\left[\delta\int_{t_2}^{t_1}\rho(\dot{\mathbf{u}})^{\mathrm{T}}\dot{\mathbf{u}}\mathrm{d}t\right]\mathrm{d}\Omega \tag{3.220}$$

Again, since this change of the order of integral operations does not affect the results, the chain rule of variation and the scalar property of time integration in Eq. (3.220) yield

$$\int_{t_1}^{t_2}\delta\left(\rho(\dot{\mathbf{u}})^{\mathrm{T}}\dot{\mathbf{u}}\right)\mathrm{d}t = \rho\int_{t_1}^{t_2}\left(\delta(\dot{\mathbf{u}})^{\mathrm{T}}\dot{\mathbf{u}} + (\dot{\mathbf{u}})^{\mathrm{T}}\delta\dot{\mathbf{u}}\right)\mathrm{d}t = 2\rho\int_{t_1}^{t_2}\delta(\dot{\mathbf{u}})^{\mathrm{T}}\dot{\mathbf{u}}\mathrm{d}t \tag{3.221}$$

Note that the variation and time differentiation are also exchangeable, namely

$$\int_{t_1}^{t_2}\left(\delta(\dot{\mathbf{u}})^{\mathrm{T}}\dot{\mathbf{u}}\right)\mathrm{d}t = \int_{t_1}^{t_2}\left(\frac{\partial\delta\mathbf{u}^{\mathrm{T}}}{\partial t}\frac{\partial\mathbf{u}^{\mathrm{T}}}{\partial t}\right)\mathrm{d}t \tag{3.222}$$

integrating by parts with respect to the time gives rise to

$$\int_{t_1}^{t_2}\left(\frac{\partial\delta\mathbf{u}^{\mathrm{T}}}{\partial t}\frac{\partial\mathbf{u}^{\mathrm{T}}}{\partial t}\right)\mathrm{d}t = \int_{t_1}^{t_2}-\delta\mathbf{u}^{\mathrm{T}}\ddot{\mathbf{u}}\mathrm{d}t + \delta\mathbf{u}^{\mathrm{T}}(\dot{\mathbf{u}})^{\mathrm{T}}\Big|_{t_1}^{t_2} \tag{3.223}$$

Invoking the initial conditions (3.184) at initial time (t_1) and final time (t_2), it may be confirmed that $\delta\mathbf{u}^{\mathrm{T}}$ must be zero at t_1 and t_2 because no variation is able to emerge for any given constant value. Therefore, the last right hand side term in Eq. (3.223) vanishes, i.e.,

$$\int_{t_1}^{t_2}\left(\frac{\partial\delta\mathbf{u}^{\mathrm{T}}}{\partial t}\frac{\partial\mathbf{u}^{\mathrm{T}}}{\partial t}\right)\mathrm{d}t = \int_{t_1}^{t_2}-\delta\mathbf{u}^{\mathrm{T}}\ddot{\mathbf{u}}\mathrm{d}t \tag{3.224}$$

and Eq. (3.220) becomes

$$\frac{1}{2}\delta\int_{t_1}^{t_2}\left[\iiint_{\Omega}\rho(\dot{\mathbf{u}})^{\mathrm{T}}\dot{\mathbf{u}}\mathrm{d}\Omega\right]\mathrm{d}t=\iiint_{\Omega}\left[\rho\int_{t_1}^{t_2}-\delta\mathbf{u}^{\mathrm{T}}\ddot{\mathbf{u}}\mathrm{d}t\right]\mathrm{d}\Omega \tag{3.225}$$

Now, switch the order of the integration in the right hand side item of Eq. (3.225), we get

$$\delta\int_{t_1}^{t_2}\left[\frac{1}{2}\iiint_{\Omega}\rho(\dot{\mathbf{u}})^{\mathrm{T}}\dot{\mathbf{u}}\mathrm{d}\Omega\right]\mathrm{d}t=-\int_{t_1}^{t_2}\left[\rho\iiint_{\Omega}\delta\mathbf{u}^{\mathrm{T}}\ddot{\mathbf{u}}\mathrm{d}\Omega\right]\mathrm{d}t \tag{3.226}$$

Use is made of Eqs. (3.219) and (3.226), Eq. (3.215) becomes

$$\begin{aligned}\int_{t_2}^{t_1}\Bigg[&-\iiint_{\Omega}\delta\boldsymbol{\varepsilon}^{\mathrm{T}}\boldsymbol{\sigma}\mathrm{d}\Omega+\iiint_{\Omega}\delta\mathbf{u}^{\mathrm{T}}\mathbf{b}\mathrm{d}\Omega+\iint_{\Gamma_t}\delta\mathbf{u}^{\mathrm{T}}\bar{\mathbf{t}}\mathrm{d}\Gamma+\sum_{\zeta=1}^{n_\zeta}\delta\mathbf{u}^{\mathrm{T}}(\mathbf{x}_\zeta)\mathbf{T}_\zeta(\mathbf{x}_\zeta)\\&-\iiint_{\Omega}\rho\delta\mathbf{u}^{\mathrm{T}}\ddot{\mathbf{u}}\mathrm{d}\Omega\Bigg]\mathrm{d}t=0\end{aligned} \tag{3.227}$$

To satisfy the condition that for all possible choices of **u**, the time integration in Eq. (3.227) have to vanish, the well-known Galerkin weak form is finally obtained as

$$\begin{aligned}&\iiint_{\Omega}\delta\boldsymbol{\varepsilon}^{\mathrm{T}}\boldsymbol{\sigma}\mathrm{d}\Omega-\iiint_{\Omega}\delta\mathbf{u}^{\mathrm{T}}\mathbf{b}\mathrm{d}\Omega-\iint_{\Gamma_t}\delta\mathbf{u}^{\mathrm{T}}\bar{\mathbf{t}}\mathrm{d}\Gamma-\sum_{\zeta=1}^{n_\zeta}\delta\mathbf{u}^{\mathrm{T}}(\mathbf{x}_\zeta)\mathbf{T}_\zeta(\mathbf{x}_\zeta)\\&\quad+\iiint_{\Omega}\rho\delta\mathbf{u}^{\mathrm{T}}\ddot{\mathbf{u}}\mathrm{d}\Omega=0\end{aligned} \tag{3.228}$$

which is identical to Eq. (3.199) that is derived from the principle of virtual work.

By using the stress-strain relation (3.103) and the strain-displacement relation (3.85), Eq. (3.228) can be explicitly expressed in terms of displacements (essential variables) **u**,

$$\begin{aligned}&\iiint_{\Omega}\delta(\mathbf{L}\mathbf{u})^{\mathrm{T}}\mathbf{D}(\mathbf{L}\mathbf{u})\mathrm{d}\Omega-\iiint_{\Omega}\delta\mathbf{u}^{\mathrm{T}}\mathbf{b}\mathrm{d}\Omega-\iint_{\Gamma_t}\delta\mathbf{u}^{\mathrm{T}}\bar{\mathbf{t}}\mathrm{d}\Gamma\\&\quad-\sum_{\zeta=1}^{n_\zeta}\delta\mathbf{u}^{\mathrm{T}}(\mathbf{x}_\zeta)\mathbf{T}_\zeta(\mathbf{x}_\zeta)+\iiint_{\Omega}\rho\delta\mathbf{u}^{\mathrm{T}}\ddot{\mathbf{u}}\mathrm{d}\Omega=0\end{aligned} \tag{3.229}$$

For static problems, it reduces to,

$$\iiint_{\Omega} \delta(\mathbf{Lu})^{\mathrm{T}}\mathbf{D}(\mathbf{Lu})\mathrm{d}\Omega - \iiint_{\Omega} \delta\mathbf{u}^{\mathrm{T}}\mathbf{b}\mathrm{d}\Omega - \iint_{\Gamma_t} \delta\mathbf{u}^{\mathrm{T}}\bar{\mathbf{t}}\mathrm{d}\Gamma - \sum_{\zeta=1}^{n_\zeta} \delta\mathbf{u}^{\mathrm{T}}(\mathbf{x}_\zeta)\mathbf{T}_\zeta(\mathbf{x}_\zeta) = 0 \tag{3.230}$$

It is notable that we are not necessary to know the strong form of system equation in Galerkin procedure, thus it retains a wider range of applicability. In addition, Galerkin weak form is very handy in the application to solid dynamics, because one does not need to perform integration by parts. The discretized equation set can be derived very easily using approximated displacements.

2. **Constrained formulation**

Where the approximated field function does not satisfy boundary condition in Eq. (3.185), or the problem domain are embedded with parts of internal boundaries, a constrained Galerkin weak form need to be derived in a manner similar to that of Hamilton's principle, but without the modification of Lagrangian function.

(1) **Galerkin weak form with Lagrange multipliers**

For dynamic problems,

$$\begin{aligned} &\iiint_{\Omega} \delta(\mathbf{Lu})^{\mathrm{T}}\mathbf{D}(\mathbf{Lu})\mathrm{d}\Omega - \iiint_{\Omega} \delta\mathbf{u}^{\mathrm{T}}\mathbf{b}\mathrm{d}\Omega - \iint_{\Gamma_t} \delta\mathbf{u}^{\mathrm{T}}\bar{\mathbf{t}}\mathrm{d}\Gamma - \sum_{\zeta=1}^{n_\zeta} \delta\mathbf{u}^{\mathrm{T}}(\mathbf{x}_\zeta)\mathbf{T}_\zeta(\mathbf{x}_\zeta) \\ &- \iiint_{\Omega} \delta\boldsymbol{\lambda}^{\mathrm{T}}\mathbf{C}(\mathbf{u})\mathrm{d}\Omega - \iiint_{\Omega} \boldsymbol{\lambda}^{\mathrm{T}}\delta\mathbf{C}(\mathbf{u})\mathrm{d}\Omega + \iiint_{\Omega} \rho\delta\mathbf{u}^{\mathrm{T}}\ddot{\mathbf{u}}\mathrm{d}\Omega = 0 \end{aligned} \tag{3.231}$$

For static problems, it is simply reduced to

$$\begin{aligned} &\iiint_{\Omega} \delta(\mathbf{Lu})^{\mathrm{T}}\mathbf{D}(\mathbf{Lu})\mathrm{d}\Omega - \iiint_{\Omega} \delta\mathbf{u}^{\mathrm{T}}\mathbf{b}\mathrm{d}\Omega - \iint_{\Gamma_t} \delta\mathbf{u}^{\mathrm{T}}\bar{\mathbf{t}}\mathrm{d}\Gamma - \sum_{\zeta=1}^{n_\zeta} \delta\mathbf{u}^{\mathrm{T}}(\mathbf{x}_\zeta)\mathbf{T}_\zeta(\mathbf{x}_\zeta) \\ &- \iiint_{\Omega} \delta\boldsymbol{\lambda}^{\mathrm{T}}\mathbf{C}(\mathbf{u})\mathrm{d}\Omega - \iiint_{\Omega} \boldsymbol{\lambda}^{\mathrm{T}}\delta\mathbf{C}(\mathbf{u})\mathrm{d}\Omega = 0 \end{aligned} \tag{3.232}$$

(2) **Galerkin weak form with penalty factors**

For dynamic problems,

$$\iiint_{\Omega} \delta(\mathbf{Lu})^{\mathrm{T}}\mathbf{D}(\mathbf{Lu})\mathrm{d}\Omega - \iiint_{\Omega} \delta\mathbf{u}^{\mathrm{T}}\mathbf{b}\mathrm{d}\Omega - \iint_{\Gamma_t} \delta\mathbf{u}^{\mathrm{T}}\bar{\mathbf{t}}\mathrm{d}\Gamma - \sum_{\zeta=1}^{n_\zeta} \delta\mathbf{u}^{\mathrm{T}}(\mathbf{x}_\zeta)\mathbf{T}_\zeta(\mathbf{x}_\zeta)$$

$$- \iiint_{\Omega} \delta \mathbf{C}(\mathbf{u})^{\mathrm{T}} \alpha \mathbf{C}(\mathbf{u}) \mathrm{d}\Omega + \iiint_{\Omega} \rho \delta \mathbf{u}^{\mathrm{T}} \ddot{\mathbf{u}} \mathrm{d}\Omega = 0 \tag{3.233}$$

For static problems, it is simply reduced to

$$\iiint_{\Omega} \delta(\mathbf{Lu})^{\mathrm{T}} \mathbf{D}(\mathbf{Lu}) \mathrm{d}\Omega - \iiint_{\Omega} \delta \mathbf{u}^{\mathrm{T}} \mathbf{b} \mathrm{d}\Omega - \iint_{\Gamma_t} \delta \mathbf{u}^{\mathrm{T}} \bar{\mathbf{t}} \mathrm{d}\Gamma$$

$$- \sum_{\zeta=1}^{n_\zeta} \mathbf{u}^{\mathrm{T}}(\mathbf{x}_\zeta) \mathbf{T}_\zeta(\mathbf{x}_\zeta) - \iiint_{\Omega} \delta \mathbf{C}(\mathbf{u})^{\mathrm{T}} \alpha \mathbf{C}(\mathbf{u}) \mathrm{d}\Omega = 0 \tag{3.234}$$

3.6.5 Weighted Residual Method

1. **Concept**

Polynomial series expansion is extensively exercised in mathematics and engineering to yield meaningful solutions. Among them, "weighted residual methods (WRMs)" are powerful tools due to their simplicity and high accuracy. Methods dubbed with terms of "collocation", "Galerkin", "least square", and "Rayleigh-Ritz" may be looked at as special cases of WRMs.

Suppose a differential operator A that acts on a function $u(x)$ to produce another function $p(x)$

$$\mathrm{A}(u(x)) = p(x) \tag{3.235}$$

A series of independent function basis $u_1(x, y), \ldots, u_n(x, y)$ are used to construct the trial function $\tilde{u}$ that is a linear combination to approximate u such that

$$u \cong \tilde{u} = \sum_{i=1}^{n} c_i u_i(x) \tag{3.236}$$

The error or residual may be calculated by

$$R(x) = \mathrm{A}(\tilde{u}) - p(x) \neq 0 \tag{3.237}$$

This residual should be forced to zero in average sense over the domain with the help of weighting functions $W_i(x) (i = 1, 2, \ldots n)$, for example

$$\int R(x) W_i(x) \mathrm{d}x = 0 \quad (i = 1, 2, \ldots, n) \tag{3.238}$$

WRM is a very simple but powerful mathematical tool to straightforwardly transfer the most PDEs that govern the solid and fluid dynamics into approximated weak form. It is notable that the number of weighting functions $W_i(x)$ is exactly equal to the number of unknown constants c_i in $\tilde{u}$, hence the result of Eq. (3.238) is a set of n algebraic equations for $c_i(i = 1, 2, \ldots, n)$. The main difference between the descendent methods originated from Eq. (3.238) is the selection of $W_i(x)$, which gives rise to different precisions and computation efforts.

It is also necessary to emphasize that if approximated Eq. (3.236) (trial function) contains the exact solution of the problem, WRMs will produce exact solution as long as there is no numerical error in computation. This characteristic is useful in the validation of "mesh-free methods (MFMs)" based on WRMs.

2. **Formulation**

Let us consider the PDEs for solid dynamics written in a general concise functional form

$$\mathbf{Au} = \begin{Bmatrix} A_1\mathbf{u} \\ A_2\mathbf{u} \\ \cdots \\ A_{m_A}\mathbf{u} \end{Bmatrix} = 0 \tag{3.239}$$

We seek such an unknown variable function $\mathbf{u}$ from Eq. (3.239) in a domain Ω with certain boundary conditions on the domain boundary Γ

$$\mathbf{Cu} = \begin{Bmatrix} C_1\mathbf{u} \\ C_2\mathbf{u} \\ \cdots \\ C_{m_c}\mathbf{u} \end{Bmatrix} = 0 \tag{3.240}$$

If the PDEs are linear, Eqs. (3.239) and (3.240) may be specified as follows

$$\begin{cases} \mathbf{A}(\mathbf{u}) = 0 & \text{in } \Omega \\ \mathbf{C}(\mathbf{u}) = \begin{cases} \mathbf{Mu} + \bar{\mathbf{t}} = 0 & \text{on } \Gamma_t \\ \mathbf{u} - \bar{\mathbf{u}} = 0 & \text{on } \Gamma_u \end{cases} \end{cases} \tag{3.241}$$

Because the PDEs (3.239) have to be zero at each point of the domain Ω, it follows that

$$\iiint\limits_{\Omega} \mathbf{W}^{\mathrm{T}}\mathbf{Au}\mathrm{d}\Omega = \iiint\limits_{\Omega} \left[W_1\, A_1(\mathbf{u})\; W_2A_2(\mathbf{u})\; \ldots\; W_{m_A}A_{m_A}(\mathbf{u})\right]\mathrm{d}\Omega = 0 \tag{3.242}$$

in which

$$\mathbf{W} = \left[W_1\;\; W_2\; \ldots\; W_{m_A}\right]^{\mathrm{T}} \tag{3.243}$$

is a set of arbitrary functions equal to the amount of equation (or components of **u**) involved.

It may be further asserted that if Eq. (3.241) is satisfied by variable function **u** in domain, then **u** must satisfy Eq. (3.242) at all points of the domain. This statement is, however, very powerful in the exploration of CMs.

If boundary condition (3.240) is simultaneously satisfied, namely

$$\iint\limits_{\Gamma} \overline{\mathbf{W}}^{\mathrm{T}} \mathbf{C}\mathbf{u} \mathrm{d}\Gamma = \iint\limits_{\Gamma} \left[\overline{W}_1 C_1(\mathbf{u}) \; \overline{W}_2 C_2(\mathbf{u}) \ldots \overline{W}_{m_c} C_{m_c}(\mathbf{u}) \right] \mathrm{d}\Gamma = 0 \tag{3.244}$$

Then the integral equation

$$\iiint\limits_{\Omega} \mathbf{W}^{\mathrm{T}} \mathbf{A}\mathbf{u} \mathrm{d}\Omega + \iint\limits_{\Gamma} \overline{\mathbf{W}}^{\mathrm{T}} \mathbf{C}\mathbf{u} \mathrm{d}\Gamma = 0 \tag{3.245}$$

is equivalent to PDEs (3.239) and boundary conditions (3.240), subject to implicit postulation that the integrals in Eq. (3.245) are capable of being evaluated. This imposes certain restrictions on the possible families to which functions $\mathbf{W}$, $\overline{\mathbf{W}}$, and **u** must belong. Generally, $\mathbf{W}$ and $\overline{\mathbf{W}}$ are preferably bounded functions such as in the formulation of EFGM (Chap. 9) and SPH (Chap. 10).

On many occasions, it is possible to perform integration by parts on Eq. (3.245) to get an alternative expression

$$\iiint\limits_{\Omega} (\mathbf{G}\mathbf{W}^{\mathrm{T}})(\mathbf{E}\mathbf{u}) \mathrm{d}\Omega + \iint\limits_{\Gamma} (\mathbf{H}\overline{\mathbf{W}})^{\mathrm{T}} (\mathbf{F}\mathbf{u}) \mathrm{d}\Gamma = 0 \tag{3.246}$$

in which the new operators **E** and **F** should contain lower order derivatives than those **A** and **C** occurring in original operators. It follows that a lower order of continuity is paid off in the choice of functions **u** at a price of higher continuity of $\mathbf{W}$ and $\overline{\mathbf{W}}$ subject to the additional differential operators **G** and **F**. To be specific, if $2k$th-order derivatives occur in any term of **A** or **C**, then only $(2k - 1)$th-order continuous w.r.t. the derivatives of **u** is demanded (C_{2k-1} continuity). This will be detailed in the formulation of SPH (Chap. 10).

Equation (3.246) is now more "permissive" than the original problem posed by Eq. (3.241) and is called a counterpart weak form. It is rather interesting that this weak form is often more physically realistic than its original strong form, because the latter imposes an excessive "smoothness" on the true solution that is not demanded in the real physical world. It is notable that to use the WRM, the strong form of system equation needs to be known.

Clearly, almost any set of independent functions $\mathbf{W}_i$ and $\overline{\mathbf{W}}_i$ could be used for the purpose of weighting and, according to the choice of weighting function, a different name can be dubbed. In addition, in many advances CMs $\mathbf{W}_i = \overline{\mathbf{W}}_i$ is customarily postulated.

(1) **Point collocation method**

In this method the function family of Dirac Delta is employed, namely,

$$\mathbf{W}_i(\mathbf{x}) = \delta(\mathbf{x} - \mathbf{x}_i)\mathbf{n} \tag{3.247}$$

where

$$\delta(\mathbf{x} - \mathbf{x}_i) = \begin{cases} 1 \text{ if } \mathbf{x} = \mathbf{x}_i (i = 1, 2, \ldots n) \\ 0 \text{ otherwise} \end{cases} \tag{3.248}$$

and **n** is the unit vector correspondent to component variables. Taking the deformation problem in 3-D body for example, we have

$$\mathbf{n} = \begin{bmatrix} 1 & 1 & 1 \end{bmatrix}^{\mathrm{T}} \tag{3.249}$$

In the collocation method, function (4.247) must be forced zero at n points within the domain and the integration is nominal. FDM is a particular case of this weighting.

(2) **Subdomain collocation method**

The choice of $\mathbf{W}_i = \mathbf{n}$ in subdomain Ω_i and zero elsewhere leads to the subdomain collocation method. As one of the many finite volume methods, it essentially makes the zero-error for the integral over the specified subdomain.

(3) **Galerkin method**

Galerkin weak form can be formulated in many other ways, and one alternative is by WRMs. Let $\mathbf{W}_i = \mathbf{N}_i (i = 1, 2, \ldots n)$, we get the Galerkin method in which the shape (or basis) functions $\mathbf{N}_i (i = 1, 2, \ldots n)$ (viz. Chap. 4) are directly used as weighting functions. This method frequently (but not always) leads to symmetric matrices. In many literatures, the names of Petrov and Galerkin are often associated this method.

3.6.6 Specifications of Weak Form in Water Permeability

Permeability study relies on the solution of hydraulic potential (or water head) function $\phi(x, y, x)$. The steady state permeability theory for pervious continuum comes from the direct analogue of fluid dynamics in which the equation of continuity is specified as

$$\frac{\partial v_x}{\partial x} + \frac{\partial v_y}{\partial y} + \frac{\partial v_z}{\partial z} + q_0 = 0 \tag{3.250}$$

in which q_0 is the rate of inner source.

Darcy's law is additionally postulated that is expressed in the global coordinate system

$$\mathbf{v} = -\mathbf{k}\nabla\phi \tag{3.251}$$

It is always possible to determine a locally coordinate system in which the permeability tensor $\mathbf{k}$ becomes diagonal, namely

$$\mathbf{k} = \begin{bmatrix} k_x & 0 & 0 \\ 0 & k_y & 0 \\ 0 & 0 & k_z \end{bmatrix} \tag{3.252}$$

Introducing Eq. (2.251) into Eq. (3.250), the governing equation of the seepage flow in perfectly saturated continuum under the assumption that both material grains and pore water are incompressible, is associated with three orthogonal local axes as

$$-\frac{\partial}{\partial x}(k_x\frac{\partial\phi}{\partial x}) - \frac{\partial}{\partial y}(k_y\frac{\partial\phi}{\partial y}) - \frac{\partial}{\partial z}(k_z\frac{\partial\phi}{\partial z}) + q_0 = 0 \quad \mathbf{x} \in \Omega \tag{3.253}$$

There is one type of boundary Γ_ϕ on which the hydraulic potential $\phi_0(x, y, z)$ is specified (first type or Dirichlet boundary)

$$\phi(x, y, z)|_{\Gamma_\phi} = \phi_0(x, y, z) \tag{3.254}$$

There is another boundary Γ_q on which the normal gradient of hydraulic potential (or flow rate) $\overline{q}(x, y, z)$ is specified (second type or Neumann boundary)

$$\begin{cases} -k_n \cdot \left.\frac{\partial\phi}{\partial n}\right|_{\Gamma_q} = \overline{q}(x, y, z) \\ \frac{\partial\phi}{\partial n} = \frac{\partial\phi}{\partial x}n_x + \frac{\partial\phi}{\partial y}n_y + \frac{\partial\phi}{\partial z}n_z \end{cases} \tag{3.255}$$

where: $\mathbf{n} = [n_x \; n_y \; n_z]^{\mathrm{T}}$—unit outside normal vector to the boundary surface; $\overline{q}(x, y, z)$—flow rate per unit area on the boundary surface, $\overline{q}(x, y, z) = 0$ means the boundary is impervious/isolated.

Now, PDE (3.253) and boundary conditions (3.254)–(3.255) are re-written in the similar form of Eqs. (3.239)–(3.241)

$$\mathrm{A}(\phi) = -\nabla(\mathbf{k}\nabla)\phi + q_0 = 0 \quad \text{in } \mathbf{x} \in \Omega \tag{3.256}$$

$$\mathbf{C}(\phi) = \begin{cases} \phi - \phi_0(x, y, z) = 0 & \text{on } \mathbf{x} \in \Gamma_\phi \\ k_n \cdot \frac{\partial\phi}{\partial n} + \overline{q}(x, y, z) = 0 & \text{on } \mathbf{x} \in \Gamma_q \end{cases} \tag{3.257}$$

The variable function ϕ will vanish in the integrals taken along the boundary Γ_ϕ if $\phi - \overline{\phi} = 0$ is able to enforced in the discretization scheme (e.g. FEM), this is why the first type or Dirichlet boundary is often called "forced" or "essential" one. In CMs, this is realized by the construction of test (shape) functions possessing Kronecker delta property. Under such circumstances, the integral of Eq. (3.245) becomes

$$\iiint_{\Omega} W\left[-\frac{\partial}{\partial x}(k_x\frac{\partial \phi}{\partial x}) - \frac{\partial}{\partial y}(k_y\frac{\partial \phi}{\partial y}) - \frac{\partial}{\partial z}(k_z\frac{\partial \phi}{\partial z}) + q_0\right]\mathrm{d}\Omega$$
$$+ \iint_{\Gamma_q} \overline{W}\left[k_n \cdot \frac{\partial \phi}{\partial n} + \bar{q}(x, y, z)\right]\mathrm{d}\Gamma = 0 \tag{3.258}$$

As the first step towards the weak form solution, use is made of integration by parts

$$\begin{cases} \iiint_{\Omega} W\frac{\partial}{\partial x}(k_x\frac{\partial \phi}{\partial x})\mathrm{d}\Omega = -\iiint_{\Omega} \frac{\partial W}{\partial x}(k_x\frac{\partial \phi}{\partial x})\mathrm{d}\Omega + \oint_{\Gamma} W(k_x\frac{\partial \phi}{\partial x})n_x\mathrm{d}\Gamma \\ \iiint_{\Omega} W\frac{\partial}{\partial y}(k_y\frac{\partial \phi}{\partial y})\mathrm{d}\Omega = -\iiint_{\Omega} \frac{\partial W}{\partial y}(k_y\frac{\partial \phi}{\partial y})\mathrm{d}\Omega + \oint_{\Gamma} W(k_y\frac{\partial \phi}{\partial y})n_y\mathrm{d}\Gamma \\ \iiint_{\Omega} W\frac{\partial}{\partial z}(k_z\frac{\partial \phi}{\partial z})\mathrm{d}\Omega = -\iiint_{\Omega} \frac{\partial W}{\partial z}(k_z\frac{\partial \phi}{\partial z})\mathrm{d}\Omega + \oint_{\Gamma} W(k_z\frac{\partial \phi}{\partial z})n_z\mathrm{d}\Gamma \end{cases} \tag{3.259}$$

Equation (3.258) becomes

$$\iiint_{\Omega} \left[\frac{\partial W}{\partial x}(k_x\frac{\partial \phi}{\partial x}) + \frac{\partial W}{\partial y}(k_y\frac{\partial \phi}{\partial y}) + \frac{\partial W}{\partial z}(k_z\frac{\partial \phi}{\partial z}) + Wq_0\right]\mathrm{d}\Omega$$
$$= \oiint_{\Gamma} W\left[k_x\frac{\partial \phi}{\partial x}n_x + k_y\frac{\partial \phi}{\partial y}n_y + k_z\frac{\partial \phi}{\partial z}n_z\right]\mathrm{d}\Gamma$$
$$- \iint_{\Gamma_q} \overline{W}\left[k_n \cdot \frac{\partial \phi}{\partial n} + q(x, y, z)\right]\mathrm{d}\Gamma$$
$$= \oiint_{\Gamma} W\left[k_n . \frac{\partial \phi}{\partial n}\right]\mathrm{d}\Gamma$$
$$- \iint_{\Gamma_q} \overline{W}\left[k_n \cdot \frac{\partial \phi}{\partial n} + q(x, y, z)\right]\mathrm{d}\Gamma \tag{3.260}$$

Further, if we let

$$\overline{W} = W \text{ on } \Gamma \tag{3.261}$$

from Eq. (3.360), it follows that

$$\iiint_{\Omega} (\nabla W)^{\mathrm{T}} \mathbf{k} (\nabla \phi) \mathrm{d}\Omega = - \iiint_{\Omega} W q_0 \mathrm{d}\Omega - \iint_{\Gamma_q} W \overline{q} \mathrm{d}\Gamma \tag{3.262}$$

in which ∇ is the Hamilton operator.

It is noted that if $\overline{q} = 0$ on boundary Γ_q, the last term in Eq. (3.262) vanish. This is why the second type or Neumann boundary condition is known as "natural boundary condition".

Equation (3.262) is the weak form of the water permeability equivalent to its strong form counterpart. It admits discontinuous permeability matrix $\mathbf{k}$ and discontinuous first derivatives of ϕ, which is not easily admitted in the strong (differential) form.

3.7 Concluding Remarks

3.7.1 Non-linear Constitutive Relations for the Dynamics of Continuum

In the elaboration of governing equations for continuum dynamics hereinafter, the matrix $\mathbf{D} = [D]$ may be uniformly used, which could be elastic $\mathbf{D}^e$ or elasto-plastic $\mathbf{D}^{ep}$, dependent on the specific constitutive relation postulated. The well-posed problem by either strong form or weak form holds these constitutive relations. Take the Galerkin method of static problems for example, Eq. (3.230) may be simply specified with elasto-plastic performance as follows

$$\begin{aligned} &\iiint_{\Omega} \delta(\mathbf{L}\Delta\mathbf{u})^{\mathrm{T}} \mathbf{D}^{ep} (\mathbf{L}\Delta\mathbf{u}) \mathrm{d}\Omega - \iiint_{\Omega} \delta\Delta\mathbf{u}^{\mathrm{T}} \Delta\mathbf{b} \mathrm{d}\Omega \\ &- \iint_{\Gamma_t} \delta\Delta\mathbf{u}^{\mathrm{T}} \Delta\bar{\mathbf{t}} \mathrm{d}\Gamma - \sum_{\zeta=1}^{n_\zeta} \delta\mathbf{u}^{\mathrm{T}}(\mathbf{x}_\zeta) \mathbf{T}_\zeta(\mathbf{x}_\zeta) = 0 \end{aligned} \tag{3.263}$$

In the most CMs, Eq. (3.263) leads to incremental-iterative schemes such as the "variable stiffness method". If we prefer to use elasto-viscoplastic constitutive Eq. (3.138) with explicit time-stepping scheme, Eq. (3.230) may be alternatively expressed by

$$\iiint_{\Omega} \delta(\mathbf{L}\Delta\mathbf{u})^{\mathrm{T}} \mathbf{D}^{e} (\mathbf{L}\Delta\mathbf{u}) \mathrm{d}\Omega - \iiint_{\Omega} \delta(\mathbf{L}\Delta\mathbf{u})^{\mathrm{T}} \Delta\boldsymbol{\sigma}^{vp} \mathrm{d}\Omega$$

$$-\iiint_{\Omega} \delta\Delta\mathbf{u}^{\mathrm{T}}\Delta\mathbf{b}d\Omega - \iint_{\Gamma_t} \delta\Delta\mathbf{u}^{\mathrm{T}}\Delta\bar{\mathbf{t}}d\Gamma - \sum_{\zeta=1}^{n_\zeta} \delta\mathbf{u}^{\mathrm{T}}(\mathbf{x}_\zeta)\mathbf{T}_\zeta(\mathbf{x}_\zeta) = 0 \quad (3.264)$$

In the most CMs, Eq. (3.264) leads to incremental-iterative schemes, such as the "initial stress method" and "initial strain method".

3.7.2 Non-linear Constitutive Relations for the Dynamics of Particle Cluster

Basically, all continua may be simulated by particle-based methods in sufficiently small scale (e.g. nano-or molecular-scale), not to mention often encountered macro-scale mechanical scenarios such as the landslide composed of mixed granular rock and soil (Gary et al. 2014). Such a particle system may be directly simulated using the classical mechanics of particles illustrated in Sect. 3.2. If damping is considered, Eq. (3.1) becomes

$$\mathbf{F}_i + \mathbf{R}_i = m_i\ddot{\mathbf{x}}_i + C\dot{\mathbf{x}}_i \quad (i = 1, 2, 3, \ldots, n) \quad (3.265a)$$

or

$$\mathbf{F}_i + \mathbf{R}_i = m_i\ddot{\mathbf{x}}_i + \alpha m_i\dot{\mathbf{x}}_i \quad (i = 1, 2, 3, \ldots, n) \quad (3.265b)$$

in which C_i is the damping factor and α is the damping coefficient. The active force $\mathbf{F}_i$ and restraint force $\mathbf{R}_i$ on particle i are dependent on the boundary conditions and inter-particle actions. Use is made of Eq. (3.39), it follows that

$$\mathbf{F}_i + \mathbf{R}_i = m_i\ddot{\mathbf{u}}_i + C_i\dot{\mathbf{u}}_i \quad (i = 1, 2, 3, \ldots, n) \quad (3.266a)$$

or

$$\mathbf{F}_i + \mathbf{R}_i = m_i\ddot{\mathbf{u}}_i + \alpha m_i\dot{\mathbf{u}}_i \quad (i = 1, 2, 3, \ldots, n) \quad (3.266b)$$

where $\mathbf{u}$ is the displacement vector. With the help of updated Lagrange scheme, a corresponding incremental formulation is further get

$$\Delta\mathbf{F}_i = m_i\Delta\ddot{\mathbf{u}}_i + C_i\Delta\dot{\mathbf{u}}_i \quad (i = 1, 2, 3, \ldots, n) \quad (3.267a)$$

or

$$\Delta\mathbf{F}_i = m_i\Delta\ddot{\mathbf{u}}_i + \alpha m_i\Delta\dot{\mathbf{u}}_i \quad (i = 1, 2, 3, \ldots, n) \quad (3.267b)$$

Suppose the restraint force $\mathbf{R}_i$ is in proportional to the displacement increment but in opposite direction, we have

$$\Delta \mathbf{F}_i = k\Delta \mathbf{u}_i + C_i \Delta \dot{\mathbf{u}}_i + m_i \Delta \ddot{\mathbf{u}}_i \quad (i = 1, 2, 3, \ldots, n) \tag{3.268a}$$

or

$$\Delta \mathbf{F}_i = m_i \Delta \ddot{\mathbf{u}}_i + k\Delta \mathbf{u}_i + \alpha m_i \Delta \dot{\mathbf{u}}_i \quad (i = 1, 2, 3, \ldots, n) \tag{3.268b}$$

in which k is the stiffness coefficient.

A good example is the "discrete element methods (DEMs)" that will be detailed in Chap. 8. In the following paragraphs, only the essential issues, particularly the constitutive laws that related to the "particle-based methods (PBMs)" for continuum mechanics will be further discussed.

At the first glance of Eq. (3.268), it is easy to realize that the constitutive laws elaborated in this chapter for the substance parcel restrained by REV are not directly applicable to PBMs. Since the constitutive laws for particle dynamics simulation depends on the assigned particle properties and inter-particle action laws, it is not so well defined a priori in the similar way to classical continuum mechanics. However, the constitutive behaviors described previously can be generally replicated within the particle assemblage using the inter-particle actions of "soft-spheres". In fact, critical states and stress-dependent elasto-plastic behaviors in strain hardening/softening responses might be a natural outgrowth of the particle nature of material. These behaviors reflect the changes in granular packing and dilation accompanying compression and shear actions.

A full specification for the constitutive relation of simulated granular material has not yet been completed, although various characterization studies have been conducted under specific loading path, such as granular shear (Morgan 1999, 2004) and Mohr-Coulomb failure criterion (Potyondy and Cundall 2004). One prevalent approach to overcome the lack of certainty about the constitutive response of geo-material is to assign constitutive properties (i.e., Mohr-Coulomb parameters) in advance, and then fine-tunes inter-particle stresses to attain the expected response using the observational data of, for example, a small part of real-world engineering structure. As a demonstration, bi-axial or tri-axial test conducted on a confined particle assemblage might exhibit elastic strain from initial loading until reaching a peak strength, at this point discrete bonds begin to break, causing progressive strain softening until a through-going fracture develops. Simulations of such a numerical material under different stress paths and loading conditions will provide valuable insights into the yield and constitutive relation of geo-material (Coetzee 2017). Such an approach offers more exact matching of desired constitutive properties, meanwhile retains the ability towards the simulation of discontinuity initiation and evolution (Egholm 2007).

References

Abraham R, Marsden JE. Foundations of mechanics. London (UK): Benjamin-Cummings; 1978.

Airy GB. On the strains in the interior of beams. Philos Trans Roy Soc Lond. 1864;1863:153.

Akin JE. Finite element analysis with error estimators: an introduction to the FEM and adaptive error analysis for engineering students. Amsterdam (The Netherlands): Elsevier, Butterworth-Heinemann; 2005.

Arnold VI. Mathematical methods of classical mechanics. New York (USA): Springer; 1978.

Auricchio F, Taylor RL. Two material models for cyclic plasticity: nonlinear kinematic hardening and generalized plasticity. Int J Plast. 1995;11(1):65–98.

Barton NR, Bandis S, Bakhtar K. Strength, deformation and conductivity coupling of rock joints. Int J Rock Mech Min Sci Geomech Abstr. 1985;22(3):121–40.

Batchelor GK. An introduction to fluid dynamics. Cambridge (UK): Cambridge University Press; 1967.

Bathe KJ. Finite-element-method. Berlin (Germany): Springer; 2002.

Bažant ZP, Krizek RJ. Endochronic constitutive law for liquefaction of sand. J Eng Mech Div ASCE. 1976;102(EM2):225–38.

Bažant ZP, Belytschko T, Chang TP. Continuum theory for strain-softening. J Eng Mech ASCE. 1984;110(12):1666–92.

Bennett A. Lagrangian fluid dynamics. Cambridge (UK): Cambridge University Press; 2006.

Benson DJ. Computational methods in Lagrangian and Eulerian. Comput Meth Appl Mech Eng. 1992;99(2–3):235–394.

Brebbia CA, editor. Finite element systems (a handbook). Berlin (Germany): Springer; 1985.

Bristcau MO, Glowinski R, Mantel B, Periaux J, Perrier P. Numerical methods for incompressible and compressible Navier-Stokes problems. In: Gallagher RH, Carey G, Oden JT, Zienkiewicz OC, editors. Finite elements in fluids, vol. 6. Chichester (UK): Wiley; 1985. p. 1–40.

Bristeau MO, Glowinski R, Periaux J. Numerical methods for the Navier-Stokes equations. Applications to the simulation of compressible and incompressible viscous flow. Comput Phys Rep. 1987;6(1–6):73–187.

Budd CJ, Iserles A. Geometric integration: numerical solution of differential equations on manifolds. Philos Trans Roy Soc Lond A. 1999;357:943–1133.

Cauchy AL. Sur les equations qui experiment les conditions d'equilibre on les lois du mouvement interieur. Exerc Math. 1828;3:160–87 (in French).

Cauchy AL. Sur les l'equilibre et le mouvement interieur des corps. Exerc Math. 1829;4:293–319 (in French).

Chen SH. Computational geomechanics and hydraulic structures. Berlin (Germany): Springer; 2018.

Chen WF, Baladi GY. Soil plasticity: theory and implementation. Amsterdam (The Netherlands): Elsevier; 1985.

Chen SH, Wang HR, Xiong WL. Study of the seepage characteristics of joint. J Wuhan Univ Hydraul Electr Eng (WUHEE). 1989;22(1):51–60 (in Chinese with English abstract).

Chen SH, Qin WX, Xu Q. Composite element method and application of trace simulation for strain localization bands. Chin J Rock Mechan Eng. 2007;26(6):1116–22 (in Chinese with English abstract).

Ciarlet PG. The finite element method for elliptic problems. Amsterdam (The Netherlands): North-Holland Publishing Company; 1978.

Coetzee CJ. Review: calibration of the discrete element method. Powder Technol. 2017;310:104–42.

Crisfield MA. Non-linear finite element analysis of solid and structures, vol. 1: Essentials. London (UK): Wiley; 2000.

Dahmen W. Wavelet and multiscale methods for operator equations. Acta Numer. 1997;6:55–228.

Demuth RB, Margolin LG, Nichols BD, Adams TF, Smith BW. SHALE: A computer program for solid dynamics. New Mexico (USA): Los Alamos National Laboratory; 1985.

Desai CS, Toth J. Disturbed state constitutive modeling based on stress-strain and nondestructive behavior. Int J Solids Struct. 1996;33(11):1619–50.

Donéa J, Fasoli-Stella P, Giuliani S. Finite element solution of the transient fluid-structure problem in Lagrangian coordinates. In: Proceedings of the international meeting on fast reactor safety and related physics, vol. 3, Chicago (USA); 1976, p. 1427–35.

Donéa J, Fasoli-Stella P, Giuliani S. Lagrangian and Eulerian finite-element techniques for transient fluid-structure interaction problems. San Francisco (USA): Trans SMiRT-4; 1977, Paper B1/2.

Donéa J, Giuliani S, Halleux JP. An arbitrary Lagrangian-Eulerian finite element method for transient dynamic fluid-interactions. Comput Meth Appl Mech Eng. 1982;33(1–3):689–723.

Doltsinis IS, editor. Advances in computational nonlinear mechanics (CISM). Wien (Austria): Springer; 1989.

Dragon A, Mróz Z. A model for plastic creep of rock-like materials accounting for the kinetics of fracture. Int J Rock Mech Min Sci. 1979;16(4):235–59.

Drucker DC, Prager W. Soil mechanics and plastic analysis for limit design. Q Appl Math. 1952;10(2):157–65.

Duarte F, Gormaz R, Natesan S. Arbitrary Lagrangian-Eulerian method for Navier-Stokes equations with moving boundaries. Comput Meth Appl Mech Eng. 2004;193(45–47):4819–36.

Egholm DL. A new strategy for discrete element numerical models: 1. Theory. J Geophys Res. 2007;112:B05203.

Eisenberg MA, Phillips A. A theory of plasticity with non-coincident yield and loading surfaces. Acta Mech. 1971;11(3):247–60.

Erlicher S, Point N. Endochronic theory, non-linear kinematic hardening rule and generalized plasticity: a new interpretation based on generalized normality assumption. Int J Solids Struct. 2006;43(14–15):4175–200.

Euler L. Principes generaux de l'etat d'equilibre des fluids. Memoires del'academie des sciences de Berlin 11 E225. 1757;217–73 (in French).

Fernandez M, Moubachir M. Sensitivity analysis for an incompressible aeroelastic system. Math Models Methods Appl Sci. 2002;12(8):1109–30.

Frémond M. Non-smooth thermomechanics. Berlin (Germany): Springer; 2002.

Gagarina E, Ambati VR, Nurijanyan S, van der Vegt JJW, Bokhove O. On variational and symplectic time integrators for Hamiltonian systems. J Comput Phys. 2016;306:370–89.

Gary GG, Morgan JK, Sanz PF. Overview of continuum and particle dynamics methods for mechanical modeling of contractional geologic structures. J Struct Geol. 2014;59:19–36.

Glowinski R. Numerical methods for nonlinear variational problems. New York (USA): Springer; 1984.

Goldstein H. Classical mechanics. 2nd ed. MA (USA): Addison-Wesley; 1980.

Goodman RE, Taylor R, Brekke TL. A model for the mechanics of jointed rock. J Soil Mech Found Div ASCE. 1968;94(SM3):637–59.

Griffiths DV, Gioda G. Advanced numerical applications and plasticity in geomechanics. Wien (Austria): Springer; 2000.

Gustafson K. Domain decomposition, operator trigonometry, Robin condition. Contemp Math. 1998;218:432–7.

Gustafson K, Abe T. The third boundary condition—was it Robin's? Math Intell. 1998a;20(1):63–71.

Gustafson K, Abe T. (Victor) Gustave Robin: 1855–1897. Math Intell. 1998b;20(2):47–53.

Hahn DW, Özisik MN. Heat conduction. 3rd ed. Hoboken, NJ (USA): Wiley; 2012.

Hairer E, Lubich C, Wanner G. Geometric numerical integration. 2nd ed. Springer series in computational mathematics. Berlin (Germany): Springer; 2006.

Hamilton WR. On a general method of expressing the paths of light, and of the planets, by the coefficients of a characteristic function. Dublin Univ Rev Q Mag. 1833;1:795–826.

Harlow FH, Welch JE. Numerical calculation of time-dependent viscous incompressible flow of fluid with free surface. Phys Fluids. 1965;8(12):2182–9.

Hill R. The mathematical theory of plasticity. Oxford (UK): Oxford University Press; 1998.

Hirt CW, Nichols BD. Volume of fluid (VOF) method for the dynamics of free boundaries. J Comput Phys. 1981;39(1):201–25.

Hirt CW, Amsden AA, Cook JL. An arbitrary Lagrangian-Eulerian computing method for all flow speed. J Comp Phys. 1974;14(3):227–53.

Hrennikoff AP. Solution of problems of elasticity by the framework method. J Appl Mech. 1941;8:169–75.

Hughes TJR, Liu WK, Zimmermann TK. Lagrangian-Eulerian finite-element formulation for incompressible viscous flows. Comp Meth Appl Mech Eng. 1981;29(3):329–49.

Humpheson C, Naylor DJ. The importance of the form of the failure criterion. C/R/243/75. Swansea (UK): University of Wales; 1975.

Inglis CE. Stresses in a plate due to the presence of cracks and sharp corners. Proc Inst Naval Arch. 1913;55:219–41.

Inoue T, Kitagawa H, Shima S, editors. Computational plasticity. London (USA): Elsevier; 1990.

Irwin G. Analysis of stresses and strains near the end of a crack traversing a plate. J Appl Mech. 1957;24:361–4.

Jaeger J, Cook NG, Zimmerman R. Fundamentals of rock mechanics. 4th ed. MT (USA): Blackwell; 2007.

Jirásek M, Bažant ZP. Inelastic analysis of structures. London (UK): Wiley; 2001.

Kachanov LM. A microcrack model of rock inelasticity. Part I: frictional sliding on microcracks. Mech Mater. 1982a;1(1):19–27.

Kachanov LM. A microcrack model of rock inelasticity. Part II: propagation of microcracks. Mech Mater. 1982b;1(1):29–41.

Kachanov LM. A microcrack model of rock inelasticity. Part III: time-dependent growth of microcracks. Mech Mater. 1982c;1(2):123–9.

Katz VJ. The history of Stokes's theorem. Math Mag. 1979;52(3):146–56.

Kawamoto T, Ichikawa Y, Kyoya T. Deformation and fracturing behaviour of discontinuous rock mass and damage mechanics theory. Int J Numer Anal Meth Geomech. 1988;12(1):1–30.

Kirsch G. Die theorie der elastisizät und die bedürfnisse der festigkeitslehre. Z Ver Dtsch Ing. 1898;42:797–807 (in German).

Knothe K, Wessels H. Finite Elemente, Eine Einfuhrung fur Ingenieure. Berlin (Germany): Springer; 2008. (in German).

Krajcinovic D. Continuum damage mechanics revisited: basic concepts and definitions. J Appl Mech. 1983;52(4):829–34.

Kundu PK, Cohen IC. Fluid mechanics. Pennsylvania (USA): Academic Press; 2001.

Lagrange J. Nouvelles recherches sur la nature et la propagation du son (New researches on the nature and propagation of sound). Miscellanea Taurinensia (also known as: Mélanges de Turin). 1762;2:11–172.

Landau LD, Lifshitz EM. Fluid mechanics. Oxford (UK): Pergamon; 1959.

Lemaitre J. A course on damage mechanics. 2nd ed. Berlin (German): Springer; 1990.

Lemaitre J, Chaboche JL. Mechanics of solid materials. Cambridge (UK): Cambridge University Press; 1990.

Leroy Y, Ortiz M. Finite element analysis of strain localization in frictional materials. Int J Numer Anal Meth Geomech. 1989;13(1):53–74.

Lin FB, Bažant ZP. Convexity of smooth yield surface of frictional material. J Eng Mech. 1986;112(11):1259–62.

Logan JD. Transport modeling in hydrogeochemical systems. New York (USA): Springer; 2001.

Lubliner J. Plasticity theory. NY (USA): Macmillan Publishing Company; 1990.

Lubliner J, Taylor RL, Auricchio F. A new model of generalized plasticity and its numerical implementation. Int J Solids Struct. 1993;30(22):3171–84.

Mahtab M, Goodman RE. Three dimensional analysis of joint rock slope. In: Proceedings of the 2nd ISRM congress, vol. 3. Beograd (Yugoslavia): Privredni Pregled; 1970, p. 353–60.

Marsden JE, Ratiu TS. Introduction to mechanics and symmetry: a basic exposition of classical mechanical systems, vol. 17. Berlin (Germany): Springer; 1999.

Marsden JE, West M. Discrete mechanics and variational integrators. In: Acta numerica. Cambridge (UK): Cambridge University Press; 2001, p. 357–514.

Michino MJ, Findley WN. An historical perspective of yield surface investigation for metals. Int J Non-Linear Mech. 1976;11(1):59–82.

Miyoshi T. Foundations of the numerical analysis of plasticity. Lectures notes in numerical and applied analysis, vol. 7. Tokyo (Japan): Kynokunia Company Ltd.; 1985.

Monaghan JJ. Simulating free surface flow with SPH. J Comput Phys. 1994;110(2):399–406.

Morgan JK. Numerical simulations of granular shear zones using the distinct element method: II. The effect of particle size distribution and interparticle friction on mechanical behavior. J Geophys Res. 1999;B104:2721–32.

Morgan JK. Particle dynamics simulations of rate and state dependent frictional sliding of granular fault gouge. Pure Appl Geophys. 2004;161:1877–91.

Morris JP, Fox PJ, Zhu Y. Modeling low Reynolds number incompressible flows using SPH. J Comput Phys. 1997;136(1):214–26.

Munson BR, Young DF, Okiishi TH. Fundamentals of fluid mechanics. 3rd ed. New York (USA): Wiley; 1998.

Muskhelishvili NI. Some basic problems in the theory of elasticity. Leyden (The Netherlands): Noordhoff Ltd.; 1953.

Navier CLMH. Memoire sur les lois de l'equilibre et du mouvement des corps solides elastiques. Mem Acad Sci Fr. 1824;7:375–93 (in French).

Oda M, Suzuki K, Maeshibu T. Elastic compliance for rock-like materials with random cracks. Soils Found. 1984;24(3):27–40.

Oden JT. A general theory of finite elements. Part I. Topological considerations. Int J Numer Meth Eng. 1969a;1(2):205–21.

Oden JT. A general theory of finite elements. Part II. Applications. Int J Numer Meth Eng. 1969b;1(3):247–59.

Oden JT. Finite elements of nonlinear continua. New York (USA): McGraw-Hill; 1972.

Ortiz M, Leroy Y, Needleman A. A finite element method for localized failure analysis. Comput Meth Appl Mech Eng. 1987;61(2):189–214.

Osher S, Sethian JA. Fronts propagating with curvature-dependent speed: algorithms based on Hamilton-Jacobi formulations. J Comput Phys. 1988;79(1):12–49.

Owen DRJ, Hinton E. Finite elements in plasticity: theory and practice. Swansea (UK): Pineridge Press Ltd.; 1980.

Perzyna P. The constitutive equations for rate sensitive plastic materials. Quart Appl Math. 1963;20(4):321–32.

Perzyna P. Fundamental problems in viscoplasticity. Adv Appl Mech. 1966;9(2):243–377.

Pijaudier-Cabot G, Bažant ZP. Nonlocal damage theory. J Eng Mech Div ASCE. 1987;113(10):1512–33.

Pironneau O. Finite element methods for fluids. Chichester (UK): Wiley; 1989.

Pivonka P, Lackner R, Mang HA. Shapes of loading surfaces of concrete models and their influence on the peak load and failure mode in structural analyses. Int J Eng Sci. 2003;41(13–14):1649–65.

Potyondy DO, Cundall PA. A bonded-particle model for rock. Int J Rock Mech Min Sci. 2004;41:1329–64.

Prat PC, Bažant ZP. Tangential stiffness of elastic materials with systems of growing or closing cracks. J Mech Phys Solids. 1997;45(4):611–36.

Rice JR. Inelastic constitutive relations for solids: an integral variable theory and its application to metal plasticity. J Mech Phys Solids. 1971;19(6):433–55.

Rockafellar RT. Convex analysis. Princeton (USA): Princeton University Press; 1969.

Rudnicki JW, Rice JR. Conditions for the localization of deformation in pressure-sensitive dilatant materials. J Mech Phys Solids. 1975;23(6):371–94.

Rossmanith HP. The struggle for recognition of engineering fracture mechanics. In: Rossmanith HP, editor. Fracture research in retrospect, an anniversary volume in honour of professor George R. Irwin's 90th birthday. Rotterdam (The Netherlands): AA Balkema; 1997.

Ruth RD. A canonical integration technique. IEEE Trans Nucl Sci. 1983;30(4):2669–71.

Schlatter B. Modeling fluid flow using smoothed particle hydrodynamics. Master thesis. USA: Oregon State University; 1999.
Serret JA, editor. Oeuvres de Lagrange, vol. 1. Paris (France): Gauthier-Villars; 1867 (in French).
Simo JC, Hughes TJR. Elastoplasticity and viscoplasticity—computational aspects. New York (USA): Springer; 1988.
Spiegel MR, Lipschutz S, Spellman D. Vector analysis. 2nd ed. New York (USA): McGraw Hill; 2009.
Strang G, Fix G. An analysis of the finite element method. Englewood Cliffs, NJ (USA): Prentice-Hall; 1973.
Tonti E. Variational formulation of non-linear differential equations. Bull Acad Roy Belg Classe Sci. 1969;55:137–65 (first part) and 263–78 (second part).
Turner M, Clough RW, Martin HC, Topp LJ. Stiffness and deflection analysis of complex structures. J Aeronaut Sci/Inst Aeronaut Sci. 1956;23(9):805–23.
Vainberg MM. Variational methods for the study of nonlinear operators. San Francisco, CA (USA): Holden-Day Inc.; 1964.
Valanis KC. A theory of viscoplasticity without a yield surface. Archiwum Mechaniki Stossowanej. 1971;23(4):517–51.
Watanabe O, Atluri SN. Internal time, general internal variable, and multi-yield-surface theories of plasticity and creep: a unification of concepts. Int J Plast. 1986;2(1):37–57.
Weighardt K. Uber einen Grenzubergang der Elastititatslehre und seine Anwendung auf die Statik hochgradig statisch unbestimmter Fachwerke. Verhandlungen Des Vereins Zur Beforderung Des Gewerbefleisses. 1906;85:139–76 (in German).
Westergaard HM. Bearing pressures and cracks. ASME J Appl Mech. 1939;6:A49–53.
Williams ML. On the stress distribution at the base of a stationary crack. ASME J Appl Mech. 1957;24(1):109–14.
Yeung PK. Lagrangian investigations of turbulence. Ann Rev Fluid Mech. 2002;34:115–42.
Yu MH, Li JC. Computational plasticity with emphasis on the application of the unified strength theory. Hangzhou (China): Zhejiang University Press; 2012.
Zhang JL, Chen SH. Study of time-stepping adaptivity in FEM analysis of time-dependent problems. J Wuhan Univ Hydraul Electr Eng (WUHEE). 1996;29(1):79–84 (in Chinese with English abstract).
Zhang JL, Chen SH. An implicit scheme algorithm of elastic-viscoplastic FEM with time-stepping adaptive for bolted jointed rock masses. J Hydraulic Eng. 1997;19(Suppl):168–75 (in Chinese with English abstract).
Zienkiewicz OC. The finite element method. London (UK): McGraw-Hill Book Company Limited; 1977.
Zienkiewicz OC, Cormeau IC. Viscoplasticity solution by finite element process. Arch Mech. 1972;24(5–6):873–88.
Zienkiewicz OC, Cormeau IC. Visco-plasticity and plasticity—an alternative for finite element solution of material nonlinearities. In: Computing methods in applied sciences and engineering (Part 1). Berlin (Germany): Springer; 1974, p. 259–87.
Zienkiewicz OC, Mróz Z. Generalized plasticity formulation and applications to geomechanics. In: Desai CS, Gallagher RH, editors. Mechanics of engineering materials. New York (USA): Wiley; 1984. p. 655–79.
Zienkiewicz OC, Pande GN. Some useful forms of isotropic yield surfaces for soil and rock mechanics. In: Gudehus G, editor. Finite element in geomechanics. New York (USA): Wiley; 1977. p. 179–90.
Zienkiewicz OC, Valliappan S, King IP. Elasto-plastic solutions of engineering problems "initial stress", finite element approach. Int J Numer Meth Eng. 1969;1(1):75–100.
Zienkiewicz OC, Huang M, Pastor M. Localization problems in plasticity using finite element with adaptive remeshing. Int J Numer Anal Mech Geomech. 1995;19(2):127–48.

Chapter 4
Preparative Knowledge of Numerical Analysis

Abstract This chapter is devoted to provide the basic knowledge of numerical analysis for the elaboration and discussion of following chapters concerning advanced CMs. Starting with the floating-point arithmetic, numerical stability is understood as the core in the study and application of CMs in addition to precision. The mainstream numerical techniques such as the interpolation/approximation/integration and the solution of ODEs/PDEs/IEs, are detailed. The fundamental CMs such as FDM and FEM are particularly employed to exemplify the power of numerical analysis.

4.1 General

4.1.1 Definitions

"Numerical analysis" studies the algorithms for the solution of mathematics problems involving arbitrary real or complex variables (numbers). It is not only restricted to the results of integers or rational numbers, but also concerns with the existence and correctness of the solution as well as the time and other resources (e.g. computer memory) required in the solution. It is without exaggeration to say that an advance in sciences could not be achieved, or a significant engineering project could not be developed, without the support of numerical analysis.

Numerical analysis lies at the meeting point of pure mathematics, computer sciences and application areas. It is interesting to find that the complaints from pure mathematicians, computer scientists, as well as the broad community of engineers, are very different and often contradictory. However, performed as the dual-role of researcher and practitioner in geotechnical and hydraulic engineering, the reliability of analysis results is highly concerned by us because in practice, we only ever have approximate answers blamed for the following facts:

- Some algorithms such as Newton's method may be intrinsically approximate ones that converge towards, but never reach at the desired result in a finite number of steps. How fast they converge to the acceptable solution is therefore questionable.

S. Chen, *Advanced Computational Methods and Geomechanics*, Springer Tracts in Civil Engineering, https://doi.org/10.1007/978-981-19-7427-4_4

- We approximate the "real numbers" set R by the "floating-point numbers" set F, and the arithmetic result of every elementary computational operation (+, −, ×, ÷) with real numbers is rounded to the nearest number of F on a computer. We need to understand the set F for answering the problems w.r.t. how the rounding-off errors are accumulated and controlled by good algorithms.

The corner stones of numerical analysis are "harmonic analysis", "functional analysis", "complexity theory", "probability theory and statistics", "abstract algebra", "topology/algebraic/differential geometry", and "approximation theory" (Ciarlet and Lions 2000), of which the last two are the major topics in this chapter.

Solution of nonlinear equation could be one of the earliest numerical analysis problem put forward as: suppose we have an equation $f(x) = 0$ in unknown variable x, how to find one, some, or all of the variable values satisfying this equation? Other frequently problems encountered in modern numerical analysis (e.g. FDM, FEM) are, for example, finding an eigenvalue of an $n \times n$ matrix, solving a set (system) of n linear equations in n unknown variables, minimize a linear function of n variables subject to m linear constraints, etc.

Surface fitting is an important problem in many fields, including CAD/CAM/CAE, computer graphics, and computer vision. The solution of an "ordinary differential equation (ODE)" or "partial differential equation (PDE)" is, basically to find the unknown function as a surface in space: in 2-D it is a curve and, in four-or higher-dimension it is a hyper surface. Normally, two types of surface fitting, namely, "interpolation" and "approximation" may be distinguished (Hoschek and Lasser 1993; Pieg and Tiller 1997). The former creates an interpolating surface that passes through the data points (nodes, knots); in contrast, the latter creates an approximating surface that passes near the data points subject to minimizing the deviation of the surface from the data points. For the former, the data points are customarily given in grids; whereas for the latter, the data points may be randomly distributed. Furthermore, it may be assumed that the noise of point clouds has been filtered by, for example, a "moving least squares (MLS)" method (Alexa et al. 2003). Normally, for a governing ODE/PDE (strong form), the difference technique is customarily exercised, whereas its counterpart integration equation (weak form) suggests the technique of interpolation/approximation. In the most cases (not always), interpolation and approximation techniques are used to solve the weak form equation in spatial domain meanwhile the difference techniques are used for temporal domain.

In the last decades, approximation theory has undergone a revolution on different fronts, and in the long run, this revolution underscores its relevance to computational mathematics. Whether in university courses or in research and engineering communities, there is no simple distinguish between "proper" numerical analysis and approximation theory. We all take it for granted that the bases of approximation theory should be acquired by any budding numerical analysis. Particularly, the topics mostly concerned in the representative advanced CMs elaborated in this book are:

- Fitting a function by interpolation and approximation;
- Minimize a function of several variables;

- Solve an "ordinary differential equation (ODE)";
- Solve a "partial differential equation (PDE)".

The discussion of these topics constructs the main contents of this chapter.

4.1.2 Significant Historical Events

Throughout modern history, leading mathematicians have been involved with the applications of mathematics in science and engineering. Among many other contributions, Gauss made crucial contributions to "least-squares data fitting", "linear equation system solution", "numerical quadrature", as well as "fast Fourier transform", although the last one did not become widely known until its rediscovery by Cooley and Tukey in 1965.

Around 1900, the numerical side of mathematics started to become less conspicuous in the activities of mathematicians. This was a consequence of the growth of general mathematics in the field in which, for technical reasons, mathematical rigor had to be the heart of the matter. For example, many advances in the early twentieth century sprang from new ability to reason rigorously about infinity, which is a subject far from numerical analysis.

In the 1940s, the electronic digital programmable computer was invented. From this moment numerical analysis began to explode mainly in the community of engineering specialists. Half a century on, numerical analysis has grown into one of the largest branches of mathematics. Thanks to the efforts of these specialists and mathematicians and ever more powerful computers, we have reached a point where most of the classical mathematics problems derived from science and engineering can be solved numerically to high accuracy. Most of the algorithms that make this possible were invented since the 1950s.

Numerical analysis is built on one of strong foundations: the mathematical subject of approximation theory. This field encompasses the classical questions of interpolation and series expansions associated with the names Newton, Fourier, Gauss and many others; the semi-classical problems of polynomial and rational minimax approximation associated with the names Chebyshev, Bernstein and many others; and major newer topics including "splines", "radial basis functions" and "wavelets". In almost every area of numerical analysis, it is a fact that, sooner or later, the discussion comes down to approximation theory.

The list in Table 4.1 attempts to identify some of the most significant developments in the history of numerical analysis. In each case key early figures are cited, more or less chronologically, and a milestone year is given.

Table 4.1 Most significant events in the history of numerical analysis

Year	Development	Key early figures
1671	Newton's method	Newton, Raphson, Simpson
1795	Least-squares fitting	Gauss, Legendre
1814	Gauss quadrature	Gauss, Jacobi, Christoffel, Stieltjes
1826	Gaussian elimination	Lagrange, Gauss, Jacobi
1855	Adams ODE formulas	Euler, Adams, Bashforth
1895	Runge–Kutta ODE formulas	Runge, Heun, Kutta
1910	Finite differences for PDE	Richardson, Southwell, Courant, von Neumann, Lax
1936	Floating-point arithmetic	Torres y Quevedo, Zuse, Turing
1943	Finite elements for PDE	Courant, Feng, Argyris, Clough
1946	Spline fitting	Schoenberg, de Casteljau, Bezier, de Boor
1947	Monte Carlo simulation	Ulam, von Neumann, Metropolis
1947	Simplex algorithm	Kantorovich, Dantzig
1952	Conjugate gradient iterations	Lanczos, Hestenes, Stiefel
1952	Stiff ODE solvers	Curtiss, Hirschfelder, Dahlquist, Gear
1958	Orthogonal linear algebra	Aitken, Givens, Householder, Wilkinson, Golub
1959	Quasi-Newton iterations	Davidon, Fletcher, Powell, Broyden
1961	QR algorithm for eigenvalues	Rutishauser, Kublanovskaya, Francis, Wilkinson
1965	Fast Fourier transform	Cooley, Tukey, Sande
1971	Radial basis function fitting	Hardy, Askey, Duchon, Micchell
1971	Spectral methods for PDE	Chebyshev, Lanczos, Clenshaw, Orszag, Gottlieb
1973	Multigrid iterations	Fedorenko, Bakhvalov, Brandt, Hackbusch
1976	Nonsymmetric Krylov iterations	Vinsome, Saad, van der Vorst, Sorensen
1977	IEEE arithmetic	Kahan
1977	MATLAB	Moler
1977	Preconditioned matrix iterations	van der Vorst, Meijerink
1982	Wavelets	Morlet, Grossmann, Meyer, Daubechies
1984	Interior methods in optimization	Fiacco, McCormick, Karmarkar, Megiddo
1987	Fast multipole method	Rokhlin, Greengard
1991	Automatic differentiation	Iri, Bischof, Carle, Griewank

4.1.3 Paramount Issues

Suppose we investigate a mathematical construct A that is not amenable to computation. It often makes sense to replace it by another approximate mathematical construct A_ε where ε denotes one or more built-in parameters. A_ε is amenable to computation, although it is distinct from, yet it is close to, the original construct A. It follows that some obvious pertinent questions will give rise: "How closely does A_ε approximate A?" "Can we estimate the error between A and A_ε by an appropriate measure (scale)?" "What are the relationship of parameters and the error?" To answer such questions, certain basic knowledge concerning the virtualities of computer and the demands of numerical stability are indispensable.

1. **Floating-point arithmetic**

It is often referred that the scientific revolution was set in motion when Galileo and others made it a principle that everything must be measured, which led physical laws to be mathematically expressed. Finer measurements led to refined laws, which in turn, led to better technology for still finer measurements.

Computers certainly play a key role in the narration of numerical analysis, yet there is a misunderstanding about their performance. Some people imagine that scientists and mathematicians generate formulas, and then, by inserting numbers into these formulas, computers grind out the necessary results. The reality is actually nothing like this, because the most mathematical problems cannot be solved by a finite sequence of elementary operations.

It should be aware of that computers cannot represent real or complex numbers exactly. For example, a quotient like 1/7 evaluated on a computer will normally yield an approximate result unless the computers are so designed to work in base 7. Computers approximate real numbers by a system of "floating-point arithmetic" invented by Konrad Zuse in the 1930s: each real number is represented in a digital equivalent of scientific notation, so the scale does not matter unless the number is very huge or tiny to bring about overflow or underflow. Until the 1980s, different computers possessed different arithmetic properties. In 1985, after years of discussion, the "IEEE (Institute of Electrical and Electronics Engineers) standard for binary floating-point arithmetic", or "IEEE arithmetic" for short, was successfully accepted. This standard has subsequently become universal on the computer processors of many kinds.

An IEEE double precision real number consists of a 64-bit word that is divided into 53 bits plus 11 bits, the former are for a signed fraction in base 2 whereas the latter are for a signed exponent. Since $2^{-53} \approx 1.1 \times 10^{-16}$, IEEE standard represents the number on the real line to a relative accuracy of approximate 16 digits. Since $2^{\pm 210} \approx 10^{\pm 308}$, IEEE standard works for the number with up-and low bounds about 10^{308}–10^{-308}.

Computers do not represent numbers merely, but also perform four elementary operations (e.g. addition, subtraction, multiplication, and division), and more complicated results are obtained from sequences of these elementary operations. In floating-point arithmetic, the result of each elementary operation is almost exactly correct in the sense that: if "×" is one of four operations in its ideal form and "(∗)" is the corresponding operation realized on the computer, then for any floating-point numbers x and y, we have $x \times y = (x * y)(1 + \varepsilon)$, subject to there is no underflow or overflow. Being a very tiny quantity in its absolute value, ε is smaller than a fraction known as machine epsilon ε_{mach} that measures the precision of computer. In the IEEE system, $\varepsilon_{mach} = 2^{-53} \approx 1.1 \times 10^{-16}$.

Taking the interval [1, 2] for example, on a computer it is approximated by about 10^{16} numbers. It is interesting to compare how fineness of such a discretization with that of the discretization of physics. In a handful solid substance parcel, the number of atoms in a line from one point to another is on the order of 10^8 (the cubic root of Avogadro's number). Such a parcel behaves quite like a continuum to justify customary definitions of physical quantities such as density, pressure, stress, strain, and temperature. At resent, almost no variables and parameters in physics is more than 12 or 13 digits of accuracy, thus IEEE numbers are orders of magnitude more precise than any numbers in science and engineering.

Nevertheless, the floating-point arithmetic is widely regarded as an ugly and dangerous compromise. The source of such a misunderstanding lies in numerical instability, namely, the amplification of rounding-off errors from microscopic scale to macroscopic scale by certain computation algorithms. Although this risk is very real, yet it should always bear in our mind that numerical instability could trigger the evolving distortion of a vast nonlinear system, such as the case using PFC to simulate millions particles. However, this issue is not sufficiently aware of in the community of advanced CMs, to say nothing of how to cope with. In fact, the far more interesting business for many scholars is to design a process for the execution of fast algorithms that quickly converge to "approximate" answers. Anyway, we have to admit that it seems rather reasonable insofar, if the validation examples under ideal laboratory conditions are as good as applicable to engineering problems, rounding-off error analysis is rarely the central issue, but only the bridge over the obstacle of numerical stability.

2. Stability

(1) Stability of natural model

In the study of natural system, it is often true that a small change in actions will produce just a small change in the system state. For example, if a tiny variation in the Earth's rotation period would produce a huge change in climate, the world would be a very different place for us to live in.

Scientists often build mathematical models in an attempt to understand nature. Although these models are not able to faithfully replicate all natural phenomena, yet the important characteristics of the nature need to be preserved. One of the most important characteristics to be preserved is the stability: namely, does small change produce small change?

A PDE is said to be stable if for every set of initial data (at $t = 0$) and boundary conditions, the solution remains bounded as t approaches infinity. A PDE is called strongly stable if, for every set of initial data and boundary conditions, the solution approaches zero as t approaches infinity. Unfortunately, for a stable PDE, another worry over the approximation often emerges because of the mathematical model, even though it has been greatly simplified w.r.t. the nature laws, might still be too complicated to be exactly solved on a computer.

(2) **Stability and consistency of computation model**

Suppose we have constructed a mathematical model that is stable, the next step might be the calculation on computer towards the prediction of various action effects. When computer began to be used to solve PDEs after World War II, the numerical instability—phenomenon of the greatest practical importance, emerged. For instance, mathematical models w.r.t. the motion and deformation of fluids and solids, involve the solution of PDEs given in Chap. 3. Even though these PDEs are indeed stable, it is possible that the equation set arise from the discretization running on computer is no longer stable, namely, small changes in initially input data on the computer, or small rounding-off errors, will produce very large changes in the computed outcomes. An important job of the numerical analyst is to make sure that this does not happen, such a theme of stability recurs throughout the whole study of the approximation with CMs.

As before, the term "instability" refers to the unbounded amplification of local errors in a computation process, but now the dominant local errors are usually attributable to those of discretization rather than rounding-off. Instability typically manifests itself as an oscillatory error in the computed solution that blows up exponentially as more time marching steps are taken. Germund Dahlquist, an outstanding Swedish mathematician, is among the first who noted this effect in 1956. His landmark paper is nowadays looked as one of the most significant events marking the birth of modern numerical analysis with fundamental theorem:

$$\text{consistency} + \text{stability} = \text{convergence}.$$

This theorem is based on three definitions:

- Consistency is the property that discrete formulae possess locally positive order of accuracy and thus model the right ODE;
- Stability is the property that errors introduced at one time-marching step cannot grow unboundedly at later steps;
- Convergence is the property that as $\Delta t \to 0$, in the absence of rounding-off errors, numerical solutions converge to correct answers.

4.2 Interpolation

4.2.1 Overview

In numerical analysis, approximation procedures include "least-squares" and "Chebyshev approximation" by polynomials, "orthogonal-polynomial theories", "Padé approximation", "spline theory", and possibly "simple rational approximation" (Cheney 1966). However, it is normally recognized that one of the easiest procedures for approximating a function $u = f(x)$ is "interpolation" or "data fitting".

It is notable that data fitting algorithms do not take into account of the fact that we are trying to solve a PDE (strong form) or IE (weak form). However, data fitting is much concerned with approximation problems and its ideas can be used in the design of test spaces (Franke 1982; Schaback 1995).

Suppose the values u_i $(i = 0, 1, 2, \ldots, n)$ of function $f(x)$ at several distinct points $(x_0, x_1, \ldots, x_n)$ are known, the task of interpolation is to determine the free parameters $(a_{0,}, a_1, \ldots, a_n)$ in an interpolated function $F(x)$ in order to meet the condition

$$F(x_i) = u_i = f(x_i) \quad (i = 0, 1, 2, \ldots, n) \tag{4.1}$$

If the values of the first derivative of $f(x)$ at some points x_i are also known, they are also useful to determine the free parameters in $F(x)$.

Let E be a vector space of $n + 1$ dimension. The general interpolation problem is to find $F \subset E$ such that (Davis 1963; Brezinski 1992)

$$l_i(F) = u_i = f(x_i) \quad (i = 0, 1, \ldots, n) \tag{4.2}$$

in which l_i are linear functionals on E.

If $(\phi_0, \phi_1, \ldots, \phi_n)$ form a basis of E, then F can be written as a trial function $F = a_0\phi_0 + a_1\phi_1 + \cdots + a_n\phi_n$ and the preceding interpolation conditions lead to

$$a_0 l_0(\phi_0) + a_1 l_1(\phi_1) + \cdots + a_n l_n(\phi_n) = u_i \quad (i = 0, 1, \ldots, n) \tag{4.3}$$

which is a set of $n + 1$ linear equations in the $n + 1$ unknowns $a_0, a_1, \ldots, a_n$.

1. **Polynomial interpolation**

With a long history and a large amount of theoretical results accumulated, the approximation by means of polynomials is the base of almost all modern numerical methods.

Obviously, the easiest F in Eq. (4.1) is polynomial function. A scheme for quadratic interpolation was also used much earlier by Arabic mathematician Al-Birfini (973–1048) and certainly by many of his followers. It is vocable due to Harriot in 1611, Briggs in 1624 and Wallis in 1656, by constructing the table of logarithms that assumed equidistant interpolation points and made use of first and second order finite differences. It corresponds to the interpolation by a polynomial of degree two.

Interpolation by a polynomial with arbitrary degrees (still on equidistant points) is dubbed as the "Gregory-Newton formula" (Newton 1687). It seems that Leibniz also obtained similar results independently around 1672 (Hofmann 1974). The formula was written in a more suitable form by Cotes in 1708. Taylor (1715) made the great contribution by the expansion of a function into infinite polynomial series. Interpolation on the points of geometric progression led to a simplification similar to the arithmetical progression attributable to Stirling and Schelibach, their process was rediscovered by Runge (1891).

In 1794, Lagrange gave his famous interpolation formula but he attributed it to Newton although some commentators claimed that the formula goes back to Euler or Waring. The error for Lagrange interpolation polynomial was first given by Cauchy (1840). The interpolation problem was treated in its full generality by Hermite (1878), and today the interpolation with some unnecessarily imposed consecutive derivatives is called "Hermite-Birkhoff interpolation".

The first algorithm of interpolation polynomials at $(x_i, \ldots, x_{i+k})$ recursively computed for varying i and k was proposed by Aitken and further improved by Neville (1934). This "Neville-Aitken scheme" was extended by Molbach for the interpolation by any family of functions forming Chebyshev system. These two formulae led to "Richardson's scheme" for convergence acceleration by extrapolation (Brezinski and Zaglia 1991).

The early examples w.r.t. the non-convergence of polynomials at equidistant interpolation points were provided by Méray (1884, 1896), Runge (1901) and Borel (1897). The problem was then handled theoretically by Faber (1914) and Bernstein (1912, 1918), they proved that when any family of interpolation points is given in a closed interval, it is always possible to find a continuous function such that the sequence of interpolation polynomials at these points does not converge uniformly to the function.

Generalization of Lagrange, Newton and Neville-Aitken schemes were obtained by Gasca and his co-workers (Gasca and Carmona 1982; Gasca and Maetzu 1982; Gasca and Lebrón 1983; Gasca and Ramirez 1984; Gasca and Lebrón 1987). An overview of these topics can be found in the literatures of, for example, Gasca (1990). In the paper of Lorentz et al. (1983), a history of interpolation and approximation since the 1960s also can be found.

2. **Rational interpolation**

After the first attempt by Cauchy and Jacobi, Padé treated interpolation problem by a rational fraction. It is much more powerful for the approximation of non-smooth functions with bounded variation (Newman 1964; Popov 1977; Padé 1982). For the recent developments of this problem, including the rational Hermite interpolation and the functions of several variables, readers are referred to the works by Cuyt and Wuytack (1987), and Cuyt (1992).

3. **Spline interpolation**

Cardinal interpolation refers to the interpolation at the integer points on the real line. The first survey on this problem might trace back to Whittaker (1935). The subject was also connected to cardinal series by Higgins (1985).

Spline interpolation consists of piecewise polynomial interpolation whose values and some of its first derivatives coincide at the interpolation points. The subject had been discussed as early as 1906 by Birkhoff, although splines constructed on equidistant nodes were already used before that time by Runge (1901) (quadratic splines). Afterwards, Quade and Collatz (1938) used arbitrary degree splines for the approximation of the Fourier coefficients of periodic functions. In 1946, Schoenberg proposed a landmark work of cardinal splines.

The history of splines can be divided into two periods: Before 1964, papers were mostly devoted to the properties of piecewise polynomials; after that year the introduction of functional analysis allowed for the generalizations in Hilbert spaces and the studies on their minimization properties (Laurent 1972).

An important class of spline functions is the B-splines. These splines became widely exercised after a recurrence relation for their computation obtained by Cox (1972) and De Boor (1972). The cases of two variables began to be handled in the late 1960s by a tensor product approach. However, many curves and surfaces are badly represented by polynomials, and the introduction of rational curves and surfaces is demanded.

Spline functions have many applications in "computer aided design (CAD)". As recorded by Farin (1991), the story began in 1958 when the French automobile company "Citroën" asked De Casteljau to develop a design system for helping engineers. He had the idea of using Bernstein polynomials for this purpose but, due to secrecy, his result was not allowed to publish. Bezier, an engineer at another French automobile company "Renault", heard that Citroën was developing such a project. He initiated a new approximation to curves and surfaces, consisting of piecewise polynomials controlled by polygonal lines. Such curves and surfaces are now dubbed with the name of Bezier (1972).

4. **Radial basis interpolation**

Radial basis functions can be looked at as a generalization of splines in multi-dimensions. A linear spline function interpolating the function $f(x)$ at points $(x_0 < x_1 < \ldots < x_n)$ can be written as

$$F(x) = \sum_{0}^{n} \lambda_i |x - x_i| \tag{4.4}$$

in which the coefficients λ_i are defined by the interpolation conditions.

An extension to the multivariate case may be achieved by replacing the absolute value in Eq. (4.4) by the Euclidean norm, thus $F(x)$ becomes a linear combination of functions spherically symmetric w.r.t. the origin—exactly a characteristic of radial basis function approximation

$$F(x) = \sum_{0}^{n} \lambda_i \Phi(|x - x_i|) \tag{4.5}$$

in which Φ is a given function. Nowadays very much in favor, is the use of wavelets (Daubechies 1992). A general exposition of this subject can be found in the papers of Powell (1987) and Light (1992).

5. **Multivariate interpolation**

Early attempts to study the multivariate polynomial interpolation are attributable to Narumi (1920) and Steffensen (1927).

The first difficulty with multivariate interpolation comes from the nonexistence of Chebyshev systems, thus that the uni-solvent of interpolation problem depends on the relative locations of interpolation points; the second difficulty arises from the generalization of divided differences to several variables (Thacher and Milne 1960; Guenther and Roetman 1970). In the 1960s, these problems became very tractable thanks to the development of FEM. For example, the problem of interpolation on the subsets of grids was solved by Werner (1980) and Hakopian (1982), the Hermite-Birkhoff interpolation in multi-variables was handled by Gasca and Maeztu (1982).

As pointed out by Ciarlet and Raviart (1972), there are basically two approaches for the interpolation problem with multi-variables. The first approach is classical one that consists of prescribing points in the domain and the correspondent function values at these points, then the Lagrange or Hermite interpolation polynomial is sought. Generally, the error bounds are obtained as the sup-norm of some higher-order derivatives of the function (Coatmélec 1966; Ciarlet and Wagschal 1971). The second approach is proposed by Ciarlet and Raviart (1972) that consists of approximating the function in the Sobolev space with the sole assumption that some polynomials remain invariant in the approximation procedure. Such an approximation is not necessarily defined as an interpolation problem but it contains Lagrange or Hermite interpolation as special cases.

For the comprehensive review, readers are referred to the papers by Oden (1991).

4.2.2 Smooth Interpolation

In spite of the fact that splines and other advanced interpolations are widely exercised to approximate functions, polynomials preserve their basic significant role in CMs. Hence the interpolation theory by means of polynomials will be further detailed hereinafter. To start with, the smooth interpolation will be handled first.

The determination of an interpolation function that passes through a certain number of data points (viz. Fig. 4.1) is equivalent to the solution of a number of equations comprising identical number of data points and coefficients in the function. Typical data fitting algorithm produces a function $F(x)$ in the form of

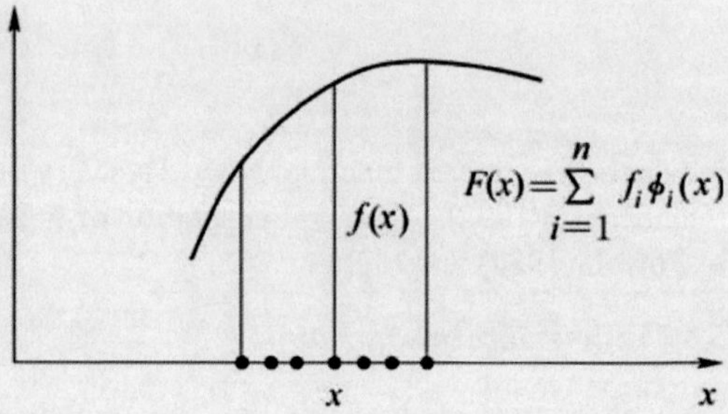

Fig. 4.1 Smoothing interpolation of C^{n+1} continuity

$$F(x) = \sum_{i=0}^{n} f_i \phi_i(x) \tag{4.6}$$

in which the values f_i are the member of data set (e.g. function values, derivatives, etc.) and the functions $\phi_i(x)$ are termed as "basis functions" or "shape functions". At this point it is necessary to specify what "good" shape functions are expected:

- Their connectivity must be evaluated in a bounded computing time of order n;
- They should rely on the node positions only;
- Their integrals and derivatives must be evaluated in a simple manner (e.g. with a few Gauss points);
- They must allow for the representation of the discontinuity in its derivatives;
- They must enable to construct nodal interpolation that take a unit value at nodes.

All of above five conditions may be found in both mesh-based or mesh-free methods. A good comment on this issue was provided by Idelsohn and Oñate (2006).

Let $F(x)$ be a set of functions and $l_0, l_1, \ldots, l_n$ be the linear functionals defined for every $f \in F$. Suppose $F_n (F_n \subset F)$ is an $n+1$ dimensional subset of F generated by the shape functions $\{\phi_i\}$, $(i = 0, 1, \ldots, n)$, i.e.

$$F_n = (\phi \in F_n : \phi = a_0\phi_0 + a_1\phi_1 + \cdots + a_n\phi_n) \tag{4.7}$$

in which $a_0, a_1, \ldots, a_n$ are the coefficient components in a $n+1$ dimensional vector $\{a_i\}$.

The general interpolation problem can be formulated such that: for a given function $f \in F$, a function $\phi \in F_n$ needs to be found such that

$$l_i(\phi) = l_i(f) \quad (i = 0, 1, 2, \ldots, n) \tag{4.8}$$

Recalling $\phi = \sum_{i=0}^{n} a_i\phi_i$ and the linearity of l_i, the condition (4.8) can be written as

$$l_i(\phi) = l_i(\sum_{j=0}^{n} a_j\phi_j) = (\sum_{j=0}^{n} a_j l_i(\phi_j)) = l_i(f) \quad (i = 0, 1, 2, \ldots, n) \tag{4.9}$$

So the general interpolation problem (4.7) becomes the equivalent problem to solve $n + 1$ coefficient vector $\{a_i\}$ of the linear equation set as follows

$$\sum_{j=0}^{n} a_j l_i(\phi_j) = l_i(f) \quad (i = 0, 1, 2, ..., n) \tag{4.10}$$

The formula (4.10) has a unique solution if the following condition holds

$$D = \begin{vmatrix} l_0(\phi_0) & l_0(\phi_1) & \dots & l_0(\phi_n) \\ l_1(\phi_0) & l_1(\phi_1) & \dots & l_1(\phi_n) \\ \dots & \dots & \dots & \dots \\ l_n(\phi_0) & l_n(\phi_1) & \dots & l_n(\phi_n) \end{vmatrix} \neq 0 \tag{4.11}$$

and the solution can be found via Cramer's formula that

$$\phi = \sum_{i=0}^{n} \Delta_i \phi_i / D \tag{4.12}$$

in which Δ_i is the determinant obtained from the determinant D by replacing its ith column with the vector $\{l_0(f)\, l_1(f) \dots\, l_n(f)\}$

$$\Delta_i = \sum_{j=0}^{n} l_j(f) \Delta_{ji} \quad (i = 0, 1, \dots, n) \tag{4.13}$$

in which Δ_{ji} are the cofactors of the elements in the ith column of D.

The unique function (4.12) derived under the condition (4.10) is called the "generalized interpolation polynomial" of the function f over the system $\{\phi_i\}(i = 0, 1, \dots, n)$. It follows that

$$\phi = \sum_{i=0}^{n} l_i(f) \Phi_i \tag{4.14}$$

in which

$$\Phi_i = \sum_{j=0}^{n} \Delta_{ij} \phi_j / D \tag{4.15}$$

Φ_i belongs to F but does not depend on the function f. Equation (4.14) is usually called the "Lagrange interpolation formula" at the basic Lagrange polynomials that possess "Kronecker Delta property"

$$l_i(\Phi_j) = \delta_{ij} = \begin{cases} 1 \text{ for } i = j \\ 0 \text{ for } i \neq j \end{cases} \tag{4.16}$$

Depending on the choice of $\{\phi_i\}(i = 0, 1, \ldots, n)$ and the sequence of $\{l_i\}(i = 0, 1, \ldots, n)$, different interpolations can be formulated.

1. **Chebyshev's interpolation**

A system of functions $\{\phi_i\}$ $(i = (0, 1, \ldots, n)$ is dubbed as "Chebyshev's system" on the set Δ if every function $\phi = a_0\phi_0 + a_1\phi_1 + \cdots + a_n\phi_n$ has at most n zero points on Δ, where $\{a_i\}$ comprises arbitrary real numbers meeting $\sum_0^n |a_i| > 0$.

In the case when the functionals l_i defined in Eq. (4.8) are linear and subject to $l_i(f) = f(x_i)$ $(i = 0, 1, 2, \ldots, n)$, for the interpolation problem $\phi = \sum_0^n a_i\phi_i$ subject to $\phi(x_i) = f(x_i)$, a necessary and sufficient condition to have a unique solution for every function f defined on the set Δ and for every choice of points $\{x_i\}$ $(i = 0, 1, 2, \ldots, n)$ $(x_i \in \Delta, x_i \neq x_j$ for $i \neq j)$, is that, $\{\phi_i\}$ $(i = 0, 1, 2, \ldots, n)$ is a Chebyshev system.

2. **Lagrange's interpolation polynomial**

The algebraic polynomial of nth degree

$$L_n(f, x) = \sum_{i=0}^{n} f(x_i) \frac{(x - x_0)(x - x_1) \ldots (x - x_{i-1})(x - x_{i+1}) \ldots (x - x_n)}{(x_i - x_0)(x_i - x_1) \ldots (x_i - x_{i-1})(x_i - x_{i+1}) \ldots (x_i - x_n)} \tag{4.17}$$

satisfies the condition $L_n(f, x_i) = f(x_i)$ $(i = 0, 1, 2, \ldots, n)$. This polynomial exists for every function f when the points of interpolation satisfy $x_i \neq x_j$ for $i \neq j$. Equation (4.17) can be rewritten as

$$\begin{cases} L_n(f, x) = \sum_{i=0}^{n} f(x_i) \dfrac{\omega(x)}{(x - x_i)\omega'(x_i)} \\ \omega(x) = \prod_0^n (x - x_i) \\ \omega'(x) = \prod_0^n (x_i - x_j) \quad \text{subject to } (i \neq j) \end{cases} \tag{4.18}$$

in which the polynomials of nth degree

$$L_{n,i}(x) = \frac{\omega(x)}{(x - x_i)\omega'(x_i)} \tag{4.19}$$

are called as the "basic Lagrange polynomials" that are constructed on points $\{x_i\}$ $(i = 0, 1, 2, \ldots, n)$ and possess Kronecker Delta property.

$$L_{n,i}(x_j) = \begin{cases} 1 \text{ for } i = j \\ 0 \text{ for } i \neq j \end{cases} \tag{4.20}$$

Hence L_n can be used as an approximant of function f,

$$f(x) = L_n(f, x) + R_n(f, x) \tag{4.21}$$

in which $R_n(f, x)$ is the remainder term. For a given function $f(x) \in C^{n+1}$ and points $\{x_i\}$ $(i = 0, 1, 2, \ldots, n)$, R_n is given as

$$R_n(f, x) = f^{(n+1)}(\xi)\frac{\omega(x)}{(n+1)!} \tag{4.22}$$

in which ξ is a number between x_0 and x_n.

4.2.3 Segment (Patch) Interpolation

There are many other ideas in data fitting. For example, a prevalent technique in CMs such as FEM is the piecewise polynomial interpolation (viz. Fig. 4.2).

After the creation of FE mesh for a specified domain (e.g. concrete dam), basic unknown variables (e.g. displacement, hydraulic potential, temperature, etc.) will be defined at nodal points. A set of "shape" (also called "basis" or occasionally "interpolation") functions is chosen on reference (parent) element to uniquely interpolate the nodal variables onto the mapped elements. To determine a one-to-one correspondence between the Cartesian coordinates in reference element and the curvilinear coordinates in mapped elements, the so-called isoparametric element is customarily employed which conducts coordinate mapping (transforming) between the reference element and mapped element. With the help of the principle of virtual work (or functional variation), an equation set w.r.t. basic variables that pronounces "the

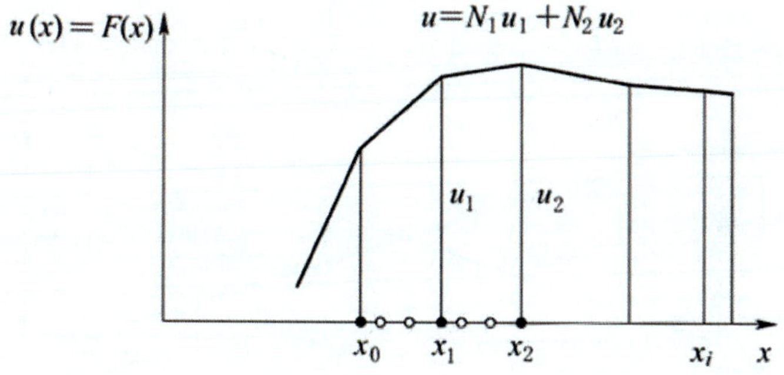

Fig. 4.2 Segment interpolation of C_0 continuity

equivalent nodal forces are equilibrated by the boundary loads", is established, which may be algebraically solved to get the basic variables tied to nodes. These solved basic variables are further employed to calculate their gradient related variables such as strain/stress, seepage/heat flow rate.

Traditional nodal shape functions of C_0 continuity as "standard" ones are prevalent in the most FE algorithms, which possess a well-known property termed as "partition of unity (PU)" that each nodal shape function is equal to 1 at its binding node whereas at the other nodes it is equal to 0, and the summation of all the nodal shape functions in an element is equal to 1.

A serious drawback with these standard basis functions exists, however, when the element order is updated, the new shape functions have to be regencrated totally and hence all the calculations are repeated. Hence it is attractive to avoid this drawback by constructing the variable interpolation as a series in which the nodal shape functions do not depend on the number of nodes in the element. This idea gives rise to "hierarchical" shape functions (Babuška et al. 1981).

1. **One-dimensional elements**

Take the bar element illustrated in Fig. 4.3 for example. The traditional 1-D element of C_0 continuity entails the displacement interpolation along the element by

$$u = N_1\hat{u}_1 + N_2\hat{u}_2 \tag{4.23}$$

in which the standard shape (basis) functions of isoparametric element are

$$\begin{cases} N_1 = \phi_1 = -\frac{1}{2}(\xi - 1) \\ N_2 = \phi_2 = \frac{1}{2}(\xi + 1) \end{cases} \quad \xi \in [-1, 1] \tag{4.24}$$

The normalized (parametric, simplex) coordinate ξ is a dimensionless geometrical variable with its origin at the middle of element (viz. Fig. 4.3) and is calculated by

$$\xi = \frac{2x - x_2 - x_1}{x_2 - x_1} \tag{4.25}$$

where: x_1—global coordinate of the left end; x_2—global coordinate of the right end.

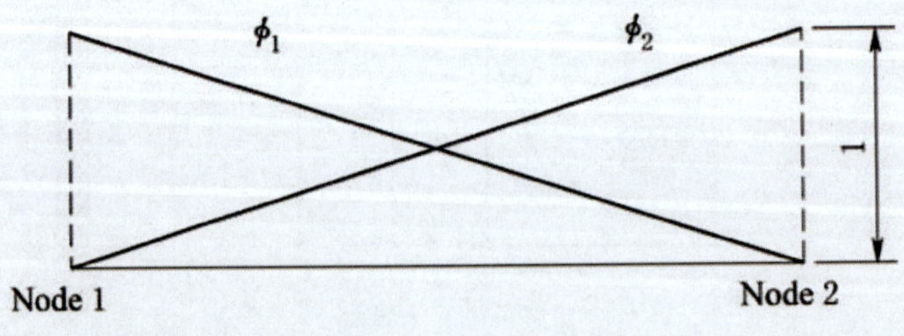

Fig. 4.3 1-D element and basis functions

A progressively increasing number of nodes will improve the interpolation accuracy. To generate a set of standard shape functions using the 2nd order polynomial for the purpose of higher order interpolation, we can write

$$u = N_1\hat{u}_1 + N_2\hat{u}_2 + N_3\hat{u}_3 \tag{4.26}$$

in which

$$\begin{cases} N_1 = \dfrac{1}{2}\xi(\xi - 1) \\ N_2 = \dfrac{1}{2}\xi(\xi + 1) \\ N_3 = (1 - \xi)(1 + \xi) \end{cases} \tag{4.27}$$

$\hat{u}_1$, $\hat{u}_2$ and $\hat{u}_3$ are exactly the basic (essential) variables at nodes, namely, the nodal displacements tied to the points of left end, right end and middle, respectively. The standard shape functions N_1, N_2 and N_3 defined by the normalized coordinate ξ possess the property of PU in Lagrange interpolation, i.e. at the binding node it is equal to 1 whereas at the other node it is equal to 0, and the summation of all the shape functions of the element is equal to 1. Cross reference Eq. (4.24) with Eq. (4.27), it is clear that the shape functions of different order will not be overlapped at all, this will result in the unpleasant consequence that the stiffness matrix of a lower order element is not the sub-matrix of higher order ones. For a fixed mesh, if we try to use higher order elements to improve the computation accuracy, the whole computation procedure should be re-started from the beginning, which results in low computation efficiency.

The hierarchical concept also may be well illustrated by the 1-D problem in Fig. 4.3. It keeps the linear interpolation (4.23) unchanged using the standard shape functions N_1 and N_2 defined in Eq. (4.24), meanwhile adds a polynomial basis function ϕ_3 correspondent to the virtual node represented by the round spot in Fig. 4.2 that is always designed to have zero values at the ends of element. The corresponding quadratic (2nd-order) interpolation is now in the form of

$$u = N_1u_1 + N_2u_2 + \phi_3u_3 \tag{4.28}$$

in which

$$\phi_3 = (1 + \xi)(1 - \xi) \tag{4.29}$$

Contrast Eqs. (4.26) and (4.27) with Eqs. (4.28) and (4.29), the new variables $\hat{u}_1$, $\hat{u}_2$, $\hat{u}_3$ may be interpreted by

$$\begin{cases} u_1 = \hat{u}_1 \\ u_2 = \hat{u}_2 \\ u_3 = \hat{u}_3 - (\hat{u}_1 + \hat{u}_2)/2 \end{cases} \tag{4.30}$$

These new essentials variables are now not ones explicitly tied to nodes (e.g. local displacements). However, the final solution is identical because the interpolation of u within the element is unchanged.

As we have already noted, the above basis function ϕ_3 is not obligated to be uniquely defined, and many possibilities exist in the quadratic expression subject to the null valuc restraint at the end nodes of element.

For a cubic (3rd-order) element, we further add an item ϕ_4 to Eq. (4.28) where ϕ_4 is any cubic (3rd-order) polynomial which is null at $\xi = \pm 1$ (i.e. at nodes 1 and 2). Again, a selective polynomial set is available. For example, we can simply write

$$\phi_4 = \xi\left(1 - \xi^2\right) \tag{4.31}$$

The coordinate mapping (transforming) between the reference element and real world element holds by the standard shape functions identical to Eqs. (4.23)–(4.25).

Following a similar manner, we can define much higher-order hierarchical elements by adding basis function series step by step.

2. **Shape functions using orthogonal polynomials**

Use is made of orthogonal polynomial series, a much better choice of hierarchical basis functions may be get to profit from their advantage that the derived stiffness matrix is nearly diagonal (Cheng and Chen 1999; Chen 2018). This is important for the purpose of saving computer memory and accelerating computation speed. In addition, the stiffness matrix employing orthogonal polynomial series exhibits better condition number of numerical stability. Legendre and Jacobian polynomials are prevalent orthogonal polynomials used for many purposes. Take the former for example, they are expressed by

$$\phi_p = \int_{-1}^{\xi} P_{p-2}(x)\mathrm{d}x \quad (p = 3, 4, \ldots) \tag{4.32}$$

in which P_k is the k-order normalized Legendre polynomial

$$P_k(x) = \sqrt{\frac{2k+1}{2}} \frac{1}{2^{k-1}(k-1)!} \frac{\mathrm{d}^k}{\mathrm{d}x^k}\left(x^2 - 1\right)^k \quad (k = 1, 2, \ldots) \tag{4.33}$$

Several front items of Eq. (4.32) may be specified as follows

$$\begin{cases} \phi_3 = \sqrt{\dfrac{3}{2}} \times \dfrac{1}{2}(\xi^2 - 1) \\ \phi_4 = \sqrt{\dfrac{5}{2}} \times \dfrac{1}{2}(\xi^3 - \xi) \\ \phi_5 = \sqrt{\dfrac{7}{2}} \times \dfrac{1}{8}(5\xi^4 - 6\xi^2 + 1) \\ \phi_6 = \sqrt{\dfrac{9}{2}} \times \dfrac{1}{8}(7\xi^5 - 10\xi^3 + 3\xi) \\ \phi_7 = \sqrt{\dfrac{11}{2}} \times \dfrac{1}{16}(21\xi^6 - 35\xi^4 + 15\xi^2 - 1) \end{cases} \quad (4.34)$$

The corresponding curves are plotted in Fig. 4.4.

3. **Two-and three-dimensional elements**

For 2- and 3-D FEM, a standard element possesses nodal shape functions associated with its nodes which meet the condition of PU. For hierarchical element, two/three additional groups of basis functions are demanded, that are associated with the edge, face and body of the element. For details, readers are referred to the literatures (e.g., Chen 2018).

4.3 Approximation

4.3.1 Overview

Although interpolation is prevalent for approximating a given continuous function f, yet it could present non-convergence even in the polynomial case. The theorem proved by Weierstrass (1885) using analytic functions states that: a sequence of polynomials converging uniformly to f on a bounded interval exists, but these polynomials cannot be obtained by interpolation at a fixed set of points. The proofs of "Weierstrass theorem" were further given by Picard (1891), Volterra (1897) and Lerch (1903) using Fourier expansions, more elementary proofs were given by Lebesgue (1898) and Mittad-Leffler (1900). Weierstrass theorem was generalized to a compact set (instead of an interval) by Mfjntz (Stone 1948).

A general "approximation" is hence more appealing because it determines a function that passes near a certain number of data points subject to minimizing the deviation of this function from the data points. For the problems about the existence, uniqueness, convergence, effectiveness, etc., of such an approximation solution, they are coped with by two essential works: the choice of approximating function F and

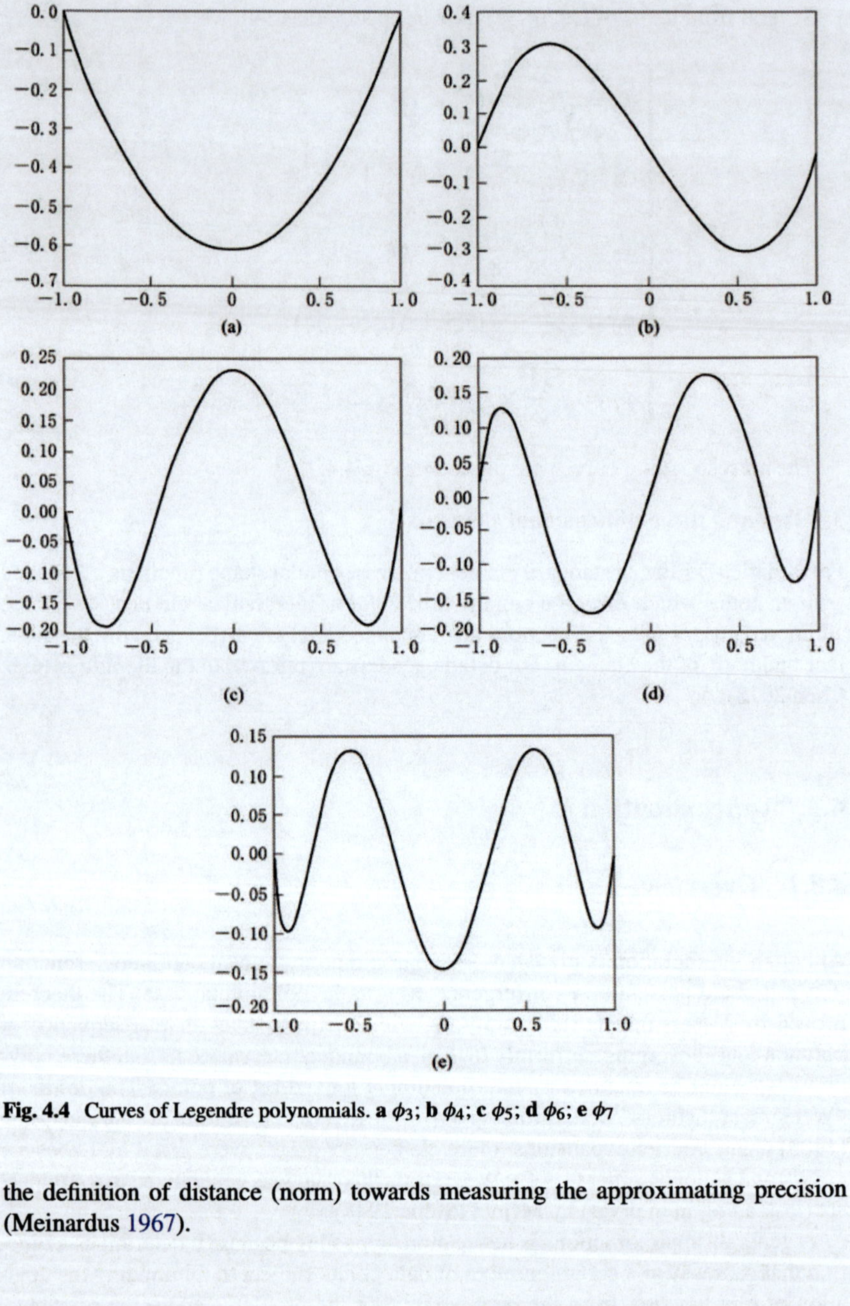

Fig. 4.4 Curves of Legendre polynomials. **a** ϕ_3; **b** ϕ_4; **c** ϕ_5; **d** ϕ_6; **e** ϕ_7

the definition of distance (norm) towards measuring the approximating precision (Meinardus 1967).

Let's define the L_p-norm and $L_\infty(f)$-norm for the function f on interval $[a, b]$ as follows

$$L_p(f) = \left[\int_a^b |f(x)|^p \mathrm{d}x \right]^{1/p} \tag{4.35}$$

$$L_\infty(f) = \sup_{a \le x \le b} |f(x)| \left[\int_a^b \sup|f(x)|^p \mathrm{d}x \right]^{1/p} \tag{4.36}$$

Equation (4.36) is called as "Chebyshev-norm", too.

Where $p = 2$ in Eq. (4.35), one get the "Euclidean-norm" that may be denoted as $L_2(.)$ or simply $\|.\|_2$. The best approximant problem consists of finding F in some set of functions such that norm $L(f - F)$ is a minimum. Chebyshev-norm was used for the first time by Laplace in 1799 for approximating the solution of an inconsistent linear equation system, but it was completely impracticable for hand computations. In 1806, Laplace handled the same problem by L_2-norm that leading to the discovery of least squares approximation. L_1-norm was studied for the first time by Markov but the corresponding computational technique was only obtained much later by Wiener in 1933.

In advanced CMs, most types of basis functions possess good approximation properties. Hence, the approximation properties of various schemes discussed hereinafter can be understood by interpreting function $f(x)$ with an approximate Dirac Delta function plus weighted average. For a broad survey about approximation theory, our readers are referred to the literature, among many others, of Tikhomirov (1990).

1. **Linear approximation**

By linear approximation, it means that the function $f(x)$ is approximated by

$$F(x) = \sum_{i=0}^{n} a_i \phi_i(x) \tag{4.37}$$

in which $\phi_i(x)$ $(i = 0, 1, 2, \ldots, n)$ is the given function. Obviously, the easiest case corresponds to polynomials $\phi_i(x) = x^i$ $(i = 0, 1, 2, \ldots, n)$.

The first systematic study of polynomial approximation is attributable to Chebyshev (1854). The proofs of the existence, uniqueness and continuity of the best uniform approximant polynomial was given by Borel (1905). He also gave the alternation theorem stating that the error of the best uniform approximant polynomial of degree n achieves its maximum in at least $n+1$ points with alternating signs. The error convergent rate of the best uniform polynomial approximation was given by Jackson (1911). The geometric character of the convergence for functions was proved by Bernstein (1912). Since the best approximant polynomial in Chebyshev sense could be difficult to obtain in practice, Hornecker (1958, 1959) studied how to obtain near-best approximants. Extensive tabulations of best approximants to elementary functions can be found in Hastings (1955), Abramowitz and Stegun (1964), Hart et al. (1965). Orthogonal polynomials w.r.t. a general linear functional is nowadays very much in favor (Maroni 1991; Brezinski and Sadok 1993) such as in the hierarchical FEM illustrated previously.

More results and historical references are referred to the works of Natanson (1964) and Cheney (1966).

2. **Nonlinear approximation**

The simplest case of approximation in a nonlinear family is the approximation by rational functions. The subject was first discussed by Chebyshev in 1859

$$F(x) = \sum_{i=0}^{n} a_i r^{t_i x} \tag{4.38}$$

in which a_i $(i = 0, 1, 2, \ldots, n)$ and t_i $(i = 0, 1, 2, \ldots, n)$ have to be determined. Chebyshev proved that the error of best approximant must reach its maximum value in at least n points. The existence of best rational approximant was studied by Walsh (1931), and the uniqueness was proved by Achieser (1947).

For an exposition of modern nonlinear approximation, our readers are referred to the works of Braess (1986), Petrushev and Popov (1987).

3. **Multivariate approximation**

The theorem of Weierstrass was extended to multi-variables by Picard (1891) and that the uniqueness for linear functions of two variables was proved by Collatz (1956). Linear projection operators are very useful since they allow multivariate approximants to be built from univariate functions by means of tensor products (Light and Cheney 1985).

There are many differences between univariate and multivariate approximation. One is the occurrence of infinite-dimensional subspaces for the approximants, and of unusual Banach spaces and norms. Moreover, the geometry of the domain for approximation plays a crucial role. In addition, the existence of best approximants becomes difficult question in multivariate approximation.

All these questions had been comprehensively reviewed by Cheney (1966).

4.3.2 Least-Squares Approximation

Interpolation can only be used for smooth regular data points (nodes, knots), such as in the solution of PDEs by FDM. However, we are frequently encountered with irregularities in handling experimental data. Under such circumstances, it is wise to approximate the data by functions that contain lower degrees of freedom, for example, lower-degree polynomials, such that they do not pass through every point.

The best fitting in least-squares sense minimizes the sum of squared residuals that is defined as the difference between observed value and fitted value. Thc first concise exposition of the "least-squares" analysis was published by Legendre in 1805. The idea was also independently formulated by Adrain in 1808. In 1822, Gauss stated that the least-squares approach to regression analysis is optimal in the following sense: the error with a mean of zero is uncorrelated and has equal variances, which is known as the "Gauss-Markov theorem". In the next two centuries, researchers found many different ways of implementing least squares analysis (Aldrich 1998).

There are a variety of linear least squares problems. As the special cases of general optimization problem in R^n, nonlinear least squares problems are closely related to nonlinear equation systems. Basic methods for solving such problems are iterative, each iteration step usually requires the solution of a related linear least squares problem. An excellent survey was given by Dennis (1977).

1. **Linear least-squares approximation**

The "linear least-squares (LLS)" problem is primary importance in many applications. Suppose, for example, that we want to fit a linear model to the given data set. In order to reduce the influence of errors in the data set, we can use a larger number m of measurements than the number n of unknowns, namely, $m > n$. The resulting problem is to solve an overdetermined linear system: to find a vector $\mathbf{x} \in R^n$ such that $\mathbf{Ax}$ is the best approximation to the known vector $\mathbf{b} \in R^m$, where $\mathbf{A} \in R^{m \times n}$. $\mathbf{x}$ is the linear least squares solution of the algebraic equation set

$$\mathbf{Ax} = \mathbf{b} \tag{4.39}$$

There are many ways of defining the best solution. A possible choice is to let $\mathbf{x}$ be a solution of the minimization problem such that

$$\min \|\mathbf{Ax} - \mathbf{b}\|_2 \quad \text{for } \mathbf{A} \in R^{m \times n}, \mathbf{x} \in R^n, \mathbf{b} \in R^m \tag{4.40}$$

in which $\|.\|_2$ denotes the Euclidean-norm.

One important source of LLS problems is linear statistics. Suppose that m observation data $\mathbf{b} = (b_1, b_2, \ldots, b_m)$ are related to n unknown parameters $\mathbf{x} = (x_1, x_2, \ldots, x_n)$, if $\mathbf{A}$ has rank n, Gauss showed that in case $\mathbf{A}$ has full rank, the least squares estimation of $\mathbf{x}$ exhibits the smallest variance in the class of estimation methods fulfilling two conditions:

- No systematic errors (no bias) in the estimation; and
- The estimation function is linear w.r.t. $\mathbf{b}$.

Note that this property does not depend on any assumed probability distribution laws of the error.

2. **Moving least-squares approximation**

The idea of the "moving least-squares (MLS)" is a simple generalization of LLS (Lancaster and Salkauskas 1981). Suppose a function is defined at a finite set of points, and the approximant value at point $x*$ is looked for. One may use all available nodes, but for the purpose of computational efficiency, it is reasonable to select some small subset of nodes x_i $(i = 1, 2, \ldots, n)$ (called the stencil or star), preferably in the neighborhood of $x*$.

Suppose that the function $f(x)$ is sufficiently smooth to be approximated by some well-defined function basis

$$f(x) = \sum_{k=1}^{m} a_k(x*) \times p_k(x) = \mathbf{p}^{\mathrm{T}}(x)\mathbf{a}(x*) \tag{4.41}$$

Usually, $\mathbf{p}_k(x)$ is a set of complete polynomial basis of arbitrary order, in 1-D cases for example

$$\mathbf{p}(x) = 1 \quad \text{for } m = 1 \text{ (constant base)} \tag{4.42}$$

$$\mathbf{p}(x) = \begin{bmatrix} 1 & x \end{bmatrix}^{\mathrm{T}} \quad \text{for } m = 2 \text{ (linear base)} \tag{4.43}$$

$$\mathbf{p}(x) = [1 \quad x \quad x^2/2]^{\mathrm{T}} \quad \text{for } m = 3 \text{ (quadric base)} \tag{4.44}$$

Equation (4.41) can be normally interpreted as a Taylor expansion of the unknown function $f(x)$ around the point $x = x*$. With this choice, the coefficient $a_1 = a_1(x*)$ can be understood as the combination of unknown function and its derivatives at $x*$, whereas the coefficients $a_k = a_k(x*)(k = 2, \ldots, m)$ are the fractions of the $k - 1$ order derivatives of unknown function at $x*$, respectively (viz. Fig. 4.5).

However, Eq. (4.41) also may be alternatively interpreted that $\mathbf{a} = \mathbf{a}(x*)$ is arbitrary function with "moving" $x*$, this extends its application because the derivatives of unknown function are no more concerned. The function $\mathbf{a} = \mathbf{a}(x*)$ is such evaluated that the approximation through n nodes meets the condition

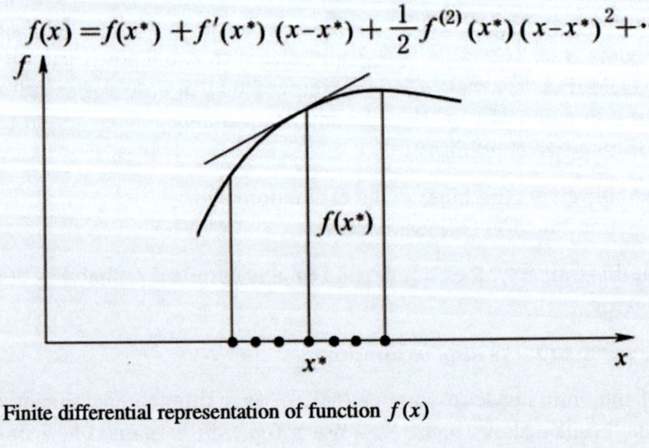

Fig. 4.5 Finite differential representation of function $f(x)$

$$f(x_i) = \mathbf{p}^{\mathrm{T}}(x_i)\mathbf{a}(x) = u_i \quad (i = 1, 2, \ldots, n) \tag{4.45}$$

in which u_i denotes the value at nodes i and the fixed point $x*$ is replaced by moving point x hereinafter.

Normally, the positions of nodes should satisfy certain geometrical requirements, so that the linear equation set (4.45) is not singular. However, instead of trying to define and satisfy these requirements, one selects more nodes than equations, namely, the number n of nodes should be not smaller than the number m of terms in Eq. (4.41)

$$n \geq m \tag{4.46}$$

Thus Eq. (4.45) becomes an overdetermined linear equation set, and by weighting each equation with appropriate weight W_i, the traditional least-squares method may be employed for the solution. The weight function may, in general, depends on the location of node x_i $(i = 1, 2, \ldots, n)$ within a compact domain termed as "support domain" w.r.t. x. The simple and isotropic weight function dependent on the distance from central point is most prevalent

$$W_i = W(r_i) \text{ where } r_i = |x - x_i| (i = 1, 2, \ldots, n) \tag{4.47}$$

At minimum, the weight function should be decaying with the argument of distance, i.e.

$$W(r_1) \geq W(r_2) \Leftrightarrow r_1 < r_2 \tag{4.48}$$

Now we denote the approximant (4.45) as

$$u^{\mathrm{h}}(x, x_i) = \mathbf{p}^{\mathrm{T}}(x_i)\mathbf{a}(x) \tag{4.49}$$

which is explicitly dependent on the position x_i merely. Towards the "best fitting" $u^{\mathrm{h}}(x, x_i)$ from Eq. (4.49) w.r.t. all the scattered points x_i $(i = 1, 2, \ldots, n)$, a functional of weighted residual is constructed in the form of

$$J = \sum_{i=1}^{n} W(x - x_i)[u^{\mathrm{h}}(x, x_i) - u(x_i)]^2 = \sum_{i=1}^{n} W(x - x_i)[\mathbf{p}^{\mathrm{T}}(x_i)\mathbf{a}(x) - u(x_i)]^2 \tag{4.50}$$

in which $u(x_i)$ is the observation value at point x_i, and $\mathbf{a}(x)$ should be so chosen to minimize the weighted residual defined in Eq. (4.50) for all the value of x, which is realized by the stationary condition of functional J

$$\frac{\partial J}{\partial \mathbf{a}} = 0 \tag{4.51}$$

Equation (4.51) results in

$$\mathbf{A}(x)\mathbf{a}(x) = \mathbf{B}(x)\mathbf{u} \tag{4.52}$$

where $\mathbf{A}(x)$ is called the "weighted moment matrix"

$$\mathbf{A}(x) = \sum_{i=1}^{n} W(x - x_i)\mathbf{p}(x_i)\mathbf{p}^{\mathrm{T}}(x_i) \tag{4.53}$$

and

$$\mathbf{B}(x) = [\mathbf{B}_1 \ldots \mathbf{B}_i \ldots \mathbf{B}_n] \tag{4.54}$$

$$\mathbf{B}_i(x) = W(x - x_i)\mathbf{p}(x_i) \quad (i = 1, 2, \ldots, n) \tag{4.55}$$

$$\mathbf{u} = [u(x_1) \ldots u(x_i) \ldots u(x_n)]^{\mathrm{T}} \tag{4.56}$$

The solution of Eq. (4.52) gives rise to

$$\mathbf{a}(x) = \mathbf{A}(x)^{-1}\mathbf{B}(x)\mathbf{u} \tag{4.57}$$

On substitution into Eq. (4.49) in which the individual node x_i has been smeared within the support domain, it follows that

$$u^h(x) = \sum_{i=1}^{n} \mathbf{p}^{\mathrm{T}}(x_i)(\mathbf{A}(x))^{-1}\mathbf{B}_i(x)u(x_i) \tag{4.58}$$

or simply

$$\begin{cases} u^h(x) = \sum_{i=1}^{n} \phi_i(x)u_i \\ \phi_i(x) = \mathbf{p}^{\mathrm{T}}(x_i)(\mathbf{A}(x))^{-1}\mathbf{B}_i(x) \end{cases} \tag{4.59}$$

These shape functions $\phi_i (i = 1, 2, \ldots, n)$ may be used to construct trial function and as test functions towards the formulation of MFMs, for example, the "element-free Galerkin method (EFGM)".

Equation (4.59) can be written in the form of

$$u^h(x) = \Phi(x)\mathbf{u} \tag{4.60}$$

$$\Phi(x) = [\phi_1(x) \ldots \phi_i(x) \ldots \phi_n(x)]. \tag{4.61}$$

$\Phi(x)$ is the "shape function matrix" of MLS corresponding to n nodes in the support domain.

A glance of special case $m = 1$ would be interesting. Under such circumstances, we have

$$\mathbf{p}(x_i) = 1 \tag{4.62}$$

$$\mathbf{A}(x) = \sum_{i=1}^{n} W(x - x_i) \tag{4.63}$$

$$\mathbf{B}(x) = [W(x - x_1) \dots W(x - x_i) \dots W(x - x_n)] \tag{4.64}$$

Then, the shape function (4.59) is simplified as

$$\phi_i = \frac{W(x - x_i)}{\sum_j^n W(x - x_j)} \quad (i = 1, 2, \dots, n) \tag{4.65}$$

This version of MLS is sufficient to implement the analog of FDM on an irregular nodal grids, more precisely, it allows for solving PDEs using the "nodal collocation method". The main drawback of this approach is that the approximant is defined only at data nodes, and nothing is known about the solution between these nodes.

Figure 4.6 exposits a 1-D example of the MLS approximant for displacement $u^h(x_i)$ at node i: it depends not only on the nodal parameter u_i but also on the nodal parameters corresponding to all the nodes within the support domain of node i. This makes the imposition of essential boundary conditions more complicated than that, for example, using standard shape functions in FEM.

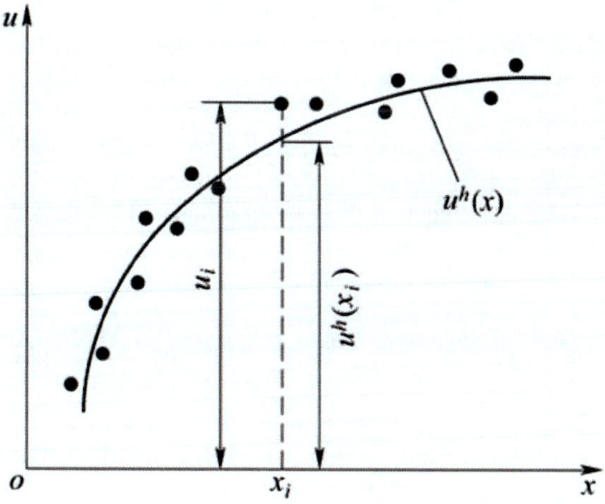

Fig. 4.6 Function $u^h(x)$ in the MLS approximant

In general, the MLS approximated function $f(x)$ (4.41) is continuous. More exactly, $f(x) \subset C^{\min(m,l)}$ subject to

① Basis functions $p_i \subset C^m$ $(i = 1, 2, \ldots, n)$;
② Weighting function $W_i(r) \subset C^l$ $(i = 1, 2, \ldots, n)$;
③ All data nodes from the whole domain are included in the stencil.

It is often assumed that the weight function has a support domain entailed by a finite ball with radius R, namely

$$W_i(r) = 0 \Leftrightarrow r > R \tag{4.66}$$

Then condition ③ may be modified as:

④ All nodes from the local support of weight function are included in a finite ball with radius R.

An example of weight function meeting condition (4.66) is

$$W_i(r) = \begin{cases} \sqrt{4/\pi}(1 - (r/R)^2)^4 & \text{where } r < R \\ 0 & \text{where } r \geq R \end{cases} \tag{4.67}$$

Noting that the weight function also depends on the position of central point x in addition to radius r, it should be continuous with respect to both arguments, i.e.,

$$W(r) = W(r, x) \tag{4.68}$$

One can easily modify above procedure to define a dense approximant—pointwise approximant. This has been, with reasonable success, applied to the solution of various PDEs using Galerkin type of weak form. However, it should be noted that although MLS shape functions satisfy PU, i.e. $\sum_{i=1}^{n} \phi_i = 1$, that can be easily found from Eq. (4.65), yet they do not possess the Kronecker Delta property, i.e. $\phi_i(x_j) \neq \delta_{ij}$, hence $u^h(x_j) \neq u_j$. Therefore, Eq. (4.60) is not the interpolant, but rather the approximant of function. The lack of Kronecker Delta property results significant difficulties in handling essential boundary conditions.

We will see later that MLS shape functions can be used in various weak formulations such as the EFGM (Chap. 9) for the solution of PDEs.

4.3.3 Inverse Distance Weighted Methods

One class of data fitting algorithms is based on the so-called "inverse distance weighted methods" (Shephard 1968) in which the scattered data set of $f(\mathbf{x}_i)$ is interpolated by a function $F(\mathbf{x})$ such that

$$F(\mathbf{x}) = \sum_{i=1}^{n} f(\mathbf{x}_i) W_i(\mathbf{x}) / \sum_{j=1}^{n} W_j(\mathbf{x}) \tag{4.69}$$

where weights are typically chosen as decaying functions of the distance between $\mathbf{x}$ and $\mathbf{x}_i$. The basic shape functions are constructed as

$$\phi_i(\mathbf{x}) = \frac{W_i(\mathbf{x})}{\sum_{j=1}^{n} W_j(\mathbf{x})} \tag{4.70}$$

Note that these shape functions are actually identical to Eq. (4.65) derived by the MLS approximation where only the constant item in complete polynomial is kept (i.e. $m = 1$), hence it is clear that they satisfy the condition of PU.

One possible generalization is to seek F in the form of

$$F(\mathbf{x}) = \widetilde{F}(\mathbf{x}, a_0, a_1, \ldots, a_m) \tag{4.71}$$

in which the parameters $(a_0, a_1, \ldots, a_m)$ are determined by the stationary condition (for each fixed $\mathbf{x}$)

$$\min . \left[\sum_{i=1}^{n} \left(f(\mathbf{x}_i) - \widetilde{F}(\mathbf{x}, a_0, a_1, \ldots, a_m) \right)^2 W_i(\mathbf{x}) \right] \tag{4.72}$$

Suppose that the function $\widetilde{F}(\mathbf{x}, a_0, a_1, \ldots, a_m)$ is a polynomial in $\mathbf{x}$ and $(a_0, a_1, \ldots, a_m)$ are the coefficients defining the polynomial, Eq. (4.72) exactly leads to the MLS method. If one carries out the minimization operation, the global approximation F takes the form of

$$F(\mathbf{x}) = \sum_{i=1}^{n} f(\mathbf{x}_i) \tilde{\phi}_i(\mathbf{x}) \tag{4.73}$$

in which $\tilde{\phi}_i(\mathbf{x})$ are the desired shape functions.

Another generalization of Shephard method uses the fact that functions $\phi_i(\mathbf{x})$ form a PU. For example, if the values and derivatives of the function are given at nodes $\mathbf{x}_i$, then one can construct

$$F(\mathbf{x}) = \sum_{i=1}^{n} \phi_i(\mathbf{x}) L_i(\mathbf{x}) \tag{4.74}$$

in which $L_i(\mathbf{x})$ are the Taylor polynomials about nodes $\mathbf{x}_i$. This is the starting point of the "partition of unity method (PUM)" (Babuška and Melenk 1997) and is today also known as the "generalized finite element method (GFEM)" (Strouboulis et al. 2001). The core ideas of PUM are, first, the construction of spaces with local approximation properties and, second, the conformity of these spaces. This is the cornerstone of MFMs such as EFGM (Belytschko et al. 1994), GFDM (Liszka and Orkisz 1980) and SPH (Gingold and Monaghan 1977; Lucy 1977) that use the shape functions for constructing trial function as well as for test functions. The hp-clouds method (Duarte and Oden 1995) can be looked at as a mixture of EFGM and PUM if the local approximation is realized by polynomials.

For the detailed comparisons and connections between these various methods, readers are referred to the overview by Duarte (1995).

4.3.4 Radial Basis Functions

Another set of approximation methods use "radial basis functions (RBFs)" (Hardy 1971; Franke 1982; Powell 1987; Schaback 1995), also known as "multi-quadrics". In the present sense, the function $f(\mathbf{x})$ is approximated as a linear combination of radial basis functions. It is emphasized here that these methods are, in their original form, non-local and therefore do not lead to sparse stiffness matrices.

RBFs belong to a class of functions where the Euclidean distance r of a fixed point (sometimes referred to as a "knot") from a variable field point is used as independent variable. The advantage of RBFs is that they involve a single independent variable regardless of the dimension of the problem. First being applied to geodesic approximation, RBFs have since been widely exercised in a number of fields such as multivariate interpolation, neural networks, kinetic modeling, solution of differential and integral equations, etc.

Function ϕ is dubbed with "radial" if it is a univariate function centered at the coordinate origin in terms of variable r, namely,

$$\phi(\mathbf{x}) = \phi(r), \quad \phi : R^n \to R \tag{4.75}$$

in which the real-valued variable defined by

$$r = \|\mathbf{x}\| \in R \tag{4.76}$$

is usually an Euclidean norm on R^n. Obviously, the definition (4.75) implies that for a radial function ϕ, the relation

$$\phi(\mathbf{x}_1) = \phi(\mathbf{x}_2) \quad \mathbf{x}_1, \mathbf{x}_2 \in R^n \tag{4.77}$$

holds only if

$$\|\mathbf{x}_1\| = \|\mathbf{x}_2\| \tag{4.78}$$

Take the Gaussian radial function centered at the origin on R^2 (2-D space) for example, it can be written as

$$\phi(\mathbf{x}) = \phi(r) = e^{-cr^2} \text{ for } \mathbf{x} = (x, y) \in R^2 \text{ and } r = \sqrt{x^2 + y^2} \in R \tag{4.79}$$

in which $c > 0$ is the shape parameter.

Figure 4.7 shows two graphs of Gaussian radial functions, one with the shape parameter $c = 1$ and another with $c = 4$. It is found that a smaller c causes flatter surface, whereas a larger c leads to sharper and localized surface. Additionally, the first-order derivatives of the Gaussian radial function can be obtained using the chain rule of differentiation as follows

$$\frac{\partial \phi}{\partial \mathbf{x}} = \frac{\partial \phi}{\partial r}\frac{\partial r}{\partial \mathbf{x}} = -2c\mathbf{x}e^{-r^2} \tag{4.80}$$

More generally, a smooth radial function can be so defined that its value depends only on the Euclidean norm of a vector from a central point $\tilde{\mathbf{x}}$—reference point, to the field point $\mathbf{x}$ (viz. Fig. 4.8), namely

$$\phi(\mathbf{x},\tilde{\mathbf{x}}) = \phi(r) \tag{4.81}$$

in which

$$r = \|\mathbf{x} - \tilde{\mathbf{x}}\| \in R \tag{4.82}$$

Thus, the radial function with simple shape can be easily applied in multi-dimensional space without causing significant extra effort. In addition, the evaluation of its derivatives is simpler than that of MLS approximation. Some typical radial functions are collected below:

$$\text{Power spline (PS) } \phi = r^{2m} \quad (m = 1, 2, 3, \ldots) \tag{4.83}$$

$$\text{Thin plate spline (TPS) } \phi = r^{2m} \ln r \quad (m = 1, 2, 3, \ldots) \tag{4.84}$$

$$\text{Multi-quadric (MQ) } \phi = \sqrt{r^2 + c^2} \tag{4.85}$$

$$\text{Gaussian (GS) } \phi = e^{-cr^2} \tag{4.86}$$

Since the radial functions are only Euclidean norm dependent, they have distinctive properties of being invariant under all Euclidean transformations (i.e., translations, rotations, reflections), and being insensitive to the dimensions of the space. This brings about the advantage in generating a large number of linearly independent basis functions, which enables RBFs to straightforwardly solve the scattered data interpolation problem (Larsson and Fornberg 2003).

In the study of CMs, a function can be interpolated by the linear combination of RBFs centered at a series of points $\{\mathbf{x}_i\}(i = 1, 2, \ldots, n)$

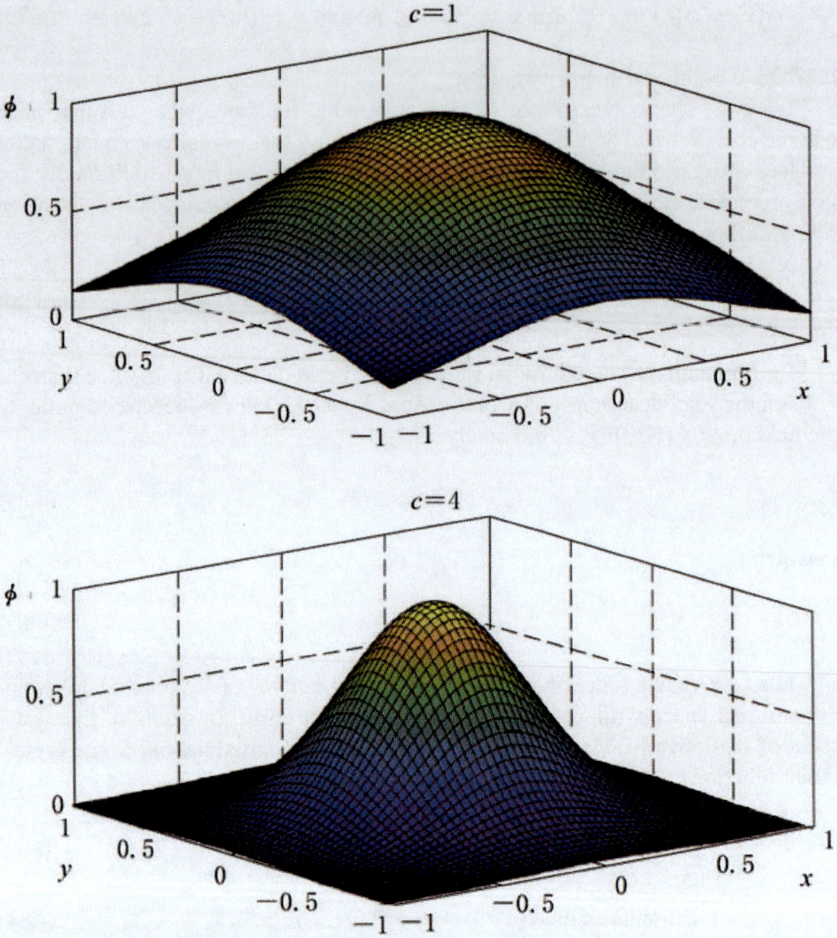

Fig. 4.7 Gaussian radial functions with $c = 1$ and $c = 4$ centered at origin

Fig. 4.8 Definition of Euclidean distance between center point and field point

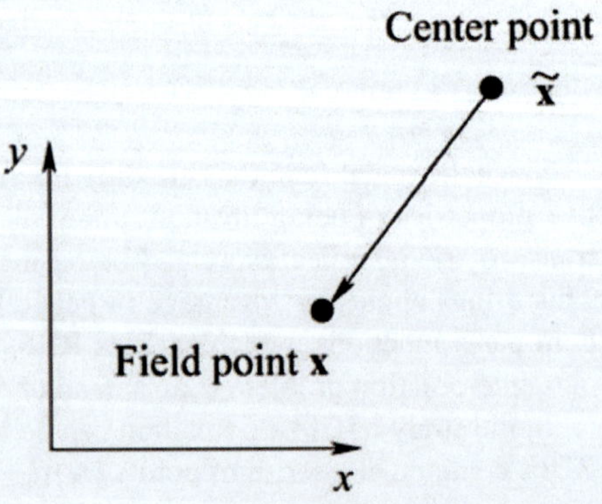

$$f(\mathbf{x}) \approx F(\mathbf{x}) = \sum_{i=1}^{n} a_i \phi(\mathbf{x} - \mathbf{x}_i) \tag{4.87}$$

in which $f(\mathbf{x})$ is the target function to be approximated, a_i is the coefficients, and n is the number of center points $\mathbf{x}_i$ (Kythe 1996). If there are exactly the same number of observation values (target function values) $f\{\mathbf{x}_i\}(i = 1, 2, \ldots, n)$, Eq. (4.87) can be expressed in matrix form

$$\begin{bmatrix} \phi(\mathbf{x}_1 - \mathbf{x}_1) & \phi(\mathbf{x}_1 - \mathbf{x}_2) & \ldots & \phi(\mathbf{x}_1 - \mathbf{x}_n) \\ \phi(\mathbf{x}_2 - \mathbf{x}_1) & \phi(\mathbf{x}_2 - \mathbf{x}_2) & \ldots & \phi(\mathbf{x}_2 - \mathbf{x}_n) \\ \ldots & \ldots & \ldots & \ldots \\ \phi(\mathbf{x}_n - \mathbf{x}_1) & \phi(\mathbf{x}_n - \mathbf{x}_2) & \ldots & \phi(\mathbf{x}_n - \mathbf{x}_n) \end{bmatrix} \begin{Bmatrix} a_1 \\ a_2 \\ \ldots \\ a_n \end{Bmatrix} = \begin{Bmatrix} f(\mathbf{x}_1) \\ f(\mathbf{x}_2) \\ \ldots \\ f(\mathbf{x}_n) \end{Bmatrix} \tag{4.88}$$

or

$$\mathbf{\Phi a} = \mathbf{f} \tag{4.89}$$

by solving this linear system of equations, one can determine all unknown coefficients.

$$\mathbf{a} = \mathbf{\Phi}^{-1}\mathbf{f} \tag{4.90}$$

For geotechnical and hydraulic structures, the simulation of ground surfaces and faults that are normally curved and irregular, is crucial in the pre-process procedure to create block assemble system or even further to discretize into elements and nodes. Assume that the coordinates of the n points on one ground (or fault) surface are arranged in the following manner.

$$\begin{bmatrix} x_1 & x_2 & \ldots & x_i & \ldots & x_n \\ y_1 & y_2 & \ldots & y_i & \ldots & y_n \\ z_1 & z_2 & \ldots & z_i & \ldots & z_n \end{bmatrix} \tag{4.91}$$

Then the surface formula may be constructed as the combination of RBFs $r_i^2 \ln r_i^2 (i = 1, 2, 3, \ldots, n)$ plus a plane function (Chen 2018)

$$Z(x, y) = a_0 + a_1 x + a_2 y + \sum_{i=1}^{n} t_i r_i^2 \ln(r_i^2 + \varepsilon) \tag{4.92}$$

in which

$$r_i = \sqrt{(x - x_i)^2 + (y - y_i)^2} \tag{4.93}$$

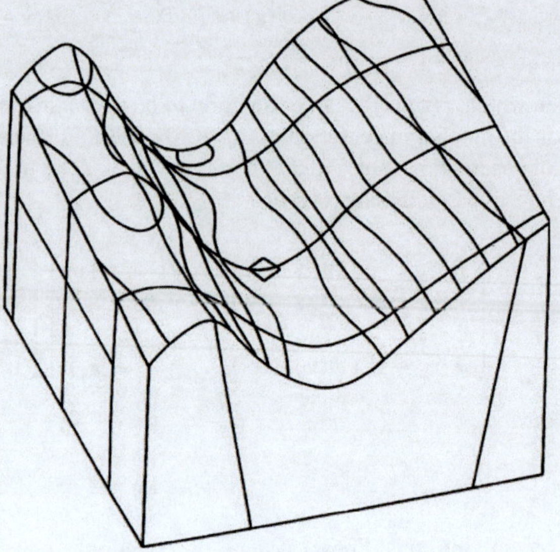

Fig. 4.9 Axonometric drawing of an integrated block system

and ε is a small fraction (e.g. $\varepsilon = 0.001$). On inserting the coordinates of n points into Eq. (4.92), n linear equations comprising $n+3$ coefficients $(a_0, a_1, a_2, t_1, t_2, \ldots, t_n)$ can be obtained. After being supplemented by three conditions

$$\begin{cases} \sum_{i=1}^{n} t_i = 0 \\ \sum_{i=1}^{n} t_i x_i = 0 \\ \sum_{i=1}^{n} t_i y_i = 0 \end{cases} \tag{4.94}$$

$n+3$ equations for solving $n+3$ coefficients are established.

Figure 4.9 shows an approximated valley segment which contains one ground surface and six faults using Eqs. (4.92)–(4.94).

4.3.5 Manifolds

A manifold is a topological space that locally resembles Euclid space near each point. More precisely, each point of a n-dimensional manifold has a neighborhood that is homeomorphic to the n-dimensional Euclid space.

Suppose a group M meets:

- M is a Hausdorff topology space;
- For a point at M, there exists open group $U \subset M$ homeomorphism to R^n or a sub-set of R^n;
- M possesses enumerable base of a topology.

Then, M is a "topology manifold" of dimension n.

One-dimensional manifolds include straight lines and curves. For a simple curve it can be expressed in equation $y = f(x)$, whereas for complex curve we have to express it in partially inter-overlapped segments. It is naturally for each segment a different coordinate system is employed for convenience. The consequence is that a point in the curve may be interpreted in several different coordinate systems. With these coincident points, we may cruise the curve form one coordinate system to another.

Figure 4.10 shows a polyline $A_3 - A_4 - B_4 - B_5 - C_4 - C_5$. Topology ignores bending/breaking, so the broken segments $A_3 - B_4$, $A_4 - C_4$ and $B_5 - C_5$ are treated exactly the same as small pieces of straight line. Take $A_3 - B_4$ for instance, any point in this segment can be uniquely described by its abscissa coordinate. So, its projection $A_1 - A_2$ is continuous and invertible and called as "chart". In Fig. 4.10, three segments $A_3 - B_4$, $A_4 - C_4$ and $B_5 - C_5$ form three charts $A_1 - A_2$, $B_1 - B_2$, $C_1 - C_2$. They cover the whole polyline and form the "atlas" of the polyline. Charts in an atlas may be partially overlapped and a manifold may be represented in several charts. If two charts are partially overlapped, their overlapped portion representing the same region of the manifold is termed as the "manifold element" in NMM.

2-D manifolds are surfaces such as planes and spheres. They can all be embedded in the 3-D real space.

The concept of manifold is central to many topics of modern geometry and mathematical physics because it allows complicated structures to be described and understood in term of simpler local topological properties in Euclidean space. Manifolds are naturally looked at as the solution of equation set and as the graph of function.

Gauss might be the first to consider abstract spaces as mathematical objects in their own right. He gave a calculation method for the curvature of a surface without considering the ambient space in which the surface lies. Such a surface would, in modern terminology, be a manifold. Riemann was the first to generalize the idea of a real-world surface to higher dimensions.

Another important mathematical source of manifold was the analytical mechanics developed in the nineteenth century by Poisson, Jacobi and Hamilton. The possible states of a mechanical system are thought to be the points in an abstract space, namely phase space. Geometrical and topological aspects of classical mechanics were emphasized by Poincaré, too, and his notion of a chain of manifolds is a precursor to the modern notion of atlas.

Shi employed the basic concept of topology manifold in his formulation of "numerical manifold method (NMM)" (Chap. 6). He defined the concept of mathematics cover (chart) for each finite element node that is a set of finite elements surrounding the node. Each cover is overlapped partially with its adjacent covers.

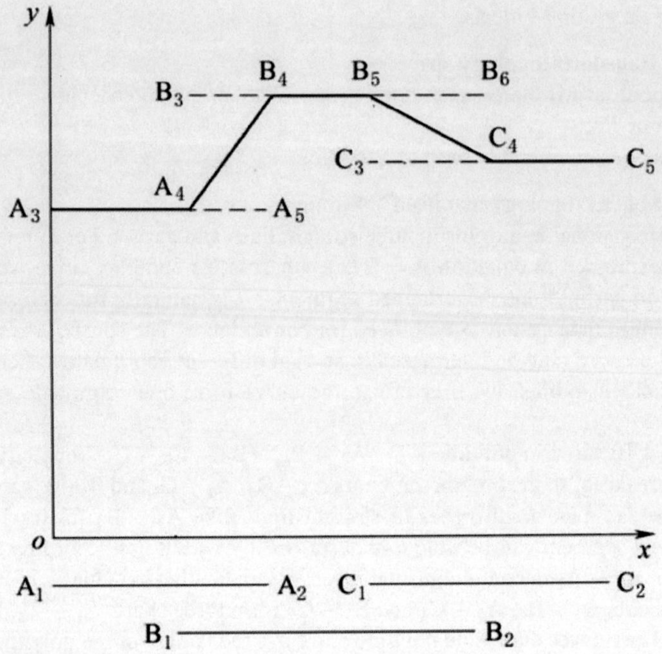

Fig. 4.10 Polyline $A_3 - A_4 - B_4 - B_5 - C_4 - C_5$ and charts

4.4 Numerical Integration (Quadrature)

4.4.1 Concept

The calculation of integrals and derivatives of function is the elementary operation in CMs. It is fortunate that, as the former belongs smoothing operator, its results are much more robust than the differentiation, whereas the latter is well-known being sensitive to "noisy". However, it is unfortunate that, while one can almost always differentiate functions by hand (if the derivative exists at all), most functions cannot be integrated by hand in closed-form. Instead, they must be integrated approximately by a process known as "numerical integration" or "numerical quadrature". Quadrature is very old problem appearing in the records of many ancient civilizations because it is in fact related to the geometrical problem—the determination of farmland areas. To give an example, the work of Archimedes on the area calculation of a circle related to the value of π is well known.

Although the term quadrature was a historical synonym of integration in general that converts areas into equivalent squares, yet in modern usage this term almost exclusively refers to algorithms for function integration. In particular, suppose we demand following definite integral over an interval $[0, \pi]$ (viz. Fig. 4.11)

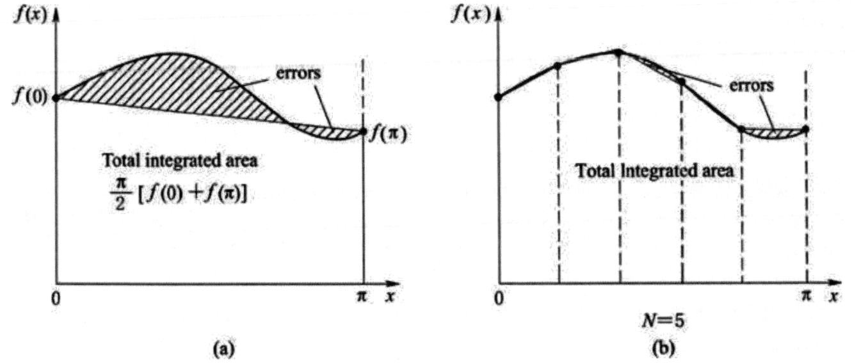

Fig. 4.11 Illustration of integrating on [0, π]. **a** Trapezoidal rule; **b** composite trapezoidal rule

$$I = \int_0^{\pi} f(x)\mathrm{d}x \tag{4.95}$$

In numerical analysis, we turn to approximate this integral I by a sum I_n for $n+1$ quadrature points (nodes, knots) $x_i \in [0, \pi]$ and corresponding quadrature weights W_i

$$I = \int_0^{\pi} f(x)\mathrm{d}x = \sum_{i=0}^{n} W_i f(x_i) + R_n(f) \approx \sum_{i=0}^{n} W_i f(x_i) = I_n \tag{4.96}$$

Then the central problem becomes: how can we choose the points (knots) x_i and weights W_i so that the error $|I_n - I|$ goes to zero as rapidly as possible when we increase n? This problem has a long history related to big names Newton–Cotes, Runge, Gaussian, Chebyshev, Clenshaw–Curtis, et al. It is closely related to other important numerical algorithms such as numerical linear algebra (e.g. Lanczos iterations), approximation theory, and fast Fourier transforms (FFTs). For higher-dimension integration (cubature), the story becomes even more intricate, ranging from statistics (Monte-Carlo integration) and number theory (low-discrepancy sequences and quasi-Monte–Carlo methods) to fractals (sparse grids).

In general, numerical quadrature is purposed to obtain an approximant value of a definite integral for $f(x)$ in $[a, b]$

$$I(f) = \int_a^b f(x)\mathrm{d}x \tag{4.97}$$

A wise idea is to replace the function f by an interpolation polynomial P_{n-1} of degree $n-1$, then an approximate value $I_n(f)$ of $I(f)$ may be obtained under the form of

$$I_n(f) = \sum_{i=1}^{n} A_i^{(n)} f(x_i) \tag{4.98}$$

in which $x_i (i = 1, 2, \ldots, n)$ are the interpolation points, also called as the "knots", of quadrature formula (4.98). Correspondingly, the coefficients $A_i^{(n)}$ are called as the "weights". According to the terminology due to Radau (1880), formula (4.98) is said to have a "degree of exactness (DOE)" $n - 1$.

The knots $x_i (i = 1, 2, \ldots, n)$ can be chosen arbitrarily. If, for simplicity, they are taken equidistant in $[a, b]$, the corresponding quadrature formula is named after Newton and Cotes. If they are taken in an optimal way to achieve the highest possible DOE, the quadrature formula is named after Gauss.

It seems that Cavalieri was the first to propose a formula for numerical quadrature now called "Simpson's formula" that also appeared in Newton's book (1687). Cotes took up Newton's work and gave the quadrature formulae up to 11 knots of interpolation. These formulae are now called "Newton–Cotes quadrature formulae". The case $n = 1$ gives trapezoidal rule (viz. Fig. 4.11a) and the case $n = 2$ is usually called "Simpson's rule" because Atwood (1798) said that it was obtained by Simpson. Actually, Simpson (1743) himself gave the rules up to $n = 6$ knots.

Towards the general convergence theorem of Eq. (4.99), the sufficiency was proved by Steklov while the necessity was given by Pelya. The theorem states that: a necessary and sufficient condition for a quadrature formula to be convergent for every continuous function in $[a, b]$ is that it converges for polynomials and that there exists a number M such that for all n, $\sum_{i=1}^{n} \left| A_i^{(n)} \right| \leq M$.

As a direct conclusion, interpolatory quadrature formulae with positive weights are convergent since they exactly integrate polynomials of degree zero at least. Particularly, the trapezoidal rule is convergent for every continuous function f in $[a, b]$ although its convergence can be quite slow. The works to accelerate convergence had been conducted by many scholars through the whole twentieth century (Richardson 1910; Romberg 1955; Bauer 1961; Laurent 1963; Goldstine 1977; Gautschi 1981; Gowing 1983).

4.4.2 General Formula

Let's build a general quadrature formula as follows

$$\int_a^b f(x)\mathrm{d}x = \sum_{k=0}^{m} \sum_{\lambda=0}^{r_k} A_{k\lambda} f^{(\lambda)}(x_k) + R_n(f) \tag{4.99}$$

in which $n = r_0 + r_1 + \cdots + r_m + m + 1$.

Where the coefficients $A_{k\lambda}$ do not depend on the function, the knots $\{x_k\}$ $(k = 0, 1, \ldots, m)$ are different within the interval $[a, b]$, and to every knot x_k a positive number r_k is assigned. In addition, if $f(x)$ is an algebraic polynomial of degree $n-1$, $R_n(f) = 0$.

The quadrature formula (4.99) always exists and is unique, namely, the coefficients $A_{k\lambda}$ are defined in a unique way. Let P_{n-1} be the Hermite polynomial of degree $n-1$ that satisfies interpolation conditions

$$P_{n-1}^{(\lambda)}(x_k) = f^{(\lambda)}(x_k) \quad \text{for } (k = 0, 1, \ldots, m) \text{ and } (\lambda = 0, 1, \ldots, r_k) \tag{4.100}$$

This polynomial can be written in the form of

$$P_{n-1}(x) = \sum_{k=0}^{m} \sum_{\lambda=0}^{r_k} f^{(\lambda)}(x_k) L_{k\lambda}(x) \tag{4.101}$$

in which $L_{k\lambda}$ are polynomials of degree $n-1$ subject to following restraints

$$L_{k\lambda}^{(j)}(x_i) = 0 \quad \text{for } (k \neq j) \text{ and } (j = 0, 1, \ldots, r_k) \tag{4.102}$$

$$L_{k\lambda}^{(j)}(x_i) = \begin{cases} 1 \text{ where } \lambda = j, \\ 0 \text{ where } \lambda \neq j \end{cases} \tag{4.103}$$

Hence it is evident that the quadrature

$$\int_a^b f(x)\mathrm{d}x \approx \int_a^b P_{n-1}(x)\mathrm{d}x \tag{4.104}$$

belongs the type in Eq. (4.99) with coefficients $A_{k\lambda} = \int_a^b L_{k\lambda}(x)\mathrm{d}x$. It is exact for the polynomials of degree $n-1$. Vice versa, if Eq. (4.104) is exact for the polynomials of degree lower than n, then

$$\int_a^b L_{k\lambda}(x)\mathrm{d}x = \sum_{i=0}^{m} \sum_{j=0}^{r_i} A_{ij} L_{k\lambda}^{(j)}(x_i) \tag{4.105}$$

In the case of equidistant knots where $x_k = a + k(b-a)/m$ and $r_k = 0$ $(k = 0, 1, \ldots, m)$, Eq. (4.105) is called the "Newton–Cotes formula". Other examples of simplification are:

- Rectangle formula $(m = 0)$

$$\int_a^b f(x)\mathrm{d}x = (b-a) f\left(\frac{b-a}{2}\right) + \frac{(b-a)^3}{24} f''(\xi) \quad \text{for } \xi \in (a, b) \tag{4.106}$$

- Trapezoidal rule ($m = 1$)

$$\int_a^b f(x)\mathrm{d}x = \frac{(b-a)}{2}(f(b) - f(a)) - \frac{(b-a)^3}{12} f''(\xi) \quad \text{for } \xi \in (a, b) \tag{4.107}$$

- Simpson's rule ($m = 3$)

$$\int_a^b f(x)\mathrm{d}x = \frac{(b-a)}{6}[f(a) + 4f(\frac{b+a}{2}) + f(b)] - \frac{(b-a)^5}{2880} f^{(4)}(\xi) \quad \text{for } \xi \in (a, b) \tag{4.108}$$

4.4.3 Gauss–Legendre Integration

An important question is put forward that: if we let the knots vary freely and then compute the weights, what maximum "algebraic degree of exactness (ADE)" can be achieved and how can these optimal knots be obtained?

In 1814, Gauss presented a paper to give an integral formula with weights. Scholars who took up the work of Gauss and made significant contribution are Jacobi (1826), Christoffel (1858, 1877), Mehler (1864), Posse (1875), Darboux (1878), Radau (1880), Stieltjes (1884), Markov (1885), Bernstein (1918), Davis (1953). A detailed representation of Gauss quadrature formula can be found, for example, in the work of Ralston (1965), where the knots and weights are given for different values of n.

Take the Gauss–Legendre quadrature for example, the abscissas for n-point rule are the roots of the Legendre function of degree n. The first five Legendre polynomials in Eq. (4.33) are listed below

$$\begin{cases} P_1(x) = x \\ P_2(x) = \frac{1}{2}(3x^2 - 1) \\ P_3(x) = \frac{1}{2}(5x^3 - 3x) \\ P_4(x) = \frac{1}{8}(35x^4 - 30x^2 + 3) \\ P_5(x) = \frac{1}{8}(63x^5 - 70x^3 + 15x) \end{cases} \tag{4.109}$$

There are a number of algorithms that can be used to find the roots of above polynomials, such as the iterative method proposed by Newton.

1. **2-point rule**

The Gauss–Legendre integration rule of two-point is expressed by

$$\int_{-1}^{1} f(x)\mathrm{d}x \approx W_1 f(x_1) + W_2 f(x_2) \tag{4.110}$$

in which x_1, x_2 are the abscissas and w_1, w_2 are the weights. The roots of equation $P_2(x) = 0$ are easily get as follows

$$\begin{cases} x_1 = \dfrac{1}{\sqrt{3}} \\ x_2 = -\dfrac{1}{\sqrt{3}} \end{cases} \tag{4.111}$$

To find the weights W_1 and W_2, one need two simultaneous equations. Use is made of our knowledge concerning the definite integration of 1 and x, we get

$$\begin{cases} \int_{-1}^{1} \mathrm{d}x = 2 = W_1 + W_2 \Rightarrow 2 = W_1 + W_2 \\ \int_{-1}^{1} x\mathrm{d}x = 0 = W_1 f(\frac{1}{\sqrt{3}}) + W_2 f(-\frac{1}{\sqrt{3}}) = W_1 \frac{1}{\sqrt{3}} - W_2 \frac{1}{\sqrt{3}} \Rightarrow W_1 = W_2 = 1 \end{cases} \tag{4.112}$$

Inserting Eqs. (4.111) and (4.112) into Eq. (4.110) gives rise to

$$\int_{-1}^{1} f(x)\mathrm{d}x \approx f\left(\frac{1}{\sqrt{3}}\right) + f\left(-\frac{1}{\sqrt{3}}\right) \tag{4.113}$$

2. **4-point rule**

With the roots of equation $P_4(x) = 0$ and definite integration of $1, x, x^2$ and x^3, the abscissas and weights for 4-point rule are

$$\begin{cases} x_1 = -0.339981043584856 \\ x_2 = -0.861136311594053 \\ x_3 = -0.339981043584856 \\ x_4 = -0.861136311594053 \end{cases} \tag{4.114}$$

$$\begin{cases} W_1 = 0.652145154862546 \\ W_2 = 0.347854845137454 \\ W_3 = 0.652145154862546 \\ W_4 = 0.347854845137454 \end{cases} \tag{4.115}$$

3. **n-point rule**

A generalization for a higher order integration rule is

$$\int_{-1}^{1} f(x)\mathrm{d}x \approx W_1 f(x_1) + W_2 f(x_2) + \cdots + W_n f(x_n) \tag{4.116}$$

in which x and W are the abscissas and the weights applicable for n-point rule. A table for higher order Gauss–Legendre rule is available in relevant handbooks (Abramowitz and Stegun 1964).

Any definite integral in the interval [a, b] can be transformed to an integral in the interval $[-1, 1]$ using the formula

$$\int_{a}^{b} f(x)\mathrm{d}x = \frac{b-a}{2}\int_{-1}^{1} f\left(\frac{b-a}{2}x + \frac{b+a}{2}\right)\mathrm{d}x \tag{4.117}$$

hence we get

$$\int_{a}^{b} f(x)\mathrm{d}x \approx \frac{b-a}{2}\sum_{i=1}^{n} W_i f\left(\frac{b-a}{2}x_i + \frac{b+a}{2}\right) \tag{4.118}$$

in which x and W are the abscissas and the weights applicable for n point rule.

If there are singularities in the interval $[a, b]$, the fixed point Gauss–Legendre rule may lead to incorrect results and hence must be avoided. A good example is provided by the quadrature of $\int_{-1}^{1} \frac{1}{x^2}\mathrm{d}x$: its exact solution is -2 whereas the two point rule leads to the answer of 6.

4.4.4 Cubature

Numerical procedures for approximating double-or triple-integrals are particularly called "cubature" although the term "quadrature" still may be generally used. The first method was proposed by Maxwell (1877) by constructing a formula of degree 7 with 27 knots for the cube $[-1, +1]^3$. The first mathematician who investigated the connection between orthogonal polynomials in two variables and cubature was

Appell in 1890. Radon (1948) constructed formulas for the square, circle and triangle using orthogonal polynomials. Cubature formula on sphere was studied by Sobolev (1962). These are very useful towards the advanced CMs such as the MLPG that demands the integrals over circles and sphere.

The difficulties encountered with the irregular regions and the necessity of economizing the number of function evaluations led to the "Monte-Carlo methods" (Metropolis and Ulam 1949; Metropolis 1989; Eckhardt 1989). Recent developments on the subject can be found in the works of Keast and Fairweather (1987), Espelid and Genz (1992).

Gauss–Legendre product rule is the simplest solution to extend the conventional 1-D Gauss–Legendre quadrature into cubature problems by a change of coordinates either from global Cartesian to local Cartesian (Zienkiewicz et al. 2005) or from global Cartesian to polar (De and Bathe 2001). The former is dependent on the transformation of a geometrical manifold (polygon/circle, polyhedron/sphere) into standard square/cubic, in FEM such a transformation is implemented with the help of isoparametric element concept. Whereas the latter fully makes use the advantage of radial basis characteristics: a higher degree rule for radial integral and a midpoint rule for angular integral. Today, Gauss–Legendre product rule for cubature is prevalent in the literatures of CMs (Mazzia and Pini 2010).

With Gauss–Legendre product rule, 1-D quadrature schemes discussed previously can be directly generalized to higher dimension. As the dimension increases, the number of quadrature points grows. For the 3-D integral, we can conduct the integration along x-axis first according to Eq. (4.116) such that

$$\iiint\limits_{e} f(x, y, z)\mathrm{d}x\mathrm{d}y\mathrm{d}z \approx \sum_{i}^{n_i} W_i \iint\limits_{s} f(x_i, y, z) \tag{4.119}$$

in which e is the integration domain (element, cell), s is the integration section perpendicular to x-axis at x_i.

Next, we approximate the integral along y-axis as

$$\iiint\limits_{e} f(x, y, z)\mathrm{d}x\mathrm{d}y\mathrm{d}z \approx \sum_{i}^{n_i}\sum_{j}^{n_j} W_i W_j \int\limits_{l} f(x_i, y_j, z) \tag{4.120}$$

in which l is integration line perpendicular to x-axis and y-axis at (x_i, y_i).

Finally, the Gauss–Legendre product rule for cubature is obtained as

$$\iiint\limits_{e} f(x, y, z)\mathrm{d}x\mathrm{d}y\mathrm{d}z \approx \sum_{i}^{n_i}\sum_{j}^{n_j}\sum_{k}^{n_k} W_i W_j W_k f(x_i, y_j, z_k) \tag{4.121}$$

Note that the knot numbers n_i, n_j and n_k are not necessarily identical. Since it is not so easy to get the position (x_i, y_j, z_k) for nonstandard regions, a transformation similar to Eq. (4.117) is demanded. This transformation with the combination of sub-divisions and shape functions of isoparametric finite element may solve a very large number of cubature issues derived from CMs.

1. **Cubature on arbitrary polygon**

Figure 4.12 shows an arbitrary polygon whose vertex coordinates are known. The polygon is firstly divided into a number of triangles. For each triangle, we draw lines from its centroid to the middle points of its edges, the polygon is divided into n_e quadrilaterals. Then the Gaussian-Legendre product rule for cubature leads to

$$\iint_S f(x, y)\mathrm{d}x\mathrm{d}y \approx \sum_e^{n_e} \int_{-1}^{1}\int_{-1}^{1} [f(\xi, \eta)|J|]_e \mathrm{d}\xi \mathrm{d}\eta = \sum_e^{n_e} \left[\sum_i^{n_i} \sum_j^{n_j} W_i W_j f(\xi_i, \eta_j)|J|_{ij} \right]_e \tag{4.122}$$

in which the knot positions ξ_i and η_j are independently recurred using Eqs. (4.110)–(4.116) according to the precision desired; $|J|$ is the Jacobian determinant derived from the transformation using the standard shape functions of isoparametric finite element (Zienkiewicz et al. 2005).

2. **Cubature in arbitrary polyhedron**

Firstly, the arbitrary polyhedron is divided into a number of tetrahedrons. For each tetrahedron the following sub-divide are conducted (see Fig. 4.13): drawing lines from tetrahedron centroid to the centroids of its surfaces, and drawing lines from surface centroids to the middle points of edges. In this manner one tetrahedron will be definitely sub-divided into four hexahedra. After coordinate transform the 3-D Gaussian-Legendre product rule for cubature can be implemented for any function $f(x, y, x)$ defined in the polyhedron as

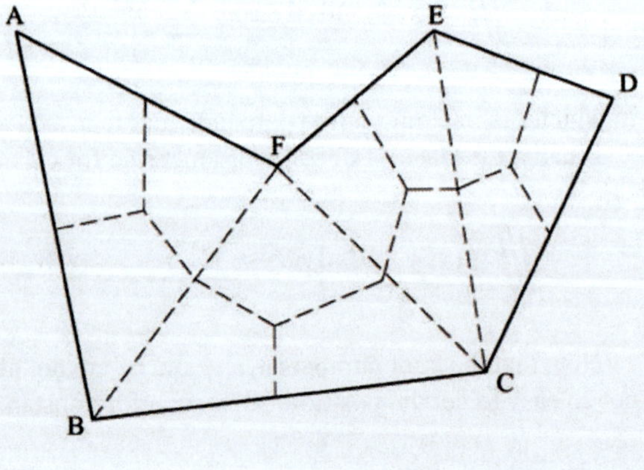

Fig. 4.12 Element dividing on arbitrary polygon towards cubature

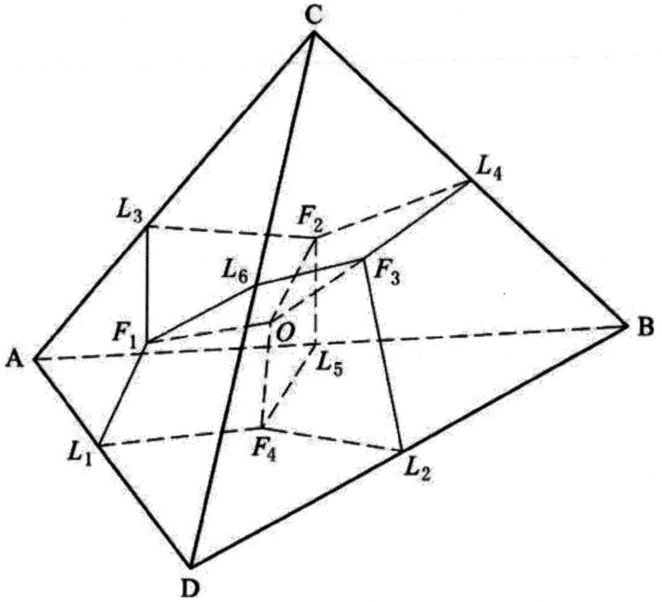

Fig. 4.13 Element dividing in tetrahedron towards cubature

$$\iiint_{\Omega} f(x, y, z)\mathrm{d}x\mathrm{d}y\mathrm{d}z \approx \sum_{e}^{n_e} \int_{-1}^{1}\int_{-1}^{1}\int_{-1}^{1} [f(\xi, \zeta, \eta)|J|]_e \mathrm{d}\xi \mathrm{d}\zeta \mathrm{d}\eta$$
$$= \sum_{e}^{n_e} \left[\sum_{i}^{n_i}\sum_{j}^{n_j}\sum_{k}^{n_k} W_i W_j W_k f(\xi_i, \zeta_j, \eta_k)|J|_{ijk} \right]_e \tag{4.123}$$

in which n_e is the amount of integration elements produced in the polyhedron.

4.5 Solution of Ordinary Differential Equations

4.5.1 Finite Differences

A difference equation is an equation of unknown function sequence, that is often derived from the operation of numerical discretization for ODE or PDE using, for example, FDM. Naturally, the computer is able to provide the values of unknown function only at a discrete set of points, which is implemented by first to replace the given differential equation using a difference equation that approximates it, then to calculate the successive approximant values of the desired function from the difference equation.

In time domain, suppose we know that a certain sequence $(u_0, u_1, u_2, \ldots)$ dependent on time t satisfies the following equations with initial condition u_0 and u_1,

$$t_i = i\Delta t \tag{4.124}$$

$$u_{i+2} + au_{i+1} + bu_i = 0 \quad (i = 0, 1, \ldots, \infty) \tag{4.125}$$

in which i is the time-marching step sequence and Δt is the time-marching step interval (length). Such equations are frequently encountered when an ODE or PDE is solved on computers.

In space domain, let the function $u(x)$ defined at the data points $\{x_i\}(i = 0, 1, \ldots, n)$ that form an equidistant set with a step size h, e.g. $x_i = x_0 + ih$ $(i = 0, 1, 2, \ldots, n)$, and $u_i = u(x_i)$. The formulae $\Delta u_i = u_{i+1} - u_i$ for $(i = 0, 1, 2, \ldots, n)$ are called "forward finite differences" of the first order. Inductively, forward finite differences of order $k + 1$ are given by

$$\Delta^{(k+1)} u_i = \Delta^{(k)} u_{i+1} - \Delta^{(k)} u_i \quad (k = 0, 1, 2, \ldots) \tag{4.126}$$

Other notations are also used in finite differences

$$\begin{cases} \Delta u_i = \nabla u_{i+1} = \delta u_{i+1/2} = u^{(1)}_{i+1/2} = u_{i+1} - u_i \\ \nabla^{(k+1)} u_i = \nabla^{(k)} u_i - \nabla^{(k)} u_{i-1} \\ \delta^{(k+1)} u_i = \delta^{(k)} u_{i+1/2} - \delta^{(k)} u_{i-1/2} \\ u_i^{(k+1)} = u^{(k)}_{i+1/2} - u^{(k)}_{i-1/2} \end{cases} \quad (i = 0, 1, 2, \ldots, n) \tag{4.127}$$

in which $\nabla^{(k)} u_i$ is the "backward finite difference" of order k, and $\delta^{(k)} u_i$ is the "central finite difference" of order k.

The main properties of the finite differences are summarized below

$$\begin{cases} \Delta^{(k)} u_i = \sum_{j=0}^{k} (-1)^{k+j} \binom{k}{j} u_{i+j} \\ \nabla^{(k)} u_i = \Delta^{(k)} u_{i-k} \\ u_i^k = \Delta^{(k)} u_{i-k/2} \\ \Delta^{(k)} u_i = u^{(k)}_{i+k/2} \\ \nabla^{(k)} u_i = h^k k! u(x_i, x_{i+1}, \ldots, x_{i+k}) \end{cases} \tag{4.128}$$

4.5.2 *Finite Differential Methods*

1. **Concept**

The modern study on natural phenomena mathematically described by both ODEs and PDEs usually requires a significant application of computation effort. Differential equation contains various derivatives of $u(x)$ and various known functions of x. The order of a differential equation is the order of the highest derivative that appears in it.

Around 1850, numerical solution of ODEs began to get attention (Shampine 1994). The Adams formulae are based on the polynomial interpolation in equidistant points (fewer than ten in practice), which are now called multistep methods for ODEs.

Let's consider a simple initial value problem below

$$\frac{\mathrm{d}u}{\mathrm{d}t} = \dot{u} = u^{(1)} = f(t, u) \tag{4.129}$$

with independent time variable $t > 0$.

Pick a small time-marching step length $\Delta t > 0$ and consider a finite set of time points

$$t_i = i\Delta t \quad \text{for } i \geq 0 \tag{4.130}$$

the ODE may be replaced by an algebraic approximant that enables us to calculate a succession of approximant values

$$u_i \approx u(t_i) \quad \text{for } i \geq 0 \tag{4.131}$$

The simplest approximation is given by the "forward Euler scheme" (viz. Fig. 4.14)

$$u^*_{i+1} = u_i + \Delta t f(t_i, u_i) \tag{4.132}$$

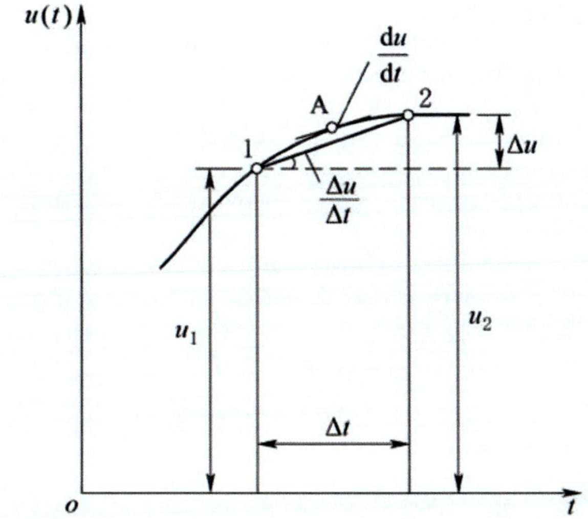

Fig. 4.14 Finite difference approximation to first derivative

Using the abbreviation $f_i = f(t_i, u_i)$, Eq. (4.132) becomes

$$u_{i+1} \approx u_{i+1}^* = u_i + \Delta t_i f_i \tag{4.133}$$

It is well-known that the forward Euler scheme is tightly restrained by numerical stability in the selection of step length Δt_i. On the contrary, the "backward Euler scheme"

$$u_i^* = u_{i-1} + \Delta t f(t_i, u_{i-1}) \tag{4.134}$$

is unconditionally stable.

Using $f(t_i, u_i)$ only for the gradient in the time interval, in Euler's scheme there was no possibility to include the information from time t_{i+1} at which the solution will be obtained, hence it is only first-order accurate.

The easiest extension of forward Euler method is known as the "improved Euler method" or "Heun's method". It is obtained by using forward Euler's scheme (4.132) before applying the trapezoidal rule such that

$$\begin{cases} u_{i+1}^* = u_i + \Delta t_i f(t_i, u_i) \\ u_{i+1} \approx u_i + \dfrac{1}{2}\Delta t_i \left[f(t_i, u_i) + f(t_{i+1}, u_{i+1}^*) \right] \end{cases} \tag{4.135}$$

This scheme is shown in Fig. 4.15. The broken line possesses a gradient $f(t_i, u_i)$ giving the point marked by an open circle corresponding to backward Euler's scheme (4.134). At this point the gradient is calculated from the differential equation, giving $f(t_{i+1}, u_{i+1}^*)$. Afterwards, the mean of two gradients is used to get u_{i+1} by Eq. (4.135). In Fig. 4.15 this has been shown by transferring a broken line of gradient $f(t_{i+1}, u_{i+1}^*)$ back to the original point, then drawing a dotted line which is the mean of two gradients to obtain u_{i+1}.

Heun's method yields a local error of $O((\Delta t)^3)$ and hence a global error of $O((\Delta t)^2)$. For hyperbolic conservation laws, the spectrum of spatial differential operator is purely imaginary. Since the absolute stability region of forward Euler method and Heun's method intersects the imaginary axis only at the origin, they are typically not stable for the evolution of hyperbolic conservation laws. However, they are one-step, one-stage, one-derivative methods and so serve as the basis for many other methods.

Another approach to raising the order of accuracy is the inclusion of higher order time derivatives. Take the ODE (4.129) for example, the truncated second-order Taylor series is

$$u_{i+1} = u_i + \Delta t f(u_i) + \frac{1}{2}\Delta t^2 f^{(1)}(u_i) \tag{4.136}$$

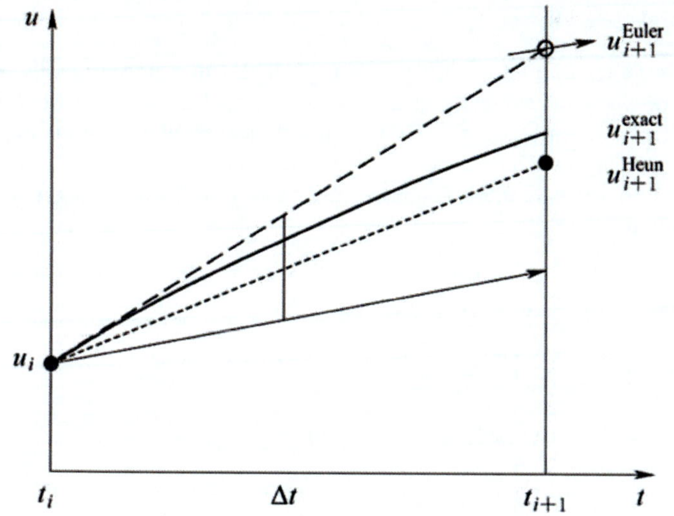

Fig. 4.15 One step scheme of Heun

The challenge in using Taylor series is the increasingly difficulty in the computation of derivatives, so it is rarely exercised in practice. However, the idea behind it appears in various other schemes.

A huge amount of effort over the past seven decades (and even earlier) has been made to numerically solving differential equations, which gives rise to many ingenious algorithms and associated codes. Most algorithms are based upon the desire of discretizing ODEs in a good manner to keep the local truncation errors as small as possible, to ensure their stability (the local errors do not grow), and to constrain errors within specified tolerance levels. The resulting discrete equation set is then solved with carefully designed linear and nonlinear solvers often dealing with very large amount of unknowns. With the help of effective (a-priori and a-posteriori) error control strategies, these algorithms can often lead to very accurate solutions.

However, existing methods based on the analysis of local truncation errors do not respect, or even take into account, the qualitative and global characteristics of ODEs. The geometric integration has, in contrast, enabled to systematically incorporate the qualitative information of the underlying problem in a natural way. Such algorithms (e.g. Runge–Kutta method, Störmer-Verlet leapfrog method, etc.) turn out to have excellent qualitative properties. As a side benefit to incorporate qualitative characteristics, the resulting numerical schemes are often more efficient than other schemes and easier to analyze, too, because one may exploit the qualitative theory of the underlying differential equations (such as its Hamiltonian structure or group invariance) by using the powerful backward error analysis methods developed by Hairer, Reich and their co-workers (Hairer and Lubich 1997, 1999; Reich 1999).

The methods discussed hereinafter are classical, and many have been applied by default to hyperbolic PDEs over the years.

2. **Multiphase methods**

Hereinafter, the subscript i of time t_i is neglected, and even further, the whole t_i could be dropped in the relevant quantities, subject to there is no risk of misleading.

In one step scheme, using more intermediate stages allows for the augment in both the order and the size of stability region, which is particularly important for the evolution of hyperbolic PDEs (Jameson et al. 1981). In the following, two of them are presented.

- Three-stage third-order method

$$\begin{cases} u^{(1)} = u_i \\ u^{(2)} = u_i + \Delta t f(u^{(1)}) \\ u^{(3)} = u_i + \dfrac{1}{2}\Delta t\left[f(u^{(1)}) + f(u^{(2)})\right] \\ u_{i+1} = u_i + \dfrac{1}{2}\Delta t\left[f(u^{(1)}) + f(u^{(3)})\right] \end{cases} \tag{4.137}$$

- Four-stage fourth-order method

$$\begin{cases} u^{(1)} = u_i \\ u^{(2)} = u_i + \dfrac{1}{2}\Delta t f(u^{(1)}) \\ u^{(3)} = u_i + \dfrac{1}{2}\Delta t f(u^{(2)}) \\ u^{(4)} = u_i + \Delta t f(u^{(3)}) \\ u_{i+1} = u_i + \dfrac{1}{6}\Delta t\left[f(u^{(1)}) + 2f(u^{(2)}) + 2f(u^{(3)}) + f(u^{(4)})\right] \end{cases} \tag{4.138}$$

It is notable that in Eqs. (4.137), (4.138), Δt_i is presumably constant hence its subscript corresponding to time-marching step sequence is neglected.

The term "fourth-order" gives rise a new concept in the numerical solution of ODE and PDE: the appearance of convergence as $\Delta t \to 0$. To be specific, Eq. (4.138) is of fourth-order in the sense that it will normally converge at rate $O((\Delta t)^4)$. Most often, the order in a range of 3–6 enables excellent accuracy for all kinds of computation task. Higher order formulae are occasionally used when still higher accuracy is demanded.

While the increase in stages leads to better accuracy and larger allowable time-marching step length in term of absolute stability, larger computation effort is demanded, too. This increased effort is usually more than offset by the fact that larger time-marching steps can be taken whether accuracy or stability is the dominated concern.

3. **Multistep methods**

Using more steps also allows for the augment in both the order and size of stability region. k-step methods for Eq. (4.129) are typically written in the form of

$$\sum_{j=0}^{k} \alpha_j u_{i+1-j} = \sum_{j=0}^{k} \beta_j f(u_{i+1-j}) \tag{4.139}$$

in which the coefficients α_i and β_i are chosen to optimize the accuracy and stability properties. The explicit ones of this type are known as "Adams–Bashforth methods" whereas the implicit ones are called "Adams–Moulton methods". In the following we list just a few of them.

- Two-step second-order "Adams–Bashforth method"

$$u_{i+1} = u_i + \frac{1}{2}\Delta t\left[3f(u_i) - f(u_{i-1})\right] \tag{4.140}$$

- Three-step third-order "Adams–Bashforth method"

$$u_{i+1} = u_i + \frac{1}{12}\Delta t\left[23f(u_i) - 16f(u_{i-1}) + 5f(u_{i-2})\right] \tag{4.141}$$

- One-step second-order "Adams–Moulton method" (also known as "Crank–Nicolson method")

$$u_{i+1} = u_i + \frac{1}{2}\Delta t\left[f(u_i) + f(u_{i+1})\right] \tag{4.142}$$

Another approach to multistep methods is to evaluate the function f only at the right end point, i.e. $\beta_j = 0$ whenever $j \neq 0$. These are known as "backward differentiation formulae (BDF)". The second-order BDF is

$$u_{i+1} = \frac{4}{3}u_i - \frac{1}{3}f(u_{i-1}) + \frac{2}{3}\Delta t.f(u_{i+1}) \tag{4.143}$$

The advantage of multistep methods is that the computation cost is steady per time-marching step, but the storage requirement is increased. The problem of starting values for k initial steps can be estimated by using a one-step method with very small step length.

It is notable that high-order explicit multistep methods tend to have smaller stability regions than high-order explicit Runge–Kutta methods.

4. **Multistage-multistep methods**

Multistep methods require little computation per time-marching step but need more storage capacity, whereas multistage methods require more computation per time-marching step but can be designed to have lower storage capacity. The use of additional derivatives may further alleviate the storage requirement, but these derivatives are not always easy or cheap to get.

Combining multiple stages and steps may provide higher-order methods while balancing the computational effort and storage requirements (Vos et al. 2011). These methods are sometimes called "hybrid methods", "general linear methods", or pseudo "Runge–Kutta (R-K)" methods. This class of methods was originally proposed by Butcher (1965), Gear (1965) and Gragg (1964) and has been studied extensively in the works of Butcher (1996) and Jackiewicz (2009).

4.5.3 Runge–Kutta Methods

Heun's method in Eq. (4.135) is actually the second-order member belongs to a large family termed as the "Runge–Kutta (R-K)" methods—a class of one-step algorithms that make use of Taylor series expansion indirectly in a manner of

$$u_{i+1} = u_i + \Delta t.\Delta(t, u_i, \Delta t) \tag{4.144}$$

$$\Delta(t, u_i, \Delta t) = \dot{u}_i + \frac{\Delta t}{2}\ddot{u}_i + \frac{\Delta t^2}{3!}\dddot{u}_i + \cdots + O(\Delta t^p)\left[\frac{\Delta t^p}{(p+1)!}\overset{p+1}{u}(\xi_n)\right] \tag{4.145}$$

in which $O(\Delta t^p)$ is truncation error.

The general form of explicit R-K method for Eq. (4.129) is

$$\begin{cases} u(t_{i+1}) = u(t_i) + \Delta t.\Phi(t, u_i, \Delta t) \\ \Phi(t, u, \Delta t) = \sum_{r=1}^{m} c_r k_r \\ k_1 = f(t, u) \\ k_r = f(u + \Delta t.a_r, u + \Delta t \sum_{s=1}^{r-1} b_{rs} k_s) \quad (r = 2, 3, \ldots, m) \end{cases} \tag{4.146}$$

Equation (4.146) requires the ability to calculate the right hand side of the differential equation at a number of points intermediate between t_i and t_{i+1}. More coefficients k_s mean higher order of convergence by adjusting these coefficients, but they need be iteratively solved. The prevalent explicit versions can be found in many books on numerical methods. In the following just only three of them are presented.

- 2-order

$$\begin{cases} u_{i+1} = u_i + k_1 \\ k_1 = \Delta t.f(t + \frac{1}{2}\Delta t, u_i + \frac{1}{2}k_1) \end{cases} \tag{4.147}$$

- 4-order

$$\begin{cases} u_{i+1} = u_i + \frac{1}{2}\Delta t(k_1 + k_2) \\ k_1 = \Delta t.f\left(t + \frac{3-\sqrt{3}}{6}\Delta t, u_i + \frac{k_1}{4} + \frac{3-2\sqrt{3}}{12}k_2\right) \\ k_2 = \Delta t.f\left(t + \frac{3+\sqrt{3}}{6}\Delta t, u_i + \frac{k_2}{4} + \frac{3+2\sqrt{3}}{12}k_2\right) \end{cases} \tag{4.148}$$

- 6-order

$$\begin{cases} u_{i+1} = u_i + \frac{5}{18}k_1 + \frac{4}{9}k_2 + \frac{5}{18}k_3 \\ k_1 = \Delta t.f\left(t + \frac{5-\sqrt{15}}{10}\Delta t, u_i + \frac{5}{36}k_1 + \frac{10-3\sqrt{15}}{45}k_2 + \frac{25-6\sqrt{15}}{180}k_3\right) \\ k_2 = \Delta t.f\left(t + \frac{\Delta t}{2}, u_i + \frac{10+3\sqrt{15}}{72}k_1 + \frac{2}{9}k_2 + \frac{10-3\sqrt{15}}{12}k_3\right) \\ k_3 = \Delta t.f\left(t + \frac{5+\sqrt{15}}{10}\Delta t, u_i + \frac{5}{36}k_3 + \frac{10+3\sqrt{15}}{45}k_2 + \frac{25+6\sqrt{15}}{180}k_1\right) \end{cases} \tag{4.149}$$

Although R-K methods are normally easier to implement, yet sometimes they are harder for error analysis than multistep formulas. For any number s, it is a trivial matter to derive the coefficients of the s-step Adams–Bashforth formula, which has order of accuracy $p = s$. For R-K methods, by contrast, there is no simple relationship between the number of "stages" (i.e. function evaluations per step) and the attainable order of accuracy. The methods with $s = 1, 2, 3, 4$ were known to have order $p = s$, but it was not until 1963 that s = 6 stages are validated to achieve order $p = 5$. To achieve orders $p = 6, 7, 8$ the minimal numbers of stages are $s = 7, 9, 11$ respectively. While for $p > 8$, exact minimal stages are still not found. Fortunately, these higher orders are rarely needed for practical purposes. The error analysis involves beautiful mathematical tools (e.g. graph theory), and the key figure in this area since the 1960s is Butcher.

For problems with time-dependent boundary conditions, caution is needed that the natural method of imposing these conditions in R-K methods reduces the accuracy to first-order locally, and second-order globally, regardless of the spatial operator (Carpenter et al. 1995). One approach to dealing with this problem is to impose the exact boundary conditions only at each step instead of at the intermediate stages; another approach is to impose boundary conditions derived from the physical boundary condition and its derivatives at each intermediate stage. The former retains accuracy, but requires a much smaller time-marching step length to maintain stability; whereas the latter retains the time-marching step length required for stability, but if the hyperbolic PDE is nonlinear it only works for R-K methods with order $p = 3$.

Since R-K methods are difficult to give truncation error, the R-K-F (Fehlberg) scheme using nesting technology is very practical. It gives two sets of R-K formulae with m-order and $m + 1$-order accuracy, the difference in these two sets provide an estimator of error.

4.5.4 *Stable Considerations*

The primary consideration when integrating any initial-value problem with ODE is the computational effort towards a given terminal (collapse) time. Clearly then, the goal is to permit the largest allowable time-marching step length.

1. **Absolute stability (linear stability)**

Most hyperbolic problems are solved by explicit methods, where the famous stability condition budding "Courant-Friedrichs-Lewy (CFL)" requires that the step length must be so taken that is approximately proportional to the spatial mesh size. Absolute stability typically demands an even smaller step length than that with the CFL condition. For smooth problems, the step length dictated by absolute stability is both necessary and sufficient to guarantee convergence. However, in the presence of shocks or sharp gradients, additional nonlinear stability properties are needed to ensure convergence.

2. **Robustness (nonlinear stability)**

By robustness it means that the numerical approximation produces outcomes meeting certain physical or mathematical constraints. For example, in many problems it is demanded that certain quantities such as density or pressure remain positive. In addition, spurious oscillations should be avoided, too. To satisfy such constraints, step length restriction is often tighter than those from absolute stability.

Nonlinear stability conditions become critical for the convergence in the presence of shocks or sharp gradients. Available techniques involve limiters to ensure, for example, total variation bounds, entropy stability, maximum principle preservation, or positivity.

3. **Accuracy**

Given that the time-marching step length is generally the size of space grid, accuracy considerations typically dictate that the time-marching step length will be approximately the same order as the space discretization size. For the moment however, beyond this mild requirement, accuracy w.r.t. time discretization is not well tackled in the advanced CMs for geotechnical and hydraulic structures. Unlike the case for other areas, the time-marching step control based on error estimation is not widely exercised, and the accuracy is not generally considered a restraint factor in choosing time integration schemes. Typically, guidelines for choosing a time-marching step length that ensures accuracy, stability, and robustness are only obtained by the analysis of simple (e.g. 1-D and/or scalar) problems, then the practical step lengths for multi-dimensional systems are adjusted mainly dependent on the experiences of practitioners.

This situation would impose potential risk w.r.t. the reliability in the simulation of real-world scenarios, particularly when the object is discretized into millions of particles/nodes with nonlinear performances. In such a nonlinear and vast system, accuracy is no longer the "accuracy" only, because a small perturbation due to computation error may trigger the unrealistic phenomena such as bifurcation, mutation, etc.

4.5.5 Geometric Integration

It is worth to indicate that in modern computation, small rounding-off errors introduce non-symplectic perturbations to Hamiltonian problems which can be an important influential factor in long term tracing integrations (such as satellite orbit calculation) (Earn and Tremaine 1992). Geometric integration with symplectic methods allows us for a chance to examine the existing prevalent methods in a new light, maybe to improve or generalize them, or maybe to help explain their behaviors and performances not noticed before. It is still a relatively new subject area and much work have yet to be done.

The history of symplectic methods is interesting. Early work in the mathematician community specifically aimed to construct symplectic methods for ODEs is given by De Vogelaere (1956), Ruth (1983), Feng (1985a, b). However, symplectic integrators themselves have been unconsciously exercised for much longer than this. A lot of well-established and very effective numerical methods possess successful precision because they are symplectic even though this fact was not recognized when they were first constructed, examples may be given by Gauss–Legendre methods and Störmer-Verlet (leapfrog) methods (Hildebrand 1987).

The details of geometric integration with symplectic methods can be found in the reviews of Sanz-Serna (1997), Budd and Iserles (1999), Budd and Piggott (2001), Mclachlan and Quispel (2001), Hairer et al. (2002).

1. **Hamiltonian ODEs and symplectic geometry**

Probably the first significant area where geometric ideas were introduced is the symplectic integration for Hamilton ODEs.

The initial problem of Hamilton canonical equation elaborated in Chap. 3 (viz Eq. (3.19)) is

$$\dot{u} = \begin{Bmatrix} \dot{q} \\ \dot{p} \end{Bmatrix} = \begin{Bmatrix} \partial H/\partial \mathbf{p} \\ -\partial H/\partial \mathbf{q} \end{Bmatrix} = \mathbf{J}\frac{\partial H}{\partial \mathbf{u}} \quad \text{for } \mathbf{v}(0) = \mathbf{v}_0 \tag{4.150}$$

in which $H = H(\mathbf{p}, \mathbf{q})$ is Hamiltonian function. In addition to the conservation of Hamiltonian function, a key characteristic of a Hamiltonian system is the symplectic feature of its phase flow.

Although there are various discretization algorithms for ODE set, yet they do not take into account of the fact that the solution of Hamiltonian equation $\mathbf{u}(0) \rightarrow \mathbf{u}(t)$ is actually a canonical transformation. Hence as the time going on, the approximate solution would exhibit larger and larger deviation from original Hamiltonian system. An algorithm that although still produces error yet holds canonical transformation, is tagged with "symplectic". Such an approximant solution may possess the conservation of energy and other symplectic invariables. To begin with, we rewrite Eq. (4.129) in a vector form

$$\dot{u} = \mathbf{f}(\mathbf{u}) \tag{4.151}$$

this system is canonically Hamiltonian if

$$\mathbf{f}(\mathbf{u}) = \mathbf{J}^{-1}\nabla H \tag{4.152}$$

in which ∇ is the Hamilton operator and $\mathbf{J}$ is the skew-symmetric matrix (viz. Eq. (3.21)).

The phase flow construct a single-parameter differential homeomorphism group $\mathbf{g}(t) : R^{2n} \rightarrow R^{2n}$. Hence the solution of Eq. (4.151) may be expressed as

$$\mathbf{u}(t) = \mathbf{g}(t)\mathbf{u}(0) \tag{4.153}$$

in which $\mathbf{g}(t)$ is a symplectic mapping.

According to Eqs. (3.30)–(3.32), the condition that the canonical transformation will not change original Eq. (4.153) is entailed by

$$\left(\frac{\partial \mathbf{g}}{\partial \mathbf{u}}\right)^{\mathrm{T}} \mathbf{J} \left(\frac{\partial \mathbf{g}}{\partial \mathbf{u}}\right) = \mathbf{S}^{\mathrm{T}}\mathbf{J}\mathbf{S} = \mathbf{J} \tag{4.154}$$

in which $\mathbf{S}$ is the symplectic matrix defined as

$$\mathbf{S} = \left(\frac{\partial \mathbf{g}}{\partial \mathbf{u}}\right) \tag{4.155}$$

The symplectic differential algorithm is constructed according condition (4.154). For the explicit differential scheme of single step

$$\mathbf{u}_{i+1} = \mathbf{g}(\mathbf{u}_i) \tag{4.156}$$

let

$$\mathbf{S} = \left(\frac{\partial \mathbf{g}}{\partial \mathbf{u}_i}\right) = \left(\frac{\partial \mathbf{u}_{i+1}}{\partial \mathbf{u}_i}\right) \tag{4.157}$$

if Eq. (4.154) is met, the scheme (4.157) is symplectic conservative. Symplectic mapping possesses group characteristics, namely, if $\mathbf{g}_i$ $(i = 1, 2, \ldots, m)$ is a symplectic mapping, the composite mapping

$$\mathbf{g} = \mathbf{g}_m \cdot \mathbf{g}_{m-1} \cdot \ldots \cdot \mathbf{g}_1 \tag{4.158}$$

is symplectic, too. Accordingly, we can construct a higher order symplectic algorithm using lower order ones.

Suppose that a one-step method of constant step length Δt is applied to approximate the solution $\mathbf{u}(t)$ in Eq. (4.151) at time $t = i\Delta t$ by the vector $\mathbf{u}_n \in R^{2d}$. This will induce a discrete flow mapping $\mathbf{g}_{\Delta t}$ on R^{2d} which is an approximation to the continuous flow mapping $\mathbf{g}(\Delta t)$. The mapping $\mathbf{g}_{\Delta t}$ is "symplectic" if it satisfies the identity (4.154), which is a requirement of natural geometric property.

We will now able to construct numerical methods with this property. They retain many of the other features (in particular ergodic properties) of their continuous counterparts. However, it is notable that if the step length Δt is not constant, such as in an adaptive algorithm, particular cares have to be taken with this definition and much of the advantage of using a symplectic scheme will be lost (Sanz-Serna and Calvo 1994) unless the equation defining Δt at each time step also preserves the geometric structure in some way (Hairer 1997).

2. Symplectic Runge–Kutta methods

It is fortunate that an important class of Runge–Kutta methods turn out to be symplectic (Butcher 1987; Forest and Ruth 1990; Suris 1990). The implicit mid-point rule considered earlier is a member of this class.

We illustrate here a simple but important example of a symplectic method. This is the second-order implicit, (symmetric) midpoint R-K method. With the help of Eqs. (4.152), (4.147) takes the form

$$u_{i+1} = u_i + \Delta t.f\left(\frac{u_i + u_{i+1}}{2}\right) = u_i + \Delta t.\mathbf{J}^{-1}\nabla H\left(\frac{u_i + u_{i+1}}{2}\right) \tag{4.159}$$

For a sufficiently small time-marching step length Δt, the implicit mid-point rule defines a diffeomorphism. Differentiating Eq. (4.159) gives rise to the Cayley transformation

$$\mathbf{S} = \frac{\partial u_{i+1}}{\partial u_i} = \mathbf{I} + \frac{\Delta t}{2}\mathbf{J}^{-1}\mathbf{H}''\left(\frac{u_i + u_{i+1}}{2}\right)\left(\mathbf{I} + \frac{\partial u_{i+1}}{\partial u_i}\right) \tag{4.160}$$

$$\mathbf{H}'' = \begin{bmatrix} H_{pp} & H_{pq} \\ H_{qp} & H_{qq} \end{bmatrix} \tag{4.161}$$

hence th Jacobian matrix of transformation $u_i \to u_{i+1}$ is given by

$$\mathbf{S} = \frac{\partial u_{i+1}}{\partial u_i} = \left[\mathbf{I} - \frac{\Delta t}{2}\mathbf{J}^{-1}\mathbf{H}''\left(\frac{u_i + u_{i+1}}{2}\right)\right]^{-1} \times \left[\mathbf{I} + \frac{\Delta t}{2}\mathbf{J}^{-1}\mathbf{H}''\left(\frac{u_i + u_{i+1}}{2}\right)\right] \tag{4.162}$$

We have already known that Eq. (4.154) is the condition to verify a mapping to be symplectic, on the substitution of Eq. (4.162) into Eq. (4.154), we only need to check the condition

$$\begin{aligned}&\left[\mathbf{I} + \frac{\Delta t}{2}\mathbf{J}^{-1}\mathbf{H}''\left(\frac{u_i + u_{i+1}}{2}\right)\right]\mathbf{J}\left[\mathbf{I} + \frac{\Delta t}{2}\mathbf{J}^{-1}\mathbf{H}''\left(\frac{u_i + u_{i+1}}{2}\right)\right]^{\mathrm{T}} \\ &= \left[\mathbf{I} - \frac{\Delta t}{2}\mathbf{J}^{-1}\mathbf{H}''\left(\frac{u_i + u_{i+1}}{2}\right)\right]\mathbf{J}\left[\mathbf{I} - \frac{\Delta t}{2}\mathbf{J}^{-1}\mathbf{H}''\left(\frac{u_i + u_{i+1}}{2}\right)\right]^{\mathrm{T}}\end{aligned} \tag{4.163}$$

The conclusion now follows immediately from the symmetry of Hessian matrix and the fact that $\mathbf{J}^{\mathrm{T}} = \mathbf{J}^{-1} = -\mathbf{J}$.

4.5.6 Higher-Order Differential Equations

Suppose that we want to solve a single ODE of the second order, say

$$m\ddot{u} + c\dot{u} + ku = f \tag{4.164}$$

The first strategy is to split the differential operator for transforming the single second-order equation into a pair of simultaneous first-order equations that can be handled as before. To do this, one chooses two unknown functions u and v, where the function v is the first-order derivative of u. Then u and v satisfy two simultaneous first-order differential equations as follows

$$\begin{cases}\dot{u} = v \\ \dot{v} = (-cv - ku + f)/m\end{cases} \tag{4.165}$$

The second strategy is to directly solve Eq. (4.164) using the discretization scheme of higher-order differences that is the extension of first-order ones, either multiphase or multistep techniques, or the combination of the both. For example, the two-step and second-order central difference scheme that repeatedly uses the backward Euler scheme is prevalent in the advanced CMs

$$\begin{aligned}\ddot{u}_i &= \frac{1}{\Delta t}(\dot{u}_i - \dot{u}_{i-1})\\ &= \frac{1}{\Delta t^2}((u_i - u_{i-1}) - (u_{i-1} - u_{i-2}))\\ &= \frac{1}{\Delta t^2}(u_i - 2u_{i-1} + u_{i-2})\end{aligned} \tag{4.166}$$

In the dynamic response issue of geotechnical and hydraulic structures to earthquake actions, normally FEM is employed to semi-discretize the spatial domain of the structure (Newmark 1959; Severn 1978; Chopra and Chakrabarti 1981; Paulay and Priestley 1992; Wilson 1998; ICOLD 2001, 2002; Clough and Penzien 2003), by this procedure the ODE set of 2-order is obtained

$$\mathbf{M}\ddot{\widehat{\mathbf{U}}}(t) + \mathbf{C}\dot{\widehat{\mathbf{U}}}(t) + \mathbf{K}\widehat{\mathbf{U}}(t) = \mathbf{F} \tag{4.167}$$

where: **M**—mass matrix; **C**—damping matrix; **K**—stiffness matrix; $\widehat{\mathbf{U}}(t)$, $\dot{\widehat{\mathbf{U}}}(t) = \frac{\partial \widehat{\mathbf{U}}(t)}{\partial t}$, $\ddot{\widehat{\mathbf{U}}}(t) = \frac{\partial^2 \widehat{\mathbf{U}}(t)}{\partial t^2}$—vectors of nodal displacement, velocity, and acceleration, respectively.

In Eq. (4.167), **F** is the dynamic force vector, in the case of earthquake we have

$$\mathbf{F} = -\mathbf{M}\ddot{\widehat{\mathbf{U}}}_g(t) \tag{4.168}$$

where: $\ddot{\widehat{\mathbf{U}}}_g(t) = \frac{\partial^2 \widehat{\mathbf{U}}_g(t)}{\partial t^2}$—earthquake acceleration of ground motions.

The mass matrix **M** of structure system is assembled using $\mathbf{m}^e$ of the element e computed by

$$\mathbf{m}^e = \iiint\limits_{\Omega_e} \mathbf{N}^{\mathrm{T}} \rho \mathbf{N} \mathrm{d}\Omega \tag{4.169}$$

where ρ—volumetric density of the material, kg/m^3.

In the structural analysis w.r.t. dynamic response, the proportional damping (Rayleigh damping) model is commonly stipulated (Caughey and Kelly 1965; Bathe 1982)

$$\mathbf{C} = \alpha_0 \mathbf{M} + \alpha_1 \mathbf{K} \tag{4.170}$$

in which the damping coefficients α_0 and α_1 are found by the damping ratio through the formula

$$\begin{cases} \zeta_1 = \dfrac{a_0}{2\omega_1} + \dfrac{a_1\omega_1}{2} \\ \zeta_2 = \dfrac{a_0}{2\omega_2} + \dfrac{a_1\omega_2}{2} \end{cases} \tag{4.171}$$

in which ω_1 and ω_2 are the first and second natural frequencies.

The "time-history analysis" is then implemented by the direct time integration step-by-step. It enables the designer to determine the number of cycles w.r.t. nonlinear behavior, the magnitude of excursion into the nonlinear range, and the time the structure remains nonlinear. Towards the direct solution of Eq. (4.167), Newmark-β algorithm and Wilson-θ algorithm (Newmark 1959; Bathe and Wilson 1972) basically belong to the particular case of one-step scheme using quadratic truncated Taylor series expansion, are most prevalent. The remarkable advantage of these algorithms in comparison with the other linear counterparts using the assumption of linear accelerate within the time interval $[t, t + \Delta t]$, is their good stability in time-marching process.

1. **Newmark-β method**

The nodal displacements, velocities, and accelerations can be expressed as

$$\begin{cases} \widehat{\mathbf{U}}_{i+1} = \widehat{\mathbf{U}}_i + \Delta t\dot{\widehat{\mathbf{U}}}_i + \left(\dfrac{1}{2} - \beta\right)\Delta t^2\ddot{\widehat{\mathbf{U}}}_i + \beta\Delta t^2\ddot{\widehat{\mathbf{U}}}_{i+1} \\ \dot{\widehat{\mathbf{U}}}_{i+1} = \widehat{\mathbf{U}}_i + \Delta t\dot{\widehat{\mathbf{U}}}_i + \left(\dfrac{1}{2} - \gamma\right)\Delta t\ddot{\widehat{\mathbf{U}}}_i + \gamma\Delta t\ddot{\widehat{\mathbf{U}}}_{i+1} \end{cases} \tag{4.172}$$

in which β and γ are dimensionless parameters.

If $\beta = 1/6$ and $\gamma = 1/2$ are pre-defined in Eq. (4.172), we obtain a conditionally stable and widely used integration scheme. In such an approach the acceleration at nodes is assumed to vary linearly during the time interval $[t, t + \Delta t]$. The relations between the nodal displacements, velocities and accelerations are

$$\dot{\widehat{\mathbf{U}}}_{i+1} = \frac{3}{\Delta t}\widehat{\mathbf{U}}_{i+1} - \left(\frac{3}{\Delta t}\widehat{\mathbf{U}}_i + 2\dot{\widehat{\mathbf{U}}}_i + \frac{\Delta t}{2}\ddot{\widehat{\mathbf{U}}}_i\right) \tag{4.173}$$

and

$$\ddot{\widehat{\mathbf{U}}}_{i+1} = \frac{6}{(\Delta t)^2}\widehat{\mathbf{U}}_{i+1} - \left(\frac{6}{(\Delta t)^2}\widehat{\mathbf{U}}_i + \frac{6}{\Delta t}\dot{\widehat{\mathbf{U}}}_i + 2\ddot{\widehat{\mathbf{U}}}_i\right) \tag{4.174}$$

On the substitution of $\dot{\widehat{\mathbf{U}}}_{i+1}$ and $\ddot{\widehat{\mathbf{U}}}_{i+1}$ in Eq. (4.167), it follows that

$$\begin{cases} \widehat{\mathbf{U}}_{i+1} = (\overline{\mathbf{K}})^{-1}\overline{\mathbf{F}}_i \\ \overline{\mathbf{K}} = \mathbf{K} + \dfrac{6}{(\Delta t)^2}\mathbf{M} + \dfrac{3}{\Delta t}\mathbf{C} \\ \overline{\mathbf{F}}_i = \mathbf{F}_i + \mathbf{M}\left(\dfrac{6}{(\Delta t)^2}\widehat{\mathbf{U}}_i + \dfrac{6}{\Delta t}\dot{\widehat{\mathbf{U}}}_i + 2\ddot{\widehat{\mathbf{U}}}_i\right) + \mathbf{C}\left(\dfrac{3}{\Delta t}\widehat{\mathbf{U}}_i + 2\dot{\widehat{\mathbf{U}}}_i + \dfrac{\Delta t}{2}\ddot{\widehat{\mathbf{U}}}_i\right) \end{cases} \tag{4.175}$$

2. **Wilson-θ method**

By introduce a control parameter θ ($\theta \geq 1$), the Wilson-θ algorithm assumes that the acceleration is linear within the interval $[t, t + \theta\Delta t]$, and it is proved that the unconditional stability will be guaranteed when $\theta \geq 1.37$. In practical computation $\theta = 1.4$ is normally exercised.

At time $t + \tau$, or time step denotes as $i + \tau$, the acceleration may be interpolated by

$$\ddot{\hat{\mathbf{U}}}_{i+\tau} = \ddot{\hat{\mathbf{U}}}_i + \frac{\tau}{\theta\Delta t}\left(\ddot{\hat{\mathbf{U}}}_{i+\theta\Delta t} - \ddot{\hat{\mathbf{U}}}_i\right) \quad (0 \leq \tau \leq \theta\Delta t) \tag{4.176}$$

On the integral and let $\tau = \theta\Delta t$, the velocity and displacement are obtained as follows

$$\begin{cases} \dot{\hat{\mathbf{U}}}_{i+\theta\Delta t} = \dot{\hat{\mathbf{U}}}_i + \dfrac{\theta\Delta t}{2}\left(\ddot{\hat{\mathbf{U}}}_{i+\theta\Delta t} + \ddot{\hat{\mathbf{U}}}_i\right) \\ \hat{\mathbf{U}}_{i+\theta\Delta t} = \hat{\mathbf{U}}_i + \theta\Delta t\dot{\hat{\mathbf{U}}}_i + \dfrac{(\theta\Delta t)^2}{6}\left(\ddot{\hat{\mathbf{U}}}_{i+\theta\Delta t} + 2\ddot{\hat{\mathbf{U}}}_i\right) \end{cases} \tag{4.177}$$

which further provides the acceleration and velocity at time $\tau + \theta\Delta t$

$$\begin{cases} \ddot{\hat{\mathbf{U}}}_{i+\theta\Delta t} = \dfrac{6}{(\theta\Delta t)^2}\left(\hat{\mathbf{U}}_{i+\theta\Delta t} - \hat{\mathbf{U}}_i\right) - \dfrac{6}{\theta\Delta t}\dot{\hat{\mathbf{U}}}_i - 2\ddot{\hat{\mathbf{U}}}_i \\ \dot{\hat{\mathbf{U}}}_{i+\theta\Delta t} = \dfrac{3}{\theta\Delta t}\left(\hat{\mathbf{U}}_{i+\theta\Delta t} - \hat{\mathbf{U}}_i\right) - 2\dot{\hat{\mathbf{U}}}_i - \dfrac{\theta\Delta t}{2}\ddot{\hat{\mathbf{U}}}_i \end{cases} \tag{4.178}$$

Equation (4.178) is now introduced into Eq. (4.177) to give displacement $\hat{\mathbf{U}}_{i+\theta\Delta t}$. Since the displacement $\hat{\mathbf{U}}_i$ and velocity $\dot{\hat{\mathbf{U}}}_i$ and acceleration $\ddot{\hat{\mathbf{U}}}_i$ at time t, as well as the displacement $\hat{\mathbf{U}}_{i+\theta\Delta t}$ at time $t + \theta\Delta t$, are all known, the displacement $\hat{\mathbf{U}}_{i+1}$, velocity $\dot{\hat{\mathbf{U}}}_{i+1}$ and acceleration $\ddot{\hat{\mathbf{U}}}_{i+1}$ at time $t + \Delta t$ or time step $i + 1$, may be recursively calculated. To summarize, following steps are specified for the implementation of Wilson-θ algorithm:

① Calculation of system stiffness matrix $\mathbf{K}$, damping matrix $\mathbf{C}$ and mass matrix $\mathbf{M}$;

② For the initial values of $\hat{\mathbf{U}}_0$, $\dot{\hat{\mathbf{U}}}_0$, and $\mathbf{F}_0$, calculate $\ddot{\hat{\mathbf{U}}}_0$ using the governing equation;

③ Let $\theta = 1.4$, calculate the constants

$$\begin{aligned} &a_0 = \frac{6}{(\theta\Delta t)^2}, a_1 = \frac{3}{\theta\Delta t}, a_2 = 2a_1, a_3 = \frac{\theta\Delta t}{2}, a_4 = \frac{a_0}{\theta}, \\ &a_5 = -\frac{a_2}{\theta}, a_6 = 1 - \frac{3}{\theta}, a_7 = \frac{\Delta t}{2}, a_8 = \frac{\Delta t^2}{6} \end{aligned} \tag{4.179}$$

④ Assemble the equivalent stiffness matrix

$$\overline{\mathbf{K}} = \mathbf{K} + a_0\mathbf{M} + a_1\mathbf{C} \tag{4.180}$$

⑤ Calculate the equivalent load at time $t + \theta\Delta t$ or time step $i + \theta\Delta t$

$$\overline{\mathbf{F}}_{i+\theta\Delta t} = \mathbf{F}_{i+\theta\Delta t} + \mathbf{M}\left(a_0\widehat{\mathbf{U}}_i + a_2\dot{\widehat{\mathbf{U}}}_i + 2\ddot{\widehat{\mathbf{U}}}_i\right) + \mathbf{C}\left(a_1\widehat{\mathbf{U}}_i + 2\dot{\widehat{\mathbf{U}}}_i + a_3\ddot{\widehat{\mathbf{U}}}_i\right) \tag{4.181}$$

⑥ Calculate the displacement at time $t + \theta\Delta t$

$$\widehat{\mathbf{U}}_{i+\theta\Delta t} = (\overline{\mathbf{K}})^{-1}\overline{\mathbf{F}}_{i+\theta\Delta t} \tag{4.182}$$

⑦ Calculate the acceleration, velocity, and displacement at time $t + \Delta t$

$$\begin{cases} \ddot{\widehat{\mathbf{U}}}_{i+1} = a_4\left(\widehat{\mathbf{U}}_{i+1} - \widehat{\mathbf{U}}_i\right) + a_5\dot{\widehat{\mathbf{U}}}_i + a_6\ddot{\widehat{\mathbf{U}}}_i \\ \dot{\widehat{\mathbf{U}}}_{i+1} = \dot{\widehat{\mathbf{U}}}_i + a_7\left(\ddot{\widehat{\mathbf{U}}}_{i+1} + \ddot{\widehat{\mathbf{U}}}_i\right) \\ \widehat{\mathbf{U}}_{i+1} = \widehat{\mathbf{U}}_i + \Delta t\dot{\widehat{\mathbf{U}}}_i + a_8\left(\ddot{\widehat{\mathbf{U}}}_{i+1} + 2\ddot{\widehat{\mathbf{U}}}_i\right) \end{cases} \tag{4.183}$$

4.6 Solution of Partial Differential Equations

4.6.1 Concept

Roughly speaking, the principle of the "finite difference methods (FDMs)" for ODEs is, given a number of discretization points (nodes) defined in a mesh (grid), to assign one discrete unknown variable per one discretization point, and to write one equation per the discretization point. At each discretization point, the derivatives of the unknown variables are replaced by finite differences through the use of Taylor expansions.

A large family of FDM for ODEs has been constructed with correspondent efficient algorithms in a wide field of application-oriented problems. The stability conditions of the fundamental FDM have been well-investigated and the relevant accuracy estimators have been constructed. The demands of electronics and mechanics give strong impetus to the development of a broad class of methods for solving stiff equation systems. Problems arise from control theory, biology, and medicine also play important roles for the progress in FDM.

The achievements of FDM for PDEs are even more impressive (O'Brien et al. 1951). Finite difference counterparts of main differential operators for continuum mechanics have been well-constructed, including those that with conservation properties. The elegant theory of approximation, stability, and convergence of FDM has been established.

Take the linear wave equation for example, the correspondent PDE is

$$\frac{\partial v}{\partial t} = \frac{\partial v}{\partial x} \tag{4.184}$$

Numerically, we first pick the space and time steps Δx and Δt for a regular grid

$$x_j = j\Delta x, t_i = i\Delta t \quad \text{for } j, i \geq 0 \tag{4.185}$$

Then replace the PDE by algebraic formulae that compute a succession of approximate values

$$v_i^j \approx v(t_i, x_j) \quad \text{for } j, i \geq 0 \tag{4.186}$$

A well-known discretization scheme is given by the Lax-Wendroff formula

$$v_{i+1}^j = v_i^j + \frac{1}{2}\lambda\left(v_i^{j+1} - v_i^{j-1}\right) + \frac{1}{2}\lambda^2\left(v_i^{j+1} - v_i^j + v_i^{j-1}\right) \tag{4.187}$$

in which $\lambda = \Delta t / \Delta x$.

The fundamental FDM for PDEs was invented around 1910 by Richardson for stress analysis and meteorology. Afterwards, solution scheme of mathematical physics was published by Courant, Friedrichs and Lewy in 1928. Although their works became famous today, the impact of their ideas was limited before the invention and growth of modern computers. In World War II, powerful computers gave a strong impetus to the development and application of FDM for solving PDEs. The group led by von Neumann at the Los Alamos laboratory made considerable theoretical progress so that the large scale practical applications of FDM became possible with the aid of computer (O'Brian et al. 1951). For parabolic equations, a highlight of early theory was the milestone paper by John (1952). FDM then entered its golden age during the 1950s and 1960s, and major contributions were given by Douglas, Kreiss, Lees, Samarskii, Widlund and many others. At the end of this period the theory of FDM for PDEs had been well-developed. The Lax-Wendroff formula (4.187) and its relatives had grown into a breathtakingly powerful subject known as "computational fluid dynamics (CFD)". Early treatments of linear and nonlinear equations in 1-D soon moved to 2-D, and eventually, to 3-D. It is now a routine to solve problems involving millions of variables on discrete grids with thousands of points in each of three directions.

Important landmark in the progress of FDM is the development of the "alternating direction implicit method", "fractional steps method", and "splitting method". Today, many complicated problems in physics and mechanics, such as the electrodynamics, solid mechanics, fluid mechanics, theory of particle and radiation transfer, atmosphere and ocean dynamics, plasma physics, etc., may be solved by DFM.

However, for multi-dimensional problems in general domains, the situation is less satisfactory, partly because FDM employ the solution values at the points (nodes) on uniform mesh (grid), which does not necessarily fit the domain. The new research outcomes on parabolic and hyperbolic equations using fundamental FDM after 1970s are relative few.

Von Neumann and his colleagues discovered that some numerical methods for PDEs were subject to catastrophic instabilities. Again, take Eq. (4.187) for example, if λ is held fixed at a value smaller than or equal to 1, the method will converge to the correct solution following Δt, $\Delta x \to 0$ (ignoring rounding-off errors). Whereas, if λ is greater than 1, the method will blowup. Soon, the Lax equivalence theorem published by Lax and Richtmyer in 1956. Dahlquist (1963) gave rigor to this observation.

4.6.2 Pure Initial Value Problems

For the simple homogeneous heat equation in 1-D space, the pure initial value problem of parabolic type is governed by

$$\begin{cases} \frac{\partial u}{\partial t} = \frac{\partial^2 u}{\partial x^2} & \text{for } x \in R \text{ and } t \geq 0 \\ u(x,0) = u_0(x) & \text{for } x \in R \end{cases} \tag{4.188}$$

in which R denotes the real axis and $u_0(x)$ is a prescribed smooth bounded function. It is well known that this problem permits a closed-form solution that provides rigorous benchmark for numerical algorithms.

$$u(x,t) = \frac{1}{2\sqrt{\pi t}} \int_{-\infty}^{+\infty} e^{-(x-\xi)^2/(4t)} u_0(\xi) \mathrm{d}\xi = g(t) u_0(x) \tag{4.189}$$

Towards the numerical solution of Eq. (4.188) by FDM, a grid of points (nodes) $(x, t) = (jh, i\Delta t)$ in Fig. 4.16 is constructed where $h = \Delta x$ and Δt are grid parameters that are sufficiently small, j and i are integers, $i \geq 0$.

At these grid nodes, one look for an approximant u_i^j of Eq. (4.188) by solving a problem in which the derivatives have been replaced by finite difference quotients. Suppose that forward and backward difference quotients are employed, the simplest finite difference equation derived from Eq. (4.188) is

$$\frac{u_{i+1}^j - u_i^j}{\Delta t} = \frac{u_i^{j+1} - 2u_i^j + u_i^{j-1}}{h^2} \quad \text{for } -\infty \leq j \leq +\infty \text{ and } i \geq 0 \tag{4.190}$$

or

$$u_{i+1}^j = \lambda u_i^{j-1} + (1 - 2\lambda) u_i^j + \lambda u_i^{j+1} \equiv (E^{\Delta t h} u_i)_j \quad \text{for } -\infty \leq j \leq +\infty \text{ and } i \geq 0 \tag{4.191}$$

in which $\lambda = \Delta t / h^2$.

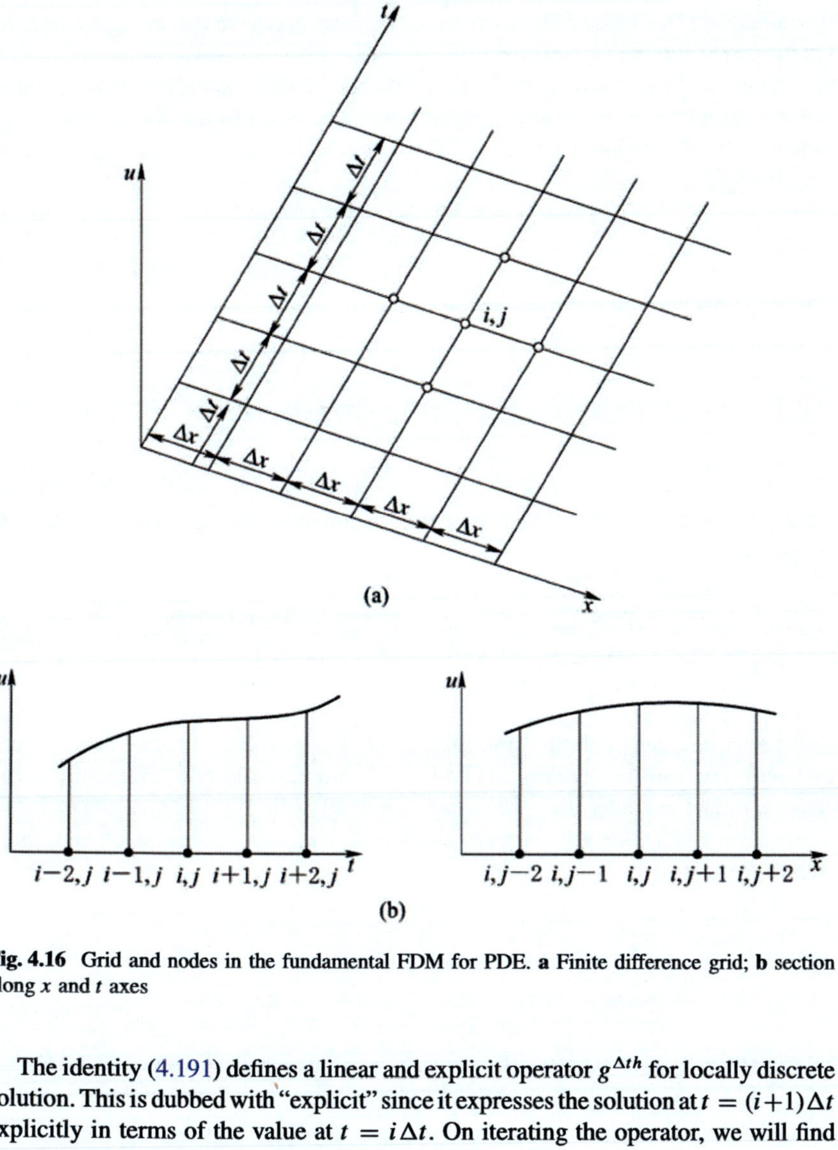

Fig. 4.16 Grid and nodes in the fundamental FDM for PDE. **a** Finite difference grid; **b** section along x and t axes

The identity (4.191) defines a linear and explicit operator $g^{\Delta th}$ for locally discrete solution. This is dubbed with "explicit" since it expresses the solution at $t = (i+1)\Delta t$ explicitly in terms of the value at $t = i\Delta t$. On iterating the operator, we will find that the solution of the discrete problem is

$$u_i^j = (g_i^{kh} u_0)_j \tag{4.192}$$

where $\lambda \leq 1/2$, the boundedness of the discrete solution in terms of the discrete initial data is referred to as stability. On the contrary, the method is unstable for $\lambda > 1/2$, namely, even though the initial data are very small, the discrete solution tends to infinity as $i \to \infty$. This may be interpreted that very small perturbations of the initial data (for instance by rounding-off errors) may give rise to big changes in the discrete solution at later times.

Denote $C^{m,n}$ the function set of (x, t) with bounded derivatives of orders at most m and n, respectively w.r.t x and t, it has been proved that (Lax and Richtmyer 1956): assume $u \in C^{m,n}$ and $\lambda = \Delta t/h^2 \leq 1/2$, there is a constant $C = C(u, T)$ such that

$$\left\| u(x, t) - u^i \right\| \leq Ch^2 \quad \text{for } i\Delta t \leq T \tag{4.193}$$

This also states that the scheme in Eq. (4.191) possesses first-order accuracy in time and second-order accuracy in space.

4.6.3 Mixed Initial Boundary Value Problems

IN real-world physical and mechanical scenarios, one is required to solve the parabolic equation on a finite interval with boundary values given at its endpoints for positive time, i.e.

$$\begin{cases} \frac{\partial u}{\partial t} = \frac{\partial^2 u}{\partial x^2} & \text{for } x \in [0, 1] \text{ and } t \geq 0 \\ u(0, t) = u(1, t) = 0 & \\ u(x, 0) = v(x) & \text{for } x \in [0, 1] \end{cases} \tag{4.194}$$

For the approximate solution one may again discretize the domain using a grid of nodes, but now by dividing interval [0, 1] into n sub-intervals of equidistance $h = 1/n$, and consider the grid nodes $(jh, i\Delta t)$ with $j = 1, 2, \ldots, n$ and $i = 1, 2, \ldots, \infty$. Let u_i^j the approximate solution at $(jh, i\Delta t)$, the natural explicit finite difference scheme is

$$\begin{cases} \partial_t u_i^j = \partial_x \bar{\partial}_x u_i^j & \text{for } (j = 1, 2, \ldots, n) \text{ and } i \geq 0 \\ u_i^0 = u_i^n = 0 & \text{for } i > 0 \\ u_0^j = v^j = v(jh) & \text{for } (j = 1, 2, \ldots, n) \end{cases} \tag{4.195}$$

where ∂_t and ∂_x are the finite differences w.r.t. time and space, respectively. For u_i^j $(j = 1, 2, \ldots, n)$, the solution of u_{i+1}^j is given

$$\begin{cases} u_{i+1}^j = \lambda(u_i^{j-1} + u_i^{j+1}) + (1 - 2\lambda)u_i^j & \text{for } (j = 1, 2, \ldots, n) \text{ and } i \geq 0 \\ u_i^0 = u_i^n = 0 & \text{for } i > 0 \end{cases} \tag{4.196}$$

This scheme is referred to as the "forward Euler scheme" with maximum-norm stable subject to $\lambda = \Delta t/h^2 \leq 1/2$ or $\Delta t \leq h^2/2$. This stability requirement is quite restrictive in practice hence we are naturally motivated to find methods being able to use h and Δt of the same order of magnitude. For this purpose, one may define an approximate solution implicitly by the following backward equation

$$\begin{cases} \overline{\partial}_t u^j_{i+1} = \partial_x \overline{\partial}_x u^j_{i+1} & \text{for } (j = 1, 2, \ldots, n) \text{ and } i \geq 0 \\ u^0_{i+1} = u^n_{i+1} = 0 & \text{for } i > 0 \\ u^j_o = v^j = v(jh) & \text{for } (j = 1, 2, \ldots, n) \end{cases} \tag{4.197}$$

For u^j_i $(j = 1, 2, \ldots, n)$, the solution of u^j_{i+1} is given

$$\begin{cases} (1 + 2\lambda)u^j_{i+1} - \lambda(u^{j-1}_{i+1} + u^{j+1}_{i+1}) = u^j_i & \text{for } (j = 1, 2, \ldots, n) \text{ and } i \geq 0 \\ u^0_{i+1} = u^n_{i+1} = 0 & \text{for } i > 0 \end{cases} \tag{4.198}$$

In matrix notation, Eq. (4.198) may be written as

$$\mathbf{B}\widehat{\mathbf{U}}^{i+1} = \widehat{\mathbf{U}}^{i} \tag{4.199}$$

in which $\widehat{\mathbf{U}}^{i+1}$ and $\widehat{\mathbf{U}}^{i}$ are the vectors with $(n - 1)$ components, $\mathbf{B}$ is the matrix of diagonally dominant, symmetric, and tridiagonal

$$\mathbf{B} = \begin{bmatrix} 1+2\lambda & -\lambda & 0 & \ldots & 0 & 0 \\ -\lambda & 1+2\lambda & -\lambda & \ldots & 0 & 0 \\ 0 & -\lambda & 1+2\lambda & \ldots & 0 & 0 \\ \ldots & \ldots & \ldots & \ldots & \ldots & \ldots \\ 0 & 0 & 0 & \ldots & -\lambda & 1+2\lambda \end{bmatrix} \tag{4.200}$$

Equation (4.198) is referred to as the "backward Euler scheme" which was firstly proposed by Laasonen (1949). Although it does not impose restraints on the mesh ratio $\lambda = \Delta t/h^2$, yet since it is only first-order accurate in time, the time discretization error will dominate unless Δt is much smaller than h.

It is noted that there are a number of methods with second-order accurate in time, such as the prevalent method credited to Crank and Nicolson (1947).

4.6.4 Example—FLAC

Merits of the well-known code "FLAC (Fast Lagrangian Analysis of Continue)" (Marti and Cundall 1982) that enable it being widely accepted in the simulation of geomaterials are:

- FDM is undertaken in spatial–temporal domain by mixed discretization, to transfer the PDEs governing the motion of an engineering structure into discrete system dominated by the Newton's second law at nodes;
- Explicit scheme of finite difference w.r.t. space and time is undertaken to calculate the derivatives;

- Artificial damping is employed for the purpose of quickly reaching static status, hence static problem may be solved by dynamic algorithm;
- Global stiffness matrix is not demanded;
- For each iterative loop, the nodal coordinates are updated, hence the finite deformation issue may be handled effectively.

1. **Governing equations for solid motion**

Denote **v** the velocity of an infinitesimal parcel (cell), the momentum Eq. (3.185) with tensor notation is given as follows

$$\sigma_{\beta\alpha,\beta} + b_\alpha = \rho \frac{\mathrm{d}v_\alpha}{\mathrm{d}t} \quad (\alpha, \beta = 1, 2, 3) \tag{4.201}$$

and the traction (stress) boundary condition on Γ_t becomes

$$\bar{t}_\alpha = \sigma_{\beta\alpha} n_\beta \quad (\alpha, \beta = 1, 2, 3) \tag{4.202}$$

The Cauchy strain rate is derived from Eq. (3.80),

$$\dot{\varepsilon}_{\alpha\beta} = \frac{1}{2}(v_{\alpha,\beta} + v_{\beta,\alpha}) \quad (\alpha, \beta = 1, 2, 3) \tag{4.203}$$

In which the Greek superscripts α and β are employed to indicate the coordinate directions for the simplification of notations $x = x_1, y = x_2$ and $z = x_3$.

In FLAC, the material (Lagrangian) cell undergoes transitional and rotational movement of rigid body, too. Under such finite deformation circumstance, the "stress increment in co-rotational coordinate system" $\Delta\overline{\sigma}_{\varepsilon\beta}$ revised by the angular velocity should be employed for in the elasto-plastic constitutive relation (3.137), i.e.

$$\Delta\overline{\sigma}_{\alpha\beta} = H_{\alpha\beta}(\sigma_{\alpha\beta}, \dot{\varepsilon}_{\alpha\beta}\Delta t, K) \tag{4.204}$$

However, for the facilitation of understanding FLAC algorithm, the movement of rigid body will be neglected in the following discussion, i.e.

$$\Delta\overline{\sigma}_{\alpha\beta} = \Delta\sigma_{\alpha\beta} \tag{4.205}$$

2. **Spatial discretization**

In Fig. 4.17, a tunnel segment with two intersected faults and an irregular ground surface is discretized into a grid composed of tetrahedrons.

For the representative tetrahedral element e with $n = 4$ nodes shown in Fig. 4.18, we denote the superscript l (1, 2, ..., 4) as the local node sequence. If necessary, f is used as the local sequence of face, and (l) is the local face sequence that is not connected to node l.

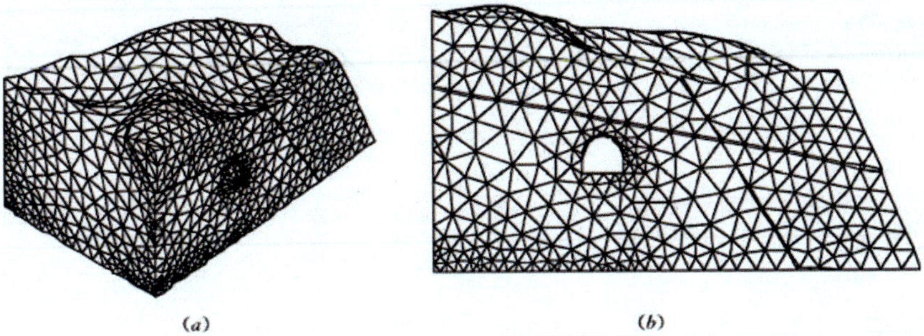

Fig. 4.17 There-dimensional cavern discretized by tetrahedral elements. **a** Axonometric view; **b** front view

Fig. 4.18 Tetrahedral element *e*

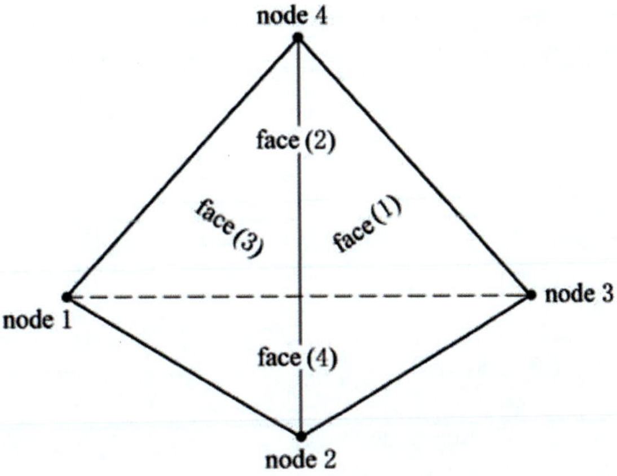

With a mapping to the standard (parent) tetrahedron element in Fig. 4.19, any variable component $q_\alpha(x_1^P, x_2^P, x_3^P)(\alpha = 1, 2, 3)$ at point P within the element may be interpolated by nodal basis functions N_l $(l = 1, 2, 3, 4)$ such that:

$$\begin{cases} q_\alpha = \sum_{l=1}^{n} \hat{q}_\alpha^l N_l \ (\alpha = 1, 2, 3) \\ N_l = \xi_l \qquad (l = 1, \ldots, 4) \end{cases} \tag{4.206}$$

in which ξ is the local coordinate of point P

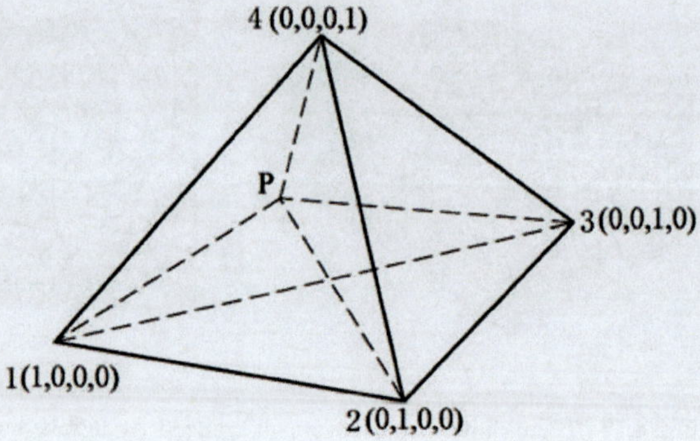

Fig. 4.19 Standard tetrahedral element

$$\begin{cases} \xi_1 = \dfrac{V_{\mathrm{P}234}}{V_{1234}} \\ \xi_2 = \dfrac{V_{1\mathrm{P}34}}{V_{1234}} \\ \xi_3 = \dfrac{V_{12\mathrm{P}4}}{V_{1234}} \\ \xi_4 = \dfrac{V_{123\mathrm{P}}}{V_{1234}} \end{cases} \tag{4.207}$$

In Eq. (4.206), $\hat{q}_\alpha^l$ ($\alpha = 1, 2, 3$) may be interpreted as the nodal values of displacement $\hat{u}_\alpha^l$ and velocity $\hat{v}_\alpha^l$. It entails linear velocity and displacement distribution as well as constant strain and stress rate distribution in tetrahedral element. In addition, nodal basis functions $N_l(l = 1, \ldots, 4)$ possess Kronecker Delta property

$$N^l(x_1^{\mathrm{P}}, x_2^{\mathrm{P}}, x_3^{\mathrm{P}}) = \delta_{l\mathrm{P}} \quad \text{for } (l, \mathrm{P} = 1, \ldots, 4) \tag{4.208}$$

Hence, the essential boundary conditions may be exactly met. $N_l(l = 1, \ldots, 4)$ also possess the property of PU, i.e.

$$\sum_{l=1}^{4} N_l = 1 \tag{4.209}$$

According to Gauss theory (3.147), we have

$$\iiint\limits_{\Omega_e} \nabla \mathbf{v} \mathrm{d}\Omega = \oiint\limits_{\Gamma_e} \mathbf{v}.\mathbf{n} \mathrm{d}\Gamma \tag{4.210}$$

in which $\mathbf{n}$ is the outward unit normal vector on the tetrahedral face.

Since the velocity is linearly distributed, from Eq. (4.210) we get

$$V v_{\alpha,\beta} = \sum_{f=1}^{4} \overline{v}_\alpha^f n_\beta^f S^f \quad (\alpha, \beta = 1, 2, 3) \tag{4.211}$$

in which S^f is the area of face f, V is the volume of tetrahedral body, the average velocity $\overline{v}_\alpha^f$ on face f is

$$\overline{v}_\alpha^f = \frac{1}{3} \sum_{l=1,(l)\neq f}^{4} v_\alpha^l \quad (\alpha = 1, 2, 3) \tag{4.212}$$

on inserting into Eq. (4.211) it follows that

$$\mathbf{v}_{,\beta} = \frac{1}{3V} \sum_{l=1}^{4} \mathbf{v}^l . \mathbf{n}^{(l)} . S^{(l)} \quad (\beta = 1, 2, 3) \tag{4.213}$$

on inserting into Eq. (4.203), the strain rate strain increment are expressed by

$$\dot{\varepsilon}_{\alpha\beta} = \frac{1}{6V} \sum_{l=1}^{4} (v_\alpha^l n_\beta^{(l)} + v_\beta^l n_\alpha^{(l)}) S^{(l)} \quad (\alpha, \beta = 1, 2, 3) \tag{4.214}$$

$$\Delta\varepsilon_{\alpha\beta} = \frac{\Delta t}{6V} \sum_{l=1}^{4} (v_\alpha^l n_\beta^{(l)} + v_\beta^l n_\alpha^{(l)}) S^{(l)} \quad (\alpha, \beta = 1, 2, 3) \tag{4.215}$$

3. **Nodal motion equations**

(1) **Element analysis**

Denote $F_\alpha^l(t)$ as the unbalance nodal force at node l derived from Eq. (4.201) such that

$$F_\alpha^l(t) = \frac{T_\alpha^l}{3} + \frac{b_\alpha V}{4} - m^l \left(\frac{\mathrm{d}v_\alpha}{\mathrm{d}t} \right)^l \quad \text{for } (l = 1, \ldots, 4) \text{ and } (\alpha = 1, 2, 3) \tag{4.216}$$

where

$$T_\alpha^l = \frac{1}{V} \sigma_{\alpha\beta} n_\beta^{(l)} S^{(l)} \quad \text{for } (l = 1, \ldots, 4) \text{ and } (\alpha, \beta = 1, 2, 3) \tag{4.217}$$

$$m^l = \frac{\alpha_1 V}{4} \quad \text{for } (l = 1, \ldots, 4) \tag{4.218}$$

in which $\alpha_1 = \rho$ for dynamical problems, for static problems α_1 is a fraction of density ρ to get better numerical stability.

(2) System assemblage

Denote $\langle l \rangle$ as the overall sequence of representative node l in the whole discrete system, the contribution from all the tetrahedral elements surrounding node $\langle l \rangle$ can be assembled in an identical manner of FEM, which gives rise to

$$F_\alpha^{\langle l \rangle}\left(t, (v_\alpha^{\langle k \rangle}, \ldots, v_\alpha^{\langle p \rangle})^{\langle l \rangle}, K\right) = M^{\langle l \rangle}\left(\frac{\mathrm{d}v_\alpha}{\mathrm{d}t}\right)^{\langle l \rangle} \quad \text{for } (\alpha = 1, 2, 3) \text{ and } (\langle l \rangle = 1, \ldots, n_n) \tag{4.219}$$

where: v_α—velocity component at the coordinate axis direction α; $\langle k \rangle$, …, $\langle p \rangle$—nodes surrounding the representative node $\langle l \rangle$; n_n—amount of nodes in the whole discrete system; $M^{\langle l \rangle}$—mass at $\langle l \rangle$ contributed from its surrounding nodes; $F_\alpha^{\langle l \rangle}$—unbalance force component at $\langle l \rangle$ contributed from its surrounding nodes.

According to Eq. (4.219), the ODE w.r.t. nodal acceleration is obtained as follows

$$\left(\frac{\mathrm{d}v_\alpha}{\mathrm{d}t}\right)^{\langle l \rangle} = \frac{1}{M^{\langle l \rangle}} F_\alpha^{\langle l \rangle} \quad \text{for } (\alpha = 1, 2, 3) \text{ and } (\langle l \rangle = 1, \ldots, n_n) \tag{4.220}$$

If the static state is reached, $F_i^{\langle l \rangle} = 0$, otherwise the continuous flow will approach to a non-zero constant state.

(3) Damping

Damping is employed to decay the vibration of system induced by unbalance force, in this manner to quickly reach the static state solution. It is presumably proportional to the unbalance force for the global node $\langle l \rangle$, namely

$$Fd_\alpha^{\langle l \rangle}(t) = -C\left|F_\alpha^{\langle l \rangle}(t)\right|\mathrm{sign}\left(v_\alpha^{\langle l \rangle}\right) \quad \text{for } (\alpha = 1, 2, 3) \text{ and } (\langle l \rangle = 1, \ldots, n_n) \tag{4.221}$$

in which C is the damping coefficient, normally $C = 0.8$ is defaulted in FLAC.

$Fd_\alpha^{\langle l \rangle}(t)$ is looked at as a revision of the unbalance force $F_\alpha^{\langle l \rangle}$ in Eqs. (4.219) and (4.220), i.e.

$$F_\alpha^{\langle l \rangle} \to F_\alpha^{\langle l \rangle} + Fd_\alpha^{\langle l \rangle} \quad \text{for } (\alpha = 1, 2, 3) \text{ and } (\langle l \rangle = 1, \ldots, n_n) \tag{4.222}$$

4. Solution procedure

For each node with global sequence $\langle l \rangle$, there are altogether 15 equations available: 3 momentum equations + 6 strain rate equations + 6 constitutive equations. Correspondingly, there are altogether 15 unknown variables: 3 velocity components + 6 stress rate components + 6 strain rate components. In FLAC, the central difference scheme is applied to Eq. (4.220) for the iterative solution of non-linear problem, namely

$$v_\alpha^{\langle l \rangle}(t + \frac{\Delta t}{2}) = v_\alpha^{\langle l \rangle}(t - \frac{\Delta t}{2}) + \frac{\Delta t}{M^{\langle l \rangle}} F_\alpha^{\langle l \rangle}\left(t, (v_\alpha^{\langle k \rangle}, \ldots, v_\alpha^{\langle p \rangle})^{\langle l \rangle}, K\right)$$

$$\text{for } (\alpha = 1, 2, 3) \text{ and } (\langle l \rangle = 1, \ldots, n_n) \tag{4.223}$$

The velocities and coordinates of nodes are revised as is common in the updated Lagrange (UL) formalism,

$$u_\alpha^{\langle l \rangle}(t + \Delta t) = u_\alpha^{\langle l \rangle}(t) + \Delta t v_\alpha^{\langle l \rangle}(t + \frac{\Delta t}{2}) \text{ for } (\alpha = 1, 2, 3) \text{ and } (\langle l \rangle = 1, \ldots, n_n) \tag{4.224}$$

$$x_\alpha^{\langle l \rangle}(t + \Delta t) = x_\alpha^{\langle l \rangle}(t) + \Delta t v_\alpha^{\langle l \rangle}(t + \frac{\Delta t}{2}) \text{ for } (\alpha = 1, 2, 3) \text{ and } (\langle l \rangle = 1, \ldots, n_n) \tag{4.225}$$

The strain rate and strain increment as well as stress increment may be calculated using Eqs. (4.212)–(4.215) and (4.203)–(4.205), afterwards, strain and stress are updated as follows

$$\varepsilon_{\alpha\beta}(t + \Delta t) = \varepsilon_{\alpha\beta}(t) + \Delta\varepsilon_{\alpha\beta}(t + \frac{\Delta t}{2}) \quad (\alpha, \beta = 1, 2, 3) \tag{4.226}$$

$$\sigma_{\alpha\beta}(t + \Delta t) = \sigma_{\alpha\beta}(t) + \Delta\sigma_{\alpha\beta}(t + \frac{\Delta t}{2}) \quad (\alpha, \beta = 1, 2, 3) \tag{4.227}$$

For more details, such as the mixed discretization exploiting both the advantages of tetrahedral element and hexahedral element, our readers are referred to the manual of FLAC (ITASCA 2002).

4.7 Solution for Weak Form Equations

4.7.1 Ritz Method

Also termed as "Rayleigh–Ritz Method (RRM)", this method employs the variational counterpart of differential problem: after the trial function being introduced into functional $J(c_0, c_1, c_2, \ldots, c_n)$, the minimization operator is undertaken to get stationary solution of unknown coefficients $(c_0, c_1, c_2, \ldots, c_n)$ such that

$$\frac{\partial J(c_0, c_1, c_2, \ldots, c_n)}{\partial c_i} = 0 \quad (i = 0, 1, 2, \ldots, n) \tag{4.228}$$

Symmetric matrices will arise whenever a variational principle exists, which is one of the most important merits of variational approaches. However, it is worthwhile to indicate that frequently such a symmetry will be derived directly from Galerkin method and weighted residual method, which leads to know (previously unknown) the existence of variational principles (Tonti 1969). Generally, the difficulty with RRM is the determination of corresponding functional for variational operation.

Take the Poisson equation for example,

$$\begin{cases} -\nabla^2 u = -\frac{\partial^2 u}{\partial x^2} - \frac{\partial u}{\partial y^2} = f(x, y) \ (x, y) \in G \\ u|_\Gamma = \phi(x, y) \\ u \in M \end{cases} \tag{4.229}$$

in which ∇^2 is the Laplace operator. The varational solution u being exactly the solution of PDE (4.229) may be get by the minimum of functional $F[u]$

$$F[u] = \iiint\limits_\Omega \left[\left(\frac{\partial u}{\partial x}\right)^2 + \left(\frac{\partial u}{\partial y}\right)^2\right] d\Omega - 2 \iint\limits_\Gamma f u \, d\Gamma \tag{4.230}$$

Independent function basis $u_0(x, y), u_1(x, y), \ldots, u_n(x, y)$ are so constructed that meet

$$\begin{cases} u_i(x, y) \in C^2(\Omega + \Gamma) \\ u_0(x, y)|_\Gamma = \phi(x, y) \\ u_i(x, y)|_\Gamma = 0 \quad (i = 1, \ldots, n) \end{cases} \tag{4.231}$$

The linear combination of these basis functions constructs the trial function

$$\begin{cases} u \cong \tilde{u}(x, y) = \sum\limits_{i=0}^{n} c_i u_i(x, y) \\ \tilde{u}(x, y) \in M \\ c_0 = 1 \end{cases} \tag{4.232}$$

On inserting into Eq. (4.230), we have

$$\begin{aligned} F[\tilde{u}] &= J(c_0, c_1, c_2, \ldots, c_n) \\ &= \iiint\limits_\Omega \left[\left(\sum_{i=0}^{n} c_i \frac{\partial u_i(x, y)}{\partial x}\right)^2 + \left(\sum_{i=0}^{n} c_i \frac{\partial u_i(x, y)}{\partial y}\right)^2\right] d\Omega \\ &\quad - 2 \iint\limits_\Gamma f(x, y) u_0(x, y) d\Gamma \end{aligned} \tag{4.233}$$

Condition of stationary solution leads to

$$\frac{\partial J}{\partial c_j} = 2 \iiint_{\Omega} \left[\left(\sum_{i=0}^{n} c_i \frac{\partial u_i}{\partial x} \right) \frac{\partial u_j}{\partial x} + \left(\sum_{i=0}^{n} c_i \frac{\partial u_i}{\partial y} \right) \frac{\partial u_j}{\partial y} \right] \mathrm{d}\Omega = 0 \quad (j = 1, 2, \ldots, n) \tag{4.234}$$

Denote

$$[u_i, u_j] = \iiint_{\Omega} \left[\frac{\partial u_i}{\partial x} \frac{\partial u_j}{\partial x} + \frac{\partial u_i}{\partial y} \frac{\partial u_j}{\partial y} \right] \mathrm{d}\Omega \tag{4.235}$$

we get the equation set for the solution of n coefficients ($c_1, c_2, ..., c_n$)

$$\begin{cases} [u_0, u_1] + c_1[u_1, u_1] + \ldots + c_n[u_n, u_1] = 0 \\ [u_0, u_2] + c_1[u_1, u_2] + \ldots + c_n[u_n, u_2] = 0 \\ \ldots \\ [u_0, u_n] + c_1[u_1, u_n] + \ldots + c_n[u_n, u_n] = 0 \end{cases} \tag{4.236}$$

It is emphasized that independent function basis should be defined in the entire domain and restrained by all boundary conditions. This demand is not so easy to meet before the appearance of FEM using piecewise interpolation in sub-domains.

4.7.2 *Finite Element Method*

1. **Concept**

The "finite element methods (FEM)" is a significant success for the solution of PDEs from diverse roots in engineering and mathematics. Instead of approximating a differential operator by a difference quotient in FDM, FEM approximates the solution by trial functions that can be broken up into simple pieces. For instance, one may discretize the domain into elementary geometrical piece sets such as triangles or tetrahedrons. The restriction of trial functions to each piece is a polynomial of low degree. Taking the advantage of functional analysis and solving a weak form system equation within the corresponding finite-dimensional sub-space, there is often a guarantee that the computed solution is optimal within that sub-space.

FEM for solid dynamics may be equally formulated using the principles of variational and weighted residual (Chap. 3), among them the virtual work principle (3.199) that may be derived from Hamilton's principle and actually equivalent to Galerkin method is most prevalent. The major procedures may be implemented as follows:

① Construct shape functions. They are employed to construct the trial function for the approximation of essential variable—displacement or its increment with a certain order of consistency.

② Substitute the trial function into Eq. (3.199). This gives rise to a set of "ordinary differential equations (ODEs)" w.r.t. time only. For static problems, a set of algebraic equations is obtained.
③ Solve the equation set. Normally, standard FDM is employed for the solution of ODEs to get dynamic process, whereas a number of standard algebraic equation solvers are available for static field.

2. **Basic formulation for static elasticity**

We should construct a FE mesh in which the whole structure domain is divided into a number of discrete sub-domains, namely "clcments", connected at discrete points called "nodes". Certain of these nodes will have fixed displacements, and others will have prescribed loads.

Figure 4.20 shows the discrete FE mesh for a gravity dam with its foundation, and Fig. 4.21 establishes the coordinate transformation using standard shape functions identical to that for the unknown function transformation, namely, shape functions defining the element geometry and the field function are the same. Such a type of element comes under the term of "isoparametric element".

Actually, the gravity dam in Fig. 4.20 may be well-simulated by the 2-D (plane strain) formulation of FEM. Very often (although not always), the quadrilateral isoparametric element with four nodes is exercised, and its coordinate transformation using standard shape functions is shown in Fig. 4.22.

For two-dimensional problem, the global node sequence is denoted as i $(1, 2, \ldots, n_n)$ where n_n is the amount of nodes. For representative element $fe(e)$ (or briefly e) $(1, 2, \ldots, n_e)$, a local sequence of node is assigned as $l(1, \ldots, n)$ where a link list $i = \langle l \rangle$ is preset and n is the amount of nodes defining element e. For the 2-D linear element shown in Fig. 4.22, $n = 4$. The nodal coordinate $\mathbf{x}_e$ and the interior coordinate $\mathbf{x}_P$ at any position P are defined as

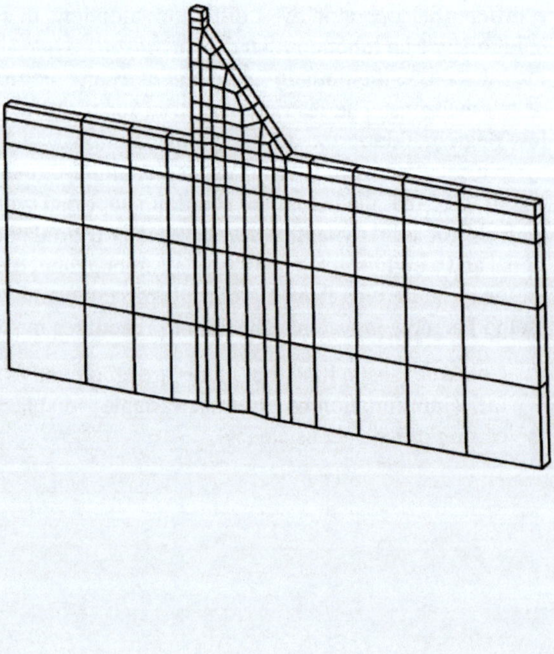

Fig. 4.20 Finite element mesh of a gravity dam

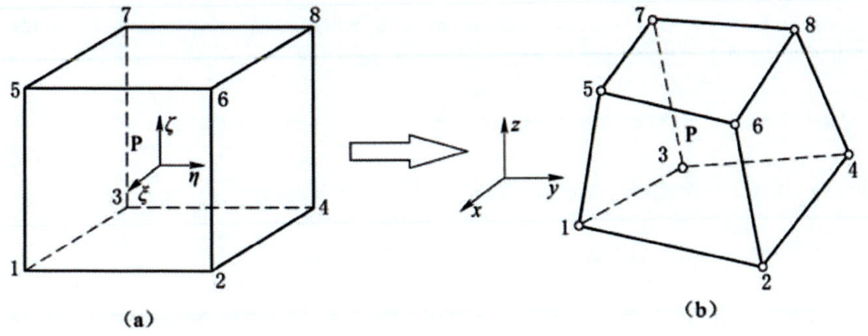

Fig. 4.21 Hexahedral element. **a** Parent element; **b** real object element after mapping

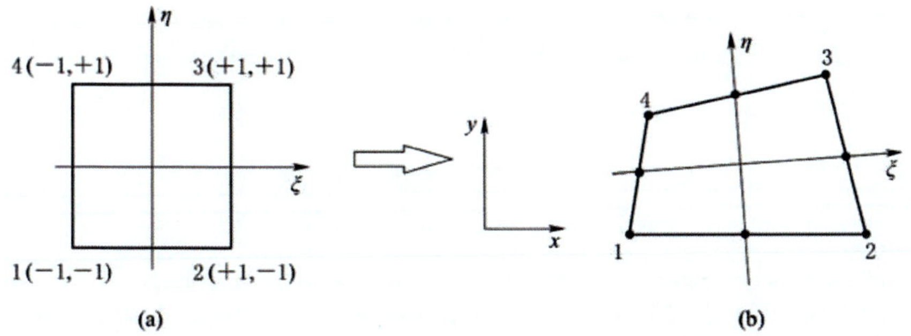

Fig. 4.22 Quadrilateral element. **a** Parent element; **b** real object element after mapping

$$\begin{cases} \mathbf{x}_e = [\mathbf{x}_1 \ldots \mathbf{x}_l \ldots \mathbf{x}_n]^{\mathrm{T}} & (e = 1, 2, \ldots, n_e) \\ \mathbf{x}_l = [x_l \quad y_l]^{\mathrm{T}} & (l = 1, \ldots, n) \\ \mathbf{x}_{\mathrm{P}} = [x_{\mathrm{P}} \quad y_{\mathrm{P}}]^{\mathrm{T}} & \end{cases} \tag{4.237}$$

The quadrilateral element in Fig. 4.22b may be looked as a mapping from the standard squre element term as "parent element" in Fig. 4.22a, the coordinate $\mathbf{x}_{\mathrm{P}}$ of any point P within the element may be interpolated by the nodal basis functions $N_l(l = 1, \ldots, n)$ such that:

$$\mathbf{x}_{\mathrm{P}} = \mathbf{N}\mathbf{x}_e \tag{4.238}$$

in which $\mathbf{N}$ is the "shape function matrix",

$$\begin{cases} \mathbf{N} = [\mathbf{N}_1 \ldots \mathbf{N}_l \ldots \mathbf{N}_n] = [N_1\mathbf{I} \ldots N_l\mathbf{I} \ldots N_n\mathbf{I}] \\ \mathbf{I} = \begin{bmatrix} 1 & 0 \\ 0 & 1 \end{bmatrix} \end{cases} \tag{4.239}$$

in which $\mathbf{I}$ stands for 2×2 unit matrix and N_l is the basis function at node l. For the linear quadrilateral element in Fig. 4.22, we have

$$\begin{cases} N_l = \frac{1}{4}(1+\xi_0\xi)(1+\eta_0\eta) \\ \xi_0 = (-1)^l \\ \eta_0 = (-1)^{[l/2+0.5]} \end{cases} \quad (l = 1, \ldots, n = 4) \tag{4.240}$$

where ξ_0, η_0 are the normalized coordinates of node l.

It is notable that the transformation (mapping) (4.238) is revertible, namely, if the coordinate $\mathbf{x}_P$ in real object element is specified, the correspondent local coordinate (ξ, η) may be uniquely calculated.

Now we denote the nodal displacement increment $\Delta\hat{\mathbf{u}}_e$ as

$$\begin{cases} \Delta\hat{\mathbf{u}}_e = [\Delta\hat{\mathbf{u}}_1 \ldots \Delta\hat{\mathbf{u}}_l \ldots \Delta\hat{\mathbf{u}}_n]^{\mathrm{T}} & (e = 1, 2, \ldots, n_e) \\ \Delta\hat{\mathbf{u}}_l = [\Delta\hat{u}_{xl} \quad \Delta\hat{u}_{yl}]^{\mathrm{T}} & (l = 1, \ldots, n) \end{cases} \tag{4.241}$$

There also exists nodal force increment

$$\begin{cases} \Delta\mathbf{f}_e = [\Delta\mathbf{f}_1 \ldots \Delta\mathbf{f}_l \ldots \Delta\mathbf{f}_n]^{\mathrm{T}} & (e = 1, 2, \ldots, n_e) \\ \Delta\mathbf{f}_l = [\Delta f_{xl} \quad \Delta f_{yl}]^{\mathrm{T}} & (l = 1, \ldots, n) \end{cases} \tag{4.242}$$

Displacement approximation (trial function) within an isoparametric element is the interpolation of nodal displacements (as yet unknown) via shape functions identical to that used in the coordinate transformation

$$\Delta\mathbf{u}_e = \mathbf{N}\Delta\hat{\mathbf{u}}_e \quad (e = 1, 2, \ldots, n_e) \tag{4.243}$$

For small deformation problems, on inserting Eq. (4.243) into Eqs. (3.85) and (3.86), Cauchy strain increment is calculated by

$$\Delta\varepsilon_e = \mathbf{B}\Delta\hat{\mathbf{u}}_e \quad (e = 1, 2, \ldots, n_e) \tag{4.244}$$

in which $[B]$ is the "strain matrix"

$$\mathbf{B} = [B] = [\mathbf{B}_1 \ldots \mathbf{B}_l \ldots \mathbf{B}_n] \tag{4.245}$$

$$\mathbf{B}_l = \mathbf{L}N_l = \begin{bmatrix} \frac{\partial N_l}{\partial x} & 0 \\ 0 & \frac{\partial N_l}{\partial y} \\ \frac{\partial N_l}{\partial y} & \frac{\partial N_l}{\partial x} \end{bmatrix} \quad (l = 1, \ldots, n) \tag{4.246}$$

According to the Hooke's law we have elastic constitutive relation

$$\Delta\sigma_e = \mathbf{D}\Delta\varepsilon_e \tag{4.247}$$

in which $\Delta\sigma_e$ is the stress incremental vector, $\Delta\varepsilon_e$ is the strain incremental vector, **D** is the elastic matrix.

On the substitution of Eq. (4.244) into Eq. (4.247), it follows that

$$\Delta\sigma_e = \mathbf{S}\Delta\hat{\mathbf{u}}_e \tag{4.248}$$

in which **S** is the "stress matrix"

$$\mathbf{S} = \mathbf{DB} \tag{4.249}$$

Suppose a small, virtual displacement $\delta\Delta\hat{\mathbf{u}}_e$ occurs to the element e that in turn produces a virtual strain $\delta\Delta\varepsilon_e$ according to Eq. (4.248), the virtual work principle in Eq. (3.199) will give rise to

$$(\delta\Delta\hat{\mathbf{u}}_e)^{\mathrm{T}}\Delta\mathbf{f}_e = \iiint\limits_{\Omega_e} (\delta\Delta\varepsilon_e)^{\mathrm{T}}\Delta\sigma_e \mathrm{d}\Omega \tag{4.250}$$

$$\Delta\mathbf{f}_e = \iiint\limits_{\Omega_e} \mathbf{N}^T\Delta\mathbf{b}\mathrm{d}\Omega + \iint\limits_{\Gamma_e} \mathbf{N}^T\Delta\bar{\mathbf{t}}\mathrm{d}\Gamma + \mathbf{N}^{\mathrm{T}}\Delta\mathbf{T} \tag{4.251}$$

in which $\Delta\mathbf{b}$, $\Delta\bar{\mathbf{t}}$ and $\Delta\mathbf{T}$ are the nodal force incremental vectors corresponding to volumetric load, surface load, and concentrated load within the element or on the element edges.

Denoting

$$\delta\Delta\varepsilon_e = \mathbf{B}\delta\Delta\hat{\mathbf{u}}_e \tag{4.252}$$

and use is made of Eqs. (4.247), (4.250) may be specified as

$$(\delta\Delta\hat{\mathbf{u}}_e)^{\mathrm{T}}\Delta\mathbf{f}_e = \iiint\limits_{\Omega_e} (\delta\Delta\hat{\mathbf{u}}_e)^{\mathrm{T}}\mathbf{B}^{\mathrm{T}}\mathbf{DB}\Delta\hat{\mathbf{u}}_e \mathrm{d}\Omega \tag{4.253}$$

Since the virtual displacement and strain are arbitrary, the governing (equilibrium) equation of the representative element $e(1, 2, \ldots, n_e)$ is derived

$$\mathbf{k}_e\Delta\hat{\mathbf{u}}_e = \Delta\mathbf{f}_e \tag{4.254}$$

$$\mathbf{k}_e = \iiint\limits_{\Omega_e} \mathbf{B}^{\mathrm{T}}\mathbf{DB}\mathrm{d}\Omega \tag{4.255}$$

where

$$\mathbf{k}_e = \begin{bmatrix} \mathbf{k}_{1,1} & \dots & \mathbf{k}_{1,l} & \dots & \mathbf{k}_{1,k} & \dots & \mathbf{k}_{n,n} \\ \dots & \dots & \dots & \dots & \dots & \dots & \dots \\ \mathbf{k}_{l,1} & \dots & \mathbf{k}_{l,l} & \dots & \mathbf{k}_{l,k} & \dots & \mathbf{k}_{l,n} \\ \dots & \dots & \dots & \dots & \dots & \dots & \dots \\ \mathbf{k}_{k,1} & \dots & \mathbf{k}_{k,l} & \dots & \mathbf{k}_{k,k} & \dots & \mathbf{k}_{k,n} \\ \dots & \dots & \dots & \dots & \dots & \dots & \dots \\ \mathbf{k}_{n,1} & \dots & \mathbf{k}_{n,l} & \dots & \mathbf{k}_{n,k} & \dots & \mathbf{k}_{n,n} \end{bmatrix} \tag{4.256}$$

in which the subscripts l and k are the local node sequences (viz. Fig. 4.22), and

$$\mathbf{k}_{l,k} = \iiint_{\Omega_e} \mathbf{B}_l^{\mathrm{T}} \mathbf{D}_e \mathbf{B}_k \mathrm{d}\Omega \quad (l, k = 1, 2, \dots, n) \tag{4.257}$$

$$\Delta \mathbf{f}_l = \iiint_{\Omega_e} \mathbf{N}_l^{\mathrm{T}} \Delta \mathbf{b} \mathrm{d}\Omega + \iint_{\Gamma_e} \mathbf{N}_l^{\mathrm{T}} \Delta \bar{\mathbf{t}} \mathrm{d}\Gamma + \mathbf{N}_l^{\mathrm{T}} \Delta \mathbf{T} \quad (l = 1, 2, \dots, n) \tag{4.258}$$

For the integrals in Eqs. (4.257) and (4.258), Gaussian quadrature is normally employed with the help of isoparametric element concept (Zienkiewicz et al. 2005).

By looping over each element, the global stiffness matrix $\mathbf{K}$ and force vector $\Delta\mathbf{F}$ of the structure system are assembled in a way similar to the "matrix truss method", and the governing (equilibrium) equation of the whole structure system is expressed by

$$\mathbf{K}\Delta\widehat{\mathbf{U}} = \Delta\mathbf{F} \tag{4.259}$$

in which $\mathbf{K}$ is the global (system) stiffness matrix, $\Delta\mathbf{F}$ is the incremental load of the structure system at the current time step. For the structural domain discretized into a mesh containing n_e elements and n_n node, they are specified as follows:

$$\mathbf{K} = \begin{bmatrix} \mathbf{K}_{1,1} & \dots & \mathbf{K}_{1,i} & \dots & \mathbf{K}_{1,j} & \dots & \mathbf{K}_{1,n_n} \\ \dots & \dots & \dots & \dots & \dots & \dots & \dots \\ \mathbf{K}_{i,1} & \dots & \mathbf{K}_{i,i} & \dots & \mathbf{K}_{i,j} & \dots & \mathbf{K}_{i,n_n} \\ \dots & \dots & \dots & \dots & \dots & \dots & \dots \\ \mathbf{K}_{j,1} & \dots & \mathbf{K}_{j,i} & \dots & \mathbf{K}_{j,j} & \dots & \mathbf{K}_{j,n_n} \\ \dots & \dots & \dots & \dots & \dots & \dots & \dots \\ \mathbf{K}_{n_n,1} & \dots & \mathbf{K}_{n_n,i} & \dots & \mathbf{K}_{n_n,j} & \dots & \mathbf{K}_{n_n,n_n} \end{bmatrix} \quad (i, j = 1, 2, \dots, n_n) \tag{4.260}$$

in which the subscripts i and j are the global node sequences, the 2×2 sub-matrix $\mathbf{K}_{i,i}$ depends on the node i only, whereas the 2×2 sub-matrix $\mathbf{K}_{i,j}$ $(i \neq j)$ depends on the nodes i and j. Similarly, we have

$$\Delta\widehat{\mathbf{U}} = \left[\Delta\hat{\mathbf{u}}_1 \dots \Delta\hat{\mathbf{u}}_i \dots \Delta\hat{\mathbf{u}}_j \dots \Delta\hat{\mathbf{u}}_{n_n}\right]^{\mathrm{T}} \tag{4.261}$$

$$\Delta\mathbf{F} = \left[\Delta\mathbf{F}_1 \ldots \Delta\mathbf{F}_i \ldots \Delta\mathbf{F}_j \ldots \Delta\mathbf{F}_{n_n}\right]^{\mathrm{T}} \tag{4.262}$$

in which $\Delta\hat{\mathbf{u}}_i$ and $\Delta\mathbf{F}_i$ are 2×1 sub-vectors. Suppose there is a link list w.r.t. local node sequence (l, k) and global node sequence (i, j), namely, $i = \langle l \rangle$ and $j = \langle k \rangle$, the assemblage operation is conducted such that

$$\mathbf{k}_{l,k} \to \mathbf{K}_{i,j} \tag{4.263}$$

$$\Delta\mathbf{f}_l \to \Delta\mathbf{F}_i \tag{4.264}$$

The arrow " $\to$ " means the operation of assemble.

After the solution of displacement increment $\Delta\widehat{\mathbf{U}}$, the strain increment $\Delta\varepsilon_e$ and stress increment $\Delta\sigma_e$ for each element may be computed individually using Eqs. (4.243) and (4.244).

3. **Extended formulation for elasto-viscoplasticity**

The constitutive relation of incremental form (Chap. 3) may be implemented in FEM by the initial strain algorithm. For example, on inserting the elasto-viscoplastic constitutive Eq. (3.138) into Eqs. (4.250), (4.259) becomes

$$\mathbf{K}\Delta\widehat{\mathbf{U}} = \Delta\mathbf{F} + \Delta\mathbf{F}^{vp} \tag{4.265}$$

where the equivalent (initial) nodal load increment $\Delta\mathbf{F}^{vp}$ attributable to viscoplastic deformation is assembled by $\Delta\mathbf{f}_e^{vp}$ that is calculated by

$$\Delta\mathbf{f}_e^{vp} = \iiint\limits_{\Omega_e} \mathbf{B}^{\mathrm{T}}\mathbf{D}\Delta\varepsilon_e^{vp}\mathrm{d}\Omega \quad (e = 1, 2, \ldots, n_e) \tag{4.266}$$

$$\Delta\mathbf{f}_e^{vp} = \left[\Delta\mathbf{f}_1^{vp} \ldots \Delta\mathbf{f}_l^{vp} \ldots \Delta\mathbf{f}_n^{vp}\right]^{\mathrm{T}} \tag{4.267}$$

$$\Delta\mathbf{f}_l^{vp} = \iiint\limits_{\Omega_e} \mathbf{B}_l^{\mathrm{T}}\mathbf{D}\Delta\varepsilon_e^{vp}\mathrm{d}\Omega \quad (l = 1, 2, \ldots, n) \tag{4.268}$$

According to the link list $i = \langle l \rangle$ between local node sequence l and global node sequence i, the assemblage operation is conducted such that

$$\Delta\mathbf{f}_l^{vp} \to \Delta\mathbf{F}_i^{vp} \tag{4.269}$$

4. **Extended formulation for dynamics**

Dynamic solution with FEM customarily makes use of weak form in space domain but strong form in time domain. By the virtual work principle and D'Alembert's principle, the discrete equation set of FEM for the dynamic response under the exciting of earthquake is formulated in Eq. (4.167), where the global mass matrix **M** is detailed as

$$\mathbf{M} = \begin{bmatrix} \mathbf{M}_{1,1} & \ldots & \mathbf{M}_{1,i} & \ldots & \mathbf{M}_{1,j} & \ldots & \mathbf{M}_{1,n_n} \\ \ldots & \ldots & \ldots & \ldots & \ldots & \ldots & \ldots \\ \mathbf{M}_{i,1} & \ldots & \mathbf{M}_{i,i} & \ldots & \mathbf{M}_{i,j} & \ldots & \mathbf{M}_{i,n_n} \\ \ldots & \ldots & \ldots & \ldots & \ldots & \ldots & \ldots \\ \mathbf{M}_{j,1} & \ldots & \mathbf{M}_{j,i} & \ldots & \mathbf{M}_{j,j} & \ldots & \mathbf{M}_{j,n_n} \\ \ldots & \ldots & \ldots & \ldots & \ldots & \ldots & \ldots \\ \mathbf{M}_{n_n,1} & \ldots & \mathbf{M}_{n_n,i} & \ldots & \mathbf{M}_{n_n,j} & \ldots & \mathbf{M}_{n_n,n_n} \end{bmatrix} \quad (i, j = 1, 2, \ldots, n_n) \tag{4.270}$$

This global mass matrix is assembled by $\mathbf{m}_e$ of each element

$$\mathbf{m}_e = \iint\limits_{\Omega_e} \mathbf{N}^{\mathrm{T}} \rho \mathbf{N} \mathrm{d}x \mathrm{d}y \quad (e = 1, 2, \ldots, n_e) \tag{4.271}$$

$$\mathbf{m}_e = \begin{bmatrix} \mathbf{m}_{1,1} & \ldots & \mathbf{m}_{1,l} & \ldots & \mathbf{m}_{1,k} & \ldots & \mathbf{m}_{n,n} \\ \ldots & \ldots & \ldots & \ldots & \ldots & \ldots & \ldots \\ \mathbf{m}_{l,1} & \ldots & \mathbf{m}_{l,l} & \ldots & \mathbf{m}_{l,k} & \ldots & \mathbf{m}_{l,n} \\ \ldots & \ldots & \ldots & \ldots & \ldots & \ldots & \ldots \\ \mathbf{m}_{k,1} & \ldots & \mathbf{m}_{k,l} & \ldots & \mathbf{m}_{k,k} & \ldots & \mathbf{m}_{k,n} \\ \ldots & \ldots & \ldots & \ldots & \ldots & \ldots & \ldots \\ \mathbf{m}_{n,1} & \ldots & \mathbf{m}_{n,l} & \ldots & \mathbf{m}_{n,k} & \ldots & \mathbf{m}_{n,n} \end{bmatrix} \tag{4.272}$$

$$\mathbf{m}_{l,k} = \iint\limits_{\Omega_{me}} \mathbf{N}_l^{\mathrm{T}} \rho \mathbf{N}_k \mathrm{d}x \mathrm{d}y \quad (l, k = 1, 2, \ldots, n) \tag{4.273}$$

According to the link lists $i = \langle l \rangle$ and $j = \langle k \rangle$ w.r.t. local node sequences l and k and global node sequences i and j, the assemblage operation is conducted such that

$$\mathbf{m}_{l,k} \to \mathbf{M}_{i,j} \tag{4.274}$$

5. **Solution of matrix equation**

The solution of Eqs. (4.175), (4.182), (4.259), (4.265) relies on the algorithms for matrix equation

$$[A]\{x\} = \{b\} \tag{4.275}$$

or

$$\mathbf{Ax} = \mathbf{b} \tag{4.276}$$

in which $\mathbf{x}$ and $\mathbf{b}$ are N-dimensional vectors representing the forces and unknown variables (for 2-D displacement problem, $N = 2 \times n_n$), $\mathbf{A}$ is an $N \times N$ non-singular, sparse, and positive definite matrix representing the discrete form of differential operator.

Solution algorithms to obtain $\mathbf{x}$ fall into two basic classes: direct methods and iterative methods. Readers are referred to the books of Golub and Loan (1996) and Saad (2003) in which these algorithms are discussed in more detail.

(1) Direct methods

Direct methods demand the calculation of $\mathbf{A}^{-1}\mathbf{b}$ directly. The first consideration is the possibility to assemble matrix $\mathbf{A}^{-1}$ and perform the foregoing matrix–vector product explicitly. Nevertheless, this is seldom ever advocated. Efficient direct solution methods factorize the original matrix $\mathbf{A}$ into several component matrices, for which the operation of inverse applied to a vector is trivial. For example, the LU decomposition factorizes $\mathbf{A}$ in a form of $\mathbf{A} = \mathbf{LU}$, where $\mathbf{L}$ and $\mathbf{U}$ are lower and upper triangular matrices of same dimension. Following such a factorization, the solution may be obtained via first performing $\mathbf{y} = \mathbf{L}^{-1}\mathbf{b}$ (forward), and then computing $\mathbf{x} = \mathbf{U}^{-1}\mathbf{y}$ (backward).

Modern prevalent factorization techniques can further exploit the sparsity in FE operator $\mathbf{A}$. To perform the factorization, sparse direct methods require $O(N^{3/2})$ floating point operations in 2-D problems, and $O(N^2)$ operations in 3-D ones. In addition, the forward and backward procedures demand $O(N \log N)$ and $O(N^{4/3})$ floating point operations in 2- and 3-dimensional problems, respectively (Li and Widlund 2006).

The memory usage and CPU time for the solution using factorization techniques can become prohibitively expensive when the number of unknown variables becomes large. In addition, high-resolution is not always guaranteed, even with the usage of double-precision and super computer.

(2) Iterative methods

Iterative methods pursue updates to the solution via a sequence of iterative operations. A simple iterative method (e.g. Richardson's method) is constructed below to explain their features.

① Set the iteration counter $k = 0$ and choose the initial guess for the solution $\mathbf{x}^0 = 0$.
② Update the solution

$$\mathbf{x}^{k+1} = \mathbf{x}^k + (\mathbf{b} - \mathbf{A}\mathbf{x}^k) \tag{4.277}$$

③ Update the counter $k = k + 1$ then return to step ②.

At each iteration step, a collapse condition is prescribed to terminate the iterative sequence when $\mathbf{x}^k$ is sufficiently accurate. Typically, it may be related to the residual in a manner of

$$\mathbf{r} = \mathbf{b} - \mathbf{A}\mathbf{x}^k \tag{4.278}$$

Iterative methods exhibit an attractive property that only matrix–vector and vector-vector product operations are employed to obtain the solution. Thus in terms of memory usage, they could be much cheaper compared to direct methods.

Nowadays, a large number of robust iterative methods are available. We are able to choose any appropriate one according to the properties of matrix, e.g. symmetric or non-symmetric, positive definite or non-positive definite, etc. For an implementation "catalog" of different iterative methods, the work by Barrett and his co-workers is recommended (Barrett et al. 1994).

In all cases, the number of iterations needed to reach a convergent solution could be rather large, and will generally increase as the resolution of FE mesh is refined. To accelerate the convergence of iterative procedure, use is prevalently made of pre-conditioners. In the example of aforementioned Richardson's method, the iterative equation (4.277) is reformed by a pre-conditioner matrix **B**:

$$\mathbf{x}^{k+1} = \mathbf{x}^k + \mathbf{B}^{-1}(\mathbf{b} - \mathbf{A}\mathbf{x}^k) \tag{4.279}$$

Intuitively, it may be inferred that **B** should be a close approximation to **A**, and the cost (memory usage and CPU time) should be much cheaper to undertake the action of its inverse operation than that of **A**.

An optimal pre-conditioner is one by which the number of iterations to reach convergence is independent of mesh resolution. The pre-conditioners which possess this property are called multi-level pre-conditioners. An excellent coverage of the theory and implementation concerning multi-level pre-conditioners can be found in the works by Wesseling (1992), Briggs et al. (2000), Trottenberg et al. (2001), and Falgout (2006).

A multi-level pre-conditioner tries to construct a hierarchical representation of **A** containing multi (two or more)-levels. At the top of this hierarchy is the operator **A**, and at each subsequent level we have a "coarser" representation of the operator from higher level. The under-lying principle of such multi-level algorithms is that we use solutions from the coarser level to accelerate the convergence at the finer levels. Information concerning the solution is passed between different levels in the hierarchy via interpolant (fine → coarse) and prolongation (coarse → fine) operators.

(3) SSOR-PCG method for general purpose

The method of "symmetric successive over-relaxation pre-conditioned conjugate gradient" (SSOR-PCG) (Lin 1997), as a well preformed iterative solver for Eq. (4.276), is presented as follows.

① Let R^N represent the real N-dimensional vector space, and suppose $\mathbf{M} = \left(\mathbf{S}^{\mathrm{T}}\mathbf{S}\right)^{-1}$ is a symmetric and positive definite matrix, Eq. (4.276) is transformed into the algorithm of "pre-conditioned conjugate gradient (PCG)"

$$\begin{cases} \mathbf{A}'\mathbf{x}' = \mathbf{b}' \\ \mathbf{A}' = \mathbf{SAS}^{\mathrm{T}} \\ \mathbf{b}' = \mathbf{Sb} \\ \mathbf{x}' = \mathbf{S}^{-\mathrm{T}}\mathbf{x} \end{cases} \tag{4.280}$$

② Towards the iterative procedure, we set the initial values

$$\begin{cases} \mathbf{x}^0 \\ \mathbf{g}^0 = \mathbf{Ax}^0 \\ \mathbf{h}^0 = \mathbf{M}^{-1}\mathbf{g}^0 \quad R : \delta = \left(\mathbf{g}^k, \mathbf{h}^k\right) \\ \mathbf{d}_0 = -\mathbf{h}^0 \\ k = 0 \end{cases} \tag{4.281}$$

③ If $\delta \le \varepsilon$, the iteration collapses, otherwise

$$\begin{cases} \tau_k = \dfrac{\left(\mathbf{g}^k, \mathbf{h}^k\right)}{\left(\mathbf{d}^k, \mathbf{Ad}^k\right)} \\ \mathbf{x}^{k+1} = \mathbf{x}^k + \tau_k\mathbf{d}^k \\ \mathbf{g}^{k+1} = \mathbf{g}^k + \tau_k\mathbf{Ad}^k \\ \mathbf{h}^{k+1} = \mathbf{M}^{-1}\mathbf{g}^{k+1} \\ \beta_k = \dfrac{\left(\mathbf{g}^{k+1}, \mathbf{h}^{k+1}\right)}{\left(\mathbf{g}^k, \mathbf{h}^k\right)} \\ \mathbf{d}^{k+1} = -\mathbf{h}^{k+1} + \beta_k\mathbf{d}^k \\ k = k + 1 \end{cases} \tag{4.282}$$

The convergent rate of equivalent Eq. (4.280) is dependent on $\mathrm{Cond}\left(\mathbf{A}'\right)$—the condition number of matrix $\mathbf{A}'$ that will be addressed later in this chapter. If $\mathbf{M}$ is a unit matrix, then Eq. (4.280) becomes the "conjugate gradient" (CG) iterative scheme with convergent rate depending on $\mathrm{Cond}(\mathbf{A})$. Normally, $\mathrm{Cond}(\mathbf{A})$ is rather large and we may expect that $\mathrm{Cond}\left(\mathbf{A}'\right) \ll \mathrm{Cond}(\mathbf{A})$. The matrix $\mathbf{M}$ so selected is termed as the "pre-conditioner" of $\mathbf{A}$. A well-performed pre-conditioner should meet the following requirements:

- $\mathrm{Cond}\left(\mathbf{A}'\right) \ll \mathrm{Cond}(\mathbf{A})$;
- Compared with $\mathbf{A}$, it does not demand too large additional memory;
- It is much easier to solve $\mathbf{Mh} = \mathbf{g}$ than to solve $\mathbf{Ax} = \mathbf{b}$.

Where the decomposition matrices of the "symmetric successive over-relaxation (SSOR) method is used to construct $\mathbf{M}$ in a manner of

$$\mathbf{M} = (2 - \Theta)^{-1}(\mathbf{D}/\Theta + \mathbf{L})(\mathbf{D}/\Theta)^{-1}(\mathbf{D}/\Theta + \mathbf{L})^{\mathrm{T}} \tag{4.283}$$

the PCG algorithm is termed as the "SSOR-PCG method". In Eq. (4.283), **D** is the diagonal matrix of **A**, **L** is the lower triangular matrix of **A**, i.e. $\mathbf{A} = \mathbf{D} + \mathbf{L} + \mathbf{L}^{\mathrm{T}}$, $0 < \Theta < 2$ is the relaxation factor.

The selection of appropriate relaxation factor Θ is still problematic because there is no theoretical support for complex engineering problems. As a result, empirical values based on computation experiences are widely exercised. For caution, Θ should be very close to 1.0, and $\Theta = 1.0$–1.1 is commonly advisable.

4.8 Concluding Remarks

This chapter provides basic knowledge for Chaps. 5–11 concerning advanced CMs. It is worthwhile to emphasize that stability is always at the core in the study and application of CMs in addition to precision.

The stability issue may be put forward at different levels. At the PDE level, it is said to be stable if for every set of initial data and boundary conditions, the solution remains bounded as the time approaches infinity. For a stable PDE however, another level of numerical stability and consistency emerges: the equation set arise from the discretization running on computer might be no longer stable, namely, small changes in initial data on the computer, or small rounding-off errors, would produce very large changes in the computed outcomes. Numerical instability is actually derived from the combined characteristics of PDEs and discretization schemes, the former is related to the "stiff of PDEs" whereas the latter is the "numerical stability" directly related to the question of "what is a good algorithm?" (Gear 1971, 1981; Hairer and Wanner 1996).

4.8.1 *Stiff Equation and Symplectic Method*

Instability typically manifests itself as an oscillatory error in the computed solution that blows up exponentially as more time-marching steps are taken. This is intrinsically due to the phenomenon of equation stiffness. Although it is not possible to provide an exact definition for the stiffness of PDEs, yet several qualitative statements can be (and mostly have been) made in an attempt to encapsulate the notion of stiffness.

In mathematics, a stiff equation is a differential equation for which certain numerical methods are numerically unstable, unless the time-marching step length is extremely small. Alternatively, the definition for stiff PDEs with any initial conditions is that the numerical method with a finite region of absolute stability enforces us to take an excessively small step length to ensure the smoothness of solution (Lambert 1992).

The term "stiff" is used because such equations correspond to tight coupling between the driver and driven in servomechanisms, although the origins of stiffness are various (Burden and Faires 2011). Transient solutions with increasing oscillation occur in a wide variety of mechanical problems, including the study of differential equations w.r.t. the motion of spring-mass system with large spring constants. These often emerge, for example, where the penalty method for restraint condition and contact adjustment are employed in CMs.

In handling stiff equation, a discretization scheme with less "excessive linearity" will lead to better stable algorithm, although it does not always guarantee the precision.

The consideration of qualitative (geometric) properties of ODEs and PDEs is important when designing time-marching step schemes with better stable performance. Traditionally, we tend to prefer the black box approach that leads to excellent results. In many situations it is able to avoid looking too far into the underlying structure of problems. However, in many modern examples such as the preservation of bounded solutions and periodic orbits for satellite and planet, the geometric integration with symplectic method has shown that: a consideration of the underlying structure of a problem may open up the new possibility to discretize the time domain such that the system structure is not totally destroyed. Such a consideration has led to a number of robust discretization schemes that are both more accurate and stable attributable to the conservation of energy and other symplectic invariables. This is understandable if we recall the space discretization using either FDM or FEM, by taking into account the space structure into account in the deduction of algebra equation set for all the nodal variables, oscillation may be alleviated or even totally avoided.

However, geometric integration with symplectic method is still a relatively new subject and much work have yet to be done. In this chapter, we just show how to examine whether an existed finite differential algorithm belongs to the symplectic method of geometric integration (viz. Eqs. (4.159)–(4.163)).

4.8.2 Error Analysis and Condition Number

In the field of numerical analysis, the "condition number" of a function measures how much the output value of the function will change for a small disturbance in the input data. In other words, condition number is used to measure how much error in the output from an error in the input. Hence, it may give us a good index w.r.t. numerical stability, particularly in the discretization and approximation of PDEs using matrix equation for nodal variables. Although condition numbers are frequently applied to the question in linear algebra, yet they can be defined for non-linear functions in several variables. A typical example is given by the linear regression problem, in which the condition number of moment matrix can be used as a diagnostic for multi-collinearity (Belsley et al. 1980; Pesaran 2015).

A problem with a low condition number is said to be well-conditioned, otherwise it is said to be ill-conditioned. For latter, the correct solution/answer to the equation becomes hard to find. Given a function $f(\mathbf{x})$ and an approximation $\tilde{f}(\mathbf{x})$ with input $\mathbf{x}$, the absolute error is defined by a norm of their difference

$$\|\delta f\| = \left\| f(\mathbf{x}) - \tilde{f}(\mathbf{x}) \right\| \tag{4.284}$$

and the relative error is calculated as follows

$$\left\| f(\mathbf{x}) - \tilde{f}(\mathbf{x}) \right\| / \| f(\mathbf{x}) \| \tag{4.285}$$

where $\|.\|$ is the norm defined in Eqs. (4.35), (4.36), and Euclidean-norm $\|.\|_2$ is customarily employed. In this context, the absolute condition number of the function f is defined as

$$\mathrm{Cond}(f) = \lim_{\varepsilon \to 0} \left(\sup_{\|\delta \mathbf{x}\| \le \varepsilon} \left(\frac{\|\delta f(\mathbf{x})\|}{\|\delta \mathbf{x}\|} \right) \right) \tag{4.286}$$

and the relative condition number is further defined as

$$\mathrm{Cond}(f) = \lim_{\varepsilon \to 0} \left(\sup_{\|\delta \mathbf{x}\| \le \varepsilon} \left(\frac{\|\delta f(\mathbf{x})\| / \| f(\mathbf{x})\|}{\|\delta \mathbf{x}\| / \|\mathbf{x}\|} \right) \right) \tag{4.287}$$

For a differentiable function $f(x)$ depending on scalar x, Eq. (4.287) is identical to

$$\mathrm{Cond}(f) = \left| \frac{x f'(x)}{f(x)} \right| \tag{4.288}$$

To clarify the definition of above formulae, suppose there is a small disturbance Δx in x, the relative change in x is $[(x + \Delta x) - x]/x = \Delta x / x$, while the relative change in $f(x)$ is $[f(x + \Delta x) - f(x)]/f(x)$. On taking their ratio it yields

$$\begin{aligned} \frac{[f(x + \Delta x) - f(x)]/f(x)}{\Delta x / x} &= \frac{x}{f(x)} \frac{f(x + \Delta x) - f(x)}{(x + \Delta x) - x} \\ &= \frac{x}{f(x)} \frac{f(x + \Delta x) - f(x)}{\Delta x} \approx \frac{x f'(x)}{f(x)} \end{aligned} \tag{4.289}$$

The prevalent application in CMs is the condition number associated with a linear equation (4.276). Note that the condition number gives a bound on how inaccurate the solution $\mathbf{x}$ will be after approximation, and is a property of the matrix $\mathbf{A}$ only, not the algorithm or floating-point accuracy.

Let $\mathbf{e}$ be the error in $\mathbf{b}$ (viz. Eq. (4.276)). Assuming that $\mathbf{A}$ is a nonsingular matrix, the error in the solution $\mathbf{A}^{-1}\mathbf{b}$ is $\mathbf{A}^{-1}\mathbf{e}$. The ratio of the relative errors is calculated by

$$\left(\frac{\|\mathbf{A}^{-1}\mathbf{e}\|}{\|\mathbf{A}^{-1}\mathbf{b}\|} \right) / \left(\frac{\|\mathbf{e}\|}{\|\mathbf{b}\|} \right) = \frac{\|\mathbf{A}^{-1}\mathbf{e}\|}{\|\mathbf{e}\|} \frac{\|\mathbf{b}\|}{\|\mathbf{A}^{-1}\mathbf{b}\|} \tag{4.290}$$

The maximum value for nonzero $\mathbf{b}$ and $\mathbf{e}$ is further defined as

$$\text{Cond}(\mathbf{A}) = \max_{\mathbf{e},\mathbf{b}\neq 0}\left(\frac{\|\mathbf{A}^{-1}\mathbf{e}\|}{\|\mathbf{e}\|}\frac{\|\mathbf{b}\|}{\|\mathbf{A}^{-1}\mathbf{b}\|}\right)$$
$$= \max_{\mathbf{e}\neq 0}\left(\frac{\|\mathbf{A}^{-1}\mathbf{e}\|}{\|\mathbf{e}\|}\right)\max_{\mathbf{x}\neq 0}\left(\frac{\|\mathbf{A}\mathbf{x}\|}{\|\mathbf{x}\|}\right) = \|\mathbf{A}^{-1}\|\,\|\mathbf{A}\| \tag{4.291}$$

It is clear that

$$\text{Cond}(\mathbf{A}) = \|\mathbf{A}^{-1}\|\,\|\mathbf{A}\| \geq \|\mathbf{A}^{-1}\mathbf{A}\| = 1 \tag{4.292}$$

If $\|.\|$ is the matrix norm induced by the standard Euclidean space norm of vectors (spectral norm) that is defined in the square-summable sequence space L_2 (viz. Eqs. (4.35), (4.36)) and denoted as $\|.\|_2$, then

$$\text{Cond}(\mathbf{A}) = \frac{|\lambda_{\max}(\mathbf{A})|}{|\lambda_{\min}(\mathbf{A})|} \tag{4.293}$$

where $\lambda_{\max}(\mathbf{A})$ and $\lambda_{\min}(\mathbf{A})$ are maximal and minimal (by moduli) eigenvalues of $\mathbf{A}$, respectively.

The condition number Cond($\mathbf{A}$) is a way of describing how well or badly the algebraic equation system (4.276) could be approximated. If $\mathbf{A}$ is a scalar multiple of a linear isometry, Cond($\mathbf{A}$) is exactly unity, then it is able to find an approximated solution whose precision is not worse than that of the input data, although it does not mean that the algorithm will converge rapidly. If Cond($\mathbf{A}$) is infinite, it implies that the problem is ill-posed, and no algorithm can be expected to reliably find a solution. If Cond($\mathbf{A}$) is not too larger than unity, $\mathbf{A}$ is well-conditioned, which means that its inverse can be get with good accuracy. Otherwise $\mathbf{A}$ is said to be ill-conditioned, such a matrix is almost singular, and its inverse computation is prone to large numerical errors.

The first use of condition number is credited to Turing (1948). By a number of successive contributions (Todd 1950; Wilkinson 1963), nowadays we know that Cond($\mathbf{A}$) can be thought of both as a measure of the sensitivity of a linear system solution to the perturbation in the input data and as a measure of the sensitivity of a matrix inverse to the perturbation in the matrix itself (Higham 1995).

In most CMs such as FEM, matrices governing the discretized system are, in general, not well-conditioned. This is blamed for a number of factors such as the large variation in element sizes due to the existence of various structural planes, the large artificial stiffness coefficients of contact faces, etc. Under such circumstances, good convergence is more possibly obtained by the iterative solver with an efficient preconditioner (Ashby et al. 1989; Axelsson 1994; Saint-Georges et al. 1996).

In the presence of linear constraints where dependent unknowns are eliminated directly or by Lagrange multipliers, the efficiency of iterative solver deteriorates meanwhile a direct solver is not affected, the sub-space projection method (Saint-Georges et al. 1998) is advisable, its origin goes back to the works in optimization methods (Fletcher and Reeves 1964; Goldfarb 1969).

These issues are however, needed further explorations for the advances CMs elaborated in this book towards their robust algorithms and standardized codes that may be convincingly accepted by practitioners.

References

Abramowitz M, Stegun IA, editors. Handbook of mathematical functions with formulas, graphs, and mathematical tables. Washington DC (USA): National Bureau of Standards; 1964.

Achieser NI. Lectures on the theory of approximation. Moscow (USSR): Gostekhizdat; 1947. (in Russian).

Aldrich J. Doing least squares: perspectives from Gauss and Yule. Int Stat Rev. 1998;66(1):61–81.

Alexa M, Behr J, Cohhen-Or D, Fleishman S, Levin D, Silva CT. Computing and rendering point set surfaces. IEEE Trans Visual Comput Graph. 2003;9(1):3–15.

Ashby SF, Manteuffel TA, Saylor PE. Adaptive polynomial preconditioning for Hermitian indefinite linear systems. BIT. 1989;29:583–609.

Atwood G. A disquisition on the stability of ships. Philos Trans Royal Soc. 1798;88(1):127–260.

Axelsson O. Iterative solution methods. Cambridge (UK): Cambridge University Press; 1994.

Babuška I, Melenk JM. The partition of unity method. Int J Numer Meth Eng. 1997;40(4):727–58.

Babuška I, Szabó BA, Katz IN. The p-version of the finite element method. SIAM J Numer Anal. 1981;18(3):515–45.

Barrett R, Berry M, Chan TF, Demmel J, Donato J, Dongarra J, Eijkhout V, Pozo R, Romine C, van der Vorst H. Templates for the solution of linear systems: building blocks for iterative methods. Philadelphia (USA): SIAM; 1994.

Bathe KJ. Finite element procedures in engineering analysis. New Jersey (USA): Prentice-Hall; 1982.

Bathe KJ, Wilson EL. Stability and accuracy analysis of direct integration methods. Earthquake Eng Struct Dyn. 1972;1(3):283–91.

Bauer EL. La méthode d'intégration numérique de Romberg. In: Colloque sur l'analyse numérique. Louvain (Belgium): Librairie Universitaire; 1961. p. 119–129 (in French).

Belsley DA, Kuh E, Welsch RE. The condition number. In: Regression diagnostics: identifying influential data and sources of collinearity. New York (USA): Wiley; 1980. p. 100–104.

Belytschko T, Lu YY, Gu L. Element-free Galerkin methods. Int J Numer Meth Eng. 1994;37(2):229–56.

Bernstein SN. Quelques remarques sur l'interpolation. Math Ann. 1918;79:1–12 (in French).

Bernstein SN. Sur la valeur asymptotique de la meilleure approximation des fonctions analytiques. C R Acad Sci Paris. 1912;155:1062–5 (in French).

Bezier P. Numerical control: mathematics and applications. New York (USA): Wiley; 1972.

Birkhoff GD. General mean value and remainder theorems with applications to mechanical differentiation and quadrature. Trans Amer Math Soc. 1906;7:107–36.

Borel E. Sur l'interpolation. C R Acad Sci Paris. 1897;104:673–6 (In French).

Borel E. Leçons sur les Fonctions de Variables Réelles. Paris (French): Gauthier-Villars; 1905. (in French).

Braess D. Nonlinear approximation theory. Berlin (Germany): Springer; 1986.

Brezinski C. Biorthogonality and its applications to numerical analysis. New York (USA): CRC Press; 1992.

Brezinski C, Sadok H. Lanczos-type algorithms for solving systems of linear equations. Appl Numer Math. 1993;11(6):443–73.

Brezinski C, Zaglia MR. Extrapolation methods. theory and practice. Amsterdam (Netherlands): Springer; 1991.

Briggs WL, Henson VE, McCormick SF. A multigrid tutorial. 2nd ed. Philadelphia (USA): Society for Industrial and Applied Mathematics; 2000.
Budd CJ, Iserles A. Geometric integration: numerical solution of differential equations on manifolds. Philos Trans Royal Soc A Math Phys Eng Sci. 1999;357(1754):945–56.
Budd CJ, Piggott MD. The geometric integration of scale invariant ordinary and partial differential equations. J Comput Appl Math. 2001;128(1–2):399–422.
Burden RL, Faires JD. Numerical analysis. 9th ed. Boston, MA (USA): Brooks/Cole; 2011.
Butcher JC. A modified multistep method for the numerical integration of ordinary differential equations. J Assoc Comput Mach. 1965;12(1):124–35.
Butcher JC. The numerical analysis of ordinary differential equations: Runge-Kutta and general linear methods. New York, NY (USA): Wiley Interscience; 1987.
Butcher JC. General linear methods. Comput Math Appl. 1996;31(4–5):105–12.
Carpenter MH, Gottlieb D, Abarbanel S, Don WS. The theoretical accuracy of Runge-Kutta time discretizations for the initial boundary value problem: a study of the boundary error. SIAM J Sci Comput. 1995;16(6):1241–52.
Cauchy AL. Sur les fonctions interpolaires. C R Acad Sci Paris. 1840;11:775–89 (in French).
Caughey TH, Kelly MEJ. Classical normal modes in damped linear dynamic systems. J Appl Mech ASME. 1965;32(3):583–8.
Chebyshev PL. Sur les questions de minima qui se rattachent à la représentation approximative des fonctions. Mm Acad Imp Sci St Ptersbourg, Ser 6, Sci Math Phys. 1854;7:199–291 (in French).
Chen SH. Computational geomechanics and hydraulic structures. Berlin (Germany): Springer; 2018.
Cheney EW. Introduction to approximation theory. New York (USA): McGraw-Hill; 1966.
Cheng Z, Chen SH. Three dimensional hierarchical finite element method for hydraulic structure. J Hydraulic Eng. 1999;21(12):53–8 (in Chinese with English abstract).
Chopra AK, Chakrabarti P. Earthquake analysis of concrete gravity dams including dam-water-foundation rock interaction. Earthquake Eng Struct Dyn. 1981;9(4):363–83.
Christoffel EB. Über die Gaußische Quadratur und eine Verallgemeinerung derselben. Journal Die Reine Und Angewandte Mathematik. 1858;55:61–82 (in German).
Christoffel EB. Sur une classe particulibre de fonctions entières et de fractions continues. Ann Mat Pur Appl. 1877;8(2):1–10 (in French).
Ciarlet PG, Lions JL, editors. Handbook of numerical analysis. Amsterdam (Netherlands): Elsevier Science, North Holland; 2000.
Ciarlet PG, Raviart PA. General Lagrange and Hermite interpolation in R^n with applications to finite element methods. Arch Rat Mech Anal. 1972;46(3):177–99.
Ciarlet PG, Wagschal C. Multipoint Taylor formulas and applications to the finite element method. Numer Math. 1971;17(1):84–100.
Clough RW, Penzien J. Dynamics of structures. 3rd ed. Berkeley (USA): Computers & Structures Inc.; 2003.
Coatmélec C. Approximation et interpolation des fonctions différentiables de plusieurs variables. Ann Scient Ecole Norm Sup. 1966;83(3):271–341 (in French).
Collatz L. Approximation von Funktionen bei einer und bei mehreren unabhingigen Vernderlichen. Zamm J Math Mech. 1956;36(5–6):198–211 (in German).
Courant R, Friedrichs KO, Lewy H. Ober die partiellen Differenzengleichungen der mathematischen Physik. Math Ann. 1928;100(2):32–74 (in German).
Cox MG. The numerical evaluation of B-splines. IMA J Appl Math. 1972;10(2):134–49.
Crank J, Nicolson P. A practical method for numerical integration of solution of partial differential equations of heat-conduction type. Proc Cambridge Philos Soc. 1947;43:50–67.
Cuyt A. Rational Hermite interpolation in one and more variables. In: Singh SP, editor. Approximation theory, spline functions and applications. Dordrecht (Netherlands): Kluwer Academic; 1992. p. 69–103.
Cuyt A, Wuytack L. Nonlinear methods in numerical analysis. Amsterdam (Netherlands): Elsevier Science; 1987.
Dahlquist G. A special stability problem for linear multistep methods. BIT. 1963;3(1):27–43.

Darboux G. Mémoire sur l'approximation des fonctions de trés grands nombres, et sur une classe étendue de développements en série. Journal De Mathématiques Pures Et Appliquées. 1878;4:5–57 (in French).

Davis PJ. Errors of numerical approximation for analytic functions. J Rat Mech Anal. 1953;2:303–13.

Davis PJ. Interpolation and approximation. Waltham (USA): Blaisdell; 1963.

De S, Bathe KJ. The method of finite spheres with improved numerical integration. Comp Struct. 2001;79(22–25):2183–96.

De Boor C. On calculating with B-splines. J Approx Theor. 1972;6(1):50–62.

De Vogelaere R. Methods of integration which preserve the contact transformation property of Hamiltonian equations. Tech Report (No. 4). Indiana (USA): Dept Math, Univ Notre Dame; 1956.

Dennis JE. Nonlinear least squares and equations. In: Jacobs DAH, editor. The state of the art in numerical analysis. New York (USA): Academic; 1977. p. 269–312.

Duarte A. A review of some meshless methods to solve partial differential equations. Austin, TX (USA): TICAM, The University of Texas at Austin; 1995. TICAM Report 95-06.

Duarte CA, Oden JT. Hp clouds—a meshless method to solve boundary-value problems. Texas(USA): Texas Institute for Computational and Applied Mechanics (TICAM), the University of Texas at Austin (USA); 1995. Technical Report 95-05.

Earn DJD, Tremaine S. Exact numerical studies of Hamiltonian maps: iterating without roundoff error. Physica D. 1992;56(1):1–22.

Eckhardt R. Stan Ulam, John von Neumann, and the Monte Carlo method. In: Cooper NG, editor. From cardinals to chaos. Cambridge (UK): Cambridge University Press; 1989. p. 131–7.

Espelid TO, Genz A. Numerical integration. Recent developments, software and applications. Dordrecht (Netherlands): Kluwer; 1992.

Faber G. Uber die interpolatorische Darstellung stetiger Funktionen. Jahresber Dtsch Math-Ver. 1914;23:192–210 (in German).

Falgout RD. An introduction to algebraic multigrid. Comput Sci Eng. 2006;8(6):24–33.

Farin G. Splines in CAD/CAM. Surv Math Ind. 1991;1:39–73.

Feng K, editor. Beijing symposium on differential geometry and differential equation. Beijing (China): Science Press; 1985a.

Feng K. On difference schemes and symplectic geometry. In: Feng K, editor. In: Proceedings of Beijing symposium on differential geometry and differential equation. Beijing (China): Science Press; 1985b. p. 42–58.

Fletcher R, Reeves CM. Function minimization by conjugate gradients. Comput J. 1964;7(2):149–54.

Forest E, Ruth RD. Fourth-order symplectic integration. Physica D. 1990;43(1):105–17.

Franke R. Scattered data interpolation: test of some methods. Math Comput. 1982;38:181–200.

Gasca M. Multivariate polynomial interpolation. In: Dahmen W, Gasca M, Micchelli CA, editors. Computation of curves and surfaces. Dordrecht (Netherlands): Kluwer; 1990. p. 215–36.

Gasca M, Carmona AL. A general recurrence interpolation formula and its applications to multivariate interpolation. J Approx Theor. 1982;34(4):361–74.

Gasca M, Lebrón E. A note on recurrence interpolation formulae for certain sets of points in R^k. In: Collatz L, Meinardus G, Werner H, editors. Numerical methods of approximation theory, vol. 7. Basel (Switzerland): Birkhuser; 1983. p. 77–85.

Gasca M, Lebrón E. On Aitken-Neville formulae for multivariate interpolation. North-Holland Math Stud. 1987;133(1):133–40.

Gasca M, Maetzu JI. On Lagrange and Hermite interpolation problems in R^k. Numer Math. 1982;39(1):1–14.

Gasca M, Ramirez V. Interpolation systems in R^k. J Approx Theor. 1984;42(1):36–51.

Gear CW. Hybrid methods for initial value problems in ordinary differential equations. Soc Ind Appl Math B Numer Anal. 1965;2(1):69–86.

Gear CW. Numerical initial-value problems in ordinary differential equations. Englewood Cliffs (USA): Prentice Hall; 1971.
Gear CW. Numerical solution of ordinary differential equations: is there anything left to do? SIAM Rev. 1981;23(1):10–24.
Gingold RA, Monaghan JJ. Smoothed particle hydrodynamics—theory and application to non-spherical stars. Mon Not R Astron Soc. 1977;181(3):375–89.
Goldfarb D. Extension of Davidon's variable metric method to maximization under linear inequality and equality constraints. SIAM J Appl Math. 1969;17(4):739–64.
Goldstine HH. A history of numerical analysis from the 16th through the 19th century. New York (USA): Springer; 1977.
Golub GH, Loan CFV. Matrix computations. Baltimore (USA): Johns Hopkins University Press; 1996.
Gowing R. Roger cotes, natural philosopher. Cambridge (UK): Cambridge University Press; 1983.
Gragg WBHJS. Generalized multistep predictor-corrector methods. J ACM. 1964;11(2):188–209.
Gregory J. Exercitationes geometricae. London (UK); 1668.
Gautschi W. A survey of Gauss-Christoffel quadrature formulae. In: Butzer PL, Fehire E, editors. Christoffel EB, the influence of his work on mathematics and the physical sciences. Basel (Switzerland): Birkhauser; 1981. p. 72–147.
Guenther RB, Roetman EL. Some observations on interpolation in higher dimensions. Math Comput. 1970;24:517–21.
Hairer E. Variable time step integration with symplectic methods. Appl Numer Math. 1997;25(2–3):219–27.
Hairer E, Lubich C. The life-span of backward error analysis for numerical integrators. Numer Math. 1997;76(4):441–62.
Hairer E, Lubich C. Asymptotic expansions and backward analysis for numerical integrators. In: Llave RDL, Petzold LR, Lorenz J, editors. Dynamics of algorithms, ima volumes in mathematics and its applications 118. New York (USA): Springer; 1999. p. 91–106.
Hairer E, Lubich C, Wanner G. Geometric numerical integration. Springer series in computational mathematics, vol. 31. Heidelberg (Germany): Springer; 2002.
Hairer E, Wanner G. Solving ordinary differential equations II: stiff and differential-algebraic problems. 2nd ed. Berlin (Germany): Springer; 1996.
Hakopian HA. Multivariate divided differences and multivariate interpolation of Lagrange and Hermite type. J Approx Theor. 1982;34(3):286–305.
Hardy RL. Multiquadric equations of topography and other irregular surface. Geophys Res. 1971;76(8):1905–15.
Hart JE, Cheney EW, Lawson CL, Maehly HJ, Mesztenyi CK, Rice JR, Thacher HG Jr, Witzgall C. Handbook of computer approximations. New York (USA): Wiley; 1965.
Hastings C. Approximations for digital computers. Princeton (USA): Princeton University Press; 1955.
Hermite C. Sur la formule d'interpolation de Lagrange. J Reine Angew Math. 1878;84:70–9 (in French).
Higgins JR. Five short stories about the cardinal series. Bull Amer Math Soc. 1985;12(1):45–89.
Higham DJ. Condition numbers and their condition numbers. Linear Algebra Appl. 1995;214:193–213.
Hildebrand FB. Introduction to numerical analysis. 2nd ed. New York (USA): Dover Publications Inc; 1987.
Hofmann JE. Leibniz in Paris, 1672–1676. Cambridge (UK): Cambridge University Press; 1974.
Hornecker G. Evaluation approcheé de la meilleure approximation polynômiale d'ordre n de f(x) sur un segment fini [a, b]. Chiffres. 1958;1:157–69 (in French).
Hornecker G. Approximations rationnelles voisines de la meilleure approximation au sens de Tchebycheff. C R Acad Sci Paris. 1959;249:939–41 (in French).
Hoschek J, Lasser D. Fundamentals of computer aided geometric design. Wellesley, MA (USA): Peters AK; 1993.

ICOLD. Computational procedures for dam engineering—reliability and applicability (Bulletin 122). Paris (France): ICOLD; 2001.

ICOLD. Seismic design and evaluation of structures appurtenant to dams (Bulletin 123). Paris (France): ICOLD; 2002.

Idelsohn SR, Oñate E. To mesh or not to mesh. That is the question ... Comput Meth Appl Mech Eng. 2006;195(37–40):4681–4696.

ITASCA. FLAC—fast lagrangian analysis of continua. User's Guide. Minneapolis (USA): Itasca Consulting Group, Inc; 2002.

Jackiewicz Z. General linear methods for ordinary differential equations. Hoboken, NJ (USA): Wiley; 2009.

Jackson D. Über die Genauigkeit der Annahrung stetiger Funktionen durch ganze rationale Funktionen. PhD Dissertation, Göttingen; 1911.

Jacobi CGJ. Über Gauß neue Methode, die Werthe der Integrale näherungsweise zu finden. J Reine Angew Math. 1826;1:301–8 (in German).

Jacobi CGJ. Über die Darstellung einer Reihe gegebener Werthe durch eine gebrochen rationale Funktion. J Reine Angew Math. 1845;30:127–56 (in German).

Jameson A, Schmidt W, Turkel E. Numerical solutions of the Euler equations by finite volume methods using Runge-Kutta time-stepping schemes. In: AIAA 14th fluid and plasma dynamic conference. Palo Alto, California (USA): American Institute of Aeronautics and asronautics; 1981. p. 1981–1259.

John F. On the integration of parabolic equations by difference methods, I: linear and quasilinear equations for the infinite interval. Comm Pure Appl Math. 1952;5(2):155–211.

Johnson WW. On Cotesian numbers: their history, computation, and values to n = 20. Quart J Pure Appl Math. 1915;46:52–65.

Keast P, Fairweather G. Numerical integration. Dordrecht (Netherlands): Reidel; 1987.

Kythe PK. Fundamental solutions for differential operators and applications. Boston (USA): Birkhauser; 1996.

Laasonen P. Über eine Methode zur Lösung der Wärmeleitungsgleichung. Acta Math. 1949;81:309–17 (in German).

Lagrange JL. Sur une methode particuliere d'approximation et d'interpolation. Nouv Mem Acad Sci Berlin. 1783;38:279–289 (in French).

Lambert JD. Numerical methods for ordinary differential systems. New York (USA): Wiley; 1992.

Lancaster P, Salkauskas K. Surfaces generated by moving least squares methods. Math Comput. 1981;37(155):141–58.

Larsson E, Fornberg B. A numerical study of some radial basis functions based solution methods for elliptic PDEs. Comput Math Appl. 2003;46(5–6):891–902.

Laurent PJ. Un théorème de convergence pour le procédé d'extrapolation de Richardson. C R Acad Sci Paris. 1963;256:1435–7 (in French).

Laurent PJ. Approximation et Optimisation. Paris (France): Hermann; 1972.

Lax PD, Richtmyer RD. Survey of the stability of linear finite difference equations. Comm Pure Appl Math. 1956;9:267–93.

Lebesgue H. Sur l'approximation des fonctions. Bull Sci Math. 1898;22:278–87 (in French).

Legendre AM. Nouvelles méthodes pour la détermination des orbites des comètes (New Methods for the Determination of the Orbits of Comets). Paris (France): Didot F; 1805 (in French).

Lerch M. Sur un point de la théorie des fonctions géneratrices d'Abel. Acta Math. 1903;27:339–52 (in French).

Li J, Widlund O. FETI-DP, BDDC and block Cholesky methods. Int J Numer Meth Eng. 2006;66(2):250–71.

Light WA, Cheney EW. Approximation theory in tensor product spaces, lecture notes in mathematics, vol. 1169. Berlin (Germany): Springer; 1985.

Light WA. Some aspects of radial basis function approximation. In: Singh SP, editor. Approximation theory, spline functions and applications. Dordrecht (Netherlands): Kluwer; 1992. p. 163–90.

Lin SZ. Improved iterative format of symmetric successive over relaxation-preconditioned conjugated radient method. J Numer Meth Comp Appl. 1997;18(4):266–70 (in Chinese with English abstract).

Liszka T, Orkisz J. The finite difference method at arbitrary irregular grids and its application in applied mechanics. Comput Struct. 1980;11(1–2):83–95.

Lorentz GG, Jetter K, Riemenschneider SD. Birkhoff Interpolation. Reading, MA (USA): Addison-Wesley; 1983.

Lucy LB. A numerical approach to the testing of the fission hypothesis. Astron J. 1977;82(12):1013–24.

Markov AA. Sur la méthode de Gauss pour le calcul approché des intégrales. Math Ann. 1885;25(3):427–43 (in French).

Maroni P. Une thorié algebrique des polynômes orthogonaux. Application aux polynômes orthogonaux semi-classiques. In: Brezinski C, Draux A, Magnus AP, Maroni P, Rou-Veaux A, editors. Orthogonal polynomials and their applications. Basel (Switzerland): Baltzer; 1991. p. 95–130 (in French).

Marti J, Cundall P. Mixed discretization procedure for accurate solution of plasticity problems. Int J Numer Anal Meth Geomech. 1982;6(1):129–39.

Maxwell JC. On approximate multiple integration between limits of summation. Proc Cambridge Philos Soc. 1877;3:39–47.

Mazzia A, Pini G. Product Gauss quadrature rules vs cubature rules in the meshless local Petrov-Galerkin method. J Complex. 2010;26(1):82–101.

Mclachlan RI, Quispel GRW. Six lectures on the geometric integration of ODEs. In: De-Vore R, Iserles A, Süli E, editors. Foundations of computational mathematics. London mathematical society lecture note series, vol. 284; 2001. p. 155–210.

Mehler EG. Bemerkungen zur Theorie der mechanischen Quadraturen. J Reine Angew Math. 1864;63:152–7 (in German).

Meinardus G. Approximation of functions: theory and numerical methods. Berlin (Germany): Springer; 1967.

Méray C. Observations sur la légitimité de l'interpolation. Ann École Norm Sup. 1884;2:165–76 (in French).

Méray C. Nouveaux exemples d'interpolation illusoires. Bull Sci Math. 1896;20:266–71 (in French).

Metropolis N. The beginning of the Monte Carlo method. In: Cooper NG, editor. From cardinals to chaos. Cambridge (UK): Cambridge University Press; 1989. p. 125–30.

Metropolis N, Ulam S. The Monte Carlo method. J Amer Stat Assoc. 1949;44(247):335–41.

Mittad-Leffler G. Sur la représentation analytique des fonctions d'une variable réelle. Rend Circ Mat Palermo. 1900;14:217–24 (in French).

Narumi S. Some formulas in the theory of interpolation of many independent variables. Tdhoku Math J. 1920;18:309–21.

Natanson IP. Constructive function theory. New York (USA): Ungar; 1964.

Neville EH. Iterative interpolation. J Indian Math Soc. 1934;20:87–120.

Newman D. Rational approximation to lxl. Michigan Math J. 1964;11:11–4.

Newmark NM. A method of computation for structural dynamics. J Eng Mech Div ASCE. 1959;85(EM3):67–94.

Newton I. Philosophiae Naturalis Principia Mathematica. London (UK); 1687.

O'Brien GG, Hyman MA, Kaplan S. A study of the numerical solution of partial differential equations. J Math Phys. 1951;29(1–4):223–51.

Oden JT. Finite elements: an introduction. In: Ciarlet PG, Lions JL, editors. Handbook of numerical analysis, vol. II. Finite element methods (Part 1). Amsterdam (Netherlands): Elsevier; 1991. p. 3–15.

Padé H. Sur la representation approcbcc d'une fonction par des fractions rationelles. Ann Ec Norm. 1982;9(Suppl):1–92.

Paulay T, Priestley MJN. Seismic design of reinforced concrete and masonry buildings. New York (USA): Wiley; 1992.

Pesaran MH. The multicollinearity problem. In: Time series and panel data econometrics. New York (USA): Oxford University Press; 2015. p. 67–72.

Petrushev PP, Popov VA. Rational approximation of real functions. Cambridge (UK): Cambridge University Press; 1987.

Picard E. Sur la représentation approcheé des functions. C R Acad Sci Paris. 1891;112:183–6 (in French).

Pieg L, Tiller W. The NURBS book. 2nd ed. New York (USA): Springer; 1997.

Popov V. Uniform rational approximation of the class V_r, and its applications. Acta Mathematica Academiae Scientiarum Hungarica. 1977;29(1–2):119–29.

Posse C. Sur les quadratures. Nouvelles Annales De Mathématiques. 1875;14(2):49–62 (in French).

Powell MJD. Radial basis functions for multivariable interpolation: a review. In: Mason JC, Cox MG, editors. Algorithms for approximation. Oxford (UK): Carendon Press; 1987. p. 143–67.

Quade W, Collatz L. Zur Interpolationstheorie der reellen periodischen Funktionen. Proc. Preuss Akad Wiss (Math Phys Kl). 1938;30:383–429 (in German).

Radau R. Etude sur les formules d'approximation qui servent à calculer la valeur numérique d'une intégrale définie. J Math Pures Appl. 1880;6(3):283–336 (in French).

Radon J. Zur mechanischen Kubatur. Monatshefte Für Mathematik. 1948;52(4):286–300 (in German).

Ralston A. A first course in numerical analysis. New York (USA): McGraw-Hill; 1965.

Reich S. Backward error analysis for numerical integrators. SIAM J Numer Anal. 1999;36(5):1549–70.

Richardson LE. The approximate arithmetic solution by finite differences of physical problems involving differential equations, with an application to the stress in a masonry dam. Phil Trans Roy Soc London Ser A. 1910;210:307–357.

Romberg W. Vereinfachte numerische integration. Kgl Norske Vid Seisk Forsk. 1955;28:30–6.

Runge C. Über eine numerische Berechnung der Argumente der cyklischen, hyperbolischen und logarithmischen Funktionen. Acta Math. 1891;15(1):221–47 (in German).

Runge C. Über empirische Funktionen und die Interpolation zwischen äquidistanten Ordinaten. Z Angew Math Phys. 1901;46:224–43 (in German).

Runge C, Willers FA. Numerische und graphische Integration. In: Encyklopiidie der mathematischen Wissenschaften, vol. 2, Sec. 3, Heft 2. Leipzig (Germany): Teubner; 1915. p. 47–17 (in German).

Ruth RD. A canonical integration technique. IEEE Trans Nucl Sci. 1983;30(4):2669–71.

Saad Y. Iterative methods for sparse linear systems. Philadelphia (USA): SIAM; 2003.

Saint-Georges P, Notay Y, Wake G. Efficient iterative solution of constrained finite element analyses. Comput Methods Appl Mech Eng. 1998;160(1–2):101–14.

Saint-Georges P, Warzée G, Notay Y, Beauwens R. High-performance PCG solver for FEM structural analyses. Int J Numer Methods Eng. 1996;39(8):1313–40.

Sanz-Serna JM. Geometric integration. In: Duff IS, Watson GA, editors. The state of the art in numerical analysis. Oxford (UK): Clarendon; 1997. p. 121–43.

Sanz-Serna JM, Calvo MP. Numerical hamiltonian problems. London (UK): Chapman and Hall; 1994.

Schaback R. Creating surfaces from scattered data using radial basis functions. In: Daehlen M, Lyche T, Schumaker LL, editors. Mathematical methods in CAGD III. Boston (USA): Academic; 1995. p. 1–21.

Severn RT. The aseismic design of concrete dams (part one). Int J Water Power Dam Constr. 1978;28(1):37–38 (Part one), 28(2):41–46 (Part two).

Shampine LF. Numerical solution of ordinary differential equations. New York (USA): Chapman & Hall; 1994.

Shephard D. A two-dimensional function for irregularly-spaced data. In: Proceedings of 23rd AMC national conference. New York (USA): AMC; 1968. p. 517–524.

Simpson T. Mathematical dissertations on a variety of physical and analytical subjects. London (UK); 1743.

Sobolev SL. The formulas of mechanical cubature on the surface of a sphere. Sib Math J. 1962;3(5):486–96.
Steffensen JF. Interpolation. New York (USA): Chelsea; 1927.
Stieltjes TJ. Variétés. Notes sur quelques formules pour l'évaluation de certaines intégrales. Bull Astr Paris. 1884;1:568–569 (in French).
Stone MH. The generalized Weierstrass approximation theorem. Math Mag. 1948;21(5):167–184, 237–254.
Strouboulis T, Copps K, Babuška I. The generalized finite element method. Comput Meth Appl Mech Eng. 2001;190(32–33):4081–193.
Suris Y. Hamiltonian methods of Runge-Kutta type and their variational interpretation. Math Model. 1990;2(4):78–87.
Ten DI. Lectures on wavelets. Philadelphia (USA): SIAM; 1992.
Taylor B. Methodus Incrementorum Directa et Inverta. London (UK); 1715 (in French).
Thacher Jr HC, Milne WE. Interpolation in several variables. J SIAM. 1960;8:33–42.
Tikhomirov VM. Approximation theory. In: Gamkrelidze RV, editor. Analysis II. Berlin (Germany): Springer; 1990. p. 93–243.
Todd J. The condition of a certain matrix. Math Proc Cambridge Philos Soc. 1950;46(1):116–8.
Tonti E. Variational formulation of non-linear differential equations. Bull Acad Roy Belg Classe Sci. 1969;55:137–165 and 263–278.
Trottenberg U, Oosterlee CW, Schüller A. Multigrid. Orlando (USA): Academic Press Inc; 2001.
Turing AM. Rounding-off errors in matrix processes. Quart J Mech Appl Math. 1948;1(1):287–308.
Volterra V. Sul principio di Dirichlet. Rendiconti Del Circolo Matematico Di Palermo. 1897;11:83–6 (in Italian).
Vos PEJ, Eskilsson C, Bolis A, Chun S, Kirby RM, Sherwin SJ. A generic framework for time-stepping partial differential equations (PDEs): eneral linear methods, object-oriented implementation and application to fluid problems. Int J Comput Fluid Dyn. 2011;25(3):107–25.
Walsh JL. The existence of rational functions of best approximation. Trans Amer Math Soc. 1931;33:668–89.
Weierstrass K. Über die analytische Darstellbarkeit sogenannter willkürlicher Funktionen einer reellen Veränderlichen. Sitzungsberichte der Königlich Preußischen Akademie der Wissenschaften zu Berlin. 1885;2:633–639 and 789–805 (in German).
Werner H. Remarks on Newton type multivariate interpolation for subsets of grids. Computing. 1980;25(2):181–91.
Wesseling P. An introduction to multigrid methods. Chichester (UK): Wiley; 1992.
Whittaker JM. Interpolatory function theory. Cambridge (UK): Cambridge University Press; 1935.
Wilkinson JH. Rounding errors in algebraic processes. Englewood Cliffs, NJ (USA): Prentice-Hall Inc.; 1963.
Wilson EL. Three dimensional static and dynamic analysis of structures: a physical approach with emphasis on earthquake engineering. 2nd ed. Berkeley, CA (USA): Computers and Structures Inc.; 1998.
Zienkiewicz OC, Taylor RL, Zhu JZ. The finite element method: its basis and fundamentals. 6th ed. Butterworth-Heinemann; 2005.

Chapter 5
General Finite Element Methods with Special Focus on XFEM

Abstract This chapter summarizes the principles related to a wide spectrum of mesh-based methods that belong to the category of "general finite element methods (GFEMs)". They are the most promising advanced CMs in explicitly handling local failure phenomena and meso-scale components (e.g. deformation localization bands, concrete cracks, structural planes, etc.) to give their detailed description using simple "finite element (FE)" mesh. Such a FE mesh may be generated beforehand to discretize the engineering structure concerned, where the deployment and size of FEs are dominated by the structure configuration and the gradient of field functions (e.g. strains and flow rates). The existence and evolution of local failure phenomena and meso-scale components are allocated within enriched elements. The basic variables within enriched element may be interpolated from the correspondent "enriched" nodal variables only (CEM), or from "enriched" nodal variables and basis functions together (XFEM). According to the virtual work or variational principle, the governing equations are established to solve these nodal variables. In this manner, less restraint is imposed on the mesh generation with considerable amount of heterogeneous components, which allows for a great simplification in the pre-process work towards the analysis for complex engineering structures. By the validation example of the 1-D discontinuous bar containing joint, it is clear that the FEM with joint element, the XFEM and the CEM with different enrichments will give identical outcomes. By the engineering example, it is shown that the GFEMs specified in this chapter already possess high ability to handle complex engineering issues, but the enriched sub-models for crack-tips are additionally demanded, which is cumbersome because of the repeated data mapping between the global mesh of arch dam and the local mesh containing crack-tip.

5.1 General

Most rock-like materials are inhomogeneous and embedded with "structural planes" (inclusive faults, joints, etc.) or "fractures" (inclusive cracks). In order to characterize such features in the numerical model, either explicitly or implicitly, various methods have been developed. Normally, fundamental mesh-based methods such as the FDM,

S. Chen, *Advanced Computational Methods and Geomechanics*, Springer Tracts in Civil Engineering, https://doi.org/10.1007/978-981-19-7427-4_5

FEM and BEM are cumbersome due to the need for regular mesh to match the geometry of structural plane/fracture. This becomes a major difficulty in problems with evolving fractures where the mesh must be regenerated at each step.

It is worthwhile to indicate that linked to the nonlinear bifurcation of continuum, localization phenomenon with intense inelastic bands is a common occurrence in frictional materials such as soil (Vardoulakis et al. 1978; Vardoulakis 1980), rock (Waversik, Brace 1971) and concrete (Van Mier 1984), which brings about the similar difficulty in fundamental CMs w.r.t. the inception and evolution of narrow localized bands. The mechanism responsible for the formation of localization bands varies widely from one material to another. Although a full understanding of post-bifurcation phenomenon is to emerge, yet experimental data suggest that localization may result in the overall softening of specimen, even the material remains hardening at all times (Read and Hegemier 1984). An explanation of this nature was first noted by Mandel (1964) and subsequently by Rudnicki and Rice (1975).

For structural problems where their characteristic length defined by the smallest structure dimension is much larger than the "representative element volume (REV)" of material, the structural planes/fractures can be implicitly modeled with a homogenization technique to obtain the equivalent properties of the material at macroscale level. A macroscopic constitutive equation is then constructed to model the material response to a given set of loads and boundary conditions. The major benefit of this equivalent approach is that large scale problems can be solved by fundamental CMs, but a major drawback emerges that it is not able to predict various complicated microscopic effects. For problems where the REV is either larger than or the same order of the characteristic length of engineering structure, the continuum hypothesis is violated and the explicit representation of structural planes/fractures is desirable.

Explicit approach for structural planes/fractures in FEM has been motivated since the late 1960s (Hansbo and Hansbo 2004). Various models dubbed with, such as "joint element" (Goodman et al. 1968; Ghaboussi et al. 1973), "fracture element" (Zienkiewicz et al. 1970), "interface element" (Katona 1983), "thin-layer element" (Desai et al. 1984), etc., have been developed. It is also possible to simulate crack growth after special procedures having been developed and implemented. The use of standard elements is possible after a minor adaptation of element mesh, and the calculation of J-integral is often a routine option. These models were applied to successfully simulate the rock failure dominated by structural plane slip and the concrete rupture due to crack growth, but it was found that the results strongly depend on the element type as well as on the mesh refinement.

5.1.1 Fundamental Finite Element Method

Finite element analysis for crack growth (as well as localization evolution) involves three major steps.

1. **Construction of mesh that matches exactly the crack geometry**

This is often cumbersome, because the geometry of crack is normally independent of the CAD model that describes the computation domain.

For 2-D fracture problems under plane stress or plane strain state, 8-node element (viz. Fig. 5.1) provides very accurate results.

Figure 5.2 shows a portion of FE mesh with 8-node elements containing a localization band or joint (structural plane) or crack segment, in the last case there is a crack-tip that is the singularity point.

For nonlinear materials where the strain localization emerges, the evolution of shear band may be described by the theory established by Oritz et al. (1987).

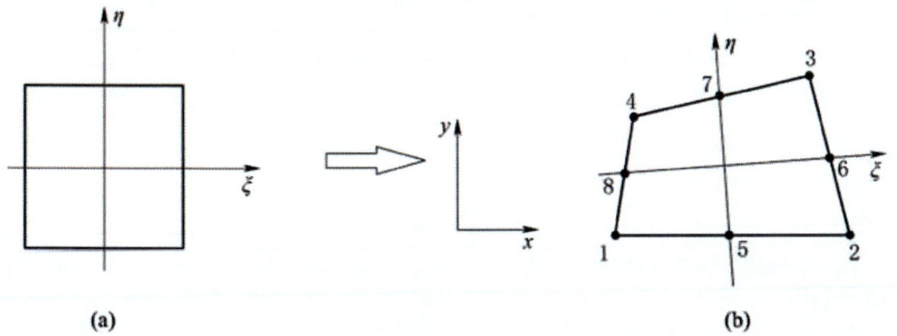

Fig. 5.1 Quadratic isoparametric element with 8 nodes. **a** Parent element; **b** real object element after mapping

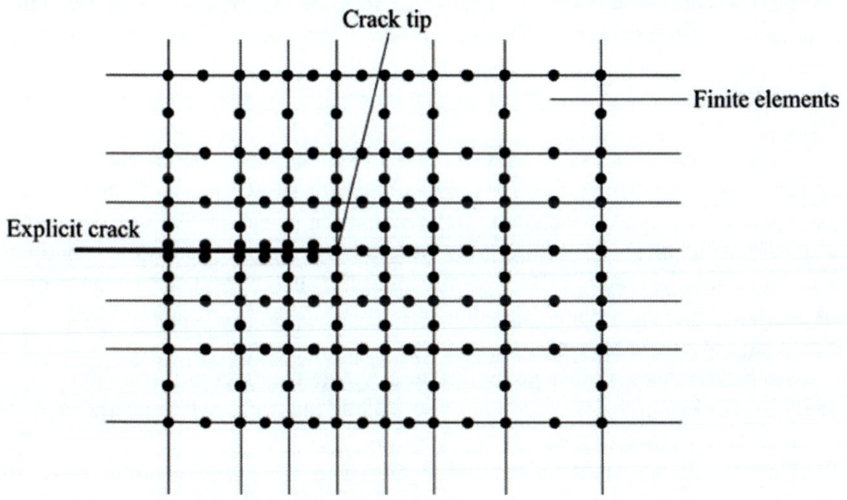

Fig. 5.2 FE mesh containing a crack segment and tip

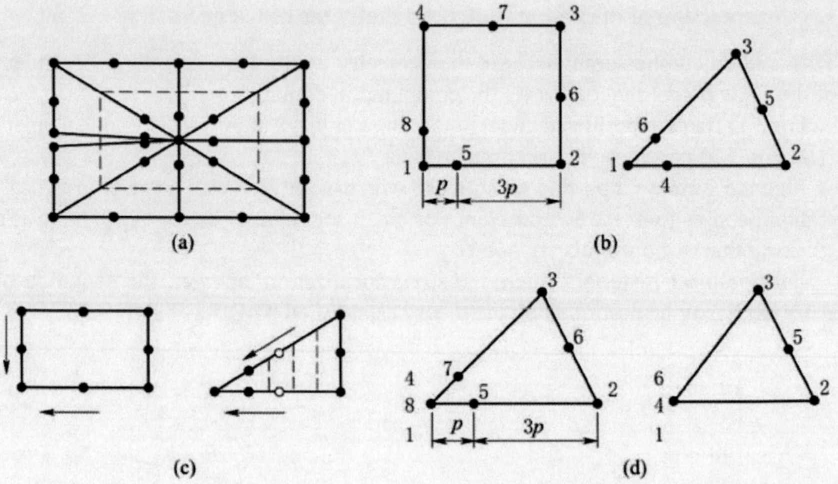

Fig. 5.3 Special element around crack-tip in 2-D FEM (Barsoum 1974, 1976). **a** Mesh around crack-tip; **b** quarter point elements; **c** collapse of elements; **d** collapsed elements

For linearly elastic materials, fracture mechanics says that the stresses at crack-tip are proportional to $1/\sqrt{r}$, with r is the distance to the crack-tip. This crack-tip singularity needs to be appropriately represented (viz. Fig. 5.3a), but, due to the fact that standard FEM is founded on piecewise differentiable polynomial approximations, it performs not very efficiently in the problems with discontinuous and/or singular solutions. We are naturally to reckon that the very steep stress gradient at crack-tip could be better described with a lot of small elements. However, it is found that refining the mesh at crack-tip will not lead to convergence, i.e. further refinement will always render different results. The reason for this unacceptable mesh dependency is that, the singularity in the stress field is not described by the element interpolation (basis) functions.

To circumvent the shortcomings of standard element in the description of crack-tip stress, one can change the interpolation function of element in such a manner that the crack-tip field is included. This results in a so-called "enriched element", which is only applicable at the crack-tip. Another solution is to modify the variational principle that is the base of FEM. The resulting element is hybrid one, where stresses are unknown nodal variables. This demands a revision of fundamental FEM codes, which are not always possible, particularly for some commercial ones.

Fortunately, there is other possibility to calculate the crack-tip stress field accurately by, for example, the adaption of so-called "quarter point elements" derived from standard 8-node elements (viz. Fig. 5.3b).

Quarter point elements are devised by Barsoum in 1976 as a simple means to describe the stress singularity at crack-tip. They are standard 8-node quadrilateral or 6-node triangular elements, where two mid-side points are simultaneously moved towards one corner node, such that they divide the element edge with a ratio of 1:3

(viz. Fig. 5.3b). It can be proved that they describe the $1/\sqrt{r}$-stress field singularity at crack-tip accurately. Quarter point elements, also referred to as "distorted quadratic" elements, can be made very easily at the crack-tip by replacing some mid-side nodes of the element. The remarkable advantage is that they allow the use of standard finite element packages.

Another prevalent solution with standard finite element packages is to use a "collapsed" technique where three nodes are situated at the same location but not connected (viz. Fig. 5.3c, d), and it appears to describe a $1/r$ singularity rather well.

The crack-tip mesh is often made as a crack-tip reset, with quarter point elements at the crack-tip. Lim et al. (1993) advised to use transition elements between quarter point elements and regular midpoint elements for better results.

2. **Stability analysis of shear band and crack**

For strain localization problems, a bifurcation analysis is used to determine the evolution of localized deformation modes, whereas for crack growth problems, the computation of "stress intensity factors (SIFs)" must be carefully undertaken. If one is concentrated on the issue of crack growth only, normally, domain integrals (e.g. J-integral) are found to be more accurate than extrapolations made at the crack-tip subject to the integration domain is sufficiently large. To further raise accuracy of domain integrals, use of special crack-tip elements may be helpful. Among them, Barsoum (1974, 1976) elements are well performed. There are few, if any, modifications of the finite element codes dealing with Barsoum elements.

3. **Update of crack front**

Problems of crack growth often involve two or more well-differentiated length scales: a global-or macro-scale commensurate with the macroscopic features of engineering structure, and a local-or meso-scale in the order of shear band thickness or crack opening. In many cases of interest, the meso-scale steadily fails and the macro-scale steadily propagates in time. Depending on the loading level, this process can be self-limiting or continue at infinitum. Examples of the former are found in solids with an intrinsic characteristic length, such as non-local and polar continua, solids with cohesive interfaces, and others. On the contrast, rate-dependent solids lacking an internal length exhibit shear bands without self-limiting.

As the crack is part of mesh, the update of mesh, as well as the data structure associated with the crack geometry, are demanded. Blamed for the time-dependent location, the shape of crack is generally not known beforehand, hence the solution methods incorporating some ability of adaptive mesh naturally suggest themselves (Ortiz and Quigley 1991).

Insofar, the explicit treatment of fractures (localization bands and cracks) and their growth remains intractable in FEM. The most difficult problem is that the FE mesh needs to coincide with fractures, which often complicates the mesh generator. When simulating the evolution of fractures, re-meshing is inevitable that makes the simulation tedious and time consuming.

5.1.2 General Finite Element Method

Over half century, various approaches have been made to overcome the shortcomings arise from inserting imperfections in FE mesh (Borja et al. 2000). The idea of adding interpolation terms to the element in order to ameliorate its deficient properties has a long standing in the literatures of FEM, and numerous "incompatible" elements have been created in a variety of application areas. Wilson and his co-workers (Wilson et al. 1973) are the pioneers who added two bending modes to an isoparametric quadrilateral for the simulation of its bending properties. When the onset of localization is detected, the element interpolation is extended by adding suitably defined shape functions that reproduce the localized deformation modes. These additional shape functions render the element incompatible and satisfy the patch test (Taylor et al. 1976). The extra "degrees of freedom (DOFs)" representing the amplitudes of localized modes are eliminated by static condensation. It should be emphasized that all these operations can be implemented at the element level, hence they do not require a mesh design to take explicit account of likely directions of localization bands (Pietruszcak and Mroz 1981).

In problems involving strong strain localization, Ortiz and his co-workers (Ortiz et al. 1987; Nacar et al. 1989; Leroy and Ortiz 1990) found that the performance of conventional isoparametric element can be substantially improved by adding mesostructure, in this manner the element is conferred ability to develop strain discontinuities, or "shocks", of arbitrary orientation and polarization. They presented an interesting example in which a plane strain sand specimen is subjected to uniaxial compression. Examples of this nature illustrate the need for specialized FEs tailored to localization problems.

Consider now the incremental FE analysis for an inelastic body expected to undergo strain localization. By virtue of the very nature of localization instabilities, the onset of localization can be detected pointwisely at, say, the reduced quadrature points within each element. The geometry of the incipient localization modes is also found readily in the same manner. For rate-independent solids, localization is detected by monitoring the eigenvalues of the acoustic tensor. For rate-dependent solids, a similar analysis based on the tangent modulus of the limiting rate-independent material yields the earliest time at which localization becomes possible (Ortiz et al. 1987).

Once localization is detected within an element $fe(e)$ (or briefly e) $(1, 2, ..., n_e)$ (viz. Fig. 5.4), incompatible shape functions possessing appropriate strain discontinuities are added to the element interpolation such that

$$\mathbf{u}_e(\mathbf{x}) = \mathbf{N}(\mathbf{x})\hat{\mathbf{u}}_b + \mathbf{M}(\mathbf{x})\hat{\mathbf{u}}_L \tag{5.1}$$

in which $\mathbf{u}_e(\mathbf{x})$ is the displacement field within the element; $\hat{\mathbf{u}}_b$ and $\mathbf{N}(\mathbf{x})$ are the basic nodal displacement vector and shape function matrix of standard finite element; $\mathbf{M}(\mathbf{x})$ is the localized mode matrix and $\hat{\mathbf{u}}_L$ is the corresponding amplitude vector.

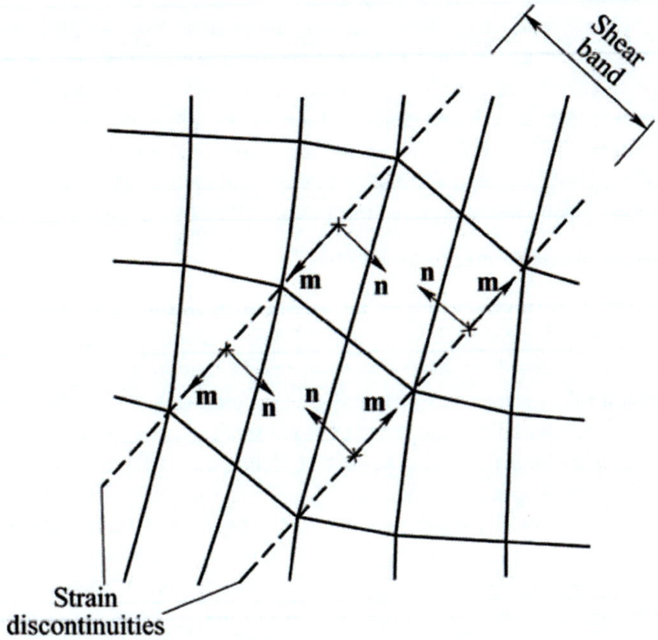

Fig. 5.4 Close-up view of the central portion of localization band

N(**x**) is typically given by the serendipity family of 4-to 9-node isoparametric elements in 2-D problems (viz. Eqs. (4.239) and (4.240)). **M**(**x**) exhibits the jump across shear bands that are constructed using the localized deformation modes fully determined by pairs of unit vectors (**m**, **n**) representing the shear displacement direction and the normal to the surface of shear band, respectively (viz. Fig. 5.4). It should be noted that the shape functions in Eq. (5.1) are incompatible, i.e., in general they do not satisfy the C^0 continuity across element boundary hence the enriched element becomes non-conforming.

The method of Ortiz overcomes the difficulties associated with standard finite elements in the problems involving strain localization and crack growth. It relies on the fact that the conditions for localization are purely local, namely, they can be ascertained in a pointwise fashion. Once the onset of localization is detected within an element, its strain interpolation may be enriched by means of additional shape functions which reproduce the localized deformation patterns. The results are thus fairly independent of the relative orientation of shear band w.r.t. mesh. This remarkable advantage actually gave great impetus towards the simulation of shear band and latterly the crack using fixed finite element mesh. Together with the other advances such as the p-refinement using hierarchical basis functions, a series of new FEM based on the "partition of unity (PU)" were timely created (Melenk and Babuška 1996; Babuška and Melenk 1997), whose comprehensive expression is the "generalized finite element methods (GFEMs)" (Strouboulis et al. 2000; Strouboulis et al. 2001). The mesh in GFEMs is totally independent of the problem domain and can always be regular. GFEMs augment the finite element space with high-order terms or handbook functions to tackle the problems with multiple reentrant corners, voids, and cracks. Three major members of GFEMs, namely, the "composite element method (CEM)", the "extended finite element method (XFEM)", and the "numerical manifold method (NMM)" that are rather important in the advanced CMs for geotechnical and hydraulic structures, will be briefly presented below.

1. **Composite element method (CEM)**

From the point view of practitioners, the crucial difficulty in the explicit simulation of structural planes such as rock joints with FEM lies in the pre-process to discretize the calculation domain. This arises from the reality that on the one hand, there are a large amount of rock joints of different shapes and sizes, and, a large amount of rock bolts/drainage holes/cooling pipes of small size (e.g. at a diameter of 3–10 cm) deployed within a small width and largely stretched zone (e.g. spaced at 1.5–3 m); on the other hand, the special elements available possess definite nodes, and some of them should be the common nodes of the host entity elements. Consequently, this, plus the complicity in the configuration and construction of engineering structures, will lead to time consuming and tedious pre-process overheads.

CEM is good at explicitly simulating meso-structures arise from joints, bolts, drainage holes, and cooling pipes within the prototype engineering structures (Chen 2018). The most attractive advantage with this method is that it may give a detailed description of the non-homogenous and discontinuous properties within an element using standard shape functions typically given by the serendipity family. In this way less restraint is imposed on the mesh generation for complicated geotechnical and hydraulic structures with considerable amount of meso-structural components, which allows for a great simplification in the pre-process work towards the engineering application.

For the strain–stress problems particularly focused on structural planes, standard FE mesh should be generated to discretize the engineering structure firstly. The deployment and sizes of elements are dependent on the structure configuration and stress gradient, cares over the existence of the discontinuity system need not to be taken. Then the algebra and geometry calculi will be conducted with the messages of the structural plane system and the FE mesh, to construct the composite element (CE) mesh. In such a CE mesh there are several standard elements containing structural plane segments, which are defined as composite elements.

Suppose a composite element $ce(e)$ (or briefly e) $(1, 2, ..., n_e)$ contains n_r structural plane segments which delimit it into n_r sub-elements allowing for different mechanical properties ($n_r = 2$ in Fig. 5.5). These sub-elements are not necessarily to be the fundamental finite elements.

The displacement $\Delta\mathbf{u}_{r_i}(\mathbf{x})$ within each sub-element is interpolated from n nodal displacements $\Delta\hat{\mathbf{u}}_{r_i} = \left[\Delta\hat{\mathbf{u}}_{r_i}^1 \ldots \Delta\hat{\mathbf{u}}_{r_i}^l \ldots \Delta\hat{\mathbf{u}}_{r_i}^k \ldots \Delta\hat{\mathbf{u}}_{r_i}^n\right]^{\mathrm{T}}$ tied to the composite element (viz. Fig. 5.6) by

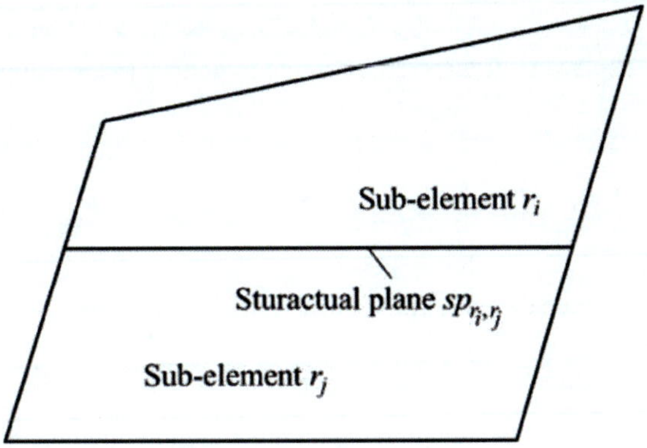

Fig. 5.5 Schematic drawing of a composite element containing one structural plane segment

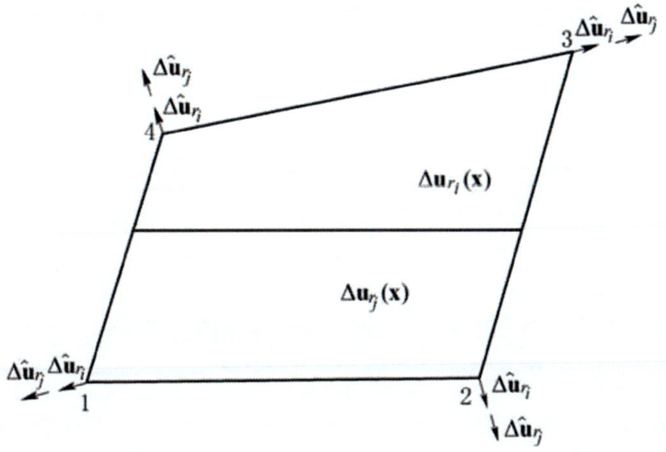

Fig. 5.6 Diagram to the displacement interpolation in a composite element containing one structural plane segment

$$\Delta\mathbf{u}_{r_i}(\mathbf{x}) = \mathbf{N}(\mathbf{x})\Delta\hat{\mathbf{u}}_{r_i} \quad (r_i = 1, \ldots, n_r) \tag{5.2}$$

in which **N** stands for the shape function matrix of FEM defined over the whole composite element; superscripts l and k stand for the local node sequences. For the 4-node element in 2-D problems (viz. Figs. 5.6 and 4.22), $n = 4$ and **N** has been given in Eqs. (4.239) and (4.240).

The strain increments of intact rock sub-elements are routinely calculated by the formula

$$\Delta\varepsilon_{r_i}(\mathbf{x}) = \mathbf{B}(\mathbf{x})\Delta\hat{\mathbf{u}}_{r_i} \quad (r_i = 1, \ldots, n_r) \tag{5.3}$$

in which $\mathbf{B}(\mathbf{x})$ is identical to that in Eqs. (4.245) and (4.246).

The displacement increments in the rock sub-elements will give rise to relative displacement increments on the structural plane segments, which will be expressed in the local Cartesian coordinate system as

$$\Delta\mathbf{u}_{sp_{r_i,r_j}}(\mathbf{x}) = \mathbf{T}_{sp_{r_i,r_j}}(\Delta\mathbf{u}_{r_i}(\mathbf{x}) - \Delta\mathbf{u}_{r_j}(\mathbf{x})) = \mathbf{T}_{sp_{r_i,r_j}}\mathbf{N}(\Delta\hat{\mathbf{u}}_{r_i} - \Delta\hat{\mathbf{u}}_{r_j}) \quad (r_i, r_j = 1, \ldots, n_r) \tag{5.4}$$

in which the coordinate transformation matrix $\mathbf{T}_{sp_{r_i,r_j}}$ has been defined in Eq. (3.125) (viz. Fig. 3.6c).

Suppose the virtual displacements manifest in the rock sub-elements

$$\delta\mathbf{u}_{r_i}(\mathbf{x}) = \mathbf{N}\delta\hat{\mathbf{u}}_{r_i} \quad (r_i = 1, \ldots, n_r) \tag{5.5}$$

The corresponding virtual strains and relative displacements will be

$$\delta\boldsymbol{\varepsilon}_{r_i} = \mathbf{B}\delta\hat{\mathbf{u}}_{r_i} \quad (r_i = 1, \ldots, n_r) \tag{5.6}$$

$$\delta\hat{\mathbf{u}}_{sp_{r_i,r_j}} = \mathbf{T}_{j_{r_i,r_j}}\mathbf{N}\left(\delta\hat{\mathbf{u}}_{r_i} - \delta\hat{\mathbf{u}}_{r_j}\right) \quad \left(r_i, r_j = 1, \ldots, n_r\right) \tag{5.7}$$

and the virtual work principle for the composite element containing structural planes will be

$$\sum_{r_i=1}^{n_r} W_{r_i} + \sum_{j_{r_i,r_j}=1}^{n_j} W_{sp_{r_i,r_j}} = \sum_{r_i=1}^{n_r} \left(\delta\hat{\mathbf{u}}_{r_i}\right)^{\mathrm{T}} \Delta\mathbf{f}_{r_i} \tag{5.8}$$

in which $\Delta\mathbf{f}_{r_i}$ are the nodal load increments at the composite element which can be transferred from the load increments at the rock sub-elements according to the identical algorithm of FEM. The component virtual works W_{r_i} and $W_{sp_{r_i,r_j}}$ contributed from rock sub-elements and structural plane segments will be elaborated as follows.

(1) **Sub-element**

$$W_{r_i} = \iint\limits_{\Omega_{r_i}} \left(\delta\boldsymbol{\varepsilon}_{r_i}\right)^{\mathrm{T}} \Delta\boldsymbol{\sigma}_{r_i} \mathrm{d}\Omega \quad (r_i = 1, \ldots, n_r) \tag{5.9}$$

Take the elasto-viscoplastic constitutive Eq. (3.138) and the strain–displacement relation (5.3) into account, the virtual work in the rock material may be expressed as

$$W_{r_i} = \iint\limits_{\Omega_{r_i}} (\delta\hat{\mathbf{u}}_{r_i})^{\mathrm{T}}\mathbf{B}^{\mathrm{T}}\mathbf{D}_{r_i}(\mathbf{B}\Delta\hat{\mathbf{u}}_{r_i} - \Delta\boldsymbol{\varepsilon}_{r_i}^{vp})\mathrm{d}\Omega \quad (r_i = 1, \ldots, n_r) \tag{5.10}$$

(2) **Structural plane segment**

$$W_{sp_{r_i,r_j}} = \int_{\Gamma_{sp_{r_i,r_j}}} \left(\delta\hat{\mathbf{u}}_{sp_{r_i,r_j}}\right)^{\mathrm{T}} \Delta\boldsymbol{\sigma}_{sp_{r_i,r_j}} \mathrm{d}\Gamma \tag{5.11}$$

Use is made of the elasto-viscoplastic constitutive relation (3.142) and the strain–displacement relation (5.4), we get

$$W_{sp_{r_i,r_j}} = \int_{\Gamma_{sp_{r_i,r_j}}} \begin{pmatrix} \left((\delta\hat{\mathbf{u}}_{r_i})^{\mathrm{T}} - (\delta\hat{\mathbf{u}}_{r_j})^{\mathrm{T}}\right)\mathbf{N}^{\mathrm{T}}\left(\mathbf{T}_{sp_{r_i,r_j}}\right)^{\mathrm{T}}\mathbf{D}_{sp_{r_i,r_j}} \\ \left(\mathbf{T}_{sp_{r_i,r_j}}\mathbf{N}\left(\Delta\hat{\mathbf{u}}_{r_i} - \Delta\hat{\mathbf{u}}_{r_j}\right) - \Delta\mathbf{u}^{vp}_{sp_{r_i,r_j}}\right) \end{pmatrix} \mathrm{d}\Gamma \tag{5.12}$$

(3) **Composite element**

Because $\delta\hat{\mathbf{u}}_{r_i}$ and $\delta\hat{\mathbf{u}}_{r_j}$ are arbitrary, hence on introducing Eqs. (5.10) and (5.12) into Eq. (5.8) and arranging the resultant equation regarding the different virtual displacements, the validity of the virtual work principle will leads to

$$\begin{bmatrix} \mathbf{k}_{r_1,r_1} & \cdots & \mathbf{k}_{r_1,r_i} & \cdots & \mathbf{k}_{r_1,r_j} & \cdots & \mathbf{k}_{r_1,n_r} \\ \cdots & \cdots & \cdots & \cdots & \cdots & \cdots & \cdots \\ \mathbf{k}_{r_i,r_1} & \cdots & \mathbf{k}_{r_i,r_i} & \cdots & \mathbf{k}_{r_i,r_j} & \cdots & \mathbf{k}_{r_i,n_r} \\ \cdots & \cdots & \cdots & \cdots & \cdots & \cdots & \cdots \\ \mathbf{k}_{r_j,r_1} & \cdots & \mathbf{k}_{r_j,r_i} & \cdots & \mathbf{k}_{r_j,r_j} & \cdots & \mathbf{k}_{r_j,n_r} \\ \cdots & \cdots & \cdots & \cdots & \cdots & \cdots & \cdots \\ \mathbf{k}_{n_r,r_1} & \cdots & \mathbf{k}_{n_r,r_i} & \cdots & \mathbf{k}_{n_r,r_j} & \cdots & \mathbf{k}_{n_r,n_r} \end{bmatrix} \begin{Bmatrix} \Delta\hat{\mathbf{u}}_{r_1} \\ \cdots \\ \Delta\hat{\mathbf{u}}_{r_i} \\ \cdots \\ \Delta\hat{\mathbf{u}}_{r_j} \\ \cdots \\ \Delta\hat{\mathbf{u}}_{n_r} \end{Bmatrix} = \begin{Bmatrix} \Delta\mathbf{f}_{r1} \\ \cdots \\ \Delta\mathbf{f}_{r_i} \\ \cdots \\ \Delta\mathbf{f}_{r_j} \\ \cdots \\ \Delta\mathbf{f}_{n_r} \end{Bmatrix} + \begin{Bmatrix} \Delta\mathbf{f}^{vp}_{r1} \\ \cdots \\ \Delta\mathbf{f}^{vp}_{r_i} \\ \cdots \\ \Delta\mathbf{f}^{vp}_{r_j} \\ \cdots \\ \Delta\mathbf{f}^{vp}_{n_r} \end{Bmatrix} \tag{5.13}$$

in which

$$\begin{cases} \mathbf{k}_{r_i,r_i} = \iint\limits_{\Omega_{r_i}} \mathbf{B}^{\mathrm{T}}\mathbf{D}_{r_i}\mathbf{B}\,\mathrm{d}\Omega + \sum\limits_{r_j \neq r_i} H_2(r_i, r_j) \int\limits_{\Gamma_{sp_{r_i,r_j}}} \mathbf{N}^{\mathrm{T}}\mathbf{T}^{\mathrm{T}}_{sp_{r_i,r_j}}\mathbf{D}_{sp_{r_i,r_j}}\mathbf{N}\mathrm{d}\Gamma, \\ H_2(r_i, r_j) = \begin{cases} 1 & \text{where } r_i \text{ and } r_j \text{ are adjacent} \\ 0 & \text{where } r_i \text{ and } r_j \text{ are not adjacent} \end{cases} \end{cases} \tag{5.14}$$

$$\begin{cases} \Delta\mathbf{f}^{vp}_{r_i} = \iint\limits_{\Omega_{r_i}} \mathbf{B}^{\mathrm{T}}\mathbf{D}_{r_i}\Delta\boldsymbol{\varepsilon}^{vp}_{r_i}\mathrm{d}\Omega + \sum\limits_{r_j \neq r_i}^{n_j} H_3(r_i, r_j) \int\limits_{\Gamma_{sp_{r_i,r_j}}} \mathbf{N}^{\mathrm{T}}\mathbf{T}^{\mathrm{T}}_{sp_{r_i,r_j}}\mathbf{D}_{sp_{r_i,r_j}}\Delta\mathbf{u}^{vp}_{sp_{r_i,r_j}}\mathrm{d}\Gamma \\ H_3(r_i, r_j) = \begin{cases} 1 & \text{where } r_i > r_j \\ -1 & \text{where } r_i < r_j \\ 0 & \text{where } r_i \text{ and } r_j \text{ are not adjacent} \end{cases} \end{cases} \tag{5.15}$$

Equations (5.13)–(5.15) are the equilibrium equation set governing the composite element containing n_r entity (rock) sub-elements. According to the link list w.r.t.

local nodal sequence and global nodal sequence, they may be assembled into the expanded equilibrium equation set of the whole structure system. Suppose only one linear isoparametric element e is transfer to the composite element comprising two sub-elements, namely $n = 4$ and $n_r = 2$, Eq. (5.13) becomes

$$\begin{bmatrix} \mathbf{k}_{r1,r1} & \mathbf{k}_{r1,r2} \\ \mathbf{k}_{r2,r1} & \mathbf{k}_{r2,r2} \end{bmatrix} \begin{Bmatrix} \Delta\hat{\mathbf{u}}_{r1} \\ \Delta\hat{\mathbf{u}}_{r2} \end{Bmatrix} = \begin{Bmatrix} \Delta\mathbf{f}_{r1} \\ \Delta\mathbf{f}_{r2} \end{Bmatrix} + \begin{Bmatrix} \Delta\mathbf{f}_{r1}^{vp} \\ \Delta\mathbf{f}_{r2}^{vp} \end{Bmatrix} \tag{5.16}$$

in which

$$\mathbf{k}_{r_i,r_j} = \begin{bmatrix} \mathbf{k}_{r_i,r_j}^{1,1} & \mathbf{k}_{r_i,r_j}^{1,2} & \mathbf{k}_{r_i,r_j}^{1,3} & \mathbf{k}_{r_i,r_j}^{1,4} \\ \mathbf{k}_{r_i,r_j}^{2,1} & \mathbf{k}_{r_i,r_j}^{2,2} & \mathbf{k}_{r_i,r_j}^{2,3} & \mathbf{k}_{r_i,r_j}^{2,4} \\ \mathbf{k}_{r_i,r_j}^{3,1} & \mathbf{k}_{r_i,r_j}^{3,2} & \mathbf{k}_{r_i,r_j}^{3,3} & \mathbf{k}_{r_i,r_j}^{3,4} \\ \mathbf{k}_{r_i,r_j}^{4,1} & \mathbf{k}_{r_i,r_j}^{4,2} & \mathbf{k}_{r_i,r_j}^{4,3} & \mathbf{k}_{r_i,r_j}^{4,4} \end{bmatrix} \quad (i, j = 1, 2) \tag{5.17}$$

$$\Delta\hat{\mathbf{u}}_{r_i} = \begin{bmatrix} \Delta\hat{\mathbf{u}}_{r_i}^1 & \Delta\hat{\mathbf{u}}_{r_i}^2 & \Delta\hat{\mathbf{u}}_{r_i}^3 & \Delta\hat{\mathbf{u}}_{r_i}^4 \end{bmatrix}^{\mathrm{T}} \quad (i = 1, 2) \tag{5.18}$$

$$\Delta\mathbf{f}_{r_i} = \begin{bmatrix} \Delta\mathbf{f}_{r_i}^1 & \Delta\mathbf{f}_{r_i}^2 & \Delta\mathbf{f}_{r_i}^3 & \Delta\mathbf{f}_{r_i}^4 \end{bmatrix}^{\mathrm{T}} \quad (i = 1, 2) \tag{5.19}$$

$$\Delta\mathbf{f}_{r_i}^{vp} = \begin{bmatrix} \Delta\mathbf{f}_{r_i}^{vp,1} & \Delta\mathbf{f}_{r_i}^{vp,2} & \Delta\mathbf{f}_{r_i}^{vp,3} & \Delta\mathbf{f}_{r_i}^{vp,4} \end{bmatrix}^{\mathrm{T}} \quad (i = 1, 2) \tag{5.20}$$

Cross-referencing Eqs. (5.16)–(5.20) to Eqs. (4.254)–(4.258), it is found that a composite element containing one structural plane segment will double the DOFs of the nodes that construct the original finite element. Hence, with a link list relating the local node sequences l and k to the expanded (doubled in this case) global node sequences i and j, namely, $i = \left\langle \begin{matrix} l \\ r_i \sim r_j \end{matrix} \right\rangle$ and $j = \left\langle \begin{matrix} k \\ r_i \sim r_j \end{matrix} \right\rangle$, the assemblage operation for the whole structure system is conducted in a similar manner of FEM illustrated in Chap. 4. Generally, if a composite element consists of n_r rock sub-elements, its DOFs will be n_r times that of the original finite element, whereas if an element does not contain structural plane segments, it will be degenerated automatically to the standard finite element. In this way composite elements and finite elements can be coupled well in one discrete mesh and there is no essential difficulty in the code compiling.

(4) Notes

The most remarkable characteristic of CEM that differentiates it from other members of GEFMs (e.g. XFEM, NMM) are (Chen 2018).

- The use of standard or hierarchical shape functions for all the sub-elements within a composite element that are uniformly given by the serendipity family in fundamental FEM;
- There are individual nodal displacement vectors for correspondent sub-elements;
- Discontinuous jump between sub-elements naturally emerges without the help of Heaviside function in the interpolation stage;
- It permits any irregular shape of sub-element arise from the existence of structural planes, bolts, holes, and pipes.

However, if the SIFs are demanded for crack growth computation, both the asymptotic shape function and corresponding nodal displacement enrichment functions for crack-tip are required in a similar manner of XFEM.

For details, readers are referred to the literatures of Chen and his co-workers (Chen 1987; Chen and Qiang 2004; Chen et al. 2004a, b; Chen and Feng 2006; Chen et al. 2008a, b; Chen and Shahrour 2008; Chen et al. 2010; 2011; 2012; 2015; Chen 2018).

2. **Numerical manifold method (NMM)**

In many aspects, NMM (Shi 1991, 1992; Chen et al. 1998) is similar to the GFEMs except for the treatment of fractures (faults, joints and cracks) and discrete blocks by the truncated shape functions in a unified form.

NMM normally uses FE mesh to construct mathematic covers (patches), each of them belongs to the serendipity family. A polygon composed of FEs surrounding one internal node defines an original physical cover that may be further sub-divided by fractures into several physical covers share identical node. The overlap operation for all the physical covers produces manifold element system that is similar to the "atlas" in topology manifold. Taking Fig. 5.7 for example, the physical cover 1 is composed of element 1 only, the physical cover 3 is composed of element 1 and element 2, the physical cover 11 is composed of elements 1, 2, 3, and 4.

The essential variables (e.g. displacements) in a manifold element may be interpolated by the complete polynomials tied to physical covers (yet unknown) and

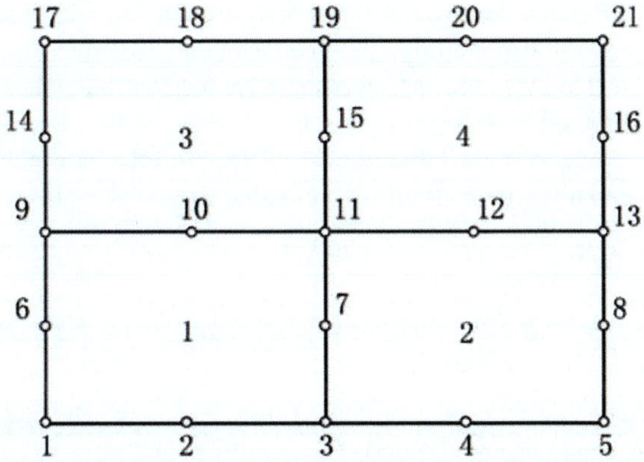

Fig. 5.7 Mathematical covers and physical covers

weighted by the shape functions. If the complete polynomials are of zero-order (constants) and there are no structural planes, NMM degenerates into fundamental FEM. The remarkable advantage of NMM is its ability to provide a solution of higher procession beneficial from complete polynomials, in addition to the potentiality to describe discontinuous jump between manifold elements. However, it is notable that for the complete polynomials with higher order (to say, higher than 2), the stability and existence of the solution have not been well-justified yet.

The principles and algorithms of NMM will be detailed in Chap. 6.

3. **Extended finite element method (XFEM)**

Based on a standard Galerkin procedure and using the concept of PU (Duarte and Oden 1995; Melenk and Babuška 1996) to accommodate internal boundaries in the discrete model, XFEM may be looked as a member of GFEMs. In XFEM, cracks and their tips are directly represented by incorporating enrichment shape functions and corresponding nodal variables, hence there is neither need for FE mesh generating to conform to fractures nor need for re-meshing to trace fracture growth (Yazid et al. 2009).

XFEM was originally proposed by presenting an enrichment strategy of FE approximation so that the crack growth problem can be solved with minimal re-meshing (Belytschko and Black 1999). Sooner, a much more elegant technique by adapting an enrichment that includes the asymptotic near tip field and a Heaviside function $H(x)$ was introduced, and the algorithm was further expanded to three-dimensional static crack modeling (Dolbow 1999; Möes et al. 1999; Dolbow et al. 2000; Sukumar et al. 2000; Belytschko et al. 2001; Sukumar and Prévost 2003). In contrast to the element enrichment scheme of Benzley (1974), the enrichment accuracy of XFEM based on PU is almost independent of element size. The cohesive crack in the framework of XFEM had been tackled (Dolbow et al. 2001; Wells and Sluys 2001; Moës and Belytschko 2002; Liu and Borja 2009), too.

It is notable that XFEM looks much more like a direct generalization of the method proposed by Ortiz et al. (1987) for the issue of localization: apart from the enrichment for crack itself using discontinuous enrichment function $H(x)$ of Heaviside type that performs similar functionality to the localized mode matrix $\mathbf{M}(\mathbf{x})$ in Eq. (5.1), XFEM undertakes further enrichment using additional asymptotic shape functions and corresponding nodal displacements for crack-tip.

XFEM has been implemented in the commercial code ABAQUS. It is increasingly adopted by other commercial software, with a few plug-ins and actual core implementations available (e.g. ANSYS, etc.). Hereinafter in this Chapter, the principles and algorithms of XFEM will elucidated in detail.

5.2 Crack-Surface Enrichment in XFEM

The essential idea in XFEM that is closely related to GFEMs (Duarte et al. 2000; Strouboulis et al. 2000; Strouboulis et al. 2001), is to add discontinuous enrichment functions to the FE approximation that meets the partition of unity. To be specific, an element containing crack segment extends the nodal displacement for those nodes that are associated with the element crossed by the crack.

5.2.1 Concept

Suppose that the solid material occupying a domain Ω is crossed by the crack Γ_c with normal vector $\mathbf{n}$ (viz. Fig. 5.8), then the domain is divided into two portions (sub-domains):

- The side approaching $\mathbf{n}$ is considered as the positive portion and noted as Ω^+;
- The side leaving $\mathbf{n}$ is considered as negative portion and noted as Ω^- ($\Omega = \Omega^+ \cup \Omega^-$).

For the 2-D FE model with one crack in Fig. 5.9, the basic node set of all elements is denoted as S_b, the enriched node set of the elements which are completely split by the crack is denoted as S_c and the enriched node set around each crack-tip is denoted as S_t.

Let's consider the case that one crack splits an entire element e and the crack-tip advances element by element each time (Fig. 5.10). Under such circumstances, the approximate displacement increment $\Delta\mathbf{u}_e(\mathbf{x})$ can be simplified without introducing the crack-tip branch functions. The displacement field is decomposed into two parts as

$$\Delta\mathbf{u}_e(\mathbf{x}) = \Delta\mathbf{u}_b(\mathbf{x}) + \Delta\mathbf{u}_c(\mathbf{x})\phi(\mathbf{x}) \quad (e = 1, 2, 3, ..., n_e) \tag{5.21}$$

in which $\Delta\mathbf{u}_b(\mathbf{x})$ and $\Delta\mathbf{u}_c(\mathbf{x})$ are continuous displacement functions. The component $\Delta\mathbf{u}_b(\mathbf{x}, t)$ is the basic displacement increment identical to that in the fundamental FEM, whereas the component $\Delta\mathbf{u}_c(\mathbf{x}, t)$ is used to represent the discontinuous displacement cause by the crack, and $\phi(\mathbf{x})$ is the enriched approximation basis function particularly designed to describe this displacement jump.

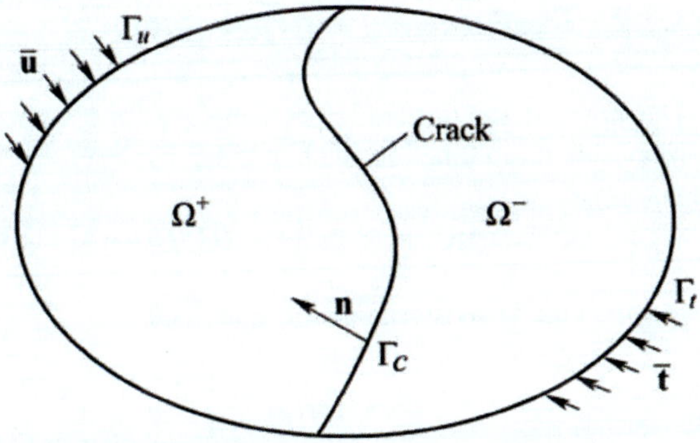

Fig. 5.8 Domain crossed by crack

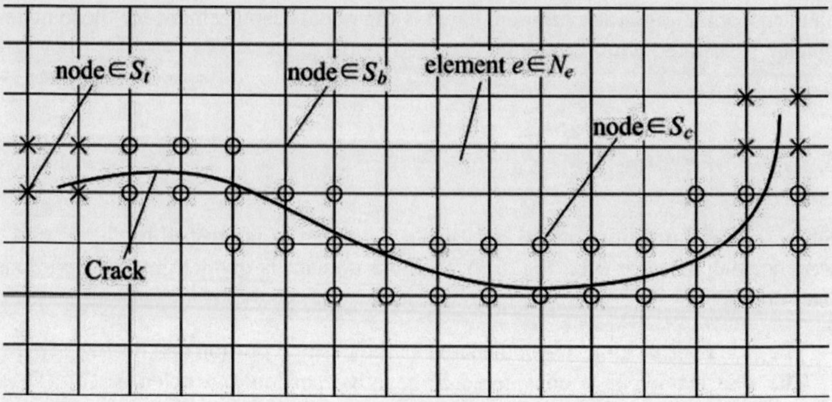

Fig. 5.9 Enriched nodes

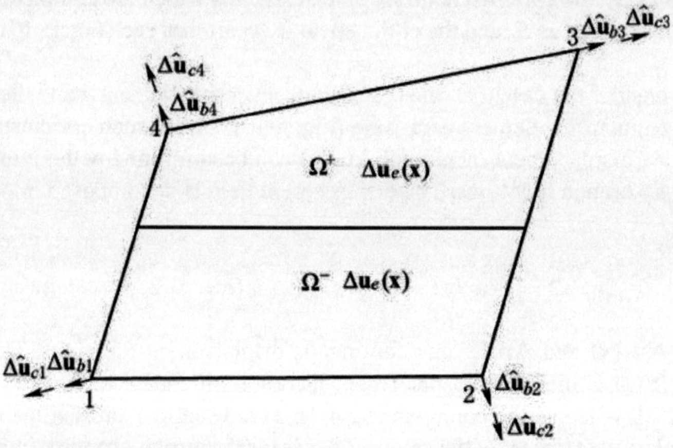

Fig. 5.10 Diagram to the displacement interpolation in an element completely split by crack

5.2.2 *Displacement Interpolation*

The simplest and most efficient technique for enriched approximation makes use of Heaviside jump function, such that the displacement filed is written as

$$\Delta\mathbf{u}_e(\mathbf{x}) = \Delta\mathbf{u}_b(\mathbf{x}) + H(\mathbf{x})\Delta\mathbf{u}_c(\mathbf{x}) \quad (e = 1, 2, 3, ..., n_e) \tag{5.22}$$

in which the Heaviside function is defined as

$$H(\mathbf{x}) = \begin{cases} +1 & \text{if } \mathbf{x} \in \Omega^{+} \\ -1 & \text{if } \mathbf{x} \in \Omega^{-1} \end{cases} \tag{5.23}$$

The displacement field $\Delta \mathbf{u}_e(\mathbf{x})$ is piecewise continuous in each Ω^+ or Ω^-, hence it is differentiable respectively in each sub-domain.

For an standard element with n nodes but without crack segment, if the displacement increment of each node is known, the displacement increment at any point in this element may be determined by using the interpolation function (viz. Eq. (4.243))

$$\Delta \mathbf{u}_b(\mathbf{x}) = \sum_{l=1}^{n} \mathbf{N}_{b,l}(\mathbf{x}) \Delta \hat{\mathbf{u}}_{b,l} = \mathbf{N}_b(\mathbf{x}) \Delta \hat{\mathbf{u}}_b \tag{5.24}$$

in which $\Delta \mathbf{u}_b(\mathbf{x})$ is the displacement at position $\mathbf{x}$, $\Delta \hat{\mathbf{u}}_{b,l}$ is the basic nodal displacement vector at node $l \in S_b$

$$\Delta \hat{\mathbf{u}}_b = \left[\Delta \hat{\mathbf{u}}_{b,1} \ldots \Delta \hat{\mathbf{u}}_{b,l} \ldots \Delta \hat{\mathbf{u}}_{b,k} \ldots \Delta \hat{\mathbf{u}}_{b,n}\right]^{\mathrm{T}} \tag{5.25}$$

$$\Delta \hat{\mathbf{u}}_{b,l} = \begin{Bmatrix} \Delta \hat{u}_{b,x,l} \\ \Delta \hat{u}_{b,y,l} \end{Bmatrix} \quad (l = 1, 2, \ldots, n) \tag{5.26}$$

and $\mathbf{N}_b(\mathbf{x})$ is the standard basis (shape) function matrix defined in Eq. (4.239).

Similarly, for a split element with enriched nodes, the displacement at any point within the element can be given as

$$\Delta \mathbf{u}_e(\mathbf{x}) = \mathbf{N}_b(\mathbf{x}) \Delta \hat{\mathbf{u}}_b + \mathbf{N}_c(\mathbf{x}) H(\mathbf{x}) \Delta \hat{\mathbf{u}}_c \quad (e = 1, 2, 3, \ldots, n_e) \tag{5.27}$$

where $\Delta \hat{\mathbf{u}}_b$ represents the basic (normal) displacement increment for the normal DOFs in the fundamental FEM, while $\Delta \hat{\mathbf{u}}_c$ is the displacement increment for the enriched DOFs.

It is worth to mention that normally $\mathbf{N}_b = \mathbf{N}_c = \mathbf{N}$ in Eq. (5.27), hence

$$\Delta \mathbf{u}_e(\mathbf{x}) = \mathbf{N}(\mathbf{x}) \Delta \hat{\mathbf{u}}_b + \mathbf{N}(\mathbf{x}) H(\mathbf{x}) \Delta \hat{\mathbf{u}}_c \quad (e = 1, 2, 3, \ldots, n_e) \tag{5.28}$$

in which $\Delta \hat{\mathbf{u}}_{c,l}$ represents the extended nodal displacement vector for those nodes $l \in S_c$ with enriched DOFs,

$$\Delta \hat{\mathbf{u}}_c = \left[\Delta \hat{\mathbf{u}}_{c,1} \ldots \Delta \hat{\mathbf{u}}_{c,l} \ldots \Delta \hat{\mathbf{u}}_{c,k} \ldots \Delta \hat{\mathbf{u}}_{c,n}\right]^{\mathrm{T}} \tag{5.29}$$

$$\Delta \hat{\mathbf{u}}_{c,l} = \begin{Bmatrix} \Delta \hat{u}_{c,x,l} \\ \Delta \hat{u}_{c,y,l} \end{Bmatrix} \quad (l = 1, 2, \ldots, n) \tag{5.30}$$

5.2.3 Strain and Stress

1. **Solids**

The strain in the enriched element entirely split by crack can be given directly

$$\Delta\boldsymbol{\varepsilon}(\mathbf{x}) = \mathbf{L}\Delta\mathbf{u}_e = \sum_{l=1}^{n}\mathbf{B}_l(\mathbf{x})\Delta\hat{\mathbf{u}}_{b,l} + \sum_{l=1}^{n}H(\mathbf{x})\mathbf{B}_l(\mathbf{x})\Delta\hat{\mathbf{u}}_{c,l} \tag{5.31}$$

for 2-D problems

$$\mathbf{B}_l = \mathbf{L}N_l = \begin{bmatrix} \frac{\partial N_l}{\partial x} & 0 \\ 0 & \frac{\partial N_l}{\partial y} \\ \frac{\partial N_l}{\partial y} & \frac{\partial N_l}{\partial x} \end{bmatrix} \tag{5.32}$$

Then, the stress field of solid can be determined by using the constitutive law in an incremental form of

$$\Delta\boldsymbol{\sigma} = \mathbf{D}\Delta\boldsymbol{\varepsilon} - \Delta\boldsymbol{\sigma}^{vp} \tag{5.33}$$

in which **D** is the constitutive matrix relating the stress and strain, that will be the elastic matrix when the Hook's law is applied; the initial stress increment $\Delta\boldsymbol{\sigma}^{vp}$ due to the viscoplastic deformation has been specified in Eqs. (3.138)–(3.141).

2. **Cracks**

Since there exists the cohesive stress $\Delta\boldsymbol{\sigma}_c$ at the internal boundary for the cohesive crack, among many others, Wells and Sluys (2001), Moës and Belytschko (2002), Khoei and Nikbakht (2007), Li and Yuan (2013) had considered the cohesive crack in the framework of the XFEM, but they need alternatively define the Heaviside function. Actually, the traction stress on cohesive crack may be considered without the change of Heaviside function in Eq. (5.23) with the help of elasto-viscoplastic constitutive relation (viz. Eq. (3.142))

$$\Delta\boldsymbol{\sigma}_c = \mathbf{D}_c\Delta\mathbf{u}_c - \Delta\boldsymbol{\sigma}_c^{vp} \tag{5.34}$$

$$\begin{aligned}\Delta\mathbf{u}_c &= \mathbf{T}_c\sum_{l=1}^{n} H_{\Omega^+}(\mathbf{x})\mathbf{N}_l(\mathbf{x})\Delta\hat{\mathbf{u}}_{c,l} - \mathbf{T}_c\sum_{l=1}^{n} H_{\Omega^-}(\mathbf{x})\mathbf{N}_l(\mathbf{x})\Delta\hat{\mathbf{u}}_{c,l} \\ &= 2\mathbf{T}_c\mathbf{N}(\mathbf{x})\Delta\hat{\mathbf{u}}_c \quad \text{where } \mathbf{x}\in\Gamma_c\end{aligned} \tag{5.35}$$

in which $\mathbf{D}_c$ is the elastic stiffness matrix of the crack defined in Eq. (3.123); the initial stress increment $\Delta\boldsymbol{\sigma}_c^{vp}$ due to the viscoplastic deformation has been specified in Eqs. (3.143–3.145); the transformation $\mathbf{T}_c$ has been defined in Eq. (3.125).

5.2.4 Governing Equation Set

The virtual work principle with internal traction boundary (cohesive crack) without taking the dynamic force into account are given as follows

$$\begin{aligned}&\iint_{\Omega}\delta\Delta\boldsymbol{\varepsilon}^{\mathrm{T}}\Delta\boldsymbol{\sigma}\mathrm{d}\Omega + \int_{\Gamma_c}\delta\Delta\mathbf{u}^{\mathrm{T}}\Delta\boldsymbol{\sigma}_c\mathrm{d}\Gamma = \iint_{\Omega}\delta\Delta\mathbf{u}^{\mathrm{T}}\Delta\mathbf{b}\mathrm{d}\Omega + \int_{\Gamma_t}\delta\Delta\mathbf{u}^{\mathrm{T}}\Delta\bar{\mathbf{t}}\,\mathrm{d}\Gamma \\ &+\sum_{\zeta=1}^{n_\zeta}\delta\Delta\mathbf{u}^{\mathrm{T}}(\mathbf{x}_\zeta)\Delta\mathbf{T}_\zeta(\mathbf{x}_\zeta)\end{aligned} \tag{5.36}$$

where: $\Delta\boldsymbol{\sigma}$—second order Cauchy stress tensor; $\Delta\mathbf{b}$—body force per unit volume; $\bar{\mathbf{t}}$—external force on the boundary Γ_t; $\mathbf{T}_\zeta(\mathbf{x}_\zeta)$—concentrated force at $\mathbf{x}_\zeta$ in the domain Ω; $\Delta\boldsymbol{\sigma}_c$—stress on the crack surface Γ_c, if $\Delta\boldsymbol{\sigma}_c$ equals zero, the crack is considered as traction free.

On the substitution of Eqs. (5.31)–(5.35) into Eq. (5.36), the linear equation set similar to that of FEM finally turns out that

$$\begin{bmatrix}\mathbf{k}_{b,b} & \mathbf{k}_{b,c}\\ \mathbf{k}_{c,b} & \mathbf{k}_{c,c}\end{bmatrix}\begin{Bmatrix}\Delta\hat{\mathbf{u}}_b\\ \Delta\hat{\mathbf{u}}_c\end{Bmatrix} = \begin{Bmatrix}\Delta\mathbf{f}_b\\ \Delta\mathbf{f}_c\end{Bmatrix} + \begin{Bmatrix}\Delta\mathbf{f}_b^{vp}\\ \Delta\mathbf{f}_c^{vp}\end{Bmatrix} \tag{5.37}$$

It is noted that Eq. (5.37) is symbolically identical to Eq. (5.16), hence the assemblage of an extended finite element containing one crack segment may be routinely undertaken with the similar procedure of a composite element containing one structural plane segment.

1. **Sub-matrices**

The sub-matrix $\mathbf{k}_{b,b}$ in Eq. (5.37) is exactly the same as that in the fundamental FEM, i.e.

$$\mathbf{k}_{b,b} = \iint_{\Omega_e}\mathbf{B}^{\mathrm{T}}\mathbf{D}\mathbf{B}\mathrm{d}\Omega \tag{5.38}$$

Other two sub-matrices considering the coupling effects between the "normal" and "enriched" DOFs, are expressed as

$$\begin{cases} \mathbf{k}_{b,c} = \iint\limits_{\Omega_e} \mathbf{B}^{\mathrm{T}}\mathbf{D}H(\mathbf{x})\mathbf{B}\mathrm{d}\Omega \\ \mathbf{k}_{c,b} = \iint\limits_{\Omega_e} H(\mathbf{x})\mathbf{B}^{\mathrm{T}}\mathbf{D}\mathbf{B}\mathrm{d}\Omega \end{cases} \tag{5.39}$$

It is easy to find their symmetry that $\mathbf{k}_{b,c} = \mathbf{k}_{c,b}$.

The fourth sub-matrix in Eq. (5.37) represents the contribution of the crack itself. For a closed crack (or structural plane), it is expressed by

$$\mathbf{k}_{c,c} = \iint\limits_{\Omega_e} \mathbf{B}^{\mathrm{T}}\mathbf{D}\mathbf{B}\mathrm{d}\Omega + 4\int\limits_{\Gamma_c} (\mathbf{T}_c\mathbf{N})^{\mathrm{T}}\mathbf{D}_c(\mathbf{T}_c\mathbf{N})\mathrm{d}\Gamma \tag{5.40}$$

For an open crack (without traction), Eq. (5.40) is simplified as

$$\mathbf{k}_{c,c} = \iint\limits_{\Omega_e} \mathbf{B}^{\mathrm{T}}\mathbf{D}\mathbf{B}\mathrm{d}\Omega \tag{5.41}$$

2. **Sub-vectors**

The right hand side of Eq. (5.37) consists of the nodal vectors correspondent to the unbalance forces for "normal" and "enriched" DOFs, in which $\Delta\mathbf{f}_b$ and $\Delta\mathbf{f}_c$ are contributed from external forces identically calculated by that of fundamental FEM

$$\begin{cases} \Delta\mathbf{f}_b = \iint\limits_{\Omega_e} \mathbf{N}^{\mathrm{T}}\Delta\mathbf{b}\mathrm{d}\Omega + \int\limits_{\Gamma_t} \mathbf{N}^{\mathrm{T}}\Delta\bar{\mathbf{t}}\mathrm{d}\Gamma + \sum\limits_{\mathbf{x}_\zeta \in e} \mathbf{N}^{\mathrm{T}}(\mathbf{x}_\zeta)\Delta\mathbf{T}_\zeta(\mathbf{x}_\zeta) \\ \Delta\mathbf{f}_c = \iint\limits_{\Omega_e} \mathbf{N}^{\mathrm{T}}H(\mathbf{x})\Delta\mathbf{b}\mathrm{d}\Omega + \int\limits_{\Gamma_t} \mathbf{N}^{\mathrm{T}}H(\mathbf{x})\Delta\bar{\mathbf{t}}\mathrm{d}\Gamma + \sum\limits_{\mathbf{x}_\zeta \in e} \mathbf{N}^{\mathrm{T}}(\mathbf{x}_\zeta)H(\mathbf{x})\Delta\mathbf{T}_\zeta(\mathbf{x}_\zeta) \end{cases} \tag{5.42}$$

where as $\Delta\mathbf{f}_b^{vp}$ and $\Delta\mathbf{f}_c^{vp}$ are respectively the nodal force vectors attributable to the initial force, or in other term, the initial stress due to viscoplastic deformation

$$\begin{cases} \Delta\mathbf{f}_b^{vp} = \iint\limits_{\Omega_e} \mathbf{B}^{\mathrm{T}}\Delta\boldsymbol{\sigma}^{vp}\mathrm{d}\Omega \\ \Delta\mathbf{f}_c^{vp} = \iint\limits_{\Omega_e} H(\mathbf{x})\mathbf{B}^{\mathrm{T}}\Delta\boldsymbol{\sigma}^{vp}\mathrm{d}\Omega + 2\int\limits_{\Gamma_c} H(\mathbf{x})(\mathbf{T}_c\mathbf{N})^{\mathrm{T}}\Delta\boldsymbol{\sigma}_c^{vp}\mathrm{d}\Gamma \end{cases} \tag{5.43}$$

Take the elasto-viscoplastic constitutive relations (3.138) and (3.142) for example, we have

$$
\begin{cases}
\Delta \boldsymbol{\sigma}^{vp} = \mathbf{D} H_1(F)\gamma \dfrac{\partial Q}{\partial \boldsymbol{\sigma}} \Delta t = \mathbf{D} H_1(F)\gamma (\nabla_\sigma Q)\Delta t \quad \text{for solid material} \\
\Delta \boldsymbol{\sigma}_c^{vp} = \mathbf{D}_c H_1(F_c)\gamma \dfrac{\partial Q_c}{\partial \boldsymbol{\sigma}_c} \Delta t = \mathbf{D}_c H_1(F_c)\gamma (\nabla_{\sigma_c} Q_c)\Delta t \quad \text{for closed and cohesive crack} \\
\Delta \boldsymbol{\sigma}_c^{vp} = 0 \quad \text{for open crack without traction}
\end{cases}
\tag{5.44}
$$

3. **Retrogression**

Where there is no crack embedded in the element, Eq. (5.37) is exactly the same as that for the fundamental FEM

$$
\mathbf{k}\Delta\hat{\mathbf{u}} = \Delta\mathbf{f} + \Delta\mathbf{f}^{vp} \tag{5.45}
$$

It is clear that although the definitions of nodal variables are different, an enriched element containing one crack segment will double the DOFs of the nodes that construct the original finite element. Hence, with a link list w.r.t. the local node sequence of the enriched element and the extended (doubled in this case) global node sequence, the assemblage operation of system is conducted in a similar manner of CEM illustrated previously in this chapter. It may be said that essentially, there are no great difference between FEM (with joint element), CEM and XFEM.

5.3 Crack-Tip Enrichment in XFEM

The use of crack-tip enrichment serves two major purposes:

- If the crack is terminated in the interior of an element, then enriching the crack-tip element with the Heaviside function solely would be not sufficiently accurate. The crack-tip enrichment functions may ensure that the crack terminates precisely at the crack-tip.
- The use of linear elastic asymptotic crack-tip fields serve as suitable enrichment functions because they possess the correct near-tip behavior even with relatively coarse FE mesh (Möes et al. 1999; Daux et al. 2000; Sukumar et al. 2000; Sukumar and Prévost 2003).

5.3.1 Displacement Interpolation

Equation (5.37) is sufficient for the element containing crack segment that completely penetrates through the element. However, if an element with n nodes belonging to

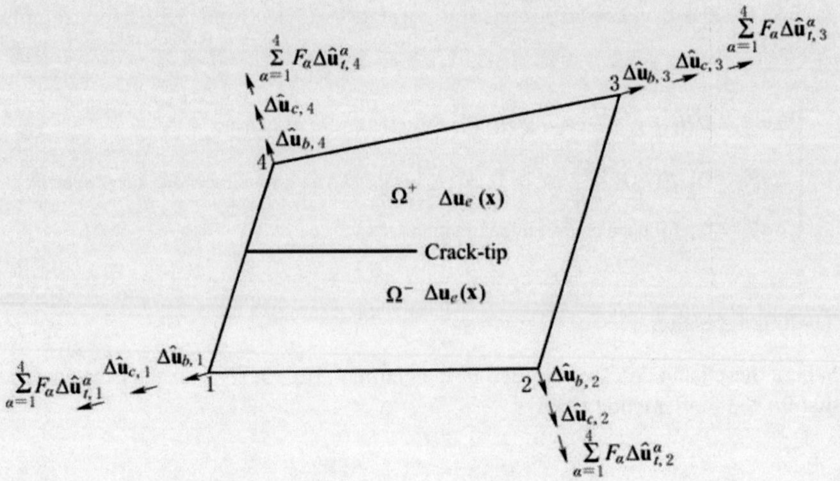

Fig. 5.11 Diagram to the displacement interpolation in element containing crack-tip

set S_t (viz. Fig. 5.11) contains a segment of crack with its tip inside, the enrichment should be extended to take the influence of crack-tip in a manner of

$$\Delta\mathbf{u}_e(\mathbf{x}) = \sum_{l=1}^{n} \mathbf{N}_l(\mathbf{x})\Delta\hat{\mathbf{u}}_{b,l} + \sum_{l=1}^{n} \mathbf{N}_l(\mathbf{x}) \sum_{\alpha=1}^{4} F_\alpha \Delta\hat{\mathbf{u}}_{t,l}^{\alpha} = \mathbf{N}(\mathbf{x})\Delta\hat{\mathbf{u}}_b + \mathbf{N}(\mathbf{x})\Delta\hat{\mathbf{u}}_t$$
$$(e = 1, 2, 3, \ldots, n_e) \tag{5.46}$$

in which $\mathbf{N}_l(x)$ is the standard shape function matrix, and the enrichment coefficient vector $\Delta\mathbf{u}_t$ is associated with n nodes. For 2-D problems, the amount of enrichment functions for each node l around the crack-tip are 4.

$$\Delta\hat{\mathbf{u}}_t = \sum_{\alpha=1}^{4} F_\alpha \Delta\hat{\mathbf{u}}_t^{\alpha} \tag{5.47}$$

$$\Delta\hat{\mathbf{u}}_t^{\alpha} = \begin{Bmatrix} \Delta\hat{\mathbf{u}}_{t,1}^{\alpha} \\ \cdots \\ \Delta\hat{\mathbf{u}}_{t,l}^{\alpha} \\ \cdots \\ \Delta\hat{\mathbf{u}}_{t,k}^{\alpha} \\ \cdots \\ \Delta\hat{\mathbf{u}}_{t,n}^{\alpha} \end{Bmatrix} \quad (\alpha = 1, 2, 3, 4) \tag{5.48}$$

$$\Delta\hat{\mathbf{u}}_{t,l}^{\alpha} = \begin{Bmatrix} \Delta\hat{u}_{t,x,l}^{\alpha} \\ \Delta\hat{u}_{t,y,l}^{\alpha} \end{Bmatrix} \quad \text{for } (\alpha = 1, 2, 3, 4) \text{ and } (l = 1, 2, ..., n) \tag{5.49}$$

The crack-tip enrichment functions $F_{\alpha}(\alpha = 1, 2, 3, 4)$ in isotropic elasticity are derived from the asymptotic displacement fields (viz. Eq. (3.191)) (Bui 1978; Fleming et al. 1997) such that

$$\begin{aligned} \mathbf{F}(r,\theta) &= \left[\, F_1(r,\theta)\ F_2(r,\theta)\ F_3(r,\theta)\ F_4(r,\theta) \,\right]^{\mathrm{T}} \\ &= \left[\, \sqrt{r}\sin\left(\tfrac{\theta}{2}\right)\ \sqrt{r}\cos\left(\tfrac{\theta}{2}\right)\ \sqrt{r}\sin\left(\tfrac{\theta}{2}\right)\sin\theta\ \sqrt{r}\cos\left(\tfrac{\theta}{2}\right)\sin\theta \,\right]^{\mathrm{T}} \end{aligned} \tag{5.50}$$

in which (r, θ) is a polar coordinate system with origin at the crack-tip (viz. Fig. 3.13). Note that the first term in Eq. (5.50) is discontinuous across the crack surface whereas the last three terms are continuous.

Under the circumstances where the crack-tip is at or very close to the edge of element, the "Heaviside jump function" also might be employed to intrude the discontinuous property across the crack surface, namely, approximation (5.46) will be further extended in a form of

$$\Delta\mathbf{u}_e(\mathbf{x}) = \sum_{l=1}^{n} \mathbf{N}_l \Delta\hat{\mathbf{u}}_{b,l} + \sum_{l=1}^{n} H(\mathbf{x}) \mathbf{N}_l \Delta\hat{\mathbf{u}}_{c,l} + \sum_{l=1}^{n} \mathbf{N}_l(\mathbf{x}) \sum_{\alpha=1}^{4} F_{\alpha} \Delta\hat{\mathbf{u}}_{t,l}^{\alpha}$$
$$(e = 1, 2, 3, ..., n_e) \tag{5.51}$$

5.3.2 *Strain and Stress*

The strain increment in the element containing crack-tip is now given as

$$\Delta\boldsymbol{\varepsilon}(\mathbf{x}) = \mathbf{L}\Delta\mathbf{u}_e = \sum_{l=1}^{n} \mathbf{B}_l(\mathbf{x}) \Delta\,\hat{u}_{b,l} + \sum_{l=1}^{n} H(\mathbf{x}) \mathbf{B}_l(\mathbf{x}) \Delta\,\hat{u}_{c,l} + \sum_{l=1}^{n} \sum_{\alpha=1}^{4} \mathbf{B}_{t,l}^{\alpha}(\mathbf{x}) \Delta\,\hat{u}_{t,l}^{\alpha} \tag{5.52}$$

where

$$\mathbf{B}_{t,l}^{\alpha}(\mathbf{x}) = F_{\alpha}(\mathbf{x}) \mathbf{B}_l(\mathbf{x}) \quad \text{for} \quad (l = 1, 2, ..., n) \quad \text{and} \quad (\alpha = 1, 2, 3, 4) \tag{5.53}$$

Consequently, the stress increment can be determined by using an adequate constitutive law

$$\Delta\boldsymbol{\sigma} = \mathbf{D}\left(\sum_{l=1}^{n}\mathbf{B}_l(\mathbf{x})\Delta\hat{\mathbf{u}}_{b,l} + \sum_{l=1}^{n}H(\mathbf{x})\mathbf{B}_l(\mathbf{x})\Delta\hat{\mathbf{u}}_{c,l} + \sum_{l=1}^{n}\sum_{\alpha=1}^{4}\mathbf{B}_{t,l}^{\alpha}(\mathbf{x})\Delta\hat{\mathbf{u}}_{t,l}^{\alpha}\right) - \Delta\boldsymbol{\sigma}^{vp} \tag{5.54}$$

5.3.3 Governing Equation Set

Use is made of the virtual work principle (5.36), the equilibrium condition of the enriched element containing both the crack segment and crack-tip are interpreted by a set of linear equations

$$\begin{bmatrix} \mathbf{k}_{b,b} & \mathbf{k}_{b,c} & \mathbf{k}_{b,1} & \ldots & \mathbf{k}_{b,4} \\ \mathbf{k}_{c,b} & \mathbf{k}_{c,c} & \mathbf{k}_{c,1} & \ldots & \mathbf{k}_{c,4} \\ \mathbf{k}_{1,b} & \mathbf{k}_{1,c} & \mathbf{k}_{1,1} & \ldots & \mathbf{k}_{1,4} \\ \ldots & \ldots & \ldots & \ldots & \ldots \\ \mathbf{k}_{4,b} & \mathbf{k}_{4.c} & \mathbf{k}_{4,1} & \ldots & \mathbf{k}_{4,4} \end{bmatrix} \begin{Bmatrix} \Delta\hat{\mathbf{u}}_b \\ \Delta\hat{\mathbf{u}}_c \\ \Delta\hat{\mathbf{u}}_{t,1} \\ \ldots \\ \Delta\hat{\mathbf{u}}_{t,4} \end{Bmatrix} = \begin{Bmatrix} \Delta\mathbf{f}_b \\ \Delta\mathbf{f}_c \\ \Delta\mathbf{f}_{t,1} \\ \ldots \\ \Delta\mathbf{f}_{t,4} \end{Bmatrix} + \begin{Bmatrix} \Delta\mathbf{f}_b^{vp} \\ \Delta\mathbf{f}_c^{vp} \\ \Delta\mathbf{f}_{t,1}^{vp} \\ \ldots \\ \Delta\mathbf{f}_{t,4}^{vp} \end{Bmatrix} \tag{5.55}$$

in which the additional stiffness matrices considering the coupling effects between "normal" DOFs, "enriched" DOFs for crack segment, and "enriched" DOFs for crack-tip, are

$$\begin{cases} \mathbf{k}_{b,\alpha} = \mathbf{k}_{\alpha,b} = \iint_{\Omega_e}\mathbf{B}^{\mathrm{T}}\mathbf{D}\mathbf{B}_t^{\alpha}\mathrm{d}\Omega \ (\alpha = 1, \ldots, 4) \\ \mathbf{k}_{c,\alpha} = \mathbf{k}_{\alpha,c} = \iint_{\Omega_e}H(\mathbf{x})\mathbf{B}^{\mathrm{T}}\mathbf{D}\mathbf{B}_t^{\alpha}\mathrm{d}\Omega \ (\alpha = 1, \ldots, 4) \\ \mathbf{k}_{\alpha,\beta} = \mathbf{k}_{\beta,\alpha} = \iint_{\Omega_e}(\mathbf{B}_t^{\alpha})^{\mathrm{T}}\mathbf{D}\mathbf{B}_t^{\beta}\mathrm{d}\Omega \ (\alpha, \beta = 1, \ldots, 4) \end{cases} \tag{5.56}$$

The right hand side of Eq. (5.55) consists of nodal unbalance force vectors, where the additional contribution from the crack-tip is expressed as

$$\begin{cases} \Delta\mathbf{f}_{t,\alpha} = \iint_{\Omega_e}\mathbf{N}^{\mathrm{T}}F_{\alpha}\Delta\mathbf{b}\mathrm{d}\Omega + \int_{\Gamma_t}\mathbf{N}^{\mathrm{T}}F_{\alpha}\Delta\bar{\mathbf{t}}\,\mathrm{d}\Gamma + \sum_{\mathbf{x}_{\zeta}\in e}\mathbf{N}^{\mathrm{T}}(\mathbf{x}_{\zeta})F_{\alpha}\Delta\mathbf{T}_{\zeta}(\mathbf{x}_{\zeta}) \\ \Delta\mathbf{f}_{t,\alpha}^{vp} = \iint_{\Omega_e}(\mathbf{B}_t^{\alpha})^{\mathrm{T}}\Delta\boldsymbol{\sigma}^{vp}\mathrm{d}\Omega \end{cases} \quad (\alpha = 1, \ldots, 4) \tag{5.57}$$

There are two difficulties in the integration of relevant sub-matrices and sub-vectors (Sukumar and Prévost 2003): the discontinuity along the crack, and the singularity at the crack-tip. Ventura (2006) showed how the standard Gauss quadrature could be employed in the elements with crack without splitting the elements in sub-cells or presenting any additional approximation. However, it is prevalent that the sub-domain integration scheme is usually implemented by partitioning the cracked element into standard sub-elements (viz. Fig. 4.12) first, then the Gaussian-Legendre product rule (viz. Eq. (4.122)) will be exercised.

Comparing Eqs. (5.55) and (5.37), it is evident that if an element need be simultaneously enriched by both the crack plane and crack-tip (not always), the element DOFs are further extended by six times w.r.t. original homogeneous element. With the help of well-designed link list concerning local node (inclusive basic and enriched) sequences and global node sequences, namely, $i = \left\langle \begin{matrix} l \\ b, c, t \end{matrix} \right\rangle$ and $j = \left\langle \begin{matrix} k \\ b, c, t \end{matrix} \right\rangle$, the sub-matrices (5.56) and sub-vectors (5.57) may be assembled into the extended global equilibrium equation set of the structure system in a similar manner of CEM illustrated previously in this chapter.

5.3.4 *Crack-Tip Enrichment Strategy*

The crack-tip enrichment described above is a conventional approach dubbed as "topological enrichment" on the element containing crack-tips. The nodes belonging to such an element is enriched only, the enriched area therefore directly depends on the mesh density (viz. Fig. 5.12a). Béchet et al. (2005) improved the robustness of XFEM for the stress analysis around cracks by introducing a different enrichment scheme in which a given domain is enriched regardless of mesh density (viz. Fig. 5.12b). Similar geometrical enrichment schemes in square zone for quadrilateral element mesh are also proposed (Xiao and Karihaloo 2006). It is normally said that one more layer of elements surrounding the crack-tip leads to a dramatic reduction of compotation error by at least 90%.

The geometrical enrichment is particularly useful where the crack-tip is ended at the element node or edge as shown in Fig. 5.13.

5.3.5 *Numerical Implementation*

1. **Simulation of crack growth**

Theoretically, a FE mesh may be crossed by any curved crack (represent by the continuous line in Fig. 5.14). Hence in calculation the crack should firstly be discretized. A broken line in each cracked element (viz. Fig. 5.14) is customarily used to replace the real crack.

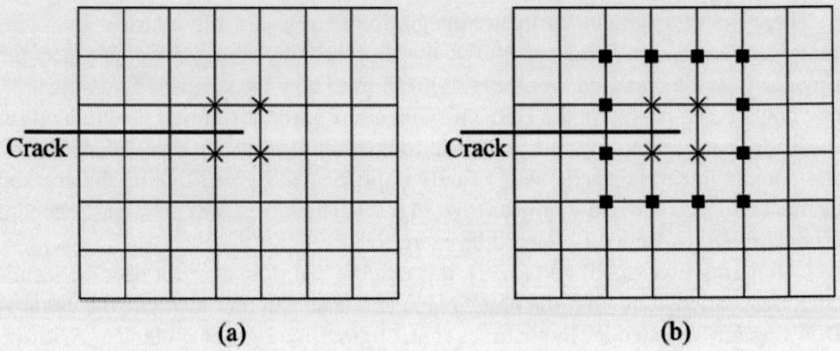

Fig. 5.12 Selected elements for the enrichment. a Topological enrichment; b geometrical enrichment

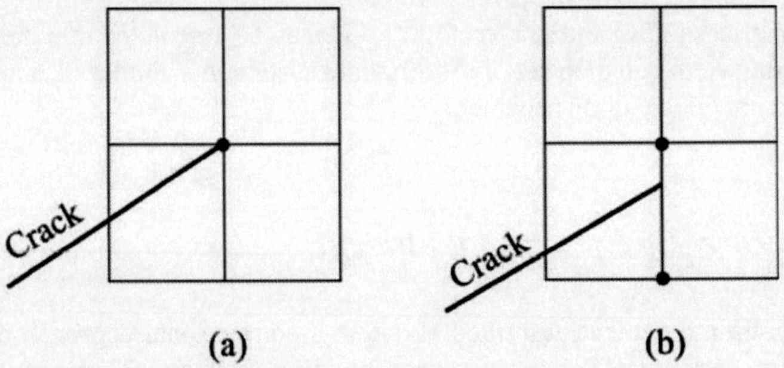

Fig. 5.13 Position of crack-tip. a At node; b at edge

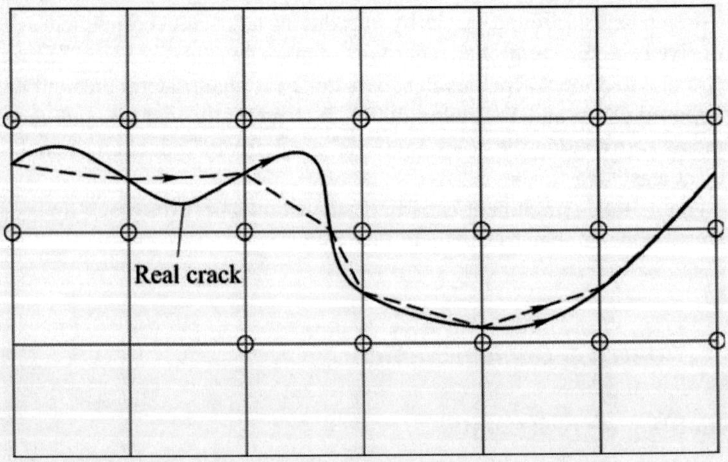

Fig. 5.14 Finite element mesh crossed by a crack

After defining the crack position, certain nodes should be chosen for the DOF enrichment. The rules for picking these nodes are simple and clear: all the nodes associated with the elements crossed by crack should be enriched. For each of these nodes, "additional" DOFs are added for representing the displacement jump or/and stress singularity caused by the crack. Furthermore, the crack has its growth direction, as illustrated in Fig. 5.14 from left to right.

There is an important problem that the stress state of a newly split element needs to be adjusted. There are basically two methods for this purpose: one is to integrate the traction on the cracking segment, then it is looked at as a distributed force exerted on the newly created crack segment; another is to make use of balance condition to get the unbalance nodal force at the newly cracked element. The latter will be implemented by the "restraint-relaxation" procedure described as follows.

① Calculate the additional stress before crack growth. Figure 5.15a shows an element in which $\Delta \mathbf{f}_t$ is the nodal force vector balanced by the enriched (singular) stress $\Delta \boldsymbol{\sigma}_t$. Under the circumstances of linear elastic state, from Eq. (5.52) one gets

$$\Delta \boldsymbol{\sigma}_t = \mathbf{D} \sum_{l=1}^{n} \sum_{\alpha=1}^{4} \mathbf{B}_{t,l}^{\alpha}(\mathbf{x}) \Delta \hat{\mathbf{u}}_{t,l}^{\alpha} \tag{5.58}$$

② Calculate the unbalanced nodal force. Suppose the element is fully clamped, the stress $\Delta \boldsymbol{\sigma}_t$ may be looked at as the initial stress

$$\Delta \boldsymbol{\sigma}_t^0 = \Delta \boldsymbol{\sigma}_t \tag{5.59}$$

Since the imagined clamp does not actually exist, the unbalanced nodal force before and after the cracking will cause the strain and stress adjustment. Figure 5.15b shows the element before (in broken line) and after (in solid line) the adjustment where the unbalanced nodal forces $\Delta \mathbf{f}_b^t$ and $\Delta \mathbf{f}_c^t$ to release the element singularity are calculated by

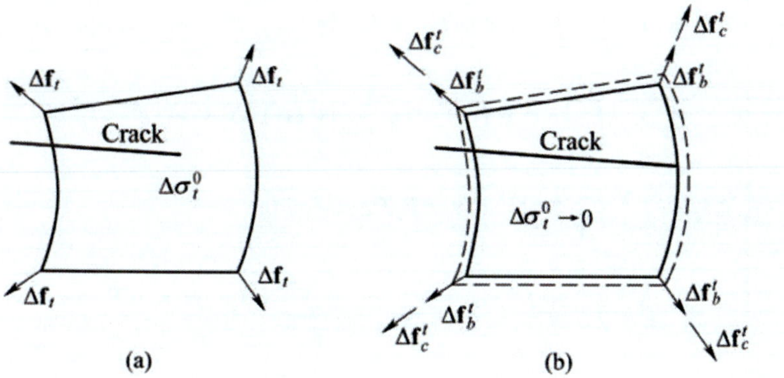

Fig. 5.15 Diagram to the calculation of unbalanced force. **a** Before cracking; **b** after cracking

$$\begin{cases} \Delta \mathbf{f}_b^t = \iint\limits_{\Omega_e} \mathbf{B}^{\mathrm{T}} \Delta \boldsymbol{\sigma}_t^0 \mathrm{d}\Omega \\ \Delta \mathbf{f}_c^t = \iint\limits_{\Omega_e} H(\mathbf{x}) \mathbf{B}^{\mathrm{T}} \Delta \boldsymbol{\sigma}_t^0 \mathrm{d}\Omega \end{cases} \tag{5.60}$$

These unbalanced forces will be employed to compute the incremental displacement, strain and stress due to the crack growth, namely, Eq. (5.51) will be replaced by Eq. (5.28), and the update operation is undertaken before the next step of crack growth iteration as follows

$$\begin{cases} \Delta \mathbf{f}_b^{vp} \to \Delta \mathbf{f}_b^{vp} + \Delta \mathbf{f}_b^t \\ \Delta \mathbf{f}_c^{vp} \to \Delta \mathbf{f}_c^{vp} + \Delta \mathbf{f}_c^t \end{cases} \tag{5.61}$$

2. **Quadrature**

For a crack crossed element or/and crack-tip embedded element, both the elemental stiffness matrices and nodal force vectors (viz. Eqs. (5.38)–(5.43) and (5.56)–(5.57) and (5.60)) should be integrated over the whole element. However, not as that for the fundamental FEM and CEM, the function in the integrant is no longer continuous, because the Heaviside function gives different values in sub-domains Ω^+ and Ω^-. Therefore, the integration should be performed respectively in each sub-domain.

In 2-D case, the author of this book divides element into a set of sub-triangles, then each triangle is further divided into sub-quadrilaterals in the sequence of i_{Ω_e}, as illustrated in Fig. 4.12. Simultaneously, the crack segments and boundary segments (if any) are cut into sub-line segments in the sequence of j_{Γ_c} or/and j_{Γ_t}. Taking Eq. (5.42) for example, it can be additively calculated using the Gaussian quadrature on the normalized coordinates ξ and η of every sub-divided quadrilateral element i_{Ω_e} and the normalized coordinates ξ of every sub-divided line (bar) element j_{Γ_t}

$$\begin{cases} \Delta \mathbf{f}_b = \sum\limits_{i_{\Omega_e}} \int_{-1}^{1} \int_{-1}^{1} \mathbf{N}^{\mathrm{T}} \Delta \mathbf{b} |J(\xi, \eta)| \mathrm{d}\xi \mathrm{d}\eta \\ \qquad + \sum\limits_{j_{\Gamma_t}} \int_{-1}^{1} \mathbf{N}^{\mathrm{T}} \Delta \bar{\mathbf{t}} |J(\xi)| \mathrm{d}\xi + \sum\limits_{\mathbf{x}_\zeta \in e} \mathbf{N}^{\mathrm{T}}(\mathbf{x}_\zeta) \Delta \mathbf{T}_\zeta(\mathbf{x}_\zeta) \\ \Delta \mathbf{f}_c = \sum\limits_{i_{\Omega_e}} \int_{-1}^{1} \int_{-1}^{1} \mathbf{N}^{\mathrm{T}} H(\mathbf{x}) \Delta \mathbf{b} |J(\xi, \eta)| \mathrm{d}\xi \mathrm{d}\eta \\ \qquad + \sum\limits_{j_{\Gamma_t}} \int_{-1}^{1} \mathbf{N}^{\mathrm{T}} H(\mathbf{x}) \Delta \bar{\mathbf{t}} |J(\xi)| \mathrm{d}\xi + \sum\limits_{\mathbf{x}_\zeta \in e} \mathbf{N}^{\mathrm{T}}(\mathbf{x}_\zeta) H(\mathbf{x}) \Delta \mathbf{T}_\zeta(\mathbf{x}_\zeta) \end{cases} \tag{5.62}$$

in which $|J|$ is the Jacobian determinant.

5.4 Crack Growth

When the crack growth (propagation) should be considered, two important problems emerge. The first one concerns the determination of macroscopic crack inception, i.e., the onset criterion for the transition from diffuse damage to macroscopic and localized crack. The second is related to the determination of the crack growth direction.

In the fundamental FEM, there are a number of well-known crack growth criteria to calibrate crack onset and growth direction according to fracture mechanics. One example is the "maximum hoop (circumferential) stress criterion" (Erdogan and Sih 1963; Song et al. 2006) that makes use the averaged stresses on the nearby integration points and the material strength. Another important example is the "SIFs based criteria" that may be further distinguished as direct and indirect types. Direct criteria use the singularity of the elements at crack-tip to get SIFs straightforwardly, whereas indirect criteria comprising "displacement method", "stress method" and "maximum energy release rate criterion" (Nuismer 1975) employ the stress or displacement obtained by the computation to undertake further analysis for SIFs. In addition, third example may be given by the special criteria such as the "maximum strain energy density criterion" (Sih 1974), the "strip-yield model" according to Dugdale (1960), the "cohesive force model" according to Barenblatt (1962), and the "J-integral method". It is worthwhile to note that direct criteria require a FEM mesh with very small elements close to the crack-tip, which limits its applicability to complex problems, whereas indirect and other special criteria mentioned above permit the use of larger elements. It is interesting to indicate that all these criteria are known to give coherent results: they predict slightly different angles for the initial kink, but they all predict that once the kink is initiated, the crack trajectory is such that $K_{II} = 0$ (principle of local symmetry).

SIF criterion has been the most prevalent in the fundamental FEM analysis for concrete structures (e.g. dams). In this criterion, the stress near crack-tip theoretically approaches infinity at the crack-tip according to the expression $\sigma = K/\sqrt{2\pi r}$, where r is the distance from the crack-tip and K is the SIF depending on the load and crack dimensions, etc. When K reaches a critical value K_c, the crack propagates.

In this book, the crack criterion and crack growth direction are handled by "linear elastic fracture mechanics (LEFM)" that has been addressed in **Chaps.** 2 and 3, where the stress and displacement methods for the SIFs by least square fitting and the J-integral method are employed. These methods are widely exercised in the fundamental FEM hence there is no essential difference in the application to XFEM and CEM in this chapter, and NMM in **Chap.** 6 as well.

5.4.1 Cracking Criteria

The factor SIF K_a is a parameter depending on the load acting on elements, the dimensions of the crack and the sizes of the specimen (Chan et al. 1970; Linsbauer et al. 1989a, b). In the simplest and most frequent case, a crack belongs Mode I fracture and the index a will be equal to I. Whereas if the crack evolves in a complex stress state, the index a will be equal to II or III, respectively.

In XFEM, due to the enrichment for near-tip field, the SIFs can be directly expressed as a linear combination of enriched coefficients. In 2-D problem, if the crack-tip is located inside an element e, then for a pure mode I crack, K_I may be directly calculated by (Sukumar and Prévost 2003):

$$K_I = \frac{E\sqrt{2\pi}}{4(1-\nu^2)} \sum_{l=1}^{n} N_l(\mathbf{x}) u_{t,y,l}^1 \quad \text{where } \mathbf{x} \in \Omega_e \tag{5.63}$$

in which $N_l(\mathbf{x})$ is the finite element shape function for local node l, and $u_{t,y,l}^1$ is the enriched coefficients tied to near-tip function F_1 and the displacement approximation in the y-direction for node l.

However, such a direct extraction of SIFs is not sufficiently accurate, hence extrapolation (displacement or stress) techniques or crack-tip flux integrals are normally employed to indirectly get SIFs. In general, the latter leads to higher accuracy.

1. **Stress extrapolation method for SIFs**

According to the asymptotic stress function (3.191) around a crack-tip, a general expression when $\theta = 0$ may be deduced

$$\sigma = K/\sqrt{2\pi r} \tag{5.64}$$

Specifically

$$\begin{cases} K_I = \lim\limits_{r\to 0}[\sigma_y(r, \theta = 0)\sqrt{2\pi r}] \\ K_{II} = \lim\limits_{r\to 0}[\tau_{xy}(r, \theta = 0)\sqrt{2\pi r}] \end{cases} \tag{5.65}$$

Take K_I for example and suppose that a set of data (K_{Ii}, r_i), $(i = 1, 2, ..., N)$ has been obtain by the XFEM computation in which $K_{Ii} = \sigma_{yi}\sqrt{2\pi r_i}$, the asymptotic characteristic may be elucidated in Fig. 5.16.

We construct the linear fitting formula such that

$$\bar{K} = Ar + B \tag{5.66}$$

in which $r = 0$, $\bar{K}_I = K_I = B$. The normal equation of least-squares method is established

$$S = \sum_{i}^{N} (\bar{K}_{Ii} - K_{Ii})^2 = \sum_{i}^{N} (Ar_i + B - K_{Ii})^2 = \min. \tag{5.67}$$

The solution of Eq. (5.67) gives rise to

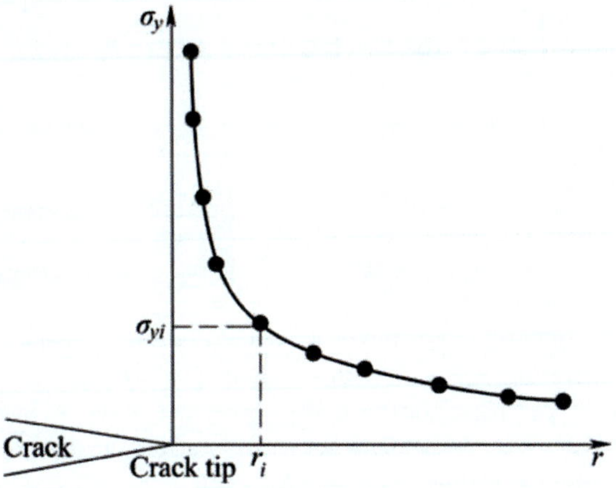

Fig. 5.16 Stress distribution near the crack surface ($\theta = 0$)

$$\begin{cases} A = \dfrac{\sum\limits_i^N r_i \sum\limits_i^N K_{Ii} - N \sum\limits_i^N r_i K_{Ii}}{\left(\sum\limits_i^N r_i\right)^2 - N \sum\limits_i^N r_i^2} \\ K_I \approx B = \dfrac{\sum\limits_i^N r_i \sum\limits_i^N r_i K_{Ii} - \sum\limits_i^N r_i^2 \sum\limits_i^N K_{Ii}}{\left(\sum\limits_i^N r_i\right)^2 - N \sum\limits_i^N r_i^2} \end{cases} \tag{5.68}$$

The computation of K_{II} may be identically undertaken to get

$$\begin{cases} A = \dfrac{\sum\limits_i^N r_i \sum\limits_i^N K_{IIi} - N \sum\limits_i^N r_i K_{IIi}}{\left(\sum\limits_i^N r_i\right)^2 - N \sum\limits_i^N r_i^2} \\ K_{II} \approx B = \dfrac{\sum\limits_i^N r_i \sum\limits_i^N r_i K_{IIi} - \sum\limits_i^N r_i^2 \sum\limits_i^N K_{IIi}}{\left(\sum\limits_i^N r_i\right)^2 - N \sum\limits_i^N r_i^2} \end{cases} \tag{5.69}$$

2. **Displacement extrapolation method for SIFs**

If $\theta = \pm\pi$, we obtain the relative displacement of the crack walls according to Eq. (3.191) such that

$$\begin{cases} \Delta u_x = u_x(\theta = +\pi, r) - u_x(\theta = -\pi, r) = K_{II}\sqrt{\dfrac{r}{2\pi}}(1+\chi)/G \\ \Delta u_y = u_y(\theta = +\pi, r) - u_y(\theta = -\pi, r) = K_{I}\sqrt{\dfrac{r}{2\pi}}(1+\chi)/G \end{cases} \tag{5.70}$$

$$\chi = \begin{cases} \dfrac{3-\nu}{1+\nu} & \text{plane stress} \\ 3-4\nu & \text{plane strain} \end{cases} \tag{5.71}$$

In the computation of Δu_{xi} and Δu_{yi}, a number of node pairs should be arranged according to Fig. 5.17 in a manner of (4,2), (5,3), etc.

The basic algorithm of displacement extrapolation for SIFs K_I and K_{II} is identical to the stress extrapolation, as has been illustrated in Eqs. (5.66)–(5.69). This displacement extrapolation method has been implemented in the commercial software such as ANSYS.

To the knowledge of author, 3-D SIF computation has not been satisfactorily handled yet. It differs from 2-D case because of extra geometric complexity, plus the eigenvalue problem to determine the singularity exponents and the associated 3-D primal and dual modes (Doitrand et al. 2020).

3. **Path independent *J*-integral**

The most accurate method for SIFs that allows to get rid of the mixed modes is based on a path independent integral. Rice and his co-workers developed a new toughness measure without the restraint of linear-elastic approximation ahead of crack-tip. Their analysis leads to the design of *J*-integral (Rice 1968; Rice and Rosengren 1968).

In 3-D problems, *J*-integral is actually a vector **J** with three components mathematically defined as (viz. Fig. 5.18).

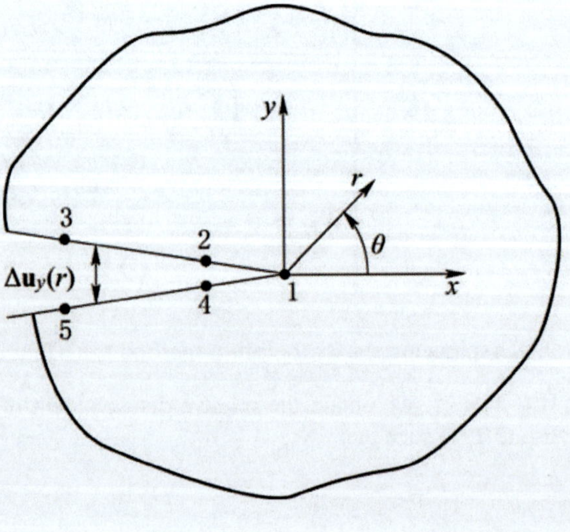

Fig. 5.17 Nodal pairs for the path definition in displacement extrapolation

$$J_\alpha = \int_\Gamma \left(w.n_\alpha - t_\beta \frac{\partial u_\beta}{\partial x_\alpha} \right) \mathrm{d}\Gamma \text{ with } w = \int_0^\varepsilon \boldsymbol{\sigma}^\mathrm{T} \mathrm{d}\boldsymbol{\varepsilon} \quad \text{for} \quad (\alpha, \beta = 1, 2, 3) \quad (5.72)$$

where: Γ—arbitrary clockwise path around crack-tip; w—density of strain energy; t_β—components of traction vectors; u_β—components of displacement vectors; $\mathrm{d}\Gamma$—incremental length along the path.

What is customarily called the "J-integral" in fracture mechanics, is the first component of vector **J**, namely, the component in x-direction. Suppose a crack is now introduced along the x-axis in Fig. 5.19. The J-integral is calculated along a closed path, which runs around the crack-tip and along both the walls of crack surface. The integral can be written as a summation of four parts. Obviously the calculated value is zero for a domain without crack.

Fig. 5.18 Closed curve with unit normal vector and stress vector

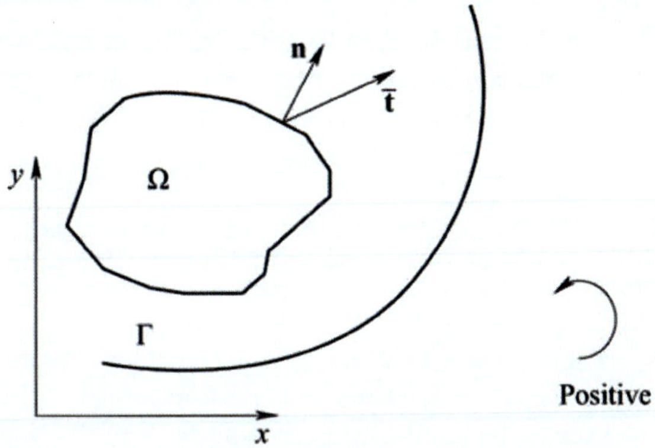

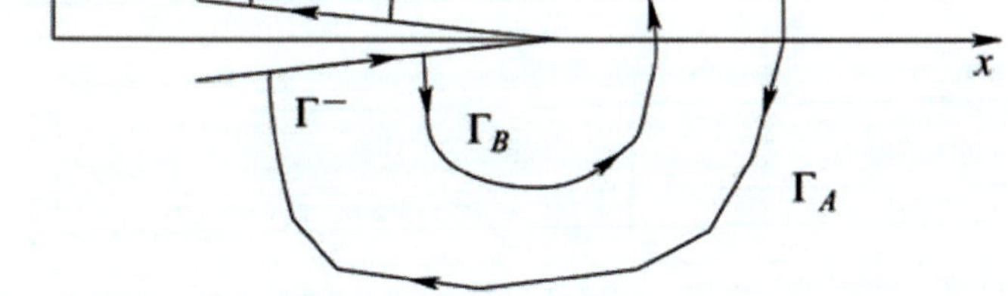

Fig. 5.19 Closed curve around crack

It is notable that if the crack is not loaded, the traction is null ($t_\beta = 0$), otherwise the stress components of $\boldsymbol{\sigma}_c$ that is the accumulation of Eq. (5.34) should be employed

$$\begin{cases} t_x = \tau_{c,yx} \\ t_y = \sigma_{c,y} \end{cases} \tag{5.73}$$

On the crack surface, the unit normal vector is pointing in y-direction, so its first component n_x is zero. This results in the integrand along the crack walls being null, and the J-integral is the sum of two terms representing the integrand along two lines around the crack-tip. When the movement over the path is anti-clockwise, we assign the value a positive value, otherwise the result is negative. With this sign convention, it appears that the value of both integrals has to be equal. This important result means that the integral over a non-closed path around the crack-tip, starting at one wall of crack surface and ending at the other, always has the same value and so the J-integral is path independent. This character makes use of additional postulations that the material is homogeneous, the domain is not subjected to volume loads, the acceleration is zero, the deformation and rotations are small. With above declarations, we get

$$J = J_1 = J_x = \int_\Gamma \left(w.n_x - t_x \frac{\partial u_x}{\partial x} - t_y \frac{\partial u_y}{\partial x} \right) \mathrm{d}\Gamma \tag{5.74}$$

J-integral represents the change rate of net potential energy with respect to crack advance (per unit thickness of crack front). Also can be thought of as the energy flow into the crack-tip, J-integral is a measure of the singularity strength at the crack-tip. For the Mode I crack, it follows immediately that J-integral is equivalent to the energy release rate G. This means that crack growth criteria can be constructed with J-integral similar to that with K and G.

It is reminded that J-integral reduces to the Griffith theory for linear-elastic body. Specially, formulae (2.33) and (2.35) can be used to infer an equivalent K_{Ic} value from J_{Ic} for high toughness and ductile solids where the validity of K_{Ic} will requires unreasonably large test specimens. In 2-D cases, J-integral is related to the mixed mode SIFs through

$$\begin{cases} J = \dfrac{K_I^2}{E^*} + \dfrac{K_{II}^2}{E^*} \\ E^* = \begin{cases} E & \text{plane stress} \\ E/(1-\nu^2) & \text{plane strain} \end{cases} \end{cases} \tag{5.75}$$

in which E and ν are the Young's modulus and Poisson's ratio, respectively.

J-integral also may be related to the crack-tip opening in CTOD-based criterion. Based on empirical observations, a practical relation is suggested by ASTM, including the influences of specimen dimensions and ultimate strength σ_u.

J-integral has been implemented in a number of commercial FEM codes such as ABAQUS. It is emphasized however that, above analyses are limited to situations where plastic deformation at the crack-tip does not extend to the furthest edge of loaded portion. In addition, although the conventional J-integral defined in Eq. (5.74) may be employed to construct the crack growth criterion with material property J_c, yet it is impossible to separate K_I from K_{II} directly (viz. Eq. (5.75)). Since the cracking direction is dependent on the ratio of K_I/K_{II} using, for example, the "maximum tangential (hoop) stress (MTS)" criterion, the calculation of K_I and K_{II} individually is nontrivial.

In the context of 2-D LEFM, and by the decomposition of displacement field into symmetric and antisymmetric components w.r.t. local crack-tip coordinate system, Xie et al. (1995) developed a practical approach for the issue of mixed-mode crack growth: A VCE technique is used to calculate Mode-I and Mode-II energy release rates from these decomposed displacement fields; the corresponding SIFs are then calculated from the energy release rates. This method may be directly used in XFEM (CEM and NMM as well).

4. **Others**

(1) **Energy balance approach**

In the energy balance approach, it is postulated that a certain amount of energy G_c is absorbed by the formation of a unit area of crack surface. When the crack propagates, a certain amount of stored energy is released, and this released energy is equal to or greater than the absorbed energy. The FEM with rather large elements may be used to determine the energy release rate in the energy balance approach (Salah EI-Din and EI-Adawy Nassef 1975), but the formation of crack cannot be explained.

Hillerborg, Modéer and Petersson (1976) proposed a model to establish the relation of G_c and other parameters, in which the crack is assumed to propagate when the stress at the crack-tip reaches tensile strength σ_T. When the crack opens, the stress will not fall to zero immediately, but will be reduced gradually to zero following the increase of crack width w until critical value w_c (viz. Fig. 5.20). For that part of crack where $w < w_c$, the "crack" in reality corresponds to a micro-cracked zone with some remaining strength for stress transfer. As there is a resistant stress to be overcome in opening the crack, energy needs to be absorbed. To open the crack from zero to or beyond w_c, the amount of energy absorption per unit crack area is the area between the curve and the coordinate axes in Fig. 5.20, namely

$$\int_0^{w_c} \sigma \mathrm{d}w = G_c \tag{5.76}$$

There are a number of simplifications w.r.t. the curves in Fig. 5.20, among them the straight line (viz. Fig. 5.21) seems an acceptable choice, particularly in the engineering application it is normally assumed that the concrete is linear-elastic until σ_T is reached (Evans and Marathe 1968). It follows that Eq. (5.76) gives rise to

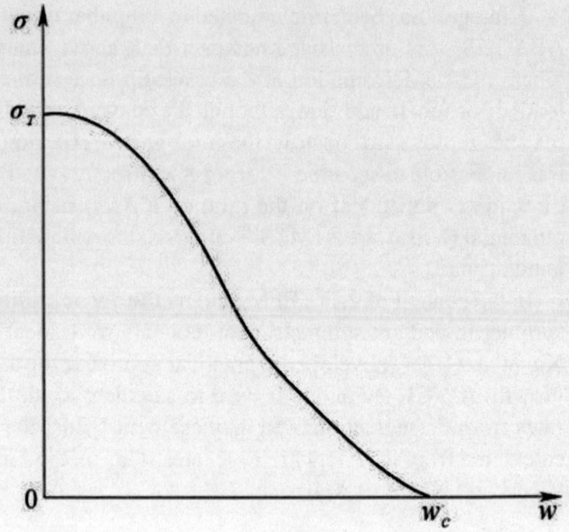

Fig. 5.20 Variation of stress σ with crack width w: general case

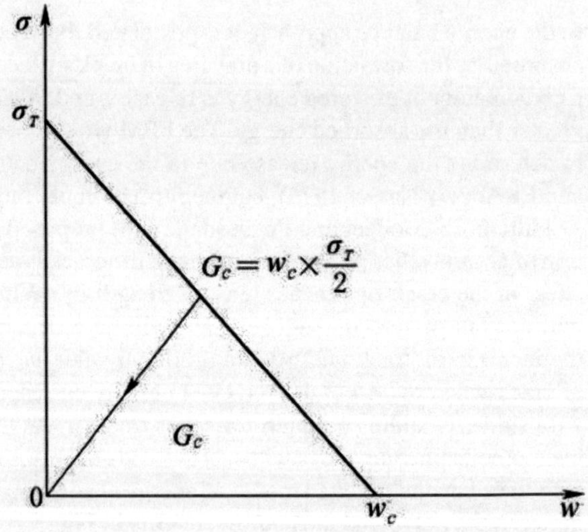

Fig. 5.21 Variation of stress σ with crack width w: linear cohesive law

$$\int_0^{w_c} \sigma \,\mathrm{d}w = \sigma_T w_c/2 = G_c \tag{5.77}$$

or

$$w_c = 2G_c/\sigma_T \tag{5.78}$$

Take the ordinary concrete for example, G_c/σ_T is of the order 0.005–0.01 mm (Welch and Haisman 1969), hence w_c is of the order 0.01– 0.02 mm.

(2) **Domain form of interaction integral**

An improved J-integral termed as "interaction integral method" (Yau et al. 1980) to get K_I and K_{II} individually may be formulated on the basis of conservation integral.

Conservation laws in elasticity have received considerable attention for their applications to science and engineering problems. A general mathematical form of conservation laws states that: the integral of a function of certain field variables over the closed surface surrounding a sub-region in an elastically deformed body, vanishes.

Applications of conservation laws to fracture mechanics leads to the so-called path-independent integrals, among them J-integral is well-known. The value of J-integral is equivalent to Irwin's energy release rate, and may be related to crack-tip SIFs in both linear and nonlinear elastic solids, too. However, one drawback in the application of J-integral approach is the difficulty in dealing with mixed-mode fracture problems, because the SIF solution for each individual mode cannot be directly and explicitly separated.

Fortunately, the domain form of interaction integral (Yau et al. 1980; Moran and Shih 1987; Doitrand et al. 2020) provides a well-established technique to determine mixed-mode SIFs with FEM that may be of course applied to XFEM (CEM and NMM as well) without essential difficulties. In this technique, strain auxiliary fields are introduced and superposed on the actual fields. By judicious choice of the auxiliary fields, the interaction integral can be directly related to the mixed-mode SIFs, and the domain form of interaction integral is computed (Duarte et al. 2001) by the least squares fit method to minimize the errors among the stresses calculated numerically and their asymptotic values. Xiao and Karihaloo (2003) enhanced the accuracy of local fields and determined the SIFs directly without extra post-processing. Liu and his co-workers (Liu et al. 2004) extended this technique to directly evaluate the mixed mode SIFs in homogeneous and bi-materials.

5.4.2 *Crack Growth Direction*

The determination of crack growth orientation is one of the difficult issues in modelling macroscopic cracks. In the past few decades, a series of fracture criteria have been proposed and widely exercised that can be divided into three categories: stress-based criteria, strain-based criteria and energy-based criteria. The most prevalent ones make use of the "maximum tangential (hoop) stress (MTS)", "maximum tangential strain (MTSN)", "maximum energy release rate (MERR)", and "minimum strain energy density (MSED)".

One of the remarkable advantages of stress-based criteria over energy based ones is the simple and clear role of each stress field parameter in the estimation of fracture onset. However, energy-based criteria employ more realistic assumptions taking into account of the energy dissipation mechanism during crack growth. Anyway, both the stress and energy based criteria are subject to the modifications for the better explanation of crack onset under different loading and boundary conditions (Berto and Lazzarin 2014; Lazzarin et al. 2014; Torabi and Pirhadi 2015).

A literature review reveals that, in some materials, strain based criteria provide more realistic description than stress and energy based counterparts (Maiti and Smith 1984; Wu and Li 1989; Nalla et al. 2003). As early as in 1974, Wu pointed out that a meaningful description of concrete fracture must be based on strain. Inspired by his work, Chang (1981) developed the MTSN criterion to study the mixed mode fracture in angled crack problem, and its generalized form seams rather promising in the engineering practices nowadays (Mirsayar 2015).

1. **MTS criterion**

MTS constructs a stress-based criterion on the hypothesis that: a crack-tip will move in the direction where the highest value of tangential stress component is found. Under the polar coordinate system in Fig. 5.22, this can be expressed mathematically by first and second derivatives of the tangential stress component σ_{tt} w.r.t. the angle θ (Erdogan and Sih 1963), namely, the formula

$$\frac{\partial \sigma_{tt}}{\partial \theta} = 0 \tag{5.79}$$

leads to the equation for the solution of crack growth direction angle θ_c as follows

$$K_I \sin\theta_c + K_{II}(3\cos\theta_c - 1) = 0 \tag{5.80}$$

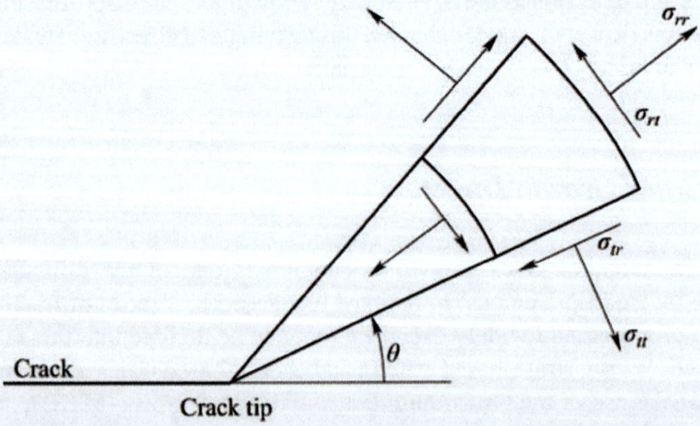

Fig. 5.22 Cylindrical stress components at the crack-tip

Equation (5.80) normally gives two solutions

$$\theta_c = 2\tan^{-1}\frac{1}{4}\left(\frac{K_I}{K_{II}} \pm \sqrt{\left(\frac{K_I}{K_{II}}\right)^2 + 8}\right) \tag{5.81}$$

but the condition

$$\frac{\partial^2 \sigma_{tt}}{\partial \theta^2} < 0 \tag{5.82}$$

or

$$\frac{3}{4}\frac{K_I}{\sqrt{2\pi r}}\left(-\frac{1}{4}\cos\frac{\theta_c}{2} - \frac{3}{4}\cos\frac{3\theta_c}{2}\right) + \frac{3}{4}\frac{K_{II}}{\sqrt{2\pi r}}\left(\frac{1}{4}\sin\frac{\theta_c}{2} + \frac{9}{4}\sin\frac{3\theta_c}{2}\right) < 0 \tag{5.83}$$

help to select a correct θ_c. If $K_{II} = 0$, then $\theta_c = 0$ (pure mode I). It is notable that if $K_{II} > 0$, the crack growth angle $\theta_c < 0$; whereas if $K_{II} < 0$, $\theta_c > 0$.

Once θ_c is fixed, we have

$$\sigma_{tt}(\theta = \theta_c) = \frac{K_{Ic}}{\sqrt{2\pi r}} \tag{5.84}$$

Hence MTS criterion of crack onset may be obtained as

$$\frac{1}{4}\frac{K_I}{K_{1c}}\left(3\cos\frac{\theta_c}{2} + \cos\frac{3\theta_c}{2}\right) + \frac{1}{4}\frac{K_{II}}{K_{1c}}\left(-3\sin\frac{\theta_c}{2} - 3\sin\frac{3\theta_c}{2}\right) = 1 \tag{5.85}$$

A computationally more amenable expression for θ_c is

$$\theta_c = 2\tan^{-1}\left[\frac{-2K_{II}/K_I}{1 + \sqrt{1 + 8(K_{II}/K_I)^2}}\right] \tag{5.86}$$

2. **MTSN criterion**

MTSN criterion states that: a crack will propagate in the direction θ_c where the tangential strain $\varepsilon_{tt}(\theta = \theta_c)$ reaches its maximum value ε_c at a critical distance r_c from the crack-tip. The criterion could simply be obtained by the stationary condition

$$\frac{\partial \varepsilon_{tt}}{\partial \theta} = 0 \tag{5.87}$$

and additional condition for the unique solution of θ_c

$$\frac{\partial^2 \varepsilon_{tt}}{\partial \theta^2} < 0 \tag{5.88}$$

Once θ_c is available, with the linear elastic relationship between the tangential strain and stress components, MTSN criterion of crack onset may be expressed as

$$\begin{cases} \varepsilon_{tt}(\theta = \theta_c, r = r_c) = \varepsilon_c = \sigma_T / E \\ \varepsilon_{tt} = \dfrac{1}{E}(\sigma_{tt} - \nu\sigma_{rr}) \end{cases} \tag{5.89}$$

In which E, ν, σ_T are the Young's modulus, Poisson's ratio, and tensile strength of the material, respectively; ε_{tt} and σ_{rr} are the tangential and radial components of the stress field in cylindrical coordinates with origin at the crack-tip (viz. Fig. 5.22), respectively; critical distance r_c may be regarded as the length of the damaged zone around crack-tip which is a material property (Taylor et al. 2004; Taylor 2008).

For more details, readers are referred to the work of Chang (1981).

3. **SED criterion**

This is an energy-based criterion in which SED U is the stored elastic energy per unit volume. For linear elastic material, this specific energy is easily calculated around crack-tip region and appears to be inversely proportional to the distance to the crack-tip (r). Hence, the strain energy density factor S defined as the product $S = r.U$ is postulated as r-independent.

SED criterion states that: at crack growth, the crack-tip will move toward the point where S is the minimum. For details, readers are referred to the relevant literatures (Sih 1974).

4. **MERR criterion**

Also called as G criterion, MERR criterion is an energy-based criterion, too. The basic consideration is that: the magnitude of the energy release rate at any point will vary with the inclination of crack extension; at some angle w.r.t. the local crack plane, this energy release rate will reach the maximum.

For the application in CMs (e.g. XFEM, CEM, NMM), a set of virtual crack extensions w.r.t. crack front will give different local energy release rates. From these values and their corresponding directions, the nodal maximum energy release rate G and corresponding direction can be found. For details, readers are referred to the relevant literatures (Nuismer 1975; Claydon 1992).

5.4.3 Implementation

The computation procedure for the crack growth issue in elastic body is flowcharted in Fig. 5.23.

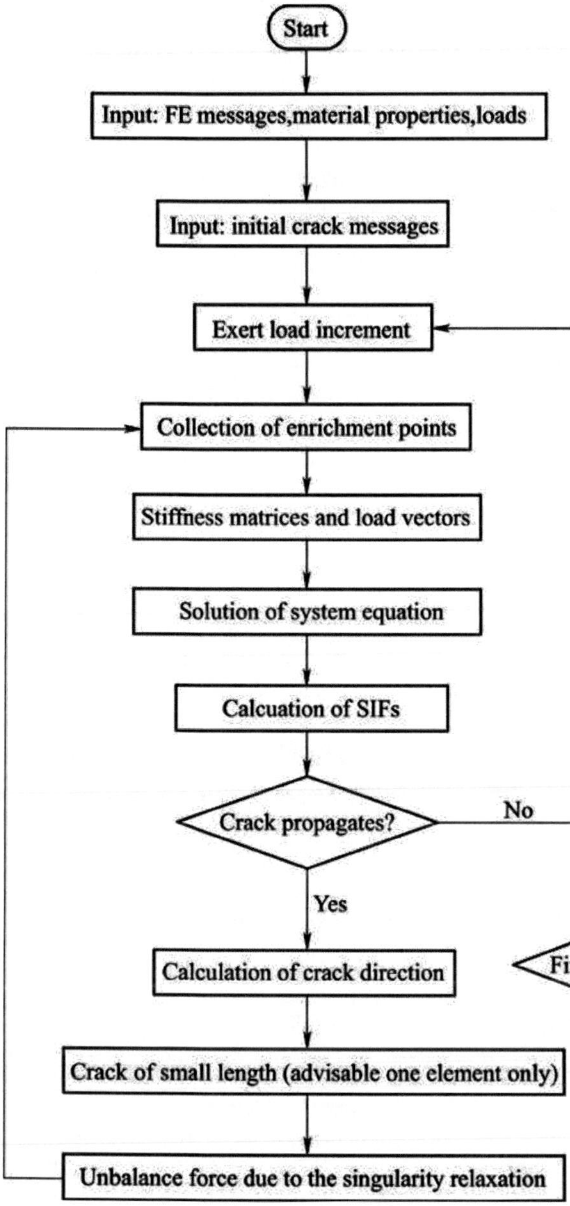

Fig. 5.23 Flowchart for the computation using XFEM

5.5 Validations and Applications

5.5.1 *Elastic and Discontinuous Problem of 1-D bar*

A 1-D elastic and discontinuous bar under the action of tension in Fig. 5.24 is studied for the verification of the principle and basic algorithm of XFEM, and more important, for the justification that the basic algorithms of XFEM and fundamental FEM (with joint element) as well as CEM provide identical results subject to the neglect of the influence of mesh density.

The bar with a length 2L is divided into two equal parts Ω^+ and Ω^- by the traction interface Γ_c with zero initial opening (aperture), and is submitted to an imposed displacement $\bar{u}$ at its right end. The Young's modulus of the bar is E and the stiffness of the interface is k_n.

The virtual work principle (5.36) may be simplified for this bar problem as below

$$\int_{\Omega} \delta\varepsilon\sigma \mathrm{d}\Omega + \delta u_c \sigma_c\big|_{\Gamma_c} = \sum_{k=1}^{n_k} \delta u(x_k) T_k(x_k) \tag{5.90}$$

1. **Fundamental FEM**

In Fig. 5.25a, each bar sub-part (Ω^- and Ω^+) is meshed with one linear finite element only ($e1$ and $e2$), separated by an traction interface on which no boundary conditions are prescribed. On each node, only one degree of freedom (DOF) is considered, u_i representing the nodal displacement in the bar direction. Using classical 2-nodal joint element $sp_{e1,e2}$ for the traction interface, where node 2 corresponds to Γ_c^-, and node 3 to Γ_c^+, on these two nodes the displacements are unknown and will be solved finally.

The displacements imposed on nodes 1 and 4 are 0.0 and $\bar{u}$, respectively. On each half bar, strain and stress are constant and calculated individually by

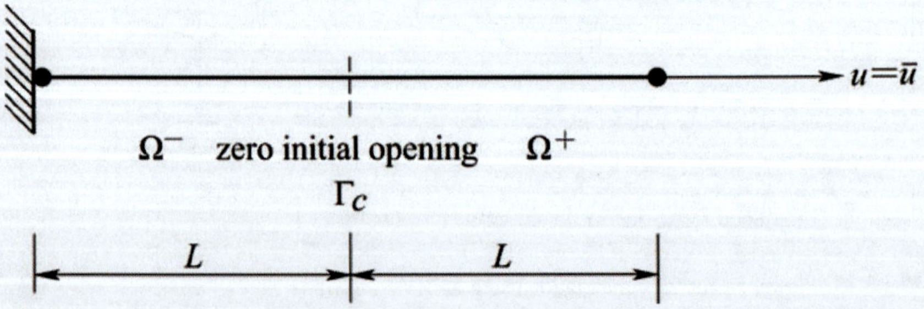

Fig. 5.24 A1-D elastic and discontinuous bar under the tensile action

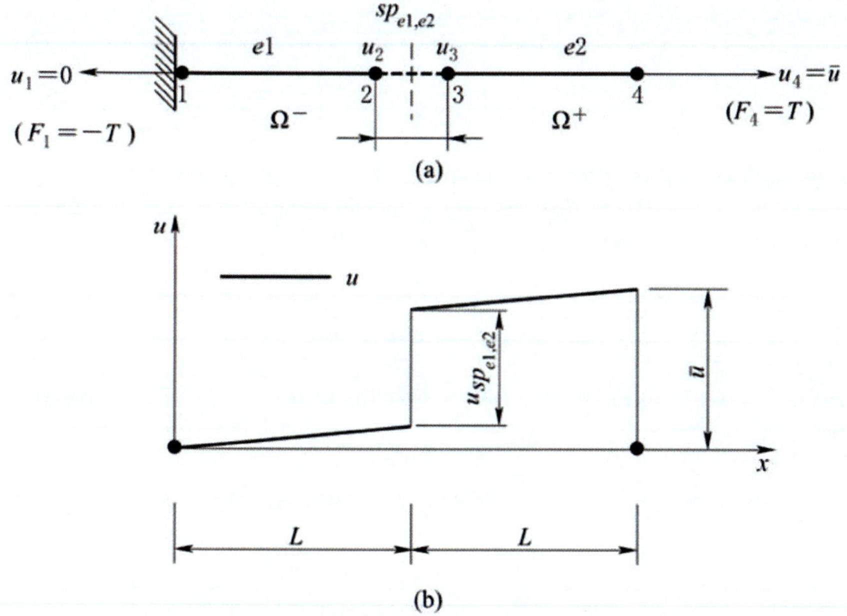

Fig. 5.25 Weak formulation of FEM. **a** Diagram to the calculation of discontinuous bar by FEM; **b** displacement distribution along the bar

$$\begin{cases} \varepsilon_{e1} = \dfrac{u_2 - u_1}{L}, \ \sigma_{e1} = E\varepsilon_{e1} \ \text{where } e1 \in \Omega^- \\ \varepsilon_{e2} = \dfrac{u_4 - u_3}{L}, \ \sigma_{e2} = E\varepsilon_{e2} \ \text{where } e2 \in \Omega^+ \end{cases} \tag{5.91}$$

The contribution of Ω^- to the virtual work in Eq. (5.90) is specified by

$$\int\limits_{\Omega^{-1}} \delta\varepsilon_{e1}\sigma_{e1}\mathrm{d}\Omega = \frac{E}{L^2}(\delta u_2 - \delta u_1)(u_2 - u_1) = \{\delta u_{e1}\}[K_{e1}]\{u_{e1}\} \tag{5.92}$$

where

$$[K_{e1}] = \frac{E}{L}\begin{bmatrix} 1 & -1 \\ -1 & 1 \end{bmatrix} \tag{5.93}$$

is the stiffness matrix of the left half-bar.

The contribution of Ω^+ to the virtual work in Eq. (5.90) is specified by

$$\int\limits_{\Omega^{+1}} \delta\varepsilon_{e2}\sigma_{e2}\mathrm{d}\Omega = \frac{E}{L}(\delta u_4 - \delta u_3)(u_4 - u_3) = \{\delta u_{e2}\}[K_{e2}]\{u_{e2}\} \tag{5.94}$$

where

$$[K_{e2}] = \frac{E}{L}\begin{bmatrix} 1 & -1 \\ -1 & 1 \end{bmatrix} \tag{5.95}$$

is the stiffness matrix of the right half-bar.

On interface Γ_c, the average stress $\sigma_{sp_{e1,e2}}$ and the opening $u_{sp_{e1,e2}}$ can be expressed as

$$\begin{cases} u_{sp_{e1,e2}} = (u_3 - u_2) \\ \sigma_{sp_{e1,e2}} = k_n(u_3 - u_2) \end{cases} \tag{5.96}$$

and consequently, its contribution to the weak formulation in Eq. (5.90) is specified by

$$\delta u_{sp_{e1,e2}} \sigma_{sp_{e1,e2}} = (\delta u_3 - \delta u_2) k_n (u_3 - u_2) = \{\delta u_{sp_{e1,e2}}\}[K_c]\{u_{sp_{e1,e2}}\} \tag{5.97}$$

where

$$[K_c] = k_n\begin{bmatrix} 1 & -1 \\ -1 & 1 \end{bmatrix} \tag{5.98}$$

is the stiffness matrix of interface Γ_c.

The global equilibrium equation set is assembled from Eqs. (5.92), (5.94) and (5.97) as follows

$$[K]\{u\} = \frac{E}{L}\begin{bmatrix} 1 & -1 & 0 & 0 \\ -1 & 1+\frac{Lk_n}{E} & -\frac{Lk_n}{E} & 0 \\ 0 & -\frac{Lk_n}{E} & 1+\frac{Lk_n}{E} & -1 \\ 0 & 0 & -1 & 1 \end{bmatrix}\begin{Bmatrix} u_1 \\ u_2 \\ u_3 \\ u_4 \end{Bmatrix} = \begin{Bmatrix} F_1 \\ F_2 \\ F_3 \\ F_4 \end{Bmatrix} \tag{5.99}$$

Using the boundary conditions ($u_1 = 0$ and $u_4 = \bar{u}$, $F_1 = -F_4$ and $F_3 = F_2 = 0$), it is reduced to

$$\begin{cases} -\dfrac{E}{L}u_2 = F_1 & \text{①} \\ \dfrac{E}{L}\left(1 + \dfrac{Lk_n}{E}\right)u_2 - k_n u_3 = 0 & \text{②} \\ -k_n u_2 + \dfrac{E}{L}\left(1 + \dfrac{Lk_n}{E}\right)u_3 - \bar{u} = 0 & \text{③} \\ \dfrac{E}{L}(-u_3 + \bar{u}) = F_4 & \text{④} \end{cases} \tag{5.100}$$

Its solution gives rise to

$$\begin{cases} u_2 = \bar{u}\dfrac{\frac{Lk_n}{E}}{1+2\frac{Lk_n}{E}} = \bar{u}\dfrac{1}{2+\frac{E}{Lk_n}} \\ u_3 = \bar{u}\dfrac{1+\frac{Lk_n}{E}}{1+2\frac{Lk_n}{E}} = \bar{u}\dfrac{1+\frac{E}{Lk_n}}{2+\frac{E}{Lk_n}} \end{cases} \tag{5.101}$$

$$\begin{cases} F_1 = -\bar{u}\dfrac{\frac{E}{L}}{2+\frac{E}{Lk_n}} \\ F_4 = \dfrac{E}{L}(\bar{u} - u_3) = \bar{u}\dfrac{\frac{E}{L}}{2+\frac{E}{Lk_n}} \end{cases} \tag{5.102}$$

The corresponding interfacial opening u_c is calculated by

$$u_{sp_{e1,e2}} = u_3 - u_2 = \bar{u}\dfrac{\frac{E}{Lk_n}}{2+\frac{E}{Lk_n}} = \bar{u}\dfrac{1}{1+2\frac{Lk_n}{E}} \tag{5.103}$$

Figure 5.26b exposits the displacement distribution along the bar directly by FEM without the help of intermediate variables

2. **CEM**

In Fig. 5.26a, only one composite element with two nodes (1, 4) is employed for the whole bar. The sub-parts (Ω^- and Ω^+) are corresponding to the sub-elements (r_1 and r_2) separated by the traction interface $sp_{r1,r2}$ on which no boundary conditions are prescribed. At each node, there are two degrees of freedom (DOFs) corresponding to $\{u_{r1}\} = [u_{r1,1}\ u_{r1,4}]^{\mathrm{T}}$ and $\{u_{r2}\} = [u_{r2,1}\ u_{r2,4}]^{\mathrm{T}}$ towards the displacement interpolation for sub-elements in the bar direction. On the traction interface $sp_{r1,r2}$, the relative displacement is unknown, but is computed later.

The displacements imposed at nodes 1 and 4 are 0.0 and $\bar{u}$, respectively. On each half bar, strain and stress are constant and calculated individually by

$$\begin{cases} \varepsilon_{r1} = \dfrac{u_{r1,4} - u_{r1,1}}{2L},\ \sigma_{r1} = E\varepsilon_{r1} \text{ where } r1 \in \Omega^- \\ \varepsilon_{r2} = \dfrac{u_{r2,4} - u_{r2,1}}{2L},\ \sigma_{r2} = E\varepsilon_{r2} \text{ where } r2 \in \Omega^+ \end{cases} \tag{5.104}$$

The contribution of r_1 to the virtual work in Eq. (5.90) is specified by

$$\int\limits_{r1} \delta\varepsilon_{r1}\sigma_{r1}\mathrm{d}\Omega = L\frac{E}{4L^2}(\delta u_{r1,4} - \delta u_{r1,1})(u_{r1,4} - u_{r1,1}) = \{\delta u_{r1}\}[K_{r1}]\{u_{r1}\} \tag{5.105}$$

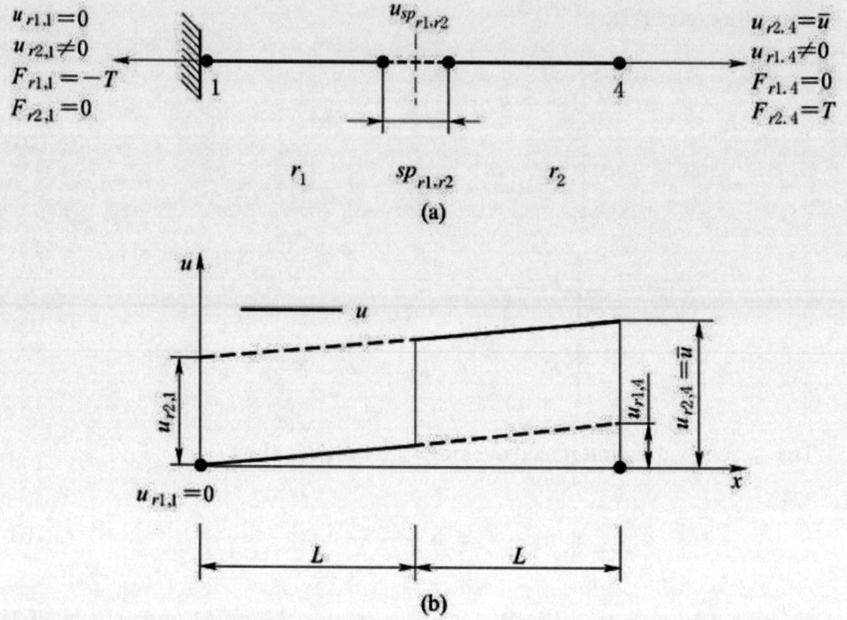

Fig. 5.26 Weak formulation of CEM. **a** Diagram to the calculation of discontinuous bar by CEM; **b** displacement distribution along the bar

where

$$[K_{r1}] = \frac{E}{4L}\begin{bmatrix} 1 & -1 \\ -1 & 1 \end{bmatrix} \tag{5.106}$$

is the stiffness matrix of the left half-bar.

For the right half-bar r_2, we get

$$\int_{r2} \delta\varepsilon_{r2}\sigma_{r2}\mathrm{d}\Omega = L\frac{E}{4L^2}(\delta u_{r2,4} - \delta u_{r2,1})(u_{r2,4} - u_{r2,1}) = \{\delta u_{r2}\}[K_{r2}]\{u_{r2}\} \tag{5.107}$$

where

$$[K_{r1}] = \frac{E}{4L}\begin{bmatrix} 1 & -1 \\ -1 & 1 \end{bmatrix} \tag{5.108}$$

is the stiffness matrix of the right half-bar.

On the interface $sp_{r1,r2}$, the average stress $\sigma_{sp_{r1,r2}}$ and the opening $u_{sp_{r1,r2}}$ can be written as

$$
\begin{cases}
\sigma_{sp_{r1,r2}} = k_n\left(\dfrac{u_{r2,4}+u_{r2,1}}{2} - \dfrac{u_{r1,4}+u_{r1,1}}{2}\right) \\
u_{sp_{r1,r2}} = \left(\dfrac{u_{r2,4}+u_{r2,1}}{2} - \dfrac{u_{r1,4}+u_{r1,1}}{2}\right)
\end{cases}
\tag{5.109}
$$

Consequently, its contribution to the virtual work in Eq. (5.90) is

$$
\begin{aligned}
\delta u_{sp_{r1,r2}}\sigma_{sp_{r1,r2}} &= k_n\left(\frac{\delta u_{r4,4}+\delta u_{r4,1}}{2} - \frac{\delta u_{r1,4}+\delta u_{r1,1}}{2}\right)\left(\frac{u_{r4,4}+u_{r4,1}}{2} - \frac{u_{r1,4}+u_{r1,1}}{2}\right) \\
&= \frac{1}{4}k_n(\delta u_{r4,4}+\delta u_{r4,1}-\delta u_{r1,4}-\delta u_{r1,1})(u_{r4,4}+u_{r4,1}-u_{r1,4}-u_{r1,1}) \\
&= \{\delta u_{sp_{r1,r2}}\}[K_{sp_{r1,r2}}]\{u_{sp_{r1,r2}}\}
\end{aligned}
\tag{5.110}
$$

where

$$
[K_{sp_{r1,r2}}] = \frac{1}{4}k_n\begin{bmatrix} 1 & 1 & -1 & -1 \\ 1 & 1 & -1 & -1 \\ -1 & -1 & 1 & 1 \\ -1 & -1 & 1 & 1 \end{bmatrix}
\tag{5.111}
$$

is the stiffness matrix of the interface $sp_{r1,r2}$.

The global equilibrium equation set of bar is assembled from Eqs. (5.105), (5.107) and (5.110) as follows

$$
[K]\begin{Bmatrix} u_{r1,1} \\ u_{r1,4} \\ u_{r2,1} \\ u_{r2,4} \end{Bmatrix} = \begin{Bmatrix} F_{r1,1} \\ F_{r1,4} \\ F_{r2,1} \\ F_{r2,4} \end{Bmatrix}
\tag{5.112}
$$

in which the global stiffness matrix is assembled in a manner of

$$
\begin{aligned}
[K] &= \frac{1}{4}k_n\begin{bmatrix} 1 & 1 & -1 & -1 \\ 1 & 1 & -1 & -1 \\ -1 & -1 & 1 & 1 \\ -1 & -1 & 1 & 1 \end{bmatrix} + \frac{E}{4L}\begin{bmatrix} 1 & -1 & 0 & 0 \\ -1 & 1 & 0 & 0 \\ 0 & 0 & 0 & 0 \\ 0 & 0 & 0 & 0 \end{bmatrix} + \frac{E}{4L}\begin{bmatrix} 0 & 0 & 0 & 0 \\ 0 & 0 & 0 & 0 \\ 0 & 0 & 1 & -1 \\ 0 & 0 & -1 & 1 \end{bmatrix} \\
&= \frac{E}{4L}\begin{bmatrix} 1+\frac{Lk_n}{E} & -(1-\frac{Lk_n}{E}) & -\frac{Lk_n}{E} & -\frac{Lk_n}{E} \\ -(1+\frac{Lk_n}{E}) & 1+\frac{Lk_n}{E} & -\frac{Lk_n}{E} & -\frac{Lk_n}{E} \\ -\frac{Lk_n}{E} & -\frac{Lk_n}{E} & 1+\frac{Lk_n}{E} & -(1-\frac{Lk_n}{E}) \\ -\frac{Lk_n}{E} & -\frac{Lk_n}{E} & -(1-\frac{Lk_n}{E}) & 1+\frac{Lk_n}{E} \end{bmatrix}
\end{aligned}
\tag{5.113}
$$

Use is made of boundary conditions

$$\begin{cases} u_{r1,1} = 0 \\ u_{r2,4} = \bar{u} \end{cases} \tag{5.114}$$

and,

$$\begin{cases} F_{r1,4} = 0 \\ F_{r2,1} = 0 \\ F_{r1,1} = F_1 = -F_{r2,4} = -F_4 = -T \end{cases} \tag{5.115}$$

equation set (5.112) becomes

$$\begin{cases} \dfrac{E}{4L}\left[-(1-\dfrac{Lk_n}{E})u_{r1,4} - \dfrac{Lk_n}{E}u_{r2,1} - \dfrac{Lk_n}{E}\,\bar{u}\right] = -T & ① \\ \dfrac{E}{4L}\left[(1+\dfrac{Lk_n}{E})u_{r1,4} - \dfrac{Lk_n}{E}u_{r2,1} - \dfrac{Lk_n}{E}\,\bar{u}\right] = 0 & ② \\ \dfrac{E}{4L}\left[-\dfrac{Lk_n}{E}u_{r1,4} + (1+\dfrac{Lk_n}{E})u_{r2,1} - (1-\dfrac{Lk_n}{E})\,\bar{u}\right] = 0 & ③ \\ \dfrac{E}{4L}\left[-\dfrac{Lk_n}{E}u_{r1,4} - (1-\dfrac{Lk_n}{E})u_{r2,1} + (1+\dfrac{Lk_n}{E})\,\bar{u}\right] = T & ④ \end{cases} \tag{5.116}$$

The combination of formulae ① and ④, or the combination of formulae ② and ③ leads to

$$u_{r1,4} + u_{r2,1} - \bar{u} = 0 \tag{5.117}$$

or

$$u_{r1,4} = \bar{u} - u_{r2,1} \tag{5.118}$$

On inserting Eq. (5.118) into the formula ③ of Eq. (5.114), we obtain

$$-\frac{Lk_n}{E}(-u_{r2,1} + \bar{u}) + (1+\frac{Lk_n}{E})u_{r2,1} - (1-\frac{Lk_n}{E})\,\bar{u} = 0 \tag{5.119}$$

which gives rise to

$$\begin{cases} u_{r2,1} = \bar{u}\dfrac{1}{1+\frac{2Lk_n}{E}} \\ u_{r1,4} = \bar{u}\left(1 - \dfrac{1}{1+\frac{2Lk_n}{E}}\right) \end{cases} \tag{5.120}$$

Used is made of Eqs. (5.109), (5.114) and (5.120), the opening of $sp_{r1,r2}$ may be calculated by

$$u_{sp_{r1,r2}} = \bar{u}\frac{1}{1+\frac{2Lk_n}{E}} \tag{5.121}$$

which is identical to Eq. (5.103). In addition, by the formulae ① and ④ in Eq. (5.116), we may get the restraint forces F_1 and F_2 that are identical to Eq. (5.102).

Figure 5.26b exposits the displacement distribution along the bar by CEM with the help of intermediate displacement variables u_{r1} and u_{r2}

3. **XFEM**

In Fig. 5.27a, only one enriched element with two nodes (1,4) is employed for the whole bar. At each node, there are two degrees of freedom (DOFs) corresponding to $\{u_b\} = [u_{b,1}\ u_{b,4}]^{\mathrm{T}}$ and $\{u_c\} = [u_{c,1}\ u_{c,4}]^{\mathrm{T}}$ for the displacement interpolation. At the cohesive (traction) interface (crack) c, the relative displacement is unknown but will be computed later using $\{u_c\} = [u_{c,1}\ u_{c,4}]^{\mathrm{T}}$ and Heaviside function (viz. Eq. (5.21)).

The displacements imposed at nodes 1 and 4 are 0.0 and $\bar{u}$, respectively. On each half bar, the strain and stress are constant and superimposed using

$$\begin{cases} \varepsilon_b = \dfrac{u_{b,4} - u_{b,1}}{2L} & \text{for the whole bar } \Omega \\ \varepsilon_c^- = -\dfrac{u_{c,4} - u_{c,1}}{2L} & \text{for the left half - bar } \Omega^- \\ \varepsilon_c^+ = \dfrac{u_{c,4} - u_{c,1}}{2L} & \text{for the right half - bar } \Omega^+ \end{cases} \tag{5.122}$$

$$\sigma = E(\varepsilon_b + \varepsilon_c) \tag{5.123}$$

On the interface c, the stress σ_c and the opening u_c can be calculated by

$$\begin{cases} u_c = (\dfrac{u_{c,4} + u_{c,1}}{2}) - (-\dfrac{u_{c,4} + u_{c,1}}{2}) = (u_{c,4} + u_{c,1}) \\ \sigma_c = k_n(u_{c,4} + u_{c,1}) \end{cases} \tag{5.124}$$

At the ends of bar, there are restraint forces

$$\{f_b\} = \begin{Bmatrix} F_{b,1} \\ F_{b,2} \end{Bmatrix} \tag{5.125}$$

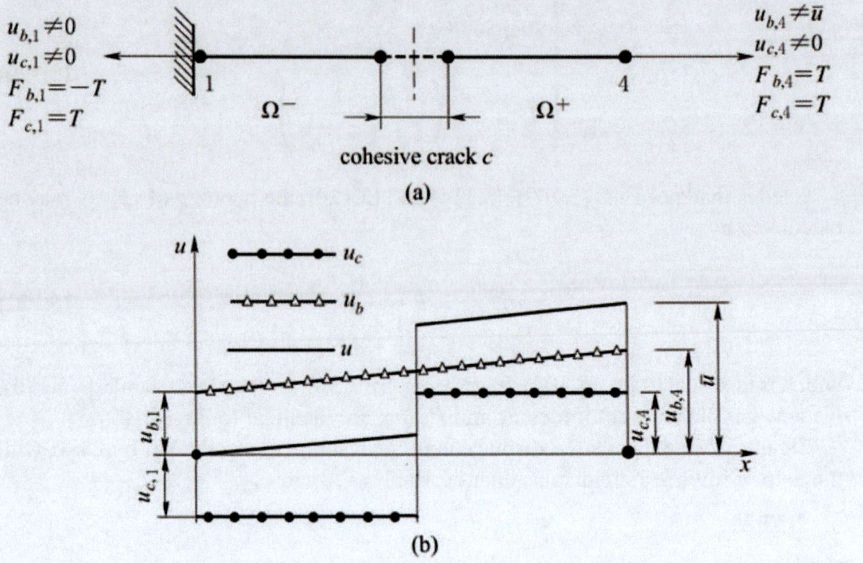

Fig. 5.27 Weak formulation of XFEM. **a** Diagram to the calculation of discontinuous bar by XFEM; **b** displacement distribution along the bar

$$\{f_c\} = \begin{Bmatrix} F_{c,1} \\ F_{c,2} \end{Bmatrix} \tag{5.126}$$

The virtual work in Eq. (5.90) may be specifically expressed by

$$\begin{aligned}
\int_{\Omega} \delta\varepsilon\sigma \mathrm{d}\Omega + \delta u_c \sigma_c &= \int_{\Omega^-} \delta(\varepsilon_b + \varepsilon_c^-) E (\varepsilon_b + \varepsilon_c^-) \mathrm{d}\Omega \\
&+ \int_{\Omega^+} \delta(\varepsilon_b + \varepsilon_c^+) E (\varepsilon_b + \varepsilon_c^+) \mathrm{d}\Omega + k_n(\delta u_{c,4} + \delta u_{c,1})(u_{c,4} + u_{c,1}) \\
&= \delta u_{b,1} F_{b,1} + \delta u_{b,2} F_{b,2} + u_{c,1} F_{c,1} + \delta u_{c,2} F_{c,2}
\end{aligned} \tag{5.127}$$

Use is made of Eqs. (5.122)–(5.124), the global equilibrium equation set may be obtained as follows

$$[K] \begin{Bmatrix} u_{b,1} \\ u_{b,4} \\ u_{c,1} \\ u_{c,4} \end{Bmatrix} = \begin{Bmatrix} F_{b,1} \\ F_{b,4} \\ F_{c,1} \\ F_{c,4} \end{Bmatrix} \tag{5.128}$$

in which the system stiffness matrix is

$$[K] = \frac{E}{2L}\begin{bmatrix} 1 & -1 & 0 & 0 \\ -1 & 1 & 0 & 0 \\ 0 & 0 & 1+\frac{2Lk_n}{E} & -(1-\frac{2Lk_n}{E}) \\ 0 & 0 & -(1-\frac{2Lk_n}{E}) & 1+\frac{2Lk_n}{E} \end{bmatrix} \tag{5.129}$$

Noting that the restraint forces meet the conditions

$$F_{b,4} = -F_{b,1} = F_{c,1} = F_{c,4} = T \tag{5.130}$$

Equation (5.128) becomes

$$\begin{cases} \frac{E}{2L}\left[u_{b,1} - u_{b,4}\right] = -T & ① \\ \frac{E}{4L}\left[-u_{b,1} + u_{b,4}\right] = T & ② \\ \frac{E}{4L}\left[(1+\frac{2Lk_n}{E})u_{c,1} - (1-\frac{2Lk_n}{E})u_{c,4}\right] = T & ③ \\ \frac{E}{4L}\left[-(1-\frac{2Lk_n}{E})u_{c,1} + (1+\frac{2Lk_n}{E})u_{c,4}\right] = T & ④ \end{cases} \tag{5.131}$$

At the first glance, it would be despairing because the formulae in Eq. (5.131) are not independent. However, it may be easily solved by the appropriate use of these formulae plus following boundary conditions

$$\begin{cases} u_{b,1} - u_{c,1} = 0 \\ u_{b,4} + u_{c,4} = \bar{u} \end{cases} \tag{5.132}$$

Firstly, formula ③ reduced by ④ leads to

$$u_{c,1} - u_{c,4} = 0 \tag{5.133}$$

From Eqs. (5.132) and (5.133) one get an important relation

$$u_{b,1} = u_{c,1} = u_{c,4} \tag{5.134}$$

Then, the combination of formulae ② and ③ leads to

$$-u_{b,1} + u_{b,4} = (1+\frac{2Lk_n}{E})u_{c,1} - (1-\frac{2Lk_n}{E})u_{c,4}$$

or

$$u_{b,4} = \frac{4Lk_n}{E}u_{c,1} + u_{b,1} = (1+\frac{4Lk_n}{E})u_{b,1} \tag{5.135}$$

Afterwards, use is made of Eqs. (5.134) and (5.132), it follows that

$$u_{b,1}(2+\frac{4Lk_n}{E})=\bar{u} \tag{5.136}$$

or

$$u_{b,1}=\frac{\bar{u}}{2}\frac{1}{1+\frac{2Lk_n}{E}} \tag{5.137}$$

On inserting Eqs. (5.137) into (5.135), we further obtain

$$u_{b,4}=\frac{\bar{u}}{2}\frac{1+\frac{4Lk_n}{E}}{1+\frac{2Lk_n}{E}} \tag{5.138}$$

Finally, the opening of crack is calculated by

$$u_c=u_{c,1}+u_{c,2}=2u_{b,1}=\bar{u}\frac{\frac{E}{Lk_n}}{2+\frac{E}{Lk_n}}=\bar{u}\frac{1}{1+2\frac{Lk_n}{E}} \tag{5.139}$$

which is identical to Eq. (5.103) (by FEM) and Eq. (5.121) (by CEM). The same applies to the restraint forces at the ends of bar using the formula ① of Eq. (5.131).

Figure 5.27b exposits the displacement distribution along the bar by CEM with the help of intermediate displacement variables u_b and u_c

5.5.2 Cracking Growth in Concrete Arch Dam

As a kind of slim shell structure flexibly supported by its deformable abutments and foundation, high concrete arch dam exhibits remarkable spatial-time characteristics. For nearly 20 years, the author had been engaged in the consultant works for Xiaowan Hydropower Project (XHP) that accommodates a high double curvature arch dam. During this period, one of special missions was to realize a four-dimensional (spatial-time) numerical simulation procedure by the dynamic input of the parameters tightly related to the structural geometry and material properties (e.g. thermal, permeability, deformation/stress) following the whole construction progress and service operation. This procedure has been working well in tracing the real work status of the arch dam. A series of difficult technical issues in the construction of Xiaowan Dam, such as the dam body cracking, are successfully solved.

1. **Presentation of the project**

At a height of 294.5 m, Xiaowan Arch Dam was one of the most challenging hydropower projects in China facing crucial difficulties (e.g. complicated geological conditions) beyond the existing design codes/specifications and the experiences in many aspects. Attributable to its special leading position and current technology situation, explorations and studies for Xiaowan Dam had been continuously conducted throughout the whole process from the early design phases to the implementation and operation periods. Particularly, in the initial period of construction, the information-based dynamic design concept was well accepted and the systematic instrumentation was arranged for the dam. Therefore, this dam is an ideal and precious object for the study on the spatial-time characteristics of high double-curvature arch dams, because it is able to provide abundant, detailed and accurate field data for the verification (Zou et al. 2016).

The main hydraulic structures layout in the project are the double curvature arch dam, plunge pool, auxiliary weir, flood release and diversion tunnels, and underground power plant system (viz. Figs. 5.28 and 5.29). The river bed is at the EL.950.5 m and the dam crest is at the EL. 1245 m. The dam thickness varies from 12 m at the crest to 72.9 m at the base of the crown cantilever.

The project is located in a V-shaped valley. Through the optimal study on the concrete composition (cement, flying ash, aggregate, admixture), the dam body was casted in three concrete zones of A ($C_{180}40$), B ($C_{180}35$), C ($C_{180}30$). For more details, readers are referred to the literature by Zou, the author and co-workers (Zou et al. 2016).

2. **Characteristics of the computation**

To handle the multi-field issues involved in the growth of dam, three computation meshes were generated w.r.t. the dynamic fields of temperature, seepage, and strain/stress, respectively (viz. Fig. 5.30).

The physical and mechanical parameters involved in the computation and the observation points in the instrumentation system are referred to the literature by Zou, the author and co-workers (Zou et al. 2016).

3. **Temperature and thermal stresses**

The temperature of concrete varies along with the concrete placement and the implementation of temperature control measures. At the initial stage of dam concrete placement, the temperature control measures were determined in accordance with the routine method specified in the existing design codes available. However, the actual temperature pickup was excessively high in the period between the end of cooling phase I and the beginning of cooling phase II cooling. In addition, the range and rate of temperature drop were very sharp during cooling phase II. These phenomena had been revealed by both the computation and the field observation (viz. Fig. 5.31).

This spatial gradient and temporal variation of temperature resulted in high tensile stress up to 1.2–2.2 MPa in the early construction period (see the stress envelope below the EL. 1110 m in Fig. 5.32).

After the timely adjustment of temperature control measures in light of the actual spatial-time evolution of dam body temperature and the cracking mechanism, the temperature field had been controlled effectively. To be specific, the temperature basically remained unchanged when the cooling phase I was over and drops steadily

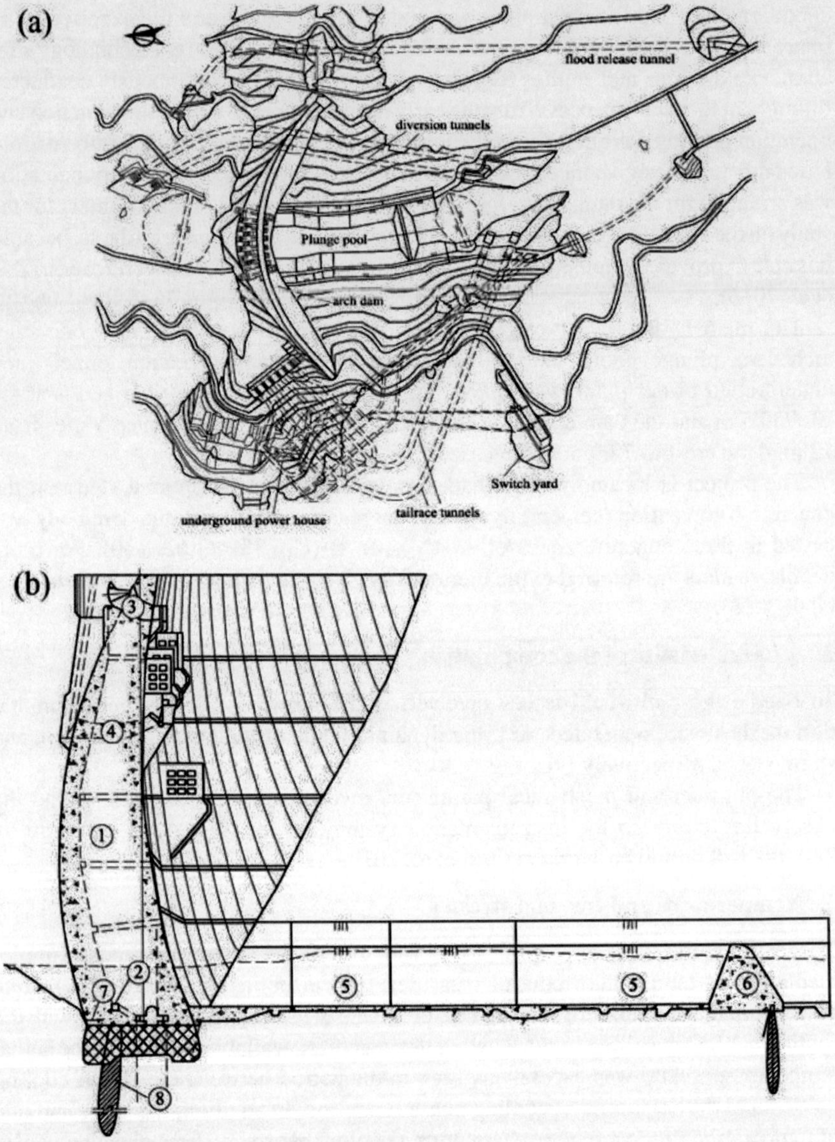

Fig. 5.28 Xiaowan Project (Diagram). **a** Plan; **b** cantilever profile. ①–arch dam; ②–foundation grouting gallery; ③–overflow spillway; ④–orifice spillway; ⑤–plunge pool; ⑥–auxiliary weir; ⑦–bottom inducing joint; ⑧–grouting curtain.

Fig. 5.29 Xiaowan Project (photo)

along with the cooling phase II, although the temperature picked up at a certain degree when the cooling phase II had been over, yet the final stabled temperature was reached around 19 °C (viz. Fig. 5.33).

As a result, the principal tensile stress during the late construction period was limited within 0.5–0.8 MPa at the most locations (see the stress envelope above the EL. 1110 m in Fig. 5.32).

4. **Temperature-induced cracks**

In the early batch of dam concrete designed according to the conventional specifications, the spatial gradient and temporal variation of temperature caused high tensile stress (1.2–2.2 MPa, viz. Fig. 5.32). Since the actual tensile strength of dam concrete is merely about 1.4 MPa, therefore temperature-induced cracks manifested and extended significantly during the early phases of dam concrete placement.

The chemical grouting treatment was undertaken for these temperature-induced cracks immediately after their detection, meanwhile the new, high-quality post-cooling scheme was analyzed and feedback designed for the later batches of dam concrete.

To estimate the risks of further crack growth during the later phase construction, reservoir impoundment, and service under NSL, a series sub-models containing crack-tips were cut from the dam body for the refined analyses with the help of SIFs (K_{I}, K_{II}, K_{III}). Our CEM code with asymptotic enrichment for the element containing crack-tip that is actually identical to XFEM, was employed. Figure 5.34 shows such a sub-model cut from the dam monolith 21# embedded with crack-tips.

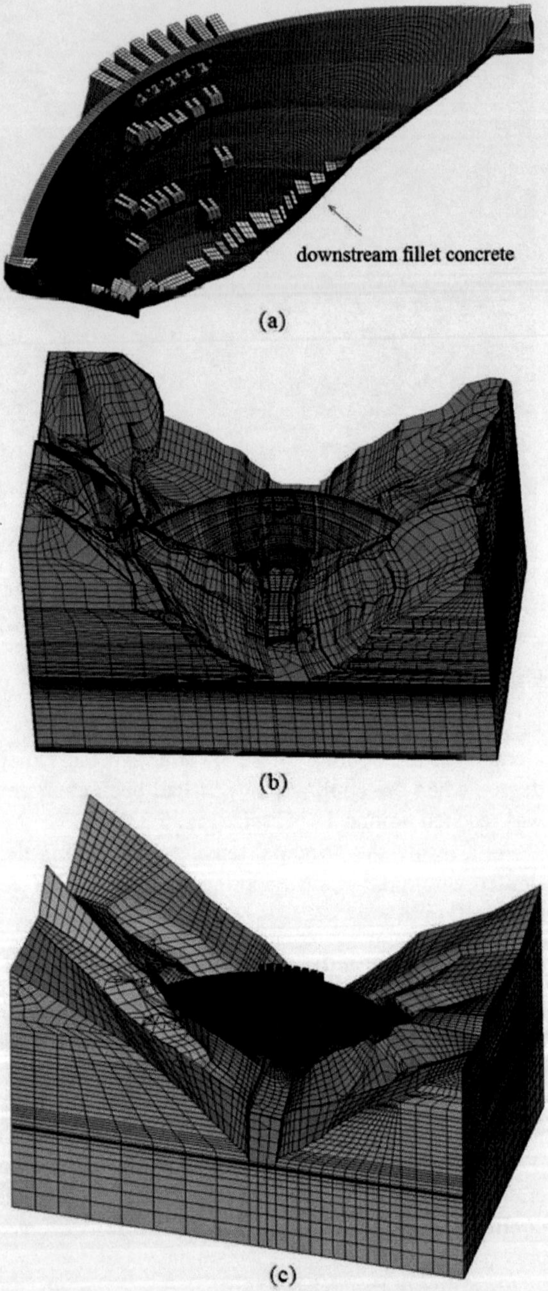

Fig. 5.30 FE meshes of the dam and foundation/abutments: construction phase. **a** Temperature analysis (448,966 elements; 505,958 nodes); **b** seepage analysis (1,058,610 elements; 791,986 nodes); **c** stress/strain analysis (1,163,662 elements; 942,608 nodes)

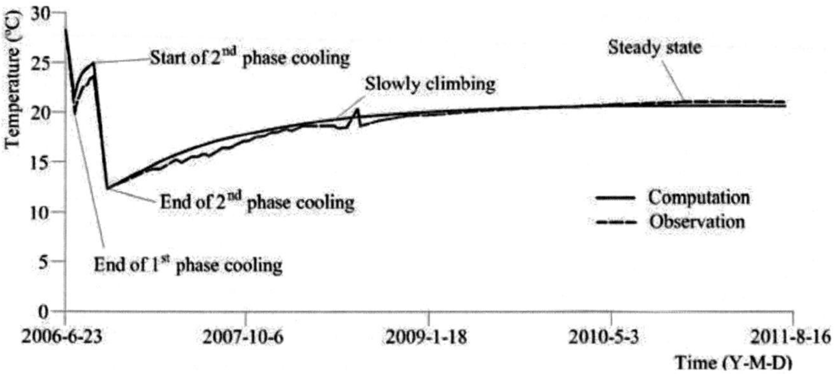

Fig. 5.31 Typical temperature history: routine design of temperature control measures

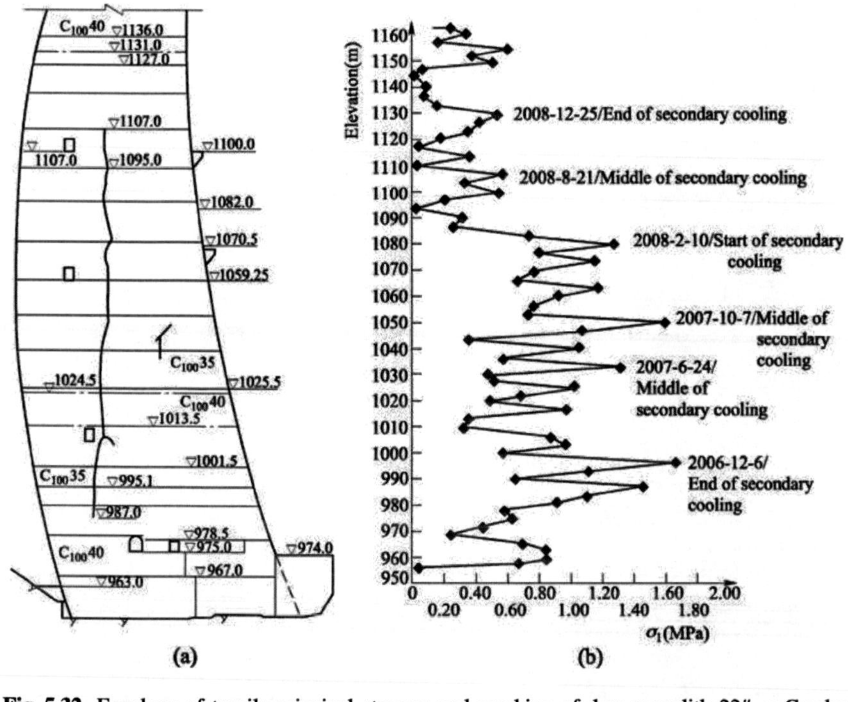

Fig. 5.32 Envelope of tensile principal stresses and cracking of dam monolith 22#. **a** Cracks; **b** envelope of tensile principal stresses

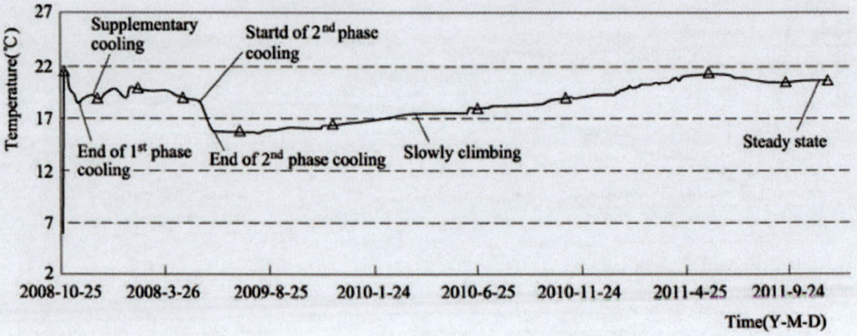

Fig. 5.33 Typical temperature history: after the adjustment of temperature control measures

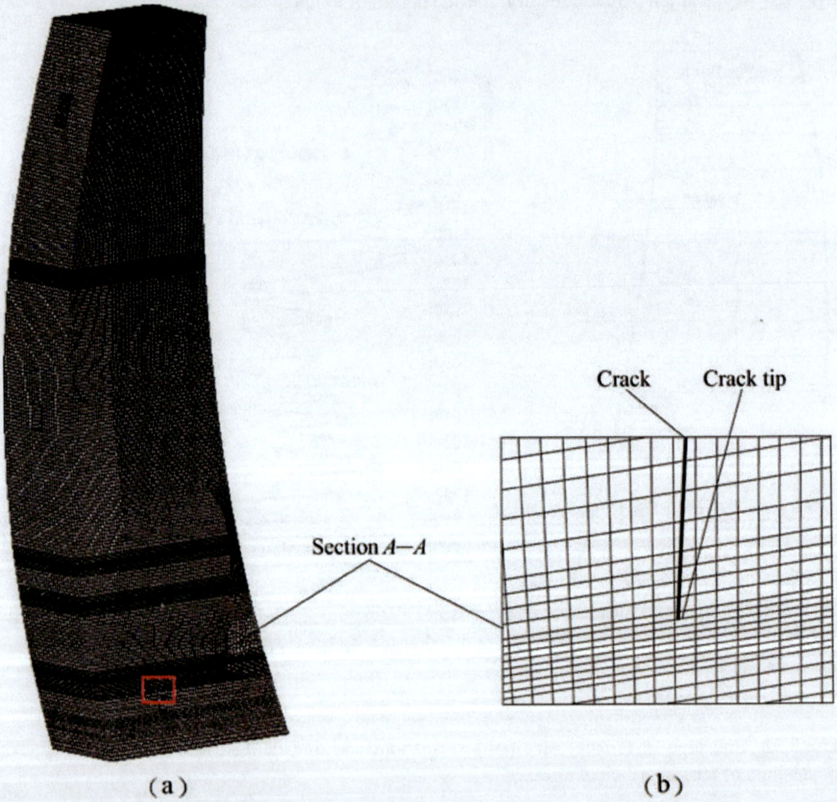

Fig. 5.34 Sub-model cut from the dam monolith 21#. **a** Refined sub-model mesh; **b** local section containing lower crack-tip

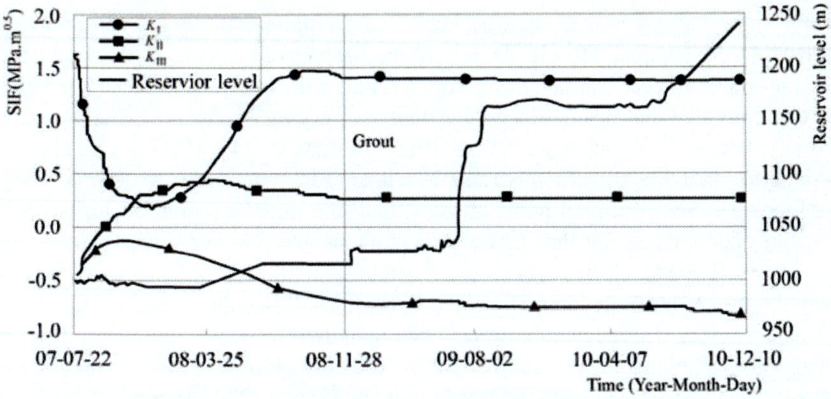

Fig. 5.35 History of SIFs at the lower crack-tip (EL.1010 m) in dam monolith 21#

This sub-model evolved following the whole process from the construction to the operation of dam. During this process the temperature and external force actions, as well as the boundary displacements that should be interpolated from the aforementioned overall simulation models, were input or adjusted at each time-marching step.

Figure 5.35 shows the evolution of SIFs at the lower crack-tip (EL.1010 m) in the dam monolith 21#. K_{I} exhibits a sharp drop shortly after the tip element cracks, afterwards it rebounds gradually. However, when the dam is poured to the EL. 1150 m, K_{I} begins to drop again. After the chemical grouting, K_{I} continues to drop slowly. The history of K_{II} is similar to that of K_{I} before the dam is poured to the EL. 1150 m, but after that K_{II} continues to mount. The chemical grouting slows down the rise of K_{II} remarkably. The evolution of K_{III} is opposite to that of K_{II}, and during the reservoir impounding K_{III} is much smaller than K_{I} and K_{II}. Based on above analysis, it is confirmed that this crack will be stable during the whole reservoir impoundment and service periods.

According to all of the analyses concerning the SIFs, the most significant conclusion was made that: the thermal cracks in Xiaowan Arch Dam are stable. This has been verified by the monitored data during the reservoir impounding and the normal operation since December 16, 2008 (the first day of reservoir impounding).

5.6 Concluding Remarks

The purpose of this chapter is to show how a fundamental FEM algorithm can be used to create XFEM (CEM as well).

5.6.1 Implementation of XFEM

There was a period when the distinctions among PUM, GFEM and XFEM were rather confused, and, as a consequence, they could be considered almost identical methods.

Three important advantages can be identified for these mesh-based formulations when they are compared to MFMs. Firstly, the strategy of extrinsically enriching the PU consumes a lower computation effort. Secondly, the interpolation characteristics of approximate functions produce the solution satisfying Kronecker Delta property. Finally, the similarity between PUM/GFEM /XFEM/CEM and FEM can allow a straightforward migration for them and the reuse of FEM code-structure (Borja 2008).

From the computational point of view, both the migration and reusing aspects depend, of course, on how the FEM code is implemented. The viability of extending a fundamental FEM code by the enrichment strategy has already been demonstrated by Strouboulis and his co-workers (Strouboulis et al. 2000; Strouboulis et al. 2001) using an object-oriented strategy, this strategy has been used by many later works (Bordas et al. 2007). Beyond the general ideas about the migration from FEM to other CMs such as PUM/GFEM/XFEM/CEM, Sukumar and Prévost (2003) described details of its implementation within a general-purposed FE code written in Fortran. Advances about this topic can be found in the overview presented by Fries and Belytschko (2010).

XFEM has been implemented in a number of sophisticated codes such as ABAQUS on OOP paradigm, in addition to a number of original open sources (Dunant et al. 2007).

It needs to be noted that from the author's point view of this book, however, the implementation of CEM for geotechnical and hydraulic engineering is well worth trying. This is because by the assembling procedure w.r.t. sub-elements, the appearance of rock fractures, rock bolts, drain holes and cooling pipes within the standard finite element may be easily and uniformly handled. The first impression of CEM ability has been given in this chapter, for more details readers are referred to the precursor book (Chen 2018).

5.6.2 Advances in XFEM

XFEM has obtained so promising results that we have immediately found the opportunities for many kinds of problems. Although these subjects are, however, out of the scope of this book, yet it is worthwhile to give a brief summary for the further reading.

- **Precondition.** It is notable that the enrichment strategy of XFEM can produce a badly conditioned stiffness matrix, which in turn leads to severe loss of accuracy in the numerical solution. Under such circumstances, preconditioners should be undertaken (Béchet et al. 2005). Babuška and Banerjee (2012) presented a new approach, named stable GFEM (SGFEM), which is able to overcome such problems.
- **Numerical integration.** If the enrichment function is polynomial, the integration of such terms can be exactly undertaken by Gauss quadrature. However, when the handbook function is singular or even discontinuous, Gauss quadrature

can produce inaccurate solutions. Under such circumstances, different techniques should be selectively used that include (incomplete list): subdomain quadrature (Möes et al. 1999); adaptive quadrature (Strouboulis et al. 2000); replacement of the discontinuous/non-differentiable functions that enrich the PU by equivalent polynomials (Ventura 2006); integration technique with a conformal mapping of the element domain (or its sub-cells) onto a unit disk (Natarajan et al. 2009); equivalent contour integrals in the place of domain integrals (Belytschko et al. 2009).

- **Adaptive strategy and error estimations**. The literatures related to this issue for mesh-based PUM are still limited. One of the first works to address such problem was proposed by Strouboulis and his co-worker (Strouboulis et al. 2001) by the use of recovery-based error measure for h-adaptive meshes. This kind of error measure is also the subject of the most error estimators that have so far been reported for the PUMs (Fries and Belytschko 2010). In the context of the implicit residual-based error strategies under GFEM/XFEM approach, the works of Barros, Bordas and their co-workers can be referred (Barros et al. 2004; Bordas and Duflot 2007). Other important work is the automated meshing that explores the OOP technique with integrated mesh generator (Dunant et al. 2007). This architecture produces advantages in very complex enrichment schemes with adaptive mesh refinement towards the solution of engineering problems related to the fracture mechanics.
- **Finite strain problems**. Legrain et al. (2005) studied the finite strain problems of XFEM. Béchet et al. (2005) improved the robustness for the stress analysis around cracks by introducing a different enrichment scheme, developing new integration quadrature for asymptotic functions, and implementing a preconditioning scheme adapted to the enriched functions. Xiao and Karihaloo (2006) improved the accuracy of crack-tip fields using higher order quadrature. Dumstorff and Meschke (2007) numerically studied the performance of different crack propagation criteria in the framework of XFEM.
- **Multiple cracks and crack nucleation.** Due to the fact that $H(x)$ has only two values of $+1$ and -1, it is insufficient for modelling complex crack surface patterns. To handle the holes and complex cracks with multiple branches, Daux and his co-workers (Daux et al. 2000) introduced a junction function $J(x)$ as the supplement to $H(x)$. Their work enables to model the interactions between cracks and holes, or the systems with cracks emanating from holes. Similar methods can also be found in a number of literatures (Wells and Sluys 2001; Borst et al. 2004; Hansbo and Hansbo 2004; Gasser and Holzapfel, 2005; Fries and Belytschko 2006). But it is still not easy with XFEM to determine enrichment functions for arbitrarily branched or intersecting cracks. Recently, based on PU method, a GFEM for the poly-crystals with discontinuous grain boundaries was proposed by incorporating different discontinuous function (Simone et al. 2006), and it was extended for crack problems with branches (Duarte et al. 2007). However, its enrichment functions are linearly dependent, and it is only suitable for the polycrystalline aggregate with crack-tips stopping at element edges.
- **Graded materials.** Comi and Mariani (2007) further proposed a XFEM simulation of quasi-brittle fracture in functionally graded materials.

- **Dynamic modeling.** Belytschko et al. (2003) presented a new enrichment technique to avoid the difficulties encountered with the original XFEM in time-dependent problems. Réthoré et al. (2005a, b) proposed an energy-conserving scheme for dynamic crack growth. Menouillard et al. (2006) formulated an efficient explicit time-marching scheme. Prabel et al. (2007) demonstrated that, the modeling of crack growth in a dynamic elastic–plastic media can be done using linear functions approximation. Song and Belytschko (2009) elaborated a "cracking node method" for dynamic fracture, where the growth of actual crack is approximated with contiguous discrete crack segments that lie on the element nodes and span only two adjacent elements.
- **Elastic–plastic fracture.** Elguedj et al. (2006) used the well-known Hutchinson-Rice-Rosengren (HRR) (Hutchinson 1968; Rice and Rosengren, 1968) fields to represent the singularities in elastic–plastic fracture mechanics. Their method enables to carry out the fatigue fracture analysis with mixed mode cracks in homogeneous, isotropic and elastic–plastic body. Based on the Fourier analysis, several strategies of enrichment were compared and a six-enrichment-functions basis was proposed. Their enrichment basis for crack-tip was coupled with a Newton like iterative scheme and a radial return method for plastic flow.
- **Intrinsic XFEM.** Fries and Belytschko (2006) proposed an intrinsic XFEM for treating arbitrary discontinuities. Unlike the standard XFEM, no additional unknowns are introduccd at the nodes whose supports are crossed by discontinuities. The method constructs an approximation space consisting of mesh-based and enriched-MLS functions near discontinuities.
- **Superposed approaches.** Lee et al. (2004) exposed a combination of XFEM and mesh superposition method (s-version FEM) previously proposed by Fish (1992). The near-tip field is modeled by superimposed quarter point elements on an overlaid mesh and the rest of crack is implicitly described by a step function on PU. Xiao and Karihaloo (2007) proposed an implementation of the "hybrid crack element (HCE)" on a general FE mesh and in combination with XFEM. Nakasumi et al. (2008) presented a connection of mesh superposition technique and XFEM to analyze the crack propagation in large scale or complicated engineering structures. Wyart and his co-workers (Wyart et al. 2007; 2008) proposed a possible approach to implement XFEM in general purpose FEM software. It is actually a sub-structure approach to decompose the cracked component into safe sub-domains and cracked sub-domains, then they are separately analyzed by the general-purposed FEM code and special-designed XFEM code.
- **Others.** XFEM has also been utilized in a variety of areas such as fluid mechanics, phase transformations, material sciences, solidification problems (Wagner et al. 2001; Merle and Dolbow 2002; Chessa and Belytschko 2003a, b; Gerstenberger and Wall 2008; Faivre et al. 2016; Paul et al. 2018).

It is also necessary to remind that there are a number of CMs to simulate crack growth without using enrichment functions (Wells and Sluys 2001; Borst et al. 2004; Gasser and Holzapfel 2005; Fries and Belytschko 2006). In addition, to handle rock block rotation and complete detachment as well as large scale opening of fractures, there are also a number of more competent methods available such as the "numerical manifold method (NMM)" (Shi 1992), "discrete element methods (DEMs)" (Cundall and Hart 1992), etc. Some of representative ones will be addressed in the **Chaps. 6–11** of this book.

References

Babuška I, Banerjee U. Stable generalized finite element method (SGFEM). Comput Methods Appl Mech Eng. 2012;201–204:91–111.

Babuška I, Melenk JM. Partition of unity method. Int J Numer Meth Eng. 1997;40(4):727–58.

Barenblatt G. The mathematical theory of equilibrium cracks in brittle fracture. Adv Appl Mech. 1962;7:55–129.

Barros FB, Proença SPB, de Barcellos CS. On error estimator and shape p-adaptivity in the generalized finite element method. Int J Numer Methods Eng. 2004;60(14):2373–98.

Barsoum RS. Application of quadratic isoparametric elements in linear fracture mechanics. Int J Fract. 1974;10(1):603–5.

Barsoum RS. On the use of isoparametric finite elements in linear fracture mechanics. Int J Numer Methods Eng. 1976;10(1):25–37.

Béchet E, Minnebo H, Möes N, Burgardt B. Improved implementation and robustness study of the X-FEM for stress analysis around cracks. Int J Numer Methods Eng. 2005;64(8):1033–56.

Belytschko T, Black T. Elastic crack growth in finite elements with minimal remeshing. Int J Numer Methods Eng. 1999;45(5):601–20.

Belytschko T, Möes N, Usui S, Parimi C. Arbitrary discontinuities in finite elements. Int J Numer Methods Eng. 2001;50(4):993–1013.

Belytschko T, Chen H, Xu J, Zi G. Dynamic crack propagation based on loss of hyperbolicity and a new discontinuous enrichment. Int J Numer Methods Eng. 2003;58(12):1873–905.

Belytschko T, Gracie R, Ventura G. A review of extended/generalized finite element methods for material modeling. Modell Simul Mater Sci Eng. 2009;17(4): 043001.

Benzley S. Representation of singularities with isoparametric finite elements. Int J Numer Methods Eng. 1974;8(3):537–45.

Berto F, Lazzarin P. Recent developments in brittle and quasi-brittle failure assessment of engineering materials by means of local approaches. Mater Sci Eng R: Rep. 2014;75:1–48.

Bordas S, Duflot M. Derivative recovery and a posteriori error estimate for extended finite elements. Comput Methods Appl Mech Eng. 2007;196:3381–99.

Bordas S, Nguyen PV, Dunant C, Guidoum A, Nguyen-Dang H. An extended finite element library. Int J Numer Meth Eng. 2007;71(6):703–32.

Borja RI. Assumed enhanced strain and the extended finite element methods: a unification of concepts. Comput Methods Appl Mech Eng. 2008;197:2789–803.

Borja RI, Regueiro RA, Lai TY. FE modeling of strain localization in soft rock. J Geotech Geoenviron Eng. 2000;126:335–43.

Bui HD. Mécanique de la rupture fragile. Paris (France): Masson; 1978. (in French).

Chan SK, Tuba IS, Wilson WK. On the finite element method in linear fracture mechanics. Eng Fract Mech. 1970;2(1):1–17.

Chang KJ. On the maximum strain criterion—a new approach to the angled crack problem. Eng Fract Mech. 1981;14(1):107–24.

Chen SH. Computational geomechanics and hydraulic structures. Berlin (Germany): Springer; 2018.

Chen SH, Qiang S. Composite element model for discontinuous rock masses. Int J Rock Mech Min Sci Geomech Abstr. 2004;41(7):865–70.

Chen SH, Shahrour I. Composite element method for the bolted discontinuous rock masses and its application. Int J Rock Mech Min Sci. 2008;45(3):384–96.

Chen G, Ohnishi Y, Ito T. Development of high-order manifold method. Int J Numer Methods Eng. 1998;43(4):685–712.

Chen SH, Qiang S, Chen SF, Egger P. Composite element model of the fully grouted rock bolt. Rock Mech Rock Eng. 2004a;37(3):193–212.

Chen SH, Feng XM, Shahrour I. Numerical estimation of REV and permeability tensor for fractured rock masses by composite element method. Int J Numer Anal Meth in Geomech. 2008a;32(12):1459–77.

Chen SH, Qiang S, Shahrour I, Egger P. Composite element analysis of gravity dam on a complicated rock foundation. Int J Geomech, ASCE. 2008b;8(5):275–84.

Chen SH, Xue LL, Xu GS, Shahrour I. Composite element method for the seepage analysis of rock masses containing fractures and drainage holes. Int J Rock Mech Min Sci. 2010;47(5):762–70.

Chen SH, Su PF, Shahrour I. Composite element algorithm for the thermal analysis of mass concrete: simulation of lift joint. Finite Elem Anal Des. 2011;47(5):536–42.

Chen SH, He J, Shahrour I. Estimation of elastic compliance matrix for fractured rock masses by composite element method. Int J Rock Mech Min Sci. 2012;49(1):156–64.

Chen SH, Zhang X, Shahrour I. Composite element model for the bonded anchorage head of stranded wire cable in tension. Int J Numer Anal Meth Geomech. 2015;39(12):1352–68.

Chen SH, Feng XM. Composite element model for rock mass seepage flow. J Hydrodyn (Ser. B). 2006; 18(2):219–24.

Chen SH, Xu Q, Hu J. Composite element method for seepage analysis of geo-technical structures with drainage hole array. J Hydrodyn (Ser. B). 2004b; 16(3):260–66.

Chen SH. The Elasto-plastic and Elasto-viscoplastic analysis of the discontinuous rock masses. PhD Thesis. WUHEE, Wuhan (China); 1987 (in Chinese with English summary).

Chessa J, Belytschko T. An extended finite element method for two-phase fluids. ASME J Appl Mech. 2003a;70(1):10–7.

Chessa J, Belytschko T. An enriched finite element method and level sets for axisymmetric two-phase flow with surface tension. Int J Numer Methods Eng. 2003b;58(13):2041–64.

Claydon PW. Maximum energy release rate distribution from a generalized 3D virtual crack extension method. Eng Fract Mech. 1992;42(6):961–9.

Comi C, Mariani S. Extended finite element simulation of quasi-brittle fracture in functionally graded materials. Comput Methods Appl Mech Eng. 2007;196(41–44):4013–26.

Cundall PA, Hart DH. Numerical modelling of discontinua. Eng Comput. 1992;9(2):101–13.

Daux C, Möes N, Dolbow J, Sukumar N, Belytschko T. Arbitrary branched and intersecting cracks with the extended finite element method. Int J Numer Meth Eng. 2000;48(12):1741–60.

de Borst R, Remmers JJC, Needleman A, Abellan MA. Discrete vs smeared crack models for concrete fracture: bridging the gap. Int J Numer Meth Eng. 2004;28(7–8):583–607.

Desai CS, Zamman MM, Lightner JG, Siriwardane HJ. Thin layer element for interfaces and joints. Int J Numer Anal Meth Geomech. 1984;8(1):19–43.

Doitrand A, Leguillon D, Martin E. Computation of generalized stress intensity factors of 3D singularities. Int J Solids Struct. 2020;190:271–80.

Dolbow J, Moës N, Belytschko T. Discontinuous enrichment in finite elements with a partition of unity method. Finite Elem Anal Des. 2000;36(3–4):235–60.

Dolbow J, Moës N, Belytschko T. An extended finite element method for modeling crack growth with frictional contact. Comput Methods Appl Mech Eng. 2001;190(51–52):6825–46.

Dolbow J. An extended finite element method with discontinuous enrichment for applied mechanics. PhD Thesis. Northwestern University (USA); 1999.

Duarte CA, Babuška I, Oden J. Generalized finite element methods for three dimensional structural mechanics problems. Comput Struct. 2000;77(2):215–32.

Duarte CA, Hamzeh O, Liszka T, Tworzyd W. A generalized finite element method for the simulation of three-dimensional dynamic crack propagation. Comput Methods Appl Mech Eng. 2001;190(15–17):2227–62.

Duarte CA, Reno LG, Simone A. A high-order generalized FEM for through-the-thickness branched cracks. Int J Numer Meth Eng. 2007;72(3):325–51.

Duarte CA, Oden JT. Hp clouds–a meshless method to solve boundary-value problems. Technical Report 95–05. Texas(USA): Texas Institute for Computational and Applied Mechanics (TICAM), the University of Texas at Austin (USA); 1995.

Dugdale DS. Yielding of steel sheets containing slits. J Mech Phys Solids. 1960;8(2):100–4.

Dumstorff P, Meschke G. Crack propagation criteria in the framework of X-FEM-based structural analyses. Int J Numer Anal Methods Geomech. 2007;31(2):239–59.

Dunant C, Nguen VP, Belgasmia M, Bordas S, Guidoum A. Architecture tradeoffs of integrating a mesh generator to partition of unity enriched object-oriented finite element software. Revue Européenne De Mécanique Numérique. 2007;16(2):237–58.

Elguedj T, Gravouil A, Combescure A. Appropriate extended functions for X-FEM simulation of plastic fracture mechanics. Comput Methods Appl Mech Eng. 2006;195(7–8):501–15.

Erdogan F, Sih GC. On the crack extension in plates under plane loading and transverse shear. J Basic Eng. 1963;85(4):519–27.

Evans PH, Marathe MS. Microcracking and stress-strain curves for concrete in tension. Mater Struct (RILEM).1968; 1(1):61–64.

Faivre M, Paul B, Golfier F, Giot R, Massin P, Colombo D. 2D coupled HM-XFEM modeling with cohesive zone model and applications to fluid-driven fracture network. Eng Fract Mech. 2016;159:115–43.

Fish J. The s-version of the finite element method. Comput Struct. 1992;43:539–47.

Fleming M, Chu YA, Moran B, Belytschko T. Enriched element-free Galerkin methods for crack tip fields. Int J Numer Methods Eng. 1997;40(8):1483–504.

Fries T, Belytschko T. The intrinsic XFEM: a method for arbitrary discontinuities without additional unknowns. Int J Numer Methods Eng. 2006;68(13):1358–85.

Fries TP, Belytschko T. The extended/generalized finite element method: an overview of the method and its applications. Int J Numer Methods Eng. 2010;84:253–304.

Gasser TC, Holzapfel GA. Modeling 3D crack propagation in unreinforced concrete using PUFEM. Comput Meth Appl Mech Eng. 2005;194(25–26):2859–96.

Gerstenberger A, Wall WA. An extended finite element method/Lagrange multiplier based approach for fluid-structure interaction. Comput Methods Appl Mech Eng. 2008;197(19–20):1699–714.

Ghaboussi J, Wilson EL, Isenberg J. Finite element for rock joints and interfaces. J Soil Mech Found, ASCE.1973; 99(10): 849–62.

Goodman RE, Taylor RL, Brekke TL. A model for the mechanics of jointed rock. J Soil Mech Found Div, ASCE.1968; 94(SM3):637–59.

Hansbo A, Hansbo P. A finite element method for the simulation of strong and weak discontinuities in solid mechanics. Comput Meth Appl Mech Eng. 2004;193(33–35):3523–40.

Hillerborg A, Modéer M, Petersson PE. Analysis of crack formation and crack growth in concrete by means of fracture mechanics and finite elements. Cem Concr Res. 1976;6(6):773–82.

Hutchinson JW. Plastic stress and strain fields at a crack tip. J Mech Phys Solids. 1968;16:337–47.

Katona MG. A simple contact-friction interface element with applications to buried culverts. Int J Numer Anal Meth Geomech. 1983;7(3):371–84.

Khoei AR, Nikbakht M. An enriched finite element algorithm for numerical computation of contact friction problems. Int J Mech Sci. 2007;49(2):183–99.

Lazzarin P, Campagnolo A, Berto F. A comparison among some recent energy-and stress-based criteria for the fracture assessment of sharp V-notched components under Mode I loading. Theor Appl Fract Mech. 2014; 71(SI):21–30.

Lee S, Soon JH, Yoon YC, Zi G, Belytschko T. Combined extended and superimposed finite element method for cracks. Int J Numer Methods Eng. 2004;59(8):1119–36.

Legrain G, Moës N, Verron E. Stress analysis around crack tips in finite strain problems using the extended finite element method. Int J Numer Methods Eng. 2005;63(2):290–314.

Leroy Y, Ortiz M. Finite element analysis of transient strain localization phenomena in frictional solids. Int J Numer Anal Methods Geomech. 1990;14(2):93–124.

Li H, Yuan H. Cohesive zone modelling of low cycle fatigue cracks in cracked and notched specimens. Fatigue Fract Eng Mater Struct. 2013;36(12):1246–57.

Lim IL, Johnston IW, Choi SK. Application of singular quadratic distorted isoparametric elements in linear fracture mechanics. Int J Numer Methods Eng. 1993;36(14):2473–99.

Linsbauerr HN, Ingraffea AR, Rossmanith HP, Wawrzynek PA. Simulation of cracking in large arch dam: Part 1. J Struc Eng, ASCE. 1989a;115(7):1599–615.

Linsbauer HN, Ingraffea AR, Rossmanith HP, Wawrzynek PA. Simulation of cracking in large arch dam: Part 2. J Struct Eng, ASCE. 1989b;115(7):1616–30.

Liu F, Borja RI. An extended finite element framework for slow-rate frictional faulting with bulk plasticity and variable friction. Int J Numeri Anal Methods Geomech. 2009;33(13):1535–60.

Liu XY, Xiao QZ, Karihaloo BL. XFEM for direct evaluation of mixed mode SIFs in homogeneous and bi-materials. Int J Numer Methods Eng. 2004;59(8):1103–18.

Maiti SK, Smith RA. Criteria for brittle fracture in biaxial tension. Eng Fract Mech. 1984;19(5):793–804.

Mandel J. Conditions de stabilite et postulat de Drucker. In: Proceeding IUTAM Symposium on rheology and soil mechanics. Berlin (Germany): Springer-Verlag; 1964. p. 58–68.

Melenk JM, Babuška I. The partition of unity finite element method: basic theory and applications. Comput Meth Appl Mech Eng. 1996;29(1–4):289–314.

Menouillard T, Réthoré J, Combescure A, Bung H. Efficient explicit time stepping for the extended finite element method (X-FEM). Int J Numer Methods Eng. 2006;68(9):911–39.

Merle R, Dolbow J. Solving thermal and phase change problems with the extended finite element method. Comput Mech. 2002;28(5):339–50.

Van Mier JGM. Strain-softening of concrete under multiaxial loading conditions. PhD Thesis. Technische Hogeschool Eindhoven (Germany); 1984.

Mirsayar MM. Mixed mode fracture analysis using extended maximum tangential strain criterion. Mater Design. 2015;86:941–7.

Möes N, Dolbow J, Belytschko T. A finite element method for crack growth without remeshing. Int J Numer Meth Eng. 1999;46(1):131–50.

Moës N, Belytschko T. Extended finite element method for cohesive crack growth. Eng Fract Mech. 2002;69(7):813–33.

Moran B, Shih CF. Crack tip and associated domain integrals from momentum and energy balance. Eng Fract Mech. 1987;27(6):615–41.

Nacar A, Needleman A, Ortiz M. A finite element method for analyzing localization in rate dependent solids at finite strains. Comp Meth Appl Mech Eng. 1989;73(3):235–58.

Nakasumi S, Suzuki K, Ohtsubo H. Crack growth analysis using mesh superposition technique and X-FEM. Int J Numer Methods Eng. 2008;75(3):291–304.

Nalla RK, Kinney JH, Ritchie RO. Mechanistic fracture criteria for the failure of human cortical bone. Nat Mater. 2003;2(3):164–8.

Natarajan S, Bordas S, Mahapatra DR. Numerical integration over arbitrary polygonal domains based on Schwarz-Christoffel conformal mapping. Int J Numer Meth Eng. 2009;80(1):103–34.

Nuismer R. An energy release rate criterion for mixed mode fracture. Int J Fract. 1975;11(2):245–50.

Ortiz M, Quigley JJ. Adaptive mesh refinement in strain localization problems. Comput Methods Appl Mech Eng. 1991;90(1–3):781–804.

Ortiz M, Leroy Y, Needleman A. A finite element method for localized failure analysis. Comput Methods Appl Mech Eng. 1987;61(2):189–124.

Paul B, Faivre M, Massin P, Giot R, Colombo D, Golfier F, Martin A. 3D coupled HM-XFEM modeling with cohesive zone model and applications to nonplanar hydraulic fracture propagation and multiple hydraulic fractures interference. Comput Methods Appl Mech Eng. 2018;342:321–53.

Pietruszczak ST, Mroz Z. Finite element analysis of deformation of strain softening materials. Int J Numer Methods Eng. 1981;17(3):327–34.

Prabel B, Combescure A, Gravouil A, Marie S. Level set XFEM non-matching meshes: application to dynamic crack propagation in elastic–plastic media. Int J Numer Methods Eng. 2007;69(8):1553–69.

Read HE, Hegemier GA. Strain softening of rock, soil and concrete-a review article. Mech Mater. 1984;3(4):271–94.

Réthoré J, Gravouil A, Combescure A. An energy-conserving scheme for dynamic crack growth using the extended finite element method. Int J Numer Methods Eng. 2005a;63(5):631–59.

Réthoré J, Gravouil A, Combescure A. A combined space-time extended finite element method. Int J Numer Methods Eng. 2005b;64(2):260–84.

Rice JR. A path independent integral and the approximate analysis of strain concentration by notches and cracks. J Appl Mech. 1968;35(2):379–86.

Rice J, Rosengren G. Plane strain deformation near a crack tip in a power-law hardening material. J Mech Phys Solids. 1968;16(1):1–12.

Rudnicki JW, Rice JR. Conditions for the localization of deformation in pressure-sensitive dilatant materials. J Mech Phys Solids. 1975;23(6):371–94.

Salah EI-Din AS, EI-Adawy Nassef MM. A modified approach for estimating the cracking moment of reinforced concrete beams. ACl J. 1975; (7):356–60.

Shi G. Modeling rock joints and blocks by manifold method. In: Tillerson JR, Wawersik WR, editors. Proceeding 33rd US Symposium on rock mechanics, New Mexico (USA). Rotterdam (Netherlands): AA Balkema; 1992. p. 639–48.

Shi G. Manifold method of material analysis. In: Transactions of the 9th army conference on applied mathematics and computing. Report No. 92–1. Minneapolis (USA): US Army Research Office; 1991. p. 57–262.

Sih GC. Strain energy density factor applied to mixed mode crack problems. Int J Fract. 1974;10(3):305–21.

Simone A, Duarte CA, Van der Giessen A. A generalized finite element method for polycrystals with discontinuous grain boundaries. Int J Numer Methods Eng. 2006;67(8):1122–45.

Song JH, Belytschko T. Cracking node method for dynamic fracture with finite elements. Int J Numer Meth Eng. 2009;77(3):360–85.

Song SH, Paulino GH, Buttlar WG. Simulation of crack propagation in asphalt concrete using an intrinsic cohesive zone model. J Eng Mech. 2006;132(11):1215–23.

Strouboulis T, Babuška I, Copps K. The design and analysis of the generalized finite element method. Comput Methods Appl Mech Eng. 2000;181(1–3):43–69.

Strouboulis T, Copps KI, Babuška I. The generalized finite element method. Comput Methods Appl Mech Eng. 2001;190(32–33):4081–193.

Sukumar N, Prévost JH. Modeling quasi-static crack growth with the extended finite element method Part I: computer implementation. Int J Solids Struct. 2003;40(26):7513–37.

Sukumar N, Moës N, Moran B, Belytschko T. Extended finite element method for three-dimensional crack modelling. Int J Numer Methods Eng. 2000;48(11):1549–70.

Taylor D. The theory of critical distances. Eng Fract Mech. 2008;75(7):1696–705.

Taylor RL, Beresford PJ, Wilson EL. A non-conforming element for stress analysis. Int J Numer Meth Eng. 1976;10(6):1211–9.

Taylor D, Merlo M, Pegley R, Cavatorta MP. The effect of stress concentrations on the fracture strength of polymethylmethacrylate. Mater Sci Eng. 2004;A382(1–2):288–94.

Torabi AR, Pirhadi E. Stress-based criteria for brittle fracture in key-hole notches under mixed mode loading. Eur J Mech, A/Solids. 2015;49:1–12.

Vardoulakis I. Shear band inclination and shear modulus of sand in biaxial tests. Int J Numer Anal Methods Geomech. 1980;4(2):103–19.

Vardoulakis I, Goldscheider M, Gudehus G. Formation of shear bands in sand bodies as a bifurcation problem. Int J Numer Anal Methods Geomech. 1978;2(2):99–128.

Ventura G. On elimination of quadrature subcells for discontinuous functions in the extended finite element method. Int J Numer Methods Eng. 2006;66(5):761–95.

Wagner GJ, Möes N, Liu WK, Belytschko T. The extended finite element method for rigid particles in Stokes flow. Int J Numer Methods Eng. 2001;51(3):293–313.

Waversik WR, Brace WF. Post-failure behavior of a granite and a diabase. Rock Mech. 1971;3(2):61–85.

Welch GB, Haisman B. Fracture toughness measurements of concrete. Report No R 42, University of New South Wales, Sydney (Australia): Uniciv; 1969.

Wells GN, Sluys LJ. A new method for modeling cohesive cracks using finite elements. Int J Numer Mech Eng. 2001;50(12):2667–82.

Wilson EL, Taylor RL, Doherty WP, Ghaboussi J. Incompatible displacement models. In: Fenves SJ, Perrone N, Robinson AR, Schnobrich WC, editors. Numerical and computer methods in structural mechanics. New York (USA): Academic Press; 1973. p. 43–57.

Wu X, Li X. Analysis and modification of fracture criteria for mixed-mode crack. Eng Fract Mech. 1989;34(1):55–64.

Wu HC. Dual failure criterion for plane concrete. J Eng Mech Div, ASCE. 1974; 100(6):1167–181.

Wyart E, Duflot M, Coulon D, Martiny P, Pardoen T, Remacle JF, Lani F. Substructuring FE-XFE approaches applied to three-dimensional crack propagation. J Comput Appl Math. 2008;215:626–38.

Wyart E, Coulon D, Duflot M, Pardoen T, Remacle JF, Lani F. A substructured FE-shell/XFE-3D method for crack analysis in thin walled structures. Int J Numer Methods Eng. 2007; 72:757–79.

Xiao Q, Karihaloo B. Direct evaluation of accurate coefficients of the linear elastic crack tip asymptotic field. Fatigue Fract Eng Mater Struct. 2003;26(8):719–29.

Xiao Q, Karihaloo B. Improving the accuracy of XFEM crack tip fields using higher order quadrature and statically admissible stress recovery. Int J Numer Methods Eng. 2006;66(9):1378–410.

Xiao Q, Karihaloo B. Implementation of hybrid crack element on a general finite element mesh and in combination with XFEM. Comput Methods Appl Mech Eng. 2007;196(13–16):1864–73.

Xie M, Gerstle W, Rahulkumar R. Energy-based automatic mixed mode crack-propagation modeling. J Eng Mech ASCE. 1995;121(8):914–23.

Yau JF, Wang SS, Corten HT. A mixed-mode crack analysis of isotropic solids using conservation laws of elasticity. J Appl Mech. 1980;47(2):335–41.

Yazid A, Abdelkader N, Abdelmadjid H. State-of-the-art review of the X-FEM for computational fracture mechanics. Appl Math Model. 2009;33(12):4269–82.

Zienkiewicz OC, Best B, Dullage C, Stagg K. Analysis of nonlinear problems in rock mechanics with particular reference to jointed rock systems. In: Proceeding Second International Congress on rock mechanics. Belgrade (Yugoslavia); 1970. p. 8–14.

Zou LC, Chen SH, Wang GJ, Fu SJ, Tang XL, Xu Q, Zhao ZY, Wang WM, Xie M, He J. Simulation for the spatial-time characteristics of high arch dam. Geotech Geol Eng. 2016;34(4):949–80.

Chapter 6
General Finite Element Methods with Special Focus on NMM

Abstract This chapter summarizes the principle and algorithm related to another member of mesh-based method, namely, the NMM that belongs to the category of "general finite element methods (GFEMs)". The unique characteristic of NMM is, instead of conventional "node" in the fundamental FEM, it defines a new core object termed as "cover" that is composed of the finite element patches surrounding the node. Polynomial cover functions replacing the constant nodal variables (i.e. displacements) are used to interpolate the field functions within manifold elements created by the union of covers. The coefficients in the polynomial cover function are solved by the algebraic equation set according to the virtual work or variational principle. In this manner, less restraint is imposed on the mesh generation with considerable amount of structural planes, which allows for a great simplification in the pre-process work towards the computation for complex geotechnical and hydraulic structures.

6.1 General

Named after the mathematical notion of manifold, the "numerical manifold method (NMM)" was initially proposed and further developed by Shi (1991, 1992, 1995, 1996a, b, 1997, 2012) on the base of finite cover approximation.

NMM feasibly simulates structural planes in jointed rocks using elements of any shape truncated by structural planes (Ma et al. 2010; An et al. 2014; Li and Zhao 2019). This is mainly attributable to its dual-cover system inclusive "mathematical covers" and "physical covers".

The mathematical covers overlay the whole space occupied by the engineering structure domain. It is usually generated from several kinds of FE meshes, the FEs that share one common node form a "mathematical cover (MC)". The creation of mathematical covers is supposed to have a slight influence on simulation results (Zhang et al. 2010b).

The "physical cover (PC)" is formed by the intersection of mathematical cover and the modeling domain including interior and exterior boundaries. Specifically, "physical patches" are formed by cutting a mathematical cover with structural planes,

S. Chen, *Advanced Computational Methods and Geomechanics*, Springer Tracts in Civil Engineering, https://doi.org/10.1007/978-981-19-7427-4_6

such as physical boundaries, material interfaces, joints and faults. Then, the union of physical patches constitutes physical covers. The overlap portion of neighboring physical covers is called a "manifold element (ME)" that is the basic computation unit in NMM.

NMM straightforwardly models structural planes by splitting a mathematical cover into several physical covers completely cut by the structural planes, then an independent local function is tied to each physical cover for the interpolation of field variable within the manifold element union-operated by physical cover set. The major advantage with dual-cover systems in NMM is that arbitrary boundaries and structural planes in thc domain may be explicitly permitted without conforming mesh. Therefore, the mesh generation in NMM is greatly simplified and the fracture propagation can be simulated without re-meshing.

A simplex integration method is customarily employed in NMM that was initially proposed for DDA (viz. Chap. 8). This enables us to conveniently evaluate the integration over elements intersected by interior structural planes and external boundaries. For the 2-D ME of general shape, this scheme will first partition the element into several sub-triangles, then the quadrature will be conducted on each sub-triangle. However, another integration scheme elaborated in CEM and XFEM is conveniently feasible, too, in which each sub-triangle is further refined by three quadrilateral sub-elements, afterwards the standard Gauss quadrature is conducted on these sub-elements (viz. Chap. 5).

It needs to be noted that the fundamental FEM may be looked as a special case of NMM when the following conditions are met:

- MCs in NMM are generated from the standard FE mesh;
- Weight functions defined on MCs are identical to the shape functions in FEM;
- Cover functions defined on PCs are constants;
- Physical features such as structural planes (e.g. joints, faults, etc.) and external boundaries do not intersect MEs.

In this context, NMM can be looked at as a member of GFEMs.

6.2 Covers and Manifold Elements

6.2.1 Conceptual Illustration of One-Dimensional Problems

The portrait of a physical problem including the domain in which the problem is defined, and all the physical features such as the structural planes (e.g. cracks, joints, faults, material interfaces, holes, etc.) and the boundaries on which certain conditions are prescribed. Such a well-defined domain is referred to as "physical domain", whereas a domain independent of but completely cover the physical domain is called the "mathematical domain".

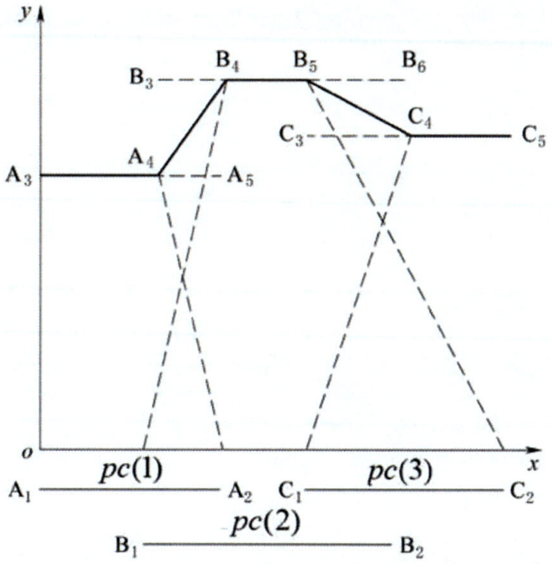

Fig. 6.1 Covers for one-dimensional bar

Shi is the first who employed the basic concept of topology manifold to establish NMM. His formulation is based on a number of mathematical covers (charts), each of them is a set of finite elements surrounding one node. Each mathematical cover is overlapped partially with its adjacent ones.

With reference to the schematic diagram in Fig. 6.1, the definitions of basic components in NMM, namely the mathematical cover, the physical cover, the manifold element, the cover function, and the weight function, are generally exposited. There are three arbitrary mathematical covers $A_1 - A_2$, $B_1 - B_2$, $C_1 - C_2$ along 1-D continuous bar, since they are not intersected by material boundaries and discontinuous point, they are identical to physical covers denoted as $pc(1)$, $pc(2)$, $pc(3)$.

Cover function may be constant, linear, or polynomials. Suppose it is constant, then on every physical cover the cover function is defined as

$$\begin{cases} u_1(x) = A_3 - A_5 \ x \in pc(1) \\ u_2(x) = B_3 - B_6 \ x \in pc(2) \\ u_3(x) = C_3 - C_5 \ x \in pc(3) \end{cases} \tag{6.1}$$

The cover functions are linked each other via weight functions $W_i(x)(i = 1, 2, 3)$ over the whole bar. The weight function may be defined as

$$\begin{cases} W_i(x) \geq 0 \ x \in pc(i) \\ W_i(x) = 0 \ x \notin pc(i) \end{cases} \tag{6.2}$$

For a manifold element that is the portion overlapped by several PCs, it is demanded that

$$\sum_{x\in pc(j)} W_j(x) = 1 \tag{6.3}$$

Suppose the weight function is linear, then according to the above definitions one gets

$$\begin{cases} W_1(x)u_1(x) = \mathrm{A}_3 - \mathrm{A}_4 - \mathrm{A}_2 \\ W_2(x)u_2(x) = \mathrm{B}_1 - \mathrm{B}_4 - \mathrm{B}_5 - \mathrm{B}_2 \\ W_3(x)u_3(x) = \mathrm{C}_1 - \mathrm{C}_4 - \mathrm{C}_5 \end{cases} \tag{6.4}$$

The weighted cover function over the whole physical domain gives a polyline distribution of $u(x)$

$$u(x) = \sum_{i=1}^{3} W_i(x)u_i(x) = \mathrm{A}_3 - \mathrm{A}_4 - \mathrm{B}_4 - \mathrm{B}_5 - \mathrm{C}_4 - \mathrm{C}_5 \tag{6.5}$$

6.2.2 Two-Dimensional Extension

For two-dimensional problems, a mathematical cover can be constructed as a union of small plane patches that is consecutively denoted as $mc(i)(i = 1, 2, \ldots n_m)$. The $mc(i)$ could be, in turn, partitioned into several physical covers denoted as $pc(i_j)(j = 1, 2, \ldots, n_i)$ $(pc(i_j) \subset \Omega_i)$, or directly denoted as i_j, where j is the sequence of physical cover originated from the mathematical cover $mc(i)$, and n_i is the amount of physical covers created from $mc(i)$. In other words, physical covers are in fact the sub-division of mathematical covers by the physical domain.

1. **Mathematical covers using triangular finite element mesh**

(1) **Homogeneous domain**

Figure 6.2a shows a triangular FE mesh for a homogeneous domain, the corresponding $n_m = 6$ mathematical covers are listed in Table 6.1. It is noted that the amount of mathematical covers is always equal to the amount of nodes n_n, namely, $n_m = n_n$.

PCs are the intersection of mathematical covers and physical domain. In Fig. 6.2a the intersection of mathematical cover $mc(1)$ and material domain is a portion of physical domain, hence it is a physical cover denoted as 1_1 $(pc(1_1))$. All the other physical covers are listed in Table 6.2. It is clear that for a normal FE mesh of continuous and homogeneous domain, each mathematical cover is uniquely related to a physical cover. In other words, the amount of physical covers n_p is always equal

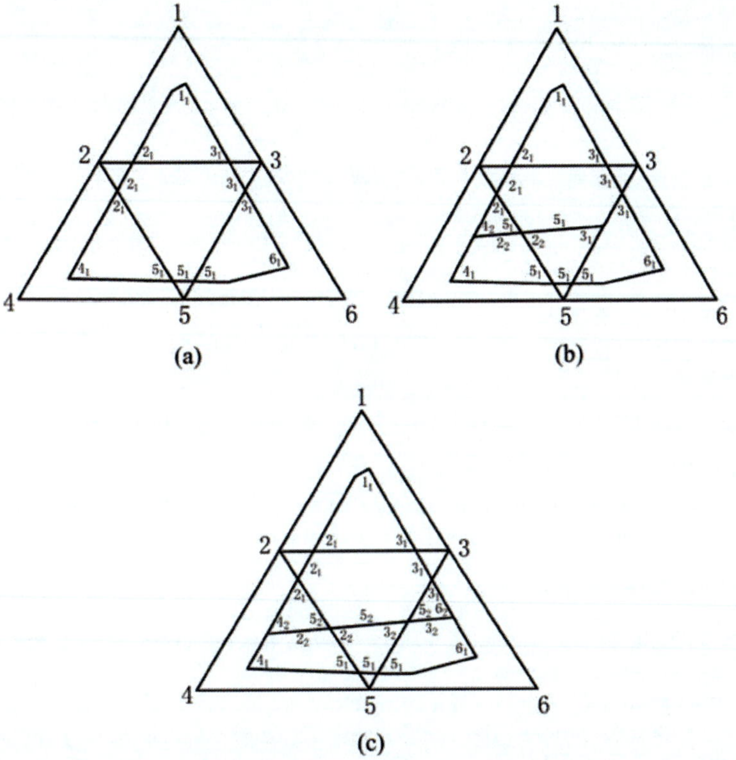

Fig. 6.2 Triangular element mesh based covers. **a** Homogeneous domain; **b** crack tapered at the element edge within the domain; **c** crack penetrating the whole domain

Table 6.1 Relation of node and mathematical cover (viz. Fig. 6.2a)

Node	1	2	3	4	5	6
$mc(i)$	1(Δ123)	2 (□1453)	3(□1256)	4(Δ245)	5(□4632)	6(Δ563)

to the amount of mathematical covers n_m, namely, $n_p = n_m$. In addition, the amount of manifold elements n_{me} is equal to the amount of finite element n_{fe}.

The manifold element is defined as the common region shared by several physical covers. For triangular FE mesh, any point in the domain is located in the common portion or boundary of three physical covers, this common portion is defined as "manifold element (ME)". The manifold elements correspondent on Fig. 6.2a are

Table 6.2 Relation of mathematical covers and physical covers (viz. Fig. 6.2a)

$mc(i)$	1	2	3	4	5	6
$pc(i_j)$	1_1	2_1	3_1	4_1	5_1	6_1

Table 6.3 Manifold elements embedded in finite elements (viz. Fig. 6.2a)

Finite element	Δ123	Δ245	Δ253	Δ356
Embedded manifold element	$1_1 2_1 3_1$	$2_1 4_1 5_1$	$2_1 5_1 3_1$	$3_1 5_1 6_1$

Table 6.4 Relation of mathematical covers and physical covers (viz. Fig. 6.2b)

$mc(i)$	1	2	3	4	5	6
$pc(i_j)$	1_1	$2_1, 2_2$	3_1	$4_1, 4_2$	5_1	6_1

Table 6.5 Manifold elements embedded in finite elements (viz. Fig. 6.2b)

Finite element	Δ123	Δ245	Δ253	Δ356
Embedded manifold element	$1_1 2_1 3_1$	$2_1 4_2 5_1, 2_2 4_1 5_1$	$2_1 5_1 3_1, 2_2 5_1 3_1$	$3_1 5_1 6_1$

summarized in Table 6.3. It is also clear that for a normal FE mesh of continuous and homogeneous domain, each ME is identically overlapped with FE.

(2) **Crack tapered at element edge**

Figure 6.2b shows the triangular FE mesh with tapered crack (joint), the corresponding six mathematical covers are identically listed in Table 6.1.

The mathematical cover $mc(4)$ is completely cut into three sub-regions and two of them are within the problem domain, so two physical covers, i.e. $pc(4_1)$ and $pc(4_2)$ (4_1 and 4_2) are created. Similarly, $mc(2)$ also forms two physical covers, i.e. $pc(2_1)$ and $pc(2_2)$ (2_1 and 2_2). Since $mc(5)$ is only partially intersected by joint, hence there is one physical cover $pc(5_1)(5_1)$ solely. Table 6.4 lists the relation of mathematical covers and physical covers, and Table 6.5 lists the manifold elements embedded in finite elements.

It is noted that where structural planes (joints, cracks) are embedded in mathematical covers, the amount of mathematical covers is still equal to the amount of nodes, but the amount of physical covers will be larger than the amount of mathematical covers, namely, $n_p > n_m$. In addition, the amount of manifold elements n_{me} will be larger than the amount of finite element n_{fe}, namely, $n_{me} > n_{fe}$.

(3) **Crack penetrating the whole domain**

In Fig. 6.2c there is a crack (joint) goes completely through the whole physical domain, and correspondingly, eleven physical covers $1_1, 2_1, 2_2, 3_1, 3_2, 4_1, 4_2, 5_1, 5_2, 6_1, 6_2$ are created and listed in Table 6.6. Altogether, seven manifold elements are found in Table 6.7.

Table 6.8 exposes basic differences in the fundamental FEM and NMM. Particularly, the important characteristics of triangular mesh-based NMM are summarized as follows:

Table 6.6 Relation of mathematical covers and physical covers (viz. Fig. 6.2c)

$mc(i)$	1	2	3	4	5	6
$pc(i_j)$	1_1	$2_1, 2_2$	$3_1, 3_2$	$4_1, 4_2$	$5_1, 5_2$	$6_1, 6_2$

Table 6.7 Manifold elements embedded in finite elements (viz. Fig. 6.2c)

Finite element	$\Delta 123$	$\Delta 245$	$\Delta 253$	$\Delta 356$
Embedded manifold element	$1_1 2_1 3_1$	$2_1 4_2 5_2, 2_2 4_1 5_1$	$2_1 5_2 3_1, 2_2 5_1 3_2$	$3_1 5_2 6_2, 3_2 5_1 6_1$

Table 6.8 Comparison of finite element method and numerical manifold method

FEM	Node	FE defined by nodes	Edge defined by nodes
NMM	Physical cover	ME defined by the common region PCs	Boundary/intersection of PCs

- Shape of manifold element may be arbitrary;
- Three physical covers that form a manifold element may be looked at as three general "nodes" of the manifold element;
- Along the common edge, the adjacent FEs share identical physical cover or general "node".
- The MEs derived from one FE due to crack will possess different physical covers or general "nodes".

2. **Mathematical covers using quadrilateral finite element mesh**

In Fig. 6.3, a quadrilateral FE mesh completely overlaying the physical domain is employed to define mathematical covers, in which a tapered crack (joint) is embedded and propagated in sequence. Tables 6.9, 6.10 and 6.11 give the relationship of MEs and FEs.

6.3 Kinematics Aspects

6.3.1 *Approximation of Displacement Field*

For displacement field, the NMM approximation is made over the physical covers. Each $pc(i_j)$ is tied by local approximation functions termed as "cover functions". These cover functions for the interpolation of displacement $\mathbf{u}$ in a manifold element can be constructed by any functions that are able to describe the local characteristics of solution towards a high-accuracy global approximation. Particularly, the linear combination of mutually independent functions with given order and constant coefficients are prevalent.

In 2-D problems, on each mathematical cover $mc(i)(i = 1, 2, \ldots n_m)$ a piecewise weight function $W_i(x, y)$ is defined that satisfies the following conditions

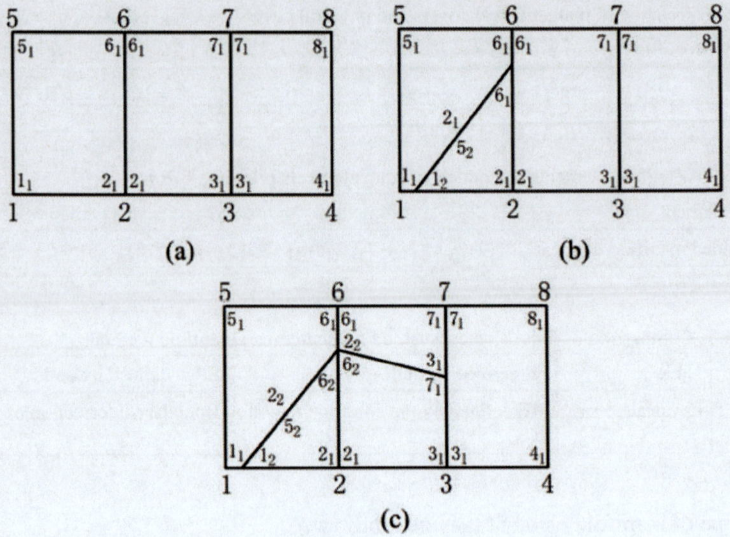

Fig. 6.3 Quadrilateral element mesh based covers. **a** Homogeneous domain; **b** crack splits quadrilateral element 1 2 6 5; **c** crack splits quadrilateral elements 1 2 6 5 and 2 3 7 6

Table 6.9 Finite elements and corresponding manifold elements in Fig. 6.3a

Finite element	1 2 6 5	2 3 7 6	3 4 8 7
Manifold element	$1_1 2_1 6_1 5_1$	$2_1 3_1 7_1 6_1$	$3_1 4_1 8_1 7_1$

Table 6.10 Finite elements and corresponding manifold elements in Fig. 6.3b

Finite element	1 2 6 5	2 3 7 6	3 4 8 7
Manifold element	$1_2 2_1 6_1 5_2$, $1_1 2_1 6_1 5_1$	$2_1 3_1 7_1 6_1$	$3_1 4_1 8_1 7_1$

Table 6.11 Finite elements and corresponding manifold elements in Fig. 6.3c

Finite element	1 2 6 5	2 3 7 6	3 4 8 7
Manifold element	$1_2 2_1 6_2 5_2$, $1_1 2_2 6_1 5_1$	$2_1 3_1 7_1 6_2$, $2_2 3_1 7_1 6_1$	$3_1 4_1 7_1 8_1$

$$\begin{cases} W_i(x, y) \geq 0 & (x, y) \in mc(i) \\ W_i(x, y) = 0 & (x, y) \notin mc(i) \\ \sum_{l=1}^{n} W_l(x, y) = 1 & (x, y) \in mc(i), l \in fe(e) \end{cases} \tag{6.6}$$

where $fe(e)$ is the FE sequence that is defined by n nodes with local sequence $l(1, 2, \ldots, n)$ (inclusive node i) to compose $mc(i)$.

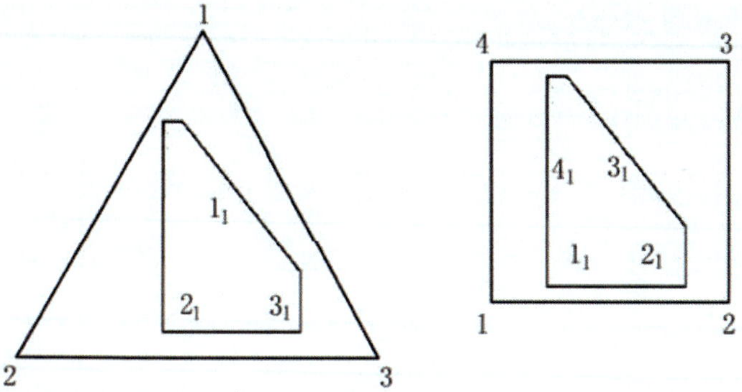

(a) (b)

Fig. 6.4 Manifold element for two-dimensional problems. **a** Manifold element in triangular finite element mesh; **b** manifold element in quadrilateral finite element mesh

According to Eq. (6.6), the weight function set $W_i(x, y)$ meets the "partition of unity (PU)". In addition, it is continuous over the mathematical cover $mc(i)$, and has non-zero value only on its corresponding mathematical cover $mc(i)$. These weight functions can be employed to paste all the cover functions together to give a global approximation over each manifold element.

Let's taking the 2-D problem for example. In Fig. 6.4a, there are three mathematical covers 1, 2, 3 related to the triangular element, the common region of three physical covers $1_1, 2_1, 3_1$ forms a manifold element $me(1_1 2_1 3_1)$ sequenced as $me(e)(e = 1, 2, \ldots, n_{me})$ in the whole domain, where n_{me} is the amount of manifold elements. In Fig. 6.4b, four mathematical covers (1, 2, 3, 4) based physical covers $(1_1, 2_1, 3_1, 4_1)$ form a manifold element $me(1_1 2_1 3_1 4_1)$ sequenced as $me(e)$, too. The interpolation of displacement function for the manifold element is generally expressed as:

- Triangular mesh-based

$$\mathbf{u}_{me(e)} = \begin{Bmatrix} u_x(x, y) \\ u_y(x, y) \end{Bmatrix}$$

$$= \begin{bmatrix} W_1(x, y) & 0 & W_2(x, y) & 0 & W_3(x, y) & 0 \\ 0 & W_1(x, y) & 0 & W_2(x, y) & 0 & W_3(x, y) \end{bmatrix} \begin{Bmatrix} \hat{u}_{x,pc(1_1)} \\ \hat{u}_{y,pc(1_1)} \\ \hat{u}_{x,pc(2_1)} \\ \hat{u}_{y,pc(2_1)} \\ \hat{u}_{x,pc(3_1)} \\ \hat{u}_{y,pc(3_1)} \end{Bmatrix} \tag{6.7}$$

- Quadrilateral mesh-based

$$\mathbf{u}_{me(e)} = \begin{Bmatrix} u_x(x, y) \\ u_y(x, y) \end{Bmatrix}$$

$$= \begin{bmatrix} W_1(x, y) & 0 & W_2(x, y) & 0 & W_3(x, y) & 0 & W_4(x, y) & 0 \\ 0 & W_1(x, y) & 0 & W_2(x, y) & 0 & W_3(x, y) & 0 & W_4(x, y) \end{bmatrix} \begin{Bmatrix} \hat{u}_{x,pc(1_1)} \\ \hat{u}_{y,pc(1_1)} \\ \hat{u}_{x,pc(2_1)} \\ \hat{u}_{y,pc(2_1)} \\ \hat{u}_{x,pc(3_1)} \\ \hat{u}_{y,pc(3_1)} \\ \hat{u}_{x,pc(4_1)} \\ \hat{u}_{y,pc(4_1)} \end{Bmatrix} \tag{6.8}$$

Denote n as the amount of physical covers towards the construction of a manifold element that is identical to the amount of nodes defining a finite element, and we have established a link list $i = \langle l \rangle$ relation the local physical cover sequence $l = pc(i_j)$ and global physical cover sequence $i (i = 1, 2, \ldots, n_{pc})$, then with the definition of

$$\begin{cases} \mathbf{W} = [\mathbf{W}_1 \ldots \mathbf{W}_l \ldots \mathbf{W}_n] \\ \mathbf{W}_l = \begin{bmatrix} W_l(x, y) & 0 \\ 0 & W_l(x, y) \end{bmatrix} \quad (l = 1, 2, \ldots, n) \end{cases} \tag{6.9}$$

$$\hat{\mathbf{u}}_{me(e)} = \left[\hat{\mathbf{u}}_1 \ldots \hat{\mathbf{u}}_l \ldots \hat{\mathbf{u}}_n\right]^{\mathrm{T}} \tag{6.10}$$

$$\hat{\mathbf{u}}_l = \begin{Bmatrix} \hat{u}_{x,l} \\ \hat{u}_{y,l} \end{Bmatrix} \quad (l = 1, 2, \ldots, n) \tag{6.11}$$

Equations (6.7) and (6.8) may be uniformly written as

$$\mathbf{u}_{me(e)}(x, y) = \mathbf{W}\hat{\mathbf{u}}_{me(e)} \tag{6.12}$$

It can be easily proved that the approximation (6.12) can reproduce the cover functions (or local approximation functions), in which the constant displacement functions $\hat{\mathbf{u}}_{me(e)}$ are defined at n physical covers (viz. Eq. 6.10).

Towards a higher computational accuracy, higher-order isoparametric finite element, even with p-version of refinement, may be exercised (Chen 2018), but Shi proposed another way by using polynomial "cover function" for displacement in a manner of

$$\hat{\mathbf{u}}_l(x, y) = \begin{Bmatrix} \hat{u}_{x,l}(x, y) \\ \hat{u}_{y,l}(x, y) \end{Bmatrix} = \sum_{k=1}^{m} \begin{bmatrix} f_{l,k}(x, y) & 0 \\ 0 & f_{l,k}(x, y) \end{bmatrix} \begin{Bmatrix} \hat{u}_{l,2k-1} \\ \hat{u}_{l,2k} \end{Bmatrix} \quad (l = 1, 2, \ldots, n) \tag{6.13}$$

in which m is the amount of complete polynomial items.

Denote

$$\begin{cases} \mathbf{fun}_l = \begin{bmatrix} f_{l,1}(x,y) & 0 & \ldots\ldots & f_{l,m}(x,y) & 0 \\ 0 & f_{l,1}(x,y) & \ldots\ldots & 0 & f_{l,m}(x,y) \end{bmatrix} \\ \hat{\mathbf{u}}_l = \begin{bmatrix} \hat{u}_{l,1} & \hat{u}_{l,2} & \ldots\ldots & \hat{u}_{l,2m-1} & \hat{u}_{l,2m} \end{bmatrix}^{\mathrm{T}} \end{cases} \quad (l = 1, 2, \ldots, n) \tag{6.14}$$

where $\hat{\mathbf{u}}_l$ is the coefficient vector of the polynomial cover function at the physical cover l that constructs the manifold element *me*. Instead of Eq. (6.11), we now get

$$\hat{\mathbf{u}}_l(x, y) = \mathbf{fun}_l \hat{\mathbf{u}}_l \tag{6.15}$$

Specifically, a constant cover function $f_{l,1}$ $(m = 1)$ may be simply obtained

$$\hat{\mathbf{u}}_l(x, y) = \begin{bmatrix} 1 & 0 \\ 0 & 1 \end{bmatrix} \begin{Bmatrix} \hat{u}_{l,1} \\ \hat{u}_{l,2} \end{Bmatrix} = \begin{Bmatrix} \hat{u}_{x,l} \\ \hat{u}_{y,l} \end{Bmatrix} = \hat{\mathbf{u}}_l = \text{constant nodal vector} \tag{6.16}$$

in which $\hat{u}_{x,l}$ and $\hat{u}_{y,l}$ are constant and equal to the displacement components tied to the physical cover l.

A complete linear cover function $f_{l,k}$ $(k = 1, \ldots, m)$ $(m = 3)$ of a manifold element leads to

$$\hat{\mathbf{u}}_l(x, y) = \begin{bmatrix} 1 & 0 & x & 0 & y & 0 \\ 0 & 1 & 0 & x & 0 & y \end{bmatrix} \begin{Bmatrix} \hat{u}_{l,1} \\ \hat{u}_{l,2} \\ \hat{u}_{l,3} \\ \hat{u}_{l,4} \\ \hat{u}_{l,5} \\ \hat{u}_{l,6} \end{Bmatrix} \quad (l = 1, 2, \ldots, n) \tag{6.17}$$

It is notable that for polynomial functions with order higher than two, the convergence of NMM is still questionable for the moment. If we further denote the shape function matrix $\mathbf{N}_{me(e)}(e = 1, 2, \ldots, n_{me})$ as

$$\mathbf{N}_{me(e)} = \begin{bmatrix} \mathbf{N}_1 & \ldots & \mathbf{N}_l & \ldots & \mathbf{N}_n \end{bmatrix} = \begin{bmatrix} \mathbf{W}_1.\mathbf{fun}_1 & \ldots & \mathbf{W}_l.\mathbf{fun}_l & \ldots & \mathbf{W}_n.\mathbf{fun}_n \end{bmatrix} \tag{6.18}$$

$$\hat{\mathbf{u}}_{me(e)} = \begin{bmatrix} \hat{\mathbf{u}}_1 & \ldots & \hat{\mathbf{u}}_l & \ldots & \hat{\mathbf{u}}_n \end{bmatrix}^{\mathrm{T}} \tag{6.19}$$

then instead of Eq. (6.12), the displacement function within the manifold element is expressed by its parameter vector $\hat{\mathbf{u}}_{me(e)}(e = 1, 2, \ldots, n_{me})$

$$\mathbf{u}_{me(e)}(x, y) = \begin{Bmatrix} u_x(x, y) \\ u_y(x, y) \end{Bmatrix} = \mathbf{N}_{me(e)} \hat{\mathbf{u}}_{me(e)} \tag{6.20}$$

Actually, for any manifold element $me(e)$ that is formed by n physical covers, Eq. (6.20) gives a general interpolation scheme subject to the appropriate selection of weight functions.

It is emphasized again that in NMM, the mathematical covers are constructed totally independent of the boundaries and material interfaces. Unless the constant cover function ($m = 1$) is used (viz. Eq. 6.16), the essential boundary conditions cannot be accomplished by directly enforcing the degrees of freedom like in FEM, but should be imposed by the methods of Lagrange multipliers or penalty factor, or artificial cohesive boundaries. The details will be exposited later in Sect. 6.4.

6.3.2 *Specification of Weight Functions and Cover Functions*

1. **Three-nodal triangle element (constant nodal cover m = 1)**

For the three-nodal triangle element with area A in Fig. 6.5, three sub-triangles ΔP23, ΔP31 and ΔP12 with areas A_1, A_2 and A_3 respectively, may be used to define the area coordinates of point P(x, y) as follows

$$\begin{cases} L_1 = \frac{A_1}{A} = \frac{(a_1+b_1x+c_1y)}{2A} \\ L_2 = \frac{A_2}{A} = \frac{(a_2+b_2x+c_2y)}{2A} \\ L_3 = \frac{A_3}{A} = \frac{(a_3+b_3x+c_3y)}{2A} \end{cases} \tag{6.21}$$

in which (x_1, y_1), (x_2, y_2), (x_3, y_3) are the coordinates of nodes.

The standard shape functions in FEM may be further defined such that

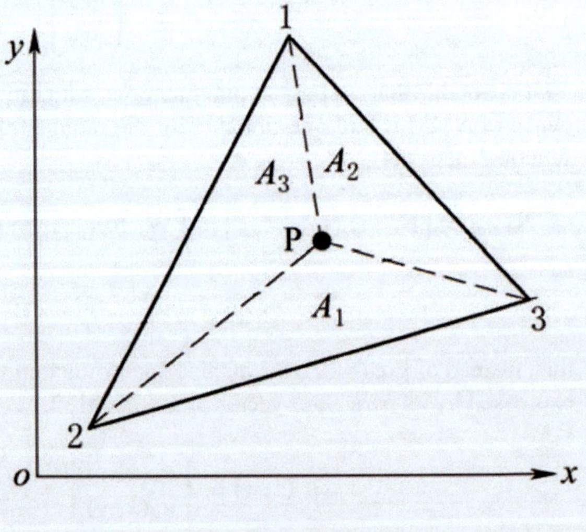

Fig. 6.5 Area coordinates

$$\begin{cases} N_1 = L_1 \\ N_2 = L_2 \\ N_3 = L_3 \end{cases} \tag{6.22}$$

At x-y plan, the coordinate interpolation is given by

$$\begin{cases} x = \sum_{l=1}^{3} N_l x_l \\ y = \sum_{l=1}^{3} N_l y_l \end{cases} \tag{6.23}$$

Let

$$\begin{cases} W_1(x, y) = N_1 \\ W_2(x, y) = N_2 \\ W_3(x, y) = N_3 \end{cases} \tag{6.24}$$

we have

$$W_1(x, y) + W_2(x, y) + W_3(x, y) = \sum_{l=1}^{3} N_l = 1 \tag{6.25}$$

and

$$W_l(x, y) \geq 0 \quad (l = 1, 2, 3) \tag{6.26}$$

For a finite element that is not intersected by joint, it is identical to the manifold element with same displacement (cover) functions. The only difference is in the terms "shape function" and "weight function".

(1) **Case of** Fig. 6.2b

Now we look at the triangular mesh in Fig. 6.2b containing 6 mathematical covers, 8 physical covers and 6 manifold elements, in which the physical covers $pc(4_1)$ and $pc(4_2)$, $pc(2_1)$ and $pc(2_2)$ are formed by the intersection of structural plane (crack) through the mathematical covers 4 and 2, respectively. All the covers and manifold elements are collected in Tables 6.4 and 6.5.

Suppose the cover function for each physical cover is constant, then for the manifold elements $me(2_1 4_2 5_1)$ and $me(2_2 4_1 5_1)$, the weight functions of physical covers $pc(2_1)$ and $pc(2_2)$ are identical to the shape function related to node 2 in the finite element $\triangle 245$ and denoted as $W_1(x, y)$, the weight functions of physical cover $pc(4_1)$ and $pc(4_2)$ are identical to the same shape function related to node 4 in the finite element $\triangle 245$ and denoted as $W_2(x, y)$, the weight function of physical cover $pc(5_1)$ is exactly the shape function related to node 5 in the finite element $\triangle 245$ and denoted as $W_3(x, y)$. The constant cover functions (6.10) are grouped as:

- $me(2_1 4_2 5_1)$.

$$\hat{\mathbf{u}}_{me(e)} = [\,\hat{u}_{x,2_1}\ \hat{u}_{y,2_1}\ \hat{u}_{x,4_2}\ \hat{u}_{y,4_2}\ \hat{u}_{x,5_1}\ \hat{u}_{y,5_1}\,]^{\mathrm{T}} \tag{6.27}$$

- $me(2_2 4_1 5_1)$.

$$\hat{\mathbf{u}}_{me(e)} = [\,\hat{u}_{x,2_2}\ \hat{u}_{y,2_2}\ \hat{u}_{x,4_1}\ \hat{u}_{y,4_1}\ \hat{u}_{x,5_1}\ \hat{u}_{y,5_1}\,]^{\mathrm{T}} \tag{6.28}$$

Similarly, for the manifold elements $me(2_1 5_1 3_1)$ and $me(2_2 5_1 3_1)$, we have

- $me(2_1 5_1 3_1)$.

$$\hat{\mathbf{u}}_{me(e)} = [\,\hat{u}_{x,2_1}\ \hat{u}_{y,2_1}\ \hat{u}_{x,5_1}\ \hat{u}_{y,5_1}\ \hat{u}_{x,3_1}\ \hat{u}_{y,3_1}\,]^{\mathrm{T}} \tag{6.29}$$

- $me(2_2 5_1 3_1)$.

$$\hat{\mathbf{u}}_{me(e)} = [\,\hat{u}_{x,2_2}\ \hat{u}_{y,2_2}\ \hat{u}_{x,5_1}\ \hat{u}_{y,5_1}\ \hat{u}_{x,3_1}\ \hat{u}_{y,3_1}\,]^{\mathrm{T}} \tag{6.30}$$

The manifold elements $me(3_1 5_1 6_1)$ and $me(1_1 2_1 3_1)$ without structural plane (crack) are composed of original physical covers ($pc(3_1)$, $pc(5_1)$, $pc(6_1)$) and ($pc(1_1)$, $pc(2_1)$, $pc(3_1)$) with corresponding weight functions ($W_1(x, y)$, $W_2(x, y)$, $W_3(x, y)$) that are identical to the shape functions related to the finite elements $\Delta 356$ and $\Delta 123$, respectively. Hence.

- $me(3_1 5_1 6_1)$.

$$\hat{\mathbf{u}}_{me(e)} = [\,\hat{u}_{x,3_1}\ \hat{u}_{y,3_1}\ \hat{u}_{x,5_1}\ \hat{u}_{y,5_1}\ \hat{u}_{x,6_1}\ \hat{u}_{y,6_1}\,]^{\mathrm{T}} \tag{6.31}$$

- $me(1_1 2_1 3_1)$.

$$\hat{\mathbf{u}}_{me(e)} = [\,\hat{u}_{x,1_1}\ \hat{u}_{y,1_1}\ \hat{u}_{x,2_1}\ \hat{u}_{y,2_1}\ \hat{u}_{x,3_1}\ \hat{u}_{y,3_1}\,]^{\mathrm{T}} \tag{6.32}$$

(2) **Case of** Fig. 6.2c

For the case with seven manifold elements in Fig. 6.2c, the displacement interpolations are given as follows

- $me(2_1 4_2 5_1)$.

It is identical to Eq. (6.27).

- $me(2_2 4_1 5_1)$.

It is identical to Eq. (6.28).

- $me(2_1 5_2 3_1)$.

$$\hat{\mathbf{u}}_{me(e)} = [\,\hat{u}_{x,2_1}\ \hat{u}_{y,2_1}\ \hat{u}_{x,5_2}\ \hat{u}_{y,5_2}\ \hat{u}_{x,3_1}\ \hat{u}_{y,3_1}\,]^{\mathrm{T}} \tag{6.33}$$

- $me(2_2 5_1 3_2)$.

$$\hat{\mathbf{u}}_{me(e)} = [\,\hat{u}_{x,2_2}\ \hat{u}_{y,2_2}\ \hat{u}_{x,5_1}\ \hat{u}_{y,5_1}\ \hat{u}_{x,3_2}\ \hat{u}_{y,3_2}\,]^{\mathrm{T}} \tag{6.34}$$

- $me(3_1 5_2 6_2)$.

$$\hat{\mathbf{u}}_{me(e)} = [\,\hat{u}_{x,3_1}\ \hat{u}_{y,3_1}\ \hat{u}_{x,5_2}\ \hat{u}_{y,5_2}\ \hat{u}_{x,6_2}\ \hat{u}_{y,6_2}\,]^{\mathrm{T}} \tag{6.35}$$

- $me(3_2 5_1 6_1)$.

$$\hat{\mathbf{u}}_{me(e)} = [\,\hat{u}_{x,3_2}\ \hat{u}_{y,3_2}\ \hat{u}_{x,5_1}\ \hat{u}_{y,5_1}\ \hat{u}_{x,6_1}\ \hat{u}_{y,6_1}\,]^{\mathrm{T}} \tag{6.36}$$

The manifold element $me(1_1 2_1 3_1)$ without structural plane (crack) is composed of original physical covers ($pc(1_1)$, $pc(2_1)$, $pc(3_1)$) with corresponding the weight functions ($W_1(x, y)$, $W_2(x, y)$, $W_3(x, y)$) that are identical to the shape functions related to the finite element $\triangle 123$. Hence

- $me(1_1 2_1 3_1)$.

It is identical to Eq. (6.32).

Above constructed cover function possesses characteristics that the displacement along the common edge of elements are continuous attributable to the use of standard FE shape functions. However, by the assignment of different physical covers to the manifold elements that is actually the sub-elements of a finite element cut through by structural planes (cracks, joints), the discontinuity at the structural plane is not kept, in this manner the open and slip of structural planes may be naturally simulated.

2. **Four-nodal quadrilateral element (constant nodal cover m = 1)**

Use is made of the shape functions defined in Eqs. (4.237)–(4.239), the coordinate interpolation at plane xoy is given by

$$\begin{cases} x = \sum\limits_{l=1}^{n} N_l(\xi, \eta) x_l \\ y = \sum\limits_{l=1}^{n} N_l(\xi, \eta) y_l \end{cases} \tag{6.37}$$

in which $n = 4$. It follows that we can get the reverse transformation $\xi = \xi(x, y)$, $\eta = \eta(x, y)$, hence the shape functions can be implicitly expressed as

$$N_l(\xi, \eta) = N_l(\xi(x, y), \eta(x, y)) \tag{6.38}$$

If we let

$$\begin{cases} W_1(x, y) = N_1(\xi, \eta) \\ W_2(x, y) = N_2(\xi, \eta) \\ W_3(x, y) = N_3(\xi, \eta) \\ W_4(x, y) = N_4(\xi, \eta) \end{cases} \tag{6.39}$$

it is obviously that for any point $P(x, y)$ within the manifold element, PU is valid for the weight functions

$$W_1(x, y) + W_2(x, y) + W_3(x, y) + W_4(x, y) = \sum_{l=1}^{n} N_l(\xi, \eta) = 1 \tag{6.40}$$

(1) **Case of** Fig. 6.3b

For the manifold elements $me(1_2 2_1 6_1 5_2)$ and $me(1_1 2_1 6_1 5_1)$ in Fig. 6.3b, the weight functions of physical covers $pc(1_1)$ and $pc(1_2)$ are identical to the shape function of the node 1 in finite element □1265 and denoted as $W_1(x, y)$, the weight functions of physical covers $pc(2_1)$ and $pc(6_1)$ are identical to the shape functions of nodes 2 and 6 in finite element □1265 and denoted as $W_2(x, y)$ and $W_3(x, y)$, the weight functions of physical covers $pc(5_1)$ and $pc(5_2)$ are identical to the shape function of node 5 in finite element □1265 and denoted as $W_4(x, y)$. In this manner the constant cover functions in Eq. (6.10) are grouped as:

- $me(1_2 2_1 6_1 5_2)$.

$$\hat{\mathbf{u}}_{me(e)} = [\,\hat{u}_{x,1_2}\ \hat{u}_{y,1_2}\ \hat{u}_{x,2_1}\ \hat{u}_{y,2_1}\ \hat{u}_{x,6_1}\ \hat{u}_{y,6_1}\ \hat{u}_{x,5_2}\ \hat{u}_{y,5_2}\,]^{\mathrm{T}} \tag{6.41}$$

- $me(1_1 2_1 6_1 5_1)$.

$$\hat{\mathbf{u}}_{me(e)} = [\,\hat{u}_{x,1_1}\ \hat{u}_{y,1_1}\ \hat{u}_{x,2_1}\ \hat{u}_{y,2_1}\ \hat{u}_{x,6_1}\ \hat{u}_{y,6_1}\ \hat{u}_{x,5_1}\ \hat{u}_{y,5_1}\,]^{\mathrm{T}} \tag{6.42}$$

The adjacent manifold element $me(2_1 3_1 7_1 6_1)$ is identical to the case without structural plane, namely, it is composed of four original physical covers ($pc(2_1)$, $pc(3_1)$, $pc(7_1)$, $pc(6_1)$) with corresponding weight functions ($W_1(x, y)$, $W_2(x, y)$, $W_3(x, y)$, $W_4(x, y)$) identical to the shape functions tie to nodes (2, 3, 7, 6) of the finite element □2376. The same applies to the manifold element $me(3_1 4_1 8_1 7_1)$. Hence

- $me(2_1 3_1 7_1 6_1)$.

$$\hat{\mathbf{u}}_{me(e)} = [\,\hat{u}_{x,2_1}\ \hat{u}_{y,2_1}\ \hat{u}_{x,3_1}\ \hat{u}_{y,3_1}\ \hat{u}_{x,7_1}\ \hat{u}_{y,7_1}\ \hat{u}_{x,6_1}\ \hat{u}_{y,6_1}\,]^{\mathrm{T}} \tag{6.43}$$

- $me(3_1 4_1 8_1 7_1)$.

$$\hat{\mathbf{u}}_{me(e)} = [\,\hat{u}_{x,3_1}\ \hat{u}_{y,3_1}\ \hat{u}_{x,4_1}\ \hat{u}_{y,4_1}\ \hat{u}_{x,8_1}\ \hat{u}_{y,8_1}\ \hat{u}_{x,7_1}\ \hat{u}_{y,7_1}\,]^{\mathrm{T}} \tag{6.44}$$

For other manifold elements uniquely related with finite elements, their weight functions and interpolations are identical to fundamental FEM.

(2) **Case of** Fig. 6.3c

For the five manifold elements in Fig. 6.3c, the displacement interpolations are given below:

- $me(1_2 2_1 6_2 5_2)$.

$$\hat{\mathbf{u}}_{me(e)} = [\,\hat{u}_{x,1_2}\ \hat{u}_{y,1_2}\ \hat{u}_{x,2_1}\ \hat{u}_{y,2_1}\ \hat{u}_{x,6_1}\ \hat{u}_{y,6_1}\ \hat{u}_{x,5_2}\ \hat{u}_{y,5_2}\,]^{\mathrm{T}} \tag{6.45}$$

- $me(1_1 2_2 6_1 5_1)$.

$$\hat{\mathbf{u}}_{me(e)} = [\,\hat{u}_{x,1_1}\ \hat{u}_{y,1_1}\ \hat{u}_{x,2_2}\ \hat{u}_{y,2_2}\ \hat{u}_{x,6_1}\ \hat{u}_{y,6_1}\ \hat{u}_{x,5_1}\ \hat{u}_{y,5_1}\,]^{\mathrm{T}} \tag{6.46}$$

- $me(2_1 3_1 7_1 6_2)$.

$$\hat{\mathbf{u}}_{me(e)} = [\,\hat{u}_{x,2_1}\ \hat{u}_{y,2_1}\ \hat{u}_{x,3_1}\ \hat{u}_{y,3_1}\ \hat{u}_{x,7_1}\ \hat{u}_{y,7_1}\ \hat{u}_{x,6_2}\ \hat{u}_{y,6_2}\,]^{\mathrm{T}} \tag{6.47}$$

- $me(2_2 3_1 7_1 6_1)$.

$$\hat{\mathbf{u}}_{me(e)} = [\,\hat{u}_{x,2_2}\ \hat{u}_{y,2_2}\ \hat{u}_{x,3_1}\ \hat{u}_{y,3_1}\ \hat{u}_{x,7_1}\ \hat{u}_{y,7_1}\ \hat{u}_{x,6_1}\ \hat{u}_{y,6_1}\,]^{\mathrm{T}} \tag{6.48}$$

- $me(3_1 4_1 8_1 7_1)$.

  It is identical to Eq. (6.44).

6.3.3 *Strain*

In the following content, the quadrilateral mesh-based NMM will be specially elaborated in details. Under such circumstances, each manifold element is constructed by four physical covers, namely, $n = 4$. It is reminded that, in addition to the overall sequence of the physical covers $pc(i)(i = 1, 2, \ldots, n_{pc})$,or briefly i, of the whole structure domain, the auxiliary use of local consequence of physical covers $l(1, 2, \ldots, n)$ will facilitate the deduction of governing equation set for structural systems. In addition, the incremental formulation will be implemented hereinafter for the purpose of non-linear (i.e. elasto-viscoplastic) analysis.

1. **Manifold elements**

For a manifold element $me(e)(e = 1, 2, \ldots, n_{me})$ composed of n (= 4) physical covers, the displacement interpolation is the weighted summation of these n physical covers (viz. Eqs. 6.19 and 6.20). In an incremental approach, the time domain is divided into a finite number of time-marching steps. Suppose the variables such as the displacement, velocity and acceleration are known at the beginning time t of the current step, and the increments of these variables at the end time $t + \Delta t$ need to be solved.

For the manifold elements $me(e)(e = 1, 2, \ldots, n_{me})$ based on quadrilateral FE mesh, at time t the strain increment $\Delta\boldsymbol{\varepsilon}_{me(e)} = [\Delta\varepsilon_x\ \Delta\varepsilon_y\ \Delta\varepsilon_z]^T$ is given by

$$\Delta\boldsymbol{\varepsilon}_{me(e)} = \mathbf{B}_{me(e)}\Delta\hat{\mathbf{u}}_{me(e)} \tag{6.49}$$

in which $\Delta\hat{\mathbf{u}}_{me(e)}$ is the nodal parameter vector defined in Eq. (6.19), and

$$\mathbf{B}_{me(e)} = \mathbf{L}\mathbf{N}_{me(e)} \tag{6.50}$$

or

$$\mathbf{B}_{me(e)} = \left[\mathbf{B}_1 \ldots \mathbf{B}_l \ldots \mathbf{B}_n\right] = \left[\mathbf{L}\mathbf{N}_1 \ldots \mathbf{L}\mathbf{N}_l \ldots \mathbf{L}\mathbf{N}_n\right] \tag{6.51}$$

where $\mathbf{L}$ is a differential operator matrix simplified from Eq. (3.86) for 2-D problems

$$\mathbf{L} = \begin{bmatrix} \frac{\partial}{\partial x} & 0 \\ 0 & \frac{\partial}{\partial y} \\ \frac{\partial}{\partial y} & \frac{\partial}{\partial x} \end{bmatrix} \tag{6.52}$$

and $\mathbf{N}_{me(e)}$ is the matrix of general shape function given in Eq. (6.18).

2. **Joint element between manifold elements**

In Fig. 6.6a, the finite element $fe(1265)$ is cut into two manifold elements $me(1_1 2_2 6_1 5_1)$ and $me(1_2 2_1 6_2 5_2)$ whose global consequence numbers are simply denote as $me(e_1)$ and $me(e_2)$ (vis. Fig. 6.6b).

There are a number of prevalent techniques to enforce displacement compatibility conditions between manifold elements, such as the Lagrange multiplier method,

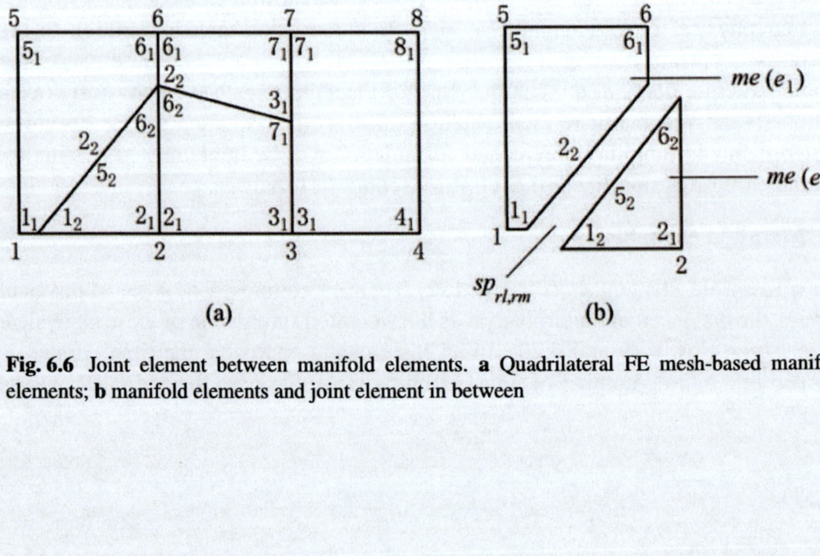

Fig. 6.6 Joint element between manifold elements. **a** Quadrilateral FE mesh-based manifold elements; **b** manifold elements and joint element in between

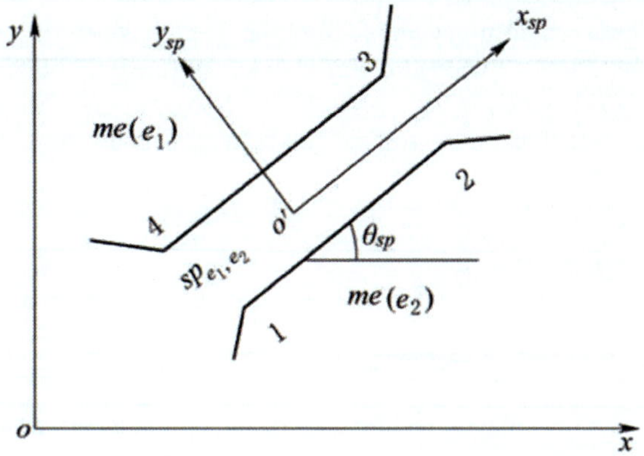

Fig. 6.7 Joint element for cohesive contact face

the penalty factor method, or the augmented Lagrange multiplier method (Wang et al. 1998; Tsay et al. 1999; Terada et al. 2003). In this book, the joint element $sp_{me(e_1),me(e_2)}$ abbreviated as *sp* is employed to simulate the cohesive contact between manifold elements. For this joint element, the local contact nodal pairs (1, 4) and (2, 3) are deployed (viz. Fig. 6.7), where the 1-D local coordinate system is used in a way similar to Fig. 3.6.

The relative displacements along the joint element are calculated by

$$\Delta \mathbf{u}_{sp} = [\, \Delta u_{x,sp} \; \Delta u_{y,sp} \,]^{\mathrm{T}} \tag{6.53}$$

On invoking Eq. (6.20), Eq. (6.53) becomes

$$\Delta \mathbf{u}_{sp} = \mathbf{T}_{sp}(\mathbf{N}_{me(e_1)}\Delta\hat{\mathbf{u}}_{me(e_1)} - \mathbf{N}_{me(e_2)}\Delta\hat{\mathbf{u}}_{me(e_2)}) \tag{6.54}$$

where the coordinate transformation matrix $\mathbf{T}_{sp}$ has been defined in Eq. (3.125).

3. **Point-face contact: hard spring elements**

It is unique that the point-edge contact between manifold elements may be particularly considered towards the potential simulation of finite deformation related to the rotation/dislocation of rock blocks, because NMM aims at solving discontinuous problems even with rigid motion (Shi 1991, 1992, 1995, 1996a, b, 1997, 2012).

When intersected with structural planes, each mathematical cover forms several physical covers associated with individual cover functions. The adjacent manifold elements formed by these physical covers may be independent on each other in the framework of NMM. The displacement field across manifold elements may be discontinuous, but one manifold element cannot penetrate into another. Such a constraint is normally termed as "non-penetration" condition.

In NMM, the contact detection technique is identical to DDA. In addition, from the point view of engineers, for the simulation of finite deformation related to the rotation/dislocation of rock blocks, DDA and DEM are sufficient w.r.t. the balance between accuracy and facilitation. Hence the contact detection technique of NMM will not be included in this chapter. If necessary however, the original works of Shi (1991, 1992, 1995, 1996a, b, 1997, 2012) concerning the penetration and slip, the sub-stiffness matrices and sub-load vectors due to finite deformation, are referred.

6.3.4 Stress

1. **Manifold elements**

The constitutive relation (3.138) with explicit time-marching scheme may be written as

$$\Delta\boldsymbol{\sigma}_{me(e)}(t) = \mathbf{D}_{me(e)}(\Delta\boldsymbol{\varepsilon}_{me(e)}(t) - \dot{\boldsymbol{\varepsilon}}^{vp}_{me(e)}(t)\Delta t) = \mathbf{D}_{me(e)}\Delta\boldsymbol{\varepsilon}_{me(e)}(t) - \Delta\boldsymbol{\sigma}^{vp}_{me(e)}(t) \tag{6.55}$$

2. **Joint element between manifold elements**

The constitutive relation (3.142) with explicit time-marching scheme may be written as

$$\Delta\boldsymbol{\sigma}_{sp}(t) = \mathbf{D}_{sp}(\Delta\mathbf{u}_{sp}(t) - \dot{\mathbf{u}}^{vp}_{sp}(t)\Delta t) = \mathbf{D}_{sp}(\Delta\mathbf{u}_{sp}(t) - \Delta\mathbf{u}^{vp}_{sp}(t)) \quad = \mathbf{D}_{sp}\Delta\mathbf{u}_{sp}(t) - \Delta\boldsymbol{\sigma}^{vp}_{sp}(t) \tag{6.56}$$

6.4 Dynamics Aspects: Basic Formulation

6.4.1 Concept

1. **Governing equations**

In the formulation of a computational method for solid mechanics, boundary conditions are the constituent parts (Fig. 6.8). In general, the boundary conditions are categorized into two types: essential boundary condition on Γ_u and natural boundary condition on Γ_t, the latter is for the imposition of tractions $\bar{\mathbf{t}}$, whereas the former is for the imposition displacement restraints $\mathbf{u} = \bar{\mathbf{u}}$. In addition, the structural planes (joint, fault, crack) $\Gamma_{sp.}$ which may be looked at as internal boundaries, frequently emerges in the domain to separate it into two or more parts of sub-domains.

In NMM, mathematical covers are constructed totally independent of boundaries and structural planes. The essential boundary conditions thus cannot be accomplished by directly enforcing the DOFs like in FEM, but is usually prescribed by Lagrange multipliers or penalty factors in the weak form of governing equation.

Fig. 6.8 Boundaries and structural planes

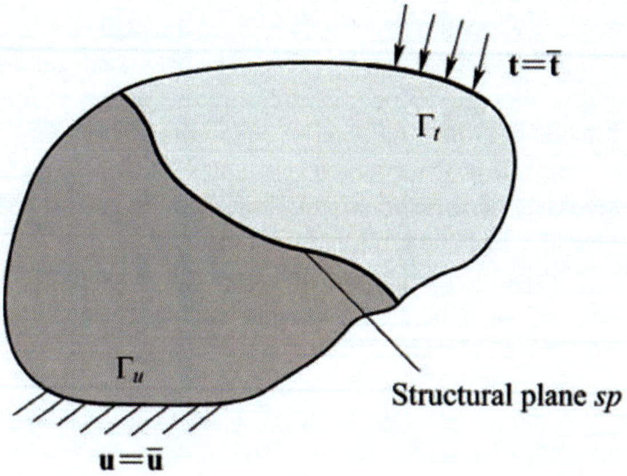

The original formulation of NMM was based on the minimization principle of potential energy, and penalty factors in the form of stiff springs for structural planes were employed to prevent the penetration between manifold elements. For a detailed formulation using the minimization principle of potential energy, readers are referred to the literatures of Shi and his co-workers (Shi 1991, 1992, 1995, 1996a, b, 1997, 2012).

In this chapter, the principle of virtual work is used to derive the governing equation for NMM. According to Eq. (3.198) and taking into account of the internal traction boundary (cohesive crack, structural plane) as well as the dynamic force, we get

$$\begin{aligned}
&\sum_{e}^{n_{me}} \iint_{\Omega_{me(e)}} \delta\Delta\boldsymbol{\varepsilon}_{me(e)}^{\mathrm{T}} \Delta\boldsymbol{\sigma}_{me(e)} \mathrm{d}\Omega + \sum_{e}^{n_{me}} \iint_{\Omega_{me(e)}} \delta\Delta\mathbf{u}_{me(e)}^{\mathrm{T}} \rho \Delta\ddot{\mathbf{u}}_{me(e)} \mathrm{d}\Omega \\
&+ \sum_{sp=1}^{n_{sp}} \int_{\Gamma_{sp}} \delta\Delta\mathbf{u}_{sp}^{\mathrm{T}} \Delta\boldsymbol{\sigma}_{sp} \mathrm{d}\Gamma + \int_{\Gamma_u} \delta\Delta\mathbf{u}^{\mathrm{T}} \alpha(\Delta\mathbf{u} - \Delta\bar{\mathbf{u}}) \mathrm{d}\Gamma \\
&= \sum_{e}^{n_{me}} \iint_{\Omega_{me(e)}} \delta\Delta\mathbf{u}_{me(e)}^{\mathrm{T}} \Delta\mathbf{b}_{me(e)} \mathrm{d}\Omega + \int_{\Gamma_t} \delta\Delta\mathbf{u}_t^{\mathrm{T}} \Delta\bar{\mathbf{t}} \mathrm{d}\Gamma + \sum_{\zeta=1}^{n_\zeta} \delta\Delta\mathbf{u}_\zeta^{\mathrm{T}}(\mathbf{x}_\zeta) \Delta\mathbf{T}_\zeta(\mathbf{x}_\zeta)
\end{aligned} \tag{6.57}$$

where: $\Delta\boldsymbol{\sigma}$—second order Cauchy stress tensor; $\Delta\mathbf{b}$—body force per unit volume; $\Delta\bar{\mathbf{t}}$—external force on the boundary Γ_t; $\Delta\mathbf{T}_\zeta(\mathbf{x}_\zeta)$—point (concentrated) force exerted on $\mathbf{x}_\zeta$; $\Delta\bar{\mathbf{u}}$—prescribed displacements on the boundary Γ_u; α—penalty factor; $\Delta\boldsymbol{\sigma}_{sp}$—stress on structural plane Γ_{sp}, if $\Delta\boldsymbol{\sigma}_{sp}$ is equal to zero, it is considered as traction free.

It is noted that some items originally appear in Eq. (6.57), such as the virtual works attributable to constant initial geo-stress and vertex-to-edge contact, are neglected. The major reason is the considerations on the requirements in geotechnical and hydraulic structures focused on structural plane (or crack growth) problem: the initial geo-stress is not constant and need be back analyzed by adjusting boundary tractions and volumetric actions; in addition, the vertex-to-edge contact is normally accompanied with rotational motion due to the finite deformation under the circumstances of landslide process, which is troublesome for the code implementation and not very necessary because where rotational motion should be handled, other advanced CMs such as DEM and DDA exhibit more suitability.

Similar to FEM, the assemble operation from the corresponding sub-matrices and sub-vectors of manifold elements will finally turn out one solvable algebraic equation set

$$\mathbf{K}\Delta\hat{\mathbf{U}} + \mathbf{M}\Delta\ddot{\hat{\mathbf{U}}} = \Delta\mathbf{F} + \Delta\mathbf{F}^{vp} \tag{6.58}$$

in which $\mathbf{K}$ is the system stiffness matrix, $\mathbf{M}$ is the system mass matrix, $\Delta\mathbf{F}$ is the incremental load vector at the current time step, $\Delta\mathbf{F}^{vp}$ is the incremental load vector due to viscoplastic deformation. For each physical cover with global sequence $i\,(i = 1, 2, \ldots, n_{pc})$ is assigned where n_{pc} is the amount of physical covers, they are specified as follows:

$$\mathbf{K} = \begin{bmatrix} \mathbf{K}_{1,1} & \ldots & \mathbf{K}_{1,i} & \ldots & \mathbf{K}_{1,j} & \ldots & \mathbf{K}_{1,n_{pc}} \\ \ldots & \ldots & \ldots & \ldots & \ldots & \ldots & \ldots \\ \mathbf{K}_{i,1} & \ldots & \mathbf{K}_{i,i} & \ldots & \mathbf{K}_{i,j} & \ldots & \mathbf{K}_{i,n_{pc}} \\ \ldots & \ldots & \ldots & \ldots & \ldots & \ldots & \ldots \\ \mathbf{K}_{j,1} & \ldots & \mathbf{K}_{j,i} & \ldots & \mathbf{K}_{j,j} & \ldots & \mathbf{K}_{j,n_{pc}} \\ \ldots & \ldots & \ldots & \ldots & \ldots & \ldots & \ldots \\ \mathbf{K}_{n_{pc},1} & \ldots & \mathbf{K}_{n_{pc},i} & \ldots & \mathbf{K}_{n_{pc},j} & \ldots & \mathbf{K}_{n_{pc},n_{pc}} \end{bmatrix} \tag{6.59}$$

in which the sub-matrix of stiffness $\mathbf{K}_{i,i}$ depends on the physical cover i only, whereas $\mathbf{K}_{i,j}\,(i \neq j)$ depends on both the physical covers i and j. They are all $2m \times 2m$ sub-matrices where m is the amount of items in the polynomial displacement cover function.

$$\mathbf{M} = \begin{bmatrix} \mathbf{M}_{1,1} & \ldots & \mathbf{M}_{1,i} & \ldots & \mathbf{M}_{1,j} & \ldots & \mathbf{M}_{1,n_{pc}} \\ \ldots & \ldots & \ldots & \ldots & \ldots & \ldots & \ldots \\ \mathbf{M}_{i,1} & \ldots & \mathbf{M}_{i,i} & \ldots & \mathbf{M}_{i,j} & \ldots & \mathbf{M}_{i,n_{pc}} \\ \ldots & \ldots & \ldots & \ldots & \ldots & \ldots & \ldots \\ \mathbf{M}_{j,1} & \ldots & \mathbf{M}_{j,i} & \ldots & \mathbf{M}_{j,j} & \ldots & \mathbf{M}_{j,n_{pc}} \\ \ldots & \ldots & \ldots & \ldots & \ldots & \ldots & \ldots \\ \mathbf{M}_{n_{pc},1} & \ldots & \mathbf{M}_{n_{pc},i} & \ldots & \mathbf{M}_{n_{pc},j} & \ldots & \mathbf{M}_{n_{pc},n_{pc}} \end{bmatrix} \tag{6.60}$$

in which the sub-matrix of mass $\mathbf{M}_{i,i}$ depends on the physical cover i only, whereas $\mathbf{M}_{i,j}(i \neq j)$ depends on both the physical covers i and j. They are all $2m \times 2m$ sub-matrices.

$$\Delta\widehat{\mathbf{U}} = \left[\, \Delta\hat{\mathbf{u}}_1 \ldots \Delta\hat{\mathbf{u}}_i \ldots \Delta\hat{\mathbf{u}}_j \ldots \Delta\hat{\mathbf{u}}_{n_{pc}} \,\right]^{\mathrm{T}} \tag{6.61}$$

$$\Delta\mathbf{F} = \left[\, \Delta\mathbf{F}_1 \ldots \Delta\mathbf{F}_i \ldots \Delta\mathbf{F}_j \ldots \Delta\mathbf{F}_{n_{pc}} \,\right]^{\mathrm{T}} \tag{6.62}$$

$$\Delta\mathbf{F}^{vp} = \left[\, \Delta\mathbf{F}_1^{vp} \ldots \Delta\mathbf{F}_i^{vp} \ldots \Delta\mathbf{F}_j^{vp} \ldots \Delta\mathbf{F}_{n_{pc}}^{vp} \,\right]^{\mathrm{T}} \tag{6.63}$$

In Eqs. (6.61)–(6.63) $\Delta\hat{\mathbf{u}}_i$, $\Delta\mathbf{F}_i$ and $\Delta\mathbf{F}_i^{vp}$ are $2m \times 1$ sub-vectors.

It is notable that where the traditional structural dynamics is handled and the structural response is the major concern, the Rayleigh damping that partially contributed from viscoelasticity, is commonly stipulated (viz. Eq. 4.170). As a result, Eq. (6.58) may be symbolically extended to the incremental Eq. (4.167), namely, instead of Eq. (6.58), we have

$$\mathbf{K}\Delta\widehat{\mathbf{U}} + \mathbf{C}\Delta\dot{\widehat{\mathbf{U}}} + \mathbf{M}\Delta\ddot{\widehat{\mathbf{U}}} = \Delta\mathbf{F} + \Delta\mathbf{F}^{vp} = \Delta\mathbf{P} \tag{6.64}$$

2. **Solution schemes**

Equation (6.64) may be solved by the Newmark-β or Wilson-θ method routinely (viz. Chap. 4). However, for incremental form, these algorithms should be implemented in different style. Taking the Newmark-β method for example, we first write

$$\begin{cases} \Delta\ddot{\widehat{\mathbf{U}}}(t) = \frac{\Delta\widehat{\mathbf{U}}(t)}{\alpha\Delta t^2} - \frac{\dot{\widehat{\mathbf{U}}}(t)}{\alpha\Delta t} - \frac{\ddot{\widehat{\mathbf{U}}}(t)}{2\alpha} \\ \Delta\dot{\widehat{\mathbf{U}}}(t) = \left[(1-\beta)\ddot{\widehat{\mathbf{U}}}(t) + \beta\ddot{\widehat{\mathbf{U}}}(t+\Delta t)\right]\Delta t = \Delta t\ddot{\widehat{\mathbf{U}}}(t) + \beta\Delta t\Delta\ddot{\widehat{\mathbf{U}}}(t) \end{cases} \tag{6.65}$$

the substitution into Eq. (6.64) gives rise to the solution algorithm detailed below:

① Calculation of system stiffness matrix **K**, damping matrix **C** and mass matrix **M**;

② Calculation of constants

$$\begin{cases} \alpha \geq 0.25(0.5+\beta)^2 \\ \beta \geq 0.5 \\ a_0 = \frac{1}{\alpha\Delta t^2} \\ a_1 = \frac{\beta}{\alpha\Delta t} \\ a_2 = \frac{1}{\alpha\Delta t} \\ a_3 = \frac{1}{2\alpha} - 1 \\ a_4 = \frac{\beta}{\alpha} - 1 \\ a_5 = \frac{\Delta t}{2}(\frac{\beta}{\alpha} - 2) \\ a_6 = (1-\beta)\Delta t \\ a_7 = \beta\Delta t \end{cases} \tag{6.66}$$

③ Assemble of the equivalent stiffness matrix

$$\overline{\mathbf{K}}(t) = \mathbf{K}(t) + a_0\mathbf{M}(t) + a_1\mathbf{C}(t) \tag{6.67}$$

④ Calculation of the equivalent load vector

$$\Delta\overline{\mathbf{P}}(t) = \Delta\mathbf{P}(t) + \mathbf{M}(t)\left(a_2\dot{\widehat{\mathbf{U}}}(t) + (a_3+1)\ddot{\widehat{\mathbf{U}}}(t)\right) + \mathbf{C}(t)\left((a_4+1)\dot{\widehat{\mathbf{U}}}(t) + a_5\ddot{\widehat{\mathbf{U}}}(t)\right) \tag{6.68}$$

⑤ Solution of the displacement increment

$$\Delta\widehat{\mathbf{U}}(t) = \overline{\mathbf{K}}^{-1}\Delta\overline{\mathbf{P}}(t) \tag{6.69}$$

⑥ Calculation of the acceleration, velocity, and displacement at time $t + \Delta t$

$$\begin{cases} \ddot{\widehat{\mathbf{U}}}(t+\Delta t) = a_0\Delta\widehat{\mathbf{U}}(t) - a_2\dot{\widehat{\mathbf{U}}}(t) - a_3\ddot{\widehat{\mathbf{U}}}(t) \\ \dot{\widehat{\mathbf{U}}}(t+\Delta t) = \dot{\widehat{\mathbf{U}}}(t) + a_6\ddot{\widehat{\mathbf{U}}}(t) + a_7\ddot{\widehat{\mathbf{U}}}(t+\Delta t) \\ \widehat{\mathbf{U}}(t+\Delta t) = \widehat{\mathbf{U}}(t) + \Delta\widehat{\mathbf{U}}(t) \end{cases} \tag{6.70}$$

It is notable that the time-marching step length should be small enough to guarantee the accuracy, because in one time interval Δt, the matrices of stiffness and mass and damping are all constant. In addition, where a crack growth emerges in Δt, these matrices need to be reformed, which accounts for considerable computation efforts. This is also why in this book we do not take into account of point-to-edge contact due to the finite deformation and rigid rotation declared in many literatures on NMM, because under such circumstances, DDA (Chap. 8) or DEM (Chap. 7) would be more convenient and sufficient accurate from the point view of practitioners.

6.4.2 Sub-matrices of Stiffness and Sub-vectors of Load

1. Sub-matrix and sub-vector contributed from the strain of manifold element

The virtual work of manifold element $me(e)(e = 1, 2, \ldots, n_{me})$ due to strain is

$$\iint\limits_{\Omega_{me(e)}} \delta\Delta\boldsymbol{\varepsilon}_{me(e)}^{\mathrm{T}}\Delta\boldsymbol{\sigma}_{me(e)}\mathrm{d}\Omega = \delta\hat{\mathbf{u}}_{me(e)}^{\mathrm{T}}\left(\iint\limits_{\Omega_{me(e)}} \mathbf{B}_{me(e)}^{\mathrm{T}}\mathbf{D}_{me(e)}\mathbf{B}_{me(e)}\mathrm{d}\Omega\hat{\mathbf{u}}_{me(e)} - \iint\limits_{\Omega_{me(e)}} \mathbf{B}_{me(e)}^{\mathrm{T}}\boldsymbol{\sigma}_{me(e)}^{vp}\mathrm{d}\Omega\right) \tag{6.71}$$

in which the integrant is over the entire manifold element $me(e)$. For the n physical covers with local sequence $l \supset me(e)$, the arbitrary property of $\delta\hat{\mathbf{u}}_{me(e)}$ gives rise to

$$\mathbf{k}_{l,k} = \iint\limits_{\Omega_{me(e)}} \mathbf{B}_l^{\mathrm{T}} \mathbf{D}_{me(e)} \mathbf{B}_k \mathrm{d}\Omega \quad \text{for } l, k \supset me(e) \tag{6.72}$$

$$\Delta \mathbf{f}_l^{vp} = \iint\limits_{\Omega_{me(e)}} \mathbf{B}_l^{\mathrm{T}} \boldsymbol{\sigma}_{me(e)}^{vp} \mathrm{d}\Omega \quad \text{for } l \supset me(e) \tag{6.73}$$

With the help of link list that relates local physical cover sequences (l, k) and global physical sequences (i, j) in the form of $i = \langle l \rangle$ and $j = \langle k \rangle$, they are assembled in a manner similar to the fundamental FEM such that

$$\mathbf{k}_{l,k} \rightarrow \mathbf{K}_{i,j} \tag{6.74}$$

and

$$\Delta \mathbf{f}_l^{vp} \rightarrow \Delta \mathbf{F}_i^{vp} \tag{6.75}$$

The arrow "→" means the operation to be assembled to $\mathbf{K}_{i,j}$ and $\Delta \mathbf{F}_i^{vp}$ in Eqs. (6.59) and (6.63).

2. **Sub-matrix and sub-vector contributed from the inertia of manifold element**

The virtual work of manifold element $me(e)(e = 1, 2, \ldots, n_{me})$ due to inertia is

$$\iint\limits_{\Omega_{me(e)}} \delta \Delta \mathbf{u}_{me(e)}^{\mathrm{T}} (\rho \Delta \ddot{\mathbf{u}}_{me(e)}) \mathrm{d}\Omega = \delta \hat{\mathbf{u}}_{me(e)}^{\mathrm{T}} \iint\limits_{\Omega_{me(e)}} \mathbf{N}_{me(e)}^{\mathrm{T}} \rho \mathbf{N}_{me(e)} \mathrm{d}x \mathrm{d}y \ddot{\hat{\mathbf{u}}}_{me(e)} \tag{6.76}$$

For the n physical covers $l \supset me(e)$, Eq. (6.76) gives rise to

$$\mathbf{m}_{l,k} = \iint\limits_{\Omega_{me(e)}} \mathbf{N}_l^{\mathrm{T}} \rho \mathbf{N}_k \mathrm{d}\Omega \quad \text{for } l, k \supset me(e) \tag{6.77}$$

with the help of $i = \langle l \rangle$ and $j = \langle k \rangle$ that relate local physical cover sequences and global physical sequences, $2m \times 2m$ sub-matrix $\mathbf{m}_{l,k}$ is assembled into the physical covers

$$\mathbf{m}_{l,k} \rightarrow \mathbf{M}_{i,j} \tag{6.78}$$

3. **Sub-matrix and sub-vector contributed from the relative displacement of joint element**

The virtual work of the joint element representing a segment of structural plane (crack surface) between manifold element $me(e_1)(e_1 = 1, 2, \ldots, n_{me})$ and manifold element $me(e_2)(e_2 = 1, 2, \ldots, n_{me})$ is

$$\int_{\Gamma_{sp}} \delta\Delta\mathbf{u}_{sp}^{\mathrm{T}}\Delta\boldsymbol{\sigma}_{sp}\mathrm{d}\Gamma = \int_{\Gamma_{sp}} (\delta\Delta\hat{\mathbf{u}}_{me(e_1)}^{\mathrm{T}}\mathbf{N}_{me(e_1)}^{\mathrm{T}} - \delta\Delta\hat{\mathbf{u}}_{me(e_2)}^{\mathrm{T}}\mathbf{N}_{me(e_2)}^{T})\mathbf{T}_{sp}^{\mathrm{T}}\mathbf{D}_{sp}\mathbf{T}_{sp}$$

$$(\mathbf{N}_{me(e_1)}\Delta\hat{\mathbf{u}}_{me(e_1)} - \mathbf{N}_{me(e_2)}\Delta\hat{\mathbf{u}}_{me(e_2)})\mathrm{d}\Gamma - \int_{\Gamma_{sp}} (\delta\Delta\hat{\mathbf{u}}_{me(e_1)}^{\mathrm{T}}\mathbf{N}_{me(e_1)}^{\mathrm{T}}$$

$$-\delta\Delta\hat{\mathbf{u}}_{me(e_2)}^{\mathrm{T}}\mathbf{N}_{me(e_2)}^{\mathrm{T}})\mathbf{T}_{sp}^{\mathrm{T}}\Delta\boldsymbol{\sigma}_{sp}^{vp}(t)\mathrm{d}\Gamma \tag{6.79}$$

For the n physical covers constructing manifold element $me(e_1)$ and other n physical covers constructing manifold element $me(e_2)$, Eq. (6.79) leads to

$$\mathbf{k}_{l,k} = \int_{\Gamma_{sp}} \mathbf{N}_l^{\mathrm{T}}\mathbf{T}_{sp}^{\mathrm{T}}\mathbf{D}_{sp}\mathbf{T}_{sp}\mathbf{N}_k\mathrm{d}\Gamma \quad \text{for } l, k \supset me(e_1) \tag{6.80}$$

$$\mathbf{k}_{l,k} = \int_{\Gamma_{sp}} \mathbf{N}_l^{\mathrm{T}}\mathbf{T}_{sp}^{\mathrm{T}}\mathbf{D}_{sp}\mathbf{T}_{sp}\mathbf{N}_k\mathrm{d}\Gamma \quad \text{for } l, k \supset me(e_2) \tag{6.81}$$

$$\mathbf{k}_{l,k} = -\int_{\Gamma_{sp}} \mathbf{N}_l^{\mathrm{T}}\mathbf{T}_{sp}^{\mathrm{T}}\mathbf{D}_{sp}\mathbf{T}_{sp}\mathbf{N}_k\mathrm{d}\Gamma \quad \text{for } l \supset me(e_1), k \supset me(e_2) \tag{6.82}$$

$$\mathbf{k}_{k,l} = -\int_{\Gamma_{sp}} \mathbf{N}_k^{\mathrm{T}}\mathbf{T}_{sp}^{\mathrm{T}}\mathbf{D}_{sp}\mathbf{T}_{sp}\mathbf{N}_l\mathrm{d}\Gamma \quad \text{for } k \supset me(e_1), l \supset me(e_2) \tag{6.83}$$

$$\Delta\mathbf{f}_l^{vp} = \int_{\Gamma_{sp}} \mathbf{N}_l^{\mathrm{T}}\mathbf{T}_{sp}^{\mathrm{T}}\Delta\boldsymbol{\sigma}_{sp}^{vp}\mathrm{d}\Gamma \quad \text{for } l \supset me(e_1) \tag{6.84}$$

$$\Delta\mathbf{f}_k^{vp} = -\int_{\Gamma_{sp}} \mathbf{N}_k^{\mathrm{T}}\mathbf{T}_{sp}^{\mathrm{T}}\Delta\boldsymbol{\sigma}_{sp}^{vp}\mathrm{d}\Gamma \quad \text{for } k \supset me(e_2) \tag{6.85}$$

With the help of link list that relates local physical cover sequences and global physical sequences, they are assembled to the global counterparts

$$\mathbf{k}_{l,k} \rightarrow \mathbf{K}_{i,j} \tag{6.86}$$

$$\Delta\mathbf{f}_l^{vp} \rightarrow \Delta\mathbf{F}_i^{vp} \tag{6.87}$$

$$\Delta\mathbf{f}_k^{vp} \rightarrow \Delta\mathbf{F}_j^{vp} \tag{6.88}$$

4. **Sub-matrix and sub-vector contributed from the prescribed boundary displacement**

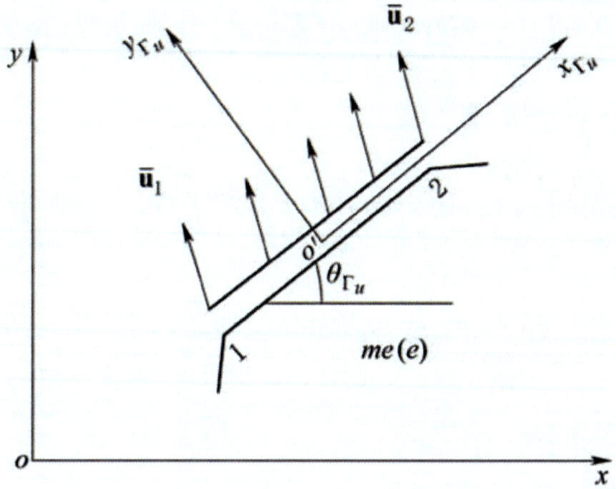

Fig. 6.9 Boundary segment with prescribed displacement

Suppose a boundary segment of manifold element $me(e)(e = 1, 2, \ldots, n_{me})$ is straight and defined by the vertex points 1 and 2 in Fig. 6.9.

The virtual work of the prescribed displacement between manifold element $me(e)$ and boundary wall is

$$\int_{\Gamma_u} \delta \Delta \mathbf{u}^{\mathrm{T}} \alpha (\Delta \mathbf{u} - \bar{\mathbf{u}}) \mathrm{d}\Gamma = \int_{\Gamma_u} \delta \Delta \hat{\mathbf{u}}_{me(e)}^{\mathrm{T}} \mathbf{N}_{me(e)}^{\mathrm{T}} \alpha (\mathbf{N}_{me(e)} \hat{\mathbf{u}}_{me(e)} - \bar{\mathbf{u}}) \mathrm{d}\Gamma$$
$$= \delta \Delta \hat{\mathbf{u}}_{me(e)}^{\mathrm{T}} \int_{\Gamma_u} \mathbf{N}_{me(e)}^{\mathrm{T}} \alpha \mathbf{N}_{me(e)} \hat{\mathbf{u}}_{me(e)} \mathrm{d}\Gamma$$
$$- \delta \Delta \hat{\mathbf{u}}_{me(e)}^{\mathrm{T}} \int_{\Gamma_u} \mathbf{N}_{me(e)}^{\mathrm{T}} \alpha \bar{\mathbf{u}} \mathrm{d}\Gamma \tag{6.89}$$

For n physical covers constructing manifold element $me(e)$, Eq. (6.89) gives rise to

$$\mathbf{k}_{l,k} = \alpha \int_{\Gamma_u} \mathbf{N}_l^{\mathrm{T}} \mathbf{N}_k \mathrm{d}\Gamma \quad \text{for } l, k \supset me(e) \tag{6.90}$$

$$\Delta \mathbf{f}_l = \alpha \int_{\Gamma_u} \mathbf{N}_l^{\mathrm{T}} \bar{\mathbf{u}} \mathrm{d}\Gamma \quad \text{for } l \supset me(e) \tag{6.91}$$

They are assembled to the global counterparts routinely

$$\mathbf{k}_{l,k} \to \mathbf{K}_{i,j} \tag{6.92}$$

$$\Delta \mathbf{f}_l \to \Delta \mathbf{F}_i \tag{6.93}$$

6.4.3 Sub-vectors of Load Attributable to External Forces

1. **Body loads**

Assuming that $\Delta\mathbf{b} = \begin{bmatrix} \Delta b_x & \Delta b_y \end{bmatrix}^{\mathrm{T}}$ is the constant volumetric force exerting on the manifold element $me(e)(e = 1, 2, \ldots, n_{me})$, then its virtual work should be equal to that of the load vector $\Delta\mathbf{f}_l$ tied to the physical covers constructing $me(e)$, namely

$$\iint\limits_{\Omega_{me(e)}} \delta\Delta\mathbf{u}^{\mathrm{T}}_{me(e)}\Delta\mathbf{b}_{me(e)}\mathrm{d}\Omega = \delta\hat{\mathbf{u}}^{\mathrm{T}}_{me(e)} \iint\limits_{\Omega_{me}(e)} \mathbf{N}^{\mathrm{T}}_{me(e)}\Delta\mathbf{b}\mathrm{d}\Omega = \delta\hat{\mathbf{u}}^{\mathrm{T}}_{me(e)}\Delta\mathbf{f}_l \tag{6.94}$$

hence

$$\Delta\mathbf{f}_l = \iint\limits_{\Omega_{me(e)}} \mathbf{N}^{\mathrm{T}}_l\Delta\mathbf{b}\mathrm{d}\Omega \quad \text{for } l \supset me(e) \tag{6.95}$$

The assemblage to the system load vector is under taken as follows

$$\Delta\mathbf{f}_l \rightarrow \Delta\mathbf{F}_i \tag{6.96}$$

2. **Boundary traction loads**

Suppose a boundary segment of manifold element $me(e)(e = 1, 2, \ldots, n_{me})$ is straight and defined by the vertex points 1 and 2 in Fig. 6.10.

Assuming that $\Delta\bar{\mathbf{t}} = \begin{bmatrix} \Delta\bar{t}_x & \Delta\bar{t}_y \end{bmatrix}^{\mathrm{T}}$ is the distributed boundary traction acting on Γ_t that are partially overlapped with an edge of manifold element $me(e)$, then its virtual work should be equal to that of the load vector $\Delta\mathbf{f}_l$ tied to the physical covers constructing $me(e)$, namely

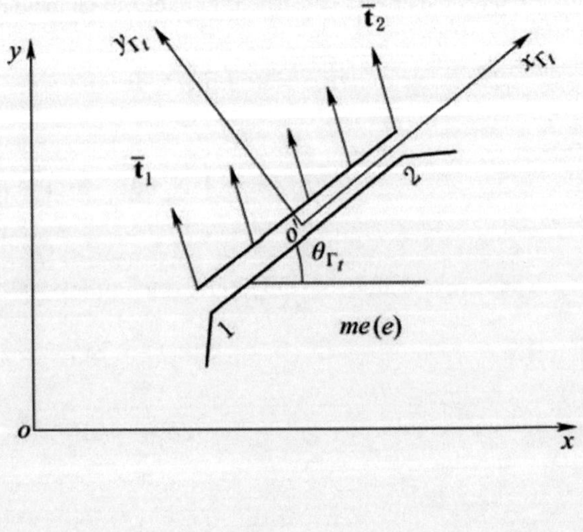

Fig. 6.10 Boundary segment exerted by traction

$$\int_{\Gamma_t} \delta\Delta\mathbf{u}_t^{\mathrm{T}}\Delta\bar{\mathbf{t}}\mathrm{d}\Gamma = \delta\hat{\mathbf{u}}_{me(e)}^{\mathrm{T}}\int_{\Gamma_t} \mathbf{N}_{me(e)}^{\mathrm{T}}\Delta\bar{\mathbf{t}}\mathrm{d}\Gamma = \delta\hat{\mathbf{u}}_{me(e)}^{\mathrm{T}}\Delta\mathbf{f}_l \tag{6.97}$$

For n physical covers constructing the manifold element $me(e)$, we get

$$\Delta\mathbf{f}_l = \int_{\Gamma_t} \mathbf{N}_{me(e)}^{\mathrm{T}}\Delta\bar{\mathbf{t}}\mathrm{d}\Gamma \quad \text{for } l \supset me(e) \tag{6.98}$$

This sub-vectors $\Delta\mathbf{f}_l$ are assembled to the global counterpart, too (viz. Eq. 6.96).

(3) **Point loads**

The concentrated loading force $\Delta\mathbf{T}_\zeta = \left[\Delta T_x \ \Delta T_y\right]^{\mathrm{T}}$ acts on the point (x_ζ, y_ζ) of manifold element $me(e)(e = 1, 2, \ldots, n_{me})$ simply produces the equality w.r.t. virtual works as

$$\delta\Delta\mathbf{u}_\zeta^{\mathrm{T}}(\mathbf{x}_\zeta)\Delta\mathbf{T}_\zeta(\mathbf{x}_\zeta) = \delta\Delta\hat{\mathbf{u}}_{me(e)}^{\mathrm{T}}\mathbf{N}_{me(e)}(x_\zeta, y_\zeta)^{\mathrm{T}}\Delta\mathbf{T}_\zeta(x_\zeta, y_\zeta) = \delta\hat{\mathbf{u}}_{me(e)}^{\mathrm{T}}\Delta\mathbf{f}_l \tag{6.99}$$

hence this concentrated force is transferred to the n physical covers constructing the manifold element $me(e)$

$$\Delta\mathbf{f}_l = \mathbf{N}_i(x_\zeta, y_\zeta)^{\mathrm{T}}\Delta\mathbf{T}_\zeta(\mathbf{x}_\zeta) \quad \text{for } l \supset me(e) \tag{6.100}$$

This sub-vectors $\Delta\mathbf{f}_l$ are identically assembled as illustrated in Eq. (6.96).

6.5 Dynamics Aspects: Crack Growth Problems

6.5.1 Concept

NMM simulates structural planes and cracks (cohesive and open) by splitting a mathematical cover into several physical covers tied by individual cover functions, in this manner the discontinuity can be well represented with displacement compatibility conditions along the joint element model elaborated previously (Sect. 6.4).

Figure 6.11 shows the 2-D domain containing a crack, where the regular mathematical covers are generated from rectangular FE mesh. For a mathematical cover completely split by the crack surface into two isolated regions, two individual physical covers are created (marked with circle). For a mathematical cover is partially cut by the crack surface, only one physical cover is created (marked with cross) and termed as "singular physical covers" because it contains crack-tip in contrast to other conventional physical covers. The crack will lead to a displacement discontinuity in elements ①–⑦.

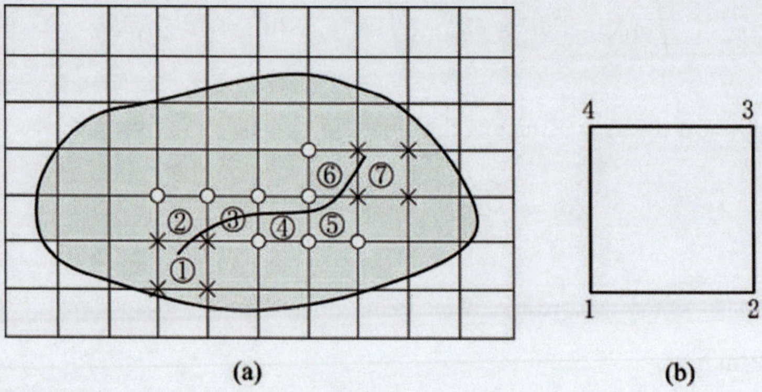

Fig. 6.11 Modeling a two-dimensional crack with the NMM. **a** NMM model; **b** local sequence for the four nodes associated with a finite element

For clarity, we first investigate the displacement field in element ②. There are four mathematical covers associated with element ②, i.e. $mc(1), mc(2), mc(3)$ and $mc(4)$, where $l\,(1, 2, 3, 4)$ is the local sequence of each mathematical cover associated with the local sequence of node shown in Fig. 6.11b. Mathematical covers $mc(1)$ and $mc(2)$ are partially cut by the crack and thus each forms only one physical cover $pc(1_1)$ and $pc(2_1)$, respectively, while mathematical covers $mc(3)$, and $mc(4)$ are completely split by the crack surface and thus each forms two individual physical covers $pc(3_1)$ and $pc(3_2)$, $pc(4_1)$ and $pc(4_2)$, respectively.

6.5.2 Displacement Enrichment

Now we turn to examine the displacement field in elements ① and ⑦, and element ① will be taken as a representative extended manifold element (Fig. 6.12).

Note that all the four mathematical covers associated with element ①, i.e. $mc(1)$, $mc(2)$, $mc(3)$ and $mc(4)$, are only partially cut by the crack surface, and thus each forms one singular physical cover, $pc(1_1)$, $pc(2_1)$, $pc(3_1)$ and $pc(4_1)$, respectively. For simplicity, in Figs. 6.11b and 6.12, the local physical cover sequence is given as $l = 1, 2, 3, 4$ that is identical to mathematical cover sequence. In order to seamlessly represent the displacement discontinuity across the crack surface and to accurately describe the crack-tip singularity, the singular physical covers are enriched with the asymptotic crack-tip functions derived from linear-elastic fracture mechanics

$$\mathbf{F} = \begin{bmatrix} F_1 & F_2 & F_3 & F_4 \end{bmatrix} = \begin{bmatrix} \sqrt{r}\sin\frac{\theta}{2} & \sqrt{r}\cos\frac{\theta}{2} & \sqrt{r}\sin\theta\sin\frac{\theta}{2} & \sqrt{r}\sin\theta\cos\frac{\theta}{2} \end{bmatrix} \tag{6.101}$$

in which r and θ are polar coordinates in the local crack-tip coordinate system (viz. Fig. 3.13). These four basis functions are able to construct arbitrary displacement

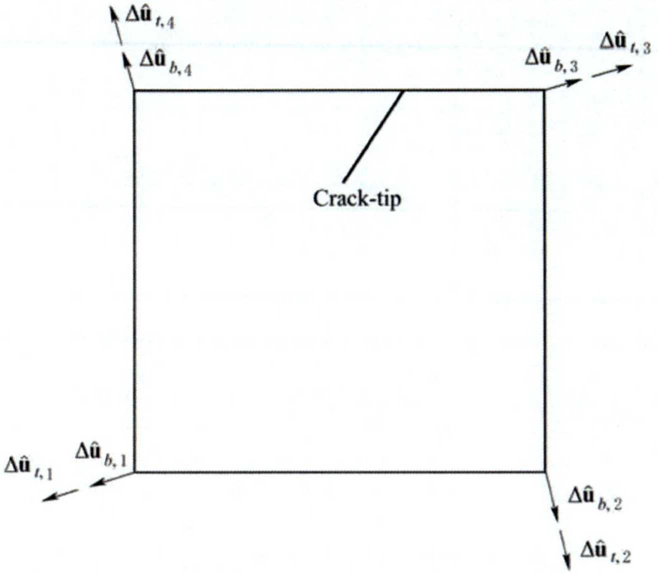

Fig. 6.12 Displacement interpolation of enriched manifold element ①

field near the crack-tip, and moreover, the first term is discontinuous across the crack surface (i.e. $\theta = \pm\pi$), which allows for the displacement discontinuity across crack surfaces in partially cut manifold elements (e.g. $me(e) = 1$ and 7 in Fig. 6.11a). As a consequence, instead of Eq. (6.20) we get the extended cover function for the enriched manifold element containing crack-tip

$$\Delta\mathbf{u}_{me(e)} = \mathbf{N}_{me(e)}\Delta\hat{\mathbf{u}}_{b,me(e)} + \mathbf{N}_{me(e)}\Delta\hat{\mathbf{u}}_{t,me(e)} \tag{6.102}$$

in which $\Delta\hat{\mathbf{u}}_{b,me(e)}$ is the coefficient vector of the basic cover function (viz. Eq. 6.19); the matrix of enriched basis function matrix and corresponding coefficient vector of cover function are expressed by

$$\begin{cases} \Delta\hat{\mathbf{u}}_{t,me(e)} = \sum\limits_{\alpha=1}^{4} F_\alpha \Delta\hat{\mathbf{u}}_t^\alpha \\ \Delta\hat{\mathbf{u}}_t^\alpha = \begin{Bmatrix} \Delta\hat{\mathbf{u}}_{t,1}^\alpha \\ \ldots \\ \Delta\hat{\mathbf{u}}_{t,l}^\alpha \\ \ldots \\ \Delta\hat{\mathbf{u}}_{t,k}^\alpha \\ \ldots \\ \Delta\hat{\mathbf{u}}_{t,n}^\alpha \end{Bmatrix} \quad (\alpha = 1, 2, 3, 4) \end{cases} \tag{6.103}$$

$$\begin{cases} \Delta\hat{\mathbf{u}}_{t,l}^{\alpha} = \begin{Bmatrix} \hat{u}_{t,x,l}^{\alpha} \\ \hat{u}_{t,y,l}^{\alpha} \end{Bmatrix} & \text{(constant cover function } m = 1) \\ \Delta\hat{\mathbf{u}}_{t,l}^{\alpha} = \begin{Bmatrix} \hat{u}_{t,l,1}^{\alpha} \\ \hat{u}_{t,l,2}^{\alpha} \\ \hat{u}_{t,l,3}^{\alpha} \\ \hat{u}_{t,l,4}^{\alpha} \\ \hat{u}_{t,l,5}^{\alpha} \\ \hat{u}_{t,l,6}^{\alpha} \end{Bmatrix} & \text{(linear cover function } m = 3) \end{cases} \tag{6.104}$$

6.5.3 Strain and Stress

The strain increment in the enriched manifold element is calculated by

$$\begin{aligned} \Delta\boldsymbol{\varepsilon}_{me(e)}(\mathbf{x}) &= \mathbf{L}\Delta\mathbf{u}_{me(e)} = \Delta\boldsymbol{\varepsilon}_{b,me(e)} + \Delta\boldsymbol{\varepsilon}_{t,me(e)} \\ &= \mathbf{B}_{b,me(e)}(\mathbf{x})\Delta\hat{\mathbf{u}}_{b,me(e)} + \mathbf{B}_{t,me(e)}(\mathbf{x})\Delta\hat{\mathbf{u}}_{t,me(e)} \\ &= \sum_{l=1}^{n}\mathbf{B}_{b,l}(\mathbf{x})\Delta\hat{\mathbf{u}}_{b,l} + \sum_{l=1}^{n}\sum_{\alpha=1}^{4}\mathbf{B}_{t,l}^{\alpha}(\mathbf{x})\Delta\hat{\mathbf{u}}_{t,l}^{\alpha} \end{aligned} \tag{6.105}$$

where $\mathbf{B}_{b,l}$ is the strain matrix identical to $\mathbf{B}_l$ in Eq. (6.51), and

$$\mathbf{B}_{t,l}^{\alpha}(\mathbf{x}) = F_{\alpha}(\mathbf{x})\mathbf{B}_{b,l}(\mathbf{x}) \quad \text{for } (l = 1, 2, \ldots, n) \text{ and } (\alpha = 1, 2, 3, 4) \tag{6.106}$$

Then, the stress increment can be determined by using the elasto-viscoplastic constitutive equation

$$\begin{aligned} \Delta\boldsymbol{\sigma}_{me(e)} &= \mathbf{D}_{me(e)}\left(\Delta\boldsymbol{\varepsilon}_{b,me(e)} + \Delta\boldsymbol{\varepsilon}_{t,me(e)}\right) - \Delta\boldsymbol{\sigma}_{me(e)}^{vp} \\ &= \mathbf{D}_{me(e)}\left(\sum_{l=1}^{n}\mathbf{B}_{b,l}(\mathbf{x})\Delta\hat{\mathbf{u}}_{b,l} + \sum_{l=1}^{n}\sum_{\alpha=1}^{4}\mathbf{B}_{t,l}^{\alpha}(\mathbf{x})\Delta\hat{\mathbf{u}}_{t,l}^{\alpha}\right) - \Delta\boldsymbol{\sigma}^{vp} \end{aligned} \tag{6.107}$$

6.5.4 Governing Equation Set

The virtual work of an enriched manifold element containing crack-tip becomes

$$\iint\limits_{\Omega_{me(e)}} \delta\Delta\boldsymbol{\varepsilon}_{me(e)}^{\mathrm{T}}\Delta\boldsymbol{\sigma}_{me(e)}\mathrm{d}\Omega = \iint\limits_{\Omega_{me(e)}} \left(\delta\Delta\boldsymbol{\varepsilon}_{b,me(e)} + \delta\Delta\boldsymbol{\varepsilon}_{t,me(e)}\right)^{\mathrm{T}}$$

$$
\begin{aligned}
&\mathbf{D}_{me(e)}\left(\Delta\boldsymbol{\varepsilon}_{b,me(e)} + \Delta\boldsymbol{\varepsilon}_{t,me(e)}\right)\mathrm{d}\Omega \\
&- \iint\limits_{\Omega_{me(e)}} \left(\delta\Delta\boldsymbol{\varepsilon}_{b,me(e)} + \delta\Delta\boldsymbol{\varepsilon}_{t,me(e)}\right)^{\mathrm{T}} \boldsymbol{\sigma}_{me(e)}^{vp}\mathrm{d}\Omega \\
&= \iint\limits_{\Omega_{me(e)}} \left(\delta\Delta\boldsymbol{\varepsilon}_{b,me(e)}\right)^{\mathrm{T}}\mathbf{D}_{me(e)}\left(\Delta\boldsymbol{\varepsilon}_{b,me(e)}\right)\mathrm{d}\Omega \\
&+ \iint\limits_{\Omega_{me(e)}} \left(\delta\Delta\boldsymbol{\varepsilon}_{b,me(e)}\right)^{\mathrm{T}}\mathbf{D}_{me(e)}\left(\Delta\boldsymbol{\varepsilon}_{t,me(e)}\right)\mathrm{d}\Omega \\
&+ \iint\limits_{\Omega_{me(e)}} \left(\delta\Delta\boldsymbol{\varepsilon}_{t,me(e)}\right)^{\mathrm{T}}\mathbf{D}_{me(e)}\left(\Delta\boldsymbol{\varepsilon}_{b,me(e)}\right)\mathrm{d}\Omega \\
&+ \iint\limits_{\Omega_{me(e)}} \left(\delta\Delta\boldsymbol{\varepsilon}_{t,me(e)}\right)^{\mathrm{T}}\mathbf{D}_{me(e)}\left(\Delta\boldsymbol{\varepsilon}_{t,me(e)}\right)\mathrm{d}\Omega \\
&- \iint\limits_{\Omega_{me(e)}} \left(\delta\Delta\boldsymbol{\varepsilon}_{b,me(e)}\right)^{\mathrm{T}}\boldsymbol{\sigma}_{me(e)}^{vp}\mathrm{d}\Omega \\
&- \iint\limits_{\Omega_{me(e)}} \left(\delta\Delta\boldsymbol{\varepsilon}_{t,me(e)}\right)^{\mathrm{T}}\boldsymbol{\sigma}_{me(e)}^{vp}\mathrm{d}\Omega
\end{aligned}
\tag{6.108}
$$

which may give rise to a set of linear equations

$$
\begin{bmatrix} \mathbf{k}^{b,b} & \mathbf{k}^{b,t} \\ \mathbf{k}^{t,b} & \mathbf{k}^{t,t} \end{bmatrix}
\begin{Bmatrix} \Delta\hat{\mathbf{u}}_b \\ \Delta\hat{\mathbf{u}}_t \end{Bmatrix}
= \begin{Bmatrix} \Delta\mathbf{f}_b \\ \Delta\mathbf{f}_t \end{Bmatrix}
+ \begin{Bmatrix} \Delta\mathbf{f}_b^{vp} \\ \Delta\mathbf{f}_t^{vp} \end{Bmatrix}
\tag{6.109}
$$

or in more detailed form

$$
\begin{bmatrix} \mathbf{k}_{b,b} & \mathbf{k}_{b,t_1} & \cdots & \mathbf{k}_{b,t_4} \\ \mathbf{k}_{t_1,b} & \mathbf{k}_{t_1,t_1} & \cdots & \mathbf{k}_{t_1,t_4} \\ \cdots & \cdots & \cdots & \cdots \\ \mathbf{k}_{t_4,b} & \mathbf{k}_{t_4,t_1} & \cdots & \mathbf{k}_{t_4,t_4} \end{bmatrix}
\begin{Bmatrix} \Delta\hat{\mathbf{u}}_b \\ \Delta\hat{\mathbf{u}}_{t,1} \\ \cdots \\ \Delta\hat{\mathbf{u}}_{t,4} \end{Bmatrix}
= \begin{Bmatrix} \Delta\mathbf{f}_b \\ \Delta\mathbf{f}_{t,1} \\ \cdots \\ \Delta\mathbf{f}_{t,4} \end{Bmatrix}
+ \begin{Bmatrix} \Delta\mathbf{f}_b^{vp} \\ \Delta\mathbf{f}_{t,1}^{vp} \\ \cdots \\ \Delta\mathbf{f}_{t,4}^{vp} \end{Bmatrix}
\tag{6.110}
$$

in which $\mathbf{k}_{b,b}$, $\Delta\mathbf{f}_b$ and $\Delta\mathbf{f}_b^{vp}$ are the basic stiffness matrix, nodal load vector attributable to external force, and nodal load vector attributable to viscoplasticity. They are assembled from the sub-matrices and sub-vectors $\mathbf{k}_{l,k}$, $\Delta\mathbf{f}_l$ and $\Delta\mathbf{f}_l^{vp}$ correspondent on the n physical covers constructing the manifold element in a similar manner of the fundamental NMM. The additional stiffness matrices considering the coupling effects between the "normal" or "basic" DOFs and "enriched" DOFs for crack-tip are calculated by

$$\begin{cases} \mathbf{k}_{b,\alpha} = \mathbf{k}_{\alpha,b} = \iint\limits_{\Omega_e} (\mathbf{B}_b)^{\mathrm{T}} \mathbf{D}_{me(e)} \mathbf{B}_t^{\alpha} \mathrm{d}\Omega \ \ (\alpha = 1, \ldots, 4) \\ \mathbf{k}_{\alpha,\beta} = \mathbf{k}_{\beta,\alpha} = \iint\limits_{\Omega_e} (\mathbf{B}_t^{\alpha})^{\mathrm{T}} \mathbf{D}_{me(e)} \mathbf{B}_t^{\beta} \mathrm{d}\Omega \ (\alpha, \beta = 1, \ldots, 4) \end{cases} \tag{6.111}$$

The right hand side of Eq. (6.110) is composed of nodal force vectors, where the additional contribution from the crack-tip is expressed as

$$\begin{cases} \Delta \mathbf{f}_{t,\alpha} = \iint\limits_{\Omega_{me(e)}} \mathbf{N}_{me(e)}^{\mathrm{T}} F_{\alpha} \Delta \mathbf{b} \mathrm{d}\Omega + \int\limits_{\Gamma_t} \mathbf{N}_{me(e)}^{\mathrm{T}} F_{\alpha} \Delta \bar{\mathbf{t}} \mathrm{d}\Gamma + \sum\limits_{\mathbf{x}_{\zeta} \in me(e)} \mathbf{N}_{me(e)}^{\mathrm{T}} (\mathbf{x}_{\zeta}) F_{\alpha} \Delta \mathbf{T}_{\zeta} (\mathbf{x}_{\zeta}) \\ \Delta \mathbf{f}_{t,\alpha}^{vp} = \iint\limits_{\Omega_{me(e)}} (\mathbf{B}_t^{\alpha})^{\mathrm{T}} \Delta \boldsymbol{\sigma}^{vp} \mathrm{d}\Omega \end{cases}$$
$$(\alpha = 1, \ldots, 4) \tag{6.112}$$

Equations (6.109)–(6.112) may be directly employed to assemble the system equation set. Suppose the physical cover $i = \langle l \rangle$ is enriched. It is evident that if a manifold element need be enriched by crack-tip, its DOFs are extended by five times the original homogeneous manifold element. With the help of well-designed link list concerning local physical cover (inclusive basic and enriched) sequences and global cover sequences, the sub-matrices (6.111) and sub-vectors (6.112) may be assembled in a similar manner of CEM and EFEM illustrated previously in Chap. 5. Symbolically, such an extended global equilibrium equation set of the structure system may be expressed as

$$\mathbf{K} \Delta \hat{\mathbf{U}} = \Delta \mathbf{F} + \Delta \mathbf{F}^{vp} \tag{6.113}$$

where

$$\mathbf{K} = \begin{bmatrix} \mathbf{K}_{1,1} & \ldots & \mathbf{K}_{1,i} & \mathbf{K}_{1,i}^{b,t} & \ldots & \mathbf{K}_{1,j} & \ldots & \mathbf{K}_{1,n_{pc}} \\ \ldots & \ldots & \ldots & \ldots & \ldots & \ldots & \ldots & \ldots \\ \mathbf{K}_{i,1} & \ldots & \mathbf{K}_{i,i}^{b,b} & \mathbf{K}_{i,i}^{b,t} & \ldots & \mathbf{K}_{i,j} & \ldots & \mathbf{K}_{i,n_{pc}} \\ \mathbf{K}_{i,1}^{t,b} & \ldots & \mathbf{K}_{i,i}^{t,b} & \mathbf{K}_{i,i}^{t,t} & \ldots & \mathbf{K}_{i,j}^{t,b} & \ldots & \mathbf{K}_{i,n_{pc}}^{t,b} \\ \ldots & \ldots & \ldots & \ldots & \ldots & \ldots & \ldots & \ldots \\ \mathbf{K}_{j,1} & \ldots & \mathbf{K}_{j,i} & \mathbf{K}_{j,i}^{b,t} & \ldots & \mathbf{K}_{j,j} & \ldots & \mathbf{K}_{j,n_{pc}} \\ \ldots & \ldots & \ldots & \ldots & \ldots & \ldots & \ldots & \ldots \\ \mathbf{K}_{n_{pc},1} & \ldots & \mathbf{K}_{n_{pc},i} & \mathbf{K}_{n_{pc},i}^{b,t} & \ldots & \mathbf{K}_{n_{pc},j} & \ldots & \mathbf{K}_{n_{pc},n_{pc}} \end{bmatrix} \tag{6.114}$$

$$\Delta \hat{\mathbf{U}} = \left[\Delta \hat{\mathbf{u}}_1 \ldots \Delta \hat{\mathbf{u}}_{b,i} \ \Delta \hat{\mathbf{u}}_{t,i} \ldots \Delta \hat{\mathbf{u}}_j \ldots \Delta \hat{\mathbf{u}}_{n_{pc}} \right]^{\mathrm{T}} \tag{6.115}$$

$$\Delta \mathbf{F} = \left[\Delta \mathbf{F}_1 \ldots \Delta \mathbf{F}_{b,i} \ \Delta \mathbf{F}_{t,i} \ldots \Delta \mathbf{F}_j \ldots \Delta \mathbf{F}_{n_{pc}} \right]^{\mathrm{T}} \tag{6.116a}$$

$$\Delta \mathbf{F}^{vp} = \left[\Delta \mathbf{F}_1^{vp} \ldots \Delta \mathbf{F}_{b,i}^{vp} \ \Delta \mathbf{F}_{t,i}^{vp} \ldots \Delta \mathbf{F}_j^{vp} \ldots \Delta \mathbf{F}_{n_{pc}}^{vp} \right]^{\mathrm{T}} \tag{6.116b}$$

in which $\Delta\hat{\mathbf{u}}_i$, $\Delta\mathbf{F}_i$ and $\Delta\mathbf{F}^{vp}$ are $2m \times 1$ sub-vectors. Normally, $\mathbf{K}_{j,i}^{b,t} = \mathbf{K}_{i,j}^{b,t} = 0$ if the physical cover j does not construct the same extended manifold element $me(e)$ with the physical cover i (i.e. topological enrichment).

In summary, NMM models cracks by splitting a mathematical cover completely into several physical covers attached with individual cover functions, plus enriching singular functions where the crack-tip is tapered within. Further details for modeling complex problems such as branched and intersecting cracks can be found in relevant literatures (Ma et al. 2009; Zhang et al. 2010a, b).

6.5.5 Crack Onset and Growth Criteria

It is reminded that the crack onset and growth criteria in NMM are identical to that in XFEM (viz. Chap. 5).

6.5.6 Integration Schemes

In 2-D problems, although mathematical covers are usually constructed on triangular or quadrilateral FE meshes, the manifold element may have general shape because of the intersection by structural planes and cracks. Thus, direct integration scheme over the whole manifold element is not always possible. Fortunately, the existing simplex integration scheme and the isoparametric integral scheme are both practical for the quadrature of the matrices and vectors over the manifold elements of arbitrary shapes, subject to appropriate sub-domain partition.

1. **Simplex integration**

Considering the fact that all the polygonal manifold elements can be partitioned into several triangles, and the weight functions and cover functions are simple functions (e.g. polynomials), the accurate integration can be carried out by virtue of simplex integration scheme.

In order to undertake the simplex integration, an arbitrary manifold element $me(\mathrm{V}_1\mathrm{V}_2\mathrm{V}_3\mathrm{V}_4\mathrm{V}_5\ldots\mathrm{V}_{n+1})$ needs to be discretized into a set of oriented triangles (i.e. triangulation) depicted in Fig. 6.13, whose oriented boundary is composed of the ordered edges $\mathrm{V}_1-\mathrm{V}_2$, $\mathrm{V}_2-\mathrm{V}_3$, $\mathrm{V}_3-\mathrm{V}_4$, $\mathrm{V}_4-\mathrm{V}_5$, ..., $\mathrm{V}_n-\mathrm{V}_{n+1}$ and $\mathrm{V}_{n+1}-\mathrm{V}_1$. A boundary preserved triangulation can be easily achieved by connecting each edge to a fixed point, e.g. O(0, 0) for convenience. In this way, a set of oriented triangles are obtained, each of them with three ordered vertices is a simplex, and the quadrature over the manifold element becomes a sum of simplex integration analytically. Detailed description about the simplex integration can be found in relevant literatures (Shi 1996a, b; Li et al. 2005; Li and Cheng 2005).

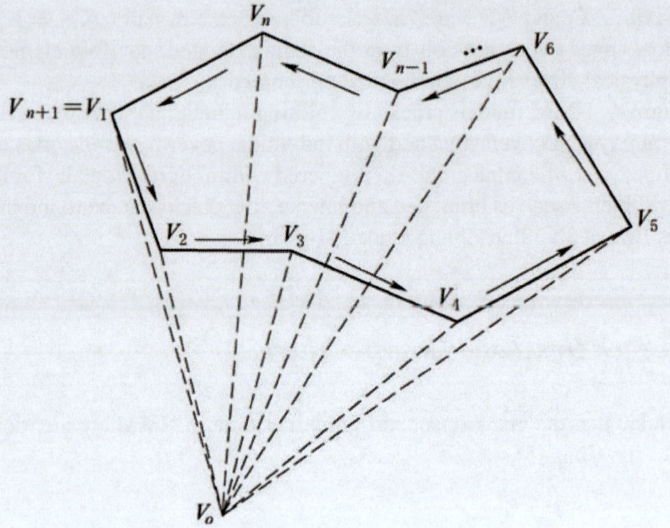

Fig. 6.13 Arbitrary manifold element and its triangulation

2. **Gaussian quadrature**

The simplex integration scheme is limited to the cases where the cover functions are polynomials. If non-polynomial functions are adopted, we may turn to other methods such the Gaussian quadrature. The manifold element of general shape is first partitioned into several sub-triangles (Sukumar and Prévost 2003). The quadrature is carried out on each sub-triangle, and then summation is carried out to obtain the integration over the whole manifold element.

This idea may be further expanded to get higher accurate quadrature by sub-dividing each triangle into three quadrilateral elements. Afterwards, the Gaussian quadrature scheme for the isoparametric element may be undertaken routinely. This algorithm has been illustrated in Chap. 4 (viz. Fig. 4.12 and Eq. 4.122).

6.5.7 Solution Procedure for Crack Growth Problems

The solution procedure for the crack growth problems with NMM is flow charted as follows (viz. Fig. 6.14):

① Input the structural model and generate mathematical covers. The text file that need be imported includes the geometry information, material properties, and some other setting variables (e.g. body forces and boundary conditions).

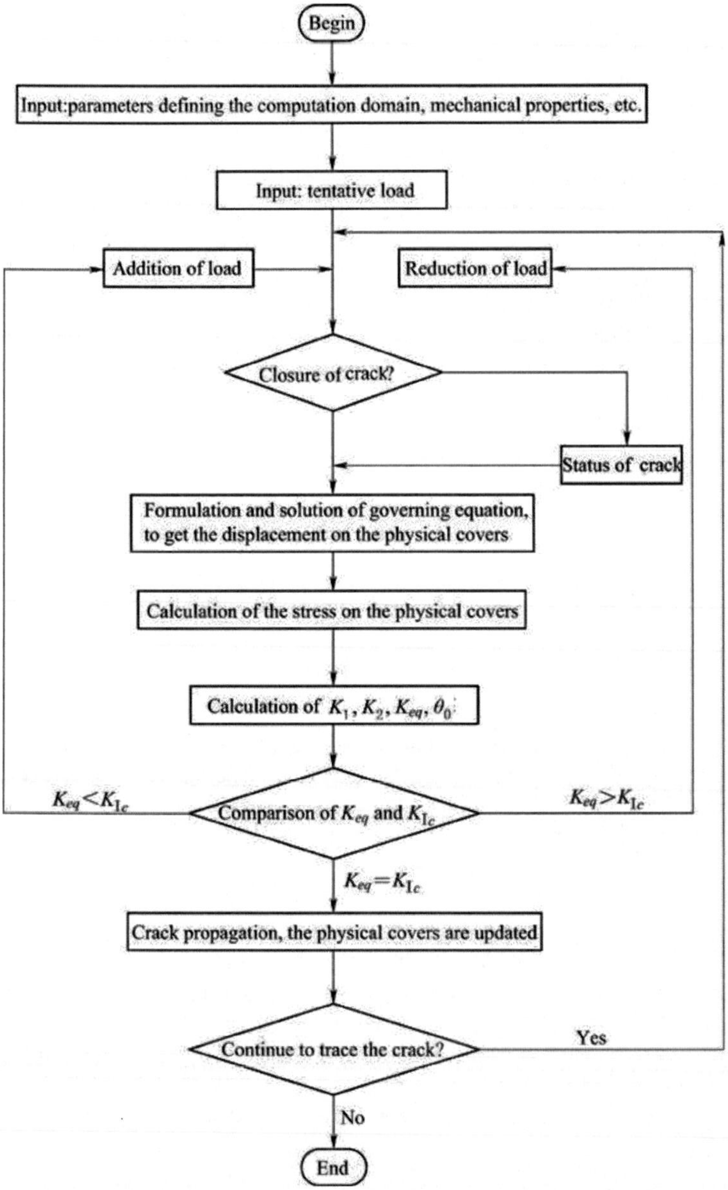

Fig. 6.14 Flow chart to the computation of crack growth with NMM

② Generate manifold elements. This is based on the topological operation between the imported structural planes and mathematical covers.
③ Generate link lists. The contact detection algorithm is implemented to detect the pairs of neighboring elements and their links are generated accordingly.
④ Detect the contact of node pairs between manifold elements.
⑤ Computation of element stiffness matrices and load vectors. These matrices and vectors are assembled to get the global equilibrium equation set, from which the global vector of unknown coefficients in cover functions are solved.
⑥ The mechanical fields (displacements, strains, stresses) of a manifold element may be calculated straightforwardly.
⑦ Once the failure criteria w.r.t. the slip and open of structural plane, the growth of crack, etc., need be met, the status of element link lists and node pairs should be reset.
⑧ Remove failed links to generate new manifold element. Additional manifold elements need to be generated when at least a crack has completed cut through the whole element.

6.6 Validations

6.6.1 One-Dimensional Bar Problem

A 1-D elastic and discontinuous bar under the action of tension in Fig. 5.24 is studied for the verification of the philosophy and basic algorithm of NMM. Only one finite element with two nodes (1 and 4) is employed for the solution. Each bar sub-part (Ω^- and Ω^+) is one manifold element separated by the interface $sp_{me(1),me(2)}$. On each node, two physical covers $pc(1_1)$, $pc(1_2)$ and $pc(4_1)$, $pc(4_2)$ are created to interpolate the displacements of manifold elements $me(1)$ and me(2). On $sp_{me(1),me(2)}$, the relative displacement is unknown but will be calculated lately.

Since it is clear that the bar is under the state of constant strain and stress, hence the constant cover function is employed (i.e. $m = 1$). As a result, the shape functions are identical to that in the FEM meeting both the conditions of PU and Kronecker Delta properties. Consequently, essential boundary conditions due to the displacement restraints at the ends of bar may be imposed directly without the help of penalty technique. Or in other words, displacements equal to 0 and $\overline{u}$ are imposed on nodes 1 and 4, respectively.

On each manifold element, deformation and stress are equal to

$$\begin{cases} \varepsilon_{me(1)} = \frac{u_{pc(4_2)} - u_{pc(1_1)}}{2L} \quad \text{and} \quad \sigma_{me(1)} = E\varepsilon_{me(1)} \text{ for } me(1) \in \Omega^- \\ \varepsilon_{me(2)} = \frac{u_{pc(4_1)} - u_{pc(1_2)}}{2L} \quad \text{and} \quad \sigma_{me(2)} = E\varepsilon_{me(2)} \text{ for } me(2) \in \Omega^+ \end{cases} \tag{6.117}$$

Their contribution to the virtual work is individually calculated as follows:

- Left half-bar.

$$\int\limits_{me(1)} \delta\varepsilon_{me(1)}\sigma_{me(1)}\mathrm{d}\Omega = L\frac{E}{4L^2}(\delta u_{pc(4_2)} - \delta u_{pc(1_1)})(u_{pc(4_2)} - u_{pc(1_1)})$$

$$= \{\delta u_{me(1)}\}[K_{me(1)}]\{u_{me(1)}\} \tag{6.118}$$

where

$$[K_{me(1)}] = \frac{E}{4L}\begin{bmatrix} 1 & -1 \\ -1 & 1 \end{bmatrix} \tag{6.119}$$

is the stiffness matrix of the left half-bar.

- Right half-bar.

$$\int\limits_{me(2)} \delta\varepsilon_{me(2)}\sigma_{me(2)}\mathrm{d}\Omega = L\frac{E}{4L^2}(\delta u_{pc(4_1)} - \delta u_{pc(1_2)})(u_{pc(4_1)} - u_{pc(1_2)})$$

$$= \{\delta u_{me(2)}\}[K_{me(2)}]\{u_{me(2)}\} \tag{6.120}$$

and

$$[K_{me(2)}] = \frac{E}{4L}\begin{bmatrix} 1 & -1 \\ -1 & 1 \end{bmatrix} \tag{6.121}$$

is the stiffness matrix of the right half-bar.

- Interface. The average stress $\sigma_{sp_{me(1),me(2)}}$ and the aperture $u_{sp_{me(1),me(2)}}$ can be written as

$$\begin{cases} \sigma_{sp_{me(1),me(2)}} = k_n\left(\frac{u_{pc(4_1)}+u_{pc(2_2)}}{2} - \frac{u_{pc(1_1)}+u_{pc(4_2)}}{2}\right) \\ u_{sp_{me(1),me(2)}} = \left(\frac{u_{pc(4_1)}+u_{pc(2_2)}}{2} - \frac{u_{pc(1_1)}+u_{pc(4_2)}}{2}\right) \end{cases} \tag{6.122}$$

Consequently, the contribution of interface to the virtual work is calculated by

$$\begin{aligned} &\delta u_{sp_{me(1),me(2)}} \cdot \sigma_{sp_{me(1),me(2)}} \\ &= k_n\left(\frac{\delta u_{pc(4_1)} + \delta u_{pc(2_2)}}{2} - \frac{\delta u_{pc(1_1)} + \delta u_{pc(4_2)}}{2}\right)\left(\frac{u_{pc(4_1)} + u_{pc(2_2)}}{2} - \frac{u_{pc(1_1)} + u_{pc(4_2)}}{2}\right) \\ &= \frac{1}{4}k_n(\delta u_{pc(4_1)} + \delta u_{pc(2_2)} - \delta u_{pc(1_1)} - \delta u_{pc(4_2)})(u_{pc(4_1)} + u_{pc(2_2)} - u_{pc(1_1)} - u_{pc(4_2)}) \\ &= \{\delta u_{sp_{me(1),me(2)}}\}[K_{sp_{me(1),me(2)}}]\{u_{sp_{me(1),me(2)}}\} \end{aligned} \tag{6.123}$$

where

$$[K_{sp_{me(1),me(2)}}] = \frac{1}{4}k_n \begin{bmatrix} 1 & 1 & -1 & -1 \\ 1 & 1 & -1 & -1 \\ -1 & -1 & 1 & 1 \\ -1 & -1 & 1 & 1 \end{bmatrix} \quad (6.124)$$

is the stiffness matrix corresponding to $sp_{me(1),me(2)}$.

The global equilibrium condition of the whole bar is derived from above analyses w.r.t. two subparts of bar and the interface in between, that

$$[K]\begin{Bmatrix} u_{pc(1_1)} \\ u_{pc(4_2)} \\ u_{pc(1_2)} \\ u_{pc(4_1)} \end{Bmatrix} = \begin{Bmatrix} F_{pc(1_1)} \\ F_{pc(4_2)} \\ F_{pc(1_2)} \\ F_{pc(4_1)} \end{Bmatrix} \quad (6.125)$$

in which the global stiffness matrix is assembled in a manner of

$$\begin{aligned} [K] &= \frac{1}{4}k_n \begin{bmatrix} 1 & 1 & -1 & -1 \\ 1 & 1 & -1 & -1 \\ -1 & -1 & 1 & 1 \\ -1 & -1 & 1 & 1 \end{bmatrix} + \frac{E}{4L}\begin{bmatrix} 1 & -1 & 0 & 0 \\ -1 & 1 & 0 & 0 \\ 0 & 0 & 0 & 0 \\ 0 & 0 & 0 & 0 \end{bmatrix} \\ &+ \frac{E}{4L}\begin{bmatrix} 0 & 0 & 0 & 0 \\ 0 & 0 & 0 & 0 \\ 0 & 0 & 1 & -1 \\ 0 & 0 & -1 & 1 \end{bmatrix} \\ &= \frac{E}{4L}\begin{bmatrix} 1+\frac{Lk_n}{E} & -(1-\frac{Lk_n}{E}) & -\frac{Lk_n}{E} & -\frac{Lk_n}{E} \\ -(1-\frac{Lk_n}{E}) & 1+\frac{Lk_n}{E} & -\frac{Lk_n}{E} & -\frac{Lk_n}{E} \\ -\frac{Lk_n}{E} & -\frac{Lk_n}{E} & 1+\frac{Lk_n}{E} & -(1-\frac{Lk_n}{E}) \\ -\frac{Lk_n}{E} & -\frac{Lk_n}{E} & -(1-\frac{Lk_n}{E}) & 1+\frac{Lk_n}{E} \end{bmatrix} \end{aligned} \quad (6.126)$$

On invoking the boundary conditions

$$\begin{cases} u_{pc(1_1)} = 0 \\ u_{pc(4_1)} = \overline{u} \end{cases} \quad (6.127)$$

and,

$$\begin{cases} F_{pc(1_2)} = 0 \\ F_{pc(4_2)} = 0 \\ F_{pc(1_1)} = F_1 = -F_{pc(4_1)} = -F_4 = -T \end{cases} \quad (6.128)$$

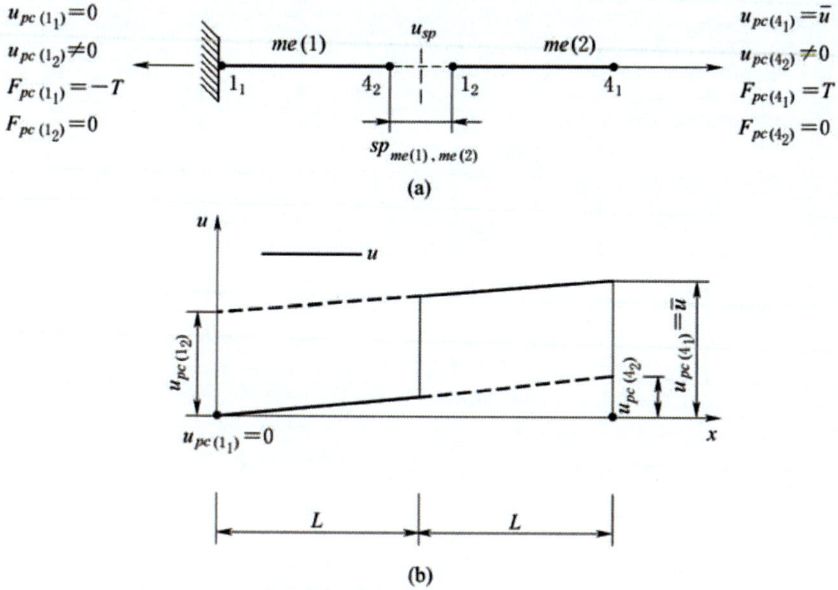

Fig. 6.15 Weak formulation of NMM. **a** Diagram to the calculation of discontinuous bar by NMM; **b** displacement distribution along the bar

the global equilibrium condition (6.125) becomes

$$\begin{cases} \frac{E}{4L}\left[-(1-\frac{Lk_n}{E})u_{pc(4_2)}-\frac{Lk_n}{E}u_{pc(1_2)}-\frac{Lk_n}{E}\bar{u}\right]=-T \\ \frac{E}{4L}\left[(1+\frac{Lk_n}{E})u_{pc(4_2)}-\frac{Lk_n}{E}u_{pc(1_2)}-\frac{Lk_n}{E}\bar{u}\right]=0 \\ \frac{E}{4L}\left[-\frac{Lk_n}{E}u_{pc(4_2)}+(1+\frac{Lk_n}{E})u_{pc(1_2)}-(1-\frac{Lk_n}{E})\bar{u}\right]=0 \\ \frac{E}{4L}\left[-\frac{Lk_n}{E}u_{pc(4_2)}-(1-\frac{Lk_n}{E})u_{pc(1_2)}+(1+\frac{Lk_n}{E})\bar{u}\right]=T \end{cases} \tag{6.129}$$

This equation set is identical to that of CEM (viz. Eq. 5.116), if the following nomenclature replacement is conducted

$$\begin{cases} u_{pc(1_1)}=u_{r1,1} \\ u_{pc(1_2)}=u_{r1,4} \\ u_{pc(4_1)}=u_{r2,4} \\ u_{pc(4_2)}=u_{r2,1} \end{cases} \tag{6.130}$$

This example confirms that for the elastic bar problem, the solutions by FEM, CEM, XFEM and NMM are identical. The displacement distribution by NMM is plotted in Fig. 6.15b.

6.6.2 Computation of SIF

A series of laboratory tests (Tasdemir et al. 1990) that represent the well-documented testing program on the growth of crack have been widely employed by many

Fig. 6.16 Sample size

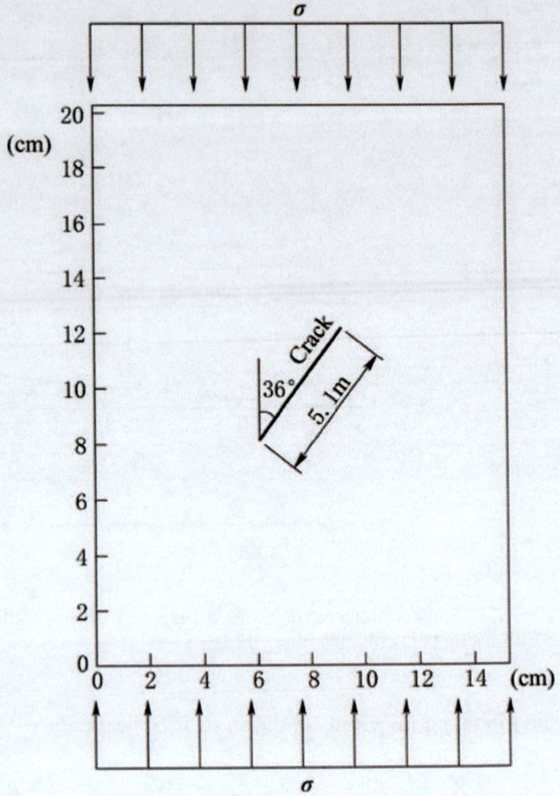

researchers in the field of CMs for algorithm validations (Scavia 1995; Wang et al. 1999). The tests were carried out on the prismatic mortar specimens with a notch cut at 36° to the vertical and subjected to a vertical compressive load σ (viz. Fig. 6.16).

The analysis is carried out by applying a vertical pressure $\sigma = 10$ MPa loaded in the increment of 0.25 MPa. The specimen size is 15.3 cm (width) × 20.3 (height) × 7.6 cm (thickness), the length of crack is $2a = 5.1$ cm, the Young's modulus is $E = 27$ GPa, the Poisson's ratio is $\nu = 0.2$, the friction angle on crack faces is $\varphi = 14°$, the toughness is $K_{1c} = 0.52\ \text{MPa}\sqrt{\text{m}}$.

Table 6.12 gives the relation of crack growth length L versus exerted pressure σ, in which the data in column E are from experiment, in C1 are computed by Scavia (1995), in C2 are computed by Wang et al. (1999), in C3 are computed by the author. Figure 6.17 exhibits a phase of crack growth trajectory, where the open of crack is amplitude 1000 times.

It can be concluded from this example that the agreement between NMM calculation and test outcomes is satisfactory. However, checking is limited to the crack growth trajectory, as there is no sufficient and reliable information on the tested data.

Table 6.12 Comparison of experimental and computed values: applied loads σ versus grown crack length L

	$L = 0$ cm				$L = 1.27$ cm				$L = 2.5$ cm			
Load σ (MPa)	E	C1	C2	C3	E	C1	C2	C3	E	C1	C2	C3
	3.78	3.35	3.72	3.76	6.77	6.50	6.43	6.53	8.26	8.50	8.15	8.17

Fig. 6.17 Crack growth trajectory

6.7 Concluding Remarks

6.7.1 Relation with Other Prevalent CMs

1. **Relation with fundamental FEM**

In FEM, the problem domain is divided into a collection of standard elements with fixed number of nodes at vertices and/or on sides. Trial functions, usually of polynomial, are employed to approximate the unknown field variables at the element level where nodes serve as interpolation points (knots) for unknown variables to generate local algebraic equations. The elemental equations are then assembled into a global equation set whose solution produces unknown variables at all the nodes, subject to properly defined initial and boundary conditions. On the contrary, in NMM the approximation is established on physical covers to which cover functions containing unknown coefficients are tied. If the cover function is constant and equivalent to the unknown variables of each cover, the weight functions are so chosen as the nodal shape functions in FEM, the structural planes and external boundaries do not intersect the manifold elements, NMM will exactly retrogress to the fundamental FEM. In other words, FEM is a special case of NMM.

2. **Relation with GFEMs**

Several modifications to the fundamental FEM have been made since 1990s. One typical example is the creation of GFEMs, which embed local and/or analytical solutions into the FEM approximation. They have been successfully employed to tackle the problems with multiple voids and cracks. Where special functions such as the asymptotic crack-tip functions are incorporated to enrich the local approximation, NMM is able to give identical results of XFEM and CEM that belong to GFEMs. Since they are all based on PU, it is clear that the mathematical and technical developments in GFEM can be partially used in NMM without major modifications. However, the linear dependence problem stemming from the multiplication of PU functions and local approximation functions in GFEM also may emerge in extended NMM.

A typical comparison may be given by XFEM (see Chap. 5), in which discontinuous and asymptotic crack-tip functions are incorporated to account for the crack arbitrarily aligned with FE mesh. In XFEM, the generalized Heaviside function $H(x)$ is demanded to describe the discontinuity across the crack surface. On the contrary, owing to the introduction and consecutively augment of covers, NMM accommodates crack surface without the help of the Heaviside function in XFEM, or the discontinuous function in GFEM.

CEM was formulated step-by-step by the author since the beginning of this century (Chen 2018). It gives another modification w.r.t. the problems of complex fracture pattern in elements and linear dependence stemming from the multiplication of PU functions and local approximation functions. This is archived by the standard or/and hierarchical shape functions of FEM whereas assign totally different nodal variables for each sub-element explicitly delimited by joints, cracks, and interfaces of any shapes (even closed sub-domains such as bolt and pipe) within the finite element. It is easy to covert an existing FEM code into CEM one and to include many mature FEM techniques such as the h-or p-version adaptive refinements.

3. **Relation with DDA**

DDA was originally developed to solve the problems of jointed rock mass. Since rock blocks generally undergo small deformation, DDA simulates each discrete element as a simple deformable block with only six degrees of freedom (DOFs) in a 2-D setting. Since NMM employs higher-order cover functions to raise the DOFs towards more accurate description for the displacement field and stress field in each block, it may be looked at as a modification of DDA because NMM preserves all the characteristics of the discrete element such as the kinematics constraints and contact detection. In other words, if one discrete block is one manifold element with linear displacement field, then NMM is identical to DDA.

4. **Relation with DEM**

DEM is an explicit version of "discrete element methods (DEMs)", while DDA is an implicit version of DEMs. Compared with DEM, NMM possesses following four basic advantages (Jing 2003):

- The equilibrium condition is automatically satisfied for quasi-static problems without using excessive iteration cycles;
- The length of time-marching step can be larger without introducing numerical instability;
- Closed-form integrations can be performed without the help of Gaussian quadrature techniques;
- It is possible to covert an existing FEM code into NMM code and to include many mature FEM techniques.

In NMM however, matrix equation set is produced and need be solved. On the contrary, DEM does not need to follow suit.

6.7.2 Remarkable Developments and Applications

Since the initiation of NMM, various developments and applications have been achieved during the last three decades. Most of the publications are included in the series proceedings of the "International Conference on Analysis of Discontinuous Deformation (ICADD) symposia". The main developments and applications of NMM are summarized as follows.

1. **General extension**

This extension is mainly focused on the improvement of approximation accuracy. In the original version of NMM, the simplest triangular manifold element with constant cover function was adopted for 2-D issues, but the criteria for crack inception and propagation were neglected. Since then, various improvements have been conducted. Shyu and Salami (1995) implemented quadrilateral isoparametric elements. Chiou et al. (2002) studied the mixed-mode fracture growth by combining NMM with the virtual crack extension model. Cheng et al. (2002) incorporated Wilson nonconforming elements in NMM. Chen et al. (1998) developed a higher-order cover functions, afterwards Su et al. (2003) proposed a subroutine in the commercial software Mathematica to automatically produce the expressions for high-order NMM. Terada et al. (2003) introduced the "finite cover method (FCM)" as an alias of NMM. In rapid succession, Lin et al. (2005), Cheng and Zhang (2008), Jiang et al. (2009), He and Ma (2010), He et al. (2013) gave the formulations of 3-D NMM with high-order cover functions using tetrahedral or/and hexahedral FE meshes. An et al. (2011) investigated the linear dependence problem of approximation space using finite element covers and polynomial local functions at both elemental and global levels. Cai et al. (2013), Zheng and Xu (2014) put forward the new strategies for the simulation of crack growth, including the rank deficiency induced by the high-order cover functions, the integrals with singularity, and the kinked cracks. Ghasemzadeh et al. (2014) proposed a high-order NMM towards dynamic problems. Later on, He and Zhang (2015), Yang and Zheng (2016), Xu et al. (2017) established a high-order local approximation function and alleviated/eliminated the linear dependence on the triangular and quadrilateral mathematical meshes, respectively. Their works are rather prevalent in fracture analysis as well as in free and forced vibration analysis.

Zheng et al. (2014) adopted the influencing domains of nodes in the "moving least squares (MLS)" interpolation when constructing mathematical covers and proposed the MLS-based NMM to better address the problems with large deformation/movement. They further proposed a so-called projection-contraction algorithm to tackle the complementarity problem in the static growth of multiple fractures (Zheng et al. 2015). Recently, Liu et al. (2018a, b), studied the micro/macro failure of rock by the Voronoi element-based NMM.

For the stress wave propagation issues, Zhao et al. (2014) improved the performance of NMM in rock masses by importing Newmark solver, edge-to-edge contact scheme and non-reflection boundary condition. Wu and Fan (2014) developed a time dependent absorbing boundary condition. Wei et al. (2018) also proposed some new boundary conditions to study the seismic response. Yang et al. (2018) developed boundary settings for the seismic dynamic response analysis of rock masses.

Zhang et al. (2014, 2017, 2018) studied the thermo-mechanical fractures and transient heat conduction problems using NMM.

For the simulation of finite deformation problem, the fundamental NMM in Lagrangian formalism may be improved in a manner similar to its FEM counterpart (Wei and Jiang 2017).

2. **Extension of NMM for complicated interface and crack-growth problems**

Tsay et al. (1999) applied NMM with a local mesh refinement and auto-remeshing schemes to simulate crack growth. Zhang et al. (1999b), Wang et al. (2003) extended NMM to model shear failure.

Terada et al. (2003) formulated the "finite cover method (FCM)" for the static equilibrium state of engineering structure with arbitrary physical boundaries including material interfaces and the discontinuities caused by the rupture of material interface. The displacement compatibility conditions on the material interface are enforced by Lagrange multipliers. FCM may result in generalized elements that are partially within physical domains. Terada and Kurumatani (2004, 2005) assessed that the performance of such elements is equivalent to and is not inferior to FEM, and further extended FCM to elasto-plastic problems, in which the interface elements are introduced to impose the essential boundary conditions and the displacement compatibility conditions on material interfaces. Terada et al. (2007) extended FCM for the analyses of progressive failure processes involving cohesive zone fracture. Kurumatani and Terada (2009) extended FCM towards the simulation of cracks in quasi-brittle heterogeneous solids by introducing a method termed as "multi-cover-layer modeling", in order to capture both weak and strong discontinuities.

3. **Extension of NMM in Eulerian formalism**

The PU-based approximation strategy of the NMM is almost the same as that of the PU-FEM and has been inherited to the "finite cover method (FCM)" that introduced by Terada and his co-workers (Terada et al., 2003; Kurumatani and Terada 2005, 2009) independently of the PU-FEM. Su et al. (2003) suggested a new version of NMM with fixed mathematical covers to compute large displacements. After each time-marching step, if the deformed domain boundaries intersect the fixed mathematical covers, new physical covers and new manifold elements are created. Terada et al. (2007) gave the Eulerian FCM for the hyperelastic bodies undergoing large deformation. In order to realize the implicit solution scheme for quasi-static equilibrium problems, they used both the Lagrangian and Eulerian mathematical meshes simultaneously, and proposed a numerical algorithm to repeatedly perform re-covering and mapping over generalized elements. Okazawa et al. (2007) developed another Eulerian FCM for solid dynamics, which is able to consider the kinematic and dynamic boundary conditions by incorporating the approximation of FCM into the existing Eulerian FEM.

4. **Hybrid methods**

To serve different purposes, it is wise to couple NMM with other member of CMs.

(1) **With BEM**

To overcome the drawback that NMM cannot simulate infinitely large domains, coupling of NMM for near-field analysis with BEM (Zhang et al. 1999a), EFGM (Zhou et al. 1999), and others for far-field analysis, had been explored.

(2) **With XFEM**

Ma et al. (2009) incorporated the enriched functions of XFEM/GFEM into the fundamental NMM to simulate complex crack patterns. Li et al. (2005), Gao and Cheng (2011) developed an enriched meshless manifold method for 2-D crack modeling. In their works, the influence domains of nodes form the mathematical covers with the weight functions, while the physical covers are formed by the sub-divisions of mathematical covers by the problem domain. Recently, Xu et al. (2019) proposed an extended NMM to handle multiple crack growth problem in which the crack-tip enrichment is identical to that of XFEM.

(3) **With particle-based methods**

Many hybrid methods by the coupling of, for example, NMM-EFGM (Zhou et al. 1999), NMM-DDA (Miki et al. 2010), etc., were explored. Among them, a model originally entitled "particle manifold method (PMM)" and an improvement version termed as PNMM (Zhao et al. 2012; Sun et al. 2013, 2014; Li et al. 2017; Li and Zhao 2019), is remarkable.

PNMM is defined as the combination of mathematical covers and particles. A relationship termed as "link" is defined to describe the continuous state between neighboring particles. The basic motivation of this study is to reduce the complexity of topological operations between polygons, especially in 3-D NMM. The generation of manifold elements forms the first discretization level in PNMM. Then an extra level of discretization is incorporated by dividing each manifold element into a group of particles. As the governing equation and DOFs are defined on manifold elements, introduction of particles does not bring about additional DOFs and does not affect the form of governing equation. The material properties, boundary conditions and body forces are all assigned to particles individually. Using particle integration scheme, the global matrices and vectors in the governing equation set may be assembled (Sun 2012). The most notable improvement of PNMM compared to original NMM is that the polygon-to-polygon contact is simplified to particle-to-particle contact that is easier to be carried out. In addition, the particle integration scheme is suitable for any integrands on both the convex and concave domains.

More details of PNMM can be found in recent literatures (Li et al. 2017).

5. Others

Li et al. (2005), Li and Cheng (2005) derived the governing equation set of NMM from the method of weighted residual (MWR), which enables to extend it to problems such as heat conduction and potential fluid flow.

Today, NMM has been applied in various simulating problems derived from geotechnical and hydraulic structures, such as the rock masses containing joints of different scales (Lin et al. 1999), saturated-unsaturated unsteady groundwater flow in porous media (Ohnishi et al. 1999), crack in dam and foundation (Zhang et al. 1999a, b), dynamic and nonlinear analysis of saturated porous media (Zhang and Zhou 2006), shear response of heterogeneous rock joint and dynamic friction mechanism of rock mass (Ma et al. 2007), fluid-solid interaction analysis (Su and Huang 2007), toppling failure of rock slope (Zhang et al. 2010a), progressive failure in rock slope (Wong and Wu 2014), etc.

6.7.3 Further Developments

Further developments with large room on NMM are briefly mentioned, among many others, as follows.

1. **Linear dependence problem**

Similar to other CMs based on the PU property of weight functions, any analytically known functions (handbook functions) or numerically obtained solutions that reflect the local characteristics of boundary value problem, can be incorporated into the local approximation functions termed as "cover functions" to improve the efficiency and accuracy of NMM. However, NMM shape functions determined from the product of weight functions and basis of cover functions may be linearly dependent, thus lead to a singular global stiffness matrix. The linear dependence problem involved in

NMM need be fully investigated and possible ways for avoidance should be figured out rigorously.

2. **Generation of mathematical covers and construction of their weight functions**

For convenience, NMM usually adopts FE meshes to generate mathematical covers and construct associated weight functions. However, other possibilities that are able to fully explore the advantages of NMM need be worked out. For example, instead of FE mesh, some earlier researchers used the influence domain of node as the mathematical cover, which resulted in the so-called "meshless manifold method".

3. **Fracture analysis**

As mentioned earlier, NMM seems promising in the simulation of fracture growth and transition from continuum to discontinuum, such as the rock blasting process where an explosive is detonated in a borehole to break rock, the rock crushing operation where the rock is broken by a crushing machine to achieve a desired size distribution, the structural demolition work where the tall building is demolished by carefully placed explosive, the rock or soil slope failure scenario where a slip body is detached under comprehensive actions (gravitation, seepage, earthquake, etc.).

Towards a satisfactorily handling aforementioned problems, a sound toolkit must be capable of:

- Calculating fracture onset point;
- Calculating subsequent fracturing path (direction and length);
- Treating the energy loss caused by fracturing process;
- Treating the discontinuous interactions along fractures.

Ma et al. (2009) studied multiply branched and intersected cracks. In their works, singular physical covers are enriched with asymptotic crack-tip functions, SIFs are evaluated by virtue of the domain form of interaction integral. Zhang et al. (2010a) made the further extension to simulate the growth of complex cracks, with the maximum circumferential stress criterion for crack growth.

However, such capabilities still need to be further improved, and practical exercises with field and laboratory verifications are highly demanded.

Taking the rock blasting process for example, except the stress wave effect, the detonation gas also plays an important role in extending the stress wave induced fractures. So, for a realistic rock blasting simulation, the equation of state for the expansion of detonation gas, together with a model for gas flow through fracturing solid, should be well implemented into NMM.

4. **Complex domain problems**

Since the mathematical covers in NMM can be completely independent of the geometry of problem domain, it is especially suitable for engineering structures with complex domains including hundreds of voids, cracks, and/or inclusions which may be expensive or practically impossible to be solved using fundamental FEM. The applications of NMM on this field need to be especially explored by the strategies of, for example, CEM (2018).

5. **Large-scale computational capabilities**

Problems involving millions of elements (particles) with irregular shapes are of practical importance in many branches of science and engineering. Today, the most urgent challenges are how to efficiently detect and represent the contact interactions between them, how to include the effects of realistic-shaped elements (particles), and how to display results in a manageable CPU time. All these are crucial obstacle that should be overcome in order to successfully apply NMM.

6. **Eulerian NMM**

For large deformation problems, the Lagrangian NMM may suffer from highly distorted mesh and thus cannot retain desired accuracy without appropriate but complicated improvement. Under such circumstances, the Eulerian NMM in which Eulerian mesh is fixed to the space may bring about an advantage. Although a number of works have been done, yet there remain unsolved technical difficulties in the strategy to accurately and stably map field variables (mass, velocity, and constitutive relation) into Eulerian mesh. In addition, the issues of computational efficiency and robustness need be further clarified, too.

References

An XM, Li LX, Ma GW, Zhang HH. Prediction of rank deficiency in partition of unity based methods with plane triangular or quadrilateral meshes. Comput Methods Appl Mech Eng. 2011;200(5–8):665–74.

An XM, Ning Y, Ma GW, He L. Modeling progressive failures in rock slopes with nonpersistent joints using the numerical manifold method. Int J Numer Anal Methods Geomech. 2014;38(7):679–701.

Cai Y, Zhuang X, Zhu H. A generalized and efficient method for finite cover generation in the numerical manifold method. Int J Comput Methods. 2013;10(5):1350028.

Chen SH. Computational geomechanics and hydraulic structures. Berlin (Germany): Springer; 2018.

Chen G, Ohnishi Y, Ito T. Development of high-order manifold method. Int J Numer Methods Eng. 1998;43(4):685–712.

Cheng YM, Zhang YH. Formulation of a three-dimensional numerical manifold method with tetrahedron and hexahedron elements. Rock Mech Rock Eng. 2008;41(4):601–28.

Cheng YM, Zhang YH, Chen WS. Wilson non-conforming element in numerical manifold method. Int J Numer Methods Biomed Eng. 2002;18(12):877–84.

Chiou YJ, Lee YM, Tsay RJ. Mixed mode fracture propagation by manifold method. Int J Fract. 2002;114(4):327–47.

Gao HF, Cheng YM. A complex variable meshless manifold method for fracture problems. Int J Comput Methods. 2011;17(1):55–81.

Ghasemzadeh H, Ramezanpour MA, Bodaghpour S. Dynamic high order numerical manifold method based on weighted residual method. Int J Numer Methods Eng. 2014;100(8):596–619.

He L, Ma GW. Development of 3D numerical manifold method. Int J Comput Methods. 2010;7(1):107–29.

He L, Zhang QB. Numerical investigation of arching mechanism to underground excavation in jointed rock mass. Tunn Undergr Space Technol. 2015;50:54–67.

He L, An XM, Ma GW, Zhao ZY. Development of three-dimensional numerical manifold method for jointed rock slope stability analysis. Int J Rock Mech Min Sci. 2013;64:22–35.

Jiang QH, Zhou CB, Li DQ. A three-dimensional numerical manifold method based on tetrahedral meshes. Comput Struct. 2009;87(13–14):880–9.

Jing L. A review of techniques, advances and outstanding issues in numerical modelling for rock mechanics and rock engineering. Int J Rock Mech Min Sci. 2003;40(3):283–353.

Kurumatani M, Terada K. Finite cover method with mortar elements for elastoplasticity problems. Comput Mech. 2005;36(1):45–61.

Kurumatani M, Terada K. Finite cover method with multi-cover layers for the analysis of evolving discontinuities in heterogeneous media. Int J Numer Meth Eng. 2009;79(1):1–24.

Li SC, Cheng YM. Enriched meshless manifold method for two-dimensional crack modeling. Theoret Appl Fract Mech. 2005;44(3):234–48.

Li X, Zhao J. An overvicw of particle-based numerical manifold method and its application to dynamic rock fracturing. J Rock Mech Geotech Eng. 2019;11(3):684–700.

Li SC, Cheng YM, Wu YF. Numerical manifold method based on the method of weighted residuals. Comput Mech. 2005;35(6):470–80.

Li X, Zhang QB, He L, Zhao J. Particle-based numerical manifold method to model dynamic fracture process in rock blasting. Int J Geomech. 2017;17(5):E4016014.

Lin JS, Koo CY, Chern JC. Manifold method analysis of rock masses containing joints of two different scales. In: Amadei B, editor. ICADD-3: third international conference on analysis of discontinuous deformation, Vail, Colorado (USA). Rotterdam (The Netherlands): American Rock Mechanics Association, AA Balkema; 1999, p. 231–42.

Lin SZ, Qi Y, Su HD. Formulation of high-order numerical manifold method and fast simplex integration based on special matrix operations. In: MacLaughlin MM, Sitar N, editors. Proceedings of the 7th international conference on the analysis of discontinuous deformation (ICADD-7), Honolulu (USA); 2005, p. 183–90.

Liu Q, Jiang Y, Wu Z, He J. A Voronoi element based-numerical manifold method (VE-NMM) for investigating micro/macro-mechanical properties of intact rocks. Eng Fract Mech. 2018a;199:71–85.

Liu Q, Jiang Y, Wu Z, Qian Z, Xu X. Numerical modeling of acoustic emission during rock failure process using a Voronoi element based-explicit numerical manifold method. Tunn Undergr Space Technol. 2018b;79:175–89.

Ma GW, An XM, Wang MY. Numerical studies on dynamic friction mechanism of blocky rock system using manifold method. In: Proceedings of the eight international conference on the analysis of discontinuous deformation, Beijing, China; 2007, p. 221–6.

Ma GW, An XM, Zhang HH, Li LX. Modeling complex crack problems with numerical manifold method. Int J Fract. 2009;156(1):21–35.

Ma GW, An ZM, He L. The numerical manifold method: a review. Int J Comput Methods. 2010;7(1):1–32.

Miki S, Sasaki T, Koyama T, Nishiyama S, Ohnishi Y. Development of coupled discontinuous deformation analysis and numerical manifold method (NMM-DDA). Int J Comput Methods. 2010;7(1):131–50.

Ohnishi Y, Tanaka M, Koyama T. Manifold method in saturated unsaturated unsteady groundwater flow analysis. In: Amadei B, editor. ICADD-3: third international conference on analysis of discontinuous deformation, Vail, Colorado (USA). Rotterdam (The Netherlands): American Rock Mechanics Association, AA Balkema; 1999, p. 221–30.

Okazawa S, Kashiyama K, Kaneko Y. Eulerian formulation using stabilized finite element method for large deformation solid dynamics. Int J Numer Methods Eng. 2007;72:1544–59.

Scavia C. A method for the study of crack propagation in rock structures. Geotechnique. 1995;45(3):447–63.

Shi GH. Manifold method of material analysis. In: Trans 9th army conf appl math comput, Minneapolis, Minnesota (USA); 1991, p. 57–76.

Shi GH. Modeling rock joints and blocks by manifold method. In: Proceedings of 33rd US symposium on rock mechanics, Santa Fe, New Mexico (USA); 1992, p. 639–48.
Shi GH. Numerical manifold method. In: Li JC, Wang CY, Sheng J, editors. Proceedings of the first international conference on analysis of discontinuous deformation. Chungli (Taiwan): National Central University; 1995, p. 187–222.
Shi GH. Manifold method. In: Salami MR, Banks D, editors. Proceedings of the first international forum on discontinuous deformation analysis (DDA) and simulations of discontinuous media. Albuquerque, NM (USA): TSI Press; 1996a, p. 52–204.
Shi GH. Simplex integration for manifold method, FEM, DDA and analytical analysis. In: Salami MR, Banks D, editors. Proceedings of the first international forum on discontinuous deformation analysis (DDA) and simulations of discontinuous media. Albuquerque, NM (USA): TSI Press; 1996b, p. 205–62.
Shi GH. Numerical manifold method. In: Ohnishi Y, editor. Proceedings of the second international conference on analysis of discontinuous deformation. Kyoto (Japan): Japan Institute of Systems Research; 1997, p. 1–35.
Shi GH. Rock block stability analysis of slopes and underground power houses. In: Zhao J, Ohnishi Y, Zhao GF, Sasaki T, editors. Advances in discontinuous numerical methods and applications in geomechanics and geoengineering. CRC Press; 2012. p. 3–16.
Shyu K, Salami MR. Manifold with four-node isoparametric finite element method. In: First working forum on the manifold method of material analysis, California (USA); 1995, p. 165–82.
Su HD, Huang YY. Application of numerical manifold method in fluid solid interaction harmonic analysis. In: Proceedings of the eighth international conference on analysis of discontinuous deformation (ICADD-8), Beijing (China); 2007, p. 227–34.
Su HD, Xie XL, Liang QY. Automatic programming for high-order numerical manifold method. In: Lu M, editor. Proceedings of the sixth international conference on analysis of discontinuous deformation (ICADD-6). Trondheim (Norway): CRC Press; 2003, p. 153–60.
Sukumar N, Prévost JH. Modeling quasi-static crack growth with the extended finite element method. Part I: Computer implementation. Int J Solids Struct. 2003;40(26):7513–37.
Sun L. Particle manifold method (PMM) for multiscale continuous–discontinuous analysis. Ph.D. thesis, Ecole Polytechnique Federale de Lausanne (Switzerland); 2012.
Sun L, Zhao GF, Zhao J. Particle manifold method (PMM): a new continuum discontinuum numerical model for geomechanics. Int J Numer Anal Methods Geomech. 2013;37(11):1711–36.
Sun L, Zhao GF, Zhao J. A multiscale manifold method using particle representations of the physical domain. Geomech Geoeng. 2014;9(2):124–32.
Terada K, Kurumatani M. Performance assessment of generalized elements in the finite cover method. Finite Elem Anal Des. 2004;41(2):111–32.
Terada K, Kurumatani M. An integrated procedure for three-dimensional structural analysis with the finite cover method. Int J Numer Meth Eng. 2005;63(15):2102–23.
Tasdemir MA, Maji AK, Shah SP. Crack propagation in concrete under compression. J Eng Mech. 1990;116(5):1058–76.
Terada K, Asai M, Yamagishi M. Finite cover method for linear and non-linear analyses of heterogeneous solids. Int J Numer Methods Eng. 2003;58(9):1321–46.
Terada K, Ishii T, Kyoya T, Kishino Y. Finite cover method for progressive failure with cohesive zone fracture in heterogeneous solids and structures. Comput Mech. 2007;39(2):191–210.
Tsay RJ, Chiou YJ, Chuang WL. Crack growth prediction by manifold method. J Eng Mech. 1999;125(8):884–90.
Wang SL, Deng JH, Ge XR. Application of augmented Lagrangian method in frictional contact problem. Chin J Geotech Eng. 1998;17(5):67–70 (in Chinese with English abstract).
Wang SL, Ge XR, Zhang G. Numerical analysis of crack propagation under compression. Chin J Rock Mechan Eng. 1999;18(6):671–5 (in Chinese with English abstract).
Wang SL, Feng XT, Zheng H. Modeling strong discontinuities by manifold method. In: Lu M, editor. Proceedings of the sixth international conference on analysis of discontinuous deformation (ICADD-6), Trondheim (Norway); 2003, p. 141–4.
Wei W, Jiang QH. A modified numerical manifold method for simulation of finite deformation problem. Appl Math Model. 2017;48:673–87.

Wei W, Zhao Q, Jiang QH, Grasselli G. Three new boundary conditions for the seismic response analysis of geomechanics problems using the numerical manifold method. Int J Rock Mech Min Sci. 2018;105:110–22.

Wong LNY, Wu Z. Application of the numerical manifold method to model progressive failure in rock slopes. Eng Fract Mech. 2014;119:1–20.

Wu Z, Fan L. The numerical manifold method for elastic wave propagation in rock with time-dependent absorbing boundary conditions. Eng Anal Bound Elem. 2014;46:41–50.

Xu DD, Yang YT, Zheng H, Wu AQ. A high order local approximation free from linear dependency with quadrilateral mesh as mathematical cover and applications to linear elastic fractures. Comput Struct. 2017;178:1–16.

Xu DD, Wu AQ, Li C. A linearly-independent higher-order extended numerical manifold method and its application to multiple crack growth simulation. J Rock Mech Geotech Eng. 2019;11(6):1256–63.

Yang YT, Zheng H. A three-node triangular element fitted to numerical manifold method with continuous nodal stress for crack analysis. Eng Fract Mech. 2016;162:51–75.

Yang YT, Guo HW, Fu XD, Zheng H. Boundary settings for the seismic dynamic response analysis of rock masses using the numerical manifold method. Int J Numer Anal Methods Geomech. 2018;42(6):1095–122.

Zhang HW, Zhou L. Numerical manifold method for dynamic nonlinear analysis of saturated porous media. Int J Numer Anal Methods Geomech. 2006;30(9):927–51.

Zhang GX, Sugiura Y, Hasegawa H. Crack propagation by manifold and boundary element method. In: Amadei B, editor. ICADD-3: third international conference on analysis of discontinuous deformation, Vail, Colorado (USA). Rotterdam (The Netherlands): American Rock Mechanics Association, AA Balkema; 1999a, p. 273–82.

Zhang GX, Sugiura Y, Saito K. Application of manifold method to jointed dam foundation. In: Amadei B, editor. ICADD-3: third international conference on analysis of discontinuous deformation, Vail, Colorado (USA). Rotterdam (The Netherlands): American Rock Mechanics Association, AA Balkema; 1999b, p. 211–20.

Zhang GX, Zhao Y, Peng XC. Toppling failure simulation of rock slope by numerical manifold method. Int J Comput Methods. 2010a;7(1):167–89.

Zhang HH, Li LX, An XM, Ma GW. Numerical analysis of 2D crack propagation problems using the numerical manifold method. Eng Anal Bound Elem. 2010b;34(1):41–50.

Zhang HH, Ma GW, Ren F. Implementation of the numerical manifold method for thermo-mechanical fracture of planar solids. Eng Anal Bound Elem. 2014;44:45–54.

Zhang HH, Ma GW, Fan LF. Thermal shock analysis of 2D cracked solids using the numerical manifold method and precise time integration. Eng Anal Bound Elem. 2017;75:46–56.

Zhang HH, Han SY, Fan LF, Huang D. The numerical manifold method for 2D transient heat conduction problems in functionally graded materials. Eng Anal Bound Elem. 2018;88:145–55.

Zhao J, Sun L, Zhu JB. Modelling P-wave transmission across rock fractures by particle manifold method (PMM). Geomech Geoeng. 2012;7(3):175–81.

Zhao GF, Zhao XB, Zhu JB. Application of the numerical manifold method for stress wave propagation across rock masses. Int J Numer Anal Methods Geomech. 2014;38(1):92–110.

Zheng H, Xu DD. New strategies for some issues of numerical manifold method in simulation of crack propagation. Int J Numer Methods Eng. 2014;97(13):986–1010.

Zheng H, Liu F, Li C. The MLS-based numerical manifold method with applications to crack analysis. Int J Fract. 2014;190(1–2):147–66.

Zheng H, Liu F, Du X. Complementarity problem arising from static growth of multiple cracks and MLS-based numerical manifold method. Comput Methods Appl Mech Eng. 2015;295:150–71.

Zhou W, Kuo X, Yang R. Crack propagation using manifold method coupled with element free method. In: Amadei B, editor. ICADD-3: third international conference on analysis of discontinuous deformation, Vail, Colorado (USA). Rotterdam (The Netherlands): American Rock Mechanics Association, AA Balkema; 1999, p. 283–9.

Chapter 7
Discrete Element Methods with Special Focus on DEM

Abstract In this chapter, the dynamic relaxation algorithm of "distinct element method (DEM)" belonging to the family of particle-or block-based CMs, is elucidated towards the simulation of rock mass as a particle assemblage with large deformation. A distinct element represents the particle or block discretized naturally (e.g. aggregate grain, rock-fill due to crushing/blasting process, and rock block due to joint intersection) or artificially (e.g. equivalently simulated landslide body). The interaction between particles or blocks is captured by compliant contact models that accommodate virtual "inter-penetrations" to get normal and tangential contact forces related to postulated stiffness coefficients. These contact forces are in turn, applied to solve the problem iteratively. In addition, a viscous damping parameter may be introduced to reduce dynamic effects for the simulation of quasi-static states.

7.1 General

In order to predict and optimize the behavior and motion of granular matter in engineering devices, numerical simulation tools are increasingly created (Cleary and Prakash 2004), of which the "discrete element methods (DEMs)" form an important family of CMs for the motion of particle assemblage. After the laying of cornerstone in 1905 by Klein and Weighardt, the framework of DEMs was first explained by investigating equilateral triangles and square framework patterns (Weighardt 1906). Riedel (1927) worked with square patterns and shown that a transition from framework model to continuum provides reasonable results. In 1941, Hrennikoff investigated patterns with equilateral triangles and proved that this analogy is allowed for. In 1943, McHenry gave good results using square patterns. Lie (1944) made a further enhancement by considering beams for plate modeling. These early pioneering works indicated that the stress and strain fields inside continuum can be described with discrete elements by using either triangles or square patterns.

The numerical simulation for natural discrete particles was first tackled in the works of Born and Huang (1954) and Maradudin (1958), where atoms are described via the concentrated masses and contact forces due to atom interaction. Eisenstadt (1971) further modeled bonds between atoms by springs. It is notable that even

S. Chen, *Advanced Computational Methods and Geomechanics*, Springer Tracts in Civil Engineering, https://doi.org/10.1007/978-981-19-7427-4_7

if DEMs are very closely related to molecular dynamics (MD), the methods are generally distinguished by their inclusion of rotational DOFs, and often (although not always), the complicated particle geometries (including polyhedrons).

In DEMs, Lagrangian method is applied to the moving assemblage that is regarded as a system of particles with given shapes and material properties. The state of particles is obtained by time domain integration for the dynamics equations derived from the classical Newton's second law w.r.t. the motion (translational and rotational) of each particle. All the forces and moments exerting on particles are updated at every time-marching step.

There exist two basic algorithms for the solution of dynamics equations with DEMs, namely, static relaxation (Stewart and Brown 1984) and dynamic relaxation (Otter et al. 1966).

Static relaxation method solves the state of particles under equilibrium condition using unbalanced forces. It is an implicit iterative scheme intended to eliminate the residual forces and momentums.

Dynamic relaxation method solves the motion equation by using explicit scheme with which the state of particles at the next time step can be calculated from the state of particles at the current and/or previous time steps. This simplifies the solution procedure because the new particle positions may be explicitly expressed, hence it is not necessary to solve a large matrix equation. However, the time-marching step length and the damping to absorb the kinetic energy of system are pre-set to ensure the numerical stability of solution. If the state of particles at next step is calculated not only from current and previous steps but also from the next step, the equation of motion will be implicit in new positions. Generally, such an implicit scheme is computationally more expensive for the solution of equation system at each time-marching step, but relatively larger time-marching step length may be used due to its higher numerical stability.

In this chapter, the dynamic relaxation algorithm based on the distinct element method, also abbreviated as DEM that was initiated by Cundall (1971), will be elucidated towards the simulation of rock mass (Cundall and Strack 1979; Potyondy and Cundall 2004). The basic computation procedure of DEM can be implemented in following steps:

① Contacts between discrete elements (particles/blocks) are identified/updated through a detection technique;
② Contact forces between discrete elements are computed based on their relative positions;
③ Accelerations of discrete elements driven by unbalanced forces are calculated using the Newton's second law;
④ Velocities and displacements are further calculated by time integration; and
⑤ New positions of discrete elements are updated every time step.

Any empirical joint constitutive laws derived from laboratory experiments (viz. Chap. 2) can be implemented towards the interaction calculation (Bandis et al. 1983; Saeb and Amadei 1990, 1992). In addition, a viscous damping coefficient may be introduced to reduce dynamic effects for the simulation of quasi-static state.

Due to the complexity in the behavior of granular materials, substantial computational effort is normally required for DEM computations, although this becomes less and less serious attributable to the advances in computer sciences and technologies.

7.2 Contact Detection

Accounting for a significant portion of computer resources, it is widely recognized that the paramount issue in the application of DEM is to get higher efficient contact algorithm for the construction of neighborhood link list w.r.t. discrete element assembly (Nezami et al. 2004).

The most commonly used particle shape in DEM simulations is disc (2-D) and sphere (3-D), which is conceptually and computationally simple (Lin and Ng 1997). The large majority of commercial codes such as BALL (Strack and Cundall 1978), TRUBAL (Strack and Cundall 1984), and PFC (Itasca 2018), adopt spheres as the basic shape of particular element. Polygonal and polyhedral block elements also have been implemented in the commercial codes UDEC and 3DEC for solving 2-D and 3-D problems. Since 2-D problem will be detailed in this book where elements are mainly disc-shaped and polygon-shaped, the contact detection schemes and corresponding algorithms for two-dimensional "disc elements" and "polygonal elements" will be particularly addressed. In addition, the former will be briefly termed as "discs" whereas the latter will be termed as "blocks".

7.2.1 Detection Schemes

Regardless of rapid advances in computer hardware and parallel computation techniques, the huge computation effort required to update contact forces is still a major obstacle in the large scale simulations with DEM. For complex distinct element geometries, such as 3-D polyhedrons, contact detection subroutines can easily take up to 80% of the total computation time. Hence, the applicability of DEM code is directly related to the efficiency of contact detection scheme, which is usually performed by two independent stages:

- First stage. Being referred to as "neighbor search", this preliminary detection undertakes a rough search merely that aims to rule out the elements without possibility of contact and to provide a link list for all possible elements in contact. Among available algorithms, the most prevalent ones include "No Binary Search (NBS)" algorithm (Munjiza and Andrews 1998) and DESS algorithm (Perkins and Williams 2001). A comprehensive review w.r.t. neighborhood search is available in the literature of Williams and O'Connor (1995).

- Second stage. Being referred to as "geometric resolution", this refined detection pinpoints the contact positions for the calculation of contact forces. Pairs of neighboring elements obtained from the first stage are examined in more detail to find the contact points or contact areas for the calculation of correspondent contact forces.

Efficiency and geometric resolution strongly depends on the complexity of the geometric representation of distinct element. For example, if the boundaries of element are implicitly represented by a single function $f(x, y, z) = 0$, a closed-form solution is likely to be feasible (Ting et al. 1993). Otherwise, such as in the cases of polygon or polyhedron elements, the contact detection would be quite cumbersome. Barbosa (1990) introduces a simple algorithm for the contact detection between polyhedrons, which needs to compare all the vertexes of one element to all the faces of another and vice versa. This algorithm is rather similar to the original version of DDA (Shi 1993) and possesses a high computational complexity of order $O(N^2)$, where N is the number of polyhedral vertexes.

Nowadays, two approaches are commonly exercised to handle the contact detection: "discrete function representation (DFR)" (Williams and O'Connor 1995, 1999) and "continuous function representation (CFR)" (Hogue 1998).

The DFR algorithm is based on discretizing the surface of block element into a set of nodes, the contact between elements is calculated by considering the interaction between the bounding boxes of block elements. It achieves a computational complexity of order $O(N)$. Krishnasamy and Jakiela (1995), Feng and Owen (2002) introduced energy-based methods for finding contact forces, in which a potential energy function is defined for each contact point using overlap area. It was first derived as the cornerstone of combined "finite-discrete element method (FEM-DEM, FDEM)" in dealing with discontinuous problems (Munjiza 2004; Mahabadi et al. 2010; Lisjak et al. 2014; Yan et al. 2018). This is similar to the sub-meshing technique to enhance block deformability in DDA (Clatworthy and Scheele 1999).

The CFR algorithm detects inter-element contacts through the direct manipulation of the equations that mathematically describe the shape of element. It has two sub-types: the analytical CFR algorithm to resolve the contact between polyhedral block elements (Feng and Owen 2004; Nezami et al. 2007) and the numerical CFR algorithm to resolve the contact between elements with curved surfaces (Mustoe and Miyata 2001; Andrade et al. 2012; Lu et al. 2012). CFR algorithm tends to be applicable to distinct elements only of specific shapes, such as cylinders (Feng et al. 2017), ellipsoids (Xu and Chen 2012; Zhu et al. 2019), and convex polygons (Houlsby 2009; Boon et al. 2013). A CFR approach for the contact resolution of convex super-quadrics has been implemented (Podlozhnyuk et al. 2017), too. The code PFC allows for the simulation of mixed spheres and polyhedrons using a variant of the contact detection algorithm proposed by Nezami and his coworkers (2006).

As a member of analytical CFR methods, the "common plane (CP)" method (Cundall 1988) for detecting the contact between polyhedrons is prevalent and has been implemented in the code 3DEC (Houlsby 2009). In some sense, the CP is a plane that bisects the space between two elements. As a result, the expensive element-to-element contact detection problem reduces to a much faster plane-to-element contact detection problem. Once the CP is established between two elements, the normal vector to CP defines the direction of contact, which in turn, defines the direction of the contact forces between two elements. This is especially advantageous for vertex-to-vertex or edge-to-vertex contacts, where the definition of contact normal vector is non-trivial. The method has a complexity of order $O(N)$.

7.2.2 Contact Detection for Disc Elements

1. **Preliminary detection**

For preliminary detection, a background grid should be created that is schematically shown in Fig. 7.1. At the present time, the red neighbor discs are further detected because they are possible to contact with our target disc in blue. The other green discs are not necessarily to be further detected.

2. **Refined detection**

Denote d_i the representative disc element and d_j its contact disc element, there are two types of contact for d_i: disc-to-disc contact (Fig. 7.2a) and disc-to-wall contact (Fig. 7.2b).

Disc-to-disc contact is easily detected by the comparison using their radii, and the normal vector $\mathbf{n}_{d_i,d_j}$ is directed along the line between disc centers or disc center

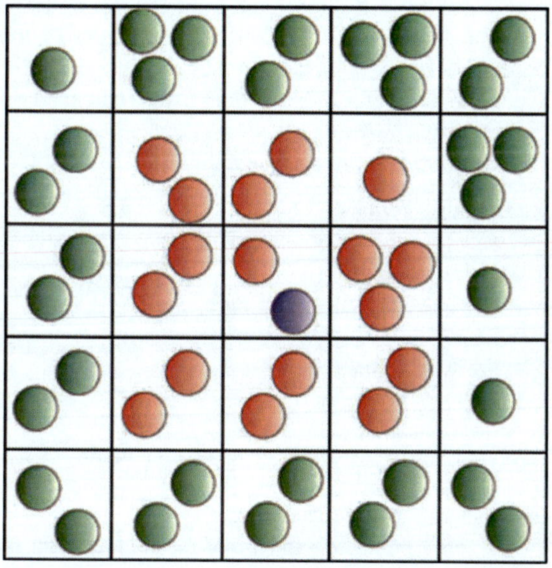

Fig. 7.1 Sketch of background grid and disc system

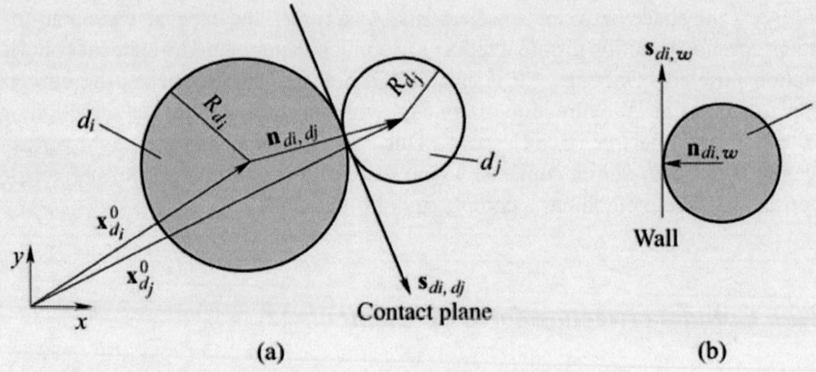

Fig. 7.2 Contact of discs. **a** Disc-to-disc contact; **b** disc-to-wall contact

and wall, namely

$$\begin{cases} \mathbf{n}_{d_i,d_j} = (\mathbf{x}^0_{d_j} - \mathbf{x}^0_{d_i})/L_{d_i,d_j} \\ L_{d_i,d_j} = \left|\mathbf{x}^0_{d_j} - \mathbf{x}^0_{d_i}\right| \end{cases} \tag{7.1a}$$

$$\begin{cases} \mathbf{n}_{d_i,w} = (\mathbf{x}^0_{w} - \mathbf{x}^0_{d_i})/L_{d_i,w} \\ L_{d_i,w} = \left|\mathbf{x}^0_{w} - \mathbf{x}^0_{d_i}\right| \end{cases} \tag{7.1b}$$

where: L_{d_i,d_j}—distance between the disc centers; $\mathbf{x}^0_{d_i}$ and $\mathbf{x}^0_{d_j}$—center coordinates of discs d_i and d_j.

For disc-to-wall contact, the normal vector $\mathbf{n}_{d_i,w}$ is directed along the line defining the shortest distance $L_{d_i,w}$ from the disc center $\mathbf{x}^0_{d_i}$ to $\mathbf{x}^0_w$ on the wall. The tangential (shear) vectors $\mathbf{s}_{d_i,d_j}$ or $\mathbf{s}_{d_i,w}$ in the "contact plane" abbreviated as "CP", are perpendicular to $\mathbf{n}_{d_i,d_j}$ or $\mathbf{n}_{d_i,w}$ according to the right hand rule, respectively. It is noted that $\mathbf{n}_{d_i,d_j} = -\mathbf{n}_{d_j,d_i}$ and $\mathbf{s}_{d_i,d_j} = -\mathbf{s}_{d_j,d_i}$.

The overlap defined as the relative contact displacement is given by:

- Disc-disc contact

$$u^n_{d_i,d_j} = R_{d_i} + R_{d_j} - L_{d_i,d_j} \tag{7.2a}$$

- Disc-wall contact

$$u^n_{d_i,w} = R_{d_i} - L_{d_i,w} \tag{7.2b}$$

in which R is the radius of disc.

The coordinates $\mathbf{x}^c_{d_i}$ and $\mathbf{x}^c_{d_j}$ of the contact point is given by

$$\mathbf{x}_{d_i}^c = \mathbf{x}_{d_i} + \left(R_{d_i} - u_{d_i,d_j}^n/2\right)\mathbf{n}_{d_i,d_j} \tag{7.3}$$

$$\mathbf{x}_{d_j}^c = \mathbf{x}_{d_j} + \left(R_{d_j} - u_{d_j,d_i}^n/2\right)\mathbf{n}_{d_j,d_i} \tag{7.4}$$

It is evident that for disc-to-disc contact, we have

$$\mathbf{x}_{d_i}^c = \mathbf{x}_{d_j}^c \tag{7.5}$$

7.2.3 Contact Detection for Polygonal Block Elements

In DEM, adjacent rock block may contact each other through bonds that are typically assigned with normal stiffness, shear stiffness, and friction coefficient (even plus cohesiveness). Crack nucleation and fracture propagation are simulated by the breakage of bonds. Consequently, elements of arbitrary shapes can be formed as a result of the simulated fracturing process.

In this chapter, 2-D blocks are all assumed to be convex polygons, because any concave ones can be simulated as a combination of several convex polygons attached to each other.

1. **Preliminarily detection**

Figure 7.3 maps distinct elements to a rectangle grid, for example, element r_i is mapped into rectangles (7,8,12,13), element r_j into rectangles (8,9,10), element r_k into rectangles (13,14,18,19); since r_i and r_j share a common rectangle 8, r_i and r_k share a common rectangle 13, but r_j and r_k share no common rectangle, hence r_i and r_j, r_i and r_k are neighbor elements, but r_j and r_k are not neighbor elements.

2. **Refined detection**

(1) **Common plane (CP) method**

In 3-D space, a CP is generally defined as a plane (viz. Fig. 7.4) perpendicular to the line $\mathrm{V}_{r_i,k} - \mathrm{V}_{r_j,l}$ and goes through the point M, where the subscripts $l(1, 2, \ldots, n)$ and $k(1, 2, \ldots, m)$ are the vertex sequences of r_i and r_j, respectively. In 2-D cases, a CP is actually a line that may be perturbed around M in both negative and positive sense (viz. Fig. 7.5).

A CP is identified by its unit normal vector $\mathbf{n}$, and any point $\mathbf{V}_0$ on it (viz. Fig. 7.6). For any two blocky distinct elements r_i and r_j, their CP should meet the following conditions:

- Condition 1: Centroids of r_i and r_j are located on the opposite sides of CP. Suppose the centroid of r_i is located on the negative side and the centroid of r_j on the positive side, denote dr_i and dr_j as the distances from $V_{r_i,k}$ and $V_{r_j,l}$ to the CP, the gap $dr_j - dr_i$ is only the function of direction vector $\mathbf{n}$ of the CP.

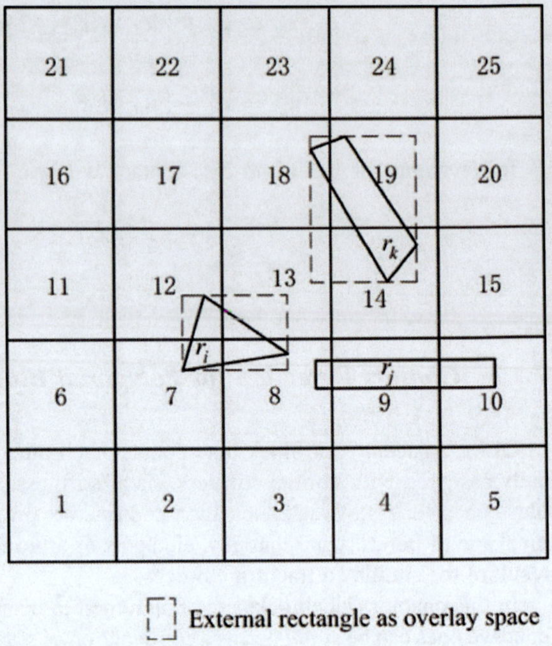

Fig. 7.3 Sketch of background grid and block system

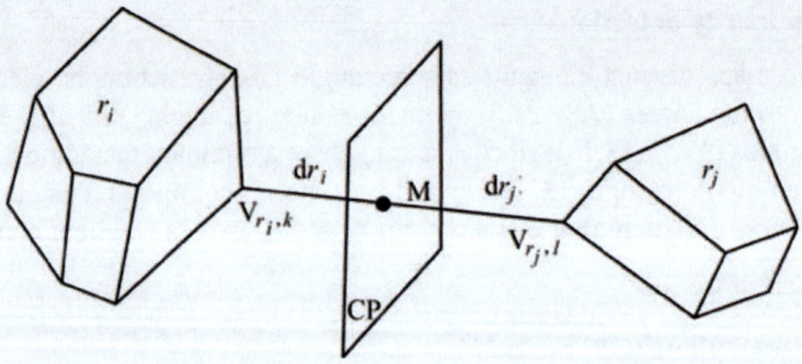

Fig. 7.4 Common plane of two polygonal blocks

- Condition 2: The gap defined as $dr_j - dr_i$ is maximum. This condition identifies the direction vector **n** by maximizing the gap.
- Condition 3: $|dr_i| = |dr_j|$. This condition specifies the location of the CP. For separated elements, the gap is always positive ($dr_i < 0$ and $dr_j > 0$); whereas for elements in contact, the gap is always negative ($dr_i > 0$ and $dr_j < 0$).

Cundall (1988) suggested a two-staged iterative procedure for finding CP. The iteration process starts from an initial guess of any arbitrary plane that might be either the CP from the previous time-marching step or the perpendicular bisector of

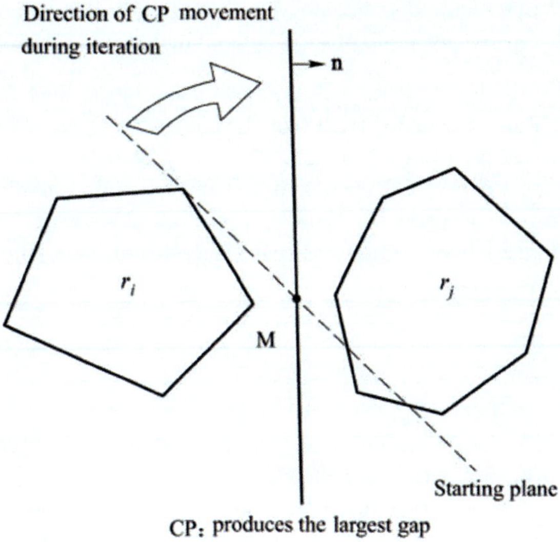

Fig. 7.5 Conventional iterative procedure for finding common plane

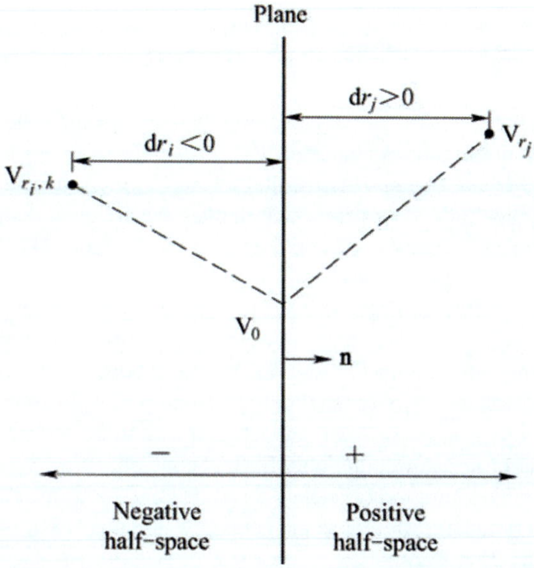

Fig. 7.6 Distance definition from point to plane

the line connecting the centers of polygonal blocks, and continues until the direction of the CP is found with reasonable accuracy:

- On the tentative CP, specify one point referred to as the reference point (i.e. point M in Fig. 7.5);

– The normal vector **n** corresponding to the maximum gap is found by iteratively rotating the tentative CP around the reference point M.

If any perturbed CP produces a gap larger than that of the current CP, the new CP replaces the current one, the closest vertexes and the reference point is updated based on the newly found CP. If all the perturbations produce smaller gaps than that of the current CP, the next iteration starts with a smaller perturbation. To be specific, for any vertexes $V_{r_i,k}$ and $V_{r_j,l}$ of block elements r_i and r_j, the distance dr_i and dr_j to the tentative CP in the space is defined as (viz. Fig. 7.6)

$$\begin{cases} dr_i = \mathbf{n} \cdot (\mathbf{V}_{r_i,k} - \mathbf{V}_0) \\ dr_j = \mathbf{n} \cdot (\mathbf{V}_{r_j,l} - \mathbf{V}_0) \end{cases} \tag{7.6}$$

It divides the space into positive and negative half-spaces, with points in positive half-space have positive distances and points in negative half-space have negative distances to the tentative CP.

The quantities $\min(dr_i)$ and $\max(dr_i)$ representing the minimum and maximum values are further calculated, respectively, taken over all the vertexes $k(1, 2, \ldots, m)$ of r_i in a manner of

$$dV_{r_i} = \begin{cases} \min(dr_i) \text{ for } (k = 1, 2, \ldots, m), \text{ if } dr_i^c < 0 \\ \max(dr_i) \text{ for } (k = 1, 2, \ldots, m), \text{ if } dr_i^c > 0 \end{cases} \tag{7.7}$$

where: m—vertex amount of r_i; dV_{r_i}—smallest distance from r_i to CP; dr_i^c—distance from the centroid of r_i to CP.

If an edge of r_i is parallel to the CP, then more than one vertex can be used to define dV_{r_i}. The vertex that defines the smallest distance in Eq. (7.7) is called the "closest vertex" of the polygon to the CP, and dV_{r_i} is back value-assigned to dr_i, namely, $dV_{r_i} \rightarrow dr_i$.

Whenever $dr_j - dr_i > d_{tol}$, where d_{tol} is a small positive pre-set tolerance, then block elements r_i and r_j are recognized as not in contact, and no CP is created. A "potential contact" indicated by the condition $0 < dr_j - dr_i < d_{tol}$ is a case for which the elements are likely to develop new contacts in the next few time-marching steps, whereas a "real contact" is a case indicated by $dr_j - dr_i < 0$. Contact points and forces are found only for real contacts.

The iteration efficiency depends on a good initial guess for tentative CP. Where it is rather different to the actual CP, the algorithm could require a large number of iteration steps.

(2) **Fast common plane (FCP) method**

For 2-D problems, Nezami and his co-workers (2004) showed that the number of iteration steps to find the correct CP between two polygons is depended on the initial guess. They proposed the "fast common plane (FCP)" method that will be detailed in this book. In Fig. 7.7, the closet vertexes of r_i and r_j are $V_{r_i,k}$ and $V_{r_j,l}$,

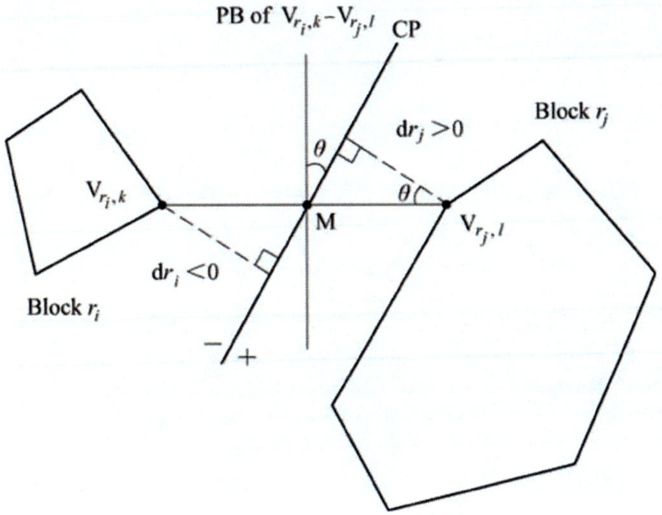

Fig. 7.7 PB of segment $V_{r_i,k} - V_{r_j,l}$ and CP

respectively, the middle point of their link-line is M. According to three conditions stated previously, one plane of the following types is pinpointed as the CP.

① It passes through the midpoint M of line segment $V_{r_i,k} - V_{r_j,l}$. Let θ measure the angle between the CP and the "perpendicular bisector (PB)" (viz. Fig. 7.7), then

$$\begin{cases} dr_i = -\left|\mathbf{M} - \mathbf{V}_{r_i,k}\right| \cos\theta \\ dr_j = \left|\mathbf{M} - \mathbf{V}_{r_j,l}\right| \cos\theta \end{cases} \tag{7.8}$$

② It is completely located within the space S (shaded area in Fig. 7.8) that is the area formed by rays $M - M_1$, $M - M_2$, $M - M_3$ or $M - M_4$. Each of them is a part of straight line L starting from the midpoint M and unlimitedly extending in outer direction, and is parallel to edges $V_{r_i,k} - V_{r_i,k+1}$, $V_{r_j,l} - V_{r_j,l+1}$, $V_{r_i,k} - V_{r_i,k-1}$, $V_{r_j,l} - V_{r_j,l-1}$, respectively.

Assume that line L, portion of which is located outside S, is a tentative CP (Fig. 7.8). Then, vertex $V_{r_j,l-1}$ is closer to L than vertex $V_{r_j,l}$. This implies that $V_{r_i,k}$ and $V_{r_j,l-1}$, but not $V_{r_i,k}$ and $V_{r_j,l}$, are the closest vertexes. Therefore, line L cannot be a candidate CP. Similarly, all lines located partially or completely outside S cannot be candidate CPs.

③ It should produce the smallest angle with PB. From Fig. 7.7, we have $dr_j - dr_i = \left|\mathbf{V}_{r_j,l} - \mathbf{V}_{r_i,k}\right| \cos\theta$. According to the Condition 2 stated previously, $dr_j - dr_i$ is the maximum, so angle θ is the minimum.

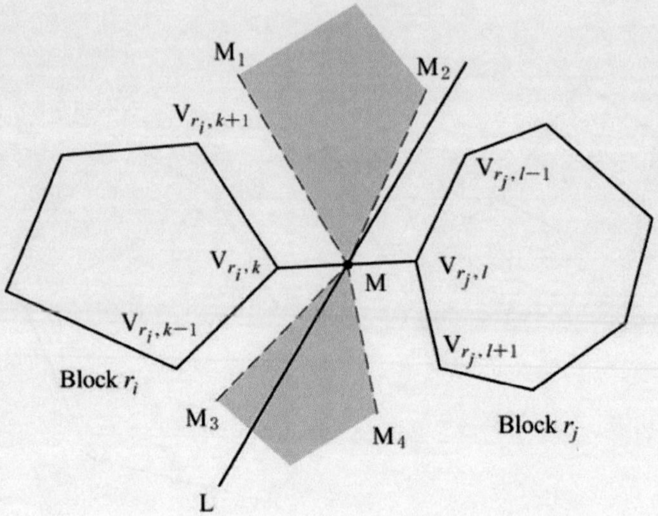

Fig. 7.8 Shaded space defined by rays $\mathrm{M}-\mathrm{M}_1$, $\mathrm{M}-\mathrm{M}_2$, $\mathrm{M}-\mathrm{M}_3$ or $\mathrm{M}-\mathrm{M}_4$

④ It is one of five candidate planes.

- If the PB is completely located in the shaded area (Fig. 7.9), from ③ the line L that makes the smallest possible angle θ with PB is the PB itself;
- If the PB is not located completely inside the shaded area (Fig. 7.8), PB is not CP. Under such circumstances, CP is one of boundary rays $\mathrm{M}-\mathrm{M}_1$, $\mathrm{M}-\mathrm{M}_2$, $\mathrm{M}-\mathrm{M}_3$ or $\mathrm{M}-\mathrm{M}_4$.

⑤ In conclusion, since any line L inside S other than on the boundaries will make a larger angle θ, hence CP is one of the following candidates:

- The perpendicular bisector (PB) of line segment $\mathrm{V}_{r_i,k}-\mathrm{V}_{r_j,l}$;
- The lines passing through the midpoint M of $\mathrm{V}_{r_i,k}-\mathrm{V}_{r_j,l}$ and being parallel to the edges $\mathrm{V}_{r_i,k}-\mathrm{V}_{r_i,k+1}$, $\mathrm{V}_{r_i,k}-\mathrm{V}_{r_i,k-1}$ of element r_i, or being parallel to the edges $\mathrm{V}_{r_j,j}-\mathrm{V}_{r_j,j+1}$ and $\mathrm{V}_{r_i,k}-\mathrm{V}_{r_i,k-1}$ of element r_j.

It is reminded that unlike Cundall's algorithm, Nezami's algorithm is mainly for separated blocks where their gap $d = dr_j - dr_i > d_{tol}$. If the gap is equal to or smaller than zero, a real contact is detected and an additional step is required by translating both blocks apart to achieve a positive and small gap.

Revised Nezami algorithms were elaborated by a number of scholars (Chen et al. 2004a, 2004b; Wang et al. 2006; Gen 2008; Descantes et al. 2019). The basic idea is to move r_i and r_j in the opposite direction of connecting line to get positive maximum distance for the CP search, afterwards these blocks are moved back to their original positions.

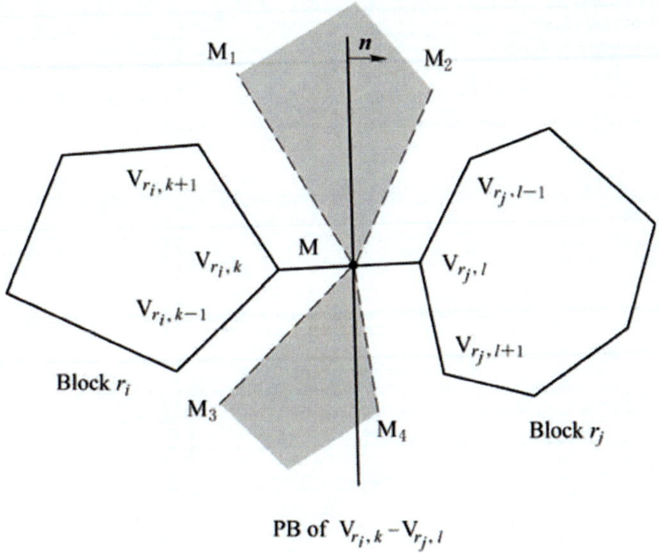

Fig. 7.9 Selection of CP by Nezami's algorithm

Figure 7.10 shows a possible initial CP (CP1) by original Nezami's algorithm and the correct CP between two interpenetrated blocks. Since CP1 is inclined so the maximum distance is negative. To overcome such a mistake, minor modifications are demanded w.r.t. Nezami's algorithm without translating elements: in case of a negative gap, the closest vertexes are simply re-defined as those that are most deeply buried into CP1. Since the final CP minimizes the overlap, hence it maximizes the distance between the sets of most deeply buried vertexes.

After the position of CP is fixed, the contact type of two polygonal block elements may be distinguished according to Table 7.1, for this purpose the number of vertexes that contact the CP is employed.

(3) **Improved FCP (IFCP)**

To accelerate the searching iteration, one may check if two block elements are edge-to-edge contacted firstly, if yes, Nezami's algorithm is not necessarily undertaken. Then, one check if two block elements are intersected, if yes, the Nezami's algorithm is not necessarily undertaken, too. These precedent steps may make use of the judgement algorithm for the penetration w.r.t. convex polygons similar to that of DDA (Chen et al. 2004a, 2004b). If all the vertexes of one block element are at the outside of all the edges of another element, these two elements are separated, and the FCP algorithm is conducted as described above.

The algorithm of IFCP is schematically flow charted in Fig. 7.11.

Fig. 7.10 Revision example of contact detection

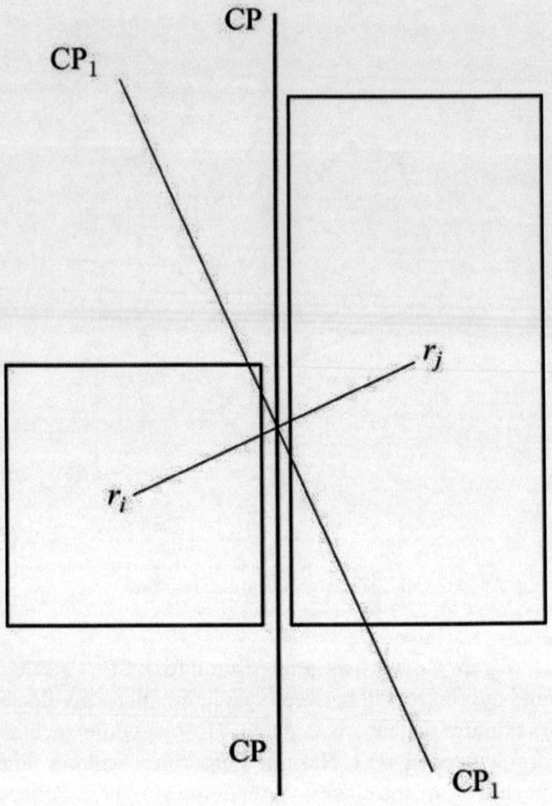

Table 7.1 Contact type of polygonal block elements

Number of vertexes on the CP		Contact type
Block r_i	Block r_j	
0	0	No contact
1	1	Vertex-to-vertex
1	2	Vertex-to-edge
2	1	Edge-to-vertex
2	2	Edge-to-edge

(4) Others

Contact detection between 2-D polygons or 3-D polyhedrons is very complex and time-consuming. In contact detection algorithms, the contact type is important

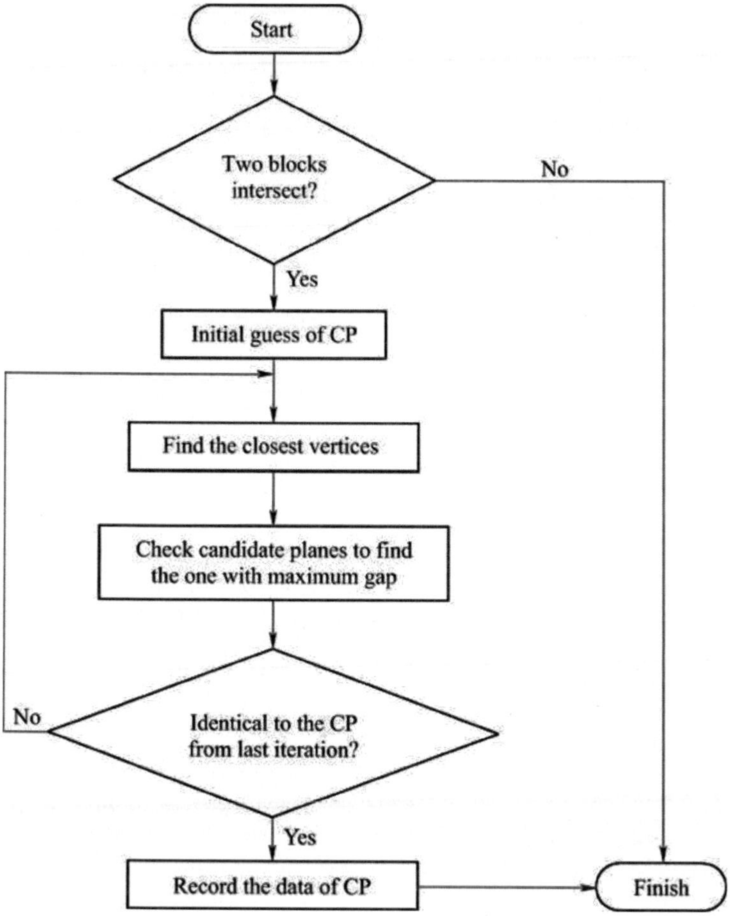

Fig. 7.11 Flow chart of the IFCP

because it determines the mechanical response between block elements. There are basically three types of contact for 2-D polygonal block elements: vertex-to-vertex, vertex-to-edge, edge-to-edge. The simplest approach is to test all possibilities, but this is low efficiency and unnecessary. Fortunately, there are internal links between blocks, vertexes and edges. These links can be utilized to reduce the search space and to improve the search efficiency significantly. However, even with these improvements, the algorithms of FCP and IFCP are still rather laborious. Hence further endeavors to improve the detection algorithms are expected.

In coupling continuous-discontinuous materials, hybrid methods such as DEM-NMM, block elements are initially neatly stacked, forming a pseudo-continuous medium. In addition, vertexes are shared by neighboring edges. Under such circumstances, it is difficult to identify contact types of vertex-to-vertex or edge-to-edge. In

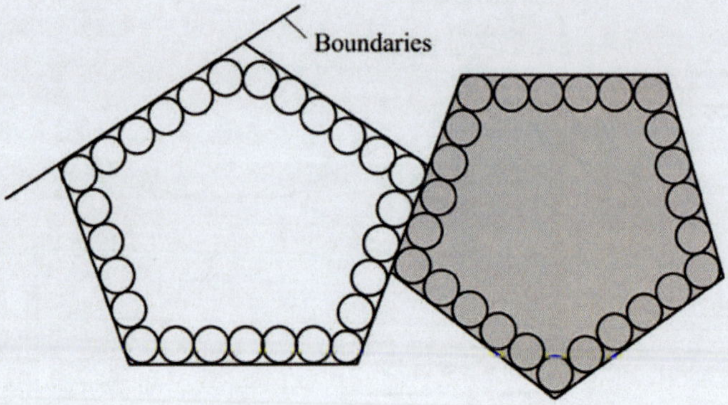

Fig. 7.12 Sketch of contact detection between blocks or between block and boundary via edge-attached discs

some cases, the contact type cannot be determined correctly. To handle such a difficult situation, Wang et al. (2015) proposed a "shrunken edge algorithm" by indenting the vertexes and edges of the block element to calculate the contact force in deformable DEM.

Alternatively, the polygonal block element can be represented by a set of discs filling the region enclosed by the element edges (Favier et al. 1999; Ren et al. 2014; Qiu et al. 2017). The relative position of those discs within the block element keeps unchanged and moves as a rigid body, hence no contact detection is conducted between these internal discs, but only calculated if the discs belonging to two different block elements overlap with each other. It facilitates the computation procedure because the contact detection between complex-shaped blocks can be simplified as that between simple-shaped discs. This idea is employed in this book for 2-D cases, where the arbitrary polygonal block element is composed of a number of discs, then the contact detection may be implemented by the multi-disc contact detection. For the sake of saving computation time, discs are only placed along the edges of element. Contact forces are only calculated when the discs belonging to different elements overlap with each other or the discs overlap with boundary wall (viz. Fig. 7.12).

7.3 Formulation of DEM with Disc Elements

A DEM simulation is started by generating a distinct (disc/polygon particles in 2-D problems) element model (assemblage) firstly, which results in all spatially orienting particles assigned with initial velocity. Then, the model executes step by step and, at each time-marching step, the Newton's second law (force = mass × acceleration) is integrated for each particle to provide updated velocities and new positions. Based on these new particle positions, contact forces are updated according to the relative displacements between particles. These updated contact forces are, in turn, used to compute the new positions and velocities of all particles for the next time-marching step. This recursion is collapsed when the simulation gives rise to a standstill (static) state.

The typical forces involved are contact and frictional forces between particles as well as the gravity. In the prevalent works, a soft-contact approach is customarily employed in which the particles are treated as rigid bodies and allowed to overlap at contact points. The magnitude of inter-particle forces depends on the overlap degree between particles. In addition to particles, various DEMs codes also include "walls" that allow one to apply velocity boundary conditions to the particle assemblage for the purpose of compaction and confinement. The particles and walls interact with one another via the forces arise from contact points.

The general algorithm for DEM is schematically flow charted in Fig. 7.13.

Despite the fact that the general DEM allows for the simulation of particles with complex shape, in a large portion of practical works, sphere (3-D problems) or disc (2-D problems) particles are widely exercised attributable to their simple algorithms for contact detection. Moreover, it is also possible to create particles with arbitrary shape by attaching two or more sphere/disc particles together termed as "clump", each clump is not allowed to break out and acts as an autonomous object serving as super-particles of general shape. It is also possible to model a brittle solid by bonding every particle to its neighbors. The resulting assemblage can be regarded as a "solid" that possesses elastic properties and is capable of "fracturing" when bonds break down in a progressive manner.

7.3.1 Generation of Disc Element Assemblage

This is implemented by the identification algorithm for particle assemblage (viz. Chap. 2).

7.3.2 Geometrical Parameters of Disc Elements

The force–displacement law of contact point lying on the CP can be described in terms of the unit normal vectors $\mathbf{n}_{d_i,d_j}$ for disc-to-disc contact (Fig. 7.2a) and $\mathbf{n}_{d_i,w}$ for disc-to-wall contact (Fig. 7.2b), respectively. With the help of these normal vectors, the contact force is decomposed into a normal component along the direction of normal vector and a shear component on the CP, respectively. The force–displacement law (constitutive equation) relates these two components of force to the corresponding components of relative displacements (overlap and shear) via the normal stiffness and shear stiffness. In both cases, the overlap $u^n_{d_i,d_j}$ or $u^n_{d_i,w}$ defined as the relative contact displacement in the normal direction, has been given by Eqs. (7.2a–7.2b).

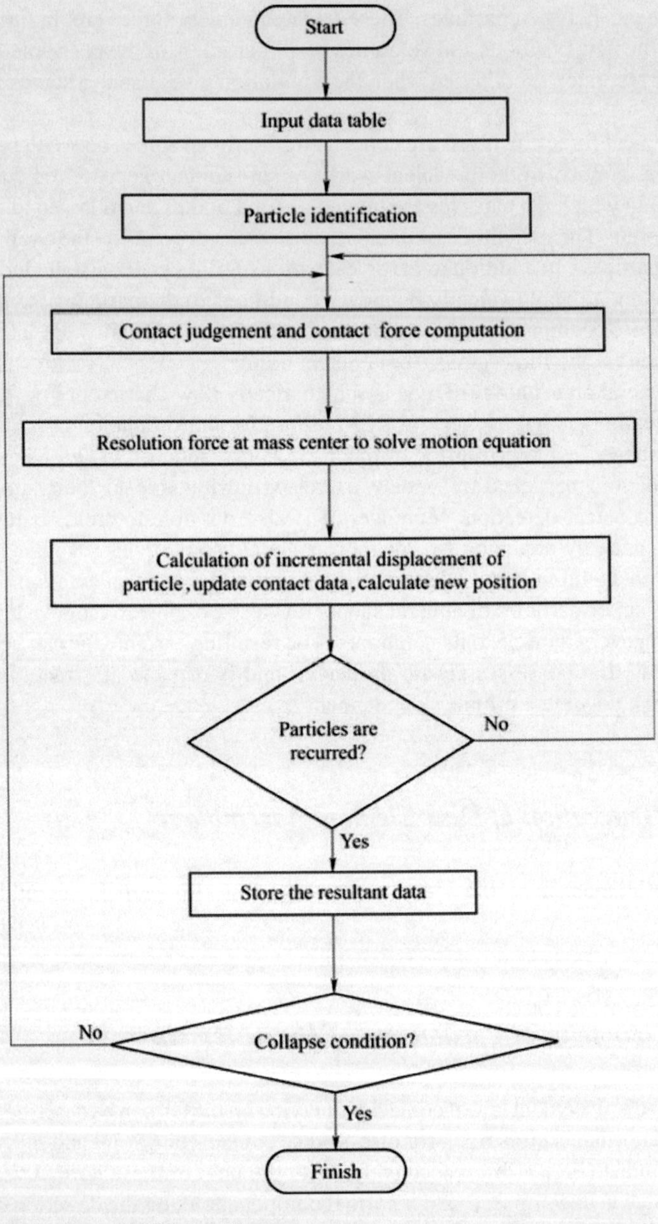

Fig. 7.13 Flow chart of DEM algorithm

7.3.3 Momentum Conservation

The motion of a single rigid disc is determined by the resultant force and moment acting upon it, and can be described in terms of translational motion and rotational motion w.r.t. the center of its mass. The translational motion of disc center is described in terms of its position $\mathbf{x}_{d_i}$, velocity $\frac{d\mathbf{x}_{d_i}}{dt}$, and acceleration $\frac{d^2\mathbf{x}_{d_i}}{dt^2}$. The rotational motion around disc center is described in terms of its angle θ_{d_i}, angular velocity $\frac{d\theta_{d_i}}{dt}$, and angular acceleration $\frac{d^2\theta_{d_i}}{dt^2}$. It is noted that the superscript (°) indicating disc center in these variables is ignored hereinafter. The motion is governed by two equations: one relates the resultant force to the translational acceleration, whereas the other relates the resultant moment to the rotational acceleration. Their positive directions are defined in Fig. 7.14.

1. **Transitional motion**

The equation for translational motion can be written in the vector form:

$$\begin{cases} m_{d_i}\dfrac{d^2\mathbf{x}_{d_i}}{dt^2} + \alpha_u m_{d_i}\dfrac{d\mathbf{x}_{d_i}}{dt} = \mathbf{F}_{d_i} \\ \mathbf{F}_{d_i} = \mathbf{f}_{d_i} + m_{d_i}\mathbf{g} \end{cases} \quad (d_i = 1, 2, \ldots, n_d) \tag{7.9}$$

where: n_d—amount of discs in the particle system; $\mathbf{f}_{d_i}$—resultant force that is the summation of all external forces acting on the disc exclusive gravity action; m_{d_i}—mass of disc; $\mathbf{g}$—gravity acceleration; α_u—transitional damping coefficient.

2. **Rotational motion**

For a disc with radius R and uniformly distributed mass, its center of mass coincides with the geometry center. Thus,

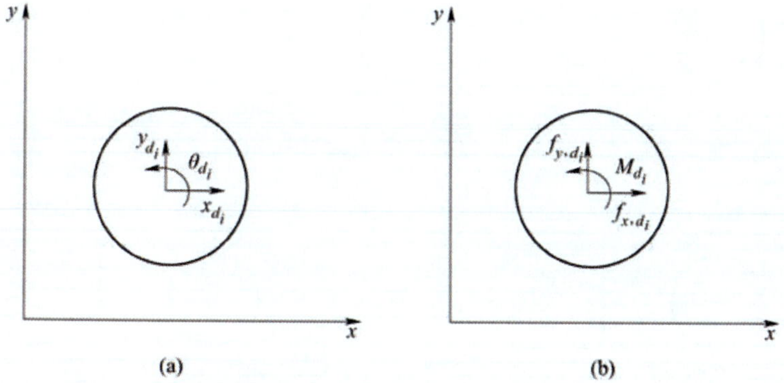

Fig. 7.14 Definition of positive directions of disc element. **a** Motion; **b** action

$$\begin{cases} I_{d_i}\dfrac{\mathrm{d}^2\theta_{d_i}}{\mathrm{d}t^2} + \alpha_\theta I_{d_i}\dfrac{\mathrm{d}\theta_{d_i}}{\mathrm{d}t} = \mathbf{M}_{d_i} \\ I_{d_i} = \dfrac{2}{5}mR^2 \\ \mathbf{M}_{d_i} = \mathbf{Mc}_{d_i} + \mathbf{Mr}_{d_i} \end{cases} \quad (d_i = 1, 2, \ldots, n_d) \tag{7.10}$$

where: $\mathbf{Mc}_{d_i}$—resultant contact moment; $\mathbf{Mr}_{d_i}$—resultant rotational friction moment; I_{d_i}—moment of inertia; α_θ—rotational damping coefficient, normally, $\alpha_\theta = \alpha_u = \alpha$.

3. **Contact force**

Force–displacement law links the relative displacements between two contact entities (disc-to-disc, disc-to-wall) to the force acting on the entities. For disc-to-disc contact, an additional force and moment arise from the deformation of "cement material" representing bonds, also can be applied to each disc (Fig. 7.15).

The contact force vector can be resolved into normal and shear components w.r.t. the local coordinate system tied to CP (viz. Fig. 7.16), as

$$\mathbf{fc}_{d_i} = \mathbf{fc}^n_{d_i} + \mathbf{fc}^s_{d_i} \quad (d_i = 1, 2, \ldots, n_d) \tag{7.11}$$

in which the superscripts n and s denote the normal and shear components, respectively.

(1) **Normal component**

For disc-to-disc contact, the unit normal vector $\mathbf{n}_{d_i,d_j}$ and tangential vector $\mathbf{s}_{d_i,d_j}$ that define the CP is given by Eqs. (7.1a–7.1b).

The overlap $u^n_{d_i,d_j}$ that is the relative contact displacement in the normal direction, is given by Eqs. (7.2a–7.2b). The location of the contact point is given by Eqs.

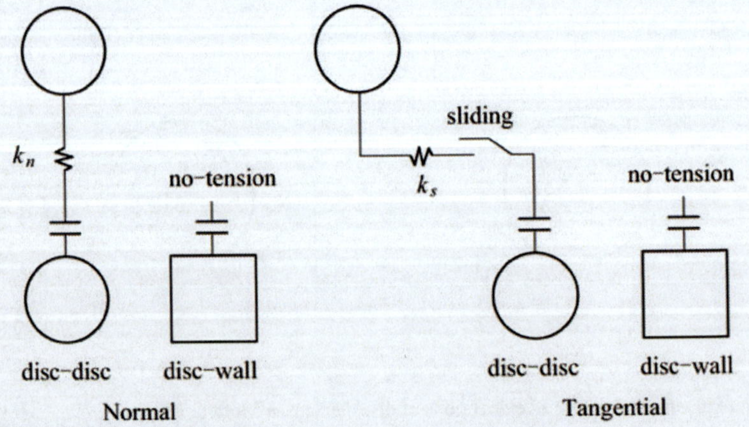

Fig. 7.15 Sketch of normal and shear contact forces

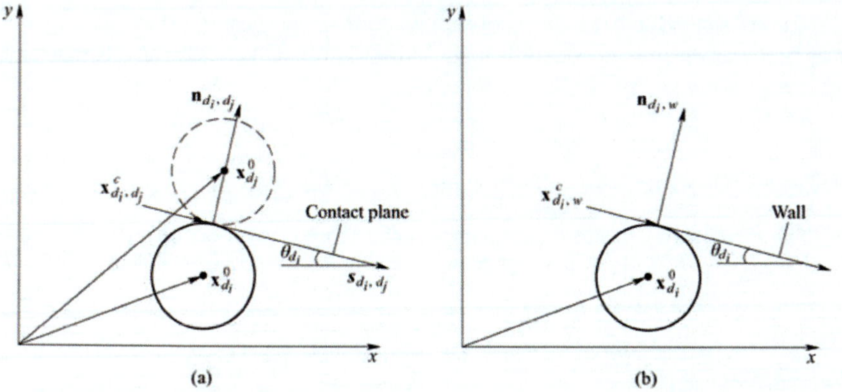

Fig. 7.16 Diagram to the calculation of contact force and moment **a** Disc-to-disc contact; **b** disc-to-wall contact

(7.3–7.5). The normal contact force is calculated by

$$\mathbf{fc}^n_{d_i} = k_n u^n_{d_i,d_j} \mathbf{n}_{d_i,d_j} \quad (d_i, d_j = 1, 2, \ldots, n_d) \tag{7.12}$$

in which k_n is the normal stiffness at the contact.

(2) **Shear component**

The shear contact force $\mathbf{fc}^s_{d_i}$ is computed in an incremental manner. When the contact emerges, the total shear contact force is initialized (to zero). Then the subsequent relative shear increment results in the incremental elastic shear force. The motion of the contact must be considered during this procedure.

In every time-marching step, the shear force $\mathbf{fc}^s_{d_i}$ is updated in two rotations: the first one concerns the new normal direction only; whereas the second one concerns the rotation from old CP to new CP.

- First rotation. The relative motion at the contact, or the contact velocity $\frac{\mathrm{d}\mathbf{u}^c_{d_i,d_j}}{\mathrm{d}t}$ that is defined as the velocity of disc d_i relative to disc d_j at the contact point C on the CP, is given by:

$$\frac{\mathrm{d}\mathbf{u}^c_{d_i,d_j}}{\mathrm{d}t} = \frac{\mathrm{d}\mathbf{x}^c_{d_j}}{\mathrm{d}t} - \frac{\mathrm{d}\mathbf{x}^c_{d_i}}{\mathrm{d}t} \tag{7.13}$$

or

$$\frac{\mathrm{d}\mathbf{u}^c_{d_i,d_j}}{\mathrm{d}t} = \left[\frac{\mathrm{d}\mathbf{x}_{d_j}}{\mathrm{d}t} + \frac{\mathrm{d}\theta_{d_j}}{\mathrm{d}t}(\mathbf{x}^c - \mathbf{x}_{d_j})\right] - \left[\frac{\mathrm{d}\mathbf{x}_{d_i}}{\mathrm{d}t} + \frac{\mathrm{d}\theta_{d_i}}{\mathrm{d}t}(\mathbf{x}^c - \mathbf{x}_{d_i})\right] \tag{7.14}$$

in which $\frac{\mathrm{d}\mathbf{x}_{d_i}}{\mathrm{d}t}$ and $\frac{\mathrm{d}\mathbf{x}_{d_j}}{\mathrm{d}t}$ are the translational velocities of d_i and d_j, respectively.

The contact velocity can be resolved into normal and shear components w.r.t. the CP. Denoting these components by $\mathbf{u}^n_{d_i,d_j}$ and $\mathbf{u}^s_{d_i,d_j}$ respectively, the shear component of the contact velocity can be calculated by

$$\frac{\mathrm{d}\mathbf{u}^s_{d_i,d_j}}{\mathrm{d}t} = \frac{\mathrm{d}\mathbf{u}^c_{d_i,d_j}}{\mathrm{d}t} - \frac{\mathrm{d}\mathbf{u}^n_{d_i,d_j}}{\mathrm{d}t} \quad (d_i, d_j = 1, 2, \ldots, n_d) \tag{7.15}$$

The incremental elastic shear component and force vector occurring in the time-marching step of Δt is further calculated by

$$\Delta\mathbf{u}^s_{d_i,d_j} = \frac{\mathrm{d}\mathbf{u}^s_{d_i,d_j}}{\mathrm{d}t}\Delta t \quad (d_i, d_j = 1, 2, \ldots, n_d) \tag{7.16}$$

$$\Delta\mathbf{fc}^s_{d_i} = k_s \Delta\mathbf{u}^s_{d_i,d_j} \quad (d_i, d_j = 1, 2, \ldots, n_d) \tag{7.17}$$

in which k_s is the shear stiffness at the contact point.

- Second rotation. The new shear contact force is found by the addition of old shear force at the start of time-marching step (should be rotated to account for the motion of CP) and the new elastic shear force increment

$$\mathbf{fc}^s_{d_i}(t + \Delta t) = \mathbf{fc}^s_{d_i}(t) + \Delta\mathbf{fc}^s_{d_i} \quad (d_i = 1, 2, \ldots, n_d) \tag{7.18}$$

(3) **Contact forces in global coordinate system**

According to the inclination angle from the global X-axis to the local S-axis defining CP and use is made of Eqs. (3.124, 3.125), the contact force vector in global coordinate system is calculated as follows

$$\mathbf{Fc}_{d_i} = \mathbf{T}^{\mathrm{T}}_{d_i}(\mathbf{fc}^n_{d_i} + \mathbf{fc}^s_{d_i}) \quad (d_i = 1, 2, \ldots, n_d) \tag{7.19}$$

The contact force is updated to get all the active force exerting on the disc

$$\mathbf{F}_{d_i} \rightarrow \mathbf{F}_{d_i} + \mathbf{Fc}_{d_i} \quad (d_i = 1, 2, \ldots, n_d) \tag{7.20}$$

(4) **Contact moment**

The contact moment is calculated by

$$\mathbf{Mc}_{d_i} = R_{d_i}\mathbf{fc}^s_{d_i} \quad (d_i = 1, 2, \ldots, n_d) \tag{7.21}$$

$$\mathbf{M}_{d_i} \rightarrow \mathbf{M}_{d_i} + \mathbf{Mc}_{d_i} \quad (d_i = 1, 2, \ldots, n_d) \tag{7.22}$$

(5) Non-linear adjustment

The values of normal and shear contact forces determined above are subject to the adjustment for meeting the nonlinear contact constitutive relation.

The criterion of non-normal tensile strength is enforced by checking whether the overlap is smaller than or equal to zero. If the answer is yes, both the normal and shear contact forces are set to zero. Slip behavior is enforced by a relation between shear force and normal force, such that two contacting discs may slip relative to one another. These treatments provide non-normal strength in tension, and allow for slip by limiting the shear force.

Normally, the slip onset is solely determined by the friction coefficient $\mu = \tan\varphi$ at the contact point, where μ is the friction coefficient and φ is the friction angle. The contact point is checked by the maximum allowable shear contact force

$$fc^s_{\max} = \mu\left|\mathbf{fc}^n_{d_i}\right| \quad (d_i = 1, 2, \ldots, n_d) \tag{7.23}$$

If $\left|\mathbf{fc}^s_{d_i}\right| > fc^s_{\max}$, then slip emerges (during the next calculation cycle) and consequently, the magnitude of $\mathbf{fc}^s_{d_i}$ is "pulled back" to $fc^s_{\max}$, namely, Eq. (7.18) is adjusted in a manner of

$$\mathbf{fc}^s_{d_i} \to \mathbf{fc}^s_{d_i}\left(fc^s_{\max}/\left|\mathbf{fc}^s_{d_i}\right|\right) \quad (d_i = 1, 2, \ldots, n_d) \tag{7.24}$$

Since the equations of motion (7.9) and (7.10) are established at the centroid of disc, hence afore-calculated forces in the forms of normal bond and shear bond need to be transmitted to the centroid of disc.

The contact of disc-to-wall is similar to the contact of disc-to-disc (viz. Fig. 7.14).

(6) Rotational friction moment

Fundamental PFC only takes into the consideration of transitional friction but neglects rotational friction. Nevertheless, it has been verified that the latter is not ignorable in the issues of shear band and particle motion (Zhu et al. 2007). Hence, to restrain the exaggerated rotation of disc, the rotational friction moment is introduced as (Zhou et al. 1999)

$$\mathbf{Mr}_{d_i} = -\mu_r Fc^n_{d_i}\bar{R}\frac{\mathrm{d}\theta_{d_i}}{\mathrm{d}t}/|\frac{\mathrm{d}\theta_{d_i}}{\mathrm{d}t}| \quad (d_i = 1, 2, \ldots, n_d) \tag{7.25}$$

$$\mathbf{M}_{d_i} \to \mathbf{M}_{d_i} + \mathbf{Mr}_{d_i} \quad (d_i = 1, 2, \ldots, n_d) \tag{7.26}$$

in which μ_r is the rotational friction coefficient between discs and $\bar{R} = \frac{R_{d_i}R_{d_j}}{R_{d_i}+R_{d_j}}$.

7.3.4 Integration of Governing Equations

1. **Velocity**

The equations of motion (7.9) and (7.10), are customarily integrated using a central finite difference procedure involving time-marching step length Δt. The translational motion velocity $\frac{\mathrm{d}\mathbf{x}_{d_i}}{\mathrm{d}t}$ and rotational motion velocity $\frac{\mathrm{d}\theta_{d_i}}{\mathrm{d}t}$ are calculated at the mid-intervals $t \pm n\Delta t/2$, while the translational motion acceleration $\frac{\mathrm{d}^2\mathbf{x}_{d_i}}{\mathrm{d}t^2}$, rotational motion acceleration $\frac{\mathrm{d}^2\theta_{d_i}}{\mathrm{d}t^2}$, force $\mathbf{Fc}_{d_i}$ and moments $\mathbf{Mc}_{d_i}$ and $\mathbf{Mr}_{d_i}$ are calculated at the primary intervals of $t \pm n\Delta t$. It follows that

$$\begin{cases} \dfrac{\mathrm{d}^2\mathbf{x}_{d_i}}{\mathrm{d}t^2} = \dfrac{1}{\Delta t}\left(\dfrac{\mathrm{d}\mathbf{x}_{d_i}}{\mathrm{d}t}(t+\Delta t/2) - \dfrac{\mathrm{d}\mathbf{x}_{d_i}}{\mathrm{d}t}(t-\Delta t/2)\right) \\ \dfrac{\mathrm{d}^2\theta_{d_i}}{\mathrm{d}t^2} = \dfrac{1}{\Delta t}\left(\dfrac{\mathrm{d}\theta_{d_i}}{\mathrm{d}t}(t+\Delta t/2) - \dfrac{\mathrm{d}\theta_{d_i}}{\mathrm{d}t}(t-\Delta t/2)\right) \end{cases} \quad (d_i = 1, 2, \ldots, n_d) \tag{7.27}$$

Substitution of Eq. (7.27) into Eqs. (7.9) and (7.10) gives rise to the solution formulae for the velocities at time $(t + \Delta t/2)$

$$\begin{cases} \dfrac{\mathrm{d}\mathbf{x}_{d_i}(t+\Delta t/2)}{\mathrm{d}t} = D_2\left[D_1\mathbf{x}_{d_i}(t-\Delta t/2) + \dfrac{\mathbf{F}_{d_i}}{m_{d_i}}\Delta t\right] \\ \dfrac{\mathrm{d}\theta_{d_i}(t+\Delta t/2)}{\mathrm{d}t} = D_2\left[D_1\theta_{d_i}(t-\Delta t/2) + \dfrac{\mathbf{M}_{d_i}}{I_{d_i}}\Delta t\right] \end{cases} \quad (d_i = 1, 2, \ldots, n_d) \tag{7.28}$$

in which $D_1 = 1 - (\alpha\Delta t/2)$, $D_2 = 1/[1 + (\alpha\Delta t/2)]$, Δt is the time-marching step length.

2. **Deflection increment**

The velocities from Eq. (7.28) is used to update the position of disc center according to

$$\begin{cases} \mathbf{x}_{d_i}(t+\Delta t/2) = \mathbf{x}_{d_i}(t-\Delta t/2) + \dfrac{\mathrm{d}\mathbf{x}_{d_i}(t+\Delta t/2)}{\mathrm{d}t}\Delta t \\ \theta_{d_i}(t+\Delta t/2) = \theta_{d_i}(t-\Delta t/2) + \dfrac{\mathrm{d}\theta_{d_i}(t+\Delta t/2)}{\mathrm{d}t}\Delta t \end{cases} \quad (d_i = 1, 2, \ldots, n_d) \tag{7.29}$$

Normally, the rotation angle is not necessary updated unless it takes some concerns in the computation task.

3. **Iterative update procedure**

The mainstream procedure for the motion of disc assemblage can be summarized as follows:

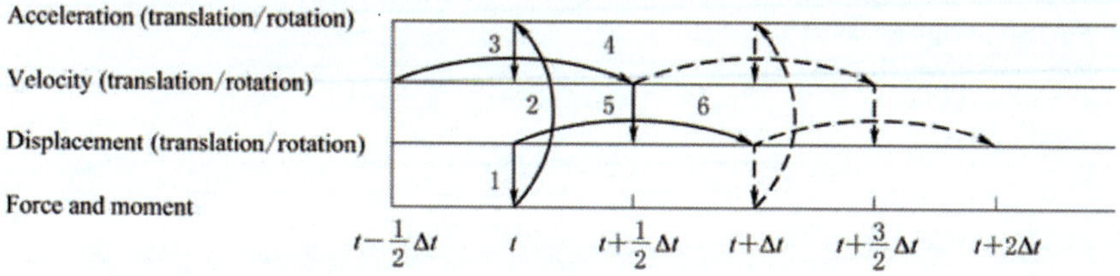

Fig. 7.17 Sketch of central difference method

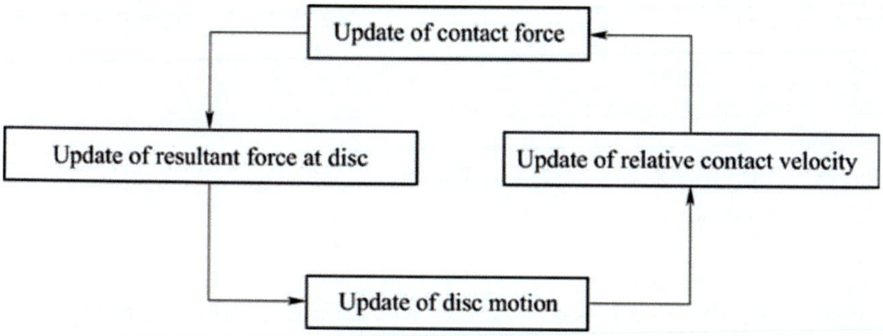

Fig. 7.18 Flow chart to the iterative procedure (disc assemblage)

- Given $\frac{d\mathbf{x}_{d_i}(t-\Delta t/2)}{dt}$, $\frac{d\theta_{d_i}(t-\Delta t/2)}{dt}$, $\mathbf{x}_{d_i}$, $\mathbf{Fc}_{d_i}(t)$, $\mathbf{Mc}_{d_i}(t)$ and $\mathbf{Mr}_{d_i}(t)$, Eq. (7.28) is employed to calculate $\frac{d\mathbf{x}_{d_i}(t+\Delta t/2)}{dt}$ and $\frac{d\theta_{d_i}(t+\Delta t/2)}{dt}$;
- Then, Eq. (7.29) is used to obtain $\mathbf{x}_{d_i}(t+\Delta t/2)$;
- Afterwards, $\mathbf{Fc}_{d_i}(t+\Delta t)$, $\mathbf{Mc}_{d_i}(t+\Delta t)$ and $\mathbf{Mr}_{d_i}(t+\Delta t)$ are updated by Eqs. (7.12, 7.18), which drive the next iterative recursion.

The algorithm is elucidated in Fig. 7.17, in which the digit indicates the sequence of solution. The flow chart is schematically shown in Fig. 7.18.

7.4 Formulation of DEM with Polygonal Elements

7.4.1 Generation of Block Element Assemblage

This may be implemented by the identification algorithm of block assemblage (viz. Chap. 2).

7.4.2 Geometrical Parameters of Block Elements

Suppose a 2-D distinct element that represents the rock block r_i $(i = 1, 2, ..., n_r)$ whose vertexes are $V_{r_i,k}$ $(k = 1, 2, ..., m)$, in which $V_{r_i,m} = V_{r_i,1}$ (viz. Fig. 7.19).

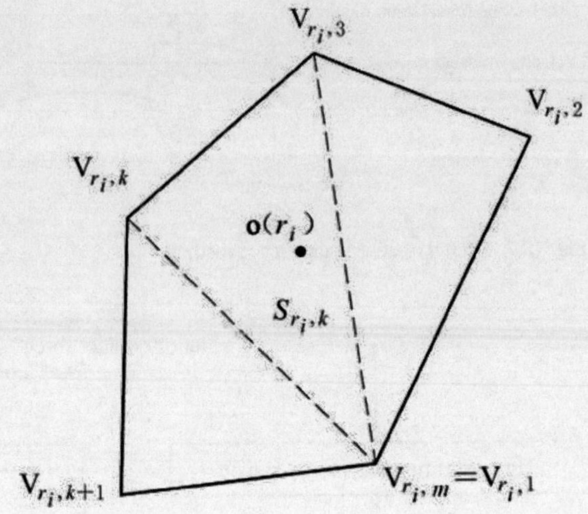

Fig. 7.19 Triangularization of polygonal block element r_i—diagram to area calculation

This block r_i is divided into $m-3$ triangles $S_{r_i,k}(k = 1, 2, ..., m-3)$ sharing a common vertex $V_{r_i,1}$, and the relevant geometrical characteristics may be calculated as follows:

- Area of triangle $S_{r_i,k}$

$$S_{r_i,k} = \frac{1}{2}\left\|(\mathbf{V}_{r_i,k+1} - \mathbf{V}_{r_i,1}) \times (\mathbf{V}_{r_i,k+2} - \mathbf{V}_{r_i,1})\right\| \quad (k = 1, 2, \ldots, m-3) \quad (7.30)$$

- Center of triangle $S_{r_i,k}$

$$\mathbf{o}(S_{r_i,k}) = \frac{1}{3}(\mathbf{V}_{r_i,k+2} + \mathbf{V}_{r_i,k+1} + \mathbf{V}_{r_i,1}) \quad (k = 1, 2, \ldots, m-3) \quad (7.31)$$

- Area of block r_i

$$A_{r_i} = \sum_{k=2}^{m-2} S_{r_i,k} \quad (7.32)$$

- Center of block r_i

$$\mathbf{x}_{r_i}^0 = \mathbf{o}(r_i) = \frac{\sum_{k=1}^{m-3} S_{r_i,k}\mathbf{o}(S_{r_i,k})}{A_{r_i}} \quad (7.33)$$

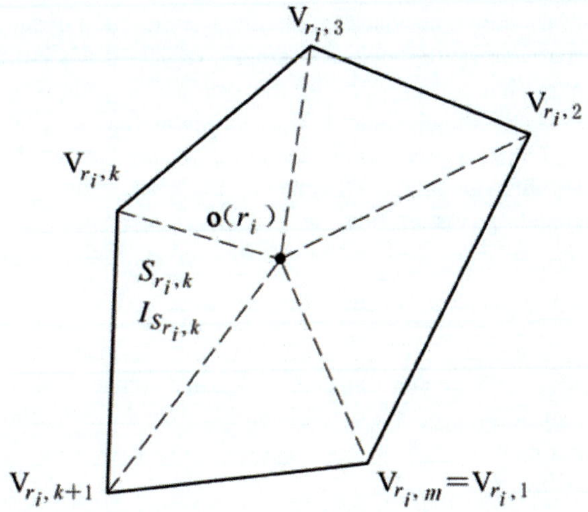

Fig. 7.20 Triangularization of polygonal element r_i—diagram to inertia calculation

The moment of inertia of block r_i may be calculated as follows (viz. Fig. 7.20):

- The center $\mathbf{o}_{r_i}$ of r_i is regarded as the origin in triangular resolution operation;
- The face is resolute into triangles $S_{r_i,k}(k = 1, 2, ..., m)$;
- The inertia tensor $I_{S_{r_i,k}}$ of triangle $S_{r_i,k}$ w.r.t. $\mathbf{o}_{r_i}$ is calculated;
- The inertia moment I_{r_i} of r_i is the summation of $I_{S_{r_i,k}}$ contributed from all triangles.

7.4.3 Momentum Conservation

1. **Translation motion**

(1) **Conventional polygonal block elements**

$$\begin{cases} m_{r_i}\dfrac{\mathrm{d}^2\mathbf{x}_{r_i}}{\mathrm{d}t^2} + \alpha m_{r_i}\dfrac{\mathrm{d}\mathbf{x}_{r_i}}{\mathrm{d}t} = \mathbf{F}_{r_i} \\ \mathbf{F}_{r_i} = \mathbf{f}_{r_i} + m_{r_i}\mathbf{g} \end{cases} \quad (r_i = 1, 2, \ldots, n_r) \tag{7.34}$$

where: $\frac{\mathrm{d}\mathbf{x}_{r_i}}{\mathrm{d}t}$—translational velocity; $\frac{\mathrm{d}^2\mathbf{x}_{r_i}}{\mathrm{d}t^2}$—translational acceleration; $\mathbf{f}_{r_i}$—resultant force, i.e., the sum of all externally forces acting on the block element exclusive gravity; m_{r_i}—total mass of block element; $\mathbf{g}$—gravity acceleration; α_u—transitional damping coefficient.

α_u should be tested for the issue of structural dynamics, but where the final static state is only the main concern, $\alpha_u = 0.7\text{–}0.8$ is customarily suggested subject to numerical test to guarantee the steady state well. It is noted that the superscript (^{o}) indicating block center is ignored hereinafter.

The polygonal block element is widely used in the simulation of landslide and toppling scenarios. The contact detection, vertex penetration and slip, and corresponding contact forces w.r.t. edge-to-edge contact and vertex-to-edge contact may be similarly calculated by that for BEA and DDA (Chap. 8).

(2) **Special polygonal block element with attached-discs serving as basic units**

Use of special polygonal element with attached-discs as basic units (viz. Fig. 7.12) may facilitate the contact detection and contact force calculation between blocks. In addition, it is particularly advantageous in the solid–fluid coupling algorithm because the interaction forces between solid blocks and fluid particles are more easy to get (Chap. 11).

The force in the momentum Eq. (7.34) may be specified as

$$\begin{cases} \mathbf{Fc}_{r_i} = \sum\limits_{k \in r_i} \sum\limits_{l \in r_j} \mathbf{Fc}_{r_i,k,l} \\ \mathbf{F}_{r_i} \rightarrow \mathbf{F}_{r_i} + \mathbf{Fc}_{r_i} \end{cases} \quad (r_i = 1, 2, \ldots, n_r) \tag{7.35}$$

in which subscript k denotes the disc d_k attached on the surface of block r_i, subscript l denotes the disc d_l attached on the surface of block r_j,$\mathbf{Fc}_{r_i,k,l}$ is the contact force calculated by Eqs. (7.19, 7.24).

2. **Rotational motion**

(1) **Conventional polygonal block elements**

$$I_{ri}\frac{\mathrm{d}^2\theta_{r_i}}{\mathrm{d}t^2} + \alpha_\theta I_{ri}\frac{\mathrm{d}\theta_{r_i}}{\mathrm{d}t} = \mathbf{M}_{r_i} \quad (r_i = 1, 2, \ldots, n_r) \tag{7.36}$$

where: $\frac{\mathrm{d}\theta_{r_i}}{\mathrm{d}t}$—angular velocity; $\frac{\mathrm{d}^2\theta_{r_i}}{\mathrm{d}t^2}$—angular acceleration;$I_{r_i}$—moment of inertia; $\mathbf{M}_{r_i}$—moment of force; α_θ—rotational damping coefficient, normally $\alpha_\theta = \alpha_u = \alpha$.

Equation (7.36) is valid in the local coordinate system with origin at the block element center of mass. The contact detection, vertex penetration and slip, and corresponding contact moments may be similarly calculated by that for BEA and DDA (Chap. 8).

(2) **Special polygonal block element with attached-discs serve as basic units**

The moment of force in Eq. (7.36) may be specified as

$$\begin{cases} \mathbf{Mc}_{r_i} = \sum\limits_{k \in r_i} \sum\limits_{l \in r_j} (\mathbf{x}_{c,k,l} - \mathbf{x}_{r_i}) \times \mathbf{Fc}_{r_i,k,l} \\ \mathbf{M}_{r_i} \rightarrow \mathbf{M}_{r_i} + \mathbf{Mc}_{r_i} \end{cases} \quad (r_i = 1, 2, \ldots, n_r) \tag{7.37}$$

where: $x_{c,k,l}$—position vector of the contact point; $\mathbf{x}_{r_i}$—position vector of the block centroid calculated by Eq. (7.33) (with superscript ignored); $\mathbf{Fc}_{r_i,k,l}$—contact force calculated by Eqs. (7.19, 7.24).

7.4.4 *Integration of Equations*

Use is made of the central differential scheme, the moment Eqs. (7.34) and (7.36) are integrated in time-matching step Δt as follows.

1. **Velocity**

The central differential scheme of Eq. (7.34) leads to

$$\frac{\mathrm{d}\mathbf{x}_{r_i}(t+\Delta t/2)}{\mathrm{d}t} = D_2\left[D_1\frac{\mathrm{d}\mathbf{x}_{r_i}(t-\Delta t/2)}{\mathrm{d}t}) + \frac{\mathbf{F}_{r_i}(t)}{m_{r_i}}\Delta t\right] \quad (r_i = 1, 2, \ldots, n_r) \tag{7.38}$$

in which $D_1 = 1 - (\alpha\Delta t/2)$, $D_2 = 1/[1 + (\alpha\Delta t/2)]$, m_{r_i} is the mass of block element r_i, Δt is the time-marching step length.

The central differential scheme of Eq. (7.36) gives rise to

$$\frac{\mathrm{d}\theta_{r_i}(t+\Delta t/2)}{\mathrm{d}t} = D_2\left[D_1\frac{\mathrm{d}\theta_{r_i}(t-\Delta t/2)}{\mathrm{d}t} + \frac{\mathbf{M}_{r_i}(t)}{I_{r_i}}\Delta t\right] \quad (r_i = 1, 2, \ldots, n_r) \tag{7.39}$$

in which $D_1 = 1 - (\alpha\Delta t/2)$ and $D_2 = 1/[1 + (\alpha\Delta t/2)]$.

2. **Deflection increment**

Transitional and rotational deflection increments is calculated by

$$\Delta\mathbf{x}_{r_i} = \frac{\mathrm{d}\mathbf{x}_{r_i}(t+\Delta t/2)}{\mathrm{d}t}\Delta t \quad (r_i = 1, 2, \ldots, n_r) \tag{7.40}$$

$$\Delta\theta_{r_i} = \frac{\mathrm{d}\theta_{r_i}(t+\Delta t/2)}{\mathrm{d}t}\Delta t \quad (r_i = 1, 2, \ldots, n_r) \tag{7.41}$$

3. **Iterative update procedure**

In each incremental step, the updates are undertaken as follows:

– Coordinates of block center

$$\mathbf{x}_{r_i}(t+\Delta t) = \mathbf{x}_{r_i}(t) + \Delta\mathbf{x}_{r_i} \quad (r_i = 1, 2, \ldots, n_r) \tag{7.42}$$

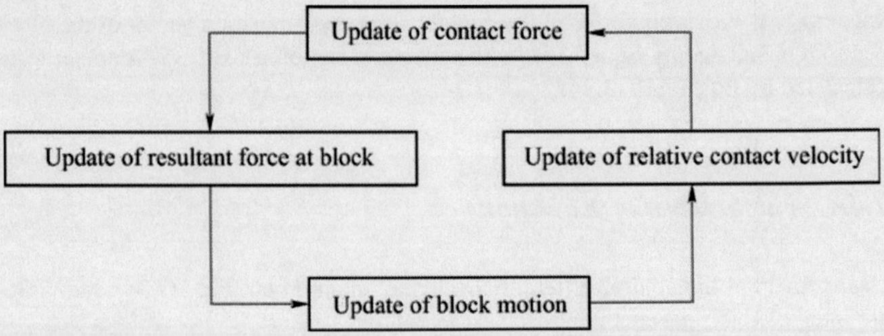

Fig. 7.21 Flow chart to the iterative procedure (block assemblage)

– Coordinates of block vertex

$$\mathbf{x}_{r_i,k}(t+\Delta t) = \mathbf{x}_{r_i,k}(t)\Delta \mathbf{x}_{r_i} + \Delta\theta_{r_i}\left[\mathbf{x}_{r_i,k}(t) - \mathbf{x}_{r_i}(t)\right]$$
$$\text{for } (r_i = 1, 2, \ldots, n_r) \text{ and } (k = 1, 2, \ldots, m) \tag{7.43}$$

The flow chart of iterative update procedure is illustrated in Fig. 7.21.

7.5 Issues Related to Applications

7.5.1 Stiffness

Stiffness influences the computation result significantly: too small of stiffness leads to large particle overlap and low computation accuracy; whereas too large stiffness leads to exaggeratedly large contact force and subsequent numerical instability characterized by irrational particle motion and increasing oscillation.

For the joints between polygonal block elements, the stiffness coefficient is desirable to be tested (viz. Chap. 2). For weak seams such as faults and interlayers, their stiffness coefficients are preferably to be tested, too. For the latter, however, normally only the Young's modulus E and Poisson's ratio ν and thickness T are provided, then the stiffness coefficients may be estimated by:

$$\begin{cases} k_n = \dfrac{E}{T} \\ k_s = \dfrac{E}{2(1+\nu)T} \end{cases} \tag{7.44}$$

For the point contact between disc elements, the customarily employed models are either the linear contact or simplified Hertz-Mindlin contact.

1. **Linear contact model**

$$\begin{cases} k_n = \dfrac{k_{n,d_i} k_{n,d_j}}{k_{n,d_i} + k_{n,d_j}} \\ k_s = \dfrac{k_{s,d_i} k_{s,d_j}}{k_{s,d_i} + k_{s,d_j}} \end{cases} \tag{7.45}$$

in which subscripts d_i and d_j denote two contacted entities (disc-to-disc, disc-to-wall).

This is simple and prevalent model, but these stiffness coefficients should be evaluated by, for example, the bi-axial compressive test.

2. **Simplified Hertz-Mindlin contact model**

It demands the Young's modulus and Poisson's ratio of disc or wall and is dependent on the disc radium such that

$$\begin{cases} k_n = \left[\dfrac{2\overline{G}\sqrt{2\overline{R}}}{3(1-\overline{\nu})} \right] \sqrt{u^n} \\ k_s = \left\{ \dfrac{2\left[\overline{G}^2 3(1-\overline{\nu})\overline{R} \right]^{1/3}}{2-\overline{\nu}} \right\} \left| Fc_{d_i}^n \right|^{1/3} \end{cases} \tag{7.46}$$

– For disc-to-disc contact

$$\begin{cases} \overline{R} = \dfrac{2R_{d_i} R_{d_j}}{R_{d_i} + R_{d_j}} \\ \overline{G} = \dfrac{1}{2}(G_{d_i} + G_{d_j}) \\ \overline{\mu} = \dfrac{1}{2}(\nu_{d_i} + \nu_{d_j}) \end{cases} \tag{7.47}$$

– For disc-to-wall contact

$$\begin{cases} \overline{R} = R_{d_i} \\ \overline{G} = G_{d_i} \\ \overline{\nu} = \nu_{d_i} \end{cases} \tag{7.48}$$

7.5.2 Damping

Damping is very complex that is still considered in somewhat an unknown or grey area (Arora 2014), in other words, the effects of damping are clear, but the characterization of damping is a puzzle waiting for further solution. This is mainly blamed for the fact that, in contrast with the inertia and stiffness forces, it is not very clear which state variables are relevant to the damping forces. Obviously, the appropriate choice of damping model depends on its detailed mechanisms. Unfortunately, these mechanisms are more varied and less well-understood than the physical mechanisms governing the stiffness and inertia. In the structural dynamics with FEM, a commonly used damping model—a kind of global damping model that was originated by Rayleigh (1897) assumes that instantaneous velocities are the only variables. The Taylor expansion then leads to a model, which encapsulates damping behavior in a dissipation matrix, directly analogous to the mass and stiffness matrices. However, it is important to avoid the misconception that, this is the only model of damping because it is rather possible for the damping forces to depend upon other quantities. Any model, which guarantees that the energy dissipation rate is nonnegative, can be a potential candidate to represent the damping effect for a given engineering structure.

In broad terms, damping mechanisms can be divided into three classes: energy dissipated throughout the bulk material in engineering structure that is also called as "material damping", dissipation of energy associated with the junctions or interfaces between structure components that is generally called as "boundary damping", and dissipation of energy associated with the fluid in contact with the engineering structure which is also called as "viscous damping". Material damping can arise from a variety of micro-structure mechanisms (Bert 1973), for small strain deformation it is often adequately represented by an equivalent linear viscoelastic model. When damping is gauged on an engineering structure, it is commonly found out to be at least an order of magnitude higher than the intrinsic material damping of the main structure components. This difference is attributed to the effects such as the frictional micro-slipping at joints and the air pumping in riveted seams, namely, the boundary damping and viscous damping.

In DEMs, boundary damping should be dominant in the description of decay scenarios. Energy supplied to the particle system is mainly dissipated through frictional sliding, namely, the friction damping occurs during the sliding adjustment entailed by Eqs. (7.23–7.24).

It is normally postulated that the viscous damping in the shear direction is not applied when sliding occurs, under such circumstances, friction damping operates solely. Where frictional sliding does not occur, two adaptive viscous damping models, one is the contact damping and the other is the global damping, may be exercised (Cundall 1987; Cundall and Hart 1992). Contact damping—a kind of local damping, operates on the relative velocities at contacts and may be envisioned as the dashpots acting in the normal and shear directions, in which the energy dissipation per cycle is independent of the rate at which the cycle is executed. These damping forces are controlled by the damping constant α and can be specified separately for each particle. In the DEM literature (Potyondy and Cundall 2004), an identical value of $\alpha = 0.7$–0.8 is commonly used.

Although these treatments are not physically true, yet both the contact damping and global damping provide similar curves of vibration decay if they are used separately. In other words, they do reduce the number of calculation cycles to reach equilibrium state for quasi static problems, and the damping does not affect the equilibrium state itself (Cundall and Strack 1979). Hence, global damping that is similar to the mass damping portion in the Rayleigh damping model, is employed in this book

$$\mathbf{fd} = -\alpha_u m \frac{\mathrm{d}\mathbf{u}}{\mathrm{d}t} \tag{7.49}$$

$$\mathbf{fd}_\theta = -\alpha_\theta I \frac{\mathrm{d}\theta}{\mathrm{d}t} \tag{7.50}$$

where: α_u and α_θ–global damping coefficients, normally an identical value α is input in the computation, but the postulation $\alpha_\theta = 0$ is also acceptable where the rotational friction moment is taken into account [viz. Eqs. (7.25–7.26)]; m—condensed mass at particle center; $\frac{\mathrm{d}\mathbf{u}}{\mathrm{d}t}$ and $\frac{\mathrm{d}\theta}{\mathrm{d}t}$–transitional and rotational velocities of the particle center.

The global damping employed in this book [viz. Eqs. (7.9, 7.10, 7.34 and 7.36)] affects the absolute velocities of discs, and may be envisioned as the effect of dashpots connecting each particle to ground. These dashpots operate both on the transitional velocity and rotational velocity.

It is notable that α is actually not constant and should be adjusted according to the ratio of its absorbed energy to the system energy. Its initial value may be estimated by the critical damping of spring oscillator

$$\alpha < 2\sqrt{km} \tag{7.51}$$

in which $k = \min(k_n, k_s)$ and $m = \frac{m_{d_i} m_{d_j}}{m_{d_i} + m_{d_j}}$ for the contact between discs with masses of m_{d_i} and m_{d_j}, respectively.

Defining

$$R = \frac{\sum E_D}{E_K} \tag{7.52}$$

in which E_D is the energy absorbed by damping and E_K is the rate of energy variation of structural system.

Numerical tests show that, when R is around 0.5, damping effect is well performed; when $R > 0.5$, the damping coefficient α is reduced by 1% for the next iterative step; when $R < 0.5$, the damping coefficient α is raised by 5% for the next iterative step.

7.5.3 Critical Time-Marching Step Length

The central difference method is customarily employed in the dynamic relaxation method, but it is conditionally stable restricted by a critical step length, namely, the computational solution will remain stable only if the time-marching step length does not exceed a critical value that is related to the minimum eigen period of the engineering structure.

First, we consider a one-dimensional mass-spring oscillator described by a point mass m, vertical spring with stiffness k, and damping coefficient α (viz. Fig. 7.22). The motion of the point mass is governed by

$$m\frac{\mathrm{d}^2u}{\mathrm{d}t^2}+\alpha\frac{\mathrm{d}u}{\mathrm{d}t}+ku=0 \tag{7.53}$$

The second order finite difference scheme towards the solution is specified as

$$\frac{\mathrm{d}u(t)}{\mathrm{d}t}=\frac{u(t+\Delta t)-u(t-\Delta t)}{2\Delta t} \tag{7.54}$$

$$\frac{\mathrm{d}^2u(t)}{\mathrm{d}t^2}=\frac{u(t+\Delta t)-2u(t)+u(t-\Delta t)}{\Delta t^2} \tag{7.55}$$

The substitution of Eqs. (7.54) and (7.55) into Eq. (7.53) gives rise to

$$(2m+\alpha\Delta t)u(t+\Delta t)+2(k\Delta t^2-2m)u(t)+(2m-\alpha\Delta t)u(t-\Delta t)=0 \tag{7.56}$$

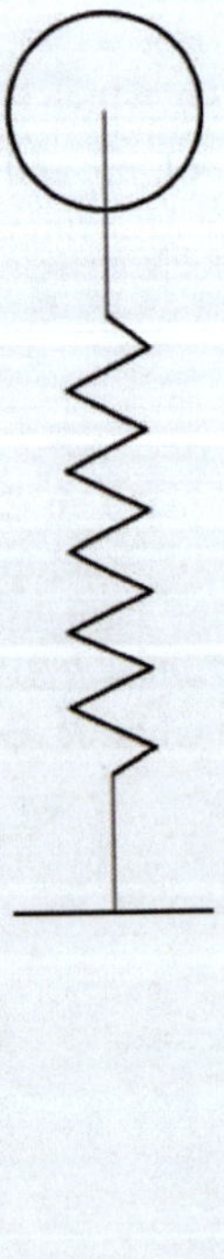

Fig. 7.22 Oscillator of single mass-spring system

From which the correspondent eigenvalue λ is the solution of

$$(2m + \alpha\Delta t)\lambda^2 + 2(k\Delta t^2 - 2m)\lambda + (2m - \alpha\Delta t) = 0 \tag{7.57}$$

This equation provides two solutions

$$\lambda_1, \lambda_2 = \left[(2m - k\Delta t^2) \pm \sqrt{k^2\Delta t^4 - 4mk\Delta t^2 + \alpha^2\Delta t^2}\right]/(2m + \alpha\Delta t) \tag{7.58}$$

To guarantee the bounded displacement u when $t \to \infty$, it is demanded that

$$\max(|\lambda_1|, |\lambda_2|) \le 1 \tag{7.59}$$

namely,

$$\Delta t \le 2\sqrt{\frac{m}{k}}(\sqrt{1+\xi^2} - \xi) \tag{7.60}$$

in which $\xi = \frac{\alpha}{2\sqrt{mk}}$.

Where $\alpha = 0$, Eq. (7.60) becomes

$$\Delta t \le 2\sqrt{\frac{m}{k}} \tag{7.61}$$

Equation (7.61) is only available for single freedom system. Since global eigenvalue analyses are impractical for complex structural system, therefore a simplified procedure is customarily implemented to estimate the critical step length at the start of each cycle.

For a multi-freedom system of discrete elements, the actual time-marching step length used in any cycle may be taken as a fraction of the critical value estimated by Eq. (7.61). If particles are connected with certain spring stiffness, the time-marching step length may be limited by the following conditions

$$\Delta t \le 2\gamma\sqrt{\frac{m_{\min}}{k_{\max}^{tra}}} \text{ and} \tag{7.62}$$

$$\Delta t \le 2\gamma\sqrt{\frac{I_{\min}}{k_{\max}^{rot}}} \tag{7.63}$$

where: $m_{\min}$—mass of the smallest particle; $I_{\min}$—inertia moment of the smallest particle; k^{tra} and k^{rot}—translational stiffness and rotational stiffness, respectively; γ—constant, normally $\gamma = 0.1$.

7.5.4 Velocity-Weakening Friction Law

Large scale landslide is one of the most catastrophic geo-hazards that significantly threats human lives and properties, particularly when it exhibits enhanced mobility with extreme velocity and long runout (travel distance). For example, on March 7th 1983, the Saleshan landslide with a volume of $3100 \times 10^4 m^3$ in the north bank of Baxie river (China) exhibited a slide velocity of 20 m/s and a runout of 1050 m (Liao and Jin 1994). The famous Lituya Bay landslide event in 1958 hinted a mean impact velocity 110 m/s (Fritz et al. 2001).

Observations show that the manifestation of enhanced mobility depends strongly on the volume (normally larger than 10^6 m^3) in addition to different but not being generally accepted mechanisms. A widely accepted opinion is, the larger the landslide, the smaller the frictional resistance. Based on the well-documented data for landslides with volume being larger than $10^5 m^3$, Scheidegger (1973) established an empirical formula stating that the apparent coefficient of friction (ratio of the fall height to the runout distance) reduces with the argument of landslide volume. Pudasaini and Miller (2013) explicitly derived landslide mobility as a function of the landslide volume in addition to other physical and geometrical parameters. They found that the landslide mobility is in proportion to the cube root of landslide volume. This volume effect also had been evidenced by Campbell et al. (1995), Davies and McSaveney (1999), Legros (2002).

Although many hypotheses have been put forward to reveal the driving mechanism of enhanced landslide mobility, yet the universally accepted physical mechanism is still a long standing problem. Anyway, it is commonly accepted that the enhanced landslide mobility is tightly related to the reduced basal friction (Dong et al. 2013). Influential factors of friction reduction comprise the dynamic friction degradation of joints and sliding surfaces, lubrication effect of water, liquefaction effect of pore water pressure, aerodynamic effect and impact mechanism of ejected mass, etc.

In the absence of water, Melosh (1979) proposed an acoustic fluidization hypothesis that sliding preferentially occurs when the overburden pressure is relieved by sufficiently strong acoustic waves. Davies and McSaveney (1999) emphasized that fragmentation is an important factor for the enhanced mobility. They proposed a fragmentation-spreading model that the increased overburden pressure in large landslides crushes the sliding mass into high velocity fragments. This fragmentation mechanism causes the rear portion of slip (fall) body to slow down and the front portion to accelerate (Wang and Tonon 2011a, 2011b). More recently, Johnson et al. (2016) numerically verified these hypotheses by DEM.

In the presence of water, Habib (1975) proposed a self-lubrication mechanism for deep seated landslides: if shear rate becomes sufficiently high, frictional heating on the slip surface can transform pore water into vapor, which forms a vapor cushion being able to sustain the sliding mass and facilitate the motion. Goren and Aharonov (2007) further suggested a thermo-poro-elastic mechanism with postulation that the frictional heating at basal shear zone is not high enough to make pore water vaporization but only increasing the pore water pressure. This pressure climbs dramatically to bring down the shear resistance because of the reduced effective normal stress. For the interpretation of landslide volume effect, a depth-dependent permeability law in which the permeability decreases with the depth was claimed that the pore water pressure in large landslides is more difficult to diffuse, as a result, long duration of landslide liquefaction leads to longer runout. Iverson (2005) pointed out that shear contraction is responsible for landslide liquefaction because it reduces the porosity at basal shear zone and consequently raises the pore water pressure. This is equally to infer that the landslide bifurcating behavior mainly depends on the initial soil porosity (Iverson et al. 2000). Use was made of the thermo-poro-elastic model with constant porosity, Goren and Aharonov (2009) emphasized that the landslide bifurcation is resulted from slight changes in permeability that dominates the diffusion rate of pore water pressure.

Since the friction coefficient is assumed to be constant in fundamental DEM, hence it is not able to satisfactorily reproduce the high velocity and long runout behaviors of landslide. According to the state-of-the-arts and the availability of computational tools, the author of this book is apt to attribute all aforementioned mechanism in a phenomenological way, namely, the dependence of friction coefficient on velocity. This is similar to the total stress approach for the landslides and embankment slopes in geotechnical and hydraulic structures, which takes the mixture of landslide mass and interstitial water as the object of study (Chen 2015). The major advantages of this approach are simple, straightforward, and applicable (Liu et al. 2016).

As a pioneering work, McDougall and Hungr made use of the frictional rheology that bulk basal friction coefficient dynamically changes with pore water pressure (McDougall and Hungr 2004). Their frictional rheology model contains two extreme scenarios: bulk basal friction coefficient becomes zero if pore water pressure equals to total bed-normal stress (i.e. full liquefaction), whereas it turns into dynamic basal friction coefficient in the absence of water (i.e. dry landslide). Jop et al. (2006) proposed a $\mu(I)$ rheology model for dense granular flows in which the granular material is postulated as a visco-plastic fluid and the coefficient of friction is dependent on the inertial number I. Although their $\mu(I)$ rheology model can well reproduce the runout distance, it failed to reproduce the observed landslide deposits. Togo et al. (2011) introduced a velocity-displacement dependent friction law that is able to account for shear displacement weakening phenomenon observed in the experiments. Pudasaini and Miller (2013) proposed a mechanical hypermobility model that uses physical and rheological arguments to get a single relationship between landslide volume or inundation area and mobility. Lucas et al. (2014) proposed a velocity-weakening friction law that the coefficient of friction dynamically changes with the sliding velocity. Since this phenomenological law is universal and can be applied to many landslides from small volume to large volume in the solar system, hence it is incorporated into our DEM code such that

$$\mu(v) = \frac{\mu_0 - \mu_w}{1 + v/v_w} + \mu_w \tag{7.64}$$

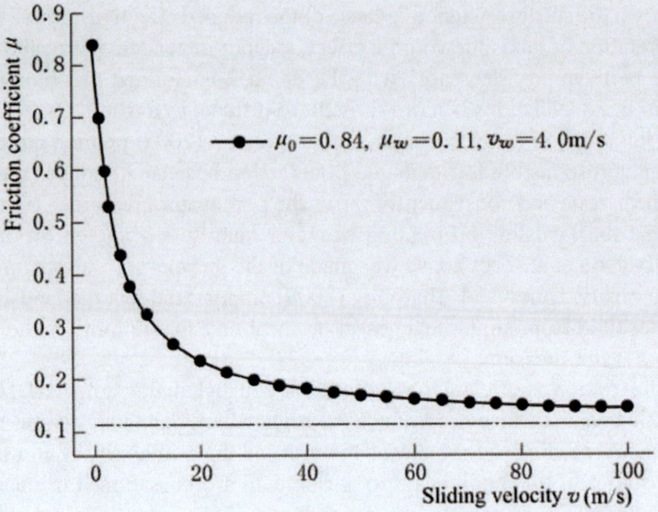

Fig. 7.23 Relationship between sliding velocity and friction coefficient (Lucas et al. 2014)

where: μ—block's friction coefficient; v—sliding velocity; μ_0 and μ_w—static and weakened friction coefficients, respectively; v_w—characteristic velocity.

In Fig. 7.23, a special case of Eq. (7.64) is visualized.

7.6 Validations

7.6.1 Single Block Vibration

A cubic block with size $1 \times 1 \times 1$ m is placed on the horizontal plane ($z = 0$ in Fig. 7.24). At the initial time, the cubic weight is suddenly exerted. The boundary (block-wall contact) stiffness $k_n = 10^6 \text{N/m}^3$, the density of cubic material is $\rho = 2 \times 10^3 \text{g/m}^3$, the damping coefficient α is selectively varied for three cases in Table 7.2.

Figure 7.25 clearly validates that the damping coefficient remarkably influences the collapse time to reach equilibrium state for pseudo-static problems.

7.6.2 Triangular Block Slide

A rigid block slide test is conducted and cross-referenced by the closed-form solution. This triangular block (right angle edges 1.5×1.5 m) slides down along a 45° inclined

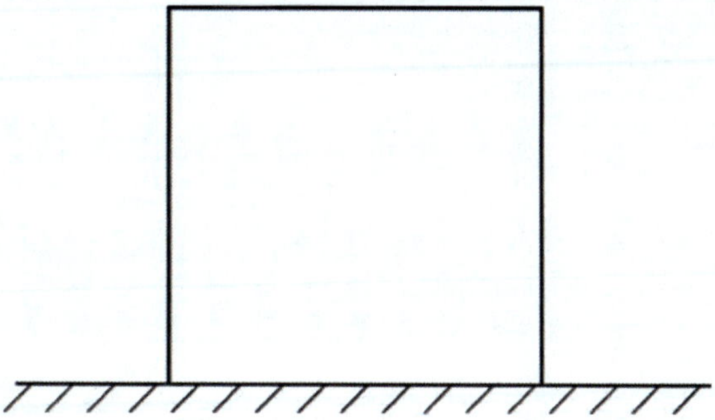

Fig. 7.24 Vibrating block model

Table 7.2 Cases with different damping coefficients

Damping coefficient	Case 1	Case 2	Case 3
α(N·s/m·g)	0.0	1.0	5.0

plane under the action of gravity only. The initial position of the triangular block is depicted in Fig. 7.26 and all the relevant DEM parameters are listed in Table 7.3.

In order to conduct a comparison with the closed-form solution, only the block's motion before hitting the horizontal bottom is presented.

1. **Closed-form solution**

The acceleration of block along the inclined plane is generally expressed by

$$\frac{\mathrm{d}v}{\mathrm{d}t} = g(\sin\theta - \cos\theta\mu(v)) \tag{7.65}$$

in which $\mu(v)$ is given in Eq. (7.64), namely

$$\begin{aligned}\frac{\mathrm{d}v}{\mathrm{d}t} &= g\left\{\sin\theta - \cos\theta\left[\frac{\mu_0 - \mu_w}{1 + v/v_w} + \mu_w\right]\mu(v)\right\}\\ &= g\sin\theta - g\cos\theta\mu_w - \frac{g\cos\theta v_w(\mu_0 - \mu_w)}{v + v_w}\end{aligned} \tag{7.66}$$

Let $A = g\sin\theta - g\cos\theta\mu_w$, $B = g\cos\theta v_w(\mu_0 - \mu_w)$, it may be simply written as:

$$\frac{\mathrm{d}v}{\mathrm{d}t} = A - \frac{B}{v + v_w} \tag{7.67}$$

Separation of variables leads to

$$\mathrm{d}t = \frac{v + v_w}{Av + Av_w - B}\mathrm{d}v = \left(\frac{1}{A} + \frac{B}{A^2}\frac{1}{v + v_w - B/A}\right)\mathrm{d}v \tag{7.68}$$

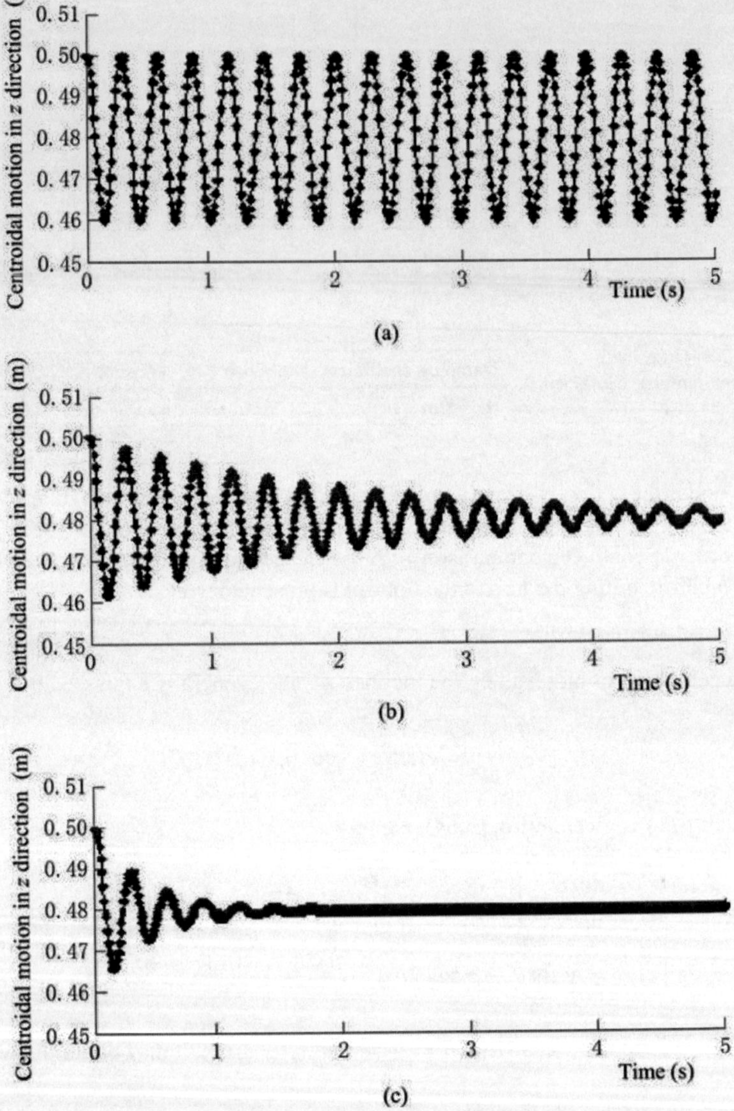

Fig. 7.25 Computation results with different damping coefficients **a** Case 1, $\alpha = 0$; **b** Case 2, $\alpha = 1.0$; **c** Case 3, $\alpha = 5.0$

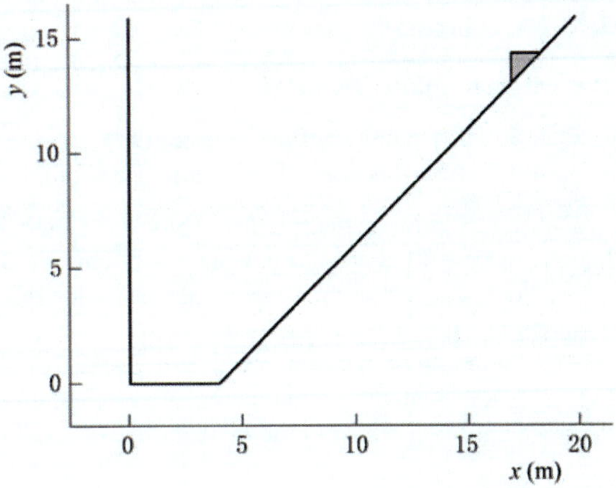

Fig. 7.26 Initial model of the triangular block sliding along the inclined ramp

Table 7.3 Parameters in the problem of triangular block slide

Basic parameters	
Block density (kg/m^3)	2000.0
Block-wall friction coefficient (°)	tan (20/25/30°)
DEM time-marching step length (s)	1.0×10^{-5}
Parameters for the linear contact model	
Block-wall normal stiffness (kN/m)	7.0×10^5
Block-wall shear stiffness (kN/m)	7.0×10^5
Parameters for the friction coefficient related to slip velocity	
μ_0	tan (20°/25°/30°)
μ_w	tan (5°)
v_w (m/s)	4.0

Multiple the two sides of Eq. (7.68) by A^2, we get

$$\left(A + B\frac{1}{v + v_w - B/A}\right)\mathrm{d}v = A^2\mathrm{d}t \tag{7.69}$$

Integration of Eq. (7.69) results in the equation towards the solution of sliding velocity

$$Av + B\ln|v + v_w - B/A| = A^2t + C \tag{7.70}$$

Taking into account the initial condition $v = 0$ when $t = 0$, the integration constant C in Eq. (7.73) is specified as

$$C = B\ln|v_w - B/A| \tag{7.71}$$

2. **DEM computation**

Two cases are studied by DEM:

- Case 1. The friction coefficient is constant.
- Case 2. The friction coefficient is related to sliding velocity according to Eq. (7.64).

Figure 7.27 exhibits the calculated block's velocity in the slope direction. It varies linearly in Case 1 but nonlinearly in Case 2. This is attributed to the fact that in Case 1 the block's acceleration keeps constant. After the activation of Eq. (7.64), the variation of friction force comes into effect.

The variation of friction coefficient against sliding velocity is illustrated in Fig. 7.28. It is evident that as the increase of sliding velocity, the friction coefficient decrease until it approaches the residual friction coefficient.

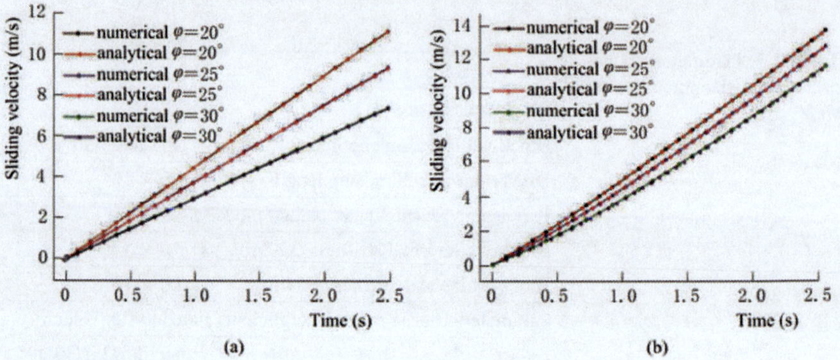

Fig. 7.27 Comparison of landslide velocities between DEM computation and closed-form solution. **a** Case 1; **b** Case 2

Fig. 7.28 Evolution of slide block friction angle in Case 2 (DEM computation)

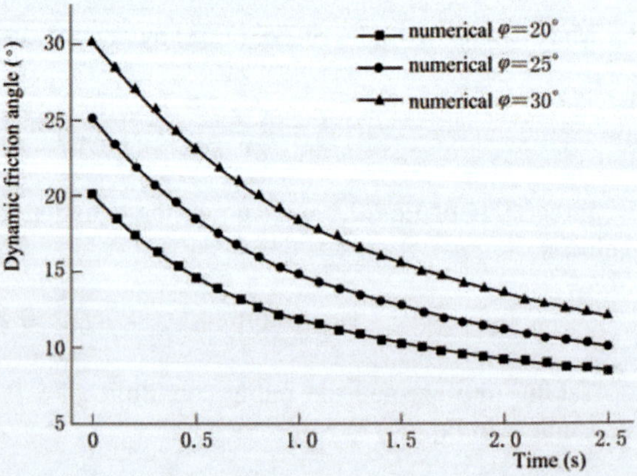

7.6.3 Disc Compaction Test

Figure 7.29 is a photo-elastic analysis model of nine-disc assembly (Cundall and Strack 1979). These discs are initially loosely packed in a square, then four rigid walls uniformly move towards each other compressing the disc assemblage, to investigate the variation of the normal force at contact point C.

The relevant parameters for the DEM computation are listed in Table 7.4. It is emphasized that the contact normal stiffness and shear stiffness are two times that of Cundall and Strack (1979) because the linear contact model is used in our study (Tan and Chen 2017).

High loading speed (Case 1) and low loading speed (Case 2) are investigated. For the former the loading speed is 0.36 m/s and the loading process stops after 40 cycles, whereas for the latter the loading speed is 0.12 m/s and the loading process stops after 120 cycles.

The normal force at contact point C is plotted in Fig. 7.30. Our results show a good agreement with that of Cundall and Strack (1979):

- The normal force at contact point C increases continuously until the end of loading, then it oscillates and finally reaches the equilibrium state;
- The normal force increasing rate (i.e. the slope of curve) in high loading speed Case 1 is much greater than that in low loading speed Case 2, hence the former exhibits larger oscillations at the end of loading cycles;
- Despite the loading speeds in these two cases are different, the final equilibrium forces are identical;

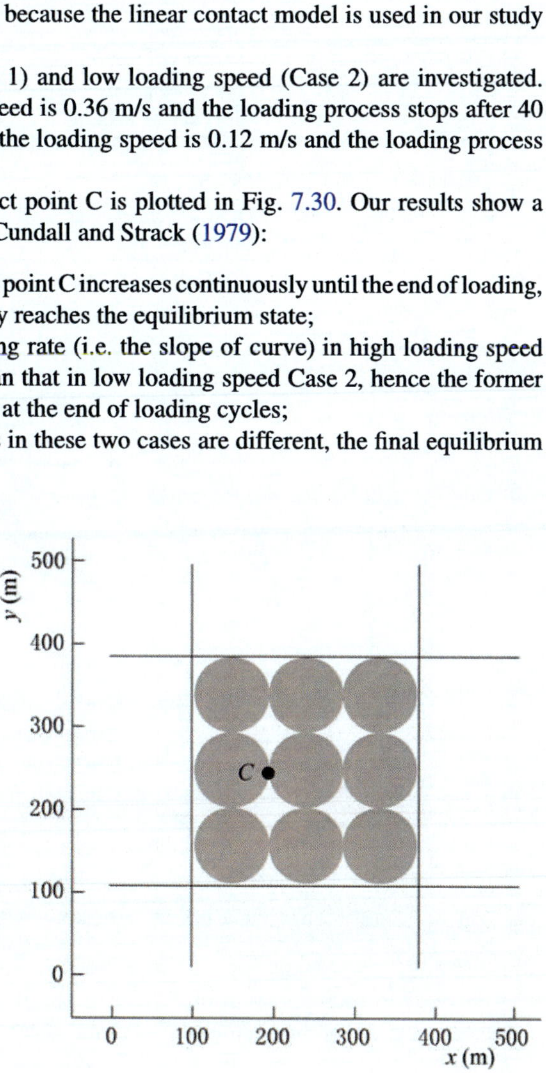

Fig. 7.29 Initial configuration of nine-disc assemblage model

Table 7.4 DEM parameters for the nine-disc test

DEM parameters	Author	Cundall and Strack (1979)
Density (kg/m^3)	1000.0	1000.0
Disc radius (m)	50.0	50.0
Disc-to-disc normal stiffness $(10^5$ kN/m)	3.0	1.5
Disc-to-disc shear stiffness $(10^5$ kN/m)	3.0	1.5
Disc-to-wall normal stiffness $(10^5$ kN/m)	7.0	/
Disc-to-wall shear stiffness $(10^5$ kN/m)	7.0	/
Damping coefficient (N·s/m·g)	2.0	/
DEM time-marching step length (s)	0.04576	0.04576

N.B. symbol "/" denotes that the corresponding parameter is not given in the paper of Cundall and Strack (1979)

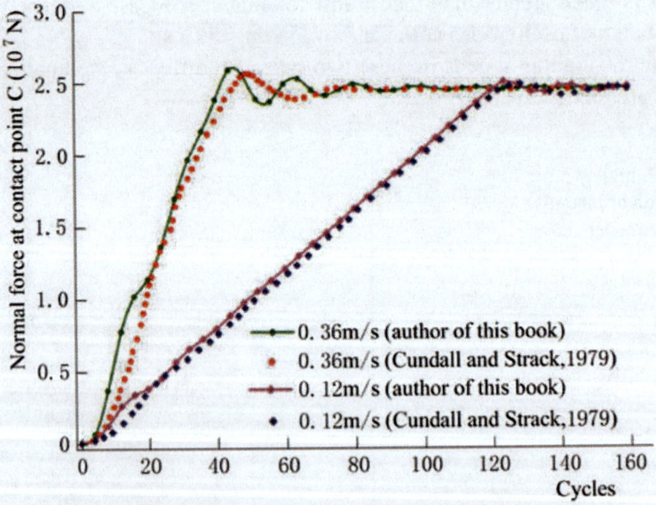

Fig. 7.30 Numerical results for nine-disc test with different load speeds

– The slightly discrepancy in high loading speed Case 1 can be explained from two aspects. First, the normal stiffness and shear stiffness of walls had not been provided in the benchmark literature (Cundall and Strack 1979), hence their values have to be postulated in our study. Second, as an important effect on the energy dissipation of granular assemblage, the damping coefficient had not been given in the benchmark literature, too.

7.6.4 Cylinder Collapse Test

The six-layer collapse test (viz. Fig. 7.31) was conducted using solid aluminum cylinders regularly deposited in an acrylic resin tank with the size of 26 cm(length) × 10 cm (width) × 26 cm (height) (Zhang et al. 2009). These cylinders possess the density of 2700 kg/m^3, the diameter of 1.0 cm, and the length of 9.9 cm. The initial conditions were imposed by a vertical valve 6 cm from the left wall of the tank, which is removed at the beginning of test allowing solid cylinders to collapse under the action of gravity.

This physical test was previously adopted to validate DEM code by Canelas et al. (2016) in which the cylinders are represented by solid discs and the tank is modeled by walls. All the relevant parameters for the computation using DEM are listed in Table 7.5. It is noted that the simplified Hertz-Mindlin contact model (viz. Eqs. (7.46–7.48) is utilized in our study.

Suppose that the discs are initially loosely packed and there are no contact forces existed on. In order to reach a state of equilibrium, the valve is removed at 1.0 s after the beginning of the calculation to give enough time for discs to settle down under the action of gravity. Afterwards the time is re-initialized as $t = 0.0$ s at the instant when the valve is removed.

The evolution of disc deposition pattern is exposited in Fig. 7.31. In general, numerical results are in good accordance with the experiment data, indicating that the collapse of solid cylinders can be well replicated by our DEM code.

The positions of the disc assemblage can be calculated by

$$\begin{cases} x = \dfrac{1}{n_d} \sum\limits_{i=1}^{n_d} x_{d_i}^0 \\ y = \dfrac{1}{n_d} \sum\limits_{i=1}^{n_d} y_{d_i}^0 \end{cases} \tag{7.72}$$

where n_d is the amount of discs and $(x_{d_i}^0, y_{d_i}^0)$ is the mass center position of disc d_i. These mass center positions are quantitatively cross-referenced by the experimental data in Fig. 7.32. It is clear that the computational results agree well with the experimental data. More specifically, the mass center position of the disc system in the horizontal direction gradually increases meanwhile that in the vertical direction gradually decreases. This is not surprising and can be explained from the fact that after the removal of valve, the solid cylinders start to collapse under the action of gravity and continue to move toward the right wall of tank along the horizontal tank bottom (viz. Fig. 7.31).

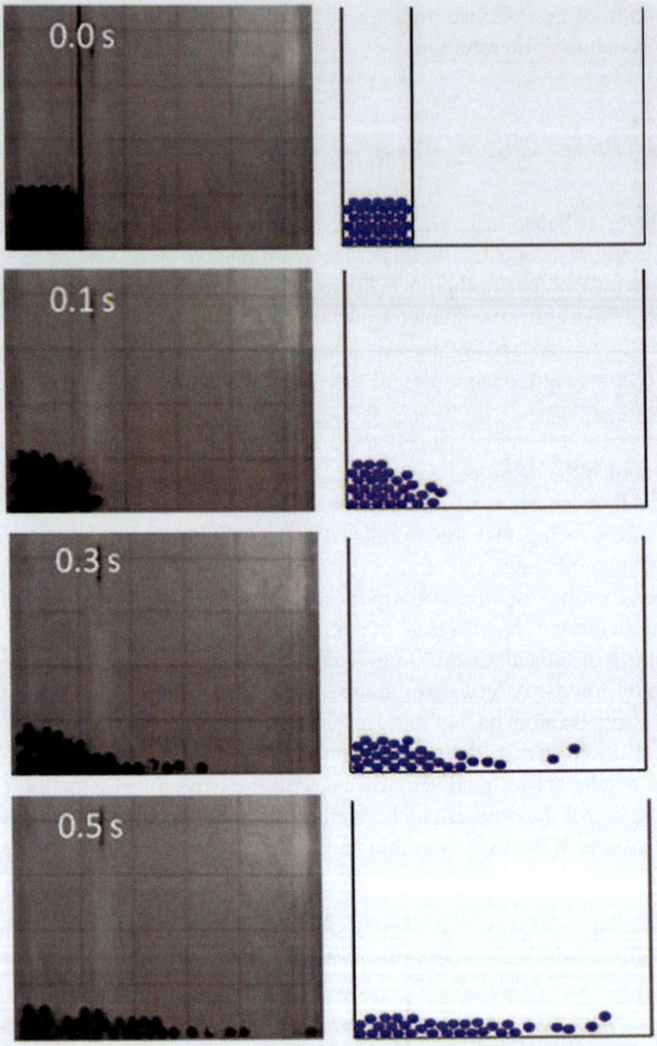

Fig. 7.31 Disc deposition patterns at $t = 0.0$, 0.1, 0.3 and 0.5 s. Left column: experiment (Zhang et al. 2009); Right column: DEM computation (Tan and Chen 2017).

The kinetic energy and potential energy of disc assemblage are extracted to give a deeper insight into the collapse of solid cylinders from an energy perspective. In our study, both the kinetic energy and potential energy are calculated under the assumption that the cylinders are discs with unit width in 2-D DEM framework. According to Shan and Zhao (2014), the dissipated energy E_d in the dry granular flow can be calculated by

Table 7.5 DEM parameters for the test of cylinder layer collapse (Canelas et al. 2016)

DEM parameters	Values
Density (kg/m^3)	2700.0
Disc radius (m)	0.005
Disc Young's modulus (kN/m^2)	69×10^6
Disc Poisson's ratio	0.30
Disc-to-disc friction coefficient	0.45
Wall Young's modulus (kN/m^2)	30×10^5
Wall Poisson's ratio	0.30
Disc-to-wall friction coefficient	0.45
Damping coefficient (N·s/m·g)	0.0
DEM time-marching step length (s)	1.0×10^{-5}

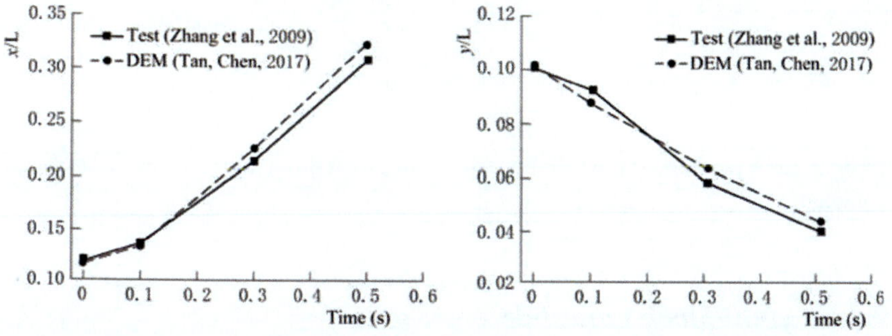

Fig. 7.32 Comparison of the mass center positions between experiment and DEM outcomes (*L* is the length of tank). **a** *x*-direction; **b** *y*-direction

$$E_d = \Delta E_{p0}^s - E_{ki}^s, \tag{7.73}$$

in which ΔE_{p0}^s is the potential energy change and E_{ki}^s is the total kinetic energy of the system.

The energy evolution in the test is demonstrated in Fig. 7.33. The total potential energy decays steadily after the removal of valve due to the collapse of solid-cylinder under the action of gravity and subsequently, the potential energy is turned into kinetic energy allowing the solid cylinders to move toward the right wall of tank. The total kinetic energy climbs slightly at the initial stage. Afterwards, it gradually decays to zero at about 1.2 s corresponds to the moment when the solid cylinders are stopped on the horizontal tank bottom. Meanwhile, the dissipated energy increases gradually until reaching a stable value. This indicates that the energy is totally dissipated out through the friction and collision between disc-to-disc and disc-to-wall contacts in the process of collapsing and moving. Another important aspect should not be neglected is that the total kinetic energy is much smaller compared to the dissipated energy. This means that only a small portion of potential energy is turned into kinetic energy

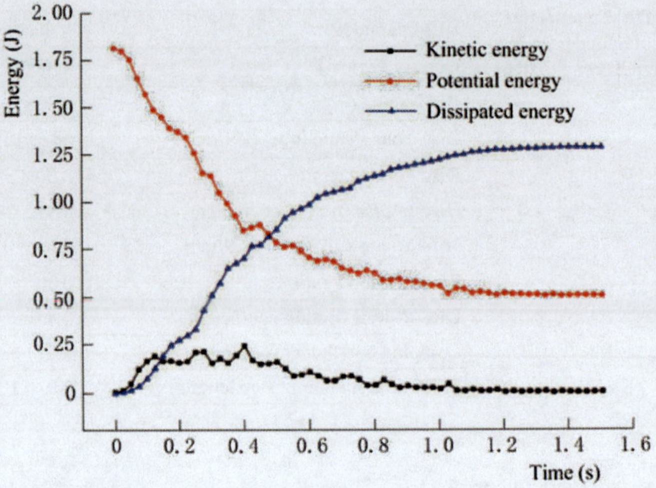

Fig. 7.33 Energy evolution in six-cylinder-layer collapse test

while a large portion of potential energy is dissipated out through the friction and collision.

7.6.5 Multi-block Landslide

Slope failure (sliding or toppling) that subsequently falls into reservoir is one of the destructive geological hazards in hydraulic engineering. In order to alleviate its adverse consequences, an effective tool to replicate the entire failure process and take into account of the landslide-water interaction is demanded. The actual landslide is hard to be well observed and only limited in-situ field data are available, numerical simulation, nevertheless, may provide an alternative way to this target.

The initial configuration of DEM model to study a landslide with multi-blocks is depicted in Fig. 7.34. All the relevant DEM parameters are listed in Table 7.6. The sliding mass is discretized into 12 blocks, the rightmost block is marked as Block 1 and the leftmost block is marked as Block 12. Two cases (scenarios) are investigated for this landslide.

Case 1: Basic strength parameters.
Case 2: Sliding velocity and friction coefficient is related by Eq. (7.64).

In the study, the coupling action of block/water is not taken into account, but it will be further addressed in the Chap. 11 of this book.

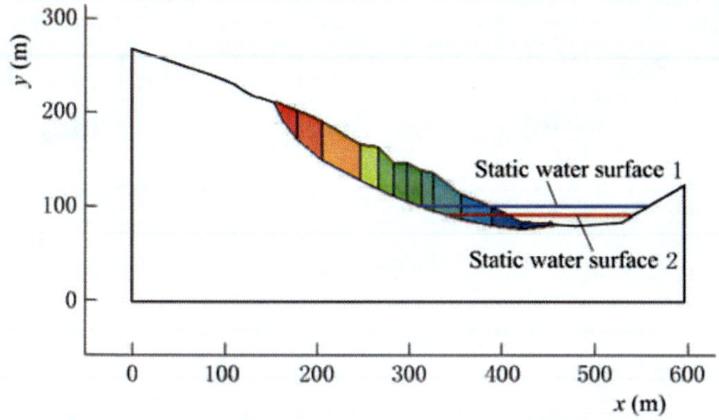

Fig. 7.34 Model set up of the multi-blocks landslide

Table 7.6 DEM parameters for the multi-blocks landslide

Basic parameters	
Block density (kg/m^3)	2000.0
Block-to-block friction coefficient	tan (20°)
Block-to-wall friction coefficient	tan (20°)
DEM time-marching step length (s)	1.0×10^{-4}
Parameters for the simplified Hertz-Mindlin contact model	
Young's modulus (kPa)	700,000.0
Poisson's ratio	0.3
Parameters for the velocity-weakening friction law	
μ_0	tan (20°)
μ_w	tan (10°)
v_w (m/s)	4.0
Parameters for the linearly distributed hydrostatic pressure calculation	
Water density (kg/m^3)	1000.0
Static water surface elevation (m)	100.0/90.0

1. **Case 1**

The landslide moves along the slip surface under the action of gravity only. The landslide deposits at different instants are illustrated in Fig. 7.35 and there is no excessive penetration between blocks or boundaries. Each block slides and rotates freely and thus allows for being disengaged from each other. At the instant of 40.0 s, the front portion of the landslide mass has not reached the opposite river bank yet.

Block runout is extracted in Fig. 7.36. In general, the runout of right block is greater than that of left block. This means that the front portion of the landslide mass drags the rear portion to move downwards due to the newly created exposure face.

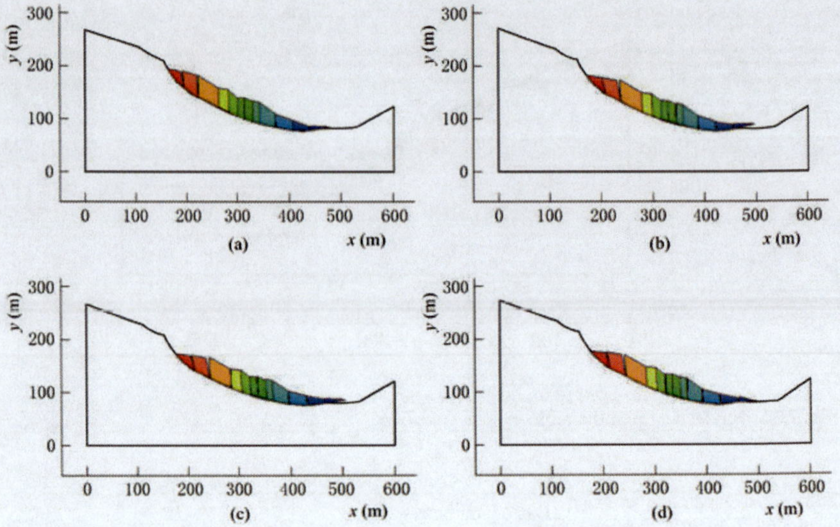

Fig. 7.35 Evolution of the landslide deposits (Case 1). **a** $t = 10.0$ s; **b** $t = 20.0$ s; **c** $t = 30.0$ s; **d** $t = 40.0$ s

The block runout increases rapidly first, then it slows down and tends to stop at the river bed in accordance with the fact that the slip surface becomes more and more flat.

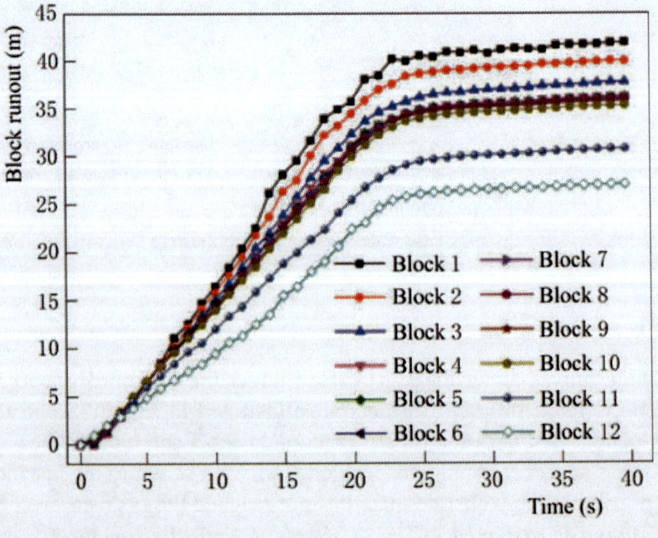

Fig. 7.36 Block runout versus time (Case 1)

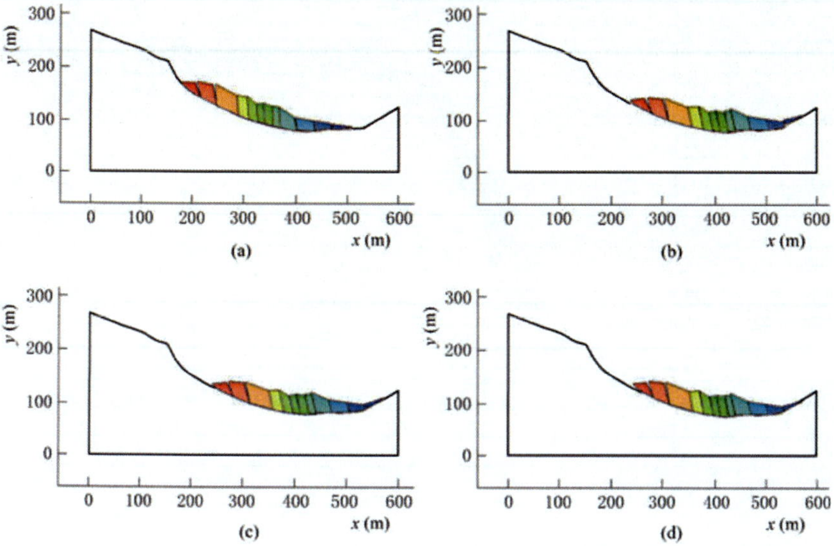

Fig. 7.37 Evolution of the landslide deposits (Case 2). **a** $t = 10.0$ s; **b** $t = 20.0$ s; **c** $t = 30.0$ s; **d** $t = 40.0$ s

2. **Case 2**

Due to the reduced friction on the slip surface and less dissipated energy, the landslide moves faster and shows certain hypermobility characteristics. In Fig. 7.37, the front portion of the landslide mass hits and subsequently climbs onto the opposite bank. Afterwards, the landslide decelerates and slows down due to the obstacles of the opposite river bank.

Block runout in Case 2 is much greater than that in Case 1. It is clear that a higher velocity and longer runout of the landslide can be well replicated by the activation of velocity-weakening friction law. The block's runout is almost unchanged after about 20.0 s, which indicates that the landslide mass has been settled down. This also can be confirmed from the landslide deposits pattern illustrated in Fig. 7.38.

The variation of block friction angle is elucidated in Fig. 7.39. In general, the block friction angle decays after the landslide onset until reaches minimum. Afterwards, it increases steadily to the static value. According to the relationship between sliding velocity and friction coefficient, it is evident that the landslide initially accelerates down to the riverbed under the action of gravity, then it decelerates because the local topography hinders its motion, especially after the front portion of the sliding mass hits the opposite river bank.

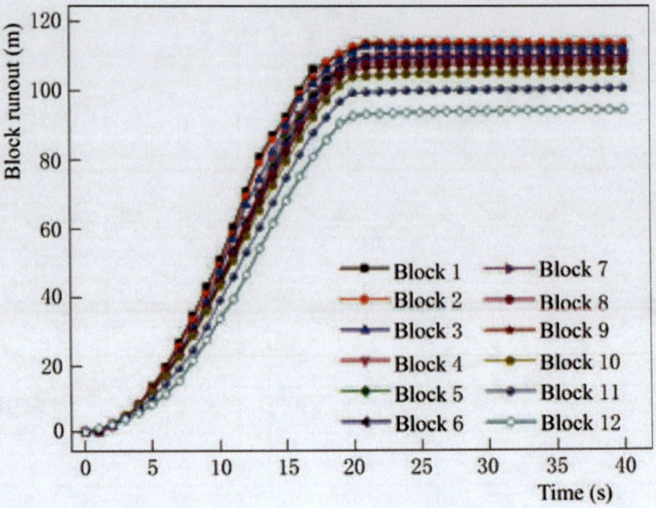

Fig. 7.38 Block runout versus time (Case 2)

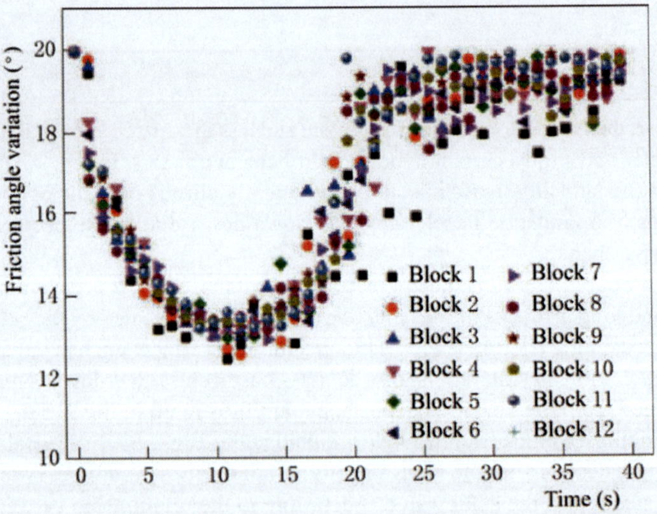

Fig. 7.39 Evolution of friction angle (Case 2)

7.6.6 Multi-particle Landslide

The initial configuration of the landslide in Fig. 7.34 is employed to establish a 2-D PFC model that comprises 3396 discs with uniform radium: first, the landslide is discretized into 12 blocks (viz. Fig. 7.34), then each block is filled with discs according to the specified porosity of slide body. The parameters used in the computation are listed in Table 7.7. Three cases (scenarios) are investigated.

Case 1: Normal case without taking into account of rotation friction and velocity related friction.
Case 2: Rotation friction is considered [viz. Eqs. (7.25–7.26)].
Case 3: High-speed and long-runout landslide due to the consideration of velocity dependent friction (viz. Eq. (7.64)).

1. **Case 1**

The velocity distribution is shown in Fig. 7.40. In the first phase of acceleration when $t \leq 20.0$ s, the leading edge has reached the opposite river bank that is much farther than the multi-block counterpart (cf. Fig. 7.35). This is attributable to the fact that discs are easier to rotate than polygons and in Case 1 the rotation friction is not taken into account (i.e. rotation friction coefficient is null).

In Fig. 7.41 the configurations of landslide at different time are presented. It is found that they only exhibit negligible change during $t = 15.0$ s and $t = 20.0$ s, indicating that the landslide approaches to stand still.

Table 7.7 Computational parameters for granular flow landslide

Basic parameters	
Density (kg/m^3)	2000.0
Disc radium (m)	0.8
Porosity	0.2
Disc-to-disc friction coefficient	tan (20°)
Disc-to-wall friction coefficient	tan (20°)
Amount of discs	3396
Damping coefficient (N·s/m·g)	0.0
Rotation friction coefficient	0.0/0.2
Time-marching step length (s)	1.0×10^{-3}
Linear contact parameters	
Disc-to-disc normal stiffness (10^5kN/m)	3.0
Disc-to-disc shear stiffness (10^5kN/m)	3.0
Disc-to-wall normal stiffness (10^5kN/m)	7.0
Disc-to-wall shear stiffness (10^5kN/m)	7.0
Parameters relating friction and velocity	
μ_0	tan (20°)
μ_w	tan (10°)
v_w(m/s)	2.0

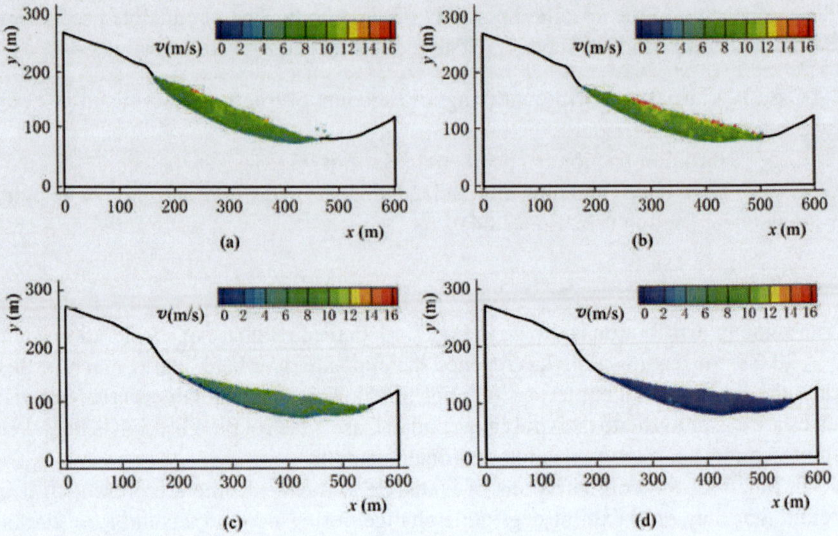

Fig. 7.40 Evolution of landslide velocity (Case 1). **a** $t = 5.0$ s; **b** $t = 10.0$ s; **c** $t = 15.0$ s; **d** $t = 20.0$ s

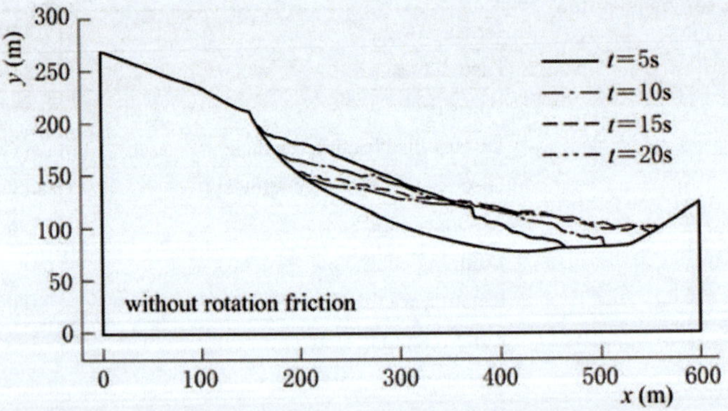

Fig. 7.41 Evolution of landslide deposits (Case 1)

2. **Case 2**

Let the rotation friction coefficient be 0.2. According to Fig. 7.42, after the consideration of rotation friction, the slip velocity and runout distance are both reduce. The leading edge is deposited at the river bed, similar to the multi-block landslide (cf. Fig. 7.35).

Figure 7.43 shows that after the consideration of rotation friction, the configurations of the landslide at different time are rather similar.

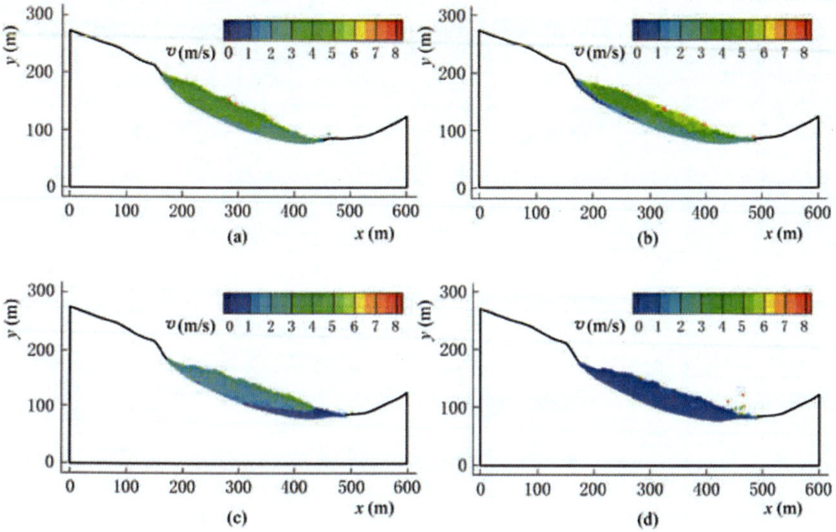

Fig. 7.42 Evolution of landslide velocity (Case 2). **a** $t = 5.0$ s; **b** $t = 10.0$ s; **c** $t = 15.0$ s; **d** $t = 20.0$ s

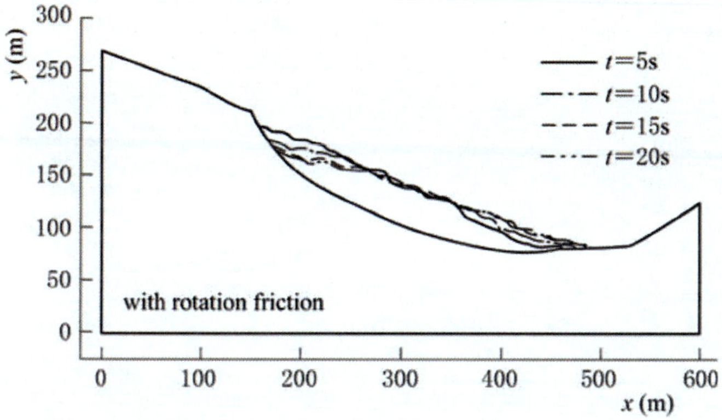

Fig. 7.43 Evolution of landslide deposits (Case 2)

3. **Case 3**

The velocity dependent friction in Eq. (7.64) is activated. The comparison between Figs. 7.44 and 7.40 validates that Case 3 exhibits higher slip velocity and longer runout distance. The leading edge climbs up to the opposite river bank.

In Fig. 7.45, we find that the configurations of the landslide at $t = 15.0$ s and $t = 20.0$ s are similar. After the stop of motion, the deposit of landslide is rather uniformly spread out on the river bed with nearly horizontal surface.

Figure 7.46 exposits the variation of friction angle at different times.

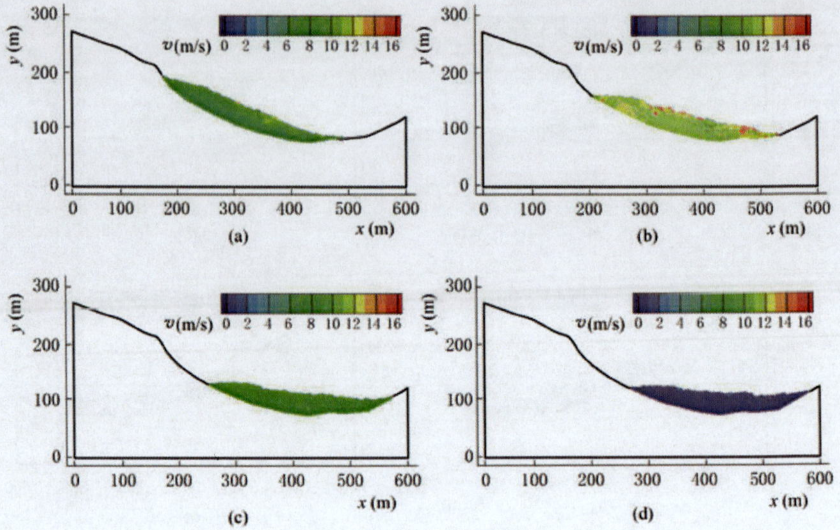

Fig. 7.44 Evolution of landslide velocity (Case 3). **a** $t = 5.0$ s; **b** $t = 10.0$ s; **c** $t = 15.0$ s; **d** $t = 20.0$ s

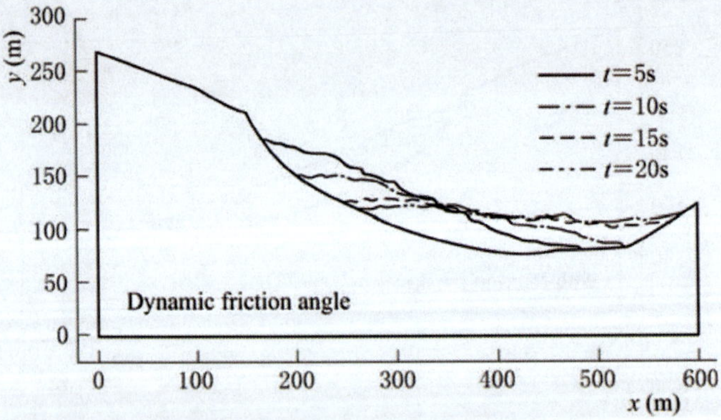

Fig. 7.45 Evolution of landslide deposits (Case 3)

4. **Deposit of landslide body**

Figure 7.47 collects the deposits of three cases at different times.

7.7 Concluding Remarks

As an important family of advanced CMs, the "discrete element methods (DEMs)" treat geomaterial as an assembly of separate particles or blocks. In contrast to the

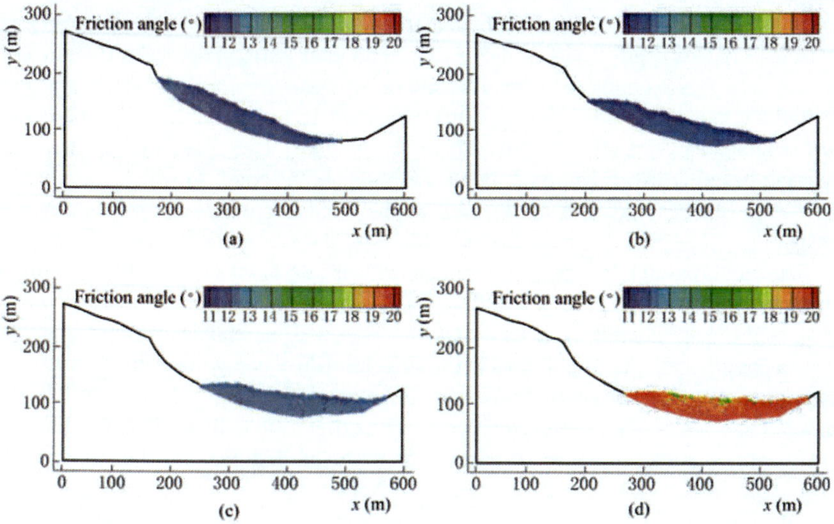

Fig. 7.46 Evolution of friction angle (Case 3). **a** $t = 5.0$ s; **b** $t = 10.0$ s; **c** $t = 15.0$ s; **d** $t = 20.0$ s

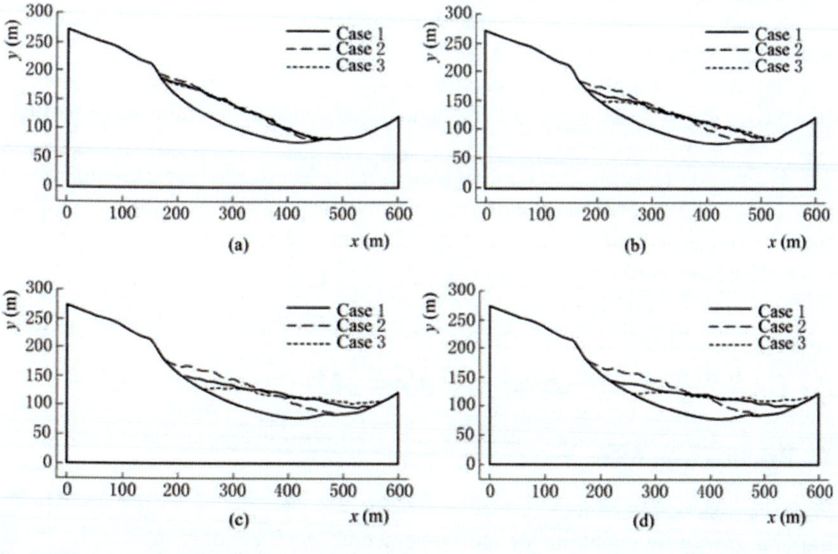

Fig. 7.47 Evolution of landslide deposits. **a** $t = 5.0$ s; **b** $t = 10.0$ s; **c** $t = 15.0$ s; **d** $t = 20.0$ s

continuum approach, DEMs are straight forward: the particles are directly handled as individual elements, they interact through local and pairwise contacts and observe the Newton's laws of motion (Kozicki and Donzé 2009; Gray et al. 2014; Coetzee 2017).

The particle-based "distinct element method (DEM)" (Cundall 1971; Cundall and Strack 1979; Potyondy and Cundall 2004; Zhao et al. 2015) is prevalently adopted to simulate geomaterials with granular micro-or meso-structures in which adjacent particles contact or bond with each other. Crack nucleation and fracture propagation are simulated by breaking bonds. There are typically two types of bonds in the particle-based DEM: the contact bond and the parallel bond, to which the normal and shear stiffness coefficients as well as the friction coefficient are customarily assigned.

The block-based DEM discretizes geomaterials into blocks using a finite number of structural planes (e.g. joints, faults). Each block may be internally sub-divided. Adjacent blocks interact in the normal direction with an elastic stiffness together with a possible tensile strength and in the tangential direction with an elastic stiffness together with a shear strength. In the standard block-based DEM, the failure of block assemblage can only be captured either in terms of plastic yield or displacements of pre-existing structural planes. Therefore, new fractures cannot be driven within the continuous rock block (Lisjak and Grasselli 2014).

DEM employs the explicit time-marching step scheme identical to that used in the explicit DFM for continuum analysis, hence it possesses a remarkable characteristic that the time-marching step length must be so small that, during a single time-marching step, disturbances due to, for example, contact actions, are sufficiently small. It is particularly important in simulating the nonlinear performance of assemblage system containing a large number of particles.

Although DEM is relatively computationally intensive, yet it possesses a remarkable advantage of parallel processing capability to scale up the number of particles or the length of simulation.

It is also easy to formulate coupling models using, such as hybrid DEM-FDM (Cai et al. 2007), FEM-DEM (Munjiza 2004; Latham et al. 2008; Ma et al. 2014; Mahabadi et al. 2010; Lisjak et al. 2014) and DEM-SPH, the last one will be addressed in the Chap. 11 of this book.

7.7.1 Relation with Other Prevalent CMs

1. **Relation with BEA**

With the assumption of face-to-face contact between blocks, in spatial domain, the implicit governing equations of the fundamental block element method or "block element analysis (BEA)" are formulated by the consideration of force and moment equilibrium condition, deformation compatibility condition, and constitutive relation on structural planes (Chen 1987). In the time domain, the equation set is implicitly and step-wisely solved to reproduce quasi-static/dynamic and non-linear performance of the whole block assemblage system. With the overlay technique using finite elements of p-refinement, BEA can tackle the complex deformation pattern in block system rather well (Chen et al. 2004a, b).

BEA is able to easily handle various practical issues such as the coupling of fluid flow across rock joints, the simulation of reinforcement mechanism, the structural dynamic response under the action of seismic shake, etc. It is also beneficial from the clear and accessible mechanical parameters which permit the experimental evaluation for a specific engineering case. Take the structural dynamics for example, the elastic viscous damping, in addition of friction damping, may be easily introduced and assigned in a similar manner of fundamental FEM.

2. **Relation with RBSM**

In the "rigid body spring model (RBSM)" (Kawai 1978), a domain is discretized into polyhedral elements by structural planes. These elements are rigid bodies interconnected through their contact faces by putting some springs between them to describe the shear or tension of structural planes. In the other words, the main assumption of RBSM is that a continuum medium can be represented by a collection of nodal masses interconnected by spring elements. When a Voronoi tessellation is adopted for both the mesh design and the scaling of spring stiffness, homogeneous isotropic materials may be simulated rather well, too.

RBSN may be extended to simulate rock fracturing. But the rotation/detachment process accompanied by other contact types that is routine in DEM, is not very well handled insofar.

3. **Relation with DDA**

The "discontinuous deformation analysis (DDA)" (Shi and Goodman 1985, 1989) for the deformation and motion of multi-block system in which rock blocks are postulated as "simply deformable", namely, polynomials are employed for the approximation of displacement field within rock blocks. The implicit governing equation set w.r.t. space is solved by taking into account of the interaction between blocks via various possible contacts.

In contrast to DEM that treats the kinematics of each block separately using an explicit time-marching scheme, DDA is basically implicit. It uses displacements as basic variables solved by a set of linearized equations that should be updated following time-marching step.

In addition to artificial damping term to dissipate energy for quasi-static problems, DDA is able to incorporate other mechanisms for energy loss easily.

4. **Relation with MD**

Even though DEM is very closely related to "molecular dynamics (MD)" (Alder and Wainright 1957; Gibson et al. 1960; Rahman 1964; Stillinger and Rahman 1974), they are generally distinguished by the inclusion of rotational DOF and complicated particle geometries.

7.7.2 Remarkable Developments and Applications

1. **State-of-the-arts in applications**

During the past decades, DEM has been proved to be a reliable tool for the study on the behavior of granular materials in micro-scale, meso-scale and even macro-scale.

Today, DEM is widely accepted as an effective computational tool for addressing engineering problems with granular and discontinuous materials, such as granular flows (Fillot et al. 2007) and rock failure motions (Fakhimi et al. 2002; Diederichs and Kaiser et al. 2004; Shimizu et al. 2010). Particularly, DEM simulations have been widely used in large scale geophysical applications such as landslides (Cleary and Campbell 1993; Campbell et al. 1995; Tan and Chen 2017; Tan et al. 2018a, 2018b). In addition, particle dampers are seen to be the hot point towards the quantitative description of dynamic process of architectural structure, because they present appealing dissipation properties and can be used in harsh environments (Ibrahim 2009; Matuttis and Chen 2014; Gagnon et al. 2019).

The application of PFC to geomaterials is still undergoing substantial development. The initial application of PFC (using discs or balls) to geomaterials was in soil mechanics w.r.t. granular flows (Cundall and Strack 1979; Walton 1984). Afterwards, it has been used to examine a variety of geologic processes inclusive the deformation of fault gouge (Morgan and Boettcher 1999; Aharonov and Sparks 2004; Guo and Morgan 2006, 2007; Abe and Mair 2005), clay smear (Egholm et al. 2008), volcanic spreading (Morgan and McGovern 2005), caldera collapse (Hardy 2008), large-scale geodynamic modeling for complex tectonic systems (Strayer and Suppe 2002; Vietor 2003; Finch et al. 2004; Strayer et al. 2004; Cardozo et al. 2005; Naylor et al. 2005; Benesh et al. 2007; Hardy et al. 2009; Miyakawa et al. 2010), etc.

To handle the issue of multi-parametrical evaluation, the "bulk calibration approach (BCA)" is prevalently employed, which is actually derived from the concept of "numerical materials" (e.g. the numerical concrete and numerical rock in Chap. 2). It does not necessarily assume a strong link between the DEM parameters and bulk material properties. On the contrary, the parameters are rather treated as adjustment factors even though the models were developed by giving the parameters certain physical meanings (Marigo and Stitt 2015). In the first step of this approach, field or laboratory experiments are performed to measure specific bulk properties. Then, the experiments are repeated numerically, following the same setup and procedures as closely as possible to the laboratory experiments, by which the model parameters are iteratively adjusted until the computed bulk behavior is matched to laboratory data with certain degree of accuracy.

The disadvantage of BCA is that more than one set of parameters will produce the same bulk response for a given laboratory experiment (Combarros et al. 2014). It is also possible that when the parameter set is calibrated using a specific experiment, the same parameter set might not provide the correct bulk behavior in modeling another experiment or application (Benvenuti et al. 2016). Hence more laboratory experiment should be conducted, and each experiment should isolate a single parameter for which its value can be determined (Jensen et al. 2014; Simons et al. 2015), or the combined results from more than one experiment should provide a unique set of parameter values. Such calibrated parameter values will compensate for other aspects (e.g. particle size and shape, contact model, etc.) not being accurately modelled (Yoon 2007; Markauskas and Kacianauskas 2011).

BCA is often used in combination with the “direct measuring approach (DMA)”. Parameters such as particle size and density might be measured directly while other parameters such as the particle-to-particle friction coefficient might be determined using a reverse calibration process by BCA. For example, it is impossible to link the rotational friction parameter to the non-sphericity of particles and therefore this is a purely empirical parameter that cannot be determined by direct measurement (Benvenuti et al. 2016).

2. **Important issues for further exploration**

If the constitutive laws match the desired objective, we are able to set up one or more particle-based models. Among the key issues to be addressed in the preparation of model, the first consideration is the model resolution. In addition, the way w.r.t. how the structural planes (e.g. faults and joints) are represented in geomaterial models, and how the boundary conditions are imposed, etc., are also important.

(1) **Model resolution and computation cost**

A basic tenet of CMs is that an inherent discontinuum at small scale can be modeled as a continuum when viewed at sufficiently large scale. The difficulty lies in determining which properties can be safely averaged and which properties must be explicitly specified in the model. Typically, this is achieved through trial-and-error (sensitivity) studies to determine if the included properties produce the expected behavior. Practically speaking, the particles in the model need to be at least one-third characteristic size expected to be resolved. This constraint is attributable to the requirement of REV and can result in a large computation effort for each time-marching step, leading to long run times. This highlights the inherent struggle between model resolution and run time.

Today, DEM plays leading role in simulating the dynamics of granular media. However, run-time efficiency is still a decisive restraint factor in large scale applications, therefore the research and investigation for alternative algorithms attaining good balance between costs and benefits, are of paramount importance.

(2) **Geometrical description of elementary particles**

It is notable that in reality, many complex-shaped particles with highly interlocked grain structures are poorly described by discs/spheres. This often leads us failing to capture the essential aspects of mechanical behavior. As a result, after 1990, efforts have been made to incorporate complex-shaped particles in the DEM implementation (Lu et al. 2015). Such as the method illustrated in this book (cf. Fig. 7.12), where a 2-D block is represented by a set of discs/spheres surrounding the edges of block (Favier et al. 1999; Kruggel-Emden et al. 2008a, 2008b; Ren 2014; Marigo and Stitt 2015; Qiu et al. 2017). However, many discs/spheres may be required to obtain a reasonable approximation of a real particle’s shape.

Nowadays, the analytical particle shapes available include: ellipses (Ting et al. 1993; Lin and Ng 1995; Lin and Ng 1997; Shodja and Nezami 2003), super-quadratics (Williams and Pentland 1992), polygons (Barbosa 1990), polyhedrons (Barbosa 1990), connected circular segments (Potapov and Campbell 1998). The “bonded particle model (BPM)" (Potyondy and Cundall 2004) or "clumped particle model (CPM)" (Cho et al. 2007) is actually an important advancement in which a group of particles are glued together as a single rigid body (Weng and Li 2012). Clumped particles emerge an irregular shape but move as a deformable and non-breakable body. In the commercial code PFC, spheres are clumped together to create irregularly shaped clusters (Markauskas et al. 2010). Apart from discs/spheres, particles in other shapes have been adopted in the development of BPM, too (Ferelleca and McDowell 2010; Feng et al. 2012; Fang et al. 2015; Smeets et al. 2015; Yan et al. 2015).

One appealing aspect of PFC-based BMP is that the discrete nature of particle assemblage effectively reproduces the phenomenology of natural or laboratory geomaterials. For example, specific configurations and load paths can be generated to replicate many constitutive behaviors such as elasticity and plasticity accompanied with dilation (changes in volume). In this manner, BMP can provide unique insights into the micro-and meso-mechanics associated with observed behaviors (Trent et al. 1987; Morgan and Boettcher 1999; Morgan 1999, 2004; Aharonov and Sparks 2004; Potyondy and Cundall 2004; Hadjigeorgiou et al. 2009; Schöpfer et al. 2009; Esmaieli et al. 2010; Zhang et al. 2011; Scholtès and Donzé 2012; Ma et al. 2014; Vallejos et al. 2016; Ma et al. 2021).

BPM is also able to introduce unphysical asperities on joint surfaces for an additional frictional resistance against sliding. For example, a "smooth joint-contact model (SJM)" was proposed to simulate joint behavior with actual geometry and morphology. It was found be able to capture the shear and dilation behavior of natural joint that exhibits significant scale effects (Ivars et al. 2011; Bahaaddini et al. 2014; Lambert and Coll 2014).

By simulating the intact rock matrix using BPM and discrete joints using SJM, a "synthetic rock mass (SRM)" model has been developed to characterize the mechanical properties of jointed rock mass including the peak strength, damage, fragmentation, brittleness, anisotropy and scale effect. Compared to the Hoek–Brown empirical approach for presumed isotropic rock masses, SRM equipped with explicit DFNs is able to derive the orientation-specific strength of naturally jointed rock, in which the influence of joint length distribution and connectivity may be taken into account (Poulsen et al. 2015).

(3) **Boundary conditions**

Test cases show that DEMs can describe the behavior of continuum quite excellently. Even with a coarse mesh, exact results can be obtained, provided that the boundaries are correctly modelled. This advantage has been explored in the study of grain material properties such as the constitutive relations of rock-fill material.

Once the model geometry of domain is designed, boundary conditions should be defined that will restrain/drive the deformation/motion of the model. Typically, in solid dynamics, prescribed boundary displacements are employed to force deformation into the rest of the domain. The boundary condition may be approximated by some far-field displacement field, in such a case the model should be free to deform against the boundary, for example, by introducing frictionless contact between the model material and the boundary face. An alternative boundary condition relies on prescribed forces rather than displacements. However, given the tendency for the material to undergo changes in strength and stiffness during deformation/motion, force-controlled boundary conditions tend to be less stable, and unpredictable.

(4) **Contact resolution and efficiency**

While there are established detection approaches for polyhedron-to-polyhedron (PH–PH) and superquadrics-to-superquadrics (SQ-SQ) contacts, an open problem is the mixed polyhedron-to-superquadrics (PH-SQ) contact detection. This is desirable as many real particle systems contain a broad range of particle shapes. Recently, Peng and Hanley (2019) proposed a novel contact detection algorithm for convex polyhedron and superquadrics which is efficient and reasonably simple.

There are also other methods which, in principle, can be applied to any shape of particle. A typical one is the Gilbert-Johnson-Keerthi (GJK) algorithm (Gilbert et al. 1988) that has been prevalent in computer graphics for detecting the penetration or finding the minimum distance between two objects with specific shapes such as polyhedrons (Ong and Gilbert 1997; den Bergen 1999; Cameron 1997; Cameron and Enhancing 1997). However, GJK algorithm is well-performed only for particles with flat surfaces, for strongly curved surfaces, pre-discretization is demanded. Other typical one need to be mentioned is that determines contact orientation using a potential field related to particle shape (Feng et al. 2012). In addition, Dong et al. (2015) proposed a contact resolution method based on the orientation discretization: the contact region is divided into cells and each cell is categorized as inside or outside of the particle intersection. Their specific approach requires a pre-computed database.

(5) **Implicit scheme in time domain**

The common practice in DEM simulations is the explicit updating methods and the use of implicit methods is very limited. Kruggel-Emden et al. (2008a, 2008b) discussed the selection of an appropriate time integration scheme for DEM. Schäfer and Negrut (2010) evaluated the potential of implicit integration methods in the MD simulation for biological molecules. Although they reported good energy conservation response by implicit methods, the elongation in the time-marching step length was limited due to the loss of convergence derived from iterative process. Tuley et al. (2010) investigated the performance of different integration schemes including implicit schemes w.r.t. the motion of a particle during its normal collision with a wall. This work was further improved by Jasion et al. (2011) to handle the response of the same integration schemes w.r.t. tangential collision. Their works showed that, more simple integrators, such as the first order symplectic Euler scheme, are more suitable than costly and conditionally stable integrators for the particle system containing thousands of particles. Liu and Qiang (2018) discussed a symplectic tracking for multi-particle system on GPUs.

Overall, the experience with implicit scheme in DEM has been limited and reportedly mixed (i.e. both positive and negative results). Despite several investigatory studies, very few have drawn their conclusions out of the results. This could be, among other reasons, due to the fact that implicit schemes are in general harder to be implemented than the explicit counterparts. Therefore, further research is encouraged to reliably implement an implicit scheme for granular matter and to quantitatively assess its costs and benefits.

The best known member of implicit DEMs is probably the "discontinuous deformation analysis (DDA)" (Shi and Goodman 1985) (Chap. 8). This method is claimed to solve assemblage systems with few thousands of particles in "reasonable times", though no comparison with explicit counterpart was presented. Another member of implicit DEMs is the "block element analysis (BEA)" (Chen 2018) (Chap. 8). Under the situations of serviceability state where the structure configuration will not be topologically changed, this method may be used to solve the multi-blocky structural system rather accurately cross-referenced with FEM and laboratory experiment.

(6) Structural planes and crack growth

One advantage of DEM is that it is able to predict the behavior of both continuum and discontinuum (Hahn et al. 2010). Therefore, it possesses practical importance in modelling structural planes and cracks.

The first question is how to pre-define these structural planes in the DEM model, or to have them emerge when and where the conditions dictate. It is possible to pre-define zones of weakness or heterogeneity that localize structural planes in particle-based models, although it is theoretically unnecessary to do so because the process of cracking inception and strain localization can emerge most completely (Ma et al. 2021), through the linkage of microscopic cracking and crushing and other heterogeneities. Insofar nevertheless, it is still impractical to track the whole localization process within the large domain of interest, this is also partially blamed for the lack of well-defined transition from a cohesive strength condition to frictional sliding on a newly formed structural planes. It is notable that in this research area, the hybrid of FEM-DEM or briefly termed as FDEM proposed by Munjiza (2004) is very active (Mahabadi et al. 2010; Lisjak et al. 2014; Yan et al. 2018; Sun et al. 2020).

The second question that has not been well-hackled is the demonstrated dependence of friction on slip velocity, maturity of contact, rock mass stiffness, and temperature.

(7) Parameters

There are a number of parameters in the DEM model such as the particle shape, size, and contact model. Unfortunately, they are not easily measured or/and tested. Take the most important parameters related to contact models for example, various commercial and open source DEM codes implement contact models differently. Ucgul et al. (2014) compared the Hertz contact model to the hysteretic spring contact model (Walton and Braun 1986) and showed that the latter was more accurate in modelling soil-tool interaction. In a review article, Boac and his co-worker (2014) showed that both the linear and nonlinear (Hertz-Mindlin) contact models can be effective to study the grain handling and processing operations. For the flow of material from a conveyor belt and impacting on a plate, the linear spring-dashpot and Hertz-Mindlin models were investigated by Grima and Wypych (2011), they concluded that on macroscopic or bulk level there was no significant difference in the outcomes. Gröeger and Katterfeld (2006) also modelled conveyor belts and transfer chutes, they concluded that if spherical particles are used, the Hertz contact model would be more appropriate. However, if more complex particle shapes are encountered, the contact parameters should be carefully calibrated.

For this issue, the review by Coetzee (2017) is recommended.

References

Abe S, Mair K. Grain fracture in 3D numerical simulations of granular shear. Geophys Res Lett. 2005;32(5):L05305.

Aharonov E, Sparks D. Stick-slip motion in simulated granular layers. J Geophys Res. 2004;109(B9):B09306.

Alder BJ, Wainright TE. Phase transition for a hard sphere system. J Chem Phys. 1957;27(5):1208–9.

Andrade JE, Lim KW, Avila CF, Vlahinić I. Granular element method for computational particle mechanics. Comput Methods Appl Mech Eng. 2012;241–244:262–74.

Arora A. Structural damping identification method using normal FRFs. Int J Solids Struct. 2014;51(1):133–43.

Bahaaddini M, Hagan PC, Mitra R, Hebblewhite BK. Scale effect on the shear behaviour of rock joints based on a numerical study. Eng Geol. 2014;181:212–23.

Bandis S, Lumsden AC, Barton NR. Fundamentals of rock joint deformation. Int J Rock Mech Min Sci and Geomech Abstr. 1983;20(6):249–68.

Barbosa RE. Discrete element models for granular materials and rock masses. PhD Thesis. University of Illinois at Urbana-Champaign (USA):1990.

Benesh NP, Plesch A, Shaw JH, Frost EK. Investigation of growth fault bend folding using discrete element modeling: implications for signatures of active folding above blind thrust faults. J Geophys Res. 2007;112(B3):B03S04.

Benvenuti L, Kloss C, Pirker S. Identification of DEM simulation parameters by artificial neural networks and bulk experiments. Powder Technol. 2016;291:456–65.

Bert CW. Material damping: an introduction review of mathematical models, measures and experimental techniques. J Sound Vib. 1973;29(2):129–53.

Boac JM, Ambrose RPK, Casada ME, Maghirang RG, Maier DE. Applications of discrete element method in modeling of grain postharvest operations. Food Eng Rev. 2014;6(4):128–49.

Boon CW, Houlsby GT, Utili S. A new contact detection algorithm for three-dimensional nonspherical particles. Powder Technol. 2013;248:94−102. https://www.sciencedirect.com/science/article/pii/S0032591019305741-bbb0115

Born M, Huang K. Dynamic theory of crystal lattice. Oxford (UK): Oxford University Press; 1954.

Cai M, Kaiser PK, Morioka H, Minami M, Maejima T, Tasaka Y, Kurose H. FLAC/PFC coupled numerical simulation of AE in large-scale underground excavations. Int J Rock Mech Min Sci. 2007;44(4):550–64.

Cameron S. A comparison of two fast algorithms for computing the distance between convex polyhedral. IEEE Trans Robot Autom. 1997;13(6):915–20.

Cameron S. Enhancing GJK. Computing minimum and penetration distances between convex polyhedral. In: Proceedings international conference on robotics and automation, vol. 4. IEEE;1997. p. 3112−3117.

Campbell CS, Cleary PW, Hopkins MA. Large-scale landslide simulations: global deformation, velocities, and basal friction. J Geophys Res Solid Earth. 1995;100(B5):8267–83.

Canelas RB, Crespo AJC, Domínguez JM, Ferreira RML, Gomez-Gesteira M. SPH-DCDEM model for arbitrary geometries in free surface solid-fluid flows. Comput Phys Commun. 2016;202:131–40.

Cardozo N, Allmendinger RW, Morgan JK. Influence of mechanical stratigraphy and initial stress state on the formation of two fault propagation folds. J Struct Geol. 2005;27(11):1954–72.
Chen SH. The elasto-plastic and elasto-viscoplastic analysis of the discontinuous rock masses. PhD Thesis. WUHEE, Wuhan (China), 1987 (in Chinese with English summary).
Chen SH. Hydraulic structures. Berlin (Germany): Springer; 2015.
Chen SH. Computational geomechanics and hydraulic structures. Berlin (Germany): Springer; 2018.
Chen SH, Li YM, Wang WM, Shahrour I. Analysis of gravity dam on a complicated rock foundation using an adaptive block element method. J Geotech Geoenviron Eng, ASCE. 2004a;130(7):759–63.
Chen WS, Zheng H, Cheng YM, Ge XR. Detection of 3D rock block contacts by penetration edges. Chin J Rock Mechan Eng. 2004b;23(4):565–71 (in Chinese with English abstract).
Cho N, Martin CD, Sego DC. A clumped particle model for rock. Int J Rock Mech Min Sci. 2007;44(7):997–1010.
Clatworthy D, Scheele F. A method of sub-meshing in discontinuous deformation analysis (DDA). In: Amadei B, editors. ICADD-3: Third international conference on analysis of discontinuous deformation, Vail, Colorado (USA). Rotterdam (Netherlands): American Rock Mechanics Association, AA Balkema;1999. p. 85−96.
Cleary PW, Campbell CS. Self-lubrication for long run-out landslides: examination by computer simulation. J Geophys Res Solid Earth. 1993;98(B12):21911–24.
Cleary PW, Prakash M. Discrete-element modelling and smoothed particle hydrodynamics: potential in the environmental sciences. Philos Trans-Royal Soc London Series a Mathemat Phys Eng Sci. 2004;362:2003–30.
Coetzee CJ. Review: calibration of the discrete element method. Powder Technol. 2017;310:104–42.
Combarros M, Feise HJ, Zetzener H, Kwade A. Segregation of particulate solids: experiments and DEM simulations. Particuology. 2014;12(1):25–32.
Cundall PA. A computer model for simulating progressive large scale movements in blocky rock systems. In: Proceedings symposium ISRM (Vol. 1). Nancy (France):1971, Paper No. II-8.
Cundall PA, Strack ODL. A discrete numerical model for granular assemblies. Geotechnique. 1979;29(1):47–65.
Cundall PA. Distinct element models of rock and soil structure. In: Brown ET, editors. Analytical and computational methods in engineering rock mechanics. London (UK): Allen & Unwin;1987, Chap. 4. p. 129−163.
Cundall PA. Formulation of a three-dimensional distinct element model. Part I. A scheme to detect and represent contacts in a system composed of many polyhedral blocks. Int J Rock Mech Min Sci Geomech Abst. 1988;25(3):107−116.
Cundall PA, Hart R. Numerical modeling of discontinua. J Eng Comput. 1992;9(2):101–13.
Davies TR, McSaveney MJ. Runout of dry granular avalanches. Can Geotech J. 1999;36(2):313–20.
Davies TR, McSaveney MJ, Hodgson KA. A fragmentation-spreading model for long-runout rock avalanches. Can Geotech J. 1999;36(6):1096–110.
den Bergen GV. A fast and robust GJK implementation for collision detection of convex objects. J Graph Tools. 1999;4(2):7–25.
Descantes Y, Tricoire F, Richard P. Classical contact detection algorithms for 3D DEM simulations: drawbacks and solutions. Comput Geotech. 2019;114: 103134.
Diederichs MS, Kaiser PK, Eberhardt E. Damage initiation and propagation in hard rock during tunneling and the influence of near-face stress rotation. Int J Rock Mech Min Sci. 2004;41(5):785–812.
Dong JJ, Yang CM, Yu WL, Lee CT, Miyamoto Y, Shimamoto T. Velocity-displacement dependent friction coefficient and the kinematics of giant landslide. In: Ugai K, Yagi H, Wakai A, editors. Earthquake-induced landslides. Berlin (Germany): Springer; 2013. p. 397–403.
Dong K, Wang C, Yu A. A novel method based on orientation discretization for discrete element modeling of non-spherical particles. Chem Eng Sci. 2015;126:500–16.
Egholm DL, Clausen OR, Sandiford M, Kristensen MB, Korstgård JA. The mechanics of clay smearing along faults. Geology. 2008;36(10):787–90.

Eisenstadt MM. Introduction to mechanical properties of materials. New York (USA): MacMillan; 1971.

Esmaieli K, Hadjigeorgiou J, Grenon M. Estimating geometrical and mechanical REV based on synthetic rock mass models at Brunswick mine. Int J Rock Mech Min Sci. 2010;47(6):915–26.

Fakhimi A, Carvalho F, Ishida T, Labuz JF. Simulation of failure around a circular opening in rock. Int J Rock Mech Min Sci. 2002;39(4):507–15.

Fang ZQ, Hu GM, Du J, Fan Z, Liu J. A contact detection algorithm for multi-sphere particles by means of two-level-grid-searching in DEM simulations. Int J Numer Meth Eng. 2015;102(13):1869–93.

Favier JF, Abbaspour-Fard MH, Kremmer M, Raji AO. Shape representation of axi-symmetrical, non-spherical particles in discrete element simulation using multi-element model particles. Eng Comput. 1999;16(4):467–80.

Feng YT, Han K, Owen DRJ. Energy-conserving contact interaction models for arbitrarily shaped discrete elements. Comput Methods Appl Mech Eng. 2012;205–208:169–77.

Feng YT, Han K, Owen DRJ. A generic contact detection framework for cylindrical particles in discrete element modelling. Comput Methods Appl Mech Eng. 2017;315:632–51.

Feng YT, Owen DRJ. An energy based corner to contact algorithm. In: Cook BK, Jensen RP, editors. Discrete element methods, numerical modeling of discontinuation. Santa Fe, New Mexico (USA):2002. p. 32–37.

Feng YT, Owen DRJ. A 2D polygon/polygon contact model: algorithmic aspects. Eng Comput. 2004;21(2–4):265–277.

Ferellec JF, McDowell GR. A method to model realistic particle shape and inertia in DEM. Granular Matter. 2010;12(5):459–67.

Fillot N, Iordanoff I, Berthier Y. Modelling third body flows with a discrete element method- a tool for understanding wear with adhesive particles. Tribol Int. 2007;40(6):973–81.

Finch E, Hardy S, Gawthorpe R. Discrete-element modelling of extensional fault-propagation folding above rigid basement fault blocks. Basin Res. 2004a;16(4):467–88.

Finch E, Hardy S, Gawthorpe R. Discrete-element modelling of contractional fault-propagation folding above rigid basement fault blocks. J Structural Geology. 2004b;25(4):515–28.

Fritz HM, Hager WH, Minor HE. Lituya Bay case: rockslide impact and wave run-up. Sci Tsunami Haz. 2001;19(1):3–22.

Gagnon L, Morandini M, Ghiringhelli GL. A review of particle damping modeling and testing. J Sound Vib. 2019;459:114865.

Gen QD. Improvement on common plane detection algorithm and three-dimensional discrete particles generation algorithm. PhD Thesis of Dalian University of Technology (China), 2008 (in Chinese with English Summary).

Gilbert EH, Johnson DW, Keerthi SS. A fast procedure for computing the distance between complex objects in three-dimensional space. IEEE J Robot Autom. 1988;4(2):193–203.

Gibson JB, Goland AN, Milgram M, Vineyard GH. Dynamics of radiation damage. Phys Rev. 1960;120:1229–53.

Grima AP, Wypych PW. Investigation into calibration of discrete element model parameters for scale-up and validation of particle-structure interactions under impact conditions. Powder Technol. 2011;212(1):198–209.

Goren L, Aharonov E. Long runout landslides: the role of frictional heating and hydraulic diffusivity. Geophys Res Lett. 2007;34(7):L07301.

Goren L, Aharonov E. On the stability of landslides: a thermo-poro-elastic approach. Earth Planet Sci Lett. 2009;277(3–4):365–72.

Gray GG, Morgan JK, Pablo F. Sanz PF. Overview of continuum and particle dynamics methods for mechanical modeling of contractional geologic structures. J Struct Geol 2014;59:19–36.

Gröeger T, Katterfeld A. On the numerical calibration of discrete element models for the simulation of bulk solids. Comput Aided Chem Eng. 2006;21(A):533–538.

Guo Y, Morgan JK. The frictional and micromechanical effects of grain comminution in fault gouge from distinct element simulations. J Geophys Res Solid Earth. 2006;111(B12):B12406.

Guo Y, Morgan JK. Fault gouge evolution and its dependence on normal stress and rock strength–results of discrete element simulations: gouge zone properties. J Geophys Res Solid Earth. 2007;112(B10):B10403.

Habib P. Production of gaseous pore pressure during rock slides. Rock Mech Rock Eng. 1975;7(4):193–7.

Hadjigeorgiou J, Esmaieli K, Grenon M. Stability analysis of vertical excavations in hard rock by integrating a fracture system into a PFC model. Tunn Undergr Space Technol. 2009;24(3):296–308.

Hahn M, Wallmersperger T, Kröplin BH. Discrete element representation of continua: proof of concept and determination of the material parameters. Comput Mater Sci. 2010;50(2):391–402.

Hardy S. Structural evolution of calderas: insights from two-dimensional discrete element simulations. Geology. 2008;36(12):927–30.

Hardy S, McClay K, Muñoz JA. Deformation and fault activity in space and time in high-resolution numerical models of doubly vergent thrust wedges. Mar Petrol Geol. 2009;26(2):232–48.

Hogue C. Shape representation and contact detection for discrete element simulations of arbitrary geometries. Eng Comput. 1998;15(3):374–90.

Houlsby GT. Potential particles: a method for modelling non-circular particles in DEM. Comput Geotech. 2009;36(6):953–9.

Hrennikoff AP. Plane stress and bending of plates by method of articulated framework. PhD Thesis. MIT, Boston, Massachusetts (USA), 1941.

Ibrahim R. In: Vibro-impact dynamics: modeling, mapping and applications, vol. 43. Berlin, Heidelberg (Germany):Springer;2009.

Itasca Consulting Group, Inc. PFC—Particle Flow Code. User's Manual (Ver. 6.0). Minneapolis (USA), 2018.

Itasca Consulting Group, Inc. 3DEC—Three-Dimensional Distinct Element Code. User's Manual (Ver.5.2). Minneapolis (USA), 2016.

Ivars DM, Pierce M, Darcel C, Reyes-Montes J, Potyondy D, Young P, Cundall P. The synthetic rock mass approach for jointed rock mass modeling. Int J Rock Mech Min Sci. 2011;48(2):219–44.

Iverson RM. Regulation of landslide motion by dilatancy and pore pressure feedback. J Geophys Res Earth Surf. 2005;110:F02015.

Iverson RM, Reid ME, Iverson NR, LaHusen RG, Logan M, Mann JE, Brien DL. Acute sensitivity of landslide rates to initial soil porosity. Science. 2000;290(5491):513–6.

Jasion G, Shrimpton J, Danby M, Takeda K. Performance of numerical integrators on tangential motion of DEM within implicit flow solvers. Comput Chem Eng. 2011;35(11):2218–26.

Jensen A, Fraser K, Laird G. Improving the precision of discrete element simulations through calibration models. In: Proceedings 13th international LS-DYNA users conference. Detroit, MI (USA):2014. p. 1−12.

Johnson BC, Campbell CS, Melosh HJ. The reduction of friction in long runout landslides as an emergent phenomenon. J Geophys Res Earth Surf. 2016;121(5):881–9.

Jop P, Forterre Y, Pouliquen O. A constitutive law for dense granular flows. Nature. 2006;441(7094):727–30.

Kawai T. New discrete models and their application to seismic response analysis of structures. Nucl Eng Des. 1978;48(1):207–29.

Klein F, Weighardt K. Über Spannungsflachen und reziproke Diagramme, mit besonderer Beruck-sichtigung der Maxwellschen Arbeiten. Archiv der Mathematik und Physik.1905, 3 Reihe 7 and 8:1~10 and 95~119 (in German).

Kozicki J, Donzé FV. Yade-open DEM: an open-source software using a discrete element merhod to simulate granular material. Eng Comput. 2009;26(7):786–805.

Krishnasamy J, Jakiela MJ. A method to resolve ambiguities in corner-corner interactions polygons in the context of motion simulation. Eng Comput. 1995;12(2):135–44.

Kruggel-Emden H, Sturm M, Wirtz S, Scherer V. Selection of an appropriate time integration scheme for the discrete element method (DEM). Comput Chem Eng. 2008a;32(10):2263–79.

Kruggel-Emden H, Rickelt S, Wirtz S, Scherer V. A study on the validity of the multi-sphere discrete element method. Powder Technol. 2008b;188(2):153–65.

Lambert C, Coll C. Discrete modeling of rock joints with a smooth-joint contact model. J Rock Mech Geotech Eng. 2014;6(1):1–12.

Latham JP, Munjiza A, Garcia X, Xiang JS, Guises R. Three-dimensional particle shape acquisition and use of shape library for DEM and FEM/DEM simulation. Miner Eng. 2008;21(11):797–805.

Legros F. The mobility of long-runout landslides. Eng Geol. 2002;63(3–4):301–31.

Lie KH. Berechnung der Fachwerke und ihrer verwandten Systeme auf neuem Wege. Der Stahlbau. 1944;8(9):41–50 (in German).

Liao X, Jin P. Study on the correlation between the volume of high-speed landslide mass and the runout distance. Subgrade Eng. 1994;6:9–12 (in Chinese with English abstract).

Lin XS, Ng TT. Contact detection algorithms for 3-dimensional ellipsoids in discrete element modelling. Int J Numer Anal Methods Geomech. 1995;19(9):653–9.

Lin XS, Ng TT. A three-dimensional discrete element model using arrays of ellipsoids. Geotechnique. 1997;47(2):319–29.

Lisjak A, Tatone BSA, Grasselli G, Vietor T. Numerical modelling of the anisotropic mechanical behaviour of opalinus clay at the laboratory-scale using FEM/DEM. Rock Mech Rock Eng. 2014;47(1):187–206. https://www.sciencedirect.com/journal/journal-of-rock-mechanics-and-geotechnical-engineering/vol/11/issue/6

Lisjak A, Grasselli G. A review of discrete modeling techniques for fracturing processes in discontinuous rock masses. J Rock Mech Geotech Eng. 2014;6(4):301–14.

Liu W, He SM, Li XP, Xu Q. Two-dimensional landslide dynamic simulation based on a velocity-weakening friction law. Landslides. 2016;13(5):957–65.

Liu ZC, Qiang J. Symplectic multi-particle tracking on GPUs. Comput Phys Commun. 2018;226:10–7.

Lu G, Third JR, Müller CR. Critical assessment of two approaches for evaluating contacts between super-quadric shaped particles in DEM simulations. Chem Eng Sci. 2012;78:226–35.

Lu G, Third JR, Müller CR. Discrete element models for non-spherical particle systems: from theoretical developments to applications. Chem Eng Sci. 2015;127:425–65.

Lucas A, Mangeney A, Ampuero JP. Frictional velocity-weakening in landslides on Earth and on other planetary bodies. Nat Commun. 2014;5:3417.

Ma G, Zou YX, Chen Y, Tang LW, Ng TT, Zhou W. Spatial correlation and temporal evolution of plastic heterogeneity in sheared granular materials. Powder Technology. 2021;378(Part A):263–273.

Ma G, Zhou W, Chang XL, Yuan W. Combined FEM/DEM modeling of triaxial compression tests for rockfills with polyhedral particles. Int J Geomech. 2014;14(4):04014014.

Mahabadi OK, Cottrell BE, Grasselli G. An example of realistic modelling of rock dynamics problems: FEM/DEM simulation of dynamic Brazilian test on Barre granite. Rock Mech Rock Eng. 2010;43(6):707–16.

Maradudin AA. Screw dislocations and discrete elastic theory. J Phys Chem Solid. 1958;9(1):1–20.

Markauskas D, Kacianauskas R, Džiugys A, Navakas R. Investigation of adequacy of multi-sphere approximation of elliptical particles for DEM simulations. Granul Matter. 2010;12(1):107–23.

Markauskas D, Kacianauskas R. Investigation of rice grain flow by multi-sphere particle model with rolling resistance. Granul Matter. 2011;13(2):143–8.

Marigo M, Stitt EH. Discrete element method (DEM) for industrial applications: comments on calibration and validation for the modelling of cylindrical pellets. Kona Powder and Particle J. 2015;32:236–52.

Matuttis HG, Chen JF. Understanding the discrete element method: simulation of non-spherical particles for granular and multi-body system. Singapore: John Wiley & Sons, Ltd, 2014.

McDougall S, Hungr O. A model for the analysis of rapid landslide motion across three-dimensional terrain. Can Geotech J. 2004;41(6):1084–97.

McHenry D. A lattice analogy for the solution of stress problems. J Institut of Civil Eng. 1943;21(1):59–82.

Melosh HJ. Acoustic fluidization: a new geologic process? J Geophys Res: Solid Earth. 1979;84(B13):7513–20.

Miyakawa A, Yamada Y, Matsuoka T. Effect of increased shear stress along a plate boundary fault on the formation of an out-of-sequence thrust and a break in surface slope within an accretionary wedge, based on numerical simulations. Tectonophysics. 2010;484(1–4):127–38.

Morgan JK. Numerical simulations of granular shear zones using the distinct element method: II. The effect of particle size distribution and interparticle friction on mechanical behavior. J Geophys Res Solid Earth. 1999;104(B2):2721−2732.

Morgan JK. Particle dynamics simulations of rate and state dependent frictional sliding of granular fault gouge. Pure Appl Geophys. 2004;161(9–10):1877–91.

Morgan JK, Boettcher MS. Numerical simulations of granular shear zones using the distinct element method: I. Shear zone kinematics and the micromechanics of localization. J Geophys Res Solid Earth. 1999;104(B2):2703−2719.

Morgan JK, McGovern PJ. Discrete element simulations of gravitational volcanic deformation: 1. Deformation structures and geometries. J Geophys Res Solid Earth. 2005a;110(B5):B05402.

Morgan JK, McGovern PJ. Discrete element simulations of gravitational volcanic deformation: 2. Mechanical analysis. J Geophys Res Solid Earth. 2005b;110(B5):B05403.

Munjiza A. The combined finite-discrete element method. New York (USA): Wiley; 2004.

Munjiza A, Andrews KRF. NBS contact detection algorithm for bodies of similar size. Int J Numer Meth Eng. 1998;43(1):131–49.

Mustoe GGW, Miyata M. Material flow analyses of noncircular-shaped granular media using discrete element methods. J Eng Mech. 2001;127(10):1017–26.

Naylor M, Sinclair HD, Willett S, Cowie PA. A discrete element model for orogenesis and accretionary wedge growth. J Geophys Res Solid Earth. 2005;110(B12):B12404.

Nezami EG, Hashash YMA, Zhao D, Ghaboussi J. A fast contact detection algorithm for 3-D discrete element method. Comput Geotech. 2004;31(7):575–87.

Nezami EG, Hashash YMA, Zhao D, Ghaboussi J. Shortest link method for contact detection in discrete element method. Int J Numer Anal Methods Geomech. 2006;30(8):783–801.

Nezami EG, Hashash YMA, Zhao D, Ghaboussi J. Simulation of front end loader bucket-soil interaction using discrete element method. Int J Numer Anal Methods Geomech. 2007;31(9):1147–62.

Otter JRH, Cassell AC, Hobbs RE. Dynamic relaxation. ICE Proceed. 1966;35(4):633–56.

Ong CJ, Gilbert EG. The Gilbert-Johnson-Keerthi distance algorithm: a fast version for incremental motions. In: Proceedings international conference on robotics and automation, vol. 2. Albuquerque, NM (USA): IEEE Press;1997.1183−1189.

Peng D, Hanley KJ. Contact detection between convex polyhedra and superquadrics in discrete element codes. Powder Technol. 2019;356:11–20.

Perkins E, Williams JR. A fast contact detection algorithm insensetive to object size. Eng Comput. 2001;18(1–2):48–61.

Podlozhnyuk A, Pirker S, Kloss C. Efficient implementation of superquadric particles in discrete element method within an open-source framework. Comput Particle Mech. 2017;4(1):101–18.

Potapov AV, Campbell CS. A fast model for the simulation of non-round particles. Granular Matter. 1998;1(1):9–14.

Potyondy DO, Cundall PA. A bonded-particle model for rock. Int J Rock Mech Min Sci. 2004;41(8):1329–64.

Poulsen BA, Adhikary DP, Elmouttie MK, Wilkins A. Convergence of synthetic rock mass modelling and the Hoek-Brown strength criterion. Int J Rock Mech Min Sci. 2015;80:171–80.

Pudasaini SP, Miller SA. The hypermobility of huge landslides and avalanches. Eng Geol. 2013;157:124–32.

Qiu LC, Jin F, Lin PZ, Liu Y, Han Y. Numerical simulation of submarine landslide tsunamis using particle based methods. J Hydrodyn. 2017;29(4):542–51.

Rahman A. Correlations in the motion of atoms in liquid Argon. Phys Rev. 1964;136(2A):405–11.

Rayleigh L. Theory of sound (Two Volumes). New York (USA): Dover Publications; 1897.

Ren B, Jin Z, Gao R, Wang YX, Xu ZL. SPH-DEM modeling of the hydraulic stability of 2D blocks on a slope. J Waterway Port Coastal and Ocean Eng ASCE. 2014;140(6):04014022.

Riedel W. Beiträge zur Lösung des ebenen Problems eines elastischen Körpers mittels der Airyschen Spannungsfunktion. Zeitschrift fur Angewandte Mathematik und Mechanik.1927, 7(Heft 3):169–188 (in German).

Saeb S, Amadei B. Modelling joint response under constant or variable normal stiffness boundary conditions. Int J Rock Mech Min Sci and Geomech Abstr. 1990;27(3):213–7.

Saeb S, Amadei B. Modelling rock joints under shear and normal loading. Int J Rock Mech Min Sci and Geomech Abstr. 1992;29(3):267–78.

Scholtès L, Donzé FV. Modeling progressive failure in fractured rock masses using a 3D discrete element method. Int J Rock Mech Min Sci. 2012;52:18–30.

Schöpfer MPJ, Abe S, Childs C, Walsh JJ. The impact of porosity and crack density on the elasticity, strength and friction of cohesive granular materials: Insights from DEM Modelling. Int J Rock Mech Min Sci. 2009;46(2):250–61.

Schäfer N, Negrut D. A quantitative assessment of the potential of implicit integration methods for molecular dynamics simulation. J Comput Nonlinear Dyn. 2010;5(3): 031012.

Scheidegger AE. On the prediction of the reach and velocity of catastrophic landslides. Rock Mech and Rock Eng. 1973;5(4):231–6.

Shan T, Zhao J. A coupled CFD-DEM analysis of granular flow impacting on a water reservoir. Acta Mech. 2014;225(8):2449–70.

Shi GH. Block system modeling by discontinuous deformation analysis. Southampton (UK): Computational Mechanics Publications; 1993.

Shi GH, Goodman RE. Two dimensional discontinuous deformation analysis. Int J Numer Anal Meth Geomech. 1985;9(6):541–56.

Shi GH, Goodman RE. Generalization of two-dimensional discontinuous deformation analysis for forward modelling. Int J Numer Anal Meth Geomech. 1989;13(4):359–80.

Shimizu H, Koyama T, Ishida T, Chijimatsu M, Fujita T, Nakama S. Distinct element analysis for class II behavior of rocks under uniaxial compression. Int J Rock Mech Min Sci. 2010;47(2):323–33.

Shodja HM, Nezami EG. A micromechanical study of rolling and sliding contacts in assemblies of oval granules. Int J Numer Anal Meth Geomech. 2003;27(5):403–24.

Simons TAH, Weiler R, Strege S, Bensmann S, Schilling M, Kwade A. A ring shear tester as calibration experiment for DEM simulations in agitated mixers—a sensitivity study. Procedia Eng. 2015;102:741–748.

Smeets B, Odenthal T, Vanmaercke S, Ramon H. Polygon-based contact description for modeling arbitrary polyhedra in the discrete element method. Comput Methods Appl Mech Eng. 2015;290:277–89.

Stewart IJ, Brown ET. A static relaxation method for the analysis of excavation in discontinuous rock. In: Proceedings ISRM symposium on design and performance of underground excavations. Cambridge (UK): British Geotechnical Society; 1984. 149–154.

Stillinger FH, Rahman A. Improved simulation of liquid water by molecular dynamics. J Chem Phys. 1974;60(4):1545–57.

Strayer LM, Suppe J. Out of plane motion of a thrust sheet during along strike propagation of a thrust ramp: a distinct element approach. J Struct Geol. 2002;24(4):637–50.

Strayer LM, Erickson SG, Suppe J. Influence of growth strata on the evolution of fault-related folds—distinct-element models. In: McClay KR, editors. Thrust tectonics and hydrocarbon systems, vol. 82. American Association of Petroleum Geologists Memoir; 2004:p. 413–37.

Strack ODL, Cundall PA. The distinct element method as a tool for research in granular media, Part 1. Report to NSF, University of Minnesota (USA); 1978.

Strack ODL, Cundall PA. Fundamental studies of fabric in granular materials. Report to NSF, University of Minnesota (USA); 1984.

Sun L, Liu QS, Grasselli G, Tang XH. Simulation of thermal cracking in anisotropic shale formations using the combined finite-discrete element method. Comput Geotech. 117:103237. https://www.sciencedirect.com/journal/engineering-fracture-mechanics/vol/266/suppl/C

Tan H, Chen S. A hybrid DEM-SPH model for deformable landslide and its generated surge waves. Adv Water Resour. 2017;108:256–76.

Tan H, Xu Q, Chen S. Subaerial rigid landslide-tsunamis: insights from a block DEM-SPH model. Eng Anal Boundary Elem. 2018a;95:297–314.

Tan H, Ruffini G, Heller V, Chen SH. A numerical landslide-tsunami hazard assessment technique applied on hypothetical scenarios at Es Vedrà, offshore Ibiza. J Marine Sci Eng. 2018b;6(4):111.

Ting JM, Khwaja M, Meahum LR, Rowell HD. An ellipsed-based discrete element model for granular materials. Int J Numer Anal Meth Geomech. 1993;17(9):603–23.

Togo T, Shimamoto T, Ma SL, Hirose T. High-velocity frictional behavior of Longmenshan fault gouge from Hongkou outcrop and its implications for dynamic weakening of fault during the 2008 Wenchuan earthquake. Earthq Sci. 2011;24(3):267–81.

Trent BC, Margolin LG, Cundall PA, Gaffney ES. The micromechanics of cemented granular material. In: Constitutive laws for engineering materials: theory and applications. Amsterdam (UK): Elsevier, 1987. p. 795−802.

Tuley R, Danby M, Shrimpton J, Palmer M. On the optimal numerical time integration for Lagrangian DEM within implicit flow solvers. Comput Chem Eng. 2010;34(10):886–99.

Ucgul M, Fielke JM, Saunders C. Three-dimensional discrete element modelling of tillage: determination of a suitable contact model and parameters for a cohesionless soil. Biosyst Eng. 2014;121:105–17.

Vallejos JA, Suzuki K, Brzovic A, Mas ID. Application of synthetic rock mass modeling to veined core-size samples. Int J Rock Mech Min Sci. 2016;81:47–61.

Vietor T. Numerical simulation of collisional orogeny using the distinct element technique. In: Konietzky H, editor. Numerical modeling in micromechanics via particle methods. Lisse (Netherlands): AA Balkema; 2003. p. 295–301.

Walton OR. Numerical simulation of inelastic, frictional particle-particle interactions. In: Roco MC, editor. Particulate two-phase flow. Boston (USA): Butterworth-Heinemann; 1984. p. 884–911.

Walton OR, Braun RL. Viscosity, granular-temperature, and stress calculations for shearing assemblies of inelastic, frictional disks. J Rheol. 1986;30(5):949–80.

Wang Y, Tonon F. Discrete element modeling of rock fragmentation upon impact in rock fall analysis. Rock Mech Rock Eng. 2011a;44(1):23–35.

Wang Y, Tonon F. Dynamic validation of a discrete element code in modeling rock fragmentation. Int J Rock Mech Min Sci. 2011b;48(4):535–45.

Wang JQ, Lin G, Liu J. Analysis and improvement of scheme to detect contacts in system composed of 3D polyhedral blocks. Chin J Rock Mechan Eng. 2006;25(11):2247–57 (in Chinese with English abstract).

Wang J, Li SH, Feng C. A shrunken edge algorithm for contact detection between convex polyhedral blocks. Comput Geotech. 2015;63:315–30.

Weng MC, Li HH. Relationship between the deformation characteristics and microscopic properties of sandstone explored by the bonded-particle model. Int J Rock Mech Min Sci. 2012;56:34–43.

Weighardt K. Über einen Grenzubergang der Elastititatslehre und seine Anwendung auf die Statik hochgradig statisch unbestimmter Fachwerke. Verhandlungen Des Vereins Zur Beforderung Des Gewerbefleisses. 1906;85:139–76 (in German).

Williams JR, O'Connor R. A linear complexity intersection algorithm for discrete element simulations of arbitrary geometries. Int J CAE-Eng Comput Special Ed Discrete Element Meth. 1995;12(2):185–201.

Williams JR, O'Connor R. Discrete element simulation and the contact problem. Arch Comput Methods Eng. 1999;6(4):279–304.

Williams JR, Pentland AP. Superquadrics and modal dynamics for discrete elements in interactive design. Eng Comput. 1992;9(2):115–27.

Xu W, Chen H. Mesostructural characterization of particulate composites via a contact detection algorithm of ellipsoidal particles. Powder Technol. 2012;221:296–305.

Yoon JS. Application of experimental design and optimization to PFC model calibration in uniaxial compression simulation. Int J Rock Mech Min Sci. 2007;44(6):871–89.

Yan C, Jiao YY, Zheng H. A fully coupled three-dimensional hydro-mechanical finite discrete element approach with real porous seepage for simulating 3D hydraulic fracturing. Comput Geotech. 2018;96:73–89.

Yan Z, Wilkinson SK, Stitt EH, Marigo M. Discrete element modelling (DEM) input parameters: understanding their impact on model predictions using statistical analysis. Comput Particle Mech. 2015;2(3):283–99.

Zhang Q, Zhu H, Zhang L, Ding X. Study of scale effect on intact rock strength using particle flow modeling. Int J Rock Mech Min Sci. 2011;48(8):1320–8.

Zhang S, Kuwabara S, Suzuki T, Kawano Y, Morita K, Fukuda K. Simulation of solid-fluid mixture flow using moving particle methods. J Comput Phys. 2009;228(7):2552–65.

Zhao T, Dai F, Xu NW, Liu Y, Xu Y. A composite particle model for non-spherical particles in DEM simulations. Granul Matter. 2015;17(6):763–74.

Zhou YC, Wright BD, Yang RY, Xu BH, Yu AB. Rolling friction in the dynamic simulation of sandpile formation. Phys A. 1999;269(2–4):536–53.

Zhu HP, Zhou ZY, Yang RY, Yu AB. Discrete particle simulation of particulate systems: theoretical developments. Chem Eng Sci. 2007;62(13):3378–96.

Zhu ZG, Xu WX, Chen HS. The fraction of overlapping interphase around 2D and 3D polydisperse non-spherical particles: theoretical and numerical models. Comput Methods Appl Mech Eng. 2019;345:728–47.

Chapter 8
Discrete Element Methods with Special Focus on DDA

Abstract In this chapter, the implicit (static) relaxation algorithm of "discontinuous deformation analysis (DDA)" belonging to the family of block-based CMs, is elucidated towards the simulation of rock mass as an assemblage of rock blocks exhibiting large deformation and separation. The block system may be discretized naturally due to the DFN in rock mass or artificially in equivalently homogeneous geomaterial. The force–displacement relation of blocks under equilibrium state is approached by an implicit iterative scheme intended to eliminate the residual or unbalanced forces and momentums. Another static relaxation algorithm of "block element analysis (BEA)" belonging the same CM family towards the simulation of rock mass with small (infinitesimal) deformation is expounded in parallel, for a better understanding of the specific merits and drawbacks of each member in block-based CMs.

8.1 General

In DEMs, the finite differential schemes used for the motion of particle assemblage may be basically distinguished into two types: dynamic relaxation and static relaxation (Otter et al. 1966).

Dynamic relaxation scheme implemented in, for example, the "distinct element method (DEM)" handles non-linear static problem using dynamic equation that is integrated with critical pumping step by step. To ensure the accuracy of solution, damping is employed to absorb the kinetic energy of the particle assemblage system. Due to its explicit characteristic, it is not necessary to solve a large matrix equation, but the damping coefficient and rigorous time-marching step length are pre-set.

Static relaxation scheme (Stewart and Brown 1984) solves the particle assemblage system under equilibrium state using iteration procedure w.r.t. unbalanced forces. It is an implicit scheme intended to eliminate the residual force and momentum. The methods of "discontinuous deformation analysis (DDA)" (Shi 1988, 1992, 1993) and "block element analysis (BEA)" (Chen 1987, 1992) belong to static relaxation scheme. In essence similar to FEM, they employ a type of mesh (matrix) where all the elements are real isolated rock blocks of any shape bounded by existing structural planes. When the blocks are in contact, a tensile cut-off Mohr–Coulomb's law is

S. Chen, *Advanced Computational Methods and Geomechanics*, Springer Tracts in Civil Engineering, https://doi.org/10.1007/978-981-19-7427-4_8

applied to the contacts (boundaries as well) for the adjustment of superfluous forces. The simultaneous equilibrium equation set is formulated and solved for each loading or/and time increment. In FEM, the number of unknowns is decided by the sum of the DOFs of all nodes, whereas in DDA and BEA, the number of unknowns is the sum of the DOFs of all blocks, hence they also may be regarded as a kind of NMM belonging to GFEMs.

It should be emphasized that the static relaxation scheme considers both statics and dynamics with an implicit formulation in the space domain and a step marching algorithm either implicitly or explicitly in the time domain. The difference between static analysis and dynamic analysis is that the former pre-sets zero velocity in the beginning of each time-marching step, while the latter inherits the velocity of the previous time-marching step.

The formulation of static relaxation may be based on either the minimization of potential energy or the virtual work principle. Numerical penalties in the form of stiff springs are applied at the block contacts to either prevent penetration or to release tension between them. When the system converges to an equilibrium state, however, there are inevitable penetration energies at each contact, which balance the contact forces. Thus, the energy of penetration (the deformation of the springs) can be used to calculate the normal and shear contact forces.

In this book, 2-D polygonal particles, termed as "block elements" or simply "blocks", will be handled to elucidate the principles and algorithms of DEMs with static relaxation scheme. The basic ideas are first introduced using BEA under the state of infinitesimal deformation, afterwards DDA subjected to finite block rotation and detachment will be elaborated in detail.

8.2 Introduction to BEA

8.2.1 Concept

The analysis with the contact assumption of face-to-face (3-D) or edge-to-edge (2-D) between blocks merely is the major peculiarity of fundamental BEA. Its governing equations are formulated by three pillar conditions: the force and moment equilibrium condition for rigid block, the deformation compatibility condition between blocks, and the elasto-viscoplastic constitutive relation of structural plane (Chen 1987; Chen et al. 1988; Chen 1992, 1993). The methodology has been further expanded into the areas of reinforcement analysis, stochastic analysis, seepage analysis, coupling analysis, and dynamic analysis (Chen et al. 1994; Chen et al. 2000; Wang and Chen 2001; Chen et al. 2010). A more delicate BEA may be implemented, too, by taking into account of complicated block deformations using, for example, the hierarchical refinement of p-version, in which each deformable block is circumscribed by an overlapped finite element (standard or hierarchical) (Chen et al. 2004). If the asymptotic enrichment near crack-tip is employed for the overlapped finite element that is similar to XFEM, the crack growth within a deformable block also may be simulated routinely.

BEA has been recommended by the DL/T 5353-2006《Design specification for engineering slopes in water resources and hydropower project》.

8.2.2 *Formulation of BEA with Rigid Block Elements*

1. **Coordinate systems**

In Fig. 8.1, r_i $(i = 1, 2, ..., n_r)$ is taken as a representative block element where n_r is the amount of block elements, the boundaries of the block are composed of structural planes denoted by sp_{r_i,r_j}, the neighboring block contacting block r_i through sp_{r_i,r_j} is denoted by $r_j (j = 1, 2, ..., n_r)$.

The polygonal block element in Fig. 8.2 is delimited by three structural planes, their strikes, dips and mean points can be decided through DFN. For each structural plane patch, a local coordinate system is also needed to facilitate the formulation. The origin of the local coordinate system related to structural plane sp_{r_i,r_j} is at the plane's centroid, $y_{sp_{r_i,r_j}}$-axis is perpendicular to the plane, the direction of $x_{sp_{r_i,r_j}}$-axis is coincident with the dip direction. This coordinate system divides the space into positive and negative half-spaces. Take the representative block r_i in Fig. 8.2 for example, it is located in the negative half-space of structural plane sp_{r_i,r_j} that plays the similar role of CP in DEM (Chap. 7).

The global coordinate of the origin of local coordinate system is denoted as $(x^0_{sp_{r_i,r_j}}, y^0_{sp_{r_i,r_j}})$, the global coordinate of the geometry center of block r_i is denoted as $(x^0_{r_i}, y^0_{r_i})$. The coordinate transformation from global coordinate system to local coordinate system is defined by matrix $\mathbf{T}_{sp_{r_i,r_j}}$

$$\mathbf{x}_{sp_{r_i,r_j}} = \mathbf{T}_{sp_{r_i,r_j}} \mathbf{x} \quad (i, j = 1, 2, ..., n_r) \tag{8.1}$$

in which n_r is the amount of block elements, $\mathbf{T}_{sp_{r_i,r_j}}$ has been defined in Eq. (3.125). At any point $(x_{sp_{r_i,r_j}}, y_{sp_{r_i,r_j}})$ on sp_{r_i,r_j}, the stress and deformation (relative displacement)

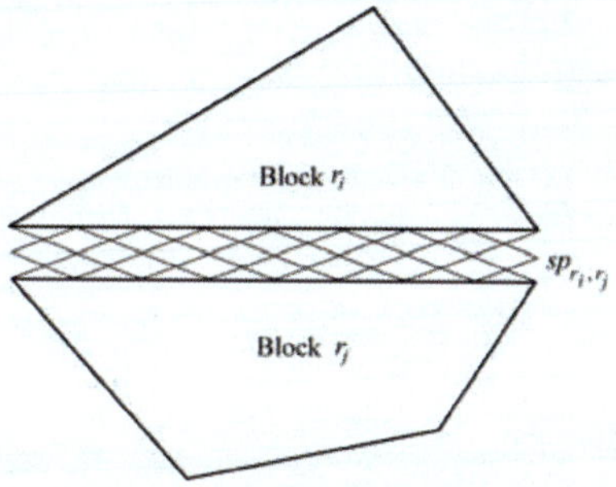

Fig. 8.1 Contact of block elements r_i and r_j

Fig. 8.2 Global and local coordinate systems of the representative block r_i

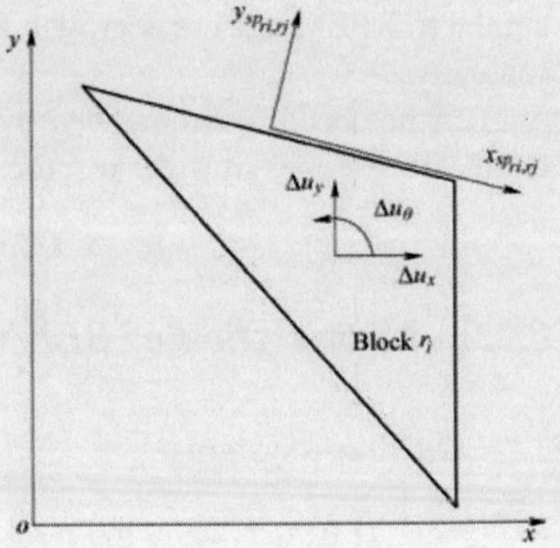

increments are denoted as

$$\begin{cases} \Delta\boldsymbol{\sigma}_{sp_{r_i,r_j}} = \left[\Delta\tau_{yx} \quad \Delta\sigma_y\right]^{\mathrm{T}}_{sp_{r_i,r_j}} \\ \Delta\mathbf{u}_{sp_{r_i,r_j}} = \left[\Delta u_x \quad \Delta u_y\right]^{\mathrm{T}}_{sp_{r_i,r_j}} \end{cases} \tag{8.2}$$

It is notable that tensile stress $\Delta\sigma_y$ is defined as positive one in the formulation of BEA.

All the resultant load increments w.r.t. external forces that exert on the centroid of block r_i are denoted in the global coordinate system as (viz. Fig. 8.3)

$$\Delta\mathbf{f}_{r_i} = [\Delta f_x \;\; \Delta f_y \;\; \Delta f_\theta]^{\mathrm{T}}_{r_i} \quad (i = 1, 2, ..., n_r) \tag{8.3}$$

in which Δf_θ is the momentum at the center of r_i.

Fig. 8.3 Diagram to the equilibrium of representative block r_i

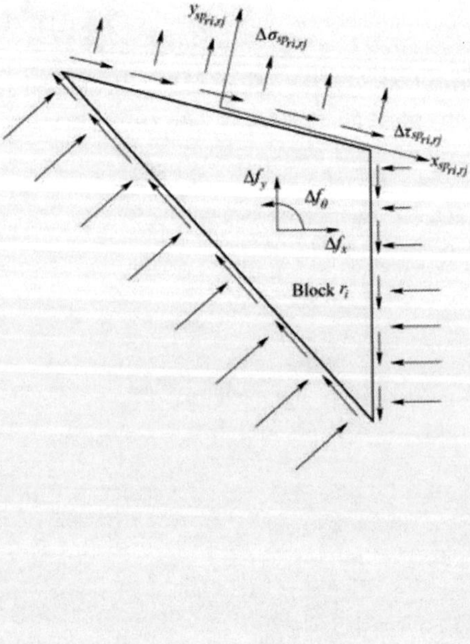

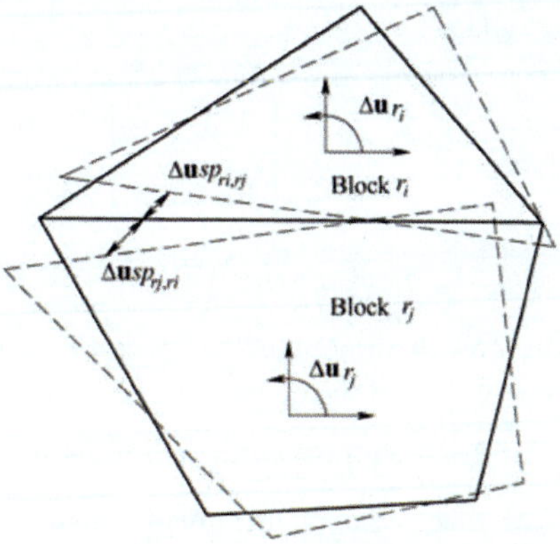

Fig. 8.4 Deformation of sp_{r_i,r_j} attributable to adjacent r_i and r_j

The deflection increments at the centroid of block r_i are also denoted in the global coordinate system as (viz. Figs. 8.2 and 8.4)

$$\Delta \mathbf{u}_{r_i} = \begin{bmatrix} \Delta u_x & \Delta u_y & \Delta u_\theta \end{bmatrix}_{r_i}^{\mathrm{T}} \quad (i = 1, 2, ..., n_r) \tag{8.4}$$

in which Δu_θ is the angular deflection at the center of r_i.

2. **Governing equations**

(1) **Force and moment equilibrium equations**

Integrating the stress increments on each sp_{r_i,r_j} into the resultant force with respect to the centroid of block r_i (viz. Fig. 8.3), we get the equilibrium equation

$$\Delta \mathbf{F}_{r_i} - \sum_{sp_{r_i,r_j} \subset r_i} H(sp_{r_i,r_j}) \int_{\Gamma_{sp_{r_i,r_j}}} \mathbf{P}_{sp_{r_i,r_j}} \Delta \boldsymbol{\sigma}_{sp_{r_i,r_j}} \mathrm{d}x_{sp_{r_i,r_j}} = 0 \quad (i = 1, 2, ..., n_r) \tag{8.5}$$

in which the Heaviside function is defined as

$$H(sp_{r_i,r_j}) = \begin{cases} 1 \text{ for } r_i \text{ in the positive half} - \text{space of } sp_{r_i,r_j} \\ -1 \text{ for } r_i \text{ in the negative half} - \text{space of } sp_{r_i,r_j} \end{cases} \tag{8.6}$$

and $[P]_{sp_{r_i,r_j}}$ is the 3×2 matrix defined as

$$\mathbf{P}_{sp_{r_i,r_j}} = \begin{bmatrix} \mathbf{T}_{sp_{r_i,r_j}}^{\mathrm{T}} \\ \mathbf{A}_{sp_{r_i,r_j}} + \mathbf{B}_{sp_{r_i,r_j}} \mathbf{T}_{sp_{r_i,r_j}}^{\mathrm{T}} \end{bmatrix} \tag{8.7}$$

in which

$$\begin{cases} \mathbf{A}_{sp_{r_i,r_j}} = \begin{bmatrix} 0 & -x_{sp_{r_i,r_j}} \end{bmatrix} \\ \mathbf{B}_{sp_{r_i,r_j}} = \begin{bmatrix} -(y^0_{sp_{r_i,r_j}} - y^0_{r_i}) & (x^0_{sp_{r_i,r_j}} - x^0_{r_i}) \end{bmatrix} \end{cases} \tag{8.8}$$

In Eq. (8.8), $\mathbf{x}^0_{sp_{r_i,r_j}} = [x^0_{sp_{r_i,r_j}} \;\; y^0_{sp_{r_i,r_j}}]^{\mathrm{T}}$ is the global coordinate of the mid-point of the structural plane sp_{r_i,r_j} on block $P_0(x_0, y_0)$, and $\mathbf{x}^0_{r_i} = [x^0_{r_i} \;\; y^0_{r_i}]^{\mathrm{T}}$ is the global coordinate of the centroid of block P_1.

(2) **Deformation compatibility equations**

The displacement increments of blocks P_2 and P_3 will give rise to deformation increment $\Delta\mathbf{u}_{p_{r_i,r_j}}$ on their contacting structural plane sp_{r_i,r_j} (viz. Fig. 8.4). According to kinematics they may be related by the formula

$$\Delta\mathbf{u}_{sp_{r_i,r_j}} = H(sp_{r_i,r_j})\mathbf{T}_{sp_{r_i,r_j}}(\mathbf{Q}_{r_i}\Delta\mathbf{u}_{r_i} - \mathbf{Q}_{r_j}\Delta\mathbf{u}_{r_j}) \quad (i, j = 1, 2, ..., n_r) \tag{8.9}$$

in which

$$\mathbf{Q}_{r_i} = \begin{bmatrix} 1 & 0 & -(y - y^0_{r_i}) \\ 0 & 1 & (x - x^0_{r_i}) \end{bmatrix} \quad (i = 1, 2, ..., n_r) \tag{8.10}$$

$\mathbf{Q}_{r_i}$ actually plays the role of shape function matrix that interpolates the displacement at any point within the rigid body using the deflections defined at centroid (cf. Fig. 8.2). Particularly, for the point on structural plane sp_{r_i,r_j}, the coordinate in Eq. (8.10) is calculated by

$$\mathbf{x} = \mathbf{T}^{\mathrm{T}}_{sp_{r_i,r_j}}\mathbf{x}_{sp_{r_i,r_j}} + \mathbf{x}^0_{r_i} \tag{8.11}$$

(3) **Constitutive equations**

According to the potential theory of viscoplasticity (Owen and Hinton 1980), the deformation and stress increments at any point $x_{j_{r_i,r_j}}$ on the structural plane sp_{r_i,r_j} shall observe the elasto-viscoplastic constitutive relation (3.142)–(3.145).

(4) **Equilibrium equations of the whole block system: static problems**

Invoking Eqs. (3.142) and (8.9), the equilibrium equation of block r_i Eq. (8.5) becomes

$$\mathbf{k}_{r_i,r_i}\Delta\mathbf{u}_{r_i} + \sum_{r_j}\mathbf{k}_{r_i,r_j}\Delta\mathbf{u}_{r_j} = \Delta\mathbf{f}_{r_i} + \Delta\mathbf{f}^{vp}_{r_i} \quad \text{for } (i = 1, 2, ..., n_r) \tag{8.12}$$

where r_j loops over all the neighboring blocks surrounding block r_i in corresponding to the structural plane $sp_{r_i,r_j} \subset r_i$, and

$$\begin{cases} \mathbf{k}_{r_i,r_i} = \sum\limits_{sp_{ri,rj} \subset r_i} \int\limits_{\Gamma_{sp_{ri,rj}}} \mathbf{P}_{sp_{r_i,r_j}} \mathbf{D}_{sp_{r_i,r_j}} \mathbf{T}_{sp_{r_i,r_j}} \mathbf{Q}_{r_i} \mathrm{d}x_{sp_{r_i,r_j}} \\ \mathbf{k}_{r_i,r_j} = -\int\limits_{\Gamma_{sp_{ri,rj}}} \mathbf{P}_{sp_{r_i,r_j}} \mathbf{D}_{sp_{r_i,r_j}} \mathbf{T}_{sp_{r_i,r_j}} \mathbf{Q}_{r_j} \mathrm{d}x_{sp_{r_i,r_j}} \qquad (i,j = 1,2,\ldots,n_r) \\ \Delta \mathbf{f}_{r_i}^{vp} = \sum\limits_{sp_{ri,rj} \subset r_i} H(sp_{r_i,r_j}) \int\limits_{\Gamma_{sp_{r_i,r_j}}} \mathbf{P}_{sp_{r_i,r_j}} \boldsymbol{\sigma}_{sp_{r_i,r_j}}^{vp} \mathrm{d}x_{sp_{r_i,r_j}} \end{cases} \tag{8.13}$$

For the other blocks, there also exist foregoing equations. Assembly all of such equations in a similar way to FEM, the equilibrium equation set of the block system is finally constructed

$$\mathbf{K}\Delta\mathbf{U} = \Delta\mathbf{F} + \Delta\mathbf{F}^{vp} \tag{8.14}$$

in which

$$\mathbf{K} = \begin{bmatrix} \mathbf{k}_{r_1,r_1} & \ldots\ \mathbf{k}_{r_1,r_i} & \ldots\ \mathbf{k}_{r_1,r_j} & \ldots\ \mathbf{k}_{r_1,r_{nr}} \\ \ldots & \ldots\ \ldots & \ldots\ \ldots & \ldots\ \ldots \\ \mathbf{k}_{r_i,r_1} & \ldots\ \mathbf{k}_{r_i,r_i} & \ldots\ \mathbf{k}_{r_i,r_j} & \ldots\ \mathbf{k}_{r_i,r_{nr}} \\ \ldots & \ldots\ \ldots & \ldots\ \ldots & \ldots\ \ldots \\ \mathbf{k}_{r_j,r_1} & \ldots\ \mathbf{k}_{r_j,r_i} & \ldots\ \mathbf{k}_{r_j,r_j} & \ldots\ \mathbf{k}_{r_j,r_{nr}} \\ \ldots & \ldots\ \ldots & \ldots\ \ldots & \ldots\ \ldots \\ \mathbf{k}_{r_{nr},r_1} & \ldots\ \mathbf{k}_{r_{nr},r_i} & \ldots\ \mathbf{k}_{r_{nr},r_j} & \ldots\ \mathbf{k}_{r_{nr},r_{nr}} \end{bmatrix} \tag{8.15}$$

$$\Delta\mathbf{U} = \left[\Delta\mathbf{u}_{r_1} \ldots \Delta\mathbf{u}_{r_i} \ldots \Delta\mathbf{u}_{r_j} \ldots \Delta\mathbf{u}_{r_{nr}}\right]^{\mathrm{T}} \tag{8.16}$$

$$\Delta\mathbf{F} = \left[\Delta\mathbf{f}_{r_1} \ldots \Delta\mathbf{f}_{r_i} \ldots \Delta\mathbf{f}_{r_j} \ldots \Delta\mathbf{f}_{r_{nr}}\right]^{\mathrm{T}} \tag{8.17}$$

$$\Delta\mathbf{F}^{vp} = \left[\Delta\mathbf{f}_{r_1}^{vp} \ldots \Delta\mathbf{f}_{r_i}^{vp} \ldots \Delta\mathbf{f}_{r_j}^{vp} \ldots \Delta\mathbf{f}_{r_{nr}}^{vp}\right]^{\mathrm{T}} \tag{8.18}$$

are the stiffness matrix, displacement increment, load increment, and equivalent (initial) force increment due to viscoplastic deformation, respectively.

At any time-marching step, we solve Eq. (8.14) firstly. Then use is made of the displacement increment $\Delta\mathbf{U}$ of the block system, the deformation and stress increments are calculated by Eqs. (8.9) and (3.142). Next, identify yield points on the structural plane: if yield occurs, viscoplastic strain/stress increments and corresponding equivalent load increments are calculated by Eqs. (3.142)–(3.145) and Eq. (8.13). Afterwards, the procedure goes to the next time-marching step.

(5) Equilibrium equations of the whole block system: dynamic problems

The force and moment equilibrium Eq. (8.5) in terms of stress and force are changed by taking the inertia of block element into account such that

$$-\iint\limits_{\Omega_{r_i}} \mathbf{Q}_{r_i}^T \Delta \mathbf{f}_{r_i}^{in} \mathrm{d}\Omega + \sum_{sp_{r_i,r_j} \subset r_i} H(sp_{r_i,r_j}) \int\limits_{\Gamma_{sp_{r_i,r_j}}} \mathbf{P}_{sp_{r_i,r_j}} \Delta \boldsymbol{\sigma}_{sp_{r_i,r_j}} \mathrm{d}x_{sp_{r_i,r_j}} = \Delta \mathbf{f}_{r_i}(t) \quad (i = 1, 2, \cdots, n_r) \tag{8.19}$$

in which $\Delta \mathbf{f}_{r_i}(t)$ and $\Delta \mathbf{f}_{r_i}^{in}$ are the external load increment and volumetric inertia increment acting on the centroid of block element r_i, respectively; $\mathbf{Q}_{r_i}$ has been defined in Eq. (8.10).

According to D'Alembert principle, we get

$$\Delta \mathbf{f}_{r_i}^{in} = -\rho_{r_i} \mathbf{Q}_{r_i} \frac{\mathrm{d}^2 \Delta \mathbf{u}_{r_i}}{\mathrm{d}t^2} \quad (i = 1, 2, ..., n_r) \tag{8.20}$$

On the substitution of Eqs. (3.142), (8.9) and (8.20) into Eq. (8.19), the dynamic equilibrium equation of block element r_i becomes

$$\mathbf{m}_{r_i,r_i} \frac{\mathrm{d}^2 \Delta \mathbf{u}_{r_i}}{\mathrm{d}t^2} + \mathbf{k}_{r_i,r_i} \Delta \mathbf{u}_{r_i} + \sum_{r_j} \mathbf{k}_{r_i,r_j} \Delta \mathbf{u}_{r_j} = \Delta \mathbf{f}_{r_i}(t) + \Delta \mathbf{f}_{r_i}^{vp}(t) \quad (i = 1, 2, \cdots, n_r) \tag{8.21}$$

where r_j runs over all the neighboring block elements around block element r_i in corresponding to structural plane sp_{r_i,r_j}, and

$$\mathbf{m}_{r_i} = \iint\limits_{\Omega_{r_i}} \mathbf{Q}_{r_i}^T \rho_{r_i} \mathbf{Q}_{r_i} \mathrm{d}\Omega \quad (i = 1, 2, \cdots, n_r) \tag{8.22}$$

Assemble all the block elements governed by Eq. (8.21) in a similar way to the dynamic FEM, it follows that

$$\mathbf{M} \frac{\mathrm{d}^2 \Delta \mathbf{U}}{\mathrm{d}t^2} + \mathbf{K} \Delta \mathbf{U} = \Delta \mathbf{F}(t) + \Delta \mathbf{F}^{vp}(t) \tag{8.23}$$

in which

$$\mathbf{M} = \begin{bmatrix} \mathbf{m}_{r_1,r_1} & ... \; 0 & ... \; 0 & ... \; 0 \\ ... & ... \; ... & ... \; ... & ... \; ... \\ 0 & ... \; \mathbf{m}_{r_i,r_i} & ... \; 0 & ... \; 0 \\ ... & ... \; ... & ... \; ... & ... \; ... \\ 0 & ... \; 0 & ... \; \mathbf{m}_{r_j,r_j} & ... \; 0 \\ ... & ... \; ... & ... \; ... & ... \; ... \\ 0 & ... \; 0 & ... \; 0 & ... \; \mathbf{m}_{r_{nr},r_{nr}} \end{bmatrix} \tag{8.24}$$

$$\frac{d^2\Delta\mathbf{U}}{dt^2} = \left[\frac{d^2\Delta\mathbf{u}_{r_1}}{dt^2} \cdots \frac{d^2\Delta\mathbf{u}_{r_i}}{dt^2} \cdots \frac{d^2\Delta\mathbf{u}_{r_j}}{dt^2} \cdots \frac{d^2\Delta\mathbf{u}_{r_{nr}}}{dt^2}\right]^{\mathrm{T}} \tag{8.25}$$

Damping is normally demanded in the practical applications. Similar to the dynamic FEM, Rayleigh damping model (Bathe 1982) is employed for the dynamic response of engineering structure in BEA

$$\mathbf{C} = \alpha_0\mathbf{M} + \alpha_1\mathbf{K} \tag{8.26}$$

in which α_0 and α_1 are defined in Eq. (4.171).

The momentum equation set (8.23) is therefore becomes

$$\mathbf{M}\frac{d^2\Delta\mathbf{U}}{dt^2} + \mathbf{C}\frac{d\Delta\mathbf{U}}{dt} + \mathbf{K}\Delta\mathbf{U} = \Delta\mathbf{F}(t) + \Delta\mathbf{F}^{vp}(t) \tag{8.27}$$

where

$$\frac{d\Delta\mathbf{U}}{dt} = \left[\frac{d^2\Delta\mathbf{u}_{r_1}}{dt} \cdots \frac{d\Delta\mathbf{u}_{r_i}}{dt} \cdots \frac{d\Delta\mathbf{u}_{r_j}}{dt} \cdots \frac{d\Delta\mathbf{u}_{r_{nr}}}{dt}\right]^{T} \tag{8.28}$$

is the velocity of rock block system.

For the seismic response problem, Eq. (8.27) can be further expressed as

$$\mathbf{M}\frac{d^2\Delta\mathbf{U}}{dt^2} + \mathbf{C}\frac{d\Delta\mathbf{U}}{dt} + \mathbf{K}\Delta\mathbf{U} = -\mathbf{M}\mathbf{a} + \Delta\mathbf{F}^{vp}(t) \tag{8.29}$$

in which $\mathbf{a}$ is the earthquake acceleration of ground.

Equation (8.27) or Eq. (8.29) is formalistly similar to that of the dynamic FEM. Therefore, the Wilson-θ or Newmark-β methods prevalently exerted in the dynamic FEM can be applied to solve them without any difficulties (viz. Chap. 4).

8.2.3 Formulation of BEA with Deformable Block Elements

1. **Concept of overlay elements**

A preliminary attempt to introduce block deformation into BEA was made by the complete polynomial interpolation of displacement field within blocks (Chen 1994). Later on, the displacement in rock block was revolutionarily interpolated from the nodal displacements of its overlay finite element that circumscribes the block, to attain much better numerical stability, and to allow for the adaptive refinement (Chen et al. 2004).

We choose r_i as the representative block, and denote its neighboring block as block r_j connecting r_i through the structural plane sp_{r_i,r_j}. The deformation and stress increments on sp_{r_i,r_j} are denoted as

$$\Delta \mathbf{u}_{sp_{r_i,r_j}} = \begin{bmatrix} \Delta u_x & \Delta u_y \end{bmatrix}^{\mathrm{T}}_{sp_{r_i,r_j}} \tag{8.30}$$

$$\Delta \boldsymbol{\sigma}_{sp_{r_i,r_j}} = \begin{bmatrix} \Delta \tau_{yx} & \Delta \sigma_y \end{bmatrix}^{\mathrm{T}}_{sp_{r_i,r_j}} \tag{8.31}$$

The displacement, strain, and stress increments in block r_i are denoted as

$$\Delta \mathbf{u}_{r_i} = \begin{bmatrix} \Delta u_x & \Delta u_y \end{bmatrix}^{\mathrm{T}}_{r_i} \quad \text{for } (i = 1, 2, \ldots, n_r) \tag{8.32}$$

$$\Delta \boldsymbol{\varepsilon}_{r_i} = \begin{bmatrix} \Delta \varepsilon_x & \Delta \varepsilon_y & \Delta \gamma_{xy} \end{bmatrix}^{\mathrm{T}}_{r_i} \quad \text{for } (i = 1, 2, \cdots, n_r) \tag{8.33}$$

$$\Delta \boldsymbol{\sigma}_{r_i} = \begin{bmatrix} \Delta \sigma_x & \Delta \sigma_y & \Delta \tau_{xy} \end{bmatrix}^{\mathrm{T}}_{r_i} \quad \text{for } (i = 1, 2, \cdots, n_r) \tag{8.34}$$

In Fig. 8.5, a finite element (standard or hierarchically refined) with n nodes ($n = 4$) is endued to circumscribe block r_i. This FE is termed as "overlay finite element (OFE)". The OFEs corresponding to different blocks can be inter-overlapped. The displacements $\Delta \mathbf{u}_{r_i}$ in block r_i are interpolated from the nodal displacements $\Delta \hat{\mathbf{u}}_{r_i}$ tied to the OFE by the formula

$$\Delta \mathbf{u}_{r_i} = \mathbf{N}_{r_i} \Delta \hat{\mathbf{u}}_{r_i} \quad (i = 1, 2, \cdots, n_r) \tag{8.35}$$

in which

$$\Delta \hat{\mathbf{u}}_{r_i} = \begin{bmatrix} \Delta \hat{\mathbf{u}}_1 \ldots \Delta \hat{\mathbf{u}}_l \ldots \Delta \hat{\mathbf{u}}_n \end{bmatrix}^{\mathrm{T}}_{r_i} \quad (i = 1, 2, \cdots, n_r) \tag{8.36}$$

$$\Delta \hat{\mathbf{u}}_l = \begin{bmatrix} \Delta \hat{u}_{x,l} & \Delta \hat{u}_{y,l} \end{bmatrix}^{\mathrm{T}} \tag{8.37}$$

and $\mathbf{N}_{r_i}$ is the shape function matrix of the OFE

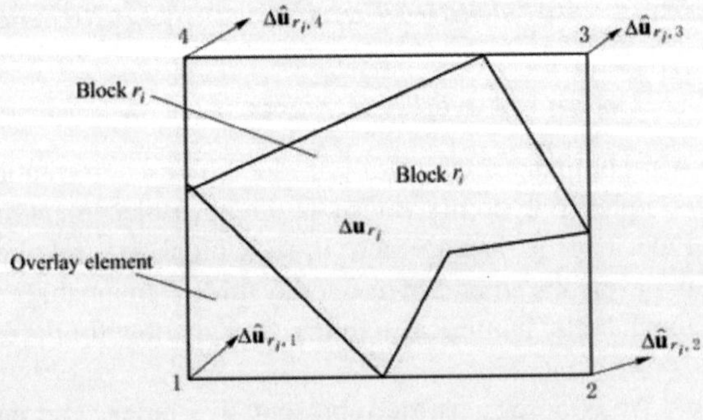

Fig. 8.5 Block r_i circumscribed by overlay finite element

$$\mathbf{N}_{r_i} = \left[N_1\mathbf{I} \ldots N_l\mathbf{I} \ldots N_n\mathbf{I} \right] \tag{8.38}$$

where $\mathbf{I}$ stands for 2×2 unit matrix for 2-D problems.

In Eqs. (8.35)–(8.38), $\mathbf{N}_{r_i}$ is defined in the whole OFE, $\Delta\hat{\mathbf{u}}_{r_i}$ is the general displacement vector corresponding to the general DOFs of OFE, $N_l(l = 1, ..., n)$ is the nodal shape function (viz. Chap. 4). It should be emphasized that the interpolation expressed in Eq. (8.35) is only valid within each circumscribed block element.

2. **Deformation compatibility equations**

Within r_i we have

$$\Delta\boldsymbol{\varepsilon}_{r_i} = \mathbf{B}_{r_i}\Delta\hat{\mathbf{u}}_{r_i} \quad (i = 1, 2, \cdots, n_r) \tag{8.39}$$

in which $\mathbf{B}_{r_i}$ is the strain matrix

$$\mathbf{B}_{r_i} = \left[\mathbf{B}_1 \ldots \mathbf{B}_l \ldots \mathbf{B}_n \right]_{r_i} \quad (i = 1, 2, \cdots, n_r) \tag{8.40}$$

$$\mathbf{B}_l = \begin{bmatrix} \frac{\partial N_l}{\partial x} & 0 \\ 0 & \frac{\partial N_l}{\partial y} \\ \frac{\partial N_l}{\partial y} & \frac{\partial N_l}{\partial x} \end{bmatrix} \quad (l = 1, \ldots, n) \tag{8.41}$$

The relative displacement increment on sp_{r_i,r_j} is

$$\Delta\mathbf{u}_{sp_{r_i,r_j}} = H(sp_{r_i,r_j})\mathbf{T}_{sp_{r_i,r_j}}\left(\mathbf{N}_{r_i}\Delta\hat{\mathbf{u}}_{r_i} - \mathbf{N}_{r_j}\Delta\hat{\mathbf{u}}_{r_j}\right) \quad (r_i, r_j = 1, 2, \cdots, n_r) \tag{8.42}$$

in which the matrix $\mathbf{T}_{sp_{r_i,r_j}}$ has been defined in Eq. (3.125), and $H(sp_{r_i,r_j})$ has been defined in Eq. (8.6).

3. **Constitutive equations**

Use is made of the potential elasto-viscoplastic theory (Owen and Hinton 1980), the stress and strain increments in block r_i can be related by Eq. (3.138) such that

$$\Delta\boldsymbol{\sigma}_{r_i} = \mathbf{D}_{r_i}\Delta\boldsymbol{\varepsilon}_{r_i} - \Delta\boldsymbol{\sigma}_{r_i}^{vp} \quad (i = 1, 2, \cdots, n_r) \tag{8.43}$$

in which elastic matrix $\mathbf{D}_{r_i}$ is the function of Young's modulus and Poisson's ratio; the viscoplastic stress increment $\Delta\boldsymbol{\sigma}_{r_i}^{vp}$ is given according to the associated flow rule with Drucker-Prager yield criterion.

Meanwhile, the deformation and stress increments at any point on sp_{r_i,r_j} will observe Eq. (3.142) which is restated as

$$\Delta\boldsymbol{\sigma}_{sp_{r_i,r_j}} = \mathbf{D}_{sp_{r_i,r_j}}\Delta\mathbf{u}_{sp_{r_i,r_j}} - \Delta\boldsymbol{\sigma}_{sp_{r_i,r_j}}^{vp} \tag{8.44}$$

in which the elastic matrix $\mathbf{D}_{sp_{r_i,r_j}}$ is the function of the normal and tangential stiffness coefficients [viz. Eq. (3.123)]; the viscoplastic stress increment $\Delta\boldsymbol{\sigma}^{vp}_{sp_{r_i,r_j}}$ is given according to the non-associated flow rule with the Mohr–Coulomb yield criterion.

4. **Governing equations**
(1) **Virtual work principle**

The virtual work principle for a discrete system containing blocks n_r is constructed as

$$\sum_{i=1}^{n_r} W_{r_i} + \sum_{sp_{r_i,r_j}=1}^{n_{sp}} W_{sp_{r_i,r_j}} = W_f \tag{8.45}$$

in which W_{r_i} is the virtual work contributed from r_i, $W_{sp_{r_i,r_j}}$ is the virtual work contributed from sp_{r_i,r_j}, W_f is the virtual work of external forces, n_r and n_{sp} are the amounts of blocks and structural plane patches in the whole rock block system, respectively.

If blocks r_i and r_j demonstrate virtual nodal displacements $\delta\Delta\hat{\mathbf{u}}_{r_i}$ and $\delta\Delta\hat{\mathbf{u}}_{r_j}$, then the virtual displacements inside the blocks are $\delta\Delta\mathbf{u}_{r_i}$ and $\delta\Delta\mathbf{u}_{r_j}$, the corresponding virtual strains are $\delta\Delta\boldsymbol{\varepsilon}_{r_i}$ and $\delta\Delta\boldsymbol{\varepsilon}_{r_j}$, the virtual deformation on sp_{r_i,r_j} is $\delta\Delta\mathbf{u}_{sp_{r_i,r_j}}$. They must observe following restraints

$$\begin{cases} \delta\Delta\mathbf{u}_{r_i} = \mathbf{N}_{r_i}\delta\Delta\hat{\mathbf{u}}_{r_i} \\ \delta\Delta\mathbf{u}_{r_j} = \mathbf{N}_{r_j}\delta\Delta\hat{\mathbf{u}}_{r_j} \end{cases} \tag{8.46}$$

$$\begin{cases} \delta\Delta\boldsymbol{\varepsilon}_{r_i} = \mathbf{B}_{r_i}\delta\Delta\hat{\mathbf{u}}_{r_i} \\ \delta\Delta\boldsymbol{\varepsilon}_{r_j} = \mathbf{B}_{r_j}\delta\Delta\hat{\mathbf{u}}_{r_j} \end{cases} \tag{8.47}$$

$$\delta\mathbf{u}_{sp_{r_i,r_j}} = H(sp_{r_i,r_j})\mathbf{T}_{sp_{r_i,r_j}}\left(\mathbf{N}_{r_i}\delta\hat{\mathbf{u}}_{r_i} - \mathbf{N}_{r_j}\delta\hat{\mathbf{u}}_{r_j}\right) \quad (i, j = 1, 2, \cdots, n_r) \tag{8.48}$$

- Virtual work contributed from block r_i

$$W_{r_i} = \iint\limits_{\Omega_{r_i}} \delta\boldsymbol{\varepsilon}^{\mathrm{T}}_{r_i}\Delta\boldsymbol{\sigma}_{r_i}d\Omega \quad (i = 1, 2, \cdots, n_r) \tag{8.49}$$

- Virtual work contributed from structural plane sp_{r_i,r_j}

$$W_{sp_{r_i,r_j}} = \int\limits_{\Gamma_{sp_{r_i,r_j}}} \delta\Delta\mathbf{u}^{\mathrm{T}}_{sp_{r_i,r_j}}\Delta\boldsymbol{\sigma}_{sp_{r_i,r_j}}dx_{sp_{r_i,r_j}} \quad (i, j = 1, 2, \cdots, n_r) \tag{8.50}$$

- Virtual work of external forces

$$W_f = \sum_{i=1}^{n_r} \delta \Delta \hat{\mathbf{u}}_{r_i}^{\mathrm{T}} \Delta \mathbf{f}_{r_i} \tag{8.51}$$

$\Delta \mathbf{f}_{r_i}$ is the general load vector tied to the OFE that is transferred from the forces exerting on block element r_i (will be detailed later).

(2) **Equilibrium equations**

On the substitution of Eqs. (8.35)–(8.44) and (8.46)–(8.51) into Eq. (8.45), and recall that $\delta \Delta \hat{\mathbf{u}}_{r_i}$ is arbitrary, it follows that

$$\begin{aligned}\Delta \mathbf{f}_{r_i} = & \iint_{\Omega_{r_i}} \mathbf{B}_{r_i}^{\mathrm{T}} \mathbf{D}_{r_i} \mathbf{B}_{r_i} \Delta \hat{\mathbf{u}}_{r_i} \mathrm{d}\Omega - \iint_{\Omega_{r_i}} \mathbf{B}_{r_i}^{\mathrm{T}} \Delta \boldsymbol{\sigma}_{r_i}^{vp} \mathrm{d}\Omega \\ & + \sum_{sp_{r_i,r_j} \subset r_i} \int_{\Gamma_{sp_{r_i,r_j}}} \Big(\mathbf{N}_{r_i}^{\mathrm{T}} \mathbf{T}_{sp_{r_i,r_j}}^{\mathrm{T}} \mathbf{D}_{sp_{r_i,r_j}} \mathbf{T}_{sp_{r_i,r_j}} \mathbf{N}_{r_i} \Delta \hat{\mathbf{u}}_{r_i} \\ & - \mathbf{N}_{r_i}^{\mathrm{T}} \mathbf{T}_{sp_{r_i,r_j}}^{\mathrm{T}} \mathbf{D}_{sp_{r_i,r_j}} \mathbf{T}_{sp_{r_i,r_j}} \mathbf{N}_{r_j} \Delta \hat{\mathbf{u}}_{r_j} - H(sp_{r_i,r_j}) \mathbf{N}_{r_i}^{\mathrm{T}} \mathbf{T}_{sp_{r_i,r_j}}^{\mathrm{T}} \Delta \boldsymbol{\sigma}_{sp_{r_i,r_j}}^{vp} \Big) \mathrm{d}x_{sp_{r_i,r_j}}\end{aligned} \tag{8.52}$$

in which the subscript sp_{r_i,r_j} runs over all the structural planes surrounding r_i, and the subscript r_j runs over all the neighboring blocks of block r_i corresponding to sp_{r_i,r_j}. Equation (8.52) may be simply expressed as

$$\mathbf{k}_{r_i,r_i} \Delta \hat{\mathbf{u}}_{r_i} + \sum_{sp_{r_i,r_j}} \mathbf{k}_{r_i,r_j} \Delta \hat{\mathbf{u}}_{r_j} = \Delta \mathbf{f}_{r_i} + \Delta \mathbf{f}_{r_i}^{vp} \tag{8.53}$$

in which

$$\begin{cases} \mathbf{k}_{r_i,r_i} = \iint_{\Omega_{r_i}} \mathbf{B}_{r_i}^{\mathrm{T}} \mathbf{D}_{r_i} \mathbf{B}_{r_i} \mathrm{d}\Omega + \sum_{sp_{r_i,r_j} \subset r_i} \int_{\Gamma_{sp_{r_i,r_j}}} \mathbf{N}_{r_i}^{\mathrm{T}} \mathbf{T}_{sp_{r_i,r_j}}^{\mathrm{T}} \mathbf{D}_{sp_{r_i,r_j}} \mathbf{T}_{sp_{r_i,r_j}} \mathbf{N}_{r_j} \mathrm{d}x_{sp_{r_i,r_j}} \\ \mathbf{k}_{r_i,r_j} = \mathbf{k}_{r_j,r_i} = -\int_{\Gamma_{sp_{r_i,r_j}}} \mathbf{N}_{r_i}^{\mathrm{T}} \mathbf{T}_{sp_{r_i,r_j}}^{\mathrm{T}} \mathbf{D}_{sp_{r_i,r_j}} \mathbf{T}_{sp_{r_i,r_j}} \mathbf{N}_{r_j} \mathrm{d}x_{sp_{r_i,r_j}} \quad \text{for } \Gamma_{sp_{r_i,r_j}} \subset r_i \qquad (i, j = 1, 2, \cdots, n_r) \\ \Delta \mathbf{f}_{r_i}^{vp} = \iint_{\Omega_{r_i}} \mathbf{B}_{r_i}^{\mathrm{T}} \Delta \boldsymbol{\sigma}_{r_i}^{vp} \mathrm{d}\Omega + \sum_{sp_{r_i,r_j} \subset r_i} \int_{\Gamma_{sp_{r_i,r_j}}} \mathbf{N}_{r_i}^{\mathrm{T}} \mathbf{T}_{sp_{r_i,r_j}}^{\mathrm{T}} \Delta \boldsymbol{\sigma}_{sp_{r_i,r_j}}^{vp} \mathrm{d}x_{sp_{r_i,r_j}} \end{cases} \tag{8.54}$$

Assemble Eq. (8.53) into the equilibrium equation set of the whole block system following the similar routine of FEM, we get

$$\begin{bmatrix} \mathbf{k}_{r_1,r_1} & \ldots & \mathbf{k}_{r_1,r_i} & \ldots & \mathbf{k}_{r_1,,r_j} & \ldots & \mathbf{k}_{r_1,r_{nr}} \\ \ldots & & \ldots\ldots & & \ldots\ldots & & \ldots\ldots \\ \mathbf{k}_{r_i,r_1} & \ldots & \mathbf{k}_{r_i,r_i} & \ldots & \mathbf{k}_{r_i,r_j} & \ldots & \mathbf{k}_{r_i,r_{nr}} \\ \ldots & & \ldots\ldots & & \ldots\ldots & & \ldots\ldots \\ \mathbf{k}_{r_j,r_1} & \ldots & \mathbf{k}_{r_j,r_i} & \ldots & \mathbf{k}_{r_j,r_j} & \ldots & \mathbf{k}_{r_j,r_{nr}} \\ \ldots & & \ldots\ldots & & \ldots\ldots & & \ldots\ldots \\ \mathbf{k}_{r_{nr},r_1} & \ldots & \mathbf{k}_{r_{nr},r_i} & \ldots & \mathbf{k}_{r_{nr},r_j} & \ldots & \mathbf{k}_{r_{nr},r_{nr}} \end{bmatrix} \begin{Bmatrix} \Delta\hat{\mathbf{u}}_{r_1} \\ \ldots \\ \Delta\hat{\mathbf{u}}_{r_i} \\ \ldots \\ \Delta\hat{\mathbf{u}}_{r_j} \\ \ldots \\ \Delta\hat{\mathbf{u}}_{r_{nr}} \end{Bmatrix} = \begin{Bmatrix} \Delta\mathbf{f}_{r_1} \\ \ldots \\ \Delta\mathbf{f}_{r_i} \\ \ldots \\ \Delta\mathbf{f}_{r_j} \\ \ldots \\ \Delta\mathbf{f}_{r_{nr}} \end{Bmatrix} + \begin{Bmatrix} \Delta\mathbf{f}^{vp}_{r_1} \\ \ldots \\ \Delta\mathbf{f}^{vp}_{r_i} \\ \ldots \\ \Delta\mathbf{f}^{vp}_{r_j} \\ \ldots \\ \Delta\mathbf{f}^{vp}_{r_{nr}} \end{Bmatrix} \tag{8.55}$$

or

$$\mathbf{K}\Delta\widehat{\mathbf{U}} = \Delta\mathbf{F} + \Delta\mathbf{F}^{vp} \tag{8.56}$$

In Eq. (8.56), $\mathbf{K}$ is the global stiffness matrix assembled from the elementary sub-matrices and $\Delta\widehat{\mathbf{U}}$ is the global general displacement increments vector

$$\Delta\widehat{\mathbf{U}} = \left[\Delta\hat{\mathbf{u}}_{r_1} \ldots \Delta\hat{\mathbf{u}}_{r_i} \ldots \Delta\hat{\mathbf{u}}_{r_j} \ldots \Delta\hat{\mathbf{u}}_{r_{nr}}\right]^{\mathrm{T}} \tag{8.57}$$

$\Delta\mathbf{F}$ and $\Delta\mathbf{F}^{vp}$ are the correspondent global general load vectors

$$\Delta\mathbf{F} = \left[\Delta\mathbf{f}_{r_1} \ldots \Delta\mathbf{f}_{r_i} \ldots \Delta\mathbf{f}_{r_j} \ldots \Delta\mathbf{f}_{r_{nr}}\right]^{\mathrm{T}} \tag{8.58}$$

$$\Delta\mathbf{F}^{vp} = \left[\Delta\mathbf{f}^{vp}_{r_1} \ldots \Delta\mathbf{f}^{vp}_{r_i} \ldots \Delta\mathbf{f}^{vp}_{r_j} \ldots \Delta\mathbf{f}^{vp}_{r_{nr}}\right]^{\mathrm{T}} \tag{8.59}$$

With the solved nodal displacements $\Delta\widehat{\mathbf{U}}$ from Eq. (8.56), the displacements/strains/stresses within block r_i and on structural plane sp_{r_i,r_j} can be calculated by Eqs. (8.35, 8.39, 8.42, 8.43, 8.44).

(3) **Nodal loads**

There are two types of nodal load increment in Eq. (10.53). One is the equivalent load $\Delta\mathbf{f}^{vp}_{r_i}$ due to viscoplastic deformation given in Eq. (10.54). Another is the equivalent load $\Delta\mathbf{f}_{r_i}$ due to actions (forces), which can be accumulated from the different types of actions by the virtual work consideration w.r.t. balance condition, normally $\Delta\mathbf{f}_{r_i} = \Delta\mathbf{f}_{r_i,T} + \Delta\mathbf{f}_{\overline{r_i,t}} + \Delta\mathbf{f}_{r_i,b}$.

- Concentrated force.

  If a concentrated (point) force $\Delta\mathbf{T}_\zeta = \left[\Delta T_{x_\zeta}\ \Delta T_{y_\zeta}\right]^{\mathrm{T}}$ exerts on block r_i, and the coordinate of acting point is $\mathbf{x}_\zeta = \left(x_\zeta\ y_\zeta\right)$, then its virtual work is

$$\delta\mathbf{u}^{\mathrm{T}}_{r_i}\Delta\mathbf{T}_\zeta = \delta\hat{\mathbf{u}}^{\mathrm{T}}_{r_i}\mathbf{N}^{\mathrm{T}}_{r_i}\big|_{\mathbf{x}_\zeta}\Delta\mathbf{T}_\zeta \quad \text{for } \mathbf{x}_\zeta \in r_i \tag{8.60}$$

It must be equal to the virtual work due to external load $\Delta\mathbf{f}_{r_i,T}$ such that

$$\delta\hat{\mathbf{u}}_{r_i}^{\mathrm{T}}\Delta\mathbf{f}_{r_i,T} = \delta\hat{\mathbf{u}}_{r_i}^{\mathrm{T}}\mathbf{N}_{r_i}^{\mathrm{T}}\big|_{\left(x_\zeta\ y_\zeta\right)}\Delta\mathbf{T}_\zeta \quad \text{for } \mathbf{x}_\zeta \in r_i \tag{8.61}$$

Whence it follows that

$$\Delta\mathbf{f}_{r_i,T} = \mathbf{N}_{r_i}^{\mathrm{T}}\big|_{\left(x_\zeta\ y_\zeta\right)}\Delta\mathbf{T}_\zeta \quad \text{for } \mathbf{x}_\zeta \in r_i \tag{8.62}$$

- Surface distributed force.

  The surface distributed force $\Delta\bar{\mathbf{t}} = \left[\Delta\bar{t}_x\ \Delta\bar{t}_y\right]^{\mathrm{T}}$ exerting on the structural plane (or domain boundary) of block r_i can be transferred into the load vector in the same way of point force as

$$\Delta\mathbf{f}_{r_i,\bar{t}} = \int_{\Gamma_{spr_i,r_j}} \mathbf{N}_{r_i}^{\mathrm{T}}\Delta\bar{\mathbf{t}}\mathrm{d}x_{spr_i,r_j} \quad \text{for } \Gamma_{spr_i,r_j} \subset r_i \tag{8.63}$$

- Volumetric force.

  The volumetric force $\Delta\mathbf{b} = \left[\Delta b_x\ \Delta b_y\right]^{\mathrm{T}}$ of block r_i can also be balanced by the corresponding nodal load vector as follows

$$\Delta\mathbf{f}_{r_i,b} = \iint_{\Omega_{r_i}} \mathbf{N}_{r_i}^{\mathrm{T}}\Delta\mathbf{b}\mathrm{d}\Omega \tag{8.64}$$

5. **Refinement strategies**

(1) **Refinement with regard to the deformation pattern in large and complex blocks**

Towards the formulation of a practical p-refinement BEA, the error estimation and the order-upgrade of shape function may be carried out on the block element individually. For details, readers are referred to literatures (Chen 2018).

(2) **Refinement with regard to the crack-tip within blocks**

In the block system identification for a computation domain, a field measured or statistically created "discrete fracture network (DFN)" is demanded. Figure 8.6a shows the DFN in a sampling window (10 × 10 m), and there are 48 block elements completely delimited by joints. The tapered joint within a block element are normally neglected as an invalid edge (Fig. 8.6b), if the main concern is the macro-scale deformation and the motion of block system. However, where the crack growth is of importance, such a tapered joint should be kept and simulated using the asymptotic enrichment near tip field within the framework of OFE (Fig. 8.6c), which is similar to that in XFEM.

Take 2-D problems for example, the displacement interpolation (8.35) is extended in a manner of

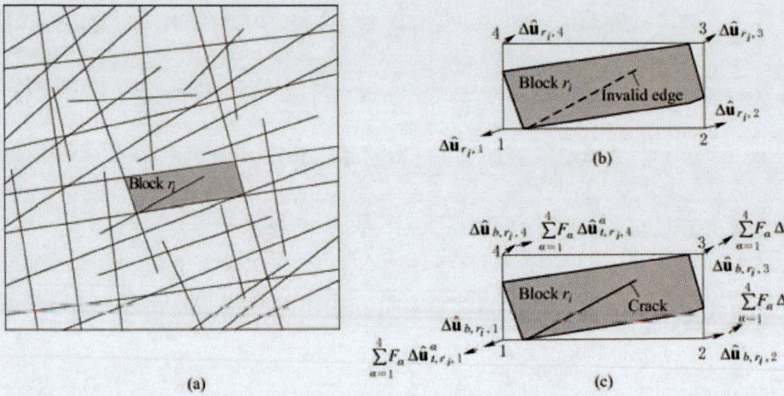

Fig. 8.6 Handling of joints in deformable block element with OFE. **a** Stochastic DFN at $x - y$ plane (sample size = 10 m). **b** Invalid edge for tapered joint. **c** Crack-tip for tapered joint

$$\Delta\mathbf{u}_{r_i} = \mathbf{N}_{r_i}\Delta\hat{\mathbf{u}}_{b,r_i} + \mathbf{N}_{r_i}\Delta\hat{\mathbf{u}}_{t,r_i} \quad (i = 1, 2, \cdots, n_r) \tag{8.65}$$

In which $\mathbf{N}_{r_i}$ and $\Delta\hat{\mathbf{u}}_{b,r_i}$ are identical to that in Eqs. (8.36 and 8.37), specifically

$$\begin{cases} \Delta\hat{\mathbf{u}}_{b,r_i} = \left[\Delta\hat{\mathbf{u}}_{b,r_i,1} \ldots \Delta\hat{\mathbf{u}}_{b,r_i,l} \ldots \Delta\hat{\mathbf{u}}_{b,r_i,n}\right]^{\mathrm{T}} \\ \Delta\hat{\mathbf{u}}_{b,r_i,l} = \begin{Bmatrix} \Delta\hat{u}_{b,r_i,x,l} \\ \Delta\hat{u}_{b,r_i,y,l} \end{Bmatrix} \end{cases}$$
$$\text{for } (r_i = 1, 2, \cdots, n_r) \text{ and } (l = 1, 2, \ldots, n) \tag{8.66}$$

is the nodal displacement increment for the normal DOF identical to FEM, and the nodal enrichment coefficients $\Delta\hat{\mathbf{u}}_{t,r_i}$ for the OFE containing crack-tip are given by

$$\Delta\hat{\mathbf{u}}_{t,r_i} = \sum_{\alpha=1}^{4} F_\alpha \Delta\hat{\mathbf{u}}^{\alpha}_{t,r_i} \tag{8.67}$$

$$\Delta\hat{\mathbf{u}}^{\alpha}_{t,r_i} = \left[\Delta\hat{\mathbf{u}}^{\alpha}_{t,r_i,1} \ldots \Delta\hat{\mathbf{u}}^{\alpha}_{t,r_i,l} \ldots \Delta\hat{\mathbf{u}}^{\alpha}_{t,r_i,n}\right]^{\mathrm{T}} \quad (\alpha = 1, 2, 3, 4) \tag{8.68}$$

$$\Delta\hat{\mathbf{u}}^{\alpha}_{t,r_i,l} = \begin{Bmatrix} \Delta\hat{u}^{\alpha}_{t,r_i,x,l} \\ \Delta\hat{u}^{\alpha}_{t,r_i,y,l} \end{Bmatrix} \quad \text{for } (\alpha = 1,2,3,4) \text{ and } (l = 1, 2, \ldots, n) \tag{8.69}$$

for the standard linear 2-D finite element, $n = 4$ in Eqs. (8.66)–(8.69).

The crack-tip enrichment functions s in isotropic elasticity are derived from the asymptotic displacement fields [viz. Equation (3.191)], which use four asymptotic functions to span the Westergaard solution at crack-tip in the development expansion of crack-tip displacement field (Bui 1978; Fleming et al. 1997)

$$\begin{aligned}\mathbf{F}(r,\theta) = \left[\,F_1(r,\theta)\;F_2(r,\theta)\;F_3(r,\theta)\;F_4(r,\theta)\,\right]^{\mathrm{T}} = \\ \left[\,\sqrt{r}\sin\left(\tfrac{\theta}{2}\right)\;\sqrt{r}\cos\left(\tfrac{\theta}{2}\right)\;\sqrt{r}\sin\left(\tfrac{\theta}{2}\right)\sin\theta\;\sqrt{r}\cos\left(\tfrac{\theta}{2}\right)\sin\theta\,\right]^{\mathrm{T}}\end{aligned} \tag{8.70}$$

in which (r, θ) is a polar coordinate system with origin at the crack-tip (viz. Fig. 3.13). Note that the first function in Eq. (8.70) is discontinuous across the crack whereas the last three functions are continuous.

The strain in the enriched element is now extended as

$$\boldsymbol{\varepsilon}_{r_i} = \mathbf{B}_{b,r_i}\Delta\hat{\mathbf{u}}_{b,r_i} + \sum_{\alpha=1}^{4}\mathbf{B}^{\alpha}_{t,r_i}\Delta\overset{\wedge\alpha}{\mathbf{u}}_{t,r_i} \quad (i = 1, 2, \cdots, n_r) \tag{8.71}$$

where $\mathbf{B}_{b,r_i}$ is identical to that in Eq. (8.40)

$$\begin{cases}\mathbf{B}_{b,r_i} = \mathbf{B}_{r_i} = \left[\,\mathbf{B}_1 \ldots \mathbf{B}_l \ldots \mathbf{B}_n\,\right] \; (i = 1, 2, \cdots, n_r) \\ \mathbf{B}_l = \begin{bmatrix}\frac{\partial N_l}{\partial x} & 0 \\ 0 & \frac{\partial N_l}{\partial y} \\ \frac{\partial N_l}{\partial y} & \frac{\partial N_l}{\partial x}\end{bmatrix} \qquad (l = 1, 2, \ldots, n)\end{cases} \tag{8.72}$$

and

$$\begin{cases}\mathbf{B}^{\alpha}_{t,r_i} = \left[\,\mathbf{B}^{\alpha}_{t,r_i,1} \ldots \mathbf{B}^{\alpha}_{t,r_i,l} \ldots \mathbf{B}^{\alpha}_{t,r_i,n}\,\right] & (i = 1, 2, \cdots, n_r) \\ \mathbf{B}^{\alpha}_{t,r_i,l} = \mathbf{B}_l F_\alpha & (\alpha = 1, 2, 3, 4)\end{cases} \tag{8.73}$$

Then, the stress field can be determined by using an adequate constitutive law such that

$$\boldsymbol{\sigma}_{r_i} = \mathbf{D}\boldsymbol{\varepsilon}_{r_i} \quad (i = 1, 2, \cdots, n_r) \tag{8.74}$$

Invoking the virtual work principle, the governing equation set of the block element r_i containing crack-tip is obtained

$$\begin{bmatrix}\mathbf{k}_{b,b} & \mathbf{k}_{b,1} & \ldots & \mathbf{k}_{b,4} \\ \mathbf{k}_{1,b} & \mathbf{k}_{1,1} & \ldots & \mathbf{k}_{1,4} \\ \ldots & \ldots & \ldots & \ldots \\ \mathbf{k}_{4,b} & \mathbf{k}_{4,1} & \ldots & \mathbf{k}_{4,4}\end{bmatrix}\begin{Bmatrix}\Delta\hat{\mathbf{u}}_{b,r_i} \\ \Delta\hat{\mathbf{u}}^{1}_{t,r_i} \\ \ldots \\ \Delta\hat{\mathbf{u}}^{4}_{t,r_i}\end{Bmatrix} = \begin{Bmatrix}\Delta\mathbf{f}_{b,r_i} \\ \Delta\mathbf{f}^{1}_{t,r_i} \\ \ldots \\ \Delta\mathbf{f}^{4}_{t,r_i}\end{Bmatrix} + \begin{Bmatrix}\Delta\mathbf{f}^{vp}_{b,r_i} \\ \Delta\mathbf{f}^{vp,1}_{t,r_i} \\ \ldots \\ \Delta\mathbf{f}^{vp,4}_{t,r_i}\end{Bmatrix} \quad (i = 1, 2, \cdots, n_r) \tag{8.75}$$

The stiffness matrix considering the coupling effects between "normal" and additional "enriched" DOFs, are

$$\begin{cases} \mathbf{k}_{b,b} = \iint\limits_{\Omega_{r_i}} (\mathbf{B}_{b,r_i})^{\mathrm{T}} \mathbf{D}_{r_i} \mathbf{B}_{b,r_i} \mathrm{d}\Omega \\ \mathbf{k}_{b,\alpha} = \mathbf{k}_{\alpha,b} = \iint\limits_{\Omega_e} (\mathbf{B}_{b,r_i})^{\mathrm{T}} \mathbf{D}_{r_i} \mathbf{B}^{\alpha}_{t,r_i} \mathrm{d}\Omega \quad (i = 1, 2, \cdots, n_r), (\alpha, \beta = 1, \ldots, 4) \\ \mathbf{k}_{\alpha,\beta} = \mathbf{k}_{\beta,\alpha} = \iint\limits_{\Omega_e} (\mathbf{B}^{\alpha}_{t,r_i})^{\mathrm{T}} \mathbf{D}_{r_i} \mathbf{B}^{\beta}_{t,r_i} \mathrm{d}\Omega \end{cases} \tag{8.76}$$

In the right hand side of Eq. (8.75), load vectors $\Delta\mathbf{f}_{r_i}$ due to external force and $\Delta\mathbf{f}^{vp}_{r_i}$ due to viscoplastic deformation can be calculated by the similar routine in XFEM (Chap. 5) w.r.t. the "normal" and additional "enriched" DOFs, respectively

$$\begin{cases} \Delta\mathbf{f}_{b,r_i} = \iint\limits_{\Omega_{r_i}} (\mathbf{N}_{r_i})^{\mathrm{T}} \Delta\mathbf{b} \mathrm{d}\Omega + \int\limits_{\Gamma_t} (\mathbf{N}_{r_i})^{\mathrm{T}} \Delta\bar{\mathbf{t}} \mathrm{d}\Gamma + \sum\limits_{\mathbf{x}_\zeta \in r_i} (\mathbf{N}_{r_i})^{\mathrm{T}} (\mathbf{x}_\zeta) \Delta\mathbf{T}_\zeta (\mathbf{x}_\zeta) \\ \Delta\mathbf{f}^{vp}_{b,r_i} = \iint\limits_{\Omega_{r_i}} (\mathbf{B}_{b,r_i})^{\mathrm{T}} \Delta\boldsymbol{\sigma}^{vp} \mathrm{d}\Omega \qquad (i = 1, 2, \cdots, n_r) \end{cases} \tag{8.77}$$

in which

$$\begin{cases} \Delta\mathbf{f}^{\alpha}_{t,r_i} = \iint\limits_{\Omega_{r_i}} (\mathbf{N}_{r_i})^{\mathrm{T}} F_\alpha \Delta\mathbf{b} \mathrm{d}\Omega + \int\limits_{\Gamma_t} (\mathbf{N}_{r_i})^{\mathrm{T}} F_\alpha \Delta\bar{\mathbf{t}} \mathrm{d}\Gamma + \sum\limits_{\mathbf{x}_\zeta \in r_i} (\mathbf{N}_{r_i})^{\mathrm{T}} F_\alpha \Delta\mathbf{T}_\zeta (\mathbf{x}_\zeta) \\ \Delta\mathbf{f}^{vp,\alpha}_{t,r_i} = \iint\limits_{\Omega_{r_i}} (\mathbf{B}^{\alpha}_{t,r_i})^{\mathrm{T}} \Delta\boldsymbol{\sigma}^{vp} \mathrm{d}\Omega \qquad (\alpha = 1, ..., 4), (i = 1, 2, \cdots, n_r) \end{cases} \tag{8.78}$$

If the cracking criterion and direction as well as the stress release w.r.t. crack growth are handled, the methods and algorithms are identical to that of XFEM, too.

8.2.4 Key Algorithms

1. **Boundary condition arise from truncation foundation**

For many geotechnical and hydraulic structures, the calculation domain merely comprises a portion of foundation or surrounding rock that is trancated from the Earth crust. The condition on such an artificial boundary need to be carefully handled to guarantee the computation presicion.

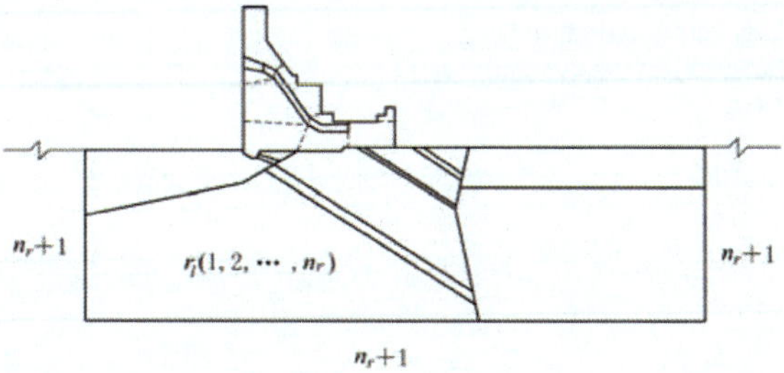

Fig. 8.7 Baozhusi Gravity dam with $n_r = 28$ movable blocks and one fixed boundary block

Take the static analysis of gravity dam for example, the exterior portion of dam foundation (viz. Fig. 8.7) may be simply looked at as a fixed and monolithic block (numbered as block $n_r + 1$) in the half space. The blocks at its vicinity are restrained via face-to-face (3-D) or edge-to-edge (2-D) contact but always keep $\Delta\hat{\mathbf{u}}_{n_r+1} = 0$ in the governing equations.

However, for seismic problem, thus an artificial boundary need be particular treatment, because the seismic wave will be completely reflected on the artificial boundary if it is not appropriately handled, which will lead to erroneous results. Therefore, it is necessary to introduce restraints at the artificial boundary (Clayton and Engquist1980; Higdon 1986), for the purpose of eliminating unreal boundary reflection.

Various artificial boundary conditions are available based on the seminal work of Lysmer and Kuhlemeyer (1969). In BEA, the visco-elastic artificial boundary condition proposed by Deeks and Randolph (1994) is adopted for the dynamic analysis, which is the development of viscous boundary condition. The general form of such condition in the dynamic FEM is expressed as

$$\sigma_{l,\alpha}(t) = -K^{\alpha} \times u_{l,\alpha}(t) - C^{\alpha} \times \frac{\mathrm{d}u_{l,\alpha}(t)}{\mathrm{d}t} \tag{8.79}$$

in which $\sigma_{l,\alpha}(t)$, $u_{l,\alpha}(t)$ and $\frac{\mathrm{d}u_{l,\alpha}(t)}{\mathrm{d}t}$ are the stress, displacement, and velocity of node l along the direction α at time t; K^{α} and C^{α} are the elastic modulus and damping factor along the direction α. They are the functions of material characteristics.

The physical meaning of visco-elastic boundary is that at every point along each direction of the boundary, an element containing spring and damping is designed, which is used to simulate the absorption and recovery of the artificial boundary.

For the dynamic response using BEA, the spring-damping element is put at the artificial boundary as illustrated in Fig. 8.8.

For the rigid block r_i contacted to artificial boundary sp_b, the dynamic governing Eq. (8.27) written in the element level is

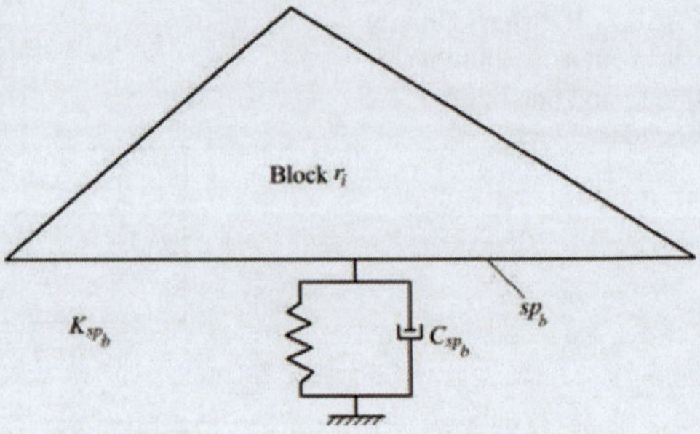

Fig. 8.8 Visco-elastic boundary in the dynamic BEA

$$\mathbf{m}_{r_i,r_i}\frac{\mathrm{d}^2\Delta\mathbf{u}_{r_i}}{\mathrm{d}t^2}+\mathbf{c}_{r_i,r_i}\frac{\mathrm{d}\Delta\mathbf{u}_{r_i}}{\mathrm{d}t}+\sum_{r_j}\mathbf{c}_{r_i,r_j}\frac{\mathrm{d}\Delta\mathbf{u}_{r_j}}{\mathrm{d}t}+\mathbf{k}_{r_i,r_i}\Delta\mathbf{u}_{r_i}+\sum_{r_j}\mathbf{k}_{r_i,r_j}\Delta\mathbf{u}_{r_i}$$
$$=\Delta\mathbf{f}(t)_{r_i}+\Delta\mathbf{f}^{vp}(t)_{r_i}+\Delta\mathbf{p}_{r_i}\quad(i=1,2,\cdots,n_r)\tag{8.80}$$

in which $\Delta\mathbf{p}_{r_i}$ is integrated from the visco-elastic stress $\Delta\boldsymbol{\sigma}_{sp_b}$ at the artificial boundary sp_b

$$\Delta\mathbf{p}_{r_i}=-\sum_{sp_b\subset r_i}H(sp_b)\int_{\Gamma_{sp_b}}\mathbf{P}_{sp_b}\Delta\boldsymbol{\sigma}_{sp_b}\mathrm{d}x_{sp_b}\ (i=1,2,\cdots,n_r)\tag{8.81}$$

Invoking the general concept of visco-elastic boundary in Eq. (8.79), it follows that

$$\Delta\boldsymbol{\sigma}_{sp_b}=-\mathbf{D}_{sp_b}\Delta\mathbf{u}_{sp_b}-\mathbf{C}_{sp_b}\frac{\mathrm{d}\Delta\mathbf{u}_{sp_b}}{\mathrm{d}t}\tag{8.82}$$

in which $\Delta\mathbf{u}_{sp_b}$ and $\frac{\mathrm{d}\Delta\mathbf{u}_{sp_b}}{\mathrm{d}t}$ are the displacement and displacement rates at the artificial boundary sp_b.

According to the deformation compatibility condition (8.9), for the block element r_i contacted to the artificial boundary sp_b, we have

$$\begin{cases}\Delta\mathbf{u}_{sp_b}=H(sp_b)\mathbf{T}_{sp_b}\mathbf{Q}_{sp_b}\Delta\mathbf{u}_{r_i}\\ \dfrac{\mathrm{d}\Delta\mathbf{u}_{sp_b}}{\mathrm{d}t}=H(sp_b)\mathbf{T}_{sp_b}\mathbf{Q}_{sp_b}\dfrac{\mathrm{d}\Delta\mathbf{u}_{r_i}}{\mathrm{d}t}\end{cases}(i=1,2,\cdots,n_r)\tag{8.83}$$

and

$$\mathbf{D}_{sp_b}=\begin{bmatrix}k_{s,sp_b}&0&0\\0&k_{s,sp_b}&0\\0&0&k_{n,sp_b}\end{bmatrix}\tag{8.84}$$

$$\mathbf{C}_{sp_b} = \begin{bmatrix} C_{s,sp_b} & 0 & 0 \\ 0 & C_{s,sp_b} & 0 \\ 0 & 0 & C_{n,sp_b} \end{bmatrix} \tag{8.85}$$

where: k_{n,sp_b} and k_{s,sp_b}—normal stiffness and shear stiffness; C_{n,sp_b} and C_{s,sp_b}—normal damping factor and shear damping factor.

Invoking Eqs. (8.81)–(8.85), Eq. (8.80) becomes

$$\begin{aligned} &\mathbf{m}_{r_i,r_i}\frac{d^2\Delta\mathbf{u}_{r_i}}{dt^2} + \left(\mathbf{c}_{r_i,r_i} + \mathbf{c}_{r_i}\right)\frac{\mathrm{d}\Delta\mathbf{u}_{r_i}}{\mathrm{d}t} \\ &\quad + \sum_{r_j}\mathbf{c}_{r_i,r_j}\frac{\mathrm{d}\Delta\mathbf{u}_{r_j}}{\mathrm{d}t} + \left(\mathbf{k}_{r_i,r_i} + \mathbf{k}_{r_i}\right)\Delta\mathbf{u}_{r_i} + \sum_{r_j}\mathbf{k}_{r_i,r_j}\Delta\mathbf{u}_{r_j} \\ &= \Delta\mathbf{f}(t)_{r_i} + \Delta\mathbf{f}^{vp}(t)_{r_i} \quad (i = 1, 2, \cdots, n_r) \end{aligned} \tag{8.86}$$

in which

$$\mathbf{k}_{r_i} = \sum_{sp_b \subset r_i}\int_{\Gamma_{sp_b}} \mathbf{Q}_{r_i}^{\mathrm{T}}\mathbf{T}_{sp_b}^{\mathrm{T}}\mathbf{D}_{sp_b}\mathbf{T}_{sp_b}\mathbf{Q}_{r_i}\mathrm{d}x_{sp_b} \quad (i = 1, 2, \cdots, n_r) \tag{8.87}$$

$$\mathbf{c}_{r_i} = \sum_{sp_b \subset r_i}\int_{\Gamma_{sp_b}} \mathbf{Q}_{r_i}^{\mathrm{T}}\mathbf{T}_{sp_b}^{\mathrm{T}}\mathbf{C}_{sp_b}\mathbf{T}_{sp_b}\mathbf{Q}_{r_i}\mathrm{d}x_{sp_b} \quad (i = 1, 2, \cdots, n_r) \tag{8.88}$$

It is evident that the introduction of visco-elastic artifical boundary condition only changes the diagonal elements of the stiffness and damping matrices.

2. **Numerical integration on DFN**

Calculation of the matrices and vectors in the formulation of BEA requires the numerical integration (quadrature) on structural planes (joint, fault, crack) and in entire entity blocks. A Gaussian quadrature method on the irregular structural plane as well as in the irregular block entity may be employed for such purpose (Chap. 4).

3. **Searching for safety factors**

A potential failure case (mode) is defined as a portion of rock mass formed by one block or the combination of several blocks whose "factor of safety (FOS)" is comparatively lower than that stipulated in the design specifications. If a domain contains a large amount of blocks, theoretically there will emerge substantial failure cases, and the search for their FOS will account for huge computer resources. To minimize the search effort, a heuristic search strategy called "intelligent search method" had been proposed. The search strategy is characterized that only the possible block combinations enlightened by the previously calibrated block combinations, are to be considered in the successive search (Xu et al. 2000).

8.3 Kinematics of DDA

Where the separation and dislocation of block elements emerge as the consequence of large deformation, the fundamental BEA elaborated in Sect. 8.2 are subject to significant reconstruction of governing equations, particularly the contact evolution from edge-to-edge (2-D) only to mixed vertex-to-edge and edge-to-edge, as well as their update during the motion of block assemblage. In doing so, the formulation will be too delicate, because it results in complex algorithm with the help of, for example, the FEM of finite deformation.

Under such circumstances, DDA demonstrates its strong merit that is able to simulate the motion of rock mass accompanied by block separation and dislocation related to the post failure, meanwhile to give a certain description of strain/stress within blocks. Hereinafter, the rest content will focus on the principle and algorithm of DDA.

8.3.1 *Deformation Patterns*

In the basic formulation of DDA, the rock mass is considered to be an assemblage of blocks separated by structural planes, particularly joints. Each block is arbitrary in geometry and the displacement within block observes complete first order polynomial distribution. For the 2-D case, a block possesses six DOFs, and the displacement (u_x, u_y) at any point (x, y) in block element r_i may be expressed as:

$$\mathbf{u}_{r_i} = \begin{Bmatrix} u_{r_i,x} \\ u_{r_i,y} \end{Bmatrix} = \mathbf{N}_{r_i}\hat{\mathbf{u}}_{r_i} \quad (i = 1, 2, \cdots, n_r) \tag{8.89}$$

in which the general displacement vector $\hat{\mathbf{u}}_{r_i}$ comprises six variables:

$$\hat{\mathbf{u}}_{r_i} = \begin{bmatrix} u_x^0 \; u_y^0 \; u_\theta^0 \; \varepsilon_x \; \varepsilon_y \; \gamma_{xy} \end{bmatrix}^{\mathrm{T}} \tag{8.90}$$

and $\mathbf{N}_{r_i}$ is the general shape matrix

$$\mathbf{N}_{r_i} = \begin{bmatrix} 1 \; 0 \; -(y-y^0) \; (x-x^0) \; 0 & (y-y^0)/2 \\ 0 \; 1 \; (x-x^0) \quad 0 & -(y-y^0) \; (x-x^0)/2 \end{bmatrix} \tag{8.91}$$

where $\left(u_x^0 \; u_y^0 \; u_\theta^0 \right)$ are the transition and rotation at the geometrical centroid (x^0, y^0) of the block element, $(\varepsilon_x, \varepsilon_y, \gamma_{xy})$ are the constant normal and shear strains in the whole block element. Equations (8.90) and (8.91) may be abbreviated in the form of

$$\begin{cases} \mathbf{N}_{r_i} = \begin{bmatrix} t_{11} & t_{12} & t_{13} & t_{14} & t_{15} & t_{16} \\ t_{21} & t_{22} & t_{23} & t_{24} & t_{25} & t_{26} \end{bmatrix} \\ \hat{\mathbf{u}}_{r_i} = [\hat{u}_1 \quad \hat{u}_2 \quad \hat{u}_3 \quad \hat{u}_4 \quad \hat{u}_5 \quad \hat{u}_6]^{\mathrm{T}}_{r_i} \end{cases} \tag{8.92}$$

It is notable that if the rotation u_θ^0 is large, Eqs. (8.89)–(8.91) is used to get $\hat{\mathbf{u}}_{r_i}$ firstly, afterwards the transitional motion of each vertex should be revised by

$$\begin{cases} u_x = (x - x^0)(\cos u_\theta^0 - 1) - (y - y^0))\sin u_\theta^0 \\ u_y = (y - y^0)(\cos u_\theta^0 - 1) + (x - x^0)\sin u_\theta^0 \end{cases} \tag{8.93}$$

This is to avoid the distortion of block area due to the first order polynomial shape matrix (8.91).

8.3.2 *Types and Detection Techniques of Contact*

Similar to other particle-based CMs (e.g. DEM, BEA), few simple and robust algorithms could provide sophisticated and efficient solutions of contact positions and contact directions. Of cause, the common plane (CP) and its improved versions initially for DEM may be used in the contact detection for DDA. Contact potential energy that is the cornerstone of FEM-DEM in dealing with the discontinuous problems may be successfully employed in the DDA, too (Xu et al. 2020). In this book, the initial version of contact detection algorithm proposed by Shi for DDA will be elucidated. By the distinction of "convex angle" and "concave angle", the detection algorithm may be implemented without the combination of convex polygons into one concave polygon. It is a simple algorithm that needs to compare all the vertexes of one element to all the faces of another and vice versa, and is rather similar to the algorithm for the contact detection between polyhedrons in DEM (Barbosa 1990). In Shi's work, the term "vertex" is replaced by the term "angle".

1. **Types of contact**

When blocks move, they tend to penetrate each other giving rise to contact forces. The first step of contact detection is to identify the types of contact between arbitrarily polygonal blocks. Shi distinguished three different kinds of contact showing in Fig. 8.9.

① Contact between one edge and another angle (vertex) (viz. Fig. 8.9a): the inter-penetration will occur if and only if angle (vertex) V_1 passes line (edge) $V_2 - V_3$ that is the reference line.

② Contact between one acute (convex) angle (vertex) and one obtuse (concave) angle (viz. Fig. 8.9b): the inter-penetration will occur if one of the following two conditions is met:

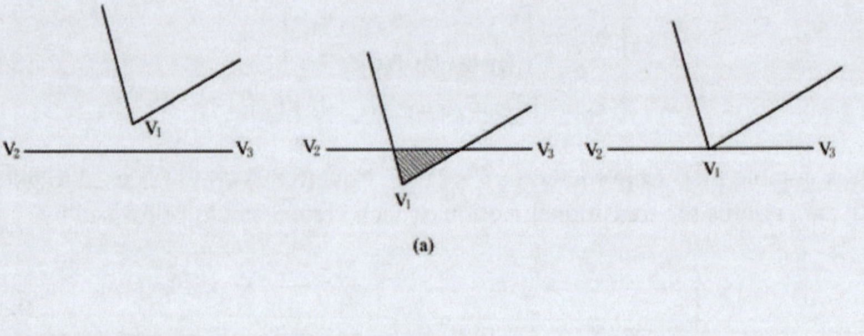

Fig. 8.9 Types of contact. **a** Angle-to-edge. **b** Convex angle-to-concave angle. **c** Convex angle-to-convex angle

Vertex V_1 passes line $V_2 - V_3$;
Point V_1 passes line $V_2 - V_4$ where $V_2 - V_3$ and $V_2 - V_4$ are two reference lines.
③ Contact between two acute angles (vertexes) (viz. Fig. 8.9c): Inter-penetration will occur if each of the two reference lines of one vertex is crossed by another vertex. The reference lines for the contact detection between two convex angles are selected using Table 8.1 according to the included angles α and β of four edges.

Table 8.1 Reference lines for the contact detection of two convex angles

α	β	First reference line	Second reference line
$\le 180°$	$\le 180°$	O-E3	O-E2
$\le 180°$	$> 180°$	O-E3	O-E4
$> 180°$	$\le 180°$	O-E1	O-E2
$> 180°$	$> 180°$	O-E1	O-E4

It is important to mention that in DDA, all types of contact will be equivalently simulated as one or two vertex-to-edge contacts.

2. **Neighbor searching**

Neighbor searching is the first step in contact detection. This is identical to the mapping from block elements circumscribed by a rectangle box (frame) to rectangle grid (viz. Fig. 7.3).

3. **Penetration detection**

In the second step called as "geometric resolution", pairs of contact blocks obtained from the first step are examined in more detail to find the possible contact points towards the calculation of contact forces.

We define a rectangle frame for each block element schematically shown in Fig. 8.10. An incremental detection approach may be adopted with the postulations that:

- The contact state at the beginning of current time t is known, and the contact state at the end of current step $t = t + \Delta t$ is to be solved;
- The time increment Δt for each time step is small enough so that the displacements of all the points within the problem domain are smaller than a pre-defined maximum tolerance d_0, which is used as the criterion for contact detection (viz. Fig. 8.11).

The surface point of one block element may contact any portion of the surface of another, hence the contact areas are not known a-priori and may change considerably within a time-marching step. So, the algorithm for detecting contact at the beginning of each time-marching step is implemented by two sub-steps:

- A global search is first carried out to find the blocks that might possibly come into contact, namely, if the distance between the frame boxes of any two blocks is smaller than $2d_0$, they are likely to contact each other within this time-marching step;
- Then, a local search is followed to figure out all the vertex-to-edge contact pairs.

Fig. 8.10 Definition of rectangle frame

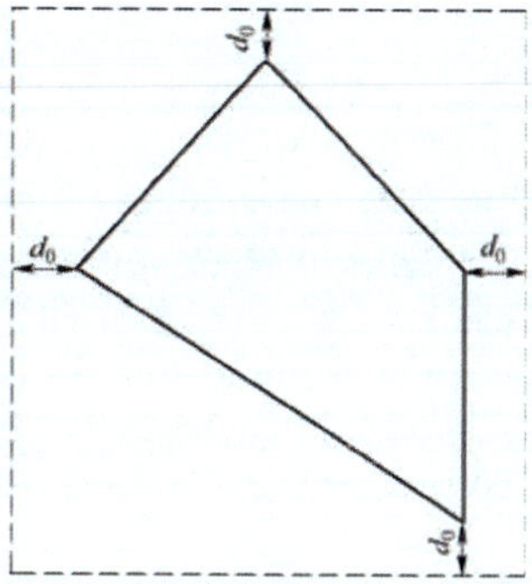

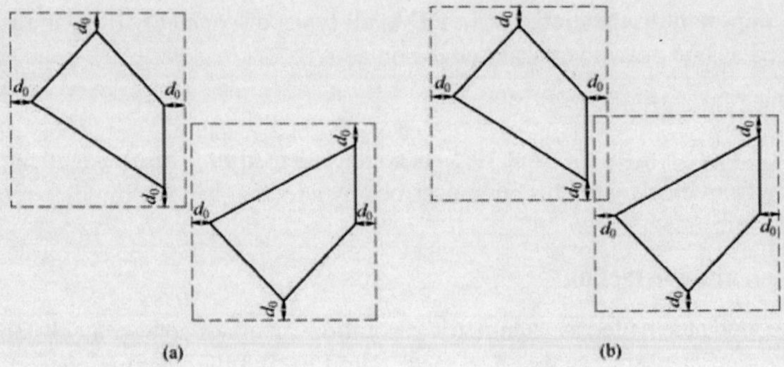

Fig. 8.11 Preliminary detection. **a** Non-adjacent blocks. **b** Adjacent blocks

8.4 Dynamics of DDA

8.4.1 Equilibrium Equation

Suppose there are n_r block elements constituting the block system, the simultaneous equilibrium equation set is written in the matrix form as follows

$$\begin{bmatrix} \mathbf{k}_{r_1,r_1} & \cdots & \mathbf{k}_{r_1,r_i} & \cdots & \mathbf{k}_{r_1,r_j} & \cdots & \mathbf{k}_{r_1,r_{nr}} \\ \cdots & \cdots & \cdots & \cdots & \cdots & \cdots & \cdots \\ \mathbf{k}_{r_i,r_1} & \cdots & \mathbf{k}_{r_i,r_i} & \cdots & \mathbf{k}_{r_i,r_j} & \cdots & \mathbf{k}_{r_i,r_{nr}} \\ \cdots & \cdots & \cdots & \cdots & \cdots & \cdots & \cdots \\ \mathbf{k}_{r_j,r_1} & \cdots & \mathbf{k}_{r_j,r_i} & \cdots & \mathbf{k}_{r_j,r_j} & \cdots & \mathbf{k}_{r_j,r_{nr}} \\ \cdots & \cdots & \cdots & \cdots & \cdots & \cdots & \cdots \\ \mathbf{k}_{r_{nr},r_1} & \cdots & \mathbf{k}_{r_{nr},r_i} & \cdots & \mathbf{k}_{r_{nr},r_j} & \cdots & \mathbf{k}_{r_{nr},r_{nr}} \end{bmatrix} \begin{Bmatrix} \hat{\mathbf{u}}_{r_1} \\ \cdots \\ \hat{\mathbf{u}}_{r_i} \\ \cdots \\ \hat{\mathbf{u}}_{r_j} \\ \cdots \\ \hat{\mathbf{u}}_{r_{nr}} \end{Bmatrix} = \begin{Bmatrix} \mathbf{f}_{r_1} \\ \cdots \\ \mathbf{f}_{r_i} \\ \cdots \\ \mathbf{f}_{r_j} \\ \cdots \\ \mathbf{f}_{r_{nr}} \end{Bmatrix} \tag{8.94a}$$

or briefly

$$\mathbf{K}\widehat{\mathbf{U}} = \mathbf{F} \tag{8.94b}$$

in which each coefficient sub-matrix $\mathbf{k}_{r_i,r_j}$ $(i \neq j)$ is defined by the contacts between blocks r_i and r_j only, whereas the sub-matrix $\mathbf{k}_{r_i,r_i}$ depends on both the material properties of block r_i itself plus the contacts between blocks r_i and r_j. Since each block r_i possesses six DOFs defined by the components of $\hat{\mathbf{u}}_{r_i}$ in Eq. (8.90), $\mathbf{k}_{r_i,r_j}$ is a 6×6 sub-matrix, $\hat{\mathbf{u}}_{r_i}$ and $\mathbf{f}_{r_i}$ are 6×1 sub-vectors. These sub-matrices and sub-vectors may be derived by, the virtual work principle or the minimization of the potential energy, the latter was employed in the initial works of Shi.

The total potential energy Π is contributed from the block strain Π_s, inertia force Π_{in}, load Π_f, initial stress Π_σ, displacement constraint Π_{sp}, deformation due to the hard spring at normal penetration Π_{np}, tangential slip deformation Π_{ts}, frictional deformation Π_{sc}, etc., namely

$$\Pi = \Pi_s + \Pi_{in} + \Pi_f + \Pi_\sigma + \Pi_{sp} + \Pi_{np} + \Pi_{ts} + \Pi_{sc} \tag{8.95}$$

The basic operation to derive the sub-matrices and sub-vectors is to invoke the stationary condition w.r.t. correspondent energy components in Eq. (8.95). Generally, the sub-matrix $\mathbf{k}_{r_i,r_j}$ and the sub-vector $\mathbf{f}_{r_i}$ are obtained by demanding

$$\begin{cases} \mathbf{k}_{r_i,r_j} = \dfrac{\partial^2 \Pi}{\partial \hat{\mathbf{u}}_{r_i} \partial \hat{\mathbf{u}}_{r_j}} \\ \mathbf{f}_{r_i} = -\dfrac{\partial \Pi(0)}{\partial \hat{\mathbf{u}}_{r_i}} \end{cases} \quad (i, j = 1, 2, ..., n_r) \tag{8.96}$$

8.4.2 *Sub-matrices of Stiffness and Sub-vectors of Strains/Stresses/Loads*

1. **Elastic strain**

The strain energy Π_s of block element r_i is

$$\begin{aligned} \Pi_s &= \frac{1}{2} \iint\limits_{\Omega_{r_i}} \boldsymbol{\varepsilon}_{r_i}^{\mathrm{T}} \boldsymbol{\sigma}_{r_i} \mathrm{d}\Omega = \frac{1}{2} \iint\limits_{\Omega_{r_i}} \boldsymbol{\varepsilon}_{r_i}^{\mathrm{T}} \mathbf{D}_{r_i} \boldsymbol{\varepsilon}_{r_i} \mathrm{d}\Omega \\ &= \frac{1}{2} \iint\limits_{\Omega_{r_i}} \hat{\mathbf{u}}_{r_i}^{\mathrm{T}} \mathbf{D}_{r_i} \hat{\mathbf{u}}_{r_i} \mathrm{d}\Omega = \frac{S}{2} \hat{\mathbf{u}}_{r_i}^{\mathrm{T}} \mathbf{D}_{r_i} \hat{\mathbf{u}}_{r_i} \quad (i = 1, 2, ..., n_r) \end{aligned} \tag{8.97}$$

in which S is the area of block element r_i.

For each time-marching step, assume the block is linear-elastic and under plane stress state, it follows that

$$\mathbf{D}_{r_i} = \frac{E}{1 - \nu^2} \begin{bmatrix} 0\,0\,0\,0\,0\,0 \\ 0\,0\,0\,0\,0\,0 \\ 0\,0\,0\,0\,0\,0 \\ 0\,0\,0\,1\,\nu\,0 \\ 0\,0\,0\,\nu\,1\,0 \\ 0\,0\,0\,0\,0\,\frac{1-\nu}{2} \end{bmatrix} \quad (i = 1, 2, ..., n_r) \tag{8.98}$$

in which E is the Young's modulus and ν is the Poisson' ratio. For plane strain problem, $E \to \frac{E}{1-\nu^2}$ and $\nu \to \frac{\nu}{1-\nu}$ are employed to replace the E and ν in Eq. (8.98).

Invoking the stationary condition with regard to Π_e, it follows that

$$\frac{\partial^2 \Pi_s}{\partial \hat{\mathbf{u}}_{r_i} \hat{\mathbf{u}}_{r_i}} = \frac{S}{2} \frac{\partial^2}{\partial \hat{\mathbf{u}}_{r_i} \partial \hat{\mathbf{u}}_{r_i}} (\hat{\mathbf{u}}_{r_i}^{\mathrm{T}} \mathbf{D}_{r_i} \hat{\mathbf{u}}_{r_i}) = S\mathbf{D}_{r_i} \quad (i = 1, 2, ..., n_r) \tag{8.99}$$

Assembling operation to the sub-matrix $\mathbf{k}_{r_i, r_i}$ is carried out such that

$$S\mathbf{D}_{r_i} \rightarrow \mathbf{k}_{r_i, r_i} \quad (i = 1, 2, ..., n_r) \tag{8.100}$$

where the arrow " $\rightarrow$ " means the operation to assemble.

We note that the 6 × 6 sub-matrix (8.100) has only null elements in the first three columns and first three rows. Therefore, there are no resistances to rigid body movement (u_0, v_0, r_0). In order to provide an adequate resistance to prevent the block element from going too far when it detaches from others, an artificial resistance matrix is constructed as

$$\delta \mathbf{D}_{r_i} = \begin{bmatrix} \delta_x & 0 & 0 & 0 & 0 & 0 \\ 0 & \delta_y & 0 & 0 & 0 & 0 \\ 0 & 0 & \delta_\theta & 0 & 0 & 0 \\ 0 & 0 & 0 & 0 & 0 & 0 \\ 0 & 0 & 0 & 0 & 0 & 0 \\ 0 & 0 & 0 & 0 & 0 & 0 \end{bmatrix} \quad (i = 1, 2, ..., n_r) \tag{8.101}$$

which will be assembled to the sub-matrix $\mathbf{k}_{r_i, r_i}$, too

$$\delta \mathbf{D}_{r_i} \rightarrow \mathbf{k}_{r_i, r_i} \quad (i = 1, 2, ..., n_r) \tag{8.102}$$

2. **Initial stress**

The potential energy in block element r_i due to initial stress $\boldsymbol{\sigma}^0 = \left[0\ 0\ 0\ \sigma_x^0\ \sigma_y^0\ \tau_{xy}^0 \right]^{\mathrm{T}}$ that is postulated as constant within the block element may be expressed as

$$\Pi_\sigma = -\iint_{\Omega_i} \boldsymbol{\varepsilon}_{r_i}^{\mathrm{T}} \boldsymbol{\sigma}^0 \mathrm{d}\Omega = -S \hat{\mathbf{u}}_{r_i}^{\mathrm{T}} \boldsymbol{\sigma}^0 \quad (i = 1, 2, ..., n_r) \tag{8.103}$$

Invoking the stationary condition w.r.t. Π_σ, we get

$$-\frac{\partial \Pi_\sigma(0)}{\partial \hat{\mathbf{u}}_{r_i}} = S \frac{\partial \hat{\mathbf{u}}_{r_i}^{\mathrm{T}}}{\partial \hat{\mathbf{u}}_{r_i}} \boldsymbol{\sigma}^0 = S\, \boldsymbol{\sigma}^0 \quad (i = 1, 2, ..., n_r) \tag{8.104}$$

This 6 × 1 sub-matrix $S\boldsymbol{\sigma}^0$ is assembled to $\mathbf{f}_{r_i}$ in Eq. (8.94), namely

$$S\boldsymbol{\sigma}_0 \rightarrow \mathbf{f}_{r_i} \tag{8.105}$$

3. **Point force**

The potential energy due to a point (concentrated) force $\mathbf{T}_\zeta = \left[T_{x_\zeta} \; T_{y_\zeta} \right]^{\mathrm{T}}$ exerted on the location $\mathbf{x}_\zeta = \left(x_\zeta \; y_\zeta \right)$ of block element r_i is simply given by

$$\Pi_p = -(f_x u_x + f_y u_y) = -\hat{\mathbf{u}}_{r_i}^{\mathrm{T}} \mathbf{N}_{r_i}^{\mathrm{T}} \Big|_{\mathbf{x}_\zeta} \mathbf{T}_\zeta \quad \text{for } \mathbf{x}_\zeta \in r_i \tag{8.106}$$

in which (u_x, u_y) is the displacement at point (x, y). To minimize Π_p, its derivative is calculated to get

$$-\frac{\partial \Pi_p(0)}{\partial \hat{\mathbf{u}}_{r_i}} = \mathbf{N}_{r_i}^{\mathrm{T}} \Big|_{\mathbf{x}_\zeta} \mathbf{T}_\zeta \quad \text{for } \mathbf{x}_\zeta \in r_i \tag{8.107}$$

which is the product of 6×2 matrix and 2×1 vector, and is assembled into the vector $\mathbf{f}_{r_i}$ in Eq. (8.94)

$$\mathbf{N}_{r_i}^{\mathrm{T}} \Big|_{\mathbf{x}_\zeta} \mathbf{T}_\zeta \to \mathbf{f}_{r_i} \tag{8.108}$$

4. **Body force**

Denote $\mathbf{b} = \left[b_x \; b_y \right]^{\mathrm{T}}$ as the volumetric load exerting on (x_0, y_0) that is the centroid of block element r_i calculated by

$$\begin{cases} x^0 = \iint\limits_{\Omega_{r_i}} x \mathrm{d}\Omega \Big/ \iint\limits_{\Omega_{r_i}} \mathrm{d}\Omega \\ y^0 = \iint\limits_{\Omega_{r_i}} y \mathrm{d}\Omega \Big/ \iint\limits_{\Omega_{r_i}} \mathrm{d}\Omega \end{cases} \tag{8.109}$$

The potential energy due to this body force is

$$\begin{aligned} \Pi_b &= -\iint\limits_{\Omega_{r_i}} (b_x u_x + b_y u_y) \mathrm{d}\Omega = -\iint\limits_{\Omega_{r_i}} \hat{\mathbf{u}}_{r_i}^{\mathrm{T}} \mathbf{N}_{r_i}^{\mathrm{T}} \mathbf{b} \mathrm{d}\Omega \\ &= -S \hat{\mathbf{u}}_{r_i}^{\mathrm{T}} \mathbf{b} \quad (i = 1, 2, \ldots, n_r) \end{aligned} \tag{8.110}$$

in which (u_x, u_y) is the displacement at the point (x, y) of block element r_i. The derivative of Π_b is undertaken to get corresponding sub-vector

$$-\frac{\partial \Pi_b(0)}{\partial \hat{\mathbf{u}}_{r_i}} = S\mathbf{b} \quad (i = 1, 2, \ldots, n_r) \tag{8.111}$$

This 6×1 sub-vector is assembled into the load vector $\mathbf{f}_{r_i}$ in Eq. (8.94), i.e.

$$S\mathbf{b} \to \mathbf{f}_{r_i} \tag{8.112}$$

in which S is the area of block element r_i.

5. **Inertia force**

Denote $(u_x(t), u_y(t))$ as the time dependent displacement at any point (x, y) of block element r_i with density ρ. The force of inertia is

$$\mathbf{f}^{in} = \begin{Bmatrix} f_x \\ f_y \end{Bmatrix} = -\rho \begin{Bmatrix} \dfrac{\mathrm{d}^2 u_x(t)}{\mathrm{d}t^2} \\ \dfrac{\mathrm{d}^2 u_y(t)}{\mathrm{d}t^2} \end{Bmatrix} \tag{8.113}$$

and the correspondent potential energy is given by

$$\Pi_{in} = -\iint_{\Omega_{r_i}} \mathbf{u}_{r_i}^{\mathrm{T}} \mathbf{f}^{in} \mathrm{d}\Omega = \iint_{\Omega_{r_i}} \rho \hat{\mathbf{u}}_{r_i}^{\mathrm{T}} \mathbf{N}_{r_i}^{\mathrm{T}} \mathbf{N}_{r_i} \frac{\mathrm{d}^2 \hat{\mathbf{u}}_{r_i}}{\mathrm{d}t^2} \mathrm{d}\Omega \quad (i = 1, 2, ..., n_r) \tag{8.114}$$

in which

$$\frac{\mathrm{d}^2 \hat{\mathbf{u}}_{r_i}}{\mathrm{d}t^2} = \left[\frac{\mathrm{d}^2 u_x^0}{\mathrm{d}t^2} \ \frac{\mathrm{d}^2 u_y^0}{\mathrm{d}t^2} \ \frac{\mathrm{d}^2 u_\theta^0}{\mathrm{d}t^2} \ \frac{\mathrm{d}^2 \varepsilon_x}{\mathrm{d}t^2} \ \frac{\mathrm{d}^2 \varepsilon_y}{\mathrm{d}t^2} \ \frac{\mathrm{d}^2 \gamma_{xy}}{\mathrm{d}t^2} \right]_{r_i}^{\mathrm{T}} \tag{8.115}$$

The central Euler scheme may be used for time discretization

$$\begin{aligned} \frac{\mathrm{d}^2 \hat{\mathbf{u}}_{r_i}(t)}{\mathrm{d}t^2} &= \frac{1}{\Delta t^2}(\hat{\mathbf{u}}_{r_i}(t) - 2\hat{\mathbf{u}}_{r_i}(t - \Delta t) + \hat{\mathbf{u}}_{r_i}(t - 2\Delta t)) \\ &= \frac{1}{\Delta t^2}((\hat{\mathbf{u}}_{r_i}(t) - \hat{\mathbf{u}}_{r_i}(t - \Delta t)) - (\hat{\mathbf{u}}_{r_i}(t - \Delta t) - \hat{\mathbf{u}}_{r_i}(t - 2\Delta t))) \\ &= \frac{1}{\Delta t^2}(\Delta \hat{\mathbf{u}}_{r_i}(t) - \Delta \hat{\mathbf{u}}_{r_i}(t - \Delta t)) \end{aligned} \tag{8.116}$$

where: $\Delta \hat{\mathbf{u}}_{r_i}(t)$—displacement increment of present time step; $\Delta \hat{\mathbf{u}}_{r_i}(t - \Delta t)$—displacement increment of the last time step; Δt—time-marching step length.

Equation (8.114) becomes

$$\Pi_{in} = \frac{\rho}{\Delta t^2} \iint_{\Omega_i} \hat{\mathbf{u}}_{r_i}^{\mathrm{T}} \mathbf{N}_{r_i}^{\mathrm{T}} \mathbf{N}_{r_i} (\Delta \hat{\mathbf{u}}_{r_i}(t) - \Delta \hat{\mathbf{u}}_{r_i}(t - \Delta t)) \mathrm{d}\Omega \quad (i = 1, 2, ..., n_r) \tag{8.117}$$

For each temporary equilibrium state at the end of time-marching step, the derivatives of Π_{in} w.r.t. block displacement variables are undertaken as follows

$$\frac{\partial^2 \Pi_{in}}{\partial \hat{\mathbf{u}}_{r_i} \partial \hat{\mathbf{u}}_{r_i}} = \frac{2\rho}{\Delta t^2} \iint_{\Omega_i} \mathbf{N}_{r_i}^{\mathrm{T}} \mathbf{N}_{r_i} \mathrm{d}\Omega \quad (i = 1, 2, ..., n_r) \tag{8.118}$$

$$-\frac{\partial \Pi_{in}(0)}{\partial \hat{\mathbf{u}}_{r_i}} = \frac{\rho}{\Delta t^2} \iint_{\Omega_i} \mathbf{N}_{r_i}^{\mathrm{T}} \mathbf{N}_{r_i} \Delta \hat{\mathbf{u}}_{r_i}(t - \Delta t) \mathrm{d}\Omega \quad (i = 1, 2, ..., n_r) \tag{8.119}$$

The stationary condition of Π_{in} leads to a 6 × 6 sub-matrix to be assembled into stiffness matrix $\mathbf{k}_{r_i, r_i}$

$$\frac{2\rho}{\Delta t^2} \iint_{\Omega_i} \mathbf{N}_{r_i}^{\mathrm{T}} \mathbf{N}_{r_i} \mathrm{d}\Omega \rightarrow \mathbf{k}_{r_i, r_i} \tag{8.120}$$

and a 6 × 1 sub-vector to be assembled into the load vector $\mathbf{f}_{r_i}$ in Eq. (8.94)

$$\frac{\rho}{\Delta t^2} \iint_{\Omega_i} \mathbf{N}_{r_i}^{\mathrm{T}} \mathbf{N}_{r_i} \Delta \hat{\mathbf{u}}_{r_i}(t - \Delta t) \mathrm{d}\Omega \rightarrow \mathbf{f}_{r_i} \tag{8.121}$$

It is notable that this original DDA does not consider damping, which is identically to postulate the zero velocity at the start moment of each time-marching step, namely, $\frac{\partial \hat{\mathbf{u}}_{r_i}(0)}{\partial t} = 0$. However, it is possible that, by assuming global damping forces proportional to absolute velocities such as in DEM and BEA, general approaches may be established in which viscous damping matrix is assembled to the global governing equations of DDA. Such a treatment enables DDA to be more adequate in the dynamic response analysis of geotechnical and hydraulic structures under seismic actions.

6. **Excavation load**

Basically, the load on an excavation exposed surface needs to be firstly calculated by the initial stress on the exposure face in opposite direction, namely, in the formulation of excavation load vector using Eq. (8.105), the initial stress $\boldsymbol{\sigma}^0$ is replaced by $-\boldsymbol{\sigma}^0$ straightforwardly.

In Fig. 8.12, a compressed beam (length L_0, height h_0, unity width = 1) is fixed at two ends. Suppose there is initial stress field $\boldsymbol{\sigma}^0 = \left[\sigma_x^0 \; 0 \; 0\right]^{\mathrm{T}}$.

Now, the right end constraint is partially removed to simulate stress release. From original position B to new position C, the elongation of beam is $\Delta e = \eta \Delta L (0 \leq \eta \leq 1)$, where ΔL is the displacement of beam after the total removal of right end constraint. Two methods are available to get the partially released state at C.

(1) **Reverse of force**

In Fig. 8.13a, at the right end of beam, force P_1 is exerted as

$$P_1 = E \frac{\eta \Delta L}{L_0} h_0 = -\eta \sigma_x^0 h_0 \tag{8.122}$$

Fig. 8.12 Partially removed restraint at the right end of beam

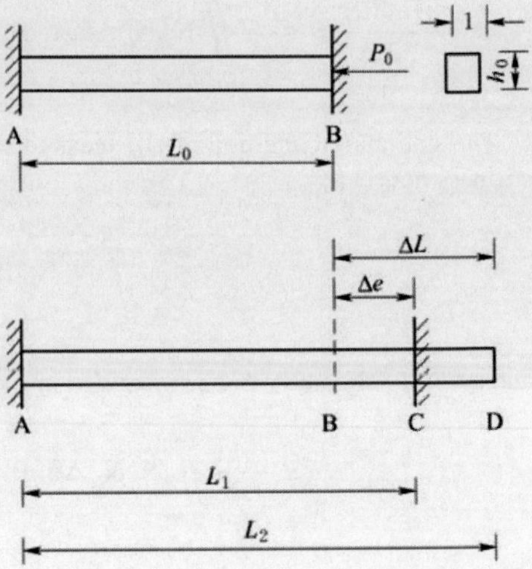

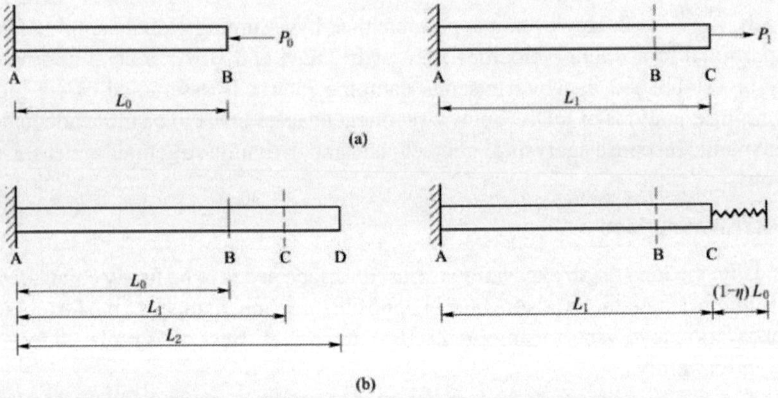

Fig. 8.13 Two methods for the partial release effects of beam. **a** Reverse load at the right end. **b** Inverse initial stress-boundary spring

According to the equilibrium condition, P_1 gives rise to stress increment $\Delta\boldsymbol{\sigma}^1 = \left[-\eta\sigma_x^0\ 0\ 0\right]^{\mathrm{T}}$. When the right end moves to C, the resultant stress is $\boldsymbol{\sigma} = \boldsymbol{\sigma}^0 + \Delta\boldsymbol{\sigma}^1 = \left[(1-\eta)\sigma_x^0\ 0\ 0\right]^{\mathrm{T}}$.

(2) **Reverse of initial stress**

In Fig. 8.13b, the initial stress is looked at as a general force. After the removal of restraint, under the action of $-\sigma_x^0$ the resulted axial displacement is $\Delta L = -\frac{\sigma_x^0}{E}L_0$. Attributable to this displacement increment, the position of right end is shifted from B to D. The boundary condition of this position does not meet the new boundary condition (at C). A spring is therefore added its compression deformation is $(1-\eta)\Delta L$.

After the iterative procedure the new position is at C, where the actual deformation is $\eta \Delta L$, that gives rise to stress increment $\Delta \boldsymbol{\sigma}^2 = \left[-\eta \sigma_x^0 \; 0 \; 0 \right]^{\mathrm{T}}$. As a result, $\boldsymbol{\sigma} = \boldsymbol{\sigma}^0 + \Delta \boldsymbol{\sigma}^2 = \left[(1-\eta)\sigma_x^0 \; 0 \; 0 \right]^{\mathrm{T}}$.

This example validates that the reverse of force and the reverse of initial stress present identical outcomes.

8.4.3 Sub-matrices of Stiffness and Sub-vectors of Displacement Constraints

As a boundary condition, the displacement δ of certain block may be pre-defined in specific directions. δ at a point can be specified as an input displacement (or pre-tension distance) by using a very stiff spring. Denote a directional vector $\mathbf{l} = [l_x \;\; l_y]^{\mathrm{T}}$ $(l_x + l_y = 1)$ for the displacement direction of δ, the compression d of spring is given by

$$\begin{aligned} d =& \delta - (l_x u_x + l_y u_y) = \delta - (\mathbf{N}_{r_i}(x, y)\hat{\mathbf{u}}_{r_i})^{\mathrm{T}} \mathbf{l} \\ =& \delta - \mathbf{l}^{\mathrm{T}} \mathbf{N}_{r_i}(x, y)\hat{\mathbf{u}}_{r_i} \quad (i = 1, 2, \ldots, n_r) \end{aligned} \tag{8.123}$$

The strain energy of this spring is calculated by

$$\begin{aligned} \Pi_{sp} &= \frac{p}{2} d^2 = \frac{p}{2}((\mathbf{N}_{r_i}(x, y)\hat{\mathbf{u}}_{r_i})^{\mathrm{T}} \mathbf{l} - \delta)^2 \\ &= \frac{p}{2}(\hat{\mathbf{u}}_{r_i}^{\mathrm{T}} \mathbf{N}_{r_i}^{\mathrm{T}}(x, y)\mathbf{l}.\mathbf{l}^{\mathrm{T}} \mathbf{N}_{r_i}(x, y)\hat{\mathbf{u}}_{r_i} - 2\delta \hat{\mathbf{u}}_{r_i}^{\mathrm{T}} \mathbf{N}_{r_i}^{\mathrm{T}}(x, y)\mathbf{l} + \delta^2) \quad (i = 1, 2, \ldots, n_r) \\ &= \frac{p}{2} \hat{\mathbf{u}}_{r_i} \mathbf{C}_{r_i} \mathbf{C}_{r_i}^{\mathrm{T}} \hat{\mathbf{u}}_{r_i} - p\delta \hat{\mathbf{u}}_{r_i}^{\mathrm{T}} \mathbf{C}_{r_i} + \frac{p}{2}\delta^2 \end{aligned} \tag{8.124}$$

$$\mathbf{C}_{r_i} = \mathbf{N}_{r_i}^{\mathrm{T}} \mathbf{l} \quad (i = 1, 2, \ldots, n_r) \tag{8.125}$$

in which p is the stiffness of spring that should be sufficiently large, normally ranged from $10 \times E$ to $1000 \times E$ to guarantee the displacement of spring only accounts for a fraction 10^{-3} to 10^{-4} of total displacement.

Invoking the stationary condition (8.96) w.r.t. the potential energy Π_{sp} in Eq. (8.124), the second-order derivative gives rise sub-matrix

$$\frac{\partial^2 \Pi_{sp}}{\partial \hat{\mathbf{u}}_{r_i} \partial \hat{\mathbf{u}}_{r_i}} = p\mathbf{C}_{r_i} \mathbf{C}_{r_i}^{\mathrm{T}} \quad (i = 1, 2, \ldots, n_r) \tag{8.126}$$

This 6×6 sub-matrix must be assembled to the sub-matrix $\mathbf{k}_{r_i, r_i}$ in Eq. (8.94), namely

$$p\mathbf{C}_{r_i}\mathbf{C}_{r_i}^{\mathrm{T}} \rightarrow \mathbf{k}_{r_i,r_i} \quad (i = 1, 2, ..., n_r) \tag{8.127}$$

In addition, the first-order derivatives of Π_{sp} gives rise to sub-vector

$$-\frac{\partial \Pi_m(0)}{\partial \hat{\mathbf{u}}_{r_i}} = p\delta\mathbf{C}_{r_i} \quad (i = 1, 2, ..., n_r) \tag{8.128}$$

This 6×1 sub-vector also must be assembled to the sub-vector $\mathbf{f}_{r_i}$ in Eq. (8.94), namely

$$p\delta\mathbf{C}_{r_i} \rightarrow \mathbf{f}_{r_i} \tag{8.129}$$

Under the circumstances where there is no movement along the direction (l_x, l_y), $\delta = 0$ is pre-set in Eq. (8.128), and only the 6×6 sub-matrix (8.127) need be calculated and assembled.

8.4.4 Compatibility Conditions at Block Contact Point

It is unique that the vertex-to-edge contact between block elements is particularly considered towards the simulation of finite deformation related rotation/dislocation for rock blocks.

1. **Block penetration and slip**

When block contacts are involved, it is necessary to ensure that:

- No inter-penetration occurs between two block elements;
- No tensile force exists between two block elements;
- No violation for shear strength criterion between two block elements.

The application of stiff spring depends on the three types of contact discussed previously:

- Contact between an acute angle (vertex) and an edge (viz. Fig. 8.9a): the stiff spring is applied when the vertex V_1 passes the reference line.
- Contact between an acute angle and an obtuse angle (viz. Fig. 8.9b): if the vertex V_1 passes reference line $V_2 - V_3$, a stiff spring between vertex V_1 and line $V_2 - V_3$ is applied; if the vertex V_1 passes the reference line $V_2 - V_4$, a stiff spring between vertex V_1 and line $V_2 - V_4$ is applied. It is possible to use two stiff springs in the contact position.
- Contact between two acute (convex) angles (viz. Fig. 8.9c): there are two reference lines in this case. If these two reference lines are passed by the vertex simultaneously, inter-penetration takes place. The normal distances from the reference lines to the contact vertex are d_{v1}, and d_{v2}. Assume $d_{v1} < d_{v2}$, then the stiff spring is attached between the vertex and its reference line with distance d_{v1}. For this type

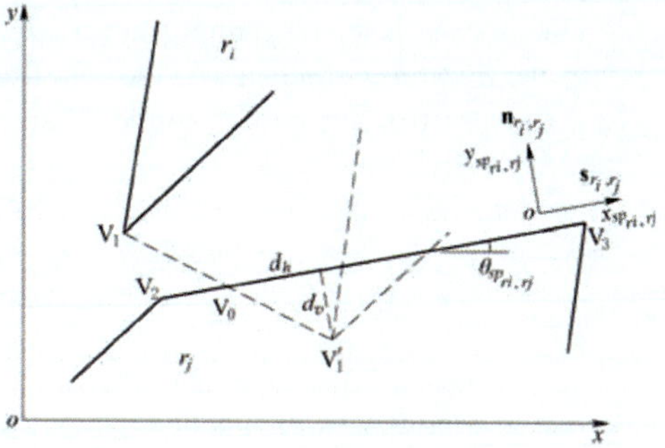

Fig. 8.14 Normal inter-penetration and tangential slip

of contact, only one stiff spring is applied. The physical function of stiff spring is to push the invaded vertex back along the shortest path.

When both the inter-penetration and slip occur at a given position, two stiff springs are applied. Similar to the penalty method, the non-tension and non-penetration conditions of blocks are enforced by add or remove hard springs p at the contact point, normally $p = (1 \sim 1000)E$ in which E is the Young's modulus of block element. These springs start from the contact point and lies in the directions normal and parallel to the reference line. For each vertex-to-edge contact pair, there are three possible contact modes: opening, sliding, and sticking. At the beginning of each time-marching step, the contact modes for all the vertex-to-edge contacts are assumed as sticking except those contact pairs inherited from the last time-marching step. For a sticking contact pair, a normal spring is applied to push the vertex away from the edge in the normal direction and another shear spring is applied to avoid the tangential displacement between vertex and edge.

In Fig. 8.14, the coordinates and displacements of vertexes $V_k (k = 1, 2, 3)$ are (x_k, y_k) and $(u_{x,k}, u_{y,k})$, where V_1 belongs to block element r_i but V_2 and V_3 belong to block element r_j. P_0 is termed as the "first entrance point", and the reference line $V_2 - V_3$ is actually the structural plane sp_{r_i,r_j}.

After the application of load increment at current time-marching step, the equilibrium equation is solved and the field variables such as displacements, stresses, etc., are obtained. The entrance distance between vertex V_1 and reference line $V_2 - V_3$ will be updated as follows:

- If the entrance line of a contact pair is positive, its contact mode changes to open, and all the springs and frictional forces are removed.
- If the entrance line of a contact pair is negative, the numerical penetration will violate the physical law, hence a normal spring is applied to push the vertex away from the entrance.
- If the tangential contact traction is smaller than its maximum value, i.e.,

$$F < N\mu + c \tag{8.130}$$

in which F is the tangential contact traction, N is the normal contact traction, μ is the friction coefficient, c is the cohesion. The sticking mode is assigned, and a shear spring is added to avoid the relative tangential displacement between V_1 and $V_2 - V_3$.

- If the tangential resultant contact traction exceeds its maximum value, i.e.

$$F \geq N\mu + c \tag{8.131}$$

the vertex will slide along the entrance line (reference line $V_2 - V_3$), as a result, a pair of frictional force is added.

After above contact detection and contact pair adjustment, Eq. (8.94) is updated accordingly and solved again. This iteration procedure is repeated until all the possible vertex-to-edge contact pairs satisfy the contact conditions, afterwards the computation proceeds to the next time step.

2. **Penetration**

(1) **Normal hard spring**

Suppose before deformation/motion of block system, V_1 and $V_2 - V_3$ are not contacted (viz. Fig. 8.15a), where vertexes V_1, V_2 and V_3 form an anti-clockwise triangle. After the deformation, V_1 penetrates into $V_2 - V_3$ with a normal distance d_v, where vertexes V_1, V_2 and V_3 form a clockwise triangle (viz. Fig. 8.15b). We denote:

$$S_0 = \begin{vmatrix} 1 & x_1 & y_1 \\ 1 & x_2 & y_2 \\ 1 & x_3 & y_3 \end{vmatrix} \tag{8.132}$$

$$\Delta = \begin{vmatrix} 1 & x_1 + u_{x,1} & y_1 + u_{y,1} \\ 1 & x_2 + u_{x,2} & y_2 + u_{y,2} \\ 1 & x_3 + u_{x,3} & y_3 + u_{y,3} \end{vmatrix} \tag{8.133}$$

where S_0 and Δ is the twice areas of triangle $\Delta V_1 V_2 V_3$ before and after deformation, respectively. Accordingly, the relation of the vertex V_1 and line $V_2 - V_3$ before and after deformation may be judged. If Δ is negative, the normal penetration of block

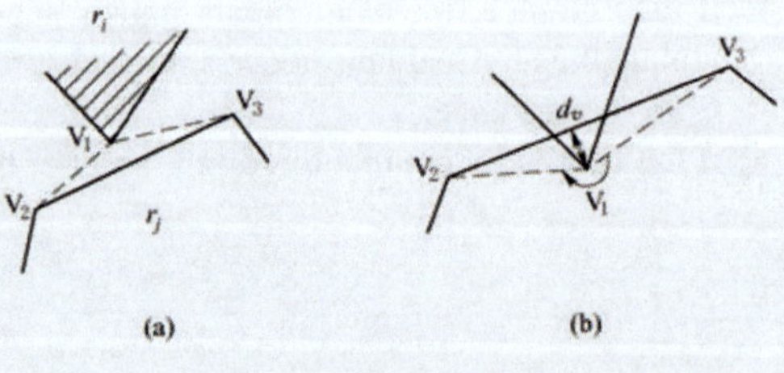

Fig. 8.15 Acute angle (vertex) V_1 of block element r_i contacts line $V_2 - V_3$ (sp_{r_i,r_j}). **a** Before contact. **b** After contact

r_i and r_j will occur with the distance of

$$\begin{cases} d_v = \dfrac{\Delta}{\sqrt{(x_2 - x_3)^2 + (y_2 - y_3)^2}} \\ \Delta = \dfrac{1}{L}\begin{vmatrix} 1 & x_1 + u_{x,1} & y_1 + u_{y,1} \\ 1 & x_2 + u_{x,2} & y_2 + u_{y2} \\ 1 & x_3 + u_{x,3} & y_3 + u_{y,3} \end{vmatrix} \end{cases} \tag{8.134}$$

in which L is the length of $\mathbf{V}_2 - \mathbf{V}_3$.

The expansion of determinant (8.134) followed by the neglect of the second order infinitesimal lead to

$$d_v = \frac{1}{L}\left(S_0 + \begin{Bmatrix} y_2 - y_3 \\ x_3 - x_2 \end{Bmatrix}^{\mathrm{T}} \begin{Bmatrix} u_{x,1} \\ u_{y,1} \end{Bmatrix} + \begin{Bmatrix} y_3 - y_1 \\ x_1 - x_3 \end{Bmatrix}^{\mathrm{T}} \begin{Bmatrix} u_{x,2} \\ u_{y,2} \end{Bmatrix} + \begin{Bmatrix} y_1 - y_2 \\ x_2 - x_1 \end{Bmatrix}^{\mathrm{T}} \begin{Bmatrix} u_{x,3} \\ u_{y,3} \end{Bmatrix}\right) \tag{8.135}$$

or

$$d_v = \frac{S_0}{L} + \mathbf{G}_{v,r_i}\hat{\mathbf{u}}_{r_i} + \mathbf{H}_{v,r_j}\hat{\mathbf{u}}_{r_j} = \frac{S_0}{L} + \hat{\mathbf{u}}_{r_i}^{\mathrm{T}}\mathbf{G}_{v,r_i}^{\mathrm{T}} + \hat{\mathbf{u}}_{r_j}^{\mathrm{T}}\mathbf{H}_{v,r_j}^{\mathrm{T}} \quad (i, j = 1, 2, ..., n_r) \tag{8.136}$$

$\mathbf{G}_{v,r_i}$ and $\mathbf{H}_{v,r_j}$ are 1×6 vectors calculated by

$$\mathbf{G}_{v,r_i} = \frac{1}{L}\begin{Bmatrix} y_2 - y_3 \\ x_3 - x_2 \end{Bmatrix}^{\mathrm{T}} \mathbf{N}_{r_i}(x_1, y_1) \quad (i = 1, 2, ..., n_r) \tag{8.137}$$

$$\begin{aligned} \mathbf{H}_{v,r_j} =& \frac{1}{L}\begin{Bmatrix} y_3 - y_1 \\ x_1 - x_3 \end{Bmatrix}^{\mathrm{T}} \mathbf{N}_{r_j}(x_2, y_2) \\ &+ \frac{1}{L}\begin{Bmatrix} y_1 - y_2 \\ x_2 - x_1 \end{Bmatrix}^{\mathrm{T}} \mathbf{N}_{r_j}(x_3, y_3) \quad (j = 1, 2, ..., n_r) \end{aligned} \tag{8.138}$$

(2) **Potential energy and sub-matrices**

The deformation energy due to the hard spring p_v at normal penetration is calculated by

$$\begin{aligned} \Pi_{np} =& \frac{p_v}{2} d_v^2 = \frac{p_v}{2}\left(\frac{S_0}{L} + \mathbf{G}_{v,r_i}\hat{\mathbf{u}}_{r_i} + \mathbf{H}_{v,r_j}\hat{\mathbf{u}}_{r_j}\right)^2 \\ =& \frac{p_v}{2}(\hat{\mathbf{u}}_{r_i}^{\mathrm{T}}\mathbf{G}_{v,r_i}^{\mathrm{T}}\mathbf{G}_{v,r_i}\hat{\mathbf{u}}_{r_i} + \hat{\mathbf{u}}_{r_j}^{\mathrm{T}}\mathbf{H}_{v,r_j}^{\mathrm{T}}\mathbf{H}_{v,r_j}\hat{\mathbf{u}}_{r_j} + 2\hat{\mathbf{u}}_{r_j}^{\mathrm{T}}\mathbf{H}_{v,r_j}^{\mathrm{T}}\mathbf{G}_{v,r_i}\hat{\mathbf{u}}_{r_i} \\ &+ \frac{2S_0}{L}\mathbf{G}_{v,r_i}\hat{\mathbf{u}}_{r_i} + \frac{2S_0}{L}\mathbf{H}_{v,r_j}\hat{\mathbf{u}}_{r_j} + \frac{S_0^2}{L^2}) \end{aligned} \tag{8.139}$$

Invoking the stationary condition (8.96) w.r.t. the potential energy in Eq. (8.139), four 6×6 sub-matrixes and two 6×1 sub-vectors are obtained. They are assembled into $\mathbf{k}_{r_i,r_i}$, $\mathbf{k}_{r_j,r_i}$, $\mathbf{k}_{r_i,r_j}$, $\mathbf{k}_{r_j,r_j}$, $\mathbf{f}_{r_i}$ and $\mathbf{f}_{r_j}$ in Eq. (8.94), respectively

$$\frac{\partial^2 \Pi_{np}}{\partial \hat{\mathbf{u}}_{r_i} \partial \hat{\mathbf{u}}_{r_i}} = p_v \mathbf{G}_{v,r_i}^{\mathrm{T}} \mathbf{G}_{v,r_i} \rightarrow \mathbf{k}_{r_i,r_i} \quad (i = 1, 2, ..., n_r) \tag{8.140}$$

$$\frac{\partial^2 \Pi_{np}}{\partial \hat{\mathbf{u}}_{r_j} \partial \hat{\mathbf{u}}_{r_j}} = p_v \mathbf{H}_{v,r_j}^{\mathrm{T}} \mathbf{H}_{v,r_j} \rightarrow \mathbf{k}_{r_j,r_j} \quad (j = 1, 2, ..., n_r) \tag{8.141}$$

$$\frac{\partial^2 \Pi_{np}}{\partial \hat{\mathbf{u}}_{r_i} \partial \hat{\mathbf{u}}_{r_j}} = p_v \mathbf{H}_{v,r_i}^{\mathrm{T}} \mathbf{G}_{v,r_j} \rightarrow \mathbf{k}_{r_i,r_j} \quad (i, j = 1, 2, ..., n_r) \tag{8.142}$$

$$\frac{\partial^2 \Pi_{np}}{\partial \hat{\mathbf{u}}_{r_j} \partial \hat{\mathbf{u}}_{r_i}} = p_v \mathbf{H}_{v,r_j}^{\mathrm{T}} \mathbf{G}_{v,r_i} \rightarrow \mathbf{k}_{r_j,r_i} \quad (i, j = 1, 2, ..., n_r) \tag{8.143}$$

$$-\frac{\partial \Pi_{np}(0)}{\partial \hat{\mathbf{u}}_{r_i}} = -p_v \frac{S_0}{L} \mathbf{G}_{v,r_i}^{\mathrm{T}} \rightarrow \mathbf{f}_{r_i} \quad (i = 1, 2, ..., n_r) \tag{8.144}$$

$$-\frac{\partial \Pi_{np}(0)}{\partial \hat{\mathbf{u}}_{r_j}} = -p_v \frac{S_0}{L} \mathbf{H}_{v,r_j}^{\mathrm{T}} \rightarrow \mathbf{f}_{r_j} \quad (j = 1, 2, ..., n_r) \tag{8.145}$$

3. **Tangential slip**
(1) **Tangential hard spring**

In Fig. 8.14, point $\mathrm{P}_0(x_0, y_0)$ is the initial contact position (first entrance point), then the slip distance d_h of vertex V_1 along edge $\mathrm{V}_2 - \mathrm{V}_3$ is calculated by

$$d_h = \frac{1}{|\mathbf{V}_3 - \mathbf{V}_2|} (\mathbf{V}_1' - \mathbf{V}_0).(\mathbf{V}_3 - \mathbf{V}_2) \tag{8.146}$$

The expansion of Eq. (8.146) and the neglect of the second order infinitesimal lead to

$$\begin{aligned} d_h &= \frac{S_0}{L} + \mathbf{G}_{h,r_i}^{\mathrm{T}} \hat{\mathbf{u}}_{r_i} + \mathbf{H}_{h,r_j}^{\mathrm{T}} \hat{\mathbf{u}}_{r_j} \\ &= \frac{S_0}{L} + \hat{\mathbf{u}}_{r_i}^{\mathrm{T}} \mathbf{G}_{h,r_i} + \hat{\mathbf{u}}_{r_j}^{\mathrm{T}} \mathbf{H}_{h,r_j} \quad (i, j = 1, 2, ..., n_r) \end{aligned} \tag{8.147}$$

in which

$$\mathbf{G}_{h,r_i} = \frac{1}{L} \mathbf{N}_{r_i}^{\mathrm{T}}(x_1, y_1) \begin{Bmatrix} x_3 - x_2 \\ y_3 - y_2 \end{Bmatrix} \tag{8.148}$$

$$\mathbf{H}_{h,r_j} = \frac{1}{L} \mathbf{N}_{r_j}^{\mathrm{T}}(x_0, y_0) \begin{Bmatrix} x_2 - x_3 \\ y_2 - y_3 \end{Bmatrix} \tag{8.149}$$

(2) **Potential energy and sub-matrices**

The potential energy due to tangential slip deformation is calculated by

$$\Pi_{ts} = \frac{p_h}{2} d_h^2 = \frac{p_h}{2}\left(\frac{S_0}{L} + \mathbf{G}_{h,r_i}^{\mathrm{T}} \hat{\mathbf{u}}_{r_i} + \mathbf{H}_{h,r_j}^{\mathrm{T}} \hat{\mathbf{u}}_{r_j}\right)^2$$
$$= \frac{p_h}{2}[(S_0/L)^2 + \hat{\mathbf{u}}_{r_i}^{\mathrm{T}} \mathbf{G}_{h,r_i}^{\mathrm{T}} \mathbf{G}_{h,r_i} \hat{\mathbf{u}}_{r_i} + \hat{\mathbf{u}}_{r_j}^{\mathrm{T}} \mathbf{H}_{h,r_j}^{\mathrm{T}} \mathbf{H}_{h,r_j} \hat{\mathbf{u}}_{r_j} \quad (i, j = 1, 2, ..., n_r)$$
$$+ 2\hat{\mathbf{u}}_{r_i}^{\mathrm{T}} \mathbf{G}_{h,r_i}^{\mathrm{T}} \mathbf{H}_{h,r_j} \hat{\mathbf{u}}_{r_j} + \frac{2S_0}{L} \hat{\mathbf{u}}_{r_i}^{\mathrm{T}} \mathbf{G}_{h,r_i} + \frac{2S_0}{L} \hat{\mathbf{u}}_{r_j}^{\mathrm{T}} \mathbf{H}_{h,r_j}] \tag{8.150}$$

Invoking the stationary condition (8.96) w.r.t. the potential energy Π_{ts}, the sub-matrices and sub-vectors are obtained and assembled by the operations identical to normal penetration

$$\frac{\partial^2 \Pi_{ts}}{\partial \hat{\mathbf{u}}_{r_i} \partial \hat{\mathbf{u}}_{r_i}} = p_h \mathbf{G}_{h,r_i}^{\mathrm{T}} \mathbf{G}_{h,r_i} \to \mathbf{k}_{r_i,r_i} \quad (i = 1, 2, ..., n_r) \tag{8.151}$$

$$\frac{\partial^2 \Pi_{ts}}{\partial \hat{\mathbf{u}}_{r_j} \partial \hat{\mathbf{u}}_{r_j}} = p_h \mathbf{H}_{h,r_j}^{\mathrm{T}} \mathbf{H}_{h,r_j} \to \mathbf{k}_{r_j,r_j} \quad (j = 1, 2, ..., n_r) \tag{8.152}$$

$$\frac{\partial^2 \Pi_{ts}}{\partial \hat{\mathbf{u}}_{r_i} \partial \hat{\mathbf{u}}_{r_j}} = p_h \mathbf{G}_{h,r_i}^{\mathrm{T}} \mathbf{H}_{h,r_j} \to \mathbf{k}_{r_i,r_j} \quad (i, j = 1, 2, ..., n_r) \tag{8.153}$$

$$\frac{\partial^2 \Pi_{ts}}{\partial \hat{\mathbf{u}}_{r_j} \partial \hat{\mathbf{u}}_{r_i}} = p_h \mathbf{H}_{h,r_j}^{\mathrm{T}} \mathbf{G}_{h,r_i} \to \mathbf{k}_{r_j,r_i} \quad (i, j = 1, 2, ..., n_r) \tag{8.154}$$

$$-\frac{\partial \Pi_{ts}}{\partial \hat{\mathbf{u}}_{r_i}} = -p_h \frac{S_0}{L} \mathbf{G}_{h,r_i} \to \mathbf{f}_{r_i} \quad (i = 1, 2, ..., n_r) \tag{8.155}$$

$$-\frac{\partial \Pi_{ts}}{\partial \hat{\mathbf{u}}_{r_j}} = -p_h \frac{S_0}{L} \mathbf{H}_{h,r_j} \to \mathbf{f}_{r_j} \quad (j = 1, 2, ..., n_r) \tag{8.156}$$

4. **Friction corresponding to tangential slip**

According to the Mohr–Coulomb criterion, the friction resistance force is calculated by

$$F_s = p_v d_v \,.\, \tan\varphi \,.sign[(\mathbf{V}_1' - \mathbf{V}_0).(\mathbf{V}_3 - \mathbf{V}_2)] \tag{8.157}$$

The directional vector of friction resistance force may be defined by the unit vector **s** along the positive direction of local coordinate axis x_{sp} (viz. Fig. 8.14) as follows

$$\mathbf{s} = \frac{1}{L}\begin{Bmatrix} x_3 - x_2 \\ y_3 - y_2 \end{Bmatrix} \tag{8.158}$$

- Frictional deformation energy of block element r_i due to the frictional slip

$$\Pi_{sc} = F_s \mathbf{u}^{\mathrm{T}}(x_1, y_1)\mathbf{s} = F_s \hat{\mathbf{u}}_{r_i}^{\mathrm{T}} \mathbf{N}_{r_i}^{\mathrm{T}}(x_1, y_1)\mathbf{s} \tag{8.159}$$

Invoking the stationary operation (8.96) for Π_{sc} w.r.t. general displacement vector, the frictional load exerting on block element r_i is obtained

$$-\frac{\partial \Pi_{sc}(0)}{\partial \hat{\mathbf{u}}_{r_i}} = -F_s \mathbf{N}_{r_i}^{\mathrm{T}}(x_1, y_1)\mathbf{s} \rightarrow \mathbf{f}_{r_i} \quad (i = 1, 2, ..., n_r) \tag{8.160}$$

- Frictional deformation energy of block element r_j due to the frictional slip

$$\Pi_{sc} = -F_s \mathbf{u}^{\mathrm{T}}(x_0, y_0)\mathbf{s} = -F_s \hat{\mathbf{u}}_{r_j}^{\mathrm{T}} \mathbf{N}_{r_j}^{\mathrm{T}}(x_0, y_0)\mathbf{s} \tag{8.161}$$

$$-\frac{\partial \Pi_{sc}(0)}{\partial \hat{\mathbf{u}}_{r_j}} = F_s \mathbf{N}_{r_j}^{\mathrm{T}}(x_0, y_0)\mathbf{s} \rightarrow \mathbf{f}_{r_j} \quad (j = 1, 2, ..., n_r) \tag{8.162}$$

8.5 Key Issues and Algorithms

The control input parameters concerning computation efforts for DDA are the amount of time-marching steps, the upper limit of step length in each step, the assumed maximum displacement ratio, the penalty factor or contact spring stiffness, and the energy dissipation parameter (sometimes). Hatzor and Feintuch (2001) explored the relationships between the time-marching step length and the maximum displacement ratio by closed-form solutions for dynamic problems (block on an incline subjected to dynamic load). Doolin and Sitar (2002), Tsesarsky et al. (2002) further studied the role of time-marching step length and penalty factor by closed-form solutions and data from shaking table experiments, respectively.

8.5.1 Time-Marching Step Length

Time-marching step length (size, interval) significantly affects the accuracy and efficiency of DDA. For problems where closed-form solutions exist, the optimal step length can be determined accurately. Numerical solutions, however, have to be used where closed-form solutions do not exist. Under such circumstances, the control parameters must be determined in advance on the basis of previous experience and engineering judgment. One possible way to estimate the suitability of the selected step length in DDA is to check the average number of iterations per time-marching step required for penetration adjustment.

The system of equilibrium equations is solved for the displacement variables using "open-close" iterations for penetration adjustment within each time-marching step: first, the solution is checked to see how well "no tension and no penetration" constraints are met; if tension or penetration is found along any contact, the constraints are adjusted by selecting new blocks and constraining positions, and the modified versions of stiffness matrix **K** as well as load vector **F** are formed, from which a new solution is obtained; this iteration process is repeated until no tension and no penetration is guaranteed for all of the blocks; afterwards, the next time-marching step is conducted.

8.5.2 Damping

Before we attempt to apply DDA to a full scale problem of jointed rock mass, it is necessary to check whether the dynamic DDA displacements are matched by closed-form solutions. Fundamental DDA formulation is completely linear-elastic without energy dissipation mechanism other than the mechanical energy required for the deformation of contact springs and intact block material. The frictional sliding along structural planes is the main source of energy consumption, because no "artificial" damping was introduced in the initial formulation of DDA. Although being an honest mathematical approach, yet it is not entirely realistic because irreversible processes such as crushing of block material at contact points, or temporary resistance to sliding offered by asperities, are not modeled. Such energy dissipation mechanism, loosely referred to as "damping", must be activated during block system deformation. Otherwise DDA should be expected to provide exaggerated displacements, or long last vibration before a static state may be acceptable.

Hatzor and Feintuch (2001) demonstrated the validity of DDA results for the fully dynamic analysis of a single block that is on an incline and subjected to dynamic action. They run the new version of DDA (Shi 2002) without the introduction of any damping, as in the closed-form solution by MacLaughlin (1997). The difference in the closed-form and DDA solutions is within 1– 2%. Hatzor and Feintuch further investigated three different sinusoidal functions of increasing complexity for the dynamic action input, and checked the agreement between the DDA and closed-form solutions. A good agreement was obtained in all cases. It is notable that both the closed-form solution and DDA solution neglected damping.

Towards the fully dynamic analysis for the geotechnical and hydraulic structures under seismic actions, however, it is necessary to introduce certain damping for additional energy dissipation mechanism apart from contact friction, if a realistic displacement prediction is sought (Ishikawa et al. 2002). Examples for such energy dissipation mechanism may be the block fracture at contact points, the contact surface damage during slip, etc.

Use was made of the shaking table experiment at the Earthquake Engineering Laboratory, UC Berkeley (USA), Wartman et al. (2003) studied the dynamic displacement problem of rock block on an incline. Tsesarsky et al. (2002) conducted identical numerical tests using DDA. For 0% energy dissipation, the velocity of each block at the end of a time-marching step is completely transferred to the beginning of next time-marching step. For 1.5% energy dissipation, the initial velocity of a time step is 1.5% smaller than the terminal velocity of the previous time step. These outcomes show that with zero energy dissipation, DDA tends to overestimate the displacements by as much as 80%. With as little as 2% energy dissipation however, DDA displacements match the physical test results within the error of 5%.

Use was made of a slope with a stepped base consisting of 50 blocks, McBride and Scheele (2001) studied the multi-block toppling problem. Their conclusion was that as much as 20% energy dissipation was required in order to obtain realistic agreement between the physical model and DDA model.

These findings suggest that a realistic application of dynamic DDA must incorporate certain form of energy dissipation such as the "kinetic coefficient" (Hatzor et al. 2002a) in order to account for additional energy loss mechanism that is not modeled previously. Although kinetic coefficient is applicable in some cases like rock fall, yet sometimes damping effect described by kinetic coefficient alone is not adequate, this is mainly blamed for the fact that kinetic damping is not a physically valid damping mechanism.

Introduction of physical damping mechanism, for example a "spring + dashpot model" at contact points would most certainly provide more accurate results. The commonly used viscous damping force may be assumed to be proportional to velocity of each block. A dimensionless damping ratio also may be defined to determine whether the system will oscillate by comparing to a critical damping ratio. In simple DDA model, kinetic damping may be approximately transferred to the viscous damping (Fu et al. 2015a). Although these treatments are not physically ideal, yet they may provide similar scenarios of vibration decay, subject to contact and global damping are used separately. In other words, they do reduce the amount of calculation cycles to reach the equilibrium state for quasi static problems, and the damping does not affect the equilibrium state itself.

However, the correct value of damping coefficient still remains an open question. As has been explained in the Chap. 7 of this book, damping mechanism can be derived from three sources: material damping, boundary (contact) damping, and viscous damping. In DDA, contact damping should be dominant in the description of decay scenarios. It is actually a kind of local damping operates on the relative velocity at contact point and may be envisioned as resulting from dashpots acting in the normal and shear directions on contact springs (Yu et al. 2019). Sasaki et al. (2005) also introduced a formulation of viscous damping, and a seismic landslide case was studied using their proposed method.

From the point view of application, the author tends to use Rayleigh damping model as global damping model for DDA that assumes instantaneous velocity the only variable. In doing so, the deduction concerning inertia force from Eq. (8.113) to Eq. (8.121) should be revised, and the incremental form of general displacement vector $\Delta\hat{\mathbf{u}}_{r_i}$ should be used to replace the general displacement vector $\hat{\mathbf{u}}_{r_i}$. Hence from Eq. (8.114), the incremental potential energy due to inertia is

$$\Pi_{in} = -\iint\limits_{\Omega_{r_i}} \Delta\mathbf{u}_{r_i}^{\mathrm{T}}\Delta\mathbf{f}^{in}\mathrm{d}\Omega = \iint\limits_{\Omega_{r_i}} \rho\Delta\hat{\mathbf{u}}_{r_i}^{\mathrm{T}}\mathbf{N}_{r_i}^{\mathrm{T}}\mathbf{N}_{r_i}\frac{\partial^2\Delta\hat{\mathbf{u}}_{r_i}}{\partial t^2}\mathrm{d}\Omega \quad (i = 1, 2, ..., n_r) \tag{8.163}$$

Invoking the stationary operation for Π_{in} w.r.t. general displacement vector, the inertia load exerting on block element r_i is obtained

$$\begin{aligned} -\frac{\partial\Pi_{in}(0)}{\partial\Delta\hat{\mathbf{u}}_{r_i}} &= -\left(\iint\limits_{\Omega_{r_i}} \rho\Delta\hat{\mathbf{u}}_{r_i}^{\mathrm{T}}\mathbf{N}_{r_i}^{\mathrm{T}}\mathbf{N}_{r_i}\mathrm{d}\Omega\right)\frac{\mathrm{d}^2\Delta\hat{\mathbf{u}}_{r_i}}{\mathrm{d}t^2} \\ &= -\,\mathbf{m}_{r_i,r_i}\frac{\mathrm{d}^2\Delta\hat{\mathbf{u}}_{r_i}}{\mathrm{d}t^2} \quad (i = 1, 2, ..., n_r) \end{aligned} \tag{8.164}$$

Being independently assembled in the equilibrium condition (8.94) in the incremental form such that $-\mathbf{m}_{r_i,r_i}\frac{\mathrm{d}^2\Delta\hat{\mathbf{u}}_{r_i}}{\mathrm{d}t^2} \to \Delta\mathbf{f}_{r_i}$, we get

$$\mathbf{M}\frac{\mathrm{d}^2\Delta\hat{\mathbf{U}}}{\mathrm{d}t^2} + \mathbf{K}\Delta\hat{\mathbf{U}} = \Delta\mathbf{F} \tag{8.165}$$

where

$$\mathbf{M} = \begin{bmatrix} \mathbf{m}_{r_1,r_1} & ...\ 0 & ...\ 0 & ...\ 0 \\ ... & ...\ ... & ...\ ... & ...\ ... \\ 0 & ...\ \mathbf{m}_{r_i,r_i} & ...\ 0 & ...\ 0 \\ ... & ...\ ... & ...\ ... & ...\ ... \\ 0 & ...\ 0 & ...\ \mathbf{m}_{r_j,r_j} & ...\ 0 \\ ... & ...\ ... & ...\ ... & ...\ ... \\ 0 & ...\ 0 & ...\ 0 & ...\ \mathbf{m}_{r_{nr},r_{nr}} \end{bmatrix} \tag{8.166}$$

$$\frac{\mathrm{d}^2\Delta\hat{\mathbf{U}}}{\mathrm{d}t^2} = \left[\frac{\mathrm{d}^2\Delta\hat{\mathbf{u}}_{r_1}}{\mathrm{d}t^2} \cdots \frac{\mathrm{d}^2\Delta\hat{\mathbf{u}}_{r_i}}{\mathrm{d}t^2} \cdots \frac{\mathrm{d}^2\Delta\hat{\mathbf{u}}_{r_j}}{\mathrm{d}t^2} \cdots \frac{\mathrm{d}^2\Delta\hat{\mathbf{u}}_{r_{nr}}}{\mathrm{d}t^2}\right]^{\mathrm{T}} \tag{8.167}$$

Similar to Eq. (8.26), we postulate

$$\mathbf{C} = \alpha_0\mathbf{M} + \alpha_1\mathbf{K} \tag{8.168}$$

in which α_0 and α_1 are defined in Eq. (4.171). The dynamic governing equation set of the block element system with global damping is

$$\mathbf{M}\frac{\mathrm{d}^2\Delta\hat{\mathbf{U}}}{\mathrm{d}t^2} + \mathbf{C}\frac{\mathrm{d}\Delta\hat{\mathbf{U}}}{\mathrm{d}t} + \mathbf{K}\Delta\hat{\mathbf{U}} = \Delta\mathbf{F}(t) \tag{8.169}$$

where

$$\frac{\mathrm{d}\Delta\widehat{\mathbf{U}}}{\mathrm{d}t} = \left[\frac{\mathrm{d}\Delta\hat{\mathbf{u}}_{r_1}}{\mathrm{d}t} \ldots \frac{\mathrm{d}\Delta\hat{\mathbf{u}}_{r_i}}{\mathrm{d}t} \ldots \frac{\mathrm{d}\Delta\hat{\mathbf{u}}_{r_j}}{\mathrm{d}t} \ldots \frac{\mathrm{d}\Delta\hat{\mathbf{u}}_{r_{nr}}}{\mathrm{d}t}\right]^{\mathrm{T}} \tag{8.170}$$

The "time-history analysis" is then implemented by the direct time integration step-by-step, for this purpose the Newmark-β algorithm and Wilson-θ algorithm (Newmark 1959; Bathe and Wilson 1972) basically belonging to the particular case of one-step schemes using quadratic truncated Taylor series expansion, may be employed.

Again, it is emphasized that within each time-marching step, the iteration (geometrical) should be conducted for contact adjustment followed by the revision for the matrices of mass, damping and stiffness in Eq. (8.169).

In addition, since the incremental equation set (8.169) is employed, non-linear constitutive relations (e.g. elasto-plasticity or elasto-viscoplasticity) for both the block entities and contact planes may be implemented with possibly more rigorous and expansive non-linear iteration (physical) within one time-marching step. This is, however, not very necessary because on the one hand, higher order distribution of strain in blocks should be competently employed for this purpose, on the other hand, the unique advantage of DDA could be undermined considerably in doing so.

8.5.3 *Velocity-Weakening Friction Law*

In handling the large scale landslide exhibiting enhanced mobility with extreme velocity and long runout, the velocity-weakening friction law (7.64) that the coefficient of friction dynamically changes with the sliding velocity (Lucas et al. 2014) may be incorporated into DDA.

8.5.4 *Solution and Iteration of the System Equation*

After the solution of Eq. (8.94) or Eq. (8.169), the block position may be updated using general displacement or its increment. With the help of non-penetration and non-tension conditions, the hard springs are added or/and removed, in this way a revised equilibrium equation is get. The iterative computation is collapsed until certain convergent criteria, such as the allowable displacement increment or residual load increment, are met (viz. Fig. 8.16).

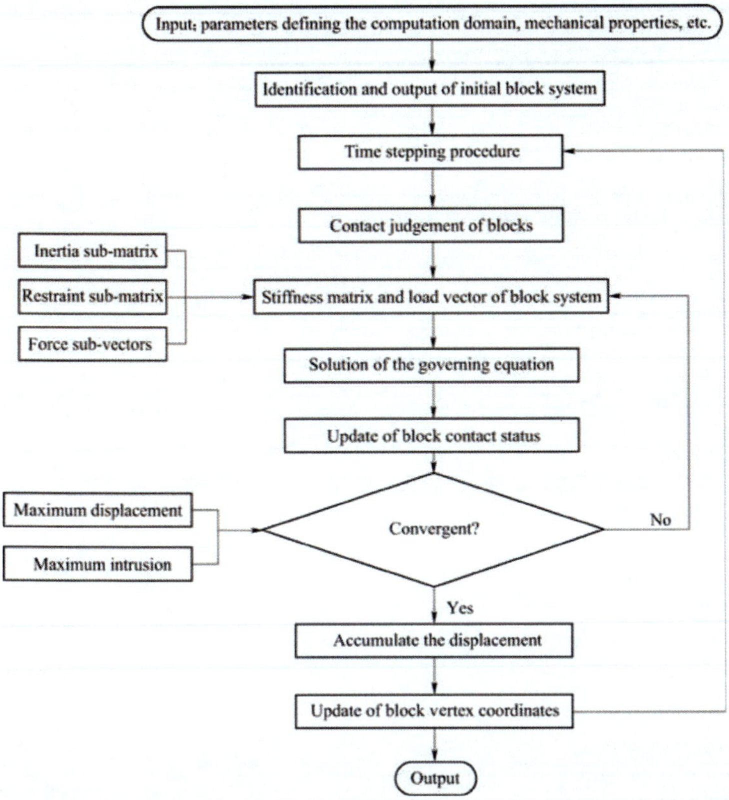

Fig. 8.16 Flowchart of DDA program

8.6 Validations

8.6.1 Cantilever Beam Bending

In Fig. 8.17, an elastic cantilever beam (length $L = 6.0$ m; width $b = 0.5$ m; height $h = 1.0$ m) exerted by a concentrated force $P = 0.5$ MPa at its free (right) end is exposited. The elastic parameters used in the study are $E = 10000$ MPa and $\nu = 0.17$.

Both the p-refinement BEA and the fundamental FEM are employed to solve this problem. In the former the cantilever is looked at as one block element, whereas in the latter it is divided uniformly along the length and height into 12 sections and 4 layers (i.e. 48 finite elements).

Figure 8.18 shows the computed and analytically solved flexural curves of the beam axis, and Fig. 8.19 shows the normal stress distribution on the fixed (left) end section. In BEA, the order p of the shape functions along the width is fixed to 1, meanwhile along both the length and height directions it is adaptively upgraded from

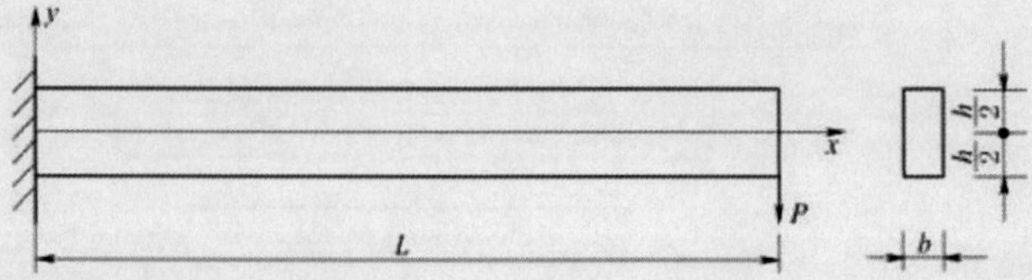

Fig. 8.17 Diagram of elastic cantilever beam

1 until to 4. The DOFs, energy norm, relative error, and the flexure are summarized in Table 8.2.

Fig. 8.18 Flexural curve of beam axis (I = sectional inertia moment)

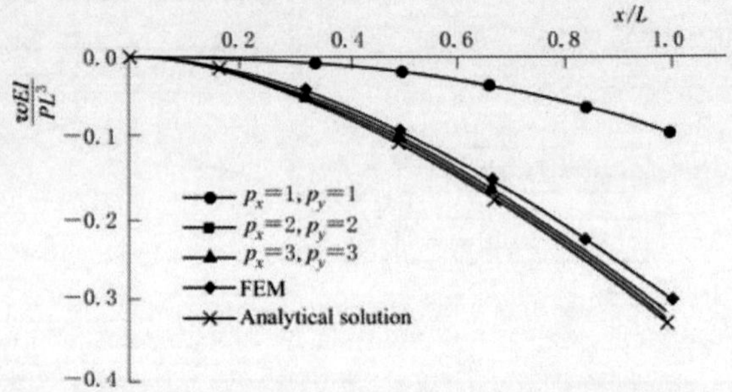

Fig. 8.19 Normal stress distribution on the fixed (left) end section (I = sectional inertia moment)

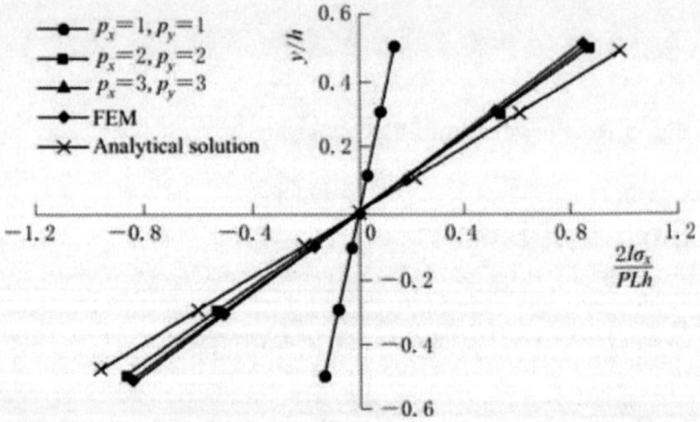

Table 8.2 Variation of p-refinement indices of the cantilever beam

Adaptive step	Order of shape function p	System DOFs	Energy norm	Relative error%	Flexure (m)
1	1	24	0.0679009	96.68	0.004610
2	2	48	0.2658887	44.12	0.070694
3	3	72	0.2962777	0.35	0.087778
4	4	102	0.2962795	–	0.087782

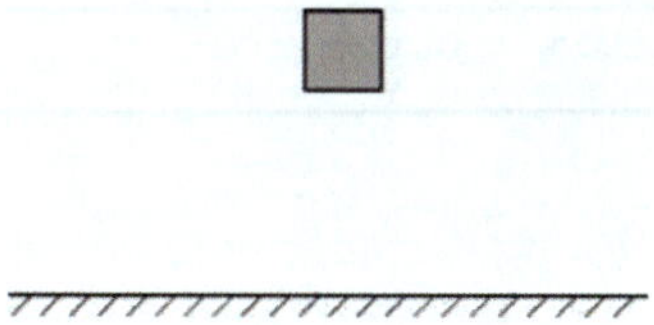

Fig. 8.20 Diagram of free falling body

Table 8.3 Computation of free falling body

Time (s)	DDA		Analytical	
	Velocity (m/s)	Displacement (m)	Velocity (m/s)	Displacement (m)
0.0	0.0	0.0	0	0
0.15	– 1.47	0.1102	– 1.47	0.11035
0.25	– 2.45	0.3062	– 2.45	0.30625
0.50	– 4.9	1.225	– 4.9	1.225
0.75	– 7.35	2.7562	– 7.35	2.75625
1.0	– 9.8	4.90	– 9.8	4.90
1.15	– 11.27	6.48	– 11.27	6.48025
1.25	– 12.25	7.6562	– 12.25	7.65625
1.35	– 13.23	8.9302	– 13.23	8.93025
1.392	– 13.6416	9.4946	– 13.6416	9.5

8.6.2 *Block Free Falling*

Figure 8.20 shows a free falling block (1×1 m). The center of block is 10 m above the ground plane hence its lower bottom is 9.5 m high. The time step length is $\Delta t =$ 0.0001 s. Table 8.3 lists the results from the DDA computation and direct calculation using Newton's second law, which identically give landing time 1.392 s.

8.6.3 *Block Sliding*

Figure 8.21 shows a block sliding along the slope with two inclination angles: the upper portion AB with a length of $L_1 = 8m$ and an angle of $\alpha_1 = 30°$, the lower portion BC with an angle of $\alpha_2 = 15°$. The friction angle is fixed but φ_1 is variable, different friction angles $\varphi_1 = 23°, 25°, 28°, 29.7°$ and $30.1°$ are tested.

Figures 8.22 and 8.23 exposit the velocity and runout distance versus time. According to Fig. 8.23, if the runout distance is longer than 8 m, the block is going in the lower portion of the slope. The computation results indicate that:

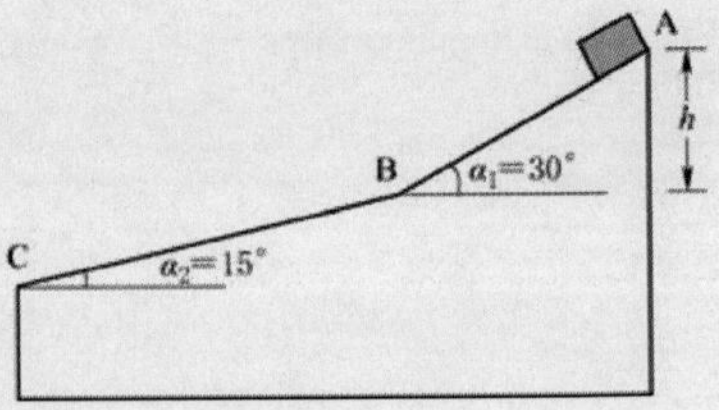

Fig. 8.21 Diagram of sliding block

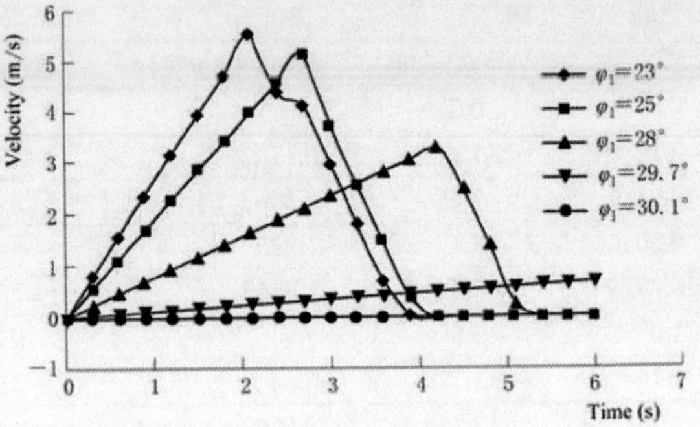

Fig. 8.22 Velocity versus time

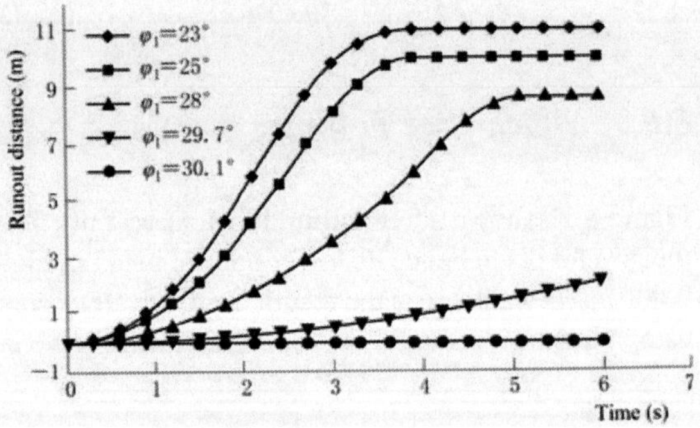

Fig. 8.23 Runout distance versus time

① Where $\varphi_1 = 30.1°$, the block is stable with zero velocity, but the velocity starts to increase when $\varphi_1 = 29.7°$. Hence the computation accuracy of block stability using DDA is validated;
② The smaller of φ_1, the larger of the slip velocity. When φ_1 is smaller than 22°, the block is not able to rest on the lower portion BC.

8.7 Engineering Applications

8.7.1 *Gravity Dam Stability: Baozhusi Project, China*

1. **Presentation of the project**

Baozhusi Project (viz. Fig. 8.24) is located on the Bailongjiang River, China. The 132 m tall and 524.48 m long concrete gravity dam creates a 2,550 million m^3 reservoir. On either side of power station at the dam's base, there are two gate-controlled chute spillways. In addition, there are two pairs of orifice spillways. Below the left orifice there are two bottom outlets.

2. **Characteristics of the computation**

The 17th dam monolith of 21.5 m thickness and 131 m high is shown in Fig. 8.25, in which a penstock is installed. The width of dam's base is 92 m (exclusive the powerhouse). The rock masses under the dam base is Ordovician sand stone and at the downstream of dam toe is Silurian shale (S_1). Major large-scale structural planes are faults (F_4, F_2) and argillic intercalated layers (D_5, $D_{7,8}$, D_1, D_3, D_6).

Preliminary studies showed that the safety of dam foundation against sliding was insufficient. The designer then revised the structural design by replacing the open and permanent joint between dam and powerhouse with a bonded temporary joint, in this manner the dam and powerhouse were supposed to work together resisting the reservoir water thrust.

The computation study was carried out for the revised design to analyze the deformation and stress of dam/foundation system, which is expected to answer the questions such as "what is the failure mechanism of the structure (sliding within the foundation or cracking in the dam body)"? "Is the safety of the dam sufficient"? A parallel laboratory physical test was carried out, too. The main prototype mechanical parameters used in the computation and test are listed in Tables 8.4 and 8.5.

According to the equality principle with respect to the relative displacement of fault walls, the stiffness coefficients k_n and k_s input in the BEA computation are related to the Young's modulus E and Poisson's ratio ν of the filler in fault or argillic intercalated layer by Eq. (7.44).

Figure 8.25 shows the block assemblage system consisting of 28 block elements. The block elements in the foundation are exactly defined by the structural planes, whereas the dam body and powerhouse are discretized by dummy structural planes. In the discretization of dam, the penstock and the dam/powerhouse joint are taken into account. Similar to the penalty method for contact issues, the stiffness coefficients and strength of dummy structural planes in the dam body and power house should be high enough to ensure the computation precision. The error tolerance e_t in the adaptive upgrade of the order p of shape functions is stipulated as $e_t = 5\%$.

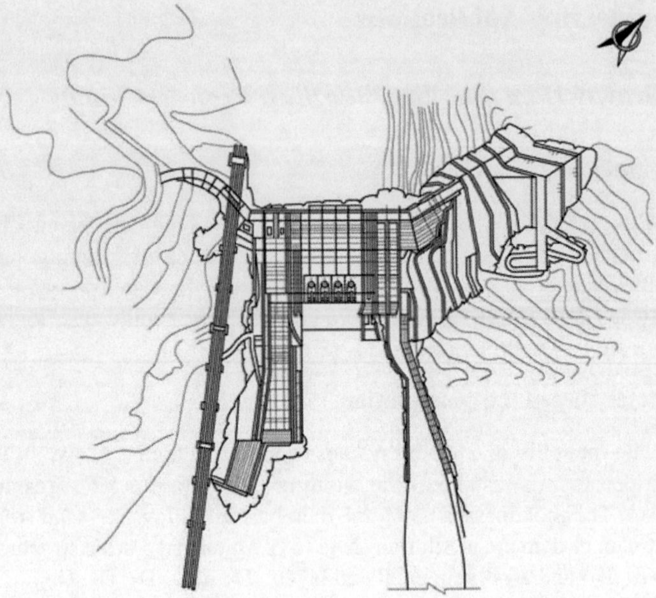

(a)

(b)

Fig. 8.24 Baozhusi project, China. **a** Plan. **b** Left-downstream view

3. **Test configuration and procedure**

The physical test model in Fig. 8.26 is made of geo-mechanical material with high specific weight and low elasticity and strength, which is ideal for the study of the displacement and failure mechanism of dam/foundation system. However, the stress in this kind of physical model cannot be well gauged by the conventional instruments.

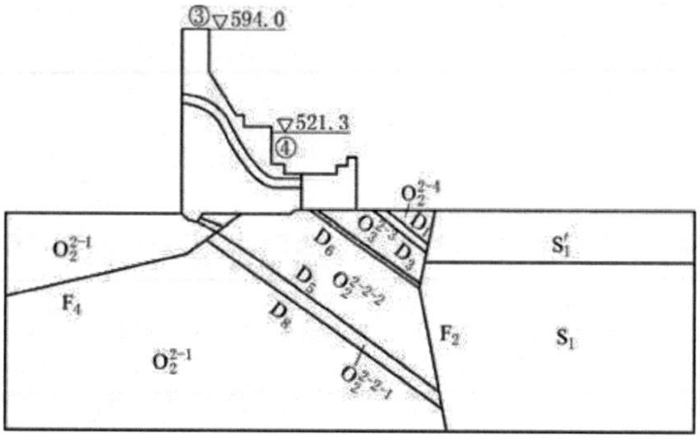

Fig. 8.25 Dam monolith 17: Baozhusi Project, China. ③ and ④—displacement gauge (object) points in laboratory physical test

Table 8.4 Prototype mechanical parameters of the rock masses and concrete

Rock	Unit weight γ_r (kN/m³)	Friction angle φ (°)	Cohesion c(MPa)	Deformation modulus E(GPa)	Poisson's ratio ν	Tensile strength σ_t (kPa)
O_2^{2-1}	23.52	45.0	0.6	7.53	0.27	10.0
O_2^{2-2-1}	25.97	45.0	0.6	2.89	0.30	6.6
O_2^{2-2-2}, O_2^{2-3}	25.48	45.0	1.0	18.70	0.18	47.6
O_2^{2-4}	25.87	45.0	0.8	11.90	0.25	29.0
S_1	25.48	26.6	0.4	4.00	0.30	22.0

Table 8.5 Prototype mechanical parameters of structural planes

Structural plane	Deformation modulus E(MPa)	Poisson's ratio ν	Friction angle φ(°)	Cohesion c(MPa)	Thickness T (m)	Remark
D_1, D_3	30.0	0.40	20.8	0.01	0.01	–
D_6	50.0	0.40	19.3	0.02	0.01	–
$D_{7,8}$	50.0	0.40	21.8	0.01	0.01	Upstream side of F_4
$D_{7,8}$	300.0	0.30	21.8	0.02	0.01	Downstream side of F_4
D_5	40.0	0.40	14.0	0.01	0.01	Upstream side of F_4
D_5	300.0	0.30	16.7	0.02	0.01	Downstream side of F_4
F_4	1400.0	0.35	26.6	0.02	0.50	–
F_2	3000.0	0.35	24.2	0.05	1.00	–

Fig. 8.26 Gravity dam model made of geo-mechanical materials

The coefficients of scaling (Chen 2018) in the physical model test are designed as follows:

- Geometry, 170;
- Young's modulus, 170;
- Volumetric weight, 1;
- Displacement, 17;
- Friction coefficient, 1.

The static hydraulic pressure of the reservoir is equalized by a series of jacks arranged along the upstream dam face, the output of jacks can be controlled to simulate the reservoir water level fluctuation. The overloading process is implemented in 7 steps (viz. Table 8.6) to extract the failure mechanism and correspondent "factor of safety (FOS)" of the dam, where the overload increments and overload factors are calculated by the formulas

$$P_i = \frac{1}{2}\gamma(2H_i - H)H \tag{8.171}$$

$$Fs_i = 2\frac{H_i}{H} - 1 \tag{8.172}$$

in which H is the upstream normal water depth at the beginning of overloading, H_i is the overloading water depth at the ith overloading step.

Altogether 7 displacement object points are layout on the dam and powerhouse. In addition, a number of strain gauges are deployed on the upstream dam face to detect dam cracking.

4. **Computation results**

(1) **Displacement**

Table 8.6 Overloading steps

Overloading step	Upstream water level (m)	Upstream water depth H_i(m)	Overload increment P_i(MN)	Overload factor Fs_i	Remark
1	594.70	130.70	–	1.00	–
2	610.38	146.38	103.90	1.24	–
3	626.07	162.07	124.01	1.48	–
4	646.98	182.98	150.82	1.80	–
5	654.82	190.82	160.88	1.92	–
6	659.50	195.50	166.87	1.99	–
7	660.05	196.05	156.83	2.00	Upstream cracking near penstock

Figure 8.27 shows the calculated displacement increments when the overload factor Fs is equal to 2.0. The displacement increments against Fs at the gauge points ③ and ④ (viz. Fig. 8.25) are plotted in Fig. 8.28.

Fig. 8.27 Accumulated displacements (computed)

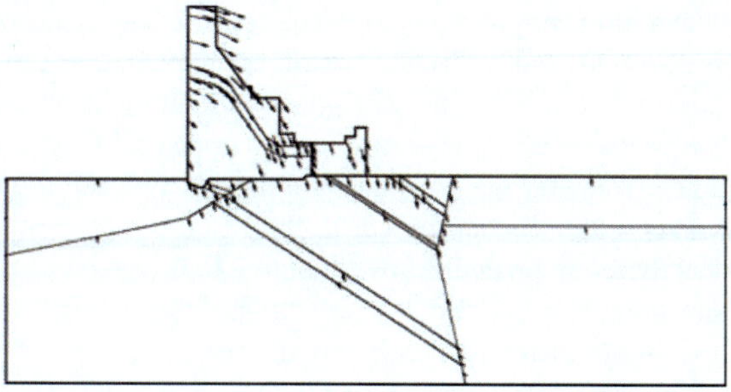

Fig. 8.28 Displacements at the gauge points versus overload factor Fs

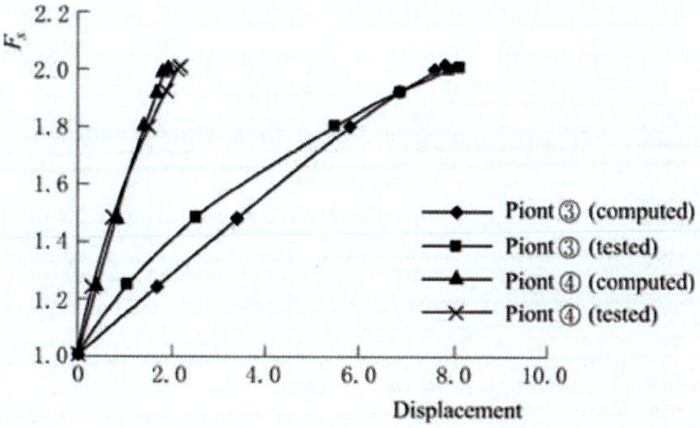

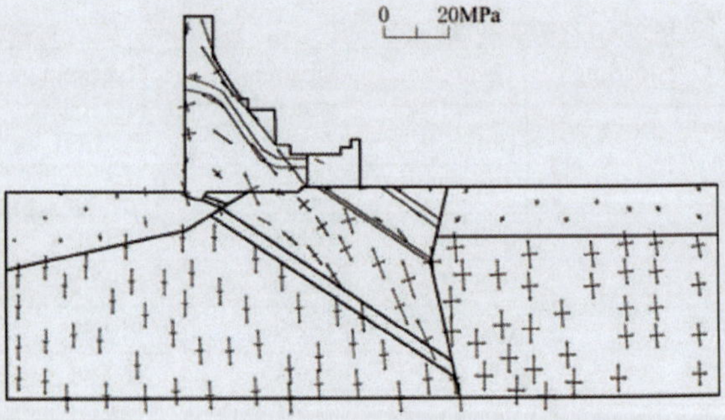

Fig. 8.29 Principal stresses (computed)

(2) **Stress**

Figure 8.29 gives the calculated principal stresses when the overloading factor Fs is equal to 2.0. Unfortunately, in the geo-mechanical material the strain cannot be measured credibly by conventional instruments, hence there is no cross-reference between the computation and test with regard to strains and stresses.

(3) **Failure mechanism and factor of safety**

The overloading factor Fs can be regarded as a FOS of dams. In the physical model test, it was found that the first crack appeared at the upstream dam body near the penstock (EL. 558.4 m) when the overloading factor $Fs = 2.0$. After the overloading factor exceeded 2.0, several cracks appeared between the penstock and dam heel. The crack near the penstock propagated fast and the upstream jack pressure in the test could not be sustained, leading to the failure of dam. The corresponding FOS can therefore be defined as equal to 2.0. It can therefore be confirmed that the failure mechanism is dam body cracking, and the stability against foundation sliding is no longer dominant consideration after the revision of design.

Figure 8.30 shows the tensile yield zones when $Fs = 2.0$ by BEA, which are the areas enclosed by the contour of point safety factor equal to unit and the upstream boundary. It is clear that the failure mechanism can be well revealed by BEA. The computation gives a factor of safety FOS = 2.3, which is higher than that from the physical model test. This is mainly attributable to the perfect plasticity (no hardening and softening) and associated flow rule postulated in the BEA computation. The computation also indicates that on the upstream riverbed there is a tensile cracking zone. It will propagate along with the overloading process, too, but much slower than that occurs in the dam body. Therefore, it is not the dominate factor entailing the safety of dam.

8.7.2 *Landslide Accident*

1. **Presentation of the landslide**

The Qianjiangping Landslide (viz. Figs. 8.31 and 8.32) is located at the Qinggan River that is a tributary of the Yangtze River, China. The massive rock of volume

Fig. 8.30 Tensile yield zones under the overload factor $Fs = 2$ (computed)

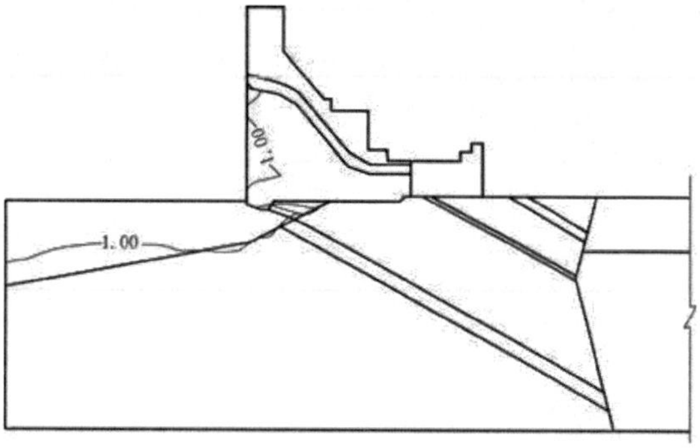

$20.4 \times 10^6 m^3$ failed on 13 July, 2003. Other geometrical features of the landslide body are: longitudinal length 1150 m, surface area 6.8×10^5 m^2, average thickness 30 m.

It is a typical single dip-out landslide. Before the failure, its crest is at EL.400–500 m and its foot is near the river bed (EL.93–95 m); its front portion is inclined at an angle of 15–20° and its rear portion is inclined at an angle of 25–30°. At the south-western slope there is a cliff with relative height of 80–305 m and length of 800 m. This cliff provides the third exposure face for the landslide.

After the failure, its rear elevation is lowered down to EL.350–405 m; the width of front deposit portion is 600 m and the width of rear deposit portion is 380 m.

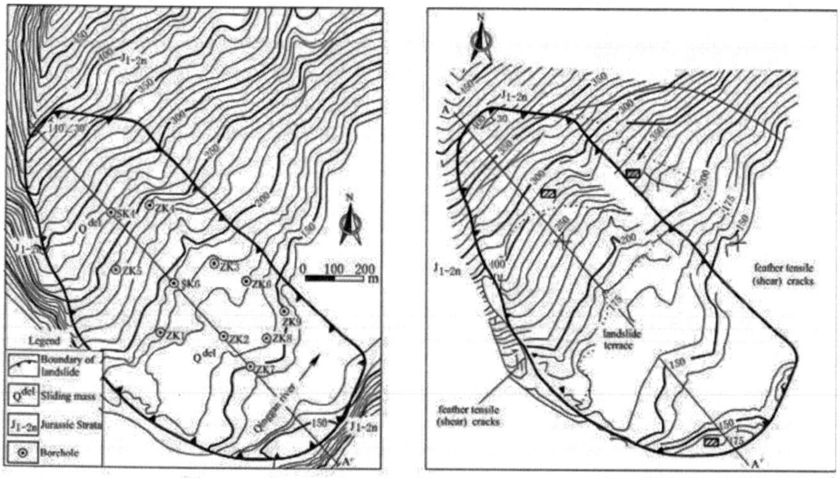

(a) (b)

Fig. 8.31 Topography of Qianjiangping Landslide, China. a Before failure. b After failure

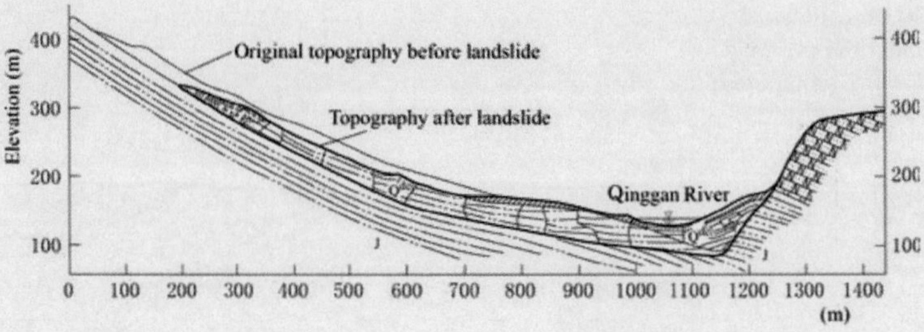

Fig. 8.32 Geologic section

(1) Topographic and geologic features

The rock stratum exposures in the landslide area are Jurassic System (J1-2n), Quaternary eluvium, slope deposit (Qdl + el) and Riverbed alluvium (Qal). Jurassic System (J1-2n) is composed of thick and green-grey feldspathic quartz sandstone, siltstone intercalated with purple silty clay rock, and mudstone. The attribute of rock layer is 130°∠15°–33°, the dip angle varies from 33° at the rear portion to 15° at near the riverbed. There are two joint sets that dominate the rear tensile cracking and flank shear-tensile cracking during the landslide:

- Set 1. Nearly strikes SN, the trace length on ground surface is 8–10 m, its attribute is 270°∠70°, and its space is 0.5–0.8 m;
- Set 2. Nearly EW and nearly vertical, the surface exposure trace length is 3–5 m, its attribute is 36^0∠70°, the joint space is ≈ 1 m.

(2) Landslide scenarios

① 15–18 days before the failure. Phenomena observed were: on 27 June, at EL.350 m and the north-eastern side of the rear portion, the wall of a residence house cracked; at EL. 280 m and the south-western side of the rear portion, the highway was damaged with a cracking of 3–6 cm wide, this cracking continued to 10 cm wide on 4 July, and a scarp of 10 cm high was recorded.

② 14 h before the failure (July 12). At EL.140 m, a 0.5 cm wide crack emerged; a bit later, another 3–5 cm wide crack emerged at EL.182 m.

③ At 0:20am of 13 July. Landslide occurred.

(3) Morphological features of the landslide

The slip surface is in a shape of armchair with its rear and front boundaries similar to arcs, the side boundary is fan-expanded. Figure 8.33 shows the slope after failure.

According to the observation on rear wall scratches, the slip direction is nearly coincident to the bedrock dip direction.

According to the observation on the abrasion of reservoir bank, the maximum surge wave height is 23–25 m.

Fig. 8.33 Landslide after failure. **a** Upstream view. **b** Opposite bank view

2. **DDA model**

(1) **Block element discretization**

The landslide is discretized into 395 block elements (viz. Fig. 8.34). The whole bedrock is handled as a fixed monolithic block element. In the thick direction of the landslide body, 4–7 layers of block elements are discretized, namely, the thickness of each layer is 4–5 m. The width of block is 7–10 m that is delimited by the EW oriented joint set.

(2) **Mechanical parameters**

There are three geomaterial zones in the upper portion with different mechanical parameters: one is below the river water level 135 m, another is above the river water level (135 m), the third zone is for bedrock. The mechanical parameters are listed in Table 8.7, in which line (A) and line (B) corresponding to the normal situation and the situation of rainfall + reservoir impounding where the strength is reduced. The time-marching step length is $\Delta t = 0.01\text{s}$.

3. **Landslide scenarios**

(1) **Driven towards the slipping onset**

Fig. 8.34 DDA model

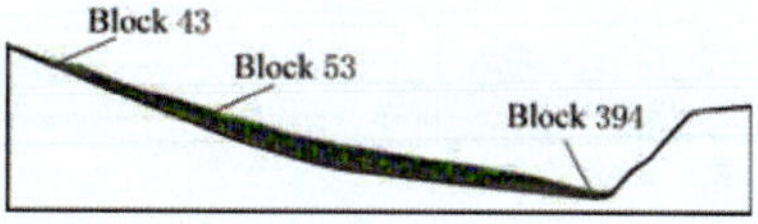

Table 8.7 Mechanical parameters of the landslide

Case	Deformation		Joint strength		Slip surface strength	
	E(GPa)	ν	φ(°)	c(MPa)	φ(°)	c(MPa)
Parameter (A)	0.5–1.0	0.35	18–23	0.03–0.06	18–23	0.02–0.05
Parameter (B)	0.5–1.0	0.35	18–20	0.03–0.05	12–15	0.02–0.03
Bedrock	2.0	0.25	/	/	/	/

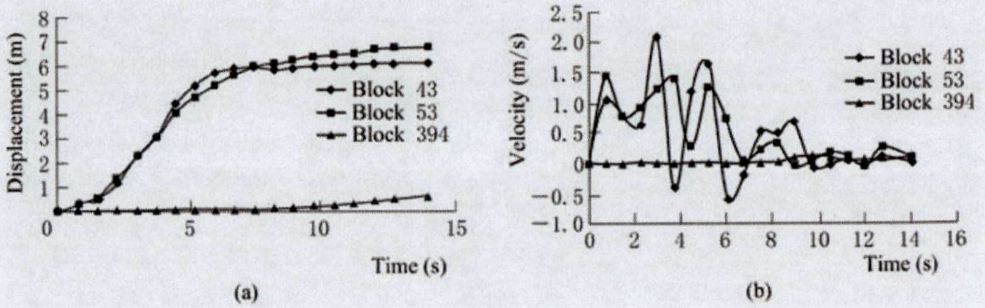

Fig. 8.35 Initial history of horizontal displacement and velocity: case 1. **a** Displacement. **b** Velocity

Three monitor points on block elements 43, 53 and 394 (viz. Fig. 8.34) are selected to capture the onset of slope failure. DDA analysis is undertaken for three cases.

Case 1: No rainfall and no reservoir impoundment. The parameters in the line (A) of Table 8.7 are employed. According to Fig. 8.35, the slope is stable. The upper portion (block elements 43 and 53) exhibits a certain deformation but the displacement and velocity of lower portion (block element 394) are negligibly small.

Case 2: Reservoir level 135 m. The parameters are the same as in Case 1. Under such circumstances, buoyancy is activated for the block elements below EL.135 m. The monitored displacement and velocity are shown in Fig. 8.36. It is clear that under this case, the slope keeps stable, too.

Case 3. Reservoir impoundment 135 m + rainfall induced strength deterioration. The parameters in the line (B) of Table 8.7 are input. According to Fig. 8.37, at initial period (OA section in the curves), block elements 43 and 53 get perceivable acceleration, but the displacement and acceleration of block element 394 is smaller, hence no overall failure will occur in this period; with the ongoing of time passing the initial period OA, block element 394 exhibits larger displacement and acceleration. Afterwards, the displacement and velocity of all three block elements are simultaneously varied, indicating overall failure.

By DDA analysis it is highlighted that the mechanism of this landslide is the combined action of reservoir impoundment and heavy rain fall.

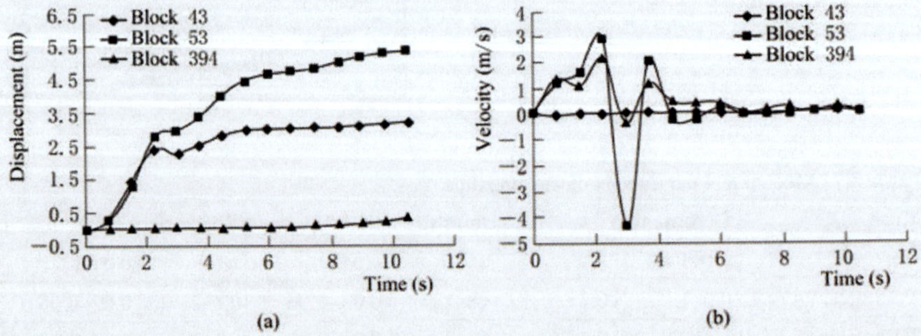

Fig. 8.36 Initial history of horizontal displacement and velocity: case 2.**a** Displacement. **b** Velocity

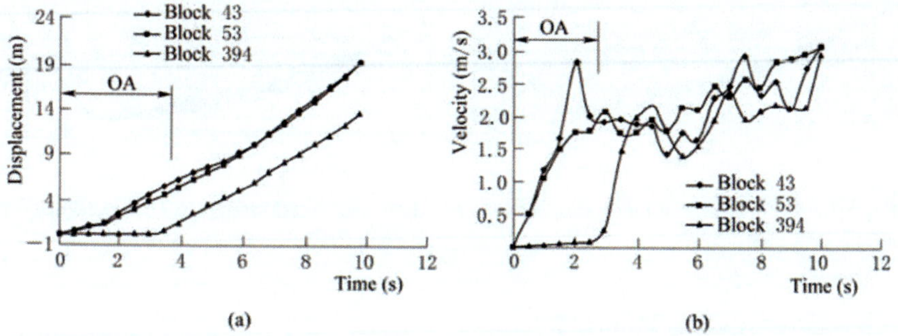

Fig. 8.37 Initial history of horizontal displacement and velocity: case 3. **a** Displacement. **b** Velocity

(2) **Slip process**

According to DDA computation, three periods of the landslide may be distinguished.

① Initiation. It is represented by the curve section OA in Fig. 8.38. Rear and upper portions exhibit remarkable deformation; shear phenomena between layers, particularly in surface layers, are obvious; the foot of landslide is compressed with high stress concentration.

② Acceleration. Following the evolution of shear deformation, the compressive failure occurs at the foot of landslide body in a manner of locally upwards squeezing; the overall slip down starts with the overall shear failure, as showed by the curve section AB in Fig. 8.38; the maximum velocity is 5 m/s.

③ Deceleration. Following the reduction of potential energy and the hinder of opposite reservoir bank, the velocity slows down until standstill, as shown by the curve section BC in Fig. 8.38; the duration of slip is about 40 s, the maximum dropdown height is 105 m; the deformation patterns of the landslide at $t = 15$s and $t = 50$s are exhibited in Fig. 8.39.

Fig. 8.38 Whole history of horizontal velocity

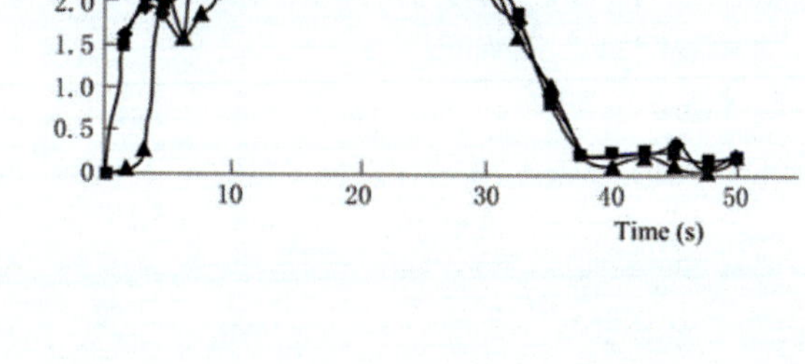

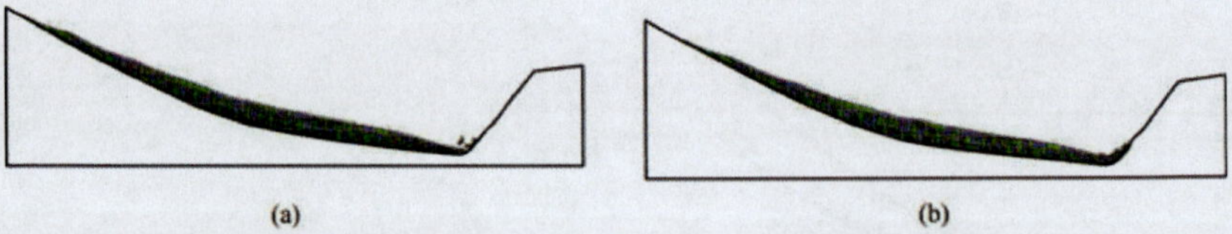

Fig. 8.39 Deformation of block system. a Deformation after 15,000 steps (15 s). b Deformation after 50,000 steps (50 s)

8.8 Concluding Remarks

We briefly summarize the state-of-the arts and expectations of DDA to close this chapter, for more details, the book by Hatzor and Shi (2017) is recommended.

The author is of the opinion that, the outmost merits of original version of DDA is that it possesses a good balance in its double simulation abilities for large block transition/rotation and simple deformation within block itself. In the former respect it out-performs the original version of BEA, whereas in the latter respect it out-performs the original version of DEM. It seems a human nature that all these referred methods have been under the improvements in an attempt to get the merits of others. Of course all these efforts are praiseworthy, but the risk of personality loss of individual method needs to be aware of, because just the personality enables the CM member occupies a unique role in straightforwardly solving the special engineering issue.

8.8.1 Validations and Applications

Over the last decades, researchers in the DDA community have dedicated a great deal of effort to document the accuracy of DDA by validation and application studies, that is particularly important since in many projects we should consider the safety margin concerning human life.

DDA has been successfully employed in the simulation of cut slope and landslide (MacLaughlin et al. 2001; Hatzor et al. 2004; Ning and Zhao 2013), rock fall (Wu et al. 2005; Ma et al. 2011), rock cavern (Hatzor et al. 2002b; Hatzor and Bakun-Mazor 2011), rock blast (Zhao et al. 2007; Ning et al. 2010), rock burst (Hatzor et al. 2017),etc. These applications have been discussed in various comprehensive reviews, to list only a few, Jing (1998, 2003) for essentials, MacLaughlin and Doolin (2006), Peng et al. (2019a, b) for static validation, Yagoda-Biran and Hatzor (2016) for dynamic verifications.

1. **Dams**

This is particularly concerned by Chinese scholars and engineers due to their extensive exercises in dam engineering since the 1990s. Recently, Zhang and Shi (2021) introduced the load sub-matrix of water pressure and described the calculation procedures of hydro-mechanical coupled analysis in the framework of DDA, which was further applied to evaluate the stability of a gravity dam foundation.

2. **Landslides and cut slopes**

A DDA analysis for Vajont landslide with nine vertical slices specified by Hendron and Patton (1985) is an important advancement (MacLaughlin 1997; Sitar et al. 2005).

The dynamic response of earthquake-induced landslides has received significant attention. The initial studies were mainly focused on the mobility of landslides (Huang et al. 2016; Song et al. 2016; Peng et al. 2019b). To model the attenuation in high-speed long-runout landslides triggered by earthquakes, the back analysis of Donghekou landslide taking into account of joint roughness degradation was conducted with DDA (Wang et al. 2021). The Masada National Monument in Israel subjected to dynamic earthquake action calibrated by DDA was well documented (Hatzor 1999; Hatzor et al. 2002a, 2004).

Bao et al. (2014), Fu et al. (2015a, b), Zhang et al. (2015a, b) reported the studies on the seismic response of rock mass with particular concern on boundary setting problems.

The scenarios of rock fall procedure from slope have been extensively studied, among which a rather large portion of works made use of DDA (Sasaki et al. 2005; Zheng et al. 2014).

Chen et al. (1995) described the use of DDA to aid the design of road cut and tunnel for a highway project. The right bank slope of Dagangshan Hydropower Project (China) was analyzed by 3-D DDA, the computational results claimed good agreement with monitoring results (Ma et al. 2020).

The slope failure process includes crack onset, growth and coalescence during the formation of a slip surface (small deformation stage) and block movement, rotation and fragmentation during the sliding process (large deformation stage). For a better understanding of the entire process of slope failure, a model that combines the fundamental CMs (e.g. FEM) and DDA is desirable: the slope models are progressively destabilized by the critical gravity approach, in this manner both the failure onset and collapse process are analyzed (Tang et al. 2017).

3. **Tunnels**

Several software enhancements have been implemented to facilitate block generation, development of in situ stresses, excavation operation, and incorporation of cable support elements. Dong et al. (1996) presented an excellent case study involving a tunnel and entrance slope excavation in blocky rock mass. The tunnel is actually an instrumented research edit hence the material properties and in situ stresses are available. The displacement predictions by DDA and FEM were cross-referenced at 8 observation points around the tunnel, and it was found that FEM predicts larger displacements than DDA.

Hatzor and Benary (1998) examined a roof failure of an ancient underground water reservoir constructed in the chalk in Israel.

Wu et al. (2004) employed DDA to investigate the mechanical behaviors of jointed rock masses with different-dipping angles during tunneling excavation, the results were well cross-referenced by experimental data. Do et al. (2017) used DDA to predict the sloping-surface subsidence induced by inclined mining stopes. Do and Wu (2020) used the DDA to simulate a tunnel excavation operation under various conditions with special focus on the ground subsidence and stress distribution, they showed that DDA is able to simulate the mechanical behaviors of the jointed rock strata in tunnel excavation.

Fu et al. (2017) discussed the applicability of existing input methods for DDA towards the seismic response study of the rock cavern in Dagangshan Hydropower Project (China).

4. **Macro-scale properties of block/particle assembly**

Ishikawa and Ohnishi (1999) examined the applicability of DDA for the cyclic plastic deformation of railway ballast in order to better understand track deformation. They declared that the cyclic deformation of ballast is resulted from particle realignment during loading, and given correct density. In 2001, they further studied the time-dependent behaviors of railroad ballast using DDA for the simulation of tri-axial tests (Ishikawa and Ohnishi 2001). Their results indicated that: DDA is capable of simulating time-dependent deformation of granular material without a specific constitutive model for viscosity; time dependency is more pronounced at lower confining stresses; and strain is rate-dependent (slower loading induces large strain).

A modified DDA algorithm was proposed to simulate the failure behavior of jointed rock mass (Zhang et al. 2008). In their proposed algorithm, by using the Monte-Carlo technique, DFN is created in the domain of interest. The triangular DDA block system is automatically generated by the advanced front method. In the process of generating blocks, numerous artificial joints come into being, and once the stress states at some artificial joints reach the failure criterion given beforehand, artificial joints will turn into real joints. In this way, the whole fragmentation process of rock mass can be replicated. Recently, cohesive crack model also has been employed to extend DDA (Gong et al. 2022).

8.8.2 Improvements

1. **Higher-order deflection and deformation**

In the original DDA, rock blocks are simply deformable, namely, the stress and strain is identical throughout the block element. In the most problems of rock slope, attributable to the "low stress" and high stiff of geomaterial, this assumption is acceptable. However, this simplification becomes problematic in the cases characterized by high stress gradient, soft material and irregular rock block shape. Improvements are continued to acquire higher accuracy (Wu 2015; Yu et al. 2020).

MacLaughlin and his co-workers (MacLaughlin and Sitar, 1996; MacLaughlin 1997) proposed a second order formulation of DDA by truncating Taylor series for sine and cosine at the quadratic terms. The resulting expression is used to derive modified potential energy, including the terms for inertia, point force, body force, etc. Many of which are identical to the original terms proposed by Shi.

Cheng and Zhang (2000) proposed two modifications for DDA to account for large rotation and large strain. New potential energy terms and corresponding mass matrix were derived, and the rotation value was iterated. To resolve the variations of stress and strain in rock block, they used constant strain triangular sub-blocks, between them nodal connectivity was enforced by springs. With a negligible performance penalty, the rotations were found considerably more accurate than the usual linear scheme, even than the second order scheme proposed by MacLaughlin and Sitar (1996). Incorporation of a higher order displacement function in DDA has been extensively validated for various beam bending problems (Koo et al. 1995; Koo and Chern 1996; Ma et al. 1996; Hsiung 2001).

In the study of stress wave propagation in jointed rock mass, Liu et al. (2018) discretized continuous rock blocks by flat-top PU mesh, which was generated independently of the problem domain. High-order polynomials were employed as local displacement approximation to improve the numerical accuracy.

Towards a refined simulation of jointed rock mass, multi-spring model was presented for edge-to-edge contact, too (Zhang et al. 2020).

2. **Time integration**

When contact constraints are enforced using penetration-based penalty method, very rigorous restraint on time-marching step length may enhance the perturbation in contact forces (Doolin and Sitar 2002).

Based on the stress wave propagation through continuous media, Wang et al. (1995) proposed a method for determining time-marching step length and penalty factor. They conducted the analysis of single elastic block under step loading, which exhibits strong damping for large step length. The analysis for penalty factor was based on the conservation of linear momentum on either side of propagating shock wave. Coaxial impact of two elastic rods was further used for deriving the relationship of penalty factor and step length.

Wang et al. (1996) showed that the time integration scheme in DDA is similar to the GN22 member of Newmark family (Zienkiewicz and Taylor 2000). They conducted a spectral investigation to ascertain the properties of DDA scheme, and found out that the spectrum of DDA scheme bifurcates from conjugate pair roots to two distinct real roots at a certain time step/period ratio. In addition, DDA scheme induces the most damping, and converges to the static solution rather fast. Doolin and Sitar (2004) continued the work by examining DDA scheme in greater detail. A geometrically non-linear algorithm for the integration of DDA scheme was derived. For constant stiffness, it produces results identical to Newmark algorithm.

3. **Crack growth**

Ke (1993, 1997) proposed an extended DDA code to incorporate "artificial" joints for simulating crack growth. The formulation is implemented by extending the original

Mohr–Coulomb contact law (friction + cohesion) to allow a tensile force to exist at the boundary of each sub-block. Validations included the cases of cantilever beam under bending, sample under uniaxial compression, and stress concentration near crack-tip.

Chang (1994) proposed a sub-meshing technique to enhance block deformability. The "nodal-based discontinuous deformation analysis (NDDA)" that is similar to FDEM (Munjiza 2004) greatly improves the stress accuracy within each DDA block by generating a local FE mesh inside the block meanwhile inherits the unique block kinematics of standard DDA. Each FE edge inside the block is treated as a potential crack, which enables the transformation of block material from continuum to discontinuum through the tensile and shear fracture mechanism (Shyu 1993; Bao 2010). Clatworthy and Scheele (1999) gave validations by the comparison to FEM solutions for the deflection of cantilever beam and the stress distribution of specimen under uniaxial compression. They declared that the results are identical "within the range of rounding-off errors". In addition, three numerical examples were examined to demonstrate the capability of NDDA in capturing brittle fracture process (Tian et al. 2014).

4. **Coupling with water**

Kim, Amadei and Pan (1999a, b) presented an DDA version incorporating water pressure, sequenced excavation, shotcrete lining and bolt support for the purpose to analyze the Unju (high speed rail) tunnel in Korea. Their implementation was divided into a finite element part for flows, and an interaction part with DDA blocks. The flow part was checked against the experimental work performed by Grenoble (1989), whereas the interaction part was validated by the solution of Sneddon and Lowengrub (1969).

Other important fields are the hydraulic fracturing of rock (Choo et al. 2016), the interaction of unsaturated soil–water (Guo et al. 2019), the landslides with induced surge waves (Wang et al. 2016; Wang et al. 2019b; Peng et al. 2021), etc., involving solid–fluid interaction.

5. **Three-dimensional problems**

A comparison of 2-D and 3-D DDA may be found in the work by Chen and his co-workers (Chen et al. 2013). The 3-D DDA established by Shi (2001a, b) reported good accuracy (relative error less than 0.2%) for two examples of block sliding. Yeung et al. (2002) constructed two plaster models of tetrahedral blocks that satisfy the similitude for physical quantities, to study rock wedge failures. The failure modes from the laboratory tests were compared with simulations using 3-D DDA code. It was noted that accurately releasing the physical blocks is a technical challenge, and sliding contacts are known to exhibit initial perturbation (Doolin and Sitar 2002).

Contact detection is always essential in DDA, and it is still a crucial obstacle in the 3-D DDA blamed for very large amount of computation efforts (Shi 2015; Wang et al. 2019a; Zheng et sal. 2018). This is why the most interest in particulate media are directed to the contact between spherical elements by a number of authors inclusive (only cited a few) Ke and Bray (1995), Thomas et al. (1996), Thomas and Bray

(1999), O'Sullivan and Bray (2001), Balden et al. (2001). The contact detection of high efficiency for polyhedral block system is still the most concern and an essential breakthrough is expected (Wu et al. 2014; Zhang et al. 2015a, 2016a, b, c, 2018a, b).

References

Balden V, Scheele F, Nurick G. Discontinuous deformation analysis in ball milling. In: Bicanic N, editor. Proceedings of the fourth international conference on analysis of discontinuous deformation. Glasgow (UK): University of Glasgow; 2001. p. 337–48.

Bao HR. Nodal based discontinuous deformation analysis. PhD Thesis. Civil and Environmental Engineering, Nanyang Technology University; 2010.

Bao HR, Yagoda-Biran G, Hatzor YH. Site response analysis with two-dimensional numerical discontinuous deformation analysis method. Earthq Eng Struct Dyn. 2014;43(2):225–46.

Barbosa RE. Discrete element models for granular materials and rock masses. PhD Thesis. University of Illinois at Urbana-Champaign (USA); 1990.

Bathe KJ. Finite element procedures in engineering analysis. New Jersey (USA): Prentice- Hall; 1982.

Bathe KJ, Wilson EL. Stability and accuracy analysis of direct integration methods. Earthquake Eng Struct Dynam. 1972;1(3):283–91.

Bui HD. Mécanique de la rupture fragile. Paris (France): Masson; 1978. (in French).

Chang CT. Nonlinear dynamic discontinuous deformation analysis with finite element meshed block system. PhD Thesis. University of California at Berkeley (USA); 1994.

Chen G, Zheng L, Zhang Y, Wu J. Numerical simulation in rockfall analysis: a close comparison of 2-D and 3-D DDA. Rock Mech Rock Eng. 2013;46(3):527–41.

Chen S, Chern JC, Koo CY. Study on performance of tunnel near slope by DDA. In: Li JC, Wang CY, Sheng J, editors. Proceedings of the first international conference on analysis of discontinuous deformation. Chungli, Taiwan: National Central University; 1995. p. 109–23.

Chen SH. The Elasto-plastic and Elasto-viscoplastic Analysis of the Discontinuous Rock Masses. PhD Thesis. WUHEE, Wuhan (China), 1987 (in Chinese with English summary).

Chen SH. Block-fault model and its application in the elasto-viscoplatic analysis of rock slope. J Hydraulic Eng. 1992;14(7):9–15 (in Chinese with English abstract).

Chen SH. Analysis of reinforced rock foundation using elastic-viscoplastic block theory. In: Ribeiro e Sousa L, Grossmann NF, editors. Proceedings 1993 ISRM international symposium—EUROCK 93. Rotterdam (Netherlands): AA Balkema; 1993. p. 45–51.

Chen SH. A new development in elastic-viscoplastic block theory of rock masses. In: Siriwardane HJ, Zaman MM, editors. Proceeding 8th international conference compounds method advance geomechanical. Rotterdam (Netherlands): AA Balkema; 1994. p. 2235–40.

Chen SH. Computational geomechanics and hydraulic structures. Berlin (Germany): Springer; 2018.

Chen SH, Li YM, Wang WM, Shahrour I. Analysis of gravity dam on a complicated rock foundation using an adaptive block element method. J Geotech Geoenviron Eng, ASCE. 2004;130(7):759–63.

Chen SH, Shen BK, Huang MH. Stochastic elastic-viscoplastic analysis for discontinuous rock masses. Int J Numer Meth Eng. 1994;37(14):2429–44.

Chen SH, Wang HR, Xiong WL. Elasto-plastic and elasto-viscoplastic block element theory of jointed rock masses. In: Yuan JX, editor. Proceeding 3rd Chinese conference numbers analytic method geotechnical engineering. Zhuhai (China);1988. p. 684–701 (in Chinese with English abstract).

Chen SH, Wang WM, She CX, Xu MY. Unconfined seepage analysis of discontinuous rock slope. J Hydrodyn. 2000;12(3):75–86.

Chen SH, Wang WM, Zheng HF, Shahrour I. Block element method for the seismic stability of rock slopes. Int J Geotech Geoenviron Eng, ASCE. 2010;136(12):1610–7.

Cheng YM, Zhang YH. Rigid body rotation and block internal discretization in DDA analysis. Int J Numer Anal Meth Geomech. 2000;24(6):567–78.

Choo LQ, Zhao Z, Chen H, Tian Q. Hydraulic fracturing modeling using the discontinuous deformation analysis (DDA) method. Comput Geotech. 2016;76:12–22.

Clayton RW, Engquist B. Absorbing boundary conditions for wave equation migration. Geophysics. 1980;45(5):895–904.

Clatworthy D, Scheele F. A method of sub-meshing in discontinuous deformation analysis (DDA). In: Amadei B, editor. ICADD-3: third international conference on analysis of discontinuous deformation, vail, Colorado (USA). Rotterdam (Netherlands): American Rock Mechanics Association, AA Balkema; 1999. p. 85–96.

Deeks AJ, Randolph MF. Axisymmetric time-domain transmitting boundaries. J Eng Mech, ASCE. 1994;120(1):25–42.

Do TN, Wu JH, Lin HM. Investigation of sloped surface subsidence during inclined seam extraction in a jointed rock mass using discontinuous deformation analysis. Int J Geomech. 2017;17(8):04017021.

Do TN, J Wu JH. Simulation of the inclined jointed rock mass behaviors in a mountain tunnel excavation using DDA. Comput Geotech. 2020;117:103249.

Doolin DM, Sitar N. Displacement accuracy of discontinuous deformation analysis method applied to sliding block. J Eng Mech. 2002;128(11):1158–68.

Doolin DM, Sitar N. Time integration in discontinuous deformation analysis. J Eng Mech. 2004;130(3):249–58.

Dong XC, Wu AQ, Ren F. A preliminary application of discontinuous deformation analysis (DDA) to the Three Gorges Dam project. In: Salami MR, Banks D, editors. Proceedings of the first international forum on discontinuous deformation analysis (DDA) and simulations of discontinuous media. Albuquerque, NM (USA): TSI Press; 1996. p. 310–17.

Fleming M, Chu YA, Moran B, Belytschko T. Enriched element-free Galerkin methods for crack tip fields. Int J Numer Meth Eng. 1997;40(8):1483–504.

Fu XD, Qian YH, Chen J, Leng XL. Extension of the discontinuous deformation analysis method to simulate seismic response of a large rock cavern complex. Int J Geomech. 2017;17(5):E4016008.

Fu XD, Sheng Q, Zhang YH, Chen J. Application of the discontinuous deformation analysis method to stress wave propagation through a one-dimensional rock mass. Int J Rock Mech Min. 2015a;80:155–70.

Fu XD, Sheng Q, Zhang YH, Zhou YQ, Dai F. Boundary setting method for the seismic dynamic response analysis of engineering rock mass structures using the discontinuous deformation analysis method. Int J Numer Anal Methods Geomech. 2015b;39(15):1693–712.

Gong SL, Hu CB, Guo LX, Ling DS, Chen GQ, Zhang XL. Extended DDA with rotation remedies and cohesive crack model for simulation of the dynamic seismic landslide. Eng Fract Mech. 2022;266:108395.

Grenoble BA. Influence of geology on seepage and uplift in concrete gravity dam foundations. PhD Thesis. University of Colorado, Boulder (USA); 1989.

Guo L, Li T, Chen G, Yu P, Peng X, Yang D. A method for microscopic unsaturated soil-water interaction analysis based on DDA. Comput Geotech. 2019;108:143–51.

Hatzor YH. Dynamic rock slope stability analysis at Masada national monument using block theory and DDA. In: Amadei B, Kranz R, Scott GA, Smeallie PH, editors. Rock mechanics for industry, proceedings of the 37th US rock mechanics symposium. Rotterdam (Netherlands): AA Balkema; 1999. p. 63–70.

Hatzor YH, Arzi AA, Tsesarsky M. Realistic dynamic analysis of jointed rock slopes using DDA. In: Hatzor YH, editor. Stability of rock structures: proceedings of the fifth international conference of analysis of discontinuous deformation. Lisse (Netherlands): AA Balkema; 2002a. p. 47–56.

Hatzor YH, Arzi AA, Zaslavsky Y, Shapira A. Dynamic stability analysis of jointed rock slopes using the DDA method: King Herod's palace, Masada, Israel. Int J Rock Mech Min Sci. 2004;41(5):813–32.

Hatzor Y, Bakun-Mazor D. Modelling dynamic deformation in natural rock slopes and underground openings with DDA: review of recent results. Geomech Geoeng. 2011;6(4):283–92.
Hatzor YH, Benary R. The stability of a laminated Voussoir beam: back analysis of a historic roof collapse using DDA. Int J Rock Mech Min Sci Geomech & Abstr. 1998;35(2):165–81.
Hatzor YH, Feintuch A. The validity of dynamic displacement prediction using DDA. Int J Rock Mech Min Sci & Geomech Abstr. 2001;38(4):599–606.
Hatzor YH, He BG, Feng XT. Scaling rockburst hazard using the DDA and GSI methods. Tunn Undergr Space Technol. 2017;70:343–62.
Hatzor YH, Shi GH. Discontinuous deformation analysis in rock mechanics practice. CRC Press; 2017.
Hatzor YH, Talesnick M, Tsesarsky M. Continuous and discontinuous stability analysis of the bell-shaped caverns at Bet Guvrin, Israel. Int J Rock Mech Min Sci. 2002b;39(7):867–86.
Hendron AJ, Patton FD. The Vaiont slide, a geotechnical analysis based on new geologic observations of the failure surface. Report GL-85-5. Vicksburg, MS (USA): US Army Corps of Engineers Waterways Experiment Station; 1985.
Higdon RL. Absorbing boundary conditions for difference approximations to the multi-dimensional wave equation. Math Compt. 1986;47(176):437–59.
Huang D, Song YX, Cen DF, Fu GY. Numerical modeling of earthquake-induced landslide using an improved discontinuous deformation analysis considering dynamic friction degradation of joints. Rock Mech Rock Eng. 2016;49(12):4767–86.
Hsiung SM. Discontinuous deformation analysis (DDA) with nth order polynomial displacement functions. In: Elsworth D, Tinucci JP, Heasley KA, editors. Proceedings of the 38th US rock mechanics symposium. Washington DC (USA). Rotterdam (Netherlands): AA Balkema; 2001. p. 1437–44.
Ishikawa T, Ohnishi Y. Analysis of cyclic plastic deformation of railroad ballast by DDA. In: Amadei B, editor. Third international conference on analysis of discontinuous deformation, Vail, Colorado (USA). Rotterdam (Netherlands): AA Balkema; 1999. p. 107–16.
Ishikawa T, Ohnishi Y. A study of time dependency of granular materials with DDA. In: Bicanic N, editor. Proceedings of the fourth international conference on analysis of discontinuous deformation. Glasgow (UK): University of Glasgow; 2001. p. 271–80.
Ishikawa T, Sekine E, Ohnishi Y. Shaking table tests of coarse granular materials with discontinuous deformation analysis. In: Hatzor YH, editor. Proceedings of the fifth international conference on analysis of discontinuous deformation. Abingdon (UK): AA Balkema; 2002. p. 181–188.
Jing L. Formulation of discontinuous deformation analysis (DDA)—an implicit discrete element model for block systems. Eng Geol. 1998;49(3–4):371–81.
Jing L. A review of techniques, advances and outstanding issues in numerical modeling for rock mechanics and rock engineering. Int J Rock Mech Min Sci. 2003;40(3):283–353.
Ke TC. Simulated testing of two-dimensional heterogeneous and discontinuous rock masses using discontinuous deformation analysis. PhD Thesis. University of California, Berkeley (USA); 1993.
Ke TC, Bray J. Modeling of particulate media using discontinuous deformation analysis. J Eng Mech. 1995;121(11):1234–43.
Ke TC. Application of DDA to simulate fracture propagation in solid. In: Ohnishi Y, editor. Proceedings of the second international conference on analysis of discontinuous deformation. Kyoto (Japan): Japan Institute of Systems Research; 1997. p. 155–85.
Kim YI, Amadei B, Pan E. Modeling the effect of water, excavation sequence and rock reinforcement with discontinuous deformation analysis. Int J Rock Mech Min Sci & Geomech Abstr. 1999a;36(7):949–70.
Kim YI, Amadei B, Pan E. Modeling the effect of geology on uplift in concrete gravity dam foundations. In: Amadei B, Kranz RL, Scott GA, Smeallie PH, editors. Rock mechanics for industry, Proceedings of the 37th US rock mechanics symposium. Vail, Colorado (USA): USRM; 1999b. p. 527–534.
Koo CY, Chern JC. The development of DDA with third order displacement function. In: Salami R, Banks D, editors. Proceeding of the 1st international forum on discontinuous deformation

analysis (dda) and simulations of discontinuous media. Albuquerque (USA): TSI Press; 1996. p. 342–9.

Koo CY, Chern JC, Chen S. Development of second order displacement function for DDA. In: Li JC, Wang CY, Sheng J, editors. Proceedings of the first international conference on analysis of discontinuous deformation. Chungli, Taiwan: National Central University; 1995. p. 91–108.

Liu XY, Zhao ZY, Ma AQ. Simulating stress wave with flat-top partition of unity based high-order discontinuous deformation analysis. Eng Anal Boundary Elem. 2018;91:110–23.

Lucas A, Mangeney A, Ampuero JP. Frictional velocity-weakening in landslides on Earth and on other planetary bodies. Nat Commun. 2014;5:3417.

Lysmer J, Kuhlemeyer RL. Finite dynamic model for infinite media. J Eng Mech Div, ASCE. 1969;95(4):859–77.

Ma K, Liu GY, Guo LJ, Zhuang DY, Collins DS. Deformation and stability of a discontinuity-controlled rock slope at Dagangshan hydropower station using three-dimensional discontinuous deformation analysis. Int J Rock Mech Min Sci. 2020;130:104313.

Ma G, Matsuyama H, Nishiyama S, Ohnishi Y. Practical studies on rockfall simulation by DDA. J Rock Mech Geotech Eng. 2011;3(1):57–63.

Ma MY, Zaman M, Zhou JH. Discontinuous deformation analysis using the third-order displacement function. In: Salami MR, Banks D, editors. Proceedings of the first international forum on discontinuous deformation analysis (dda) and simulations of discontinuous media, Berkeley, CA (USA). Albuquerque, NM (USA): TSI Press; 1996. p. 383–94.

MacLaughlin MM. Discontinuous deformation analysis of the kinematics of landslides. PhD Thesis. University of California, Berkeley (USA); 1997.

MacLaughlin MM, Doolin DM. Review of validation of the discontinuous deformation analysis (DDA) method. Int J Numer Anal Meth Geomech. 2006;30:271–305.

MacLaughlin MM, Sitar N. Rigid body rotations in DDA. In: Salami MR, Banks D, editors. Proceedings of the first international forum on discontinuous deformation analysis (DDA) and simulations of discontinuous media, Berkeley, CA (USA). Albuquerque, NM (USA): TSI Press; 1996. p. 620–36.

MacLaughlin M, Sitar N, Doolin D, Abbot T. Investigation of slope-stability kinematics using discontinuous deformation analysis. Int J Rock Mech Min Sci. 2001;38(5):753–62.

McBride AT, Scheele F. Validation of discontinuous deformation analysis using a physical model. Struct Eng Mech Comput. 2001;1:719–26.

Munjiza A. The Combined finite-discrete element method. New York (USA): Wiley; 2004.

Newmark NM. A method of computation for structural dynamics. J Eng Mech Div, ASCE. 1959;85(EM3):67–94.

Ning YJ, Yang J, An XM, Ma GW. Modelling rock fracturing and blast-induced rock mass failure via advanced discretisation within the discontinuous deformation analysis framework. Comput Geotech. 2010;38:40–9.

Ning YJ, Zhao ZY. A detailed investigation of block dynamic sliding by the discontinuous deformation analysis. Int J Numer Anal Meth Geomech. 2013;37(15):2373–93.

O'Sullivan C, Bray JD. A comparative evaluation of two approaches to discrete element modelling participate media. In: Bicanic N, editor. Proceedings of the fourth international conference on analysis of discontinuous deformation. Glasgow (UK): University of Glasgow; 2001. p. 97–110.

Otter JRH, Cassell AC, Hobbs RE. Dynamic relaxation. ICE Proc. 1966;35(4):633–56.

Owen DRJ, Hinton E. Finite elements in plasticity: theory and practice. Swansea (UK): Pineridge Press Ltd.; 1980.

Peng XY, Chen GQ, Fu HY, Yu PC, Zhang YB, Tang ZC, Zeng L, Wang W. Development of coupled DDA-SPH method for dynamic modelling of interaction problems between rock structure and soil. Int J Rock Mech Min Sci. 2021;146:104890.

Peng XY, Chen GQ, Yu PC, Zhang YB, Wang JM. Improvement of joint definition and determination in three-dimensional discontinuous deformation analysis. Comput Geotech. 2019a;110:148–60.

Peng XY, Yu PC, Chen GQ, Xia M, Zhang YB. CPU-Accelerated explicit discontinuous deformation analysis and its application to landslide analysis. Appl Math Model. 2019b;77:216–34.

Sasaki T, Hagiwara I, Sasaki K, Yoshinaka R, Nishiyama E. Earthquake response analysis of a rockfall by discontinuous deformation analysis. In: MacLaughlin M, Sitar N, editors. Proceedings of the seventh international conference on the analysis of discontinuous deformation (ICADD-7). Honolulu, Hawaii (USA); 2005. p. 137–46.

Shi GH. Block system modeling by discontinuous deformation analysis. Southampton (UK): Computational Mechanics Publications; 1993.

Shi GH. Discontinuous deformation analysis: a new numerical model for the statics and dynamics of block systems. PhD Thesis. University of California, Berkeley (USA); 1988.

Shi GH. Single and multiple block limit equilibrium of keyblock method and discontinuous deformation analysis. In: Hatzor YH, editor. Stability of rock structures: proceedings of the fifth international conference of analysis of discontinuous deformation. Lisse (Netherlands): AA Balkema; 2002. p. 3–43.

Shi GH. Discontinuous deformation analysis: a new numerical model for the statics and dynamics of deformable block structures. Eng Comput. 1992;9(2):157–68.

Shi GH. Three dimensional discontinuous deformation analysis. In: Bicanic N, editor. Proceedings of the fourth international conference on analysis of discontinuous deformation. Glasgow (UK): University of Glasgow; 2001a. p. 1–22.

Shi GH. Three dimensional discontinuous deformation analysis. In: Elsworth D, Tinucci JP, Heasley KA, editors. Proceedings of the 38th US rock mechanics symposium, American Rock Mechanics Association, Washington DC (USA). Rotterdam (Netherlands): AA Balkema; 2001b. p. 1421–28.

Shi G. Contact theory. Sci China (technol Sci). 2015;58(9):1450–96.

Shyu K. Nodal based discontinuous deformation analysis. PhD Thesis. University of California at Berkeley (USA); 1993.

Sitar N, MacLaughlin MM, Doolin DM. Influence of kinematics on landslide mobility and failure mode. J Geotech Geoenviron Eng. 2005;131(6):716–28.

Sneddon IN, Lowengrub M. Crack problems in the classical theory of elasticity. New York (USA): Wiley & Sons; 1969.

Song YX, Huang D, Cen DF. Numerical modelling of the 2008 Wenchuan earthquake-triggered Daguangbao landslide using a velocity and displacement dependent friction law. Eng Geol. 2016;215:50–68.

Stewart IJ, Brown ET. A Static relaxation method for the analysis of excavation in discontinuous rock. In: Proceeding ISRM symposium on design and performance of underground excavations. Cambridge (UK): British Geotechnical Society; 1984. p. 149–54.

Tang SB, Huang RQ, Tang CA, Liang ZZ, Heap MJ. The failure processes analysis of rock slope using numerical modelling techniques. Eng Fail Anal. 2017;79:999–1016.

Thomas PA, Bray JD. Capturing nonspherical shape of granular media with disk clusters. J Geotech Geoenviron Eng, ASCE. 1999;125(3):169–78.

Thomas PA, Bray JD, Ke TC. Discontinuous deformation analysis for soil mechanics. In: Salami MR, Banks D, editors. Proceedings of the first international forum on discontinuous deformation analysis (DDA) and simulations of discontinuous media, Berkeley, CA (USA). Albuquerque, NM (USA): TSI Press; 1996. p. 454–61.

Tian Q, Zhao ZY, Bao HR. Block fracturing analysis using nodal-based discontinuous deformation analysis with the double minimization procedure. Int J Numer Anal Meth Geomech. 2014;38(9):881–902.

Tsesarsky M, Hatzor YH, Sitar N. Dynamic block displacement prediction-validation of DDA using analytical solutions and shaking table experiments. In: Hatzor YH, editor. Stability of rock structures: proceedings of the fifth international conference of analysis of discontinuous deformation. Lisse (Netherlands): AA Balkema; 2002. p. 195–203.

Wang CY, Chuang CC, Sheng J. Time integration theories for the DDA method with finite element meshes. In: Salami MR, Banks D, editors. Proceedings of the first international forum on discontinuous deformation analysis (DDA) and simulations of discontinuous media, Berkeley, CA (USA). Albuquerque, NM (USA): TSI Press; 1996. p. 263–87.

Wang WM, Chen SH. A new implementation of elastic-viscoplastic block theory for rock masses. In: Kundu T, Desai CS, editors. Proceedings 10th international conference on computer method and advances in geomechanics. Rotterdam (Netherlands): AA Balkema; 2001. p. 417–22.

Wang X, Wu W, Zhu HH, Lin JS, Zhang H. Contact detection between polygonal blocks based on a novel multi-cover system for discontinuous deformation analysis. Comput Geotech. 2019a;111:56–65.

Wang CY, Sheng J, Chen MH. Dynamic-contact analysis scheme applied in the DDA method. In: Li JC, Wang CY, Sheng J, editors. Proceedings of the first international conference on analysis of discontinuous deformation. Chungli, Taiwan: National Central University; 1995. p. 433–59.

Wang JM, Zhang YB, Chen YL, Wang QD, Xiang CL, Fu HY, Wang P, Zhao JX, Zhao LH. Back-analysis of Donghekou landslide using improved DDA considering joint roughness degradation. Landslides. 2021;18(5):1925–35.

Wang W, Chen GQ, Zhang H, Zhou SH, Liu SG, Wu YQ, Fan FS. Analysis of landslide-generated impulsive waves using a coupled DDA-SPH method. Eng Anal Bound Elem. 2016;64:267–77.

Wang W, Yin KL, Chen GQ, Chai B, Han Z, Zhou J. Practical application of the coupled DDA-SPH method in dynamic modeling for the formation of landslide dam. Landslides. 2019b;16(5):1021–32.

Wartman J, Bray JD, Seed RB. Inclined plane studies of the Newmark sliding block procedure. J Geotech Geoenviron Eng, ASCE. 2003;129(8):673–84.

Wu JH. The elastic distortion problem with large rotation in discontinuous deformation analysis. Comput Geotech. 2015;69:352–64.

Wu JH, Ohnishi Y, Nishiyama S. Simulation of the mechanical behavior of inclined jointed rock masses during tunnel construction using discontinuous deformation analysis (DDA). Int J Rock Mech Min Sci. 2004;41(5):731–43.

Wu JH, Ohnishi Y, Nishiyama S. A development of the discontinuous deformation analysis for rock fall analysis. Int J Numer Anal Methods Geomech. 2005;29(10):971–88.

Wu W, Zhu HH, Zhuang XY, Ma GW, Cai YC. A multi-shell cover algorithm for contact detection in the three dimensional discontinuous deformation analysis. Theor Appl Fract Mech. 2014;72:136–49.

Xu DD, Wu AQ, Yang YT, Lu B, Liu F, Zheng H. A new contact potential based three-dimensional discontinuous deformation analysis method. Int J Rock Mech Min Sci. 2020;127:104206.

Xu MY, Wang WM, Chen SH. Research on the dangerous sliding-block combination of rock slopes. Rock Soil Mech. 2000;21(2):148–51 (in Chinese with English abstract).

Yagoda-Biran G, Hatzor YH. Benchmarking the numerical discontinuous deformation analysis method. Comput Geotech. 2016;71:30–46.

Yeung MR, Sun N, Jiang QH. A study of wedge stability physical models, block theory and three-dimensional discontinuous deformation analysis. In: Hatzor YH, editor. Stability of rock structures, proceedings of the fifth international conference on analysis of discontinuous deformation. Abingdon (UK): AA Balkema; 2002. p. 171–80.

Yu PC, Chen GQ, Peng XY, Guo LX, Zhang YB. Extension and application of discontinuous deformation analysis with a damped contact spring model. Int J Rock Mech Min Sci. 2019;123:104123.

Yu PC, Zhang YB, Peng XY, Chen GQ, Zhao JX. Distributed-spring edge-to-edge contact model for two-dimensional discontinuous deformation analysis. Rock Mech Rock Eng. 2020;53(1):365–82.

Zhang H, Chen GQ, Zheng L, Han Z, Zhang YB, Wu YQ, Liu SG. Detection of contacts between three-dimensional polyhedral blocks for discontinuous deformation analysis. Int J Rock Mech Min Sci. 2015a;78:57–73.

Zhang H, Liu SG, Chen GQ, Zheng L, Zhang YB, Wu YQ, Jing PD, Wang W, Han Z, Zhong GH. Extension of three-dimensional discontinuous deformation analysis to frictional-cohesive materials. Int J Rock Mech Min Sci. 2016a;86:65–79.

Zhang H, Liu SG, Han Z, Zheng L, Zhang YB, Wu YQ, Li YG, Wang W. A new algorithm to identify contact types between arbitrarily shaped polyhedral blocks for three-dimensional discontinuous deformation analysis. Comput. Geotech. 2016b;80(Dec):1–15.

Zhang H, Liu SG, Wang W, Zheng L, Zhang YB, Wu YQ, Han Z, Li YG, Chen GQ. A new DDA model for kinematic analyses of rockslides on complex 3-D terrain. Bull Eng Geol Environ. 2018a;77(2):555–71.

Zhang H, Liu SG, Zheng L, Zhong GH, Lou S, Wu YQ, Han Z. Extensions of edge-to-edge contact model in three dimensional discontinuous deformation analysis for friction analysis. Comput Geotech. 2016c;71:261–75.

Zhang H, Liu SG, Zheng L, Zhu HH, Zhuang XY, Zhang YB, Wu YQ. Method for resolving contact indeterminacy in three dimensional discontinuous deformation analysis. Int J Geomech. 2018b;18(10):04018130.

Zhang QH, Shi GH. Verification of a DDA-based hydro-mechanical model and its application to dam foundation stability analysis. Int J Rock Mech Min Sci. 2021;138:104627.

Zhang XL, Jiao YY, Zhao J. Simulation of failure process of jointed rock. J Cent South Univ Technol. 2008;15(6):888–94.

Zhang YB, Wang JM, Zhao JX, Chen GQ, Yu PC, Yang T. Multi-spring edge-to-edge contact model for discontinuous deformation analysis and its application to the tensile failure behavior of rock joints. Rock Mech Rock Eng. 2020;53(3):1243–57.

Zhang YB, Wang JM, Xu Q, Chen GQ, Zhao JX, Zheng L, Han Z, Yu PC. DDA validation of the mobility of earthquake-induced landslides. Eng Geol. 2015b;194(SI):38–51.

Zheng F, Jiao YY, Sitar N. Generalized contact model for polyhedra in three-dimensional discontinuous deformation analysis. Int J Numer Anal Meth Geomech. 2018;42(13):1471–92.

Zheng L, Chen G, Li Y, Zhang YB, Kasama K. The slope modeling method with GIS support for rockfall analysis using 3D DDA. Geomech Geoeng. 2014;9(2):142–52.

Zhao ZY, Gu J, Bao HR. Understanding fracture patterns of rock mass due to blast load—a DDA approach. In: Proceedings of the eight international conference on the analysis of discontinuous deformation, Beijing, China; 2007. p. 147–50.

Zienkiewicz OC, Taylor RL. The finite element method: solid mechanics. 5th ed., vol. 2. Oxford (UK): Butterworth Heinemann; 2000.

Chapter 9
Mesh-Free Methods with Special Focus on EFGM

Abstract TO undertake a successful computation task w.r.t. the construction and operation process of engineering structures, one of the mostly concerned obstacles is the limited or incomplete capability to automatically generate regular grid or to re-mesh retorted grid for complex structure domain containing structural planes and undergoing finite deformation. This chapter presents the principle and basic algorithm of the "element free Galerkin method (EFGM)" in structural dynamics that belongs to the family of "mesh-free methods (MFMs)". In EFGM, the basis functions no longer belong to standard PU (e.g. the shape functions of fundamental FEM), instead, they are created from the neighborhood field nodes in an "influence or support domain" of any shape by the technique of "moving least squares (MLS)" approximation. The nodal variables in the piecewise trial function may be solved from the algebraic equation set according to the virtual work principle or variational principle (weak form). In this manner, no rigorous restraint is imposed on the generation of scattered field nodes (points), which allows for a great simplification in the pre-process work of complex engineering structure. Although it has not outperformed FEMs in the routine problems of geotechnical and hydraulic structures, yet EFGM exhibits great expectation due to its flexibility in handling structural planes, crack growth, and other large deformation issues.

9.1 General

Fundamental CMs such as FDM, FVM and FEM are originally defined on grids (meshes) of data points (nodes). Each point has a fixed number of pre-defined neighboring points, and the connectivity between neighboring points can be used to define mathematical operators such as differences, derivatives and integrations. These operators are then used to construct discrete equation set to simulate PDEs (e.g. Navier-Stokes equations) directly.

Mesh (grid) generation is paramount for the pre-process in fundamental CMs and can be a very time-consuming burden for practitioners (Chen, 2018). Triangulation is the most prevalent technique to create 2-D triangular and 3-D tetrahedral mesh. The process can be fully automated for 2-D planes (curved surfaces as well) and 3-D

S. Chen, *Advanced Computational Methods and Geomechanics*, Springer Tracts in Civil Engineering, https://doi.org/10.1007/978-981-19-7427-4_9

spaces using, for example, the "advancing front technique (AFT)" and "Delaunay triangulation algorithm (DTA)". Quadrilateral mesh that possess much high computation efficiency and accuracy, can be generated either "directly" or "indirectly". The former (Zhu et al. 1991; Chen et al. 1996) converts pre-generated triangular elements into quadrilateral ones by means of splitting or merging operations, whereas the latter constructs and places quadrilateral elements directly into the domain. In the history of CMs, it is a dream to have entirely automated generators for any element types inclusive the hexahedral elements that have been proved to be superior to tetrahedral ones in terms of higher computation accuracy. Unfortunately, the existing techniques cannot generate high quality hexahedral element mesh (Cao et al. 1998; Chen et al. 1998b) for complex engineering structures. As a result, such a generator does not emerge on the software market insofar, although a number of semi-automated generators are available in most commercial software packages.

Where the material being simulated is able to move around such as in computational fluid dynamics (CFD), or exhibit large deformations such as in computational solid dynamic (CSD), the connectivity of mesh will be difficult to maintain without introducing error into the computation (Lee and Bathe 1993). Even worse, if a mesh becomes tangled or degenerated during computation, the operators defined on the mesh may no longer give correct output. Of course, the mesh may be re-created during the simulation (a process called re-meshing), but this will introduce an important source of numerical diffusion, since all the existing data points must be mapped onto new and different ones (Perić et al. 1996).

"Mesh-free methods (MFMs)" (also alternatively termed as "meshless methods") are intended to remedy the problems resulted from the retorted connectivity of computation mesh when the material exhibits strong nonlinear and discontinuous behaviors. They are particularly useful under following circumstances:

- Simulations where creating a regular mesh from a complex 3-D object may be tremendously difficult or require intensive human efforts;
- Simulations where field nodes may be created and/or element may be destroyed, such as in the process of rock fracturing and landsliding;
- Simulations where the problem domain is prone to move out of alignment w.r.t. fixed mesh, such as in the process of rock falling.

MFMs may be regarded as the means to generate a smooth surface approximation among various specified point values. They were initially introduced by Shepard (1968), and late on were extended for the same purpose by Lancaster and Salkauskas (1981). But it was not until the late-1990s when substantial and significant advances were made. Since then, much more MFM algorithms had been created, among them the "element free Galerkin method (EFGM)" (Belytschko and Lu 1995; Belytschko et al. 1996b) is an outstanding member in which the basis functions are created from the neighborhood nodes in a "domain of support (or influence)" by the "moving least squares (MLS)" approximation. This enable EFGM to exhibit great flexibility to handle the structural problem concerning crack growth in material (Belytschko et al. 2000).

In MFMs, the domain and boundary are represented (not discretized) by field nodes only, in other words, they do not need a mesh for field variables approximation.

However, some MFMs such as the EFGM that will be addressed in details in this book, require a background grid for the quadrature of system matrices and vectors. There are other MFMs such as the "meshless local Petrov-Galerkin method (MLPGM)" (Atluri and Zhu 1998, 2000) that do not require a grid for both the field variables approximation and quadrature. These types of MFMs are called "essential" ones.

9.2 Concept

9.2.1 Kernel Functions

In the fundamental FDM, the domain for 1-D simulation of function $u(x, t)$ would be represented by a set of data values $u_{i,k}$ at points x_i and times t_k, where

$$\begin{cases} x_{i+1} - x_i = h \ \ (i = 1, 2, 3, \ldots) \\ t_{k+1} - t_k = \Delta t \ (k = 1, 2, 3, \ldots) \end{cases} \tag{9.1}$$

We can define the derivatives in PDE by some finite difference formulae on this domain, for example

$$\begin{cases} \dfrac{\partial u}{\partial x} = \dfrac{u_{i+1}^k - u_{i-1}^k}{2h} \\ \dfrac{\partial u}{\partial t} = \dfrac{u_i^{k+1} - u_i^k}{\Delta t} \end{cases} \text{for } (i = 1, 2, 3, \ldots) \text{ and } (k = 1, 2, 3, \ldots) \tag{9.2}$$

Then $u(x, t)$ and its derivatives can be used to transfer PDE into an algebra equation set. In the most fundamental CMs since the 1960s, such an equation set can be alternatively created with the help of mechanical principles and shape functions that are distinguished as follows:

- The first category is based on the principles using of virtual work, Hamilton's function, and potential energy that are normally in the form of IEs (weak form). With a proper choice of test (shape) and trial (interpolation) functions, it gives rise to FEM.
- The second category is based on the residual principle that, in fact, is more general for solid deformations and fluid flows, as long as PDEs (strong form) are provided. With a proper choice of test (shape) and trial (interpolation) functions, it gives rise to FVM.
- The third category is based on the Taylor series expansion, which leads to fundamental FDM and FLAC.
- The fourth category is based on the conservation laws on each finite volume in the domain, which may be represented by FVM and SPH.

Although MFMs may be formulated using a variety of mechanical principles, yet a strong preference is perceivable: in the spatial discretization of solid domain, the first two categories are exercised more often, because the discretized equations derived on the "weak form" are normally more stable for better results. However, for the discretization of time, the third category (Taylor series) is predominant, which is normally termed as "strong form". In addition, the engineering practices also show that the last two categories could be well-performed in the spatial discretization w.r.t. fluid flow and heat transfer.

Typically MFMs start from the spatial approximation of an essential variable $u(x, t)$ via its nearest neighboring points in a manner of

$$u(x, t) = \sum_i K(\|x - x_i\|)u_i(x_i, t) \tag{9.3}$$

in which $K(\|x - x_i\|)$ is the "kernel function" that is nonlinearly dependent on the distance between field point x and neighboring data points (nodes) x_i. It operates on nearby data points for smoothness and other useful qualities such as derivatives. By linearity, the partial derivative w.r.t. the space domain may be calculated by

$$\frac{\partial u(x, t)}{\partial x} = \sum_i \frac{\partial K(\|x - x_i\|)}{\partial x} u_i(x_i, t) \tag{9.4}$$

Equations (9.3) and (9.4) are employed to transfer the PDE or IE being simulated into the ODE that may be solved by any schemes of FDM w.r.t. the time domain. If the kernel function is defined with the characteristic of "compact support", it will be the non-zero function only for nearby points within certain times "smoothing length". This is advantageous because we can save the effort by calculating the summation over only a few points. Another major advantage of MFMs is that the formulae for the calculation of function $u(x, t)$ and its derivatives use the points in any order, so it does not matter if the points move around or even exchange positions.

The major disadvantage of MFMs is that they require extra programming to determine the nearest neighbor points of a representative point, this is actually another kind of connectivity although without rigorous pattern is demanded.

In this chapter, the kernel function (9.3) is exactly the PU shape function defined by both the MLS approximation and compact weight function on the "support domain", which leads to the EFGM derived from the weak form (Galerkin) governing equations for solid dynamics. In the next chapter, we will go further, namely, the compact weight function is used to directly derive the kernel function (9.3) for the formulation of SPH from the strong form (Navier-Stokes) of the governing equations for fluid dynamics.

9.2.2 MLS Approximation and Shape Functions

In MFMs, the construction of well-performed shape functions has been and still is the major challenge, because these shape functions should be obtained with the help of pre-defined knowledge about the relationship of data points (nodes).

The currently prevalent technique for the construction of shape functions is the MLS approximation that was a common procedure for scattered data approximation by local polynomial fitting in the least squares sense. MFMs take the advantage of MLS to estimate the desired parameters by seeking a limited number of nearest points in the support domain of a representative point rather than to explore the global solution. In this way, the solution may be obtained more rapidly and there is no need for high degree polynomial. As a result, the artificial and unwanted oscillations in unconstrained regions may be avoided (Liu 2003). Being able to obtain the shape functions with higher order continuity and consistency from the basis functions with lower order continuity plus a suitable weight function, MLS possesses advantages of simple calculation, high precision and good smoothness (Huertaet al. 2004).

Although the MLS approximation is not the only means for building approximation, yet it is fundamental for the most of current MFMs (Fries and Matthies 2003). This is mainly attributable to their typically local design, namely, evaluation of a function only involves the values from nearby points that can be efficiently explored using search data structures (Bentley and Friedman 1979).

However, it need to point out that, since the shape functions constructed by MLS approximation do not possess Kronecker delta property, the essential boundary condition cannot be exactly met. Since the shape functions in EFGM are constructed concurrently with the process of assembling the global system equations, one possible solution is to impose essential boundary conditions in the stage of shape function creation or stiffness matrix calculation. By doing so, the final discretized system will not contain the DOFs for the points with specified boundary conditions. However, much work is still required to realize this idea. An alternative but promising way for the construction of shape functions with Kronecker delta property is to make use of "point interpolation method (PIM)" (Liu and Gu 2001).

Suppose a domain is discretized into scattered nodes (Fig. 9.1a and b). Each node is assigned with a compact "support domain" that is customarily circle or rectangle in 2-D problems. Support domain determines the amount of nodes used for the approximation of field variables at the representative point concerned. Actually, for any field (or quadrature) point $\mathbf{x}_Q$ marked with "•" in Fig. 9.1c, a compactly support domain may be defined, although it would be more precisely defined as "influence domain" within which a number nodes exert influence on $\mathbf{x}_Q$ (Liu 2003). Anyway, in a solid computation domain, for a node assemblage scattered without highly irregular distribution, the distinguish between support domain and influence domain in the construction of algorithm is not very necessary.

The subscript of $\mathbf{x}_Q$ is normally neglected where there is no risk of misleading. Similar to the one dimensional case, the general approximate function in a two-dimensional support domain $\Omega_{\mathbf{x}}$ is given as follows

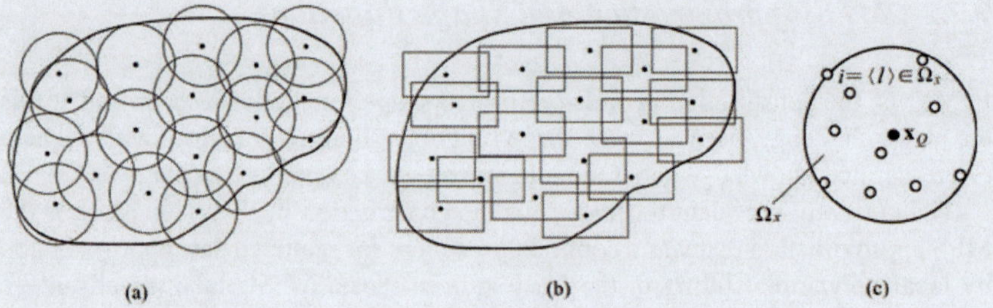

Fig. 9.1 Field nodes and field point with corresponding support domain. **a** Circle support domain; **b** rectangle support domain; **c** support domain $\Omega_{\mathbf{x}}$ of the field point $\mathbf{x}_Q$

$$u(\mathbf{x}) = \sum_{k=1}^{m} p_k(\mathbf{x})a_k(\mathbf{x}) = \mathbf{p}^{\mathrm{T}}(\mathbf{x})\mathbf{a}(\mathbf{x}) \quad \text{for } \mathbf{x} \in \Omega_{\mathbf{x}} \tag{9.5}$$

in which the time variable t is dropped for simplicity, $a_k(\mathbf{x})$ is the coefficient function of $\mathbf{x}$, $p_k(\mathbf{x})$ is the complete polynomial base function of $\mathbf{x}$.

Since the constant in the polynomial base function $p_k(\mathbf{x})$ of Eqs. (4.42)–(4.44) may be merged into the unknown coefficient $a_k(\mathbf{x})$, hence in 2-D cases following complete polynomial bases are employed

$$\mathbf{p}^{\mathrm{T}}(\mathbf{x}) = \begin{bmatrix} 1 \; x \; y \end{bmatrix} \quad \text{for } m = 3 \text{ (linear base)} \tag{9.6}$$

$$\mathbf{p}^{\mathrm{T}}(\mathbf{x}) = \begin{bmatrix} 1 \; x \; y \; x^2 \; xy \; y^2 \end{bmatrix} \quad \text{for } m = 6 \text{ (quadric base)} \tag{9.7}$$

$$\mathbf{p}^{\mathrm{T}}(\mathbf{x}) = \begin{bmatrix} 1 \; x \; y \; x^2 \; xy \; y^2 \; x^3 \; x^2y \; xy^2 \; y^3 \end{bmatrix} \quad \text{for } m = 10 \text{ (cubic base)} \tag{9.8}$$

It is notable that the coefficients $a_k(\mathbf{x})(k = 1, 2, \ldots, m)$ are coordinate dependent in MLS instead of constants in normal least squares technique.

For the solution of $a_k(\mathbf{x})(k = 1, 2, \ldots, m)$, we suppose that the function values at n nodes within $\Omega_{\mathbf{x}}$ are known as $u_i = u(x_i)(i \in \Omega_{\mathbf{x}})$, then the local approximate function at point $\mathbf{x}$ corresponding to Eq. (9.5) is degenerated as valid only surrounding these nodes, namely

$$u^h(\mathbf{x}, \mathbf{x}_i) = \sum_{k=1}^{m} p_k(\mathbf{x}_i)a_k(\mathbf{x}) = \mathbf{p}^{\mathrm{T}}(\mathbf{x}_i)\mathbf{a}(\mathbf{x}) \quad \text{for } i \in \Omega_{\mathbf{x}} \text{ and } \mathbf{x} \in \Omega_{\mathbf{x}} \tag{9.9}$$

in which superscript h denotes the size of supporting domain. Coefficient vector $\mathbf{a}(\mathbf{x})$ is decided by the weighted least square technique to minimize the overall difference in approximate function $u^h(\mathbf{x}, \mathbf{x}_i)$ and exact function u_i for all $i \in \Omega_{\mathbf{x}}$, which has been elaborated in Eq. (4.57).

Denote l as the local node sequence in the support domain $\Omega_{\mathbf{x}}$, and suppose that there is a link list relating it to the global node sequence in a manner of $i = \langle l \rangle$, the substitution of the coefficient vector $\mathbf{a}(\mathbf{x})$ in Eq. (4.57) into Eq. (9.5) yields an approximation in terms of the nodal coefficient $\hat{\mathbf{u}}$ such that

$$\mathbf{u}^h(\mathbf{x}) = \sum_{l}^{n} \phi_l(\mathbf{x}) u_l = \mathbf{\Phi}\hat{\mathbf{u}} \tag{9.10}$$

in which $\phi_l(\mathbf{x})$ is the shape function of local node l.

$$\begin{cases} \mathbf{\Phi} = [\phi_1 \ldots \phi_l \ldots \phi_n] = \mathbf{p}^{\mathrm{T}}\mathbf{A}^{-1}(\mathbf{x})\mathbf{B}(\mathbf{x}) \\ \phi_l(\mathbf{x}) = \sum_{k=1}^{m} p_k(\mathbf{x}_i)\big(\mathbf{A}^{-1}(\mathbf{x})\mathbf{B}(\mathbf{x})\big)_{kl} \end{cases} \quad i = \langle l \rangle \in \Omega_{\mathbf{x}} \tag{9.11}$$

It is notable that if $W(\|\mathbf{x} - \mathbf{x}_i\|) \in C^m$ (viz. Eq. (4.53)), then $\phi_i(\mathbf{x}) \in C^m$. Termed as "compactly supported trial function", the local approximation $u^h(\mathbf{x})$ in Eq. (9.10) is applicable to any other sub-domains for the integration in the governing equation of weak form, afterwards the nodal vector $\hat{\mathbf{u}}$ may be combined to get the global algebraic equation set over the whole domain Ω concerned.

The MLS shape functions (9.11) possess following remarkable characteristics:

- Consistency. The capacity to replicate complete order of polynomials.
- Reproduction. The capacity to replicate the basis functions used to construct shape functions

$$\sum_{k=1}^{m} \phi_l(\mathbf{x}) p_k(\mathbf{x}_l) = p_k(\mathbf{x}) \quad \text{for } i = \langle l \rangle \in \Omega_{\mathbf{x}} \tag{9.12}$$

It is notable that this characteristic is useful in handling crack-tip problem.

- Partition of unity

$$\sum_{l=1}^{n} \phi_l(\mathbf{x}) = 1 \quad \text{for } i = \langle l \rangle \in \Omega_{\mathbf{x}}. \tag{9.13}$$

- Non-Kronecker delta property. Since $\phi_i(\mathbf{x}_j) \neq \delta_{ij}$, hence $u^h(\mathbf{x}_i) \neq u_i$ which means that Eq. (9.10) is the approximation only (viz. Figs. 9.4, 9.5 and 9.6) instead of the interpolation in mesh-based methods.

Because $\mathbf{A}$ is symmetrical, Eq. (9.11) may be rewritten as

$$\phi_l(\mathbf{x}) = \mathbf{p}^{\mathrm{T}}(\mathbf{x})\mathbf{A}^{-1}(\mathbf{x})W(\mathbf{x} - \mathbf{x}_l)\mathbf{p}(\mathbf{x}_l) = \mathbf{c}^{\mathrm{T}}(\mathbf{x})\mathbf{p}(\mathbf{x}_l)W_l(\mathbf{x}) \tag{9.14}$$

$$\mathbf{c}(\mathbf{x}) = \mathbf{A}^{-1}(\mathbf{x})\mathbf{p}(\mathbf{x}) \tag{9.15}$$

The first-order derivative is the linear combination of the derivatives w.r.t. $\phi_l(\mathbf{x})$

$$\frac{\partial \phi_l(\mathbf{x})}{\partial \mathbf{x}} = \frac{\partial \mathbf{c}^{\mathrm{T}}(\mathbf{x})}{\partial \mathbf{x}}\mathbf{p}(\mathbf{x}_l)W_l(\mathbf{x}) + \mathbf{c}^{\mathrm{T}}(\mathbf{x})\mathbf{p}(\mathbf{x}_l)\frac{\partial W_l(\mathbf{x})}{\partial \mathbf{x}} \tag{9.16}$$

in which

$$\begin{aligned}\frac{\partial \mathbf{c}^{\mathrm{T}}(\mathbf{x})}{\partial \mathbf{x}} &= \frac{\partial \mathbf{A}^{-1}(\mathbf{x})}{\partial \mathbf{x}}\mathbf{p}(\mathbf{x}) + \mathbf{A}^{-1}(\mathbf{x})\frac{\partial \mathbf{p}(\mathbf{x})}{\partial \mathbf{x}} \\ &= -\mathbf{A}^{-1}(\mathbf{x})\frac{\partial \mathbf{A}(\mathbf{x})}{\partial \mathbf{x}}\mathbf{A}^{-1}(\mathbf{x})\mathbf{p}(\mathbf{x}) + \mathbf{A}^{-1}(\mathbf{x})\frac{\partial \mathbf{p}(\mathbf{x})}{\partial \mathbf{x}} \\ &= \mathbf{A}^{-1}(\mathbf{x})\left[\frac{\partial \mathbf{A}(\mathbf{x})}{\partial \mathbf{x}}\mathbf{c}^{\mathrm{T}}(\mathbf{x}) + \frac{\partial \mathbf{p}(\mathbf{x})}{\partial \mathbf{x}}\right] \\ &= \mathbf{A}^{-1}(\mathbf{x})\frac{\partial \mathbf{b}(\mathbf{x})}{\partial \mathbf{x}}\end{aligned} \tag{9.17}$$

$$\frac{\partial \mathbf{A}(\mathbf{x})}{\partial \mathbf{x}} = \sum_{l=1}^{n}\frac{\partial W_l(\mathbf{x})}{\partial \mathbf{x}}\mathbf{p}(\mathbf{x}_l)\mathbf{p}^{\mathrm{T}}(\mathbf{x}_l) \tag{9.18}$$

9.2.3 *Weight Functions*

Weight functions do influence the computation accuracy and efficiency greatly. They are compactly supported functions possessing following properties:

- Positivity;
- Maximum at $\mathbf{x}$;
- Decay following the increase of distance from $\mathbf{x}$;
- Continuity and derivability;
- Uniqueness w.r.t. coefficient $\mathbf{a}(\mathbf{x})$.

There are a number of weight functions available (viz. Chap. 4), of which the negative lag function will be employed in this chapter as follows

$$W_i(\mathbf{x}) = \begin{cases} \frac{\exp[-(d_i/d)^k]-\exp[-(d_{si}/d)^k]}{1-\exp[-(d_{si}/d)^k]} & \text{if } d_i \le d_{si} \\ 0 & \text{if } d_i \ge d_{si} \end{cases} \quad i = \langle l \rangle \in \Omega_{\mathbf{x}} \tag{9.19}$$

where: $i = \langle l \rangle$—global node sequence in the support domain; $d_i = \|\mathbf{x} - \mathbf{x}_i\|$—distance from $\mathbf{x}$ to $\mathbf{x}_i$; d_{si}—support radius of $\mathbf{x}$ that is actually the action domain of the weight function for the integration point $\mathbf{x}$; d—parameter dominating relative weight.

$$\begin{cases} d = ad_c \\ d_c = \underset{i,j \in \Omega_x}{\text{Max}} \left\| \mathbf{x}_j - \mathbf{x}_i \right\| \end{cases} \quad \text{for } i = \langle l \rangle \text{ and } j = \langle k \rangle \in \Omega_{\mathbf{x}} \tag{9.20}$$

If the nodes are uniformly distributed, d_c is the maximum space between nodes. d_c also may be defined as the average space between the nodes near the field point $\mathbf{x}$.

Too small radius d_{si} will lead to the failure of uniqueness, i.e., $\mathbf{A}^{-1}(\mathbf{x})$ in Eq. (9.11) does not exist; whereas too large radius d_{si} will cost high computation resources. In this book, $k = 2$, $d = d_c$, $d_{si} = 4d_c$ are customarily employed.

9.2.4 Support Domain and Influence Domain

"Support domain" and "influence domain" are the concepts often used in the mesh-free community to carry the same meaning in approximation but can lead to different implementation and code of MFMs. A general consideration of them may help us to construct/select more practically efficient and competently accurate weight functions. To clarify the concepts of support and influence domains, Liu and Liu (2003) gave the definition in the general context of MFMs as follows.

1. **Support domain**

The support domain for a field (quadrature) point at $\mathbf{x}_Q = (x, y, z)$ is the domain where the information for all the nodes inside this domain is used to determine the information at this field point (viz. Fig. 9.2). The weight function W is tightly related to the space of nodes (in EFGM)/particles (in SPH) and smooth length, the former

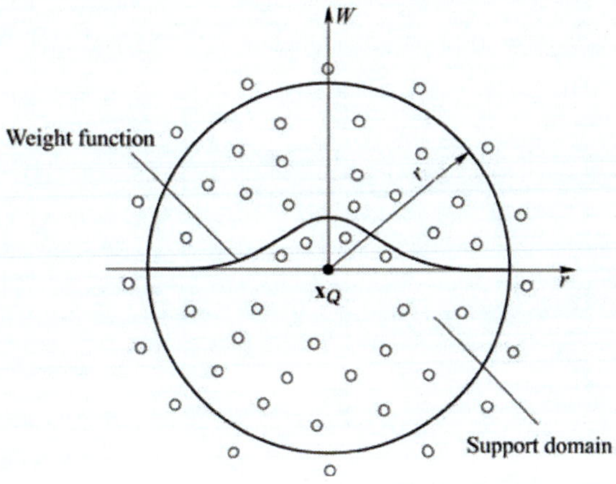

Fig. 9.2 Sketch of weight function and support domain

is used to measure the contribution of other nodes to the field point within support domain, the latter is used to define the size of support domain.

2. **Influence domain**

The concept of support domain works well if the node density does not vary too drastically in the problem domain. However, in solving practical problems where the node density can vary rather drastically, the use of support domain based on the current point of interest can lead to an unbalanced selection of nodes for constructing shape functions, which in turn, may lead to serious error. To prevent this kind of misleading, the concept of influence domain may be alternatively used to select nodes for constructing shape functions w.r.t. a representative point. In Fig. 9.3, central point i is the representative field node, other points are n_i field nodes within the circular influence domain of i.

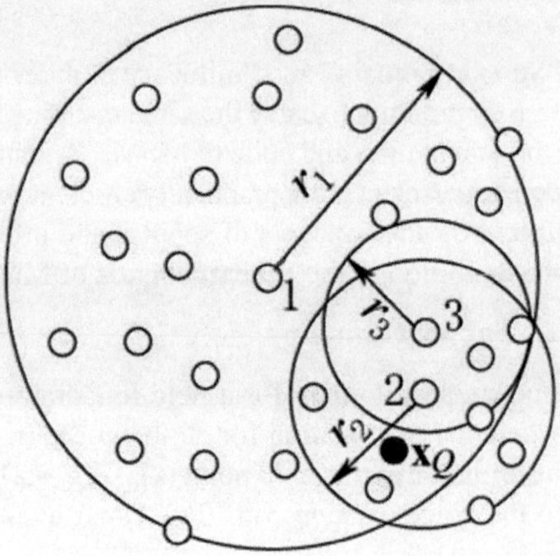

Fig. 9.3 Influence domain of field point

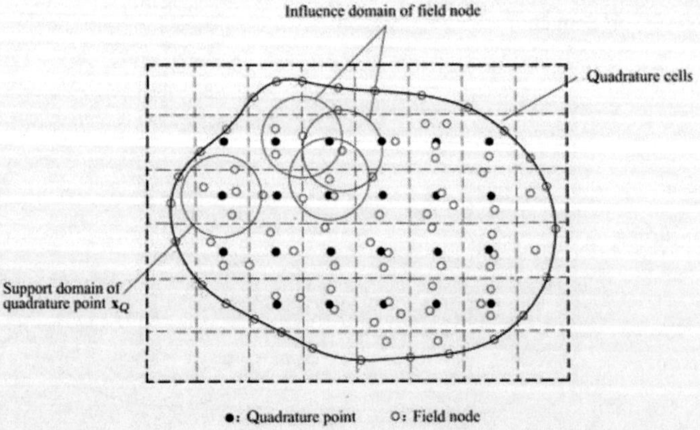

Fig. 9.4 Support domain and influence domain in background grid (quadrature cells)

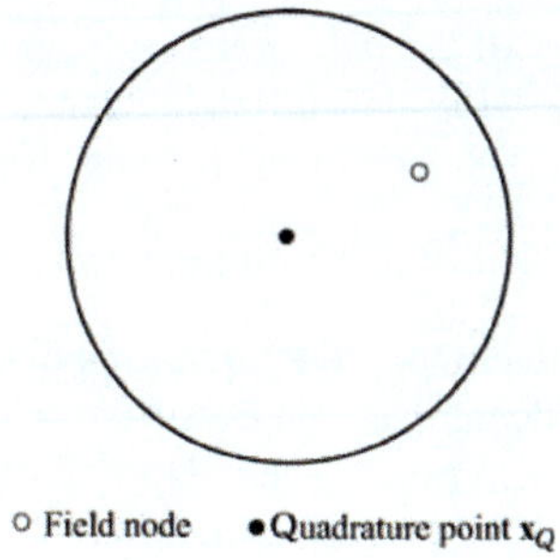

Fig. 9.5 One node domain

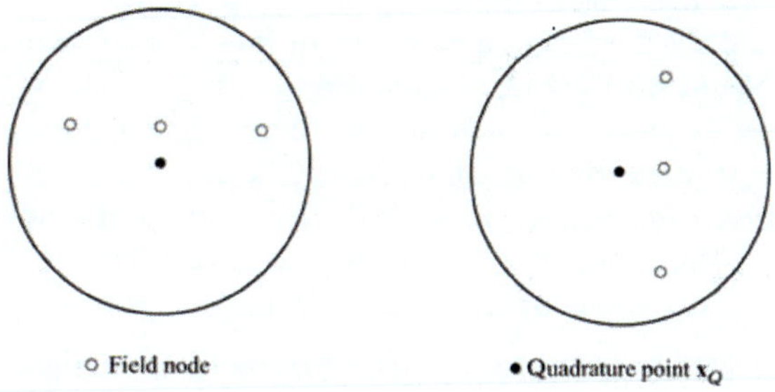

Fig. 9.6 All the nodes in a straight line; **a** in raw; **b** in column

The influence domain is so defined as a domain where a node at $\mathbf{x}_i = (x, y, z)$ exerts its influences. Hence, the influence domain is associated with a node i in the MFMs, and the support domain goes with any field point $\mathbf{x}_Q$ (e.g. quadrature point), which can be, but does not necessarily have to be, a node.

In the EFGM that will be specially handled in this chapter, the influence domain can be different from one node to another. As shown in Fig. 9.3, node 1 has an influence radius of r_1, and node 2 has an influence of r_2, etc. The fact that the dimension of influence domain can be different allows some nodes to have further influence than others and prevents unbalanced nodal distribution in constructing shape functions. Take the field (quadrature) point $\mathbf{x}_Q$ marked with black dot "•" for example, nodes 1 and 2 will be used, but node 3 will not be considered, even though node 3 is closer to $\mathbf{x}_Q$ compared with node 1.

The dimensions of influence domain can be determined using a procedure similar to that described in the support domain. Use of an influence domain provides an alternative way for the selection of nodes/particles in approximation, and it works better for highly irregular distributed discrete system. From the definition, it follows that:

- When the concept of support domain is used, the consideration is based on a field point $\mathbf{x}_Q$; when the concept of influence domain is used, the consideration is based on the node $\mathbf{x}_i$ only.
- If a node i is within the support domain of field point $\mathbf{x}_Q$, then node i exert an influence on point $\mathbf{x}_Q$, and thus $\mathbf{x}_Q$ is within the influence domain of node i.

- If the field point $\mathbf{x}_Q$ happens to be the node i, then this node will have a support domain and an influence domain simultaneously, such as the particles in the method of SPH (Chap. 10).

3. **Shape of support and influence domains**

The support domain for a field point can be a local sub-region of the entire problem domain, or can be the entire problem domain. In the latter case, the solution of that field point is related to all nodes in the problem domain. Since the importance of different nodes are different, a local support domain is usually preferred in order to reduce computational effort, in which only fewer nodes within the sub-region are used in the field variable approximation for the field point concerned. The dimension and shape of the support domain for different points may be different. The most commonly used support domains in 2-D problems are ellipse (or strictly circle) and rectangle (or strictly square) (viz. Fig. 9.1). The support domain is usually taken to be symmetric, but sometimes have to be non-symmetric, especially for the field points near boundary, or for the field points with special considerations.

The influence domain for a node can also be global as the entire problem domain or local as a sub-region in the entire problem domain with finite dimension and shape.

4. **Dimension of support and influence domains**

The accuracy of approximation in EFGM directly depends on the node amount in the support domain of the field point which is often a quadrature point $\mathbf{x}_Q$ (or the central point) of the integration cell in Fig. 9.4. A suitable support domain should be so chosen to ensure a proper area of coverage for approximation. To define the support domain for $\mathbf{x}_Q$, its dimension d_s is determined by

$$d_s = \alpha_s d_c \tag{9.21}$$

in which α_s is the dimensionless factor of the support domain and d_c is the characteristic length that relates to the average nodal spacing near the point $\mathbf{x}_Q$. For example, $\alpha_s = 3.0$ means a support domain whose radius is 3.0 times the average nodal spacing. If the nodes are uniformly distributed, d_c is simply the distance between two neighboring nodes. The actual amount of nodes n can be determined by counting all the nodes within the support domain. α_s should be pre-defined by carrying out numerical experiments for the same class of problems for which solutions already exist. Generally, $\alpha_s = 2.0 - 4.0$ may lead to good results. It is notable that, if background cells are provided, support domain can also be defined by the background cell itself.

In addition to the circle domain, the support domain may be any shapes. If the rectangular cell is used as support domain, its size may be selected such that

$$\begin{cases} d_{sx} = \alpha_{sx} d_{cx} \\ d_{sy} = \alpha_{sy} d_{cy} \end{cases} \tag{9.22}$$

in which d_{cx} and d_{cy} are the characteristic lengths related to the nodal spacing near the point $\mathbf{x}_Q$, α_{sx} and α_{sy} are the factors along the directions of x and y, respectively.

The dimension of support domain can vary both temporarily and spatially. In some cases, it also may vary in dimensions, and therefore it is a scalar in 1-D, a vector in 2-D and a tensor in 3-D.

5. **Average nodal spacing**

In the monograph by Liu (2003), the calculation of average nodal spacing has been given as follows.

(1) **1-D cases**

$$d_c = \frac{D_s}{\left(n_{D_s} - 1\right)} \tag{9.23}$$

in which D_s is an estimated d_s, n_{D_s} is the number of nodes that are covered by a known domain with the dimension D_s.

There is no essential difficulty to determine d_s for the field point $\mathbf{x}_Q$ with non-uniformly distributed nodes. The procedure is implemented as follows:

① Estimate d_s for point $\mathbf{x}_Q$, which gives D_s;
② Count nodes that covered by D_s;
③ Use Eq. (9.23) to calculate d_c;
④ Finally, calculate d_s using Eq. (9.22) for a given α_s.

(2) **2-D cases**

$$d_c = \frac{\sqrt{A_s}}{\left(\sqrt{n_{A_s}} - 1\right)} \tag{9.24}$$

in which A_s is an estimated area that is covered by the support domain of dimension d_s, n_{A_s} is the amount of nodes that are covered by the estimated domain within the area of A_s.

6. **Nodal arrangement in support and influence domains**

In addition to node amount, the node arrangement also needs to be taken care of w.r.t. the inverse operation of matrix **A** in Eq. (9.11).

Taking the linear base function (6) for example, if there is only one node $\mathbf{x}_1$ within the circular support domain of $\mathbf{x} = \mathbf{x}_Q$ (viz. Fig. 9.5), we get

$$\mathbf{A}(\mathbf{x}) = w(\mathbf{x} - \mathbf{x}_1)\begin{bmatrix} 1 & x_1 & y_1 \\ x_1 & x_1^2 & x_1 y_1 \\ y_1 & y_1 x_1 & y_1^2 \end{bmatrix} \tag{9.25}$$

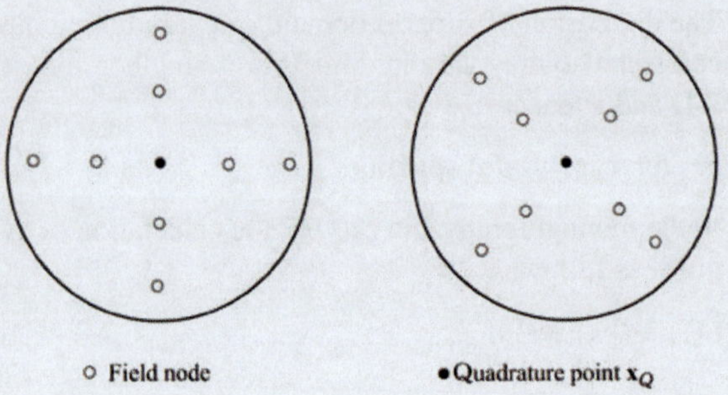

Fig. 9.7 Singularity pattern of node distribution for quadric base function

Since the second line and third line are linearly dependent, the determinant $|\mathbf{A}(\mathbf{x})|$ is zero so the inverse matrix $\mathbf{A}^{-1}$ does not exist.

Even there are a number of nodes in the support domain of $\mathbf{x} = \mathbf{x}_Q$ distributed in a straight line (viz. Fig. 9.6), since either their coordinate $x_l(l = 1, 2, \ldots, n)$ or $y_l(l = 1, 2, \ldots, n)$ are identical, the determinant $|\mathbf{A}(\mathbf{x})|$ is zero and the inverse matrix $\mathbf{A}^{-1}$ does not exist, too.

Similarly, for the quadric base function (9.7), if nodes $\mathbf{x}_l(l = 1, 2, \ldots, n)$ are arranged in two perpendicular straight lines (viz. Fig. 9.7), the singularity of $\mathbf{A}$ will give rise too.

9.3 Basic Formulation of EFGM

Hereinafter in this chapter, the interest is directed to support domain only towards the formulation of EFGM.

9.3.1 *Domain Discretization and Variable Approximation*

1. **Material entities**

A domain should be discretized in to a grid containing n_e cells and n_n nodes. In Fig. 9.8, where there is crack or boundary, the support domain of $\mathbf{x}_Q$ should not across the crack or boundary, namely, the link between the nodes in circle domain and $\mathbf{x}_Q$ is not intersected with the crack or boundary.

Suppose the support domain of $\mathbf{x} = \mathbf{x}_Q$ comprises n nodes, according to MLS, the approximate displacement $\mathbf{u}(\mathbf{x})$ in the support domain $\Omega_{\mathbf{x}_Q}$ may be approximated by the nodal displacement $\hat{\mathbf{u}}_l$ $\left(i = \langle l \rangle \in \Omega_{\mathbf{x}_Q}\right)$ such that

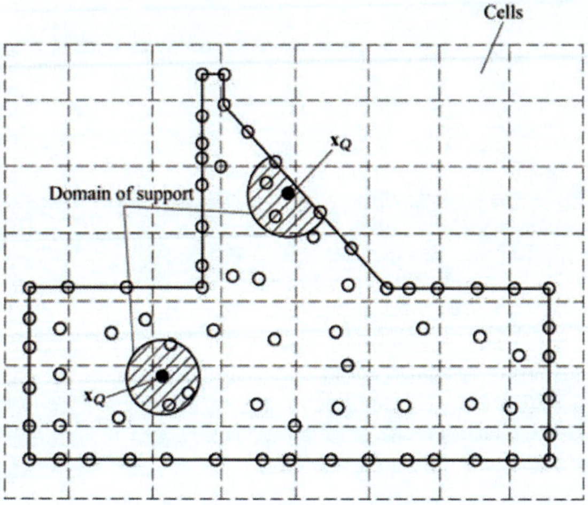

Fig. 9.8 Cell and support domain

$$\mathbf{u}(x_Q, y_Q) = \left[u_x \; u_y\right]^{\mathrm{T}} = \boldsymbol{\Phi}\hat{\mathbf{u}} \tag{9.26}$$

$$\begin{cases} \hat{\mathbf{u}} = \left[\hat{\mathbf{u}}_1 \ldots \hat{\mathbf{u}}_l \ldots \hat{\mathbf{u}}_n\right]^{\mathrm{T}} \\ \hat{\mathbf{u}}_l = \begin{Bmatrix} u_{x,l} \\ u_{y,l} \end{Bmatrix} \end{cases} \tag{9.27}$$

$$\begin{cases} \boldsymbol{\Phi} = \left[\boldsymbol{\phi}_1 \ldots \boldsymbol{\phi}_l \ldots \boldsymbol{\phi}_n\right] \\ \boldsymbol{\phi}_l = \phi_l \mathbf{I} \end{cases} \tag{9.28}$$

in which $\mathbf{I}$ is the 2×2 unit matrix.

The subscript indicating quadrature point $\mathbf{x}_Q$ is neglected where there is no risk of misleading as has been stated previously. The elastic strain and stress are derived as

$$\begin{cases} \boldsymbol{\varepsilon} = \mathbf{B}\hat{\mathbf{u}} \\ \boldsymbol{\sigma} = \mathbf{S}\hat{\mathbf{u}} \end{cases} \tag{9.29}$$

in which

$$\begin{cases} \mathbf{B} = \left[\mathbf{B}_1 \ldots \mathbf{B}_l \ldots \mathbf{B}_n\right] \\ \mathbf{B}_l = \mathbf{L}\boldsymbol{\phi}_l \end{cases} \quad \text{for } l \in \Omega_{\mathbf{x}_Q} \tag{9.30}$$

$$\mathbf{S} = \mathbf{D}\mathbf{B} \tag{9.31}$$

where: $\mathbf{L}$—differential operator matrix given in Eq. (3.86) for 3-D problems and in Eq. (5.52) for 2-D problems; $\mathbf{D}$—elastic matrix.

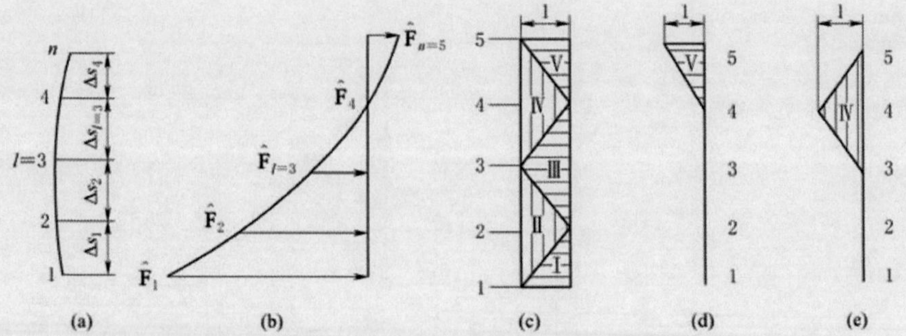

Fig. 9.9 Distribution of variables along boundary ($n = 5$). **a** Discretization of boundary curve into segments along the direction of **s**; **b** variable vector **F**; **c** mathematical covers corresponding to nodes; **d** mathematical cover of end-node n (and 1 as well); **e** mathematical cover of internal-node l

2. **Boundaries**

Boundary nodes (interior and exterior) are grouped according to exerted restraint conditions. In this chapter, n_t denotes the amount of boundary nodes with prescribed traction $\bar{\mathbf{t}}$, n_u denoted the amount of essential boundary nodes with prescribed displacement $\bar{\mathbf{u}}$, n_{sp} denotes the amount of internal nodes at the contact face or structural plane. In Fig. 9.9, a portion of specific boundary is discretized by n nodes $\left(n \in n_t \cup n_u \cup n_{sp}\right)$.

Any variable vector **F** may be interpolated using Lagrange scheme along the boundary segment.

$$\mathbf{F}(\mathbf{x}) = \sum_{l=1}^{n} \mathbf{N}_l(s)\hat{\mathbf{F}}_l \tag{9.32}$$

$$\begin{cases} \hat{\mathbf{F}} = \left[\hat{\mathbf{F}}_1 \ldots \hat{\mathbf{F}}_l \ldots \hat{\mathbf{F}}_n\right]^{\mathrm{T}} \\ \hat{\mathbf{F}}_l = \left[\hat{\mathbf{F}}_{x,l}\ \hat{\mathbf{F}}_{y,l}\right]^{\mathrm{T}} \end{cases} \tag{9.33}$$

$$\begin{cases} \mathbf{N}_l(s) = N_l(s)\mathbf{I} \\ \mathbf{I} = \begin{bmatrix} 1 & 0 \\ 0 & 1 \end{bmatrix} \\ N_l(s) = \dfrac{\|\mathbf{x} - \mathbf{x}_1\|.\|(\mathbf{x} - \mathbf{x}_2)\| \ldots \|(\mathbf{x} - \mathbf{x}_n)\|}{\|\mathbf{x}_l - \mathbf{x}_1\|.\|(\mathbf{x}_l - \mathbf{x}_2)\| \ldots \|(\mathbf{x}_l - \mathbf{x}_n)\|} \end{cases} \tag{9.34}$$

in which N_l is the interpolation function of Lagrange, s is the arc length of the boundary starting from point $\mathbf{x}_1$, $F_{x,l}$ and $F_{y,l}$ are the variable components at the boundary node l.

Another way to interpolate the variable vector $\mathbf{F}(\mathbf{x})$ makes use of segment interpolation with the help of mathematical covers defined in NMM. Suppose the boundary segment and the variable vector $\mathbf{F}(\mathbf{x})$ on any segment is linear, then the 1-D shape functions (4.24) may be employed for the interpolation to get

$$\begin{cases} \mathbf{F}(\mathbf{x}) = \sum_{l=1}^{n} \mathbf{F}_l(\mathbf{x}) \\ \mathbf{F}_l(\mathbf{x}) = \mathbf{N}_l(\xi)\hat{\mathbf{F}}_l \quad (l = 1, 2, \ldots, n) \end{cases} \tag{9.35}$$

Where

$$\mathbf{N}_l(\xi) = \begin{bmatrix} N_l(\xi) & 0 \\ 0 & N_l(\xi) \end{bmatrix} \tag{9.36}$$

$$N_l(\xi) = \begin{cases} N_2(\xi) & \text{for } \Delta s_{l-1} \\ N_1(\xi) & \text{for } \Delta s_l \end{cases} \tag{9.37}$$

The integration along the boundary is simply conducted as follows

$$\int_{\Gamma} \mathbf{F}(\mathbf{x})\mathrm{d}\Gamma = \sum_{l}^{n} \int_{\Delta s_{l-1}+\Delta s_l} \mathbf{N}_l\hat{\mathbf{F}}_l\mathrm{d}\Gamma \tag{9.38}$$

If we further postulate that all the segments are straight, then Eq. (9.38) may be directly integrated and it follows that

$$\int_{\Gamma} \mathbf{F}(\mathbf{x})\mathrm{d}\Gamma = \frac{1}{2}\begin{bmatrix} \hat{\mathbf{F}}_1\Delta s_1 + \hat{\mathbf{F}}_2(\Delta s_1 + \Delta s_2) + \ldots \\ +\hat{\mathbf{F}}_l(\Delta s_{l-1} + \Delta s_l) + \ldots + \hat{\mathbf{F}}_n\Delta s_{n-1} \end{bmatrix} \tag{9.39}$$

9.3.2 Governing Equations

For 2-D elasticity body without taking into account of structural planes and dynamic action, the Galerkin Eq. (3.234) can be written as (Belytschko et al. 1996b)

$$\iint_{\Omega} (\mathbf{L}\delta\mathbf{u})^{\mathrm{T}}\mathbf{D}(\mathbf{L}\mathbf{u})\mathrm{d}\Omega - \iint_{\Omega} \delta\mathbf{u}^{\mathrm{T}}\mathbf{b}\mathrm{d}\Omega$$

$$- \int_{\Gamma_t} \delta\mathbf{u}^{\mathrm{T}}\bar{\mathbf{t}}\mathrm{d}\Gamma - \sum_{\zeta=1}^{n_\zeta} \delta\mathbf{u}^{\mathrm{T}}(\mathbf{x}_\zeta)\mathbf{T}_\zeta(\mathbf{x}_\zeta) - \delta W_u(\mathbf{u}) = 0 \tag{9.40}$$

where: $\delta\mathbf{u}$—testing function; $\delta W_u(\mathbf{u})$—virtual work due to essential boundary conditions.

One drawback of MLS is that it is an approximation rather than the interpolant, i.e. $\phi_i(x_j) \neq \delta_{ij}$, and consequently the shape functions from the nodes on the interior of domain are nonzero on the boundary. As a result, essential boundary conditions cannot be satisfied directly. This is why a term $\delta W_u(\mathbf{u})$ should be used in Eq. (9.40) to enforce the essential boundary conditions.

It is worthwhile to mention that another prevalent method of enforcing essential boundary conditions in MFMs is by the hybrid with FEMs (Belytschko et al. 1995b; Krongauz and Belytschko 1996; Rabczuk et al. 2006). In such a hybrid method, a row of standard FEs is placed along the essential boundaries firstly, then the FE shape functions are blended with the EFGM shape functions. The boundary conditions can then be enforced by prescribing the values at boundary nodes.

1. **Lagrange multiplier method**

Suppose in Eq. (9.40) we have

$$\delta W_u(\mathbf{u}) = \int_{\Gamma_u} \delta \boldsymbol{\lambda}^{\mathrm{T}}(\mathbf{u} - \overline{\mathbf{u}})\mathrm{d}\Gamma + \int_{\Gamma_u} \delta \mathbf{u}^{\mathrm{T}} \boldsymbol{\lambda} \mathrm{d}\Gamma \tag{9.41}$$

in which $\boldsymbol{\lambda}$ is a Lagrange multiplier vector. It follows that

$$\begin{aligned} &\iint_{\Omega} (\mathbf{L}\delta\mathbf{u})^{\mathrm{T}}\mathbf{D}(\mathbf{L}\mathbf{u})\mathrm{d}\Omega - \iint_{\Omega} \delta\mathbf{u}^{\mathrm{T}}\mathbf{b}\mathrm{d}\Omega - \int_{\Gamma_t} \delta\mathbf{u}^{\mathrm{T}}\overline{\mathbf{t}}\mathrm{d}\Gamma \\ &- \sum_{\zeta=1}^{n_\zeta} \delta\mathbf{u}^{\mathrm{T}}(\mathbf{x}_\zeta)\mathbf{T}_\zeta(\mathbf{x}_\zeta) - \int_{\Gamma_u} \delta\boldsymbol{\lambda}^{\mathrm{T}}(\mathbf{u} - \overline{\mathbf{u}})\mathrm{d}\Gamma - \int_{\Gamma_u} \delta\mathbf{u}^{\mathrm{T}}\boldsymbol{\lambda}\mathrm{d}\Gamma = 0 \end{aligned} \tag{9.42}$$

$\boldsymbol{\lambda}$ may be interpolated using Eqs. (9.35)–(9.37) along the essential boundary that is discretized into n_u boundary nodes

$$\begin{cases} \boldsymbol{\lambda}(\mathbf{x}) = \sum_{l_u=1}^{n_u} \boldsymbol{\lambda}_{l_u}(\mathbf{x}) \\ \boldsymbol{\lambda}_{l_u}(\mathbf{x}) = \mathbf{N}_{l_u}(\xi)\hat{\boldsymbol{\lambda}}_{l_u} \end{cases} \tag{9.43}$$

In Eq. (9.40), the prescribed boundary traction and displacement also may be interpolated according to Eqs. (9.35)–(9.37) such that

$$\begin{cases} \overline{\mathbf{t}}(\mathbf{x}) = \sum_{l_t=1}^{n_t} \overline{\mathbf{t}}_{l_t}(\mathbf{x}) \\ \overline{\mathbf{t}}_{l_t}(\mathbf{x}) = \mathbf{N}_{l_t}(\xi)\hat{\overline{\mathbf{t}}}_{l_t} \end{cases} \tag{9.44}$$

$$\begin{cases} \overline{\mathbf{u}}(\mathbf{x}) = \sum_{l_u=1}^{n_u} \overline{\mathbf{u}}_{l_u}(\mathbf{x}) \\ \overline{\mathbf{u}}_{l_u}(\mathbf{x}) = \mathbf{N}_{l_u}(\xi)\hat{\overline{\mathbf{u}}}_{l_u} \end{cases} \tag{9.45}$$

The substitution of Eqs. (9.26)–(9.31) and (9.43)–(9.45) into Eq. (9.42) gives rise to

$$\begin{bmatrix} \mathbf{K} & \mathbf{G} \\ \mathbf{G}^T & 0 \end{bmatrix} \begin{Bmatrix} \hat{\mathbf{U}} \\ \hat{\lambda} \end{Bmatrix} = \begin{Bmatrix} \mathbf{F} \\ \mathbf{Q} \end{Bmatrix} \tag{9.46}$$

in which **K** is the system stiffness matrix, **F** is the system load vector, **G** and **Q** are the additional matrix and vector due to the Lagrange multiplier.

Suppose a 2-D domain is discretized in to the grid comprising n_e cells and n_n nodes in which n_u is the amount of essential boundary nodes with prescribed displacement, the matrices and vectors in Eq. (9.46) may be specified as follows

$$\mathbf{K} = \begin{bmatrix} \mathbf{K}_{1,1} & \dots & \mathbf{K}_{1,i} & \dots & \mathbf{K}_{1,j} & \dots & \mathbf{K}_{1,n_n} \\ \dots & \dots & \dots & \dots & \dots & \dots & \dots \\ \mathbf{K}_{i,1} & \dots & \mathbf{K}_{i,i} & \dots & \mathbf{K}_{i,j} & \dots & \mathbf{K}_{i,n_n} \\ \dots & \dots & \dots & \dots & \dots & \dots & \dots \\ \mathbf{K}_{j,1} & \dots & \mathbf{K}_{j,i} & \dots & \mathbf{K}_{j,j} & \dots & \mathbf{K}_{j,n_n} \\ \dots & \dots & \dots & \dots & \dots & \dots & \dots \\ \mathbf{K}_{n_n,1} & \dots & \mathbf{K}_{n_n,i} & \dots & \mathbf{K}_{n_n,j} & \dots & \mathbf{K}_{n_n,n_n} \end{bmatrix} \tag{9.47}$$

$$\mathbf{F} = \left[\mathbf{F}_1 \ \dots \ \mathbf{F}_i \ \dots \ \mathbf{F}_j \ \dots \ \mathbf{F}_{n_n} \right]^{\mathrm{T}} \tag{9.48}$$

$$\hat{\mathbf{U}} = \left[\hat{\mathbf{u}}_1 \ \dots \ \hat{\mathbf{u}}_i \ \dots \ \hat{\mathbf{u}}_j \ \dots \ \hat{\mathbf{u}}_{n_n} \right]^{\mathrm{T}} \tag{9.49}$$

in which $\mathbf{K}_{i,j}$ is a 2×2 sub-matrix, $\hat{\mathbf{u}}_i$ and $\mathbf{F}_i$ are 2×1 sub-vectors.

Similarly, we have

$$\mathbf{G} = \begin{bmatrix} \mathbf{G}_{1,n_n+1} & \dots & \mathbf{G}_{1,n_n+l_u} & \dots & \mathbf{G}_{1,n_n+n_u} \\ \dots & \dots & \dots & \dots & \dots \\ \mathbf{G}_{i,n_n+1} & \dots & \mathbf{G}_{i,n_n+l_u} & \dots & \mathbf{G}_{i,n_n+n_u} \\ \dots & \dots & \dots & \dots & \dots \\ \mathbf{G}_{j,n_n+1} & \dots & \mathbf{G}_{j,n_n+l_u} & \dots & \mathbf{G}_{j,n_n+n_u} \\ \dots & \dots & \dots & \dots & \dots \\ \mathbf{G}_{n_n,n_n+1} & \dots & \mathbf{G}_{n_n,n_n+l_u} & \dots & \mathbf{G}_{n_n,n_n+n_u} \end{bmatrix} \tag{9.50}$$

$$\mathbf{Q} = \left[\mathbf{Q}_{n_n+1} \ \dots \ \mathbf{Q}_{n_n+l_u} \ \dots \ \mathbf{Q}_{n_n+n_u} \right]^{\mathrm{T}} \tag{9.51}$$

in which $\mathbf{G}_{i,n_n+l_u}$ is a 2×2 sub-matrix and $\mathbf{Q}_{n_n+l_u}$ is a 2×1 sub-vector.

After the solution for the system displacement vector $\hat{\mathbf{U}}$, the strain increment $\boldsymbol{\varepsilon}$ and stress increment $\boldsymbol{\sigma}$ for each field point (quadrature point) may be calculated individually using Eq. (9.29).

2. **Penalty method**

The principle of minimum potential energy is also applicable for the formulation of EFGM. The energy function Π in which the essential boundary is enforced by the penalty factor is expressed as

$$\Pi = \iint_{\Omega} \frac{1}{2}(\mathbf{Lu})^{\mathrm{T}}\mathbf{D}(\mathbf{Lu})\mathrm{d}\Omega - \iint_{\Omega} \mathbf{u}^{\mathrm{T}}\mathbf{b}\mathrm{d}\Omega$$
$$- \iint_{\Gamma_t} \mathbf{u}^{\mathrm{T}}\,\bar{\mathbf{t}}\,\mathrm{d}\Gamma - \sum_{\zeta=1}^{n_\zeta} \mathbf{u}^{\mathrm{T}}(\mathbf{x}_\zeta)\mathbf{T}_\zeta(\mathbf{x}_\zeta) + \int_{\Gamma_u} \frac{1}{2}\alpha(\mathbf{u}-\bar{\mathbf{u}})^2\mathrm{d}\Gamma \quad (9.52)$$

in which α is the penalty factor for essential boundary condition.

Invoking Eqs. (9.26)–(9.31) and (9.43)–(9.45), the stationary operation (viz. Eq. (8.96)) w.r.t. Π leads to

$$\mathbf{K}_{eq}\hat{\mathbf{U}} = \mathbf{F}_{eq} \quad (9.53)$$

$$\begin{cases} \mathbf{K}_{eq} = \mathbf{K} + \mathbf{K}' \\ \mathbf{F}_{eq} = \mathbf{F} + \mathbf{F}' \end{cases} \quad (9.54)$$

in which $\mathbf{K}'$ and $\mathbf{F}'$ are constructed by 2×2 sub-matrices and 2×1 vectors. In addition, $\mathbf{K}$ and $\mathbf{F}$ are identical to that in Eqs. (9.47) and (9.48).

3. **Modified variational principle**

A modified variational principle in which the Lagrange multipliers possesses physical meaning, i.e., the traction (Lu et al. 1994)

$$\delta W_u(\mathbf{u}) = \int_{\Gamma_u} \delta\mathbf{t}^{\mathrm{T}}(\mathbf{u}-\bar{\mathbf{u}})\mathrm{d}\Gamma + \int_{\Gamma_u} \delta\mathbf{u}^{\mathrm{T}}\mathbf{t}\mathrm{d}\Gamma \quad (9.55)$$

plus the penalty parameter using the stiffness coefficient of contact face

$$\delta W_u(\mathbf{u}) = \frac{k}{2}\int_{\Gamma_u} \|\mathbf{u}-\bar{\mathbf{u}}\|^2\mathrm{d}\Gamma \quad (9.56)$$

in which $\mathbf{t} = \boldsymbol{\sigma}.\mathbf{n}$ is the traction at essential boundary, k is a penalty parameter defined by the stiffness coefficient of contact face (if available).

This modified variational principle is efficient for small deformation problems. In the cases of finite deformation, a method normally dubbed as "augment Lagrange multiplier" (Tsay et al. 1999) may be alternatively exercised.

9.3.3 Implementation

To solve Eqs. (9.46) or (9.53), Ω needs be divided in to integration cells that are not necessary to by tightly related to the field nodes (Fig. 9.8). Afterwards the Gaussian quadrature may be employed to undertake the integration in above formulated equations cell by cell.

1. **Domain representation**

The engineering structure is represented using a set of nodes scattered in the problem domain and on its boundary. The density of nodes depends on the accuracy requirement of the analysis and the computer resource available. The nodal distribution is usually not uniform, and a denser distribution of nodes is often demanded in the area where strain (stress) gradient is larger. This may be adaptively controlled similar to adaptive FEM using error estimator (Häussler-Combe and Korn 1998). Because these nodes will carry the value of field variables, they are often called "field nodes".

It is convenient to group the boundary nodes in different types with prescribed traction and displacement. Where there are structural planes, contact (joint) faces are discretely simulated by node pairs that will be discussed a bit of later.

2. **Displacement approximation**

The field variables such as the displacement $\mathbf{u}$ at Gaussian quadrature point $\mathbf{x} = \mathbf{x}_Q$ within a representative cell $e(1, 2, \ldots, n_e)$ is approximated using the displacements at the field nodes within the support domain of $\mathbf{x}_Q$ (viz. Eq. (9.27)).

If necessary, the displacement Lagrange multipliers $\boldsymbol{\lambda}$, pre-described boundary traction $\bar{\mathbf{t}}$ and displacement $\bar{\mathbf{u}}$, are interpolated Eqs. (9.43)–(9.45).

3. **Formation of system equation set**

The discretized equation set of EFGM is formulated by inserting the shape functions and trial (approximation) function in the weak form equation of Galerkin method. The resulted discretized equation set is written in the nodal matrix form and assembled into the entire domain, leading to the global system equation set that is an algebra equation set for static analysis, eigenvalue equation set for free-vibration analysis, and ordinary differential equation set with respect to time for dynamic analysis.

In the assemblage of system equation, the Gaussian quadrature (1-D and 2-D) for the sub-matrices and sub-vectors in Eqs. (9.47)–(9.48) and (9.50)–(9.51) are undertaken identical to the isoparametric finite element illustrated in Eqs. (9.38) and (4.122) with the help a simple background grid of triangular or quadrilateral elements (cells), which may be created by many tools in the pre-process for fundamental FEM (Chen, 2018).

(1) **Lagrange multiplier method**

Figure 9.10 shows a portion of such a grid with nine cells on which the 2×2 Gaussian quadrature is employed and the quadrature points are denoted by the black dot "•". Take the central cell e for example, there are four support domains correspondent to four Gaussian quadrature points $\mathbf{x}_Q(Q = 1, 2, 3, 4)$, the union of support domains

$\Omega_{\mathbf{x}_Q}$ is shaded by slashes in which all the field nodes will be related via a local algebraic equation set after the Gaussian quadrature on e. Suppose the link list relating local node sequence and global node sequence gives rise to $i = \langle l \rangle$ and $j = \langle k \rangle$ for each point $\mathbf{x}_Q$, the sub-matrices and sub-vectors in Eqs. (9.47)–(9.48) and (9.50)–(9.51) w.r.t. global field nodes i and j may be derived from the 2-D quadrature cell by cell and 1-D quadrature along boundary segment by segment (viz. Fig. 9.10).

$$\begin{cases} \mathbf{k}_{i,j} = \sum_{e}^{n_e} \iint_{\Omega_e} (\mathbf{L}\boldsymbol{\phi}_l)^{\mathrm{T}} \mathbf{D} (\mathbf{L}\boldsymbol{\phi}_k) \mathrm{d}\Omega \\ \quad\ = \sum_{e}^{n_e} \iint_{\Omega_e} (\mathbf{B}_l)^{\mathrm{T}} \mathbf{D} \mathbf{B}_k \mathrm{d}\Omega \\ \mathbf{f}_i = \mathbf{f}_{i,T} + \mathbf{f}_{i,\bar{t}} + \mathbf{f}_{i,b} \end{cases} \quad \text{for } i = \langle l \rangle \text{ and } j = \langle k \rangle \tag{9.57}$$

$$\begin{cases} \mathbf{g}_{i,l_u} = -\int_{\Delta s_{l_u - 1} + \Delta s_{l_u}} \boldsymbol{\phi}_l^{\mathrm{T}} \mathbf{N}_{l_u} \mathrm{d}\Gamma \\ \mathbf{q}_{l_u} = -\int_{\Delta s_{l_u - 1} + \Delta s_{l_u}} \mathbf{N}_{l_u}^{\mathrm{T}} \mathbf{N}_{l_u} \hat{\hat{\mathbf{u}}}_{l_u} \mathrm{d}\Gamma \end{cases} \quad \text{for } i = \langle l \rangle \text{ and } l_u \in \Gamma_u \tag{9.58}$$

The assemblage operation is conducted such that

$$\mathbf{k}_{i,j} \to \mathbf{K}_{i,j} \tag{9.59}$$

$$\mathbf{f}_i \to \mathbf{F}_i \tag{9.60}$$

$$\mathbf{g}_{i,l_u} \to \mathbf{G}_{i,n_n+l_u} \tag{9.61}$$

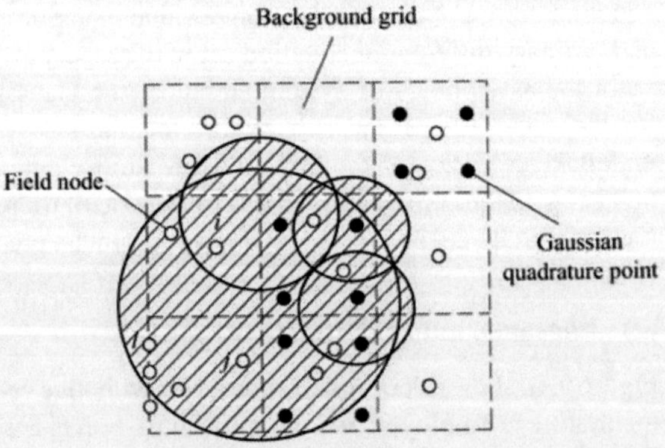

Fig. 9.10 2-D background grid for quadrature

$$\mathbf{q}_{l_u} \to \mathbf{Q}_{n_n+l_u} \tag{9.62}$$

The arrow "→" means the assembling operation.

In Eq. (9.57) the equivalent load $\mathbf{f}_i$ due to actions (forces) can be specified as follows.

- Point force.

$$\mathbf{f}_{i,T} = \sum_{\zeta=1}^{n_\varsigma} \boldsymbol{\phi}_l^{\mathrm{T}}(\mathbf{x}_\zeta)\mathbf{T}_\zeta(\mathbf{x}_\zeta) \quad \text{for } i = \langle l \rangle \in \Omega_{\mathbf{x}_\zeta} \tag{9.63}$$

- Surface distributed force.

$$\mathbf{f}_{i,\bar{t}} = \sum_{l_t=1}^{n_t} \int\limits_{\Delta s_{l_t-1}+\Delta s_{l_t}} \boldsymbol{\phi}_l^{\mathrm{T}} \mathbf{N}_{l_t} \hat{\bar{\mathbf{t}}}_{l_t} \mathrm{d}\Gamma \quad \text{for } i = \langle l \rangle \tag{9.64}$$

- Volumetric force.

$$\mathbf{f}_{i,b} = \sum_{e}^{n_e} \iint\limits_{\Omega_e} \boldsymbol{\phi}_l^{\mathrm{T}} \mathbf{b} \mathrm{d}\Omega \quad \text{for } i = \langle l \rangle \tag{9.65}$$

(2) **Penalty method**

In Eq. (9.54), the sub-matrices and sub-vectors $\mathbf{K}$ and $\mathbf{F}$ are identical to that in Eqs. (9.47)–(9.48) and calculated by Eq. (9.57), whereas $\mathbf{K}'$ and $\mathbf{F}'$ are calculated by

$$\begin{cases} \mathbf{k}'_{i,j} = \alpha \sum\limits_{l_u}^{n_u} \left(\int\limits_{\Delta s_{l_u-1}+\Delta s_{l_u}} \boldsymbol{\phi}_l^{\mathrm{T}} \boldsymbol{\phi}_k \mathrm{d}\Gamma \right) \\ \mathbf{f}'_i = \alpha \sum\limits_{l_u}^{n_u} \left(\int\limits_{\Delta s_{l_u-1}+\Delta s_{l_u}} \boldsymbol{\phi}_l^{\mathrm{T}} \mathbf{N}_{l_u} \hat{\bar{\mathbf{u}}}_{l_u} \mathrm{d}\Gamma \right) \end{cases} \quad \text{for } i = \langle l \rangle \text{ and } j = \langle k \rangle \tag{9.66a}$$

This is applicable to the boundary surface restrained both in the directions of x and y. In engineering application however, it is often (although not always) that a boundary is restrained in one direction only, for example null normal displacement and free tangential movement. Under such circumstances we should employ

$$\begin{cases} \mathbf{k}'_{i,j} = \alpha \sum_{l_u}^{n_u} \left(\int_{\Delta s_{l_u-1}+\Delta s_{l_u}} \boldsymbol{\phi}_l^{\mathrm{T}} \mathbf{I}' \boldsymbol{\phi}_k \mathrm{d}\Gamma \right) \\ \mathbf{f}'_i = \alpha \sum_{l_u}^{n_u} \left(\int_{\Delta s_{l_u-1}+\Delta s_{l_u}} \boldsymbol{\phi}_l^{\mathrm{T}} \mathbf{I}' \mathbf{N}_{l_u} \hat{\mathbf{u}}_{l_u} \mathrm{d}\Gamma \right) \end{cases} \quad \text{for } i = \langle l \rangle \text{ and } j = \langle k \rangle \tag{9.66b}$$

in which $\mathbf{I}'$ is defined as

$$\mathbf{I}' = \begin{bmatrix} s_x & 0 \\ 0 & s_y \end{bmatrix} \tag{9.67}$$

$$s_\beta = \begin{cases} 1 & \text{If the displacement at the direction } \beta \text{ is restrained} \\ 0 & \text{If the displacement at the direction } \beta \text{ is not restrained} \end{cases} \quad \beta = x, y \tag{9.68}$$

The penalty factor α is theoretically infinity to exactly meet the essential boundary condition, but too large of α will lead to the ill-conditioned system stiffness matrix. $\alpha = 2.0 \times 10^3 E$ is customarily used in which E is the Young's modulus of material.

The assemblage operation is conducted such that

$$\mathbf{k}'_{i,j} \to \mathbf{K}'_{i,j} \tag{9.69}$$

$$\mathbf{f}'_i \to \mathbf{F}'_i \tag{9.70}$$

4. **Implementation procedure**

Take Eq. (9.53) corresponding to the penalty method for example, the basic algorithm may be implemented by the following mainstream steps (viz. Fig. 9.11).

① Layout of nodes, and divide Ω into integration cells $\Omega_e (e = 1, 2, \ldots, n_e)$ and divide Γ into integration segments $\Gamma_u (l_u = 1, 2, \ldots, n_u)$, $\Gamma_t (l_t = 1, 2, \ldots, n_t)$.
② For each cell Ω_e, the following operations are undertaken:

For the Gaussian point $\mathbf{x}_Q$ in Ω_e, construct the support domain $\Omega_{\mathbf{x}_Q}$;
According to the support radius of $\mathbf{x} = \mathbf{x}_Q$, find all the nodes $\left(i = \langle l \rangle \in \Omega_{\mathbf{x}_Q}\right)$ within its support domain, calculate $\varphi_l(\mathbf{x})$ and $\frac{\partial \varphi_l(\mathbf{x})}{\partial \mathbf{x}}$ using Eqs. (9.11)–(9.16).
Use is made of Eqs. (9.57) and (9.66), calculate the sub-matrix and sub-vector contributions from the Gaussian quadrature point (knot) $\mathbf{x} = \mathbf{x}_Q$ to $i = \langle l \rangle \in \Omega_{\mathbf{x}_Q}$.

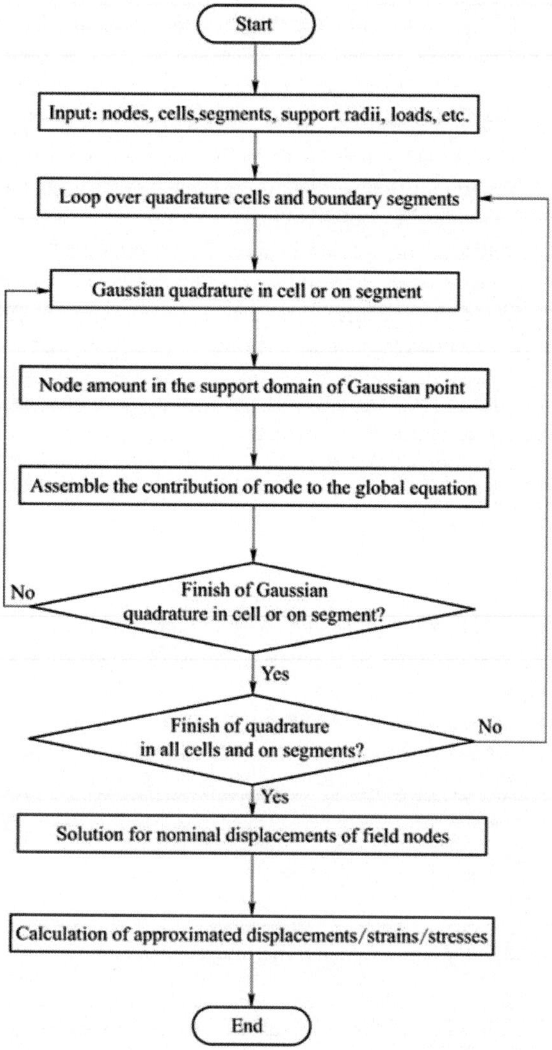

Fig. 9.11 Flowchart for the basic EFGM

With the help of link list $i = \langle l \rangle$ and $j = \langle k \rangle$, assemble the resulted sub-matrices/vectors into $\mathbf{K}_{eq}$ and $\mathbf{F}_{eq}$ as the contribution from $\mathbf{x} = \mathbf{x}_Q$ (in the cell) to the whole domain;
If the Gaussian point recursion for Ω_e has been completed, go to step ③, otherwise continue the operation (a)–(e);

③ If the recursion for all the cells is completed, go to step ④, otherwise go back to step ②;
④ For each boundary segment Δs_{l_u} or Δs_{l_t}, the following operations are undertaken:

For the Gaussian point (knot) $\mathbf{x}_Q$ on the boundary segment, construct the support domain $\Omega_{\mathbf{x}_Q}$;
According to the support radius of $\mathbf{x} = \mathbf{x}_Q$, find all the nodes $\left(i = \langle l \rangle \in \Omega_{\mathbf{x}_Q}\right)$ within its support domain, calculate $\varphi_l(\mathbf{x})$ and $\frac{\partial \varphi_l(\mathbf{x})}{\partial \mathbf{x}}$ using Eqs. (9.11)–(9.16).
Use is made of Eqs. (9.63), (9.64) and (9.66), calculate the sub-matrix and sub-vector contributions from the boundary Gaussian quadrature point.
With the help of link list $i = \langle l \rangle$ and $j = \langle k \rangle$, assemble the resulted sub-matrices/vectors into $\mathbf{K}_{eq}$ and $\mathbf{F}_{eq}$ as the contribution of $\mathbf{x} = \mathbf{x}_Q$ (on the essential boundaries with prescribed displacements) to the whole domain;
If the Gaussian point on the boundary segment Δs_{l_u} or Δs_{l_t} recursion has been completed, go to step ⑤, otherwise continue the operation (a)–(e);

⑤ If the recursion for all the boundary segments is completed, go to step ⑥, otherwise go back to step ④;
⑥ Solution of Eq. (9.53);
⑦ Calculate the strain and stress at Gaussian quadrature points according to Eqs. (9.29)–(9.31).

It is notable that:

- Nodal variable vector $\hat{\mathbf{u}}$ obtained by Eq. (9.53) is not the final solution of displacement $\mathbf{u}$, the latter should be fitted by the MLS technique according to $\hat{\mathbf{u}}$.
- If the support domain is too small, $\mathbf{A}^{-1}(\mathbf{x})$ in weight function does not exist. Hence it is necessary to let $n \geq m$, where n is the node amount in the support domain of $\mathbf{x}_Q$, m is the amount of complete basis functions in Eq. (9.5).

9.4 Dynamic Issues

9.4.1 Governing Equations

EFGM may be easily extended for the dynamic analysis. Take the penalty formulation of elastic body for example (Ouatouati and Johnson 1999), the energy function Π becomes

$$\Pi = \iint_{\Omega} \frac{1}{2}(\mathbf{L}\mathbf{u})^{\mathrm{T}} D(\mathbf{L}\mathbf{u})\mathrm{d}\Omega + \iint_{\Omega} \mathbf{u}^{\mathrm{T}} \rho \frac{\mathrm{d}^2\mathbf{u}}{\mathrm{d}t^2}\mathrm{d}\Omega$$
$$- \iint_{\Omega} \mathbf{u}^{\mathrm{T}}\mathbf{b}\mathrm{d}\Omega - \int_{\Gamma_t} \mathbf{u}^{\mathrm{T}}\bar{\mathbf{t}}\mathrm{d}\Gamma - \sum_{\zeta=1}^{n_\zeta} \mathbf{u}^{\mathrm{T}}(\mathbf{x}_\zeta)\mathbf{T}_\zeta(\mathbf{x}_\zeta)$$
$$+ \int_{\Gamma_u} \frac{1}{2}\alpha(\mathbf{u} - \bar{\mathbf{u}})^2 \mathrm{d}\Gamma = \min. \tag{9.71}$$

Following the similar deduction procedure for any weak form methods such as FEM and DDA, the derivative operation (viz. Eq. (8.96)) w.r.t. Π leads to the stationary equation

$$\mathbf{K}\hat{\mathbf{U}}(t) + \mathbf{M}\frac{\mathrm{d}^2\hat{\mathbf{U}}(t)}{\mathrm{d}t^2} = \mathbf{F} \tag{9.72}$$

in which **K** and **F** have been expressed in Eqs. (9.47)–(9.48), mass matrix **M** is constructed by 2 × 2 sub-matrix $\mathbf{m}_{i,j}$

$$\mathbf{m}_{i,j} = \sum_{e}^{n_e} \rho \iint\limits_{\Omega_e} \boldsymbol{\phi}_l^{\mathrm{T}} \boldsymbol{\phi}_k \mathrm{d}\Omega \quad \text{for } i = \langle l \rangle \text{ and } j = \langle k \rangle \tag{9.73}$$

in which ρ is the density of the material, kg/m^3.

It is notable that for free vibration analysis, the eigenvalue equation for the computation of natural frequencies neglect boundary conditions, namely, $\alpha = 0$ may be postulated.

By the postulation of Rayleigh (proportional) damping model (4.170), the governing equation of the EFGM for dynamic response is

$$\mathbf{M}\frac{\mathrm{d}^2\hat{\mathbf{U}}(t)}{\mathrm{d}t^2} + \mathbf{C}\frac{\mathrm{d}\hat{\mathbf{U}}(t)}{\mathrm{d}t} + \mathbf{K}\hat{\mathbf{U}}(t) = \mathbf{F}(t) \tag{9.74}$$

where: **M**—mass matrix; **C**—damping matrix; **K**—stiffness matrix; $\hat{\mathbf{U}}(t)$, $\frac{\mathrm{d}\hat{\mathbf{U}}(t)}{\mathrm{d}t}$, $\frac{\mathrm{d}^2\hat{\mathbf{U}}(t)}{\mathrm{d}t^2}$—vectors of displacement, velocity, and acceleration, respectively.

In the case of earthquake, it is customarily to undertake the static analysis first using all static forces, then the dynamic analysis under seismic action is carried out using dynamic force

$$\mathbf{F}(t) = -\mathbf{M}\mathbf{a}_g(t) \tag{9.75}$$

where: $\mathbf{a}_g(t)$—earthquake acceleration of ground motions.

9.4.2 Solution Techniques

The algorithm to calculate the dynamic responses of structural systems governed by Eq. (9.74) are identical to FEM elaborated in Chap. 4.

9.5 Structural Plane Issues

9.5.1 Concept

A structural plane may be looked at as a contact face between rock walls that may be well simulated by EFGM (Fleming 1997).

In Fig. 9.12a two bodies are in contact. Body A is termed as the domain Ω_{A} with a boundary Γ_{A}, while body B is termed as the domain Ω_{B} with boundary Γ_{B}. The contact structural plane, denoted as $\Gamma_{\mathrm{C}} = \Gamma_{sp}$, is an internal boundary.

We divide the structural plane with $l_{sp}(1, 2, \ldots, n_{sp})$ nodes into $n_{sp} - 1$ segments (viz. Fig. 9.9). At the quadrature point with sequence number $p(1, 2, \ldots, n_p)$ for the structural plane segment l_{sp}, there is a point pair $(p_{\mathrm{A}}, p_{\mathrm{B}})$ due to the existence of structural plane, where n_p is the amount of quadrature points on l_{sp}. Denote there are n_p^{A} and n_p^{B} field nodes in the support domain of point p that belong to Ω_A and Ω_B (Fig. 9.12b), respectively. Normally we have

$$n_p^{\mathrm{A}} + n_p^{\mathrm{B}} = n \tag{9.76}$$

To be specific, in Fig. 9.12b we have $n_p^{\mathrm{A}} = 3$ and $n_p^{\mathrm{B}} = 3$. Any node i both in Ω_{A} and the half-support domain $\Omega_{p_{\mathrm{A}}}$ of point p is particularly denoted as i_{A}; otherwise it is denoted as i_{B} if node i is simultaneously tied to Ω_{B} and $\Omega_{p_{\mathrm{B}}}$.

The inherent smoothness properties in fundamental EFGM is a two-edged sword. On one hand, it provides approximation which is sufficiently smooth. However, when a discontinuity occurs in either geometry or material aspects, this smoothness leads to consequent difficulties. Cordes and Moran (1996) treated the individual materials as separate bodies and joined them together with Lagrange multipliers. Krongauz and Belytschko (1998) used jump nodes to simulate the discontinuity in field derivatives. In this book, a contact model derived from "joint element" model is implemented

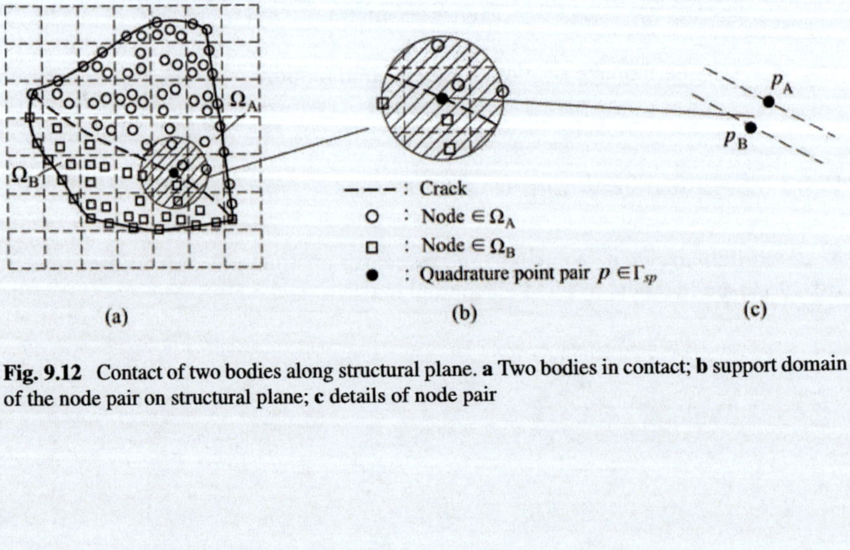

Fig. 9.12 Contact of two bodies along structural plane. **a** Two bodies in contact; **b** support domain of the node pair on structural plane; **c** details of node pair

using the constitutive relation of node pairs previously discretized along the contact face or adaptively added following the new created crack front.

9.5.2 *Elastic Contact*

The contact enforcement can be conducted via either Lagrange multipliers or penalty factors (Belytschko and Fleming 1999). We take the penalty method as an example to exhibit our contact model. For the quadrature point (knot) p on contact face, it is convenient to formulate the equation of contact in terms of the local coordinate system related to the contact face (viz. Fig. 9.13).

The relative displacement of the knot pair (p_A, p_B) at each side of contact face may be calculated in its local coordinate system by

$$\mathbf{u}_p = \begin{Bmatrix} u_{x,p} \\ u_{y,p} \end{Bmatrix} = \begin{Bmatrix} u^{A}_{x,p} \\ u^{A}_{y,p} \end{Bmatrix} - \begin{Bmatrix} u^{B}_{x,p} \\ u^{B}_{y,p} l \end{Bmatrix} \\ = \mathbf{T}_{sp}\left(\boldsymbol{\Phi}_A \hat{\mathbf{u}}_A - \boldsymbol{\Phi}_B \hat{\mathbf{u}}_B\right) \tag{9.77}$$

in which $\boldsymbol{\Phi}_A$ and $\boldsymbol{\Phi}_B$ are the shape functions of two support domains corresponding to the upper and lower portions; the coordinate transformation $\mathbf{T}_{sp}$ has been defined in Eq. (3.125).

The elastic constitutive relation of joint element model is

$$\boldsymbol{\sigma}_p = \begin{Bmatrix} \tau_p \\ \sigma_p \end{Bmatrix} = \begin{bmatrix} k_s & 0 \\ 0 & k_n \end{bmatrix} \begin{Bmatrix} \mathrm{u}_{x,p} \\ u_{y,p} \end{Bmatrix} = \mathbf{D}_p \mathbf{u}_p \tag{9.78}$$

in which k_n and k^s are the normal and tangential stiffness coefficients of the contact walls. They play similar role of the penalty factor at essential boundary but possess clear physical definition.

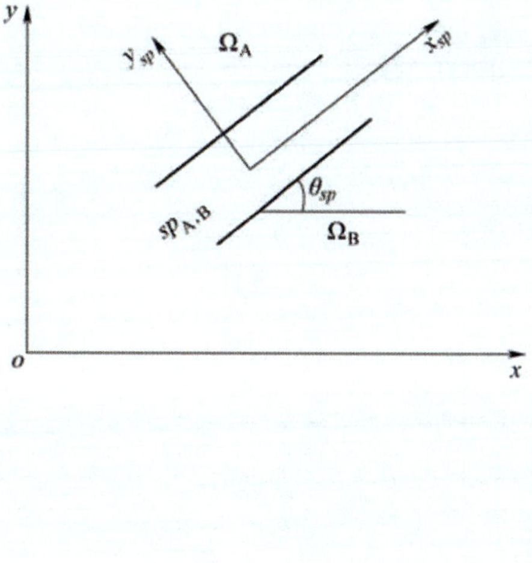

Fig. 9.13 Contact face (structural plane) and local coordinate system

The strain energy $W_u(\mathbf{u})$ correspondent to the structural plane (as interior boundary) is expressed by

$$
\begin{aligned}
W_u(\mathbf{u}) = \Pi_{sp_{A,B}} &= \frac{1}{2}\int\limits_{sp_{A,B}} (\mathbf{u}_p)^{\mathrm{T}} \mathbf{D}_{sp_{A,B}} \mathbf{u}_p \mathrm{d}\Gamma \\
&= \frac{1}{2}\int\limits_{sp_{A,B}} \left(\boldsymbol{\Phi}_{\mathrm{A}}\hat{\mathbf{u}}_{\mathrm{A}} - \boldsymbol{\Phi}_{\mathrm{B}}\hat{\mathbf{u}}_{\mathrm{B}}\right)^{\mathrm{T}} \mathbf{T}_{sp_{A,B}}^{\mathrm{T}} \mathbf{D}_{sp_{A,B}} \mathbf{T}_{sp_{A,B}} \left(\boldsymbol{\Phi}_{\mathrm{A}}\hat{\mathbf{u}}_{\mathrm{A}} - \boldsymbol{\Phi}_{\mathrm{B}}\hat{\mathbf{u}}_{\mathrm{B}}\right)\mathrm{d}\Gamma \\
&= \frac{1}{2}\int\limits_{sp_{A,B}} \left(\begin{array}{l} (\hat{\mathbf{u}}_{\mathrm{A}})^{\mathrm{T}}(\boldsymbol{\Phi}_{\mathrm{A}})^{\mathrm{T}}\mathbf{T}_{sp_{A,B}}^{\mathrm{T}}\mathbf{D}_{sp_{A,B}}\mathbf{T}_{sp_{A,B}}\boldsymbol{\Phi}_{\mathrm{A}}\hat{\mathbf{u}}_{\mathrm{A}} \\ + (\hat{\mathbf{u}}_{\mathrm{B}})^{\mathrm{T}}(\boldsymbol{\Phi}_{\mathrm{B}})^{\mathrm{T}}\mathbf{T}_{sp_{A,B}}^{\mathrm{T}}\mathbf{D}_{sp_{A,B}}\mathbf{T}_{sp_{A,B}}\boldsymbol{\Phi}_{\mathrm{B}}\hat{\mathbf{u}}_{\mathrm{B}} \\ - (\hat{\mathbf{u}}_{\mathrm{B}})^{\mathrm{T}}(\boldsymbol{\Phi}_{\mathrm{B}})^{\mathrm{T}}\mathbf{T}_{sp_{A,B}}^{\mathrm{T}}\mathbf{D}_{sp_{A,B}}\mathbf{T}_{sp_{A,B}}\boldsymbol{\Phi}_{\mathrm{A}}\hat{\mathbf{u}}_{\mathrm{A}} \\ - (\hat{\mathbf{u}}_{\mathrm{A}})^{\mathrm{T}}(\boldsymbol{\Phi}_{\mathrm{A}})^{\mathrm{T}}\mathbf{T}_{sp_{A,B}}^{\mathrm{T}}\mathbf{D}_{sp_{A,B}}\mathbf{T}_{sp_{A,B}}\boldsymbol{\Phi}_{\mathrm{B}}\hat{\mathbf{u}}_{\mathrm{B}} \end{array}\right)\mathrm{d}\Gamma
\end{aligned}
\tag{9.79}
$$

The stationary condition w.r.t. W_u demands

$$
\delta W_u(\mathbf{u}) = 0 \tag{9.80}
$$

which leads to

$$
\mathbf{k}_{sp_{A,B}} \hat{\mathbf{u}}_{sp_{A,B}} = 0 \tag{9.81}
$$

where

$$
\mathbf{k}_{sp_{A,B}} = \begin{bmatrix} \mathbf{k}_{\mathrm{A,A}} & \mathbf{k}_{\mathrm{A,B}} \\ \mathbf{k}_{\mathrm{B,A}} & \mathbf{k}_{\mathrm{B,B}} \end{bmatrix} \tag{9.82}
$$

$$
\hat{\mathbf{u}}_{sp_{A,B}} = \begin{bmatrix} \hat{\mathbf{u}}_{\mathrm{A}} & \hat{\mathbf{u}}_{B} \end{bmatrix}^{\mathrm{T}} \tag{9.83}
$$

The additional stiffness $\mathbf{k}_{sp_{A,B}}$ w.r.t. contact knot pair $(p_{\mathrm{A}}, p_{\mathrm{B}})$ is added to $\mathbf{K}$ in Eq. (9.47), which may be processed one by one according to the 2×2 component sub-matrices contributed from field nodes i_{A} and i_{B} in the domain Ω_{A} and Ω_{B}, respectively

$$\begin{cases} \mathbf{k}_{i_A,j_A} = \int\limits_{sp_{A,B}} \boldsymbol{\phi}_{l_A}^T \mathbf{T}_{sp_{A,B}}^T \mathbf{D}_{sp_{A,B}} \mathbf{T}_{sp_{A,B}} \boldsymbol{\phi}_{k_A} d\Gamma \to \mathbf{K}_{i,j} \\ \quad \text{for } i_A = \langle l_A \rangle \in \Omega_{p_A} \text{ and } j_A = \langle k_A \rangle \in \Omega_{p_A} \\ \mathbf{k}_{i_B,j_B} = \int\limits_{sp_{A,B}} \boldsymbol{\phi}_{l_B}^T \mathbf{T}_{sp_{A,B}}^T \mathbf{D}_{sp_{A,B}} \mathbf{T}_{sp_{A,B}} \boldsymbol{\phi}_{k_B} d\Gamma \to \mathbf{K}_{i,j} \\ \quad \text{for } i_B = \langle l_B \rangle \in \Omega_{p_B} \text{ and } j_B = \langle k_B \rangle \in \Omega_{p_B} \\ \mathbf{k}_{i_A,j_B} = -\int\limits_{sp_{A,B}} \boldsymbol{\phi}_{l_A}^T \mathbf{T}_{sp_{A,B}}^T \mathbf{D}_{sp_{A,B}} \mathbf{T}_{sp_{A,B}} \boldsymbol{\phi}_{k_B} d\Gamma \to \mathbf{K}_{i,j} \\ \quad \text{for } i_A = \langle l_A \rangle \in \Omega_{p_A} \text{ and } j_B = \langle k_B \rangle \in \Omega_{p_B} \\ \mathbf{k}_{i_B,j_A} = -\int\limits_{sp_{A,B}} \boldsymbol{\phi}_{l_B}^T \mathbf{T}_{sp_{A,B}}^T \mathbf{D}_{sp_{A,B}} \mathbf{T}_{sp_{A,B}} \boldsymbol{\phi}_{k_A} d\Gamma \to \mathbf{K}_{i,j} \\ \quad \text{for } i_B = \langle l_B \rangle \in \Omega_{p_B} \text{ and } j_A = \langle k_A \rangle \in \Omega_{p_A} \end{cases} \tag{9.84}$$

9.5.3 *Contact with Frictional Shear*

The friction shear strength is evaluated by Mohr-Coulomb law in Eq. (3.130). Where the yield function $F_{sp_{A,B}} > 0$, frictional slip deformation occurs. It is not difficult to formulate the energy increment related to the contact point p of Ω_A and Ω_B, the increment sub-vectors related to nodal force are calculated via the stationary operation taking into account of rigorous elasto-viscoplastic constitutive relation that

$$\Delta \mathbf{u}_p = \left(\mathbf{D}_{sp_{A,B}}\right)^{-1} \Delta \boldsymbol{\sigma}_p + \Delta \mathbf{u}_p^{vp} \tag{9.85}$$

in which the elasticity matrix $\mathbf{D}_{sp_{A,B}}$ and elasto-viscoplastic strain increment $\Delta \hat{\mathbf{u}}_p^{vp}$ are given in Eqs. (3.123) and (3.144).

Incremental form of energy is specified as

$$W_u(\Delta \mathbf{u}) = \Pi_{sp_{A,B}} = \frac{1}{2} \int\limits_{sp_{A,B}} \left(\Delta \mathbf{u}_p - \Delta \hat{\mathbf{u}}_p^{vp}\right)^T \mathbf{D}_{sp_{A,B}} \left(\Delta \mathbf{u}_p - \Delta \hat{\mathbf{u}}_p^{vp}\right) d\Gamma$$

$$
= \frac{1}{2} \int_{sp_{A,B}} \begin{pmatrix} (\Delta\hat{\mathbf{u}}_A)^T (\boldsymbol{\Phi}_A)^T \mathbf{T}^T_{sp_{A,B}} \mathbf{D}_{sp_{A,B}} \mathbf{T}_{sp_{A,B}} \boldsymbol{\Phi}_A \Delta\hat{\mathbf{u}}_A \\ +(\Delta\hat{\mathbf{u}}_B)^T (\boldsymbol{\Phi}_B)^T \mathbf{T}^T_{sp_{A,B}} \mathbf{D}_{sp_{A,B}} \mathbf{T}_{sp_{A,B}} \boldsymbol{\Phi}_B \Delta\hat{\mathbf{u}}_B \\ -(\Delta\hat{\mathbf{u}}_B)^T (\boldsymbol{\Phi}_B)^T \mathbf{T}^T_{sp_{A,B}} \mathbf{D}_{sp_{A,B}} \mathbf{T}_{sp_{A,B}} \boldsymbol{\Phi}_A \Delta\hat{\mathbf{u}}_A \\ -(\Delta\hat{\mathbf{u}}_A)^T (\boldsymbol{\Phi}_A)^T \mathbf{T}^T_{sp_{A,B}} \mathbf{D}_{sp_{A,B}} \mathbf{T}_{sp_{A,B}} \boldsymbol{\Phi}_B \Delta\hat{\mathbf{u}}_B \\ -2(\Delta\hat{\mathbf{u}}_A)^T (\boldsymbol{\Phi}_A)^T \mathbf{T}^T_{sp_{A,B}} \mathbf{D}_{sp_{A,B}} \Delta\hat{\mathbf{u}}^{vp}_p \\ +2(\Delta\hat{\mathbf{u}}_B)^T (\boldsymbol{\Phi}_B)^T \mathbf{T}^T_{sp_{A,B}} \mathbf{D}_{sp_{A,B}} \Delta\hat{\mathbf{u}}^{vp}_p + (\Delta\hat{\mathbf{u}}^{vp}_p)^T \Delta\hat{\mathbf{u}}^{vp}_p \end{pmatrix} \mathrm{d}\Gamma \tag{9.86}
$$

The stationary condition (viz. Eq. (8.96)) w.r.t. W_u leads to the additional stiffness matrix $\mathbf{k}_{sp_{A,B}}$ identical to Eq. (9.85), plus the additional load $\Delta\mathbf{f}^{vp}_{sp_{A,B}}$ arise from the viscoplastic shear slip on structural plane (joint, crack).

Equation (9.53) needs to be expressed in the incremental form of

$$
\mathbf{K}_{eq}\Delta\hat{\mathbf{U}} = \Delta\mathbf{F}_{eq} + \Delta\mathbf{F}^{vp} \tag{9.87}
$$

The additional load vector $\Delta\mathbf{F}^{vp}$ w.r.t. contact point (knot) pair (p_A, p_B) may be processed one by one according to the 2×1 component sub-vectors contributed from field nodes i_A, i_B, j_A, j_B in the domain Ω_A and Ω_B, respectively.

$$
\begin{cases} \Delta\mathbf{f}^{vp}_{i_A} = \displaystyle\int_{sp_{A,B}} \boldsymbol{\phi}^T_{l_A} \mathbf{T}^T_{sp_{A,B}} \mathbf{D}_{sp_{A,B}} \Delta\hat{\mathbf{u}}^{vp}_p \mathrm{d}\Gamma \to \Delta\mathbf{F}^{vp}_i \quad \text{for } i_A = \langle l_A \rangle \in \Omega_{p_A} \\ \Delta\mathbf{f}^{vp}_{i_B} = -\displaystyle\int_{sp_{A,B}} \boldsymbol{\phi}^T_{l_B} \mathbf{T}^T_{sp_{A,B}} \mathbf{D}_{sp_{A,B}} \Delta\hat{\mathbf{u}}^{vp}_p \mathrm{d}\Gamma \to \Delta\mathbf{F}^{vp}_i \quad \text{for } i_B = \langle l_B \rangle \in \Omega_{p_B} \\ \Delta\mathbf{f}^{vp}_{j_A} = \displaystyle\int_{sp_{A,B}} \boldsymbol{\phi}^T_{k_A} \mathbf{T}^T_{sp_{A,B}} \mathbf{D}_{sp_{A,B}} \Delta\hat{\mathbf{u}}^{vp}_p \mathrm{d}\Gamma \to \Delta\mathbf{F}^{vp}_j \quad \text{for } j_A = \langle k_A \rangle \in \Omega_{p_A} \\ \Delta\mathbf{f}^{vp}_{j_B} = -\displaystyle\int_{sp_{A,B}} \boldsymbol{\phi}^T_{k_B} \mathbf{T}^T_{sp_{A,B}} \mathbf{D}_{sp_{A,B}} \Delta\hat{\mathbf{u}}^{vp}_p \mathrm{d}\Gamma \to \Delta\mathbf{F}^{vp}_j \quad \text{for } j_B = \langle k_B \rangle \in \Omega_{p_B} \end{cases} \tag{9.88}
$$

9.5.4 Quadrature on Structural Plane and System Equation

SUppose the structural plane is discretized into n_{sp} nodes and $n_{sp} - 1$ segments previously (viz. Fig. 9.9), then for the integration of Eqs. (9.84) and (9.88), the 1-D Gaussian quadrature of n_p knots may be directly undertaken segment by segment, namely

$$\begin{cases} \mathbf{k}_{i_A,j_A} = \sum\limits_{l_{sp}}^{n_{sp}-1} \int\limits_{\Delta s_{l_{sp}}} \boldsymbol{\phi}_{l_A}^{T} \mathbf{T}_{sp_{A,B}}^{T} \mathbf{D}_{sp_{A,B}} \mathbf{T}_{sp_{A,B}} \boldsymbol{\phi}_{k_A} \mathrm{d}\Gamma \\ \quad \text{for } i_A = \langle l_A \rangle \in \Omega_{p_A} \text{ and } j_A = \langle k_A \rangle \in \Omega_{p_A} \\ \mathbf{k}_{i_B,j_B} = \sum\limits_{l_{sp}}^{n_{sp}-1} \int\limits_{\Delta s_{l_{sp}}} \boldsymbol{\phi}_{l_B}^{T} \mathbf{T}_{sp_{A,B}}^{T} \mathbf{D}_{sp_{A,B}} \mathbf{T}_{sp_{A,B}} \boldsymbol{\phi}_{k_B} \mathrm{d}\Gamma \\ \quad \text{for } i_B = \langle l_B \rangle \in \Omega_{p_B} \text{ and } j_B = \langle k_B \rangle \in \Omega_{p_B} \\ \mathbf{k}_{i_A,j_B} = -\sum\limits_{l_{sp}}^{n_{sp}-1} \int\limits_{\Delta s_{l_{sp}}} \boldsymbol{\phi}_{l_A}^{T} \mathbf{T}_{sp_{A,B}}^{T} \mathbf{D}_{sp_{A,B}} \mathbf{T}_{sp_{A,B}} \boldsymbol{\phi}_{k_B} \mathrm{d}\Gamma \\ \quad \text{for } i_A = \langle l_A \rangle \in \Omega_{p_A} \text{ and } j_B = \langle k_B \rangle \in \Omega_{p_B} \\ \mathbf{k}_{i_B,j_A} = -\sum\limits_{l_{sp}}^{n_{sp}-1} \int\limits_{\Delta s_{l_{sp}}} \boldsymbol{\phi}_{l_B}^{T} \mathbf{T}_{sp_{A,B}}^{T} \mathbf{D}_{sp_{A,B}} \mathbf{T}_{sp_{A,B}} \boldsymbol{\phi}_{k_A} \mathrm{d}\Gamma \\ \quad \text{for } i_B = \langle l_B \rangle \in \Omega_{p_B} \text{ and } j_A = \langle k_A \rangle \in \Omega_{p_A} \end{cases} \tag{9.89}$$

$$\begin{cases} \Delta\mathbf{f}_{i_A}^{vp} = \sum\limits_{l_{sp}}^{n_{sp}-1} \int\limits_{\Delta s_{l_{sp}}} \boldsymbol{\phi}_{l_A}^{T} \mathbf{T}_{sp_{A,B}}^{T} \mathbf{D}_{sp_{A,B}} \Delta\hat{\mathbf{u}}_p^{vp} \mathrm{d}\Gamma \quad \text{for } i_A = \langle l_A \rangle \in \Omega_{p_A} \\ \Delta\mathbf{f}_{i_B}^{vp} = -\sum\limits_{l_{sp}}^{n_{sp}-1} \int\limits_{\Delta s_{l_{sp}}} \boldsymbol{\phi}_{l_B}^{T} \mathbf{T}_{sp_{A,B}}^{T} \mathbf{D}_{sp_{A,B}} \Delta\hat{\mathbf{u}}_p^{vp} \mathrm{d}\Gamma \quad \text{for } i_B = \langle l_B \rangle \in \Omega_{p_B} \\ \Delta\mathbf{f}_{j_A}^{vp} = \sum\limits_{l_{sp}}^{n_{sp}-1} \int\limits_{\Delta s_{l_{sp}}} \boldsymbol{\phi}_{k_A}^{T} \mathbf{T}_{sp_{A,B}}^{T} \mathbf{D}_{sp_{A,B}} \Delta\hat{\mathbf{u}}_p^{vp} \mathrm{d}\Gamma \quad \text{for } j_A = \langle k_A \rangle \in \Omega_{p_A} \\ \Delta\mathbf{f}_{j_B}^{vp} = -\sum\limits_{l_{sp}}^{n_{sp}-1} \int\limits_{\Delta s_{l_{sp}}} \varphi_{k_B}^{T} \mathbf{T}_{sp_{A,B}}^{T} \mathbf{D}_{sp_{A,B}} \Delta\hat{\mathbf{u}}_p^{vp} \mathrm{d}\Gamma \quad \text{for } j_B = \langle k_B \rangle \in \Omega_{p_A} \end{cases} \tag{9.90}$$

After the solution of Eq. (9.87), the stress increments are calculated by Eq. (9.29) for material domain and Eq. (9.85) for structural plane, respectively. Afterwards, the accumulated quantities at field node i are obtained routinely

$$\begin{cases} \hat{\mathbf{u}}_i = \hat{\mathbf{u}}_i + \Delta\hat{\mathbf{u}}_i \\ \boldsymbol{\sigma}_i = \boldsymbol{\sigma}_i + \Delta\boldsymbol{\sigma}_i \end{cases} \quad (i = 1, 2, \ldots, n_n) \tag{9.91}$$

9.5.5 Solution Procedure

For contact problems related to structural planes, the discretized system equations are basically nonlinear, and one needs an additional iteration loop to obtain the solution. The solution procedure is implemented according to following mainstream steps:

① Normal solution of governing Eq. (9.53) without crack/joint. If there are no structural planes (or new created crack fronts), the computation is finished, otherwise go to the step ②

② Structural planes are discretized into segments forming broken contact faces, that are supposed to be initially stuck. The quadrature along the discretized contact face provides additional stiffness matrix $\mathbf{K}_{sp}$, that is assembled into the global algebraic equation set (9.87).

③ Elastic solution of Eq. (9.87) in which $\Delta\mathbf{F}^{vp} = 0$.

④ Iteration w.r.t. tension and frictional shear for the points $p = 1, 2, \ldots, n_p$ on the contact face:

(a) Yield criterion in Eq. (3.130) is employed to judge the contact state. If the yield function $F_{sp_{A,B}}(\mathbf{x}_p) \le 0$, go to (c), otherwise go to (b);
(b) Eq. (9.93) is employed to get additional load sub-vectors $\Delta\mathbf{f}_{i_A}^{vp}$ or $\Delta\mathbf{f}_{i_B}^{vp}$ and $\Delta\mathbf{f}_{j_A}^{vp}$ or $\Delta\mathbf{f}_{j_B}^{vp}$ that are assembled into $\Delta\mathbf{F}^{vp}$;
(c) If all the quadrature points p on the structural plane are over looped, go to step ⑤, otherwise return to (a).

⑤ Nodal displacement increment $\Delta\hat{\mathbf{U}}$ is solved, followed by the calculation of accumulated displacement and stress using Eq. (9.91);

⑥ If $\Delta\hat{\mathbf{U}}$ varies within tolerance, go to step ⑦, otherwise return to step ④.

⑦ Collapse of the iteration.

9.5.6 Notes

The contact model of structural plane may be improved and retrogressed easily towards the simulation of essential boundary normally or/and tangentially restrained by the prescribed displacements, and the simulation of frictional boundary contact issues.

① We regard the boundary surface as a special structural plane and the sub-domain Ω_B provides boundary displacement $\overline{\mathbf{u}}$ only. From Eq. (9.77) the relative displacement of the knot pair (p_A, p_B) at each side of boundary may be calculated by

$$\mathbf{u}_p = \begin{Bmatrix} u_{x,p} \\ u_{y,p} \end{Bmatrix} - \begin{Bmatrix} \overline{u}_{x,p} \\ \overline{u}_{y,p} \end{Bmatrix} = \mathbf{T}_{\Gamma_u}(\boldsymbol{\Phi}\hat{\mathbf{u}} - \mathbf{N}_{l_u}\hat{\overline{\mathbf{u}}}_{l_u}) \quad l_u \in \Gamma_u \tag{9.92}$$

in which $\mathbf{\Phi}$ is the shape function matrix of the support domain in $\Omega_A = \Omega$ tied to the point p on the essential boundary, and the coordinate transformation matrix $\mathbf{T}_{\Gamma_u}$ is defined in Eq. (3.125) by the inclination angle of boundary. Following the identical procedure from Eq. (9.78) to Eq. (9.84) and use is made of the numerical quadrature in Eq. (9.92), we get

$$\begin{cases} \mathbf{k}'_{i,j} = \sum\limits_{l_u}^{n_u} \int\limits_{\Delta s_{l_u-1}+\Delta s_{l_u}} \boldsymbol{\phi}_l^{\mathrm{T}} \mathbf{T}_{\Gamma_u}^{\mathrm{T}} \mathbf{D}_{\Gamma_u} \mathbf{T}_{\Gamma_u} \boldsymbol{\phi}_k \mathrm{d}\Gamma \rightarrow \mathbf{K}'_{i,j} \\ \mathbf{f}'_i = \sum\limits_{l_u}^{n_u} \left(\int\limits_{\Delta s_{l_u-1}+\Delta s_{l_u}} \boldsymbol{\phi}_l^{\mathrm{T}} \mathbf{T}_{\Gamma_u}^{\mathrm{T}} \mathbf{D}_{\Gamma_u} \mathbf{N}_{l_u} \hat{\bar{\mathbf{u}}}_{l_u} \mathrm{d}\Gamma \right) \rightarrow \mathbf{F}'_i \end{cases} \quad \text{for } i = \langle l \rangle \text{ and } j = \langle k \rangle \tag{9.93}$$

Recall that the elastic matrix $\mathbf{D}_{\Gamma_u}$ of the boundary surface is defined in Eq. (3.123). If we let the coefficients of normal stiffness and tangential (shear) stiffness of the boundary surface vary from the penalty factor α to null, Eq. (9.93) is able to simulate the restraint state from fully prescribed to totally free in the normal and tangential directions, respectively. This is actually identical to Eq. (9.66) that imposes the restraint in the directions of x-axis and y-axis.

② We may handle the frictional boundary surface as a special structural plane, too. In this case the sub-domain Ω_B provides boundary displacement $\bar{\mathbf{u}} = 0$. From Eq. (9.77) the relative displacement of the knot pair (p_A, p_B) at each side of boundary may be calculated by

$$\Delta \mathbf{u}_p = \begin{Bmatrix} \Delta u_{x,p} \\ \Delta u_{y,p} \end{Bmatrix} - \begin{Bmatrix} 0 \\ 0 \end{Bmatrix} = \mathbf{T}_{\Gamma_u} \Phi \Delta \hat{\mathbf{u}} \tag{9.94}$$

in which $\mathbf{\Phi}$ is the shape functions of the support domain in $\Omega_A = \Omega$ tied to the point p on the frictional contact boundary, and the coordinate transformation matrix $\mathbf{T}_{\Gamma_u}$ has been defined in Eq. (3.125) by the inclination angle of boundary. Following the identical procedure from Eq. (9.85) to Eq. (9.90), we get

$$\begin{cases} \mathbf{k}_{i,j} = \sum\limits_{l_u}^{n_u-1} \int\limits_{\Delta s_{l_u}} \boldsymbol{\phi}_l^{\mathrm{T}} \mathbf{T}_{\Gamma_u}^{\mathrm{T}} \mathbf{D}_{\Gamma_u} \mathbf{T}_{\Gamma_u} \boldsymbol{\phi}_k \mathrm{d}\Gamma \rightarrow \mathbf{K}_{i,j} \\ \Delta \mathbf{f}_i^{vp} = \sum\limits_{l_u}^{n_u-1} \int\limits_{\Delta s_{l_u}} \boldsymbol{\phi}_l^{\mathrm{T}} \mathbf{T}_{\Gamma_u}^{\mathrm{T}} \mathbf{D}_{\Gamma_u} \Delta \hat{\mathbf{u}}_p^{vp} \mathrm{d}\Gamma \rightarrow \Delta \mathbf{F}_i \end{cases} \quad \text{for } i = \langle l \rangle \text{ and } j = \langle k \rangle \tag{9.95}$$

③ For the opening structural plane or crack under monotonic actions, the "air element" idea may be employed. The coefficients of normal stiffness and shear stiffness in the elastic matrix $\mathbf{D}_{sp_{A,B}}$ (viz. Eqs. (9.84) and (9.89)) may be prescribed by a very small constant, for example, $k_n = k_s = (10^{-3} \sim 10^{-5})E$ is suggestible where E is the Young's modulus of the material.

9.6 EFGM Near Crack-Tips and Nonconvex Boundaries

9.6.1 Techniques for Crack Modeling

Of course the method for structural plane using contact pairs elaborated above may be employed for the simulation of the presence of crack and nonconvex boundaries within domain (Cordes and Moran 1996; Krongauz and Belytschko 1998; Zhanget al. 2004; Rabczuk and Areias 2006; Rabczuk et al. 2007; Rabczuk and Belytschko, 2007; Rabczuk and Zi 2007; Zi et al. 2007). The topological enrichment similar to XFEM (Chap. 5) may be implemented for the crack-tip nodes without essential difficulties (viz. Fig. 9.14). A unique problem due to the existence of crack-tip is, how to partially cut the support domain of a quadrature point $\mathbf{x}_Q$ near the crack-tip.

Suppose a crack is open, it may be handled as a nonconvex boundary without traction and displacement restraint, and only the field nodes in the rest portion of completely cut support domain is considered in the variable approximation for the field point, such as the point $\mathbf{x}_Q$ in Fig. 9.15a. For the partially cut domain near crack-tip, such as that of point $\mathbf{x}_Q$ in Fig. 9.15b, more variable function approximations are possible according to a number of domain cut criteria.

1. **Visibility criterion for discontinuous approximation**

The "visibility criterion" proposed by Belytschko et al. (1994a, b; 1995a) defines the support domain of a field point as the vision field at the point. All the internal

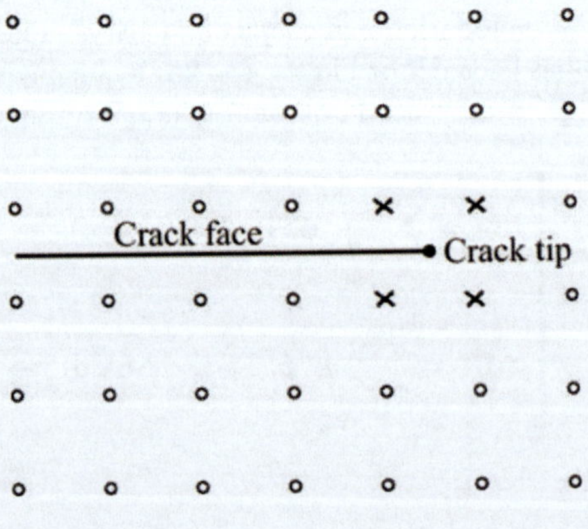

Fig. 9.14 Discrete model for near-tip crack problem with topological enrichment (node " × ")

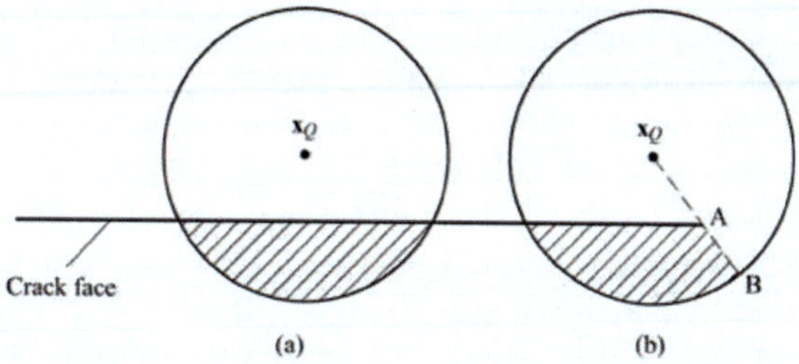

Fig. 9.15 Cut domain of support by the visibility criterion for field nodes near crack. **a** Support domain completely cut by crack; **b** Support domain partially cut by crack

and external boundaries are considered to be opaque so that the field of vision is interrupted when a boundary is encountered. For points near the crack-tip such as the point $\mathbf{x}_Q$ in Fig. 9.15b, its field of vision is cut by the crack along line A-B, and the shaded area is eliminated from the domain of support. This leads to a discontinuity in the weight function as well as the shape function along this line. The length of discontinuity depends on the nodal refinement near a nonconvex boundary or crack-tip, namely, as the nodal spacing goes to zero, the length of discontinuity tends to zero. Using this argument and the theory for nonconforming finite elements, Krysl and Belytschko (1997) showed that the discontinuous approximations generated by the visibility criterion do converge.

2. **Continuous approximation**

(1) **Diffraction and transparency methods**

Continuous and smooth approximations can be constructed near nonconvex boundaries by the "diffraction method" (Belytschko et al. 1996a). The support domain is wrapped around the crack-tip similar to the way light diffracts around sharp corner. This method, which has also been called the "wrap-around method", is quite general and can be used for cracks, smooth boundaries (e.g. interior holes) and nonconvex boundaries.

The "transparency method" (Belytschko et al. 1996b) provides another possibility for constructing continuous approximations: the transparency of crack varies so that it is completely transparent at the crack-tip and becomes completely opaque a short distance s from the crack-tip. In this way, the field of vision for a field point near the crack-tip is not abruptly truncated when it reaches the crack-tip, but rather diminishes smoothly to zero a short distance from the crack-tip.

One drawback of the transparency method is that it does not work well when field nodes are placed too close to the crack surface. To circumvent this difficulty, a restriction must be imposed on the position of nodes: they should be so placed that the normal distance from the node to the crack surface is greater than roughly $h/4$, where h is the nodal spacing.

(2) **See-through methods**

Terry (1994) proposed a "see-through method" for constructing continuous approximations near nonconvex boundaries. In his method, all or part of the boundary is made completely transparent such that discontinuity is eliminated. He found that better accuracy might be obtained for an interior hole problem when the hole boundary was not strictly enforced by the visibility criterion.

Duarte and Oden (1996), Krysl and Belytschko (1997) suggested a smoothing technique in which the crack was completely transparent if the crack-tip is within the support domain of a field point. This is also called the "continuous line criterion": if a line connecting the field node to a field point lies entirely within the support domain of the point, the node is visible. Although this technique is easy to implement and provides smooth approximations, yet it effectively shortens the crack and leads to considerable error. This method does work for cracks when the enrichment techniques are employed.

3. **Mixed criteria**

The methods described above for dealing with crack-tips and nonconvex boundaries can be used in combination. Nonconvex boundaries can be categorized as either smoothly or strongly discontinuous arise from cracks and notches. The stress concentration due to smooth boundaries can be accurately computed by increasing the nodal resolution, while the stress singularity near sharp boundaries requires infinite resolution or enrichment.
The see-through method works well for the nonconvex boundary when nodal refinement is adequate. In this case, the support domain extending across the boundary are not harmful. The diffraction method can also be used for such boundaries, but it adds complications that have been found to be superfluous.

Both the diffraction and transparency methods can be used to construct approximations which are continuous within the domain but discontinuous across the crack. The see-through method should generally not be used for cracks, but for larger wedge angles where the singularity is usually not of interest.

9.6.2 Crack-Tip Enrichment

In the work of Belytschko and Fleming (1999), methods for enhancing EFGM are distinguished into "intrinsic enrichment" where the enrichment functions are included in the EFGM basis and "extrinsic enrichment" where the approximation is enriched by adding functions externally to the EFGM basis.

1. **Overall extrinsic MLS enrichment**

In this enrichment scheme, a function closely related to the solution is added to the approximation formula (9.10) (Fleming et al. 1997). In the problem of linear elastic fracture mechanics where only one crack-tip is considered in the support domain, the near tip asymptotic field or its constituents can be added to get the approximation form such that

$$\mathbf{u}^h(\mathbf{x}) = \sum_{I\in\Omega_\mathbf{x}}^{n} \phi_I(\mathbf{x})\hat{\mathbf{u}}_{b,I} + \sum_{k=1}^{2}\sum_{I\in\Omega_\mathbf{x}}^{n} (K_I Q_I^k(\mathbf{x}_I) + K_{II} Q_{II}^k(\mathbf{x}_I)) \tag{9.96}$$

in which K_I and K_{II} are the SIFs of crack. The functions Q_{II}^k and Q_I^k describing the near-tip displacement field are given by Williams (1957)

$$\begin{cases} Q_I^1 = \dfrac{1}{2G}\sqrt{\dfrac{1}{2\pi}}\cos\dfrac{\theta}{2}\left(\kappa - 1 + 2\sin^2\dfrac{\theta}{2}\right) \\ Q_I^2 = \dfrac{1}{2G}\sqrt{\dfrac{1}{2\pi}}\sin\dfrac{\theta}{2}\left(\kappa + 1 - 2\cos^2\dfrac{\theta}{2}\right) \\ Q_{II}^1 = \dfrac{1}{2G}\sqrt{\dfrac{1}{2\pi}}\sin\dfrac{\theta}{2}\left(\kappa + 1 + 2\cos^2\dfrac{\theta}{2}\right) \\ Q_{II}^2 = \dfrac{1}{2G}\sqrt{\dfrac{1}{2\pi}}\cos\dfrac{\theta}{2}\left(\kappa - 1 - 2\sin^2\dfrac{\theta}{2}\right) \end{cases} \tag{9.97}$$

in which r is the distance from the crack-tip, θ is the angle from the tangent to the crack path (viz. Fig. 3.13), G is the shear modulus and κ is the Kolosov constant defined as

$$\kappa = \begin{cases} 3 - 4\nu & \text{plane strain problem} \\ (3-\nu)/(1+\nu) & \text{plane stress problem} \end{cases} \tag{9.98}$$

in which ν is the Poisson's ratio.

This method enables to calculate SIFs directly, but it demands the solution over the whole domain, that will lower down the computation efficiency remarkably.

2. **Extrinsic PU enrichment**

This scheme is carried out using PU methods (Belytschko et al. 1996b; Duarte and Oden 1996; Belytschko and Fleming 1999). The approximation is extrinsically augmented with enrichment functions to the existing EFGM approximation formula (9.10). It is necessary to enrich all field nodes in the support domains containing crack-tip; otherwise, the enrichment is incomplete and the solution accuracy is poor. In the work of Belytschko and Black (1999), the extrinsic basis is added only to those field nodes that are near the crack-tip.

$$\begin{aligned} \mathbf{u}(\mathbf{x}) = & \sum_{I_b \in S_b}^{n_b} \boldsymbol{\phi}_{I_b}(\mathbf{x})\hat{\mathbf{u}}_{b,I_b} + \sum_{I_c \in S_c}^{n_c} H(f(\mathbf{x}))\boldsymbol{\phi}_{I_c}(\mathbf{x})\hat{\mathbf{u}}_{c,I_c} \\ & + \sum_{I_t \in S_t}^{n_t} \boldsymbol{\phi}_{I_t}(\mathbf{x})\left(\sum_{\alpha=1}^{4} \hat{\mathbf{u}}_{t,I_t}^{\alpha} F_\alpha\right) \end{aligned} \tag{9.99}$$

in which S_b is the node set for the basic variable without crack surface (circle point in Fig. 9.16), S_c is the node set for the refinement of crack surface (point "Δ" in Fig. 9.16), S_t is the node set for the refinement of crack-tip (point "▲" in Fig. 9.16), $\mathbf{F}$ is the near-tip asymptotic displacement field function that is identical to XFEM (viz. Eq. (5.50)).

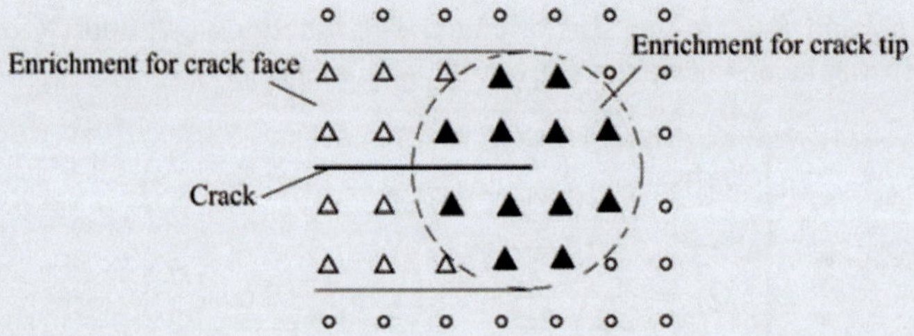

Fig. 9.16 Nodes and regions for the enrichment

The general Heaviside function in Eq. (9.99) is defined as

$$H\big(f(\mathbf{x}_Q)\big) = \begin{cases} +1 \text{ if } f(\mathbf{x}_Q) > 0 \\ -1 \text{ if } f(\mathbf{x}_Q) < 0 \end{cases} \tag{9.100}$$

in which $f(\mathbf{x}_Q)$ is defined as the distance of field point $\mathbf{x}_Q$ to the crack, when $\mathbf{x}_Q$ is above the crack H is positive, otherwise it is negative.

Suppose the support domain of field point $\mathbf{x}_Q$ is totally cut apart by the crack (viz. Fig. 15a), the refinement by the second item in Eq. (9.99) is considered only; if the support domain of field point $\mathbf{x}_Q$ is partially cut by the crack (viz. Fig. 9.15b), the refinement by the third item in Eq. (9.99) is considered only.

3. **Intrinsic basis enrichment**

One can include the asymptotic near-tip displacement field either in Eq. (9.26), or as an important ingredient, in the basis $\mathbf{p}(x)$. The choice of basis depends on the accuracy desired.

(1) **Full intrinsic enrichment**

The entire near-tip asymptotic displacement field in Eq. (5.50) may be included in the basis functions. Following some trigonometric manipulation, Fleming, Chu, Moran and Belytschko (1997) shown that all the basis functions $\mathbf{p}$ may be intrinsically enriched. Taking Eq. (9.6) for example, it may be extended as

$$\mathbf{p}(r,\theta) = \begin{bmatrix} 1 & x & y & \sqrt{r}\sin\left(\frac{\theta}{2}\right) & \sqrt{r}\cos\left(\frac{\theta}{2}\right) & \sqrt{r}\sin\left(\frac{\theta}{2}\right)\sin\theta & \sqrt{r}\cos\left(\frac{\theta}{2}\right)\sin\theta \end{bmatrix}^{\mathrm{T}} \tag{9.101}$$

The linear terms are not related to the near-tip field and are represented through the linear completeness of the EFGM, hence Eqs. (9.10) and (9.11) are unchanged but the basis polynomial is expanded. In contrast to the extrinsic methods presented previously, this method involves no additional unknowns. However, because of the increased size of basis, additional computational effort is required to invert the moment matrix $\mathbf{A}(\mathbf{x})$. In addition, the support domain must be enlarged to achieve the regularity of $\mathbf{A}(\mathbf{x})$.

Use of an enriched basis is likely to result in an ill-conditioned moment matrix $\mathbf{A}(\mathbf{x})$ that it is troublesome although this generally does not affect the final solution. It has been found that the effects of ill-conditioning can be mitigated by the diagonalization of the moment matrix via the orthogonalization technique of Gram-Schmidt (Lu et al. 1994).

When an enriched basis is used, it must be used for all nodes, or a technique described by Belytschko and Fleming (1999) must be employed to blend nodes with different basis. Simply deleting functions from the basis results in discontinuities in the approximation.

(2) **Radial enrichment**

In $\sqrt{r}$ enrichment, the basis is rather simplified

$$\mathbf{p}(r,\theta) = \begin{bmatrix} 1 & x & y & \sqrt{r} \end{bmatrix}^{\mathrm{T}} \tag{9.102}$$

in which r is the radial distance from the crack-tip. This enrichment is useful because the angular variation around the crack-tip is smooth, but the radial variation is singular in the stress. The advantage of radial enrichment is that the intrinsic basis is only expanded by one term, and inverting the moment matrix $\mathbf{A}(\mathbf{x})$ to form the shape functions is much cheaper than full enrichment. In addition, it does not seem to be necessary to use smoothing techniques because there are no discontinuities in the radial direction, but the discontinuities in the angular direction lead to a remarkable loss of accuracy when full enrichment is used. Since this enrichment does not take into account of the discontinuity behind crack-tip, its convergence is much slower.

4. **Hybrid enrichment and linear approximation**

Enriching the approximation for the entire domain is generally unnecessary because it increases computational expense without perceivable payback. This is mainly due to the fact that the crack-tip singularity is local: it extends about $0.1a$ from the crack-tip, where a is the length of crack. Hence, two local enrichment techniques may be employed to couple enriched and linear approximations, one uses a consistent coupling to maintain C^0 continuity, the other does not.

The first technique of local enrichment for crack-tip region is to use extrinsic PU enrichment (viz. Eq. (9.99)). Another technique involves coupling the approximation over a transition region that is similar to the hybrid EFGM-FEM (Belytschko et al. 1995b) (viz. Fig. 9.17). For details, our readers are referred to the work of Fleming and his co-workers (Fleming et al. 1997).

9.6.3 *Crack Growth*

The criteria and calculation algorithms of crack onset and crack growth direction are actually identical to that in XFEM (Chap. 5).

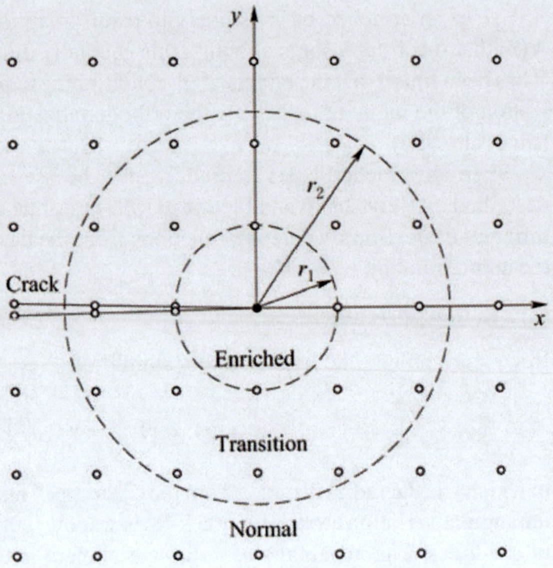

Fig. 9.17 Schematic for the hybrid of enriched and linear approximations

9.7 Validations

9.7.1 Contact Crack Within Plate

The elastic plate shown in Fig. 9.18 contains a crack of length $2a$ and is exerted by a vertical load $\sigma = -2.0 \times 10$ MPa (Li and Chen 2003a). The plate width and height is $w = 0.15$ m and $h = 0.20$ m, respectively.

The second-order polynomial base function (9.7) is employed in the construction of basis functions. In the weight function (9.19), $d_c = 0.0125$ m, $d_{si} = 4d_c$.

The uniform cells (12 × 16) are discretized for the plate. The crack and exterior boundaries w.r.t. to load and prescribed displacement are all discretized by 10 line segments, respectively.

The results presented in Figs. 9.19 and 9.20 show that the maximum computation error between exact solution and computational solution by the EFGM is within 2%.

9.7.2 Contact Block on Rigid Base

This is a classic benchmark example in the study of frictional contact problem (Simo and Laursen 1992; Li and Chen 2003b). The elastic block in Fig. 9.21 is placed on a rigid base, the loads $P_y = -200$ Pa and $P_x = 60$ Pa are exerted on the upper and

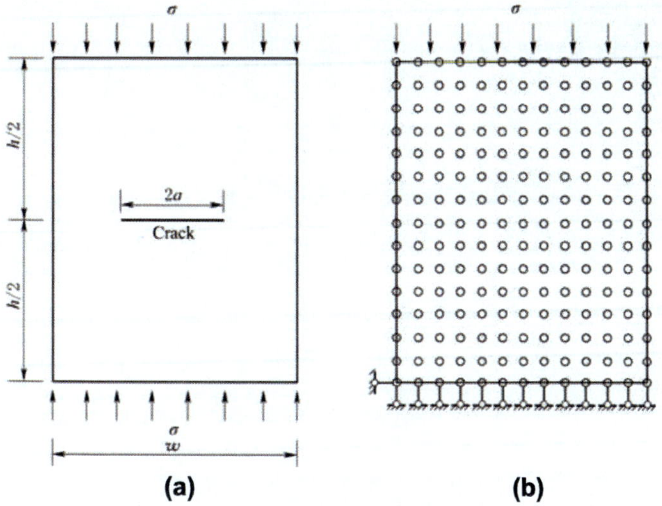

Fig. 9.18 Elastic plate containing crack. a Configuration of plate; b field nodes discretized for EFGM

Fig. 9.19 Distribution of stress σ_y along the crack ($2a = 0.05\,\text{m}$)

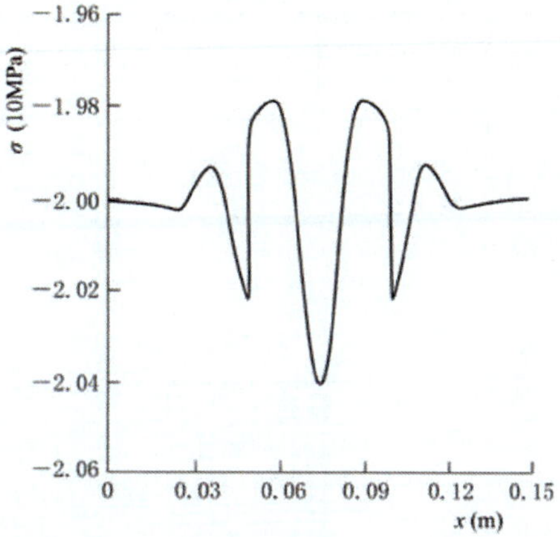

right-side edges of the block, respectively. The Young's modulus is $E = 1000\,\text{Pa}$, the Poisson's ratio is $\nu = 0.3$, the friction coefficient is $\mu = 0.5$.

The uniform cells (40×20) are discretized with 861 field nodes. The contact face and boundary faces are discretized by 10 line segments, respectively.

The second-order polynomial base function (9.7) is employed in the construction of basis functions, the parameter in the weight function (9.19) are $d_c = 0.1\,\text{m}$, $d_{si} = 4d_c$.

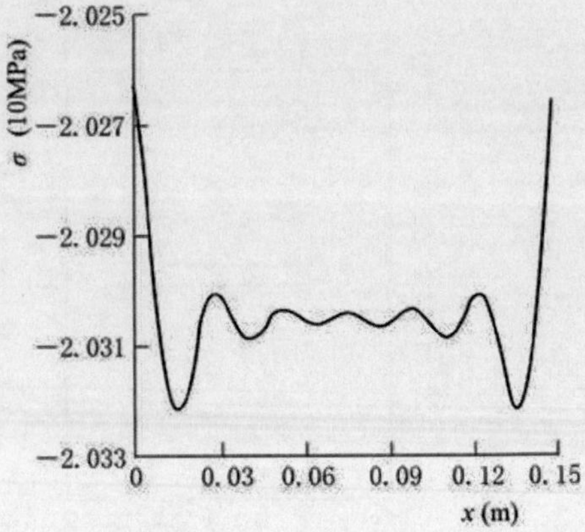

Fig. 9.20 Distribution of stress σ_y along the crack ($2a = 0.15$ m)

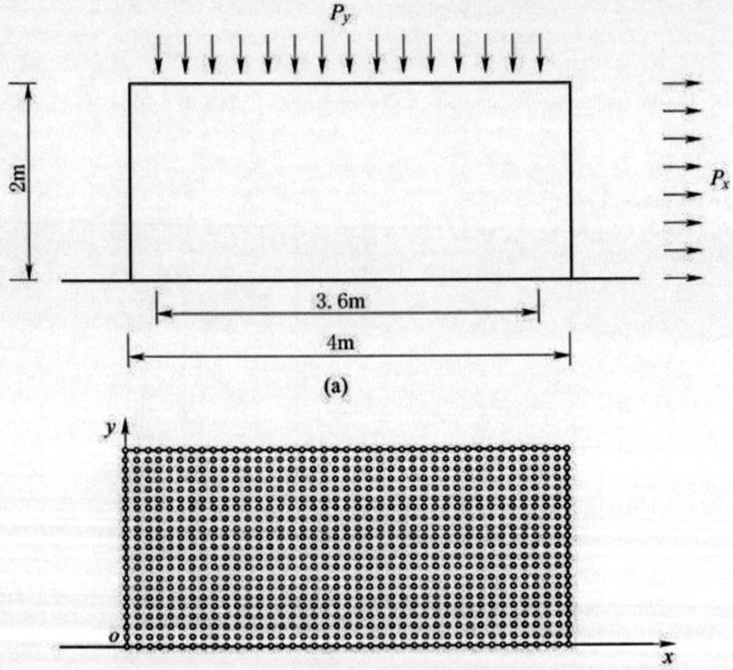

Fig. 9.21 Elastic block on rigid base. **a** Configuration of elastic block; **b** field nodes discretized for EFGM

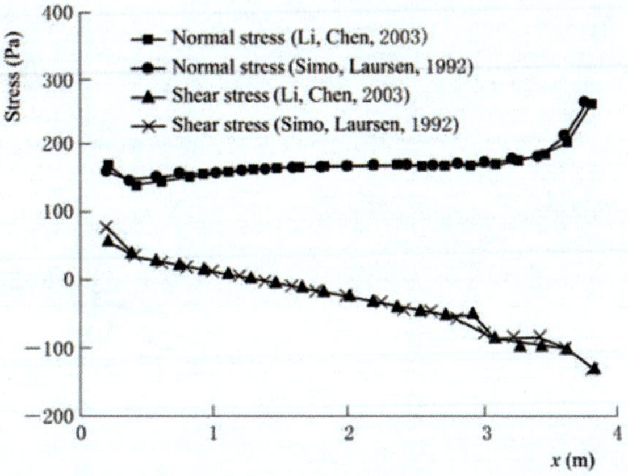

Fig. 9.22 Normal and shear stresses on contact face

Iterative collapse is controlled by the displacements difference being smaller than 1.0×10^{-10} m between two successive iteration steps. The shear and normal stresses at the contact face $3.8\,\text{m} \geq x \geq 0.2\,\text{m}$ are collected for the comparison between EFGM (Li and Chen 2003b) and FEM (Simo and Laursen 1992). Figure 9.22 indicates that the contact face segment $x \geq 3.08$ m undergoes slip state.

Figure 9.23 illustrates the elastic deformation of block. Figures 9.24 and 9.25 exposit the normal and shear contact stresses under the actions of fixed $P_y = -200$ Pa and variable $P_x = 20, 60, 100, 140, 179$ Pa, respectively. It is notable that the limit resistance against slide is $P_x = 180$ Pa.

As the rise of lateral load P_x, the normal and shear stresses at the left side of contact face decrease, whereas those that at the right side increase. Under the action of $P_x = 20$ Pa, there is no slip point on the contact face; whereas under the actions of $P_x = 60, 100, 140, 179$ Pa, the length of the contact face stays stick state are 3.08, 2.72, 2.18, 1.82 m, respectively. The computation also indicates that when $P_x = 182$ Pa, the iterative convergence cannot be maintained.

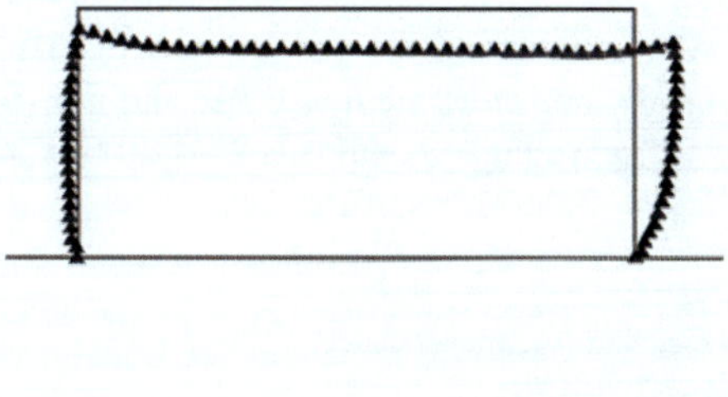

Fig. 9.23 Deformation of elastic block

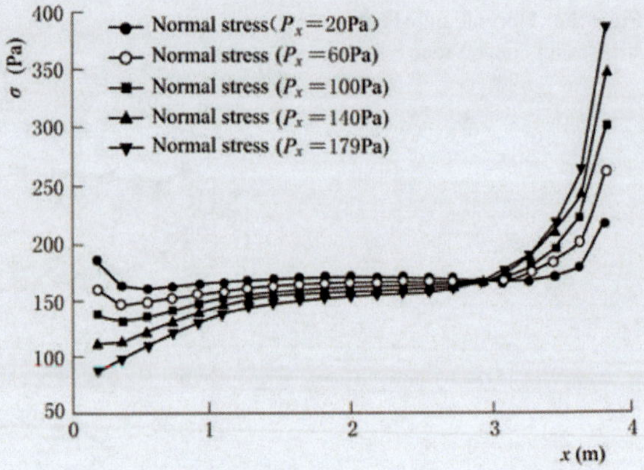

Fig. 9.24 Normal stresses on the contact face under different lateral load

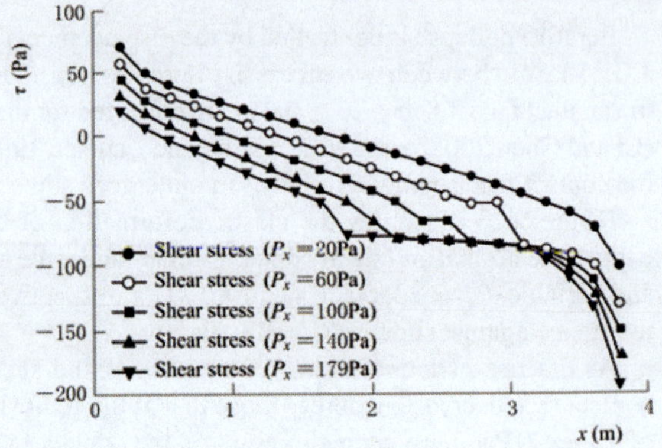

Fig. 9.25 Shear stresses on the contact face under different lateral loads

9.7.3 *Free Vibration Cantilever*

This is a classic example in the study of free vibration problem (Li and Chen 2003b). A cantilever with the length of $L = 40\,\text{m}$ is fixed at the right end, the cantilever height and width are $h = 0.8\,\text{m}$ and $w = 0.4\,\text{m}$, respectively (viz. Fig. 9.26).

The Young's modulus is $E = 20\,\text{GPa}$, the Poisson's ratio is $\nu = 0.3$, the density is $\rho = 3.0 \times 10^3\,\text{kg/m}^3$.

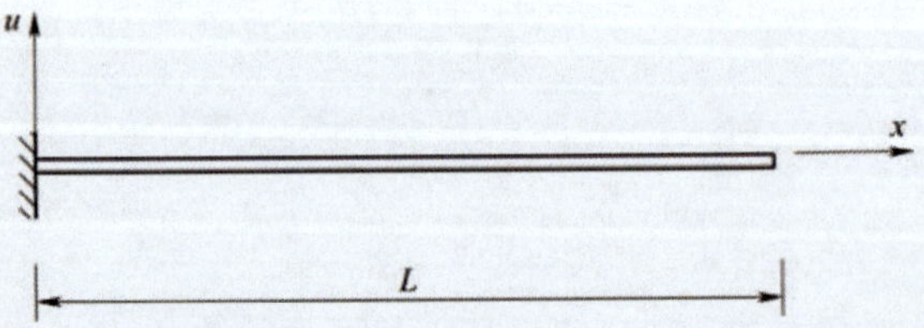

Fig. 9.26 Cantilever under free vibration

Table 9.1 Non-dimensional frequencies of cantilever (first 8 modes)

Solution method	Non-dimensional frequency parameter $\omega\sqrt{mL^4/EI}$							
Closed-form	3.516	22.039	61.695	120.903	199.859	298.556	416.991	555.165
EFGM	3.526	22.168	62.058	121.568	200.873	299.872	418.497	556.651

In the computation using EFGM, the cantilever is divided into (40 × 2) uniform cells by (41 × 3) field nodes. The displacement and force traction boundaries are all discretized into 10 line segments. The complete first-order polynomial base function (9.6) is used.

The weight function (19) is employed in this study, where the computation parameter set $k = 2$, $d_c = 1$ m, $d/d_c = 1$, $d_{si}/d_c = 4$ gives best computational solution. In Table 9.1 a non-dimensional frequency parameter $\omega\sqrt{mL^4/EI}$ is define to help the verification of computation results, where m is the mass of the unit cantilever length. The examination of these computation parameters are carried out by the following steps.

1. **For fixed value** $k = 2$

The influences of d/d_c and d_{si}/d_c on the frequencies (first 8 modes) of the cantilever are exposited in Fig. 9.27, in which the vertical coordinate axis is the non-dimensional frequency parameter $\omega\sqrt{mL^4/EI}$. It is shown that: if $d_{si}/d_c > 3.0$ and $d/d_c > 0.9$, the calculated frequencies are acceptable cross-referenced by the closed-form solution.

2. **For fixed value** $d_{si}/d_c = 4.0$ **and** $d/d_c = 1.0$

Figure 9.28 validates that k influences the computation result remarkably (first 8 modes), too. When $k = 2$ the best result is approached.

9.7.4 Simply Supported Beam Exerted by a Concentrated Impact

The size of the simply supported beam in Fig. 9.29 is identical to the cantilever in Fig. 9.26 (Li and Chen 2003b).

The concentrated impact load is $P(x, t) = P_0\delta(x - L/2)$ where $P_0 = 10$ kN/m is upwardly exerted on the middle of the beam. If the damping is neglected, the closed-form solution has been given by Clough and Penzien (1975):

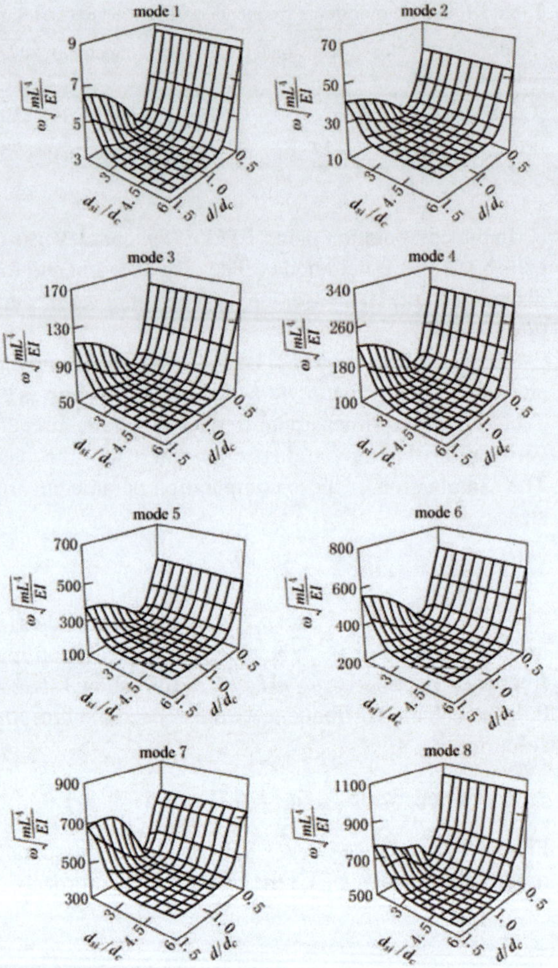

Fig. 9.27 Influences of d/d_c and d_{si}/d_c on the non-dimensional frequency parameter $\omega\sqrt{mL^4/EI}$ (with fixed $k = 2$)

$$\begin{cases} u(L/2,t) = \dfrac{2P_0L^3}{\pi^4 EI}\left(\dfrac{1-\cos\omega_1 t}{1} + \dfrac{1-\cos\omega_3 t}{81} + \dfrac{1-\cos\omega_5 t}{625} + \ldots\right) \\ \dot{u}(L/2,t) = \dfrac{2P_0L^3}{\pi^4 EI}\left(\dfrac{\omega_1\sin\omega_1 t}{1} + \dfrac{\omega_3\sin\omega_3 t}{81} + \dfrac{\omega_5\sin\omega_5 t}{625} + \ldots\right) \\ \ddot{u}(L/2,t) = \dfrac{2P_0L^3}{\pi^4 EI}\left(\dfrac{\omega_1^2\cos\omega_1 t}{1} + \dfrac{\omega_3^2\cos\omega_3 t}{81} + \dfrac{\omega_5^2\cos\omega_5 t}{625} + \ldots\right) \end{cases} \tag{9.103}$$

The dynamic EFGM using Wilson-θ algorithm is employed where $\theta = 1.4$, $\Delta t = 0.002\,\mathrm{s}$, $n = 2000$. The EFGM cells, nodes, and quadrature scheme are identical to the free vibration cantilever in Sect. 9.6.3.

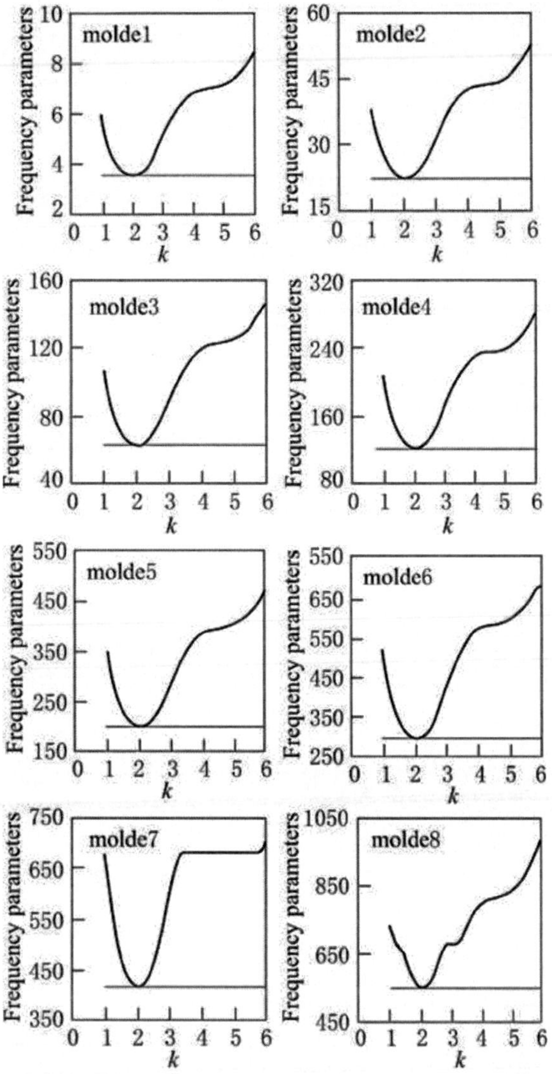

Fig. 9.28 Influences of k on the non-dimensional frequency parameter $\omega\sqrt{mL^4/EI}$ (with fixed $d/d_c = 1.0$, $d_{si}/d_c = 4.0$)

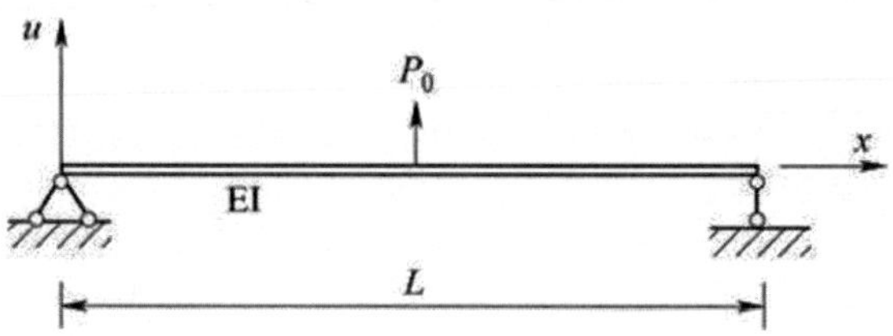

Fig. 9.29 Simply supported beam exerted by a concentrated impact. **a** Beam configuration; **b** load duration

Table 9.2 Non-dimensional frequencies of simply supported beam (first 8 modes)

Solution Method	Non-dimensional frequency parameter $\omega\sqrt{mL^4/EI}$							
Closed-form	9.870	39.478	88.826	157.914	246.740	355.306	483.611	631.655
EFGM	9.871	39.472	88.772	157.731	246.270	354.309	481.752	628.469

Table 9.2 and Figs. 9.30, 9.31 and 9.32 exhibit the comparison between closed-form solutions and EFGM solutions.

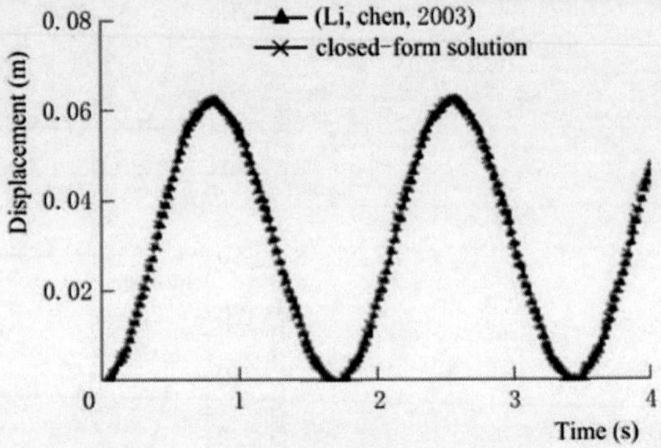

Fig. 9.30 Displacement response at the center of beam

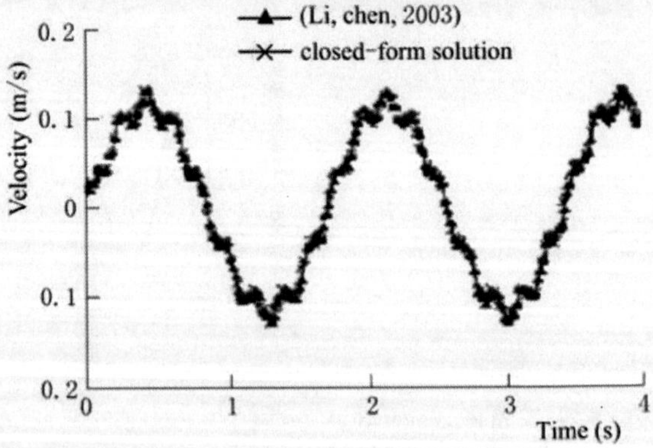

Fig. 9.31 Velocity response at the center of beam

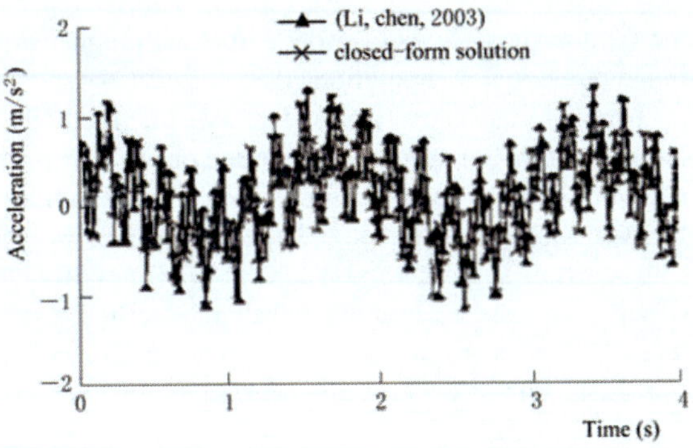

Fig. 9.32 Acceleration response at the center of beam

9.8 Concluding Remarks

9.8.1 Definition of MFMs

In this chapter, we take EFGM as a representative family member of "mesh-free methods (MFMs)" in "computation solid mechanics (CSM)", to give a general view w.r.t. the principles and algorithms of MFMs. With the name MFMs, a family of methods whose main characteristic is defined by the contraposition to FEM, in the sense that they do not need a mesh constructed by standard FEs in order to define the shape functions. In general, MFMs are the algorithms featured with following characteristics:

- The definition of shape functions depends only on the node positions;
- The evaluation of nodal connectivity is linear with the total number of nodes in the domain.

With these two statements, many of the so-called MFMs are not truly meshless insofar, because they need a correct definition of the nodal connectivity.

Among MFMs, "smooth particle hydrodynamics (SPH)" (Lucy, 1977), "element free Galerkin method (EFGM)" (Belytschko et al. 1994a), "reproducing kernel particle method (RKPM)" (Liu et al. 1995a; Chen et al. 1996a; Chen et al. 1998a), "hp-cloud method" (Duarte and Oden 1996; Liszka et al. 1996), and "partition unity method (PUM)" (Melenk and Babuška 1996; Babuška and Melenk 1997; Babuška and Zhang 1998), etc., aim to avoid re-meshing by the approximation of trial function on a set of field nodes. Such a mesh independence, however, asks us to pay higher computation effort and leads to the inability to meet the Kronecker delta property. As a result, although a variety of MFMs have been established, yet the currently available and reliable members are much more expensive and lower accurate than FEMs and FDMs because they come with various complications that affect their overall performance.

Even so, MFMs are still an important research subject in advanced CMs, because they have already exhibited some advantages in particular applications, for instance, the C_1 approximations of weak forms with higher order derivatives are convenient in the numerical computations (Liu et al. 2004).

There are basically two types of MFMs: methods based on strong form formulations (e.g. SPH) and methods based on weak form formulations (e.g. EFGM). In addition, particle-based methods such as the fundament DEM (Chap. 7), BEA and DDA (Chap. 8) may be looked at as mesh-free, too.

The strong form methods such as SPH (Lucy 1977; Gingold and Monaghan 1977) (Chap. 10) possess attractive advantages of being simple to implement and "truly" mesh-free, because no integration is required in establishing the discrete system equation set. However, they are often suffered from instability, especially when irregularly distributed field nodes are used for the PDEs with Neumann (derivative) boundary conditions, such as in the CSD with stress traction (natural) boundaries.

On the other hand, weak form methods exhibit the advantages of excellent stability and accuracy. The Neumann boundary conditions can be naturally satisfied due to the use of smoothing (integral) operators. However, these methods are said not to be "truly" mesh-free, because a background grid (local or global) is required for the quadrature.

9.8.2 Initiation and Early Developments

In the monograph of Liu (2003), the history, theory and applications of the major existing MFMs were comprehensively addressed. Other excellent reviews on the MFMs also can be found in the works by Belytschko and Lu (1995), Belytschko et al. (1996a, b), Belytschko and Fleming (1999), Li and Liu (2002), Nguyen et al. (2008), Daxini and Prajapati (2014), among many others.

Since the 1990s, much effort is concentrated on the problems to which the fundamental FDM and FEM are difficult to apply, such as the problems with free surface, deformable boundary, moving interface (for FDM), large deformation (for FEM), complex mesh generation, mesh adaptive refinement, and multi-scale resolution (for both FDM and FEM). One important goal of the initial search for MFMs is to modify the internal structure of the grid-based FDM and FEM, in this manner to get more adaptive, versatile and robust (not satisfied insofar) computation tools (Belytschko et al. 2000a, b). For this purpose, the technique of polynomial approximation that fits a number of nodes with the minimized distance between the approximated function and the value of the unknown function values, is normally employed.

Nayroles, Touzot and Viilon (1992) invoked MLS approximations in the Galerkin method to formulate the so-called "diffuse element method (DEM)". Later on, Belytschko et al. (1994a, b) made a milestone advancement by the creation of EFGM that is currently one of the most popular family member of MFMs.

Melenk and Babuška (1996) pointed out the similarities between MFMs and FEMs, and developed the so-called "partition of unity finite element method (PUFEM)", its shape functions are based on Lagrange polynomials.

Called the "finite point method (FPM)", a scheme using the Lagrangian formulation and point collocation method was proposed by Oñate and his co-workers (Oñate et al. 1996). Later on, the ideas were generalized to take into account of FE type approximations in order to obtain the computing time competence between mesh generation and node-connectivity creation. This method, called as the "meshless finite element method (MFEM)", makes use of special FE shape functions but keeps all the advantages of MFMs (Idelsohn et al. 2003). It uses the extended Delaunay tessellation to build a mesh comprising different polygonal elements (or polyhedrons in 3-D). MFEM has been successfully coupled with the particle-based method to solve CFD problems (Idelsohn et al. 2004).

In an expectation to be superior to the fundamental grid-based CMs, MFMs are particularly ambitious towards the development of next generation CMs. Their quick development in the 1990s had been recorded in many important literatures (Duarte and Oden 1995; Melenk and Babuška 1996; Gingold and Monaghan 1997).

9.8.3 *Advances Since the 2000s*

1. **Expansion of family**

Atluri and Zhu (1998) originated the "meshless local Petrov–Galerkin (MLPG)" method that requires only local background cells for quadrature. It has been applied to the analysis of beam and plate structures, fluid flows, and other mechanics areas. Detailed descriptions of MLPG and its applications can be found in the literatures by Atluri and Shen (2002), Liu and his co-workers (Liu and Chen 1995; Liu et al. 1995a, b; Liu et al. 1997; Lin and Atluri 2001). The main difference between MLPG and EFGM is that, rather than using global weak forms, in the former local weak forms are generated on overlapping sub-domains, and the quadrature is carried out in these local sub-domains merely. Atluri (2002) introduced the notion "truly meshless" since no construction of a background mesh is needed for quadrature purposes. Another well-known method mainly created for fluid mechanics is the "moving point method (MPM)" (Oñate et al. 1996).

Liu and his colleagues developed the "point interpolation method (PIM)" and some variants (Liu and Gu 2001; Gu and Liu 2001). To get an automatic computation tool, Liu further proposed the "smoothed point interpolation method (S-PIM)" by the so-called weakened weak (W2) formulation (Liu, 2019, 2010), which offers possibilities to establish various models that work well with triangular meshes that may be generated automatically and re-meshed easily.

The use of radial basis function (or together with the polynomials) well resolved the problems for both the local Petrov-Galerkin weak-form (Liu and Gu 2001) and the global Galerkin weak-form (Wang and Liu 2002). A "mesh-free weak-strong (MWS)" formulation based on the combined weak form and strong form (Liu and

Gu 2003; Liu et al. 2004; Gu and Liu 2005) was established by the joint use of MLS and radial PIM shape functions, which demands a local background mesh only for the nodes near the natural boundaries of the problem domain. In MWS, the strong form is used for all the internal nodes and the nodes on essential boundaries, whereas the local weak form (Petrov-Galerkin) is specially used for the nodes near natural boundaries. Consequently there is no need for numerical quadrature w.r.t. internal nodes and the nodes on essential boundaries. MWS is, insofar, a MFM that uses least meshing effort in the entire computation and produces most stable solution. It is one important step towards the dream of "truly meshless" meanwhile to keep the capability of producing stable and accurate solutions for solid mechanics.

MFMs also have been exercised in the solution of boundary integral equations. The "boundary mesh-free methods (BMFMs)", in which only the boundary of problem domain needs to be discretized by irregularly distributed nodes, was formulated by Mukherjee and Mukherjee (1997). A critical assessment of meshless local Petrov-Galerkin (MLPG) and local boundary integral equation (LBIE) methods was conducted by Atluri, Kim and Cho (1999a, b). The "boundary point interpolation methods (BPIM)" was further developed using polynomial PIM and radial PIM interpolations (Gu and Liu 2001; Liu and Gu 2003), which gives a much smaller discretized equation set due to the Kronecker delta properties of the PIM shape functions.

2. **Enhancing computational efficiency**

Although MFMs look quite attractive for solving a certain class of CFD and CSD problems, yet there are a number of obstacles that are tightly related to the computation efficiency, need to be overcome.

(1) **EBCs**

Mesh-free shape functions are not interpolation type because they do not possess Kronecker delta properties. Hence the imposition of "essential boundary conditions (EBCs)" consumes much computation efforts. Several techniques such as the Lagrange multiplier, penalty factor, orthogonal transformation, hybrid (with FEM), singular weighing function, boundary collocation, etc., were presented for imposing EBCs. Chen and Wang (2000) proposed two new boundary condition treatment techniques, the "mixed transformation method" and the "boundary singular kernel method", to enhance the computation efficiency for the contact problems within RKPM framework. The former is a modification of a full transformation method, while the latter introduces singularities into the kernel functions associated with the EBC nodes. Other new boundary treatment techniques developed by Ren and Liew (2002) are named as the "node interpolation method" and the "direct imposition method". In the former the shape functions associated with EBCs are constructed using node interpolation and then combined with mesh-free shape functions, while the latter rearranges the discretized system equation set and directly provides the known values of EBDs in the nodal variable vector (similar to FEM). Smoothing of the approximating functions at concave boundaries in EFGM was proposed by Belytschko and his-co-workers: by redefining a parameter governing the decay of weight function, shape functions are modified and made continuous in the domain with concave corners (Yoon et al. 2001).

A comprehensive overview of existing techniques for enforcing EBCs was presented by Fernández-Méndez and Huerta (2004).

(2) **Numerical integration**

In MFMs, for the integration of Galerkin weak form, Gauss quadrature is most commonly exercised (Dolbow and Belytschko 1999). A number of disadvantages have been reported such as its complexity, requirement of higher order quadrature rules, specialized integration zone patterns, etc. Beissel and Belytschko (1996) suggested a direct nodal integration to avoid background cells, but it leads to oscillations in the solution due to the under integration of weak form and vanishing shape functions at nodes. To overcome this drawback, a "stabilized conforming nodal integration (SCNI)" method for the elastoplastic contact analysis of metal forming processes was proposed (Chen et al. 2001; Yoon et al. 2001). In their approach, the strain smoothing stabilization was introduced to eliminate spatial instability, and the convergence was obtained by introducing an integration constraint. In an application of mesh-free formulation with SCNI, Wang and Chen (2006) proposed a locking free mesh-free formulation for curved beam with "Kirchhoff mode reproducing conditions (KRMC)". Khosravifard and Hematiyan (2010) presented a technique for the evaluation of regular domain integrals without domain discretization wherein a domain integral is transformed into a boundary integral. Their technique results in truly meshless approach with better accuracy and efficiency.

(3) **Error estimate and control**

Errors in CMs arise from a number of sources inclusive discretization resolution, quality of mathematical model, rounding off operation in computation, etc. Similar to the Z^2-estimator in h-adaptive FEM (Zienkiewicz and Zhu 1987; Chen, 1996), Chung and Belytschko (1998) proposed an estimation of local and global error based on the difference between the values of projected stress and stress given by EFGM. Effectiveness of their proposed error estimator was validated by various 1-D and 2-D problems (Gavete et al. 2001; Gavete et al. 2002; Gavete et al. 2003; Lee et al. 2003; Lee and Zhou 2004a, b).

Zhuang, Heaney and Augarde (2012) stated that the discretization error in EFGM mainly arises from not rigorously satisfying governing equations and boundary conditions. Consequently, conventional procedures for error analysis used in FEM cannot be applied straightforwardly. In FEM, it is feasible to uncouple h and p adaptive refinements but in EFGM, it is not possible because changing the density of nodes will simultaneously change both the errors e_h and e_p, the latter is related to the space of shape (test) functions. As a result, it is difficult to achieve error control and adaptive refinement in MFMs.

Kim and Atluri (2000) proposed a technique to control error and to improve solution accuracy in MLPG by adding secondary nodes in the domain where better resolution is demanded.

9.8.4 Applications

Comprehensive investigations on MFMs are closely related to their applications to complex CSD and CFD problems. Since the computational frame is based on a set of arbitrarily distributed nodes rather than on a system of pre-defined mesh/grid, MFMs are very attractive in dealing with problems that are previously difficult for fundamental mesh-based methods.

The mainstream applications of MFMs include large deformation analysis in solids (Chen et al. 1996b; Chen et al. 1998b; Li et al. 2000), structural dynamics (Liu et al. 1995b; Gu and Liu 2001; Liu et al. 2002), non-linear foundation consolidation (Wang et al. 2001), incompressible and compressible fluid flow (Lin, Atluri, 2001; Löhner et al. 2002), etc. For structural engineers in the area of geotechnical and hydraulic engineering, the most concern is the ability of MFMs to solve crack growth problems (Belytschko and Lu 1995; Belytschko and Tabbara 1996; Belytschko et al. 2000a; Carpinteri et al. 2002, 2003) in the competition with rival mesh-based methods (e.g. XFEM, CEM, NMM). Insofar, there is no unanimous conclusion w.r.t. the question which is out-performed in handling crack-growth phenomenon, although since the 2000s, remarkable advances have been archived in MFMs that will be briefly cited as below.

Duflot and Nguyen-Dang (2004a, b) worked out an enriched MFM to analyze fatigue crack growth under cyclic loading. In their method, the crack growth is modeled by successive linear extensions determined by SIFs. A fixed set of three nodes with special weight function is added at each crack-tip to accurately catch the stress singularities. Modal characteristics such as the natural frequency and mode shapes of structures are used for detecting and predicting crack-growth. In addition, Andreaus, Batra and Porfiri (2005) proposed a MLPG approach with MLS shape function for analyzing vibration of beams with multiple cracks.

In impact problems, transition in failure mode can be observed. Wang and Liu (2010) put forward an EFGM for simulating the failure transition from brittle to ductile under finite deformation state.

Under the combined thermal and hydraulic and mechanical actions, the presence of cracks will induce a strong variation in thermal/hydraulic/stress fields, which in turn, can affect the crack growth (Bouhala et al. 2012). In the structural design for concrete dams, turbines, combustion chambers, and nuclear pressure vessels, thermo-hydro-elastic fracture mechanics is widely applied. Normally, such a thermo-hydro-elastic fracture problem is practically solved by decoupling it into three separate field problems and these fields should be enriched intrinsically to represent the discontinuous temperature, fluid flow, displacement, and they traction across crack faces. For such complex engineering problems, insofar the general FEMs such as the XFEM and CEM exhibit practical out-performance attributable to their well-equipped tool-kits. To understand this personal opinion, the application example of Xiaowan arch dam (Chap. 5) may be referred.

9.8.5 Hybrid Methods

Although MFMs have found their wide applications in almost all areas of engineering structure related to solids and fluids, yet challenges in developing computationally efficient algorithms w.r.t. scalable implementation of EBCs, accurate nodal integration techniques, and adaptive error control, etc., still exist. To alleviate the difficulties, hybrid methods are looked at as a wise alternative by many researchers (Arroyo and Ortiz 2006).

A MFM can be coupled with other MFMs or fundamental MBMs to take the full advantages of each method (Fernández-Méndez and Huerta 2002). Examples include the SPH coupling with FEM (Attaway et al. 1994), the EFGM coupling with BEM (Gu and Liu 2000) and with FEM (Belytschko and Organ 1995; Hegen 1996; Liu and Gu 2000). An adaptive stress analysis package based on the mesh-free technology has been tackled (Liu and Tu 2002), too.

In 2007, Rajendran and Zhang proposed a "finite element-meshfree method (FEMM)" to synergize the advantages of both the FEM and MFM, in which the standard shape functions of isoparametric FE are employed to construct the PU, and the shape functions of MFM are used to construct the local approximations (Rajendran and Zhang 2007). It has been exercised by a number of researchers to handle the crack growth problems (Liu et al. 2018; Sun et al. 2021; Tao et al. 2021).

In 2012, EFGM combined with XFEM, named as the "extended element free Galerkin method (XEFGM)", was validated in the simulation of crack growth under thermo-mechanical actions by Bouhala, Makradi and Belouettar (2012). In their work, shape functions are constructed using MLS approximation, cracks/interfaces and crack-tips are modeled with extrinsic local enrichment. The onset of crack growth is determined by the calculated SIFs using interaction energy integral, and the crack is assumed to propagate in the direction of the maximum principal stress.

References

Andreaus U, Batra RC, Porfiri M. Vibrations of cracked Euler-Bernoulli beams using meshless local Petrov-Galerkin (MLPG) method. Comput Model Eng Sci. 2005;9(2):111–31.

Arroyo M, Ortiz M. Local maximum-entropy approximation schemes: a seamless bridge between finite elements and meshfree methods. Int J Numer Meth Eng. 2006;65(13):2167–202.

Atluri SN. The meshless Local Petrov-Galerkin (MLPG) method. Stuttgart (Germany): Tech Science Press; 2002.

Atluri SN, Cho JY, Kim HG. Analysis of thin beams using the meshless local Petrov-Galerkin method with generalized moving least squares interpolations. Comput Mech. 1999a;24(5):334–47.

Atluri SN, Kim HG, Cho JY. A critical assessment of the truly meshless local Petrov-Galerkin (MLPG) and local boundary integral equation (LBIE) methods. Comput Mech. 1999b;24(5):348–72.

Atluri SN, Shen SP. The meshless local Petrov-Galerkin (MLPG) method: a simple and less-costly alternative to the finite element and boundary element methods. CMES-Comput Model Eng Sci. 2002;3(1):11–51.

Atluri SN, Zhu TL. A new meshless local Petrov-Galerkin (MLPG) approach in computational mechanics. Comput Mech. 1998;22(2):117–27.

Atluri SN, Zhu TL. The meshless local Petrov-Galerkin (MLPG) approach for solving problems in elasto-statics. Comput Mech. 2000;25(2–3):169–79.

Attaway SW, Heinstein MW, Swegle JW. Coupling of smooth particle hydrodynamics with the finite element method. Nucl Eng Des. 1994;150(2–3):199–205.

Babuška I, Melenk JM. The partition of unity method. Int J Numer Meth Eng. 1997;40(4):727–58.

Babuška I, Zhang Z. The partition of unity method for the elastically supported beam. Comput Meth Appl Mech Eng. 1998;152(1–2):1–18.

Beissel S, Belytschko T. Nodal integration of the element-free Galerkin method. Comput Meth Appl Mech Eng. 1996;139(1–4):49–74.

Belytschko T, Black T. Elastic crack growth in finite elements with minimal remeshing. Int J Numer Meth Eng. 1999;45(5):601–20.

Belytschko T, Fleming M. Smoothing, enrichment and contact in the element-free Galerkin method. Comput Struct. 1999;71(2):173–95.

Belytschko T, Gu L, Lu YY. Fracture and crack growth by element-free Galerkin methods. Model Simul Mater Sci Eng. 1994a;2(3A):519–34.

Belytschko T, Guo T, Liu WK, Xiao SP. A unified stability analysis of meshless particle methods. Int J Numer Meth Eng. 2000a;48(9):1359–400.

Belytschko T, Krongauz Y, Fleming M, Organ D, Liu WKS. Smoothing and accelerated computations in the element free Galerkin method. J Comput Appl Math. 1996a;74(1–2):111–26.

Belytschko T, Krongauz Y, Organ D, Fleming M, Krysl P. Meshless methods: an overview and recent developments. Comp Meth Appl Mech Eng. 1996b;139(1–4):3–47.

Belytschko T, Lu YY. Element-free Galerkin methods for static and dynamic fracture. Int J Solids Struct. 1995;32(17–18):2547–70.

Belytschko T, Lu YY, Gu L. Element-free Galerkin methods. Int J Numer Meth Eng. 1994b;37(2):229–56.

Belytschko T, Lu YY, Gu L. Crack propagation by element-free Galerkin methods. Eng Fract Mech. 1995a;51(2):295–315.

Belytschko T, Organ D. Coupled finite element-element-free Galerkin method. Comput Mech. 1995;17:186–95.

Belytschko T, Organ D, Gerlach C. Element-free galerkin methods for dynamic fracture in concrete. Comp Meth Appl Mech Eng. 2000b;187(3–4):385–99.

Belytschko T, Organ D, Krongauz Y. A coupled finite element–element-free Galerkin method. Comput Mech. 1995b;17(3):186–95.

Belytschko T, Tabbara M. Dynamic fracture using element-free Galerkin methods. Int J Numer Meth Eng. 1996;39(6):923–38.

Bentley JL, Friedman JH. Data structures for range searching. ACM Comput Surv (CSUR). 1979;11(4):397–409.

Bouhala L, Makradi A, Belouettar S. Thermal and thermo-mechanical influence on crack propagation using an extended mesh free method. Eng Fract Mech. 2012;88:35–48.

Cao XH, Chen SF, Chen SH. Generation of tetrahedral meshes in 3-D domains by advancing front method. J Wuhan Univ Hydr Elec Eng (WUHEE). 1998;31(1):16–20 (in Chinese with English abstract).

Carpinteri A, Ferro G, Ventura G. The partition of unity quadrature in meshless methods. Int J Numer Meth Eng. 2002;54(7):987–1006.

Carpinteri A, Ferro G, Ventura G. The partition of unity quadrature in element-free crack modeling. Comput Struct. 2003;81(18–19):1783–94.

Chen JS, Pan C, Rogue CMOL, Wang HP. A Lagrangian reproducing kernel particle method for metal forming analysis. Comput Mech. 1998a;22(3):289–307.

Chen JS, Pan C, Wu CT, Liu WK. Reproducing kernel particle methods for large deformation analysis of nonlinear structures. Comput Meth Appl Mech Eng. 1996a;139(1–4):195–227.

Chen JS, Wu CT, Yoon S, You T. A stabilized conforming nodal integration for Galerkin mesh-free methods. Int J Numer Meth Eng. 2001;50(2):435–66.
Chen JS, Wang HP. New boundary condition treatments in meshfree computation of contact problems. Comp Meth Appl Mech Eng. 2000;187(3–4):441–68.
Chen SF, Chen SH, Cao XH. Automatic generation of unstructured hexahedron mesh for 3D complicated domain. Rock Soil Mech. 1998b;19(4):46–51 (in Chinese with English abstract).
Chen SH. Adaptive FEM analysis for two-dimensional unconfined seepage problems. J Hydrodyn. 1996;19(1):60–6.
Chen SH. Computational geomechanics and hydraulic structures. Berlin (Germany): Springer; 2018.
Chen SH, Wang JS, Zhang JL. Adaptive elasto-viscoplastic FEM analysis for hydraulic structures. J of Hydraulic Eng. 1996b;27(2):68–75 (in Chinese with English abstract).
Chung HJ, Belytschko T. An error estimate in the EFG method. Comput Mech. 1998;21(2):91–100.
Clough RW, Penzien J. Dynamics of structures. New York (USA): McGraw-Hill; 1975.
Cordes LW, Moran B. Treatment of material discontinuity in the element-free Galerkin method. Comp Meth Appl Mech Eng. 1996;139(1–4):75–89.
Daxini SD, Prajapati JM. A review on recent contribution of meshfree methods to structure and fracture mechanics applications. Sci World J. 2014;Article ID 247172.
Dolbow J, Belytschko T. Numerical integration of the Galerkin weak form in meshfree methods. Comput Mech. 1999;23(3):219–30.
Duarte CA, Oden JT. Hp clouds—a meshless method to solve boundary-value problem. Technical Report. TICAM, the University of Texas at Austin (USA); 1995.
Duarte CA, Oden JT. H-p clouds—an h–p meshless method. Numer Methods Partial Differ Equ. 1996;12(6):673–705.
Duflot M, Nguyen-Dang H. Fatigue crack growth analysis by an enriched meshless method. J Comput Appl Math. 2004a;168(1–2):155–64.
Duflot M, Nguyen-Dang H. A meshless method with enriched weight functions for fatigue crack growth. Int J Numer Meth Eng. 2004b;59(14):1945–61.
Fernández-Méndez S, Huerta A. Imposing essential boundary conditions in mesh-free methods. Comput Meth Appl Mech Eng. 2004;193(12–14):1257–75.
Fernández-Méndez S, Huerta A. Coupling finite elements and particles for adaptivity: an application to consistently stabilized convection–diffusion. In: Griebel M, Schweitzer MA, editors. Meshfree methods for partial differential equations, Lecture Notes in Computational Science and Engineering (vol. 26). Berlin (Germany): Springer; 2002. p. 117–129.
Fleming M, Chu YA, Moran B, Belytschko T. Enriched element-free Galerkin methods for crack tip fields. Int J Numer Meth Eng. 1997;40(8):1483–504.
Fleming MA. The element-free Galerkin method for fatigue and quasi-static fracture. PhD Thesis. Northwestern University (USA); 1997.
Fries TP, Matthies HG. Classification and overview of meshfree methods. Technical Report. Department of Mathematics and Computer Science, Technical University of Braunschweig (Germany); 2003.
Gavete L, Falcón S, Ruiz A. An error indicator for the element-free Galerkin method. Eur J Mech A/Solids. 2001;20(2):327–41.
Gavete L, Cuesta JL, Ruiz A. A procedure for approximation of the error in the EFG method. Int J Numer Meth Eng. 2002;53(3):677–90.
Gavete L, Gavete ML, Alonso B, Martin AJ. A posteriori error approximation in EFG method. Int J Numer Meth Eng. 2003;58(15):2239–63.
Gingold RA, Monaghan JJ. Smoothed particle hydrodynamics: theory and application to non-spherical stars. Mon Not R Astron Soc. 1977;181(3):375–89.
Gu YT, Liu GR. A coupled element free Galerkin/boundary element method for stress analysis of two-dimension solids. Comput Methods Appl Mech Eng. 2000;190:4405–19.
Gu YT, Liu GR. A local point interpolation method for static and dynamic analysis of thin beams. Comp Meth App Mech Eng. 2001;190(42):5515–28.

Gu YT, Liu GR. A meshfree weak-strong (MWS) form method for time dependent problems. Comput Mech. 2005;35(2):134–45.

Häussler-Combe U, Korn C. An adaptive approach with the element-free Galerkin method. Comput Meth Appl Mech Eng. 1998;162(1–4):203–22.

Hegen D. Element-free Galerkin methods in combination with finite element approaches. Comput Methods Appl Mech Eng. 1996;135:143–66.

Huerta A, Belytschko T, Fernández-Méndez S, Rabczuk T. Meshfree methods. In: Stein E, De Borst R, Hughes TJR, editors. Encyclopedia of computational mechanics. New York (USA): Wiley; 2004.

Idelsohn SR, Oñate E, Calvo N, Del Pin F. The meshless finite element method. Int J Numer Methods Eng. 2003;58(6):893–912.

Idelsohn SR, Oñate E, Del Pin F. The particle finite element method: a powerful tool to solve incompressible flows with free surfaces and breaking waves. Int J Numer Meth Eng. 2004;61(7):964–89.

Khosravifard A, Hematiyan MR. A new method for meshless integration in 2D and 3D Galerkin meshfree methods. Eng Anal Bound Elem. 2010;34(1):30–40.

Kim HG, Atluri SN. Arbitrary placement of secondary nodes, and error control, in the meshless local Petrov-Galerkin (MLPG) method. Comput Model Eng Sci. 2000;1(3):11–32.

Krongauz Y, Belytschko T. Enforcement of essential boundary conditions in meshless approximations using finite elements. Comp Meth Appl Mech Eng. 1996;131(1–2):133–45.

Krongauz Y, Belytschko T. EFG approximation with discontinuous derivatives. Int J Numer Meth Eng. 1998;41(7):1215–33.

Krysl P, Belytschko T. Element-free Galerkin method: convergence of the continuous and discontinuous shape functions. Comput Meth Appl Mech Eng. 1997;148(3–4):257–77.

Lancaster P, Salkauskas K. Surfaces generated by moving least squares methods. Math Comput. 1981;37(155):141–58.

Lee NS, Bathe KJ. Effects of element distortions on the performance of isoparametric elements. Int J Numer Meth Eng. 1993;36(20):3553–76.

Lee GH, Chung HJ, Choi CK. Adaptive crack propagation analysis with the element-free Galerkin method. Int J Numer Meth Eng. 2003;56(3):331–50.

Lee CK, Zhou CE. On error estimation and adaptive refinement for element free Galerkin method. Part I. Stress recovery and a posteriori error estimation. Comput Struct. 2004a;82(4–5):413–28.

Lee CK, Zhou CE. On error estimation and adaptive refinement for element free Galerkin method. Part II. Adaptive refinement. Comput Struct. 2004b;82(4–5):429–43

Li SF, Hao W, Liu WK. Mesh-free simulations of shear banding in large deformation. Int J Solids Struct. 2000;37(48–50):7185–206.

Li WD, Chen SH. Numerical modeling for frictional contact problems. Rock Soil Mech. 2003a;24(3):385–8 (in Chinese with English abstract).

Li WD, Chen SH. Vibration analysis by an effective meshless method. Chin J Comput Mech. 2003b;20(6):756–63 (in Chinese with English abstract).

Li SF, Liu WK. Meshless and particle methods and their applications. Appl Mech Rev. 2002;55(1):1–34.

Lin H, Atluri SN. The meshless local Petrov-Galerkin (MLPG) method for solving incompressible Navier-Stokes equations. CMES-Comput Model Eng Sci. 2001;2(2):117–42.

Liszka TJ, Duarte CAM, Tworzydlo WW. Hp-meshless cloud method. Comput Meth Appl Mech Eng. 1996;139(1–4):263–88.

Liu GR. Meshfree Methods: moving beyond the finite element. 2nd ed. Boca Raton (USA): CRC Press; 2003.

Liu GR. A G space theory and a weakened weak (W2) form for a unified formulation of compatible and incompatible methods: part I, theory. Int J Numer Meth Eng. 2010;81(9):1093–126.

Liu GR, Gu YT. Coupling element free Galerkin and hybrid boundary element methods using modified variational formulation. Comput Mech. 2000;26:166–73.

Liu GR, Gu YT. A point interpolation method for two-dimensional solids. Int J Numer Methods Eng. 2001;50:937–95.
Liu GR, Gu YT. A meshfree method: meshfree weak-strong (MWS) form method, for 2-D solids. Comput Mech. 2003;33(1):2–14.
Liu GR, Liu MB. Smoothed particle hydrodynamics: a meshfree particle method. New Jersey (USA): World Scientific Publishing Co., Pte. Ltd.; 2003.
Liu GR, Tu ZH. An adaptive procedure based on background cells for meshless methods. Comp Meth Appl Mech Eng. 2002;191(17–18):1923–43.
Liu GR, Wu YL, Ding H. Meshfree weak-strong (MWS) form method and its application to incompressible flow problems. Int J Numer Meth Fluids. 2004a;46(10):1025–47.
Liu L, Liu GR, Tan VBC. Element free method for static and free vibration analysis of spatial thin shell structures. Comp Meth Appl Mech Eng. 2002;191(51–52):5923–42.
Liu QS, Sun L, Tang CH, Chen L. Simulate intersecting 3D hydraulic cracks using a hybrid "FE-Meshfree" method. Eng Anal Bound Elem. 2018;91:24–43.
Liu WK, Han WM, Lu HS, Li SF, Cao J. Reproducing kernel element method. Part I. Theoretical formulation. Comput Meth Appl Mech Eng. 2004b;193(12–14):933–51.
Liu WK, Chen Y. Wavelet and multiple scale reproducing kernel method. Int J Numer Meth Fluid. 1995;21(10):901–31.
Liu WK, Jun S, Li SF, Adee J, Belytschko T. Reproducing kernel particle methods for structural dynamics. Int J Numer Meth Eng. 1995a;38(10):1655–79.
Liu WK, Jun S, Zhang YF. Reproducing kernel particle methods. Int J Numer Meth Eng. 1995b;20(8–9):1081–106.
Liu WK, Uras RA, Chen Y. Enrichment of the finite element method with reproducing kernel particle method. ASME J Appl Mech. 1997;64(4):861–70.
Löhner R, Sacco C, Oñate E, Idelsohn S. A finite point method for compressible flow. Int J Numer Meth Eng. 2002;53(8):1765–79.
Lu YY, Belytschko T, Gu L. A new implementation of the element free Galerkin method. Comp Meth Appl Mech Eng. 1994;113:397–414.
Lucy LB. A numerical approach to the testing of the fission hypothesis. Astron J. 1977;82:1013–24.
Melenk JM, Babuška I. The partition of unity finite element method: basic theory and applications. Comput Meth Appl Mech Eng. 1996;139(1–4):289–314.
Mukherjee YX, Mukherjee S. The boundary node method for potential problems. Int J Numer Meth Eng. 1997;40(5):797–815.
Nguyen VP, Rabczuk T, Bordas SPA, Duflot M. Meshless methods: a review and computer implementation aspects. Math Comput Simul. 2008;79(3):763–813.
Nayroles B, Touzot G, Viilon P. Generalizing the finite element method: diffuse approximation and diffuse elements. Comput Mech. 1992;10(5):307–18.
Oñate E, Idelsohn S, Zienkiewicz OC, Taylor RL. A finite point method in computational mechanics: applications to convective transport and fluid flow. Int J Numer Meth Eng. 1996;39(22):3839–66.
Ouatouati AE, Johnson D. A new approach for numerical modal analysis using the element-flee method. Int J Numer Meth Eng. 1999;46(1):1–27.
Perić D, Hochard CH, Dutko MD, Owen DRJ. Transfer operators for evolving meshes in small strain elasto-plasticity. Comp Meth Appl Mech Eng. 1996;137(3–4):331–44.
Rabczuk T, Areias PMA. A meshfree thin shell for arbitrary evolving cracks based on an external enrichment. Comput Model Eng Sci. 2006;16(2):115–30.
Rabczuk T, Areias PMA, Belytschko T. A simplified meshfree methods for shear bands with cohesive surfaces. Int J Numer Meth Eng. 2007;69(5):993–1021.
Rabczuk T, Belytschko T. A three dimensional large deformation meshfree method for arbitrary evolving cracks. Comput Meth Appl Mech Eng. 2007;196(29–30):2777–99.
Rabczuk T, Xiao SP, Sauer M. Coupling of meshfree methods with finite elements: basic concepts and test results. Commun Numer Meth Eng. 2006;22(10):1031–65.
Rabczuk T, Zi G. A meshfree method based on the local partition of unity for cohesive cracks. Comput Mech. 2007;39(6):743–60.

Rajendran S, Zhang BR. A FE-meshfree QUAD4 element based on partition of unity. Comput Meth Appl Mech Eng. 2007;197(1–4):128–47.
Ren J, Liew KM. Mesh-free method revisited: two new approaches for the treatment of essential boundary conditions. Int J Comput Eng Sci. 2002;3(2):219–33.
Shepard D. A two-dimensional function for irregularly spaced points. In: Proceedings of the 23rd ACM national conference. Princeton (USA): Brandon/Systems Press Inc.; 1968. p. 517–24.
Simo JS, Laursen TA. An augmented Lagrangian treatment of contact problems involving friction. Comput Struct. 1992;42(1):97–116.
Sun L, Tao SJ, Tang XH, Liu QS. Simulation of the nonplanar three-dimensional thermal cracking using the finite element-meshfree method. Appl Math Model. 2021;99:106–28.
Tao SJ, Tang XH, Rutqvist J, Liu QS, Hu MS. The influence of stress anisotropy and stress shadow on frost cracking in rock. Comput Geotech. 2021;133(4): 103967.
Terry TG. Fatigue crack propagation modeling using the element free Galerkin method. MSc Thesis. Northwestern University (USA); 1994.
Tsay RJ, Chiou YJ, Chuang WL. Crack growth prediction by manifold method. J Eng Mech. 1999;125(8):884–90.
Wang D, Chen JS. A locking-free meshfree curved beam formulation with the stabilized conforming nodal integration. Comput Mech. 2006;39(1):83–90.
Wang JD, Liu GR. A point interpolation meshless method based on radial basis functions. Int J Numer Meth Eng. 2002;54(11):1623–1548.
Wang JG, Liu GR, Wu YG. A point interpolation method for simulating dissipation process of consolidation. Comput Meth Appl Mech Eng. 2001;190:5907–22.
Wang S, Liu H. Modeling brittle-ductile failure transition with meshfree method. Int J Impact Eng. 2010;37(7):783–91.
Williams ML. On the stress distribution at the base of a stationary crack. J Appl Mech. 1957;24(1):109–14.
Yoon S, Wu CT, Wang HP, Chen JS. Efficient meshfree formulation for metal forming simulations. J Eng Mater Techn. 2001;123(4):462–7.
Zhang ZQ, Zhou JX, Wang XM, Zhang YF, Zhang L. Investigations on reproducing kernel particle method enriched by partition of unity and visibility criterion. Comput Mech. 2004;34(4):310–29.
Zhu JZ, Zienkiewicz OC, Hinton E, Wu J. A new approach to the development of automatic quadrilateral mesh generation. Int J Numer Meth Eng. 1991;32(4):849–66.
Zhuang X, Heaney C, Augarde C. On error control in the element-free Galerkin method. Eng Anal Bound Elem. 2012;36(3):351–60.
Zi G, Rabczuk T, Wall WA. Extended meshfree methods without branch enrichment for cohesive cracks. Comput Mech. 2007;40(2):367–82.
Zienkiewicz OC, Zhu JZ. A simple error estimator and adaptive procedure for practical engineering analysis. Int J Numer Meth Eng. 1987;24(2):337–57.

Chapter 10
Mesh-Free Methods with Special Focus on SPH

Abstract This chapter presents the principle and basic algorithm of the "smoothed particle method (SPH)" for fluid that belongs to the family of "mesh-free methods (MFMs)" or "mesh-free particle methods (MPMs)". As in EFGM, the trial functions are no longer created by the basis belong to standard PU, too. Instead, the fluid domain is represented by a set of arbitrarily distributed nodes without regular connectivity. Termed as "kernel approximation", at the representative node, the field function value in the governing PDEs (strong form) is replaced by the integrated function value with the help of its neighborhood field nodes in the "support domain" similar to that in EFGM. Then the representative node is further approximated by the representative particle assigned with mass, which is termed as the "particle approximation" meaning that the particle is actually the live physical objects. These two steps produce, instead of original PDEs, a set of ODEs that may be solved by routine finite difference schemes w.r.t. time-marching steps to obtain the time history of all the field variables for all the particles. The use of Lagrangian description allows SPH to be adaptive for particle distribution that evolves with the on-going of time. Consequently, it possesses high ability to handle violent "fluid–structure interaction (FSI)" problems with extremely free surface distortion.

10.1 General

In continuum mechanics, it is well known that obtaining closed-form solutions for a set of governing equations (PDEs or IEs) derived from conservation laws is usually impossible, except for very few simple cases. Therefore, efforts have to be made for numerical solutions by CMs. In doing so, one first needs to discretize the problem domain by grid where the governing equations are defined. Next, a scheme is selected to provide an approximation for the values and derivatives of field functions (variables) at grid nodes. Then the approximation is introduced in the governing equations to produce a set of ODEs comprising nodal values of field functions. Afterwards this set of discretized ODEs is solved using one of standard FDM routines with respect only to time.

S. Chen, *Advanced Computational Methods and Geomechanics*, Springer Tracts in Civil Engineering, https://doi.org/10.1007/978-981-19-7427-4_10

Problems in "computational fluid dynamics (CFD)" are customarily solved by employing the "mesh-based methods (MBMs) such as FDM, FVM and FEM. These fundamental CMs have been dominant for more than half century. An important characteristic of these fundamental CMs is that a correspondent Eulerian (for FDM and FVM) grid or Lagrangian (for FEM) mesh should be constructed as the computational framework to provide spatial discretization for PDEs or IEs. Consequently, they encounter difficulties in handling complex phenomena such as distorting free surfaces, moving material interfaces and boundaries. Motivated to find alternatives, a family of "mesh-free methods (MFMs)" or "mesh-free particle methods (MPMs)" was given birth nearly half century ago, in which the method of "smooth particle hydrodynamics (SPH)" is one of outstanding family member.

MFMs for analyzing solids and fluids share some common features, but are different in the means of function approximation and implementation process. In Chap. 9, the weak form (Galerkin principle) is solved by the construction and application of shape (basis) functions using the MLS technique and the properly selected weight function. Whereas in this chapter, the MFMs of particle type, alternatively termed as "mesh-free particle methods (MPMs)", in general employ a finite set of discrete particles to directly solve the state and motion of continuum governed by PDEs (strong form). Each particle can either be associated with one naturally generated object (e.g. rock block or granule), or artificially discretized sub-domain. These particles can possess different sizes (nano-level, micro-level, meso-level, macro-level, and even astronomical-level). Although there is no essential difficulty to formulate an algorithm without using a mesh, yet the key problem is how to ensure the stability of the numerical solution, especially when the irregularly-distributed particles in compactly supported domains are used for the problems with Neumann (derivative) boundary conditions. This is mainly blamed for the fact that without the overall restraint from weak form equation, the qualitative information of the underlying problem domain cannot be incorporated in a natural way.

SPH method, as the first member of MPMs, was originally invented for modeling astrophysical phenomena (Lucy 1977; Gingold and Monaghan 1977) because the collective movement of particle assembly is similar to the movement of a liquid or gas governed by classical Newtonian hydrodynamics. Afterwards, SPH has been extensively extended towards the study of the dynamic response with material failure as well as with fluid surface breakage. Today, SPH and its different variants have been incorporated into many commercial or open source codes (Gómez-Gesteira et al. 2012a, 2012b). Important reviews on SPH are (only limitedly cited) those by Benz (1989), Monaghan (1992), Randles and Libersky (1996), Liu and Liu (2003, 2010), Liu, Liu and Zong (2008), Basa and Lastiwka (2009), Price (2011), Bui and Nguyen (2021).

Following strategies are employed in the formulation of SPH:

① The problem domain is represented by a set of arbitrarily distributed particles. No regular connectivity for these particles is demanded (in contrast to the particle assemblage in DEMs), but problems with large deformation will require a huge number (could be millions) of particles to represent the whole domain.

② The integral function representation is used for field function approximation. This is termed as the "kernel approximation" that mathematically provides certain numerical stability attributable to its piecewise smoothing effect, subject to sufficient particles in the support domain to ensure the numerical quadrature being accurately performed.

③ The kernel approximation is further approximated using "live" particles assigned with mass, velocity and acceleration, meaning that they are identical to material particles. This operation is termed as "particle approximation" that replaces the integral w.r.t. field function and its derivatives with the summation over all the corresponding values at the neighboring particles within support domain. This produces discretized ODE system w.r.t. time.

④ The particle approximation is repeatedly performed at every time-marching step. The distribution of particles for next time-marching step depends on the current distribution of particles. Such a Lagrangian description characteristic allows SPH to possess high ability for handling problems with extremely large deformation and free surface distortion.

⑤ It is emphasized that one should figure out a proper way to determine the time-marching step sequence to ensure a stable ODE integration scheme.

10.2 Approximation in SPH

The spatial approximation in SPH is divided into two steps (Liu and Liu 2003; Violeau 2012): the first step is the integral representation or the so-called "kernel approximation" of field functions and their derivatives; the second step is the correspondent "particle approximation" with which the integral representation is approximated by the summation of the field values tied to the nearest neighboring particles.

10.2.1 Approximation of Field Functions

1. Kernel approximation

The concept of integral representation of function $f(x)$ starts from the identity

$$f(x_i) = \int_{-\infty}^{\infty} f(x)\delta(x_i - x)\mathrm{d}x \tag{10.1}$$

in which $f(x)$ is a function of position x, i is the specific position of the representative particle, and the kernel $\delta(x_i - x)$ is the Dirac Delta function (viz. Figure 10.1d) defined as

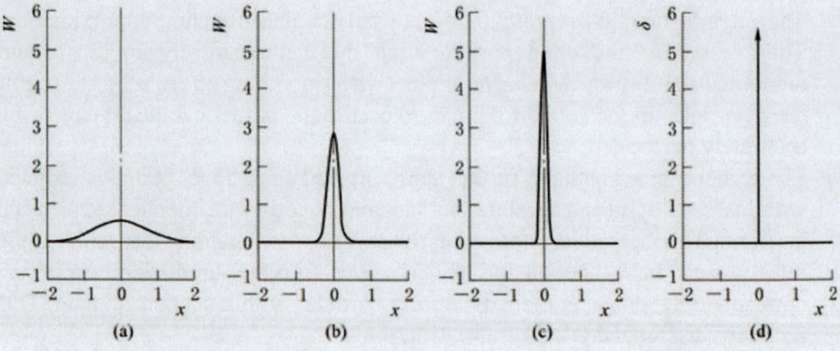

Fig. 10.1 Smoothing function $W(x_i - x, h)$ approach of Dirac Delta function $\delta(x_i - x)$ where $x_i = 0$. **a** Approach using smoothing function (flat); **b** approach using smoothing function (steep); **c** approach using smoothing function (sharp); **d** approach using exact Dirac Delta function

$$\begin{cases} \int\limits_{-\infty}^{\infty} \delta(x_i - x)\mathrm{d}x = 1 \\ \delta(x_i - x) = 0 \quad \text{for } x_i \neq x \end{cases} \tag{10.2}$$

Equation (10.1) implies that a function $f(x)$ can be rigorously represented by an integral form, as long as it is continuously defined in Ω. If $\delta(x_i - x)$ is replaced by a smoothing function $W(x_i - x, h)$ (Fig. 10.1a–c), the integral representation (10.1) may be approximated by

$$f(x_i) \approx \int\limits_{-\alpha_s h}^{\alpha_s h} f(x)W(x_i - x, h)\mathrm{d}x \tag{10.3a}$$

where: h—smooth length; α_s—dimensionless scaling factor; W—smoothing kernel function.

For 2-D problem (viz. Figure 10.2), it may be directly extended such that

$$f(\mathbf{x}_i) \approx \iint\limits_{\Omega_i} f(\mathbf{x})W(\mathbf{x}_i - \mathbf{x}, h)\mathrm{d}\Omega \tag{10.3b}$$

where: Ω_i—support domain defined by smooth length and centered at $\mathbf{x}_i$.

Equation (10.3) is a decisive thought leading to the creation of MPMs from PDEs directly, and the "smoothing kernel function" abbreviated as "smooth function" is actually a special kind of the weight function in EFGM (viz. Chap. 9). Different types of smoothing function have been used in a number of literatures (Liu and Liu 2003). The conditions or properties for the smoothing functions are summarized as follows:

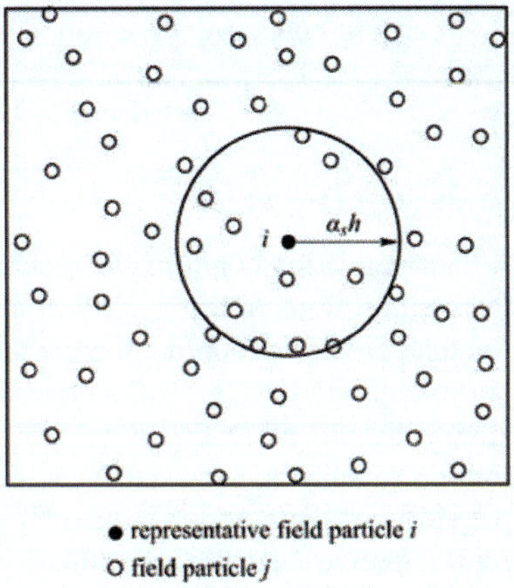

Fig. 10.2 Diagram to show the kernel approximation of 2-D domain Ω_i

① It must be normalized over the support domain of $\mathbf{x}_i$

$$\iint_{\Omega_i} W(\mathbf{x}_i - \mathbf{x}, h)\mathrm{d}\Omega = 1 \text{ (unity)} \tag{10.4}$$

② It should be compactly supported

$$W(\mathbf{x}_i - \mathbf{x}, h) = 0 \quad \text{for} \quad |\mathbf{x}_i - \mathbf{x}| > \alpha_s h \text{ (compact support)} \tag{10.5}$$

The dimension of the compact support domain is defined by the smoothing length h and dimensionless scaling factor α_s(viz. Figure 10.2). Normally, $\alpha_s = 2 \sim 3$.

③ It should be positive for any field point $\mathbf{x}$ within the support domain of particle $\mathbf{x}_i$

$$W(\mathbf{x}_i - \mathbf{x}, h) \geq 0 \quad \text{for} \quad |\mathbf{x}_i - \mathbf{x}| \leq \alpha_s h \text{ (positivity)} \tag{10.6}$$

④ It should be monotonically decaying with the increase of the distance from particle $\mathbf{x}_i$

$$|\mathbf{x}_i - \mathbf{x}_2| > |\mathbf{x}_i - \mathbf{x}_1| \rightarrow W(\mathbf{x}_i - \mathbf{x}_1, h) > W(\mathbf{x}_i - \mathbf{x}_2, h) \text{ (decay)} \tag{10.7}$$

⑤ It should approach Dirac Delta function as the smoothing length approaches to zero

$$\lim_{h \to 0} W(\mathbf{x}_i - \mathbf{x}, h) = \delta(\mathbf{x}_i - \mathbf{x}) \text{ (Dirac Delta)} \tag{10.8}$$

⑥ It should be an even function

$$W(\mathbf{x}_i - \mathbf{x}, h) = W(\mathbf{x} - \mathbf{x}_i, h) \text{ (symmetry)} \tag{10.9}$$

⑦ It should be sufficiently smooth (smoothness).

Normalization condition ① ensures that the integral of smoothing function over the support domain to be unity. It also ensures the zero-order consistency (C^0) of the integral representation of a continuum function. Condition ② transforms a SPH approximation from global operation to a local operation. This will lead to a set of sparse discretized ODE system w.r.t. time. Condition ③ states that the smoothing function should be non-negative in the support domain. Condition ④ is based on the physical consideration that a nearer particle should exert a bigger influence on the representative particle. Condition ⑤ guarantees that as the smoothing length tends to be zero, the approximation value approaches the exact one. Condition ⑥ means that particles from the same distance but different positions should exert equal action on the representative particle. Condition ⑦ aims to obtain better approximations of a function and its derivatives.

2. Particle approximation

In the SPH method, the entire domain is represented by n_p particles that carry individual mass and occupy own space. By the particle approximation. The continuous kernel approximation Eq. (10.3) is converted to the discretized forms of the summation over all the particles $j \in n_i$ in the support domain Ω_i of representative particle $i(1, 2, \ldots, n_p)$.

If the infinitesimal volume $d\mathbf{x}_j$ at the location of j is replaced by a particle with finite volume ΔV_j that is related to the particle mass by

$$m_j = \Delta V_j \rho_j \quad \text{for } j \in \Omega_i \tag{10.10a}$$

or

$$\Delta V_j = \frac{m_j}{\rho_j} \quad \text{for } j \in \Omega_i \tag{10.10b}$$

where:ρ_j—density of particle j; Ω_i—support domain of particle i containing n_i neighboring particles.

Use is made of Eq. (10.10), the continuous SPH integral representation for $f(\mathbf{x}_i)$ in Eq. (10.3) can be written in the form of discretized particle approximation as follows

$$f(\mathbf{x}_i) = \iint\limits_{\Omega_i} f(\mathbf{x})W(\mathbf{x}_i - \mathbf{x}, h)\mathrm{d}\Omega \approx \sum_{j\in\Omega_i}^{n_i} f(\mathbf{x}_j)W(\mathbf{x}_i - \mathbf{x}_j, h)\Delta V_j$$

$$= \sum_{j\in\Omega_i}^{n_i} f(\mathbf{x}_j)W(\mathbf{x}_i - \mathbf{x}_j, h)\frac{m_j}{\rho_j} \quad (i = 1, 2, \ldots, n_p) \tag{10.11}$$

or briefly

$$f(\mathbf{x}_i) = \sum_{j\in\Omega_i}^{n_i} f(\mathbf{x}_j)W_{ij}\frac{m_j}{\rho_j} \quad (i = 1, 2, \ldots, n_p) \tag{10.12}$$

Suppose $f(\mathbf{x}) = \rho$, Eq. (10.12) gives rise to

$$\rho_i = \sum_{j\in\Omega_i}^{n_i} m_j W_{ij} \quad (i = 1, 2, \ldots, n_p) \tag{10.13}$$

Equation (10.13) means that in SPH, the density of particle i is the weighted average of all its neighboring particles in the support domain. It is generally referred to as the "summation density approach" that is one of the most prevalent formulas to obtain the density.

The particle approximation is, however, related to some numerical problems such as the particle inconsistency and the tensile instability (Monaghan, 2000, 2005).

10.2.2 Approximation of Function Derivatives

In SPH, another essential role of smooth operation (10.3) is to approximate the derivatives in PDEs by the basic variables tied to randomly scattered field particles, in this manner the PDEs may be transferred into a set of linear ODEs depending on time merely. Since only the first-order derivatives w.r.t. the filed functions in N–S equations are needed in this book, hence they are detailed below.

1. Kernel approximation

The approximation of the first-order partial derivative can be obtained by replacing the function $f(\mathbf{x})$ in Eq. (10.3) with its partial derivative, namely,

$$\frac{\partial f(\mathbf{x}_i)}{\partial x_\alpha} = \iint\limits_{\Omega_i} \frac{\partial f(\mathbf{x})}{\partial x_\alpha} W(\mathbf{x}_i - \mathbf{x}, h)\mathrm{d}\Omega \quad (\alpha = 1, 2) \tag{10.14}$$

in which the Greek superscripts α is employed to indicate the coordinate direction for the simplification of notations $x = x_1$ and $y = x_2$.

Integrating by parts, we get

$$\frac{\partial f(\mathbf{x}_i)}{\partial x_\alpha} = \int_{\Gamma_i} f(\mathbf{x}) W(\mathbf{x}_i - \mathbf{x}, h) \cdot \mathbf{n}_{\Gamma_i} \mathrm{d}\Gamma - \iint_{\Omega_i} f(\mathbf{x}) \frac{\partial W(\mathbf{x}_i - \mathbf{x}, h)}{\partial x_\alpha} \mathrm{d}\Omega \quad (10.15)$$

or

$$\frac{\partial f(\mathbf{x}_i)}{\partial x_{\alpha,i}} = \int_{\Gamma_i} f(\mathbf{x}) W(\mathbf{x}_i - \mathbf{x}, h) \cdot \mathbf{n}_{\Gamma_i} \mathrm{d}\Gamma + \iint_{\Omega_i} f(\mathbf{x}) \frac{\partial W(\mathbf{x}_i - \mathbf{x}, h)}{\partial x_{\alpha,i}} \mathrm{d}\Omega \quad (10.16)$$

where $\mathbf{n}_{\Gamma_i}$ is the unit vector normal to the surface Γ_i. It is reminded that by taking the partial derivate of $W(\mathbf{x}_i - \mathbf{x}, h)$ w.r.t. to particle i, the negative sign in Eq. (10.15) is removed, which gives rise to Eq. (10.16).

Since the condition ② (viz. Equation 10.5) requires the smoothing function to vanish on the surface of support domain, which naturally leads the surface integration $\int_{\Gamma_i} f(\mathbf{x}) W(\mathbf{x}_i - \mathbf{x}, h) \cdot \mathbf{n}_{\Gamma_i} \mathrm{d}\Gamma$ to vanish for any arbitrarily function $f(\mathbf{x})$. From Eq. (10.16) it follows that

$$\frac{\partial f(\mathbf{x}_i)}{\partial x_{\alpha,i}} = \iint_{\Omega_i} f(\mathbf{x}) \frac{\partial W(\mathbf{x}_i - \mathbf{x}, h)}{\partial x_{\alpha,i}} \mathrm{d}\Omega \quad (10.17)$$

The approximation of the n-order derivatives of field function can be obtained by directly replacing the function $f(\mathbf{x})$ in Eq. (10.3) with its correspondent derivative, which gives rise to

$$\frac{\partial^n f(\mathbf{x}_i)}{\partial x_\alpha^n} = \iint_{\Omega_i} \frac{\partial^n f(\mathbf{x})}{\partial x_\alpha^n} W(\mathbf{x}_i - \mathbf{x}, h) \mathrm{d}\Omega \quad (\alpha = 1, 2) \quad (10.18)$$

Following the same procedure of the first-order partial derivatives, the conditions for producing second-or higher-order partial derivatives of a field function can be routinely obtained (Liu and Liu 2003). Since they are not employed in this book, we will not give details here.

2. Particle approximation

The particle approximation of partial derivatives for field function is further derived from Eq. (10.17) as follows

$$\frac{\partial f(\mathbf{x}_i)}{\partial x_{\alpha,i}} \approx \sum_{j \in \Omega_i}^{n_i} f(\mathbf{x}_j) \frac{\partial W(\mathbf{x}_i - \mathbf{x}_j, h)}{\partial x_{\alpha,i}} \Delta V_j$$

$$= \sum_{j\in\Omega_i}^{n_i} f(\mathbf{x}_j)\frac{\partial W(\mathbf{x}_i - \mathbf{x}_j, h)}{\partial x_{\alpha,i}}\frac{m_j}{\rho_j} = \sum_{j\in\Omega_i}^{n_i} f(\mathbf{x}_j)\frac{m_j}{\rho_j}\frac{\partial W_{ij}}{\partial x_{\alpha,i}} \quad (i = 1, 2, \ldots, n_p) \tag{10.19}$$

Equation (10.19) states that the partial derivative of a field function at representative field particle i may be approximated by the field functions at all the field particles j within its support domain that are weighted by the partial derivative of smoothing function. Now we come to understand our pioneer's wisdom that the particle approximations (10.12) and (10.19) actually convert the continuous integral representations of a field function and its first derivatives to the discretized summations over an arbitrarily set of particles. This enables SPH to abandon any background mesh for numerical quadrature. It is also notable that the particle approximation will introduce the mass and density of the particle into PDEs, which can be convenient for dynamics problems in which the density is a basic field function (variable).

10.2.3 Approximation of Function Gradients

1. Kernel approximation

USE is made of divergence theorem (viz. Equation 3.148) and following the same argument with the partial derivatives of field function, the kernel approximation of gradient operator may be directly expressed as

$$\nabla_i f(\mathbf{x}_i) = \iint_{\Omega_i} f(\mathbf{x})\nabla_i W(\mathbf{x}_i - \mathbf{x}, h)\mathrm{d}\Omega \tag{10.20}$$

By this formula, the gradient operator on a function is transmitted to a gradient operator on the smoothing function. In other words, the integral representation of the function gradient allows the spatial gradient to be determined from the values of the field function and the gradient of the smoothing function W, rather than from the gradient of the field function itself. This is very similar to that in the weak form formulations (e.g. EFGM) that reduce the consistency requirement on the unknown field functions to produce more stable solutions of PDEs.

It is reminded that if the support domain overlaps with the problem domain, the smoothing function W will be truncated by the boundary and the surface integral is no longer zero. Under such circumstances, modifications should be made to remedy the boundary effects and to keep the zero surface integration in Eq. (10.16). This issue will be discussed later on.

2. Particle approximation

The particle approximation of gradient operator may be extended from Eq. (10.17) directly

$$\begin{aligned}\nabla_i f(\mathbf{x}_i) &\approx \sum_{j\in\Omega_i}^{n_i} f(\mathbf{x}_j)\nabla_i W(\mathbf{x}_i - \mathbf{x}_j, h)\Delta V_j \\ &= \sum_{j\in\Omega_i}^{n_i} f(\mathbf{x}_j)\nabla_i W(\mathbf{x}_i - \mathbf{x}_j, h)\frac{m_j}{\rho_j} = \sum_{j\in\Omega_i}^{n_i} f(\mathbf{x}_j)\frac{m_j}{\rho_j}\nabla_i W_{ij} \quad (i = 1, 2, \ldots, n_p)\end{aligned} \tag{10.21}$$

where

$$\nabla_i W_{ij} = \frac{\mathbf{x}_i - \mathbf{x}_j}{r_{ij}}\frac{\partial W_{ij}}{\partial r_{ij}} = \frac{\mathbf{x}_{ij}}{r_{ij}}\frac{\partial W_{ij}}{\partial r_{ij}} \tag{10.22}$$

in which $r_{ij} = |\mathbf{x}_i - \mathbf{x}_j|$ is the distance between particle i and particle j.

Equation (10.21) means that the gradient of field function at particle i is approximated by the field functions at all the particles j within its support domain that are weighted by the gradient of smoothing function.

Actually, the SPH for the solution of PDEs can always be derived by a number of approaches. One approach (Benz 1989) is to multiply each term in PDEs with the smoothing function firstly, then to integrate over the volume by parts. Monaghan (1992) used a more straightforward approach termed as "golden rules" that are mathematically expressed by following identities

$$\begin{cases} \nabla f(\mathbf{x}) = \dfrac{1}{\rho}[\nabla(\rho f(\mathbf{x})) - f(\mathbf{x})\nabla\rho] \\ \dfrac{\nabla f(\mathbf{x})}{\rho} = \left[\nabla\left(\dfrac{1}{\rho}f(\mathbf{x})\right) + \dfrac{1}{\rho^2}f(\mathbf{x})\nabla\rho\right] \end{cases} \tag{10.23}$$

Applying the procedure of particle approximation to each gradient term on the RHS of Eq. (10.23), the divergence of $f(\mathbf{x})$ at particle i is obtained as

$$\begin{cases} \nabla_i f(\mathbf{x}_i) = \dfrac{1}{\rho_i}\left[\displaystyle\sum_{j\in\Omega_i}^{n_i} m_j[f(\mathbf{x}_j) - f(\mathbf{x}_i)]\nabla_i W_{ij}\right] \\ \nabla_i f(\mathbf{x}_i) = \rho_i\left[\displaystyle\sum_{j\in\Omega_i}^{n_i} m_j\left[\dfrac{f(\mathbf{x}_j)}{\rho_j^2} + \dfrac{f(\mathbf{x}_i)}{\rho_i^2}\right]\nabla_i W_{ij}\right] \end{cases} \tag{10.24}$$

One attractive characteristic of above identities is that the field function $f(\mathbf{x})$ appears in the form of paired particles. It is reminded that Eqs. (10.23) and (10.24) are valid for the first-order partial derivatives of field function, too.

10.3 Construction of Smoothing Functions

Any numerical approximation should represent the corresponding field function as closely as possible. One of the central issues for MPMs is how to effectively perform function approximation tied to a set of nodes (particles) arbitrarily scattered without using a pre-defined mesh or grid (Liu and Liu 2003).

10.3.1 Basic Requirements

In the fundamental FDM, the concept of consistency defines how well the discretized system equation set approximates PDEs. Taking the interpolation scheme for example, consistency means its ability to exactly replicate PDEs as the number of grid nodes approaches infinity and the grid size approaches zero. Namely, consistency is a prerequisite for convergence. This is attributable to Lax-Richtmyer equivalence theorem, which states that a consistent FDM scheme for well-posed PDEs is convergent if and only if it is stable.

In the fundamental FEM, consistency is closely related to the concept of completeness of its basis (shape) functions. For a FEM approximation to converge, it must approach the exact solution when the nodal distance approaches zero. It is well-known that the degree of consistency is often characterized by the order of polynomial that can be exactly replicated by the approximation. For example, if an approximation can reproduce a constant exactly, then it is said to have zero-order or C^0 consistency. In general, if an approximation can reproduce a polynomial of up to k- order exactly, it is said to have k-order or C^k consistency. The minimum consistency requirement depends on the order of PDE. For a PDE with $2k$-order, C^k consistency is demanded for Galerkin formulation. Belytschko and his co-workers (Belytschko et al. 1996, 1998) provided a detailed discussion on the consistency and completeness of CMs.

The consistency concept for the fundamental FEM may be similarly exercised in SPH, namely, for an integral representation to exactly reproduce a field function, the smoothing function should satisfy some conditions represented by the polynomial reproducibility.

1. Consistency of the kernel approximation

For a constant field function $f(\mathbf{x}) = c$ to be exactly reproduced by the SPH kernel approximation, we demand

$$f(\mathbf{x}_i) = \iint\limits_{\Omega_i} cW(\mathbf{x}_i - \mathbf{x}, h)\mathrm{d}\Omega = c \tag{10.25a}$$

or

$$\iint_{\Omega_i} W(\mathbf{x}_i - \mathbf{x}, h)\mathrm{d}\Omega = 1 \tag{10.25b}$$

It is clear that the unity condition ① in Eq. (10.4) is actually the condition for the kernel approximation to have the zero-order consistency.

For a linear function to be reproduced, we demand

$$f(\mathbf{x}_i) = \iint_{\Omega_i} (c_0 + c_1\mathbf{x}) W(\mathbf{x}_i - \mathbf{x}, h)\mathrm{d}\Omega = c_0 + c_1\mathbf{x}_i \tag{10.26}$$

Multiplying $\mathbf{x}_i$ to the both sides of Eq. (10.25), we get the identity

$$\iint_{\Omega_i} \mathbf{x}_i W(\mathbf{x}_i - \mathbf{x}, h)\mathrm{d}\Omega = \mathbf{x}_i \tag{10.27}$$

On the subtracting from Eq. (10.26), it follows that

$$\iint_{\Omega_i} (\mathbf{x} - \mathbf{x}_i) W(\mathbf{x}_i - \mathbf{x}, h)\mathrm{d}\Omega = 0 \tag{10.28}$$

This provides an insight into the shape of smoothing function: it should be symmetric so that the first moment of $W(\mathbf{x}_i - \mathbf{x}, h)$ vanishes. This actually has been defined as the condition ⑥.

2. Consistency of the particle approximation

It is notable that the kernel approximation does not ensure the consistency for the discrete form of particle approximation. The discrete counterparts of the constant and linear consistency conditions (10.25) and (10.28) are specified as

$$\sum_{j\in\Omega_i}^{n_i} W(\mathbf{x}_i - \mathbf{x}_j, h)\Delta\mathbf{x}_j = 1 \tag{10.29}$$

and

$$\sum_{j\in\Omega_i}^{n_i} \left(\mathbf{x}_i - \mathbf{x}_j\right) W(\mathbf{x}_i - \mathbf{x}_j, h)\Delta\mathbf{x}_j = 0 \tag{10.30}$$

in which n_i is the amount of particles in the support domain of the representative field particle i centered at $\mathbf{x}_i$.

These discretized consistency conditions are, however, not always satisfied, which lead to the so called "particle inconsistency" phenomenon in the discretized equation set (Morris 1996; Belytschko et al. 1996). An obvious example may be given by the particle at or near the boundary of problem domain so that its support domain is cut by the boundary.

A number of restoring processes have been proposed, but they may lead to some side-effect problems:

- The restored smoothing function might be negative in some parts of the domain, which can lead to unphysical representation of field variables (e.g. negative density or/and energy) and bring about the breakdown of computation;
- The restored smoothing function might not be monotonically decaying with the increase of particle distance;
- The restored smoothing function might not be symmetric, which may result in some serious consequences.

A famous restoring approach called "reproducing kernel particle method (RKPM)" was proposed by Liu and his co-workers (Liu and Chen 1995; Liu et al. 1995; Liu et al. 1995; You et al., 2003; Liu et al. 2004; Zhang et al., 2004). Comparing with the traditional smoothing function that is only dependent on the particle distance and applicable for all the particles, such a consistency restored smoothing function is particle-wise. It is, therefore, depends on both the distance and position of the interacting particles. The cost for this approach is additional CPU time. In addition, the particle distribution must satisfy certain conditions rather than the arbitrary distribution in the fundamental RKPM version.

The discretized inconsistency at boundaries also may be approximately overcome by proper handling the boundary condition (see later), hence in this book, only the simple forms of normal smoothing functions need to be considered.

10.3.2 Support and Influence Domains

The definition of support and influence domains for the smooth kernel function is actually identical to that of weight function in EFGM (Chap. 9). In SPH, the concept of support and influence domains for a particle is closely related to the smoothing length h of that particle, and a formula similar to Eq. (9.24) may be given as

$$r_i = \alpha_s h_i \quad (i = 1, 2, \ldots, n_p) \tag{10.31}$$

in which α_s is the dimensionless factor of the support domain, h_i is the characteristic length that relates to the particle spacing near i, n_p is the amount of particles in the problem domain. If the particles are uniformly distributed, h_i is simply the distance between two neighboring particles, i.e., identical to d_c in Eq. (9.24). The smoothing length can vary both temporarily and spatially, which normally results in non-uniform distribution. Under such circumstances, h_i can be defined as an "average" particle spacing near i.

The physical meaning of the dimensionless factor α_s is clear. For example, $\alpha_s = 2$ means a support domain whose radius is two times the average particle spacing. $\alpha_s = 2 \sim 3$ is the routine practice for most of the smoothing functions. The actual number of particles n_i tied to the representative field particle i can be simply determined by counting in the support domain.

The use of influence and support domains in the SPH method can lead to, respectively, the scatter model and gather model in approximating the field variables at particle i (Hernquist and Katz 1989). In the former, approximations on particle i are carried out on the particles whose influence domain covers particle i. In the latter, approximations on particle i are carried out on the particles within the support domain of particle i.

Since the smoothing length for two field particles may not necessarily be the same, it may happen, particularly to the scatter model, that a particle i falling within the influence domain of particle j does not influence on particle j. This is a usual case in many MPMs (inclusive SPH) that are directly derived from PDEs. They do not involve particle nature that enable us to systematically incorporate the qualitative information of the underlying problem in a natural way, and the domains are used for numerical approximation of field variables only. This unbalanced influence can lead to severe nonphysical solution (Monaghan 1992) because it obviously violates the Newton's third law. In order to overcome this problem, some kind of mean value of the smoothing length for two interacting particles is demanded (see later).

In this book, only the gather model that employs support domain concept is implemented.

10.3.3 Specification of Smoothing Functions

The smoothing function, also called as "smoothing kernel function", "smoothing kernel" or simply "kernel" in many SPH literatures, is of utmost importance because it determines the dimension and pattern and consistency of both the kernel and particle approximations. Many researchers and practitioners have tried different kinds of smoothing functions.

Normally, a smoothing function with smoother value and derivatives would yield better approximations. This is because such a smoothing function will not be too sensitive to particle disorder, hence the errors in the approximation are smaller (Monaghan 1992; Liu and Liu 2003).

The center peak value of a smoothing function is important, too, since it determines how much contribution the particle i itself to the approximation. Obviously, a smoothing function that is close to the Dirac Delta function is more accurate in terms of kernel approximations. As shown in Eq. (10.1), when W is exactly the Dirac Delta function, the integral approximation is exact.

In Table 10.1, the prevalent smoothing functions when the dimensionless scaling factor $\alpha_s = 1$ are collected. In addition, $R = r/h = |\mathbf{x}_i - \mathbf{x}|/h$ is the relative distance between two field points $\mathbf{x}_i$ and $\mathbf{x}$.

Table 10.1 Prevalent smoothing functions

Quartic smoothing function (Lucy 1977) It is used in the original version of SPH	$W(R,h) =$ $\alpha_d \begin{cases} (1-6R^2+8R^3-3R^4) & R \le 1, \\ 0 & R > 1, \end{cases}$ $\alpha_d =$ $\begin{cases} 5/(\pi h^2) & \text{two -dimension,} \\ 105/(16\pi h^3) & \text{three -dimension.} \end{cases}$
Gaussian kernel smoothing function (Gingold and Monaghan 1977) It is sufficiently smooth even for high-orders of derivatives. But it is not really compact because it never decays to zero theoretically. A larger bandwidth in the discrete system matrix may be resulted	$W(R,h) = \alpha_d e^{-R^2}$, $\alpha_d = \begin{cases} 1/(\pi h^2) & \text{two -dimension,} \\ 1/(\pi^{\frac{3}{2}} h^3) & \text{three -dimension.} \end{cases}$
Smoothing function based on the cubic spline functions (Monaghan and Lattanzio 1985) Since the second derivative of cubic spline is piecewise linear, the stability properties can be inferior	$W(R,h) =$ $\alpha_d \begin{cases} 1-\frac{3}{2}R^2+\frac{3}{4}R^3 & 0 \le R < 1, \\ \frac{1}{4}(2-R)^3 & 1 \le R < 2, \\ 0 & R \ge 2, \end{cases}$ $\alpha_d = \begin{cases} 10/(7\pi h^2) & \text{two -dimension,} \\ 1/(\pi h^3) & \text{three -dimension.} \end{cases}$
Quintic kernel function (Wendland 1995) It uses piecewise polynomial, positive definite and compactly supported radial functions of minimal degree	$W(R,h) =$ $\alpha_d \begin{cases} (1-R/2)^4(2R+1) & 0 \le R < 2, \\ 0 & R \ge 2, \end{cases}$ $\alpha_d = \begin{cases} 7/(4\pi h^2) & \text{two -dimension,} \\ 21/(16\pi h^3) & \text{three -dimension.} \end{cases}$
Dome-shaped quadratic smoothing function (Hicks and Liebrock 2000) It is reminded that this smoothing function does not have compact support for its first derivative	$W(R,h) =$ $\alpha_d \begin{cases} W(R,h) = \alpha_d(1-R^2) & R \le 1, \\ 0 & R > 1, \end{cases}$ $\alpha_d = \begin{cases} 2/(\pi h^2) & \text{two -dimension,} \\ 15/(8\pi h^3) & \text{three -dimension.} \end{cases}$

10.4 Formulation of Fundamental SPH for Weakly Compressible Fluids

The PDEs for fluid dynamics can be written as a set of well-known "Navier–Stokes (N-S)" equations, which states the conservation of mass, momentum and energy. With the particle approximation for derivatives that has been elucidated previously, now we are able to derive the SPH formulation by discretizing N-S equations spatially, leading to a set of ODEs w.r.t. particle density, pressure and velocity. This set of ODEs can be further solved via time integration using finite difference schemes.

10.4.1 Continuity Equation

Let'S restate the continuity equation in Chap. 3 as

$$\frac{\mathrm{d}\rho}{\mathrm{dt}} + \rho\left(\frac{\partial v_x}{\partial x} + \frac{\partial v_y}{\partial y} + \frac{\partial v_z}{\partial z}\right) = 0 \tag{10.32}$$

or

$$\frac{\mathrm{d}\rho}{\mathrm{dt}} + \rho \mathrm{div}(\mathbf{v}) = 0 \tag{10.33}$$

The density approximation is very important in the SPH since it basically determines the particle distribution and the smoothing length evolution. There are two approaches for density approximation in the conventional SPH.

1. Summation density

It directly applies the SPH approximation to the density itself. For a representative field particle i, the density is written in the form of

$$\rho_i = \sum_{j\in\Omega_i}^{n_i} m_j W_{ij} \quad (i = 1, 2, \ldots, n_p) \tag{10.34}$$

in which n_i is the amount of particles in the support domain of particle i, and m_j is the mass associated with particle j. W_{ij} is the smoothing function of particle i evaluated at particle j, namely

$$\begin{aligned} W_{ij} &= W\big(\mathbf{x}_i - \mathbf{x}_j, h\big) = W\big(\left|\mathbf{x}_i - \mathbf{x}_j\right|, h\big) \\ &= W\big(R_{ij}, h\big) \quad \text{for } (i = 1, 2, \ldots, n_p) \text{ and } (j \subset \Omega_i) \end{aligned} \tag{10.35}$$

in which $R_{ij} = \frac{r_{ij}}{h} = \frac{|\mathbf{x}_i - \mathbf{x}_j|}{h}$ is the relative distance of particle i and particle j, and r_{ij} is the distance between these two particles.

Recalling that W_{ij} possesses the unit of the inverse of volume, Eq. (10.34) simply approximates the density of particle i by the weighted average density of particles $j = 1, 2, \ldots, n_i$ in the support domain of i.

2. Continuity density

It makes use of the continuity equation plus some transformations, and different transformation schemes on the RHS of continuity Eq. (10.33) may provide different density approximations.

One possible way is that the SPH approximation is only applied to the velocity divergence part, namely, the density in the RHS of Eq. (10.33) is evaluated according to Eq. (10.21)

$$\frac{\mathrm{d}\rho_i}{\mathrm{d}t} = -\rho_i \sum_{j\in\Omega_i}^{n_i} \frac{m_j}{\rho_j} \mathbf{v}_j \cdot \nabla_i W_{ij} \quad (i = 1, 2, \ldots, n_p) \tag{10.36}$$

in which ∇_i is the gradient (Hamilton) operator

$$\nabla_i = \left\{ \frac{\partial}{\partial \mathbf{x}_i} \right\} \tag{10.37}$$

Use is made of the particle approximation on the gradient of the unity that is zero

$$\nabla 1 = \int_{\Omega_i} 1 \cdot \nabla_i W(\mathbf{x}_i - \mathbf{x}_j, h) \mathrm{d}\Omega_j = \sum_{j\in\Omega_i}^{n_i} \frac{m_j}{\rho_j} \nabla_i W_{ij} = 0 \tag{10.38}$$

or

$$\rho_i \mathbf{v}_i \cdot \left(\sum_{j\in\Omega_i}^{n_i} \frac{m_j}{\rho_j} \nabla_i W_{ij} \right) = \rho_i \sum_{j\in\Omega_i}^{n_i} \frac{m_j}{\rho_j} \mathbf{v}_i \cdot \nabla_i W_{ij} = 0 \tag{10.39}$$

By adding the RHS of Eq. (10.39) to the RHS of Eq. (10.36), another form of density approximation may be obtained

$$\frac{\mathrm{d}\rho_i}{\mathrm{d}t} = \rho_i \sum_{j\in\Omega_i}^{n_i} \frac{m_j}{\rho_j} \mathbf{v}_{ij} \cdot \nabla_i W_{ij} \quad (i = 1, 2, \ldots, n_p) \tag{10.40}$$

where

$$\mathbf{v}_{ij} = \mathbf{v}_i - \mathbf{v}_j \tag{10.41}$$

Equation (10.40) introduces velocity difference into the particle approximation, and is more preferable in the formulation of SPH because it accounts for the relative velocity of particle pairs in the support domain. Another benefit of Eq. (10.40) is that the use of velocity difference in anti-symmetrized form may reduce the errors arise from the particle inconsistency problem (Monaghan, 1982, 1985, 1988). In the original works of SPH with material strength, Libersky and Petschek (1990) adopted this kind of density approximation, too.

A more prevalent form of continuity density is to apply the "second golden rule of SPH" (Monaghan 1992) mathematically expressed in Eqs. (10.23) and (10.24). If we write

$$\nabla \cdot \mathbf{v} = [\nabla \cdot (\rho \mathbf{v}) - (\nabla \rho) \cdot \mathbf{v}]/\rho \tag{10.42}$$

according to the first formula in Eq. (10.24), it follows that

$$\frac{\mathrm{d}\rho_i}{\mathrm{d}t} = \sum_{j \in \Omega_i}^{n_i} m_j \mathbf{v}_{ij} \cdot \nabla_i W_{ij} \quad (i = 1, 2, \ldots, n_p) \tag{10.43}$$

It clearly shows that the density change rate of particle i is tightly related to the relative velocities between this particle and all the other surrounding particles in the support domain Ω_i, and the gradient of the smoothing function determines the contribution of these relative velocities.

3. Comments on summation density approximation and continuity density approximation

Density summation approximation meets the mass conservation rigorously since the integral of density over entire domain is exactly the total mass of all the particles, whereas the continuity density approximation does not (Monaghan 1992). However, the former exhibits edge effect, also called as "boundary particle deficiency", for the particles at the boundary of fluid domain leading to spurious results.

The summation density approximation may be modified to improve its accuracy by using, for example, boundary virtual/ghost particles or other methods. Another possible improvement is to normalize the RHS of Eq. (10.34) with the SPH summation of the smoothing function itself (Randles and Libersky 1996; Chen et al. 1999)

$$\rho_i = \frac{\sum_{j \in \Omega_i}^{n_i} m_j W_{ij}}{\sum_{j \in \Omega_i}^{n_i} \frac{m_j}{\rho_j} W_{ij}} \quad (i = 1, 2, \ldots, n_p) \tag{10.44}$$

It is well suited for simulating general fluid flow problems without discontinuities such as shock waves.

The continuity density approximation is employed in this book. This is mainly because that for simulating events with strong discontinuity (e.g. explosion, high velocity impact), it is more feasible. In addition, without demanding to calculate density before other parameters, it is more profitable in coding for the algorithms of parallel processors. However, it needs more computational efforts.

10.4.2 Momentum Equation

1. General formulation

We restate the momentum conservation equation of Lagrangian formalism (viz. Chap. 3) for 2-D problems

$$\frac{\mathrm{d}\mathbf{v}}{\mathrm{d}t} = \mathbf{F} + \mathbf{g} - \frac{1}{\rho}\nabla p + \frac{1}{\rho}\nabla\boldsymbol{\tau} \tag{10.45}$$

or

$$\frac{\mathrm{d}\mathbf{v}}{\mathrm{d}t} = \mathbf{F} + \mathbf{g} + \frac{1}{\rho}\nabla\boldsymbol{\sigma} \tag{10.46}$$

in which $\mathbf{F} = \begin{bmatrix} F_x & F_y \end{bmatrix}^{\mathrm{T}}$ is the external force per density (exclusive gravity action) exerting on fluid; $\mathbf{g} = \begin{bmatrix} g_x & g_y \end{bmatrix}^{\mathrm{T}}$ is the gravity acceleration. Normally, for the global coordinate system where the vertical axis is upright

$$\begin{cases} g_x = 0 \\ g_y = -g \end{cases} \tag{10.47}$$

The derivation of particle approximation for the momentum equation is somewhat similar to the continuity equation, which usually involves some transformations, and different transformations can provide different approximations. For example, directly applying the particle approximation to the gradient on the RHS of Eq. (10.46) yields

$$\frac{\mathrm{d}}{\mathrm{d}t}\mathbf{v}_i = \mathbf{F}_i + \mathbf{g} + \frac{1}{\rho_i}\sum_{j\in\Omega_i}^{n_i} m_j \frac{\boldsymbol{\sigma}_j}{\rho_j}\nabla_i W_{ij} \quad (i = 1, 2, \ldots, n_p) \tag{10.48}$$

Use is made of a null identity derived from the gradient operator of the unity

$$\sum_{j\in\Omega_i}^{n_i} m_j \frac{\boldsymbol{\sigma}_i}{\rho_i\rho_j}\nabla_i W_{ij} = \frac{\boldsymbol{\sigma}_i}{\rho_i}\left(\sum_{j\in\Omega_i}^{n_i}\frac{m_j}{\rho_j}\nabla_i W_{ij}\right) = 0 \tag{10.49}$$

Adding Eq. (10.49) to the LHS of Eq. (10.48) leads to

$$\frac{\mathrm{d}}{\mathrm{d}t}\mathbf{v}_i = \mathbf{F}_i + \mathbf{g} + \sum_{j\in\Omega_i}^{n_i} m_j \frac{\boldsymbol{\sigma}_i + \boldsymbol{\sigma}_j}{\rho_i\rho_j}\nabla_i W_{ij} \quad (i = 1, 2, \ldots, n_p) \tag{10.50}$$

Equation (10.50) is a frequently exercised particle approximation formulation of momentum equation. One benefit of such a symmetrized formulation is its good performance in reducing the errors arise from particle inconsistency problem (Monaghan 1982, 1985, 1988).

Considering the second formula in Eq. (10.24), Eq. (10.48) also may be approximated by

$$\frac{\mathrm{d}}{\mathrm{d}t}\mathbf{v}_i = \mathbf{F}_i + \mathbf{g} + \sum_{j\in\Omega_i}^{n_i} m_j\left(\frac{\boldsymbol{\sigma}_i}{\rho_i^2} + \frac{\boldsymbol{\sigma}_j}{\rho_j^2}\right)\nabla_i W_{ij} \quad (i = 1, 2, \ldots, n_p) \tag{10.51}$$

This is another very prevalent particle approximation formulation of momentum equation that is frequently found in the literatures and will be implemented in this book. This symmetrized formulation enables to reduce the errors arise from the particle inconsistency problem, too.

Use is made of Eqs. (3.76) and (3.99), Eq. (10.51) can be written in more detail

$$\begin{aligned}\frac{\mathrm{d}}{\mathrm{d}t}\mathbf{v}_i &= \mathbf{F}_i + \mathbf{g} - \sum_{j\in\Omega_i}^{n_i} m_j\left(\frac{p_i}{\rho_i^2} + \frac{p_j}{\rho_j^2}\right)\nabla_i W_{ij} \\ &+ \sum_{j\in\Omega_i}^{n_i} m_j\left(\frac{\mu_i\,\dot{\boldsymbol{\varepsilon}}_i}{\rho_i^2} + \frac{\mu_j\,\dot{\boldsymbol{\varepsilon}}_j}{\rho_j^2}\right)\nabla_i W_{ij} \quad (i = 1, 2, \ldots, n_p)\end{aligned} \tag{10.52}$$

On the RHS, the second term is the approximation for the pressure, and the third term is the approximation for the viscous force related to physical viscosity. They are evaluated by the state equation and the particle velocity field, respectively.

2. Physical viscosity term

Early interest of SPH is generally directed to solve the Euler equation of inviscid flows. This is partially due to the difficulty in obtaining the second derivatives in the physical viscous term of general N-S equations. Takeda et al. (1994), when treating the viscous force, directly used the second order derivatives of the smoothing function. However, there is an alternative to avoid the approximation of second derivatives that may be implemented by the so-called "nested approximations" first on the strain rate in Eq. (10.52), then on the stress. This clever idea was proposed by Flebbe and his co-workers (Flebbe et al. 1994; Riffert et al. 1995) when simulating astrophysical problems.

Take the strain rate in Eq. (3.98) for example, one gets

$$\left\{\dot{\varepsilon} = \left((\nabla\mathbf{v}^{\mathrm{T}})^{\mathrm{T}} + \nabla\mathbf{v}^{\mathrm{T}} - \frac{2}{3}\mathrm{div}(\mathbf{v})\mathbf{I}\right) = \left[\frac{\partial v_\alpha}{\partial x_\beta} + \frac{\partial v_\beta}{\partial x_\alpha}\right] - \frac{2}{3}\mathrm{div}(\mathbf{v})\mathbf{I} \quad \text{for } (\alpha,\beta = 1,2)\right. \tag{10.53}$$

in which the Greek superscripts α and β are employed to indicate the coordinate directions for the simplification of notations $x = x_1$ and $y = x_2$, the summation is taken over repeated indices similar to tensor contraction. The approximation of $\dot{\boldsymbol{\varepsilon}}_i$

for particle i can be written in the component form such that

$$\dot{\varepsilon}_{\alpha\beta,i} = \sum_{j\in\Omega_i}^{n_i} \frac{m_j}{\rho_j} v_{\beta,j} \frac{\partial W_{ij}}{\partial x_{\alpha,i}} + \sum_{j\in\Omega_i}^{n_i} \frac{m_j}{\rho_j} v_{\alpha,j} \frac{\partial W_{ij}}{\partial x_{\beta,i}} - \left(\frac{2}{3} \sum_{j\in\Omega_i}^{n_i} \frac{m_j}{\rho_j} \mathbf{v}_j \cdot \nabla_i W_{ij} \right) \delta_{\alpha\beta} \quad \text{for } (\alpha,\beta = 1,2) \text{ and } (i = 1,2,\ldots,n_p) \tag{10.54}$$

or

$$\dot{\varepsilon}_{\alpha\beta,i} = \sum_{j\in\Omega_i}^{n_i} \frac{m_j}{\rho_j} \mathbf{I}' \mathbf{v}_j \cdot \nabla_i W_{ij} - \left(\frac{2}{3} \sum_{j\in\Omega_i}^{n_i} \frac{m_j}{\rho_j} \mathbf{v}_j \cdot \nabla_i W_{ij} \right) \mathbf{I} \quad \text{for } (\alpha,\beta = 1,2) \text{ and } (i = 1,2,\ldots,n_p) \tag{10.55}$$

where

$$\begin{cases} \mathbf{I} = \begin{bmatrix} 1 & 0 \\ 0 & 1 \end{bmatrix} \\ \mathbf{I}' = \begin{bmatrix} 0 & 1 \\ 1 & 0 \end{bmatrix} \end{cases} \tag{10.56}$$

The approximation of $\dot{\varepsilon}\alpha\beta$ for particle j can be obtained in the same way. Afterwards, the calculated strain rates $\dot{\varepsilon}\alpha\beta$ for particle i and j are introduced into Eq. (10.52) for the calculation of acceleration. This approach is straightforward because the cumbersome direct approximation of the second order derivatives w.r.t. flow velocity is avoided.

Monahan (1995) employed another approach to approximate the viscous term when modeling heat conduction problems, which seems to be more acceptable in simulating low velocity flows. Morris et al. (1997) also used Monahan's approach to model low Reynolds number incompressible flows, where the momentum equations are discretized as follows

$$\begin{cases} \frac{dv_{x,i}}{dt} = F_{x,i} + g_x - \sum_{j\in\Omega_i}^{n_i} m_j\left(\frac{p_i}{\rho_i^2} + \frac{p_j}{\rho_j^2}\right)\frac{\partial W_{ij}}{\partial x_i} \\ \quad + \sum_{j\in\Omega_i}^{n_i} m_j\left(\frac{\mu_i+\mu_j}{\rho_i\rho_j}\right)v_{y,ji}\left(\frac{1}{r_{ij}}\frac{\partial W_{ij}}{\partial r_{ij}}\right) \\ \frac{dv_{y,i}}{dt} = F_{y,i} + g_y - \sum_{j\in\Omega_i}^{n_i} m_j\left(\frac{p_i}{\rho_i^2} + \frac{p_j}{\rho_j^2}\right)\frac{\partial W_{ij}}{\partial y_i} \\ \quad + \sum_{j\in\Omega_i}^{n_i} m_j\left(\frac{\mu_i+\mu_j}{\rho_i\rho_j}\right)v_{x,ji}\left(\frac{1}{r_{ij}}\frac{\partial W_{ij}}{\partial r_{ij}}\right) \end{cases} \quad (i = 1,2,\ldots,n_p) \tag{10.57}$$

or

$$\frac{d\mathbf{v}_i}{dt} = \mathbf{F}_i + \mathbf{g} - \sum_{j\in\Omega_i}^{n_i} m_j\left(\frac{p_i}{\rho_i^2} + \frac{p_j}{\rho_j^2}\right)\nabla_i W_{ij} + \sum_{j\in\Omega_i}^{n_i} m_j\left(\frac{\mu_i + \mu_j}{\rho_i\rho_j}\right)\mathbf{I}'\mathbf{v}_{ji}\left(\frac{1}{r_{ij}}\frac{\partial W_{ij}}{\partial r_{ij}}\right) \quad (i = 1, 2, \ldots, n_p) \tag{10.58}$$

in which r_{ij} is the distance between the two particles, and $\mathbf{v}_{ji} = \mathbf{v}_j - \mathbf{v}_i$.

It is emphasized that the viscosity described above is different from the artificial viscosity that is mainly for resolving shock problems.

3. State equations

In the fundamental SPH formulation for compressible flows, particle motion is driven by the pressure gradient that is calculated by the local particle density and internal energy through the state equation. However, for incompressible flows such as water, the actual state equation will lead to extremely small, even prohibitive, time-marching step length. This is also true for other fundamental CMs (e.g. FDM and FEM). Although it is possible to include the constraint of constant density into the SPH formulations, yet the resultant equations are too cumbersome (Koshizuka et al. 1998).

The real-world fact that a theoretically incompressible flow is practically compressible leads to the concept of artificial compressibility. It states that every theoretically incompressible fluid is considered as actually compressible. Therefore, it is feasible to use a quasi-incompressible state equation with "artificial compressibility" to model incompressible flows, in this manner the time derivative of pressure may be calculated easily. In this book, the Monaghan equation of state will be employed that is expressed directly as follows

$$p_i = B\left(\left(\frac{\rho_i}{\rho_0}\right)^\gamma - 1\right) \quad (i = 1, 2, \ldots, n_p) \tag{10.59}$$

where the parameters have been specified in Eq. (3.101).

Another possible choice of the state equation is

$$p_i = c^2\rho_i \quad (i = 1, 2, \ldots, n_p) \tag{10.60}$$

Morris et al. (1997) adopted the state Eq. (10.60) in the SPH for modeling low Reynolds number incompressible flows. Zhu et al. (1999) also applied this equation in a pore-scale numerical model for the flow through porous media.

For the evaluation of the sound speed in Eq. (10.60) that is a key parameter, Monaghan (1994) defined the density variation δ as

$$\delta = \frac{\Delta\rho}{\rho_0} = \frac{\rho - \rho_0}{\rho_0} = \frac{v_b^2}{c^2} = M^2 \tag{10.61}$$

in which v_b and M are the fluid bulk velocity and Mach number respectively.

If the actually sound speed (e.g. 1480 m/s for water under standard pressure and temperature) is employed, the corresponding Mach number is very small and the density variation δ will be nearly negligible. In the other words, the real-world water is nearly incompressible. Therefore, in order to approximate the water as an artificial compressible fluid, a much smaller sound speed value than the real-world one should be postulated. This sound speed value, on the one hand, must be large enough so that the behavior of the artificial compressible fluid is sufficiently close to the real-world water; on the other hand, must not be too large so that the time-marching step length relating to computation efforts is acceptable.

To keep the balance of the affordable time-marching step length and the acceptable water flow performance, there should be an optimal value for the artificial sound speed. From Eq. (10.61), it can be seen that in modeling the real-world water as an artificial compressible fluid, the artificial sound speed should be closely related to the bulk velocity of the water flow. In addition, the pressure field also needs to be well estimated with the artificial sound speed. Morris et al. (1997) argued that the square of the artificial sound speed should be comparable with the largest value of $\frac{v_b^2}{\delta}$, $\frac{\upsilon v_b}{\delta L}$, and $\frac{FL}{\delta}$, i.e.

$$c^2 = \max\left(\frac{v_b^2}{\delta}, \frac{\upsilon v_b}{\delta L}, \frac{FL}{\delta}\right) \tag{10.62}$$

where: ν/ρ—kinetic viscosity; F—magnitude of the body force; L—characteristic length, in the simulation of landslide induced tsunamis, it is the depth of water.

The calculated pressures p_i and p_j for particle i and j by Eq. (10.59) are introduced into Eq. (10.52) for the calculation of particle acceleration.

10.4.3 Energy Equation

It is notable that the combination of continuity equation, momentum equation and state equation is able to handle a large number of CFD problems such as the tsunamis (surge waves) lately shown in the examples of this book, where the evolution of energy is not the direct concern. In case the particle approximation of energy is demanded, readers are referred to literatures (Liu and Liu 2003; Hoover 2006; Violeau 2012) for details.

10.4.4 Solution Strategy

Suppose at time t, the density $\rho(t)$, pressure $p(t)$ and velocity $\mathbf{v}(t)$ are solved, then the equation set (10.43) and (10.52) may be integrated by, for example, explicit Euler scheme to obtain $\Delta\rho(t)$ and $\Delta\mathbf{v}(t)$, afterwards, the following updating is undertaken

$$\begin{cases} \rho(t+\Delta t) = \rho(t) + \Delta\rho(t) \\ \mathbf{v}(t+\Delta t) = \mathbf{v}(t) + \Delta\mathbf{v}(t) \end{cases} \tag{10.63}$$

The updated pressure may be calculated by Eq. (10.59) using $\rho(t+\Delta t)$.

In the applications of artificial compressibility for incompressible flows, the XSPH technique (where X is the unknown factor) proposed by Monaghan (1989, 1992) is useful in which the particle moves according to

$$\frac{\mathrm{d}\mathbf{x}_i}{\mathrm{d}t} = \mathbf{v}_i + \xi \sum_{j\in\Omega_i}^{n_i} m_j \left(\frac{\mathbf{v}_j - \mathbf{v}_i}{\overline{\rho}_{ij}}\right) W_{ij} \quad (i = 1, 2, \ldots, n_p) \tag{10.64}$$

in which ξ is set being equal to or larger than 0.5. The updated position $\mathbf{x}_i$ of particle i is, again, solved by the explicit Euler scheme.

According to Eq. (10.64), the fluid particle i have to move with a velocity close to the average velocity of its neighboring particles. When applied to incompressible flows, the XSPH technique can keep the moving particles more orderly; when applied to compressible flows, it can effectively reduce unphysical penetration between particles. Hence, it will be applied in the simulation of landslide induced tsunamis (surge waves) in Chap. 11.

10.5 Numerical Considerations

10.5.1 Shock Wave and δ-SPH

For the problems of shock wave and free-surface flow, the water is generally considered inviscid and incompressible, the problem is thus governed by the Euler equation in (3.169). Consequently, neglecting the viscosity approximation and only retaining the pressure one in Eq. (10.52) results in the particle approximation for the Euler momentum equation

$$\frac{\mathrm{d}}{\mathrm{d}t}\mathbf{v}_i = \mathbf{F}_i + \mathbf{g}_i - \sum_{j\in\Omega_i}^{n_i} m_j\left(\frac{p_i}{\rho_i^2} + \frac{p_j}{\rho_j^2}\right)\nabla_i W_{ij} \quad (i = 1, 2, ..., n_p) \tag{10.65}$$

In the standard SPH formulation, water is treated as weakly compressible fluid, through an equation of state which directly links the density field to the pressure field (e.g. Equation 10.59). In SPH literatures, it is well known that SPH can be affected by numerical noise on the pressure field when dealing with Euler flow. This can be very serious when a SPH solver is dynamically coupled with a structural solver. To improve the computed pressure field by weakly compressible SPH schemes without the risk of large pressure oscillations, a density diffusion term was introduced in the equation of mass conservation, which leads to "δ-SPH" (Molteni and Colagrossi 2009; Marrone et al. 2011; Antuono et al. 2012, 2015). The particle approximation of continuity equation Eq. (10.43) can be re-written as

$$\begin{aligned}\frac{\mathrm{d}\overline{\rho}_i}{\mathrm{d}t} &= \sum_{j\in\Omega_i}^{n_i} m_j \mathbf{v}_{ij}\cdot\nabla_i W_{ij} \\ &- 2\delta h c_0 \sum_{j\in\Omega_i}^{n_i} \overline{\rho}_{ij}\mathbf{x}_{ij}\cdot\nabla_i W_{ij}\frac{1}{\mathbf{x}_{ij}\cdot\mathbf{x}_{ij}}\frac{m_j}{\overline{\rho}_j} \quad (i = 1, 2, \ldots, n_p)\end{aligned} \tag{10.66}$$

$$\begin{cases}\mathbf{x}_{ij} = \mathbf{x}_i - \mathbf{x}_j \\ \mathbf{v}_{ij} = \mathbf{v}_i - \mathbf{v}_j\end{cases} \tag{10.67}$$

$$\overline{\rho}_{ij} = \overline{\rho}_i - \overline{\rho}_j \tag{10.68}$$

where: $\overline{\rho}_i, \overline{\rho}_j$—locally averaged density of fluid particle i and j, respectively; $\mathbf{v}_i, \mathbf{v}_j$—velocity of fluid particle i and j, respectively; $\mathbf{x}_i, \mathbf{x}_j$—position vector of fluid particle i and j, respectively;δ—diffusive coefficient; c_0—speed of sound at the reference density.

Anderson and Jackson (1967) firstly derived the locally averaged N-S equations. In order to transform their equations into SPH form, the locally averaged fluid density $\overline{\rho}$ needs to be defined as

$$\overline{\rho} = \varepsilon\rho \tag{10.69}$$

in which ε is the local porosity and ρ is the actual density of fluid.

On the substitution of $\overline{\rho}$ into the locally averaged N–S equations, to which the kernel approximation and particle approximation elaborated above are applied, the weakly compressible SPH where the particle density refers to the locally averaged $\overline{\rho}$ may be formulated (Sun et al. 2013; Robinson et al. 2014; Markauskas et al. 2017).

It should be kept in mind that the summation in Eq. (10.66) excludes any possible solid particles in the interaction domain of the fluid particle.

10.5.2 Artificial Viscosity

A shock wave is a very narrow transition zone. In CFD, special treatments are demanded to allow for the algorithms being capable of modeling shock waves without unphysical oscillations around the shocked region. One simple and prevalent treatment makes use of "artificial viscosity term" being added to the physical pressure term of Euler equation (viz. Equation (10.65)). This treatment may help us to diffuse sharp variations and to dissipate the energy of high frequency components. In this manner, the numerical instability caused by the sharp spatial variation (discontinuity) may be regularized.

Application of the conservation laws of mass, momentum and energy across a shock wave front requires the simulation of transformation from kinetic energy into heat energy. Physically, this energy transformation can be interpreted as a kind of viscous dissipation, which leads to the von Neumann-Richtmyer artificial viscosity. It is widely used today for removing numerical oscillations in fluid dynamic simulations with CMs (e.g. FDM, FVM, FEM, etc.). Since the 1980s, the formulation and magnitude of artificial viscosity have undergone many refinements. Today, in the selection of the viscosity for SPH, three postulations may be employed: artificial, laminar stress, and sub-particle scale for turbulent flow. Viroulet and his colleagues (Viroulet et al. 2013a; 2013b) validated that the computation results do not depend on the choice of above viscosity models.

A "Monaghan artificial viscosity" was first proposed (Monaghan and Gingold 1983) to allow for the simulation of shocks. This is the most widely accepted artificial viscosity insofar in the SPH literatures (Monaghan and Poinracic 1985; Lattanzio et al. 1986; Monaghan 1989; Liu and Liu 2003). It is reminded that there are a number of modifications/simplifications (Hernquist and Katz 1989; Morris and Monaghan 1997), of which a famous one that is actually associated with bulk viscosity (Gómez-Gesteira et al. 2012a, 2012b) will be employed in this book.

$$\Pi_{ij} = \begin{cases} -\frac{\lambda \bar{c}_{ij}\phi_{ij}}{\bar{\rho}_{ij}} & \text{for } \mathbf{v}_{ij}\cdot\mathbf{x}_{ij} < 0 \\ 0 & \text{for } \mathbf{v}_{ij}.\mathbf{x}_{ij} \geq 0 \end{cases} \tag{10.70}$$

where

$$\begin{cases} \phi_{ij} = h_{ij}\mathbf{v}_{ij}\cdot\mathbf{x}_{ij}/\left(\mathbf{v}_{ij}\cdot\mathbf{x}_{ij} + 0.01\bar{h}_{ij}^2\right) \\ \bar{c}_{ij} = \dfrac{1}{2}(c_i + c_j) \\ \bar{\rho}_{ij} = \dfrac{1}{2}(\rho_i + \rho_j) \\ \bar{h}_{ij} = \dfrac{1}{2}(h_i + h_j) \\ \mathbf{v}_{ij} = \mathbf{v}_i - \mathbf{v}_j \\ \mathbf{x}_{ij} = \mathbf{x}_i - \mathbf{x}_j \end{cases} \tag{10.71}$$

in which c_i and c_j denote the speeds of sound of fluid particle i and j, respectively, λ is the artificial viscosity coefficient that is normally in the range of 0.01–0.3. It is noted that where a uniform smooth length is adopted for the whole domain, $\overline{h}_{ij} = h$.

The SPH formulation with artificial viscosity in Eq. (10.70) will be particularly used for the Euler equation in Chap. 11 towards the simulation of landslide induced surge waves, where the artificial viscosity Π_{ij} in Eq. (10.70) is added to the pressure terms in Eq. (10.65), namely

$$\begin{cases} \dfrac{dv_{x,i}}{dt} = F_{x,i} + g_x - \sum\limits_{j\in\Omega_i}^{n_i} m_j \left(\dfrac{p_i}{\rho_i^2} + \dfrac{p_j}{\rho_j^2} + \Pi_{ij} \right) \nabla_i W_{ij} \\ \dfrac{dv_{y,i}}{dt} = F_{y,i} + g_y - \sum\limits_{j\in\Omega_i}^{n_i} m_j \left(\dfrac{p_i}{\rho_i^2} + \dfrac{p_j}{\rho_j^2} + \Pi_{ij} \right) \nabla_i W_{ij} \end{cases} \quad (i = 1, 2, \ldots, n_p) \tag{10.72}$$

10.5.3 Boundary Treatments

In the SPH method, the support domain of smoothing function will be truncated when a fluid particle approaches solid boundary. Under such circumstances, only particles inside the support domain contribute to the field variables, but without contribution comes from outside since there are no particles behind the boundary wall. This one-sided contribution will lead to incorrect solutions, because on the boundary wall, although the velocity is zero, other field variables (e.g. density, pressure) are not necessarily zero.

Early SPH researchers for CFD either did not require boundary conditions or only required simple ones such as the "non-penetrating surface" postulation. Early SPH researchers for CSD with material strength simply ignored free boundaries, too. These simplifications give rise to an approximate zero pressure boundary condition in CFD, and an approximate zero surface stress components in CSD. Although these are allowed for very simple problems with stress-free boundaries, yet they will normally lead to complacency mistakes for the problems with general boundary conditions (Campbell 1989; Takeda et al. 1994).

1. Fixed boundaries

Insofar, there are three major kinds of boundary particles, i.e., repulsive, ghost and dynamic particles, used in the studies of SPH for the reasonable simulation of fixed boundaries.

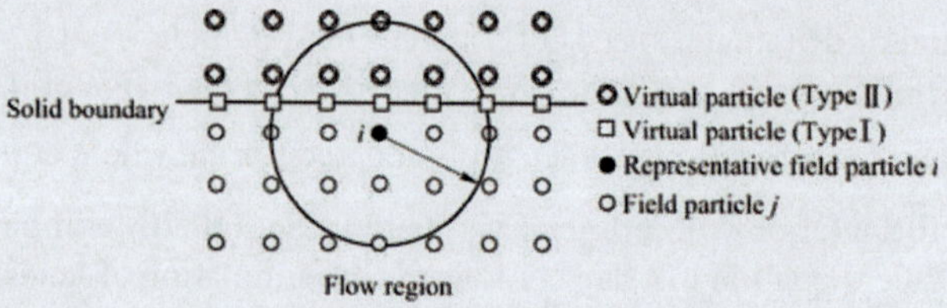

Fig. 10.3 Real particles and two types of virtual particles for solid boundary wall

(1) Repulsive boundary

Monaghan proposed a repulsive boundary condition in 1994. By exerting repulsive forces on the fluid particles that are near boundary wall, fluid particles are prevented from penetrating through boundary surface. But this boundary condition involves problem dependent parameter that needs to be adjusted according to the specific problem. In addition, these boundary particles do not contribute to the density of fluid particles.

(2) Ghost particle boundary

It is normally implemented by positioning ghost/virtual particles in the opposite side of the boundary wall to fill up the truncated region of support domain (Libersky et al. 1993; Randles and Libersky 1996; Morris et al. 1997; Dai et al. 2008). Markauskas et al. (2017) generated ghost particles instantaneously at each time-marching step instead of using pre-generated ghost particles. But this treatment may encounter difficulties in dealing with complex boundaries (Monaco et al. 2011).

In the studies by Liu and his co-workers (Liu et al. 2002; Liu and Liu, 2003), two types of virtual particles were used to treat the solid boundary conditions (viz. Figure 10.3): the first type (type I) is located right on the boundary surface and is similar to what Monaghan did in 1994, whereas the second type (type II) fills in the truncated support domain of solid region and is similar to what Libersky and his co-workers did in 1993 (Libersky et al. 1993).

For the representative particle i, all the neighboring particles within its support domain can be categorized into three groups:

- Real particles: The interior particles that are the neighbors of i;
- Virtual particles of type I: The boundary particles that are the neighbors of i;
- Virtual particles of type II: The exterior particles that are the neighbors of i.

The virtual particles of type I take part in the kernel and the particle approximations, but their position and physical variables do not evolve in the whole computation process. They are also employed to exert a repulsive boundary force to prevent the interior particles from penetrating the boundary wall. This repulsive force may be calculated by a similar approach for the molecular force of Lennard–Jones form. For the boundary particle $j \in \Omega_i$ of type I and the representative particle i approaching the boundary wall, a force is applied pair-wisely along the centerline of these two particles in a form of

$$\mathbf{F}_{ij} = \begin{cases} D\left[\left(\frac{r_0}{r_{ij}}\right)^{n_1} - \left(\frac{r_0}{r_{ij}}\right)^{n_2}\right]\frac{\mathbf{x}_{ij}}{r_{ij}^2} & \text{for } \frac{r_{ij}}{r_0} < 1 \\ 0 & \text{for } \frac{r_{ij}}{r_0} \geq 1 \end{cases} \tag{10.73}$$

where: n_1 and n_2—parameters; D—problem dependent parameter in the same scale as the square of the largest velocity; r_0—cutoff distance.

If r_0 is too large, few particles may feel the repulsive force from the virtual particles, which leads to the blowup of particle positions. On the contrary, if r_0 is too small, some particles may have already penetrated into the boundary wall before feeling the action of the repulsive force. In most practical cases, r_0 is customarily close to the initial particle spacing.

The type II virtual particles are only used to produce approximation symmetry. They can both be applied to treat solid boundaries and free surfaces. It is noted that several layers of type II virtual particles may be demanded to totally prevent the real fluid particles from escaping through boundary walls. In the validation examples of this book, three layers of type II virtual particles are normally deployed behind each boundary wall.

The numerical tests have shown that the boundary treatment using virtual/ghost particles is very stable and effective. It not only improves the accuracy of SPH approximation in the vicinity of boundary, but also prevents the particle from unphysical penetration outside the boundary wall.

(3) Dynamic boundary

Dynamic boundary condition was initially proposed by Crespo and his co-workers (Crespo et al. 2007). Boundary particles of type I in Fig. 10.3 are forced to satisfy the same governing equations of the interior fluid particles, but their velocities and positions are not updated so these boundary particles remain fixed at their original position. However, in this approach, the boundary repulsion force is generally overestimated and an unphysical gap would be formed when a thin layer of water moves along the boundary surface (Gómez-Gesteira et al. 2012a, 2012b).

Monaco and his co-workers (Monaco et al. 2011) proposed a semi-analytical approach that treats the solid boundary as a continuous material with a distribution of velocity and pressure. This approach was further developed as a unified semi-analytical wall boundary conditions for inviscid, laminar or turbulent flows in SPH (Ferrand et al. 2013; Mayrhofer et al. 2013). In the simulation of reservoir bank landslides, this boundary treatment can well replicate the irregular and complex topography of slide body and meanwhile reduces the computation effort.

Adami et al. (2012), Fraser et al. (2016) presented a generalized wall boundary condition. Sun et al. (2013) further presented a boundary force model that unifies the boundary representations for both DEM and SPH without any extra boundary particles, which saves the computation effort considerably.

In this book, the dynamic boundary condition will be implemented in the SPH code for fixed boundaries due to its simplicity.

2. Moving boundaries

For the dynamic boundary surface in the fluid–solid coupling problem using, for example, the hybrid DEM-SPH, the solid boundaries of polygonal distinct element may be represented by wall elements on which smaller discs $d_k \in \Omega_{r_i}$ attached (viz. Figure 7.12) are similar to the boundary virtual particles of type I in Fig. 10.3. They exchange information with fluid particles $i\,(1, 2, \ldots, n_p)$ through the solid disc and fluid particle interaction forces (viz. Figure 10.4)

$$\begin{cases} Fr_{x,i}^{s\to f} = m_i \sum\limits_{d_k \in \Omega_i}^{n_k} Fr_{x,k,i}^{s\to f} \\ Fr_{y,i}^{s\to f} = m_i \sum\limits_{d_k \in \Omega_i}^{n_k} Fr_{y,k,i}^{s\to f} \end{cases} \quad (i = 1, 2, \ldots, n_p) \tag{10.74}$$

$$\begin{cases} F_{x,i} \to F_{x,i} + Fr_{x,i}^{s\to f} \\ F_{y,i} \to F_{y,i} + Fr_{y,i}^{s\to f} \end{cases} \quad (i = 1, 2, \ldots, n_p) \tag{10.75}$$

where: $\mathbf{Fr}_{k,i}^{s\to f}$—repulsion force exerted on the fluid particle i by the solid particle d_k on the moving boundary of block element; m_i—mass of fluid particle i; Ω_i—support domain of fluid particle i.

The repulsion force for moving boundaries can be calculated by Qiu (2013)

$$Fr_{x,k,i}^{s\to f} = \begin{cases} \chi\left[\left(\frac{r_0}{r_{ik}}\right)^4 - \left(\frac{r_0}{r_{ik}}\right)^2\right]\frac{x_{ik}}{r_{ik}^2} & \text{for } r_{ik} \le r_0 \\ 0 & \text{for } r_{ik} > r_0 \end{cases} \tag{10.76a}$$

$$Fr_{y,k,i}^{s\to f} = \begin{cases} \chi\left[\left(\frac{r_0}{r_{ik}}\right)^4 - \left(\frac{r_0}{r_{ik}}\right)^2\right]\frac{y_{ik}}{r_{ik}^2} & \text{for } r_{ik} \le r_0 \\ 0 & \text{for } r_{ik} > r_0 \end{cases} \tag{10.76b}$$

where: $\chi = gD$, the coefficient D is chosen according to the domain configuration, and $D = H$ for open channel flow with depth; r_0—cut-off distance, normally it is set to be the initial fluid particle space; r_{ik}—distance between fluid particle i and solid particle d_k.

It is stressed that the fluid particle i will impose reaction force $\mathbf{Fr}_{k,i}^{f\to s}$ to solid particle d_k, too (viz. Figure 10.4). According to the Newton's third law we get

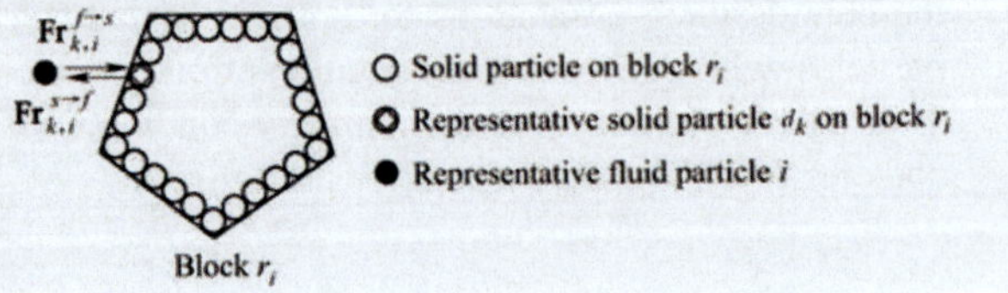

Fig. 10.4 Solid–fluid interaction forces

$$\mathbf{Fr}_{k,i}^{f \to s} = -\mathbf{Fr}_{k,i}^{s \to f} \tag{10.77}$$

This reaction force may be used to accumulate the pressure exerted at moving boundary wall, if necessary. According to Eq. (10.76), the repulsion force is zero when the distance of fluid particle i and solid particle d_k is larger than r_0, namely $d_k \notin \Omega_i$, or in other words, the moving boundary particle only exerts repulsion action on its surrounding fluid particles. It needs to be clarified that this kind of repulsive boundary condition is normally applied to the moving boundary only; for a fixed boundary condition, the dynamic boundary condition (Crespo et al. 2007) discussed previously is more reasonable. This is because with the former, the contribution of boundary particles to the density of interior particles is neglected, which will deteriorate the pressure field.

10.5.4 Variable Smoothing Length

The precision of particle approximation in SPH depends on the amount of particles within the support domain defined by $\alpha_s h$. Hence the smoothing length h directly influences the computation effort and the solution accuracy (Quinlan et al. 2006). If h is too small, there may be no sufficient particles in the support domain of the representative field particle to exert forces on it, which results in low computation accuracy. On the contrary, if h is too large, many details of particles should be smoothed out, the computation effort suffers and computation accuracy is low, too.

In the early implementations of SPH, a global particle smoothing length h was used which depends on the initial average density of the particle system. Later on, towards the solution where the fluid expands or contracts unevenly, the smoothing length is individually assigned for each particle as h_i according to the variation of its local density (Monaghan 1988, 1992). For problems that are not isotropic such as shock wave problems, the smoothing length even needs to be adaptive both in space and time (Hernquist and Katz 1989; Nelson and Papaloizou 1994). For example, use was made of the tensor form of smoothing length and ellipsoidal kernel rather than the traditional spherical kernel, Shapiro and his co-worker (Shapiro et al. 1996) developed an adaptive SPH.

There are many ways to dynamically evolve h so that the amount of neighboring particles remains relatively constant. The simplest one is to update the smoothing length according to the averaged density

$$h = h_0 \left(\frac{\rho_0}{\rho} \right)^{1/d} \tag{10.78}$$

where h_0 and ρ_0 are respectively the initial smoothing length and the initial density, d is the space dimension. Benz (1989) suggested another method to evolve the smoothing length

$$\frac{\mathrm{d}h}{\mathrm{d}t} = -\frac{1}{d}\frac{h}{\rho}\frac{\mathrm{d}\rho}{\mathrm{d}t} \tag{10.79}$$

which can be discretized using SPH approximations and calculated with the other differential equations in parallel.

10.5.5 Asymmetry of Particle Interaction

If the smoothing length varies both in time and space, each particle will possess its own smoothing length h. Hence it is natural that we endue every particle i a specific smoothing length $h = h_i$ for all the previous formulas in this chapter that contain h.

However, since h_i is not equal to h_j (often), the influence domain of particle i may cover particle j but not always vice-versa. As a result, there is the possibility that particle i exert a force on particle j but without the reaction from j, which leads to the violation of Newton's third law. This is, particularly serious in the scatter model.

The symmetry of particle interaction may be preserved by modifying smoothing length using, for example, the averaged smoothing length of the interacting particle pair i and j (Benz 1989), namely

$$h = \bar{h}_{ij} = \frac{1}{2}(h_i + h_j) \tag{10.80}$$

then one obtains the symmetric smoothing function for all the previously elaborated formulas in this chapter such that

$$W_{ij} = W(r_{ij}, \bar{h}_{ij}) \tag{10.81}$$

Other symmetry approaches that may be exercised include geometric mean, minimal value, maximal value, etc. (Liu and Liu 2003). In addition, the direct average of smoothing functions (Hernquist and Katz 1989)

$$W_{ij} = \frac{1}{2}(W(h_i) + W(h_j)) \tag{10.82}$$

may be attempted, too.

It is should be noted that above symmetry approaches are rather cumbersome in the programming, and even self-contradictory between multi-particles (more than two). As a result, insofar many codes keep to use the uniform global smooth length, and this book follows suit, although we have to pay higher computation effort due to sufficiently large h.

10.5.6 Zero-Energy Mode

It is well known that FDM suffers from a spurious zero-energy mode for which the derivative at certain regular grid point is zero: for the normal solution in which the function is constant, a solution of saw-tooth pattern exists. To handle this problem, an additional grid is normally necessary to stagger from the original grid so that the pressure and velocity may be calculated at different grid systems.

In FEM, the zero-energy mode also happens: the nodal velocities on the opposite corners of the mesh can be equal in magnitude but opposite in direction. This is also known as the "Hourglass phenomenon". It is found that this velocity field produces a spurious mode with no strain change in the mesh, and the resultant stress is zero, too, which leads to the loss of resistance to the mesh deformation of Hourglass shape. To handle this problem, a Hourglass viscosity is normally introduced to give artificial rigidity to the system. More discussions on the zero-energy mode issues can be found in the work of Swegle (1978) followed by many others.

The same spurious zero-energy mode problem also may occur in SPH when evaluating the derivatives (Belytschko et al. 2000), and an efficient remedy was proposed by Dyka and followed by others (Dyka and Ingel 1995; Dyka et al. 1997; Vignjevic et al. 2000; Randles and Libersky 2000). Their idea is somewhat similar to that associated with the staggered grids in FDM: two types of particles are used, in which the velocity particles for the momentum equation and stress particles for the pressure.

The zero-energy mode problem in SPH, however, is not as serious as that in FDM attributable to more irregular particle distribution in the former. Even the initial particle distribution is regular, it will evolve quickly to be irregular.

10.5.7 Particle Interactions

A very time consuming task for the SPH simulation is to identify all the interacting particles. The smoothing function and its derivatives are either calculated at each summation process, or obtained from other large arrays that store the previously calculated smoothing function and its derivatives. It is evident that the former is not computationally efficient and the latter demands a large storage capacity. By the pairwise interaction techniques, however, only one loop is needed. The interacting data inclusive the indices of interacting particles, their correspondent smoothing function and derivatives, are stored for only one half of all the interactions. Such a pairwise interaction technique may remarkably save storage capacity and improves computation efficiency.

Recalling that the smoothing function of representative particle i is defined at a compact support domain with dimension $\alpha_s h_i$ that contains a finite number of particles $j \subset \Omega_i$. These particles are generally referred to as the "nearest neighboring particles (NNP)". The process of finding the NNP and storing necessary data for the

SPH summation process is commonly referred to as the "nearest neighboring particle searching (NNPS)". Unlike MBMs where the position of neighboring elements/cells are well defined once the grids are generated, the NNP for SPH can vary with the ongoing of time. Today, three NNPS algorithms including "all-pair search", "linked-list search", and "tree search", are prevalent. Other efficient algorithms of NNPS such as the so-called "Bucket algorithm" are also suggestable (Liu and Liu 2003).

1. All-pair search

All-pair search is a direct and simple algorithm of NNPS. For a representative particle i, it calculates the distance r_{ij} from i to every field particle j (1,2, ..., n_p), where n_p is the amount of particles in the whole problem domain. If the distance r_{ij} is smaller than the dimension of the support domain, particle j belongs to the support domain of particle i, namely $j \in \Omega_i$. Where the symmetric smoothing lengths (inclusive uniform smoothing length) is employed, particle i is within the support domain of particle j, too. Therefore, particle i and particle j form a pair of neighboring particles. This search is undertaken for all the particles i = 1,2, ..., n_p. The complexity of the all-pair search is of order $O(n_p^2)$. Since the NNPS process is necessary for all the time-marching steps, the computation burden is very heavy.

2. Linked-list search

Monaghan (1985) proposed a linked-list search algorithm for NNPS, more details were given by Hockney and Eastwood (1988). Rhoades (1992) provided an algorithm towards NNPS that was reported to be very efficient, particularly on vector computers. Simpson (1995) described the linked-list search algorithm for 3-D SPH applied to accretion discs.

Actually, the linked-list search algorithm (viz. Figure 10.5) is rather similar to the preliminary contact detection scheme for the disc-shaped particles in DEM (cf. Figure 7.1): a temporary grid is overlaid on the problem domain, and the cell size is selected to match the dimension of support domain. For example, the cell size may be identical to the compact support length $\alpha_s h$. For a given particle i with α_s = 2, its NNP can only be in the same cell or the immediately neighboring cells. Therefore, the search is confined only over 3, 9 or 27 cells for 1-D, 2-D or 3-D

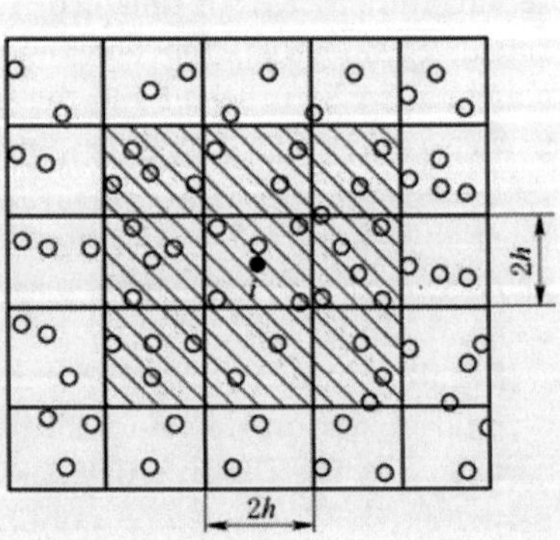

Fig. 10.5 Linked-list algorithm for NPPS in 2-D space

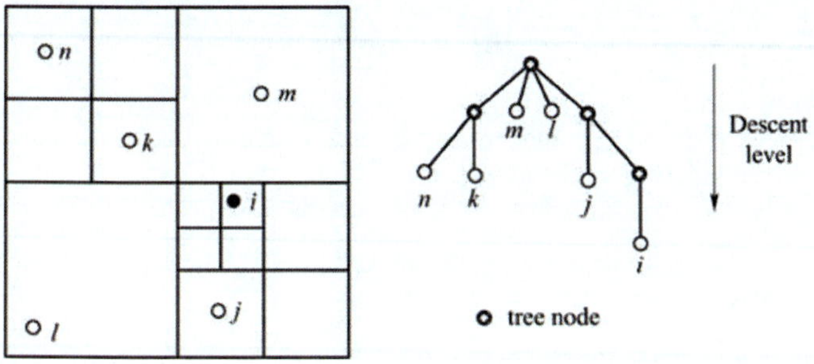

Fig. 10.6 Tree structure and tree search algorithm in 2-D space

problems, respectively. In Fig. 10.5, the dimension of the support domain is $2h$ for the particles, and the smoothing length is spatially constant.

As pointed by Monaghan and Gingold (1983), substantial savings in computational time can be achieved by using cells as a book keeping device. If the average amount of particles per cell is sufficiently small, the complexity of the linked-list search algorithm is of order $O(n_p)$.

This algorithm works well for constant smoothing length. However, when variable smoothing length is used, the cell size may not be optimal for every particle.

3. Tree search

Tree search algorithm involves creating an ordered tree structure according to the particle positions that is rather similar to the 3-D octree algorithm (2-D quadtree algorithm) in the h-adaptive FEM. For 2-D problem (viz. Figure 10.6), it recursively splits the maximal problem domain into squares (cells) as tree-branches that contain particles, until each leave on the tree contains one individual particle only.

The complexity of tree search algorithm is of order $O(n_p \log n_p)$ (Hernquist and Katz 1989; Riffert et al. 1995). Numerical tests show that the SPH combining with tree search algorithm is very efficient and robust, especially for a large amount of particles with variable smoothing lengths.

10.5.8 Time Integration

Just as other CFDs, the spatially discretized SPH equations can be temporally integrated with standard FDMs.

The explicit time integration schemes are subject to the "Courant-Friedrichs-Levy (CFL)" condition for numerical stability. This condition requires the time-marching step length being proportional to the smallest spatial particle resolution, in SPH it is equally to state that

$$\Delta t = \min\left(\frac{h}{c}\right) \tag{10.83}$$

Morris et al. (1997) gave another expression for estimating time-marching step length when considering viscous diffusion

$$\Delta t = 0.125\frac{h^2}{\upsilon} \tag{10.84}$$

in which $\upsilon = \mu/\rho$ is the kinetic viscosity.

Sometimes, where the smoothing length is too small, the time-marching step length will become prohibitively small. Under such circumstances, the Runge–Kutta (R-K) scheme with adaptive time-marching step length (Benz 1989) comes into power: the time-marching step length may be larger than that estimated by CFL condition. However, R-K scheme needs two force evaluations per step, and thus is computationally expensive. Another important consideration for the time integration in SPH is the use of individual time-steps (Hernquist and Katz 1989).

10.6 Validations

A number of numerical examples will be presented to examine the ability and efficiency of our SPH formulation and code for fluid dynamic problems. In addition, these numerical examples also provide foundations for coupling SPH with other advanced CMs such as the hybrid SPH-DEM (viz. Chap. 11).

10.6.1 Dam Break Test

Dam break test conducted by Koshizuka et al. (1995) is used to validate our SPH code. This physical experiment was previously adopted by Wu et al. (2016) for the validation of their SPH model.

The setup of the SPH model is depicted in Fig. 10.7. The initial water column is closed to the left tank wall and collapses under gravity after the start of experiment. The water column is 0.142 m wide and 0.293 m high, the water tank is 0.584 m wide and 0.350 m high. The relevant SPH parameters are listed in Table 10.2. The constant smoothing length is set to be 1.5 times the initial fluid particle spacing, and dynamic boundary conditions are employed for the tank walls. It is noted that the δ-SPH algorithm is not activated in this validation study, namely, $\delta = 0$ in Eq. (10.66).

The water flow pattern at different instants is cross-referenced by the experiment data (Koshizuka et al. 1995) in Fig. 10.8. In general, the SPH results show a satisfactory agreement with the experiment data. It is clear that at the beginning, the water column collapses under the action of gravity and moves along the tank bottom.

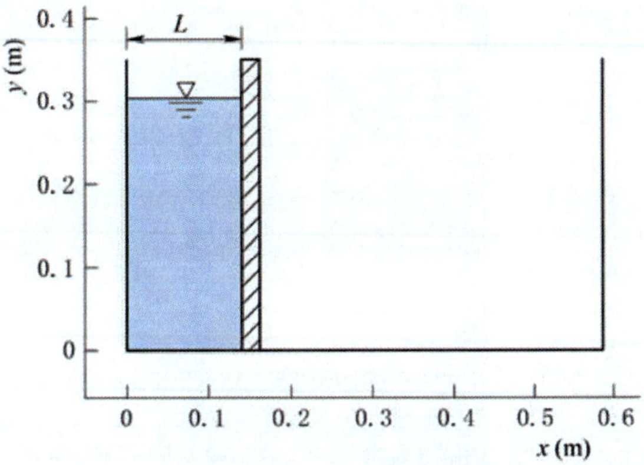

Fig. 10.7 Setup of the dam break test (Koshizuka et al. 1995)

Table 10.2 SPH parameters for the dam break test

SPH parameters	Values
Fluid density (kg/m^3)	1000.0
Fluid particle spacing (m)	0.003
Fluid particle number	4559
Boundary particle number	861
Kernel function	Wendland
Smoothing length (m)	0.0045
Boundary conditions	Dynamic BCs
Artificial viscosity coefficient η	0.3
Coefficient of sound speed λ	10
Time-marching step length(s)	1.0×10^{-5}

Subsequently, a portion of fluid particles hits the opposite flank and climbs onto the right tank wall, these fluid particles fall down again generating a plunging wave. Finally, the water flows back to the left flank.

It also can be found from Fig. 10.8 that the thin layer of water is detached from the right vertical wall during its climbing up process. Fortunately, this boundary effect can be neglected in the simulation of landslide generated surge waves because the water body in the reservoir is rather deep.

The evolution of the collapsing water front before hitting the opposite flank is quantitatively investigated and compared with the experimental data (Martin and Moyce 1952) in Fig. 10.9, they are almost identical at the beginning, whereas a slight discrepancy appears as the water proceeds to move forward.

The phenomenon that SPH fronts are shorter than the experimental data indicates that the water flow in the SPH model is slower. This slight delay is due to the limitation of the dynamic boundary condition along the tank bottom: when the water moves towards the right flank along the tank bottom, the water layer gradually becomes

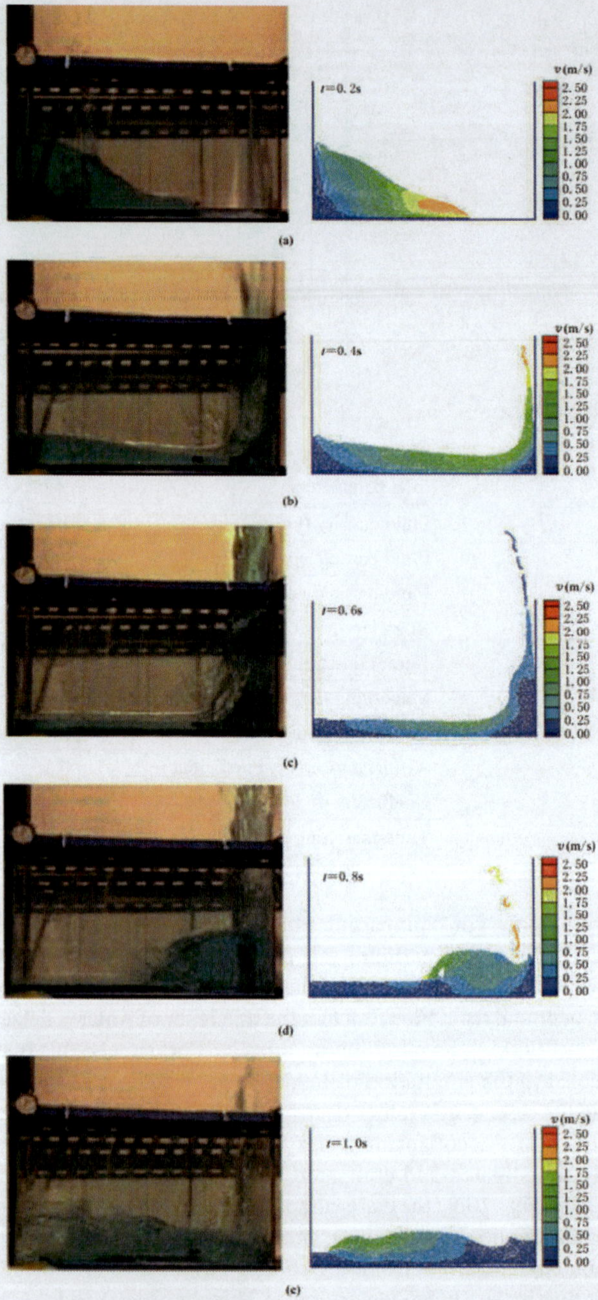

Fig. 10.8 Evolution of water flow pattern (Left column: experiment by Koshizuka et al. 1995); Right column: SPH computation by the author). **a** $t = 0.2$ s; **b** $t = 0.4$ s; **c** $t = 0.6$ s; **d** $t = 0.8$ s; **e** $t = 1.0$ s

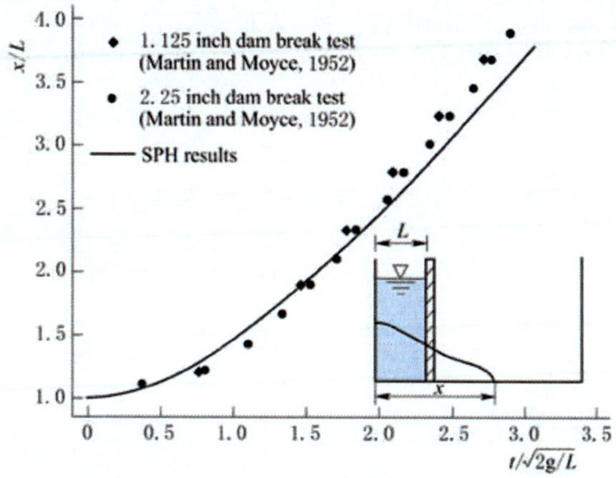

Fig. 10.9 Comparison of collapse water front between SPH results and experimental data (L is the initial water column width, x is the distance between the collapsing water front and the left flank, t is the time)

thinner and thinner, and an unrealistic physical gap is gradually formed due to the overestimated boundary repulsion force which may hinder the water motion.

10.6.2 Scott Russell's Wave Generator

A slide block being vertically pushed into the water tank is termed as "Scott Russell's wave generator" that is used to generate solitary wave (Monaghan and Kos 2000; Ataie-Ashtiani and Shobeyri 2008; Abadie et al. 2010; Yeylaghi ct al. 2017).

The model setup is depicted in Fig. 10.10 and all relevant parameters are listed in Table 10.3. The rectangular slide block with a width of 0.3 m and a height of 0.4 m is placed 0.025 m away from the left side wall of the tank for the purpose to observe the flow in gap. The length of the water tank is 2.0 m that is the same as simplified by Ataie-Ashtiani and Shobeyri (2008) for saving computation effort. The solid boundaries of moving block are replaced by dynamic boundary particles (discs) with the aim of keeping the fluid particles inside the computation domain, the reaction between fluid particles and solid discs have be elaborated in Eqs. (10.74)–(10.76).

The motion of the slide block is imposed in the manner of (Monaghan and Kos 2000):

Table 10.3 Parameters for Scott Russell's wave generator

DEM parameters	
Disc radius (m)	0.002
Number of discs attached to the block	346
DEM time-marching step length (s)	1.0×10^{-4}
SPH parameters	
Fluid particle density (kg/m^3)	1000.0
Fluid particle spacing (m)	0.005
Fixed boundary particle spacing (m)	0.0025
Number of fluid particles	16,758
Number of fixed boundary particles	5782
Kernel function	Wendland
Kernel smoothing length (m)	0.00625
Boundary conditions (BCs)	Dynamic BCs
Density filter	δ-SPH algorithm
δ-SPH coefficient	0.1
Artificial viscosity coefficient λ	0.1
Coefficient of sound speed η	10
SPH time-marching step length (s)	1.0×10^{-4}

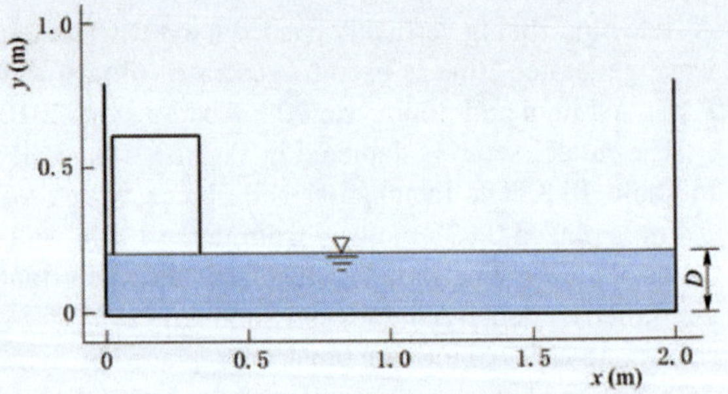

Fig. 10.10 Model setup of Scott Russell's wave generator

$$\frac{v}{\sqrt{gD}} = 1.03\frac{Y}{D}\left(1 - \frac{Y}{D}\right)^{0.5} \tag{10.85}$$

where: v—vertical velocity of the slide block; D—water depth (equal to 0.21 m in this study); Y—distance between the base of the slide block and the bottom of the tank.

The variation of v along the water depth is demonstrated in Fig. 10.11. It is clear that the slide block accelerates at the beginning and then decelerates down to zero when it arrives at the bottom of the tank.

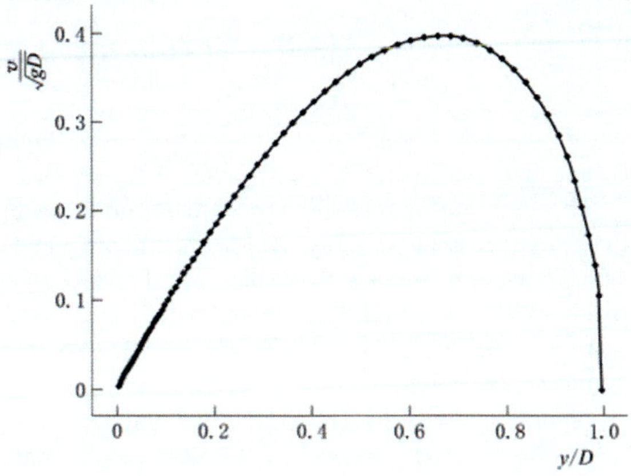

Fig. 10.11 Evolution of the vertical velocity of slide block (Monaghan and Kos 2000)

Beyond the impact zone (viz. Figure 10.12), the pressure field is stratified and nearly hydrostatic, which indicates that the δ-SPH algorithm can well eliminate spurious numerical noise on the pressure field.

After the slide block penetrating into the water body, the water displacement creates a solitary wave. It propagates outward first, then a reverse plunging wave flows back. It is also observed that an air cavity is created, meanwhile a small portion of water moves upward along the narrow gap between block and left wall of the tank that ultimately climes onto the top of the slide block.

The theoretical wave profile of the solitary wave was given by Lo and Shao (2002)

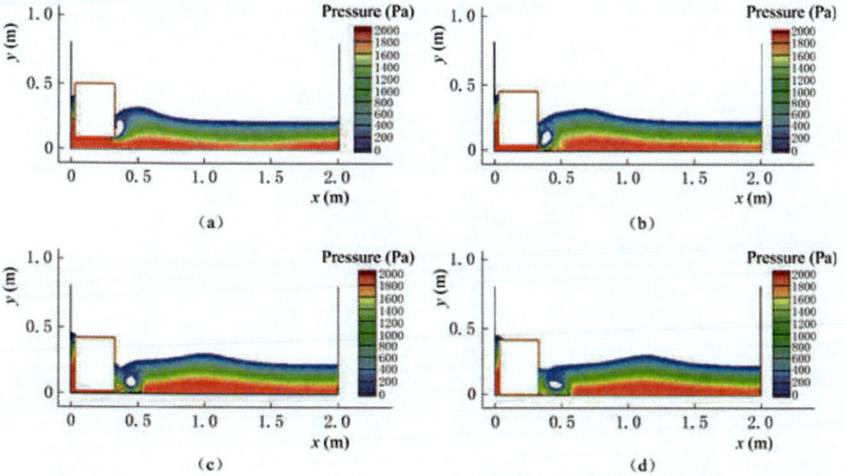

Fig. 10.12 Evolution of the pressure field of Russell's solitary wave. **a** $t = 0.27$ s; **b** $t = 0.42$ s; **c** $t = 0.61$ s; **d** $t = 0.70$ s

$$\eta(x, t) = a\text{sech}^2\left[\sqrt{\frac{3a}{4D^3}}(x - ct)\right] \tag{10.86}$$

where: η—water surface elevation; a—wave amplitude setting to be 0.092 m (Monaghan and Kos 2000); $c = \sqrt{g(D + a)}$—solitary wave velocity.

The water surface elevation predicted from the SPH model is cross-referenced by the analytical solution in Fig. 10.13. The calculated wave amplitude at 0.42 s shows a satisfactory agreement with the theoretical one. However, after 0.42 s, the calculated wave amplitude is slightly underestimated. This discrepancy is probably due to the leakage of the water in the narrow gap, which results in less water being displaced to onset solitary wave.

Figures 10.14 and 10.15 exhibit the evolution of velocity field. The flow regime in the impact zone captured by the SPH model shows a good agreement with the experiment data:

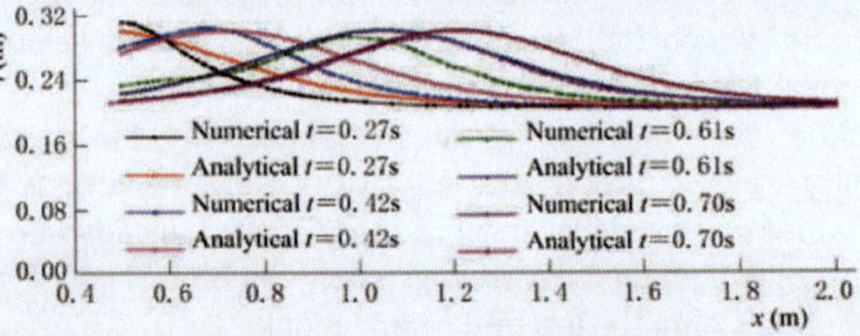

Fig. 10.13 Water surface elevation η: SPH results versus analytical solution

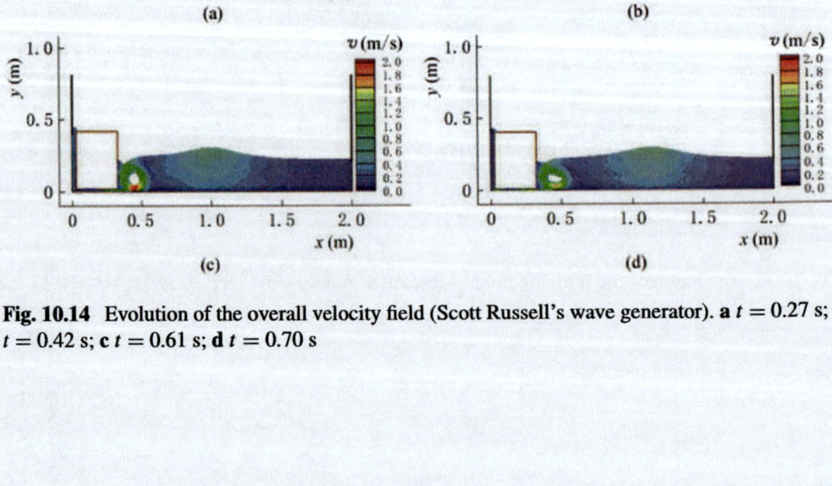

Fig. 10.14 Evolution of the overall velocity field (Scott Russell's wave generator). **a** $t = 0.27$ s; **b** $t = 0.42$ s; **c** $t = 0.61$ s; **d** $t = 0.70$ s

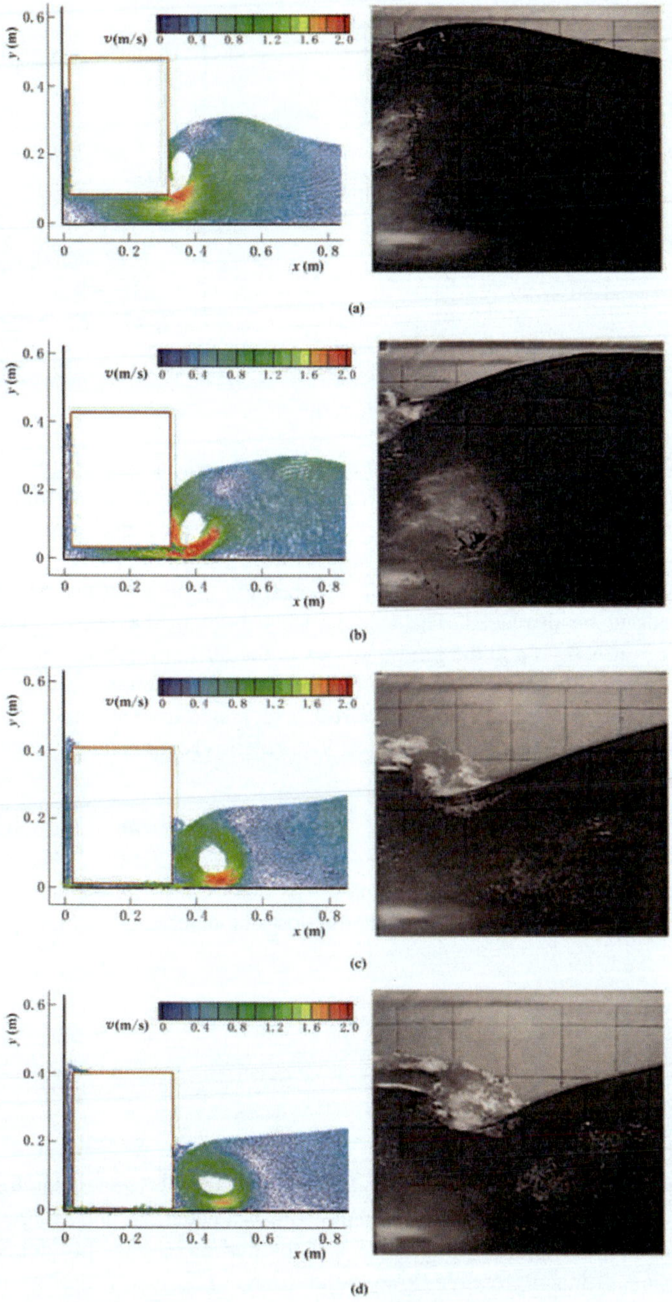

Fig. 10.15 Evolution of the velocity vector field in the impact zone. **a** $t = 0.27$ s; **b** $t = 0.42$ s; **c** $t = 0.61$ s; **c** $t = 0.70$ s (Left column: SPH by Tan and Chen 2017; Right column: experiment by Monaghan and Kos 2000)

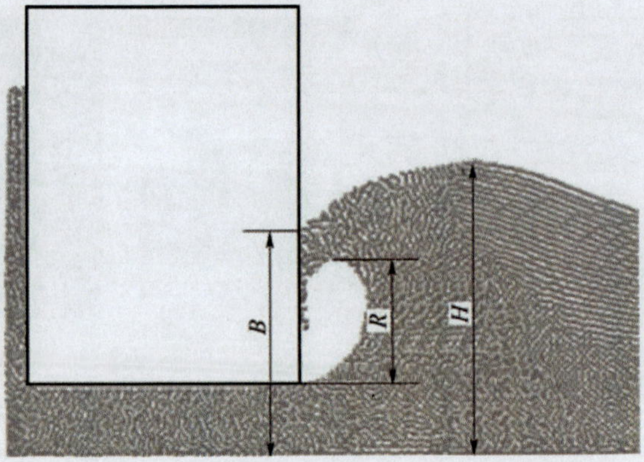

Fig. 10.16 Schematic diagram of characteristic lengths at $t = 0.285$ s

- At 0.27 s, due to the violent impact of the slide block, fluid particles in the near field gain large velocities. In this instant, the energy is transferred from the slide block to the water body. The water beneath the block moves rightward and a backwash wave reversely flows hitting the block itself.
- At 0.42 s, due to the continuous block penetration, the water beneath the block proceeds to be displaced. The solitary wave born in the impact zone propagates towards right and the fluid particles in the far field start to gain velocities, meanwhile the backwash wave plunging down forms a counterclockwise vortex.
- At the time of 0.61 s and 0.70 s, the vortex is separated from the block, and a small surge wave is generated near the block. The kinetic energy is mainly concentrated below the crest of the solitary wave.

The configuration of fluid particles at $t = 0.285$ s is shown in Fig. 10.16, from which the characteristic lengths are measured and listed in Table 10.4, together with other available data from previous studies. The quantitative comparison shows that the SPH model can accurately replicate the flow characteristics in the impact zone.

10.7 Concluding Remarks

10.7.1 Comments on MPMs

This chapter presented the fundamental algorithm of SPH—an outstanding family member of MPMs, for the CFD governed by the N-S equations and Euler equations. The resultant SPH is able to simulate various hydrodynamics problems such as compressible or incompressible flows. In addition to physical viscosity constructed from the viscous shear force, the artificial viscosity is used to model the inviscid flow problems in the simulation of shock waves. For incompressible flow problems,

Table 10.4 Comparison of characteristic lengths in Fig. 10.16 at $t = 0.285$ s

Initial fluid particle spacing (m)	Method	H(m)	R(m)	B(m)
/	Experiment[1]	0.333 ± 0.01	0.133 ± 0.02	0.303 ± 0.02
0.0105	SPH[1]	0.308	0.075	0.273
0.007	SPH[1]	0.308	0.099	0.261
0.00525	SPH[1]	0.309	0.109	0.273
0.0042	SPH[1]	0.3086	0.114	0.272
0.015	ISPH[2]	0.329	0.13	0.255
0.01	ISPH[2]	0.3301	0.146	0.268
0.008	ISPH[3]	0.3006	0.1159	0.2707
0.005	SPH [4]	0.313	0.137	0.242

N.B.: ISPH is the abbreviation of incompressible SPH. [1] by Monaghan and Kos (2000); [2] by Ataie-Ashtiani and Shobeyri (2008); [3] by Yeylaghi et al. (2017); [4] by Tan and Chen (2017)

π is applied by the special sate equation to handle them as quasi-incompressible counterparts.

The advantages of MPMs (inclusive SPH) over the conventional MBMs (inclusive the FEM and FDM in Chap. 4) can be roughly summarized as follows:

- Problem domain is discretized with particle system without a pre-fixed connectivity;
- Treatment of large deformation is relatively much easier;
- Discretization of complex geometry is relatively simpler;
- Adaptive refinement of particle distribution is much easier;
- It is easier to snapshot the state of entire physical system through tracing the motion history of all particles;
- Identification of free surfaces, moving interfaces and deformable boundaries, is relatively easier.

Many MPMs were initially developed for the system with discrete particles. Examples include the SPH for the interaction of stars in astrophysics, the molecular dynamics (MD) for the movement of atoms (Alder and Wainright 1957; Gibson and Goland 1960; Rahman, 1964; Stillinger and Rahman 1974), etc.

Sooner after the 1970s, MPMs have been modified, extended, and applied to the system of continuum media. For this purpose, an additional operation is required to generate the set of particles to represent the continuum media. Each particle represents a portion (sub-domain) of the problem domain with some attributes (e.g. mass, position, momentum and energy) that are tied to the geometric center of the sub-domain. Examples include SPH, vortex methods (Chorin 1973; Leonard 1980), "corrective smoothed particle method (CSPM)" (Chen et al. 1999, 2000), and many others.

The existing MPMs may be distinguished as strong form based and weak form based. Because no integration is required in establishing the discrete system equation set, the strong form based methods possess attractive merits: simple to implement, computationally efficient, and "truly" mesh-free. However, they are often unstable and less accurate, especially when irregularly distributed particles (nodes) are used for the PDEs with Neumann (derivative) boundary conditions, such as the CSD problems with stress (natural) boundary conditions. On the other hand, the weak form based methods (e.g. EFGM) exhibit excellent stability and accuracy, and the Neumann boundary conditions can be naturally satisfied. However, they are said not "truly" mesh-free, as a background mesh (local or global) is required for the quadrature of integrand terms.

According to the mathematical model used, MPMs also can be distinguished as deterministic or probabilistic. The former are prevalent in geotechnical and hydraulic engineering, which directly deal with the governing equations of physical law. Theoretically, once the initial and boundary conditions are weel-posed, the particle evolution in the later time can be precisely determined. The probabilistic methods are based on statistical principles, and their main representatives include MD based "Monte Carlo (MC)" method (Metropolis and Ulam 1949; Binder 1988, 1992), "direct simulation Monte Carlo (DSMC)" method (Bird 1994; Pan et al. 1999, 2000, 2002), and "lattice Bolztmann equation (LBE)" (Chen and Doolen 1998; Qian et al. 2000), etc. It needs to be noted that some MPMs can possess mixed features of deterministic or probabilistic.

Most MPMs are inherently Lagrangian, in which the particles move in the Lagrangian frame according to the internal particle interactions and external forces. However, there are also examples in which the particles are fixed in the Eulerian frame as interpolation points rather than moving objects (Laguna 1995). In addition, most MPMs use explicit scheme for the time integration. However, there are some exceptions, such as in the MPS (Koshizuka et al. 1998) that use implicit or semi-implicit schemes: the pressure term in the momentum equations is implicitly determined. It is also worthwhile to remind that even we use particles to represent a continuum domain, some kind of mesh still could be helpful to generate the initial distribution of field particles.

10.7.2 Comments on SPH

1. Computation procedure

SPH is gradually get acquainted with geotechnical and hydraulic engineers concerning macroscopic CFD and CSD, although it has been long accepted by scientists to atomistic scale simulations (Liu et al. 2002; Nitsche and Zhang 2002).

The mainstream procedure in SPH is basically similar to that in any grid-based method, except for the particle representation and approximation. As a typical member of MPMs, it is worthwhile to briefly summarized again as follows:

- Represent the problem domain with discrete particles, the information is known at the particles at an initial instant with a proper treatment on the boundary conditions;
- Represent the derivatives in the PDEs with proper particle approximations;
- From the given velocity and/or position at instant t, calculate the strain rate and/or strain, then calculate the pressure/stress of each discrete particle;
- Calculate the acceleration of each discrete particle using the afore-calculated pressure/stress;
- Use the acceleration at t to calculate new velocities and positions at instant $t + \Delta t$, where Δt is the incremental time-marching step length;
- From new velocities and/or new positions, calculate the new strain rate and/or new strain at instant $t + \Delta t$, then calculate the new pressure/stress at instant $t + \Delta t$.

Above time-marching steps are repeated until the specified collapse time.

2. Particularities

(1) Weakened operation

The weakened operation in SPH is implemented in the stage of function approximation (kernel approximation), rather than in the stage of discrete approximation as in EFGM. The application of the integral representation for field functions passes the derivative operations on the field function to the smoothing (weight) function. It reduces the requirement on the order of continuity for the field function, which leads to relative good stable for arbitrarily distributed particles for many problems with extremely large deformations (Yashiro and Okabe 2015). The accuracy of its solution naturally depends on the choice of smoothing function.

(2) Adaptive refinement

The adaptive nature of SPH is achieved at the stage of field variable approximation, and may be performed at each time-marching step. Therefore, SPH is able to handle problems with extremely large deformation.

(3) Lagrangian formalism

SPH particles carry material properties, and are allowed to move in light of the internal interactions and external forces. Functioning as both approximation points and material components, SPH particles seem to be alive. Abandon of mesh significantly simplifies the model implementation and its parallelization, even for many-core architectures (Price et al. 2006; Hegeman et al. 2006; Harada et al. 2007; Crespo et al. 2011);

Although SPH has been successful in a broad spectrum of engineering applications, yet several stumbling obstacles are to be overcome. Among which very subtle and difficult challenging issues are:

- Setting boundary conditions in SPH, such as inlets/outlets and walls of container (reservoir), is more difficult than with "mesh-based methods

(MBMs)". The logical difficulty lies in the fact that SPH was invented primarily to deal with astrophysical problems where no domain boundary exists.

- The computation cost of SPH per number of particles is significantly higher than the computation cost of MBMs per number of cells/elements when the interest is not (directly) related to density. Therefore, the simulation of constant-density flows (e.g. external aerodynamics) is less efficient with SPH.
- SPH needs large amount of particles to produce competent resolution compared with MBMs. In the typical implementations with both uniform cell/element grid and SPH particle assembly, more particles will be used to fill water body that are never rendered. As a result, the SPH for fluid simulation is more increasingly used in real-time animation and games where accuracy is not as critical as interactivity.
- SPH requires extra programming to determine NNP.

10.7.3 Improvements and Extensions

Earlier SPH suffers from tensile instabilities (Swegle et al. 1995) and lack of consistency (Morris, 1996; Belytschko et al. 1998). The application of SPH to a wide range of problems, however, have led to significant extensions and improvements. Over the past decades, different modifications or corrections have been tried to restore the consistency and tensile stability of SPH. These modifications lead to a variety of new SPH versions. The milestone works that we need to be remembered are briefly cited below (incomplete list).

Monaghan (1988, 1982, 1985) first proposed symmetrizing formulations. Later on, Johnson and Beissel (1996), Randles and Libersky (2000) gave an axis symmetry normalization formulation. Through revisiting the consistency and reproducing conditions in SPH, Liu and his co-workers proposed the "reproducing kernel particle method (RKPM)" which improves the accuracy of SPH approximation especially near the solid boundary (Liu et al. 1995). Chen et al. (1999) proposed a "corrective smoothed particle method (CSPM)" which improves the simulation accuracy both inside the problem domain and around the vicinity of boundary area. To tackle the tensile instability and zero energy mode problems, Randles and Libersky (2000) extended the "stress point method (SPM)". For a better handling of boundary, Akinci et al. (2013) introduced a versatile surface tension and two-way fluid–solid adhesion technique that allows for simulating a variety of physical effects. Mahdavi and Talebbeydokhti (2015) proposed a hybrid algorithm for the implementation of solid boundary condition and simulate the flow over a sharp crested weir. Other notable modifications or corrections of the SPH method include the "moving least square particle hydrodynamics (MLSPH)" by Dilts (1999, 2000), the integration kernel correction by Bonet and Kulasegaram (2000), and many others.

When the SPH approximation is used to create point-dependent shape functions, it can be extended to other areas of mechanics rather than classical hydrodynamics. For example, the idea of nested summation in approximating second order derivatives have allowed for more widespread application of SPH in CSD with material strength (Libersky and Petschek 1990), which is called "smoothed particle mechanics (SPM)" in some literatures (Kum et al. 1995; Posch et al. 1995). The main advantage of SPM is the possibility to deal with larger local distortion during crack growth and fragmentation process (Benz and Asphaug 1993, 1994, 1995; Rabczuk et al. 2003), impact and compact process (Randles and Libersky 1997; Hoover and Hoover 2001; Rosswog 2015), metal forming process (Bonet and Kulasegaram 2000), shock wave propagation process (Herreros and Mabssout 2011), etc.

Although SPH claims its effectiveness in modeling deformable solid aggregates (Capone et al., 2010; Chambon et al., 2011; Manenti et al., 2012, 2015; Guandalini et al., 2015), yet it generally overestimates the deformation and only suitable for modeling flow-like landslides (Rzadkiewicz et al., 1997; Shao and Lo, 2003; Ataie-Ashtiani and Shobeyri 2008; Huang and Nydal 2012; Dai et al. 2008). It is interesting to remind that, Violeau and Rogers (2016) numerically performed tri-phasic tests using N-S solver to simulate granular deformable landslide and compared with their previous experiments. They concluded that neither Newtonian nor non-Newtonian rheology is appropriate to simulate landslides.

10.7.4 Applications

For a comprehensive overview w.r.t. the industrial applications of SPH, our readers are referred to the work by Shadloo and his co-worker (Shadloo et al. 2016).

The earliest applications of SPH in engineering were mainly focused on CFD related areas. These include shock simulations (Monaghan and Gingold 1983; Monaghan 1989; Morris and Monaghan 1997), elastic flow (Swegle 1992), quasi-incompressible flows (Monaghan 1994; Morris and Monaghan 1997), gravity currents (Monaghan 1995), heat conduction (Monaghan 1995), multi-phase flows (Monaghan and Kocharyan 1995), magneto-hydrodynamics (Morris 1996), heat transfer and mass flow (Cleary 1998), ice and cohesive grains (Gutfraind and Savage 1998; Oger and Savage 1999), flow through porous media (Morris et al. 1999).

Today, SPH has been applied in many areas related to the sciences of our universe (Price 2009), such as the simulations of binary stars and stellar collisions (Benz 1988, 1989; Monaghan 1992), the evolution of the universe (Monaghan 1990), the collapse as well as the formation of galaxies (Monaghan and Lattanzio 1991), etc.

Attributable to its attractive feature in simulating large deformation and impulsive loading events, one significant application area of SPH is "high (or hyper) velocity impact (HVI)" problems concerning the effects of projectiles impacting upon space assets (satellites, space stations, shuttles). In HVI problems, shock waves propagate through the colliding bodies that behave like fluids (Zukas, 1982, 1990). Libersky and his co-workers (Libersky and Petschek 1991; Libersky et al. 1993; Libersky et al. 1995; Randles and Libersky 1996), Johnson and his co-workers (Johnson et al. 1993; Johnson et al. 1996), and many others, have made outstanding contributions in the application of SPH to impact problems.

Another important application of SPH in CSD is the explosion phenomena arising from the detonation of "high explosive (HE)". Swegle and Attaway (1995) investigated the feasibility of using SPH for underwater explosion calculations. Liu and his co-workers applied SPH to model a series of explosion phenomena (Liu et al. 2000, Liu and Liu 2003; Liu et al. 2002a, b, Liu et al. 2003a, b). For an excellent review, our readers are referred to the work of Wang and his co-workers (Wang et al. 2020).

References

Abadie S, Morichon D, Grilli S, Glockner S. Numerical simulation of waves generated by landslides using a multiple-fluid Navier-Stokes model. Coast Eng. 2010;57(9):779–94.

Adami S, Hu XY, Adams NA. A generalized wall boundary condition for smoothed particle hydrodynamics. J Comput Phys. 2012;231(21):7057–75.

Akinci N, Akinci G, Teschner M. Versatile surface tension and adhesion for SPH fluids. ACM Trans Graphics. 2013;32(6):1–8.

Alder BJ, Wainright TE. Phase transition for a hard sphere system. J Chem Phys. 1957;27(5):1208–9.

Anderson TB, Jackson R. Fluid mechanical description of fluidized beds. Equations of motion. Ind Eng Chem Fundamentals. 1967;6(4):527–39.

Antuono M, Colagrossi A, Marrone S. Numerical diffusive terms in weakly-compressible SPH schemes. Comput Phys Commun. 2012;183(12):2570–80.

Antuono M, Marrone S, Colagrossi A, Bouscasse B. Energy balance in the δ-SPH scheme. Comp Meth Appl Mech Eng. 2015;289:209–26.

Ataie-Ashtiani B, Shobeyri G. Numerical simulation of landslide impulsive waves by incompressible smoothed particle hydrodynamics. Int J Numer Meth Fluids. 2008;56(2):209–32.

Basa M, Quinlan NJ, Lastiwka M. Robustness and accuracy of SPH formulations for viscous flow. Int J Numer Meth Fluids. 2009;60(10):1127–48.

Belytschko T, Fleming M, Organ D, Krongauz Y, Liu WK. Smoothing and accelerated computations in the element free Galerkin method. J Comput Appl Math. 1996;74(1–2):111–26.

Belytschko T, Guo T, Liu WK, Xiao SP. A unified stability analysis of meshless particle methods. Int J Numer Meth Eng. 2000;48(9):1359–400.

Belytschko T, Krongauz Y, Dolbow J, Gerlach C. On the completeness of meshfree particle methods. Int J Numer Meth Eng. 1998;43(5):785–819.

Benz W. Applications of smooth particle hydrodynamics (SPH) to astrophysical problems. Comput Phys Commun. 1988;48(1):97–105.

Benz W. Smoothed particle hydrodynamics: a review. NATO Workshop. Arcs (France): Les; 1989.

Benz W, Asphaug E. Explicit 3d continuum fracture modeling with smoothed particle hydrodynamics. In: Proc 24th Lunar and Planetary Sci Conf. Lunar and Planetary Inst; 1993. P. 99–100.

Benz W, Asphaug E. Impact simulations with fracture I methods and tests. Icarus. 1994;107(1):98–116.

Benz W, Asphaug E. Simulations of brittle solids using smooth particle hydrodynamics. Comput Phys Commun. 1995;87(1–2):253–65.

Binder K. The Monte Carlo method in condensed matter physics. Berlin: Springer; 1988.

Binder K. The Monte Carlo method in statistical physics. Berlin: Springer; 1992.

Bird GA. Molecular gas dynamics and the direct simulation of gas flow. Oxford: Oxford University Press; 1994.

Bonet J, Kulasegaram S. Correction and stabilization of smooth particle hydrodynamics methods with applications in metal forming simulations. Int J Numer Meth Eng. 2000;47(6):1189–214.

Bui HH, Nguyen GD. Smoothed particle hydrodynamics (SPH) and its applications in geomechanics: from solid fracture to granular behaviour and multiphase flows in porous media. Comput Geotech. 2021;138:104315.

Campbell PM. Some new algorithms for boundary values problems in smoothed particle hydrodynamics. DNA-88-286, DNA Report, 1989.

Capone T, Panizzo A, Monaghan JJ. SPH modelling of water waves generated by submarine landslides. J Hydraul Res. 2010;48(S1):80–4.

Chambon G, Bouvarel R, Laigle D, Naaim M. Numerical simulations of granular free-surface flows using smoothed particle hydrodynamics. J Nonnewton Fluid Mech. 2011;166(12–13):698–712.

Chen JS, Yoon S, Wang HP, Liu WK. An improved reproducing kernel particle method for nearly incompressible hyperelastic solids. Comp Meth Appl Mech Eng. 2000;181:117–45.

Chen JK, Beraun JE, Carney TC. A corrective smoothed particle method for boundary value problems in heat conduction. Int J Numer Meth Eng. 1999;46(2):231–52.

Chen S, Doolen GD. Lattice Boltzmann method for fluid flows. Annu Rev Fluid Mech. 1998;30:329–46.

Chorin AJ. Discretization of a vertex sheet, with an example of roll-up. J Comput Phys. 1973;13(3):423–9.

Cleary PW. Modeling confined multi-material heat and mass flows using SPH. Appl Math Modeling. 1998;22(12):981–93.

Crespo AJC, Gómez-Gesteira M, Dalrymple RA. Boundary conditions generated by dynamic particles in SPH methods. CMC: Comput Mater Continua. 2007;5(3):173–84.

Crespo AJC, Dominguez JM, Barreiro A, Gomez-Gómez M, Rogers BD. GPUs, a new tool of acceleration in CFD: efficiency and reliability on smoothed particle hydrodynamics methods. PLoS ONE. 2011;6(6):e20685.

Dai ZL, Huang Y, Cheng HL, Xu Q. 3D numerical modeling using smoothed particle hydrodynamics of flow-like landslide propagation triggered by the 2008 Wenchuan earthquake. Eng Geol. 2014;180(SI):21–33.

Dilts GA. Moving least square particle hydrodynamics. I. Consistency and stability. Int J Numer Meth Eng. 1999;44(8):1115–55.

Dilts GA. Moving least square particle hydrodynamics. II. Conservation and boundaries. Int J Numer Meth Eng. 2000;48(10):1503–24.

Dyka CT, Ingel RP. An approach for tension instability in smoothed particle hydrodynamics (SPH). Comput Struct. 1995;57(4):573–80.

Dyka CT, Randles PW, Ingel RP. Stress points for tension instability in SPH. Int J Numer Meth Eng. 1997;40(13):2325–41.

Ferrand M, Laurence DR, Rogers BD, Violeau D, Kassiotis C. Unified semi-analytical wall boundary conditions for inviscid, laminar or turbulent flows in the meshless SPH method. Int J Numer Meth Fluids. 2013;71(4):446–72.

Flebbe O, Muzei S, Herold H, Riffert H, Ruder H. Smoothed particle hydrodynamics-physical viscosity and the simulation of accretion disks. Astrophys J. 1994;431:754–60.

Fraser K, Kiss LI, St-George L. A generalized wall boundary condition for smoothed particle hydrodynamics. In: Proceedings of 14th Int LS-DYNA Conf. 2016.

Gibson JB, Goland AN, Milgram M, Vineyard GH. Dynamics of radiation damage. Phys Rev.1960;120:1229–53.

Gingold RA, Monaghan JJ. Smoothed particle hydrodynamics: theory and application to non-spherical stars. Monthly Notices R Astron Soc. 1977;181(3):375–89.

Gómez-Gesteira M, Rogers BD, Crespo AJC, Dalrymple RA, Narayanaswamy M, Dominguez JM. SPHysics—development of a free-surface fluid solver—Part 1: theory and formulations. Comput Geosci. 2012a;48:289–99.

Gómez-Gesteira M, Crespo AJC, Rogers BD, Dalrymple RA, Dominguez JM, Barreiro A. SPHysics—development of a free-surface fluid solver—Part 2: efficiency and test cases. Comput Geosci. 2012b;48:300–7.

Guandalini R, Agate G, Manenti S, Sibilla S, Gallati M. SPH based approach toward the simulation of non-cohesive sediment removal by an innovative technique using a controlled sequence of underwater micro-explosions. Proc IUTAM. 2015;18:28–39.

Gutfraind R, Savage SB. Flow of fractured ice through wedge-shaped channels: smoothed particle hydrodynamics and discrete-element simulations. Mech Mater. 1998;29(1):1–17.

Harada T, Koshizuka S, Kawaguchi Y. Smoothed particle hydrodynamics on GPUs. Comput Graphics Int. 2007;4(4):671–91.

Hegeman K, Carr NA, Miller GSP. Particle-based fluid simulation on the GPU. In: Proceedings of International Conference on Computer Science (3994). Heidelberg: Springer; 2006.

Herreros MI, Mabssout M. A two-steps time discretization scheme using the SPH method for shock wave propagation. Comput Meth Appl Mech Eng. 2011;200(21–22):1833–45.

Hernquist L, Katz N. TreeSPH—a unification of SPH with the hierarchical tree method. Astrophys J Suppl Ser. 1989;70:419–46.

Hicks DL, Liebrock LM. Lanczos' generalized derivative: insights and applications. Appl Math Comput. 2000;112(1):63–73.

Hockney RW, Eastwood JW. Computer simulations using particles. New York (USA): Adamhilger; 1988.

Hoover WG. Smooth particle applied mechanics (The state of the art), Advanced series in nonlinear dynamics (Vol. 25). Singapore: World Scientific, 2006.

Hoover WG, Hoover CG. SPAM-based recipes for continuum simulations. Comput Sci Eng. 2001;3(2):78–85.

Huang YJ, Nydal OJ. Coupling of discrete-element method and smoothed particle hydrodynamics for liquid-solid flows. Theor Appl Mech Lett. 2012;2(1):012002.

Johnson GR, Beissel SR. Normalized smoothed functions for SPH impact computations. Int J Numer Meth Eng. 1996;39(16):2725–41.

Johnson GR, Petersen EH, Stryk RA. Incorporation of an SPH option into the EPIC code for a wide range of high velocity impact computations. Int J Impact Eng. 1993;14(1–4):385–94.

Johnson GR, Stryk RA, Beissel SR. SPH for high velocity impact computations. Comp Meth Appl Mech Eng. 1996;139:347–73.

Koshizuka S, Oka Y, Tamako H. A particle method for calculating splashing of incompressible viscous fluid. La Grange Park: American Nuclear Society Inc; 1995.

Koshizuka S, Nobe A, Oka Y. Numerical analysis of breaking waves using the moving particle semi-implicit method. Intl J Numer Meth Fluids. 1998;26(7):751–69.

Kum O, Hoover WG, Posch HA. Viscous conducting flows with smooth-particle applied mechanics. Phys Rev E. 1995;109:67–75.

Laguna P. Smoothed particle interpolation. Astrophys J. 1995;439:814–21.

Lattanzio JC, Monaghan JJ, Pongracic H, Schwartz MP. Controlling penetration. SIAM J Sci Stat Comput. 1986;7(2):591–8.

Leonard A. Vortex methods for flow simulation. J Comput Mech. 1980;37(3):289–335.

Libersky LD, Petschek AG. Smooth particle hydrodynamics with strength of materials. In: Trease HE, Fritts MF, Crowley WP, editors. Advances in the Free-Lagrange method (Vol. 248): Berlin: Springer; 1990.

Libersky LD, Petschek AG. Smooth particle hydrodynamics with strength of materials. In: Trease HE, Fritts MF, Crowley WP, editors. Advances in the Free-Lagrange method (Vol.395). Berlin: Springer; 1991. p. 248–57.

Libersky LD, Petschek AG, Carney AG, Hipp TC, Allahdadi JR, High FA. Strain Lagrangian hydrodynamics: a three-dimensional SPH code for dynamic material response. J Comput Phys. 1993;109(1):67–75.

Libersky LD, Randles PW, Carney TC. SPH calculations of fragmentation. In: Atluri SN, Yagawa G, Cruse T, editors. Proceedings of the 3rd US Cong Computational Mechanics (Computational Mechanics '95 :Theory and Applications). Dallas, TX (USA); 1995.

Liu GR, Liu MB. Smoothed particle hydrodynamics: a meshfree particle method. New Jersey: World Scientific Publishing Co., Pte. Ltd; 2003.

Liu MB, Liu GR. Smoothed Particle Hydrodynamics (SPH): an overview and recent developments. Arch Comput Meth Eng. 2010;17(1):25–76.
Liu MB, Liu GR, Lam KY. Investigations into water mitigations using a meshless particle method. Shock Waves. 2002a;12(3):181–95.
Liu MB, Liu GR, Lam KY. Coupling meshfree particle method with molecular dynamics—a novel approach for multi-scale simulations. In: Liu GR, editors. Advances in meshfree and X-FEM methods; 2002b, p. 211–6.
Liu MB, Liu GR, Lam KY. Comparative study of the real and artificial detonation models in underwater explosions. Eng Simul. 2003a;25(2):113–24.
Liu MB, Liu GR, Zong Z. An overview on smoothed particle hydrodynamics. Int J Comput Meth. 2008;5(1):135–88.
Liu MB, Liu GR, Zong Z, Lam KY. Numerical simulation of underwater explosion by SPH. In: Atluri SN, Brust FW, editors. Advances in computational engineering & science. Los Angeles; 2000, p. 1475–80.
Liu MB, Liu GR, Zong Z, Lam KY. Computer simulation of high explosive explosion using smoothed particle hydrodynamics methodology. Comput Fluids. 2003b;32(3):305–22.
Liu WK, Chen Y. Wavelet and multiple scale reproducing kernel method. Int J Numer Meth Fluid. 1995;21(10):901–31.
Liu WK, Han WM, Lu HS, Li SF, Cao J. Reproducing kernel element method. Part I. Theoretical formulation. Comput Meth Appl Mech Eng. 2004;193(12–14):933–51.
Liu WK, Jun S, Li S, Adee J, Belytschko T. Reproducing kernel particle methods for structural dynamics. Int J Numer Meth Eng. 1995a;38(10):1655–79.
Liu WK, Jun S, Zhang YF. Reproducing kernel particle methods. Int J Numer Meth Fluids. 1995b;20(8–9):1081–106.
Lo EYM, Shao S. Simulation of near-shore solitary wave mechanics by an incompressible SPH method. Appl Ocean Res. 2002;24(5):275–86.
Lucy LB. A numerical approach to the testing of the fission hypothesis. Astron J. 1977;8(12):1013–24.
Mahdavi A, Talebbeydokhti N. A hybrid solid boundary treatment algorithm for smoothed particle hydrodynamics. Scientia Iranica, Trans a, Civil Eng. 2015;22(4):1457–69.
Manenti S, Pierobon E, Gallati M, Sibilla S, D'Alpaos L, Macchi E, Todeschini S. Vajont disaster: smoothed particle hydrodynamics modeling of the post event 2D experiments. J Hydraul Eng. 2015;142(4):05015007.
Manenti S, Sibilla S, Gallati M, Agate G, Guandalini R. SPH simulation of sediment flushing induced by a rapid water flow. J Hydraul Eng. 2012;138(3):272–84.
Markauskas D, Kruggel-Emden H, Sivanesapillai R, Steeb H. Comparative study on mesh-based and mesh-less coupled CFD-DEM methods to model particle-laden flow. Powder Technol. 2017;305:78–88.
Marrone S, Antuono M, Colagrossi A, Colicchio G, Le Touzé D, Graziani G. δ-SPH model for simulating violent impact flows. Comp Meth Appl Mech Eng. 2011;200(13):1526–42.
Martin JC, Moyce WJ. An experimental study of the collapse of liquid columns on a rigid horizontal plane. Philos Trans R Soc Londn a: Math Phys Eng Sci. 1952;244(882):312–24.
Mayrhofer A, Rogers BD, Violeau D, Ferrand M. Investigation of wall bounded flows using SPH and the unified semi-analytical wall boundary conditions. Comput Phys Commun. 2013;184(11):2515–27.
Metropolis N, Ulam S. The Monte Carlo method. J Am Stat Assoc. 1949;44(247):335–41.
Molteni D, Colagrossi A. A simple procedure to improve the pressure evaluation in hydrodynamic context using the SPH. Comput Phys Commun. 2009;180(6):861–72.
Monaco AD, Manenti S, Gallati M, Sibilla S, Agate G, Guandalini R. SPH modeling of solid boundaries through a semi-analytic approach. Eng Appl Comput Fluid Mech. 2011;5(1):1–15.
Monaghan JJ. Why particle methods work. SIAM J Sci Stat Comput. 1982;3(3):422–33.
Monaghan JJ. Particle methods for hydrodynamics. Comput Phys Rep. 1985;3(2):71–124.
Monaghan JJ. An introduction to SPH. Comput Phys Commun. 1988;48(1):89–96.

Monaghan JJ. On the problem of penetration in particle methods. J Comput Phys. 1989;82(1):1–15.
Monaghan JJ. Modeling the universe. Proc Astron Soc Austr. 1990;18(3):233–7.
Monaghan JJ. Smoothed particle hydrodynamics. Annual Rev Astron Astrophys. 1992;30(1):543–74.
Monaghan JJ. Simulating free surface flow with SPH. J Comput Phys. 1994;110(2):399–406.
Monaghan JJ. Simulating gravity currents with SPH lock gates. Applied mathematics reports and preprints. Monash (Australia): Monash University, 1995.
Monaghan JJ. SPH without a tensile instability. J Comput Phys. 2000;159(2):290–311.
Monaghan JJ. Smoothed particle hydrodynamics. Rep Prog Phys. 2005;68(8):1703–59.
Monaghan JJ, Gingold RA. Shock simulation by the particle method. J Comput Phys. 1983;52(2):347–89.
Monaghan JJ, Kocharyan A. SPH simulation of multi-phase flow. Computer Physics Communication. 1995;87(1–2):225–35.
Monaghan JJ, Lattanzio JC. A refined particle method for astrophysical problems. Astron Astrophys. 1985;149(1):135–43.
Monaghan JJ, Lattanzio JC. A simulation of the collapse and fragmentation of cooling molecular clouds. Astrophys J. 1991;375(1):177–89.
Monaghan JJ, Poinracic J. Artificial viscosity for particle methods. Appl Numer Math. 1985;1(3):187–94.
Monaghan JJ, Kos A. Scott Russell's wave generator. Phys Fluids. 2000;12(3):622–30.
Morris JP. Analysis of smoothed particle hydrodynamics with applications. PhD Thesis. Monash University (Australia); 1996.
Morris JP, Fox PJ, Zhu Y. Modeling low Reynolds number incompressible flows using SPH. J Comput Phys. 1997;136(1):214–26.
Morris JP, Monaghan JJ. A switch to reduce SPH viscosity. J Comput Phys. 1997;136(1):41–50.
Morris JP, Zhu Y, Fox PJ. Parallel simulation of pore-scale flow though porous media. Comput Geotech. 1999;25(4):227–46.
Nelson RP, Papaloizou JCB. Variable smoothing lengths and energy conservation in smoothed particle hydrodynamics. Mon Not R Astron Soc. 1994;270(1):1–20.
Nitsche LC, Zhang WD. Atomistic SPH and a link between diffusion and interfacial tension. Fluid Mech Transp Phenomena. 2002;48(2):201–11.
Oger L, Savage SB. Smoothed particle hydrodynamics for cohesive grains. Comput Methods Appl Mech Eng. 1999;180(1–2):169–83.
Pan LS, Liu GR, Lam KY. Determination of slip coefficient for rarefied gas flows using direct simulation Monte Carlo. J Micromech Microeng. 1999;9(1):89–96.
Pan LS, Liu GR, Khoo BC, Song B. A modified direct simulation Monte Carlo method for low-speed microflows. J Micromech Microeng. 2000;10(1):21–7.
Pan LS, Ng TY, Xu D, Liu GR, Lam KY. Coefficient using the direct simulation Monte Carlo method. J Micromech Microeng. 2002;12(1):41–52.
Posch HA, Hoover WG, Kum O. Steady-state shear flows via nonequilibrium molecular dynamics and smoothed-particle applied mechanics. Phys Rev E. 1995;52(2):1711–9.
Price DJ. Smoothed particle hydrodynamics: things i wish my mother taught me. Adv Comput Astrophys: Methods Tools Outcomes. 2011;453:249–58.
Price DJ. Astrophysical smooth particle hydrodynamics. New Astron Rev. 2009;53(4–6):78–104.
Price DJ, Rockefeller G, Warren MS. SNSPH: A parallel 3-D smoothed particle radiation hydrodynamics code. Astrophys J. 2006;643(1):292–305.
Qian YH, Succi S, Orszag SA. Recent advances in lattice Boltzmann computing. In: Stauffer, editor. Annual reviews of computational physics (Vol. III). Singapore: World Scientific; 2000.
Qiu LC. Numerical modeling of liquid-particle flows by combining SPH and DEM. Ind Eng Chem Res. 2013;52(33):11313–8.
Quinlan NJ, Basa M, Lastiwka M. Truncation error in mesh-free particle methods. Int J Numer Meth Eng. 2006;66(13):2064–85.

Rabczuk T, Eibl J, Stempniewski L. Simulation of high velocity concrete fragmentation using SPH/MLSPH. Int J Numer Meth Eng. 2003;56(10):1421–44.
Rahman A. Correlations in the motion of atoms in liquid Argon. Phys Rev. 1964;136(2A):405–11.
Randles PW, Libersky LD. Smoothed particle hydrodynamics: some recent improvements and applications. Comp Meth Appl Mech Eng. 1996;139(1–4):375–408.
Randles PW, Libersky LD. Recent improvements in SPH modeling of hypervelocity impact. Int J Impact Eng. 1997;20(6):525–32.
Randles PW, Libersky LD. Normalized SPH with stress points. Int J Numer Meth Eng. 2000;48(10):1445–62.
Rhoades CE. A fast algorithm for calculating particle interactions in smooth particle hydrodynamic simulations. Comput Phys Commun. 1992;70(3):478–82.
Riffert H, Herold H, Flebbe O, Ruder H. Numerical aspects of the smoothed particle hydrodynamics method for simulating accretion disks. Comput Phys Commun. 1995;89(1):1–16.
Robinson M, Ramaioli M, Luding S. Fluid-particle flow simulations using two-way-coupled mesoscale SPH-DEM and validation. Int J Multiph Flow. 2014;59(1):121–34.
Rosswog S. SPH methods in the modelling of compact objects. Living Rev Comput Astrophys. 2015;1(1):1–109.
Rzadkiewicz SA, Mariotti C, Heinrich P. Numerical simulation of submarine landslides and their hydraulic effects. J Waterw Port Coast Ocean Eng. 1997;123(4):149–57.
Shadloo MS, Oger G, Touzé DL. Smoothed particle hydrodynamics method for fluid flows, towards industrial applications: motivations, current state, and challenges. Comput Fluids. 2016;136(1):11–34.
Shao S, Lo EYM. Incompressible SPH method for simulating Newtonian and non-Newtonian flows with a free surface. Adv Water Resour. 2003;26(7):787–800.
Shapiro PR, Martel EL, Villumsen JV, Owen JM. Adaptive smoothed particle hydrodynamics, with application to cosmology: methodology. Astrophys J Supplement. 1996;103(2):269–330.
Simpson JC. Numerical techniques for three-dimensional smoothed particle hydrodynamics simulations: applications to accretion disks. Astrophys J. 1995;448(2):822–31.
Stillinger FH, Rahman A. Improved simulation of liquid water by molecular dynamics. J Chem Phys. 1974;60(4):1545–57.
Sun XS, Sakai M, Yamada Y. Three-dimensional simulation of a solid-liquid flow by the DEM-SPH method. J Comput Phys. 2013;248:147–76.
Swegle JW. TOODY IV—a computer program for two-dimensional wave propagation. SAND-78-0552. Albuquerque, New Mexico: Sandia National Laboratories; 1978.
Swegle JW, Attaway SW. On the feasibility of using smoothed particle hydrodynamics for underwater explosion calculations. Comput Mech. 1995;17(1):151–68.
Swegle JW, Hicks DL, Attaway SW. Smooth particle hydrodynamics stability analysis. J Comput Phys. 1995;116(1):123–34.
Takeda H, Miyama SM, Sekiya M. Numerical simulation of viscous flow by smoothed particle hydrodynamics. Prog Theor Phys. 1994;92(5):939–59.
Tan H, Chen S. A hybrid DEM-SPH model for deformable landslide and its generated surge waves. Adv Water Resour. 2017;108:256–76.
Vignjevic R, Campbell J, Libersky L. A treatment of zero-energy modes in the smoothed particle hydrodynamics. Comp Meth Appl Mech Eng. 2000;184:67–85.
Violeau D. Fluid mechanics and the SPH method. Oxford: Oxford University Press; 2012.
Violeau D, Rogers BD. Smoothed particle hydrodynamics (SPH) for free-surface flows: past, present and future. J Hydraul Res. 2016;54(1):1–26.
Viroulet S, Cébron D, Kimmoun O, Kharif C. Shallow water waves generated by subaerial solid landslides. Geophys J Int. 2013a;193(2):747–62.
Viroulet S, Sauret A, Kimmoun O, Kharif C. Granular collapse into water: toward Tsunami landslides. J Visualization. 2013b;16(3):189–91.
Wang GH, Lu WB, Yang GD, Yan P, Chen M, Zhao XH, Li Q. A state-of-the-art review on blast resistance and protection of high dams to blast loads. Int J Impact Eng. 2020;139:103529.

Wendland H. Piecewiese polynomial, positive definite and compactly supported radial functions of minimal degree. Adv Comput Math. 1995;4(1):389–96.

Wu K, Yang D, Wright N. A coupled SPH-DEM model for fluid-structure interaction problems with free-surface flow and structural failure. Comput Struct. 2016;177:141–61.

Yashiro S, Okabe T. Smoothed particle hydrodynamics in a generalized coordinate system with a finite-deformation constitutive model. Int J Numer Meth Eng. 2015;103(11):781–97.

Yeylaghi S, Moa B, Buckham B, Oshkai P, Vasquez J, Crawford C. ISPH modelling of landslide generated waves for rigid and deformable slides in Newtonian and Non-Newtonian reservoir fluids. Adv Water Resour. 2017;107:212–32.

You Y, Chen JS, Lu H. Filters, reproducing kernel and adaptive meshfree method. Comput Mech. 2003;31(3–4):316–26.

Zhang ZQ, Zhou JX, Wang XM, Zhang YF, Zhang L. Investigations on reproducing kernel particle method enriched by partition of unity and visibility criterion. Comput Mech. 2004;34(4):310–29.

Zhu Y, Fox PJ, Morris JP. A pore-scale numerical model for flow through porous media. Int J Numer Anal Meth Geomech. 1999;23(9):881–904.

Zukas JA. Impact dynamics. New York: Wiley; 1982.

Zukas JA. High velocity impact. New York: Wiley; 1990.

Chapter 11
Hybrid Methods with Special Focus on DEM-SPH

Abstract In this chapter, the hybrid method of DEM-SHP is elaborated meant to solve the problem of "fluid–structure interaction (FSI)", with special interest being directed to the landslide and its induced surge waves (tsunamis). The landslide body is discretized by distinct elements whereas the water body is simulated by smoothed particles. On the contact surface between the landslide body and water body, a handshaking algorithm via the interaction forces including repulsion, drag and buoyant are implemented. At the end of this chapter, a series of validations and application are disclosed.

11.1 General

11.1.1 Concept

A hybrid method, in most cases, is a combination of at least two methods implicitly or explicitly, aiming to profit from the advantages of participant methods in handling the multi-phase or/and multi-scale (level) computations. The term "coupled method" also may be used, particularly (but only) meant to the interaction phenomena of multi-phase media such as the "seepage-deformation coupling" mechanism of jointed rock. The handshaking algorithm is normally demanded in hybrid methods, because it determines the compactness/consistency of the field variables around the handshaking area.

1. **One-phase issues**

In one-phase issues, hybrid methods are normally motivated to meet the different features of sub-regions with multi-scale or material/structural characteristics. There are various hybrid techniques dependent on the specific problem to be handled with (Chen 2018), of which the most useful and prevalent one is the hybrid BEM–FEM. BEM is normally used for simulating the far-field materials as equivalent elastic continua, whereas FEM is intended to tackle the non-linear and discontinuous rock-like materials at the vicinity fields where explicit representations for structural planes are demanded. The hybrid BEM–FEM was first proposed by Zienkiewicz and his

S. Chen, *Advanced Computational Methods and Geomechanics*, Springer Tracts in Civil Engineering, https://doi.org/10.1007/978-981-19-7427-4_11

co-workers (Zienkiewicz et al. 1977) then followed by other scholars (Brady and Wassyng 1981; Beer 1983; Elsworth 1986) towards strain/stress and permeability analyses. It is customarily employed for simulating underground structures (Gioda and Carini 1985; Ohkami et al. 1985; Varadarajan et al. 1985; Swoboda et al. 1987; von Estorff and Firuziaan 2000).

In the area of CFD, hybridization of the mesh-based method for far field wave propagation and the SPH for near field surge wave is an typical and prevalent exercise nowadays (Altomare et al. 2015; Tan et al. 2018b).

There are basically two handshaking algorithms for one-phase issues: in the first algorithm, the computational domain was divided into two (or more) parts without any overlap region, the nodes/elements are usually treated as common ones at the boundary between sub-regions; Alternative algorithm has been developed that employs a special region/interface, such as the joint element, which permits the discontinuous properties around the handshaking interface area.

2. **Multi-phase issues**

In multi-phase issues, particularly in the problem of "fluid–structure interaction (FSI)", the structure suffers the pressure from its surrounding fluid flow, which gives rise to the structure deflection, and conversely, the fluid flow is affected by the reaction from structure.

Violent FSI problems associated with distorted free surface flow are very common in the hydraulic engineering such as flood release flow via spillways and energy dissipation zone downstream of dams, underwater blasting and landslide generated surge waves, etc. It is also commonly observed in the nature and widely utilized in various engineering fields such as aerospace engineering (airfoil, engine blade), biomedical engineering (prosthetic heart valve, drug particles), civil engineering (bridge, skyscraper), ocean engineering (riser, offshore platform), ship and marine engineering, energy engineering (piezoelectric device, wind turbine), etc. In many cases, the FSI effect cannot be neglected because of the strong interaction between the fluid and structure, which requires careful consideration to capture the correct behavior of fluid–structure systems.

FSI problems are generally handled from three basic aspects:

- Mechanical behavior of structure;
- Free surface flow;
- Fluid–structure interaction mechanism.

Due to the complicity and limitation in analytical solution, many researchers employed experiments to understand the physical mechanism related to FSI and to maximize or minimize an objective function (e.g., increase the efficiency of energy harvesting and the suppression of structural vibration fatigue). In recent decades however, robust and efficient CMs to investigate FSI problems have been prevalent.

3. **Hybrid strategies**

(1) **Formalism consideration**

As both the fluid and solid phases are involved in FSI problems, CMs for each phase can be the same or different (Walhorn et al. 2005; Zhang et al. 2020). In general, according to the formalism frameworks, they can be dubbed with "Eulerian-Eulerian", "Eulerian–Lagrangian", and "Lagrangian-Lagrangian" (Biscarini 2010; Wu et al. 2016).

The Eulerian-Eulerian formulations are normally mesh-based (Dunne 2006; Takagi et al. 2012). It may encounter difficulties in, such as the trace of fluid free surfaces, the communication between fluid and solid domains through fluid–structure interfaces, the handling of the large displacement and deformation of structures, and the updating of the meshes for fluid and solid domains.

The Eulerian–Lagrangian formulations (Kuhl et al. 2003; Legay et al. 2006; Wang et al. 2014) may be further divided into two types, depending on whether or not the fluid mesh conforms to fluid/structure interface. "Arbitrary Lagrangian–Eulerian (ALE)" methods (Hirt et al. 1997) and "deforming-spatial-domain and space–time" procedures (Tezduyar et al. 1992) are typical ones of the former in which fluid mesh conforms to the interface. "Immersed boundary (IB)" methods (Peskin 1972, 2002; Charles 1977), "cut-cell methods" (Udaykumar et al. 1994), and "level-set methods" (Dunne 2006) belong to the latter. For the former, curvilinear and unstructured meshes are employed for an interface, but the fluid mesh should be moved or regenerated whenever the interface is changed, which increases computational efforts. For the latter, fluid mesh can be fixed regardless of the interface change. Although these methods are able to profit from the Lagrangian formulation for solid structure and Eulerian formulation for fluid flow, yet complicated and time-consuming re-meshing is still required to track large displacements.

The Lagrangian-Lagrangian formulations describe both the fluid and structure by Lagrangian formulation (Antoci et al. 2007; Franci et al. 2016; Han and Cundall 2013; Liu et al. 2014; Tang et al. 2018; Xu et al. 2019). They straightforwardly permit one to capture large displacements, to follow the motion of fluid–structure interface, and to simulate the free surface of fluid.

(2) **Discretization consideration**

Where the fluid flow involves some complexities such as deformable interfaces or interaction with structures, the numerical simulation becomes more challenging and the most effective CMs are not always available. The main reason for this is that the fundamental CMs such as the FEM for structures are best formulated in a Lagrangian frame of reference, whereas the FDM for fluids are usually formulated in a Eulerian frame of reference. As a result, the position track of moving interface requires special algorithms, which tends to degrade the accuracy and efficiency of CMs.

As the advancement in sciences and technologies, hybrid CMs constructed by grid-based methods and particle-based methods have been expanded quickly and nowadays they are viewed as rather efficient tools for the violent FSI problem. It unfolds a great expectation to provide powerful computation tools towards the challenges of giant engineering structures, particularly when encountered with high speed fluid flow and discontinuous rock-like material deformation.

Many researchers have combined FEM with SPH for the analysis of violent FSI problems (Li et al. 2015). The initial industrial application of SPH coupled with FEM was hyper-velocity impact action (Attaway et al. 1994). Such an exercise has been further expanded to simulate the FSI problems involved with impact and free surface flow (Vuyst et al. 2005; Groenenboom and Cartwright 2010). Ghost particle technique (**Chap. 10**) was introduced to impose velocity boundary conditions and to compute FSI forces (Fourey et al. 2010). A hybrid WCSPH-FEM was investigated with a Monaghan boundary condition for the simulations of sloshing and dam break (Yang et al. 2012). Recently, a particle-element contact algorithm was applied to couple ISPH and WCSPH with FEM respectively, both of them presented effective solutions (Liu et al. 2014; Long et al. 2016).

Efforts to couple DEM with a fluid solver have been achieved remarkably in recent decades. Topin and his co-workers (Topin et al. 2012) developed a hybrid DEM-CFD method to simulate the collapse of underwater granular column, by which they showed contradictory effects of the ambient water, namely, the drag effect to slow down meanwhile the lubricant effect to accelerate the granular flow. This hybrid model was also adopted by Shan and Zhao (2014) to investigate the impact behavior of a subaerial granular flow sliding into the reservoir. In addition to the drag and buoyancy, an unsteady interaction force called the "virtual mass force" was considered in their study to replicate the solid–fluid interaction.

11.1.2 Landslide and Surge Waves

Landslide and its generated surge waves (tsunamis) in water body are destructive geological disasters that may result in severe consequences on human lives and properties. For example, the landslide occurred at Lituya Bay (Alaska, USA) on the 8th of July (1958) generated the highest wave run-up of 524 m corresponding to the mean slide impact velocity of 110 m/s (Fritz et al. 2001). The surge waves induced by the Vajont landslide at northern Italy on the 9th of October (1963) overtopped the dam crest, causing downstream inundation and about 2000 casualties (Panizzo et al. 2005a; Panizzo et al. 2005b; Vacondio et al. 2013; Crosta et al. 2016; Zhao et al. 2016). In addition, reservoir bank landslides manifested in mountainous area may also give rise to additional sedimentation and cause damages to hydraulic structures.

Current approaches to investigate the landslide and generated surge waves are mainly undertaken by physical model experiments and numerical simulations.

1. **Physical experiments**

The landslide models used in the experiments can be further distinguished as rigid block slides (Heinrich 1992; Walder et al. 2003; Enet and Grilli 2007;

Sælevik et al. 2009; Viroulet et al. 2013a; Viroulet et al. 2013b; Huang et al. 2014) and granular deformable slides (Rzadkiewicz et al. 1997; Fritz et al. 2001; Fritz et al. 2003a b; Fritz and Moser 2003; Heller 2007; Mohammed and Fritz 2012; Viroulet et al. 2013b; Viroulet et al. 2014; Robbe-Saule et al. 2017). Slide impact velocity exerts, undoubtedly, a significant effect on wave characteristics (Romano et al. 2016). At high slide impact velocity, flow separation emerging on the slide shoulder creates a hydrodynamic impact crater, but there is no such flow separation at low slide impact velocity (Fritz and Moser 2003; Di Risio et al. 2009).

Using multi-blocks connected together, Sælevik, Jensen and Pedersen (2009) performed experiments to analyze surge waves generated by a potential landslide in western Norway. They found that rigid and deformable landslides may emerge different separation and collapse regimes in water.

Based on Froude similitude, Fritz et al. (2001) conducted a series of granular landslide experiments, and their data can well reproduce the slide impact at high velocity, wave generation, propagation and run-up features of the 1958 Lituya Bay event. Viroulet and his co-workers (Viroulet et al. 2013b; Viroulet et al. 2014) conducted extensive experiments to analyze the surge waves induced by granular landslide just above the water surface collapsing into the water tank at low impact velocities, and their work is pioneering to take into account of the frictional coefficient. Heller (2007) also conducted a large number of deformable landslide experiments and obtained useful empirical formulas for the maximum wave amplitude and wave height in the impact region, as well as the wave amplitude and wave height decay in the wave propagation region. The granular flow in both submarine (Rzadkiewicz et al. 1997) and subaerial (Viroulet et al. 2013b) landslide experiments deforms drastically and separates into a dense flow close to the inclined plane with a turbidity current above.

2. Numerical simulations

Since a lot of manpower and material resources are needed to conduct physical experiments, numerical simulation gradually becomes prevalent. Heinrich (1992) firstly adopted improved Nasa-Vof2D to study the surge waves generated by a triangular block sliding into a water tank. Although the computed wave profiles are very close to the experimental data, yet the motion of the triangular block needs to be pre-defined, and the assumption of the rigid landslide generally leads to an overestimated wave amplitude (Heller 2007). Ma and his co-workers (Ma et al. 2013; Ma et al. 2015) looked at the landslide body as a granular flow, and the 3D Non-Hydrostatic WAVE model was employed to study the wave generation and propagation scenarios.

With regard to numerical algorithm, the fundamental grid-based methods, such as "arbitrary Lagrangian–Eulerian (ALE)" (Hughes et al. 1981), have been widely exercised to deal with the FSI problems. Cremonesi et al. (2011) suggested a particle FEM model in which the mesh distortion is checked at each time-marching step and re-meshing is undertaken once a prescribed distortion threshold is reached. Crosta et al. (2016) presented an ALE method to model the 1963 Vajont landslide event which needs no distortion of FE mesh. Compared to the mesh-based methods, particle-based methods such as DEM and DDA free of mesh restrictions seem quite promising in the replication of landslide scenario. For more details, our readers are referred to the comprehensive review by Yavari-Ramshe and Ataie-Ashtiani (2016).

The interest of this chapter is directed to the hybrid DEM-SPH method that is particularly well suited to the problems of FSI in geotechnical and hydraulic engineering (Cleary and Prakash 2004; Ren et al. 2014; Cleary 2015; Canelas et al. 2016). The main advantage in hybrid DEM-SPH is that both the methods are particle-based, namely, particles (either solid or fluid) are driven by the forces from their neighboring ones. In addition, the underlying time integration scheme is quite similar. The only difference is that the inter-particle forces in SPH are derived from N-S equations and those in DEM are attributable to particle contacts (Huang and Nydal 2012). The engineering application is focused on the landslide process and its generated surge waves.

11.2 Interactions and Handshaking Algorithms for Solid Disc-Fluid Particle Systems

In general, either monolithic approaches or partitioned approaches may be used for the purpose of constructing hybrid methods (van Zuijlen and Bijl 2005).

In the monolithic or direct coupling approach, as the name already suggests, the flow of fluid particles and the motion of solid particles are solved simultaneously by constructing an equation set for the whole two-phase system. The direct coupling of DEM with water can be achieved using the dissipative particle dynamics (Bierwisch et al. 2011) or SPH method. Examples of these can be found in a number of literatures (Potapov et al. 2001; Huang and Nydal 2012; Qiu 2013).

In the partitioned approach, the equations for fluid flow and the equations for particle motion are solved separately with two different solvers. The solution is obtained iteratively using the result of one simulation as boundary condition for another until the solution converges. Examples of these include the coupling of DEM with FVM (Grof et al. 2009), the coupling of DEM with spherical-polyhedral particles and the Lattice Boltzmann method (Galindo-Torres 2013).

In the monolithic approach, a specialized code is needed which make the software development more complex, but it usually leads to faster code because no iterative procedure is demanded. On the contrary, the partitioned method provides certain modularity because several kinds of already existing codes can be used. In the study of this book, the partitioned method is employed for the coupling of DEM and SPH, where a handshaking approach is demanded to incorporate the interaction of fluid particles with solid particles, e.g., coupling via drag and buoyancy forces (Cleary and Morrison 2012).

11.2.1 Governing Equations of the Improved PFC for Solid Discs

1. **Momentum equations**

According to **Chap.** 7 (viz. Eqs. (7.9)–(7.10)), the translational and rotational equations with respect to the disc's centroid are determined by the Newton's second law. The general form of the momentum equations in 2-D case can be summarized as follows

$$\begin{cases} m_{d_k}\dfrac{\mathrm{d}^2 x_{d_k}}{\mathrm{d}t^2} + \alpha m_{d_k}\dfrac{\mathrm{d}x_{d_k}}{\mathrm{d}t} = F_{x,d_k} \\ m_{d_k}\dfrac{\mathrm{d}^2 y_{d_k}}{\mathrm{d}t^2} + \alpha m_{d_k}\dfrac{\mathrm{d}y_{d_k}}{\mathrm{d}t} = F_{y,d_k} \end{cases} \quad (k = 1, 2, \ldots, n_d) \tag{11.1a}$$

$$\begin{cases} F_{x,d_k} = Fc_{x,d_k}^{s\to s} + Fd_{x,d_k}^{f\to s} + Fb_{x,d_k}^{f\to s} + m_{d_k} g_x \\ F_{y,d_k} = Fc_{y,d_k}^{s\to s} + Fd_{y,d_k}^{f\to s} + Fb_{y,d_k}^{f\to s} + m_{d_k} g_y \end{cases} \quad (k = 1, 2, \ldots, n_d) \tag{11.1b}$$

$$I_{d_k}\frac{\mathrm{d}^2\theta_{d_k}}{\mathrm{d}t^2} + \alpha I_{d_k}\frac{\mathrm{d}\theta_{d_k}}{\mathrm{d}t} = M_{d_k} \quad (k = 1, 2, \ldots, n_d) \tag{11.2a}$$

$$M_{d_k} = Mc_{d_k} + Mr_{d_k} \quad (k = 1, 2, \ldots, n_d) \tag{11.2b}$$

where: $d_k = 1, 2, \ldots, n_d$—disc sequence; s—solid phase; f—fluid phase; m_k, x_{d_k} and y_{d_k}—mass and central coordinates of disc d_k; α—global damping coefficient; $\mathbf{Fc}_{d_k}^{s\to s}$—contact force between discs or disc/boundary wall (viz. Eq. (7.11)); $\mathbf{Fd}_{d_k}^{f\to s}$—drag force exerted on the solid disc d_k from fluid particle (viz. Eq. (11.7)); $\mathbf{Fb}_{d_k}^{f\to s}$—buoyancy force exerted on the solid disc d_k from fluid particle (viz. Equation (11.19)); I_{d_k}—rotational inertia; θ_{d_k}—rotational velocity (angular velocity); Mc_{d_k}—moment of contact force (viz. Eq. (11.21)); Mr_{d_k}—moment of rotational friction (viz. Eq. (7.25)).

2. **Solution schemes**

The central difference method is adopted to solve the momentum equations, which has been elucidated in Eqs. (7.27)–(7.29).

11.2.2 Governing Equations of the Improved SPH for Fluid Particles

Fluid particles' motion is solved using the locally averaged N-S equations in the SPH method (**Chap.** 10). In this method, the computational domain is discretized into a set of fluid particles carrying field variables such as density, pressure, velocity and position. There are two main steps to obtain the governing equations for SPH, i.e., kernel approximation and particle approximation (Liu and Liu 2003).

1. **Continuity equations**

The pressure field in weakly compressible SPH schemes might be noisy due to large oscillations. In this book, a density diffusion term called δ-SPH (Molteni and Colagrossi 2009; Marrone et al. 2011; Antuono et al. 2012) is added into the continuity equation to stabilize the density field and thus to improve the pressure field.

After considering the δ-SPH model, the continuity equation in **Chap.** 10 may be re-written as

$$\frac{\mathrm{d}\overline{\rho}_i}{\mathrm{d}t} = \sum_{j\in\Omega_i}^{n_i} m_j \mathbf{v}_{ij}.\nabla_i W_{ij} - 2\delta h c_0 \sum_{j\in\Omega_i}^{n_i} (\overline{\rho}_j - \overline{\rho}_i)\mathbf{x}_{ij}.\nabla_i W_{ij} \frac{1}{\mathbf{x}_{ij}.\mathbf{x}_{ij}} \frac{m_j}{\overline{\rho}_j} \quad (i = 1, 2, ..., n_p) \tag{11.3}$$

where the variables and parameters are identical to that in Eq. (10.66). However, it should be kept in mind that the summation in Eq. (11.3) excludes any possible solid particles in the interaction domain of representative fluid particle i.

2. **Momentum equations**

According to **Chap.** 10 (viz. Eq. (10.72)), the momentum equations of SPH is modified in a manner of

$$\begin{cases} \dfrac{\mathrm{d}v_{x,i}}{\mathrm{d}t} = -\displaystyle\sum_{j\in\Omega_i}^{n_i} m_j \left(\frac{p_i}{\overline{\rho}_i^2} + \frac{p_j}{\overline{\rho}_j^2} + \Pi_{ij} \right) \frac{\partial W_{ij}}{\partial x_i} + \dfrac{Fr_{x,i}^{s\to f} + Fd_{x,i}^{s\to f} + Fb_{x,i}^{s\to f}}{m_i} + g_x \\ \dfrac{\mathrm{d}v_{y,i}}{\mathrm{d}t} = -\displaystyle\sum_{j\in\Omega_i}^{n_i} m_j \left(\frac{p_i}{\overline{\rho}_i^2} + \frac{p_j}{\overline{\rho}_j^2} + \Pi_{ij} \right) \frac{\partial W_{ij}}{\partial y_i} + \dfrac{Fr_{y,i}^{s\to f} + Fd_{y,i}^{s\to f} + Fb_{y,i}^{s\to f}}{m_i} + g_y \end{cases}$$

$$(i = 1, 2, ..., n_p) \tag{11.4}$$

where: $\mathbf{g}$—gravitational acceleration; m_i—mass of fluid particle i; Π_{ij}—artificial viscosity (viz. Eq. (10.70)); $\mathbf{Fr}_i^{s\to f}$—repulsion force exerted on the fluid particle i by the vicinity solid particles (viz. Eq. (10.76)); $\mathbf{Fd}_i^{s\to f}$—drag force reacted on the fluid particle i by the vicinity solid particles (viz. Eq. (11.17)); $\mathbf{Fb}_i^{s\to f}$—buoyancy reacted on the fluid particle i by the vicinity solid particles (viz. Eq. (11.20)).

It is notable that where there is no moving boundary, the repulsion force is null (i.e. $\mathbf{Fr}_i^{s\to f} = 0$) in Eq. (11.4); On the contrary, where there are only polygonal block elements moving in fluid, such as the slip/plunge of blocky landslide into reservoir water body, the drag force $\mathbf{Fd}_i^{s\to f}$ and buoyancy $\mathbf{Fb}_i^{s\to f}$ are zero preset, because their action effects have been taken into account by the repulsion force in Eq. (11.4) already.

3. **Equation of state**

The weakly compressible equation of state (viz. Eq. (10.59)) that relates the pressure to the fluid density (Robinson et al. 2014) is used to determine the pressure term in Eq. (11.4)

$$p_i = \frac{c_0^2 \rho_0}{\gamma}\left[\left(\frac{\rho_i}{\varepsilon_i \rho_0}\right)^{\gamma} - 1\right] \quad (i = 1, 2, \ldots, n_p) \tag{11.5}$$

where: γ—often sets to be 7.0; $\rho_0 = 1000\text{kg/m}^3$—reference density; c_0—speed of sound at the reference density determined by $\eta\sqrt{gh_{\max}}$, it is usually at least 10 times the maximum flow velocity in order to keep the density fluctuation within 1%; η—coefficient of sound speed; $h_{\max}$—maximum water depth.

To simplify the formulas, hereinafter the subscript d_k indicating solid disc sequence will be frequently (not always) replaced by the subscript k indicating solid particle sequence, where there is no risk of misleading. The local porosity ε_i at the position of representative fluid particle i in Eq. (11.5) can be calculated by

$$\varepsilon_i = 1 - \sum_{k \in \Omega_i} V_k W_{ik} \quad (i = 1, 2, \ldots, n_p) \tag{11.6}$$

where: k—solid particle (i.e. disc d_k) in the support domain Ω_i; V_k—volume of solid particle k.

It is pointed out that when solid particles are in the interaction domain of a fluid particle, the local porosity decreases meanwhile pressure increases, which results in the fluid particle being pushed away from the solid particles. In order to obtain a smooth porosity field, an additional demand in the hybrid DEM-SPH model is that the smoothing length should be sufficiently larger than the solid particle diameter. Robinson et al. (2014) conducted parametric sensitivity analysis in the test of single particle sedimentation, they suggested that the smoothing length should be at least two times the solid particle diameter.

4. **Solution schemes**

The central difference method is adopted to solve the momentum equations and continuity equation with the help of state equation, which has been elucidated in Chap. 10 (viz. Eqs. (10.63)–(10.64)).

11.2.3 Solid–fluid Interactions

WE take the drag force and buoyancy as the solid–fluid handshaking forces.

1. **Drag force**

Drag force is dependent on the relative velocity of solid particle and fluid (Anderson and Jackson 1967), and a number of empirical formulas are applicable to determine its value in the framework of the locally averaged N-S equations. One of the well-known empirical formula was proposed by Sun et al. (2013) that states

$$\begin{cases} Fd_{x,d_k}^{f\to s} = \dfrac{\zeta_k}{1-\varepsilon_k}\left(\overline{v}_{x,k}^{flow} - v_{x,k}\right)V_k \\ Fd_{y,d_k}^{f\to s} = \dfrac{\zeta_k}{1-\varepsilon_k}\left(\overline{v}_{y,k}^{flow} - v_{y,k}\right)V_k \end{cases} \tag{11.7}$$

where

$$\varepsilon_k = \frac{\sum\limits_{i\in\Omega_k} \varepsilon_i \frac{m_i}{\rho_i} W_{ik}}{\sum\limits_{i\in\Omega_k} \frac{m_i}{\rho_i} W_{ik}} \tag{11.8}$$

$$\begin{cases} \overline{v}_{x,k}^{flow} = \dfrac{\sum\limits_{i\in\Omega_k} v_{x,i} \frac{m_i}{\rho_i} W_{ik}}{\sum\limits_{i\in\Omega_k} \frac{m_i}{\rho_i} W_{ik}} \\ \overline{v}_{y,k}^{flow} = \dfrac{\sum\limits_{i\in\Omega_k} v_{y,i} \frac{m_i}{\rho_i} W_{ik}}{\sum\limits_{i\in\Omega_k} \frac{m_i}{\rho_i} W_{ik}} \end{cases} \tag{11.9}$$

It is actually the combination of Ergun (1952) and Wen and Yu (1966), in which ζ_k is the interphase momentum transfer coefficient, ε_k is the local porosity, $\mathbf{v}_k$ is the transitional velocity of solid particle k (disc d_k), $\overline{\mathbf{v}}_k^{flow}$ is the average flow velocities estimated at the center position of solid particle d_k. Note that the summations (11.8) and (11.9) are only extended to neighboring fluid particles within Ω_k (recall that i denotes the fluid particle sequence and d_k denotes the solid particle sequence).

The interphase momentum transfer coefficient ζ_k is given by

$$\zeta_k = \begin{cases} 150\dfrac{(1-\varepsilon_k)^2}{\varepsilon_k}\dfrac{\mu_f}{D_k^2} + 1.75(1-\varepsilon_k)\dfrac{\rho_f}{D_k}\left|\overline{\mathbf{v}}_k^{flow} - \mathbf{v}_k\right| & \text{if } \varepsilon_k \le 0.8\,, \\ 0.75C_d\dfrac{\varepsilon_k(1-\varepsilon_k)}{D_k}\rho_f\left|\overline{\mathbf{v}}_k^{flow} - \mathbf{v}_k\right|\varepsilon_k^{-2.65} & \text{if } \varepsilon_k > 0.8 \end{cases} \tag{11.10}$$

where: μ_f—dynamic viscosity of fluid; ρ_f—density of fluid; D_k—diameter of solid particle k (disc d_k); C_d—drag coefficient.

The drag coefficient C_d is given by

$$C_d = \begin{cases} \dfrac{24}{\mathrm{Re}_k}(1 + 0.15\mathrm{Re}_k^{0.687}) & \text{if } \mathrm{Re}_k \le 1000 \\ 0.44 & \text{if } \mathrm{Re}_k > 1000 \end{cases} \tag{11.11}$$

$$\text{Re}_k = \frac{\left|\overline{\mathbf{v}}_k^{flow} - \mathbf{v}_k\right| \varepsilon_k \rho_f D_k}{\mu_f} \tag{11.12}$$

in which Re_k is the Reynolds number.

Another widely used empirical formula for the drag force term in Eq. (11.1) is a combination of the formulas by Dalla Valle (1948) and Di Felice (1994) that states

$$\begin{cases} Fd_{x,d_k}^{f\to s} = \frac{1}{8} C_d f(\varepsilon_k) \pi D_k^2 \rho_f \varepsilon_k^2 \left|\overline{\mathbf{v}}_k^{flow} - \mathbf{v}_k\right| \left(\overline{v}_{x,k}^{flow} - v_{x,k}\right) \\ Fd_{y,d_k}^{f\to s} = \frac{1}{8} C_d f(\varepsilon_k) \pi D_k^2 \rho_f \varepsilon_k^2 \left|\overline{\mathbf{v}}_k^{flow} - \mathbf{v}_k\right| \left(\overline{v}_{y,k}^{flow} - v_{y,k}\right) \end{cases} \tag{11.13}$$

where

$$C_d = \left(0.63 + \frac{4.8}{\sqrt{\text{Re}_k}}\right)^2 \tag{11.14}$$

The porosity function $f(\varepsilon_k)$ in Eq. (11.13) may be calculated by

$$f(\varepsilon_k) = \varepsilon_k^{-\zeta} \tag{11.15}$$

where

$$\zeta = 3.7 - 0.65 \exp\left[-\frac{(1.5 - \log_{10} \text{Re}_k)^2}{2}\right] \tag{11.16}$$

The reaction force $\mathbf{Fd}_k^{s\to f}$ of the drag force exerted on the fluid particle i by the solid particle k (disc d_k) needs to satisfy the Newton's third law, namely

$$\begin{cases} Fd_{x,i}^{s\to f} = -\sum_{k\in\Omega_i} \frac{1}{\Lambda_k} \frac{m_i}{\overline{\rho}_i} W_{ik} Fd_{x,d_k}^{f\to s} \\ Fd_{y,i}^{s\to f} = -\sum_{k\in\Omega_i} \frac{1}{\Lambda_k} \frac{m_i}{\overline{\rho}_i} W_{ik} Fd_{y,d_k}^{f\to s} \end{cases} \quad (i = 1, 2, \ldots, n_p) \tag{11.17}$$

in which $\mathbf{Fd}_k^{f\to s}$ is calculated by Eq. (11.7) and

$$\Lambda_k = \sum_{j\in\Omega_i} \frac{m_j}{\overline{\rho}_j} W_{jk} \tag{11.18}$$

This drag reaction is inserted in Eq. (11.4) for the SPH calculation.

2. **Buoyancy force**

The buoyancy is another fluid–solid interaction force in addition to the drag force. In Eq. (11.1), $\mathbf{Fb}_{d_k}^{f\to s}$ exerted on the solid particle k (disc d_k) by the fluid particles can be calculated by

$$\begin{cases} Fb_{x,d_k}^{f\to s} = V_k \sum\limits_{i\in\Omega_k} \dfrac{m_i p_i}{\overline{\rho}_i} \dfrac{\partial W_{ik}}{\partial x_i} \\ Fb_{y,d_k}^{f\to s} = V_k \sum\limits_{i\in\Omega_k} \dfrac{m_i p_i}{\overline{\rho}_i} \dfrac{\partial W_{ik}}{\partial y_i} \end{cases} (k = 1, 2, ..., n_d) \quad (11.19)$$

Similarly, the reaction force $\mathbf{Fb}_i^{s\to f}$ of the buoyancy exerted on the fluid particle i by the solid also needs to satisfy the Newton's third law, namely

$$\begin{cases} Fb_{x,i}^{s\to f} = -\dfrac{m_i p_i}{\overline{\rho}_i} \sum\limits_{k\in\Omega_i} V_k \dfrac{\partial W_{ik}}{\partial x_i} \\ Fb_{y,i}^{s\to f} = -\dfrac{m_i p_i}{\overline{\rho}_i} \sum\limits_{k\in\Omega_i} V_k \dfrac{\partial W_{ik}}{\partial y_i} \end{cases} (i = 1, 2, ..., n_p) \quad (11.20)$$

This buoyancy reaction is inserted in Eq. (11.4) for the SPH calculation.

It is notable that the local porosity greatly influences on the interaction between solid and fluid because it affects both drag force and buoyancy (through the locally averaged density), hence obtaining a smooth porosity field in the hybrid model is demanded in order to accurately capture the interaction mechanism.

3. **Handshaking algorithm**

As stated previously, DEM and SPH are calculated separately and they exchange information only through the solid–fluid interaction module. The detailed computational procedure of the hybrid DEM-SPH model is flowcharted in Fig. 11.1. It can be briefly divided into three modules, i.e., DEM module for the solid phase, SPH module for the fluid phase, and the solid–fluid interaction module.

An iterative process is needed to solve the hybrid DEM-SPH model. Specifically, the drag and buoyancy are exerted to the DEM module towards the solution of solid particle system, and the reaction forces of drag and buoyancy are converted to the SPH module towards the solution of fluid particle system. Afterwards the drag force and buoyancy and corresponding reaction forces are updated according to the new positions of solid and fluid particles for the next time-marching step.

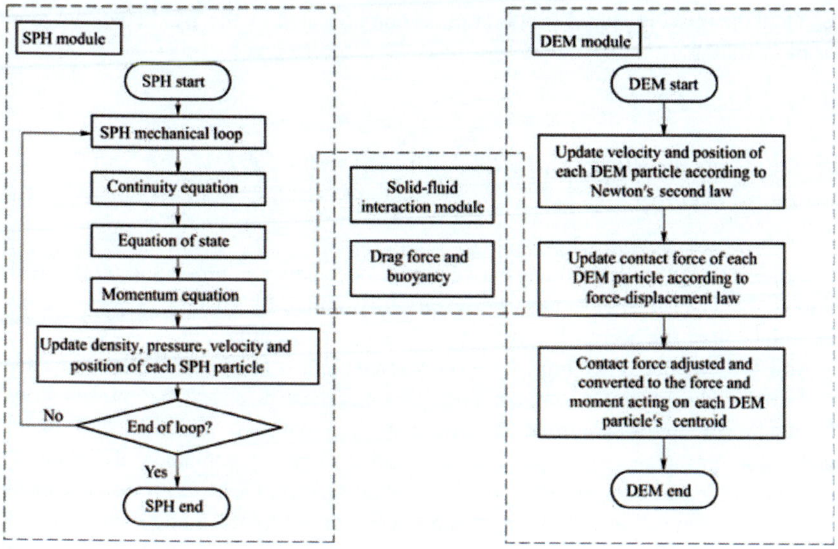

Fig. 11.1 Flowchart of the hybrid DEM-SPH model

11.3 Interactions and Hybrid Algorithms for Solid Polygon-Fluid Particle Systems

In 2-D PFC, the basic solid particles are discs. For polyhedral DEM, the solid particles coupling with SPH particles are complex polygons. For the simplicity of algorithm to simulate landslide and induced surge waves (tsunamis) with hybrid DEM-SPH, two alternatives are applicable apart from using polygonal distinct element directly.

The first one is to fill the block walls with SPH dummy particles (virtual/ghost SPH particles) to ensure the support of the kernel near wall region (Libersky et al. 1993). This method needs simple-shaped wall geometries. The ghost particles have to be recreated with every time-marching step.

The second one considers the non-vanishing surface integral when smoothing the flow quantities close to the boundary or to introduce artificial repulsion forces to prevent particles from entering the solid domain behind block wall (Adami et al. 2012). It possesses the remarkable advantage that only the wall surface needs to be described by a layer of SPH particles (Monaghan and Kajtar 2009) but it misses kernel supporting inside the wall.

Different concepts were compared by Valizadeh and Monaghan (2015), and it was found that the usage of dummy particles for the kernel support combined with a pressure repulsion from the boundary surface currently matches experimental data best. Hence it will be implemented for the coupling of SPH particles to DEM polygons in this book, where independent virtual particles for each SPH particle on the wall are used and the missing kernel support area is compensated by the 1-D kernel function for the virtual particle.

11.3.1 Governing Equations of the Improved DEM for Solid Polygons

In addition to the contact forces between blocks or between block and boundaries, in the hybrid DEM SPH model, the linearly distributed hydrostatic pressure acting on each edge of the block also needs to be taken into account.

As have been stated in **Chap.** 7, 2-D polygonal block may be potentially composed of discs. For the sake of saving computation time, these discs are only placed along the edge of the block. The relative position of discs within the same block remains unchanged during the DEM calculation. Contact force is calculated only when the discs belonging to two different blocks partially overlap with each other. Hence, the complex contact detection between blocks or between block and boundary wall can be simplified as the contact detection between discs only or between discs and boundary walls (viz. Fig. 7.12).

1. **Momentum equations**

The translational and rotational equations with respect to the block's centroid are determined by the Newton's second law and the general form of momentum equation can be expanded from Eqs. (7.34)–(7.37) as:

$$\begin{cases} m_{r_i}\dfrac{\mathrm{d}^2 x_{r_i}}{\mathrm{d}t^2} + \alpha m_{r_i}\dfrac{\mathrm{d}x_{r_i}}{\mathrm{d}t} = F_{x,r_i} \\ m_{r_i}\dfrac{\mathrm{d}^2 y_{r_i}}{\mathrm{d}t^2} + \alpha m_{r_i}\dfrac{\mathrm{d}y_{r_i}}{\mathrm{d}t} = F_{y,r_i} \end{cases} \quad (i = 1, 2, \ldots, n_r) \tag{11.21a}$$

$$\begin{cases} F_{x,r_i} = \sum\limits_{k\in r_i}\sum\limits_{l\in r_j} Fc^{s\to s}_{x,r_i,k,l} + Fvm^{f\to s}_{x,r_i} + m_{r_i} g_x + Fd^{f\to s}_{x,r_i} + Fb^{f\to s}_{x,r_i} \\ F_{y,r_i} = \sum\limits_{k\in r_i}\sum\limits_{l\in r_j} Fc^{s\to s}_{y,r_i,k,l} + Fvm^{f\to s}_{y,r_i} + m_{r_i} g_y + Fd^{f\to s}_{y,r_i} + Fb^{f\to s}_{y,r_i} \end{cases} \tag{11.21b}$$

$$I_{r_i}\frac{\mathrm{d}^2\theta_{r_i}}{\mathrm{d}t^2} + \alpha I_{r_i}\frac{\mathrm{d}\theta_{r_i}}{\mathrm{d}t} = \mathbf{M}_{r_i} \quad (i = 1, 2, \ldots, n_r) \tag{11.22a}$$

$$\mathbf{M}_{r_i} = \sum_{k\in r_i}\sum_{l\in r_j}\left(\mathbf{x}_{c,k,l} - \mathbf{x}_{r_i}\right) \times \mathbf{Fc}_{r_i,k,l} \tag{11.22b}$$

where: r_i—representative rock block sequence; k–sequence of disc d_k attached to the rock block r_i; l—sequence of disc d_l attached to the rock block r_j; notation ×—vector cross product; m_{r_i}—block's mass; I_{r_i}—moment of inertia of block

r_i; $\mathbf{Fc}_{r_i,k,l}^{s\to s}$—contact force exerted on the disc d_k of block r_i from the disc d_l of block r_j (viz. Eq. (7.35)); $\mathbf{M}_{r_i}$—moment of contact force (viz. Eq. (7.37)); $\mathbf{x}_c$—contact point position vector of disc; $\mathbf{x}_{r_i}$—centroid position vector of block r_i; α—global damping coefficient; **g**—gravitational acceleration; $\mathbf{Fd}_{r_i}^{f\to s}$—drag force (viz. Eq. (11.24)); $\mathbf{Fb}_{r_i}^{f\to s}$—buoyant force (viz. Eq. (11.25)); $\mathbf{Fvm}_{r_i}^{f\to s}$—virtual mass force (viz. Eq. (11.32)).

The mass in Eq. (11.21) can be calculated by

$$m_{r_i} = \rho_{r_i} A_{r_i} \Psi \tag{11.23}$$

where: ρ_{ri}—block's density; Ψ—block thickness, normally $\Psi = 1$ in 2-D problems; A_{ri}—block's area calculated by Eq. (7.32).

The moment of inertia in Eq. (11.22) can be calculated using the method described in **Chap.** 7 (viz. Eq. (7.33)).

2. **Solution schemes**

The central difference method is adopted to solve Eqs. (11.21)–(11.22). The integration scheme is identical to that elaborated in Eqs. (7.38)–(7.43).

In the block DEM model, the landslide mass is discretized into a set of blocks and thus the above equations are applied to all of the blocks. It is noted that the moments attributable to buoyant and drag forces are normally neglected because their resultant force passes through the geometrical center of the polygon. Furthermore, the fixed time-marching step is utilized in the block DEM calculation. In order to stabilize the explicit numerical integration scheme, the selected time-marching step length should be sufficiently small. This is actually not an added burden because the accuracy for the simulation of rock block system and fluid particle system undergoing strong configuration evolution rigorously demands small time-marching step length, too.

3. **Calculation of forces and moments**

In the DEM for multi-block system where each block boundary is attached by dummy discs, the contact force **Fc** between discs or between discs and boundary wall is calculated by the force–displacement law of, either the linear contact model or simplified Hertz-Mindlin contact model. Before being converted to the block's centroid with an additional contact moment, contact force is adjusted by Mohr–Coulomb friction law with no tensile strength cut-off. After traversing all the disc-contacts attached to block, the resultant contact force and contact moment in Eqs. (11.21)–(11.22) can be accumulated.

11.3.2 Governing Equations of the Improved SPH for Fluid Particles

Identical to Sect. 11.2.2.

11.3.3 Solid–fluid Interactions

In the hybrid DEM-SPH model, solid phase is governed by DEM and fluid phase is governed by SPH. The handshaking between solid and fluid is realized by the repulsive boundary conditions and correspondent algorithm is depicted in Fig. 11.2.

In the DEM model, the fluid → block interaction is implemented by introducing the buoyant force, drag force and virtual mass force; in the SPH model, the discs attached to the block serve as moving boundary particles exerting repulsion force on fluid particles nearby. The key feature of this handshaking algorithm is that, the discs attached to the block surface play double roles at the same time: DEM particles to calculate contact force, and SPH moving boundary particles to calculate repulsion force.

1. **Drag force**

The drag force for polygonal block is determined by (Pelinovsky and Poplavsky 1996)

$$\mathbf{Fd}_{r_i}^{f\to s} = -\tfrac{1}{2}C_d\rho A_{r_i}\mathbf{v}_{r_i}^2\frac{\mathbf{v}_{r_i}}{\left|\mathbf{v}_{r_i}\right|} \tag{11.24}$$

where: C_d—drag coefficient; ρ—water density; A_{r_i}—block main cross section perpendicular to the direction of motion.

Note that the computed drag force is opposite to the direction of the resultant velocity in the Cartesian coordinate system.

2. **Buoyant force**

It is clear that the presence of water imposes a great effect on the stability of landslide, for example, both the rainfall infiltration and reservoir impoundment are the triggering factors of reservoir bank landslide.

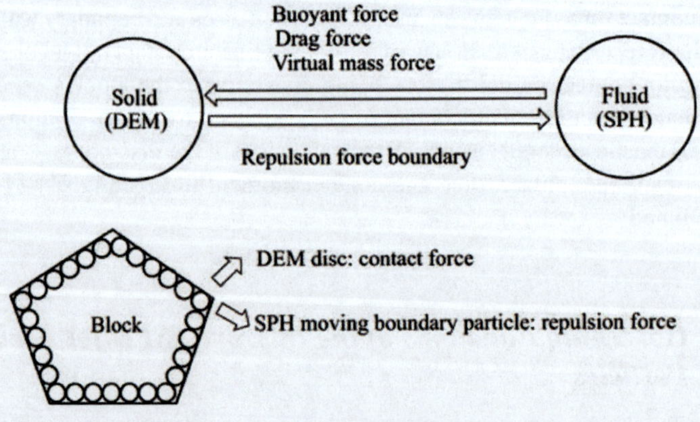

Fig. 11.2 Sketch of handshaking algorithm of DEM-SPH

(1) Distribution of hydrostatic pressure at block edges

Kim et al. (1999) and Jiao et al. (2014) considered the water pressure acting on the block as linearly distributed hydrostatic pressure in their DDA model. In our DEM model, the water → block interaction is similarly calculated according to the water pressure acting on each edge of the block and converted to the block's centroid to update its motion. Before calculating the linearly distributed hydrostatic pressure, the relative position of the edge to the static water surface should be determined with the help of following postulations (viz. Fig. 11.3):

- The vertexes attached to the block are arranged in counterclockwise direction;
- The first and second vertexes of the edge are assigned with coordinates (x_1, y_1) and (x_2, y_2), respectively;
- The vertical distances of the first and second vertexes to the static water surface are H_1 and H_2, respectively.

After the relative position of block edge to the static water surface is determined, the water pressure acting on the block edge can be calculated using a unified algorithm proposed by Kim, Amadei and Pan (1999). Taking the edge sequence $\xi(1, 2, ..., n_\xi)$ in Fig. 11.3b for example where $n_\xi = 4$. In the subroutine of water pressure calculation, it is postulated that the coordinates of the first vertex attached to the edge (referred to as point 1) is (x_1, y_1) and corresponding pressure head is H_1; the coordinates of the second vertex (referred to as point 2) is (x_2, y_2) and corresponding pressure head is H_2.

There are four situations need to be handled:

- Case 1. $H_1 \geq 0$ and $H_2 \geq 0$. The edge is totally above the static water surface (viz. edge B-C in Fig. 11.3a) and there is no need to calculate the water pressure.
- Case 2. $H_1 \leq 0$ and $H_2 \leq 0$. The edge is totally below the static water surface (viz. edge D - A in Fig. 11.3a). As a result, the water pressure distribution is in

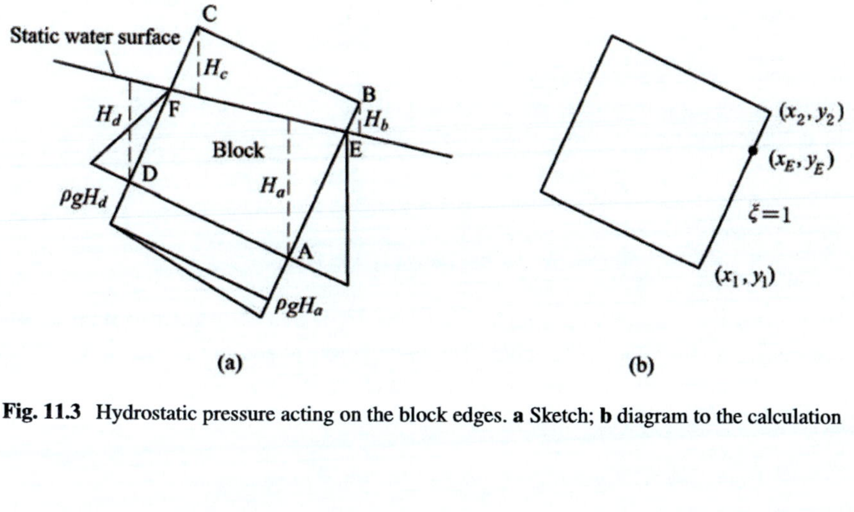

Fig. 11.3 Hydrostatic pressure acting on the block edges. **a** Sketch; **b** diagram to the calculation

trapezoid and calculated using the pressure heads corresponding to the coordinates of vertexes D and A.

- Case 3. $H_1 < 0$ and $H_2 > 0$. The edge is partially below the static water surface (viz. edge A - B in Fig. 11.3a). Since the water pressure is triangularly distributed, the coordinates $(x_3, y_3) = (x_E, y_E)$ of the intersection point E of edge A - B and static water surface is calculated first, then the water pressure is calculated using the pressure heads corresponding to the coordinates of points A and E.
- Case 4. $H_1 > 0$ and $H_2 < 0$. The edge is partially soaked below the static water surface (viz. edge C - D in Fig. 11.3a). The coordinates $(x_3, y_3) = (x_F, y_F)$ of the intersection point F of edge C - D and static water surface is calculated first, then the water pressure is calculated using the pressure heads corresponding to the coordinates of point F and D.

The above distributed hydrostatic pressures are transformed as point load acting on the corresponding block edge. Afterwards, the buoyant force is determined by the summation over each edge of the block, namely,

$$\begin{cases} Fb_{x,r_i}^{f\to s} = \sum\limits_{\xi\in r_i} Fw_{x,r_i,\xi} \\ Fb_{y,r_i}^{f\to s} = \sum\limits_{\xi\in r_i} Fw_{y,r_i,\xi} \end{cases} \tag{11.25}$$

where: ξ—edge sequence of block r_i; $Fb_{x,r_i}^{f\to s}$ and $Fb_{y,r_i}^{f\to s}$—load resulting from the water pressure acting on the block's edge.

(2) **Integration of hydrostatic pressure**

The water pressures at points 1 and 2 can be calculated as

$$\begin{cases} p_1 = \rho g|H_1| \\ p_2 = \rho g|H_2| \end{cases} \tag{11.26}$$

in which ρ is the water density.

The resultant force of the water pressure at the edge between points 1 and 2 can be calculated as

$$\begin{cases} f = \frac{1}{2}(p_1 + p_2) \times L \\ L = \|\mathbf{x}_1 - \mathbf{x}_2\| = \sqrt{(x_1 - x_2)(y_1 - y_2)} \end{cases} \tag{11.27}$$

The load f needs to be transferred to the Cartesian coordinate system before converting to the block's centroid. The calculation of the horizontal and vertical force components of point load f is undertaken as follows

$$\begin{cases} Fw_{x,r_i,\xi} = -f \sin\beta \\ Fw_{y,r_i,\xi} = f \cos\beta \\ \beta = \tan^{-1}\dfrac{y_2 - y_1}{x_2 - x_1} \end{cases} \quad \text{for} \quad x_2 > x_1 \tag{11.28}$$

$$\begin{cases} Fw_{x,r_i,\xi} = f \sin\beta \\ Fw_{y,r_i,\xi} = -f \cos\beta \\ \beta = \tan^{-1}\dfrac{y_2 - y_1}{x_2 - x_1} \end{cases} \quad \text{for} \quad x_1 > x_2 \tag{11.29}$$

$$\begin{cases} Fw_{x,r_i,\xi} = -f \\ Fw_{y,r_i,\xi} = 0 \end{cases} \quad \text{for } x_2 = x_1 \quad and \quad y_2 > y_1 \tag{11.30}$$

$$\begin{cases} Fw_{x,r_i,\xi} = f \\ Fw_{y,r_i,\xi} = 0 \end{cases} \quad \text{for } x_2 = x_1 \quad and \quad y_1 > y_2 \tag{11.31}$$

After traversing all of the edges, the buoyant load resulting from the water pressure in Eq. (11.25) can be obtained. The algorithm is flowcharted in Fig. 11.4.

3. **Virtual mass force**

Shan and Zhao (2014) considered the "virtual mass force" in their study on the FSI, this force only exhibits when the rock block element is accelerated or decelerated and is determined by

$$\begin{cases} fvm_{x,r_i}^{f\to s} = -C_m m_{0,r_i}\dfrac{\mathrm{d}^2 x_{r_i}}{\mathrm{d}t^2} \\ fvm_{y,r_i}^{f\to s} = -C_m m_{0,r_i}\dfrac{\mathrm{d}^2 y_{r_i}}{\mathrm{d}t^2} \end{cases} \tag{11.32}$$

in which C_m is the added mass coefficient that depends on the geometrical shape of block element (Watts 2000); m_{0,r_i} is the mass of the water displaced and it can be calculated by $m_{0,r_i} = \left|\mathbf{Fb}_{r_i}\right|/g$ under the assumption that the displaced water volume is equal to the block volume immersed in the water.

11.4 Validations and Applications

11.4.1 Submarine Landslide-Tsunamis

1. **Problem of Heinrich** (1992)

Heinrich (1992) conducted physical experiments to investigate the surge waves generated by a submarine rigid block sliding down along a 45° inclined plane. This experiment was widely used by researchers (Rzadkiewicz et al. 1997; Ataie-Ashtiani and

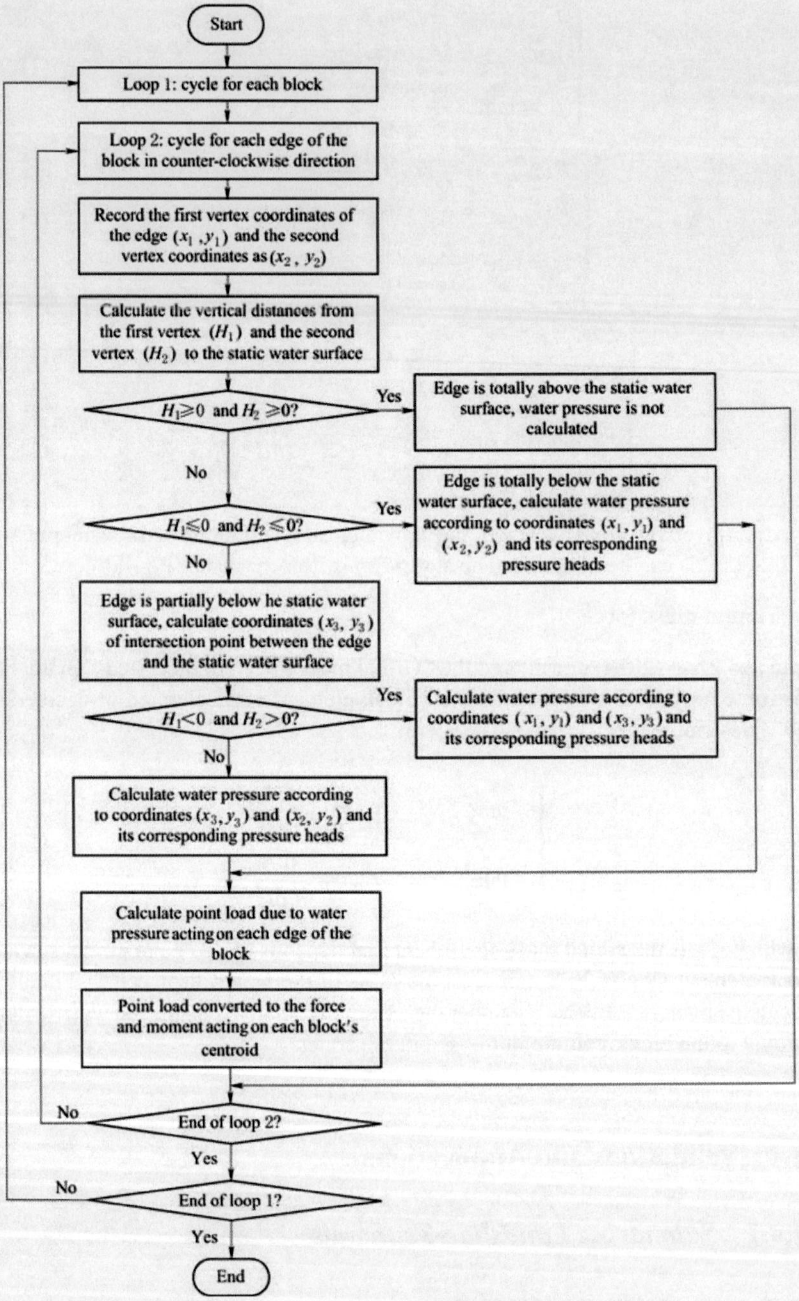

Fig. 11.4 Flowchart for the calculation of hydrostatic pressure

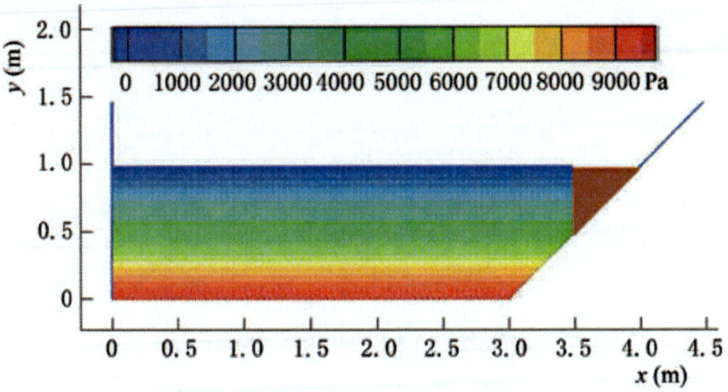

Fig. 11.5 Setup of submarine landslide model by Heinrich (1992) and corresponding hydrostatic pressure distribution in water body

Shobeyri 2008; Ataie-Ashtiani and Nik-Khah 2008; Abadie et al. 2010; Cremonesi et al. 2011; Viroulet et al. 2013b; Wang et al. 2016) for the validation of their models. The hybrid DEM-SPH model elaborated in this book just follows suit.

The initial configuration of model is depicted in Fig. 11.5. The water depth is 1.0 m and the cross section of triangular block is 0.5 × 0.5 m in both length and height. The top of the block is 0.01 m below the water surface and the fluid particles are initialized with hydrostatic pressure according to their position and density. The tank is 18.0 m long in the actual experiment, but it is truncated only 3.0 m long in numerical simulation (Ataie-Ashtiani and Shobeyri 2008; Wang et al. 2016), because SPH is very good at simulating impacts with free surfaces but is relatively poor in simulating wave propagation for long distance. In addition, this truncated tank will save considerable computation efforts by remarkably reducing the amount of SPH particles.

All the relevant parameters for the hybrid DEM-SPH computation are listed in Table 11.1. The block is surrounded by 3536 solid particles with radius of 0.003 m. Its motion is pre-defined (Grilli and Watts, 1999; Ataie-Ashtiani and Shobeyri 2008), and stops once it arrives at the bottom of tank. The vertical velocity of the block is pre-defined by

$$v(t) = \begin{cases} c_1 \tanh(c_2 t) & \text{for } t \le 0.4s \\ 0.6 & \text{for } t > 0.4s \end{cases} \tag{11.33}$$

in which $c_1 = 86$ and $c_2 = 0.0175$.

The smoothing length is set to be 1.25 times the initial fluid particle space and thus the ratio of smoothing length to solid particle diameter is 3.125 that satisfies the stipulation of smoothing length being sufficiently larger than the solid particle diameter in the hybrid DEM-SPH model (Robinson et al. 2014). The δ-SPH algorithm is adopted to reduce the spurious numerical noise and thus to stabilize the pressure field. This is in fact a one-way coupled problem because each solid particle's motion

Table 11.1 Parameters in the hybrid DEM-SPH model for submarine rigid block slide

DEM parameters	
Disc density (kg/m^3)	2500
Porosity ε	0.2
Disc radius (m)	0.003
Disc-disc normal stiffness (kN/m)	1.0×10^2
Disc-disc shear stiffness (kN/m)	1.0×10^2
Disc-disc friction coefficient	0.0875
Wall-disc normal stiffness (kN/m)	7.0×10^3
Wall-disc shear stiffness (kN/m)	7.0×10^3
Wall-disc friction coefficient	0.3640
Amount of discs	3536
Damping coefficient (N.s/m.g)	0.0
Time-marching step length (s)	1.0×10^{-4}
SPH parameters	
Particle density (kg/m^3)	1000
Particle spacing (m)	0.015
Amount of fluid particles	15,016
Amount of boundary particles	805
Kernel function	Wendland
Kernel smoothing length(m)	0.01875
Boundary condition	Dynamic
Density filter	δ-SPH algorithm
Diffusive coefficient	0.1
Artificial viscosity coefficient λ	0.3
Coefficient of sound speed η	10
Fluid viscosity (kPa·s)	1.0×10^{-6}
Time-marching step length (s)	1.0×10^{-4}

is pre-defined and the solid–fluid interaction forces need to be converted to the SPH module only to update each fluid particle's motion.

(1) **Single rigid block**

The numerical simulation was run on a computer of Intel(R) Core (TM) i7-6700 K CPU (4.00 GHz), it costed 12497 s (about 3.47 h) in the submarine rigid block slide computation for a physical test time of 3.0 s.

We first estimate the arrival time of the leading wave at the left boundary according to (Viroulet et al. 2013a)

$$t_\alpha = \frac{H - H_s}{v \sin \zeta} + \frac{x_d}{\sqrt{gH}} \tag{11.34}$$

where: H—water depth; H_s—initial submergence of block; v—solid velocity; ζ—slope angle; x_d—distance between the left boundary and the origin.

Equation (11.34) can be regarded as the moving time of the solid block slip down along the slope plus the propagation time of a linear non-dispersive wave. It is notable that the origin is located at the intersection of the slope with the horizontal bottom of tank.

According to Eq. (11.34), the estimated arrival time of the leading wave at the left boundary is 1.96 s. When the wave arrives at the left boundary, it is unable to continue propagation further due to the truncation of numerical tank. Under such circumstance, the wave is reflected from the boundary and subsequently flows back, which differs from the experiment. Therefore, in the following discussion, we mainly focused on the first two seconds of the computational outcomes.

Typical snapshots of the surge waves generated by the rigid block slide are presented in Fig. 11.6. At 0.8 s, the maximum wave amplitude above the static water surface is about 7.9 cm. It is interesting to note that this value is 7 cm given by Ataie-Ashtiani and Shobeyri (2008) and 10.1 cm given by Wang and his-co-workers (Wang et al. 2016). The block reaches the water tank bottom at approximately 1.0 s, both in our DEM-SPH model and the experiment. When the block slides down along the slope, the surrounding fluid particles gain larger velocities due to the energy transferred from the block to the water body, which gives rise to strong SFI and subsequently generated surge waves. However, the flow velocity around the block gradually decays after it stops at the tank bottom.

The velocity vectors at different times are presented in Fig. 11.7. At 0.5 s, a vortex is created in a form of clockwise flow regime near the surface of block. The velocity

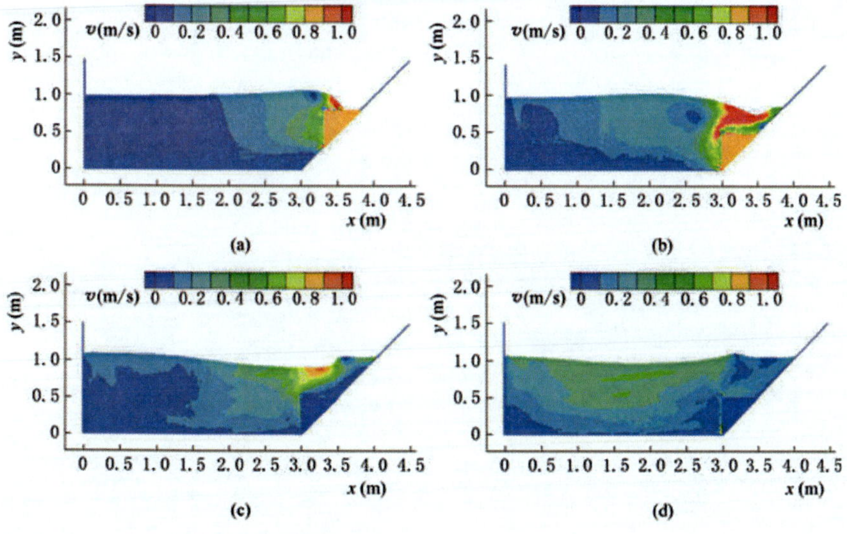

Fig. 11.6 Evolution of velocity field and wave profiles due to submarine rigid block slide. **a** $t = 0.5$ s; **b** $t = 1.0$ s; **c** $t = 1.5$ s; **d** $t = 2.0$ s

vector is dominantly directed from the block to the left ambient water, indicating that the water is pushed away by the block movement. This movement results in an elevated water surface. The fluid particles in the near field gain larger velocities while those in the far field fluid are almost undisturbed. At 1.0 s, the block arrives at the tank bottom with more fluid particles gaining velocities. The vortex is fully developed and it moves down together with the block. The fluid flows back filling the region where the block previously occupied, and another small portion of water climbs up along the slope ramp. At 1.5 s, the surge waves propagate outwards, resulting in smaller flow velocity in the near field. In addition, the vortex starts to disappear at this stage.

From Fig. 11.8 it can be found that the pressure field is correctly calculated by the δ-SPH algorithm, since it is not noisy and nearly identical to the hydrostatic field far from the block slide. The fluid particles near the block slide are subject to larger pressures, which is due to the fact that when solid particles are in the interaction domain of a fluid particle, the corresponding local porosity decreases and thus pressure increases. In this case, the ambient fluid particles are pushed away from the impermeable block slide.

Figure 11.9 exposits the water surface snapshots at different instants. In general, the computation results agree well with the experimental data. The calculated wave

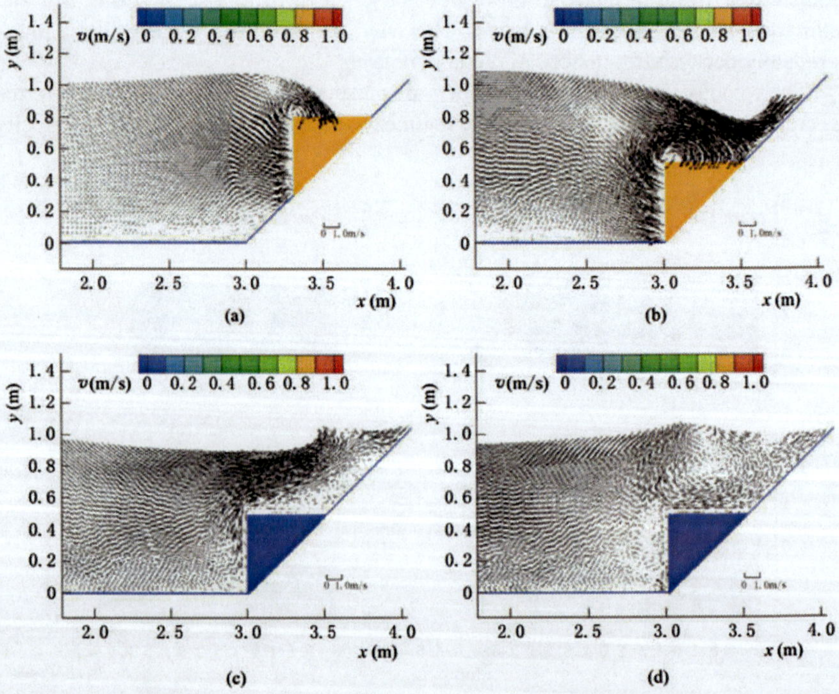

Fig. 11.7 Velocity vectors in the near field due to submarine rigid block slide. **a** $t = 0.5$ s; **b** $t = 1.0$ s; **c** $t = 1.5$ s; **d** $t = 2.0$ s

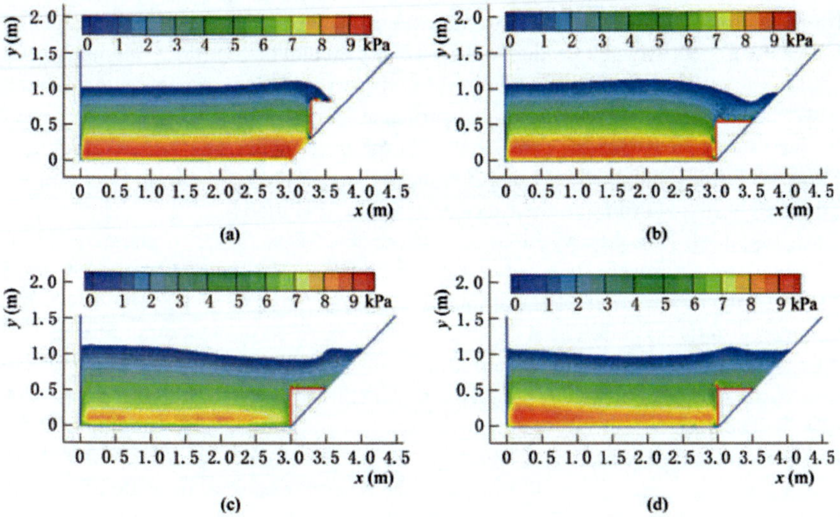

Fig. 11.8 Evolution of pressure field due to submarine rigid block slide. **a** $t = 0.5$ s; **b** $t = 1.0$ s; **c** $t = 1.5$ s; **d** $t = 2.0$ s

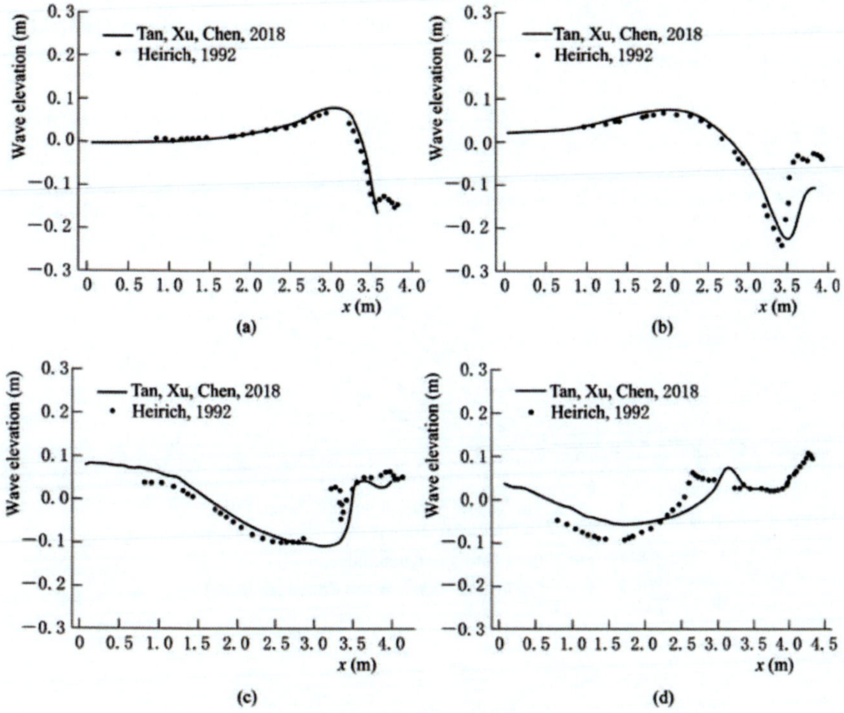

Fig. 11.9 Comparison of the water surfaces by experiment and hybrid DEM-SPH model. **a** $t = 0.5$ s; **b** $t = 1.0$ s; **c** $t = 1.5$ s; **d** $t = 2.0$ s

trough at 1.0 s is almost identical to the experimented one. However, a discrepancy increases appreciably at 2.0 s due to the mechanism that the fictitious boundary(left) wall in the truncated numerical tank reflects the wave propagation. It seems that the water waves in our DEM-SPH model propagate slightly faster than that predicted by Eq. (11.34). This is also consistent with the observation by Ataie-Ashtiani and Shobeyri (2008) that at about t = 1.5 s, the wave starts to reflect from the fictitious boundary wall, according to their incompressible SPH model.

In Fig. 11.10, the history of water surface level at $x = 1.0$ m and $x = 2.0$ m is quantitatively cross-referenced using the experimented data of Heinrich (1992) and the (incompressible) SPH computed by Ataie-Ashtiani and Shobeyri (2008). Compared to the experimented outcomes, the leading wave crest amplitude at $x =$ 1.0 m is slightly overestimated by our DEM-SPH model, but the water surface at both $x = 1.0$ m and $x = 2.0$ m matches rather well with that by Ataie-Ashtiani and Shobeyri (2008). The good agreement of the first wave crest amplitude is encouraging towards the tsunami hazard assessment using our DEM-SPH model, because it is able to predicts the maximum wave run-up along the reservoir bank or over the dam crest. In addition, the wave at $x = 2.0$ m is generated earlier than that at $x = 1.0$ m because the position $x = 2.0$ m is more close to the landslide-water interaction region. The wave amplitude decay also can be observed in the wave propagation zone.

Energy evolution analysis is carried out to give a deep insight into the landslide generated surge waves. According to Shan and Zhao (2014), the dissipated energy E_d can be calculated by

$$E_d = \Delta E^s_{po} - E^s_{ki} - E^f_{ki} - \Delta E^f_{po} \tag{11.35}$$

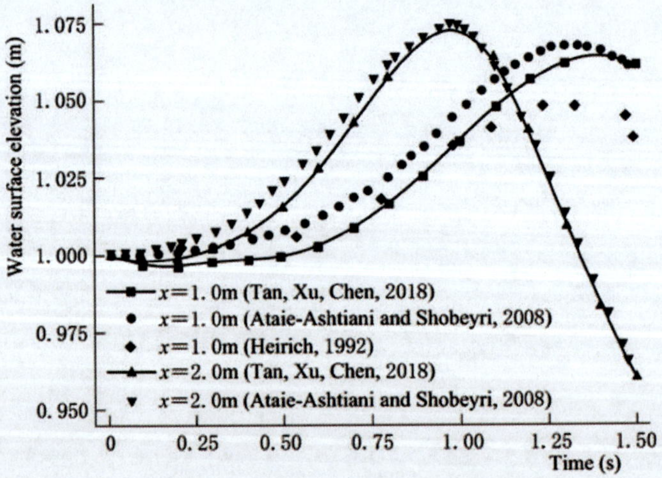

Fig. 11.10 Comparison of water surface elevation (at $x = 1.0$ m and $x = 2.0$ m)

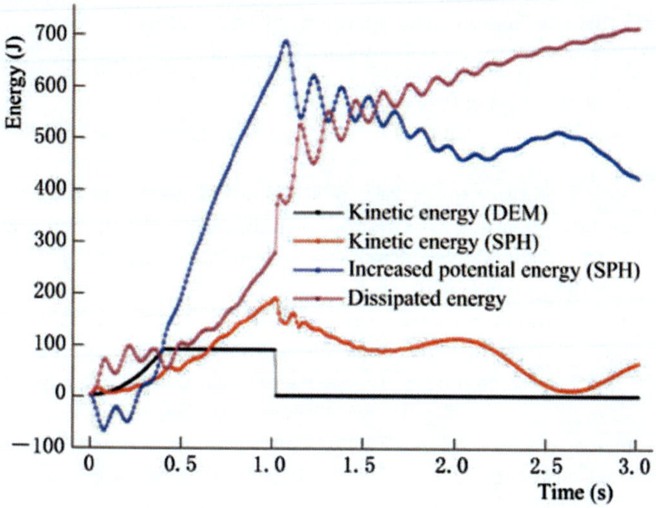

Fig. 11.11 Energy evolution of block slide and water

where: ΔE^s_{po}—potential energy change of solid system; E^s_{ki}—total kinetic energy of solid system; E^f_{ki}—total kinetic energy of fluid system; ΔE^f_{po}—potential energy change of fluid system.

From Fig. 11.11, it is clear that the kinetic energy of solid system increases steadily at the first 0.4 s, then it reaches a stable value of 90 J, and finally approaches to zero at 1.0 s indicating the stop standstill of block slide at the tank bottom. This is not surprising because the motion of each solid particle in the rigid block slide is artificially imposed.

Meanwhile, the kinetic energy of fluid particle system increases at the first 1.0 s. The fluid particles gain velocities during the strong SFI process accompanied with the energy transfer from thee block slide to the water body. The potential energy of fluid particle system decreases slightly and shortly at the very beginning, probably attributable to the mechanism that the fluid particles initially move down under gravity. Subsequently, it increases and reaches a peak at 1.0 s. This means that the ambient water displaced by the block movement raises the water surface level. Hence, surge waves are generated and propagate outwards due to the height differences between the higher water level in the near field and the almost undisturbed water level in the far field. After the block slide stops standstill, both kinetic and potential energy of the fluid particle system start to decay.

The generated surge waves periodically propagate to the left boundary and then reflected back with the process of progressively dissipated energy. It can be inferred that the surge waves will be settled down after the energy is completely dissipated out.

(2) **Deformable disc-assemblage versus single rigid block**

The triangular slide is now handled as deformable disc assemblage, to get the particularities of deformable landslide.

Typical snapshots of the surge waves generated by granular deformable slide are presented in Fig. 11.12. It is obvious that the slide no longer behaves as a rigid body but deforms significantly in the process of slip down along the inclined plane. After the slide front having hit the tank bottom, the disc mass decelerates and starts to deposit at the toe of the slope, attributable to both the presence of water and the topography that hinders the slide motion and gradually diverts the slide to move along the horizontal bottom of tank. With regard to the variation of free surface, it is less violent compared to the rigid block slide, probably due to the mechanism that a portion of energy is dissipated in the process of slide deformation, leading to less energy conversion from the granular mass to the water body.

History of the velocity vectors in the wave generation region is disclosed in Fig. 11.13. The flow regime is similar to that in the rigid block slide (cf. Fig. 11.7). But in the deformable slide, the intensity of velocity field and the wave run-up along the slope ramp are smaller than the rigid slide. At 1.5 s, the clockwise vortex still exists near the surface of sliding mass because the deformable slide has not completely stopped yet and the ambient water is still displaced by its movement.

In Fig. 11.14, the pressure field far from the deformable slide is not noisy and nearly hydrostatic, which verifies the capability of the δ-SPH algorithm in eliminating the pressure oscillations during strong landslide-water interaction.

Figure 11.15 cross-references the water surfaces between rigid slide and deformable slide. The surge waves generated by the deformable slide are smaller

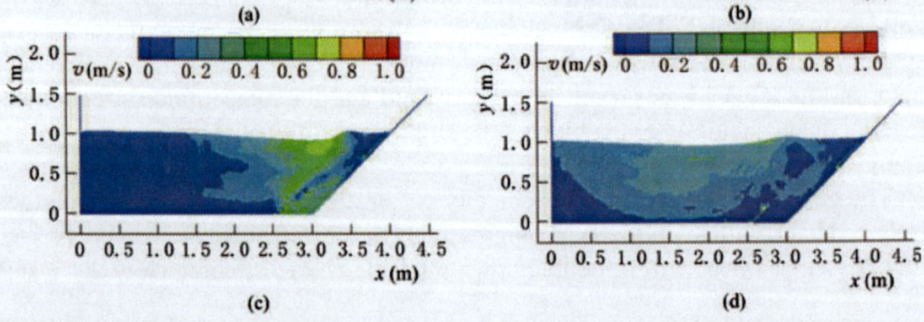

Fig. 11.12 Evolution of velocity field and wave profile due to submarine deformable slide. **a** $t = 0.5$ s; **b** $t = 1.0$ s; **c** $t = 1.5$ s; **d** $t = 2.0$ s

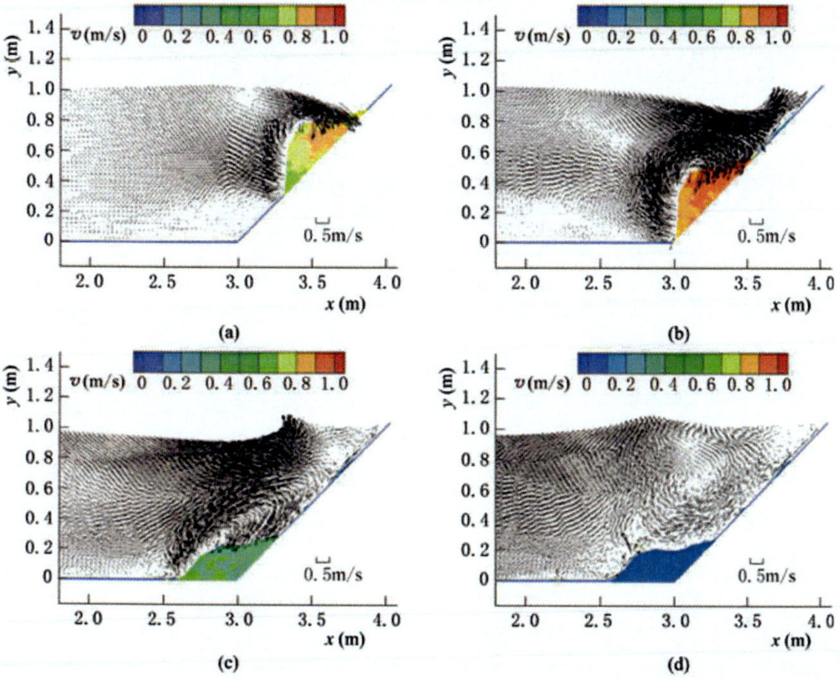

Fig. 11.13 Velocity vectors in the near field due to submarine deformable slide. **a** $t = 0.5$ s; **b** $t = 1.0$ s; **c** $t = 1.5$ s; **d** $t = 2.0$ s

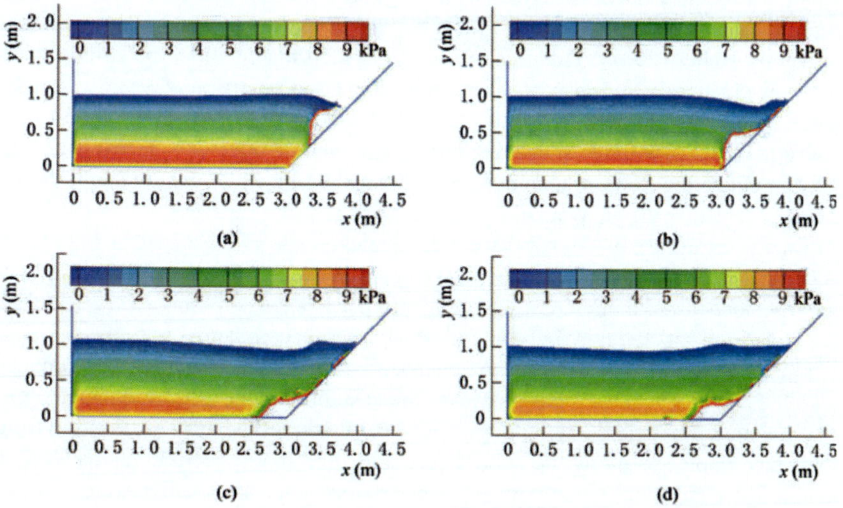

Fig. 11.14 Evolution of pressure field due to the submarine deformable slide. **a** $t = 0.5$ s; **b** $t = 1.0$ s; **c** $t = 1.5$ s; **d** $t = 2.0$ s

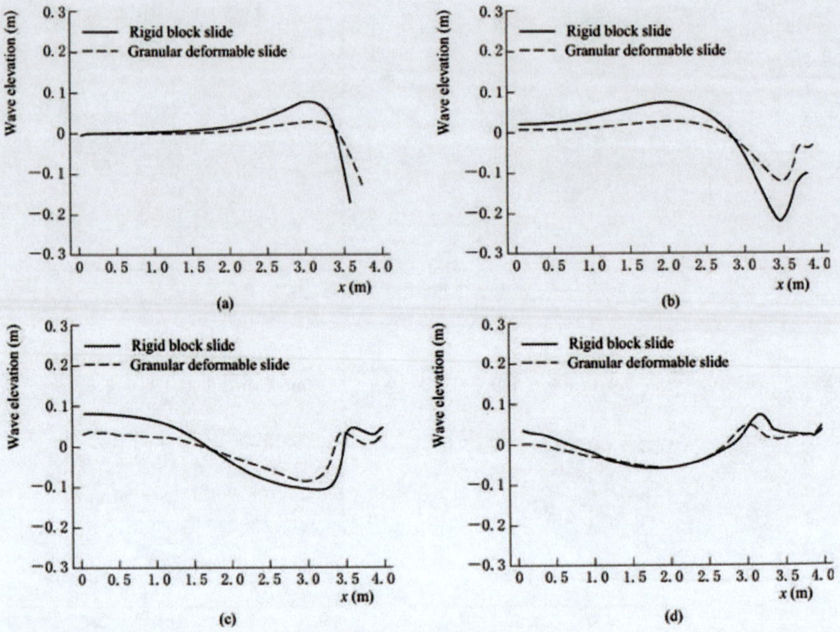

Fig. 11.15 Comparison of the water surfaces between submarine rigid and deformable slide. **a** $t = 0.5$ s; **b** $t = 1.0$ s; **c** $t = 1.5$ s; **d** $t = 2.0$ s

and less violent than that generated by the rigid block slide. The conclusion that the rigid body postulation will overestimate the wave amplitude is consistent with that drawn by Heller (2007). This is because in the deformable slide, the sliding mass deforms drastically during post-failure movement and a portion of energy is dissipated by the friction between solid particles (Mulligan and Take 2017), whereas in the rigid block slide there is no energy dissipation within the block and more energy may be converted to the water body. Therefore, the rigidity of the landslide should be emphasized in the analysis of landslide generated surge waves.

Energy evolution in the submarine deformable slide is disclosed in Fig. 11.16 according to Eq. (11.35). In the first 1.0 s, the mass slides down under the action of gravity, hence the kinetic energy of the solid particle system increases. After that, the slide front hits the tank bottom and decelerates, accompanied with the decrease of kinetic energy. At about 1.7 s, the kinetic energy approaches to zero in accordance with the fact that the deformable slide is almost stopped standstill at the toe of slope. The maximum potential energy augment of the fluid particle system resulting from the water displacement by the slide deformation and motion, is much smaller than that in the rigid block slide. Under such circumstance, the smaller water surface rise leads to reduced wave amplitude and less violent surge waves. The dissipated energy in the deformable slide is much greater than that in the rigid block slide. This validates that more energy is dissipated during deformable slide by means of

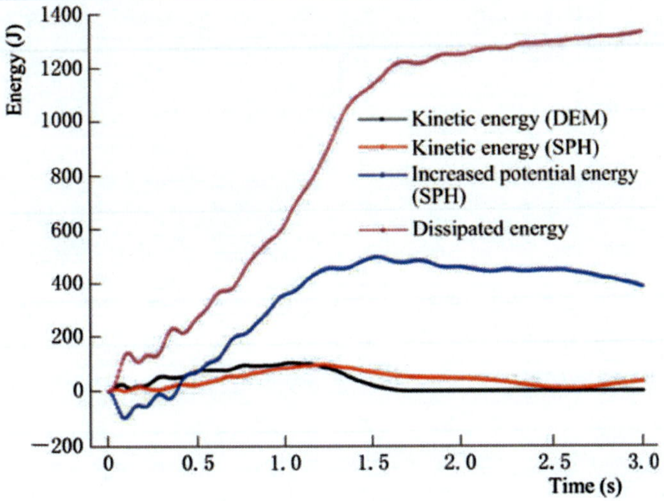

Fig. 11.16 Energy evolution in submarine granular deformable slide

inter-particle friction, and less energy is converted from the deformable slide to the water body.

It is notable that in the case of submarine granular deformable slide, the computation time is 17294 s (about 4.80 h), perceivably longer than that in the submarine rigid block slide case.

2. **Problem of Rzadkiewicz et al.** (1997)

In this validation study, the experimented data by Rzadkiewicz et al. (1997) for the submarine deformable landslide are employed. The model setup is exposed in Fig. 11.17 where the water depth is 1.60 m and the slide consists of granular particles with density 1950 kg/m^3. The triangular section of slide is 0.65×0.65 m in both length and height. It is submerged 0.10 m below the water surface and sustained by a vertical plate at its right side. The sudden removal of plate initiates the slip along the slope inclined at an angle of 45°. Dynamic boundary particles (Crespo et al. 2007) are arranged with alternate permutation to prevent the escape of fluid particles. The parameters are listed in Table 11.2 where the smoothed length is 2.5 times the diameter of disc.

In Table 11.3, the rheological models exercised previously by researchers are listed, too.

Figure 11.18 exposits the fluid pressure. It indicates that the fluid pressure is basically noise-free. In contrast, the semi-implicit algorithm MPS by Fu and Jin (2015) gives rise lower regularity of pressure. It is worthwhile to remind the difference in submarine and subaerial slide: for the former, the water surface over the slide will drop first after the onset of slip, exhibiting a wave trough; whereas for the latter, the

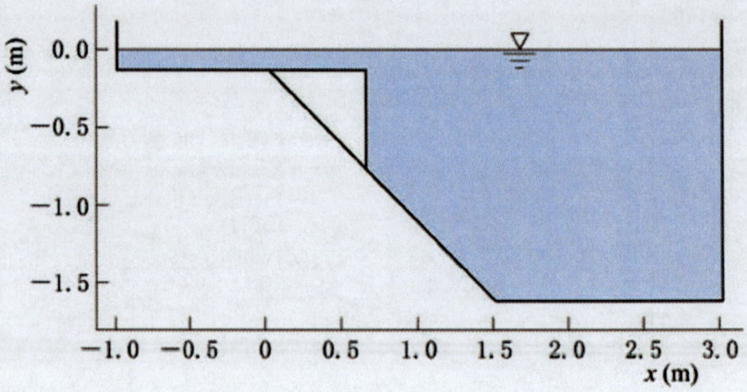

Fig. 11.17 Model setup of submarine deformable landslide

Table 11.2 Computational parameters for the hybrid DEM-SPH model for submarine deformable slide

DEM parameters	
Disc density (kg/m^3)	1950.0
Radius (m)	0.003
Porosity ε	0.2
Young's modulus (10^3 kPa)	700
Poisson's ratio	0.35
Disc-disc friction coefficient	0.00
Young's modulus of ball (10^3 kPa)	700
Poisson's ratio of wall	0.35
Disc-wall friction coefficient	0.00
Amount of discs	5977
Global damping coefficient (N.s/m.g)	0.00
Time-march step length (s)	1.0×10^{-4}
SPH parameters	
Fluid particle density (kg/m^3)	1000.0
Initial space of fluid particles (m)	0.01
Amount of fluid particles	35,595
Amount of boundary particles	4902
Smooth kernel	Wendland
Smooth length (m)	0.015
Boundary	Dynamic
Revision algorithm for pressure	δ-SPH
δ-SPH coefficient	0.1
Artificial viscosity coefficient λ	0.3
Coefficient of sound speed η	10
Dynamic viscosity of fluid (kPa·s)	1.0×10^{-6}
Time-marching step length (s)	1.0×10^{-4}

Table 11.3 Rheological model of submarine deformable landslide

Reference	Deformable landslide	Rheological model
Rzadkiewicz et al. (1997)	Non-Newton	Bingham
Ataie-Ashtiani and Shobeyri (2008)	Non-Newton	Bingham and general Cross
Capone et al. (2010)	Non-Newton	Bingham
Ma et al. (2013)	Turbulence	No Rheology
Fu and Jin (2015)	Non-Newton	Herschel-Bulkley
Qiu et al. (2017)	Non-Newton	Papanastasiou and Herschel-Bulkley
Yeylaghi et al. (2017)	Non-Newton	Bingham and general Cross

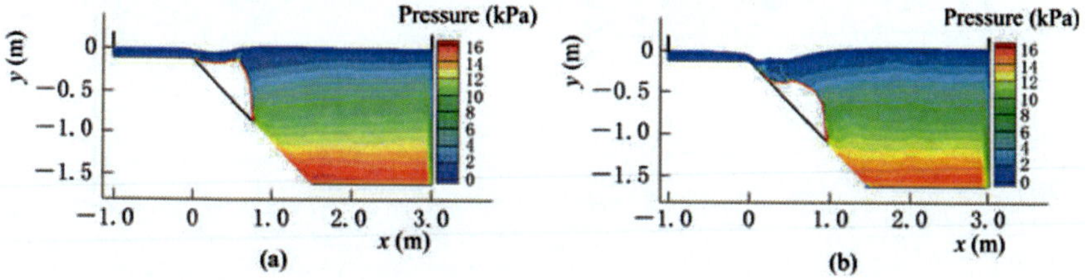

Fig. 11.18 Pressure field in water body. **a** $t = 0.4$ s; **b** $t = 0.8$ s

repulse force from the slide raises the surface, the front wave is in the form of wave crest.

Figure 11.19 shows the velocity clouds and vectors at different snapshot time. At $t = 0.4$ s, the water above the upper portion of slide moves downward, leading to water surface drop; at t=0.8 s, on the surface of slide an anti-clockwise vortex is formed, and water runs toward the center of water drop zone, which leads to local water upheave.

Figure 11.20 exhibits the snapshots of free surface at different time. The data by experiment and Nasa-Vof2D computation are all from Rzadkiewicz et al. (1997), the data by NHWAVE computation are from Ma et al. (2013), the data by ISPH computation are from Yeylaghi et al. (2017), the data by DEM-SPH computation are from the author of this book. In the Nasa-Vof2D computation (a), perfect fluid of slide is postulated; in the Nasa-Vof2D computation (b), the input parameters are $\tau_0 = 1000$ Pa, $\mu_b = 0$ Pa·s, $D = 0.004$ Pa·s.

It is verified that the computation results are rather dispersive by different methods, namely. It is evident that our DEM-SPH model provides smaller side prediction of wave trough at $t = 0.4$ s and wave crest at $t = 0.8$ s.

The deposit patterns of slide at different time are given in Fig. 11.21, in which the dada of Nasa-Vof2D are extracted from Rzadkiewicz et al. (1997), the data of

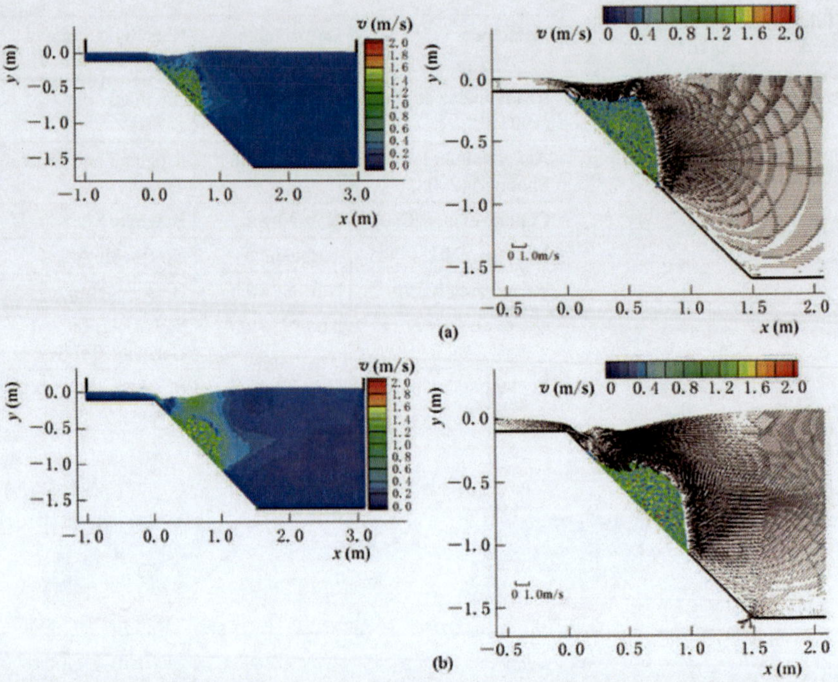

Fig. 11.19 Velocity field of water body (Left column: color cloud; Right column: vector). **a** $t = 0.4$ s; **b** $t = 0.8$ s

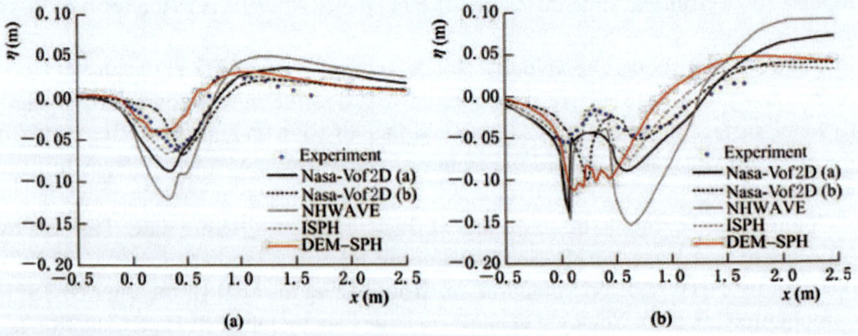

Fig. 11.20 Comparison of water surface levels. **a** $t = 0.4$ s; **b** $t = 0.8$ s

DEM-SPH are provided by the author. The computation parameters for Nasa-Vof2D (a') are $\tau_0 = 200$ Pa, $\mu_B = 0.0$, $D = 0.0$, the computation parameters for Nasa-Vof2D (b') are $\tau_0 = 1000$ Pa, $\mu_B = 0.0$, $D = 0.0$. It is understandable that the larger the yield stress, the more difficult to drive the landslide.

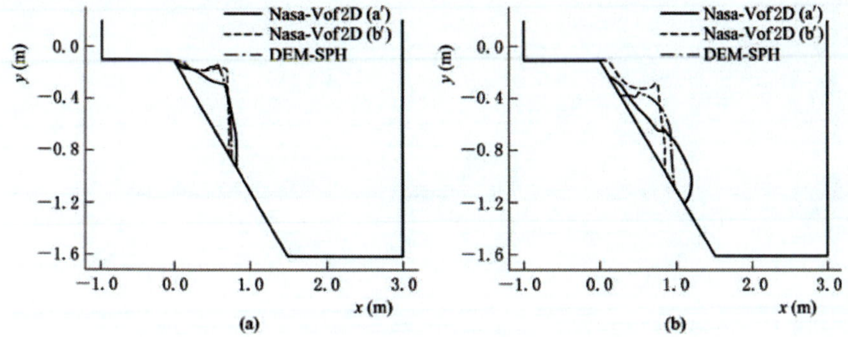

Fig. 11.21 Comparison of slide deposit patterns. **a** $t = 0.4$ s; **b** $t = 0.8$ s

As the ongoing of time, the deposit pattern exhibits larger and larger difference. The slide in Nasa-Vof2D (a') model is more similar to fluid, whereas the slide in Nasa-Vof2D (b') model is more similar to solid, the slide in our DEM-SPH model just runs in between.

11.4.2 Subaerial Landslide-Tsunamis

1. **Problem of Heinrich** (1992)

The subaerial rigid block sliding down along an inclined plane is numerically studied and cross-referenced by the experimented data of Heinrich (1992). As illustrated in Fig. 11.22, a triangular rigid block slide with cross section of 0.5×0.5 m is initially placed on the slope just above the static water surface. The slope inclination angle is 45°, the water depth is 0.40 m and other relevant parameters are listed in Table 11.4. It is noted that the linear contact model is used in this study. In addition, the horizontal length of the tank bottom is 4.0 m by the truncated simplification, in order to reduce computation efforts.

After the release of block slide, it starts to slip down and falls into the water. Subsequently, the waves are generated due to the ambient water displacement. Different from the numerical models of Heinrich (1992) and Qiu (2008) that the trajectory of the slide needs to be pre-defined, the kinematics of the slide block is calculated by DEM in our study (Tan et al. 2018a). An additional wall is placed at the toe of the ramp, being consistent with the buffer used in the experiment to abruptly stop the slide block once it arrives at the tank bottom.

In Fig. 11.23, the computed vertical slip displacement agrees well with the experiment data by Heinrich (1992), indicating that the hybrid DEM-SPH model elaborated in this book can reasonably reproduce the slide kinematics.

The whole process of landslide-tsunamis is snapshotted in Fig. 11.24. At 0.6 s, the ambient water detaches from the block slide, this is mainly attributed to the abrupt

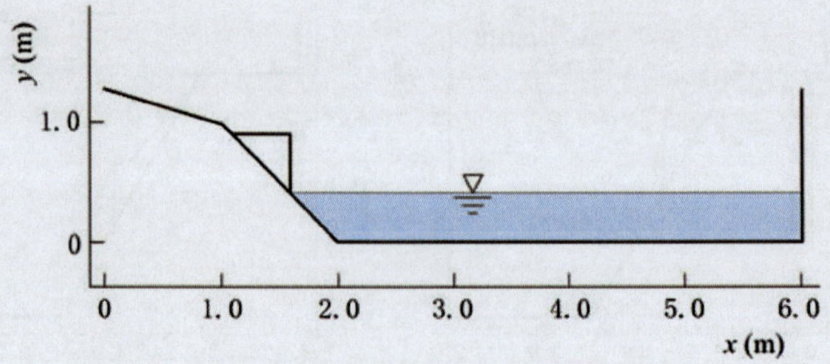

Fig. 11.22 Setup of physical experiment model for subaerial slide (Heinrich 1992)

Table 11.4 Parameters in the hybrid DEM-SPH model for subaerial rigid block slide

DEM parameters	
Disc radius (m)	0.004
Amount of discs attached to block	208
Block density (kg/m^3)	2000.0
Block friction coefficient	tan(20°)
Wall-disc friction coefficient	tan(20°)
Disc-disc normal stiffness (kN/m)	1.0×10^5
Disc-dis shear stiffness (kN/m)	1.0×10^5
Wall-disc normal stiffness (kN/m)	7.0×10^5
Wall-dis shear stiffness (kN/m)	7.0×10^5
Static water surface elevation (m)	0.40
Drag coefficient	1.00
Slide thickness (m)	0.3536
Added mass coefficient	0.76
Time-marching step length (s)	1.0×10^{-4}
SPH parameters	
Fluid particle density (kg/m^3)	1000.0
Fluid particle spacing (m)	0.01
Fixed boundary particle spacing (m)	0.005
Amount of fluid particles	16,752
Amount of fixed boundary particles	5028
Kernel function	Wendland
Kernel smoothing length (m)	0.0125
Boundary condition	Dynamic
Density filter	δ-SPH algorithm
δ-SPH coefficient	0.1
Artificial viscosity coefficient λ	0.1
Coefficient of sound speed η	10
Time-marching step length (s)	1.0×10^{-4}

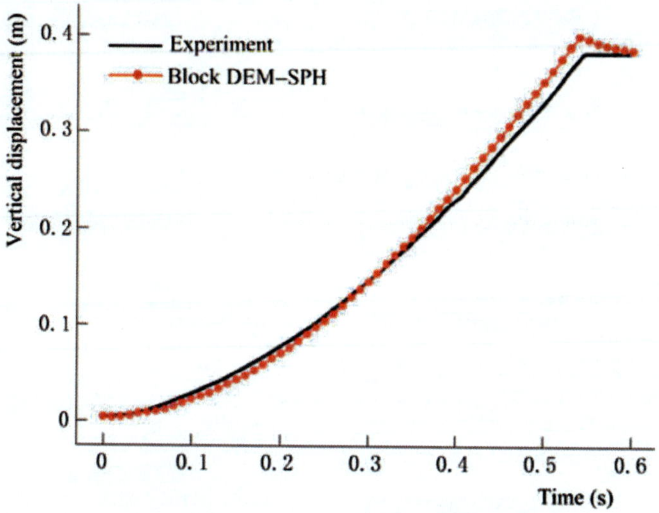

Fig. 11.23 Comparison of the vertical displacement of slide between DEM-SPH computation and experiment

stop of block. At 0.7 s, a hydrodynamic impact crater is created. At 0.8 s, the crater collapses forward forming the primary wave, a small portion of water flows back hitting the block slide and climbing up vertically. At 1.0 s, the solitary-like primary wave propagates towards right tank wall. Meanwhile, the backwash continues to run up and a jet of water is generated beyond the block slide. At 1.5 s, the water from the jet collapses and impacts into the main water body again, which dramatically disturbs the water surface and initiates a secondary wave.

The wave profiles at typical instants are plotted in Fig. 11.25 together with the experimented data. The overall agreement is good although slight discrepancies emerge. At 0.6 s, the water displaced by the block slide elevates the water surface. At 1.0 s, the water splash near the block slide is slightly overestimated by the hybrid DEM-SPH model. At 1.5 s, both the primary wave and the secondary wave can be observed.

2. **Problem of Sælevik et al.** (2009)

The numerical computations are correspondent to the experiment conducted by Sælevik et al. (2009). In their experiment setup (Fig. 11.26), the coordinate origin is defined at the intersection between the ramp and the water surface. The water depth is 0.6 m, the slide is composed of two blocks, three blocks and four blocks, respectively. It is accelerated by a conveyer belt to generate impulse waves. The ramp is inclined at an angle of 35°. Three wave probes are placed at $x = 3.25$ m, 4.24 m and 6.55 m to record wave profiles. A rounded transition is employed to smoothly divert the slide blocks from the ramp to the horizontal bottom of tank. Unfortunately, the

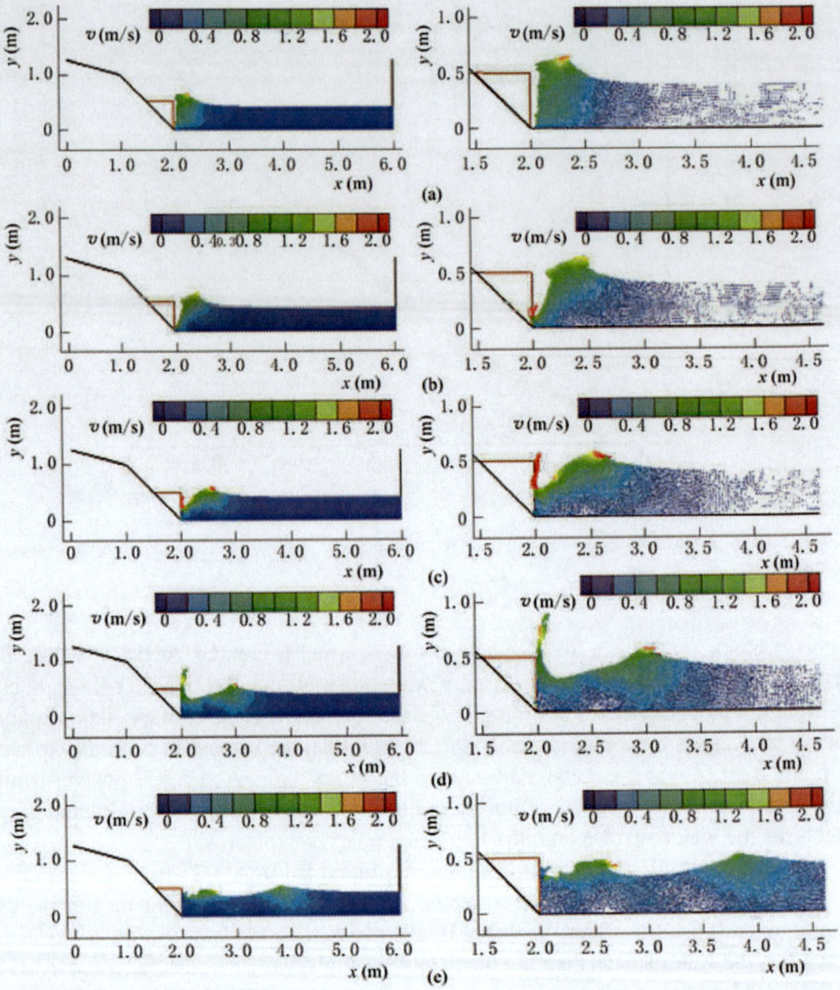

Fig. 11.24 Evolution of slide position and wave profile (Left column: velocity cloud; Right column: magnified velocity vector). **a** $t = 0.6$ s; **b** $t = 0.7$ s; **c** $t = 0.8$ s; **d** $t = 1.0$ s; **e** $t = 1.5$ s

radius of the transition is not specified in their original work. Hence in this book, a radius of 1.2 m is postulated.

In our computation, the slide blocks are placed just above the water surface and imposed a slide front impact velocity (Tan et al. 2018a). Given that no available slide kinematics can be used to tune the empirical coefficients, the drag and added mass coefficients as well as friction angle are postulated being identical to the tests of Heller and his co-workers (Heller et al. 2016). All the relevant parameters used in the computation are given in Table 11.5.

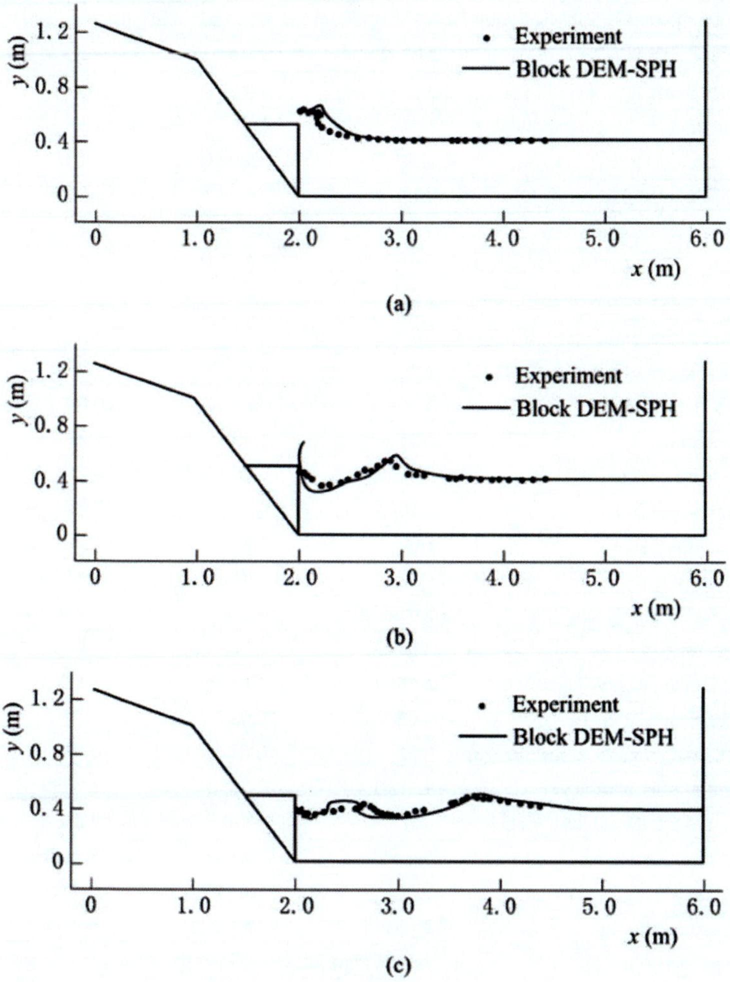

Fig. 11.25 Comparison of water surfaces between computation and experiment. **a** $t = 0.6$ s; **b** $t = 1.0$ s; **c** $t = 1.5$ s

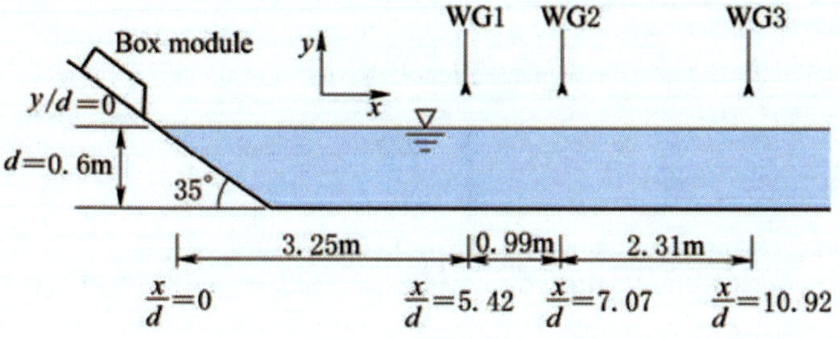

Fig. 11.26 Setup of physical experiment model for subaerial slide (Sælevik et al. 2009)

Table 11.5 Parameters in the hybrid DEM-SPH model for subaerial multi-block landslide

DEM parameters	2 blocks	3 blocks	4 blocks
Disc radius (m)	0.005	0.005	0.005
Amount of discs attached to block	247	373	503
Block density (kg/m^3)	1247.73	1269.00	1269.18
Block friction coefficient	0.45	0.45	0.45
Wall-disc friction coefficient	0.45	0.45	0.45
Disc-disc normal stiffness (10^3kN/m)	1.0	1.0	1.0
Disc-disc shear stiffness (10^3kN/m)	1.0	1.0	1.0
Wall-disc normal stiffness (10^4kN/m)	7.0	7.0	3.0
Wall-disc shear stiffness (10^4kN/m)	7.0	7.0	3.0
Initial water depth (m)	0.6	0.6	0.6
Drag coefficient	1.35	1.35	1.35
Added mass coefficient	0.76	0.76	0.76
Slide thickness (m)	0.16	0.12	0.16
Slide length (m)	1.0	1.6	2.0
Slide front impact velocity (m/s)	3.38	3.56	2.45
Time-marching step length (s)	1.0×10^{-4}	1.0×10^{-4}	1.0×10^{-4}
SPH parameters	2 blocks	3 blocks	4 blocks
Fluid particle density (kg/m^3)	1000.0	1000.0	1000.0
Fluid particle spacing (m)	0.02	0.02	0.02
Fixed boundary particle spacing (m)	0.01	0.01	0.01
Amount of fluid particles	12,890	12,890	12,890
Amount of fixed boundary particles	4571	4570	4571
Kernel function	Wendland	Wendland	Wendland
Kernel smoothing length (m)	0.025	0.025	0.025
Boundary conditions	Dynamic	Dynamic	Dynamic
Density correction	δ-SPH algorithm	δ-SPH algorithm	δ-SPH algorithm
δ-SPH coefficient	0.1	0.1	0.1
Artificial viscosity coefficient λ	0.1	0.1	0.1
Coefficient of sound speed η	10	10	10
Time-marching step length (s)	1.0×10^{-4}	1.0×10^{-4}	1.0×10^{-4}

N.B.: The linear contact model is applied; the block has a length of either 0.5 m or 0.6 m.

(1) Detachable blocks

In Fig. 11.27, the DEM model is composed of multi-blocks. More specifically, the slide possesses the thicknesses of 0.16 m, 0.12 m and 0.16 m, the slide front impact velocities of 3.38 m/s, 3.56 m/s and 2.45 m/s in the two-block case, three-block case and four-block case, respectively. Three wave probes are placed at $x =$

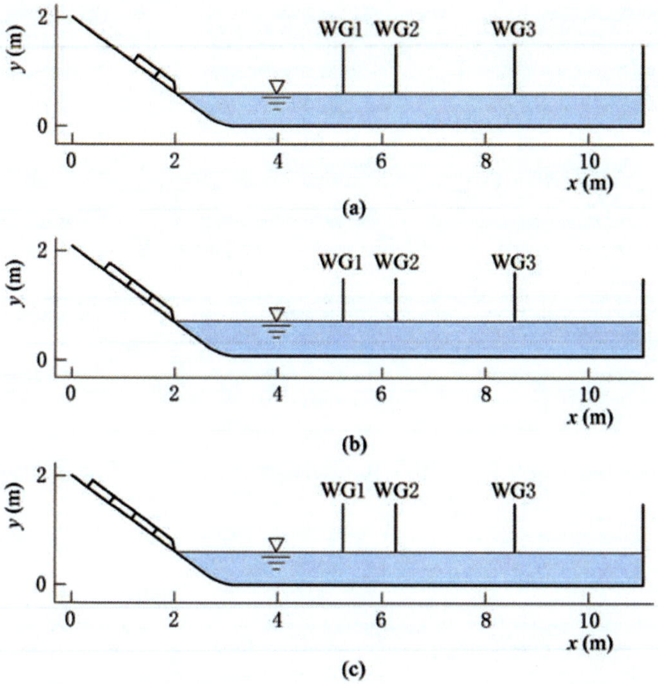

Fig. 11.27 DEM model setup for multi-block slide. **a** Two blocks; **b** three blocks; **c** four blocks

5.25 m, 6.24 m and 8.55 m in computation correspond to $x = 3.25$ m, 4.24 m and 6.55 m in experiment (viz. Fig. 11.26). The slide front angle is 45°.

It is reminded that the multi-block system may bend at the transition section. By contrast, only two points are contacted with the tank boundary walls in the case of one monolithic block (Heller et al. 2016).

The evolution of slide movement and wave propagation is snapshotted in Fig. 11.28.

The history of water surface elevation is unfolded in Fig. 11.29 along with the experimented data (Sælevik et al. 2009) and PFEM results (Salazar et al. 2016). In general, good agreements are found, particularly for the first wave. The primary wave amplitude is the largest one in the case of four-block landslide, probably due to the effect of the landslide volume. The waves by the PFEM model travel faster. The predicted arrival time of the primary wave is more accurate with the DEM-SPH model elaborated in this book. However, it is notable that the second wave is not as well resolved.

(2) **Monolithic block**

In this study, two blocks are glued together to form one monolithic (intact) block (Fig. 11.30), hence the block slide is not able to bend at the transition section (Heller

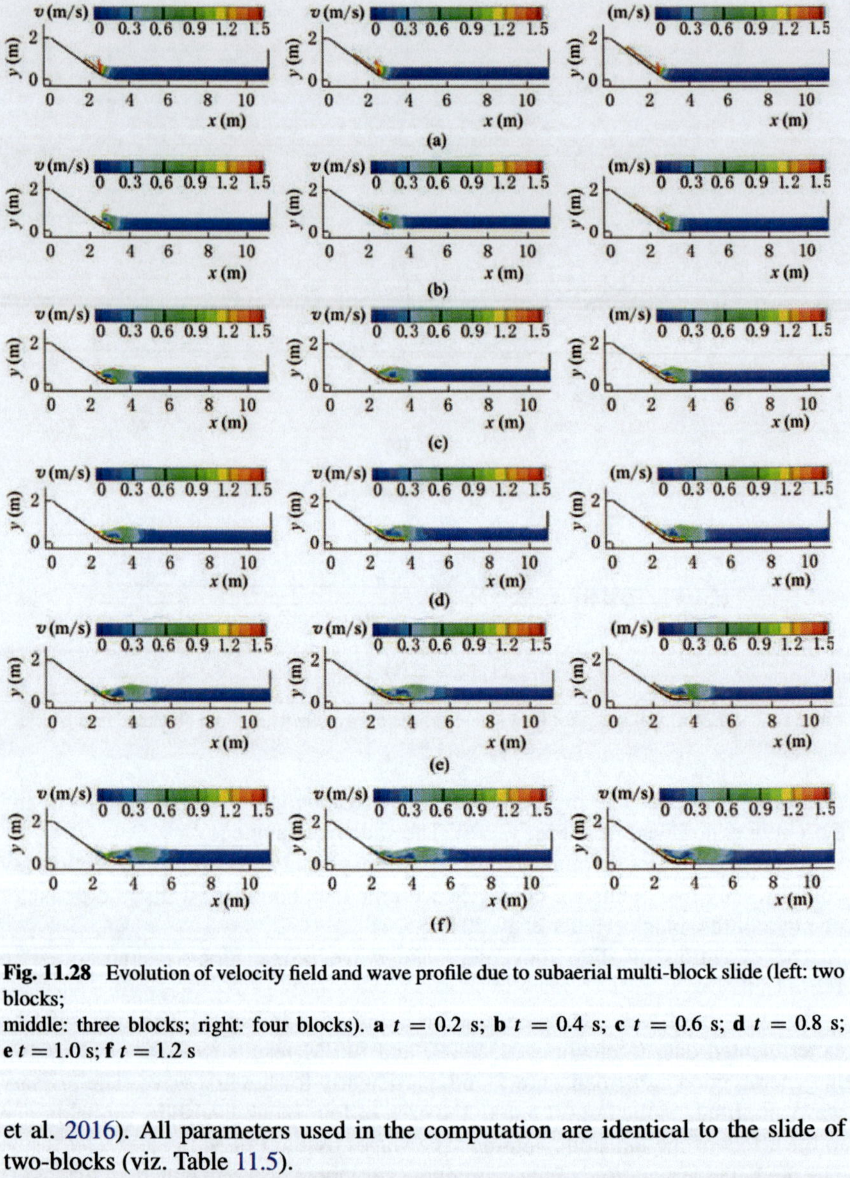

Fig. 11.28 Evolution of velocity field and wave profile due to subaerial multi-block slide (left: two blocks;
middle: three blocks; right: four blocks). **a** $t = 0.2$ s; **b** $t = 0.4$ s; **c** $t = 0.6$ s; **d** $t = 0.8$ s; **e** $t = 1.0$ s; **f** $t = 1.2$ s

et al. 2016). All parameters used in the computation are identical to the slide of two-blocks (viz. Table 11.5).

Figure 11.31 exposits the history of slide impact and wave propagation for one monolithic block slide by the computation. The runout of one monolithic block slide is longer than that of two-block slide (cf. Fig. 11.28). Hence it may be confirmed that if the landslide mass could be broken into several smaller fragments in the sliding process, a significant effect on both the runout and surge wave will be resulted in.

The wave profiles are quantitatively cross-referenced in Fig. 11.32 to validate the effect of block fragmentation. The primary wave amplitude derived by one monolithic

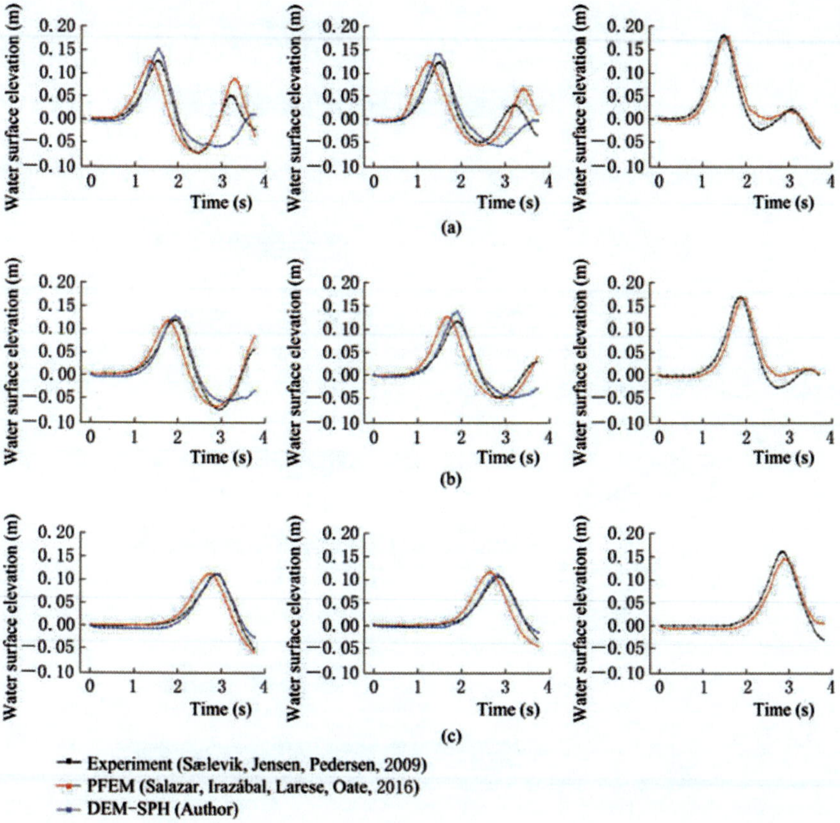

Fig. 11.29 Comparison of water surface elevation (left: two blocks; middle: three blocks; right: four blocks). **a** $x = 3.25$ m; **b** $x = 4.24$ m; **c** $x = 6.55$ m

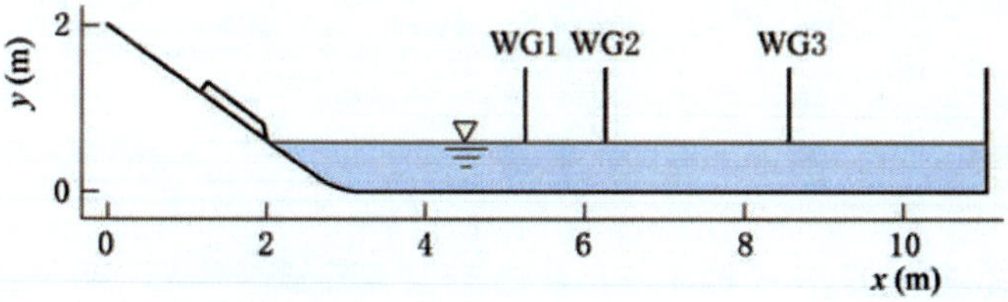

Fig. 11.30 DEM model setup for one monolithic block slide

block slide is larger than that by two separate blocks. This is because the monolithic block slide is not able to bend at the transition section, resulting in more water displacement to create primary wave. In addition, in one monolithic block case, no energy is dissipated through the inter-block friction and collision, as a consequence, more kinetic energy is transferred from the block slide to the water body.

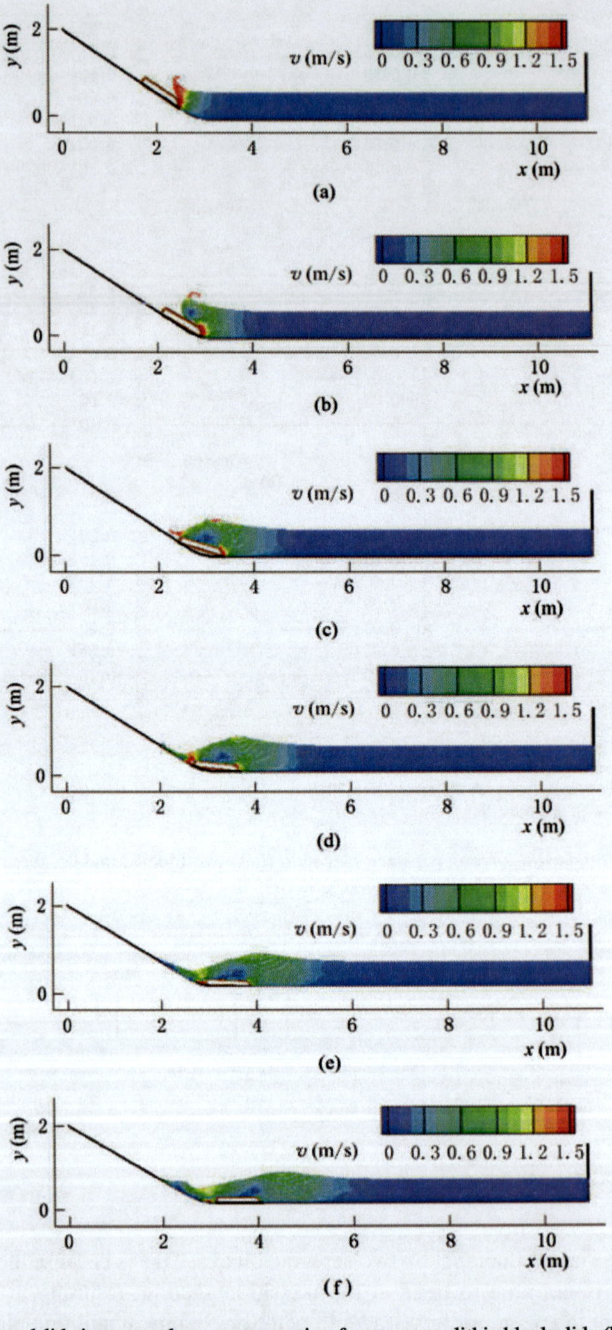

Fig. 11.31 Landslide impact and wave propagation for one monolithic block slide (by DEM-SPH computation only). **a** $t = 0.2$ s; **b** $t = 0.4$ s; **c** $t = 0.6$ s; **d** $t = 0.8$ s; **e** $t = 1.0$ s; **f** $t = 1.2$ s

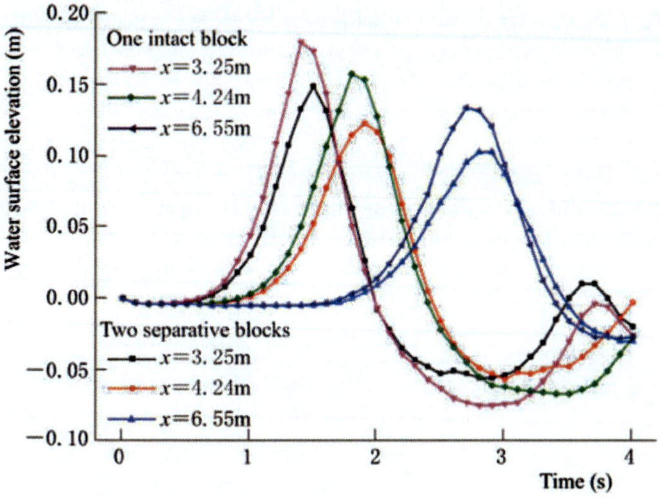

Fig. 11.32 Comparison of water surface elevations between one monolithic block slide and two-block slide

Davies et al. (1999) found that the breakage of rock may give rise to high velocity fragments in large rock avalanches. Given that the landslide body may be fragmented, the tsunami source of monolithic block slide model used in the study may overestimate the height of surge waves. Hence, the effect of block fragmentation should be taken into account in the simulation of landslide-tsunami process.

3. **Problem of Viroulet et al.** (2013a, b)

In Fig. 11.33, the water depth of tank is 0.15 m, and the slope is dipped at angle of 45°. At $x = -0.45$m, -0.75m, -1.05m and -1.35m, there are 4 wave-height gauges of which number 4 is at a distance of 0.60 m from the left tank wall. The cross section of the triangular deformable slide is 0.14×0.14 m, the slide weight is 3.00 kg. When the slip velocity is lower, the friction on the slope is negligible. To increase the friction, the glass balls are stuck on the slope to let the slope possesses identical disc-disc friction. In the experiment, the plate gate is removed suddenly to trigger the slip.

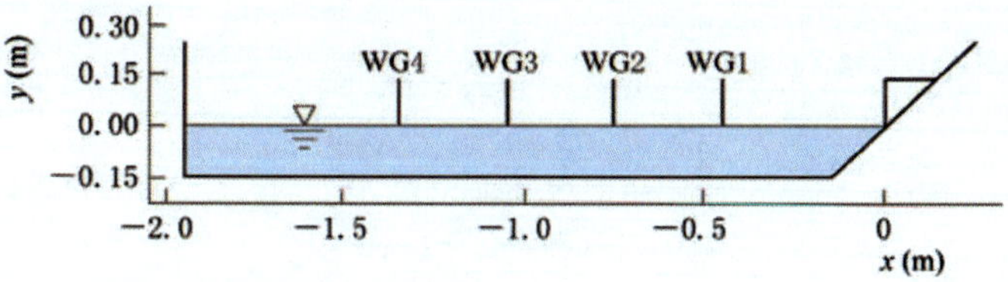

Fig. 11.33 Model setup of subaerial deformable slide (Viroulet et al. 2013b)

This experiment has been well simulated by FVM that looks at the slide block in water as two-phase model governed by N-S equations (Si et al. 2018). The parameters for the DEM-SPH computation are given in Table 11.6, in which smooth length is 1.25 times the initial space of particles, and 2.5 times the disc diameter.

In Fig. 11.34, bulb-shaped leading edge emerges that is similar to the physical experiments. This is mainly due to the action of frictional and drag forces. At t = 0.23 s, the water in front of the leading edge exhibits larger velocity, but the plunge velocity of deformable slide into water is small; under the circumstances of non-rotational friction, the water surface mounts more remarkably due the larger amount

Table 11.6 Computational parameters for the hybrid DEM-SPH model of subaerial deformable slide

DEM parameters	
Disc density (kg/m^3)	2500.0
Disc radius (m)	0.001
Porosity ε	0.2
Disc-disc normal stiffness (10^{2}kN/m)	1.0
Disc-disc tangential stiffness (10^{2}kN/m)	1.0
Disc-disc friction angle (°)	26.0
Disc-wall normal stiffness (10^{3}kN/m)	7.0
Disc-wall tangential stiffness (10^{3}kN/m)	7.0
Disc-wall friction angle (°)	26.0
Amount of discs	2495
Global damping coefficient(N.s/m.g)	0.0
Rotational friction coefficient	0.0/0.5
Time-marching step length (s)	1.0×10^{-4}
SPH parameters	
Particle density (kg/m^3)	1000.0
Initial space between particles (m)	0.004
Amount of particles	17,353
Amount of particles on fixed boundary	5212
Smooth kernel	Wendland
Smooth length (m)	0.005
Boundary condition	Dynamic
Algorithm for pressure revision	δ-SPH
δ-SPH coefficient	0.1
Artificial viscosity coefficient λ	0.3/0.01
Coefficient of sound speed η	10
Dynamic viscosity (kPa·s)	1.0×10^{-6}
Time-marching step length (s)	1.0×10^{-4}

N.B.: Linear contact model is used; disc-disc frictional coefficient is identical to Si et al. (2018)

of water displaced by the slide body. At $t = 0.41$ s, the water moves in two directions, one is outwards as head wave, another reflows backwards to the slide body, and there is a stagnation zone in between; under the circumstances of non-rotational friction, the leading edge thickens more significantly and more irregularly. At $t = 0.52$ s, the head wave propagates continuously, left the surge wave birth zone without the sight of head wave crest; under the circumstances of non-rotational friction, deposits totally submerged on the far position in the tank, whereas the activation of rotational friction keeps a part of deposits above the water surface.

Figure 11.35 indicates that the surge wave is higher when the rotational friction is neglected. In addition, the decay of wave amplitude is obvious, which observes the law of $t^{-1/3}$ (for head wave) (Viroulet et al. 2013b).

We further collect the surge wave series at wave gauges in Fig. 11.36. It is clear that the computed head wave height is overestimated compared to the experimented dada. This is probably attributable to the deficit in our present hybrid algorithm: since

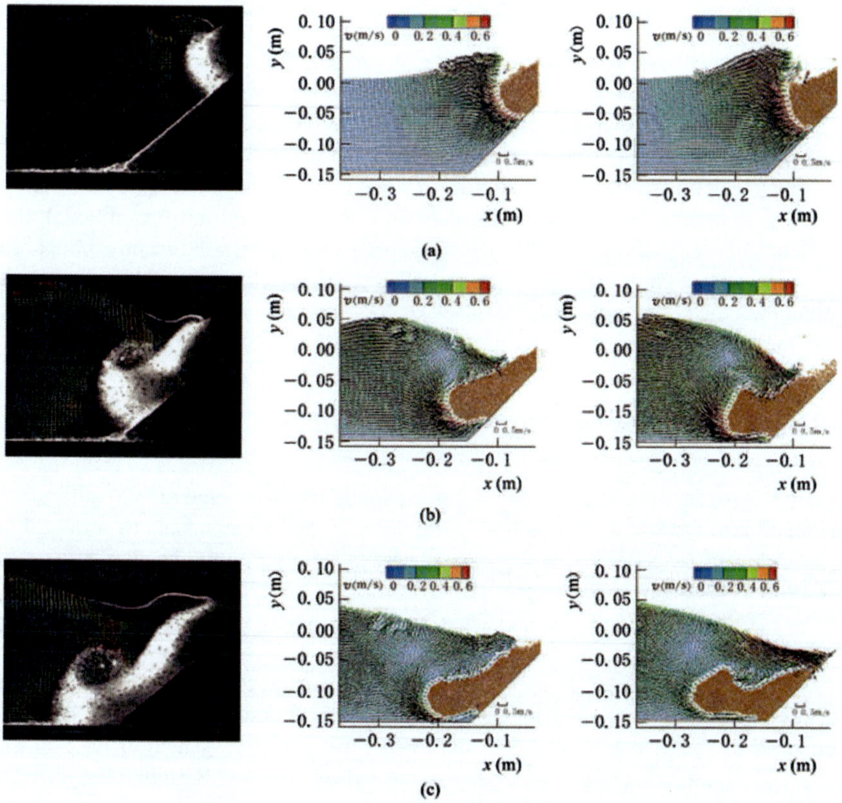

Fig. 11.34 Comparison of landslide deposits and wave profiles at near impact zone (Left column: test by Viroulet et al. 2013a, b. Mid column: $\mu_r = 0.5$ computation by the author. Right column: $\mu_r = 0.0$ computation by the author). **a** $t = 0.23$ s; **b** $t = 0.41$ s; **c** $t = 0.52$ s

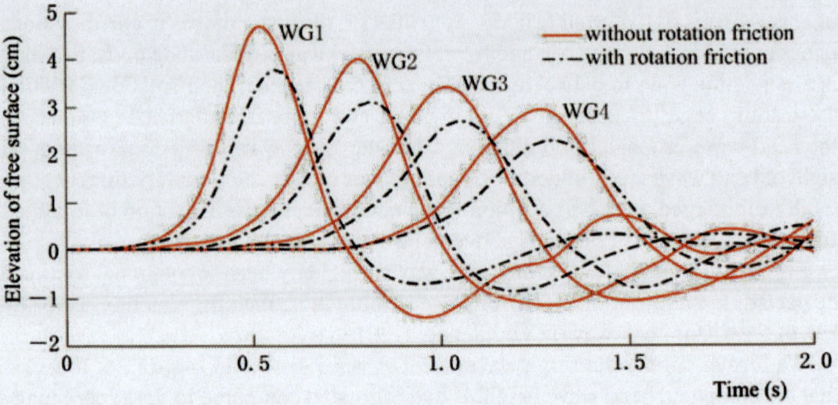

Fig. 11.35 Comparison of wave height at different wave gauges (black dotted line: $\mu_r = 0.5$; red solid line: $\mu_r = 0.0$)

the water penetrated into slide body is not simulated, the existing methods are only applicable to high speed slides or where the slide body possesses low permeability.

It is also notable that the crest of head wave by DEM-SPH and FVM (Si et al. 2018) are close, but the trough after thc head wave is not so well cross-referenced.

The prediction of the trough after head wave may be improved in DEM-SPH by employing appropriate artificial viscosity. In Fig. 11.37, we find that although the artificial viscosity has minor influence on the wave crest, yet it significantly influences on the wave trough. A smaller viscosity will weaken water viscous, this in turn, will strengthen the water surface drop.

11.4.3 Application to Lituya Bay Landslide

On the 8th of July (1958), the Lituya Bay landslide with a volume of 30.6 million m^3 plunged into the water at an approximate velocity of 110 m/s. This hypermobility landslide and its generated giant tsunami waves running on the headland up to an elevation of 524 m, is the largest ever recorded in the history. For validation, the hybrid DEM-SPH model (Tan and Chen 2017) was used to replicate this catastrophic disaster, among many other scholars.

The geometric model shown in Fig. 11.38 is identical to Shi and his co-workers (Shi et al. 2016). The water depth is 122 m and an initial velocity of 67.0 m/s is imposed to the landslide to mimic the pneumatic landslide generator used in the physical experiment (Fritz et al. 2001). In our hybrid DEM-SPH model, the granular landslide is represented by 6722 DEM discs (with a radius of 1.5 m) and the water body is represented by 4060 SPH fluid particles with an initial spacing of 6.0 m, it is noted that this value was set to be 7.5 m in Schwaiger and Higman (2007) and 6.0 m in Xenakis et al. (2017). The smoothing length is set to be 1.25 times the initial

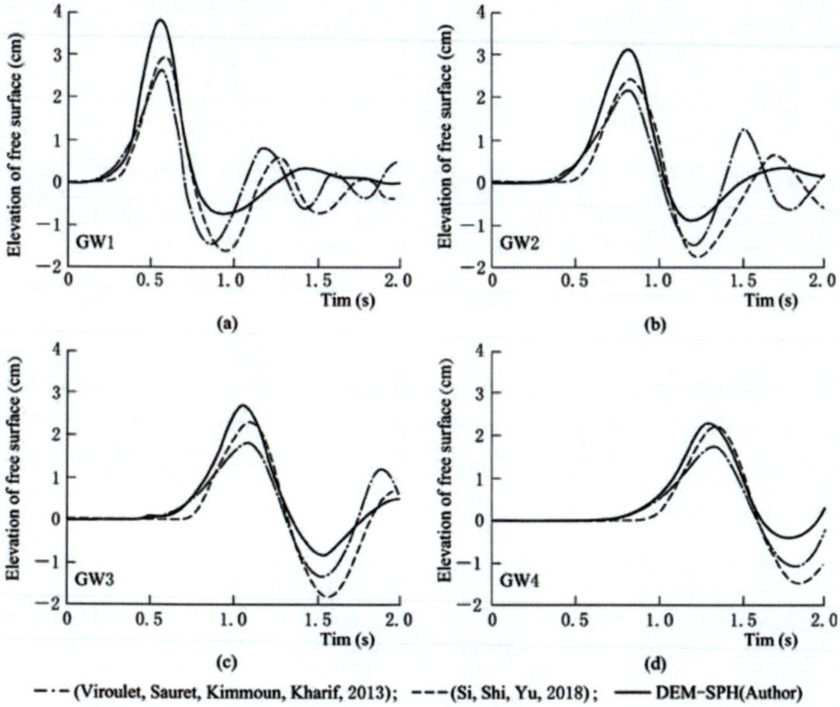

Fig. 11.36 Comparison of wave heights at different wave gauges. **a** WG 1; **b** WG 2; **c** WG 3; **d** WG 4

particle spacing. Hence, the ratio of the smoothing length to the solid disc diameter is 2.5, which satisfies the condition that the smoothing length should be at least 2.0 times the solid particle diameter (Robinson et al. 2014). All relevant parameters are given in Table 11.7, in which the disc-disc frictional coefficient is from the paper of Shi et al. (2016) and the disc-wall frictional coefficient is from the paper of Weiss et al. (2009).

The simplified Hertz-Mindlin contact model is employed in our computation, and four layer dynamic particles are used in order to prevent the fluid particles from flowing out of the boundary.

The landslide impact, wave generation, propagation and run-up are snapshotted in Fig. 11.39. In general, the DEM-SPH results agree well with both the experimented data and FEM results using generalized viscoplastic fluid model (Quecedo et al. 2004). A hydrodynamic impact crater is created due to the violent impact of the granular landslide. This crater collapses backward, resulting in a solitary-like wave with a part of water flows back to the rear of the landslide. Meanwhile, the deposits are diverted from the ramp to the horizontal bottom. The tsunami waves propagate out and climb up the headland.

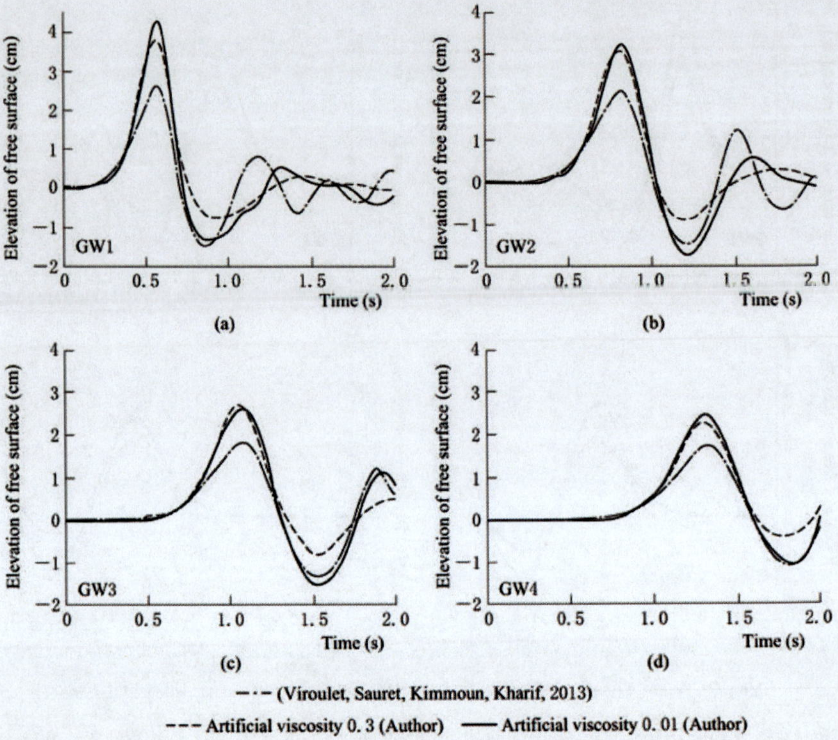

Fig. 11.37 Sensitivity analysis of wave height w.r.t. artificial viscosity coefficient. **a** WG 1; **b** WG 2; **c** WG 3; **d** WG 4

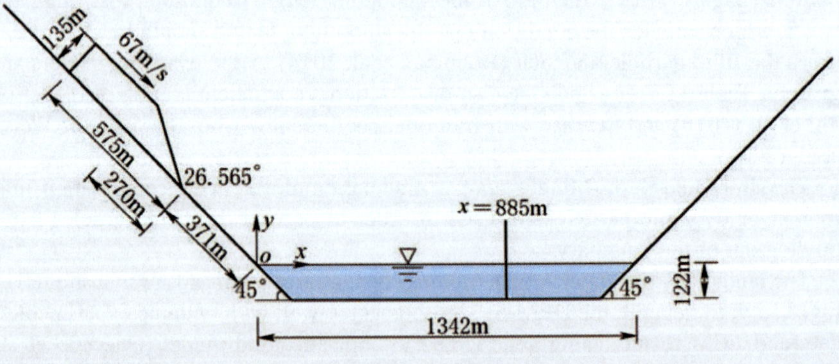

Fig. 11.38 Setup of the experiment model for Lituya Bay landslide (Shi et al. 2016)

Table 11.7 Parameters for the hybrid DEM-SPH model of Lituya Bay landslide

DEM parameters	
Initial velocity (m/s)	67.0
Solid disc density (kg/m^3)	2640.0
Solid disc radius (m)	1.5
Porosity ε	0.2
Young's modulus (10^3 kPa)	700
Poisson's ratio	0.35
Disc-disc frictional coefficient	0.9325
Wall Young's modulus (10^3 kPa)	700
Wall Poisson's ratio	0.35
Wall-disc frictional coefficient	0.40
Amount of solid discs	6722
Damping coefficient (N.s/m.g)	0.030/0.036/0.042
Time-marching step length (s)	1.0×10^{-3}
SPH parameters	
Fluid particle density (kg/m^3)	1000.0
Fluid particle spacing (m)	6.0
Amount of fluid particles	4060
Amount of boundary particles	3150
Kernel function	Wendland
Kernel smoothing length (m)	7.5
Boundary conditions	Dynamic
Density filter	δ-SPH algorithm
δ-SPH coefficient	0.1
Artificial viscosity coefficient λ	0.3
Coefficient of sound speed η	10
Fluid viscosity (kPa·s)	1.0×10^{-6}
Time-marching step length (s)	1.0×10^{-3}

The detailed process of the landslide impact and wave generation at near field is snapshotted in Fig. 11.40 along with the data from experiment and "particle FEM (PFEM)" computation (Salazar et al. 2016). It can be observed that when the landslide moves along the tank bottom, the water will be trapped beneath the landslide by both the DEM-SPH and PFEM computations, but this is not found by the experiment. Compared with the DEM-SPH model (Tan and Chen 2017), the water surface by the mesh-based PFEM is not so smooth, which indicates that re-meshing is required for this method. It is also notable that with a damping coefficient of 0.036, the landslide motion by the DEM-SPH computation is a little bit slower cross-referenced by the experiment data.

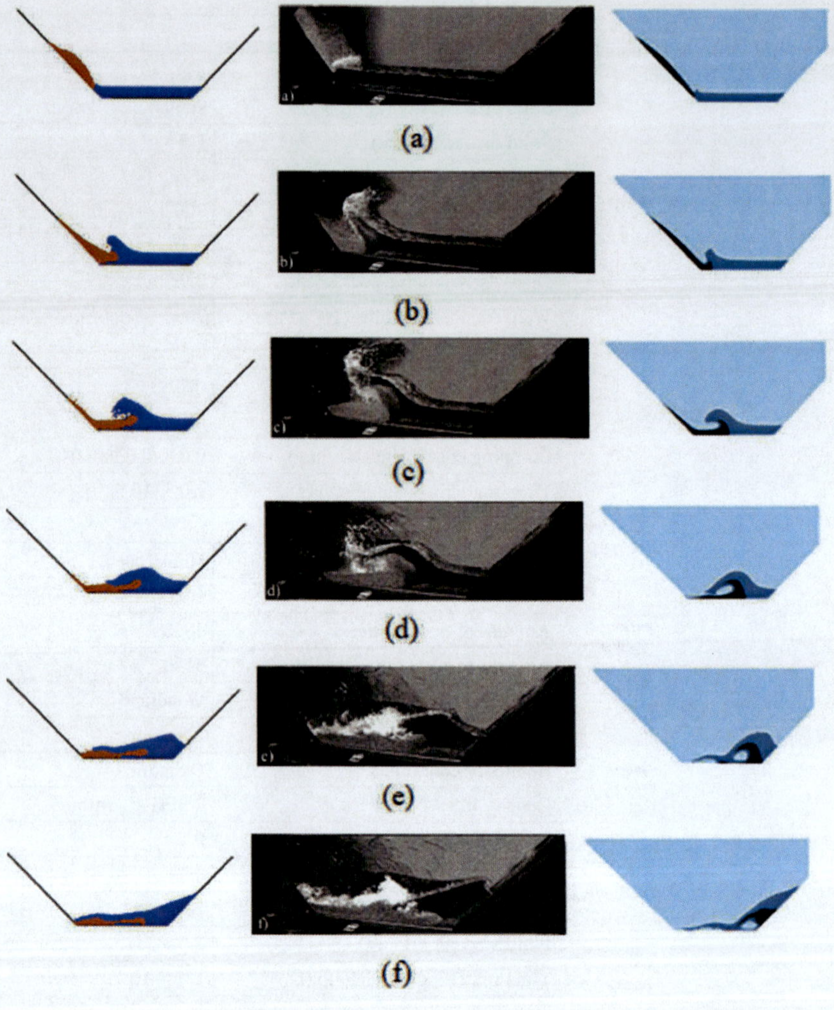

Fig. 11.39 Evolution of Lituya Bay landslide and tsunamis (Left column: DEM-SPH by Tan and Chen 2017. Middle column: experiment by Fritz et al. 2001. Right column: FEM by Quecedo et al. 2004). **a** $t = 0.76$ s; **b** $t = 5.76$ s; **c** $t = 10.76$ s; **d** $t = 15.76$ s; **e** $t = 20.76$ s; **f** $t = 25.76$ s

To explore the effect of slide kinematics on the surge waves, parametric sensitivity analysis w.r.t. damping coefficients is conducted by extracting the energy dissipation on the slide kinematics and subsequent surge waves.

Three damping coefficients, i.e., 0.030, 0.036 and 0.042, are analyzed and the results are displayed in Fig. 11.41, in which the experimental data serve as the benchmark, and the coordinate origin is set to be the intersection of the ramp with the undisturbed water level. It is clear that the smaller the damping coefficient, the faster the landslide motion, and the larger the surge waves. It seems that the damping

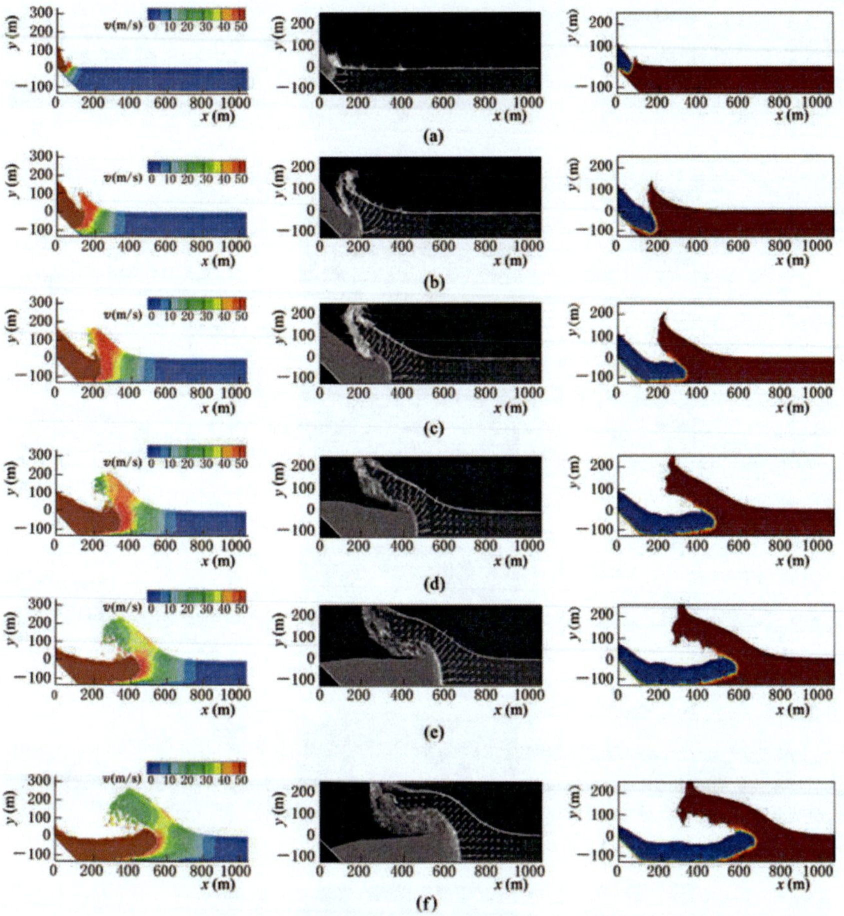

Fig. 11.40 Near field snapshots of Lituya Bay landslide and tsunamis (Left column: DEM-SPH by Tan and Chen 2017. Middle column: experiment by Fritz et al. 2009. Right column: PFEM by Salazar et al. 2016). **a** $t = 0.76$ s; **b** $t = 2.49$ s; **c** $t = 4.22$ s; **d** $t = 5.95$ s; **e** $t = 7.68$ s; **f** $t = 9.41$ s; **g** $t = 11.14$ s; **h** $t = 12.87$ s; **i** $t = 14.60$ s; **j** $t = 16.33$ s; **k** $t = 18.06$ s; **l** $t = 19.79$ s

coefficient of 0.036 best fits the experimented data w.r.t. both wave height and wave run-up at the headland.

The maximum wave height at the gauge (located at $x = 885$ m) and the maximum wave run-up at the headland are listed in Table 11.8, in which the experimented data and other computed data (as many as possible) are included. It may be found that for Lituya Bay landslide, although there are a variety of numerical models available, the maximum wave height and maximum wave run-up are not so easy to predict insofar.

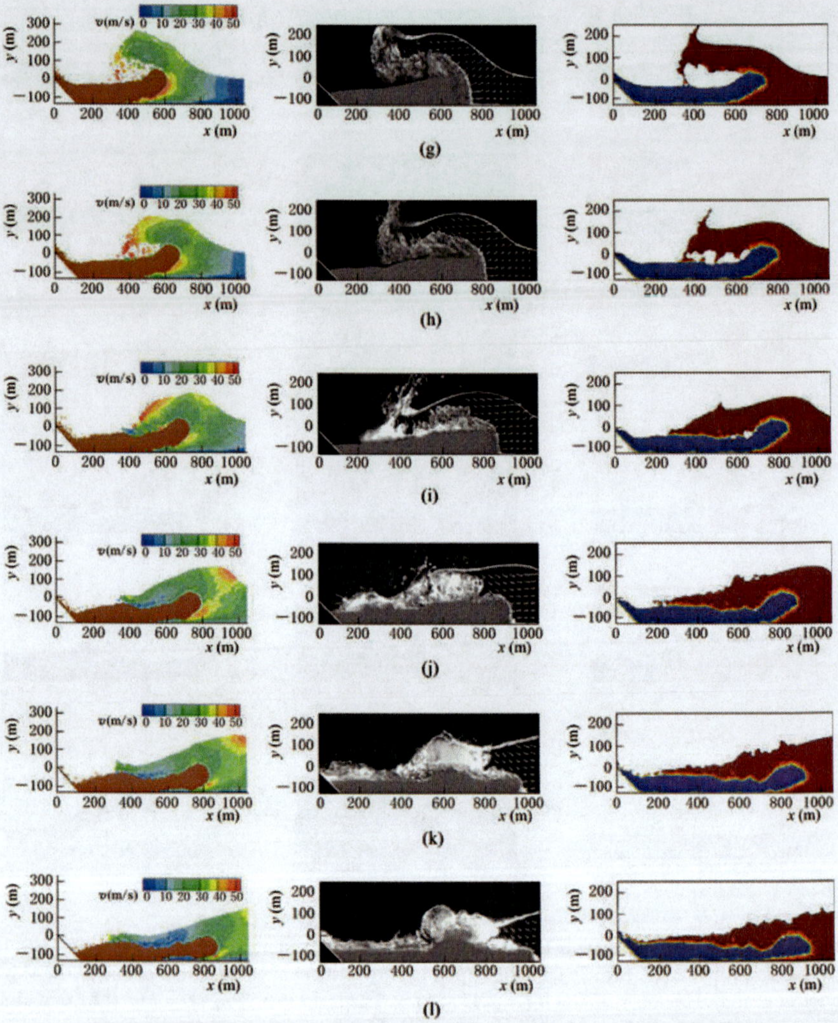

Fig. 11.40 (continued)

11.5 Concluding Remarks

DEM has been widely exercised to simulate landslide motion due to its merit in dealing with large deformation problems (Tang et al. 2009; Lin and Lin 2015). However, early works had limitations in modeling the violent FSI process blamed for the neglect of energy transfer mechanism, namely, the energy conversion from the sliding mass to the water body (Shan and Zhao 2014) and the drag force exerted on the larger landslide front edge (Viroulet et al. 2014).

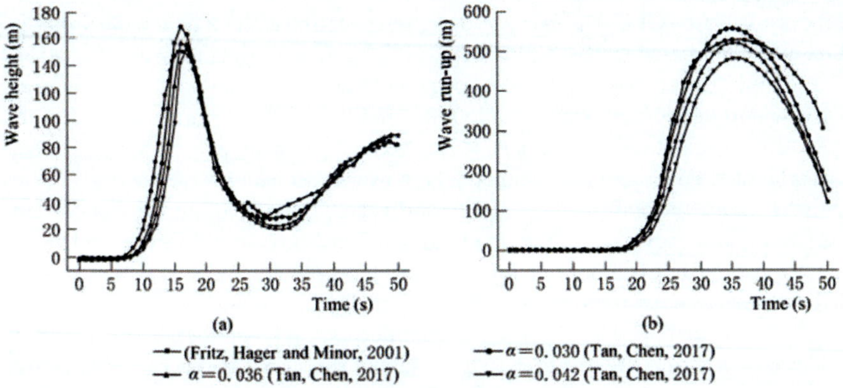

Fig. 11.41 Sensitivity analysis w.r.t. damping coefficients in terms of wave height at $x = 885$ m and wave run-up at the headland. **a** Wave height; **b** wave run-up

Table 11.8 Maximum wave height and wave run-up of the Lituya Bay landslide-induced tsunamis

Resources of data	Maximum wave height at $x =$ 885 m	Maximum wave run-up at the headland
Experimented data from Fritz et al. (2001)	152 m	526 m
Computed data from Mader and Gittings (2002)	/	580 m(+54 m)
Computed data from Quecedo et al. (2004)	227 m(+75 m)	/
Computed data from Salazar et al. (2016)	138 m(−14 m)	/
Computed data from Tan and Chen (2018)	156 m(+4 m)	519 m(−7 m)

Vacondio et al. (2013) adopted SPH to model the surge waves generated by the 1963 Vajont landslide. Although the flow process of water overtopping the dam crest had been well reproduced, yet many assumptions were demanded w.r.t. the landslide body, for example, it was assumed to be rigid and its motion was pre-defined. The force exerted by the water body on the landslide body was also neglected in their study.

Towards the solution of SFI problems such as the simulation of landslide-tsunamis, the hybrid DEM-SPH provides a promising framework. Insofar however, one of the critical restraints to the "true application" of DEM-SPH is the computation power in addition to the others such as the real-world model resolution and parameter input. This situation is more serious in 3-D problems such as the hazard assessment work for the Es Vedrà offshore landslide by the team of the author (Tan et al. 2018b). Even running both with "Central Processing Units (CPUs)" and "Graphics

Processing Units (GPUs)" (He et al. 2018), and coupling with far field wave propagation algorithm based on the depth-averaged "Non-Linear Shallow Water Equations (NLSWEs)", the outcomes are not so quantitatively accurate as the fundamental CMs within the normal work situation featured by small deformation of solid and smooth change of fluid free surface. It is rather understandable that "the good things are seldom to come in pairs", namely, the power to simulate wave break scenarios in SPH normally weakens the stability and accuracy in handling the smooth free surface of fluid. Anyway, with the continuing progress in computer power and model resolution together with more reliable parameter input, it may be expected that the hybrid methods such as the DEM-SPH illustrated in this chapter will be more extensively used and applied for the physical and mechanical scenarios that are not easily addressed with fundamental CMs. The results of such simulations can be exploited at all stages of an engineering project's life, from its early design to its operational maintenance.

It is also necessary to note that other promising Lagrangian-Lagrangian models such as the hybrid DDA-SPH method (Mikola and Sitar 2014; Wang et al. 2016; 2017; 2019; Peng et al. 2020; Yu et al. 2021) also have been well established towards the violent FSI simulations for landslide movement and generated surge waves. Many other new developments in hybrid methods, such as CFD-DEM (Shan and Zhao 2014) and NMM-SPH (Xu et al. 2019), may be found in numerous literatures, too.

References

Abadie S, Morichon D, Grilli S, Glockner S. Numerical simulation of waves generated by landslides using a multiple-fluid Navier-Stokes model. Coast Eng. 2010;57(9):779–94.

Adami S, Hu XY, Adams NA. A generalized wall boundary condition for smoothed particle hydrodynamics. J Comput Phys. 2012;231(21):7057–75.

Altomare C, Dominguez JM, Crespo AJC, Suzuki T, Caceres I, Gómez-Gesteira M. Hybridization of the wave propagation model SWASH and the meshfree particle method SPH for real coastal applications. Coast Eng J. 2015;57(4):1550024.

Anderson TB, Jackson R. Fluid mechanical description of fluidized beds. Equations of motion. Ind Eng Chem Fundam.1967; 6(4): 527–39.

Antoci C, Gallati M, Sibilla S. Numerical simulation of fluid–structure interaction by SPH. Comput Struct. 2007;85(11–14):879–90.

Antuono M, Colagrossi A, Marrone S. Numerical diffusive terms in weakly-compressible SPH schemes. Comput Phys Commun. 2012;183(12):2570–80.

Ataie-Ashtiani B, Nik-Khah A. Impulsive waves caused by subaerial landslides. Environ Fluid Mech. 2008;8(3):263–80.

Ataie-Ashtiani B, Shobeyri G. Numerical simulation of landslide impulsive waves by incompressible smoothed particle hydrodynamics. Int J Numer Meth Fluids. 2008;56(2):209–32.

Attaway SW, Heinstein MW, Swegle JW. Coupling of smooth particle hydrodynamics with the finite element method. Nucl Eng Des. 1994;150(2–3):199–205.

Beer G. Finite element, boundary element and coupled analysis of unbounded problems in elastostatics. Int J Numer Meth Eng. 1983;19(4):567–80.

Bierwisch C, Kübler R, Kleer G, Moseler M. Modelling of contact regimes in wire sawing with dissipative particle dynamics. Philos Trans Royal Soc London. Series A: Math Phys Eng Sci. 2011; 369(1945):2422–430.

Biscarini C. Computational fluid dynamics modelling of landslide generated water waves. Landslides. 2010;7(2):117–24.

Brady BHG, Wassyng A. A coupled finite element-boundary element method of stress analysis. Int J Rock Mech Min Sci Geomech Abstr. 1981;18(6):475–85.

Canelas RB, Crespo AJC, Domínguez JM, Ferreira RML, Gómez-Gesteira M. SPH–DCDEM model for arbitrary geometries in free surface solid-fluid flows. Comput Phys Commun. 2016;202:131–40.

Capone T, Panizzo A, Monaghan JJ. SPH modelling of water waves generated by submarine landslides. J Hydraul Res. 2010;48(S1):80–4.

Charles PC. Numerical analysis of blood flow in the heart. J Comput Phys. 1977;25(3):220–52.

Chen SH. Computational geomechanics and hydraulic structures. Berlin (Germany): Springer; 2018.

Cleary PW. Prediction of coupled particle and fluid flows using DEM and SPH. Miner Eng. 2015;73:85–99.

Cleary PW, Morrison RD. Prediction of 3d slurry flow within the grinding chamber and discharge from a pilot scale sag mill. Miner Eng. 2012;39:184–95.

Cleary PW, Prakash M. Discrete-element modelling and smoothed particle hydrodynamics: potential in the environmental sciences. Philos Trans Royal Soc London. Series A: Math Phys Eng Sci. 2004; 362(1822): 2003–30.

Cremonesi M, Frangi A, Perego U. A Lagrangian finite element approach for the simulation of water-waves induced by landslides. Comput Struct. 2011;89(11–12):1086–93.

Crespo AJC, Gómez-Gesteira M, Dalrymple RA. Boundary conditions generated by dynamic particles in SPH methods. Comput Mater Continua. 2007;5(3):173–84.

Crosta GB, Imposimato S, Roddeman D. Landslide spreading, impulse water waves and modelling of the Vajont rockslide. Rock Mech Rock Eng. 2016;49(6):2413–36.

Dalla Valle JM. Micromeritics: the technology of fine particles. 2nd ed. New York (USA): Pitman Publishing Corporation; 1948.

Davies TR, McSaveney MJ, Hodgson KA. A fragmentation-spreading model for long-runout rock avalanches. Can Geotech J. 1999;36(6):1096–110.

Di Felice R. The voidage function for fluid-particle interaction systems. Int J Multiph Flow. 1994;20(1):153–9.

Di Risio M, Bellotti G, Panizzo A, De Girolamo PD. Three-dimensional experiments on landslide generated waves at a sloping coast. Coast Eng. 2009;56(5–6):659–71.

Dunne T. An Eulerian approach to fluid-structure interaction and goal-oriented mesh adaptation. Int J Numer Meth Fluids. 2006;51:1017–39.

Elsworth D. A hybrid boundary-element-finite element analysis procedure for fluid flow simulation in fractured rock masses. Int J Numer Anal Meth Geomech. 1986;10(6):569–84.

Enet F, Grilli ST. Experimental study of tsunami generation by three-dimensional rigid underwater landslides. J Waterw Port Coast Ocean Eng. 2007;133(6):442–54.

Ergun S. Fluid flow through packed columns. Chem Eng Prog. 1952;48(2):89–94.

Fourey G, Oger G, Touzé DL, Alessandrini B. Violent fluid-structure interaction simulations using a coupled SPH/FEM method. In: IOP conference series: materials science and engineering, vol. 10. 2010. p. 12–41.

Franci A, Oñate E, Carbonell JM. Unified Lagrangian formulation for solid and fluid mechanics and FSI problems. Comput Meth Appl Mech Eng. 2016;298:520–47.

Fritz HM, Moser P. Pneumatic landslide generator. Int J Fluid Power. 2003;4(1):49–57.

Fritz HM, Hager WH, Minor HE. Lituya Bay case: rockslide impact and wave run-up. Sci Tsunami Haz. 2001;19(1):3–22.

Fritz HM, Mohammed F, Yoo J. Lituya Bay landslide impact generated mega-tsunami 50th anniversary. Pure Appl Geophys. 2009;166(1–2):153–75.

Fritz HM, Hager WH, Minor H E. Landslide generated impulse waves. 1. Instantaneous flow fields. Exp Fluids. 2003a; 35(6): 505–19.

Fritz HM, Hager WH, Minor HE. Landslide generated impulse waves. 2. Hydrodynamic impact craters. Exp Fluids. 2003b; 35(6):520–32.

Fu L, Jin YC. Investigation of non-deformable and deformable landslides using meshfree method. Ocean Eng. 2015;109:192–206.
Galindo-Torres SA. A coupled discrete element lattice Boltzmann method for the simulation of fluid-solid interaction with particles of general shapes. Comput Meth Appl Mech Eng. 2013;265:107–19.
Gioda G, Carini A. A combined boundary element-finite element analysis of lined openings. Rock Mech Rock Eng. 1985;18(4):293–302.
Groenenboom PHL, Cartwright BK. Hydrodynamics and fluid-structure interaction by coupled SPH-FE method. J Hydraulics Res. 2010;48(sup1):61–73.
Grof Z, Cook J, Lawrence CJ, Štěpáneka F. The interaction between small clusters of cohesive particles and laminar flow: coupled DEM/CFD approach. J Petrol Sci Eng. 2009;66(1–2):24–32.
Han Y, Cundall PA. LBM-DEM modeling of fluid-solid interaction in porous media. Int J Numer Anal Meth Geomech. 2013;37:1391–407.
He Y, Bayly AE, Hassanpour A, Muller F, Wu K, Yang DM. A GPU-based coupled SPH-DEM method for particle-fluid flow with free surfaces. Powder Technol. 2018;338:548–62.
Heinrich P. Nonlinear water waves generated by submarine and aerial landslides. J Waterw Port Coast Ocean Eng. 1992;118(3):249–66.
Heller V, Bruggemann M, Spinneken J, Rogers BD. Composite modelling of subaerial landslide-tsunamis in different water body geometries and novel insight into slide and wave kinematics. Coast Eng. 2016;109:20–41.
Heller V. Landslide generated impulse waves: prediction of near field characteristics. PhD Thesis. ETH Zurich (Switzerland); 2007.
Hirt CW, Amsden AA, Cook JL. An arbitrary Lagrangian-Eulerian computing method for all fluid speed. J Comp Phys. 1997;135:203–16.
Huang YJ, Nydal OJ. Coupling of discrete-element method and smoothed particle hydrodynamics for liquid-solid flows. Theor Appl Mech Lett. 2012;2(1): 012002.
Huang BL, Yin YP, Chen XT, Liu GN, Wang SC, Jiang ZB. Experimental modeling of tsunamis generated by subaerial landslides: two case studies of the Three Gorges Reservoir, China. Environ Earth Sci. 2014;71(9):3813–25.
Hughes TJR, Liu WK, Zimmermann TK. Lagrangian-Eulerian finite element formulation for incompressible viscous flows. Comput Meth Appl Mech. 1981;29(3):329–49.
Jiao YY, Zhang HQ, Tang HM, Zhang XL, Adoko AC, Tian HN. Simulating the process of reservoir-impoundment-induced landslide using the extended DDA method. Eng Geol, Part A. 2014;182:37–48.
Kim YI, Amadei B, Pan E. Modeling the effect of water, excavation sequence and rock reinforcement with discontinuous deformation analysis. Int J Rock Mech Min Sci. 1999;36(7):949–70.
Kuhl E, Hulshoff S, De Borst R. An arbitrary Lagrangian Eulerian finite-element approach for fluid-structure interaction phenomena. Int J Numer Meth Eng. 2003;57:117–42.
Legay A, Chessa J, Belytschko T. An Eulerian-Lagrangian method for fluid-structure interaction based on level sets. Comput Meth Appl Mech Eng. 2006;195:2070–87.
Li Z, Leduc J, Nunez-Ramirez J, Combescure A, Marongiu JC. A non-intrusive partitioned approach to couple smoothed particle hydrodynamics and finite element methods for transient fluid-structure interaction problems with large interface motion. Comput Mech. 2015;55(4):697–718.
Libersky LD, Petschek AG, Carney TC, Hipp JR, Allahdadi FA. High strain Lagrangian hydrodynamics: a three-dimensional SPH code for dynamic material response. J Comput Phys. 1993;109(1):67–75.
Lin CH, Lin ML. Evolution of the large landslide induced by Typhoon Morakot: a case study in the Butangbunasi River, Southern Taiwan using the discrete element method. Eng Geol. 2015;197:172–87.
Liu GR, Liu MB. Smoothed particle hydrodynamics: a meshfree particle method. Singapore: World Scientific Press; 2003.
Liu X, Lin P, Shao S. An ISPH simulation of coupled structure interaction with free surface flows. J Fluids Struct. 2014;48:46–61.

Long T, Hu D, Yang G, Wan D. A particle-element contact algorithm incorporated into the coupling methods of FEM-ISPH and FEM-WCSPH for FSI problems. Ocean Eng. 2016;123(1):154–63.
Ma G, Kirby JT, Shi F. Numerical simulation of tsunami waves generated by deformable submarine landslides. Ocean Model. 2013;69:146–65.
Ma G, Kirby JT, Hsu TJ, Shi F. A two-layer granular landslide model for tsunami wave generation: theory and computation. Ocean Model. 2015;93:40–55.
Mader CL, Gittings ML. Modeling the 1958 Lituya Bay mega-tsunami, II. Sci Tsunami Hazards. 2002;20(5):241–50.
Marrone S, Antuono M, Colagrossi A, Colicchio G, Le Touzé D, Graziani G. δ-SPH model for simulating violent impact flows. Comp Meth Appl Mech Eng. 2011;200(13):1526–42.
Mikola RG, Sitar N. 3D simulation of tsunami wave induced by rock slope failure using coupled DDA-SPH. In: 48th US Rock Mechanics/Geomechanics Symposium. American Rock Mechanics Association; 2014.
Mohammed F, Fritz HM. Physical modeling of tsunamis generated by three-dimensional deformable granular landslides. J Geophys Res: Oceans. 2012;117(C11):C11015.
Molteni D, Colagrossi A. A simple procedure to improve the pressure evaluation in hydrodynamic context using the SPH. Comput Phys Commun. 2009;180(6):861–72.
Monaghan JJ. Kajtar JB SPH particle boundary forces for arbitrary boundaries. Comput Phys Commun. 2009;180(10):1811–20.
Mulligan RP, Take WA. On the transfer of momentum from a granular landslide to a water wave. Coast Eng. 2017;125:16–22.
Ohkami T, Mitsui Y, Kusama T. Coupled boundary element/finite element analysis in geomechanics including body forces. Comput Geotech. 1985;1(4):263–78.
Panizzo A, De Girolamo P, Petaccia A. Forecasting impulse waves generated by subaerial landslides. J Geophys Res: Oceans. 2005a;110(C12):C12025.
Panizzo A, De Girolamo P, Di Risio M, Maistri A, Petaccia A. Great landslide events in Italian artificial reservoirs. Nat Hazards and Earth Syst Sci. 2005b;5(5):733–40.
Pelinovsky E, Poplavsky A. Simplified model of tsunami generation by submarine landslides. Phys Chem Earth. 1996;21(1–2):13–7.
Peng X, Yu P, Chen G, Xia M, Zhang Y. Development of a coupled DDA-SPH method and its application to dynamic simulation of landslides involving solid–fluid interaction. Rock Mech Rock Eng. 2020;53:113–31.
Peskin CS. The immersed boundary method. Acta Numer. 2002;11:1–39.
Potapov AV, Hunt ML, Campbell CS. Liquid-solid flows using smoothed particle hydrodynamics and the discrete element method. Powder Technol. 2001;116(2–3):204–13.
Qiu LC. Two-dimensional SPH simulations of landslide-generated water waves. J Hydraul Eng. 2008;134(5):668–71.
Qiu LC. Numerical modeling of liquid-particle flows by combining SPH and DEM. Ind Eng Chem Res. 2013;52(33):11313–8.
Qiu LC, Jin F, Lin PZ, Liu Y, Han Y. Numerical simulation of submarine landslide tsunamis using particle based methods. J Hydrodyn. 2017;29(4):542–51.
Quecedo M, Pastor M, Herreros MI. Numerical modelling of impulse wave generated by fast landslides. Int J Numer Meth Eng. 2004;59(12):1633–56.
Ren B, Jin Z, Gao R, Wang YX, Xu ZL. SPH-DEM modeling of the hydraulic stability of 2D blocks on a slope. J Waterw Port Coast Ocean Eng. 2014;140(6):04014022.
Robbe-Saule M, Morize C, Bertho Y, Sauret A, Gondret P, Radjai F, Nezamabadi S, Luding S, Delenne JY. Experimental study of wave generation by a granular collapse. Eur Phys J Conf. 2017;140(7):14007.
Robinson M, Ramaioli M, Luding S. Fluid-particle flow simulations using two-way-coupled mesoscale SPH-DEM and validation. Int J Multiph Flow. 2014;59:121–34.
Romano A, Di Risio M, Bellotti G, Molfetta MG, Damiani L, De Girolamo P. Tsunamis generated by landslides at the coast of conical islands: experimental benchmark dataset for mathematical model validation. Landslides. 2016;13(6):1379–93.

Rzadkiewicz SA, Mariotti C, Heinrich P. Numerical simulation of submarine landslides and their hydraulic effects. J Waterw Port Coast Ocean Eng. 1997;123(4):149–57.
Sælevik G, Jensen A, Pedersen G. Experimental investigation of impact generated tsunami related to a potential rock slide, Western Norway. Coastal Eng. 2009;56(9):897–906.
Salazar F, Irazábal J, Larese A, Oñate E. Numerical modelling of landslide-generated waves with the particle finite element method (PFEM) and a non-Newtonian flow model. Int J Numer Anal Meth Geomech. 2016;40(6):809–26.
Schwaiger HF, Higman B. Lagrangian hydrocode simulations of the 1958 Lituya Bay tsunamigenic rockslide. Geochem Geophys Geosyst. 2007;8(7):Q07006.
Shan T, Zhao J. A coupled CFD-DEM analysis of granular flow impacting on a water reservoir. Acta Mech. 2014;225(8):2449–70.
Shi CQ, An Y, Wu Q, Liu QQ, Cao ZX. Numerical simulation of landslide-generated waves using a soil-water coupling smoothed particle hydrodynamics model. Adv Water Resour. 2016;92:130–41.
Si PF, Shi HB, Yu XP. A general numerical model for surface waves generated by granular material intruding into a water body. Coast Eng. 2018;142:42–51.
Sun X, Sakai M, Yamada Y. Three-dimensional simulation of a solid-liquid flow by the DEM-SPH method. J Comput Phys. 2013;248:147–76.
Swoboda G, Mertz W, Beer G. Rheological analysis of tunnel excavations by means of coupled finite element (FEM)-boundary element (BEM) analysis. Int J Numer Anal Meth Geomech. 1987;11(2):115–29.
Takagi S, Sugiyama K, Li S, Matsumoto Y. A review of full Eulerian methods for fluid structure interaction problems. J Appl Mech. 2012;79: 010911.
Tan H, Chen SH. A hybrid DEM-SPH model for deformable landslide and its generated surge waves. Adv Water Resour. 2017;108:256–76.
Tan H, Xu Q, Chen SH. Subaerial rigid landslide-tsunamis: insights from a block DEM-SPH model. Eng Anal Bound Elem. 2018a;95:297–314.
Tan H, Ruffini G, Heller V, Chen SH. A numerical landslide-tsunami hazard assessment technique applied on hypothetical scenarios at Es Vedrà, offshore Ibiza. J Marine Sci Eng. 2018b;6(4):111.
Tang CL, Hu JC, Lin ML, Angelier J, Lu CY, Chan YC, Chu HT. The Tsaoling landslide triggered by the Chi-Chi earthquake, Taiwan: insights from a discrete element simulation. Eng Geol. 2009;106(1–2):1–19.
Tang Y, Jiang Q, Zhou C. A Lagrangian-based SPH-DEM model for fluid-solid interaction with free surface flow in two dimensions. Appl Math Model. 2018;62:436–60.
Tezduyar TE, Behr M, Liou J. A New strategy for finite element computations involving moving boundaries and interfaces-The deforming-spatial-domain/space-time procedure: I. The concept and the preliminary numerical tests. Comp Meth Appl Mech Eng. 1992; 94(3):339–51.
Topin V, Monerie Y, Perales F, Radjai F. Collapse dynamics and runout of dense granular materials in a fluid. Phys Rev Lett. 2012;109(18): 188001.
Udaykumar HS, Shyy W. Simulation of interfacial instabilities during solidification; Part I: conduction and capillarity effects. Numer Heat Transfer B. 1994;27:127–55.
Vacondio R, Mignosa P, Pagani S. 3D SPH numerical simulation of the wave generated by the Vajont rockslide. Adv Water Resour. 2013;59:146–56.
Valizadeh A, Monaghan JJ. A study of solid wall models for weakly compressible SPH. J Comput Phys. 2015;300:5–19.
van Zuijlen AH, Bijl H. Implicit and explicit higher order time integration schemes for structural dynamics and fluid structure interaction computations. Comput Struct. 2005;83(2–3):93–105.
Varadarajan A, Sharma KG, Singh RB. Some aspects of coupled FEBEM analysis of underground openings. Int J Numer Anal Meth Geomech. 1985;9(6):557–71.
Viroulet S, Cébron D, Kimmoun O, Kharif C. Shallow water waves generated by subaerial solid landslides. Geophys J Int. 2013a;193(2):747–62.
Viroulet S, Sauret A, Kimmoun O, Kharif C. Granular collapse into water: toward tsunami landslides. J Visual. 2013b;16(3):189–91.

Viroulet S, Sauret A, Kimmoun O. Tsunami generated by a granular collapse down a rough inclined plane. EPL (Europhysics Letters). 2014; 105(3):34004.

Von Estorff O, Firuziaan M. Coupled BEM/FEM approach for non-linear soil/structure interaction. Eng Anal Boundary Elem. 2000;24(10):715–25.

Vuyst TD, Vignjevic R, Campbell JC. Coupling between meshless and finite element methods. Int J Impact Eng. 2005;31(8):1054–64.

Walder JS, Watts P, Sorensen OE, Janssen K. Tsunamis generated by subaerial mass flows. J Geophys Res: Solid Earth. 2003;108(B5):B052236.

Walhorn E, Kölke A, Hübner B, Dinkler D. Fluid-structure coupling within a monolithic model involving free surface flows. Comput Struct. 2005;83(25–26):2100–11.

Wang GH, Zhang SR, Yu M, Li HB, Kong Y. Investigation of the shock wave propagation characteristics and cavitation effects of underwater explosion near boundaries. Appl Ocean Res. 2014;46:40–53.

Wang W, Chen GQ, Zhang H, Zhou SH, Liu SG, Wu YQ, Fan FS. Analysis of landslide-generated impulsive waves using a coupled DDA-SPH method. Eng Anal Bound Elem. 2016;64:267–77.

Wang W, Chen GQ, Zhang YB, Zheng L, Zhang H. Dynamic simulation of landslide dam behavior considering kinematic characteristics using a coupled DDA-SPH method. Eng Anal Bound Elem. 2017;80:172–83.

Wang W, Yin KL, Chen GQ, Chai B, Han Z, Zhou JW. Practical application of the coupled DDA-SPH method in dynamic modeling for the formation of landslide dam. Landslides. 2019;16(5):1021–32.

Watts P. Tsunami features of solid block underwater landslides. J Waterw Port Coast Ocean Eng. 2000;126(3):144–52.

Weiss R, Fritz HM, Wünnemann K. Hybrid modeling of the mega-tsunami runup in Lituya Bay after half a century. Geophys Res Lett. 2009;36(L9):L09602.

Wen C, Yu Y. Mechanics of fluidization. Chem Eng Prog Symp Ser. 1966;62:100–11.

Wu K, Yang DM, Wright N. A coupled SPH-DEM model for fluid-structure interaction problems with free-surface flow and structural failure. Comput Struct. 2016;177:141–61.

Xenakis AM, Lind SJ, Stansby PK, Rogers BD. Landslides and tsunamis predicted by incompressible smoothed particle hydrodynamics (SPH) with application to the 1958 Lituya Bay event and idealized experiment. Proc Royal Soc A: Math, Phys Eng Sci. 2017;473(2199):20160674.

Xu Y, Yu C, Liu F, Liu Q. A coupled NMM-SPH method for fluid-structure interaction problems. Appl Math Model. 2019;76:466–78.

Yang Q, Jones V, Mccue L. Free-surface flow interactions with deformable structures using an SPH-FEM model. Ocean Eng. 2012;55(15):136–47.

Yavari-Ramshe S, Ataie-Ashtiani B. Numerical modeling of subaerial and submarine landslide-generated tsunami waves—recent advances and future challenges. Landslides. 2016;13(6):1325–68.

Yeylaghi S, Moa B, Buckham B, Oshkai P, Vasquez J, Crawford C. ISPH modelling of landslide generated waves for rigid and deformable slides in Newtonian and Non-Newtonian reservoir fluids. Adv Water Resour. 2017;107:212–32.

Yu PC, Chen GQ, Peng XY, Zhang YB, Zhang H, Wang W. Verification and application of 2-D DDA-SPH method in solving fluid-structure interaction problems. J Fluids Struct. 2021;102: 103252.

Zhang C, Rezavand M, Hu XY. A multi-resolution SPH method for fluid-structure interactions. J Computat Phys. 2020; 110028.

Zhao T, Utili S, Crosta GB. Rockslide and impulse wave modelling in the Vajont reservoir by DEM-CFD analyses. Rock Mech Rock Eng. 2016;49(6):2437–56.

Zienkiewicz OC, Kelly DW, Bettess P. The coupling of the finite element method and boundary solution procedures. Int J Numer Meth Eng. 1977;11(2):355–75.